W9-DGF-767

Complete Solutions Manual

for

Poole's

Linear Algebra

A Modern Introduction

Second Edition

Robert Rogers
Bay State College

Australia • Brazil • Canada • Mexico • Singapore • Spain • United Kingdom • United States

© 2006 Thomson Brooks/Cole, a part of The Thomson Corporation. Thomson, the Star logo, and Brooks/Cole are trademarks used herein under license.

ALL RIGHTS RESERVED. No part of this work covered by the copyright hereon may be reproduced or used in any form or by any means—graphic, electronic, or mechanical, including photocopying, recording, taping, Web distribution, information storage and retrieval systems, or in any other manner—except as may be permitted by the license terms herein.

Printed in the United States of America

1 2 3 4 5 6 7 08 07 06 05

Printer: EPAC Technologies, Inc.

ISBN: 0-534-99859-3

For more information about our products, contact us at:
Thomson Learning Academic Resource Center
1-800-423-0563

For permission to use material from this text or product, submit a request online at
http://www.thomsonrights.com.
Any additional questions about permissions can be submitted by email to **thomsonrights@thomson.com.**

Thomson Higher Education
10 Davis Drive
Belmont, CA 94002-3098
USA

Instructors Solutions Manual
for
David Poole's
LINEAR ALGEBRA
A Modern Introduction

even- and odd-numbered exercises

2nd edition

Robert Rogers
Bay State College
Boston, Massachusetts

Contents

Chapter 1

Vectors

1.1 The Geometry and Algebra of Vectors

1.

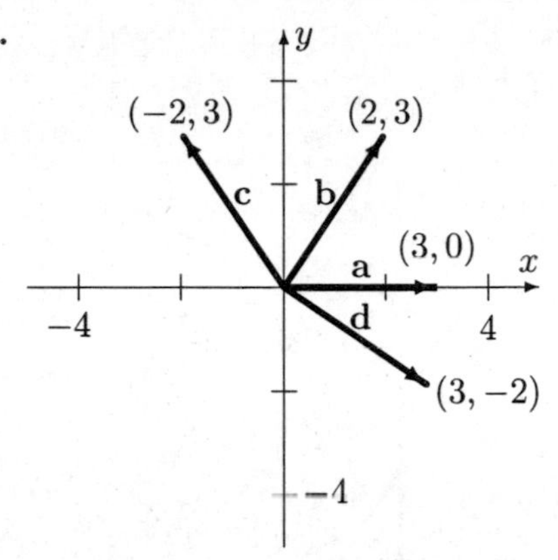

2.

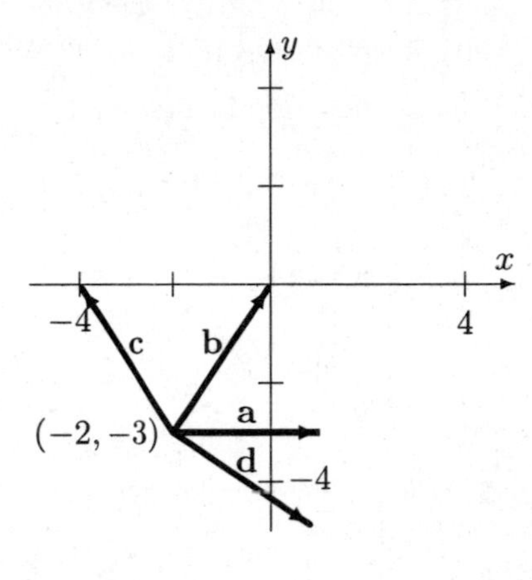

3. See Figures 1.14 and 1.15.

4. (a) Following Example 1.1, we have the following:
If $[0, 2, 0]$ is translated to $\overrightarrow{BC}$ where $C = (4, 5, 6)$,
then we must have $B = (4 - 0, 5 - 2, 6 - 0) = (4, 3, 6)$.
Note: Unlike Example 1.1, we subtract $[0, 2, 0]$ instead of adding $[0, 2, 0]$. Why?

(b) Likewise, if $[3, 2, 1]$ is translated, $B = (4 - 3, 5 - 2, 6 - 1) = (1, 3, 5)$.

(c) If $[1, -2, 1]$ is translated, $B = (4 - 1, 5 - (-2), 6 - 1) = (3, 7, 5)$.

(d) If $[-1, -1, -2]$ is translated, $B = (4 - (-1), 5 - (-1), 6 - (-2)) = (5, 6, 8)$.

5. (a) $\overrightarrow{AB} = [4 - 1, 2 - (-1)] = [3, 3]$.

(b) $[2, 1]$ (c) $[-\frac{3}{2}, \frac{3}{2}]$ (d) $[-\frac{1}{6}, \frac{1}{6}]$.

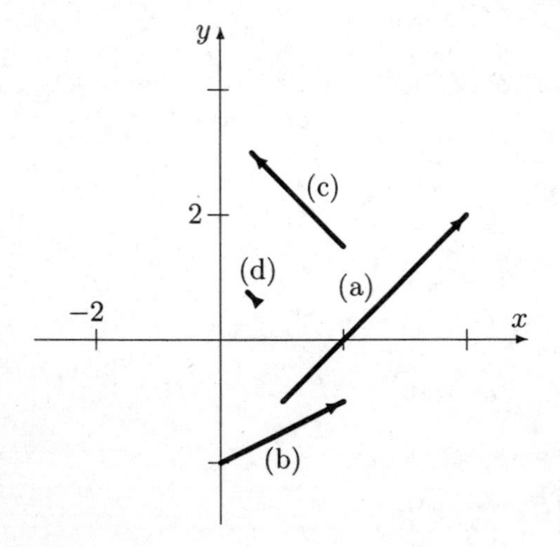

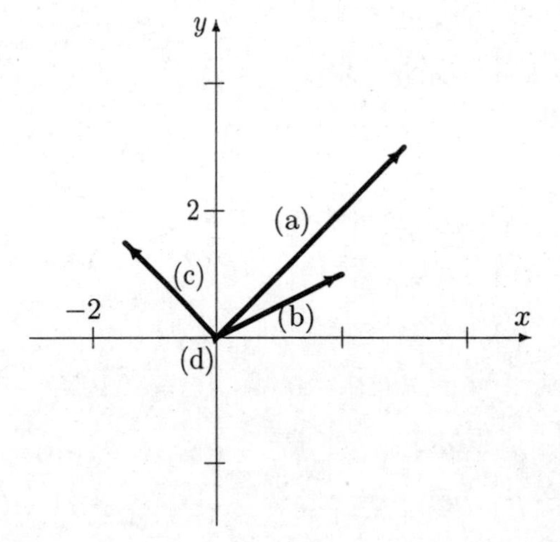

6. Recall the notation that $[a, b]$ denotes a move of a units horizontally and b units vertically.
During the first part of the walk, the hiker walks 4 km north, so $\mathbf{a} = [0, 4]$.
During the second part of the walk, the hiker walks a distance of 5 km northeast.

From the components, we get $\mathbf{b} = [5\cos 45°, 5\sin 45°] = \left[\frac{5\sqrt{2}}{2}, \frac{5\sqrt{2}}{2}\right]$.

Thus, the net displacement vector is $\mathbf{c} = \mathbf{a} + \mathbf{b} = \left[\frac{5\sqrt{2}}{2}, 4 + \frac{5\sqrt{2}}{2}\right]$.

7. $\mathbf{a} + \mathbf{b} = \begin{bmatrix} 3 \\ 0 \end{bmatrix} + \begin{bmatrix} 2 \\ 3 \end{bmatrix} = \begin{bmatrix} 3+2 \\ 0+3 \end{bmatrix} = \begin{bmatrix} 5 \\ 3 \end{bmatrix}$ **8.** $\mathbf{b} + \mathbf{c} = \begin{bmatrix} 2 \\ 3 \end{bmatrix} + \begin{bmatrix} -2 \\ 3 \end{bmatrix} = \begin{bmatrix} 0 \\ 6 \end{bmatrix}$

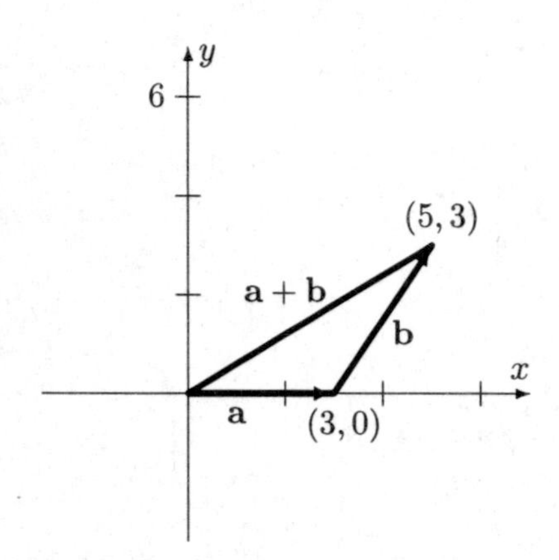

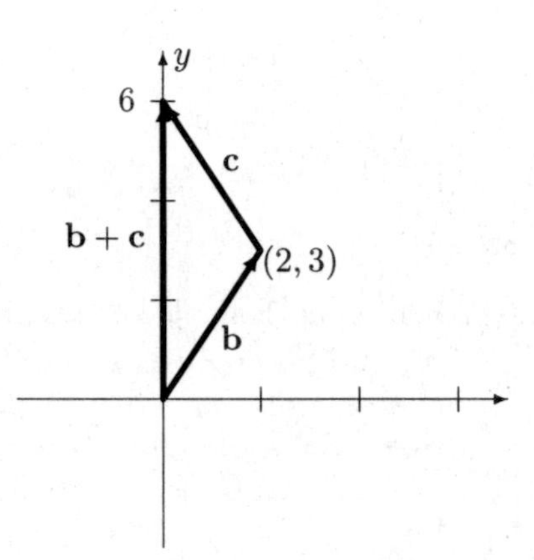

9. $\mathbf{d} - \mathbf{c} = \begin{bmatrix} 3 \\ -2 \end{bmatrix} - \begin{bmatrix} -2 \\ 3 \end{bmatrix} = \begin{bmatrix} 5 \\ -5 \end{bmatrix}$. **10.** $\mathbf{a} - \mathbf{d} = \begin{bmatrix} 3 \\ 0 \end{bmatrix} - \begin{bmatrix} 3 \\ -2 \end{bmatrix} = \begin{bmatrix} 0 \\ 2 \end{bmatrix}$.

11. $2\mathbf{a} + 3\mathbf{c} = 2[0, 2, 0] + 3[1, -2, 1] = [2 \cdot 0, 2 \cdot 2, 2 \cdot 0] + [3 \cdot 1, 3(-2), 3 \cdot 1] = [3, -2, 3]$.

12. $2\mathbf{c} - 3\mathbf{b} - \mathbf{d} = 2[1, -2, 1] - 3[3, 2, 1] - [-1, -1, -2] = [-6, -9, 1]$.

13. $\mathbf{u} = [\cos 60°, \sin 60°] = \left[\frac{1}{2}, \frac{\sqrt{3}}{2}\right]$, $\mathbf{v} = [\cos 210°, \sin 210°] = \left[-\frac{\sqrt{3}}{2}, -\frac{1}{2}\right] \Rightarrow$ (implies)

$\mathbf{u} + \mathbf{v} = \left[\frac{1}{2} - \frac{\sqrt{3}}{2}, \frac{\sqrt{3}}{2} - \frac{1}{2}\right]$, $\mathbf{u} - \mathbf{v} = \left[\frac{1}{2} + \frac{\sqrt{3}}{2}, \frac{\sqrt{3}}{2} + \frac{1}{2}\right]$.

14. (a) $\overrightarrow{AB} = \mathbf{b} - \mathbf{a}$.

(b) $\overrightarrow{BC} = \overrightarrow{OC} - \mathbf{b} = (\mathbf{b} - \mathbf{a}) - \mathbf{b} = -\mathbf{a}$.

(c) $\overrightarrow{AD} = -2\mathbf{a}$.

(d) $\overrightarrow{CF} = \overrightarrow{CB} + \overrightarrow{BA} + \overrightarrow{AF} = -\overrightarrow{BC} - \overrightarrow{AB} + \left(-\overrightarrow{AB} - \mathbf{a}\right) - \overrightarrow{AB} = 2(\mathbf{a} - \mathbf{b})$.

(e) $\overrightarrow{AC} = \overrightarrow{AB} + \overrightarrow{BC} = (\mathbf{b} - \mathbf{a}) + (-\mathbf{a}) = \mathbf{b} - 2\mathbf{a}$.

(f) $\overrightarrow{BC} + \overrightarrow{DE} + \overrightarrow{FA} = -\mathbf{a} + \left(-\overrightarrow{AB}\right) + \left(\overrightarrow{AB} + \mathbf{a}\right) = \mathbf{0}$.

15. $2(\mathbf{a}-3\mathbf{b})+3(2\mathbf{b}+\mathbf{a}) \overset{\substack{\text{property e.}\\ \text{distributivity}}}{=} (2\mathbf{a}-6\mathbf{b})+(6\mathbf{b}+3\mathbf{a}) \overset{\substack{\text{property b.}\\ \text{associativity}}}{=} (2\mathbf{a}+3\mathbf{a})+(-6\mathbf{b}+6\mathbf{b}) = 5\mathbf{a}.$

16. $-3(\mathbf{a}-\mathbf{c})+2(\mathbf{a}+2\mathbf{b})+3(\mathbf{c}-\mathbf{b}) \overset{\substack{\text{property e.}\\ \text{distributivity}}}{=} (-3\mathbf{a}+3\mathbf{c})+(2\mathbf{a}+4\mathbf{b})+(3\mathbf{c}-3\mathbf{b}) \overset{\substack{\text{property b.}\\ \text{associativity}}}{=}$
$(-3\mathbf{a}+2\mathbf{a})+(4\mathbf{b}-3\mathbf{b})+(3\mathbf{c}+3\mathbf{c}) = -\mathbf{a}+\mathbf{b}+6\mathbf{c}.$

17. $\mathbf{x}-\mathbf{a} = 2(\mathbf{x}-2\mathbf{a}) = 2\mathbf{x}-4\mathbf{a} \Rightarrow \mathbf{x}-2\mathbf{x} = \mathbf{a}-4\mathbf{a} \Rightarrow -\mathbf{x} = -3\mathbf{a} \Rightarrow \mathbf{x} = 3\mathbf{a}.$

18. $\mathbf{x}+2\mathbf{a}-\mathbf{b} = 3(\mathbf{x}+\mathbf{a})-2(2\mathbf{a}-\mathbf{b}) = 3\mathbf{x}+3\mathbf{a}-4\mathbf{a}+2\mathbf{b} \Rightarrow \mathbf{x}-3\mathbf{x} = -\mathbf{a}-2\mathbf{a}+2\mathbf{b}+\mathbf{b} \Rightarrow$
$-2\mathbf{x} = -3\mathbf{a}+3\mathbf{b} \Rightarrow \mathbf{x} = \frac{3}{2}\mathbf{a}-\frac{3}{2}\mathbf{b}.$

19.

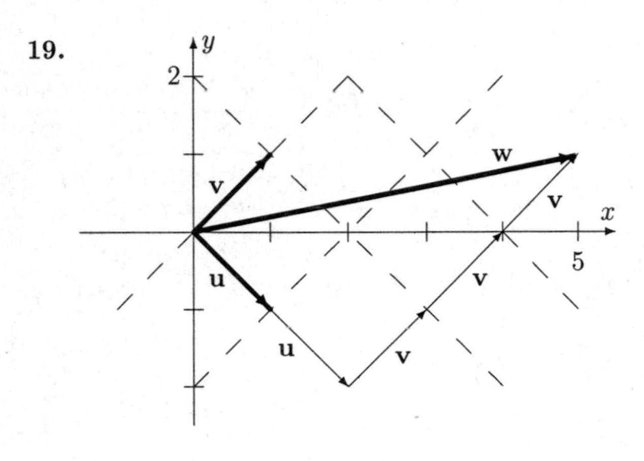

20.

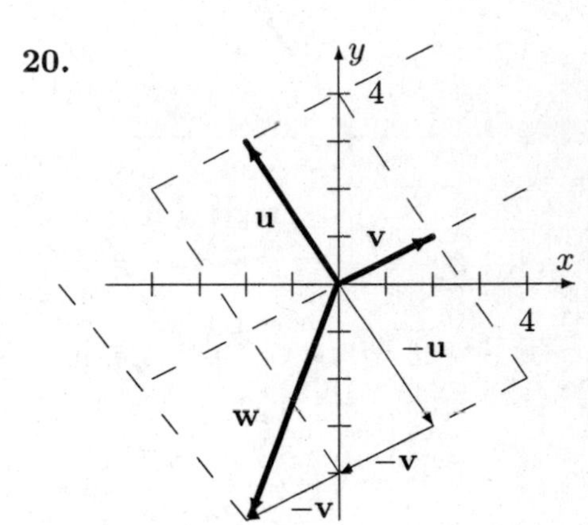

21. See Exercise 19.

22. See Exercise 20.

23. Property (d) states that $\mathbf{u} + (-\mathbf{u}) = \mathbf{0}$. The first diagram below shows $\mathbf{u}$ along with $-\mathbf{u}$. Then, as the diagonal of the parallelogram, the resultant vector is $\mathbf{0}$.

Property (e) states $c(\mathbf{u} + \mathbf{v}) = c\mathbf{u} + c\mathbf{v}$. The second figure illustrates this.

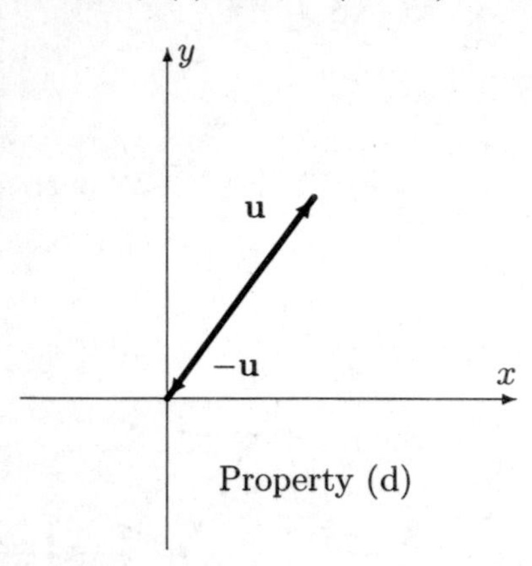

Property (d)

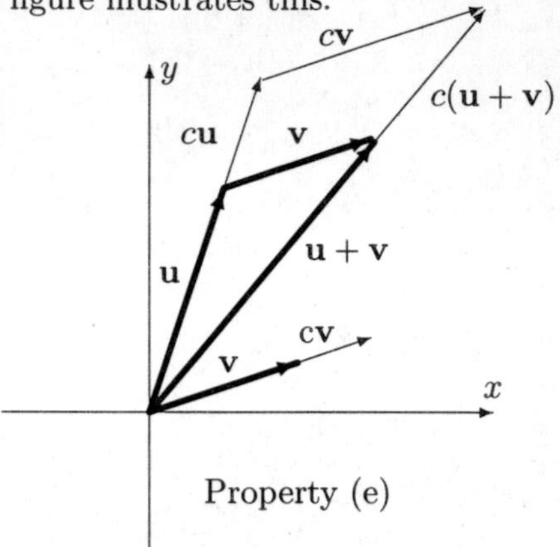

Property (e)

24. Let $\mathbf{u} = [u_1, u_2, ..., u_n]$, $\mathbf{v} = [v_1, v_2, ..., v_n]$, and let c and d be scalars in $\mathbb{R}$.
Property (d):

$$\begin{aligned} \mathbf{u} + (-\mathbf{u}) &= [u_1, u_2, ..., u_n] + (-1\,[u_1, u_2, ..., u_n]) = [u_1, u_2, ..., u_n] + [-u_1, -u_2, ..., -u_n] \\ &= [u_1 + (-u_1)\,, u_2 + (-u_2)\,, ..., u_n + (-u_n)] = [0, 0, ..., 0] = \mathbf{0} \end{aligned}$$

Property (e):

$$\begin{aligned} c\,(\mathbf{u} + \mathbf{v}) &= c\,([u_1, u_2, ..., u_n] + [v_1, v_2, ..., v_n]) = c\,([u_1 + v_1, u_2 + v_2, ..., u_n + v_n]) \\ &= [c\,(u_1 + v_1)\,, c\,(u_2 + v_2)\,, ..., c\,(u_n + v_n)] = [cu_1 + cv_1, cu_2 + cv_2, ..., cu_n + cv_n] \\ &= [cu_1, cu_2, ..., cu_n] + [cv_1, cv_2, ..., cv_n] = c\,[u_1, u_2, ..., u_n] + c\,[v_1, v_2, ..., v_n] = c\mathbf{u} + c\mathbf{v} \end{aligned}$$

Property (f):

$$\begin{aligned} (c + d)\,\mathbf{u} &= (c + d)\,[u_1, u_2, ..., u_n] = [(c + d)\,u_1, (c + d)\,u_2, ..., (c + d)\,u_n] \\ &= [cu_1 + du_1, cu_2 + du_2, ..., cu_n + du_n] = [cu_1, cu_2, ..., cu_n] + [du_1, du_2, ..., du_n] \\ &= c\,[u_1, u_2, ..., u_n] + d\,[u_1, u_2, ..., u_n] = c\mathbf{u} + d\mathbf{u} \end{aligned}$$

Property (g):

$$\begin{aligned} c\,(d\mathbf{u}) &= c\,(d\,[u_1, u_2, ..., u_n]) = c\,[du_1, du_2, ..., du_n] = [cdu_1, cdu_2, ..., cdu_n] \\ &= [(cd)\,u_1, (cd)\,u_2, ..., (cd)\,u_n] = (cd)\,[u_1, u_2, ..., u_n] = (cd)\,\mathbf{u} \end{aligned}$$

1.2 Length and Angle: The Dot Product

1. Following Example 1.8, $\mathbf{u}\cdot\mathbf{v} = \begin{bmatrix} -1 \\ 2 \end{bmatrix} \cdot \begin{bmatrix} 3 \\ 1 \end{bmatrix} = (-1)\cdot 3 + 2\cdot 1 = -3 + 2 = -1.$

2. Following Example 1.8, $\mathbf{u}\cdot\mathbf{v} = \begin{bmatrix} 3 \\ -2 \end{bmatrix} \cdot \begin{bmatrix} 4 \\ 6 \end{bmatrix} = 3\cdot 4 + (-2)\cdot 6 = 12 - 12 = 0.$

3. $\mathbf{u}\cdot\mathbf{v} = \begin{bmatrix} 1 \\ 2 \\ 3 \end{bmatrix} \cdot \begin{bmatrix} 2 \\ 3 \\ 1 \end{bmatrix} = 1\cdot 2 + 2\cdot 3 + 3\cdot 1 = 2 + 6 + 3 = 11.$

4. $\mathbf{u}\cdot\mathbf{v} = \begin{bmatrix} 3.2 \\ -0.6 \\ -1.4 \end{bmatrix} \cdot \begin{bmatrix} 1.5 \\ 4.1 \\ -0.2 \end{bmatrix} = (3.2)\cdot(1.5) + (-0.6)\cdot(4.1) + (-1.4)\cdot(-0.2) = 2.62.$

5. $\mathbf{u}\cdot\mathbf{v} = \begin{bmatrix} 1 \\ \sqrt{2} \\ \sqrt{3} \\ 0 \end{bmatrix} \cdot \begin{bmatrix} 4 \\ -\sqrt{2} \\ 0 \\ -5 \end{bmatrix} = 1\cdot 4 + (\sqrt{2})\cdot(-\sqrt{2}) + \sqrt{3}\cdot 0 + 0\cdot(-5) = 4 - 2 = 2.$

6. $\mathbf{u}\cdot\mathbf{v} = \begin{bmatrix} 1.12 \\ -3.25 \\ 2.07 \\ -1.83 \end{bmatrix} \cdot \begin{bmatrix} -2.29 \\ 1.72 \\ 4.33 \\ -1.54 \end{bmatrix} = -1.12\cdot 2.29 + 1.72\cdot 4.33 + 1.83\cdot 1.54 = 3.6265.$

7. In the remarks prior to Example 1.11, we note that finding a unit vector $\mathbf{v}$ in the same direction as a given vector $\mathbf{u}$ is called ***normalizing*** the vector $\mathbf{u}$.

Therefore, we proceed as in Example 1.12:

$\|\mathbf{u}\| = \sqrt{(-1)^2 + 2^2} = \sqrt{5}$, so a unit vector $\mathbf{v}$ in the same direction as $\mathbf{u}$ is

$$\mathbf{v} = (1/\|\mathbf{u}\|)\,\mathbf{u} = (1/\sqrt{5}) \begin{bmatrix} -1 \\ 2 \end{bmatrix} = \begin{bmatrix} -1/\sqrt{5} \\ 2/\sqrt{5} \end{bmatrix}.$$

8. Following Example 1.12, we have:

$\|\mathbf{u}\| = \sqrt{3^2 + (-2)^2} = \sqrt{13}$, so a unit vector $\mathbf{v}$ in the same direction as $\mathbf{u}$ is

$$\mathbf{v} = (1/\|\mathbf{u}\|)\,\mathbf{u} = (1/\sqrt{13}) \begin{bmatrix} 3 \\ -2 \end{bmatrix} = \begin{bmatrix} 3/\sqrt{13} \\ -2/\sqrt{13} \end{bmatrix}.$$

9. Following Example 1.12, we have:

$\|\mathbf{u}\| = \sqrt{1^2 + 2^2 + 3^2} = \sqrt{14}$, so a unit vector $\mathbf{v}$ in the same direction as $\mathbf{u}$ is

$$\mathbf{v} = (1/\|\mathbf{u}\|)\,\mathbf{u} = (1/\sqrt{14}) \begin{bmatrix} 1 \\ 2 \\ 3 \end{bmatrix} = \begin{bmatrix} 1/\sqrt{14} \\ 2/\sqrt{14} \\ 3/\sqrt{14} \end{bmatrix}.$$

10. Following Example 1.12, we have:

$\|\mathbf{u}\| = \sqrt{(3.2)^2 + (-0.6)^2 + (-1.4)^2} = \sqrt{12.56}$, so unit vector $\mathbf{v}$ in the same direction as $\mathbf{u}$

is $\mathbf{v} = (1/\|\mathbf{u}\|)\,\mathbf{u} = (1/\sqrt{12.56})\begin{bmatrix} 3.2 \\ -0.6 \\ -1.4 \end{bmatrix} = \begin{bmatrix} 3.2/\sqrt{12.56} \\ -0.6/\sqrt{12.56} \\ -1.4/\sqrt{12.56} \end{bmatrix} \approx \begin{bmatrix} 0.2548 \\ -0.0478 \\ -0.1115 \end{bmatrix}$.

11. $\|\mathbf{u}\| = \sqrt{1^2 + (\sqrt{2})^2 + (\sqrt{3})^2 + 0^2} = \sqrt{6}$, so a unit vector in the direction of $\mathbf{u}$ is

$$\mathbf{v} = (1/\|\mathbf{u}\|)\,\mathbf{u} = (1/\sqrt{6})\begin{bmatrix} 1 \\ \sqrt{2} \\ \sqrt{3} \\ 0 \end{bmatrix} = \begin{bmatrix} 1/\sqrt{6} \\ \sqrt{2}/\sqrt{6} \\ \sqrt{3}/\sqrt{6} \\ 0/\sqrt{6} \end{bmatrix} = \begin{bmatrix} 1/\sqrt{6} \\ 1/\sqrt{3} \\ 1/\sqrt{2} \\ 0 \end{bmatrix} = \begin{bmatrix} \sqrt{6}/6 \\ \sqrt{3}/3 \\ \sqrt{2}/2 \\ 0 \end{bmatrix}.$$

12. $\|\mathbf{u}\| = \sqrt{(1.12)^2 + (-3.25)^2 + (2.07)^2 + (-1.83)^2} = \sqrt{19.4507}$, so the unit vector $\mathbf{v}$ is

$$\mathbf{v} = (1/\|\mathbf{u}\|)\,\mathbf{u} = (1/\sqrt{19.4507})\begin{bmatrix} 1.12 \\ -3.25 \\ 2.07 \\ -1.83 \end{bmatrix} = \begin{bmatrix} 1.12/\sqrt{19.4507} \\ -3.25/\sqrt{19.4507} \\ 2.07/\sqrt{19.4507} \\ -1.83/\sqrt{19.4507} \end{bmatrix} \approx \begin{bmatrix} 0.2540 \\ -0.7369 \\ 0.4694 \\ -0.4149 \end{bmatrix}.$$

13. Following Example 1.13, we compute: $\mathbf{u} - \mathbf{v} = \begin{bmatrix} -1 \\ 2 \end{bmatrix} - \begin{bmatrix} 3 \\ 1 \end{bmatrix} = \begin{bmatrix} -4 \\ 1 \end{bmatrix}$, so

$d(\mathbf{u}, \mathbf{v}) = \|\mathbf{u} - \mathbf{v}\| = -\sqrt{(-4)^2 + 1^2} = \sqrt{17}$.

14. Following Example 1.13, we compute: $\mathbf{u} - \mathbf{v} = \begin{bmatrix} 3 \\ -2 \end{bmatrix} - \begin{bmatrix} 4 \\ 6 \end{bmatrix} = \begin{bmatrix} -1 \\ -8 \end{bmatrix}$, so

$d(\mathbf{u}, \mathbf{v}) = \|\mathbf{u} - \mathbf{v}\| = -\sqrt{(-1)^2 + (-8)^2} = \sqrt{65}$.

15. Following Example 1.13, we compute: $\mathbf{u} - \mathbf{v} = \begin{bmatrix} 1 \\ 2 \\ 3 \end{bmatrix} - \begin{bmatrix} 2 \\ 3 \\ 1 \end{bmatrix} = \begin{bmatrix} -1 \\ -1 \\ 2 \end{bmatrix}$, so

$d(\mathbf{u}, \mathbf{v}) = \|\mathbf{u} - \mathbf{v}\| = -\sqrt{(-1)^2 + (-1)^2 + 2^2} = \sqrt{6}$.

16. Following Example 1.13, we compute: $\mathbf{u} - \mathbf{v} = \begin{bmatrix} 3.2 \\ -0.6 \\ -1.4 \end{bmatrix} - \begin{bmatrix} 1.5 \\ 4.1 \\ -0.2 \end{bmatrix} = \begin{bmatrix} 1.7 \\ -4.7 \\ -1.2 \end{bmatrix}$, so

$d(\mathbf{u}, \mathbf{v}) = \|\mathbf{u} - \mathbf{v}\| = -\sqrt{(1.7)^2 + (-4.7)^2 + (-1.2)^2} = \sqrt{26.42} \approx 5.14$.

17. (a) $\mathbf{u} \cdot \mathbf{v}$ is a real number, so $\|\mathbf{u} \cdot \mathbf{v}\|$ is the norm of a number, which is not defined.

(b) $\mathbf{u} \cdot \mathbf{v}$ is a scalar, while $\mathbf{w}$ is a vector.
Thus, $\mathbf{u} \cdot \mathbf{v} + \mathbf{w}$ adds a scalar to a vector, which is not a defined operation.

(c) $\mathbf{u}$ is a vector, while $\mathbf{v} \cdot \mathbf{w}$ is a scalar.
Thus, $\mathbf{u} \cdot (\mathbf{v} \cdot \mathbf{w})$ is the dot product of a vector and a scalar, which is not defined.

(d) $c \cdot (\mathbf{u} + \mathbf{v})$ is the dot product of a scalar and a vector, which is not defined.

18. From trigonometry, we have:

$\cos\theta > 0 \Rightarrow \theta$ is acute, $\cos\theta < 0 \Rightarrow \theta$ is obtuse, and $\cos\theta = 0 \Rightarrow \theta$ is right.

From $\cos\theta = \dfrac{\mathbf{u} \cdot \mathbf{v}}{\|\mathbf{u}\|\,\|\mathbf{v}\|}$, we see $\mathbf{u} \cdot \mathbf{v}$ determines the sign of $\cos\theta$. Why?

Therefore, as in Example 1.14, we calculate:

$\mathbf{u} \cdot \mathbf{v} = 3 \cdot (-1) + 0 \cdot 1 = -3 < 0 \Rightarrow \cos\theta < 0 \Rightarrow \theta$ is obtuse.

19. From trigonometry, we have:

$\cos\theta > 0 \Rightarrow \theta$ is acute, $\cos\theta < 0 \Rightarrow \theta$ is obtuse, and $\cos\theta = 0 \Rightarrow \theta$ is right.

From $\cos\theta = \dfrac{\mathbf{u} \cdot \mathbf{v}}{\|\mathbf{u}\|\,\|\mathbf{v}\|}$, we see $\mathbf{u} \cdot \mathbf{v}$ determines the sign of $\cos\theta$. Why?

Therefore, as in Example 1.14, we calculate:

$\mathbf{u} \cdot \mathbf{v} = 2 \cdot 1 + (-1) \cdot (-2) + 1 \cdot (-1) = 4 > 0 \Rightarrow \cos\theta > 0 \Rightarrow \theta$ is acute.

20. Following the first step in Example 1.14, we calculate:
$\mathbf{u} \cdot \mathbf{v} = 4 \cdot 1 + 3 \cdot (-1) + (-1) \cdot 1 = 4 - 3 - 1 = 0 \Rightarrow \cos\theta = 0 \Rightarrow \theta$ is right.

21. Following the first step in Example 1.14, we calculate:
$\mathbf{u} \cdot \mathbf{v} = (0.9) \cdot (-4.5) + (2.1) \cdot (2.6) + (1.2) \cdot (-0.8) = 0.45 \Rightarrow \cos\theta > 0 \Rightarrow \theta$ is acute.

22. $\mathbf{u} \cdot \mathbf{v} = 1 \cdot (-3) + (-2) \cdot 1 + 3 \cdot (-1) + 3 \cdot 4 = -4 \Rightarrow \cos\theta < 0 \Rightarrow \theta$ is obtuse.

23. Since $\mathbf{u} \cdot \mathbf{v}$ is obviously > 0, we have $\cos\theta > 0$ which implies θ is acute.
Note: $\mathbf{u} \cdot \mathbf{v}$ is > 0 because the components of both $\mathbf{u}$ and $\mathbf{v}$ are positive.

24. As in Example 1.14, we begin by calculating $\mathbf{u} \cdot \mathbf{v}$ (if $\mathbf{u} \cdot \mathbf{v} = 0$, we're done. Why?):

$\mathbf{u} \cdot \mathbf{v} = 3 \cdot (-1) + 0 \cdot 1 = -3$, $\|\mathbf{u}\| = \sqrt{3^2 + 0^2} = \sqrt{9} = 3$, $\|\mathbf{v}\| = \sqrt{(-1)^2 + 1^2} = \sqrt{2}$.

So, $\cos\theta = \dfrac{\mathbf{u} \cdot \mathbf{v}}{\|\mathbf{u}\|\,\|\mathbf{v}\|} = \dfrac{-3}{3\sqrt{2}} = -\dfrac{\sqrt{2}}{2}$ and $\theta = \cos^{-1}\left(-\dfrac{\sqrt{2}}{2}\right) = \dfrac{3\pi}{4}$ radians or $135°$.

25. As in Example 1.14, we begin by calculating $\mathbf{u} \cdot \mathbf{v}$ (because if $\mathbf{u} \cdot \mathbf{v} = 0$ we're done. Why?):

$$\mathbf{u} \cdot \mathbf{v} = 2 \cdot 1 + (-1) \cdot (-2) + 1 \cdot (-1) = 2 + 2 - 1 = 3,$$

$$\|\mathbf{u}\| = \sqrt{2^2 + (-1)^2 + 1^2} = \sqrt{6}, \text{ and } \|\mathbf{v}\| = \sqrt{1^2 + (-2)^2 + (-1)^2} = \sqrt{6}.$$

Therefore, $\cos\theta = \dfrac{\mathbf{u} \cdot \mathbf{v}}{\|\mathbf{u}\|\,\|\mathbf{v}\|} = \dfrac{3}{\sqrt{6}\sqrt{6}} = \dfrac{1}{2}$, so $\theta = \cos^{-1}\left(\dfrac{1}{2}\right) = \dfrac{\pi}{3}$ radians or $60°$.

26. As in Example 1.14, we begin by calculating $\mathbf{u} \cdot \mathbf{v}$:

$$\mathbf{u} \cdot \mathbf{v} = 4 \cdot 1 + 3 \cdot (-1) + (-1) \cdot 1 = 4 - 3 - 1 = 0 \Rightarrow \cos\theta = 0 \Rightarrow \theta \text{ is right.}$$

If we wished to be more explicit, we could continue following Example 1.15:

$$\cos\theta = \frac{\mathbf{u} \cdot \mathbf{v}}{\|\mathbf{u}\|\,\|\mathbf{v}\|} = \frac{0}{\|\mathbf{u}\|\,\|\mathbf{v}\|} = 0, \text{ so } \theta = \cos^{-1}(0) = \frac{\pi}{2} \text{ radians or } 90°.$$

27. Following Example 1.14, we calculate:

$$\mathbf{u} \cdot \mathbf{v} = (0.9) \cdot (-4.5) + (2.1) \cdot (2.6) + (1.2) \cdot (-0.8) = 0.45,$$

$$\|\mathbf{u}\| = \sqrt{(0.9)^2 + (2.1)^2 + (1.2)^2} = \sqrt{6.66}, \text{ and}$$

$$\|\mathbf{v}\| = \sqrt{(-4.5)^2 + (2.6)^2 + (-0.8)^2} = \sqrt{27.65}.$$

Therefore, $\cos\theta = \dfrac{\mathbf{u} \cdot \mathbf{v}}{\|\mathbf{u}\|\,\|\mathbf{v}\|} = \dfrac{0.45}{\sqrt{6.66}\sqrt{27.65}} = \dfrac{0.45}{\sqrt{182.817}}$,

$$\text{so } \theta = \cos^{-1}\left(\frac{0.45}{\sqrt{182.817}}\right) \approx 1.5375 \text{ radians or } 88.09°.$$

Note: To minimize error, we do not approximate until the last step.

Since $\dfrac{0.45}{\sqrt{182.817}} \approx 0.0332816$ is a positive number close to zero,

we should expect θ to be close to but less than $90°$. Why?

28. Following Example 1.14, we calculate:

$$\mathbf{u} \cdot \mathbf{v} = 1 \cdot (-3) + (-2) \cdot 1 + 3 \cdot (-1) + 3 \cdot 4 = -4,$$

$$\|\mathbf{u}\| = \sqrt{1^2 + (-2)^2 + 3^2 + 4^2} = \sqrt{30}, \text{ and}$$

$$\|\mathbf{v}\| = \sqrt{(-3)^2 + 1^2 + (-1)^2 + 1^2} = \sqrt{12}.$$

Therefore, $\cos\theta = \dfrac{\mathbf{u} \cdot \mathbf{v}}{\|\mathbf{u}\|\,\|\mathbf{v}\|} = \dfrac{-4}{\sqrt{30}\sqrt{12}} = -\dfrac{2}{3\sqrt{10}}$,

$$\text{so } \theta = \cos^{-1}\left(-\frac{2}{3\sqrt{10}}\right) \approx 1.7832 \text{ radians or } 102.17°.$$

Note: To minimize error, we do not approximate until the last step.

Since $-\dfrac{2}{3\sqrt{10}} \approx -0.2108185$ is a negative number close to zero,

we should expect θ to be close to but greater than $90°$. Why?

29. Following Example 1.14, we calculate:

$\mathbf{u} \cdot \mathbf{v} = 1 \cdot 5 + 2 \cdot 6 + 3 \cdot \ +3 \cdot 7 + 4 \cdot 8 = 70,$

$\|\mathbf{u}\| = \sqrt{1^2 + 2^2 + 3^2 + 4^2} = \sqrt{30}$, and

$\|\mathbf{v}\| = \sqrt{5^2 + 6^2 + 7^2 + 8^2} = \sqrt{174}.$

Therefore, $\cos\theta = \dfrac{\mathbf{u} \cdot \mathbf{v}}{\|\mathbf{u}\| \, \|\mathbf{v}\|} = \dfrac{70}{\sqrt{30}\sqrt{174}} = \dfrac{35}{3\sqrt{145}},$

so $\theta = \cos^{-1}\left(\dfrac{35}{3\sqrt{145}}\right) \approx 0.2502$ radians or $14.34°$.

Note: To minimize error, we do not approximate until the last step.

Since $\dfrac{35}{3\sqrt{145}} \approx 0.9688639$ is a positive number close to 1,

we should expect θ to be close to but greater than $0°$.

30. To show $\triangle ABC$ is right, we need only show one pair of its sides meet at a right angle.

So, we let $\mathbf{u} = \overrightarrow{AB}$, $\mathbf{v} = \overrightarrow{BC}$, and $\mathbf{w} = \overrightarrow{AC}$, then by the definition of ***orthogonal*** given prior to Example 1.16, we need only show $\mathbf{u} \cdot \mathbf{v}$, or $\mathbf{u} \cdot \mathbf{w}$, or $\mathbf{v} \cdot \mathbf{w} = 0$.

Following Example 1.1 of Section 1.1, we calculate the sides of $\triangle ABC$:

$\mathbf{u} = \overrightarrow{AB} = [1 - (-3), 0 - 2] = [4, -2]$, $\mathbf{v} = \overrightarrow{BC} = [4 - 1, 6 - 0] = [3, 6]$,

$\mathbf{w} = \overrightarrow{AC} = [4 - (-3), 6 - 2] = [7, 4]$, so $\mathbf{u} \cdot \mathbf{v} = 4 \cdot 3 + (-2) \cdot 6 = 12 - 12 = 0 \Rightarrow$

The angle between $\mathbf{u} = \overrightarrow{AB}$ and $\mathbf{v} = \overrightarrow{BC}$ is $90° \Rightarrow \triangle ABC$ is a right triangle.

Note: It is obvious that $\mathbf{v}$ is not orthogonal to $\mathbf{w}$. Why?

31. To show $\triangle ABC$ is right, we need only show one pair of its sides meet at a right angle.

So, we let $\mathbf{u} = \overrightarrow{AB}$, $\mathbf{v} = \overrightarrow{BC}$, and $\mathbf{w} = \overrightarrow{AC}$, then by the definition of ***orthogonal*** given prior to Example 1.16, we need only show $\mathbf{u} \cdot \mathbf{v}$, or $\mathbf{u} \cdot \mathbf{w}$, or $\mathbf{v} \cdot \mathbf{w} = 0$.

Following Example 1.1 of Section 1.1, we calculate the sides of $\triangle ABC$:

$\mathbf{u} = \overrightarrow{AB} = [-3 - 1, 2 - 1, (-2) - (-1)] = [-4, 1, -1]$,

$\mathbf{v} = \overrightarrow{BC} = [2 - (-3), 2 - 2, (-4 - (-2)] = [5, 0, -2]$,

$\mathbf{w} = \overrightarrow{AC} = [2 - 1, 2 - 1, (-4) - (-1)] = [1, 1, -3]$.

Then $\mathbf{u} \cdot \mathbf{w} = (-4) \cdot 1 + 1 \cdot 1 + (-1) \cdot (-3) = -4 + 1 + 3 = 0 \Rightarrow$

The angle between $\mathbf{u} = \overrightarrow{AB}$ and $\mathbf{w} = \overrightarrow{AC}$ is $90° \Rightarrow \triangle ABC$ is a right triangle.

32. Following Example 1.15, we make a similar argument:

The dimensions do not matter, so we consider a cube with sides of length 1.
Also, the cube is symmetric, so we need only consider one diagonal and adjacent edge.

Orient the cube relative to the coordinate axes in $\mathbb{R}^3$, as shown in Figure 1.31.
Take the diagonal to be $[1,1,1]$ and take the adjacent edge to be $[1,0,0]$.

Then the angle θ between these two vectors satisfies:

$$\cos\theta = \frac{1\cdot 1+1\cdot 0+1\cdot 0}{\sqrt{3}\sqrt{1}} = \frac{1}{\sqrt{3}}, \text{ so } \theta = \cos^{-1}\left(\frac{1}{\sqrt{3}}\right) \approx 54.74°.$$

So, the diagonal and adjacent edge meet at 54.74°.

33. Following Example 1.15, we make a similar argument:

The dimensions do not matter, so we consider a cube with sides of length 1.
Also, the cube is symmetric, so we need only consider one pair of diagonals.

Orient the cube relative to the coordinate axes in $\mathbb{R}^3$, as shown in Figure 1.31.
Take the diagonals to be $[1,1,1]$ and $\mathbf{v} = [1,1,-1]$ (from $(1,1,0)$ to $(0,0,1)$).

Then dot product between these two vectors satisfies:

$$\mathbf{u}\cdot\mathbf{v} = 1\cdot 1+1\cdot 1+1\cdot(-1) = 1+1-1 = 1 \neq 0 \Rightarrow$$

The diagonals of a cube are not perpendicular. How might we generalize this result?

34. Following Example 1.17, we compute:

$$\mathbf{u}\cdot\mathbf{v} = \begin{bmatrix}1\\-1\end{bmatrix}\cdot\begin{bmatrix}3\\-1\end{bmatrix} = 4 \text{ and}$$

$$\mathbf{u}\cdot\mathbf{u} = \begin{bmatrix}1\\-1\end{bmatrix}\cdot\begin{bmatrix}1\\-1\end{bmatrix} = 2, \text{ so}$$

$$\text{proj}_{\mathbf{u}}(\mathbf{v}) = \left(\frac{\mathbf{u}\cdot\mathbf{v}}{\mathbf{u}\cdot\mathbf{u}}\right)\mathbf{u} = \frac{4}{2}\begin{bmatrix}1\\-1\end{bmatrix}$$

$$= \begin{bmatrix}2\\-2\end{bmatrix} = 2\mathbf{u}.$$

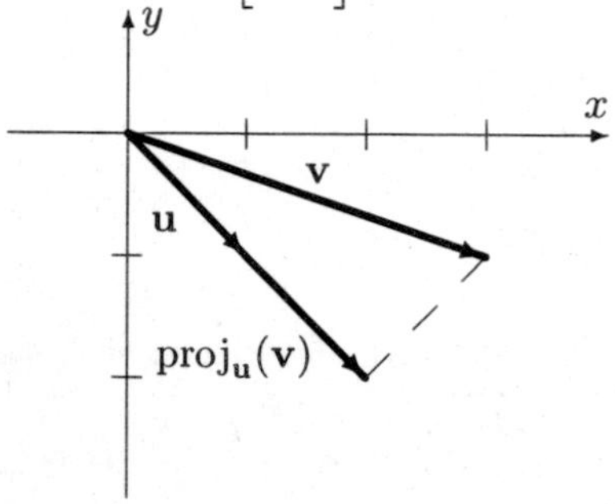

y x v u $\text{proj}_{\mathbf{u}}(\mathbf{v})$

35. Following Example 1.17, we compute:

$$\mathbf{u}\cdot\mathbf{v} = \begin{bmatrix}3/5\\-4/5\end{bmatrix}\cdot\begin{bmatrix}1\\2\end{bmatrix} = -1 \text{ and}$$

$$\mathbf{u}\cdot\mathbf{u} = \begin{bmatrix}3/5\\-4/5\end{bmatrix}\cdot\begin{bmatrix}3/5\\-4/5\end{bmatrix} = 1, \text{ so}$$

$$\text{proj}_{\mathbf{u}}(\mathbf{v}) = \left(\frac{\mathbf{u}\cdot\mathbf{v}}{\mathbf{u}\cdot\mathbf{u}}\right)\mathbf{u} = \frac{-1}{1}\begin{bmatrix}3/5\\-4/5\end{bmatrix}$$

$$= \begin{bmatrix}-3/5\\4/5\end{bmatrix} = -\mathbf{u}.$$

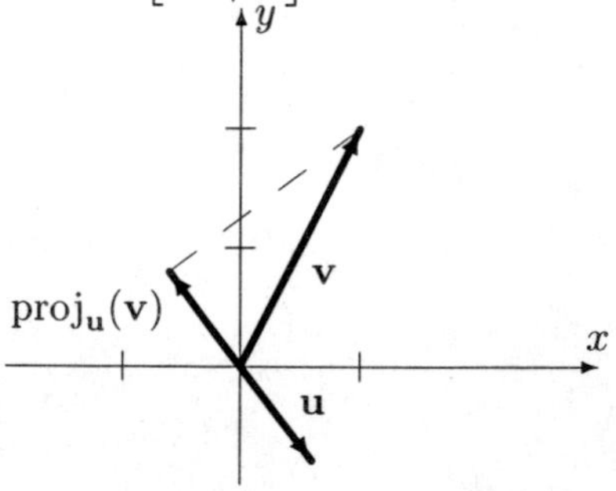

y $\text{proj}_{\mathbf{u}}(\mathbf{v})$ v x u

36. Following Example 1.17, we compute:

$$\mathbf{u}\cdot\mathbf{v} = \begin{bmatrix} 2/3 \\ -2/3 \\ -1/3 \end{bmatrix} \cdot \begin{bmatrix} 2 \\ -2 \\ 2 \end{bmatrix} = 2 \text{ and}$$

$$\mathbf{u}\cdot\mathbf{u} = \begin{bmatrix} 2/3 \\ -2/3 \\ -1/3 \end{bmatrix} \cdot \begin{bmatrix} 2/3 \\ -2/3 \\ -1/3 \end{bmatrix} = 1, \text{ so}$$

$$\operatorname{proj}_{\mathbf{u}}(\mathbf{v}) = \left(\frac{\mathbf{u}\cdot\mathbf{v}}{\mathbf{u}\cdot\mathbf{u}}\right)\mathbf{u} = \frac{2}{1}\begin{bmatrix} 2/3 \\ -2/3 \\ -1/3 \end{bmatrix}$$

$$= \begin{bmatrix} 4/3 \\ -4/3 \\ -2/3 \end{bmatrix} = 2\mathbf{u}.$$

37. Following Example 1.17, we compute:

$$\mathbf{u}\cdot\mathbf{v} = \begin{bmatrix} 1 \\ -1 \\ 1 \\ -1 \end{bmatrix} \cdot \begin{bmatrix} 2 \\ -3 \\ -1 \\ -2 \end{bmatrix} = 6 \text{ and}$$

$$\mathbf{u}\cdot\mathbf{u} = \begin{bmatrix} 1 \\ -1 \\ 1 \\ -1 \end{bmatrix} \cdot \begin{bmatrix} 1 \\ -1 \\ 1 \\ -1 \end{bmatrix} = 4, \text{ so}$$

$$\operatorname{proj}_{\mathbf{u}}(\mathbf{v}) = \left(\frac{\mathbf{u}\cdot\mathbf{v}}{\mathbf{u}\cdot\mathbf{u}}\right)\mathbf{u} = \frac{6}{4}\begin{bmatrix} 1 \\ -1 \\ 1 \\ -1 \end{bmatrix}$$

$$= \begin{bmatrix} 3/2 \\ -3/2 \\ 3/2 \\ -3/2 \end{bmatrix} = \frac{3}{2}\mathbf{u}.$$

38. Following Example 1.17, we compute:

$$\mathbf{u}\cdot\mathbf{v} = \begin{bmatrix} 0.5 \\ 1.5 \end{bmatrix} \cdot \begin{bmatrix} 2.1 \\ 1.2 \end{bmatrix} = 2.85 \text{ and}$$

$$\mathbf{u}\cdot\mathbf{u} = \begin{bmatrix} 0.5 \\ 1.5 \end{bmatrix} \cdot \begin{bmatrix} 0.5 \\ 1.5 \end{bmatrix} = 2.5, \text{ so}$$

$$\operatorname{proj}_{\mathbf{u}}(\mathbf{v}) = \left(\frac{\mathbf{u}\cdot\mathbf{v}}{\mathbf{u}\cdot\mathbf{u}}\right)\mathbf{u} = \frac{2.85}{2.5}\begin{bmatrix} 0.5 \\ 1.5 \end{bmatrix}$$

$$= \begin{bmatrix} 0.57 \\ 1.71 \end{bmatrix} = 1.14\mathbf{u}.$$

39. Following Example 1.17, we compute:

$$\mathbf{u}\cdot\mathbf{v} = \begin{bmatrix} 3.01 \\ -0.33 \\ 2.52 \end{bmatrix} \cdot \begin{bmatrix} 1.34 \\ 4.25 \\ -1.66 \end{bmatrix} = -1.5523 \text{ and}$$

$$\mathbf{u}\cdot\mathbf{u} = \begin{bmatrix} 3.01 \\ -0.33 \\ 2.52 \end{bmatrix} \cdot \begin{bmatrix} 3.01 \\ -0.33 \\ 2.52 \end{bmatrix} = 15.5194, \text{ so}$$

$$\operatorname{proj}_{\mathbf{u}}(\mathbf{v}) = \left(\frac{\mathbf{u}\cdot\mathbf{v}}{\mathbf{u}\cdot\mathbf{u}}\right)\mathbf{u} = \frac{-1.5523}{15.5194}\begin{bmatrix} 3.01 \\ -0.33 \\ 2.52 \end{bmatrix}$$

$$\approx \begin{bmatrix} -0.301 \\ 0.033 \\ -0.252 \end{bmatrix} \approx -\frac{1}{10}\mathbf{u}.$$

40. Let $\mathbf{u} = \overrightarrow{AB} = \begin{bmatrix} 2-1 \\ 2-(-1) \end{bmatrix} = \begin{bmatrix} 1 \\ 3 \end{bmatrix}$ and $\mathbf{v} = \overrightarrow{AC} = \begin{bmatrix} 4-1 \\ 0-(-1) \end{bmatrix} = \begin{bmatrix} 3 \\ 1 \end{bmatrix}$.

(a) We compute the necessary values ...

$$\mathbf{u} \cdot \mathbf{v} = \begin{bmatrix} 1 \\ 3 \end{bmatrix} \cdot \begin{bmatrix} 3 \\ 1 \end{bmatrix} = 6,$$

$$\mathbf{u} \cdot \mathbf{u} = \begin{bmatrix} 1 \\ 3 \end{bmatrix} \cdot \begin{bmatrix} 1 \\ 3 \end{bmatrix} = 10 \left(\|\mathbf{u}\| = \sqrt{10}\right),$$

$$\text{proj}_{\mathbf{u}}(\mathbf{v}) = \left(\frac{\mathbf{u} \cdot \mathbf{v}}{\mathbf{u} \cdot \mathbf{u}}\right) \mathbf{u} = \begin{bmatrix} 3/5 \\ 9/5 \end{bmatrix} \Rightarrow$$

$$\mathbf{v} - \text{proj}_{\mathbf{u}}(\mathbf{v}) = \begin{bmatrix} 12/5 \\ -4/5 \end{bmatrix} \Rightarrow$$

$$\|\mathbf{v} - \text{proj}_{\mathbf{u}}(\mathbf{v})\| = \sqrt{\left(\tfrac{12}{5}\right)^2 + \left(-\tfrac{4}{5}\right)^2} = \tfrac{4\sqrt{10}}{5}$$

... then substitute into the formula for $\mathcal{A}$:

$$\mathcal{A} = \tfrac{1}{2} \|\mathbf{u}\| \, \|\mathbf{v} - \text{proj}_{\mathbf{u}}(\mathbf{v})\|$$

$$= \tfrac{1}{2} \sqrt{10} \; \tfrac{4\sqrt{10}}{5} = 4.$$

(b) We compute the necessary values ...

$$\mathbf{u} \cdot \mathbf{v} = \begin{bmatrix} 1 \\ 3 \end{bmatrix} \cdot \begin{bmatrix} 3 \\ 1 \end{bmatrix} = 6,$$

$$\|\mathbf{u}\| = \sqrt{1^2 + 3^2} = \sqrt{10},$$

$$\|\mathbf{v}\| = \sqrt{3^2 + 1^2} = \sqrt{10} \Rightarrow$$

$$\cos\theta = \frac{\mathbf{u} \cdot \mathbf{v}}{\|\mathbf{u}\| \, \|\mathbf{v}\|} = \frac{6}{\sqrt{10}\sqrt{10}} = \frac{3}{5} \Rightarrow$$

$$\sin\theta = \sqrt{1 - \cos^2\theta} = \sqrt{1 - \left(\tfrac{3}{5}\right)^2} = \tfrac{4}{5}$$

... then substitute into the formula for $\mathcal{A}$:

$$\mathcal{A} = \tfrac{1}{2} \|\mathbf{u}\| \, \|\mathbf{v}\| \sin\theta$$

$$= \tfrac{1}{2} \sqrt{10} \sqrt{10} \; \tfrac{4}{5} = 4.$$

41. Let $\mathbf{u} = \overrightarrow{AB} = \begin{bmatrix} 4-3 \\ -2-(-1) \\ 6-4 \end{bmatrix} = \begin{bmatrix} 1 \\ -1 \\ 2 \end{bmatrix}$ and $\mathbf{v} = \overrightarrow{AC} = \begin{bmatrix} 5-3 \\ 0-(-1) \\ 2-4 \end{bmatrix} = \begin{bmatrix} 2 \\ 1 \\ -2 \end{bmatrix}$.

(a) We compute the necessary values ...

$$\mathbf{u}\cdot\mathbf{v} = \begin{bmatrix} 1 \\ -1 \\ 2 \end{bmatrix} \cdot \begin{bmatrix} 2 \\ 1 \\ -2 \end{bmatrix} = -3,$$

$$\mathbf{u}\cdot\mathbf{u} = \begin{bmatrix} 1 \\ -1 \\ 2 \end{bmatrix} \cdot \begin{bmatrix} 1 \\ -1 \\ 2 \end{bmatrix} = 6 \left(\|\mathbf{u}\| = \sqrt{6}\right),$$

$$\text{proj}_{\mathbf{u}}(\mathbf{v}) = \left(\frac{\mathbf{u}\cdot\mathbf{v}}{\mathbf{u}\cdot\mathbf{u}}\right)\mathbf{u} = \begin{bmatrix} -1/2 \\ 1/2 \\ -1 \end{bmatrix} \Rightarrow$$

$$\mathbf{v} - \text{proj}_{\mathbf{u}}(\mathbf{v}) = \begin{bmatrix} 5/2 \\ 1/2 \\ -1 \end{bmatrix} \Rightarrow$$

$$\|\mathbf{v} - \text{proj}_{\mathbf{u}}(\mathbf{v})\| = \sqrt{\left(\tfrac{5}{2}\right)^2 + \left(\tfrac{1}{2}\right)^2 + (-1)^2}$$

$$= \tfrac{\sqrt{30}}{2}$$

... then substitute into the formula for $\mathcal{A}$:

$$\mathcal{A} = \tfrac{1}{2}\|\mathbf{u}\|\,\|\mathbf{v} - \text{proj}_{\mathbf{u}}(\mathbf{v})\|$$

$$= \tfrac{1}{2}\left(\sqrt{6}\right)\left(\tfrac{\sqrt{30}}{2}\right) = \tfrac{3\sqrt{5}}{2}.$$

(b) We compute the necessary values ...

$$\mathbf{u}\cdot\mathbf{v} = \begin{bmatrix} 1 \\ -1 \\ 2 \end{bmatrix} \cdot \begin{bmatrix} 2 \\ 1 \\ -2 \end{bmatrix} = -3,$$

$$\|\mathbf{u}\| = \sqrt{1^2 + (-1)^2 + 2^2} = \sqrt{6},$$

$$\|\mathbf{v}\| = \sqrt{2^2 + 1^2 + (-2)^2} = 3 \Rightarrow$$

$$\cos\theta = \frac{\mathbf{u}\cdot\mathbf{v}}{\|\mathbf{u}\|\,\|\mathbf{v}\|} = \frac{-3}{3\sqrt{6}} = -\frac{\sqrt{6}}{6} \Rightarrow$$

$$\sin\theta = \sqrt{1-\cos^2\theta} = \sqrt{1 - \left(\tfrac{-\sqrt{6}}{6}\right)^2} = \tfrac{\sqrt{30}}{6}$$

... then substitute into the formula for $\mathcal{A}$:

$$\mathcal{A} = \tfrac{1}{2}\|\mathbf{u}\|\,\|\mathbf{v}\|\sin\theta$$

$$= \tfrac{1}{2}\left(\sqrt{6}\right)(3)\left(\tfrac{\sqrt{30}}{6}\right) = \tfrac{3\sqrt{5}}{2}.$$

42. Two vectors $\mathbf{u}$ and $\mathbf{v}$ are orthogonal ***if and only if*** [$\Leftrightarrow$] their dot product is zero. That is $\mathbf{u} \cdot \mathbf{v} = 0$. So, we set $\mathbf{u} \cdot \mathbf{v} = 0$ and solve for k:

$$\mathbf{u} \cdot \mathbf{v} = \begin{bmatrix} 2 \\ 3 \end{bmatrix} \cdot \begin{bmatrix} k+1 \\ k-1 \end{bmatrix} = 0 \Rightarrow 2(k+1) + 3(k-1) = 0 \Rightarrow 5k - 1 = 0 \Rightarrow k = \tfrac{1}{5}.$$

Substituting k back into the expression for $\mathbf{v}$ we get: $\mathbf{v} = \begin{bmatrix} \frac{1}{5}+1 \\ \frac{1}{5}-1 \end{bmatrix} = \begin{bmatrix} \frac{6}{5} \\ -\frac{4}{5} \end{bmatrix}$.

We check our answer by computing $\mathbf{u} \cdot \mathbf{v}$ (it should be zero):

$$\mathbf{u} \cdot \mathbf{v} = \begin{bmatrix} 2 \\ 3 \end{bmatrix} \cdot \begin{bmatrix} \frac{6}{5} \\ -\frac{4}{5} \end{bmatrix} = \frac{12}{5} - \frac{12}{5} = 0 \text{ as required.}$$

43. Two vectors $\mathbf{u}$ and $\mathbf{v}$ are orthogonal ***if and only if*** [$\Leftrightarrow$] their dot product is zero. That is $\mathbf{u} \cdot \mathbf{v} = 0$. So, we set $\mathbf{u} \cdot \mathbf{v} = 0$ and solve for k:

$$\mathbf{u} \cdot \mathbf{v} = \begin{bmatrix} 1 \\ -1 \\ 2 \end{bmatrix} \cdot \begin{bmatrix} k^2 \\ k \\ -3 \end{bmatrix} = 0 \Rightarrow k^2 - k - 6 = (k+2)(k-3) = 0 \Rightarrow k = -2,\ 3.$$

Substituting k back into the expression for $\mathbf{v}$ we get:

When $k = -2$, $\mathbf{v}_1 = \begin{bmatrix} (-2)^2 \\ -2 \\ -3 \end{bmatrix} = \begin{bmatrix} 4 \\ -2 \\ -3 \end{bmatrix}$. When $k = 3$, $\mathbf{v}_2 = \begin{bmatrix} 3^2 \\ 3 \\ -3 \end{bmatrix} = \begin{bmatrix} 9 \\ 3 \\ -3 \end{bmatrix}$.

We check by computing $\mathbf{u} \cdot \mathbf{v}_1$ and $\mathbf{u} \cdot \mathbf{v}_2$ (they should both be zero):

$$\mathbf{u} \cdot \mathbf{v}_1 = \begin{bmatrix} 1 \\ -1 \\ 2 \end{bmatrix} \cdot \begin{bmatrix} 4 \\ -2 \\ -3 \end{bmatrix} = 4 + 2 - 6 = 0 \text{ and } \mathbf{u} \cdot \mathbf{v_2} = \begin{bmatrix} 1 \\ -1 \\ 2 \end{bmatrix} \cdot \begin{bmatrix} 9 \\ 3 \\ -3 \end{bmatrix} = 9 - 3 - 6 = 0.$$

44. Two vectors $\mathbf{u}$ and $\mathbf{v}$ are orthogonal ***if and only if*** [$\Leftrightarrow$] their dot product is zero. That is $\mathbf{u} \cdot \mathbf{v} = 0$. So, we set $\mathbf{u} \cdot \mathbf{v} = 0$ and solve for y in terms of x:

$$\mathbf{u} \cdot \mathbf{v} = \begin{bmatrix} 3 \\ 1 \end{bmatrix} \cdot \begin{bmatrix} x \\ y \end{bmatrix} = 0 \Rightarrow 3x + y = 0 \Rightarrow y = -3x.$$

Substituting $y = -3x$ back into the expression for $\mathbf{v}$ we get: $\mathbf{v} = \begin{bmatrix} x \\ -3x \end{bmatrix} = x \begin{bmatrix} 1 \\ -3 \end{bmatrix}$.

Conclusion: Any vector orthogonal to $\begin{bmatrix} 3 \\ 1 \end{bmatrix}$ must be a multiple of $\begin{bmatrix} 1 \\ -3 \end{bmatrix}$.

Check: $\mathbf{u} \cdot \mathbf{v} = \begin{bmatrix} 3 \\ 1 \end{bmatrix} \cdot \begin{bmatrix} x \\ -3x \end{bmatrix} = 3x - 3x = 0$ for all values of x.

Note: We could also have solved for x in terms of y yielding $\mathbf{v} = \begin{bmatrix} -\frac{1}{3}y \\ y \end{bmatrix} = y \begin{bmatrix} -\frac{1}{3} \\ 1 \end{bmatrix}$.

45. As noted in the remarks just prior to Example 1.16:

The zero vector $\mathbf{0} = \begin{bmatrix} 0 \\ 0 \end{bmatrix} = \begin{bmatrix} a \\ b \end{bmatrix} \Rightarrow a = b = 0$ is orthogonal to all vectors in $\mathbb{R}^2$.
Having covered this case, we will now assume that at least one of a or $b \neq 0$.

Two vectors $\mathbf{u}$ and $\mathbf{v}$ are orthogonal ***if and only if*** $[\Leftrightarrow]$ their dot product is zero.
That is $\mathbf{u} \cdot \mathbf{v} = 0$. So, we set $\mathbf{u} \cdot \mathbf{v} = 0$ and solve for y in terms of x.

Case 1: $b \neq 0$: $\mathbf{u} \cdot \mathbf{v} = \begin{bmatrix} a \\ b \end{bmatrix} \cdot \begin{bmatrix} x \\ y \end{bmatrix} = 0 \Rightarrow ax + by = 0 \Rightarrow y = -\frac{a}{b}x.$

Substituting $y = -\frac{a}{b}x$ back into the expression for $\mathbf{v}$ we get: $\mathbf{v} = \begin{bmatrix} x \\ -\frac{a}{b}x \end{bmatrix} = x \begin{bmatrix} 1 \\ -\frac{a}{b} \end{bmatrix}.$

If we let $x = b$ we find $b \begin{bmatrix} 1 \\ -\frac{a}{b} \end{bmatrix} = \begin{bmatrix} b \\ -a \end{bmatrix}$, which clarifies the relationship between $\mathbf{u}$ and $\mathbf{v}$.

Case 2: $b = 0\,(\Rightarrow a \neq 0)$: $\mathbf{u} \cdot \mathbf{v} = \begin{bmatrix} a \\ b \end{bmatrix} \cdot \begin{bmatrix} x \\ y \end{bmatrix} = 0 \Rightarrow ax + by = 0 \Rightarrow x = -\frac{b}{a}y.$

Substituting $x = -\frac{b}{a}y$ back into the expression for $\mathbf{v}$ we get: $\mathbf{v} = \begin{bmatrix} -\frac{b}{a}y \\ y \end{bmatrix} = y \begin{bmatrix} -\frac{b}{a} \\ 1 \end{bmatrix}.$

If we let $y = -a$ we find $-a \begin{bmatrix} -\frac{b}{a} \\ 1 \end{bmatrix} = \begin{bmatrix} b \\ -a \end{bmatrix}$ exactly as in *Case 1*.

Conclusion: Any vector orthogonal to $\begin{bmatrix} a \\ b \end{bmatrix} \left(\neq \begin{bmatrix} 0 \\ 0 \end{bmatrix} \right)$ must be a multiple of $\begin{bmatrix} b \\ -a \end{bmatrix}$.

Check: $\mathbf{u} \cdot \mathbf{v} = \begin{bmatrix} a \\ b \end{bmatrix} \cdot \begin{bmatrix} bx \\ -ax \end{bmatrix} = abx - bax = 0$ for all values of x.

46. (a) The geometry of the vectors in Figure 1.26 suggests the following assertion:
For $\|\mathbf{u}+\mathbf{v}\| = \|\mathbf{u}\| + \|\mathbf{v}\|$, $\mathbf{u}$ and $\mathbf{v}$ must point in the same direction.
So, the angle θ between $\mathbf{u}$ and $\mathbf{v}$ must be $0 \Rightarrow \cos\theta = \cos 0 = 1$.

So, we have $\cos\theta = \cos 0 = 1 = \dfrac{\mathbf{u}\cdot\mathbf{v}}{\|\mathbf{u}\|\,\|\mathbf{v}\|} \Rightarrow \mathbf{u}\cdot\mathbf{v} = \|\mathbf{u}\|\|\mathbf{v}\|$.

Emulating the statement of Theorem 1.6 (Pythagoras' Theorem), we state:
For all vectors $\mathbf{u}$ and $\mathbf{v}$ in $\mathbb{R}^2$ and $\mathbb{R}^3$, $\|\mathbf{u}+\mathbf{v}\| = \|\mathbf{u}\| + \|\mathbf{v}\|$
if and only if $\mathbf{u}$ and $\mathbf{v}$ point in the same direction.

Again, we must note that pointing in same direction is equivalent to $\mathbf{u}\cdot\mathbf{v} = \|\mathbf{u}\|\|\mathbf{v}\|$.
Figuring out when to apply this condition will be the key to a successful proof.

PROOF: Proceeding as in the proof of Theorem 1.5 (The Triangle Inequality), we note:

Since both sides of the equality are nonnegative,
showing that the *square* of the left-hand side, $\|\mathbf{u}+\mathbf{v}\|^2$,
is equal to the *square* of the right hand side, $(\|\mathbf{u}\| + \|\mathbf{v}\|)^2$,
is equivalent to proving the condition.

$$\begin{aligned}
\|\mathbf{u}+\mathbf{v}\|^2 &= \mathbf{u}\cdot\mathbf{u} + 2\,(\mathbf{u}\cdot\mathbf{v}) + \mathbf{v}\cdot\mathbf{v} && \text{By Example 1.9} \\
&= \|\mathbf{u}\|^2 + 2\,(\mathbf{u}\cdot\mathbf{v}) + \|\mathbf{v}\|^2 && \text{By } \mathbf{w}\cdot\mathbf{w} = \|\mathbf{w}\|^2 \text{ for any vector } \mathbf{w} \\
&= \|\mathbf{u}\|^2 + 2\|\mathbf{u}\|\,\|\mathbf{v}\| + \|\mathbf{v}\|^2 && \text{By } \mathbf{u}\cdot\mathbf{v} = \|\mathbf{u}\|\|\mathbf{v}\| \ \textbf{(key condition)} \\
&= (\|\mathbf{u}\| + \|\mathbf{v}\|)^2. && \text{By } x^2 + 2xy + y^2 = (x+y)^2
\end{aligned}$$

(b) Since the left-hand side, $\|\mathbf{u}+\mathbf{v}\|$, is always non-negative,
we have to impose an initial condition of $\|\mathbf{u}\| \geq \|\mathbf{v}\|$.
The geometry of the vectors in Figures 1.26 and 1.30 suggests the following assertion:
For $\|\mathbf{u}+\mathbf{v}\| = \|\mathbf{u}\| - \|\mathbf{v}\|$ $\mathbf{u}$ and $\mathbf{v}$ must point in the opposite directions.
So, the angle θ between $\mathbf{u}$ and $\mathbf{v}$ must be $\pi \Rightarrow \cos\theta = \cos\pi = -1$.

$$\text{So, we have } \cos\theta = \cos\pi = -1 = \frac{\mathbf{u}\cdot\mathbf{v}}{\|\mathbf{u}\|\,\|\mathbf{v}\|} \Rightarrow \mathbf{u}\cdot\mathbf{v} = -\|\mathbf{u}\|\|\mathbf{v}\|.$$

Emulating the statement of Theorem 1.6 (Pythagoras' Theorem), we state:
For all vectors $\mathbf{u}$ and $\mathbf{v}$ in $\mathbb{R}^2$ and $\mathbb{R}^3$, $\|\mathbf{u}+\mathbf{v}\| = \|\mathbf{u}\| - \|\mathbf{v}\|$
if and only if $\mathbf{u}$ and $\mathbf{v}$ point in the opposite directions.

Again, we must note that pointing in same direction is equivalent to $\mathbf{u}\cdot\mathbf{v} = -\|\mathbf{u}\|\|\mathbf{v}\|$.
Figuring out when to apply this condition will be the key to a successful proof.

PROOF: Proceeding as in the proof of Theorem 1.5 (The Triangle Inequality), we note:

Since both sides of the equality are nonnegative,
showing that the *square* of the left-hand side, $\|\mathbf{u}+\mathbf{v}\|^2$,
is equal to the *square* of the right hand side, $(\|\mathbf{u}\| - \|\mathbf{v}\|)^2$,
is equivalent to proving the condition.

$$\begin{aligned}
\|\mathbf{u}+\mathbf{v}\|^2 &= \mathbf{u}\cdot\mathbf{u} + 2\,(\mathbf{u}\cdot\mathbf{v}) + \mathbf{v}\cdot\mathbf{v} && \text{By Example 1.9} \\
&= \|\mathbf{u}\|^2 + 2\,(\mathbf{u}\cdot\mathbf{v}) + \|\mathbf{v}\|^2 && \text{By } \mathbf{w}\cdot\mathbf{w} = \|\mathbf{w}\|^2 \text{ for any vector } \mathbf{w} \\
&= \|\mathbf{u}\|^2 - 2\|\mathbf{u}\|\,\|\mathbf{v}\| + \|\mathbf{v}\|^2 && \text{By } \mathbf{u}\cdot\mathbf{v} = -\|\mathbf{u}\|\|\mathbf{v}\| \text{ (key condition)} \\
&= (\|\mathbf{u}\| - \|\mathbf{v}\|)^2. && \text{By } x^2 - 2xy + y^2 = (x-y)^2
\end{aligned}$$

47. We prove Theorem 1.2(b) by applying the definition of the dot product.

$$\begin{aligned}\mathbf{u}\cdot(\mathbf{v}+\mathbf{w}) &= u_1(v_1+w_1)+u_2(v_2+w_2)+\cdots+u_n(v_n+w_n)\\ &= u_1v_1+u_1w_1+u_2v_2+u_2w_2+\cdots+u_nv_n+u_nw_n\\ &= (u_1v_1+u_2v_2+\cdots+u_nw_n)+(u_1w_1+u_2w_2+\cdots+u_nw_n)\\ &= \mathbf{u}\cdot\mathbf{v}+\mathbf{u}\cdot\mathbf{w}.\end{aligned}$$

48. We prove the three parts of Theorem 1.2(d)
by applying the definition of the dot product and key properties of real numbers.

Part 1: For any vector $\mathbf{u}$, we need to show $\mathbf{u}\cdot\mathbf{u}\geq 0$.
We begin by noting that for any real number x, we have $x^2\geq 0$.
So, $\mathbf{u}\cdot\mathbf{u}=u_1u_1+u_2u_2+\cdots+u_nu_n=u_1^2+u_2^2+\cdots+u_n^2\geq 0$.
Note: $u_1^2+u_2^2+\cdots+u_n^2\geq 0$ because the u_i are real numbers.

Part 2: We need to show if $\mathbf{u}=\mathbf{0}$, then $\mathbf{u}\cdot\mathbf{u}=0$.
We begin by noting that if $\mathbf{u}=\mathbf{0}$, then $u_i=0$ for all i.
If $\mathbf{u}=\mathbf{0}$, then $\mathbf{u}\cdot\mathbf{u}=\mathbf{0}\cdot\mathbf{0}=u_1^2+u_2^2+\cdots+u_n^2=0^2+0^2+\cdots+0^2=0$.

Part 3: We need to show if $\mathbf{u}\cdot\mathbf{u}=0$, then $\mathbf{u}=\mathbf{0}$.
We begin by noting that for any real number x, if $x^2=0$ then $x=0$.
If $\mathbf{u}\cdot\mathbf{u}=u_1u_1+u_2u_2+\cdots+u_nu_n=u_1^2+u_2^2+\cdots+u_n^2=0$
then $u_i^2=0$ for all i which implies $u_i=0$ because the u_i are real numbers.
Therefore, since $u_i=0$ for all i, by definition $\mathbf{u}=\mathbf{0}$.

49. We need to show $\mathrm{d}(\mathbf{u},\mathbf{v})=\|\mathbf{u}-\mathbf{v}\|=\|\mathbf{v}-\mathbf{u}\|=\mathrm{d}(\mathbf{v},\mathbf{u})$.
If we let $c=-1$ in Theorem 1.3(b), then $\|-\mathbf{w}\|=\|\mathbf{w}\|$. We use this **key fact** below.

PROOF:

$\mathrm{d}(\mathbf{u},\mathbf{v})=\|\mathbf{u}-\mathbf{v}\|$	By definition
$=\|-(\mathbf{v}-\mathbf{u})\|$	By the fact that $(x-y)=-(y-x)$
$=\|\mathbf{v}-\mathbf{u}\|$	By $\|-\mathbf{w}\|=\|\mathbf{w}\|$ **(key fact)**
$=\mathrm{d}(\mathbf{v},\mathbf{u})$.	By definition

50. We need to show $\mathrm{d}(\mathbf{u},\mathbf{w})\leq\mathrm{d}(\mathbf{u},\mathbf{v})+\mathrm{d}(\mathbf{v},\mathbf{w})$. That is, $\|\mathbf{u}-\mathbf{w}\|\leq\|\mathbf{u}-\mathbf{v}\|+\|\mathbf{v}-\mathbf{w}\|$.
This follows immediately from Theorem 1.5:

$$\|\mathbf{x}+\mathbf{y}\|\leq\|\mathbf{x}\|+\|\mathbf{y}\|\quad\text{with } \mathbf{x}=\mathbf{u}-\mathbf{v} \text{ and } \mathbf{y}=\mathbf{v}-\mathbf{w}.$$

51. We need to show $\mathrm{d}(\mathbf{u},\mathbf{v})=\|\mathbf{u}-\mathbf{v}\|=0$ if and only if $\mathbf{u}=\mathbf{v}$.
This follows immediately from Theorem 1.3(a): $\|\mathbf{w}\|=0$ if and only if $\mathbf{w}=\mathbf{0}$, with $\mathbf{w}=\mathbf{u}-\mathbf{v}$.

52. We will show $\mathbf{u}\cdot c\mathbf{v}=c(\mathbf{u}\cdot\mathbf{v})$ by applying the definitions.

$$\begin{aligned}\mathbf{u}\cdot c\mathbf{v} &= [u_1,u_2,\ldots,u_n]\cdot[cv_1,cv_2,\ldots,cv_n]=u_1cv_1+u_2cv_2+\cdots+u_ncv_n\\ &= cu_1v_1+cu_2v_2+\cdots+cu_nv_n=c(u_1v_1+u_2v_2+\cdots+u_nv_n)=c(\mathbf{u}\cdot\mathbf{v})\end{aligned}$$

53. We need to show $\|\mathbf{u}-\mathbf{v}\|\geq\|\mathbf{u}\|-\|\mathbf{v}\|$. That is, $\|\mathbf{u}\|\leq\|\mathbf{u}-\mathbf{v}\|+\|\mathbf{v}\|$.
This follows immediately from Theorem 1.5, $\|\mathbf{x}+\mathbf{y}\|\leq\|\mathbf{x}\|+\|\mathbf{y}\|$, with $\mathbf{x}=\mathbf{u}-\mathbf{v}$ and $\mathbf{y}=\mathbf{v}$.

54. If $\mathbf{u}\cdot\mathbf{v} = \mathbf{u}\cdot\mathbf{w}$ it does ***not*** follow that $\mathbf{v} = \mathbf{w}$.

An instructive counterexample is suggested by the remarks just prior to Example 1.16. Since $\mathbf{0}\cdot\mathbf{v} = 0$ for every vector $\mathbf{v}$ in $\mathbb{R}^n$, the zero vector is orthogonal to every vector. So, if $\mathbf{u} = \mathbf{0}$, we know nothing about $\mathbf{v}$ and $\mathbf{w}$ except that they are vectors in $\mathbb{R}^n$.

However, we note that $\mathbf{u}\cdot\mathbf{v} = \mathbf{u}\cdot\mathbf{w}$ implies $\mathbf{u}\cdot\mathbf{v} - \mathbf{u}\cdot\mathbf{w} = \mathbf{u}\cdot(\mathbf{v}-\mathbf{w}) = 0$. So, if $\mathbf{u}\cdot\mathbf{v} = \mathbf{u}\cdot\mathbf{w}$, it ***does*** follow that $\mathbf{v}-\mathbf{w}$ is orthogonal to $\mathbf{u}$.

55. We need to show $(\mathbf{u}+\mathbf{v})\cdot(\mathbf{u}-\mathbf{v}) = \|\mathbf{u}\|^2 - \|\mathbf{v}\|^2$ for all vectors in $\mathbb{R}^n$.
Recall, by the definitions of the dot product and the norm, $\mathbf{w}\cdot\mathbf{w} = \|\mathbf{w}\|^2$.
We apply Theorem 1.2(b) and this **key fact** to complete our *PROOF*:

$$\begin{aligned}(\mathbf{u}+\mathbf{v})\cdot(\mathbf{u}-\mathbf{v}) &= \mathbf{u}\cdot\mathbf{u} - \mathbf{u}\cdot\mathbf{v} + \mathbf{v}\cdot\mathbf{u} - \mathbf{v}\cdot\mathbf{v} && \text{By Theorem 1.2(b)}\\ &= \mathbf{u}\cdot\mathbf{u} - \mathbf{v}\cdot\mathbf{v} && \text{By the fact that } -xy + yx = 0\\ &= \|\mathbf{u}\|^2 - \|\mathbf{v}\|^2. && \text{By the fact that } \mathbf{w}\cdot\mathbf{w} = \|\mathbf{w}\|^2 \textbf{ (key fact)}\end{aligned}$$

56. (a) Let $\mathbf{u}, \mathbf{v} \in \mathbb{R}^n$. Then

$$\begin{aligned}\|\mathbf{u}+\mathbf{v}\|^2 + \|\mathbf{u}-\mathbf{v}\|^2 &= (\mathbf{u}+\mathbf{v})\cdot(\mathbf{u}+\mathbf{v}) + (\mathbf{u}-\mathbf{v})\cdot(\mathbf{u}-\mathbf{v})\\ &= (\mathbf{u}\cdot\mathbf{u} + \mathbf{v}\cdot\mathbf{v} + 2\mathbf{u}\cdot\mathbf{v}) + (\mathbf{u}\cdot\mathbf{u} + \mathbf{v}\cdot\mathbf{v} - 2\mathbf{u}\cdot\mathbf{v})\\ &= (\|\mathbf{u}\|^2 + \|\mathbf{v}\|^2) + 2\mathbf{u}\cdot\mathbf{v} + (\|\mathbf{u}\|^2 + \|\mathbf{v}\|^2) - 2\mathbf{u}\cdot\mathbf{v} = 2\|\mathbf{u}\|^2 + 2\|\mathbf{v}\|^2.\end{aligned}$$

(b)

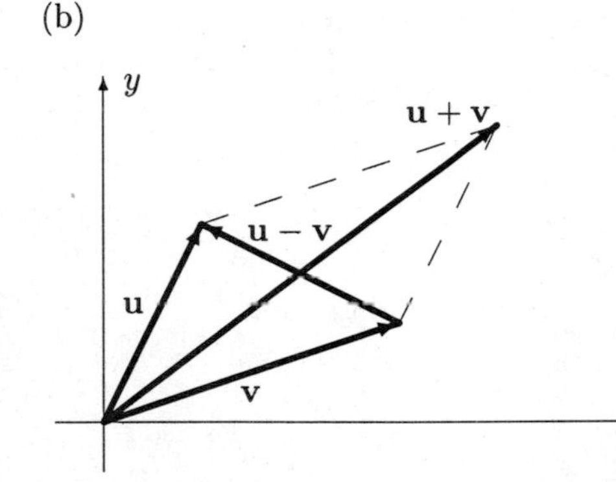

The proof in part (a) tells us that the sum of the of the squares of the lengths of the diagonals of a parallelogram is twice the sum of the squares of the lengths of its sides.

57. Let $\mathbf{u}$, $\mathbf{v} \in \mathbb{R}^n$, and consider $\frac{1}{4}\|\mathbf{u}+\mathbf{v}\|^2 - \frac{1}{4}\|\mathbf{u}-\mathbf{v}\|^2$. By definition, we have:

$$\begin{aligned}\tfrac{1}{4}\|\mathbf{u}+\mathbf{v}\|^2 - \tfrac{1}{4}\|\mathbf{u}-\mathbf{v}\|^2 &= \tfrac{1}{4}[(\mathbf{u}+\mathbf{v})\cdot(\mathbf{u}+\mathbf{v}) + (\mathbf{u}-\mathbf{v})\cdot(\mathbf{u}-\mathbf{v})]\\ &= \tfrac{1}{4}[(\mathbf{u}\cdot\mathbf{u} + \mathbf{v}\cdot\mathbf{v} + 2\mathbf{u}\cdot\mathbf{v}) - (\mathbf{u}\cdot\mathbf{u} + \mathbf{v}\cdot\mathbf{v} - 2\mathbf{u}\cdot\mathbf{v})]\\ &= \tfrac{1}{4}[(\|\mathbf{u}\|^2 - \|\mathbf{u}\|^2) + (\|\mathbf{v}\|^2 - \|\mathbf{v}\|^2) + 4\mathbf{u}\cdot\mathbf{v}] = \mathbf{u}\cdot\mathbf{v}.\end{aligned}$$

58. (a) Let $\mathbf{u}, \mathbf{v} \in \mathbb{R}^n$. Then

$$\begin{aligned}
\|\mathbf{u}+\mathbf{v}\| = \|\mathbf{u}-\mathbf{v}\| &\Leftrightarrow \|\mathbf{u}+\mathbf{v}\|^2 = \|\mathbf{u}-\mathbf{v}\|^2 \\
&\Leftrightarrow \|\mathbf{u}+\mathbf{v}\|^2 - \|\mathbf{u}-\mathbf{v}\|^2 = 0 \\
&\Leftrightarrow \tfrac{1}{4}\|\mathbf{u}+\mathbf{v}\|^2 - \tfrac{1}{4}\|\mathbf{u}-\mathbf{v}\|^2 = 0 \\
&\Leftrightarrow \mathbf{u} \text{ and } \mathbf{v} \text{ are orthogonal.}
\end{aligned}$$

(b)

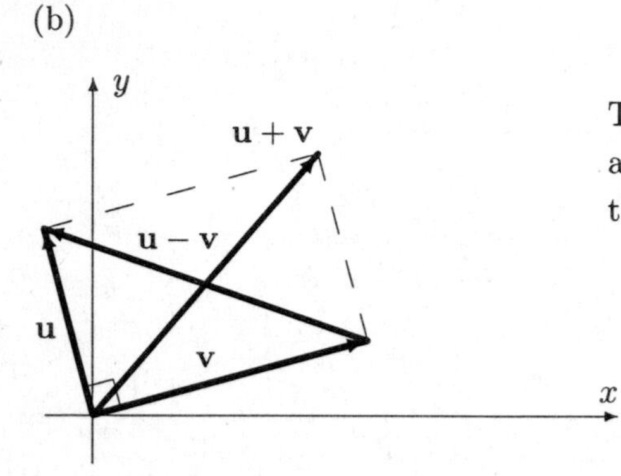

The proof in part (a) tells us that a parallelogram is a rectangle if and only if the lengths of its diagonals are equal.

59. (a) We need to show $(\mathbf{u}+\mathbf{v}) \cdot (\mathbf{u}-\mathbf{v}) = 0$ in $\mathbb{R}^n$ if and only if $\|\mathbf{u}\| = \|\mathbf{v}\|$.
By Exercise 55, $(\mathbf{u}+\mathbf{v}) \cdot (\mathbf{u}-\mathbf{v}) = \|\mathbf{u}\|^2 - \|\mathbf{v}\|^2$.
Therefore, $(\mathbf{u}+\mathbf{v}) \cdot (\mathbf{u}-\mathbf{v}) = \|\mathbf{u}\|^2 - \|\mathbf{v}\|^2 = 0$ if and only if $\|\mathbf{u}\|^2 = \|\mathbf{v}\|^2$.
It follows immediately that $\mathbf{u}+\mathbf{v}$ and $\mathbf{u}-\mathbf{v}$ are orthogonal in $\mathbb{R}^n$ if and only if $\|\mathbf{u}\| = \|\mathbf{v}\|$.

(b)

The proof in part (a) tells us that the diagonals of a parallelogram are perpendicular if and only if the lengths of its sides are equal.

60. From Example 1.9 and the fact that $\mathbf{w} \cdot \mathbf{w} = \|\mathbf{w}\|^2$, we have $\|\mathbf{u}+\mathbf{v}\|^2 = \|\mathbf{u}\|^2 + 2(\mathbf{u} \cdot \mathbf{v}) + \|\mathbf{v}\|^2$.
Taking the square root of both sides yields $\|\mathbf{u}+\mathbf{v}\| = \sqrt{\|\mathbf{u}\|^2 + 2(\mathbf{u} \cdot \mathbf{v}) + \|\mathbf{u}\|^2}$.
Substituting in the given values of $\|\mathbf{u}\| = 2$, $\|\mathbf{v}\| = \sqrt{3}$, and $\mathbf{u} \cdot \mathbf{v} = 1$
gives us $\|\mathbf{u}+\mathbf{v}\| = \sqrt{2^2 + 2(1) + \left(\sqrt{3}\right)^2} = \sqrt{4+2+3} = \sqrt{9} = 3$.

61. We need to show $\|\mathbf{u}\| = 1$ and $\|\mathbf{v}\| = 2$ imply $\mathbf{u} \cdot \mathbf{v} \neq 3$.
From Theorem 1.4 (the Cauchy-Schwarz Inequality), we have $|\mathbf{x} \cdot \mathbf{y}| \leq \|\mathbf{x}\| \, \|\mathbf{y}\|$.
Substituting in the given values of $\|\mathbf{u}\| = 1$ and $\|\mathbf{v}\| = 2$ shows $|\mathbf{u} \cdot \mathbf{v}| \leq 2$.
Therefore, $-2 \leq \mathbf{u} \cdot \mathbf{v} \leq 2$. It follows immediately that $\mathbf{u} \cdot \mathbf{v} \neq 3$.

62. (a) Assume that $\mathbf{u}$ is orthogonal to both $\mathbf{v}$ and $\mathbf{w}$, so $\mathbf{u} \cdot \mathbf{v} = \mathbf{u} \cdot \mathbf{w} = 0$.
Then $\mathbf{u} \cdot (\mathbf{v} + \mathbf{w}) = \mathbf{u} \cdot \mathbf{v} + \mathbf{u} \cdot \mathbf{w} = 0 + 0 = 0$, so $\mathbf{u}$ is orthogonal to $\mathbf{v} + \mathbf{w}$.

(b) Assume that $\mathbf{u}$ is orthogonal to both $\mathbf{v}$ and $\mathbf{w}$, so $\mathbf{u} \cdot \mathbf{v} = \mathbf{u} \cdot \mathbf{w} = 0$.
Then $\mathbf{u} \cdot (s\mathbf{v} + t\mathbf{w}) = \mathbf{u} \cdot (s\mathbf{v}) + \mathbf{u} \cdot (t\mathbf{w}) = s\,(\mathbf{u} \cdot \mathbf{v}) + t\,(\mathbf{u} \cdot \mathbf{w}) = s\,(0) + t\,(0) = 0 + 0 = 0$,
so $\mathbf{u}$ is orthogonal to $s\mathbf{v} + t\mathbf{w}$.

63. Two vectors ($\mathbf{u}$ and $\mathbf{v}$) are orthogonal if their dot product equals zero. So we evaluate:

$$\begin{aligned}
\mathbf{u} \cdot (\mathbf{v} - \operatorname{proj}_{\mathbf{u}}(\mathbf{v})) &= \mathbf{u} \cdot \left(\mathbf{v} - \left(\frac{\mathbf{u} \cdot \mathbf{v}}{\mathbf{u} \cdot \mathbf{u}}\right)\mathbf{u}\right) = \mathbf{u} \cdot \mathbf{v} - \mathbf{u} \cdot \left(\frac{\mathbf{u} \cdot \mathbf{v}}{\mathbf{u} \cdot \mathbf{u}}\right)\mathbf{u} \\
&= \mathbf{u} \cdot \mathbf{v} - \left(\frac{\mathbf{u} \cdot \mathbf{v}}{\mathbf{u} \cdot \mathbf{u}}\right)(\mathbf{u} \cdot \mathbf{u}) = \mathbf{u} \cdot \mathbf{v} - \mathbf{u} \cdot \mathbf{v} = 0.
\end{aligned}$$

64. (a) $\operatorname{proj}_{\mathbf{u}}(\operatorname{proj}_{\mathbf{u}}(\mathbf{v})) = \operatorname{proj}_{\mathbf{u}}\left(\frac{\mathbf{u} \cdot \mathbf{v}}{\mathbf{u} \cdot \mathbf{u}}\mathbf{u}\right) = \left(\frac{\mathbf{u} \cdot \mathbf{v}}{\mathbf{u} \cdot \mathbf{u}}\right)\operatorname{proj}_{\mathbf{u}}(\mathbf{u}) = \left(\frac{\mathbf{u} \cdot \mathbf{v}}{\mathbf{u} \cdot \mathbf{u}}\right)\mathbf{u} = \operatorname{proj}_{\mathbf{u}}(\mathbf{v})$.

(b)
$$\begin{aligned}
\operatorname{proj}_{\mathbf{u}}(\mathbf{v} - \operatorname{proj}_{\mathbf{u}}(\mathbf{v})) &= \operatorname{proj}_{\mathbf{u}}\left(\mathbf{v} - \frac{\mathbf{u} \cdot \mathbf{v}}{\mathbf{u} \cdot \mathbf{u}}\mathbf{u}\right) = \left(\frac{\mathbf{u} \cdot \mathbf{v}}{\mathbf{u} \cdot \mathbf{u}}\right)\mathbf{u} - \left(\frac{\mathbf{u} \cdot \mathbf{v}}{\mathbf{u} \cdot \mathbf{u}}\right)\left(\frac{\mathbf{u} \cdot \mathbf{u}}{\mathbf{u} \cdot \mathbf{u}}\right)\mathbf{u} \\
&= \left(\frac{\mathbf{u} \cdot \mathbf{v}}{\mathbf{u} \cdot \mathbf{u}}\right)\mathbf{u} - \left(\frac{\mathbf{u} \cdot \mathbf{v}}{\mathbf{u} \cdot \mathbf{u}}\right)\mathbf{u} = 0.
\end{aligned}$$

(c)

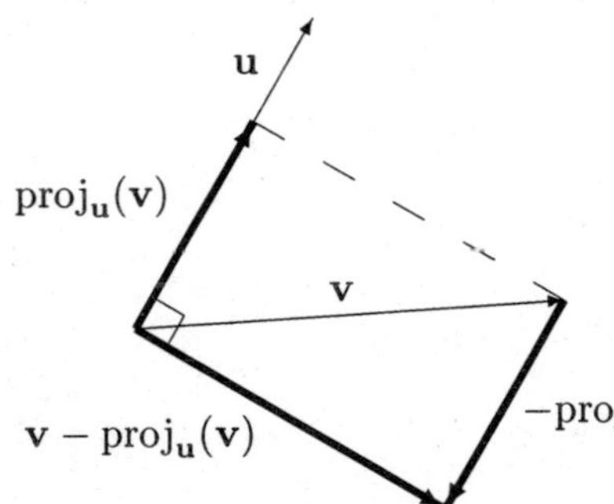

From the diagram, we see that $\operatorname{proj}_{\mathbf{u}}(\mathbf{v}) \parallel \mathbf{u}$,
so $\operatorname{proj}_{\mathbf{u}}(\operatorname{proj}_{\mathbf{u}}(\mathbf{v})) = \operatorname{proj}_{\mathbf{u}}(\mathbf{v})$.
Also, $(\mathbf{v} - \operatorname{proj}_{\mathbf{u}}(\mathbf{v})) \perp \mathbf{u}$,
so $\operatorname{proj}_{\mathbf{u}}(\mathbf{v} - \operatorname{proj}_{\mathbf{u}}(\mathbf{v})) = 0$.

65. (a) The Cauchy-Schwarz Inequality tells us $|\mathbf{u}\cdot\mathbf{v}| \le \|\mathbf{u}\|\|\mathbf{v}\|$.

Squaring both sides, we get $|\mathbf{u}\cdot\mathbf{v}|^2 \le \|\mathbf{u}\|^2\|\mathbf{v}\|^2$.

In $\mathbb{R}^2$ with $\mathbf{u} = \begin{bmatrix} u_1 \\ u_2 \end{bmatrix}$ and $\mathbf{v} = \begin{bmatrix} v_1 \\ v_2 \end{bmatrix}$, this becomes $(u_1v_1 + u_2v_2)^2 \le (u_1^2 + u_2^2)(v_1^2 + v_2^2) \Leftrightarrow$

$0 \le (u_1^2 + u_2^2)(v_1^2 + v_2^2) - (u_1v_1 + u_2v_2)^2 \Leftrightarrow 0 \le u_1^2v_2^2 + u_2^2v_1^2 - 2u_1u_2v_1v_2 \Leftrightarrow$

$0 \le \frac{1}{2}(u_1v_2 - u_2v_1)^2 + \frac{1}{2}(u_2v_1 - u_1v_2)^2$.

Since the final statement is true, all the statements are true.

(b) Let $\mathbf{u}$ and $\mathbf{v}$ be elements of $\mathbb{R}^3$. Then $|\mathbf{u}\cdot\mathbf{v}|^2 \le \|\mathbf{u}\|^2\|\mathbf{v}\|^2 \Leftrightarrow$

$(u_1v_1 + u_2v_2 + u_3v_3)^2 \le (u_1^2 + u_2^2 + u_3^2)(v_1^2 + v_2^2 + v_3^2) \Leftrightarrow$

$0 \le (u_1^2 + u_2^2 + u_3^2)(v_1^2 + v_2^2 + v_3^2) - (u_1v_1 + u_2v_2 + u_3v_3)^2 \Leftrightarrow$

$0 \le u_1^2v_2^2 + u_1^2v_3^2 + u_2^2v_1^2 + u_2^2v_3^2 + u_3^2v_1^2 + u_3^2v_2^2 - 2u_1v_1u_2v_2 - 2u_1v_1u_3v_3 - 2u_2v_2u_3v_3 \Leftrightarrow$

$$0 \le \tfrac{1}{2}(u_1v_2 - u_2v_1)^2 + \tfrac{1}{2}(u_2v_1 - u_1v_2)^2 + \tfrac{1}{2}(u_1v_3 - u_3v_1)^2 + \tfrac{1}{2}(u_3v_1 - u_1v_3)^2 + \tfrac{1}{2}(u_2v_3 - u_3v_2)^2 + \tfrac{1}{2}(u_3v_2 - u_2v_3)^2.$$

Since the final statement is true, all the statements are true.

66. $$\|\text{proj}_{\mathbf{u}}(\mathbf{v})\| = \left\|\left(\frac{\mathbf{u}\cdot\mathbf{v}}{\mathbf{u}\cdot\mathbf{u}}\right)\mathbf{u}\right\| = \left\|\frac{\|\mathbf{u}\|\|\mathbf{v}\|\cos\theta}{\|\mathbf{u}\|\mathbf{u}\|}\right\|\|\mathbf{u}\| = \frac{\|\mathbf{u}\|\|\mathbf{v}\|}{\|\mathbf{u}\|\|\mathbf{u}\|}\|\mathbf{u}\||\cos\theta|$$
$$= \|\mathbf{v}\|\,|\cos\theta| \le \|\mathbf{v}\| \text{ (since } |\cos\theta| \le 1).$$

67. We have $\text{proj}_{\mathbf{u}}(\mathbf{v}) = c\mathbf{u}$. From the figure, we see that $\cos\theta = \dfrac{c\|\mathbf{u}\|}{\|\mathbf{v}\|}$, so $\mathbf{u}\cdot\mathbf{v} = \|\mathbf{u}\|\|\mathbf{v}\|\dfrac{c\|\mathbf{u}\|}{\|\mathbf{v}\|}$ which we solve for c to get $c = \dfrac{\mathbf{u}\cdot\mathbf{v}}{\|\mathbf{u}\|^2}$. Thus, $\text{proj}_{\mathbf{u}}(\mathbf{v}) = \left(\dfrac{\mathbf{u}\cdot\mathbf{v}}{\mathbf{u}\cdot\mathbf{u}}\right)\mathbf{u}$ since $\|\mathbf{u}\|^2 = \mathbf{u}\cdot\mathbf{u}$.

68. ***Proof by Induction***

See Appendix B for further discussion and examples of *Mathematical Induction.*
Also see Chapter 3 of this Study Guide.

Using induction, we will prove the generalized Triangle Inequality:

$$\|\mathbf{v}_1 + \mathbf{v}_2 + \ldots + \mathbf{v}_n\| \leq \|\mathbf{v}_1\| + \|\mathbf{v}_2\| + \ldots + \|\mathbf{v}_n\| \text{ for all } n \geq 1.$$

1: $\|\mathbf{v}_1\| \leq \|\mathbf{v}_1\|$
This is obvious, so there is nothing to show.

k: $\|\mathbf{v}_1 + \mathbf{v}_2 + \ldots + \mathbf{v}_k\| \leq \|\mathbf{v}_1\| + \|\mathbf{v}_2\| + \ldots + \|\mathbf{v}_k\|$
This is the induction hypothesis, so there is nothing to show.

$k+1$: $\|\mathbf{v}_1 + \mathbf{v}_2 + \ldots + \mathbf{v}_k + \mathbf{v}_{k+1}\| \leq \|\mathbf{v}_1\| + \|\mathbf{v}_2\| + \ldots + \|\mathbf{v}_k\| + \|\mathbf{v}_{k+1}\|$
This is the statement we must prove using the induction hypothesis.
The Triangle inequality, $\|\mathbf{u} + \mathbf{v}\| \leq \|\mathbf{u}\| + \|\mathbf{v}\|$, will also be key.

Let $\mathbf{v}_k + \mathbf{v}_{k+1}$ by the kth vector in the induction hypothesis. Then:

$$\|\mathbf{v}_1 + \mathbf{v}_2 + \ldots + \mathbf{v}_k + \mathbf{v}_{k+1}\| \overset{\substack{\text{by}\\\text{induction}}}{\leq} \|\mathbf{v}_1\| + \|\mathbf{v}_2\| + \ldots + \|\mathbf{v}_k + \mathbf{v}_{k+1}\|$$

Applying The Triangle Inequality, we have $\|\mathbf{v}_k + \mathbf{v}_{k+1}\| \leq \|\mathbf{v}_k\| + \|\mathbf{v}_{k+1}\|$. So:

$$\|\mathbf{v}_1 + \mathbf{v}_2 + \ldots + \mathbf{v}_k + \mathbf{v}_{k+1}\| \overset{\substack{\text{by}\\\text{Tri Inq}}}{\leq} \|\mathbf{v}_1\| + \|\mathbf{v}_2\| + \ldots + \|\mathbf{v}_k\| + \|\mathbf{v}_{k+1}\|$$

We have proven (by induction) that

$$\|\mathbf{v}_1 + \mathbf{v}_2 + \ldots + \mathbf{v}_n\| \leq \|\mathbf{v}_1\| + \|\mathbf{v}_2\| + \ldots + \|\mathbf{v}_n\| \text{ for all } n \geq 1.$$

Induction can seem a little bit like *magic* at first glance.
Pay close attention to how the induction hypothesis is used to in the proof.

Exploration: Vectors and Geometry

Since Explorations are self-contained, only solutions will be provided.

1. Like Example 1.18: $\mathbf{m} - \mathbf{a} = \overrightarrow{AM} = \frac{1}{3}\overrightarrow{AB} = \frac{1}{3}(\mathbf{b} - \mathbf{a}) \Rightarrow \mathbf{m} = \mathbf{a} + \frac{1}{3}(\mathbf{b} - \mathbf{a}) = \frac{1}{3}(2\mathbf{a} + \mathbf{b})$.

In general, $\mathbf{m} - \mathbf{a} = \overrightarrow{AM} = \frac{1}{n}\overrightarrow{AB} = \frac{1}{n}(\mathbf{b} - \mathbf{a}) \Rightarrow \mathbf{m} = \mathbf{a} + \frac{1}{n}(\mathbf{b} - \mathbf{a}) = \frac{1}{n}((n-1)\mathbf{a} + \mathbf{b})$.

Note as $n \to \infty$, $\mathbf{m} \to \mathbf{a}$.

2. We use the notation that the vector beginning at the origin and ending at the point X is $\mathbf{x}$.

Therefore, from Exercise 1, we have: $\mathbf{p} = \frac{1}{2}(\mathbf{a} + \mathbf{c})$ and $\mathbf{q} = \frac{1}{2}(\mathbf{b} + \mathbf{c}) \Rightarrow$

$\overrightarrow{PQ} = \mathbf{q} - \mathbf{p} = \frac{1}{2}(\mathbf{b} + \mathbf{c}) - \frac{1}{2}(\mathbf{a} + \mathbf{c}) = \frac{1}{2}(\mathbf{b} - \mathbf{a}) = \frac{1}{2}\overrightarrow{AB}$.

3. Draw in $\overrightarrow{AC}$. Then from Exercise 2, we have: $\overrightarrow{PQ} = \frac{1}{2}\overrightarrow{AB} = \overrightarrow{SR}$.

Draw in $\overrightarrow{BD}$. Then from Exercise 2, we have: $\overrightarrow{PS} = \frac{1}{2}\overrightarrow{BD} = \overrightarrow{QR} \Rightarrow$

$PQRS$ is a parallelogram (opposite sides are parallel and congruent).

4. Following the hint, we find $\mathbf{m}$, the point that is two-thirds of the distance from A to P.

From Exercise 1, we have: $\mathbf{p} = \frac{1}{2}(\mathbf{b}+\mathbf{c}) \Rightarrow \mathbf{m} = \frac{1}{3}(2\mathbf{p}+\mathbf{a}) = \frac{1}{3}(2\cdot\frac{1}{2}(\mathbf{b}+\mathbf{c})+\mathbf{a}) = \frac{1}{3}(\mathbf{a}+\mathbf{b}+\mathbf{c})$.

Next, we find $\mathbf{m}'$, the point that is two-thirds of the distance from B to Q.

From Exercise 1, we have: $\mathbf{q} = \frac{1}{2}(\mathbf{a}+\mathbf{c}) \Rightarrow \mathbf{m}' = \frac{1}{3}(2\mathbf{q}+\mathbf{b}) = \frac{1}{3}(2\cdot\frac{1}{2}(\mathbf{a}+\mathbf{c})+\mathbf{b}) = \frac{1}{3}(\mathbf{a}+\mathbf{b}+\mathbf{c})$.

Finally, we find $\mathbf{m}''$, the point that is two-thirds of the distance from C to R.

From Exercise 1, we have: $\mathbf{r} = \frac{1}{2}(\mathbf{a}+\mathbf{b}) \Rightarrow \mathbf{m}'' = \frac{1}{3}(2\mathbf{r}+\mathbf{c}) = \frac{1}{3}(2\cdot\frac{1}{2}(\mathbf{a}+\mathbf{b})+\mathbf{c}) = \frac{1}{3}(\mathbf{a}+\mathbf{b}+\mathbf{c})$.

We have shown $\mathbf{m} = \mathbf{m}' = \mathbf{m}''$. That is, the medians intersect at the centroid, G.

5. We are given $\overrightarrow{AH}$ is orthogonal to $\overrightarrow{BC}$, that is $\overrightarrow{AH} \cdot \overrightarrow{BC} = 0$ and

$\overrightarrow{BH}$ is orthogonal to $\overrightarrow{AC}$, that is $\overrightarrow{BH} \cdot \overrightarrow{AC} = 0$.

We need to show $\overrightarrow{CH}$ is orthogonal to $\overrightarrow{AB}$, that is $\overrightarrow{CH} \cdot \overrightarrow{AB} = 0$.

$$\begin{matrix} \overrightarrow{AH} \cdot \overrightarrow{BC} = 0 \Rightarrow (\mathbf{h} - \mathbf{a}) \cdot (\mathbf{b} - \mathbf{c}) = 0 \\ \overrightarrow{BH} \cdot \overrightarrow{AC} = 0 \Rightarrow (\mathbf{h} - \mathbf{b}) \cdot (\mathbf{c} - \mathbf{a}) = 0 \end{matrix} \Rightarrow \begin{matrix} \mathbf{h}\cdot\mathbf{b} - \mathbf{h}\cdot\mathbf{c} - \mathbf{a}\cdot\mathbf{b} + \mathbf{a}\cdot\mathbf{c} = 0 \\ \mathbf{h}\cdot\mathbf{c} - \mathbf{h}\cdot\mathbf{a} - \mathbf{b}\cdot\mathbf{c} + \mathbf{a}\cdot\mathbf{b} = 0 \end{matrix} \Rightarrow$$

$0 = \mathbf{h}\cdot\mathbf{b} - \mathbf{h}\cdot\mathbf{a} - \mathbf{c}\cdot\mathbf{b} + \mathbf{a}\cdot\mathbf{c} = (\mathbf{h} - \mathbf{c}) \cdot (\mathbf{b} - \mathbf{a}) = \overrightarrow{CH} \cdot \overrightarrow{AB} = 0 \Rightarrow$

$\overrightarrow{CH}$ is orthogonal to $\overrightarrow{AB}$, so all the altitudes intersect at the orthocenter, H.

6. We are given $\overrightarrow{QK}$ is orthogonal to $\overrightarrow{AC}$, that is $\overrightarrow{QK} \cdot \overrightarrow{AC} = 0$ and
$\overrightarrow{PK}$ is orthogonal to $\overrightarrow{CB}$, that is $\overrightarrow{PK} \cdot \overrightarrow{CB} = 0$.
We need to show $\overrightarrow{RK}$ is orthogonal to $\overrightarrow{AB}$, that is $\overrightarrow{RK} \cdot \overrightarrow{AB} = 0$.
By Exercise 1, we have $\mathbf{q} = \frac{1}{2}(\mathbf{a} + \mathbf{c})$ and $\mathbf{p} = \frac{1}{2}(\mathbf{b} + \mathbf{c})$. So:

$$\left.\begin{array}{l} \overrightarrow{QK} \cdot \overrightarrow{AC} = 0 \Rightarrow (\mathbf{k} - \mathbf{q}) \cdot (\mathbf{c} - \mathbf{a}) = 0 \Rightarrow (\mathbf{k} - \frac{1}{2}(\mathbf{a} + \mathbf{c})) \cdot (\mathbf{c} - \mathbf{a}) = 0 \\ \overrightarrow{PK} \cdot \overrightarrow{CB} = 0 \Rightarrow (\mathbf{k} - \mathbf{p}) \cdot (\mathbf{b} - \mathbf{c}) = 0 \Rightarrow (\mathbf{k} - \frac{1}{2}(\mathbf{b} + \mathbf{c})) \cdot (\mathbf{b} - \mathbf{c}) = 0 \end{array}\right. \Rightarrow$$

$$\left.\begin{array}{l} \mathbf{k} \cdot \mathbf{c} - \mathbf{k} \cdot \mathbf{a} - \frac{1}{2}\mathbf{a} \cdot \mathbf{c} + \frac{1}{2}\mathbf{a} \cdot \mathbf{a} - \frac{1}{2}\mathbf{c} \cdot \mathbf{c} + \frac{1}{2}\mathbf{a} \cdot \mathbf{c} = 0 \\ \mathbf{k} \cdot \mathbf{b} - \mathbf{k} \cdot \mathbf{c} - \frac{1}{2}\mathbf{b} \cdot \mathbf{b} + \frac{1}{2}\mathbf{b} \cdot \mathbf{c} - \frac{1}{2}\mathbf{c} \cdot \mathbf{b} + \frac{1}{2}\mathbf{c} \cdot \mathbf{c} = 0 \end{array}\right. \Rightarrow$$

$$0 = \mathbf{k} \cdot \mathbf{b} - \mathbf{k} \cdot \mathbf{a} - \frac{1}{2}\mathbf{b} \cdot \mathbf{b} + \frac{1}{2}\mathbf{a} \cdot \mathbf{a} = (\mathbf{k} - \frac{1}{2}(\mathbf{b} + \mathbf{a})) \cdot (\mathbf{b} - \mathbf{a}) = (\mathbf{k} - \mathbf{r}) \cdot (\mathbf{b} - \mathbf{a}) = \overrightarrow{RK} \cdot \overrightarrow{AB} = 0 \Rightarrow$$

$\overrightarrow{RK}$ is orthogonal to $\overrightarrow{AB}$, so all perpendicular bisectors intersect at the circumcenter, K.

7. Let O be the origin, then we have: $\mathbf{b} = -\mathbf{a}$ and $\|\mathbf{a}\|^2 = \|\mathbf{c}\|^2 = r^2$, $r =$ radius of the circle.
We need to show $\overrightarrow{AC}$ is orthogonal to $\overrightarrow{BC}$, that is $\overrightarrow{AC} \cdot \overrightarrow{BC} = 0 \Rightarrow$

$$(\mathbf{c} - \mathbf{a}) \cdot (\mathbf{c} - \mathbf{b}) = (\mathbf{c} - \mathbf{a}) \cdot (\mathbf{c} + \mathbf{a}) = \|\mathbf{c}\|^2 + \mathbf{c} \cdot \mathbf{a} - \|\mathbf{a}\|^2 - \mathbf{a} \cdot \mathbf{c} = (\mathbf{a} \cdot \mathbf{c} - \mathbf{a} \cdot \mathbf{c}) + (r^2 - r^2) = 0 \Rightarrow$$

$\overrightarrow{AC}$ is orthogonal to $\overrightarrow{BC}$, so $\angle ACB$ is a right angle.

8. As in Exercise 5, we find $\mathbf{m}$, the point that is one-half of the distance from P to R.
From Exercise 1, we have: $\mathbf{p} = \frac{1}{2}(\mathbf{a} + \mathbf{b})$ and $\mathbf{r} = \frac{1}{2}(\mathbf{c} + \mathbf{d}) \Rightarrow$
$\mathbf{m} = \frac{1}{2}(\mathbf{p} + \mathbf{r}) = \frac{1}{2}(\frac{1}{2}(\mathbf{a} + \mathbf{b}) + \frac{1}{2}(\mathbf{c} + \mathbf{d})) = \frac{1}{4}(\mathbf{a} + \mathbf{b} + \mathbf{c} + \mathbf{d})$.
Next, we find $\mathbf{m}'$, the point that is one-half of the distance from Q to S.
From Exercise 1, we have: $\mathbf{q} = \frac{1}{2}(\mathbf{b} + \mathbf{c})$ and $\mathbf{s} = \frac{1}{2}(\mathbf{a} + \mathbf{d}) \Rightarrow$
$\mathbf{m}' = \frac{1}{2}(\mathbf{q} + \mathbf{s}) = \frac{1}{2}(\frac{1}{2}(\mathbf{b} + \mathbf{c}) + \frac{1}{2}(\mathbf{a} + \mathbf{d})) = \frac{1}{4}(\mathbf{a} + \mathbf{b} + \mathbf{c} + \mathbf{d})$.
We have shown $\mathbf{m} = \mathbf{m}'$.
That is, PR and QS bisect each other because they intersect at their mutual midpoint.

1.3 Lines and Planes

1. Following Example 1.20, we will:
   (a) find the normal form by substituting into $\mathbf{n}\cdot\mathbf{x}=\mathbf{n}\cdot\mathbf{p}$ and
   (b) find the general form by computing those dot products.

   (a) $\mathbf{n}=\begin{bmatrix}3\\2\end{bmatrix}$, $\mathbf{x}=\begin{bmatrix}x\\y\end{bmatrix}$, and $\mathbf{p}=\begin{bmatrix}0\\0\end{bmatrix}$ $\Rightarrow$ The normal form is $\begin{bmatrix}3\\2\end{bmatrix}\cdot\begin{bmatrix}x\\y\end{bmatrix}=\begin{bmatrix}3\\2\end{bmatrix}\cdot\begin{bmatrix}0\\0\end{bmatrix}=0$.

   (b) $\begin{bmatrix}3\\2\end{bmatrix}\cdot\begin{bmatrix}x\\y\end{bmatrix}=3x+2y$ and $\begin{bmatrix}3\\2\end{bmatrix}\cdot\begin{bmatrix}0\\0\end{bmatrix}=0\Rightarrow$ The general form is $3x+2y=0$.

2. Following Example 1.20, we will:
   (a) find the normal form by substituting into $\mathbf{n}\cdot\mathbf{x}=\mathbf{n}\cdot\mathbf{p}$ and
   (b) find the general form by computing those dot products.

   (a) $\mathbf{n}=\begin{bmatrix}5\\-3\end{bmatrix}$, $\mathbf{x}=\begin{bmatrix}x\\y\end{bmatrix}$, $\mathbf{p}=\begin{bmatrix}1\\2\end{bmatrix}$ $\Rightarrow$ Normal form is $\begin{bmatrix}5\\-3\end{bmatrix}\cdot\begin{bmatrix}x\\y\end{bmatrix}=\begin{bmatrix}5\\-3\end{bmatrix}\cdot\begin{bmatrix}1\\2\end{bmatrix}=-1$.

   (b) $\begin{bmatrix}5\\-3\end{bmatrix}\cdot\begin{bmatrix}x\\y\end{bmatrix}=5x-3y$ and $\begin{bmatrix}5\\-3\end{bmatrix}\cdot\begin{bmatrix}1\\2\end{bmatrix}=0\Rightarrow$ The general form is $5x-3y=-1$.

3. Following Example 1.21, we will:
   (a) find the vector form by substituting into $\mathbf{x}=\mathbf{p}+t\mathbf{d}$ and
   (b) find the parametric form by equating components.

   (a) $\mathbf{x}=\begin{bmatrix}x\\y\end{bmatrix}$, $\mathbf{p}=\begin{bmatrix}1\\0\end{bmatrix}$, and $\mathbf{d}=\begin{bmatrix}-1\\3\end{bmatrix}$ $\Rightarrow$ The vector form is $\begin{bmatrix}x\\y\end{bmatrix}=\begin{bmatrix}1\\0\end{bmatrix}+t\begin{bmatrix}-1\\3\end{bmatrix}$.

   (b) The vector form in (a) implies the parametric form is $\begin{matrix}x=1-t\\y=3t\end{matrix}$.

4. Following Example 1.21, we will:
   (a) find the vector form by substituting into $\mathbf{x}=\mathbf{p}+t\mathbf{d}$ and
   (b) find the parametric form by equating components.

   (a) $\mathbf{x}=\begin{bmatrix}x\\y\end{bmatrix}$, $\mathbf{p}=\begin{bmatrix}-4\\4\end{bmatrix}$, and $\mathbf{d}=\begin{bmatrix}1\\1\end{bmatrix}$ $\Rightarrow$ The vector form is $\begin{bmatrix}x\\y\end{bmatrix}=\begin{bmatrix}-4\\4\end{bmatrix}+t\begin{bmatrix}1\\1\end{bmatrix}$.

   (b) The vector form in (a) implies the parametric form is $\begin{matrix}x=-4+t\\y=\ \ 4+t\end{matrix}$.

5. Following Example 1.21, we will:
(a) find the vector form by substituting into $\mathbf{x} = \mathbf{p} + t\mathbf{d}$ and
(b) find the parametric form by equating components.

(a) $\mathbf{x} = \begin{bmatrix} x \\ y \\ z \end{bmatrix}$, $\mathbf{p} = \begin{bmatrix} 0 \\ 0 \\ 0 \end{bmatrix}$, and $\mathbf{d} = \begin{bmatrix} 1 \\ -1 \\ 4 \end{bmatrix}$ $\Rightarrow$ The vector form is $\begin{bmatrix} x \\ y \\ z \end{bmatrix} = \begin{bmatrix} 0 \\ 0 \\ 0 \end{bmatrix} + t \begin{bmatrix} 1 \\ -1 \\ 4 \end{bmatrix}$.

(b) The vector form in (a) implies the parametric form is $\begin{matrix} x = & t \\ y = & -t \\ z = & 4t \end{matrix}$.

6. Following Example 1.21, we will:
(a) find the vector form by substituting into $\mathbf{x} = \mathbf{p} + t\mathbf{d}$ and
(b) find the parametric form by equating components.

(a) $\mathbf{x} = \begin{bmatrix} x \\ y \\ z \end{bmatrix}$, $\mathbf{p} = \begin{bmatrix} 3 \\ 0 \\ -2 \end{bmatrix}$, and $\mathbf{d} = \begin{bmatrix} 0 \\ 2 \\ 5 \end{bmatrix}$ $\Rightarrow$ The vector form is $\begin{bmatrix} x \\ y \\ z \end{bmatrix} = \begin{bmatrix} 3 \\ 0 \\ -2 \end{bmatrix} + t \begin{bmatrix} 0 \\ 2 \\ 5 \end{bmatrix}$.

(b) The vector form in (a) implies the parametric form is $\begin{matrix} x = & 3 \\ y = & 2t \\ z = & -2 + 5t \end{matrix}$.

7. Following Example 1.23, we will:
(a) find the normal form by substituting into $\mathbf{n} \cdot \mathbf{x} = \mathbf{n} \cdot \mathbf{p}$ and
(b) find the general form by computing those dot products.

(a) $\mathbf{n} = \begin{bmatrix} 3 \\ 2 \\ 1 \end{bmatrix}$, $\mathbf{x} = \begin{bmatrix} x \\ y \\ z \end{bmatrix}$, $\mathbf{p} = \begin{bmatrix} 0 \\ 1 \\ 0 \end{bmatrix}$ $\Rightarrow$ The normal form is $\begin{bmatrix} 3 \\ 2 \\ 1 \end{bmatrix} \cdot \begin{bmatrix} x \\ y \\ z \end{bmatrix} = \begin{bmatrix} 3 \\ 2 \\ 1 \end{bmatrix} \cdot \begin{bmatrix} 0 \\ 1 \\ 0 \end{bmatrix} = 2$.

(b) $\begin{bmatrix} 3 \\ 2 \\ 1 \end{bmatrix} \cdot \begin{bmatrix} x \\ y \\ z \end{bmatrix} = 3x + 2y + z$ and $\begin{bmatrix} 3 \\ 2 \\ 1 \end{bmatrix} \cdot \begin{bmatrix} 0 \\ 1 \\ 0 \end{bmatrix} = 2 \Rightarrow$ The general form is $3x + 2y + z = 2$.

8. Following Example 1.23, we will:
(a) find the normal form by substituting into $\mathbf{n} \cdot \mathbf{x} = \mathbf{n} \cdot \mathbf{p}$ and
(b) find the general form by computing those dot products.

(a) $\mathbf{n} = \begin{bmatrix} 1 \\ -1 \\ 5 \end{bmatrix}$, $\mathbf{x} = \begin{bmatrix} x \\ y \\ z \end{bmatrix}$, $\mathbf{p} = \begin{bmatrix} -3 \\ 5 \\ 1 \end{bmatrix}$ $\Rightarrow$ Normal form $\begin{bmatrix} 1 \\ -1 \\ 5 \end{bmatrix} \cdot \begin{bmatrix} x \\ y \\ z \end{bmatrix} = \begin{bmatrix} 1 \\ -1 \\ 5 \end{bmatrix} \cdot \begin{bmatrix} -3 \\ 5 \\ 1 \end{bmatrix} =$
-3.

(b) $\begin{bmatrix} 1 \\ -1 \\ 5 \end{bmatrix} \cdot \begin{bmatrix} x \\ y \\ z \end{bmatrix} = x - y + 5z$, $\begin{bmatrix} 1 \\ -1 \\ 5 \end{bmatrix} \cdot \begin{bmatrix} -3 \\ 5 \\ 1 \end{bmatrix} = -3 \Rightarrow$ The general form is $x - y + 5z = -3$.

9. Following Example 1.24, we will:
(a) find the vector form by substituting into $\mathbf{x} = \mathbf{p} + s\mathbf{u} + t\mathbf{v}$ and
(b) find the parametric form by equating components.

(a) $\mathbf{x} = \begin{bmatrix} x \\ y \\ z \end{bmatrix}$, $\mathbf{p} = \begin{bmatrix} 0 \\ 0 \\ 0 \end{bmatrix}$, $\mathbf{u} = \begin{bmatrix} 2 \\ 1 \\ 2 \end{bmatrix}$, and $\mathbf{v} = \begin{bmatrix} -3 \\ 2 \\ 1 \end{bmatrix} \Rightarrow$

The vector form is $\begin{bmatrix} x \\ y \\ z \end{bmatrix} = \begin{bmatrix} 0 \\ 0 \\ 0 \end{bmatrix} + s\begin{bmatrix} 2 \\ 1 \\ 2 \end{bmatrix} + t\begin{bmatrix} -3 \\ 2 \\ 1 \end{bmatrix}$.

(b) The vector form in (a) implies the parametric form is $\begin{aligned} x &= 2s - 3t \\ y &= s + 2t \\ z &= 2s + t \end{aligned}$.

10. Following Example 1.24, we will:
(a) find the vector form by substituting into $\mathbf{x} = \mathbf{p} + s\mathbf{u} + t\mathbf{v}$ and
(b) find the parametric form by equating components.

(a) $\mathbf{x} = \begin{bmatrix} x \\ y \\ z \end{bmatrix}$, $\mathbf{p} = \begin{bmatrix} 6 \\ -4 \\ -3 \end{bmatrix}$, $\mathbf{u} = \begin{bmatrix} 0 \\ 1 \\ 1 \end{bmatrix}$, and $\mathbf{v} = \begin{bmatrix} -1 \\ 1 \\ 1 \end{bmatrix} \Rightarrow$

The vector form is $\begin{bmatrix} x \\ y \\ z \end{bmatrix} = \begin{bmatrix} 6 \\ -4 \\ -3 \end{bmatrix} + s\begin{bmatrix} 0 \\ 1 \\ 1 \end{bmatrix} + t\begin{bmatrix} -1 \\ 1 \\ 1 \end{bmatrix}$.

(b) The vector form in (a) implies the parametric form is $\begin{aligned} x &= 6 - t \\ y &= -4 + s + t \\ z &= -3 + s + t \end{aligned}$.

11. Following Example 1.24, we realize we may choose any point on ℓ,
so we will use P (Q would also be fine).

A convenient direction vector is $\mathbf{d} = \overrightarrow{PQ} = \begin{bmatrix} 2 \\ 2 \end{bmatrix}$ (or any scalar multiple of this).

Thus we obtain: $\mathbf{x} = \mathbf{p} + t\mathbf{d}$

$$= \begin{bmatrix} 1 \\ -2 \end{bmatrix} + t\begin{bmatrix} 2 \\ 2 \end{bmatrix}.$$

12. Following Example 1.24, we realize we may choose any point on ℓ,
so we will use P (Q would also be fine).

A convenient direction vector is $\mathbf{d} = \overrightarrow{PQ} = \begin{bmatrix} -2 \\ 0 \\ 4 \end{bmatrix}$ (or any scalar multiple of this).

Thus we obtain: $\mathbf{x} = \mathbf{p} + t\mathbf{d}$

$$= \begin{bmatrix} 0 \\ 1 \\ -1 \end{bmatrix} + t\begin{bmatrix} -2 \\ 0 \\ 4 \end{bmatrix}.$$

13. Following Example 1.24, we realize we need to find two direction vectors, **u** and **v**.
Since $P = (1,1,1)$, $Q = (4,0,2)$, and $R = (0,1,-1)$ lie in plane $\mathscr{P}$, we compute:

$$\mathbf{u} = \overrightarrow{PQ} = \mathbf{q} - \mathbf{p} = \begin{bmatrix} 3 \\ -1 \\ 1 \end{bmatrix} \text{ and } \mathbf{v} = \overrightarrow{PR} = \mathbf{r} - \mathbf{p} = \begin{bmatrix} -1 \\ 0 \\ -2 \end{bmatrix}.$$

Since **u** and **v** are not scalar multiples of each other, they will serve as direction vectors.
If **u** and **v** were scalar multiples of each other, we would not have a plane but simply a line.

Therefore, we have the vector equation of $\mathscr{P}$: $\begin{bmatrix} x \\ y \\ z \end{bmatrix} = \begin{bmatrix} 1 \\ 1 \\ 1 \end{bmatrix} + s \begin{bmatrix} 3 \\ -1 \\ 1 \end{bmatrix} + t \begin{bmatrix} -1 \\ 0 \\ -2 \end{bmatrix}.$

14. Following Example 1.24, we realize we need to find two direction vectors, **u** and **v**.
Since $P = (1,0,0)$, $Q = (0,1,0)$, and $R = (0,0,1)$ lie in plane $\mathscr{P}$, we compute:

$$\mathbf{u} = \overrightarrow{PQ} = \mathbf{q} - \mathbf{p} = \begin{bmatrix} -1 \\ 1 \\ 0 \end{bmatrix} \text{ and } \mathbf{v} = \overrightarrow{PR} = \mathbf{r} - \mathbf{p} = \begin{bmatrix} -1 \\ 0 \\ 1 \end{bmatrix}.$$

Since **u** and **v** are not scalar multiples of each other, they will serve as direction vectors.
If **u** and **v** were scalar multiples of each other, we would not have a plane but simply a line.

Therefore, we have the vector equation of $\mathscr{P}$: $\begin{bmatrix} x \\ y \\ z \end{bmatrix} = \begin{bmatrix} 1 \\ 0 \\ 0 \end{bmatrix} + s \begin{bmatrix} -1 \\ 1 \\ 0 \end{bmatrix} + t \begin{bmatrix} -1 \\ 0 \\ 1 \end{bmatrix}.$

15. The parametric equations and associated vector forms $\mathbf{x} = \mathbf{p} + t\mathbf{d}$ found below are ***not*** unique.

(a) As in the remarks prior to Example 1.20, we begin by letting $x = t$.
When we substitute $x = t$ into $y = 3x - 1$, we get $y = 3(t) - 1$. So, we have the following:

Parametric equations $\begin{aligned} x &= t \\ y &= -1 + 3t \end{aligned}$ and vector form $\begin{bmatrix} x \\ y \end{bmatrix} = \begin{bmatrix} 0 \\ -1 \end{bmatrix} + t \begin{bmatrix} 1 \\ 3 \end{bmatrix}.$

(b) In this case since the coefficient of y is 2, we begin by letting $x = 2t$.
When we substitute $x = 2t$ into $3x + 2y = 5$, we get $3(2t) + 2y = 5$.
Solving for y yields $y = -3t + 2.5$. So, we have the following:

Parametric equations: $\begin{aligned} x &= 2t \\ y &= 2.5 - 3t \end{aligned}$ and vector form $\begin{bmatrix} x \\ y \end{bmatrix} = \begin{bmatrix} 0 \\ 2.5 \end{bmatrix} + t \begin{bmatrix} 2 \\ -3 \end{bmatrix}.$

We discover the following pattern: if line ℓ has equation $ax + by = c$, then $\mathbf{d} = \begin{bmatrix} b \\ -a \end{bmatrix}.$

16. By convention, $\mathbf{u}$ is the vector with its tail at the origin and its head at the point U. So, when $\mathbf{x} = \mathbf{u}$, the line described by $\mathbf{x} = \mathbf{p} + t\mathbf{d}$ passes through the point U.

We note $\mathbf{x} = \mathbf{p} + t\mathbf{d}$ where $\mathbf{d} = \mathbf{q} - \mathbf{p}$ is the line that passes through P and Q, since it passes through P (when $t = 0$) in the direction of $\overrightarrow{PQ}\,(= \mathbf{q} - \mathbf{p})$.

(a) To show $\mathbf{x}$ describes the line segment $\overline{PQ}$ as t varies from 0 to 1, we need to show: when $t = 0$, $\mathbf{x} = \mathbf{p}$ which implies the line described by $\mathbf{x}$ passes through the point P and when $t = 1$, $\mathbf{x} = \mathbf{q}$ which implies the line described by $\mathbf{x}$ passes through the point Q.
When $t = 0$, $\mathbf{x} = \mathbf{p} + 0\,(\mathbf{q} - \mathbf{p}) = \mathbf{p}$.
When $t = 1$, $\mathbf{x} = \mathbf{p} + 1\,(\mathbf{q} - \mathbf{p}) = \mathbf{q}$.
Therefore, $\mathbf{x} = \mathbf{p} + t\,(\mathbf{q} - \mathbf{p})$ describes the line segment $\overline{PQ}$ as t varies from 0 to 1.

(b) As shown in ***Exploration: Vectors and Geometry*** to find the midpoint of $\overline{PQ}$, we start at P and travel half the length of $\overline{PQ}$ in the direction of the vector $\overrightarrow{PQ} = \mathbf{q} - \mathbf{p}$.
In the language of vectors, we add $\frac{1}{2}\overrightarrow{PQ} = \frac{1}{2}\,(\mathbf{q} - \mathbf{p})$ to the vector $\mathbf{p}$.
So, the vector whose head is the midpoint of $\overline{PQ}$ is $\mathbf{p} + \frac{1}{2}\,(\mathbf{q} - \mathbf{p})$.
Equating this to our expression for $\mathbf{x}$ yields: $\mathbf{p} + \frac{1}{2}\,(\mathbf{q} - \mathbf{p}) = \mathbf{p} + t\,(\mathbf{q} - \mathbf{p})$.
It follows immediately that $t = \frac{1}{2}$ and $\mathbf{x} = \mathbf{p} + \frac{1}{2}\,(\mathbf{q} - \mathbf{p}) = \frac{1}{2}\,(\mathbf{p} + \mathbf{q})$.

(c) Given $\mathbf{p} = [2, -3]$, $\mathbf{q} = [0, 1]$, and $\mathbf{x} = \frac{1}{2}\,(\mathbf{p} + \mathbf{q})$, we have:
$\mathbf{x} = \frac{1}{2}\,([0,1] + [2,-3]) = [1,-1]$. So, the midpoint of $\overline{PQ}$ is $(1, -1)$.

(d) Given $\mathbf{p} = [1,0,1]$, $\mathbf{q} = [4,1,-2]$, and $\mathbf{x} = \frac{1}{2}\,(\mathbf{p} + \mathbf{q})$, we have:
$\mathbf{x} = \frac{1}{2}\,([1,0,1] + [4,1,-2]) = \left[\frac{5}{2}, \frac{1}{2}, -\frac{1}{2}\right]$. So, the midpoint of $\overline{PQ}$ is $\left(\frac{5}{2}, \frac{1}{2}, -\frac{1}{2}\right)$.

(e) We want two points (parameterized by t_1, t_2) to split $\overline{PQ}$ into three equal segments. As in (c), the vectors whose heads are one and two-thirds of $\overline{PQ}$ from P are
$\mathbf{p} + \frac{1}{3}\,(\mathbf{q} - \mathbf{p})$ and $\mathbf{p} + \frac{2}{3}\,(\mathbf{q} - \mathbf{p})$.
Equating these to our expression for $\mathbf{x}$ yields:
$\mathbf{p} + \frac{1}{3}\,(\mathbf{q} - \mathbf{p}) = \mathbf{p} + t_1\,(\mathbf{q} - \mathbf{p})$ and $\mathbf{p} + \frac{2}{3}\,(\mathbf{q} - \mathbf{p}) = \mathbf{p} + t_2\,(\mathbf{q} - \mathbf{p})$.
It follows immediately that $t_1 = \frac{1}{3}$ and $t_2 = \frac{2}{3}$, so
$\mathbf{x}_1 = \mathbf{p} + \frac{1}{3}\,(\mathbf{q} - \mathbf{p}) = \frac{1}{3}\,(2\mathbf{p} + \mathbf{q})$ and $\mathbf{x}_2 = \mathbf{p} + \frac{2}{3}\,(\mathbf{q} - \mathbf{p}) = \frac{1}{3}\,(\mathbf{p} + 2\mathbf{q})$.
Given $\mathbf{p} = [2,-3]$, $\mathbf{q} = [0,1]$, $\mathbf{x}_1 = \frac{1}{3}\,(2\mathbf{p} + \mathbf{q})$, and $\mathbf{x}_2 = \frac{1}{3}\,(\mathbf{p} + 2\mathbf{q})$, we have:
$\mathbf{x}_1 = \frac{1}{3}\,(2\,[0,1] + [2,-3]) = \left[\frac{4}{3}, -\frac{5}{3}\right]$ and $\mathbf{x}_2 = \frac{1}{3}\,([0,1] + 2\,[2,-3]) = \left[\frac{2}{3}, -\frac{1}{3}\right]$.
So, the two points that divide $\overline{PQ}$ into three equal parts are $\left(\frac{4}{3}, -\frac{5}{3}\right)$ and $\left(\frac{2}{3}, -\frac{1}{3}\right)$.

(f) Likewise $\mathbf{p} = [1,0,-1]$, $\mathbf{q} = [4,1,-2]$, $\mathbf{x}_1 = \frac{1}{3}\,(2\mathbf{p} + \mathbf{q})$, and $\mathbf{x}_2 = \frac{1}{3}\,(\mathbf{p} + 2\mathbf{q})$, yields:
$\mathbf{x}_1 = \frac{1}{3}\,(2\,[1,0,-1] + [4,1,-2]) = \left[2, \frac{1}{3}, -\frac{4}{3}\right]$, $\mathbf{x}_2 = \frac{1}{3}\,([1,0,-1] + 2\,[4,1,-2]) = \left[3, \frac{2}{3}, -\frac{1}{3}\right]$.
So, the two points that divide $\overline{PQ}$ into three equal parts are $\left(2, \frac{1}{3}, -\frac{4}{3}\right)$ and $\left(3, \frac{2}{3}, -\frac{5}{3}\right)$.

17. Need to show ℓ_1 with slope m_1 is perpendicular to ℓ_2 with slope m_2 if and only if $m_1 m_2 = -1$. By definition, one possible form of the general equation for ℓ_1 with slope m_1 is $-m_1 x + y = b_1$.

So, the normal vector for ℓ_1 is $\mathbf{n}_1 = \begin{bmatrix} -m_1 \\ 1 \end{bmatrix}$ and the normal vector for ℓ_2 is $\mathbf{n}_2 = \begin{bmatrix} -m_2 \\ 1 \end{bmatrix}$.

Now we note ℓ_1 is perpendicular to line ℓ_2 if and only if $\mathbf{n}_1 \cdot \mathbf{n}_2 = 0$, so we have:

$\mathbf{n}_1 \cdot \mathbf{n}_2 = \begin{bmatrix} -m_1 \\ 1 \end{bmatrix} \cdot \begin{bmatrix} -m_1 \\ 1 \end{bmatrix} = m_1 m_2 + 1 = 0$ which implies $m_1 m_2 = -1$ as we were to show.

18. Given $\mathbf{d}$ is the direction vector of line ℓ and $\mathbf{n}$ is the normal vector to the plane $\mathscr{P}$, we have:
If $\mathbf{d}$ and $\mathbf{n}$ are orthogonal which implies $\mathbf{d} \cdot \mathbf{n} = 0$, then line ℓ is parallel to plane $\mathscr{P}$.
If $\mathbf{d}$ and $\mathbf{n}$ are parallel which implies $\mathbf{d} = c\mathbf{n}$ (scalar multiples), then ℓ is perpendicular to $\mathscr{P}$.

(a) Since the general form of $\mathscr{P}$ is $2x + 3y - z = 1$, its normal vector is $\mathbf{n} = \begin{bmatrix} 2 \\ 3 \\ -1 \end{bmatrix} = \mathbf{d}$.

Since $\mathbf{d} = 1\,\mathbf{n}$, ℓ is perpendicular to $\mathscr{P}$.

(b) Since the general form of $\mathscr{P}$ is $4x - y + 5z = 0$, its normal vector is $\mathbf{n} = \begin{bmatrix} 4 \\ -1 \\ 5 \end{bmatrix}$.

Since $\mathbf{d} \cdot \mathbf{n} = \begin{bmatrix} 2 \\ 3 \\ -1 \end{bmatrix} \cdot \begin{bmatrix} 4 \\ -1 \\ 5 \end{bmatrix} = 2 \cdot 4 + 3 \cdot (-1) + (-1) \cdot 5 = 0$, ℓ is parallel to $\mathscr{P}$.

(c) Since the general form of $\mathscr{P}$ is $x - y - z = 3$, its normal vector is $\mathbf{n} = \begin{bmatrix} 1 \\ -1 \\ -1 \end{bmatrix}$.

Since $\mathbf{d} \cdot \mathbf{n} = \begin{bmatrix} 2 \\ 3 \\ -1 \end{bmatrix} \cdot \begin{bmatrix} 1 \\ -1 \\ -1 \end{bmatrix} = 2 \cdot 1 + 3 \cdot (-1) + (-1) \cdot (-1) = 0$, ℓ is parallel to $\mathscr{P}$.

(d) Since the general form of $\mathscr{P}$ is $4x + 6y - 2z = 0$, its normal vector is $\mathbf{n} = \begin{bmatrix} 4 \\ 6 \\ -2 \end{bmatrix}$.

Since $\mathbf{d} = \begin{bmatrix} 2 \\ 3 \\ -1 \end{bmatrix} = \frac{1}{2} \begin{bmatrix} 4 \\ 6 \\ -2 \end{bmatrix} = \frac{1}{2}\mathbf{n}$, ℓ is perpendicular to $\mathscr{P}$.

19. Given $\mathbf{n}_1$ is the normal vector of $\mathscr{P}_1$ and $\mathbf{n}$ is the normal vector of $\mathscr{P}$, we have:
If $\mathbf{n}_1$ and $\mathbf{n}$ are orthogonal which implies $\mathbf{n}_1 \cdot \mathbf{n} = 0$, then $\mathscr{P}_1$ is perpendicular to $\mathscr{P}$.
If $\mathbf{n}_1$ and $\mathbf{n}$ are parallel which implies $\mathbf{n}_1 = c\mathbf{n}$ (scalar multiples), then $\mathscr{P}_1$ is parallel to $\mathscr{P}$.

(a) Since the general form of $\mathscr{P}$ is $2x + 3y - z = 1$, its normal vector is $\mathbf{n} = \begin{bmatrix} 2 \\ 3 \\ -1 \end{bmatrix}$.

Since $\mathbf{n}_1 \cdot \mathbf{n} = \begin{bmatrix} 4 \\ -1 \\ 5 \end{bmatrix} \cdot \begin{bmatrix} 2 \\ 3 \\ -1 \end{bmatrix} = 4 \cdot 2 + (-1) \cdot 3 + 5 \cdot (-1) = 0$, $\mathscr{P}_1$ is perpendicular to $\mathscr{P}$.

(b) Since the general form of $\mathscr{P}$ is $4x - y + 5z = 0$, its normal vector is $\mathbf{n} = \begin{bmatrix} 4 \\ -1 \\ 5 \end{bmatrix}$.

Since $\mathbf{n}_1 = 1\,\mathbf{n}$, $\mathscr{P}_1$ is parallel to $\mathscr{P}$.

(c) Since the general form of $\mathscr{P}$ is $x - y - z = 3$, its normal vector is $\mathbf{n} = \begin{bmatrix} 1 \\ -1 \\ -1 \end{bmatrix}$.

Since $\mathbf{n}_1 \cdot \mathbf{n} = \begin{bmatrix} 4 \\ -1 \\ 5 \end{bmatrix} \cdot \begin{bmatrix} 1 \\ -1 \\ -1 \end{bmatrix} = 0$, $\mathscr{P}_1$ is perpendicular to $\mathscr{P}$.

(d) Since the general form of $\mathscr{P}$ is $4x + 6y - 2z = 0$, its normal vector is $\mathbf{n} = \begin{bmatrix} 4 \\ 6 \\ -2 \end{bmatrix}$.

Since $\mathbf{n}_1 \cdot \mathbf{n} = \begin{bmatrix} 4 \\ -1 \\ 5 \end{bmatrix} \cdot \begin{bmatrix} 4 \\ 6 \\ -2 \end{bmatrix} = 0$, $\mathscr{P}_1$ is perpendicular to $\mathscr{P}$.

20. Since the vector form is $\mathbf{x} = \mathbf{p} + t\mathbf{d}$, we use the given information to determine $\mathbf{p}$ and $\mathbf{d}$.
The general equation of the given line is $2x - 3y = 1$, so its normal vector is $\mathbf{n} = \begin{bmatrix} 2 \\ -3 \end{bmatrix}$.

Our line is perpendicular to the given line, so it has direction vector $\mathbf{d} = \mathbf{n} = \begin{bmatrix} 2 \\ -3 \end{bmatrix}$.

Furthermore, since our line passes through the point $P = (2, -1)$, we have $\mathbf{p} = \begin{bmatrix} 2 \\ -1 \end{bmatrix}$.

So, the vector form of the line perpendicular to $2x - 3y = 1$ through the point $P = (2, -1)$ is

$$\begin{bmatrix} x \\ y \end{bmatrix} = \begin{bmatrix} 2 \\ -1 \end{bmatrix} + t \begin{bmatrix} 2 \\ -3 \end{bmatrix}.$$

21. Since the vector form is $\mathbf{x} = \mathbf{p} + t\mathbf{d}$, we use the given information to determine $\mathbf{p}$ and $\mathbf{d}$.

The general equation of the given line is $2x - 3y = 1$, so its normal vector is $\mathbf{n} = \begin{bmatrix} 2 \\ -3 \end{bmatrix}$.

Our line is parallel to the given line, so it has direction vector $\mathbf{d} = \begin{bmatrix} 3 \\ 2 \end{bmatrix}$.

This comes from the solution of Exercise 45 in Section 1.2: $\mathbf{n} \cdot \mathbf{d} = \begin{bmatrix} a \\ b \end{bmatrix} \cdot \begin{bmatrix} b \\ -a \end{bmatrix} = ab - ab = 0$.

Continuing, since our line passes through the point $P = (2, -1)$, we have $\mathbf{p} = \begin{bmatrix} 2 \\ -1 \end{bmatrix}$.

So, the vector form of the line parallel to $2x - 3y = 1$ through the point $P = (2, -1)$ is

$$\begin{bmatrix} x \\ y \end{bmatrix} = \begin{bmatrix} 2 \\ -1 \end{bmatrix} + t \begin{bmatrix} 3 \\ 2 \end{bmatrix}.$$

22. Since the vector form is $\mathbf{x} = \mathbf{p} + t\mathbf{d}$, we use the given information to determine $\mathbf{p}$ and $\mathbf{d}$.
A line is perpendicular to a plane if its direction vector $\mathbf{d} = \mathbf{n}$ the normal vector of the plane.

The general equation of the given plane is $x - 3y + 2z = 5$, so its normal vector is $\mathbf{n} = \begin{bmatrix} 1 \\ -3 \\ 2 \end{bmatrix}$.

Therefore, the direction vector of our line is $\mathbf{d} = \mathbf{n} = \begin{bmatrix} 1 \\ -3 \\ 2 \end{bmatrix}$.

Furthermore, since our line passes through the point $P = (-1, 0, 3)$, we have $\mathbf{p} = \begin{bmatrix} -1 \\ 0 \\ 3 \end{bmatrix}$.

So, the vector form of the line perpendicular to $x - 3y + 2z = 5$ through $P = (-1, 0, 3)$ is

$$\begin{bmatrix} x \\ y \\ z \end{bmatrix} = \begin{bmatrix} 1 \\ 0 \\ -3 \end{bmatrix} + t \begin{bmatrix} 1 \\ -3 \\ 2 \end{bmatrix}.$$

23. Since the vector form is $\mathbf{x} = \mathbf{p} + t\mathbf{d}$, we use the given information to determine $\mathbf{p}$ and $\mathbf{d}$.

A line with parametric equations $\begin{aligned} x &= a + et \\ y &= b + ft \\ z &= c + gt \end{aligned}$ has vector form $\begin{bmatrix} x \\ y \\ z \end{bmatrix} = \begin{bmatrix} a \\ b \\ c \end{bmatrix} + t \begin{bmatrix} e \\ f \\ g \end{bmatrix}$.

Therefore, its direction vector is $\mathbf{d} = \begin{bmatrix} e \\ f \\ g \end{bmatrix}$. We use this key observation below.

Since the given line has parametric equations $\begin{aligned} x &= 1 - t \\ y &= 2 + 3t \\ z &= -2 - t \end{aligned}$,

it has vector form $\begin{bmatrix} x \\ y \\ z \end{bmatrix} = \begin{bmatrix} 1 \\ 2 \\ -2 \end{bmatrix} + t \begin{bmatrix} -1 \\ 3 \\ -1 \end{bmatrix}$. So, its direction vector is $\begin{bmatrix} -1 \\ 3 \\ -1 \end{bmatrix}$.

Since our line is parallel to the given line, its direction vector is also $\mathbf{d} = \begin{bmatrix} -1 \\ 3 \\ -1 \end{bmatrix}$.

Furthermore, since our line passes through the point $P = (-1, 0, 3)$, we have $\mathbf{p} = \begin{bmatrix} -1 \\ 0 \\ 3 \end{bmatrix}$.

So, the vector form of the line parallel to the given line through $P = (-1, 0, 3)$ is

$$\begin{bmatrix} x \\ y \\ z \end{bmatrix} = \begin{bmatrix} -1 \\ 0 \\ 3 \end{bmatrix} + t \begin{bmatrix} -1 \\ 3 \\ -1 \end{bmatrix}.$$

24. Since the normal form is $\mathbf{n} \cdot \mathbf{x} = \mathbf{n} \cdot \mathbf{p}$, we use the given information to determine $\mathbf{n}$ and $\mathbf{p}$.

A plane is parallel to a given plane if their normal vectors $\mathbf{n}$ are equal.

Since the general form of the given plane is $6x - y + 2z = 3$, its normal vector is $\mathbf{n} = \begin{bmatrix} 6 \\ -1 \\ 2 \end{bmatrix}$.

Since our plane is parallel to the given plane, its normal vector is also $\mathbf{n} = \begin{bmatrix} 6 \\ -1 \\ 2 \end{bmatrix}$.

Furthermore, since our plane passes through the point $P = (0, -2, 5)$, we have $\mathbf{p} = \begin{bmatrix} 0 \\ -2 \\ 5 \end{bmatrix}$.

So, the normal form of the plane parallel to $6x - y + 2z = 3$ through $P = (0, -2, 5)$ is

$$\begin{bmatrix} 6 \\ -1 \\ 2 \end{bmatrix} \cdot \begin{bmatrix} x \\ y \\ z \end{bmatrix} = \begin{bmatrix} 6 \\ -1 \\ 2 \end{bmatrix} \cdot \begin{bmatrix} 0 \\ -2 \\ 5 \end{bmatrix} \text{ or } \begin{bmatrix} 6 \\ -1 \\ 2 \end{bmatrix} \cdot \begin{bmatrix} x \\ y \\ z \end{bmatrix} = 12.$$

25. Following Example 1.23, we will determine the general equations in two simple steps: First, we will use Figure 1.31 in Section 1.2 to find a normal vector $\mathbf{n}$ and a point vector $\mathbf{p}$. Then we will substitute into $\mathbf{n}\cdot\mathbf{x}=\mathbf{n}\cdot\mathbf{p}$ and compute the dot products to find the equations.

(a) We start with $\mathscr{P}_1$ determined by the face of the cube in the yz-plane.

It is clear that a normal vector for $\mathscr{P}_1$ is $\mathbf{n}=\begin{bmatrix}1\\0\\0\end{bmatrix}$ or any vector parallel to the x-axis.

Also we see that $\mathscr{P}_1$ passes through the origin $P=(0,0,0)$, so we set $\mathbf{p}=\begin{bmatrix}0\\0\\0\end{bmatrix}$.

Substituting into $\mathbf{n}\cdot\mathbf{x}=\mathbf{n}\cdot\mathbf{p}$ yields $\begin{bmatrix}1\\0\\0\end{bmatrix}\cdot\begin{bmatrix}x\\y\\z\end{bmatrix}=\begin{bmatrix}1\\0\\0\end{bmatrix}\cdot\begin{bmatrix}0\\0\\0\end{bmatrix}$ or $1\cdot x+0\cdot y+0\cdot z=0$.

So, the general equation for $\mathscr{P}_1$ determined by the face in the yz-plane is $x=0$.
Likewise, the general equation for $\mathscr{P}_2$ determined by the face in the xz-plane is $y=0$ and the general equation for $\mathscr{P}_3$ determined by the face in the xy-plane is $z=0$.
We have found equations for the planes that pass through the origin.
We will use this information to find equations for the planes that pass through $(1,1,1)$.
We begin with $\mathscr{P}_4$ passing through the face parallel to the face in the yz-plane.

Since $\mathscr{P}_4$ is parallel to the face in the yz-plane, its normal vector is $\mathbf{n}=\begin{bmatrix}1\\0\\0\end{bmatrix}$.

As previously noted $\mathscr{P}_4$ passes through the point $P=(1,1,1)$, so we set $\mathbf{p}=\begin{bmatrix}1\\1\\1\end{bmatrix}$.

Substituting into $\mathbf{n}\cdot\mathbf{x}=\mathbf{n}\cdot\mathbf{p}$ yields $\begin{bmatrix}1\\0\\0\end{bmatrix}\cdot\begin{bmatrix}x\\y\\z\end{bmatrix}=\begin{bmatrix}1\\1\\1\end{bmatrix}\cdot\begin{bmatrix}1\\0\\0\end{bmatrix}$ or $1\cdot x+0\cdot y+0\cdot z=1$.

So, the general equation for $\mathscr{P}_4$ is $x=1$.
Likewise, the general equations for $\mathscr{P}_5$ and $\mathscr{P}_6$ are $y=1$ and $z=1$ respectively.

(b) We will use the given information to determine $\mathbf{n}$ and $\mathbf{p}$, then compute $\mathbf{n}\cdot\mathbf{x}=\mathbf{n}\cdot\mathbf{p}$.
We begin by observing the two key facts that will enable us to find $\mathbf{n}$ and $\mathbf{p}$:
Two planes $\mathscr{P}_1$, $\mathscr{P}$ are perpendicular if their normal vectors are orthogonal, so $\mathbf{n}_1\cdot\mathbf{n}=0$.
Every vector $\mathbf{u}$ in the plane $\mathscr{P}_1$ is orthogonal to its normal vector $\mathbf{n}_1$, so $\mathbf{n}_1\cdot\mathbf{u}=0$.
Condition 1: Our plane must be perpendicular to the xy-plane, so $\mathbf{n}_1\cdot\mathbf{n}=0$. From (a),
$\mathbf{n}=\begin{bmatrix}0\\0\\1\end{bmatrix}$, so $\mathbf{n}_1\cdot\mathbf{n}=\begin{bmatrix}x\\y\\z\end{bmatrix}\cdot\begin{bmatrix}0\\0\\1\end{bmatrix}=0\Rightarrow z=0$. So, $\mathbf{n}_1$ is of the form $\begin{bmatrix}x\\y\\0\end{bmatrix}$.
Condition 2: $\mathbf{n}_1$ must be perpendicular to the vector $\mathbf{u}$ from the origin to $(1,1,1)$.
Since $\mathbf{u}=\begin{bmatrix}1-0\\1-0\\1-0\end{bmatrix}=\begin{bmatrix}1\\1\\1\end{bmatrix}$, we have $\mathbf{n}_1\cdot\mathbf{n}=\begin{bmatrix}x\\y\\0\end{bmatrix}\cdot\begin{bmatrix}1\\1\\1\end{bmatrix}=0\Rightarrow x+y=0\Rightarrow y=-x$.

So, $\mathscr{P}_1$ must be of the form $\mathbf{n}_1=\begin{bmatrix}x\\-x\\0\end{bmatrix}=x\begin{bmatrix}1\\-1\\0\end{bmatrix}$. Letting $x=1$ yields $\mathbf{n}_1=\begin{bmatrix}1\\-1\\0\end{bmatrix}$.

As previously noted $\mathscr{P}_1$ passes through the origin $P=(0,0,0)$, so we set $\mathbf{p}=\begin{bmatrix}0\\0\\0\end{bmatrix}$.

Now $\mathbf{n}\cdot\mathbf{x}=\mathbf{n}\cdot\mathbf{p}$ yields $\begin{bmatrix}1\\-1\\0\end{bmatrix}\cdot\begin{bmatrix}x\\y\\z\end{bmatrix}=\begin{bmatrix}1\\-1\\0\end{bmatrix}\cdot\begin{bmatrix}0\\0\\0\end{bmatrix}$ or $1\cdot x+(-1)\cdot y+0\cdot z=0$.
Therefore, the general equation for the plane perpendicular to the xy-plane and containing the diagonal from the origin to $(1,1,1)$ is $x-y=0$.

(c) As above, use $\mathbf{u}=[0,1,1]$ and $\mathbf{v}=[1,0,1]$ from Example 1.15 of Section 1.2 to find $\mathbf{n}$.

From $\mathbf{n}\cdot\mathbf{u}=\begin{bmatrix}x\\y\\z\end{bmatrix}\cdot\begin{bmatrix}0\\1\\1\end{bmatrix}=0\Rightarrow y+z=0\Rightarrow y=-z$.

From $\mathbf{n}\cdot\mathbf{v}=\begin{bmatrix}x\\-z\\z\end{bmatrix}\cdot\begin{bmatrix}1\\0\\1\end{bmatrix}=0\Rightarrow x+z=0\Rightarrow x=-z$.

So, the normal vector $\mathbf{n}=\begin{bmatrix}-z\\-z\\z\end{bmatrix}=z\begin{bmatrix}-1\\-1\\1\end{bmatrix}$. When $z=-1$, we have $\mathbf{n}=\begin{bmatrix}1\\1\\-1\end{bmatrix}$.

It is obvious the side diagonals pass through the origin $P=(0,0,0)$, so we set $\mathbf{p}=\begin{bmatrix}0\\0\\0\end{bmatrix}$.

Now $\mathbf{n}\cdot\mathbf{x}=\mathbf{n}\cdot\mathbf{p}$ yields $\begin{bmatrix}1\\1\\-1\end{bmatrix}\cdot\begin{bmatrix}x\\y\\z\end{bmatrix}=\begin{bmatrix}1\\1\\-1\end{bmatrix}\cdot\begin{bmatrix}0\\0\\0\end{bmatrix}$ or $1\cdot x+1\cdot y+(-1)\cdot z=0$.
The general equation for the plane containing the side diagonals is $x+y-z=0$.

26. Finding the distance between points A and B is equivalent to finding $d(\mathbf{a}, \mathbf{b})$.
Given $\mathbf{x} = [x, y, z]$, $\mathbf{p} = [1, 0, -2]$, and $\mathbf{q} = [5, 2, 4]$, we have the condition $d(\mathbf{x}, \mathbf{p}) = d(\mathbf{x}, \mathbf{q})$.
We simplify that equation to find the condition all points $X = (x, y, z)$ must satisfy.

$$d(\mathbf{x}, \mathbf{p}) = \sqrt{(x-1)^2 + (y-0)^2 + (z+2)^2} = \sqrt{(x-5)^2 + (y-2)^2 + (z-4)^2} = d(\mathbf{x}, \mathbf{q}).$$

Squaring both sides, we have: $(x-1)^2 + (y-0)^2 + (z+2)^2 = (x-5)^2 + (y-2)^2 + (z-4)^2 \Rightarrow$

$$(x^2 - 2x + 1) + y^2 + (z^2 + 4z + 4) = (x^2 - 10x + 25) + (y^2 - 4y + 4) + (z^2 - 8z + 16).$$

Noting the squares cancel and combining the other like terms, we have: $8x + 4y + 12z = 40$.
Dividing both sides by 4, we see all points $X = (x, y, z)$ lie in the plane $2x + y + 3z = 10$.

27. We will first follow Example 1.25, then use $d(Q, \ell) = \dfrac{|ax_0 + by_0 - c|}{\sqrt{a^2 + b^2}}$ and compare results.

Comparing $\begin{bmatrix} x \\ y \end{bmatrix} = \begin{bmatrix} -1 \\ 2 \end{bmatrix} + t\begin{bmatrix} 1 \\ -1 \end{bmatrix}$ to $\mathbf{x} = \mathbf{p} + t\mathbf{d}$, we see ℓ has $P = (-1, 2)$ and $\mathbf{d} = \begin{bmatrix} 1 \\ -1 \end{bmatrix}$.

As suggested by Figure 1.63, we need to calculate the length of $\overrightarrow{RQ}$,
where R is the point on ℓ at the foot of the perpendicular from Q.

Now if we let $\mathbf{v} = \overrightarrow{PQ}$, then $\overrightarrow{PR} = \text{proj}_{\mathbf{d}}(\mathbf{v})$ and $\overrightarrow{RQ} = \mathbf{v} - \text{proj}_{\mathbf{d}}(\mathbf{v})$.

Step 1. $\mathbf{v} = \overrightarrow{PQ} = \mathbf{q} - \mathbf{p} = \begin{bmatrix} 2 \\ 2 \end{bmatrix} - \begin{bmatrix} -1 \\ 2 \end{bmatrix} = \begin{bmatrix} 3 \\ 0 \end{bmatrix}$.

Step 2. $\text{proj}_{\mathbf{d}}(\mathbf{v}) = \left(\dfrac{\mathbf{d} \cdot \mathbf{v}}{\mathbf{d} \cdot \mathbf{d}}\right)\mathbf{d} = \left(\dfrac{1 \cdot 3 + (-1) \cdot 0}{1 \cdot 1 + (-1) \cdot (-1)}\right)\begin{bmatrix} 1 \\ -1 \end{bmatrix} = \dfrac{3}{2}\begin{bmatrix} 1 \\ -1 \end{bmatrix} = \begin{bmatrix} 3/2 \\ -3/2 \end{bmatrix}$.

Step 3. The vector we want is $\mathbf{v} - \text{proj}_{\mathbf{d}}(\mathbf{v}) = \begin{bmatrix} 3 \\ 0 \end{bmatrix} - \begin{bmatrix} 3/2 \\ -3/2 \end{bmatrix} = \begin{bmatrix} 3/2 \\ 3/2 \end{bmatrix}$.

Step 4. The distance $d(Q, \ell)$ from Q to ℓ is $\|\mathbf{v} - \text{proj}_{\mathbf{d}}(\mathbf{v})\| = \left\|\begin{bmatrix} 3/2 \\ 3/2 \end{bmatrix}\right\|$.

So Theorem 1.3(b) implies $\|\mathbf{v} - \text{proj}_{\mathbf{d}}(\mathbf{v})\| = \dfrac{3}{2}\left\|\begin{bmatrix} 1 \\ 1 \end{bmatrix}\right\| = \dfrac{3}{2}\sqrt{1+1} = \dfrac{3\sqrt{2}}{2}$.

Now in order to calculate $d(Q, \ell) = \dfrac{|ax_0 + by_0 - c|}{\sqrt{a^2 + b^2}}$ we need to put ℓ into general form.

If $\mathbf{d} = \begin{bmatrix} a \\ b \end{bmatrix}$, then $\mathbf{n} = \begin{bmatrix} b \\ -a \end{bmatrix}$ because $\begin{bmatrix} a \\ b \end{bmatrix} \cdot \begin{bmatrix} b \\ -a \end{bmatrix} = 0$. For ℓ, $\mathbf{d} = \begin{bmatrix} 1 \\ -1 \end{bmatrix}$ so $\mathbf{n} = \begin{bmatrix} 1 \\ 1 \end{bmatrix}$.

From $\mathbf{n} \cdot \mathbf{x} = \mathbf{n} \cdot \mathbf{p}$ we have $\begin{bmatrix} 1 \\ 1 \end{bmatrix} \cdot \begin{bmatrix} x \\ y \end{bmatrix} = \begin{bmatrix} 1 \\ 1 \end{bmatrix} \cdot \begin{bmatrix} -1 \\ 2 \end{bmatrix}$ so $x + y = 1$ and $a = b = c = 1$.

Furthermore, since $Q = (2, 2) = (x_0, y_0)$ we have $x_0 = y_0 = 2$.

So $d(Q, \ell) = \dfrac{|2 + 2 - 1|}{\sqrt{1^2 + 1^2}} = \dfrac{3}{\sqrt{2}} = \dfrac{3\sqrt{2}}{2}$ exactly as we found by following Example 1.25.

28. We will follow Example 1.25, then use $\mathrm{d}(Q,\ell) = \dfrac{|ax_0 + by_0 + cz_0 - d|}{\sqrt{a^2+b^2+c^2}}$ and compare results.

Even though the formula $\mathrm{d}(Q,\ell) = \dfrac{|ax_0 + by_0 + cz_0 - d|}{\sqrt{a^2+b^2+c^2}}$ was developed for planes, it can work for lines in $\mathbb{R}^3$ with the proper choice of $\mathbf{n} = [a, b, c]$.

Comparing $\begin{bmatrix} x \\ y \\ z \end{bmatrix} = \begin{bmatrix} 1 \\ 1 \\ 1 \end{bmatrix} + t\begin{bmatrix} -2 \\ 0 \\ 3 \end{bmatrix}$ to $\mathbf{x} = \mathbf{p} + t\mathbf{d}$, we see $P = (1,1,1)$ and $\mathbf{d} = \begin{bmatrix} -2 \\ 0 \\ 3 \end{bmatrix}$.

As suggested by Figure 1.63, we need to calculate the length of $\overrightarrow{RQ}$, where R is the point on ℓ at the foot of the perpendicular from Q.

Now if we let $\mathbf{v} = \overrightarrow{PQ}$, then $\overrightarrow{PR} = \mathrm{proj}_{\mathbf{d}}(\mathbf{v})$ and $\overrightarrow{RQ} = \mathbf{v} - \mathrm{proj}_{\mathbf{d}}(\mathbf{v})$.

Step 1. $\mathbf{v} = \overrightarrow{PQ} = \mathbf{q} - \mathbf{p} = \begin{bmatrix} 0 \\ 1 \\ 0 \end{bmatrix} - \begin{bmatrix} 1 \\ 1 \\ 1 \end{bmatrix} = \begin{bmatrix} -1 \\ 0 \\ -1 \end{bmatrix}$.

Step 2. $\mathrm{proj}_{\mathbf{d}}(\mathbf{v}) = \left(\dfrac{\mathbf{d}\cdot\mathbf{v}}{\mathbf{d}\cdot\mathbf{d}}\right)\mathbf{d} = \left(\dfrac{(-2)\cdot(-1) + (3)\cdot(-1)}{(-2)\cdot(-2) + 3\cdot 3}\right)\begin{bmatrix} -2 \\ 0 \\ 3 \end{bmatrix} = \begin{bmatrix} 2/13 \\ 0 \\ -3/13 \end{bmatrix}$.

Step 3. The vector we want is $\mathbf{v} - \mathrm{proj}_{\mathbf{d}}(\mathbf{v}) = \begin{bmatrix} -1 \\ 0 \\ -1 \end{bmatrix} - \begin{bmatrix} 2/13 \\ 0 \\ -3/13 \end{bmatrix} = \begin{bmatrix} -15/13 \\ 0 \\ -10/13 \end{bmatrix}$.

Step 4. The distance $\mathrm{d}(Q,\ell)$ from Q to ℓ is $\|\mathbf{v} - \mathrm{proj}_{\mathbf{d}}(\mathbf{v})\| = \left\|\begin{bmatrix} -15/13 \\ 0 \\ -10/13 \end{bmatrix}\right\|$.

So Theorem 1.3(b) implies $\|\mathbf{v} - \mathrm{proj}_{\mathbf{d}}(\mathbf{v})\| = \dfrac{5}{13}\left\|\begin{bmatrix} 3 \\ 0 \\ 2 \end{bmatrix}\right\| = \dfrac{5}{13}\sqrt{9+4} = \dfrac{5\sqrt{13}}{13}$.

Now in order to calculate $\mathrm{d}(Q,\ell) = \dfrac{|ax_0 + by_0 + cz_0 - d|}{\sqrt{a^2+b^2+c^2}}$ we need to put ℓ into general form.

The appropriate choice of $\mathbf{n}$ mentioned at the top follows from the following observation:

Vector $\begin{bmatrix} 3 \\ 0 \\ 2 \end{bmatrix}$ found using Theorem 1.3(b) in Step 4 is orthogonal to $\mathbf{d}$, so let $\mathbf{n} = \begin{bmatrix} 3 \\ 0 \\ 2 \end{bmatrix}$.

From $\mathbf{n}\cdot\mathbf{x} = \mathbf{n}\cdot\mathbf{p}$, $\begin{bmatrix} 3 \\ 0 \\ 2 \end{bmatrix}\cdot\begin{bmatrix} x \\ y \\ z \end{bmatrix} = \begin{bmatrix} 3 \\ 0 \\ 2 \end{bmatrix}\cdot\begin{bmatrix} 1 \\ 1 \\ 1 \end{bmatrix}$ so $3x + 2z = 5$ and $a = 3$, $b = 0$, $c = 2$, $d = 5$.

Furthermore, since $Q = (0,1,0) = (x_0, y_0, z_0)$ we have $x_0 = 0$, $y_0 = 1$, and $z_0 = 0$.

So $\mathrm{d}(Q,\ell) = \dfrac{|0+0+0-5|}{\sqrt{3^2+2^2}} = \dfrac{5}{\sqrt{13}} = \dfrac{5\sqrt{13}}{13}$ exactly as we found by following Example 1.25.

29. We will follow Example 1.26, then use $d(Q,\mathscr{P}) = \dfrac{|ax_0+by_0+cz_0-d|}{\sqrt{a^2+b^2+c^2}}$ and compare results.

By definition $ax+by+cz=d$ implies $\mathbf{n}=[a,b,c]$, so $x+y-z=0$ implies $\mathbf{n}=[1,1,-1]$.

As suggested by Figure 1.64, we need to calculate the length of $\overrightarrow{RQ} = \text{proj}_{\mathbf{n}}(\mathbf{v})$, where $\mathbf{v}=\overrightarrow{PQ}$.

Step 1. By trial and error, we find $P=(1,0,1)$ satisfies $x+y-z=0$.

Step 2. $\mathbf{v}=\overrightarrow{PQ}=\mathbf{q}-\mathbf{p}=\begin{bmatrix}2\\2\\2\end{bmatrix}-\begin{bmatrix}1\\0\\1\end{bmatrix}=\begin{bmatrix}1\\2\\1\end{bmatrix}$.

Step 3. $\text{proj}_{\mathbf{n}}(\mathbf{v})=\left(\dfrac{\mathbf{n}\cdot\mathbf{v}}{\mathbf{d}\cdot\mathbf{n}}\right)\mathbf{n}=\left(\dfrac{1\cdot1+1\cdot2-1\cdot1}{1^2+1^2+(-1)^2}\right)\begin{bmatrix}1\\1\\-1\end{bmatrix}=\dfrac{2}{3}\begin{bmatrix}1\\1\\-1\end{bmatrix}=\begin{bmatrix}2/3\\2/3\\-2/3\end{bmatrix}$.

Step 4. The distance from Q to $\mathscr{P}$ is $\|\text{proj}_{\mathbf{n}}(\mathbf{v})\| = \left\|\begin{bmatrix}2/3\\2/3\\-2/3\end{bmatrix}\right\| = \dfrac{2}{3}\left\|\begin{bmatrix}1\\1\\-1\end{bmatrix}\right\| = \dfrac{2\sqrt{3}}{3}$.

Now for $d(Q,\mathscr{P}) = \dfrac{|ax_0+by_0+cz_0-d|}{\sqrt{a^2+b^2+c^2}}$ we need identify a, b, c, d, and x_0, y_0, z_0.

Since $x+y-z=0$, $a=1$, $b=1$, $c=-1$, $d=0$. From $Q=(2,2,2)$, $x_0=y_0=z_0=2$.

So $d(Q,\mathscr{P}) = \dfrac{|2+2-2+0|}{\sqrt{1^2+1^2+(-1)^2}} = \dfrac{2}{\sqrt{3}} = \dfrac{2\sqrt{3}}{3}$ as we found by following Example 1.26.

30. We will follow Example 1.26, then use $d(Q,\mathscr{P}) = \dfrac{|ax_0+by_0+cz_0-d|}{\sqrt{a^2+b^2+c^2}}$ and compare results.

By definition $ax+by+cz=d$ implies $\mathbf{n}=[a,b,c]$, so $x-2y+2z=1$ implies $\mathbf{n}=[1,-2,2]$.

As suggested by Figure 1.64, we need to calculate the length of $\overrightarrow{RQ} = \text{proj}_{\mathbf{n}}(\mathbf{v})$, where $\mathbf{v}=\overrightarrow{PQ}$.

Step 1. By trial and error, we find $P=(1,0,0)$ satisfies $x-2y+2z=1$.

Step 2. $\mathbf{v}=\overrightarrow{PQ}=\mathbf{q}-\mathbf{p}=\begin{bmatrix}0\\0\\0\end{bmatrix}-\begin{bmatrix}1\\0\\0\end{bmatrix}=\begin{bmatrix}-1\\0\\0\end{bmatrix}$.

Step 3. $\text{proj}_{\mathbf{n}}(\mathbf{v})=\left(\dfrac{\mathbf{n}\cdot\mathbf{v}}{\mathbf{d}\cdot\mathbf{n}}\right)\mathbf{n}=\left(\dfrac{-1\cdot1+0\cdot0+0\cdot0}{1^2+(-2)^2+2^2}\right)\begin{bmatrix}1\\-2\\2\end{bmatrix}=-\dfrac{1}{9}\begin{bmatrix}1\\-2\\2\end{bmatrix}=\begin{bmatrix}-1/9\\2/9\\-2/9\end{bmatrix}$.

Step 4. The distance from Q to $\mathscr{P}$ is $\|\text{proj}_{\mathbf{n}}(\mathbf{v})\| = \left\|\begin{bmatrix}-1/9\\2/9\\-2/9\end{bmatrix}\right\| = \dfrac{1}{9}\left\|\begin{bmatrix}1\\-2\\2\end{bmatrix}\right\| = \dfrac{1}{3}$.

Now for $d(Q,\mathscr{P}) = \dfrac{|ax_0+by_0+cz_0-d|}{\sqrt{a^2+b^2+c^2}}$ we need identify a, b, c, d, and x_0, y_0, z_0.

Since $x-2y+2z=1$, $a=1$, $b=-2$, $c=2$, $d=1$. From $Q=(0,0,0)$, $x_0=y_0=z_0=0$.

So $d(Q,\mathscr{P}) = \dfrac{|0-0+0-1|}{\sqrt{1^2+(-2)^2+2^2}} = \dfrac{1}{\sqrt{9}} = \dfrac{1}{3}$ as we found by following Example 1.26.

31. Similar to Example 1.25, Figure 1.63 suggests we let $\mathbf{v} = \overrightarrow{PQ}$, then $\mathbf{w} = \overrightarrow{PR} = \text{proj}_{\mathbf{d}}(\mathbf{v})$.

Comparing $\begin{bmatrix} x \\ y \end{bmatrix} = \begin{bmatrix} -1 \\ 2 \end{bmatrix} + t\begin{bmatrix} 1 \\ -1 \end{bmatrix}$ to $\mathbf{x} = \mathbf{p} + t\mathbf{d}$, we see ℓ has $P = (-1, 2)$ and $\mathbf{d} = \begin{bmatrix} 1 \\ -1 \end{bmatrix}$.

Step 1. $\mathbf{v} = \overrightarrow{PQ} = \mathbf{q} - \mathbf{p} = \begin{bmatrix} 2 \\ 2 \end{bmatrix} - \begin{bmatrix} -1 \\ 2 \end{bmatrix} = \begin{bmatrix} 3 \\ 0 \end{bmatrix}$.

Step 2. $\mathbf{w} = \text{proj}_{\mathbf{d}}(\mathbf{v}) = \left(\dfrac{\mathbf{d}\cdot\mathbf{v}}{\mathbf{d}\cdot\mathbf{d}}\right)\mathbf{d} = \left(\dfrac{1\cdot 3 + (-1)\cdot 0}{1\cdot 1 + (-1)\cdot(-1)}\right)\begin{bmatrix} 1 \\ -1 \end{bmatrix} = \dfrac{3}{2}\begin{bmatrix} 1 \\ -1 \end{bmatrix} = \begin{bmatrix} 3/2 \\ -3/2 \end{bmatrix}$.

Step 3. So, $\mathbf{r} = \mathbf{p} + \overrightarrow{PR} = \mathbf{p} + \text{proj}_{\mathbf{d}}(\mathbf{v}) = \mathbf{p} + \mathbf{w} = \begin{bmatrix} -1 \\ 2 \end{bmatrix} + \begin{bmatrix} 3/2 \\ -3/2 \end{bmatrix} = \begin{bmatrix} 1/2 \\ 1/2 \end{bmatrix}$.

Therefore, the point R on ℓ that is closest to Q is $\left(\frac{1}{2}, \frac{1}{2}\right)$.

32. Similar to Example 1.25, Figure 1.63 suggests we let $\mathbf{v} = \overrightarrow{PQ}$, then $\mathbf{w} = \overrightarrow{PR} = \text{proj}_{\mathbf{d}}(\mathbf{v})$.

Comparing $\begin{bmatrix} x \\ y \\ z \end{bmatrix} = \begin{bmatrix} 1 \\ 1 \\ 1 \end{bmatrix} + t\begin{bmatrix} -2 \\ 0 \\ 3 \end{bmatrix}$ to $\mathbf{x} = \mathbf{p} + t\mathbf{d}$, we see ℓ has $P = (1, 1, 1)$ and $\mathbf{d} = \begin{bmatrix} -2 \\ 0 \\ 3 \end{bmatrix}$.

Step 1. $\mathbf{v} = \overrightarrow{PQ} = \mathbf{q} - \mathbf{p} = \begin{bmatrix} 0 \\ 1 \\ 0 \end{bmatrix} - \begin{bmatrix} 1 \\ 1 \\ 1 \end{bmatrix} = \begin{bmatrix} -1 \\ 0 \\ -1 \end{bmatrix}$.

Step 2. $\mathbf{w} = \text{proj}_{\mathbf{d}}(\mathbf{v}) = \left(\dfrac{\mathbf{d}\cdot\mathbf{v}}{\mathbf{d}\cdot\mathbf{d}}\right)\mathbf{d} = \left(\dfrac{(-2)\cdot(-1) + 3\cdot(-1)}{(-2)^2 + 3^2}\right)\begin{bmatrix} -2 \\ 0 \\ 3 \end{bmatrix} = \begin{bmatrix} 2/13 \\ 0 \\ -3/13 \end{bmatrix}$.

Step 3. So, $\mathbf{r} = \mathbf{p} + \overrightarrow{PR} = \mathbf{p} + \text{proj}_{\mathbf{d}}(\mathbf{v}) = \mathbf{p} + \mathbf{w} = \begin{bmatrix} 1 \\ 1 \\ 1 \end{bmatrix} + \begin{bmatrix} 2/13 \\ 0 \\ -3/13 \end{bmatrix} = \begin{bmatrix} 15/13 \\ 1 \\ 10/13 \end{bmatrix}$.

Therefore, the point R on ℓ that is closest to Q is $\left(\frac{15}{13}, 1, \frac{10}{13}\right)$.

33. Similar to Example 1.26, Figure 1.64 suggests we let $\mathbf{v} = \overrightarrow{PQ}$, then $\mathbf{w} = \overrightarrow{QR} = -\text{proj}_{\mathbf{n}}(\mathbf{v})$.

By definition $ax + by + cz = d$ implies $\mathbf{n} = [a, b, c]$, so $x + y - z = 0$ implies $\mathbf{n} = [1, 1, -1]$.

Step 1. By trial and error, we find $P = (1, 0, 1)$ satisfies $x + y - z = 0$.

Step 2. $\mathbf{v} = \overrightarrow{PQ} = \mathbf{q} - \mathbf{p} = \begin{bmatrix} 2 \\ 2 \\ 2 \end{bmatrix} - \begin{bmatrix} 1 \\ 0 \\ 1 \end{bmatrix} = \begin{bmatrix} 1 \\ 2 \\ 1 \end{bmatrix}$.

Step 3. $\mathbf{w} = \text{proj}_{\mathbf{n}}(\mathbf{v}) = \left(\frac{\mathbf{n} \cdot \mathbf{v}}{\mathbf{n} \cdot \mathbf{n}}\right)\mathbf{n} = \left(\frac{1 \cdot 1 + 1 \cdot 2 + (-1) \cdot 1}{1^2 + 1^2 + (-1)^2}\right)\begin{bmatrix} 1 \\ 1 \\ -1 \end{bmatrix} = \begin{bmatrix} 2/3 \\ 2/3 \\ -2/3 \end{bmatrix}$.

Step 4. So, $\mathbf{r} = \mathbf{p} + \overrightarrow{PQ} + \overrightarrow{QR} = \mathbf{p} + \mathbf{v} - \text{proj}_{\mathbf{n}}(\mathbf{v}) = \begin{bmatrix} 1 \\ 0 \\ 1 \end{bmatrix} + \begin{bmatrix} 1 \\ 2 \\ 1 \end{bmatrix} - \begin{bmatrix} 2/3 \\ 2/3 \\ -2/3 \end{bmatrix} = \begin{bmatrix} 4/3 \\ 4/3 \\ 8/3 \end{bmatrix}$.

Therefore, the point R in $\mathscr{P}$ that is closest to Q is $\left(\frac{4}{3}, \frac{4}{3}, \frac{8}{3}\right)$.

34. Similar to Example 1.26, Figure 1.64 suggests we let $\mathbf{v} = \overrightarrow{PQ}$, then $\mathbf{w} = \overrightarrow{QR} = \text{proj}_{\mathbf{n}}(\mathbf{v})$.

By definition $ax + by + cz = d$ implies $\mathbf{n} = [a, b, c]$, so $x - 2y + 2z = 1$ implies $\mathbf{n} = [1, -2, 2]$.

Step 1. By trial and error, we find $P = (1, 0, 0)$ satisfies $x - 2y + 2z = 1$.

Step 2. $\mathbf{v} = \overrightarrow{PQ} = \mathbf{q} - \mathbf{p} = \begin{bmatrix} 0 \\ 0 \\ 0 \end{bmatrix} - \begin{bmatrix} 1 \\ 0 \\ 0 \end{bmatrix} = \begin{bmatrix} -1 \\ 0 \\ 0 \end{bmatrix}$.

Step 3. $\mathbf{w} = \text{proj}_{\mathbf{n}}(\mathbf{v}) = \left(\frac{\mathbf{n} \cdot \mathbf{v}}{\mathbf{n} \cdot \mathbf{n}}\right)\mathbf{n} = \left(\frac{1 \cdot (-1)}{1^2 + (-2)^2 + 2^2}\right)\begin{bmatrix} 1 \\ -2 \\ 2 \end{bmatrix} = \begin{bmatrix} -1/9 \\ 2/9 \\ -2/9 \end{bmatrix}$.

Step 4. So, $\mathbf{r} = \mathbf{p} + \overrightarrow{PR} + \overrightarrow{QR} = \mathbf{p} + \mathbf{v} + \text{proj}_{\mathbf{n}}(\mathbf{v}) = \begin{bmatrix} 1 \\ 0 \\ 0 \end{bmatrix} + \begin{bmatrix} -1 \\ 0 \\ 0 \end{bmatrix} + \begin{bmatrix} -1/9 \\ 2/9 \\ -2/9 \end{bmatrix} = \begin{bmatrix} -1/9 \\ 2/9 \\ -2/9 \end{bmatrix}$.

Therefore, the point R in $\mathscr{P}$ that is closest to Q is $\left(-\frac{1}{9}, \frac{2}{9}, -\frac{2}{9}\right)$.

35. Since the given lines ℓ_1 and ℓ_2 are parallel, we can simply choose Q on ℓ_1, P on ℓ_2. Following Example 1.25, we have:

From ℓ_1, $Q = (1,1)$. From ℓ_2, we have $P = (5,4)$, $\mathbf{d} = [-2,3]$, and $\mathbf{n} = [3,2] = [a,b]$.

Step 1. $\mathbf{v} = \overrightarrow{PQ} = \mathbf{q} - \mathbf{p} = [1,1] - [5,4] = [-4,-3]$.

Step 2. $\text{proj}_{\mathbf{d}}(\mathbf{v}) = \left(\dfrac{\mathbf{d}\cdot\mathbf{v}}{\mathbf{d}\cdot\mathbf{d}}\right)\mathbf{d} = \left(\dfrac{(-2)\cdot(-4)+3\cdot(-3)}{(-2)^2+3^2}\right)\begin{bmatrix}-2\\3\end{bmatrix} = -\dfrac{1}{13}\begin{bmatrix}-2\\3\end{bmatrix} = \begin{bmatrix}2/13\\-3/13\end{bmatrix}$.

Step 3. The vector we want is $\mathbf{v} - \text{proj}_{\mathbf{d}}(\mathbf{v}) = \begin{bmatrix}-4\\-3\end{bmatrix} - \begin{bmatrix}2/13\\-3/13\end{bmatrix} = \begin{bmatrix}-54/13\\-36/13\end{bmatrix}$.

Step 4. The distance $d(Q,\ell_2)$ from ℓ_1 to ℓ_2 is $\|\mathbf{v} - \text{proj}_{\mathbf{d}}(\mathbf{v})\| = \left\|\begin{bmatrix}-54/13\\-36/13\end{bmatrix}\right\|$.

So Theorem 1.3(b) implies $\|\mathbf{v} - \text{proj}_{\mathbf{d}}(\mathbf{v})\| = \dfrac{18}{13}\left\|\begin{bmatrix}3\\2\end{bmatrix}\right\| = \dfrac{18}{13}\sqrt{4+9} = \dfrac{18\sqrt{13}}{13}$.

From $\mathbf{n}\cdot\mathbf{p} = \begin{bmatrix}3\\2\end{bmatrix}\cdot\begin{bmatrix}5\\4\end{bmatrix} = 23$, $c = 23$. Since $Q = (1,1) = (x_0, y_0)$, we have $x_0 = y_0 = 1$.

Now compare: $d(\ell_1,\ell_2) = d(Q,\ell_2) = \dfrac{|ax_0 + by_0 - c|}{\sqrt{a^2+b^2}} = \dfrac{|3+2-23|}{\sqrt{3^2+2^2}} = \dfrac{18}{\sqrt{13}} = \dfrac{18\sqrt{13}}{13}$.

36. Since the given lines ℓ_1 and ℓ_2 are parallel, we can simply choose Q on ℓ_1, P on ℓ_2. Following Example 1.25, we have:

From $[x,y,z] = [1,0,-1] + t\,[1,1,1]$, we see ℓ_1 gives us $Q = (1,0,-1)$.

From $[x,y,z] = [0,1,1] + t\,[1,1,1]$, we see ℓ_2 has $P = (0,1,1)$, $\mathbf{d} = [1,1,1]$, and $\mathbf{n} = [1,-2,1]$.

Step 1. $\mathbf{v} = \overrightarrow{PQ} = \mathbf{q} - \mathbf{p} = [1,0,-1] - [0,1,1] = [1,-1,-2]$

Step 2. $\text{proj}_{\mathbf{d}}(\mathbf{v}) = \left(\dfrac{\mathbf{d}\cdot\mathbf{v}}{\mathbf{d}\cdot\mathbf{d}}\right)\mathbf{d} = \left(\dfrac{1\cdot1-1\cdot1-1\cdot2}{1^2+1^2+1^2}\right)\begin{bmatrix}1\\1\\1\end{bmatrix} = -\dfrac{2}{3}\begin{bmatrix}1\\1\\1\end{bmatrix} = \begin{bmatrix}-2/3\\-2/3\\-2/3\end{bmatrix}$.

Step 3. The vector we want is $\mathbf{v} - \text{proj}_{\mathbf{d}}(\mathbf{v}) = \begin{bmatrix}1\\-1\\-2\end{bmatrix} - \begin{bmatrix}-2/3\\-2/3\\-2/3\end{bmatrix} = \begin{bmatrix}5/3\\-1/3\\-4/3\end{bmatrix}$.

Step 4. The distance $d(Q,\ell_2)$ from ℓ_1 to ℓ_2 is $\|\mathbf{v} - \text{proj}_{\mathbf{d}}(\mathbf{v})\| = \left\|\begin{bmatrix}5/3\\-1/3\\-4/3\end{bmatrix}\right\|$.

So Theorem 1.3(b) implies $\|\mathbf{v} - \text{proj}_{\mathbf{d}}(\mathbf{v})\| = \dfrac{1}{3}\left\|\begin{bmatrix}5\\-1\\-4\end{bmatrix}\right\| = \dfrac{1}{3}\sqrt{25+1+16} = \dfrac{\sqrt{42}}{3}$.

Set $\mathbf{n} = [5,-1,-4] = [a,b,c]$ we found in Step 4 because it is orthogonal to $\mathbf{d}$.

From $\mathbf{n}\cdot\mathbf{p} = d = [5,-1,-4]\cdot[0,1,1] = -5$. Since $Q = (1,0,-1)$, $x_0 = 1$, $y_0 = 0$, $z_0 = -1$.

Now compare: $d(Q,\ell_2) = \dfrac{|ax_0 + by_0 + cz_0 - d|}{\sqrt{a^2+b^2+c^2}} = \dfrac{|5+4+5|}{\sqrt{1^2+(-2)^2+1^2}} = \dfrac{14}{\sqrt{42}} = \dfrac{\sqrt{42}}{3}$.

37. Since the given planes $\mathscr{P}_1$ and $\mathscr{P}_2$ are parallel, we can simply choose Q in $\mathscr{P}_1$, P in $\mathscr{P}_2$. Following Example 1.26, we have:

Step 1. Since $2x + y - 2z = 0$, $Q = (0,0,0)$ is on $\mathscr{P}_1$.

Since $2x + y - 2z = 5$, $P = (0,5,0)$ is on $\mathscr{P}_2$ and $\mathbf{n} = [2,1,-2] = [a,b,c]$.

Step 2. $\mathbf{v} = \overrightarrow{PQ} = \mathbf{q} - \mathbf{p} = [0,0,0] - [0,5,0] = [0,-5,0]$

Step 3. $\text{proj}_{\mathbf{n}}(\mathbf{v}) = \left(\frac{\mathbf{n}\cdot\mathbf{v}}{\mathbf{n}\cdot\mathbf{n}}\right)\mathbf{n} = \left(\frac{2\cdot 0 - 1\cdot 5 - 2\cdot 0}{2^2 + 1^2 + (-2)^2}\right)\begin{bmatrix} 2 \\ 1 \\ -2 \end{bmatrix} = -\frac{5}{9}\begin{bmatrix} 2 \\ 1 \\ -2 \end{bmatrix} = \begin{bmatrix} -10/9 \\ -5/9 \\ 10/9 \end{bmatrix}$.

Step 4. The distance $d(Q, \mathscr{P}_2)$ from $\mathscr{P}_1$ to $\mathscr{P}_2$ is $\|\mathbf{v} - \text{proj}_{\mathbf{n}}(\mathbf{v})\| = \left\|\begin{bmatrix} -10/9 \\ -5/9 \\ 10/9 \end{bmatrix}\right\|$.

So Theorem 1.3(b) implies $\|\text{proj}_{\mathbf{n}}(\mathbf{v})\| = \frac{5}{9}\left\|\begin{bmatrix} 2 \\ 1 \\ -2 \end{bmatrix}\right\| = \frac{5}{9}\sqrt{4+1+4} = \frac{5}{3}$.

From $\mathbf{n}\cdot\mathbf{p} = d = [2,1,-2]\cdot[0,5,0] = 5$. Since $Q = (0,0,0)$, $x_0 = y_0 = z_0 = 0$.

Now compare: $d(Q, \mathscr{P}_2) = \frac{|ax_0 + by_0 + cz_0 - d|}{\sqrt{a^2+b^2+c^2}} = \frac{|0+0+0-5|}{\sqrt{2^2+1^2+(-2)^2}} = \frac{5}{\sqrt{9}} = \frac{5}{3}$.

38. Since the given planes $\mathscr{P}_1$ and $\mathscr{P}_2$ are parallel, we can simply choose Q in $\mathscr{P}_1$, P in $\mathscr{P}_2$. Following Example 1.26, we have:

Step 1. Since $x + y + z = 1$, $Q = (1,0,0)$ is on $\mathscr{P}_1$.

Since $x + y + z = 3$, $P = (3,0,0)$ is on $\mathscr{P}_2$ and $\mathbf{n} = [1,1,1] = [a,b,c]$.

Step 2. $\mathbf{v} = \overrightarrow{PQ} = \mathbf{q} - \mathbf{p} = [1,0,0] - [3,0,0] = [-2,0,0]$

Step 3. $\text{proj}_{\mathbf{n}}(\mathbf{v}) = \left(\frac{\mathbf{n}\cdot\mathbf{v}}{\mathbf{n}\cdot\mathbf{n}}\right)\mathbf{n} = \left(\frac{1\cdot(-2) + 1\cdot 0 + 1\cdot 0}{1^2+1^2+1^2}\right)\begin{bmatrix} 1 \\ 1 \\ 1 \end{bmatrix} = -\frac{2}{3}\begin{bmatrix} 1 \\ 1 \\ 1 \end{bmatrix} = \begin{bmatrix} -2/3 \\ -2/3 \\ -2/3 \end{bmatrix}$.

Step 4. The distance $d(Q, \mathscr{P}_2)$ from $\mathscr{P}_1$ to $\mathscr{P}_2$ is $\|\mathbf{v} - \text{proj}_{\mathbf{n}}(\mathbf{v})\| = \left\|\begin{bmatrix} -2/3 \\ -2/3 \\ -2/3 \end{bmatrix}\right\|$.

So Theorem 1.3(b) implies $\|\text{proj}_{\mathbf{n}}(\mathbf{v})\| = \frac{2}{3}\left\|\begin{bmatrix} 1 \\ 1 \\ 1 \end{bmatrix}\right\| = \frac{2}{3}\sqrt{1+1+1} = \frac{2\sqrt{3}}{3}$.

From $\mathbf{n}\cdot\mathbf{p} = d = [1,1,1]\cdot[3,0,0] = 3$. Since $Q = (1,0,0)$, $x_0 = 1$, $y_0 = 0$, $z_0 = 0$.

Now compare: $d(Q, \mathscr{P}_2) = \frac{|ax_0 + by_0 + cz_0 - d|}{\sqrt{a^2+b^2+c^2}} = \frac{|1+0+0-3|}{\sqrt{1^2+1^2+1^2}} = \frac{2}{\sqrt{3}} = \frac{2\sqrt{3}}{3}$.

39. Will show $d(B,\ell) = \dfrac{|ax_0 + by_0 - c|}{\sqrt{a^2+b^2}}$, where $\mathbf{n} = \begin{bmatrix} a \\ b \end{bmatrix}$, $\mathbf{n}\cdot\mathbf{a} = c$, and $B = (x_0, y_0)$.

Step 1. From Figure 1.61, we see $d(B,\ell) = \|\text{proj}_{\mathbf{n}}(\mathbf{v})\| = \left\|\left(\dfrac{\mathbf{n}\cdot\mathbf{v}}{\mathbf{n}\cdot\mathbf{n}}\right)\mathbf{n}\right\| = \dfrac{|\mathbf{n}\cdot\mathbf{v}|}{\|\mathbf{n}\|}$.

Step 2. Since $\mathbf{v} = \mathbf{b} - \mathbf{a}$, $\mathbf{n}\cdot\mathbf{v} = \mathbf{n}\cdot(\mathbf{b}-\mathbf{a}) = \mathbf{n}\cdot\mathbf{b} - \mathbf{n}\cdot\mathbf{a} = \begin{bmatrix} a \\ b \end{bmatrix}\cdot\begin{bmatrix} x_0 \\ y_0 \end{bmatrix} - c = ax_0 + by_0 - c$.

Step 3. So, $d(B,\ell) = \|\text{proj}_{\mathbf{n}}(\mathbf{v})\| = \left\|\left(\dfrac{\mathbf{n}\cdot\mathbf{v}}{\mathbf{n}\cdot\mathbf{n}}\right)\mathbf{n}\right\| = \dfrac{|\mathbf{n}\cdot\mathbf{v}|}{\|\mathbf{n}\|} = \dfrac{|ax_0+by_0-c|}{\sqrt{a^2+b^2}}$.

40. Will show $d(B,\mathscr{P}) = \dfrac{|ax_0 + by_0 + cz_0 - d|}{\sqrt{a^2+b^2++c^2}}$, where $\mathbf{n} = \begin{bmatrix} a \\ b \\ c \end{bmatrix}$, $\mathbf{n}\cdot\mathbf{a} = d$, and $B = (x_0, y_0, z_0)$.

Step 1. From Figure 1.62, we see $d(B,\mathscr{P}) = \|\text{proj}_{\mathbf{n}}(\mathbf{v})\| = \left\|\left(\dfrac{\mathbf{n}\cdot\mathbf{v}}{\mathbf{n}\cdot\mathbf{n}}\right)\mathbf{n}\right\| = \dfrac{|\mathbf{n}\cdot\mathbf{v}|}{\|\mathbf{n}\|}$.

Step 2. Since $\mathbf{v} = \mathbf{b} - \mathbf{a}$, $\mathbf{n}\cdot\mathbf{v} = \mathbf{n}\cdot(\mathbf{b}-\mathbf{a}) = \mathbf{n}\cdot\mathbf{b} - \mathbf{n}\cdot\mathbf{a} = \begin{bmatrix} a \\ b \\ c \end{bmatrix}\cdot\begin{bmatrix} x_0 \\ y_0 \\ z_0 \end{bmatrix} - d = ax_0 + by_0 + cz_0 - d$.

Step 3. So, $d(B,\mathscr{P}) = \|\text{proj}_{\mathbf{n}}(\mathbf{v})\| = \left\|\left(\dfrac{\mathbf{n}\cdot\mathbf{v}}{\mathbf{n}\cdot\mathbf{n}}\right)\mathbf{n}\right\| = \dfrac{|\mathbf{n}\cdot\mathbf{v}|}{\|\mathbf{n}\|} = \dfrac{|ax_0+by_0+cz_0-d|}{\sqrt{a^2+b^2+c^2}}$.

41. We will apply the formula from Exercise 39, $d(B,\ell) = \dfrac{|\mathbf{n}\cdot\mathbf{v}|}{\|\mathbf{n}\|}$.

Step 1. We select $B = (x_0, y_0)$ on ℓ_1 so that $\mathbf{n}\cdot\mathbf{b} = \begin{bmatrix} a \\ b \end{bmatrix}\cdot\begin{bmatrix} x_0 \\ y_0 \end{bmatrix} = ax_0 + by_0 = c_1$.

Step 2. We select A on ℓ_2 so that $\mathbf{n}\cdot\mathbf{a} = c_2$.

Step 3. Set $\mathbf{v} = \mathbf{b} - \mathbf{a}$, then $d(B,\mathscr{P}) = \dfrac{|\mathbf{n}\cdot\mathbf{v}|}{\|\mathbf{n}\|} = \dfrac{|\mathbf{n}\cdot(\mathbf{b}-\mathbf{a})|}{\|\mathbf{n}\|} = \dfrac{|\mathbf{n}\cdot\mathbf{b}-\mathbf{n}\cdot\mathbf{a}|}{\|\mathbf{n}\|} = \dfrac{|c_1-c_2|}{\|\mathbf{n}\|}$.

42. We will apply the formula from Exercise 40, $d(B,\mathscr{P}) = \dfrac{|\mathbf{n}\cdot\mathbf{v}|}{\|\mathbf{n}\|}$.

Step 1. We select $B = (x_0, y_0, z_0)$ on $\mathscr{P}_1$ so that $\mathbf{n}\cdot\mathbf{b} = \begin{bmatrix} a \\ b \\ c \end{bmatrix}\cdot\begin{bmatrix} x_0 \\ y_0 \\ z_0 \end{bmatrix} = ax_0 + by_0 + cz_0 = d_1$.

Step 2. We select A on $\mathscr{P}_2$ so that $\mathbf{n}\cdot\mathbf{a} = d_2$.

Step 3. Set $\mathbf{v} = \mathbf{b} - \mathbf{a}$, then $d(B,\mathscr{P}) = \dfrac{|\mathbf{n}\cdot\mathbf{v}|}{\|\mathbf{n}\|} = \dfrac{|\mathbf{n}\cdot(\mathbf{b}-\mathbf{a})|}{\|\mathbf{n}\|} = \dfrac{|\mathbf{n}\cdot\mathbf{b}-\mathbf{n}\cdot\mathbf{a}|}{\|\mathbf{n}\|} = \dfrac{|d_1-d_2|}{\|\mathbf{n}\|}$.

43. As in Example 1.14 of Section 1.2, we note that $\cos\theta = \dfrac{|\mathbf{u}\cdot\mathbf{v}|}{\|\mathbf{u}\|\,\|\mathbf{v}\|}$.

So, given two planes $\mathscr{P}_1$ with $\mathbf{n}_1$ and $\mathscr{P}_2$ with $\mathbf{n}_2$, we have $\cos\theta = \dfrac{|\mathbf{n}_1\cdot\mathbf{n}_2|}{\|\mathbf{n}_1\|\,\|\mathbf{n}_2\|}$.

Step 1. Since $\mathscr{P}_1$ has equation $x+y+z=0$, $\mathbf{n}_1=[1,1,1]$.
Since $\mathscr{P}_2$ has equation $2x+y-2z=0$, $\mathbf{n}_2=[2,1,-2]$.

Step 2. Therefore, $\mathbf{n}_1=[1,1,1]\cdot[2,1,-2]=1\cdot2+1\cdot1-1\cdot2=1$,
$\|\mathbf{n}_1\|=\sqrt{1^2+1^2+1^2}=\sqrt{3}$, and $\|\mathbf{n}_2\|=\sqrt{2^2+1^2+(-2)^2}=3$.

Step 3. So $\cos\theta=\dfrac{1}{3\sqrt{3}}$ and $\theta=\cos^{-1}\left(\dfrac{1}{3\sqrt{3}}\right)\approx 78.9°$.

44. As in Example 1.14 of Section 1.2, we note that $\cos\theta = \dfrac{|\mathbf{u}\cdot\mathbf{v}|}{\|\mathbf{u}\|\,\|\mathbf{v}\|}$.

So, given two planes $\mathscr{P}_1$ with $\mathbf{n}_1$ and $\mathscr{P}_2$ with $\mathbf{n}_2$, we have $\cos\theta = \dfrac{|\mathbf{n}_1\cdot\mathbf{n}_2|}{\|\mathbf{n}_1\|\,\|\mathbf{n}_2\|}$.

Step 1. Since $\mathscr{P}_1$ has equation $3x-y+2z=5$, $\mathbf{n}_1=[3,-1,2]$.
Since $\mathscr{P}_2$ has equation $x+4y-z=2$, $\mathbf{n}_2=[1,4,-1]$.

Step 2. Therefore, $\mathbf{n}_1\cdot\mathbf{n}_1=[3,-1,2]\cdot[1,4,-1]=3\cdot1-1\cdot4-2\cdot1=-3$,
$\|\mathbf{n}_1\|=\sqrt{3^2+(-1)^2+2^2}=\sqrt{14}$, and $\|\mathbf{n}_2\|=\sqrt{1^2+4^2+(-1)^2}=3\sqrt{2}$.

Step 3. So $\cos\theta=-\dfrac{3}{3\sqrt{28}}=-\dfrac{\sqrt{7}}{14}$ and $\theta=\cos^{-1}\left(-\dfrac{\sqrt{7}}{14}\right)\approx 100.9°$.
Since we need an acute angle, we take $180°-100.9°=79.1°$.

45. As in Example 1.14 of Section 1.2, we note that $\cos\theta = \dfrac{|\mathbf{u}\cdot\mathbf{v}|}{\|\mathbf{u}\|\,\|\mathbf{v}\|}$.

So, given $\mathscr{P}$ with $\mathbf{n}$ and ℓ with $\mathbf{d}$, we have $\cos\theta = \dfrac{|\mathbf{n}\cdot\mathbf{d}|}{\|\mathbf{n}\|\,\|\mathbf{d}\|}$.

Step 1. To show $\mathscr{P}$ and ℓ intersect, we note:
$x+y+2z = (2+t)+(1-2t)+2(3+t) = 9+t = 0$ implies $t=-9$.
So, $\mathscr{P}$ and ℓ intersect at the point $[2+(-9), 1-2(-9), 3+(-9)] = [-7, 19, -6]$.

Step 2. Since $\mathscr{P}$ has equation $x+y+2z=0$, $\mathbf{n}=[1,1,2]$.

Given $\begin{aligned} x &= 2+\ t \\ y &= 1-2t \\ z &= 3+\ t \end{aligned}$, ℓ satisfies $\begin{bmatrix} x \\ y \\ z \end{bmatrix} = \begin{bmatrix} 2 \\ 1 \\ 3 \end{bmatrix} + t\begin{bmatrix} 1 \\ -2 \\ 1 \end{bmatrix}$. So, $\mathbf{d} = \begin{bmatrix} 1 \\ -2 \\ 1 \end{bmatrix}$.

Step 3. Therefore, $\mathbf{n}\cdot\mathbf{d} = [1,1,2]\cdot[1,-2,1] = 1\cdot 1 - 1\cdot 2 + 2\cdot 1 = 1$,
$\|\mathbf{n}\| = \sqrt{1^2+1^2+2^2} = \sqrt{6}$, and $\|\mathbf{d}\| = \sqrt{1^2+(-2)^2+1^2} = \sqrt{6}$.

Step 4. So $\cos\theta = \dfrac{1}{\sqrt{6}\sqrt{6}} = \dfrac{1}{6}$ and $\theta = \cos^{-1}\left(\dfrac{1}{6}\right) \approx 80.4°$.

46. As in Example 1.14 of Section 1.2, we note that $\cos\theta = \dfrac{|\mathbf{u}\cdot\mathbf{v}|}{\|\mathbf{u}\|\,\|\mathbf{v}\|}$.

So, given $\mathscr{P}$ with $\mathbf{n}$ and ℓ with $\mathbf{d}$, we have $\cos\theta = \dfrac{|\mathbf{n}\cdot\mathbf{d}|}{\|\mathbf{n}\|\,\|\mathbf{d}\|}$.

Step 1. To show $\mathscr{P}$ and ℓ intersect, we note:
$4x-y-z = 4(t)-(1+2t)-(2+3t) = -3-t = 6$ implies $t=-9$.
So, $\mathscr{P}$ and ℓ intersect at the point $[(-9), 1+2(-9), 2+3(-9)] = [-9, -17, -25]$.

Step 2. Since $\mathscr{P}$ has equation $4x-y-z=0$, $\mathbf{n}=[4,-1,-1]$.

Given $\begin{aligned} x &= \quad\ t \\ y &= 1+2t \\ z &= 2+3t \end{aligned}$, ℓ satisfies $\begin{bmatrix} x \\ y \\ z \end{bmatrix} = \begin{bmatrix} 0 \\ 1 \\ 2 \end{bmatrix} + t\begin{bmatrix} 1 \\ 2 \\ 3 \end{bmatrix}$. So, $\mathbf{d} = \begin{bmatrix} 1 \\ 2 \\ 3 \end{bmatrix}$.

Step 3. Therefore, $\mathbf{n}\cdot\mathbf{d} = [4,-1,-1]\cdot[1,2,3] = 4\cdot 1 - 1\cdot 2 - 1\cdot 3 = -1$,
$\|\mathbf{n}\| = \sqrt{4^2+(-1)^2+(-1)^2} = 3\sqrt{2}$, and $\|\mathbf{d}\| = \sqrt{1^2+2^2+3^2} = \sqrt{14}$.

Step 4. So $\cos\theta = -\dfrac{1}{3\sqrt{28}} = -\dfrac{\sqrt{7}}{42}$ and $\theta = \cos^{-1}\left(-\dfrac{\sqrt{7}}{42}\right) \approx 93.6°$.

Since we need an acute angle, we take $180° - 93.6° = 86.4°$.

47. Will find an expression for $\mathbf{p}$ in terms of $\mathbf{v}$ and $\mathbf{n}$ given $\mathbf{n}$ is orthogonal to $\mathbf{p}$, that is $\mathbf{p}\cdot\mathbf{n}=0$.
Step 1. We solve for c starting from the given equation $\mathbf{p}=\mathbf{v}-c\mathbf{n}$.

$$\begin{aligned}
\mathbf{p} &= \mathbf{v}-c\mathbf{n} \quad\Rightarrow && \text{Given}\\
c\mathbf{n} &= \mathbf{v}-\mathbf{p} && \text{By } x=y-z \text{ implies } z=y-x\\
(c\mathbf{n})\cdot\mathbf{n} &= (\mathbf{v}-\mathbf{p})\cdot\mathbf{n} \quad\Rightarrow && \text{Taking the dot product of } \mathbf{n} \text{ with both sides.}\\
c(\mathbf{n}\cdot\mathbf{n}) &= \mathbf{v}\cdot\mathbf{n}-\mathbf{p}\cdot\mathbf{n} && \text{By properties of the dot product}\\
c(\mathbf{n}\cdot\mathbf{n}) &= \mathbf{n}\cdot\mathbf{v} && \text{By } \mathbf{v}\cdot\mathbf{n}=\mathbf{n}\cdot\mathbf{v} \text{ and } \mathbf{p}\cdot\mathbf{n}=0\\
c &= \frac{\mathbf{n}\cdot\mathbf{v}}{\mathbf{n}\cdot\mathbf{n}} && \text{Dividing both sides by } \mathbf{n}\cdot\mathbf{n} \text{ (a scalar)}
\end{aligned}$$

Note: Figure 1.66 also shows $c\mathbf{n}=\text{proj}_{\mathbf{n}}(\mathbf{v})=\left(\frac{\mathbf{n}\cdot\mathbf{v}}{\mathbf{n}\cdot\mathbf{n}}\right)\mathbf{n}$ which implies $c=\frac{\mathbf{n}\cdot\mathbf{v}}{\mathbf{n}\cdot\mathbf{n}}$.

Step 2. Letting $c=\frac{\mathbf{n}\cdot\mathbf{v}}{\mathbf{n}\cdot\mathbf{n}}$ in $\mathbf{p}=\mathbf{v}-c\mathbf{n}$, we have $\mathbf{p}=\mathbf{v}-\left(\frac{\mathbf{n}\cdot\mathbf{v}}{\mathbf{n}\cdot\mathbf{n}}\right)\mathbf{n}$.

48. We will find $\mathbf{p}=\text{proj}_{\mathscr{P}}(\mathbf{v})$ using $\mathbf{p}=\mathbf{v}-\left(\frac{\mathbf{n}\cdot\mathbf{v}}{\mathbf{n}\cdot\mathbf{n}}\right)\mathbf{n}$ and $\mathbf{v}=[1,0,-2]$.

(a) Step 1. Since $x+y+z=0$, $\mathbf{n}=[1,1,1]$.
So $\mathbf{n}\cdot\mathbf{v}=[1,1,1]\cdot[1,0,-2]=1\cdot1+1\cdot0-1\cdot2=-1$,
$\mathbf{n}\cdot\mathbf{n}=[1,1,1]\cdot[1,1,1]=1\cdot1+1\cdot1+1\cdot1=3$, and $c=\frac{\mathbf{n}\cdot\mathbf{v}}{\mathbf{n}\cdot\mathbf{n}}=-\frac{1}{3}$.

Step 2. So $\mathbf{p}=\mathbf{v}-\left(\frac{\mathbf{n}\cdot\mathbf{v}}{\mathbf{n}\cdot\mathbf{n}}\right)\mathbf{n}=\begin{bmatrix}1\\0\\-2\end{bmatrix}-\frac{1}{3}\begin{bmatrix}1\\1\\1\end{bmatrix}=\begin{bmatrix}1\\0\\-2\end{bmatrix}-\begin{bmatrix}1/3\\1/3\\1/3\end{bmatrix}=\begin{bmatrix}2/3\\-1/3\\-7/3\end{bmatrix}$.

(b) Step 1. Since $3x-y+z=0$, $\mathbf{n}=[3,-1,1]$.
So $\mathbf{n}\cdot\mathbf{v}=[3,-1,1]\cdot[1,0,-2]=3\cdot1-1\cdot0-1\cdot2=1$,
$\mathbf{n}\cdot\mathbf{n}=[3,-1,1]\cdot[3,-1,1]=3\cdot3+1\cdot1+1\cdot1=11$, and $c=\frac{\mathbf{n}\cdot\mathbf{v}}{\mathbf{n}\cdot\mathbf{n}}=\frac{1}{11}$.

Step 2. So $\mathbf{p}=\mathbf{v}-\left(\frac{\mathbf{n}\cdot\mathbf{v}}{\mathbf{n}\cdot\mathbf{n}}\right)\mathbf{n}=\begin{bmatrix}1\\0\\-2\end{bmatrix}-\frac{1}{11}\begin{bmatrix}1\\1\\1\end{bmatrix}=\begin{bmatrix}1\\0\\-2\end{bmatrix}-\begin{bmatrix}1/11\\1/11\\1/11\end{bmatrix}=\begin{bmatrix}10/11\\-1/11\\-23/11\end{bmatrix}$.

(c) Step 1. Since $3x-2z=0$, $\mathbf{n}=[3,0,-2]$.
So $\mathbf{n}\cdot\mathbf{v}=[3,0,-2]\cdot[1,0,-2]=3\cdot1-0\cdot0+2\cdot2=7$,
$\mathbf{n}\cdot\mathbf{n}=[3,0,-2]\cdot[3,0,-2]=3\cdot3+0\cdot0+2\cdot2=13$, and $c=\frac{\mathbf{n}\cdot\mathbf{v}}{\mathbf{n}\cdot\mathbf{n}}=\frac{7}{13}$.

Step 2. So $\mathbf{p}=\mathbf{v}-\left(\frac{\mathbf{n}\cdot\mathbf{v}}{\mathbf{n}\cdot\mathbf{n}}\right)\mathbf{n}=\begin{bmatrix}1\\0\\-2\end{bmatrix}-\frac{7}{13}\begin{bmatrix}1\\1\\1\end{bmatrix}=\begin{bmatrix}1\\0\\-2\end{bmatrix}-\begin{bmatrix}7/13\\7/13\\7/13\end{bmatrix}=\begin{bmatrix}6/13\\-7/13\\-33/13\end{bmatrix}$.

(d) Step 1. Since $2x-3y+z=0$, $\mathbf{n}=[2,-3,1]$.
So $\mathbf{n}\cdot\mathbf{v}=[2,-3,1]\cdot[1,0,-2]=2\cdot1-3\cdot0-1\cdot2=0$.

Step 2. So $\mathbf{p}=\mathbf{v}=\begin{bmatrix}1\\0\\-2\end{bmatrix}$.

Note: The projection is the vector itself because the vector is parallel to the plane. That is equivalent to the vector being orthogonal to the normal of the plane.

Exploration: The Cross Product

Since Explorations are self-contained, only solutions will be provided.

1. (a) $\mathbf{u}\times\mathbf{v}=\begin{bmatrix}1(2)-1(-1)\\1(3)-0(2)\\0(-1)-1(3)\end{bmatrix}=\begin{bmatrix}3\\3\\-3\end{bmatrix}$. (b) $\mathbf{u}\times\mathbf{v}=\begin{bmatrix}-3\\-3\\3\end{bmatrix}$.

(c) $\mathbf{u}\times\mathbf{v}=\begin{bmatrix}0\\0\\0\end{bmatrix}$. (d) $\mathbf{u}\times\mathbf{v}=\begin{bmatrix}1\\-2\\1\end{bmatrix}$.

2. $\mathbf{e}_1\times\mathbf{e}_2=\begin{bmatrix}1\\0\\0\end{bmatrix}\times\begin{bmatrix}0\\1\\0\end{bmatrix}=\begin{bmatrix}0-0(1)\\0-0\\1(1)-0\end{bmatrix}=\begin{bmatrix}0\\0\\1\end{bmatrix}=\mathbf{e}_3,$

$\mathbf{e}_2\times\mathbf{e}_3=\begin{bmatrix}0\\1\\0\end{bmatrix}\times\begin{bmatrix}0\\0\\1\end{bmatrix}=\begin{bmatrix}1\\0\\0\end{bmatrix}=\mathbf{e}_1$, and $\mathbf{e}_3\times\mathbf{e}_1=\begin{bmatrix}0\\0\\1\end{bmatrix}\times\begin{bmatrix}1\\0\\0\end{bmatrix}=\begin{bmatrix}0\\1\\0\end{bmatrix}=\mathbf{e}_2.$

3. Two vectors are orthogonal if their dot product equals zero. Check:

$$\begin{aligned}(\mathbf{u}\times\mathbf{v})\cdot\mathbf{u} &= \begin{bmatrix}u_2v_3-u_3v_2\\u_3v_1-u_1v_3\\u_1v_2-u_2v_1\end{bmatrix}\cdot\begin{bmatrix}u_1\\u_2\\u_3\end{bmatrix}\\ &= (u_2v_3-u_3v_2)\,u_1+(u_3v_1-u_1v_3)\,u_2+(u_1v_2-u_2v_1)\,u_3\\ &= (u_2v_3u_1-u_1v_3u_2)+(u_3v_1u_2-u_2v_1u_3)+(u_1v_2u_3-u_3v_2u_1)=0.\end{aligned}$$

$$\begin{aligned}(\mathbf{u}\times\mathbf{v})\cdot\mathbf{v} &= \begin{bmatrix}u_2v_3-u_3v_2\\u_3v_1-u_1v_3\\u_1v_2-u_2v_1\end{bmatrix}\cdot\begin{bmatrix}v_1\\v_2\\v_3\end{bmatrix}\\ &= (u_2v_3-u_3v_2)\,v_1+(u_3v_1-u_1v_3)\,v_2+(u_1v_2-u_2v_1)\,v_3\\ &= (u_2v_3v_1-u_2v_1v_3)+(u_3v_1v_2-u_3v_1v_2)+(u_1v_2v_3-u_1v_3v_2)=0.\end{aligned}$$

4. (a) A vector normal to the plane is $\mathbf{n} = \mathbf{u} \times \mathbf{v} = \begin{bmatrix} 0 \\ 1 \\ 1 \end{bmatrix} \times \begin{bmatrix} 3 \\ -1 \\ 2 \end{bmatrix} = \begin{bmatrix} 3 \\ 3 \\ -3 \end{bmatrix}$.

Thus, the normal form for the equation of this plane is $\mathbf{n} \cdot \mathbf{x} = \mathbf{n} \cdot \mathbf{p} \Leftrightarrow$

$$\begin{bmatrix} 3 \\ 3 \\ -3 \end{bmatrix} \cdot \begin{bmatrix} x \\ y \\ z \end{bmatrix} = \begin{bmatrix} 3 \\ 3 \\ -3 \end{bmatrix} \begin{bmatrix} 1 \\ 0 \\ -2 \end{bmatrix} \Leftrightarrow 3x + 3y - 3z = 9 \Leftrightarrow x + y - z = 3.$$

(b) Two vectors on the plane are $\mathbf{u} = \overrightarrow{PQ} = [2, 1, 1]$ and $\mathbf{v} = \overrightarrow{PR} = [1, 3, -2]$,

so a vector normal to the plane is $\mathbf{n} = \mathbf{u} \times \mathbf{v} = \begin{bmatrix} 2 \\ 1 \\ 1 \end{bmatrix} \times \begin{bmatrix} 1 \\ 3 \\ -2 \end{bmatrix} = \begin{bmatrix} -5 \\ 5 \\ 5 \end{bmatrix}$.

Thus, the normal form for the equation of this plane is $\mathbf{n} \cdot \mathbf{x} = \mathbf{n} \cdot \mathbf{p} \Leftrightarrow$

$$\begin{bmatrix} -5 \\ 5 \\ 5 \end{bmatrix} \cdot \begin{bmatrix} x \\ y \\ z \end{bmatrix} = \begin{bmatrix} -5 \\ 5 \\ 5 \end{bmatrix} \begin{bmatrix} 0 \\ -1 \\ 1 \end{bmatrix} \Leftrightarrow -5x + 5y + 5z = 0 \Leftrightarrow -x + y + z = 0.$$

5. (a) $\mathbf{v} \times \mathbf{u} = \begin{bmatrix} v_2u_3 - v_3u_2 \\ v_3u_1 - v_1u_3 \\ v_1u_2 - v_2u_1 \end{bmatrix} = - \begin{bmatrix} u_2v_3 - u_3v_2 \\ u_3v_1 - u_1v_3 \\ u_1v_2 - u_2v_1 \end{bmatrix} = -(\mathbf{u} \times \mathbf{v})$.

(b) $\mathbf{u} \times \mathbf{0} = \begin{bmatrix} u_1 \\ u_2 \\ u_3 \end{bmatrix} \times \begin{bmatrix} 0 \\ 0 \\ 0 \end{bmatrix} \begin{bmatrix} u_2(0) - u_3(0) \\ u_3(0) - u_1(0) \\ u_1(0) - u_2(0) \end{bmatrix} = \begin{bmatrix} 0 \\ 0 \\ 0 \end{bmatrix} = \mathbf{0}$.

(c) $\mathbf{u} \times \mathbf{u} = \begin{bmatrix} u_2u_3 - u_3u_2 \\ u_3u_1 - u_1u_3 \\ u_1u_2 - u_2u_1 \end{bmatrix} = \begin{bmatrix} 0 \\ 0 \\ 0 \end{bmatrix} = \mathbf{0}$.

(d) $\mathbf{u} \times k\mathbf{v} = \begin{bmatrix} u_2kv_3 - u_3kv_2 \\ u_3kv_1 - u_1kv_3 \\ u_1kv_2 - u_2kv_1 \end{bmatrix} = k \begin{bmatrix} u_2v_3 - u_3v_2 \\ u_3v_1 - u_1v_3 \\ u_1v_2 - u_2v_1 \end{bmatrix} = k(\mathbf{u} \times \mathbf{v})$.

(e) $\mathbf{u} \times k\mathbf{u} = k(\mathbf{u} \times \mathbf{u}) = k(\mathbf{0}) = \mathbf{0}$.

(f) We compute the cross product as follows:

$$\begin{aligned} \mathbf{u} \times (\mathbf{v} + \mathbf{w}) &= \begin{bmatrix} u_2(v_3 + w_3) - u_3(v_2 + w_2) \\ u_3(v_1 + w_1) - u_1(v_3 + w_3) \\ u_1(v_2 + w_2) - u_2(v_1 + w_1) \end{bmatrix} \\ &= \begin{bmatrix} (u_2v_3 - u_3v_2) + (u_2w_3 - u_3w_2) \\ (u_3v_1 - u_1v_3) + (u_3w_1 - u_1w_3) \\ (u_1v_2 - u_2v_1) + (u_1w_2 - u_2w_1) \end{bmatrix} \\ &= \begin{bmatrix} (u_2v_3 - u_3v_2) \\ (u_3v_1 - u_1v_3) \\ (u_1v_2 - u_2v_1) \end{bmatrix} + \begin{bmatrix} (u_2w_3 - u_3w_2) \\ (u_3w_1 - u_1w_3) \\ (u_1w_2 - u_2w_1) \end{bmatrix} = \mathbf{u} \times \mathbf{v} + \mathbf{u} \times \mathbf{w}. \end{aligned}$$

6. (a) We compute the cross product as follows:

$$\begin{aligned}
\mathbf{u}\cdot(\mathbf{v}\times\mathbf{w}) &= \begin{bmatrix} u_1 \\ u_2 \\ u_3 \end{bmatrix} \cdot \begin{bmatrix} v_2w_3 - v_3w_2 \\ v_3w_1 - v_1w_3 \\ v_1w_2 - v_2w_1 \end{bmatrix} \\
&= u_1v_2w_3 - u_1v_3w_2 + u_2v_3w_1 - u_2v_1w_3 + u_3v_1w_2 - u_3v_2w_1 \\
&= (u_2v_3 - u_3v_2)\,w_1 + (u_3v_1 - u_1v_3)\,w_2 + (u_1v_2 - u_2v_1)\,w_3 \\
&= (\mathbf{u}\times\mathbf{v})\cdot\mathbf{w}.
\end{aligned}$$

(b) We compute the cross product as follows:

$$\begin{aligned}
\mathbf{u}\times(\mathbf{v}\times\mathbf{w}) &= \begin{bmatrix} u_1 \\ u_2 \\ u_3 \end{bmatrix} \times \begin{bmatrix} v_2w_3 - v_3w_2 \\ v_3w_1 - v_1w_3 \\ v_1w_2 - v_2w_1 \end{bmatrix} = \begin{bmatrix} u_2(v_1w_2 - v_2w_1) - u_3(v_3w_1 - v_1w_3) \\ u_3(v_2w_3 - v_3w_2) - u_1(v_1w_2 - v_2w_1) \\ u_1(v_3w_1 - v_1w_3) - u_2(v_2w_3 - v_3w_2) \end{bmatrix} \\
&= \begin{bmatrix} (u_1w_1 + u_2w_2 + u_3w_3)\,v_1 - (u_1v_1 + u_2v_2 + u_3v_3)\,w_1 \\ (u_1w_1 + u_2w_2 + u_3w_3)\,v_2 - (u_1v_1 + u_2v_2 + u_3v_3)\,w_2 \\ (u_1w_1 + u_2w_2 + u_3w_3)\,v_3 - (u_1v_1 + u_2v_2 + u_3v_3)\,w_3 \end{bmatrix} \\
&= (u_1w_1 + u_2w_2 + u_3w_3)\begin{bmatrix} v_1 \\ v_2 \\ v_3 \end{bmatrix} - (u_1v_1 + u_2v_2 + u_3v_3)\begin{bmatrix} w_1 \\ w_2 \\ w_3 \end{bmatrix} \\
&= (\mathbf{u}\cdot\mathbf{w})\,\mathbf{v} - (\mathbf{u}\cdot\mathbf{v})\,\mathbf{w}.
\end{aligned}$$

(c) We compute the cross product as follows:

$$\begin{aligned}
\|\mathbf{u}\times\mathbf{v}\|^2 &= \left\|\begin{bmatrix} u_2v_3 - u_3v_2 \\ u_3v_1 - u_1v_3 \\ u_1v_2 - u_2v_1 \end{bmatrix}\right\|^2 = (u_2v_3 - u_3v_2)^2 + (u_1v_3 - u_3v_1)^2 + (u_1v_2 - u_2v_1)^2 \\
&= \left(u_1^2 + u_2^2 + u_3^2\right)^2\left(v_1^2 + v_2^2 + v_3^2\right)^2 - (u_1v_1 + u_2v_2 + u_3v_3)^2 \\
&= \|\mathbf{u}\|^2\,\|\mathbf{v}\|^2 - (\mathbf{u}\cdot\mathbf{v})^2.
\end{aligned}$$

7. *2:* $\mathbf{e}_1 \times (\mathbf{e}_2 \times \mathbf{e}_3) = (\mathbf{e}_1 \cdot \mathbf{e}_3)\,\mathbf{e}_2 - (\mathbf{e}_1 \cdot \mathbf{e}_2)\,\mathbf{e}_3$ [by 6(b)] $=0$ since $\mathbf{e}_i \cdot \mathbf{e}_j = 0$ for $i \neq j$.
Thus, since the $\mathbf{e}_i$ have length 1, we must have $\mathbf{e}_1 = \mathbf{e}_2 \times \mathbf{e}_3$, by 6(c).
Show $\mathbf{e}_2 = \mathbf{e}_3 \times \mathbf{e}_1$ and $\mathbf{e}_3 = \mathbf{e}_1 \times \mathbf{e}_2$ by cyclically permuting the indices.

3: $\mathbf{u}\cdot(\mathbf{u}\times\mathbf{v}) = (\mathbf{u}\times\mathbf{u})\cdot\mathbf{v}$ [by 10(a)]$= \mathbf{0}\cdot\mathbf{v}$ [by 9(c)]$= \mathbf{0}$, so $\mathbf{u}$ is orthogonal to $\mathbf{u}\times\mathbf{v}$.
Similarly, $\mathbf{v}\cdot(\mathbf{u}\times\mathbf{v}) = \mathbf{v}\cdot(-1)(\mathbf{v}\times\mathbf{u}) = (\mathbf{v}\times\mathbf{v})\cdot\mathbf{u} = \mathbf{0}$. So $\mathbf{v}$ is orthogonal to $\mathbf{u}\times\mathbf{v}$.

8. (a)
$$\begin{aligned}
\|\mathbf{u}\times\mathbf{v}\|^2 &= \|\mathbf{u}\|^2\,\|\mathbf{v}\|^2 - (\mathbf{u}\cdot\mathbf{v})^2 = \|\mathbf{u}\|^2\,\|\mathbf{v}\|^2 - \|\mathbf{u}\|^2\,\|\mathbf{v}\|^2\cos^2\theta \\
&= \|\mathbf{u}\|^2\,\|\mathbf{v}\|^2\left(1 - \cos^2\theta\right) = \|\mathbf{u}\|^2\,\|\mathbf{v}\|^2\sin^2\theta \Rightarrow \\
\|\mathbf{u}\times\mathbf{v}\| &= \|\mathbf{u}\|\,\|\mathbf{v}\|\sin\theta.
\end{aligned}$$

(b) From Figure 1, and recalling the area of a triangle is $A = \frac{1}{2}$(base)(height), we have: $A = \frac{1}{2}\left(\|\mathbf{u}\|\right)\left(\|\mathbf{v}\|\sin\theta\right)$. But $\|\mathbf{u}\times\mathbf{v}\| = \frac{1}{2}\|\mathbf{u}\|\,\|\mathbf{v}\|\sin\theta$, so $A = \frac{1}{2}\|\mathbf{u}\times\mathbf{v}\|$.

(c) Let $\mathbf{u} = \overrightarrow{AB} = [1, -1, -1]$ and $\mathbf{v} = \overrightarrow{AC} = [4, -3, 2]$. Then(b) $\Rightarrow$
The area of triangle ABC is $A = \frac{1}{2}\left\|[1, -1, -1] \times [4, -3, 2]\right\| = \frac{1}{2}\sqrt{62}$.

1.4 Code Vectors and Modular Arithmetic

1. $\mathbf{u}+\mathbf{v} = [0,1]+[1,1] = [1,0]$, $\mathbf{u}\cdot\mathbf{v} = 0+1 = 1$.

2. $\mathbf{u}+\mathbf{v} = [1,1,0]+[1,1,1] = [0,0,1]$, $\mathbf{u}\cdot\mathbf{v} = 1+1+0 = 0$.

3. $\mathbf{u}+\mathbf{v} = [1,0,1,1]+[1,1,1,1] = [0,1,0,0]$, $\mathbf{u}\cdot\mathbf{v} = 1+0+1+1 = 1$.

4. $\mathbf{u}+\mathbf{v} = [1,1,0,1,0]+[0,1,1,1,0] = [1,0,1,0,0]$, $\mathbf{u}\cdot\mathbf{v} = 0+1+0+1+0 = 0$.

5.

+	0	1	2	3
0	0	1	2	3
1	1	2	3	0
2	2	3	0	1
3	3	0	1	2

·	0	1	2	3
0	0	0	0	0
1	0	1	2	3
2	0	2	0	2
3	0	3	2	1

6.

+	0	1	2	3	3
0	0	1	2	3	4
1	1	2	3	4	0
2	2	3	4	0	1
3	3	4	0	1	2
4	4	0	1	2	3

·	0	1	2	3	4
0	0	0	0	0	0
1	0	1	2	3	4
2	0	2	4	1	3
3	0	3	1	4	2
4	0	4	3	2	1

7. $2+2+2 = 6 = 0$ in $\mathbb{Z}_3$.

8. $2\cdot 2\cdot 2 = 3\cdot 2+2 = 2$ in $\mathbb{Z}_3$.

9. $2(2+1+2) = 2(2) = 3\cdot 1+1 = 1$ in $\mathbb{Z}_3$.

10. $3+1+2+3 = 4\cdot 2+1 = 1$ in $\mathbb{Z}_4$.

11. $2\cdot 3\cdot 2 = 4\cdot 3+0 = 0$ in $\mathbb{Z}_4$.

12. $3(3+3+2) = 4\cdot 6+0 = 0$ in $\mathbb{Z}_4$.

13. $2+1+2+2+1 = 2$ in $\mathbb{Z}_3$, $2+1+2+2+1 = 0$ in $\mathbb{Z}_4$, $2+1+2+2+1 = 3$ in $\mathbb{Z}_5$.

14. $(3+4)\,(3+2+4+2) = 2\,(1) = 2$ in $\mathbb{Z}_5$.

15. $8\,(6+4+3) = 8\,(4) = 5$ in $\mathbb{Z}_9$.

16. $2^{100} = \left(2^{10}\right)^{10} = (1024)^{10} = 1^{10} = 1$ in $\mathbb{Z}_{11}$.

17. $[2,1,2]+[2,0,1] = [1,1,0]$ in $\mathbb{Z}_3^3$.

18. $[2,1,2]\cdot[2,2,1] = 1+2+2 = 2$ in $\mathbb{Z}_3^3$.

19. $[2,0,3,2]\cdot([3,1,1,2]+[3,3,2,1]) = [2,0,3,2]\cdot[2,0,3,3] = 0+0+2+1 = 3$ in $\mathbb{Z}_4^4$.
$[2,0,3,2]\cdot([3,1,1,2]+[3,3,2,1]) = [2,0,3,2]\cdot[1,4,3,3] = 2+0+4+1 = 2$ in $\mathbb{Z}_5^4$.

20. $x = 2+(-3) = 2+2 = 4$ in $\mathbb{Z}_5$.

21. $x = 1+(-5) = 1+1 = 2$ in $\mathbb{Z}_6$.

22. $x = (2)^{-1} = 2$ in $\mathbb{Z}_3$.

23. No solution. Why? Consider: $\frac{4}{2} = 2$.

24. $x = (2)^{-1} = 3$ in $\mathbb{Z}_5$.

25. $x = (3)^{-1} 4 = (2) 4 = 3$ in $\mathbb{Z}_5$.

26. No solution. Why? Consider: $\frac{6}{3} = 2$.

27. No solution. Why? $\frac{6}{2} = 3$, $\frac{8}{2} = 4$.

28. $x = (8)^{-1} 9 = (7) 4 = 6$ in $\mathbb{Z}_{11}$.

29. $x = (2)^{-1} (2 + (-3)) = (3) (2 + 2) = 2$ in $\mathbb{Z}_5$.

30. $x = (4)^{-1} (2 + (-5)) = (4)^{-1} (2 + 1)$, but $(4)^{-1}$ does not exist in $\mathbb{Z}_6$.
Therefore there is no solution.

31. Add 5 to both sides $\Rightarrow 6x = 6$, so $x = 1, 5$ (because $5 \cdot 8 = 40 = 8 \cdot 4 + 6$).

32. (a) All values. (b) All values. (c) All values.

33. (a) All $a \neq 0$ in $\mathbb{Z}_5$ have a solution because 5 is a prime number.

(b) $a = 1, 5$ because they have no common factors with 6 other than 1.

(c) a and m can have no common factors other than 1,
that is, the greatest common divisor (gcd) of a and m is 1.

34. We require that $[1, 0, 1, 1, d] \cdot [1, 1, 1, 1, 1] = 0 \Rightarrow 1 + 1 + 1 + d = 0 \Rightarrow d = 1$.
So the associated parity check code vector is $\mathbf{v} = [1, 0, 1, 1, 1]$.

35. We require that $[1, 1, 0, 1, 1, d] \cdot [1, 1, 1, 1, 1, 1] = 0 \Rightarrow 1 + 1 + 1 + 1 + d = 0 \Rightarrow d = 0$.
So the associated parity check code vector is $\mathbf{v} = [1, 1, 0, 1, 1, 0]$.

36. We check if $\mathbf{v} \cdot \mathbf{1} = \mathbf{0}$: $[1, 0, 1, 0] \cdot [1, 1, 1, 1] = 1 + 0 + 1 + 0 = 0$.
So a single error could not have occurred.

37. We check if $\mathbf{v} \cdot \mathbf{1} = \mathbf{0}$: $[1, 1, 1, 0, 1, 1] \cdot [1, 1, 1, 1, 1, 1] = 1 + 1 + 1 + 1 + 1 = 1$.
So a single error could have occurred.

38. We check if $\mathbf{v} \cdot \mathbf{1} = \mathbf{0}$: $[0, 1, 0, 1, 1, 1] \cdot [1, 1, 1, 1, 1, 1] = 1 + 1 + 1 + 1 = 0$.
So a single error could not have occurred.

39. We check if $\mathbf{v} \cdot \mathbf{1} = \mathbf{0}$: $[1, 1, 0, 1, 0, 1, 1, 1] \cdot [1, 1, 1, 1, 1, 1, 1, 1] = 1 + 1 + 1 + 1 + 1 + 1 = 0$.
So a single error could not have occurred.

40. We require that $[1, 2, 2, 2, d] \cdot [1, 1, 1, 1, 1] = 0$ in $\mathbb{Z}_3$. Thus $1 + 2 + 2 + 2 + d = 0 \Rightarrow d = 2$.

41. We require that $[3, 4, 2, 3, d] \cdot [1, 1, 1, 1, 1] = 0$in $\mathbb{Z}_5$. Thus $3 + 4 + 2 + 3 + d = 0 \Rightarrow d = 3$.

42. We require that $[1, 5, 6, 4, 5, d] \cdot [1, 1, 1, 1, 1, 1] = 0$ in $\mathbb{Z}_7$.
Thus, $1 + 5 + 6 + 4 + 5 + d = 0 \Rightarrow d = 0$.

43. We require that $[3, 0, 7, 5, 6, 8, d] \cdot [1, 1, 1, 1, 1, 1, 1] = 0$ in $\mathbb{Z}_9$.
Thus $3 + 7 + 5 + 6 + 8 + d = 0 \Rightarrow d = 7$.

44. Let $\mathbf{u}$ and $\mathbf{v}$ be arbitrary vectors in $\mathbb{Z}_m^n$ that differ in exactly one component. So: $\mathbf{u} = [u_1, u_2, \ldots, u_i, \ldots, u_n]$,$\mathbf{v} = [v_1, v_2, \ldots, v_j, \ldots, v_n]$, where $u_i = v_i$, $i \neq j$:

$$\begin{aligned}\mathbf{c} \cdot (\mathbf{u} - \mathbf{v}) &= [1, 1, \ldots, 1] \cdot ([u_1, u_2, \ldots, u_i, \ldots, u_n] - [v_1, v_2, \ldots, v_j, \ldots, v_n]) \\ &= [1, 1, \ldots, 1] \cdot ([0, 0, \ldots, u_i, \ldots, 0] - [0, 0, \ldots, v_j, \ldots, 0]) = u_i - v_j \neq 0\end{aligned}$$

so, $\mathbf{c} \cdot \mathbf{u} \neq \mathbf{c} \cdot \mathbf{v}$.

45. $\mathbf{c} \cdot \mathbf{u} = 0 \Rightarrow 3(0 + 9 + 6 + 7 + 0 + 7) + (5 + 4 + 4 + 0 + 2) + d = 0 \Rightarrow 7 + 5 + d = 0 \Rightarrow d = 8.$

46. $\mathbf{c} \cdot \mathbf{u} = 0 \Rightarrow \quad 3(0 + 4 + 1 + 1 + 4 + 2) + (1 + 0 + 4 + 8 + 1) + d = 0 \Rightarrow 6 + 4 + d = 0 \Rightarrow d = 0.$

47. Let $\mathbf{u} = [0, 4, 6, 9, 5, 6, 1, 8, 2, 0, 1, 5]$ be a UPC vector.

(a) We check if $\mathbf{c} \cdot \mathbf{u} = 0$:

$$\begin{aligned}\mathbf{c} \cdot \mathbf{u} &= [3, 1, 3, 1, 3, 1, 3, 1, 3, 1, 3, 1] \cdot [0, 4, 6, 9, 5, 6, 1, 8, 2, 0, 1, 5] \\ &= 3(0 + 6 + 5 + 1 + 2 + 1) + (4 + 9 + 6 + 8 + 0) + 5 = 5 + 7 + 5 = 7 \neq 0\end{aligned}$$

so, the UPC cannot be correct.

(b) Now $\mathbf{u} = [0, 4, u_3, 9, 5, 6, 1, 8, 2, 0, 1, 5]$, and we require that $\mathbf{c} \cdot \mathbf{u} = 0$:
$3(0 + u_3 + 5 + 1 + 2 + 1) + (4 + 9 + 6 + 8 + 0) + 5 = 0 \Leftrightarrow$
$3(u_3 + 9) + 2 = 0 \Leftrightarrow 3u_3 + 7 = 8 \Leftrightarrow 3u_3 = 1 \Leftrightarrow u_3 = 7.$
So, the correct UPC is $[0, 4, 7, 9, 5, 6, 1, 8, 2, 0, 1, 5]$.

48. Let $\mathbf{u} = [u_1, u_2, \ldots, u_{12}]$ be a correct UPC.
Let $\mathbf{u}' = \left[u_1, \ldots u_i', \ldots, u_{12}\right]$ be a UPC with an error in the ith component.
Now we check if $\mathbf{u}'$ is correct, first noting that $\mathbf{u}' = \mathbf{u} - u_i\mathbf{e}_i + u_i'\mathbf{e}_i$.
We have $\mathbf{c} \cdot \mathbf{u}' = \mathbf{c} \cdot (\mathbf{u} - u_i\mathbf{e}_i + u_i'\mathbf{e}_i) = \mathbf{c} \cdot \mathbf{u} - u_i\mathbf{c} \cdot \mathbf{e}_i + u_i'\mathbf{c} \cdot \mathbf{e}_i = \mathbf{c} \cdot \mathbf{e}_i\,(u_i' - u_i)$.
But $\mathbf{c} \cdot \mathbf{e}_i = 1$ or $3 \neq 0$ and $u_i' \neq u_i$ since the ith component is wrong.
So $\mathbf{c} \cdot \mathbf{u}' \neq 0$, and the UPC will detect the single error.

49. (a) We check if $\mathbf{c} \cdot \mathbf{u}' = 0$:

$$\begin{aligned}\mathbf{c} \cdot \mathbf{u}' &= [3,1,3,1,3,1,3,1,3,1,3,1] \cdot [0,4,7,9,2,7,0,2,0,9,4,6] \\ &= 3(0+7+2+0+0+4) + (4+9+7+2+9) + 6 = 9+1+6 = 6 \neq 0\end{aligned}$$

so the error will be detected.

(b) Transposing the third and fourth components of $\mathbf{u}$ gives $\mathbf{u}' = [0,7,9,4,2,7,0,2,0,9,4,6]$, while $\mathbf{c} \cdot \mathbf{u}' = 0$ in $\mathbb{Z}_{10}$, so the error will not be detected.

(c) Assume there is a transposition error between the ith and $(i+1)^{th}$ components and that the new UPC $= \mathbf{u}' = \mathbf{u} + \mathbf{e}_i(u_{i+1} - u_i) + \mathbf{e}_{i+1}(u_i - u_{i+1})$ satisfies $\mathbf{c} \cdot \mathbf{u}' = 0$. So:

$$\begin{aligned}0 &= \mathbf{c} \cdot (\mathbf{u} + \mathbf{e}_i(u_{i+1} - u_i) + \mathbf{e}_{i+1}(u_i - u_{i+1})) \\ &= \mathbf{c} \cdot \mathbf{u} + \mathbf{c} \cdot \mathbf{e}_i(u_{i+1} - u_i) + \mathbf{c} \cdot \mathbf{e}_{i+1}(u_i - u_{i+1}) \\ &= c_i(u_{i+1} - u_i) + c_{i+1}(u_i - u_{i+1}) = u_i(c_{i+1} - c_i) + u_{i+1}(c_i - c_{i+1})\end{aligned}$$

Case 1: If i is even then $c_i = 1$, $c_{i+1} = 3$, and the constraint becomes $0 = 2u_i + 8u_{i+1}$.
Case 2: If i is odd then $c_i = 3$, $c_{i+1} = 1$, and the constraint becomes $0 = 8u_i + 2u_{i+1}$.
So regardless of i, if u_i and u_{i+1} are transposed, the odd component of u_i and u_{i+1} is, say, u_i, and the adjacent components satisfy $8u_i + 2u_{i+1} = 0$,
the error will not be detected.
Note, the trivial solution is $u_i = u_{i+1}$, that is the adjacent components are the same.

50. $\mathbf{c} \cdot \mathbf{u} = 0 \Rightarrow [10,9,8,7,6,5,4,3,2,1] \cdot [0,3,8,7,9,7,9,9,3,d] = 1 + d \Rightarrow d = 10$.

51. $\mathbf{c} \cdot \mathbf{u} = 0 \Rightarrow [10,9,8,7,6,5,4,3,2,1] \cdot [0,3,9,4,7,5,6,8,2,d] = 4 + d \Rightarrow d = 7$.

52. Let $\mathbf{u} = [0,4,4,9,5,0,8,3,5,6]$ be an ISBN vector.

(a) $\mathbf{c} \cdot \mathbf{u} = [10,9,8,7,6,5,4,3,2,1] \cdot [0,4,4,9,5,0,8,3,5,6] = 218 = 9 \neq 0$ in $\mathbb{Z}_{11}$.
Thus the ISBN cannot be correct (because $\mathbf{c} \cdot \mathbf{u} \neq 0$).

(b) Now $\mathbf{u} = [0,4,4,9,u_5,0,8,3,5,6]$, and we require that $\mathbf{c} \cdot \mathbf{u} = 0$: $[10,9,8,7,6,5,4,3,2,1] \cdot [0,4,4,9,u_5,0,8,3,5,6] = 0 \Leftrightarrow 1 + 6u_5 = 0 \Leftrightarrow 6u_5 = 10 \Leftrightarrow u_5 = 9$.

53. (a) $\mathbf{u} = [0, 6, 7, 9, 7, 6, 2, 9, 0, 6]$ is a correct ISBN.
Consider a transposition between the fourth and fifth components.
Then $\mathbf{u}' = [0, 6, 7, 7, 9, 6, 2, 9, 0, 6]$.
Now check if $\mathbf{u}'$ is correct: $\mathbf{c} \cdot \mathbf{u}' = [10, 9, 8, 7, 6, 5, 4, 3, 2, 1] \cdot [0, 6, 7, 7, 9, 6, 2, 9, 0, 6] = 9 \neq 0$.
So, the error would be detected.

(b) In general, an ISBN vector $\mathbf{u}$ satisfies the check constraint $\sum_{i=1}^{10} (11 - i)(u_i) = 0$,
but in $\mathbb{Z}_{11}$ $(11 - i) = -i$, so the constraint becomes $\sum_{i=1}^{10} (-i)(u_i) = 0$.
Consider a transposition of any two adjacent elements (say the u_j and u_{j+1} elements).
This gives a new vector $\mathbf{u}' = \mathbf{u} + \mathbf{e}_j (u_{j+1} - u_j) + \mathbf{e}_{j+1}(u_j - u_{j+1})$.
The check constraint on $\mathbf{u}'$ becomes $\sum_{i=1}^{10} (-i)(u_i') = 0$,
but $u_i' = u_i + (\mathbf{e}_j)_i (u_{j+1} - u_j) + (\mathbf{e}_{j+1})_i (u_j - u_{j+1})$, so this becomes

$$
\begin{aligned}
\sum_{i=1}^{10} (-i)(u_i') &= = \sum_{i=1}^{10} (-i)(u_i) + \sum_{i=1}^{10} (-i)(\mathbf{e}_j)_i (u_{j+1} - u_j) + \sum_{i=1}^{10} (-i)(\mathbf{e}_{j+1})_i (u_j - u_{j+1}) \\
&= \sum_{i=1}^{10} (-i)(\mathbf{e}_j)_i (u_{j+1} - u_j) + \sum_{i=1}^{10} (-i)(\mathbf{e}_{j+1})_i (u_j - u_{j+1}) \\
&= (-j)(u_{j+1} - u_j) + (-(j+1))(u_j - u_{j+1}) \\
&= (-j)(u_{j+1} - u_j) + (-j + 10)(u_j - u_{j+1}) = 10(u_j - u_{j+1})
\end{aligned}
$$

So the constraint $\sum_{i=1}^{10} (-i)(u_i') = 0$ can only be met if $u_j = u_{j+1}$,
in which case the transposed vector is the same as the original.
So the constraint $\sum_{i=1}^{10} (-i)(u_i) = 0$ cannot be broken by transposing adjacent components.
An error made in any two adjacent components of the ISBN will be detected.

(c) See the proof in part (b).

54. Consider the ISBN $\mathbf{u} = [0, 8, 3, 7, 0, 9, 9, 0, 2, 6]$.

(a) $\mathbf{c} \cdot \mathbf{u} = [10, 9, 8, 7, 6, 5, 4, 3, 2, 1] \cdot [0, 8, 3, 7, 0, 9, 9, 0, 2, 6] = 5$ in $\mathbb{Z}_{11}$. $\mathbf{c} \cdot \mathbf{u} \neq 0$,
so this ISBN cannot be correct (because $\mathbf{c} \cdot \mathbf{u} \neq 0$).

(b) Consider the ISBN $\mathbf{u}$ with the first and second components transposed. Then:
$\mathbf{u} = [8, 0, 3, 7, 0, 9, 9, 0, 2, 6]$, $\mathbf{c} \cdot \mathbf{u} = [10, 9, 8, 7, 6, 5, 4, 3, 2, 1] \cdot [8, 0, 3, 7, 0, 9, 9, 0, 2, 6] = 0$.
So, the correct ISBN is $\mathbf{u} = [8, 0, 3, 7, 0, 9, 9, 0, 2, 6]$.

(c) Consider the ISBN $\mathbf{u} = [0, 3, 8, 7, 0, 9, 9, 0, 2, 6]$. If a transposition error occurs between the second and third components we get the vector $\mathbf{u}' = [0, 8, 3, 7, 0, 9, 9, 0, 2, 6]$,
where $\mathbf{c} \cdot \mathbf{u}' = 5 \neq 0$, so the error would be detected.
However, this error cannot be corrected because transpositions to both
$\mathbf{u} = [0, 3, 8, 7, 0, 9, 9, 0, 2, 6]$ and $\mathbf{u}'' = [8, 0, 3, 7, 0, 9, 9, 0, 2, 6]$ give proper ISBNs.

Chapter 1 Review

1. We will explain and give counter examples to justify our answers below.

(a) ***True***. Follows from the properties of $\mathbb{R}^n$ listed in Theorem 1.1 in Section 1.1:

$$\begin{aligned}
\mathbf{u} &= \mathbf{u}+\mathbf{0} && \text{Zero Property, Property (c)}\\
&= \mathbf{u}+(\mathbf{w}+(-\mathbf{w})) && \text{Additive Inverse Property, Property (d)}\\
&= (\mathbf{u}+\mathbf{w})+(-\mathbf{w}) && \text{Distributive Property, Property (b)}\\
&= (\mathbf{v}+\mathbf{w})+(-\mathbf{w}) && \text{By the given condition } \mathbf{u}+\mathbf{w}=\mathbf{v}+\mathbf{w}\\
&= \mathbf{v}+(\mathbf{w}+(-\mathbf{w})) && \text{Distributive Property, Property (b)}\\
&= \mathbf{v}+\mathbf{0} && \text{Additive Inverse Property, Property (d)}\\
&= \mathbf{v} && \text{Zero Property, Property (c)}
\end{aligned}$$

(b) ***False***. See Example 1.16 and Exercise 54 in Section 1.2. Two key counter examples:
Since $\mathbf{0}\cdot\mathbf{v}=0$ for every vector $\mathbf{v}$ in $\mathbb{R}^3$, $\mathbf{0}$ is orthogonal to every vector.
That is, if $\mathbf{u}=\mathbf{0}$, we know nothing about $\mathbf{v}$ and $\mathbf{w}$.
Let $\mathbf{u}$ and $\mathbf{v}$ be orthogonal to $\mathbf{w}$ then $\mathbf{u}\cdot\mathbf{w}=\mathbf{v}\cdot\mathbf{w}=0$.
E.g., consider $\mathbf{u}=[a,0,0]$, $\mathbf{v}=[0,b,0]$, and $\mathbf{w}=[0,0,c]$.

(c) ***False***. Note this property is ***not*** listed in Theorem 1.2 in Section 1.2.
Let $\mathbf{v}=\mathbf{0}$ then $\mathbf{u}\cdot\mathbf{0}=0$ and $\mathbf{0}\cdot\mathbf{w}=0$, but there is no restriction on $\mathbf{u}$ and $\mathbf{w}$.
Let $\mathbf{u}=\mathbf{w}$ then $\mathbf{u}\cdot\mathbf{v}=0$ and $\mathbf{v}\cdot\mathbf{u}=0$, but $\mathbf{u}\cdot\mathbf{u}\neq 0$ unless $\mathbf{u}=\mathbf{0}$.
E.g., consider $\mathbf{u}=[a,0,0]$, $\mathbf{v}=[0,b,0]$, and $\mathbf{w}=[c,0,0]$, then $\mathbf{u}\cdot\mathbf{w}=ac$.

(d) ***False***. When a line is parallel to plane then $\mathbf{d}\cdot\mathbf{n}=0$, that is $\mathbf{d}$ is ***orthogonal*** to $\mathbf{n}$.
See Figure 1.57 in Section 1.3.

(e) ***True***. Every line in plane $\mathscr{P}$ and parallel to $\mathscr{P}$ is ***orthogonal*** to its normal vector $\mathbf{n}$.
See Figure 1.62 in Section 1.3.

(f) ***True***. See the remarks following Example 1.24 in Section 1.3.

(g) ***False***. In $\mathbb{R}^3$ many non-parallel lines are ***skew*** (non-intersecting lines with $\mathbf{d}_1\neq\mathbf{d}_2$).
For example, ℓ_1 with $\mathbf{x}=t\mathbf{d}$ with $\mathbf{d}=[1,0,0]$ (the x-axis) and
ℓ_2 with $\mathbf{x}=\mathbf{p}+t\mathbf{d}$ with $\mathbf{p}=[0,0,1]$ and $\mathbf{d}=[0,1,0]$
(the line parallel to the y-axis through $[0,0,1]$).

(h) ***False***. See Examples 1.27 to 1.29 in Section 1.4.
For example, $[1,0,1]\cdot[1,0,1]=1+0+1=0$ in $\mathbb{Z}_2$.
In general, the dot product of any binary vector with an even number of 1s is 0.

(i) ***True***. See Example 1.37 in Section 1.4.
We have $\mathbf{c}\cdot\mathbf{u}=3(0+1+7+5+7+8)+(4+7+1+2+0+2)=100=0$ in $\mathbb{Z}_{10}$.

(j) ***False***. See Example 1.38 in Section 1.4.
$\mathbf{c}\cdot\mathbf{u}=[10,9,8,7,6,5,4,3,2,1]\cdot[0,5,3,2,3,4,1,7,4,8]=162\neq 0$ in $\mathbb{Z}_{11}$.

2. See Examples 1.1 and 1.5 in Section 1.1.

Let $\mathbf{w} = \begin{bmatrix} 10 \\ -10 \end{bmatrix}$ then $\mathbf{z} = 4\mathbf{u} + \mathbf{v} + \mathbf{w} = 4\begin{bmatrix} -1 \\ 5 \end{bmatrix} + \begin{bmatrix} 3 \\ 2 \end{bmatrix} + \begin{bmatrix} 10 \\ -10 \end{bmatrix} = \begin{bmatrix} 9 \\ 12 \end{bmatrix}$.

So the coordinates of the point at the head of $4\mathbf{u} + \mathbf{v}$ are $(9, 12)$.

3. See Example 1.5 in Section 1.1. ***Note***: We should do the vector arithmetic first.

$2\mathbf{x} + \mathbf{u} = 3(\mathbf{x} - \mathbf{v})$ implies $\mathbf{x} = \mathbf{u} + 3\mathbf{v} = \begin{bmatrix} -1 \\ 5 \end{bmatrix} + 3\begin{bmatrix} 3 \\ 2 \end{bmatrix} = \begin{bmatrix} 8 \\ 11 \end{bmatrix}$.

4. As in ***Exploration: Vectors and Geometry***, we have $\overrightarrow{BA} = \mathbf{b} - (-\mathbf{a}) = \mathbf{b} + \mathbf{a}$.

5. We proceed as in Example 1.14 of Section 1.2.

We have $\mathbf{u} \cdot \mathbf{v} = -1 \cdot 2 + 1 \cdot 1 - 2 \cdot 1 = -3$, $\|\mathbf{u}\| = \sqrt{(-1)^2 + 1^2 + 2^2} = \sqrt{6}$, and $\|\mathbf{v}\| = \sqrt{6}$.

Therefore, $\cos\theta = -\dfrac{3}{\sqrt{6}\sqrt{6}} = -\dfrac{1}{2}$, so $\theta = \cos^{-1}\left(-\dfrac{1}{2}\right) = \dfrac{2\pi}{3}$ radians or $120°$.

6. We proceed as in Example 1.17 of Section 1.2.
The length of $\mathbf{v}$ and the angle it makes with $\mathbf{u}$ determines how much of $\mathbf{u}$ it covers.
This is basically what $\text{proj}_{\mathbf{u}}(\mathbf{v}) = c\mathbf{u}$ says. The projection of $\mathbf{v}$ shadows a ***fraction*** of $\mathbf{u}$.

So $\mathbf{u} \cdot \mathbf{v} = \begin{bmatrix} 1 \\ -2 \\ 2 \end{bmatrix} \cdot \begin{bmatrix} 1 \\ 1 \\ 1 \end{bmatrix} = 1$, $\mathbf{u} \cdot \mathbf{u} = \begin{bmatrix} 1 \\ -2 \\ 2 \end{bmatrix} \cdot \begin{bmatrix} 1 \\ -2 \\ 2 \end{bmatrix} = 9$, $\text{proj}_{\mathbf{u}}(\mathbf{v}) = \left(\dfrac{\mathbf{u} \cdot \mathbf{v}}{\mathbf{u} \cdot \mathbf{u}}\right)\mathbf{u} = \begin{bmatrix} 1/9 \\ -2/9 \\ 2/9 \end{bmatrix}$.

In this case, the projection of $\mathbf{v}$ shadows $\frac{1}{9}$ of $\mathbf{u}$.

7. We use the given conditions to find a unit vector in the xy-plane orthogonal to $\mathbf{v} = [1, 2, 3]$.

Step 1. Figure 1.15 in Section 1.1 implies any vector in the xy-plane has a z-component of 0.
So, the vector $\mathbf{u}$ we are looking for must be of the form $\mathbf{u} = [a, b, 0]$.

Step 2. Like Exercise 42 in Section 1.2, since $\mathbf{u}$ is orthogonal to $\mathbf{v}$, we have

$$\mathbf{u} \cdot \mathbf{v} = \begin{bmatrix} 1 \\ 2 \\ 3 \end{bmatrix} \cdot \begin{bmatrix} a \\ b \\ 0 \end{bmatrix} = a + 2b = 0. \text{ So } a = -2b \text{ and } \mathbf{u} = \begin{bmatrix} -2b \\ b \\ 0 \end{bmatrix} = b\begin{bmatrix} -2 \\ 1 \\ 0 \end{bmatrix}.$$

Step 3. As in Example 1.12 of Section 1.2, we ***normalize*** $\mathbf{u}$ to create $\mathbf{w}$, the unit vector.

Letting $b = 1$ above gives us $\mathbf{u} = \begin{bmatrix} -2 \\ 1 \\ 0 \end{bmatrix}$ and $\|\mathbf{u}\| = \sqrt{(-2)^2 + 1^2 + 0^2} = \sqrt{5}$.

So ***one*** vector that works is $\mathbf{w} = \left(\dfrac{1}{\|\mathbf{u}\|}\right)\mathbf{u} = \dfrac{1}{\sqrt{5}}\begin{bmatrix} -2 \\ 1 \\ 0 \end{bmatrix} = \begin{bmatrix} -2/\sqrt{5} \\ 1/\sqrt{5} \\ 0 \end{bmatrix}$.

Note: The fact that we got to choose a value for b implies there are infinitely many solutions.

8. We begin by noting $\begin{aligned} x &= 2 - t \\ y &= 3 + 2t \\ z &= -1 + t \end{aligned}$ implies $\mathbf{x} = \begin{bmatrix} 2 \\ 3 \\ -1 \end{bmatrix} + t \begin{bmatrix} -1 \\ 2 \\ 1 \end{bmatrix}$, so $\mathbf{d} = \begin{bmatrix} -1 \\ 2 \\ 1 \end{bmatrix} = \mathbf{n}$.

From Example 1.23 of Section 1.3, we have $\mathbf{n} \cdot \mathbf{p} = \begin{bmatrix} -1 \\ 2 \\ 1 \end{bmatrix} \cdot \begin{bmatrix} 1 \\ 1 \\ 1 \end{bmatrix} = -1 \cdot 1 + 1 \cdot 2 + 1 \cdot 1 = 2$.

So the normal equation $\mathbf{n} \cdot \mathbf{x} = \mathbf{n} \cdot \mathbf{p}$ becomes the general equation $-x + 2y + z = 2$.

9. Planes that are parallel have parallel normals, so since our plane is parallel to $2x + 3y - z = 0$,

a normal to given plane and therefore our plane is $\mathbf{n} = \begin{bmatrix} 2 \\ 3 \\ -1 \end{bmatrix}$.

As in Example 1.23 of Section 1.3, we find the plane through $P = (3, 2, 5)$ with $\mathbf{n} = \begin{bmatrix} 2 \\ 3 \\ -1 \end{bmatrix}$.

With $\mathbf{p} = \begin{bmatrix} 3 \\ 2 \\ 5 \end{bmatrix}$ and $\mathbf{x} = \begin{bmatrix} x \\ y \\ z \end{bmatrix}$, we have $\mathbf{n} \cdot \mathbf{p} = 2 \cdot 3 + 3 \cdot 2 - 1 \cdot 5 = 7$.

So the normal equation $\mathbf{n} \cdot \mathbf{x} = \mathbf{n} \cdot \mathbf{p}$ becomes the general equation $2x + 3y - z = 7$.

10. As in Exercise 9, we use the given conditions to find the normal of the plane, $\mathbf{n}$.

Step 1. The normal of the plane $\mathbf{n}$ must be normal to every line in the plane.

So, we find $\mathbf{d}_1 = \overrightarrow{AB}$, $\mathbf{d}_2 = \overrightarrow{BC}$ and compute $\mathbf{n} \cdot \mathbf{d}_1 = 0$, $\mathbf{n} \cdot \mathbf{d}_2 = 0$.

$$\mathbf{d}_1 = \begin{bmatrix} 1 \\ 0 \\ 1 \end{bmatrix} - \begin{bmatrix} 1 \\ 1 \\ 0 \end{bmatrix} = \begin{bmatrix} 0 \\ -1 \\ 1 \end{bmatrix} \text{ and } \mathbf{d}_2 = \begin{bmatrix} 0 \\ 1 \\ 2 \end{bmatrix} - \begin{bmatrix} 1 \\ 0 \\ 1 \end{bmatrix} = \begin{bmatrix} -1 \\ 1 \\ 1 \end{bmatrix}.$$

$$\mathbf{n} \cdot \mathbf{d}_1 = \begin{bmatrix} a \\ b \\ c \end{bmatrix} \cdot \begin{bmatrix} 0 \\ -1 \\ 1 \end{bmatrix} = -b + c = 0 \Rightarrow c = b. \text{ So } \mathbf{n} = \begin{bmatrix} a \\ b \\ b \end{bmatrix}.$$

$$\mathbf{n} \cdot \mathbf{d}_2 = \begin{bmatrix} a \\ b \\ b \end{bmatrix} \cdot \begin{bmatrix} -1 \\ 1 \\ 1 \end{bmatrix} = -a + b + b = 0 \Rightarrow a = 2b. \text{ So } \mathbf{n} = \begin{bmatrix} 2b \\ b \\ b \end{bmatrix} = b \begin{bmatrix} 2 \\ 1 \\ 1 \end{bmatrix}.$$

Step 2. Take $P = A = (1, 1, 0)$ and compute $\mathbf{n} \cdot \mathbf{x} = \mathbf{n} \cdot \mathbf{p}$ to find the general equation.

Since $\mathbf{n} \cdot \mathbf{p} = 2 \cdot 1 + 1 \cdot 1 + 1 \cdot 0 = 3$, we have the general equation $2x + y + z = 3$.

Note: We could also have taken $\mathbf{p} = \mathbf{b}$ or $\mathbf{c}$ since $\mathbf{n} \cdot \mathbf{b} = \mathbf{n} \cdot \mathbf{c} = 3$. Why?

11. We proceed as in Exercise 41 of Section 1.2. See the notes prior to Exercise 41.

Let $\mathbf{u} = \overrightarrow{AB} = \begin{bmatrix} 1-1 \\ 0-1 \\ 1-0 \end{bmatrix} = \begin{bmatrix} 0 \\ -1 \\ 1 \end{bmatrix}$ and $\mathbf{v} = \overrightarrow{AC} = \begin{bmatrix} 0-1 \\ 1-1 \\ 2-0 \end{bmatrix} = \begin{bmatrix} -1 \\ 0 \\ 2 \end{bmatrix}$.

(a) We compute the necessary values ...

$$\mathbf{u} \cdot \mathbf{v} = \begin{bmatrix} 0 \\ -1 \\ 1 \end{bmatrix} \cdot \begin{bmatrix} -1 \\ 0 \\ 2 \end{bmatrix} = 2,$$

$$\mathbf{u} \cdot \mathbf{u} = \begin{bmatrix} 0 \\ -1 \\ 1 \end{bmatrix} \cdot \begin{bmatrix} 0 \\ -1 \\ 1 \end{bmatrix} = 2 \left(\|\mathbf{u}\| = \sqrt{2}\right),$$

$$\text{proj}_{\mathbf{u}}(\mathbf{v}) = \left(\frac{\mathbf{u} \cdot \mathbf{v}}{\mathbf{u} \cdot \mathbf{u}}\right) \mathbf{u} = \begin{bmatrix} 0 \\ -1 \\ 1 \end{bmatrix} \Rightarrow$$

$$\mathbf{v} - \text{proj}_{\mathbf{u}}(\mathbf{v}) = \begin{bmatrix} -1 \\ 1 \\ 1 \end{bmatrix} \Rightarrow$$

$$\|\mathbf{v} - \text{proj}_{\mathbf{u}}(\mathbf{v})\| = \sqrt{(-1)^2 + 1^2 + 1^2}$$

$$= \sqrt{3}$$

... then substitute into the formula for $\mathcal{A}$:

$$\mathcal{A} = \tfrac{1}{2} \|\mathbf{u}\| \, \|\mathbf{v} - \text{proj}_{\mathbf{u}}(\mathbf{v})\|$$

$$= \tfrac{1}{2} \sqrt{2} \, \sqrt{3} = \tfrac{\sqrt{6}}{2}.$$

(b) We compute the necessary values ...

$$\mathbf{u} \cdot \mathbf{v} = \begin{bmatrix} 0 \\ -1 \\ 1 \end{bmatrix} \cdot \begin{bmatrix} -1 \\ 0 \\ 2 \end{bmatrix} = 2,$$

$$\|\mathbf{u}\| = \sqrt{0^2 + (-1)^2 + 1^2} = \sqrt{2},$$

$$\|\mathbf{v}\| = \sqrt{(-1)^2 + 0^2 + +2^2} = \sqrt{5} \Rightarrow$$

$$\cos\theta = \frac{\mathbf{u} \cdot \mathbf{v}}{\|\mathbf{u}\| \, \|\mathbf{v}\|} = \frac{2}{\sqrt{10}} = \frac{\sqrt{10}}{5} \Rightarrow$$

$$\sin\theta = \sqrt{1 - \cos^2\theta} = \sqrt{1 - \left(\tfrac{\sqrt{10}}{5}\right)^2} = \tfrac{\sqrt{15}}{5}$$

... then substitute into the formula for $\mathcal{A}$:

$$\mathcal{A} = \tfrac{1}{2} \|\mathbf{u}\| \, \|\mathbf{v}\| \sin\theta$$

$$= \tfrac{1}{2} \sqrt{2} \, \sqrt{5} \, \tfrac{\sqrt{15}}{5} = \tfrac{\sqrt{6}}{2}.$$

12. From Example 1 in ***Exploration: Vectors and Geometry***, we have $\mathbf{m} = \frac{1}{2}(\mathbf{a} + \mathbf{b})$.

So $\mathbf{m} = \frac{1}{2}(\mathbf{a} + \mathbf{b}) = \frac{1}{2}\left(\begin{bmatrix} 5 \\ 1 \\ -2 \end{bmatrix} + \begin{bmatrix} 3 \\ -7 \\ 0 \end{bmatrix}\right) = \begin{bmatrix} 4 \\ -3 \\ -1 \end{bmatrix}$.

Therefore, the midpoint of the line segment between A and B is $(4, -3, -1)$.

13. We proceed as in Exercise 61 from Section 1.2.
We need to show $\|\mathbf{u}\| = 2$ and $\|\mathbf{v}\| = 3$ imply $\mathbf{u} \cdot \mathbf{v} \neq -7$.
From Theorem 1.4 (the Cauchy-Schwarz Inequality), we have $|\mathbf{x} \cdot \mathbf{y}| \leq \|\mathbf{x}\| \, \|\mathbf{y}\|$.
Substituting in the given values of $\|\mathbf{u}\| = 2$ and $\|\mathbf{v}\| = 3$ shows $|\mathbf{u} \cdot \mathbf{v}| \leq 6$.
Therefore, $-6 \leq \mathbf{u} \cdot \mathbf{v} \leq 6$. It follows immediately that $\mathbf{u} \cdot \mathbf{v} \neq -7$.

14. We will follow Example 1.26 in Section 1.3, then use $d(Q, \mathscr{P}) = \dfrac{|ax_0 + by_0 + cz_0 - d|}{\sqrt{a^2+b^2+c^2}}$.

By definition $ax + by + cz = d$ implies $\mathbf{n} = [a, b, c]$, so $2x + 3y - z = 0$ implies $\mathbf{n} = [2, 3, -1]$.

By Figure 1.62 in Section 1.3, we calculate the length of $\overrightarrow{RQ} = \text{proj}_{\mathbf{n}}(\mathbf{v})$, where $\mathbf{v} = \overrightarrow{PQ}$.

Step 1. By trial and error, we find $P = (0, 0, 0)$ satisfies $2x + 3y - z = 0$.

Step 2. $\mathbf{v} = \overrightarrow{PQ} = \mathbf{q} - \mathbf{p} = [3, 2, 5] - [0, 0, 0] = [3, 2, 5]$.

Step 3. $\text{proj}_{\mathbf{n}}(\mathbf{v}) = \left(\dfrac{\mathbf{n}\cdot\mathbf{v}}{\mathbf{d}\cdot\mathbf{n}}\right)\mathbf{n} = \left(\dfrac{2\cdot 3 + 3\cdot 2 - 1\cdot 5}{2^2 + 3^2 + (-1)^2}\right)\begin{bmatrix} 2 \\ 3 \\ -1 \end{bmatrix} = \dfrac{1}{2}\begin{bmatrix} 2 \\ 3 \\ -1 \end{bmatrix} = \begin{bmatrix} 1 \\ 3/2 \\ -1/2 \end{bmatrix}$.

Step 4. The distance from Q to $\mathscr{P}$ is $\|\text{proj}_{\mathbf{n}}(\mathbf{v})\| = \left\|\begin{bmatrix} 1 \\ 3/2 \\ -1/2 \end{bmatrix}\right\| = \dfrac{1}{2}\left\|\begin{bmatrix} 2 \\ 3 \\ -1 \end{bmatrix}\right\| = \dfrac{\sqrt{14}}{2}$.

Now for $d(Q, \mathscr{P}) = \dfrac{|ax_0 + by_0 + cz_0 - d|}{\sqrt{a^2+b^2+c^2}}$ we need identify a, b, c, d, and x_0, y_0, z_0.

Since $2x + 3y - z = 0$, $a = 1$, $b = 3$, $c = -1$, $d = 0$. From $Q = (3, 2, 5)$, $x_0 = 3, y_0 = 2, z_0 = 5$.

So $d(Q, \mathscr{P}) = \dfrac{|2\cdot 3 + 3\cdot 2 - 1\cdot 5 - 0|}{\sqrt{2^2 + 3^2 + (-1)^2}} = \dfrac{7}{\sqrt{14}} = \dfrac{\sqrt{14}}{2}$ as in Example 1.26 in Section 1.3.

15. We follow Example 1.27 in Section 1.3, then use $d(Q, \ell) = \dfrac{|ax_0 + by_0 + cz_0 - d|}{\sqrt{a^2+b^2+c^2}}$.

Comparing $\begin{bmatrix} x \\ y \\ z \end{bmatrix} = \begin{bmatrix} 0 \\ 1 \\ 2 \end{bmatrix} + t\begin{bmatrix} 1 \\ 1 \\ 1 \end{bmatrix}$ to $\mathbf{x} = \mathbf{p} + t\mathbf{d}$, we see ℓ has $P = (0, 1, 2)$ and $\mathbf{d} = \begin{bmatrix} 1 \\ 1 \\ 1 \end{bmatrix}$.

Now if we let $\mathbf{v} = \overrightarrow{PQ}$, then $\overrightarrow{PR} = \text{proj}_{\mathbf{d}}(\mathbf{v})$ and $\overrightarrow{RQ} = \mathbf{v} - \text{proj}_{\mathbf{d}}(\mathbf{v})$.

Step 1. $\mathbf{v} = \overrightarrow{PQ} = \mathbf{q} - \mathbf{p} = [3, 2, 5] - [0, 1, 2] = [3, 1, 3]$.

Step 2. $\text{proj}_{\mathbf{d}}(\mathbf{v}) = \left(\dfrac{\mathbf{d}\cdot\mathbf{v}}{\mathbf{d}\cdot\mathbf{d}}\right)\mathbf{d} = \left(\dfrac{1\cdot 3 + 1\cdot 1 + 1\cdot 3}{1\cdot 1 + 1\cdot 1 + 1\cdot 1}\right)\begin{bmatrix} 1 \\ 1 \\ 1 \end{bmatrix} = \dfrac{7}{3}\begin{bmatrix} 1 \\ 1 \\ 1 \end{bmatrix} = \begin{bmatrix} 7/3 \\ 7/3 \\ 7/3 \end{bmatrix}$.

Step 3. The vector we want is $\mathbf{v} - \text{proj}_{\mathbf{d}}(\mathbf{v}) = \begin{bmatrix} 3 \\ 1 \\ 3 \end{bmatrix} - \begin{bmatrix} 7/3 \\ 7/3 \\ 7/3 \end{bmatrix} = \begin{bmatrix} 2/3 \\ -4/3 \\ 2/3 \end{bmatrix} = -\dfrac{2}{3}\begin{bmatrix} 1 \\ -2 \\ 1 \end{bmatrix}$.

Step 4. The distance $d(Q, \ell)$ is $\|\mathbf{v} - \text{proj}_{\mathbf{d}}(\mathbf{v})\| = \left\|\begin{bmatrix} 2/3 \\ -4/3 \\ 2/3 \end{bmatrix}\right\| = \dfrac{2}{3}\left\|\begin{bmatrix} 1 \\ -2 \\ 1 \end{bmatrix}\right\| = \dfrac{2\sqrt{6}}{3}$.

$\mathbf{n}\cdot\mathbf{x} = \mathbf{n}\cdot\mathbf{p}$, $\begin{bmatrix} 1 \\ -2 \\ 1 \end{bmatrix}\cdot\begin{bmatrix} x \\ y \\ z \end{bmatrix} = \begin{bmatrix} 1 \\ -2 \\ 1 \end{bmatrix}\cdot\begin{bmatrix} 0 \\ 1 \\ 2 \end{bmatrix}$ so $x - 2y + z = 0$ and $a = 1$, $b = -2$, $c = 1$, $d = 0$.

So $d(Q, \ell) = \dfrac{|3 - 4 + 5 - 0|}{\sqrt{1^2 + (-2)^2 + 1^2}} = \dfrac{4}{\sqrt{6}} = \dfrac{2\sqrt{6}}{3}$ as in Example 1.27 of Section 1.3.

16. We follow Example 1.34 in Section 1.4.

$3 - (2+4)^3(4+3)^2 = 3 - (1)^3(2)^2 = 3 - (1)(4) = -1 = 4.$

Note: $-1 = 4$ in $\mathbb{Z}_5$ since $1 + 4 = 5 = 0 = 1 + (-1)$.

17. We begin by noting that $5 \cdot 3 = 15 = 1$ in $\mathbb{Z}_7$ so $3^{-1} = 5$ in $\mathbb{Z}_7$ since $5 \cdot 3 = 1 = 3 \cdot 3^{-1}$.

$3(x+2) = 5 \Rightarrow$	We multiply both sides by 5
$5 \cdot 3(x+2) = 5 \cdot 5 \Rightarrow$	because $3^{-1} = 5$ in $\mathbb{Z}_7$.
$x + 2 = 25 = 4 \Rightarrow$	$25 = 21 + 4 = 7 \cdot 3 + 4 = 4$ in $\mathbb{Z}_7$.
$x = 2.$	Simply subtract 2 from both sides.

Note: We should check our answer: $3(2+2) = 12 = 7 \cdot 1 + 5 = 5$ in $\mathbb{Z}_7$.

18. It is ***impossible*** to solve $3(x+2) = 5$ in $\mathbb{Z}_9$ since 3 does ***not*** have a multiplicative inverse.

Note: Any number that shares a common factor with the ***base***,
like in this case 3 and 9, will ***not*** have a multiplicative inverse. Why?

19. To find the check digit d, we follow the remarks after Example 1.37 in Section 1.4.
We find the value of d that makes $\mathbf{c} \cdot \mathbf{u} = 0$ performing all calculations in $\mathbb{Z}_{10}$.

Recall the check vector $\mathbf{c}$ for UPC code is $\mathbf{c} = [3,1,3,1,3,1,3,1,3,1,3,1]$ so we have:

$\mathbf{c} \cdot \mathbf{u} = 3 \cdot (7+3+6+7+3+7) + 1 \cdot (3+9+1+0+1) + d = 3(3) + 1(4) + d = 3 + d = 0 \Rightarrow d = 7.$

So, the check digit in UPC$[7,3,3,9,6,1,7,0,3,1,7,d]$ is 7.

Check: $\mathbf{c} \cdot \mathbf{u} = 3 \cdot (7+3+6+7+3+7) + 1 \cdot (3+9+1+0+1+7) = 13 + 7 = 0$ in $\mathbb{Z}_{10}$.

20. To find the check digit d, we follow Example 1.38 in Section 1.4.
We find the value of d that makes $\mathbf{c} \cdot \mathbf{u} = 0$ performing all calculations in $\mathbb{Z}_{11}$.

Recall the check vector $\mathbf{c}$ for ISBN code is $\mathbf{c} = [10,9,8,7,6,5,4,3,2,1]$ so we have:

$\mathbf{c} \cdot \mathbf{u} = 10 \cdot 0 + 9 \cdot 7 + 8 \cdot 6 + 7 \cdot 7 + 6 \cdot 0 + 5 \cdot 3 + 4 \cdot 9 + 3 \cdot 4 + 2 \cdot 2 + 1 \cdot d = 8+4+5+4+3+1+4+d = 0 \Rightarrow d = 4.$

So, the check digit in ISBN$[0,7,6,7,0,3,9,4,2,d]$ is 4.

Ck: $\mathbf{c} \cdot \mathbf{u} = 10 \cdot 0 + 9 \cdot 7 + 8 \cdot 6 + 7 \cdot 7 + 6 \cdot 0 + 5 \cdot 3 + 4 \cdot 9 + 3 \cdot 4 + 2 \cdot 2 + 1 \cdot 7 = 29 + 4 = 0$ in $\mathbb{Z}_{11}$.

Chapter 2

Systems of Linear Equations

2.1 Introduction to Systems of Linear Equations

1. We follow Example 2.1 and justify our assertion by applying the definition of ***linear***.
$x - \pi y + \left(\sqrt[3]{5}\right) z = 0$ ***is*** linear ***because*** power of z is 1 and π, $\sqrt[3]{5}$ are constants.

2. We follow Example 2.1 and justify our assertion by applying the definition of ***linear***.
$x^2 + y^2 + z^2 = 1$ is ***not*** linear ***because*** x, y, z occur to the power 2.

3. $x^{-1} + 7y + z = \sin\frac{\pi}{9}$ is ***not*** linear ***because*** x occurs to the power -1.

4. $2x - xy - 5z = 0$ is ***not*** linear ***because*** the product xy is of degree 2.

5. $3\cos x - 4y + z = \sqrt{3}$ is ***not*** linear ***because*** $\cos x$ is not linear.

6. $(\cos 3)\,x - 4y + z = \sqrt{3}$ ***is*** linear ***because*** $\cos 3$ and $\sqrt{3}$ are constants.

7. As in Section 1.3, we put the equation of this line into general form $ax + by = c$.
$2x + y = 7 - 3y$ is equivalent to $2x + 4y = 7$ after adding $3y$ to both sides.
Note: When the equation is ***linear*** there is no restriction on x and y. Why?

8. We begin by determining the restrictions on the variables x and y.
Typical sources are 1) division, 2) square roots, and 3) domains (like $\log x \Rightarrow x > 0$).

Step 1. Determine restriction ***type***. With $\dfrac{x^2 - y^2}{x - y} = 1$, it is division.

Step 2. Set the denominator equal to zero to determine the restriction.
We have $x - y = 0 \Rightarrow x = y$. So, the ***restriction*** is $x \neq y$.

Step 3. Simplify the given equation using algebra.

$$\frac{x^2 - y^2}{x - y} = 1 \overset{\text{factor}}{\Rightarrow} \frac{(x-y)(x+y)}{x - y} = 1 \overset{\text{cancel}}{\Rightarrow} x + y = 1.$$

Note: This tells us the given function is equivalent to the line $x + y = 1$ provided $x \neq y$.

9. We begin by determining the restrictions on the variables x and y.
Typical sources are 1) division, 2) square roots, and 3) domains (like $\log x \Rightarrow x > 0$).

Step 1. Determine restriction ***type***. With $\dfrac{1}{x} + \dfrac{1}{y} = \dfrac{4}{xy}$, it is division.

Step 2. Set the denominators equal to zero to determine the restriction.
We have $x = 0$, $y = 0$, and $xy = 0$. So, the ***restriction*** is $x, y \neq 0$.

Step 3. Simplify the given equation using algebra.

$$\frac{1}{x} + \frac{1}{y} = \frac{4}{xy} \overset{\text{common denominator}}{\Rightarrow} \frac{y}{xy} + \frac{x}{xy} = \frac{4}{xy} \overset{\text{multiply both sides by xy}}{\Rightarrow} x + y = 4.$$

Note: This tells us the given function is equivalent to the line $x + y = 4$ provided $x, y \neq 0$.

10. We begin by determining the restrictions on the variables x and y.
Typical sources are 1) division, 2) square roots, and 3) domains (like $\log x \Rightarrow x > 0$).

Step 1. Determine restriction ***type***. With $\log_{10} x - \log_{10} y = 2$, it is domains.

Step 2. Apply the domain restrictions to determine the overall restriction.
In this case, we have the overall restriction of $x > 0$ and $y > 0$.

Step 3. Simplify the given equation using algebra.

$$\log_{10} x - \log_{10} y = 2 \overset{\text{properties of logarithms}}{\Rightarrow} \log_{10} \frac{x}{y} = 2 \overset{\text{treat as exponents}}{\Rightarrow}$$

$$10^{\log_{10} \frac{x}{y}} = 10^2 \overset{\text{cancel and simplify}}{\Rightarrow} \frac{x}{y} = 100 \overset{\text{put in general form}}{\Rightarrow} x - 100y = 0.$$

Note: This tells us the given function is equivalent to the line $x - 100y = 0$ provided $x, y > 0$.

11. As in Example 2.2(a), we set $x = t$ and solve for y.
Setting $x = t$ in $3x - 6y = 0$ gives us $3t - 6y = 0$. Solving for y yields $6y = 3t \Rightarrow y = \frac{1}{2}t$.
So, we see the complete set of solutions can be written in the parametric form $[t, \frac{1}{2}t]$.

Note: We could have set $y = t$ to get $3x - 6t = 0$ and solved for x so $x = 2t$ and $[2t, t]$.

12. As in Example 2.2(a), we set $x_1 = t$ and solve for x_2.
Setting $x_1 = t$ yields $2t + 3x_2 = 5$. Solving for x_2 yields $3x_2 = 5 - 2t \Rightarrow x_2 = \frac{5}{3} - \frac{2}{3}t$.
So, a complete set of solutions written in parametric form is $[t, \frac{5}{3} - \frac{2}{3}t]$.

Note: We could have set $x_2 = t$ and solved for x_1 to get the parametric form $[\frac{5}{2} - \frac{3}{2}t, t]$.

13. As in Example 2.2(b), we set $y = s$, $z = t$ and solve for x. (Why is this a good choice?)
This substitution yields $x + 2s + 3t = 4$. Solving for x yields $x = 4 - 2s - 3t$.
So, a complete set of solutions written in parametric form is $[4 - 2s - 3t, s, t]$.

14. As in Example 2.2(b), we set $x_1 = s$, $x_2 = t$ and solve for x_3.

This substitution yields $4s + 3t + 2x_3 = 1$. Solving for x_3 yields $x_3 = \frac{1}{2} - 2s - \frac{3}{2}t$.

So, a complete set of solutions written in parametric form is $[s, t, \frac{1}{2} - 2s - \frac{3}{2}t]$.

15. The lines intersect at $(3, -3)$,
so the unique solution is $[3, -3]$.
To solve, subtract 2nd from $1^{\text{st}} \Rightarrow$
$-x = -3 \Leftrightarrow x = 3$,
so substitution $\Rightarrow y = -3$.

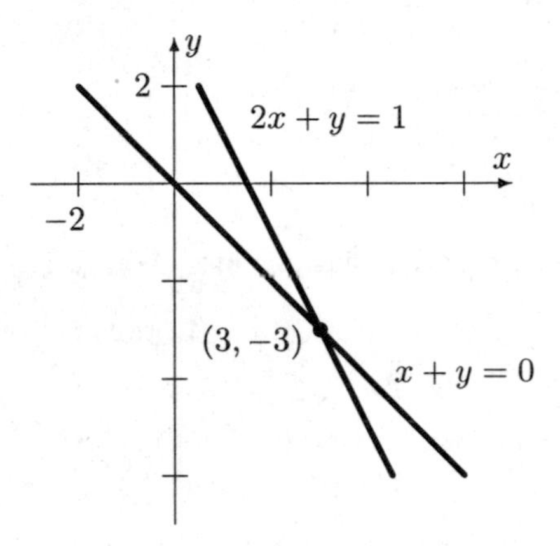

16. The lines intersect at $(3, -2)$,
so the unique solution is $[3, -2]$.
To solve, subtract $3 \times 1^{\text{st}}$ from $2^{\text{nd}} \Rightarrow$
$7y = -14 \Leftrightarrow y = -2$,
so substitution $\Rightarrow x = 3$.

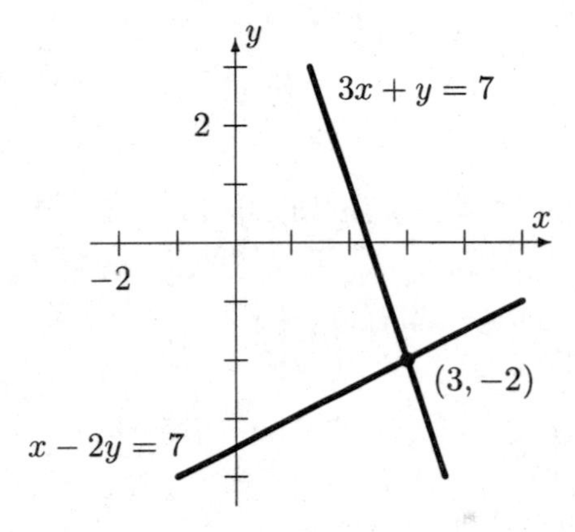

17. The lines are parallel $\Rightarrow$ no solution.
This system is inconsistent.
Add $3 \times 2^{\text{nd}}$ to $1^{\text{st}} \Rightarrow 0 = 6$.

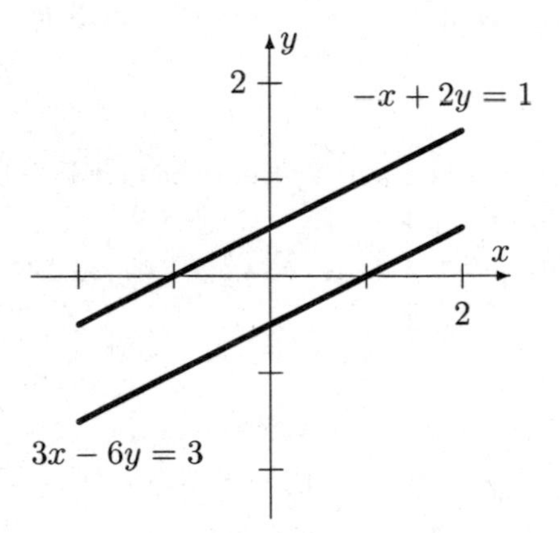

18. The graphs intersect in the line $2x - y = 4 \Rightarrow$
There are infinitely many solutions, $[t, 2t - 4]$.
The 2nd is $-\frac{3}{5} \times 1^{\text{st}}$.

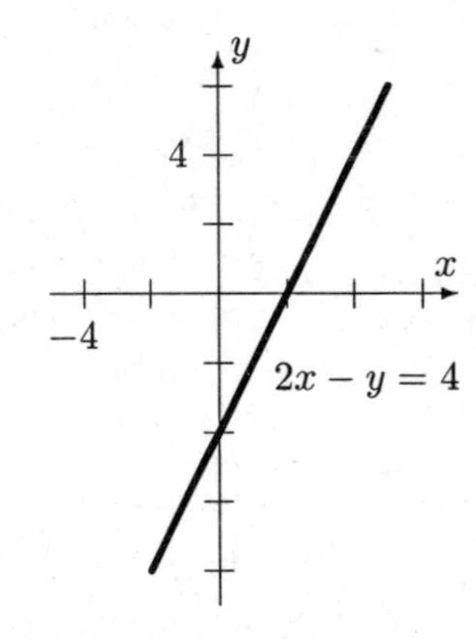

19. As in Example 2.5, we start from the last equation and work backward. We find successively $y = 3$ and $x = 1 + 2(3) - 7$. So, the unique solution is $[x, y] = [7, 3]$.

20. As in Example 2.5, we start from the last equation and work backward. We find successively $v = 3$ and $u = \frac{5}{2} + \frac{3}{2}(3) = 7$. So, the unique solution is $[u, v] = [7, 3]$.

21. We find the solution $[x, y, z] = \left[\frac{2}{3}, \frac{1}{3}, -\frac{1}{3}\right]$ using back substitution. Details below.

$$\begin{aligned} 3z &= -1 \Rightarrow & z &= -\tfrac{1}{3} \\ 2y - z &= 1 \Rightarrow 2y = 1 + z \Rightarrow y = \tfrac{1}{2} + \tfrac{1}{2}z \Rightarrow & y &= \tfrac{1}{2} + \tfrac{1}{2}\left(-\tfrac{1}{3}\right) = \tfrac{1}{3} \\ x - y + z &= 0 \Rightarrow \quad x = y - z \Rightarrow & x &= \left(\tfrac{1}{3}\right) - \left(-\tfrac{1}{3}\right) = \tfrac{2}{3} \end{aligned}$$

22. We find the solution $[x_1, x_2, x_3] = [0, 0, 0]$ using back substitution.
Note: This follows immediately from the fact that all three equations are equal to zero.

23. We find the solution $[x_1, x_2, x_3, x_4] = [5, -2, 1, 1]$ using back substitution. Details below. We find $x_3 = x_4 = 1$, $x_2 = -1 - 1 = -2$, and $x_1 = 1 - (-2) + 1 + 1 = 5$.

24. We combine the techniques of Examples 2 and 5 to find $[x, y, z] = [2 + 5t, -1 + 2t, t]$. Details: We let $z = t$ to get $y = -1 + 2t$, so $x = 5 + 3(-1 + 2t) - t = 2 + 5t$.

25. Working forward, we find $x = 2$, $y = -3 - 2(2) = -7$, and $z = -10 + 4(-7) + 3(2) = -32$. So the unique solution to the system is $[x, y, z] = [2, -7, -32]$.

26. Working forward, we find $x_1 = -1$, $x_2 = 5 + \frac{1}{2}(-1) = \frac{9}{2}$, and $x_3 = 7 - 2\left(\frac{9}{2}\right) - \frac{3}{2}(-1) = -\frac{1}{2}$. So the unique solution is $[x_1, x_2, x_3] = \left[-1, \frac{9}{2}, -\frac{1}{2}\right]$.

27. As in the solution to Example 2.6, we create the augmented matrix from the coefficients.

The system $\begin{aligned} x - y &= 0 \\ 2x + y &= 3 \end{aligned}$ has $\left[\begin{array}{cc|c} 1 & -1 & 0 \\ 2 & 1 & 3 \end{array}\right]$ as its augmented matrix.

28. As in the solution to Example 2.6, we create the augmented matrix from the coefficients.

The system $\begin{aligned} 2x_1 + 3x_2 - x_3 &= 1 \\ x_1 + x_3 &= 0 \\ -x_1 + 2x_2 - 2x_3 &= 0 \end{aligned}$ has $\left[\begin{array}{ccc|c} 2 & 3 & -1 & 1 \\ 1 & 0 & 1 & 0 \\ -1 & 2 & -2 & 0 \end{array}\right]$ as its augmented matrix.

29. The system $\begin{aligned} x + 5y &= -1 \\ -x + y &= -5 \\ 2x + 4y &= 4 \end{aligned}$ has $\left[\begin{array}{cc|c} 1 & 5 & -1 \\ -1 & 1 & -5 \\ 2 & 4 & 4 \end{array}\right]$ as its augmented matrix.

30. The system $\begin{aligned} a - 2b + d &= 2 \\ -a + b - c - 3d &= 1 \end{aligned}$ has $\left[\begin{array}{cccc|c} 1 & -2 & 0 & 1 & 2 \\ -1 & 1 & -1 & -3 & 1 \end{array}\right]$ as its augmented matrix.

31. The augmented matrix $\left[\begin{array}{ccc|c} 0 & 1 & 1 & 1 \\ 1 & -1 & 0 & 1 \\ 2 & -1 & 1 & 1 \end{array}\right]$ becomes $\begin{aligned} y + z &= 1 \\ x - y &= 1 \\ 2x - y + z &= 1 \end{aligned}$ as a system.

32. The augmented matrix $\left[\begin{array}{rrrrr|r} 1 & -1 & 0 & 3 & 1 & 2 \\ 1 & 1 & 2 & 1 & -1 & 4 \\ 0 & 1 & 0 & 2 & 3 & 0 \end{array}\right]$ becomes $\begin{array}{rl} a-b+ \quad\quad 3d+\ e & =2 \\ a+b+2c+\ d-\ e & =4 \\ b+ \quad\quad 2d+3e & =0 \end{array}$.

33. As in Example 2.4(a), we add $(x-y)+(2x+y)=0+3$ to get $3x=3 \Rightarrow x=1$ and $y=1$. A quick check confirms that $[1,1]$ is indeed the unique solution of the system.

34. As shown after Example 2.6, we row reduce the augmented matrix from Exercise 28.

$$\left[\begin{array}{rrr|r} 2 & 3 & -1 & 1 \\ 1 & 0 & 1 & 0 \\ -1 & 2 & -2 & 0 \end{array}\right] \xrightarrow{R_1 \leftrightarrow R_2} \left[\begin{array}{rrr|r} 1 & 0 & 1 & 0 \\ 2 & 3 & -1 & 1 \\ -1 & 2 & -2 & 0 \end{array}\right] \xrightarrow[R_2+R_1]{R_2-2R_1} \left[\begin{array}{rrr|r} 1 & 0 & 1 & 0 \\ 0 & 3 & -3 & 1 \\ 0 & 2 & -1 & 0 \end{array}\right] \xrightarrow{\frac{1}{3}R_2} \left[\begin{array}{rrr|r} 1 & 0 & 1 & 0 \\ 0 & 1 & -1 & 1/3 \\ 0 & 2 & -1 & 0 \end{array}\right]$$

$$\left[\begin{array}{rrr|r} 1 & 0 & 1 & 0 \\ 0 & 1 & -1 & 1/3 \\ 0 & 2 & -1 & 0 \end{array}\right] \xrightarrow{R_3-2R_2} \left[\begin{array}{rrr|r} 1 & 0 & 1 & 0 \\ 0 & 1 & -1 & 1/3 \\ 0 & 0 & 1 & -2/3 \end{array}\right] \xrightarrow[R_2+R_3]{R_1-R_3} \left[\begin{array}{rrr|r} 1 & 0 & 0 & 2/3 \\ 0 & 1 & 0 & -1/3 \\ 0 & 0 & 1 & -2/3 \end{array}\right] \Rightarrow$$

$x_1=\frac{2}{3}$, $x_2=-\frac{1}{3}$, and $x_3=-\frac{2}{3}$. So the solution is $[x_1,x_2,x_3]=\left[\frac{2}{3},-\frac{1}{3},-\frac{2}{3}\right]$.

35. As shown after Example 2.6, we row reduce the augmented matrix from Exercise 29.

$$\left[\begin{array}{rr|r} 1 & 5 & -1 \\ -1 & 1 & -5 \\ 2 & 4 & 4 \end{array}\right] \xrightarrow[R_3-2R_1]{R_2+R_1} \left[\begin{array}{rr|r} 1 & 5 & -1 \\ 0 & 6 & -6 \\ 0 & -6 & 6 \end{array}\right] \xrightarrow{R_3+R_2} \left[\begin{array}{rr|r} 1 & 5 & -1 \\ 0 & 6 & -6 \\ 0 & 0 & 0 \end{array}\right] \Rightarrow$$

$y=-1$ and $x=-1-5(-1)=4$, so the solution is $[x,y]=[4,-1]$.

36. As shown after Example 2.6, we row reduce the augmented matrix from Exercise 30.

$$\left[\begin{array}{rrrr|r} 1 & -2 & 0 & 1 & 2 \\ -1 & 1 & -1 & -3 & 1 \end{array}\right] \xrightarrow{R_2+R_1} \left[\begin{array}{rrrr|r} 1 & -2 & 0 & 1 & 2 \\ 0 & -1 & -1 & -2 & 3 \end{array}\right] \Rightarrow$$

$d=t$, $c=s$, $b=-3-s-2t$, and $a=2+2(-3-s-2t)-t=-4-2s-5t$, so the solution is $[a,b,c,d]=[-4-2s-5t,-3-s-2t,s,t]$.

37. As shown after Example 2.6, we row reduce the augmented matrix from Exercise 31.

$$\left[\begin{array}{rrr|r} 0 & 1 & 1 & 1 \\ 1 & -1 & 0 & 1 \\ 2 & -1 & 1 & 1 \end{array}\right] \xrightarrow[2R_2]{R_1 \leftrightarrow R_3} \left[\begin{array}{rrr|r} 2 & -1 & 1 & 1 \\ 2 & -2 & 0 & 2 \\ 0 & 1 & 1 & 1 \end{array}\right] \xrightarrow{R_2-R_1} \left[\begin{array}{rrr|r} 2 & -1 & 1 & 1 \\ 0 & -1 & -1 & 1 \\ 0 & 1 & 1 & 1 \end{array}\right] \xrightarrow{R_3+R_2} \left[\begin{array}{rrr|r} 2 & -1 & 1 & 1 \\ 0 & -1 & -1 & 1 \\ 0 & 0 & 0 & 2 \end{array}\right]$$

$\Rightarrow 0=2 \Rightarrow$ No solution.

38. As shown after Example 2.6, we row reduce the augmented matrix from Exercise 32.

$$\left[\begin{array}{rrrrr|r} 1 & -1 & 0 & 3 & 1 & 2 \\ 1 & 1 & 2 & 1 & -1 & 4 \\ 0 & 1 & 0 & 2 & 3 & 0 \end{array}\right] \xrightarrow[2R_3]{R_2-R_1} \left[\begin{array}{rrrrr|r} 1 & -1 & 0 & 3 & 1 & 2 \\ 0 & 2 & 2 & -2 & -2 & 2 \\ 0 & 2 & 0 & 4 & 6 & 0 \end{array}\right] \xrightarrow{R_3-R_2} \left[\begin{array}{rrrrr|r} 1 & -1 & 0 & 3 & 1 & 2 \\ 0 & 2 & 2 & -2 & -2 & 2 \\ 0 & 0 & -2 & 6 & 8 & -2 \end{array}\right]$$

Using back substitution, we get: $e=t$, $d=s$, $c=\left(-\frac{1}{2}\right)(-2-6s-8t)=1+3s+4t$.

$b=\left(\frac{1}{2}\right)(2-2(1+3s+4t)+2s+2t)=-2s-3t$, $a=2+(-2s-3t)-3s-t=2-5s-4t$.

So, the solution is $[a,b,c,d,e]=[2-5s-4t,-2s-3t,1+3s+4t,s,t]$.

39. The key to this problem is simple substitution.

(a) The fact that $x = t$ tells us that x is a free variable. What does that tell us?
The linear equations we are looking for must be multiples of each other. Why?
Substituting $t = x$ into $y = 3 - 2t$ yields $y = 3 - 2x \Rightarrow 2x + y = 3$.
Any multiple of this equation will create the system we are looking for.
For example, $2x + y = 3$ and $4x + 2y = 6$ (which is just $2\times$ the equation $2x + y = 3$).

(b) Substituting $s = y$ into $y = 3 - 2x$ yields $s = 3 - 2x \Rightarrow x = \frac{3}{2} - \frac{1}{2}s$.
The parametric solution then becomes $x = \frac{3}{2} - \frac{1}{2}s$ and $y = s$.

40. The key to this problem is simple substitution.

(a) Substituting $t = x_1$ into $x_2 = 1 + t$, $x_3 = 2 - t$ yields $x_2 = 1 + x_1$, $x_3 = 2 - x_1$.
These equations lead immediately to the system: $-x_1 + x_2 = 1$, $x_1 + x_3 = 2$.

(b) Substituting $s = x_3$ into $x_3 = 2 - x_1$ yields $s = 2 - x_1 \Rightarrow x_1 = 2 - s$.
Then substituting $2 - s = x_1$ into $x_2 = 1 + x_1$ yields $x_2 = 1 + (2 - s) \Rightarrow x_2 = 3 - s$.
The parametric solution then becomes $x_1 = 2 - s$, $x_2 = 3 - s$, and $x_3 = s$.

41. Let $u = \frac{1}{x}$, and $v = \frac{1}{y}$. Then the system of equations becomes $2u + 3v = 0$, $3u + 4v = 1$.
Solving the second equation for v gives $v = \frac{1}{4} - \frac{3}{4}u$. So, substitution $\Rightarrow 2u + 3\left(\frac{1}{4} - \frac{3}{4}u\right) = 0$.
Thus $u = 3$ and $v = \frac{1}{4} - \frac{3}{4}(3) = -2$. So, the solution is $[x, y] = \left[\frac{1}{3}, -\frac{1}{2}\right]$.

42. Let $u = x^2$, and $v = y^2$. So, the system becomes $u + 2v = 6$, $u - v = 3$.
Subtracting the second equation from the first gives $3v = 3 \Rightarrow v = 1$.
Substituting this into the second equation gives $u = 3 + 1 = 4$. Thus $u = 4$ and $v = 1 \Rightarrow$
The solution set is $[x, y] = \left[\pm\sqrt{4}, \pm\sqrt{1}\right]$. That is, $\{[2,1], [2,-1], [-2,1], [-2,-1]\}$.

43. Let $u = \tan x$, $v = \sin y$, $w = \cos z \Rightarrow u - 2v = 2$, $u - v + w = 2$, $v - w = -1$.
We form the augmented matrix and row reduce it to find the solution of the system:

$$\left[\begin{array}{ccc|c} 1 & -2 & 0 & 2 \\ 1 & -1 & 1 & 2 \\ 0 & 1 & -1 & -1 \end{array}\right] \xrightarrow{R_2 - R_1} \left[\begin{array}{ccc|c} 1 & -2 & 0 & 2 \\ 0 & 1 & 1 & 0 \\ 0 & 1 & -1 & -1 \end{array}\right] \xrightarrow{R_3 - R_2} \left[\begin{array}{ccc|c} 1 & -2 & 0 & 2 \\ 0 & 1 & 1 & 0 \\ 0 & 0 & -2 & -1 \end{array}\right].$$

Using back substitution $w = \frac{1}{2}$, $v = -\frac{1}{2}$, $u = 2 + 2\left(-\frac{1}{2}\right) = 1 \Rightarrow [u, v, w] = \left[1, -\frac{1}{2}, \frac{1}{2}\right]$.
Since $x = \tan^{-1} u$, $y = \sin^{-1} v$, $z = \cos^{-1} w$, the solution is $[x, y, z] = \left[\frac{\pi}{4}, -\frac{\pi}{6}, \frac{\pi}{3}\right]$.

44. Let $r = 2^a$, and $s = 3^b$. Then the system becomes $-r + 2s = 1$, $3r - 4s = 1$.
Adding three times the first equation to the second gives $2s = 4 \Rightarrow s = 2$.
Substituting $s = 2$ into $-r + 2s = 1$ yields $-r + 2(2) = 1 \Rightarrow r = 3 \Rightarrow [r, s] = [3, 2]$.
Since $a = \log_2 r$ and $b = \log_3 s$, the solution is $[a, b] = [\log_2 3, \log_3 2]$.

Exploration: Lies My Computer Told Me

1. $\begin{aligned} x+y &= 0 \\ x+\tfrac{801}{800}y &= 1 \end{aligned} \Rightarrow \begin{aligned} -800x-800y &= 0 \\ 800x+801y &= 800 \end{aligned} \Rightarrow \begin{aligned} x &= -800 \\ y &= 800 \end{aligned}$

2. $\begin{aligned} x+y &= 0 \\ x+1.0012y &= 1 \end{aligned} \Rightarrow \begin{aligned} -x-y &= 0 \\ x+1.0012y &= 1 \end{aligned} \Rightarrow \begin{aligned} x &= -833.33 \\ y &= 833.33 \end{aligned}$

3. $\begin{aligned} x+y &= 0 \\ x+1.00y &= 1 \end{aligned} \Rightarrow \begin{aligned} -x-y &= 0 \\ x+1.00y &= 1 \end{aligned} \Rightarrow 0=1 \Rightarrow$ No solution.

4. Even small changes in slope can cause lines to be widely divergent for large values of x.

To eight significant digits, we have:

$$\begin{aligned} 4.552x+7.083y &= 1.931 \\ 1.731x+2.693y &= 2.001 \end{aligned} \Rightarrow \begin{aligned} 7.879512x+12.260673y &= 3.342561 \\ -7.879512x-12.258536y &= -9.108552 \end{aligned} \Rightarrow$$

$$\begin{aligned} x &= 4198.8301 \\ y &= -2698.1708 \end{aligned}$$

To four significant digits, we have:

$$\begin{aligned} 4.552x+7.083y &= 1.931 \\ 1.731x+2.693y &= 2.001 \end{aligned} \Rightarrow \begin{aligned} 7.879x+12.26y &= 3.342 \\ -7.879x-12.25y &= -9.108 \end{aligned} \Rightarrow \begin{aligned} x &= 896.8 \\ y &= -576.4 \end{aligned}$$

To two significant digits, we have:

$$\begin{aligned} 4.5x+7.0y &= 1.9 \\ 1.7x+2.6y &= 2.0 \end{aligned} \Rightarrow \begin{aligned} 7.6x+11y &= 3.2 \\ -7.6x-11y &= -9 \end{aligned} \Rightarrow 0=-9 \Rightarrow \text{No solution.}$$

2.2 Direct Methods for Solving Linear Systems

1. No, this matrix is not in row echelon form. Why not? Give at least one reason.
The leading entry in row 3 appears to the left of the leading entry in row 2.

2. This matrix is in row echelon form, but not reduced row echelon form. Why not?.
There are many reasons. For example, the leading entry in row 1 is 7 not 1.

3. This matrix is in row echelon form, and also reduced row echelon form. Why is the 3 okay?
The 3 occurs in a column that does not contain a leading 1.

4. This matrix is in row echelon form, and also reduced row echelon form. Why are the 0s okay?
All three rows are zero, so no leading 1s are required.

5. No, this matrix is not in row echelon form. Why not? Give a reason.
The row of all zeroes is not at the bottom.

6. No, this matrix is not in row echelon form. Why not? Give a reason.
The leading entry in row 3 appears to the left of the leading entry in row 2.

7. No, this matrix is not in row echelon form. Why not? Give a reason.
The leading entry in row 2 appears underneath the leading entry in row 1.

8. This matrix is in row echelon form, but not reduced row echelon form. Why not?
The leading entry in row 4 is not a 1. Could we have given another reason?

9. (a) $\begin{bmatrix} 0 & 0 & 1 \\ 0 & 1 & 1 \\ 1 & 1 & 1 \end{bmatrix} \xrightarrow{R_1 \leftrightarrow R_3} \begin{bmatrix} 1 & 1 & 1 \\ 0 & 1 & 1 \\ 0 & 0 & 1 \end{bmatrix}$. (b) ... $\begin{bmatrix} 1 & 1 & 1 \\ 0 & 1 & 1 \\ 0 & 0 & 1 \end{bmatrix} \xrightarrow[R_2 - R_3]{R_1 - R_2} \begin{bmatrix} 1 & 0 & 0 \\ 0 & 1 & 0 \\ 0 & 0 & 1 \end{bmatrix}$.

10. (a) $\begin{bmatrix} 4 & 3 \\ 2 & 1 \end{bmatrix} \xrightarrow{R_2 - \frac{1}{2}R_1} \begin{bmatrix} 4 & 3 \\ 0 & -\frac{1}{2} \end{bmatrix}$ ... (b) ... $\begin{bmatrix} 4 & 3 \\ 0 & 1 \end{bmatrix} \xrightarrow{R_1 - 3R_2} \begin{bmatrix} 4 & 0 \\ 0 & 1 \end{bmatrix} \xrightarrow{\frac{1}{4}R_1} \begin{bmatrix} 1 & 0 \\ 0 & 1 \end{bmatrix}$.

11. (a) $\begin{bmatrix} 3 & 5 \\ 5 & -2 \\ 2 & 4 \end{bmatrix} \xrightarrow[R_3 - \frac{2}{3}R_1]{R_2 - \frac{5}{3}R_1} \begin{bmatrix} 3 & 5 \\ 0 & -\frac{31}{3} \\ 0 & \frac{2}{3} \end{bmatrix} \xrightarrow[\frac{3}{2}R_3]{-\frac{3}{31}R_2} \begin{bmatrix} 3 & 5 \\ 0 & 1 \\ 0 & 1 \end{bmatrix} \xrightarrow{R_3 - R_2} \begin{bmatrix} 3 & 5 \\ 0 & 1 \\ 0 & 0 \end{bmatrix}$.

(b) Continuing from (a): $\begin{bmatrix} 3 & 5 \\ 0 & 1 \\ 0 & 0 \end{bmatrix} \xrightarrow{R_1 - 5R_2} \begin{bmatrix} 3 & 0 \\ 0 & 1 \\ 0 & 0 \end{bmatrix} \xrightarrow{\frac{1}{3}R_1} \begin{bmatrix} 1 & 0 \\ 0 & 1 \\ 0 & 0 \end{bmatrix}$.

12. (a) $\begin{bmatrix} 2 & -4 & -2 & 6 \\ 3 & 1 & 6 & 6 \end{bmatrix} \xrightarrow{-2R_2} \begin{bmatrix} 2 & -4 & -2 & 6 \\ -6 & -2 & -12 & -12 \end{bmatrix} \xrightarrow{R_2 + 3R_1} \begin{bmatrix} 2 & -4 & -2 & 6 \\ 0 & -14 & -18 & 6 \end{bmatrix}$.

(b) ... from (a): $\begin{bmatrix} 2 & -4 & -2 & 6 \\ 0 & -14 & -18 & 6 \end{bmatrix} \xrightarrow[-\frac{1}{14}R_2]{\frac{1}{2}R_1} \begin{bmatrix} 1 & -2 & -1 & 3 \\ 0 & 1 & \frac{9}{7} & -\frac{3}{7} \end{bmatrix} \xrightarrow{R_1 + 2R_2} \begin{bmatrix} 1 & 0 & \frac{11}{7} & \frac{15}{7} \\ 0 & 1 & \frac{9}{7} & -\frac{3}{7} \end{bmatrix}$.

13. (a) $\begin{bmatrix} 3 & -2 & -1 \\ 2 & -1 & -1 \\ 4 & -3 & -1 \end{bmatrix} \xrightarrow[-3R_3]{-3R_2} \begin{bmatrix} 3 & -2 & -1 \\ -6 & 3 & 3 \\ -12 & 9 & 3 \end{bmatrix} \xrightarrow[R_3+4R_1]{R_2+2R_1} \begin{bmatrix} 3 & -2 & -1 \\ 0 & -1 & 1 \\ 0 & 1 & -1 \end{bmatrix} \xrightarrow{R_3+R_2} \begin{bmatrix} 3 & -2 & -1 \\ 0 & 1 & -1 \\ 0 & 0 & 0 \end{bmatrix}.$

(b) Continuing from (a): $\begin{bmatrix} 3 & -2 & -1 \\ 0 & 1 & -1 \\ 0 & 0 & 0 \end{bmatrix} \xrightarrow{R_1+2R_2} \begin{bmatrix} 3 & 0 & -3 \\ 0 & 1 & -1 \\ 0 & 0 & 0 \end{bmatrix} \xrightarrow{\frac{1}{3}R_1} \begin{bmatrix} 1 & 0 & -1 \\ 0 & 1 & -1 \\ 0 & 0 & 0 \end{bmatrix}.$

14. (a) $\begin{bmatrix} -2 & -4 & 7 \\ -3 & -6 & 10 \\ 1 & 2 & -3 \end{bmatrix} \xrightarrow[2R_3]{-2R_2} \begin{bmatrix} -2 & -4 & 7 \\ 6 & 12 & -20 \\ 2 & 4 & -6 \end{bmatrix} \xrightarrow[R_3+R_1]{R_2+3R_1} \begin{bmatrix} -2 & -4 & 7 \\ 0 & 0 & 1 \\ 0 & 0 & 1 \end{bmatrix} \xrightarrow{R_3-R_2} \begin{bmatrix} -2 & -4 & 7 \\ 0 & 0 & 1 \\ 0 & 0 & 0 \end{bmatrix}.$

(b) Continuing from (a): $\begin{bmatrix} -2 & -4 & 7 \\ 0 & 0 & 1 \\ 0 & 0 & 0 \end{bmatrix} \xrightarrow{R_1-7R_2} \begin{bmatrix} -2 & -4 & 0 \\ 0 & 0 & 1 \\ 0 & 0 & 0 \end{bmatrix} \xrightarrow{-\frac{1}{2}R_1} \begin{bmatrix} 1 & 2 & 0 \\ 0 & 0 & 1 \\ 0 & 0 & 0 \end{bmatrix}.$

15. $\begin{bmatrix} 1 & 2 & -4 & -4 & 5 \\ 0 & -1 & 10 & 9 & -5 \\ 0 & 0 & 1 & 1 & -1 \\ 0 & 0 & 0 & 0 & 24 \end{bmatrix} \xrightarrow{R_4+29R_3} \begin{bmatrix} 1 & 2 & -4 & -4 & 5 \\ 0 & -1 & 10 & 9 & -5 \\ 0 & 0 & 1 & 1 & -1 \\ 0 & 0 & 29 & 29 & -5 \end{bmatrix} \xrightarrow{8R_3} \begin{bmatrix} 1 & 2 & -4 & -4 & 5 \\ 0 & -1 & 10 & 9 & -5 \\ 0 & 0 & 8 & 8 & -8 \\ 0 & 0 & 29 & 29 & -5 \end{bmatrix}$

$\xrightarrow{R_4-3R_2} \begin{bmatrix} 1 & 2 & -4 & -4 & 5 \\ 0 & -1 & 10 & 9 & -5 \\ 0 & 0 & 8 & 8 & -8 \\ 0 & 3 & -1 & 2 & 10 \end{bmatrix} \xrightarrow{R_3\leftrightarrow R_2} \begin{bmatrix} 1 & 2 & -4 & -4 & 5 \\ 0 & 0 & 8 & 8 & -8 \\ 0 & -1 & 10 & 9 & -5 \\ 0 & 3 & -1 & 2 & 10 \end{bmatrix} \xrightarrow[\substack{R_3+2R_1 \\ R_4-R_1}]{R_2+2R_1} \begin{bmatrix} 1 & 2 & -4 & -4 & 5 \\ 2 & 4 & 0 & 0 & 2 \\ 2 & 3 & 2 & 1 & 5 \\ -1 & 1 & 3 & 6 & 5 \end{bmatrix}.$

16. $R_j \leftrightarrow R_i$ undoes $R_i \leftrightarrow R_j$, $\frac{1}{k}R_i$ undoes kR_i, $R_i - kR_j$ undoes $R_i + kR_j$.

17. $A = \begin{bmatrix} 1 & 2 \\ 3 & 4 \end{bmatrix} \xrightarrow{R_2-2R_1} \begin{bmatrix} 1 & 2 \\ 1 & 0 \end{bmatrix} \xrightarrow{-\frac{1}{2}R_1} \begin{bmatrix} -\frac{1}{2} & -1 \\ 1 & 0 \end{bmatrix} \xrightarrow{R_1+\frac{7}{2}R_2} \begin{bmatrix} 3 & -1 \\ 1 & 0 \end{bmatrix} = B.$

So A and B are row equivalent. Convert A into B by $R_2 - 2R_1$, $-\frac{1}{2}R_1$, $R_1 + \frac{7}{2}R_2$.

18. $A = \begin{bmatrix} 2 & 0 & -1 \\ 1 & 1 & 0 \\ -1 & 1 & 1 \end{bmatrix} \xrightarrow{R_1+R_2} \begin{bmatrix} 3 & 1 & -1 \\ 1 & 1 & 0 \\ -1 & 1 & 1 \end{bmatrix} \xrightarrow{R_2+2R_3} \begin{bmatrix} 3 & 1 & -1 \\ -1 & 3 & 2 \\ -1 & 1 & 1 \end{bmatrix}$

$\xrightarrow{R_3+R_1} \begin{bmatrix} 3 & 1 & -1 \\ -1 & 3 & 2 \\ 2 & 2 & 0 \end{bmatrix} \xrightarrow{R_2+R_1} \begin{bmatrix} 3 & 1 & -1 \\ 2 & 4 & 1 \\ 2 & 2 & 0 \end{bmatrix} \xrightarrow{R_2+\frac{1}{2}R_3} \begin{bmatrix} 3 & 1 & -1 \\ 3 & 5 & 1 \\ 2 & 2 & 0 \end{bmatrix} = B.$

Therefore, the matrices A and B are row equivalent.

19. Performing $R_2 + R_1$ and $R_1 + R_2$ does *not* leave rows 1 and 2 identical.
After performing $R_2 + R_1$ the second row is now $R_2' = R_2 + R_1$.
So $R_1 + R_2$ is actually $R_1 + R_2' = R_1 + (R_2 + R_1) = 2R_1 + R_2$.
Performing $R_2 + R_1$ and $R_1 + R_2$ simultaneously annuls their linearity.

20. $$\begin{bmatrix} x_1 \\ x_2 \end{bmatrix} \xrightarrow{R_2+R_1} \begin{bmatrix} x_1 \\ x_2 + x_1 \end{bmatrix} \xrightarrow{R_1-R_2} \begin{bmatrix} -x_2 \\ x_2 + x_1 \end{bmatrix} \xrightarrow{R_2+R_1} \begin{bmatrix} -x_2 \\ x_1 \end{bmatrix} \xrightarrow{-R_1} \begin{bmatrix} x_2 \\ x_1 \end{bmatrix}.$$
The net effect is to interchange the first and second rows.

21. Our first task is to show that $\begin{bmatrix} 3 & 1 \\ 2 & 4 \end{bmatrix} \xrightarrow{3R_2-2R_1} \begin{bmatrix} 3 & 1 \\ 0 & 10 \end{bmatrix}$ is *not* an elementary row operation.

Compare $3R_2 - 2R_1$ to the elementary row operations $R_i \leftrightarrow R_j$, kR_i, $R_i + kR_j$.
Clearly, $3R_2 - 2R_1$ is a combination of kR_i and $R_i + kR_j$ done at the same time.
Performing row operations simultaneously annuls their linearity.

One way to achieve the result is: $\begin{bmatrix} 3 & 1 \\ 2 & 4 \end{bmatrix} \xrightarrow{R_2-\frac{2}{3}R_1} \begin{bmatrix} 3 & 1 \\ 0 & \frac{10}{3} \end{bmatrix} \xrightarrow{3R_2} \begin{bmatrix} 3 & 1 \\ 0 & 10 \end{bmatrix}.$

22. We must show that we can create a 1 in row 1, column 1 using $R_i \leftrightarrow R_j$, kR_i, or $R_i + kR_j$.

$\begin{bmatrix} 3 & 2 \\ 1 & 4 \end{bmatrix} \xrightarrow{R_2 \leftrightarrow R_1} \begin{bmatrix} 1 & 4 \\ 3 & 2 \end{bmatrix}$. $\begin{bmatrix} 3 & 2 \\ 1 & 4 \end{bmatrix} \xrightarrow{\frac{1}{3}R_1} \begin{bmatrix} 1 & \frac{2}{3} \\ 1 & 4 \end{bmatrix}$. $\begin{bmatrix} 3 & 2 \\ 1 & 4 \end{bmatrix} \xrightarrow{R_1-2R_2} \begin{bmatrix} 1 & -6 \\ 1 & 4 \end{bmatrix}$.

$R_i \leftrightarrow R_j$ is the most direct. That is, it requires the fewest operations.
kR_i requires fewer operations than $R_i + kR_j$, but $R_i + kR_j$ gives integer results.

23. Since **rank** = the number of nonzero rows in the row echelon form of a matrix, before we answer we should put each of the matrices into row echelon form.

(1) Since this matrix A is not in its row echelon form B, we must row reduce A first.

$$\begin{bmatrix} 1 & 0 & 1 \\ 0 & 0 & 3 \\ 0 & 1 & 0 \end{bmatrix} \xrightarrow{R_2 \leftrightarrow R_3} \begin{bmatrix} 1 & 0 & 1 \\ 0 & 1 & 0 \\ 0 & 0 & 3 \end{bmatrix}.$$ So, rank A = the number of nonzero rows in $B = 3$.

(2) A is in row echelon form, so we need only count the number of its nonzero rows.

Since $\begin{bmatrix} 7 & 0 & 1 & 0 \\ 0 & 1 & -1 & 4 \\ 0 & 0 & 0 & 0 \end{bmatrix}$ has two nonzero rows, rank $A = 2$.

(3) A is in row echelon form, so we need only count the number of its nonzero rows.

Since $\begin{bmatrix} 0 & 1 & 3 & 0 \\ 0 & 0 & 0 & 1 \end{bmatrix}$ has two nonzero rows, rank $A = 2$.

(4) A is in row echelon form, so we need only count the number of its nonzero rows.

Since $\begin{bmatrix} 0 & 0 & 0 \\ 0 & 0 & 0 \\ 0 & 0 & 0 \end{bmatrix}$ has no nonzero rows, rank $A = 0$.

(5) Since this matrix A is not in its row echelon form B, we must row reduce A first.

$$\begin{bmatrix} 1 & 0 & 3 & -4 & 0 \\ 0 & 0 & 0 & 0 & 0 \\ 0 & 1 & 5 & 0 & 1 \end{bmatrix} \xrightarrow{R_2 \leftrightarrow R_3} \begin{bmatrix} 1 & 0 & 3 & -4 & 0 \\ 0 & 1 & 5 & 0 & 1 \\ 0 & 0 & 0 & 0 & 0 \end{bmatrix}.$$ So, rank $A = 2$.

(6) Since this matrix A is not in its row echelon form B, we must row reduce A first.

$$\begin{bmatrix} 0 & 0 & 1 \\ 0 & 1 & 0 \\ 1 & 0 & 0 \end{bmatrix} \xrightarrow{R_1 \leftrightarrow R_3} \begin{bmatrix} 1 & 0 & 0 \\ 0 & 1 & 0 \\ 0 & 0 & 1 \end{bmatrix}.$$ So, rank $A = 3$.

(7) Since this matrix A is not in its row echelon form B, we must row reduce A first.

$$\begin{bmatrix} 1 & 2 & 3 \\ 1 & 0 & 0 \\ 0 & 1 & 1 \\ 0 & 0 & 1 \end{bmatrix} \xrightarrow[R_4 - R_1 + R_2 + 2R_3]{\substack{R_2 - R_1 \\ R_3 - \frac{1}{2}R_1 + \frac{1}{2}R_2}} \begin{bmatrix} 1 & 2 & 3 \\ 0 & -2 & -3 \\ 0 & 0 & -\frac{1}{2} \\ 0 & 0 & 0 \end{bmatrix}.$$ So, rank $A = 3$.

(8) A is in row echelon form, so we need only count the number of its nonzero rows.

Since $\begin{bmatrix} 2 & 1 & 3 & 5 \\ 0 & 0 & 1 & -1 \\ 0 & 0 & 0 & 3 \\ 0 & 0 & 0 & 0 \end{bmatrix}$ has 3 nonzero rows, rank $A = 3$.

24. $\begin{bmatrix}1&0&0\\0&1&0\\0&0&1\end{bmatrix}, \begin{bmatrix}1&0&*\\0&1&*\\0&0&0\end{bmatrix}, \begin{bmatrix}1&*&0\\0&0&1\\0&0&0\end{bmatrix}, \begin{bmatrix}0&1&0\\0&0&1\\0&0&0\end{bmatrix}, \begin{bmatrix}0&1&*\\0&0&0\\0&0&0\end{bmatrix}, \begin{bmatrix}0&0&1\\0&0&0\\0&0&0\end{bmatrix}$, and $\begin{bmatrix}0&0&0\\0&0&0\\0&0&0\end{bmatrix}$.

25. We have the following system of equations: $\begin{bmatrix}1&2&-3\\2&-1&1\\4&-1&1\end{bmatrix}\begin{bmatrix}x_1\\x_2\\x_3\end{bmatrix} = \begin{bmatrix}9\\0\\4\end{bmatrix}$.

We form the augmented matrix and row reduce it as follows:

$$\left[\begin{array}{ccc|c}1&2&-3&9\\2&-1&1&0\\4&-1&1&4\end{array}\right] \xrightarrow[R_3-2R_2]{R_1+3R_3} \left[\begin{array}{ccc|c}13&-1&0&21\\2&-1&1&0\\0&1&-1&4\end{array}\right] \xrightarrow[-R_3]{R_1\leftrightarrow R_2} \left[\begin{array}{ccc|c}2&-1&1&0\\13&-1&0&21\\0&-1&1&-4\end{array}\right]$$

$$\xrightarrow[R_2]{R_1-R_3} \left[\begin{array}{ccc|c}2&0&0&4\\-13&1&0&-21\\0&-1&1&4\end{array}\right] \xrightarrow{\frac{1}{2}R_1} \left[\begin{array}{ccc|c}1&0&0&2\\-13&1&0&-21\\0&-1&1&-4\end{array}\right] \xrightarrow{R_2+13R_1} \left[\begin{array}{ccc|c}1&0&0&2\\0&1&0&5\\0&-1&1&-4\end{array}\right]$$

$$\xrightarrow{R_3+R_2} \left[\begin{array}{ccc|c}1&0&0&2\\0&1&0&5\\0&0&1&1\end{array}\right].\ \text{So, the solution is } \begin{bmatrix}x_1\\x_2\\x_3\end{bmatrix} = \begin{bmatrix}2\\5\\1\end{bmatrix}.$$

26. We form the augmented matrix and row reduce it as follows:

$$\left[\begin{array}{ccc|c}1&-1&1&0\\-1&3&1&5\\3&1&7&2\end{array}\right] \xrightarrow{R_3-5R_1-2R_2} \left[\begin{array}{ccc|c}1&-1&1&0\\-1&3&1&5\\0&0&0&-8\end{array}\right]$$

The third row is equivalent to the equation $0 = -8$ which clearly has no solution. Therefore, the system is inconsistent.

Does $R_3 = 5R_1 + 2R_2$ (excluding constants) cause the system to be inconsistent?

27. We form the augmented matrix and row reduce it as follows:

$$\left[\begin{array}{ccc|c}1&-3&-2&0\\-1&2&1&0\\2&4&6&0\end{array}\right] \xrightarrow{R_3+8R_1+10R_2} \left[\begin{array}{ccc|c}1&-3&-2&0\\-1&2&1&0\\0&0&0&0\end{array}\right]$$

$R_3 = -8R_1 - 10R_2$ (excluding constants) does not cause a problem here? Why?
Since the system is homogeneous (all constants = 0), the system has at least one solution.

$$\xrightarrow{R_2+R_1} \left[\begin{array}{ccc|c}1&-3&-2&0\\0&-1&-1&0\\0&0&0&0\end{array}\right] \xrightarrow{-R_2} \left[\begin{array}{ccc|c}1&-3&-2&0\\0&1&1&0\\0&0&0&0\end{array}\right] \xrightarrow{R_1+3R_2} \left[\begin{array}{ccc|c}1&0&1&0\\0&1&1&0\\0&0&0&0\end{array}\right].$$

The third row of $0 = 0$ tells us that $x_3 = t$ is a free variable.
Back substituting, we have $x_2 + t = 0 \Rightarrow x_2 = -t$ and $x_1 + t = 0 \Rightarrow x_1 = -t$.

$$\text{So, the solution is } \begin{bmatrix}x_1\\x_2\\x_3\end{bmatrix} = t\begin{bmatrix}-1\\-1\\1\end{bmatrix} \text{ or equivalently } \begin{bmatrix}x_1\\x_2\\x_3\end{bmatrix} = t\begin{bmatrix}1\\1\\-1\end{bmatrix}.$$

28. From the beginning, we know this system has infinitely many solutions. Why? Because this system has 4 variables and only 3 equations they have to satisfy.

We form the augmented matrix and row reduce it as follows:

$$\left[\begin{array}{rrrr|r} 2 & 3 & -1 & 4 & 0 \\ 3 & -1 & 0 & 1 & 1 \\ 3 & -4 & 1 & -1 & 2 \end{array}\right] \longrightarrow \left[\begin{array}{rrrr|r} 1 & 0 & 0 & 1 & \frac{1}{2} \\ 0 & 1 & 0 & 2 & \frac{1}{2} \\ 0 & 0 & 1 & 4 & \frac{1}{2} \end{array}\right].$$

The fact that $4-3=1$ tells us that $z=t$ is a free variable.
So $y+4t=\frac{1}{2} \Rightarrow y=\frac{1}{2}-4t$, $x+2t=\frac{1}{2} \Rightarrow x=\frac{1}{2}-2t$, and $w+t=\frac{1}{2} \Rightarrow w=\frac{1}{2}-t$.

So, the solution is $\begin{bmatrix} w \\ x \\ y \\ z \end{bmatrix} = \begin{bmatrix} \frac{1}{2} \\ \frac{1}{2} \\ \frac{1}{2} \\ 0 \end{bmatrix} + t \begin{bmatrix} -1 \\ -2 \\ -4 \\ 1 \end{bmatrix}$.

29. Note that there are 3 equations but only 2 variables to satisfy them.
It is helpful, therefore, to begin by noting $R_3 = 9R_1 - 4R_2$.

$$\left[\begin{array}{rr|r} 2 & 1 & 3 \\ 4 & 1 & 7 \\ 2 & 5 & -1 \end{array}\right] \xrightarrow{R_3-9R_1+4R_2} \left[\begin{array}{rr|r} 2 & 1 & 3 \\ 4 & 1 & 7 \\ 0 & 0 & 0 \end{array}\right] \xrightarrow{R_2-2R_1} \left[\begin{array}{rr|r} 2 & 1 & 3 \\ 0 & -1 & 1 \\ 0 & 0 & 0 \end{array}\right] \cdots \longrightarrow \cdots \left[\begin{array}{rr|r} 1 & 0 & 2 \\ 0 & 1 & -1 \\ 0 & 0 & 0 \end{array}\right]$$

So, the solution is $\begin{bmatrix} r \\ s \end{bmatrix} = \begin{bmatrix} 2 \\ -1 \end{bmatrix}$.

30. From the beginning, we know this system has infinitely many solutions. Why? Because this system has 4 variables and only 3 equations they have to satisfy.

We form the augmented matrix and row reduce it as follows:

$$\left[\begin{array}{rrrr|r} -1 & 3 & -2 & 4 & 0 \\ 2 & -6 & 1 & -2 & -3 \\ 1 & -3 & 4 & -8 & 2 \end{array}\right] \longrightarrow \left[\begin{array}{rrrr|r} 1 & -3 & 0 & 0 & 2 \\ 0 & 0 & 1 & -2 & -1 \\ 0 & 0 & 0 & 0 & 0 \end{array}\right]$$

Since rank $A = 2$ and $4-2=2$, we get 2 free variables: $x_2 = s$ and $x_4 = t$.
Back substituting, we get $x_4 = t$, $x_3 = 1+2t$, $x_2 = s$, and $x_1 = -2+3s$.

So, the solution is $\begin{bmatrix} x_1 \\ x_2 \\ x_3 \\ x_4 \end{bmatrix} = \begin{bmatrix} -2 \\ 0 \\ 1 \\ 0 \end{bmatrix} + s \begin{bmatrix} 3 \\ 1 \\ 0 \\ 0 \end{bmatrix} + t \begin{bmatrix} 0 \\ 0 \\ 2 \\ 1 \end{bmatrix}$.

31. From the beginning, we know this system has infinitely many solutions. Why?
Because this system has 5 variables and only 3 equations they have to satisfy.

We form the augmented matrix and row reduce it as follows:

$$\left[\begin{array}{ccccc|c} \frac{1}{2} & 1 & -1 & -6 & 0 & 2 \\ \frac{1}{6} & \frac{1}{2} & 0 & -3 & 1 & -1 \\ \frac{1}{3} & 0 & -2 & 0 & -4 & 8 \end{array}\right] \longrightarrow \left[\begin{array}{ccccc|c} 1 & 0 & -6 & 0 & -12 & 24 \\ 0 & 1 & 2 & -6 & 6 & -10 \\ 0 & 0 & 0 & 0 & 0 & 0 \end{array}\right]$$

Since rank $A = 2$ and $5 - 2 = 3$, we get 3 free variables: $x_3 = r$, $x_4 = s$, and $x_5 = t$.
Back substituting, we get $x_2 = -10 - 2r + 6s - 6t$, $x_1 = 24 + 6r + 12t$.

$$\text{So, the solution is } \begin{bmatrix} x_1 \\ x_2 \\ x_3 \\ x_4 \\ x_5 \end{bmatrix} = \begin{bmatrix} 24 \\ -10 \\ 0 \\ 0 \\ 0 \end{bmatrix} + r \begin{bmatrix} 6 \\ -2 \\ 1 \\ 0 \\ 0 \end{bmatrix} + s \begin{bmatrix} 0 \\ 6 \\ 0 \\ 1 \\ 0 \end{bmatrix} + t \begin{bmatrix} 12 \\ -6 \\ 0 \\ 0 \\ 1 \end{bmatrix}.$$

32. We should be able to see that rank $A = 3$. What does this tell us?
There are 3 equations and 3 variables. So if there is a solution, it is unique. Why?
Because the Rank Theorem (2.2) tells us there are $3 - 3 = 0$ free variables.

$$\left[\begin{array}{ccc|c} \sqrt{2} & 1 & 2 & 1 \\ 0 & \sqrt{2} & -3 & -\sqrt{2} \\ 0 & -1 & \sqrt{2} & 1 \end{array}\right] \longrightarrow \left[\begin{array}{ccc|c} 1 & 0 & 0 & \sqrt{2} \\ 0 & 1 & 0 & -1 \\ 0 & 0 & 1 & 0 \end{array}\right] \Rightarrow \text{The solution is } \begin{bmatrix} x \\ y \\ z \end{bmatrix} = \begin{bmatrix} \sqrt{2} \\ -1 \\ 0 \end{bmatrix}.$$

33. We form the augmented matrix and row reduce it as follows:

$$\left[\begin{array}{cccc|c} 1 & 1 & 2 & 1 & 1 \\ 1 & -1 & -1 & 1 & 0 \\ 0 & 1 & 1 & 0 & -1 \\ 1 & 1 & 0 & 1 & 2 \end{array}\right] \longrightarrow \left[\begin{array}{cccc|c} 1 & 0 & 0 & 1 & 0 \\ 0 & 1 & 0 & 0 & 0 \\ 0 & 0 & 1 & 0 & 0 \\ 0 & 0 & 0 & 0 & 1 \end{array}\right] \Rightarrow$$

The fourth row is equivalent to the equation $0 = 1$ which clearly has no solution.
Therefore, the system is inconsistent.

Q: Rank $A = 3$. How does that relate to the fact that there is no solution?
A: The system has no solution when A has a zero row with corresponding constant $\neq 0$.

34. When there are 4 equations and 4 variables, if the solution exists it is unique.
Why? Because the Rank Theorem (2.2) tells us there are $4 - 4 = 0$ free variables.

$$\left[\begin{array}{cccc|c} 1 & 1 & 1 & 1 & 4 \\ 1 & 2 & 3 & 4 & 10 \\ 1 & 3 & 6 & 10 & 20 \\ 1 & 4 & 10 & 20 & 35 \end{array}\right] \longrightarrow \left[\begin{array}{cccc|c} 1 & 0 & 0 & 0 & 1 \\ 0 & 1 & 0 & 0 & 1 \\ 0 & 0 & 1 & 0 & 1 \\ 0 & 0 & 0 & 1 & 1 \end{array}\right] \Rightarrow \text{The solution is } \begin{bmatrix} a \\ b \\ c \\ d \end{bmatrix} = \begin{bmatrix} 1 \\ 1 \\ 1 \\ 1 \end{bmatrix}.$$

35. Begin by thinking of this system as $[A|\mathbf{x}]$, then determine rank A by inspection.
Mentally performing $R_1 \leftrightarrow R_3$ to put matrix A into row echelon form,
makes it obvious that rank $A = 3$ (because A has 3 nonzero rows).

Since rank $A = 3$, this is a system of 3 equations and 3 variables.
Therefore, the system has a unique solution because there are $3 - 3 = 0$ free variables.

36. Begin by thinking of this system as $[A|\mathbf{x}]$, then determine rank A by inspection.
Mentally performing $R_3 - 2R_2$ implies the equation $0 = 2$.
This equation makes it obvious that this system has no solution.

Note: $R_3 = 2R_2$ implies rank $A = 2$. How does that relate to our answer?

37. Since this system has 4 variables and at most 3 equations,
it has infinitely many solutions. Why? There is at least one free variable.

38. Since this system has 5 variables and at most 3 equations,
it has infinitely many solutions. Why? There are at least two free variables.

39. We need only show that the condition $ad - bc \neq 0$ implies that rank $A = 2$. Why?
If rank $A = 2$, there are $2 - 2 = 0$ free variables so the system has a unique solution.

Case 1: $a = 0$, which implies both $b \neq 0$ and $c \neq 0$. Why? Because $0d - bc = -bc \neq 0$.

Row reduce A: $\begin{bmatrix} 0 & b \\ c & d \end{bmatrix} \xrightarrow{R_1 \leftrightarrow R_2} \begin{bmatrix} c & d \\ 0 & b \end{bmatrix}.$

A is now in row echelon form with 2 nonzero rows. Therefore, rank $A = 2$.

Case 2: $c = 0$, which implies both $a \neq 0$ and $d \neq 0$. Why? Because $ad - b0 = ad \neq 0$.

Row reduce A: $\begin{bmatrix} a & b \\ 0 & d \end{bmatrix}.$

A is now in row echelon form with 2 nonzero rows. Therefore, rank $A = 2$.

Case 3: $a \neq 0$ and $c \neq 0$.

Row reduce A: $\begin{bmatrix} a & b \\ c & d \end{bmatrix} \xrightarrow[aR_2]{cR_1} \begin{bmatrix} ac & bc \\ ac & ad \end{bmatrix} \xrightarrow{R_2 - R_1} \begin{bmatrix} ac & bc \\ 0 & ad - bc \end{bmatrix}.$

A is now in row echelon form with 2 nonzero rows. Therefore, rank $A = 2$.

40. First row reduce the system $[A|\mathbf{x}]$ and then answers parts (a), (b), and (c).

$$\left[\begin{array}{cc|c} k & 2 & 3 \\ 2 & -4 & -6 \end{array}\right] \xrightarrow{R_1 \leftrightarrow R_2} \left[\begin{array}{cc|c} 2 & -4 & -6 \\ k & 2 & 3 \end{array}\right] \xrightarrow{R_2 - \frac{1}{2}kR_1} \left[\begin{array}{cc|c} 2 & -4 & -6 \\ 0 & 2+2k & 3+3k \end{array}\right]$$

(a) There are no values of k for which this system has no solution. Why?
The system has no solution when A has a zero row with corresponding constant $\neq 0$.
$2 + 2k = 0 \Rightarrow k = -1$ is the only value of k that creates a row of all zeros.
But $k = -1 \Rightarrow$ the constant $3 + 3k = 3 - 3 = 0$. What does this imply?

(b) The system has a unique solution for $k \neq -1$. Why?
From (a), we see when $k \neq -1$ then rank $A = 2$. So, there are $2 - 2 = 0$ free variables.

(c) The only value of k for which this system has infinitely many solutions is $k = -1$.
The system has infinitely many solutions when A has a zero row with constant $= 0$.
From (a), we see this is exactly the case when $k = -1$.

41. First row reduce the system $[A|\mathbf{x}]$ and then answers parts (a), (b), and (c).

$$\left[\begin{array}{cc|c} 1 & k & 1 \\ k & 1 & 1 \end{array}\right] \xrightarrow{R_2 - kR_1} \left[\begin{array}{cc|c} 1 & k & 1 \\ 0 & 1-k^2 & 1-k \end{array}\right]$$

(a) When $k = -1$, this system has no solution. Why?
The system has no solution when A has a zero row with corresponding constant $\neq 0$.
$1 - k^2 = 0 \Rightarrow k = \pm 1$ create a zero row in A.
And $k = -1 \Rightarrow$ the constant $1 - k = 1 + 1 \neq 0$.

(b) When $k \neq \pm 1$, this system has a unique solution. Why?
From (a), we see when $k \neq \pm 1$ then rank $A = 2$. So, there are $2 - 2 = 0$ free variables.

(c) When $k = 1$, this system has infinitely many solutions.
The system has infinitely many solutions when A has a zero row with constant $= 0$.
From (a), we see this is exactly the case when $k = 1$.

42. First row reduce the system $[A|\mathbf{x}]$ and then answers parts (a), (b), and (c).

$$\left[\begin{array}{ccc|c} 1 & -2 & 3 & 2 \\ 1 & 1 & 1 & k \\ 2 & -1 & 4 & k^2 \end{array}\right] \xrightarrow{R_3 - R_1 - R_2} \left[\begin{array}{ccc|c} 1 & -2 & 3 & 2 \\ 1 & 1 & 1 & k \\ 0 & 0 & 0 & k^2 - k - 2 \end{array}\right]$$

(a) When $k \neq -1, 2$, this system has no solution. Why?
The system has no solution when A has a zero row with corresponding constant $\neq 0$.
$k^2 - k - 2 = 0 \Rightarrow k = -1, 2$ are the only values of k that make the constant 0.

(b) This system never has a unique solution because rank $A = 2$. Why?
Since rank $A = 2$, there is always at least $3 - 2 = 1$ free variable.

(c) When $k = -1, 2$, this system has infinitely many solutions.
The system has infinitely many solutions when A has a zero row with constant $= 0$.
From (a), we see this is exactly the case when $k = -1, 2$.

43. First row reduce the system $[A|\mathbf{x}]$ and then answers parts (a), (b), and (c).

$$\left[\begin{array}{ccc|c} 1 & 1 & k & 1 \\ 1 & k & 1 & 1 \\ k & 1 & 1 & -2 \end{array}\right] \longrightarrow \left[\begin{array}{ccc|c} 1 & 1 & k & 1 \\ 0 & k-1 & 1-k & 0 \\ 0 & 0 & k^2+k-2 & k+2 \end{array}\right]$$

(a) When $k = 1$, this system has no solution. Why?

The system has no solution when A has a zero row with corresponding constant $\neq 0$. $k^2 + k - 2 = 0 \Rightarrow k = 1$ makes the bottom row 0, but the constant $k + 2 = 1 + 2 = 3 \neq 0$.

(b) When $k \neq 1, -2$, this system has a unique solution. Why?

When $k \neq 1, -2$, rank $A = 3$. So, there are $3 - 3 = 0$ free variables.

(c) When $k = -2$, this system has infinitely many solutions.

The system has infinitely many solutions when A has a zero row with constant $= 0$. $k^2 + k - 2 = 0 \Rightarrow k = -2$ makes the bottom row 0 and the constant $k + 2 = -2 + 2 = 0$.

44. (a) The following system of n equations has infinitely many solutions:

$$\begin{aligned} x_1 + x_2 + \cdots + x_n &= 0 \\ 2x_1 + 2x_2 + \cdots + 2x_n &= 0 \\ &\vdots \\ nx_1 + nx_2 + \cdots + nx_n &= 0 \end{aligned}$$

Likewise, the following system of $n + 1$ equations has infinitely many solutions:

$$\begin{aligned} x_1 + x_2 + \cdots + x_n &= 0 \\ 2x_1 + 2x_2 + \cdots + 2x_n &= 0 \\ &\vdots \\ nx_1 + nx_2 + \cdots + nx_n &= 0 \\ (n+1)\,x_1 + (n+2)\,x_2 + \cdots + (n+1)\,x_n &= 0 \end{aligned}$$

(b) The system of n equations $x_1 = 0$, $x_2 = 0$, …, $x_n = 0$ has the unique solution $x_i = 0$. as does the system of $2n$ equations $x_1 = 0$, $2x_1 = 0$ …, $x_n = 0$, $2x_n = 0$.

45. As in Example 2.14, find the line of intersection of $3x + 2y + z = -1$, $2x - y + 4z = 5$.

First, observe that there will be a line of intersection. Why?
The normal vectors of the two planes, $[3, 2, 1]$ and $[2, -1, 4]$ are not parallel.

The points that lie in the intersection of the two planes correspond to the points

in the solution the system: $\begin{array}{rcrcrcr} 3x & + & 2y & + & z & = & -1 \\ 2x & - & y & + & 4z & = & 5 \end{array}$

Gauss-Jordan elimination yields: $\left[\begin{array}{rrr|r} 3 & 2 & 1 & -1 \\ 2 & -1 & 4 & 5 \end{array}\right] \longrightarrow \left[\begin{array}{rrr|r} 1 & 0 & \frac{9}{7} & \frac{9}{7} \\ 0 & 1 & -\frac{10}{7} & -\frac{17}{7} \end{array}\right]$

Replacing variables, we have: $\begin{array}{rcl} x + \frac{9}{7}z & = & \frac{9}{7} \\ y - \frac{10}{7}z & = & -\frac{17}{7} \end{array} \Rightarrow \begin{array}{rcl} z & = & 1 - \frac{7}{9}x \\ y & = & -\frac{17}{7} + \frac{10}{7}z \end{array}$

To eliminate fractions we set $x = 9t$, so $z = 1 - \frac{7}{9}(9t) = 1 - 7t$.
Substituting $z = 1 - 7t$ into $y = -\frac{17}{7} + \frac{10}{7}z$ yields: $y = -\frac{17}{7} + \frac{10}{7}(1 - 7t) = -1 - 10t$.
Summarizing, we now have $x = 9t$, $y = -1 - 10t$, and $z = 1 - 7t$.

Therefore, the line is $\begin{bmatrix} x \\ y \\ z \end{bmatrix} = \begin{bmatrix} 0 \\ -1 \\ 1 \end{bmatrix} + t \begin{bmatrix} 1 \\ -10 \\ -9 \end{bmatrix}$.

46. As in Example 2.14, find the line of intersection of $4x + y - z = 0$, $2x - y + 3z = 4$.

First, observe that there will be a line of intersection. Why?
The normal vectors of the two planes, $[4, 1, -1]$ and $[2, -1, 3]$ are not parallel.

The points that lie in the intersection of the two planes correspond to the points

in the solution the system: $\begin{array}{rcrcrcr} 4x & + & y & - & z & = & 0 \\ 2x & - & y & + & 3z & = & 4 \end{array}$

Gauss-Jordan elimination yields: $\left[\begin{array}{rrr|r} 4 & 1 & -1 & 0 \\ 2 & -1 & 3 & 4 \end{array}\right] \longrightarrow \left[\begin{array}{rrr|r} 1 & 0 & \frac{1}{3} & \frac{2}{3} \\ 0 & 1 & -\frac{7}{3} & -\frac{8}{3} \end{array}\right]$

Replacing variables, we have: $\begin{array}{rcl} x + \frac{1}{3}z & = & \frac{2}{3} \\ y - \frac{7}{3}z & = & -\frac{8}{3} \end{array} \Rightarrow \begin{array}{rcl} z & = & 2 - 3x \\ y & = & -\frac{8}{3} + \frac{7}{3}z \end{array}$

To begin we set $x = t$, so $z = 2 - 3(t) = 2 - 3t$.
Substituting $z = 2 - 3t$ into $y = -\frac{8}{3} + \frac{7}{3}z$ yields: $y = -\frac{8}{3} + \frac{7}{3}(2 - 3t) = 2 - 7t$.
Summarizing, we now have $x = t$, $y = 2 - 7t$, and $z = 2 - 3t$.

Therefore, the line is $\begin{bmatrix} x \\ y \\ z \end{bmatrix} = \begin{bmatrix} 0 \\ 2 \\ 2 \end{bmatrix} + t \begin{bmatrix} 1 \\ -7 \\ -3 \end{bmatrix}$.

47. When looking for examples, begin with familiar planes like $x = 0$, $y = 0$, and $z = 0$.

(a) Let's start with $x = 0$ and $y = 0$. These planes obviously intersect in the z-axis. Why?

As in Exercise 45: $\begin{array}{rcl} x + 0y + 0z &=& 0 \\ 0x + y + 0z &=& 0 \end{array} \Rightarrow \left[\begin{array}{ccc|c} 1 & 0 & 0 & 0 \\ 0 & 1 & 0 & 0 \end{array}\right] \Rightarrow x = 0, y = 0, z = t \Rightarrow$

The line of intersection of $x = 0$ and $y = 0$ is $\begin{bmatrix} x \\ y \\ z \end{bmatrix} = \begin{bmatrix} 0 \\ 0 \\ 0 \end{bmatrix} + t \begin{bmatrix} 0 \\ 0 \\ 1 \end{bmatrix}$, the z-axis.

All we need is one other plane that passes through the z-axis to complete our example.
It may help to sketch $\mathbb{R}^2$ and look for a line that passes through the origin.
One such line is $x = y$ which corresponds to the plane $x - y = 0$ in $\mathbb{R}^3$.

Sketch these three planes in $\mathbb{R}^3$ to confirm they intersect in the z-axis.

Q: How do we confirm these three planes intersect in the z-axis algebraically?
A: Check the intersection between $x = 0$ and $x - y = 0$. Why is that enough?

As in Exercise 45: $\begin{array}{rcl} x + 0y + 0z &=& 0 \\ x - y + 0z &=& 0 \end{array} \Rightarrow \left[\begin{array}{ccc|c} 1 & 0 & 0 & 0 \\ 1 & 1 & 0 & 1 \end{array}\right] \Rightarrow x = y = 0, \text{ and } z = t \Rightarrow$

The line of intersection of $x = 0$ and $x - y = 0$ is $\begin{bmatrix} x \\ y \\ z \end{bmatrix} = \begin{bmatrix} 0 \\ 0 \\ 0 \end{bmatrix} + t \begin{bmatrix} 0 \\ 0 \\ 1 \end{bmatrix}$.

Q: Start with $y = 0$ and $z = 0$ and then $x = 1$ and $y = 1$. Is there a pattern?

(b) Begin with $x = 0$ and $y = 0$. We need one plane that crosses across both of these.
It may help to visualize $\mathbb{R}^2$ and look for a line that cuts across the first quadrant.
It is obvious that the plane $x + y = 1$ will complete the example?

Sketch these three planes in $\mathbb{R}^3$ to confirm they intersect in pairs.
For example, $x = 0$ and $x + y = 1$ intersect in the line $[x, y, z] = [0, 1, 0] + t[0, 0, 1]$.

As in Exercise 45: $\begin{array}{rcl} x + 0y + 0z &=& 0 \\ x + y + 0z &=& 1 \end{array} \Rightarrow \left[\begin{array}{ccc|c} 1 & 0 & 0 & 0 \\ 1 & 1 & 0 & 1 \end{array}\right] \Rightarrow x = 0, \ y = 1, \text{ and } z = t \Rightarrow$

The line of intersection of $x = 0$ and $x + y = 1$ is $\begin{bmatrix} x \\ y \\ z \end{bmatrix} = \begin{bmatrix} 0 \\ 1 \\ 0 \end{bmatrix} + t \begin{bmatrix} 0 \\ 0 \\ 1 \end{bmatrix}$.

(c) An obvious example is $x = 0$, $x = 1$, and $y = 0$. Why?
The normal vector for $x = 0$ and $x = 1$ is $[1, 0, 0]$
while the normal vector for $y = 0$ is $[0, 1, 0]$.

(d) The most obvious example is $x = 0$, $y = 0$, and $z = 0$.
Note that any example of $x = a$, $y = a$, and $z = a$ will work.
Are there any other obvious pattern examples that will work?

48. As in Example 2.15, if these lines intersect we need to determine the point of intersection. As pointed out in that example, we need to change the parameter for the first line to s. We want to find an $\mathbf{x} = [x, y, z]$ that satisfies both equations simultaneously. That is, we want $\mathbf{x} = \mathbf{p} + s\mathbf{u} = \mathbf{q} + t\mathbf{v}$ or $s\mathbf{u} - t\mathbf{v} = \mathbf{q} - \mathbf{p}$.

Substituting the given $\mathbf{p}$, $\mathbf{q}$, $\mathbf{u}$, and $\mathbf{v}$ into $s\mathbf{u} - t\mathbf{v} = \mathbf{q} - \mathbf{p}$, we obtain the equations:

$$s\begin{bmatrix}1\\2\\-1\end{bmatrix} - t\begin{bmatrix}-1\\1\\0\end{bmatrix} = \begin{bmatrix}2\\2\\0\end{bmatrix} - \begin{bmatrix}-1\\2\\1\end{bmatrix} \Rightarrow \begin{array}{rcrcr} s & + & t & = & 3 \\ 2s & - & t & = & 0 \\ -s & & & = & -1 \end{array}$$

From this, the solution is easily found to be $s = 1, t = 2$.

$$\text{Therefore, the point of intersection is:} \quad \begin{bmatrix}x\\y\\z\end{bmatrix} = \begin{bmatrix}-1\\2\\1\end{bmatrix} + 1\begin{bmatrix}1\\2\\-1\end{bmatrix} = \begin{bmatrix}0\\4\\0\end{bmatrix}.$$

Check that substituting $t = 2$ into the other equation gives the same point.

49. As in Example 2.15, if these lines intersect we need to determine the point of intersection. As pointed out in that example, we need to change the parameter for the first line to s. We want to find an $\mathbf{x} = [x, y, z]$ that satisfies both equations simultaneously. That is, we want $\mathbf{x} = \mathbf{p} + s\mathbf{u} = \mathbf{q} + t\mathbf{v}$ or $s\mathbf{u} - t\mathbf{v} = \mathbf{q} - \mathbf{p}$.

Substituting the given $\mathbf{p}$, $\mathbf{q}$, $\mathbf{u}$, and $\mathbf{v}$ into $s\mathbf{u} - t\mathbf{v} = \mathbf{q} - \mathbf{p}$, we obtain the equations:

$$s\begin{bmatrix}1\\0\\1\end{bmatrix} - t\begin{bmatrix}2\\3\\1\end{bmatrix} = \begin{bmatrix}-1\\1\\-1\end{bmatrix} - \begin{bmatrix}3\\1\\0\end{bmatrix} \Rightarrow \begin{array}{rcrcr} s & - & 2t & = & -4 \\ & - & 3t & = & 0 \\ s & - & t & = & -1 \end{array}$$

From this, there is clearly no solution since $t = 0$ implies $s = -4$ and -1 at the same time. Therefore, we conclude that these lines do not intersect.

50. Similar to Example 2.15, we need to find conditions on a, b, c to create a point of intersection. As pointed out in that example, we need to change the parameter for the first line to s. We want to find an $\mathbf{x} = [x, y, z]$ that satisfies both equations simultaneously. That is, we want $\mathbf{x} = \mathbf{p} + s\mathbf{u} = \mathbf{q} + t\mathbf{v}$ or $s\mathbf{u} - t\mathbf{v} = \mathbf{q} - \mathbf{p}$.

Substituting the given $\mathbf{p}$, $\mathbf{q} = [a, b, c]$, $\mathbf{u}$, and $\mathbf{v}$ into $s\mathbf{u} - t\mathbf{v} = \mathbf{q} - \mathbf{p}$, we get:

$$s\begin{bmatrix}1\\1\\-1\end{bmatrix} - t\begin{bmatrix}2\\1\\0\end{bmatrix} = \begin{bmatrix}a\\b\\c\end{bmatrix} - \begin{bmatrix}1\\2\\3\end{bmatrix} \Rightarrow \begin{array}{rcrcl} s & - & 2t & = & a - 1 \\ s & - & t & = & b - 2 \\ -s & & & = & c - 3 \end{array}$$

From this, the conditions on a, b, c are easily found to be $Q = (s - 2t + 1, s - t + 2, -s + 3)$.

51. Recall the definition of the dot product: $\mathbf{u} \times \mathbf{v} = u_1v_1 + u_2v_2 + u_3v_3$.
So, vectors that satisfy $\mathbf{u} \cdot \mathbf{x} = \mathbf{0}$ and $\mathbf{v} \cdot \mathbf{x} = \mathbf{0}$ correspond to the system:

$$\begin{aligned} u_1x_1 + u_2x_2 + u_3x_3 &= 0 \\ v_1x_1 + v_2x_2 + v_3x_3 &= 0 \end{aligned} \quad \text{which leads to the augmented matrix} \left[\begin{array}{ccc|c} u_1 & u_2 & u_3 & 0 \\ v_1 & v_2 & v_3 & 0 \end{array}\right].$$

Gauss-Jordan elimination yields:

$$\left[\begin{array}{ccc|c} u_1 & u_2 & u_3 & 0 \\ v_1 & v_2 & v_3 & 0 \end{array}\right] \overset{u_1R_2}{\longrightarrow} \left[\begin{array}{ccc|c} u_1 & u_2 & u_3 & 0 \\ u_1v_1 & u_1v_2 & u_1v_3 & 0 \end{array}\right] \overset{R_2 - v_1R_1}{\longrightarrow} \left[\begin{array}{ccc|c} u_1 & u_2 & u_3 & 0 \\ 0 & u_1v_2 - u_2v_1 & u_1v_3 - u_3v_1 & 0 \end{array}\right]$$

The second row implies: $(u_1v_2 - u_2v_1)x_2 = (u_3v_1 - u_1v_3)x_3 \Rightarrow x_2 = \dfrac{u_3v_1 - u_1v_3}{u_1v_2 - u_2v_1}x_3$.

To clear the fraction, we let $x_3 = (u_1v_2 - u_2v_1)t \Rightarrow x_2 = (u_3v_1 - u_1v_3)t$

The first row implies: $u_1x_1 = -u_2x_2 - u_3x_3$.

So, $u_1x_1 = -u_2(u_3v_1 - u_1v_3)t - u_3(u_1v_2 - u_2v_1)t = (u_1u_2v_3 - u_1u_3v_2)t$.

Dividing both sides by u_1 yields: $x_1 = (u_2v_3 - u_3v_2)t$.

Therefore, as was to be shown: $\begin{bmatrix} x_1 \\ x_2 \\ x_3 \end{bmatrix} = \begin{bmatrix} u_2v_3 - u_3v_2 \\ u_3v_1 - u_1v_3 \\ u_1v_2 - u_2v_1 \end{bmatrix} t.$

52. This problem has two distinct parts:
1) Prove that the given nonparallel lines do not intersect, that is that they are skew.
2) Find two parallel planes, one containing each line. We will use the result of Exercise 51.

1: Similar to Example 2.15, we will show these lines, ℓ_P and ℓ_Q do not intersect.
As pointed out in that example, we need to change the parameter for the first line to s.
We want to show there is no $\mathbf{x} = [x, y, z]$ that satisfies both equations simultaneously.
So, we consider $\mathbf{x} = \mathbf{p} + s\mathbf{u} = \mathbf{q} + t\mathbf{v}$ or $s\mathbf{u} - t\mathbf{v} = \mathbf{q} - \mathbf{p}$.

Substituting the given $\mathbf{p}$, $\mathbf{q}$, $\mathbf{u}$, and $\mathbf{v}$ into $s\mathbf{u} - t\mathbf{v} = \mathbf{q} - \mathbf{p}$, we obtain the equations:

$$s\begin{bmatrix} 2 \\ -3 \\ 1 \end{bmatrix} - t\begin{bmatrix} 0 \\ 6 \\ -1 \end{bmatrix} = \begin{bmatrix} 0 \\ 1 \\ -1 \end{bmatrix} - \begin{bmatrix} 1 \\ 1 \\ 0 \end{bmatrix} \Rightarrow \begin{array}{rcrcr} 2s & & & = & -1 \\ -3s & - & 6t & = & 0 \\ s & + & t & = & -1 \end{array}$$

From this, there is clearly no solution since $s = -\frac{1}{2}$ implies $t = \frac{1}{4}$ and $-\frac{1}{2}$ at the same time.
Therefore, we conclude that these lines do not intersect.

2: From Exploration: The Cross Product in Chapter 1, $\mathbf{u} \times \mathbf{v}$ is orthogonal to both $\mathbf{u}$ and $\mathbf{v}$.
How does this help us find the parallel planes we are seeking? $\mathbf{u} \times \mathbf{v}$ can give us a normal.

From Exercise 51, taking $t = 1$ we have:

$$\mathbf{n} = \begin{bmatrix} a \\ b \\ c \end{bmatrix} = \begin{bmatrix} u_2v_3 - u_3v_2 \\ u_3v_1 - u_1v_3 \\ u_1v_2 - u_2v_1 \end{bmatrix} = \begin{bmatrix} (-3)(-1) & - & (1)(6) \\ (1)(0) & - & (2)(-1) \\ (2)(6) & - & (-3)(0) \end{bmatrix} = \begin{bmatrix} -3 \\ 2 \\ 12 \end{bmatrix}.$$

It is easy to confirm that $\mathbf{n} \cdot \mathbf{u} = 0$ and $\mathbf{n} \cdot \mathbf{v} = 0$.

So we are looking for planes of the form $-3x + 2y + 12z = d$.
One passing through $P = (1, 1, 0)$ and the other passing through $Q = (0, 1, -1)$.

Substituting $P = (1, 1, 0)$, we get $-3(1) + 2(1) + 12(0) = -1 = d$.
Substituting $Q = (0, 1, -1)$, we get $3(0) + 2(1) + 12(-1) = -10 = d$.

Therefore, $\mathscr{P}$, $-3x + 2y + 12z = -1$, and $\mathscr{Q}$, $-3x + 2y + 12z = -10$,
are parallel planes containing ℓ_P and ℓ_Q respectively.

53. Following Example 2.16, we note that we need only use addition and multiplication.
We form augmented matrix and perform Gauss-Jordan elimination in $\mathbb{Z}_3$:

$$\left[\begin{array}{cc|c} 1 & 2 & 1 \\ 1 & 1 & 2 \end{array}\right] \xrightarrow{R_2+2R_1} \left[\begin{array}{cc|c} 1 & 2 & 1 \\ 0 & 2 & 1 \end{array}\right] \xrightarrow{2R_2} \left[\begin{array}{cc|c} 1 & 2 & 1 \\ 0 & 1 & 2 \end{array}\right] \xrightarrow{R_1+R_2} \left[\begin{array}{cc|c} 1 & 0 & 0 \\ 0 & 1 & 2 \end{array}\right]$$

So the solution is $\begin{bmatrix} x \\ y \end{bmatrix} = \begin{bmatrix} 0 \\ 2 \end{bmatrix}$.

54. Following Example 2.17, we note that we need only use addition and multiplication. We form the augmented matrix and perform Gauss-Jordan elimination in $\mathbb{Z}_2$:

$$\left[\begin{array}{ccc|c} 1 & 1 & 0 & 1 \\ 0 & 1 & 1 & 0 \\ 1 & 0 & 1 & 1 \end{array}\right] \xrightarrow{R_3+R_1} \left[\begin{array}{ccc|c} 1 & 1 & 0 & 1 \\ 0 & 1 & 1 & 0 \\ 0 & 1 & 1 & 0 \end{array}\right] \xrightarrow{R_3+R_2} \left[\begin{array}{ccc|c} 1 & 1 & 0 & 1 \\ 0 & 1 & 1 & 0 \\ 0 & 0 & 0 & 0 \end{array}\right]$$

Therefore, we have: $\begin{array}{rcrcrcr} x & + & y & & & = & 1 \\ & & y & + & z & = & 0 \end{array}$

Setting the free variable $z = t$ yields: $\begin{bmatrix} x \\ y \\ z \end{bmatrix} = \begin{bmatrix} 1+t \\ t \\ t \end{bmatrix} = \begin{bmatrix} 1 \\ 0 \\ 0 \end{bmatrix} + t \begin{bmatrix} 1 \\ 1 \\ 1 \end{bmatrix}.$

Since $t = 0$ or 1, there are exactly two solutions: $\begin{bmatrix} 1 \\ 0 \\ 0 \end{bmatrix}$ and $\begin{bmatrix} 0 \\ 1 \\ 1 \end{bmatrix}$.

55. Following Example 2.17, we note that we need only use addition and multiplication. We form the augmented matrix and perform Gauss-Jordan elimination in $\mathbb{Z}_3$:

$$\left[\begin{array}{ccc|c} 1 & 1 & 0 & 1 \\ 0 & 1 & 1 & 0 \\ 1 & 0 & 1 & 1 \end{array}\right] \xrightarrow{R_3+2R_1} \left[\begin{array}{ccc|c} 1 & 1 & 0 & 1 \\ 0 & 1 & 1 & 0 \\ 0 & 2 & 1 & 0 \end{array}\right] \xrightarrow{R_3+R_2} \left[\begin{array}{ccc|c} 1 & 1 & 0 & 1 \\ 0 & 1 & 1 & 0 \\ 0 & 0 & 2 & 0 \end{array}\right] \xrightarrow{2R_3} \left[\begin{array}{ccc|c} 1 & 1 & 0 & 1 \\ 0 & 1 & 1 & 0 \\ 0 & 0 & 1 & 0 \end{array}\right]$$

$$\xrightarrow{R_2+2R_3} \left[\begin{array}{ccc|c} 1 & 1 & 0 & 1 \\ 0 & 1 & 0 & 0 \\ 0 & 0 & 1 & 0 \end{array}\right] \xrightarrow{R_1+2R_2} \left[\begin{array}{ccc|c} 1 & 0 & 0 & 1 \\ 0 & 1 & 0 & 0 \\ 0 & 0 & 1 & 0 \end{array}\right]$$

So the solution is: $\begin{bmatrix} x \\ y \\ z \end{bmatrix} = \begin{bmatrix} 1 \\ 0 \\ 0 \end{bmatrix}.$

56. Following Example 2.17, we note that we need only use addition and multiplication. We form the augmented matrix and perform Gauss-Jordan elimination in $\mathbb{Z}_5$:

$$\left[\begin{array}{cc|c} 3 & 2 & 1 \\ 1 & 4 & 1 \end{array}\right] \xrightarrow{R_2+3R_1} \left[\begin{array}{cc|c} 3 & 2 & 1 \\ 0 & 0 & 4 \end{array}\right]$$

The second row is equivalent to the equation $0 = 4$ which clearly has no solution. Therefore, the system is inconsistent.

57. Following Example 2.17, we note that we need only use addition and multiplication. We form the augmented matrix and perform Gauss-Jordan elimination in $\mathbb{Z}_7$:

$$\left[\begin{array}{cc|c} 3 & 2 & 1 \\ 1 & 4 & 1 \end{array}\right] \xrightarrow{R_2+2R_1} \left[\begin{array}{cc|c} 3 & 2 & 1 \\ 0 & 1 & 3 \end{array}\right] \xrightarrow{R_1+5R_2} \left[\begin{array}{cc|c} 3 & 0 & 2 \\ 0 & 1 & 3 \end{array}\right] \xrightarrow{5R_1} \left[\begin{array}{cc|c} 1 & 0 & 3 \\ 0 & 1 & 3 \end{array}\right]$$

So the solution is $\begin{bmatrix} x \\ y \end{bmatrix} = \begin{bmatrix} 3 \\ 3 \end{bmatrix}.$

58. Following Example 2.17, we note that we need only use addition and multiplication. We form the augmented matrix and perform Gauss-Jordan elimination in $\mathbb{Z}_5$:

$$\left[\begin{array}{cccc|c} 1&0&0&4&1\\ 1&2&4&0&3\\ 2&2&0&1&1\\ 1&0&3&0&2 \end{array}\right] \xrightarrow[R_4+4R_1]{\substack{R_2+4R_1\\ R_3+3R_1}} \left[\begin{array}{cccc|c} 1&0&0&4&1\\ 0&2&4&1&2\\ 0&2&0&3&4\\ 0&0&3&1&1 \end{array}\right] \xrightarrow{R_3+4R_2} \left[\begin{array}{cccc|c} 1&0&0&4&1\\ 0&2&4&1&2\\ 0&0&1&2&2\\ 0&0&3&1&1 \end{array}\right] \xrightarrow[R_4+2R_3]{3R_2} \left[\begin{array}{cccc|c} 1&0&0&4&1\\ 0&1&2&3&1\\ 0&0&1&2&2\\ 0&0&0&0&0 \end{array}\right]$$

Therefore, we have:
$$\begin{array}{rcrcrcl} x_1 & & & + & 4x_4 & = & 1\\ x_2 & + & 2x_3 & + & 3x_4 & = & 1\\ & & x_3 & + & 2x_4 & = & 2 \end{array}$$

Setting the free variable $x_4 = t$ yields: $\begin{bmatrix} x_1\\ x_2\\ x_3\\ x_4 \end{bmatrix} = \begin{bmatrix} 1+t\\ 2+t\\ 2+3t\\ 0+t \end{bmatrix} = \begin{bmatrix} 1\\ 2\\ 2\\ 0 \end{bmatrix} + t\begin{bmatrix} 1\\ 1\\ 3\\ 1 \end{bmatrix}.$

Since $t = 0, 1, 2, 3,$ or 4, there are 5 solutions: $\begin{bmatrix} 1\\ 2\\ 2\\ 0 \end{bmatrix}, \begin{bmatrix} 2\\ 3\\ 0\\ 1 \end{bmatrix}, \begin{bmatrix} 3\\ 4\\ 3\\ 2 \end{bmatrix}, \begin{bmatrix} 4\\ 0\\ 1\\ 3 \end{bmatrix}$, and $\begin{bmatrix} 0\\ 1\\ 4\\ 4 \end{bmatrix}$.

59. Recall the Rank Theorem (which applies to systems over $\mathbb{Z}_p$ not just $\mathbb{R}^n$) says: If the system is consistent, then: number of free variables $= n-$ rank (A).

In $\mathbb{Z}_p$, however, each free variable can only take on p different values.
If there is 1 free variable, there are $p^1 = p$ solutions as in Exercise 58.
If there are 2 free variables, there are $p \times p = p^2$ solutions.
In general, the total number of solutions is $p^{\text{number of free variables}} = p^{n-\text{rank}(A)}$.

60. Following Example 2.17, we note that we need only use addition and multiplication. Unlike all previous examples, however, 6 is not prime. Will that matter? Let's see. We form the augmented matrix and perform Gauss-Jordan elimination in $\mathbb{Z}_6$:

$$\left[\begin{array}{cc|c} 2&3&4\\ 4&3&2 \end{array}\right] \xrightarrow{R_2+R_1} \left[\begin{array}{cc|c} 2&3&4\\ 0&0&0 \end{array}\right]$$

So, we have $2x + 3y = 4$ in $\mathbb{Z}_6$.

Let's look at the multiplication table.

$\cdot$	0	1	2	3	4	5
0	0	0	0	0	0	0
1	0	1	2	3	4	5
2	0	2	4	0	2	4
3	0	3	0	3	0	3
4	0	4	2	0	4	2
5	0	5	4	3	2	1

We let $x = 0, 1, 2, 3, 4, 5$ to find all possible solutions.
When $x = 0$ or 3, we get $3y = 4$ which clearly has no solution.
When $x = 1$ or 4, we get $2 + 3y = 4 \Rightarrow 3y = 2$ which has no solution.
When $x = 2$ or 5, we get $4 + 3y = 4 \Rightarrow 3y = 0$ has three solutions $y = 0, 2, 4$.

So the six solutions are $\begin{bmatrix} 2\\ 0 \end{bmatrix}, \begin{bmatrix} 2\\ 2 \end{bmatrix}, \begin{bmatrix} 2\\ 4 \end{bmatrix}, \begin{bmatrix} 5\\ 0 \end{bmatrix}, \begin{bmatrix} 5\\ 2 \end{bmatrix}$, and $\begin{bmatrix} 5\\ 4 \end{bmatrix}$.

Exploration: Partial Pivoting

1. (a) Solving $0.00021x = 1$ to 5 significant digits, we have: $x = \dfrac{1}{0.00021} \approx 4761.9.$

(b) Solving $0.0002x = 1$ to 4 significant digits, we have: $x = \dfrac{1}{0.0002} \approx 5000.$

So, the effect of an 0.00001 error is $5000 - 4761.9 = 238.1$.

2. (a) Without partial pivoting:

Pivoting on 0.400 to 3 significant digits,
we first divide the first row by 0.400 ($\Rightarrow 1 \quad 249 \quad 250$),
then multiply it by 75.3 and subtract the result from row 2 $\Rightarrow$

$$\left[\begin{array}{cc|c} 0.4 & 99.6 & 100 \\ 75.3 & -45.3 & 30.0 \end{array}\right] \longrightarrow \left[\begin{array}{cc|c} 0.4 & 99.6 & 100 \\ 0 & 18700 & 18800 \end{array}\right] \Rightarrow y - 1.01, \text{ to 3 significant digits.}$$

Back substituting, we get $x = 250 - 249(1.01) = 250 - 251 = -1.00$.
This error was introduced because -45.3 and 30.0 were ignored
when reducing to 3 significant digits.
They were overwhelmed by 18700 and 18800 respectively.

(b) With partial pivoting:

Pivoting on 75.3 to 3 significant digits,
we first divide the first row by 75.3 ($\Rightarrow 1 \quad 0.601 \quad 0.398$),
then multiply it by 0.004 and subtract the result from row 2 $\Rightarrow$

$$\left[\begin{array}{cc|c} 75.3 & -45.3 & 30.0 \\ 0.4 & 99.6 & 100 \end{array}\right] \longrightarrow \left[\begin{array}{cc|c} 75.3 & -45.3 & 30.0 \\ 0 & 99.8 & 100 \end{array}\right] \Rightarrow y = 1.00, \text{ again to 3 significant digits.}$$

Back substituting, we get $x = \dfrac{30.0 + 45.3}{75.3} = 1.00$.
Pivoting on the largest absolute value reduces the error in our solution.

3. (a) Without partial pivoting:

Pivoting on 0.001 to 3 significant digits,
we first divide the first row by 0.001 ($\Rightarrow 1 \quad 995 \quad 1000$),
then multiply it by -10.2 and subtract the result from row 2 $\Rightarrow$

$$\left[\begin{array}{cc|c} 0.001 & 0.995 & 1.00 \\ -10.2 & 1.00 & -50.0 \end{array}\right] \longrightarrow \left[\begin{array}{cc|c} 0.001 & 0.995 & 1.00 \\ 0 & -10,100 & -10,200 \end{array}\right] \Rightarrow y = 1.01.$$

Back substituting, we get $x = \dfrac{1.00 - 1.00}{0.001} = 0$.
This error was introduced because 1.00 and -50.0 were ignored when reducing to 3 significant digits.
They were overwhelmed by $-10,100$ and $-10,200$ respectively.

With partial pivoting:

Pivoting on -10.2 to 3 significant digits,
we first divide the first row by -10.2 ($\Rightarrow 1 \quad 0.098 \quad 4.90$),
then multiply it by 0.001 and subtract the result from row 2 $\Rightarrow$

$$\left[\begin{array}{cc|c} -10.2 & 1.00 & -50.0 \\ 0.001 & 0.995 & 1.00 \end{array}\right] \longrightarrow \left[\begin{array}{cc|c} -10.2 & 1.00 & -50.0 \\ 0 & 0.995 & 1.00 \end{array}\right] \Rightarrow y = 1.00.$$

Back substituting, we get $x = \dfrac{-50.0 - 1.00}{-10.2} = \dfrac{51}{10.2} = 5$.
Pivoting on the largest absolute value reduces the error in our solution.

(b) Without partial pivoting:

Pivoting on 10 to 3 significant digits,
we first divide the first row by 10 ($\Rightarrow 1 \quad -0.7 \quad 0 \quad 0.7$),
then multiply it by 3 and add the result to row 2 and
also multiply it by 5 and subtract the result from row 3.

$$\left[\begin{array}{ccc|c} 10.0 & -7.00 & 0.00 & 7.00 \\ -3.0 & 2.09 & 6.00 & 3.91 \\ 5.0 & 1.00 & 5.00 & 6.00 \end{array}\right] \longrightarrow \left[\begin{array}{ccc|c} 10.0 & -7.00 & 0.00 & 7.00 \\ 0.0 & -0.01 & 6.00 & 6.01 \\ 0.0 & 2.50 & 5.00 & 2.50 \end{array}\right].$$

Pivoting on -0.01 to 3 significant digits,
we first divide the second row by -0.01 ($\Rightarrow 0 \quad 1 \quad -600 \quad -601$),
then multiply it by 2.5 and and subtract the result from row 3.

$$\left[\begin{array}{ccc|c} 10.0 & -7.00 & 0.00 & 7.00 \\ 0.0 & -0.01 & 6.00 & 6.01 \\ 0.0 & 2.50 & 5.00 & 2.50 \end{array}\right] \longrightarrow \left[\begin{array}{ccc|c} 10.0 & -7.00 & 0.00 & 7.00 \\ 0.0 & -0.01 & 6.00 & 6.01 \\ 0.0 & 0 & -1500 & -1500 \end{array}\right] \Rightarrow$$

Back substituting, we get $z = 1.00$, $y = 1.00$, and $x = 0.00$.
Repeat by interchanging rows 2 and 3 to make the second pivot 2.5.

Exploration: An Introduction to the Analysis of Algorithms

1. We will count the number of operations, one step at a time.

$$\left[\begin{array}{rrr|r} 2 & 4 & 6 & 8 \\ 3 & 9 & 6 & 12 \\ -1 & 1 & -1 & 1 \end{array}\right] \longrightarrow \left[\begin{array}{rrr|r} 1 & 2 & 3 & 4 \\ 3 & 9 & 6 & 12 \\ -1 & 1 & -1 & 1 \end{array}\right].$$

In this step, we performed 3 operations, $\frac{1}{2} \cdot 4 = 2$, $\frac{1}{2} \cdot 6 = 3$, and $\frac{1}{2} \cdot 8 = 4$.

Note, we don't count $\frac{1}{2} \cdot 2 = 1$ because that's automatic once we decide to multiply by $\frac{1}{2}$.

$$\left[\begin{array}{rrr|r} 1 & 2 & 3 & 4 \\ 3 & 9 & 6 & 12 \\ -1 & 1 & -1 & 1 \end{array}\right] \longrightarrow \left[\begin{array}{rrr|r} 1 & 2 & 3 & 4 \\ 0 & 3 & -3 & 0 \\ -1 & 1 & -1 & 1 \end{array}\right].$$

In this step, we performed 3 operations, $3 \cdot 2 = 6$, $3 \cdot 3 = 9$, and $3 \cdot 4 = 12$, for a total of 6.

Note, we don't count the subtractions (or additions) and again we don't count $3 \cdot 1 = 3$ that created the zero because this is automatic once we decide to multiply by 3.

$$\left[\begin{array}{rrr|r} 1 & 2 & 3 & 4 \\ 0 & 3 & -3 & 0 \\ -1 & 1 & -1 & 1 \end{array}\right] \longrightarrow \left[\begin{array}{rrr|r} 1 & 2 & 3 & 4 \\ 0 & 3 & -3 & 0 \\ 0 & 3 & 2 & 5 \end{array}\right].$$

In this step, we performed 3 operations, $1 \cdot 2 = 2$, $1 \cdot 3 = 3$, and $1 \cdot 4 = 4$, for a total of 9.

$$\left[\begin{array}{rrr|r} 1 & 2 & 3 & 4 \\ 0 & 3 & -3 & 0 \\ 0 & 3 & 2 & 5 \end{array}\right] \longrightarrow \left[\begin{array}{rrr|r} 1 & 2 & 3 & 4 \\ 0 & 1 & -1 & 0 \\ 0 & 3 & 2 & 5 \end{array}\right].$$

In this step, we performed 2 operations, $(\frac{1}{3})(-3) = -1$ and $\frac{1}{3} \cdot 0 = 0$, for a total of 11.

$$\left[\begin{array}{rrr|r} 1 & 2 & 3 & 4 \\ 0 & 1 & -1 & 0 \\ 0 & 3 & 2 & 5 \end{array}\right] \longrightarrow \left[\begin{array}{rrr|r} 1 & 2 & 3 & 4 \\ 0 & 1 & -1 & 0 \\ 0 & 0 & 5 & 5 \end{array}\right].$$

In this step, we performed 2 operations, $(-3)(-1) = 3$ and $(-3)(0) = 0$, for a total of 13.

$$\left[\begin{array}{rrr|r} 1 & 2 & 3 & 4 \\ 0 & 1 & -1 & 0 \\ 0 & 0 & 5 & 5 \end{array}\right] \longrightarrow \left[\begin{array}{rrr|r} 1 & 2 & 3 & 4 \\ 0 & 1 & -1 & 0 \\ 0 & 0 & 1 & 1 \end{array}\right].$$

In this step, we performed 1 operation, $\frac{1}{5} \cdot 5 = 5$, for a total of 14.

Finally, to complete the back substitution, we need only 3 more operations. See that? Namely, $-x_3$ to find x_2, and $2x_2$, $3x_3$ to find x_1, for a total of 17.

2. To the reduced form given below (as computed above) we have 14 operations:

$$\left[\begin{array}{ccc|c} 1 & 2 & 3 & 4 \\ 0 & 1 & -1 & 0 \\ 0 & 0 & 1 & 1 \end{array}\right].$$

Now, we create a zero above the 1 in the second column:

$$\left[\begin{array}{ccc|c} 1 & 2 & 3 & 4 \\ 0 & 1 & -1 & 0 \\ 0 & 0 & 1 & 1 \end{array}\right] \longrightarrow \left[\begin{array}{ccc|c} 1 & 0 & 5 & 4 \\ 0 & 1 & -1 & 0 \\ 0 & 0 & 1 & 1 \end{array}\right].$$

This required two operations, $(-2)(-1)$ and $(-2)(0)$, for a total of 16.

Finally, we create two zeroes above the 1 in the third column:

$$\left[\begin{array}{ccc|c} 1 & 0 & 6 & 4 \\ 0 & 1 & -1 & 0 \\ 0 & 0 & 1 & 1 \end{array}\right] \longrightarrow \left[\begin{array}{ccc|c} 1 & 0 & 0 & -1 \\ 0 & 1 & 0 & 1 \\ 0 & 0 & 1 & 1 \end{array}\right].$$

This required two operations, $(1)(1)$ and $(-5)(1)$, for a total of 18.

So, Gauss-Jordan required 1 more operation than Gaussian elimination.
Therefore, Gauss-Jordan is probably less efficient than Gaussian elimination.

3. (a) There are n operations required to create the first leading 1 because we have to divide every entry in the first row (except a_{11}) by a_{11}.
We don't have to divide a_{11} by a_{11} because the resulting 1 results from the choice itself.
There are n operations required to create the first zero in column 1 because we have to multiply every entry in the first row (except the leading 1) by a_{21}.
We don't have to multiply 1 by a_21 because the resulting 0 results from the choice itself.
There are n operations required to create each zero in column 1 because we have to multiply every entry in the first row (except the leading 1) by a_{k1}.
There are $n-1$ rows (excluding the first row), so $n+(n-1)n$ operations are required.
Recall, the first n operations created the leading 1 in row 1.

(b) As above, we see it takes $n-1$ operations to create the leading 1 in the second row and then $n-1$ operations to create the zeros in column 2.
Now there are only $n-2$ rows left to create zeroes in since rows 1 and 2 are excluded.
So, $(n-1)+(n-2)(n-1)$ operations are required to create the leading 1 in row 2 and all the zeros beneath it.
Continuing this process, we see:
$[n+(n-1)n]+[(n-1)+(n-2)(n-1)]+\cdots+[2+1\cdot 2]+1$.
This simplifies to: $n^2+(n-1)^2+\cdots+2^2+1^2$.
This simplification follows quickly from the following observation:
$n+(n-1)n = n[1+(n-1)] = n\cdot n = n^2$.

(c) There is 1 operation required to find x_{n-1} (involving x_n).
There are 2 operations required to find x_{n-2} (involving x_n and x_{n-1})
Continuing this reasoning, we get the total: $1+2+\cdots+(n-1)$.

(d) Exercises 45 and 46 in Section 2.4 and Appendix B give us the following equations:
$n^2+(n-1)^2+\cdots+2^2+1^2 = \frac{n(n+1)(2n+1)}{6} = \frac{1}{3}n^3+\frac{1}{2}n^2+\frac{1}{6}n$ and
$1+2+\cdots+(n-1) = \frac{n(n-1)}{2} = \frac{1}{2}n^2-\frac{1}{2}n$.
Adding these together we get the total number of operations is:
$\left(\frac{1}{3}n^3+\frac{1}{2}n^2+\frac{1}{6}n\right)+\left(\frac{1}{2}n^2-\frac{1}{2}n\right) = \frac{1}{3}n^3+n^2-\frac{1}{3}n$.
So, for large values of n the total number of operations required is $T(n)\approx\frac{1}{3}n^3$.

4. As we saw in Exercise 2, we get the same $n^2 + (n-1)^2 + \cdots + 2^2 + 1^2$ operations to create row echelon form.

To create the 1 zero above row 2 requires $1(n-1)$ operations.
To create the 2 zeroes above row 3 requires $2(n-2)$ operations.
Continuing with this reasoning, we see it takes $(n-1)+2(n-2)+\cdots+(n-1)(n-(n-1))$ operations to create all the necessary zeroes. Now note:

$$\begin{aligned}&(n-1) + 2(n-2) + \cdots + (n-1)(n-(n-1))\\ &\quad = (1\cdot n + 2\cdot n + \cdots + (n-1)\cdot n) - (1^2 + 2^2 + \cdots + (n-1)^2)\\ &\quad = n(1 + 2 + \cdots + (n-1)) - (1^2 + 2^2 + \cdots + (n-1)^2).\end{aligned}$$

Similar to Exercise 3, we have the following equations:
$(n-1)^2 + \cdots + 2^2 + 1^2 = \frac{n(n-1)(2n-1)}{6} = \frac{1}{3}n^3 - \frac{1}{2}n^2 + \frac{1}{6}n$ and
$1 + 2 + \cdots + (n-1) = \frac{n(n-1)}{2} = \frac{1}{2}n^2 - \frac{1}{2}n.$

Putting these together we get the total number of operations is:
$\left(\frac{1}{3}n^3 + \frac{1}{2}n^2 + \frac{1}{6}n\right) + n\left(\frac{1}{2}n^2 - \frac{1}{2}n\right) - \left(\frac{1}{3}n^3 - \frac{1}{2}n^2 + \frac{1}{6}n\right) = \frac{1}{2}n^3 + \frac{1}{2}n^2.$

So, for large values of n the total number of operations required is $T(n) \approx \frac{1}{2}n^3$.

2.3 Spanning Sets and Linear Independence

1. As in Example 2.18, we want to find scalars x and y such that:

$$x\begin{bmatrix}1\\-1\end{bmatrix}+y\begin{bmatrix}2\\-1\end{bmatrix}=\begin{bmatrix}1\\2\end{bmatrix}$$ Expanding, we obtain the system: $$\begin{aligned}x+2y&=1\\-x-y&=2\end{aligned}$$

We then row reduce the associated augmented matrix: $\left[\begin{array}{cc|c}1&2&1\\-1&-1&2\end{array}\right]\longrightarrow\left[\begin{array}{cc|c}1&0&-5\\0&1&3\end{array}\right]$

So the solution is $x=-5$, $y=2$, and the linear combination is $-5\begin{bmatrix}1\\-1\end{bmatrix}+3\begin{bmatrix}2\\-1\end{bmatrix}=\begin{bmatrix}1\\2\end{bmatrix}$.

2. As in Example 2.18, we want to find scalars x and y such that:

$$x\begin{bmatrix}4\\-2\end{bmatrix}+y\begin{bmatrix}-2\\1\end{bmatrix}=\begin{bmatrix}2\\1\end{bmatrix}$$ Expanding, we obtain the system: $$\begin{aligned}4x-2y&=2\\-2x+y&=1\end{aligned}$$

We then row reduce the associated augmented matrix: $\left[\begin{array}{cc|c}4&-2&2\\-2&1&1\end{array}\right]\longrightarrow\left[\begin{array}{cc|c}4&-2&2\\0&0&2\end{array}\right]$

Since $0\neq 2$, this system clearly has no solution. So, what do we conclude?
We conclude that $\mathbf{v}$ is not a linear combination of $\mathbf{u}_1$ and $\mathbf{u}_2$.

We could have noted $\begin{bmatrix}4\\-2\end{bmatrix}=-2\begin{bmatrix}-2\\1\end{bmatrix}$, while $\begin{bmatrix}2\\1\end{bmatrix}$ is not a multiple of $\begin{bmatrix}-2\\1\end{bmatrix}$.

3. As in Example 2.18, we want to find scalars x and y such that:

$$x\begin{bmatrix}1\\1\\0\end{bmatrix}+y\begin{bmatrix}0\\1\\1\end{bmatrix}=\begin{bmatrix}1\\2\\3\end{bmatrix}$$ Expanding, we obtain the system: $$\begin{aligned}x&=1\\x+y&=2\\y&=3\end{aligned}$$

Since $x=1$ and $y=3$ implies $x+y\neq 2$, this system clearly has no solution.
Therefore, $\mathbf{v}$ is not a linear combination of $\mathbf{u}_1$ and $\mathbf{u}_2$.

4. As in Example 2.18, we want to find scalars x and y such that:

$$x\begin{bmatrix}1\\1\\0\end{bmatrix}+y\begin{bmatrix}0\\1\\1\end{bmatrix}=\begin{bmatrix}3\\2\\-1\end{bmatrix}$$ Expanding, we obtain the system: $$\begin{aligned}x&=3\\x+y&=2\\y&=-1\end{aligned}$$

Since $x=3$ and $y=-1$ implies $x+y=2$, those values of x and y are clearly the solution.

So the solution is $x=3$, $y=-1$, and the linear combination is $3\begin{bmatrix}1\\1\\0\end{bmatrix}-1\begin{bmatrix}0\\1\\1\end{bmatrix}=\begin{bmatrix}3\\2\\-1\end{bmatrix}$.

5. Similar to Example 2.18, we want to find scalars x, y, and z such that:

$$x\begin{bmatrix}1\\1\\0\end{bmatrix}+y\begin{bmatrix}0\\1\\1\end{bmatrix}+z\begin{bmatrix}1\\0\\1\end{bmatrix}=\begin{bmatrix}1\\2\\3\end{bmatrix}$$ Expanding, we obtain the system: $$\begin{aligned}x \quad\quad + z &= 1\\ x + y \quad\quad &= 2\\ y + z &= 3\end{aligned}$$

Since $z = 1$ and $y = 2$ implies $y + z = 3$, the solution is $x = 0$, $y = 2$, $z = 1$.

Row reduce the augmented matrix to confirm that: $$\left[\begin{array}{ccc|c}1&0&1&1\\1&1&0&2\\0&1&1&3\end{array}\right]\longrightarrow\left[\begin{array}{ccc|c}1&0&0&0\\0&1&0&2\\0&0&1&1\end{array}\right]$$

So the solution is $x = 0$, $y = 2$, $z = 1$, and the linear combination is $2\begin{bmatrix}0\\1\\1\end{bmatrix}+1\begin{bmatrix}1\\0\\1\end{bmatrix}=\begin{bmatrix}1\\2\\3\end{bmatrix}$.

6. Similar to Example 2.18, we want to find scalars x, y, and z such that:

$$x\begin{bmatrix}1.0\\0.4\\4.8\end{bmatrix}+y\begin{bmatrix}3.4\\1.4\\-6.4\end{bmatrix}+z\begin{bmatrix}-1.2\\0.2\\-1.0\end{bmatrix}=\begin{bmatrix}3.2\\2.0\\-2.6\end{bmatrix}$$ The obvious solution is $x = 1$, $y = 1$, $z = 1$.

The linear combination is $\begin{bmatrix}1.0\\0.4\\4.8\end{bmatrix}+\begin{bmatrix}3.4\\1.4\\-6.4\end{bmatrix}+\begin{bmatrix}-1.2\\0.2\\-1.0\end{bmatrix}=\begin{bmatrix}3.2\\2.0\\-2.6\end{bmatrix}$.

Note: We should always look for an easy or obvious solution first.

7. Applying Theorem 2.4, we check to see if $[A|\mathbf{b}]$ is consistent. Why?
Theorem 2.4 says $[A|\mathbf{b}]$ is consistent $\Leftrightarrow$ $\mathbf{b}$ is a linear combination of the columns of A.
That is exactly what is required for $\mathbf{b}$ to be in the span of the columns of A.

So we row reduce $[A|\mathbf{b}]=\left[\begin{array}{cc|c}1&2&5\\3&4&6\end{array}\right]\longrightarrow\left[\begin{array}{cc|c}1&0&-4\\0&1&\frac{9}{2}\end{array}\right]$ to see that it is consistent.

What do we conclude? The vector $\mathbf{b}$ is in the span of the columns of A.

In particular, the solution tells us the linear combination is $-4\begin{bmatrix}1\\3\end{bmatrix}+\frac{9}{2}\begin{bmatrix}2\\4\end{bmatrix}=\begin{bmatrix}5\\6\end{bmatrix}$.

8. Applying Theorem 2.4, we check to see if $[A|\mathbf{b}]$ is consistent. Why?
Theorem 2.4 says $[A|\mathbf{b}]$ is consistent $\Leftrightarrow$ $\mathbf{b}$ is a linear combination of the columns of A.
That is exactly what is required for $\mathbf{b}$ to be in the span of the columns of A.

Row reduce $[A|\mathbf{b}]=\left[\begin{array}{ccc|c}1&2&3&10\\4&5&6&11\\7&8&9&12\end{array}\right]\longrightarrow\left[\begin{array}{ccc|c}1&0&-1&-\frac{28}{3}\\0&1&2&\frac{29}{3}\\0&0&0&0\end{array}\right]$ to see that it is consistent.

What do we conclude? The vector $\mathbf{b}$ is in the span of the columns of A.

The solution tells us one possible linear combination is $-\frac{28}{3}\begin{bmatrix}1\\4\\7\end{bmatrix}+\frac{29}{3}\begin{bmatrix}2\\5\\8\end{bmatrix}=\begin{bmatrix}10\\11\\12\end{bmatrix}$.

9. As in Example 2.19, we must show $x\begin{bmatrix}1\\1\end{bmatrix}+y\begin{bmatrix}1\\-1\end{bmatrix}=\begin{bmatrix}a\\b\end{bmatrix}$ can always be solved.

The augmented matrix is $\left[\begin{array}{cc|c}1&1&a\\1&-1&b\end{array}\right]$, and row reduction produces:

$$\left[\begin{array}{cc|c}1&1&a\\1&-1&b\end{array}\right]\xrightarrow{R_1+R_2}\left[\begin{array}{cc|c}2&0&a+b\\1&-1&b\end{array}\right]\xrightarrow[-R_2]{1/2R_1}\left[\begin{array}{cc|c}1&0&(a+b)/2\\0&1&-b\end{array}\right]\xrightarrow{R_2+R_1}\left[\begin{array}{cc|c}1&0&(a+b)/2\\-1&1&(a-b)/2\end{array}\right]$$

We see that $x=(a+b)/2$ and $y=(a-b)/2$, so for any choice of a and b we have

$$\left(\frac{a+b}{2}\right)\begin{bmatrix}1\\1\end{bmatrix}+\left(\frac{a-b}{2}\right)\begin{bmatrix}1\\-1\end{bmatrix}=\begin{bmatrix}a\\b\end{bmatrix}\qquad\text{Check this!}$$

10. As in Example 2.19, we must show $x\begin{bmatrix}3\\-2\end{bmatrix}+y\begin{bmatrix}0\\1\end{bmatrix}=\begin{bmatrix}a\\b\end{bmatrix}$ can always be solved.

The augmented matrix is $\left[\begin{array}{cc|c}3&0&a\\-2&1&b\end{array}\right]$, and row reduction produces:

$$\left[\begin{array}{cc|c}3&0&a\\-2&1&b\end{array}\right]\xrightarrow{\frac{1}{3}R_1}\left[\begin{array}{cc|c}1&0&a/3\\-2&1&b\end{array}\right]\xrightarrow{R_2+2R_1}\left[\begin{array}{cc|c}1&0&a/3\\0&1&(2a+3b)/3\end{array}\right]$$

We see that $x=a/3$ and $y=(2a+3b)/3$, so for any choice of a and b we have

$$\left(\frac{a}{3}\right)\begin{bmatrix}3\\-2\end{bmatrix}+\left(\frac{2a+3b}{3}\right)\begin{bmatrix}0\\1\end{bmatrix}=\begin{bmatrix}a\\b\end{bmatrix}\qquad\text{Check this!}$$

11. Similar to Example 2.19, we must show $x\begin{bmatrix}1\\0\\1\end{bmatrix}+y\begin{bmatrix}1\\1\\0\end{bmatrix}+z\begin{bmatrix}0\\1\\1\end{bmatrix}=\begin{bmatrix}a\\b\\c\end{bmatrix}$ can always be solved.

The augmented matrix is $\left[\begin{array}{ccc|c}1&1&0&a\\0&1&1&b\\1&0&1&c\end{array}\right]$, and row reduction produces:

$$\left[\begin{array}{ccc|c}1&1&0&a\\0&1&1&b\\1&0&1&c\end{array}\right]\xrightarrow{R_3-R_1+R_2}\left[\begin{array}{ccc|c}1&1&0&a\\0&1&1&b\\0&0&2&-a+b+c\end{array}\right]\xrightarrow{\frac{1}{2}R_3}\left[\begin{array}{ccc|c}1&1&0&a\\0&1&1&b\\0&0&1&(-a+b+c)/2\end{array}\right]$$

$$\xrightarrow{R_2-R_3}\left[\begin{array}{ccc|c}1&1&0&a\\0&1&0&(a+b-c)/2\\0&0&1&(-a+b+c)/2\end{array}\right]\xrightarrow{R_1-R_2}\left[\begin{array}{ccc|c}1&0&0&(a-b+c)/2\\0&1&0&(a+b-c)/2\\0&0&1&(-a+b+c)/2\end{array}\right]$$

We see that $x=(a-b+c)/2$, $y=(a+b-c)/2$, and $z=(-a+b+c)/2$.
So for any choice of a, b, and c we have:

$$\left(\frac{a-b+c}{2}\right)\begin{bmatrix}1\\0\\1\end{bmatrix}+\left(\frac{a+b-c}{2}\right)\begin{bmatrix}1\\1\\0\end{bmatrix}+\left(\frac{-a+b+c}{2}\right)\begin{bmatrix}0\\1\\1\end{bmatrix}=\begin{bmatrix}a\\b\\c\end{bmatrix}\qquad\text{Check this!}$$

12. Similar to Example 2.19, show $x\begin{bmatrix}1\\2\\3\end{bmatrix}+y\begin{bmatrix}-1\\-1\\0\end{bmatrix}+z\begin{bmatrix}2\\1\\-1\end{bmatrix}=\begin{bmatrix}a\\b\\c\end{bmatrix}$ can always be solved.

The augmented matrix is $\left[\begin{array}{ccc|c}1&-1&2&a\\2&-1&1&b\\3&0&-1&c\end{array}\right]$, and row reduction produces:

$$\left[\begin{array}{ccc|c}1&-1&2&a\\2&-1&1&b\\3&0&-1&c\end{array}\right]\xrightarrow[R_3-3R_1]{R_2-2R_1}\left[\begin{array}{ccc|c}1&-1&2&a\\0&1&-3&-2a+b\\0&3&-7&-3a+c\end{array}\right]\xrightarrow[R_3-3R_2]{R_1+R_2}\left[\begin{array}{ccc|c}1&0&-1&-a+b\\0&1&-3&-2a+b\\0&0&2&3a-3b+c\end{array}\right]$$

$$\xrightarrow{\frac{1}{2}R_3}\left[\begin{array}{ccc|c}1&0&-1&-a+b\\0&1&-3&-2a+b\\0&0&1&(3a-3b+c)/2\end{array}\right]\xrightarrow[R_2+3R_3]{R_1+R_3}\left[\begin{array}{ccc|c}1&0&0&(a-b+c)/2\\0&1&0&(5a-7b+3c)/2\\0&0&1&(3a-3b+c)/2\end{array}\right]$$

We see that $x=(a-b+c)/2$, $y=(5a-7b+3c)/2$, and $z=(3a-3b+c)/2$.
So for any choice of a, b, and c we have:

$$\left(\frac{a-b+c}{2}\right)\begin{bmatrix}1\\2\\3\end{bmatrix}+\left(\frac{5a-7b+3c}{2}\right)\begin{bmatrix}-1\\-1\\0\end{bmatrix}+\left(\frac{3a-3b+c}{2}\right)\begin{bmatrix}2\\1\\-1\end{bmatrix}=\begin{bmatrix}a\\b\\c\end{bmatrix}$$

13. We should describe the span of the given vectors (a) geometrically and (b) algebraically.

(a) Geometrically, we can see that the set of all linear combinations of $\begin{bmatrix}2\\-4\end{bmatrix}$ and $\begin{bmatrix}-1\\2\end{bmatrix}$ is just the line through the origin with $\begin{bmatrix}-1\\2\end{bmatrix}$ as direction vector.

Why do we not have to consider $\begin{bmatrix}2\\-4\end{bmatrix}$? Because $\begin{bmatrix}2\\-4\end{bmatrix}=-2\begin{bmatrix}-1\\2\end{bmatrix}$.

(b) Algebraically, the vector equation of this line is $\begin{bmatrix}x\\y\end{bmatrix}=\begin{bmatrix}-1\\2\end{bmatrix}t$.

That is just another way of saying that $\begin{bmatrix}x\\y\end{bmatrix}$ is in the span of $\begin{bmatrix}-1\\2\end{bmatrix}$.
Suppose we want to obtain the general equation of this line.
One method is to use the system of equations arising from the vector equation:

$$\begin{bmatrix}x\\y\end{bmatrix}=\begin{bmatrix}-1\\2\end{bmatrix}t\Rightarrow\begin{array}{rcl}x&=&-t\\y&=&2t\end{array}\quad\text{So } y=2(-x)=-2x\Rightarrow 2x+y=0.$$

14. We should describe the span of the given vectors (*a*) geometrically and (*b*) algebraically.

(a) Geometrically, we can see that the set of all linear combinations of $\begin{bmatrix} 0 \\ 0 \end{bmatrix}$ and $\begin{bmatrix} 3 \\ 4 \end{bmatrix}$ is just the line through the origin with $\begin{bmatrix} 3 \\ 4 \end{bmatrix}$ as direction vector.

Why do we not have to consider $\begin{bmatrix} 0 \\ 0 \end{bmatrix}$? Because $\begin{bmatrix} 0 \\ 0 \end{bmatrix} = 0\begin{bmatrix} 3 \\ 4 \end{bmatrix}$.

(b) Algebraically, the vector equation of this line is $\begin{bmatrix} x \\ y \end{bmatrix} = \begin{bmatrix} 3 \\ 4 \end{bmatrix}t$.

That is just another way of saying that $\begin{bmatrix} x \\ y \end{bmatrix}$ is in the span of $\begin{bmatrix} 3 \\ 4 \end{bmatrix}$.

Suppose we want to obtain the general equation of this line.
One method is to use the system of equations arising from the vector equation:

$$\begin{bmatrix} x \\ y \end{bmatrix} = \begin{bmatrix} 3 \\ 4 \end{bmatrix}t \Rightarrow \begin{array}{rcl} x &=& 3t \\ y &=& 4t \end{array} \quad \text{So } y = 4\left(\tfrac{1}{3}x\right) \Rightarrow 3y = 4x \Rightarrow 4x - 3y = 0.$$

15. We should describe the span of the given vectors (*a*) geometrically and (*b*) algebraically.

(a) Geometrically, we can see that the set of all linear combinations of $\begin{bmatrix} 1 \\ 2 \\ 0 \end{bmatrix}$ and $\begin{bmatrix} 3 \\ 2 \\ -1 \end{bmatrix}$ is just the plane through the origin with $\begin{bmatrix} 1 \\ 2 \\ 0 \end{bmatrix}$ and $\begin{bmatrix} 3 \\ 2 \\ -1 \end{bmatrix}$ as direction vectors.

(b) Algebraically, the vector equation of this plane is $\begin{bmatrix} x \\ y \\ z \end{bmatrix} = s\begin{bmatrix} 1 \\ 2 \\ 0 \end{bmatrix} + t\begin{bmatrix} 3 \\ 2 \\ -1 \end{bmatrix}$.

That is just another way of saying that $\begin{bmatrix} x \\ y \\ z \end{bmatrix}$ is in the span of $\begin{bmatrix} 1 \\ 2 \\ 0 \end{bmatrix}$ and $\begin{bmatrix} 3 \\ 2 \\ -1 \end{bmatrix}$.

Suppose we want to obtain the general equation of this plane.
One method is to use the system of equations arising from the vector equation:

$$\begin{array}{rcrcl} s &+& 3t &=& x \\ 2s &+& 2t &=& y \\ && -t &=& z \end{array} \Rightarrow \left[\begin{array}{rr|r} 1 & 3 & x \\ 2 & 2 & y \\ 0 & -1 & z \end{array}\right] \longrightarrow \left[\begin{array}{rr|c} 1 & 3 & x \\ 0 & -4 & -2x + y \\ 0 & 0 & (2x - y + 4z)/4 \end{array}\right]$$

We know this system is consistent, since $\begin{bmatrix} x \\ y \\ z \end{bmatrix}$ *is* in the span of $\begin{bmatrix} 1 \\ 2 \\ 0 \end{bmatrix}$ and $\begin{bmatrix} 3 \\ 2 \\ -1 \end{bmatrix}$.

So, we *must* have $2x - y + 4z = 0$, giving us the general equation we seek.

Note: Both $\begin{bmatrix} 1 \\ 2 \\ 0 \end{bmatrix}$ and $\begin{bmatrix} 3 \\ 2 \\ -1 \end{bmatrix}$ are orthogonal to $\begin{bmatrix} 2 \\ -1 \\ 4 \end{bmatrix}$. Should they be?

16. We should describe the span of the given vectors (*a*) geometrically and (*b*) algebraically.

(a) Geometrically, we can see that the set of all linear combinations of $\begin{bmatrix} 1 \\ 0 \\ -1 \end{bmatrix}$ and $\begin{bmatrix} -1 \\ 1 \\ 0 \end{bmatrix}$

is just the plane through the origin with $\begin{bmatrix} 1 \\ 0 \\ -1 \end{bmatrix}$ and $\begin{bmatrix} -1 \\ 1 \\ 0 \end{bmatrix}$ as direction vectors.

Q: Why do we get to ignore the third vector, $\mathbf{v}_3 = [0, -1, 1]$?
A: Since $\mathbf{v}_3 = -\mathbf{v}_1 + \mathbf{v}_2$, it does not affect the span. Why not?

(b) Algebraically, the vector equation of this plane is $\begin{bmatrix} x \\ y \\ z \end{bmatrix} = s \begin{bmatrix} 1 \\ 0 \\ -1 \end{bmatrix} + t \begin{bmatrix} -1 \\ 1 \\ 0 \end{bmatrix}$.

That is just another way of saying that $\begin{bmatrix} x \\ y \\ z \end{bmatrix}$ is in the span of $\begin{bmatrix} 1 \\ 0 \\ -1 \end{bmatrix}$ and $\begin{bmatrix} -1 \\ 1 \\ 0 \end{bmatrix}$.

Suppose we want to obtain the general equation of this plane.
One method is to use the system of equations arising from the vector equation:

$$\begin{array}{rcl} s - t &=& x \\ t &=& y \\ -s &=& z \end{array} \Rightarrow \left[\begin{array}{rr|c} 1 & -1 & x \\ 0 & 1 & y \\ -1 & 0 & z \end{array}\right] \longrightarrow \left[\begin{array}{rr|c} 1 & -1 & x \\ 0 & 1 & y \\ 0 & 0 & x+y+z \end{array}\right]$$

We know this system is consistent, since $\begin{bmatrix} x \\ y \\ z \end{bmatrix}$ *is* in the span of $\begin{bmatrix} 1 \\ 0 \\ -1 \end{bmatrix}$ and $\begin{bmatrix} -1 \\ 1 \\ 0 \end{bmatrix}$.

So, we *must* have $x + y + z = 0$, giving us the general equation we seek.

Note: Both $\begin{bmatrix} 1 \\ 0 \\ -1 \end{bmatrix}$ and $\begin{bmatrix} -1 \\ 1 \\ 0 \end{bmatrix}$ are orthogonal to $\begin{bmatrix} 1 \\ 1 \\ 1 \end{bmatrix}$. Should they be?

17. Since the three points $(1, 0, 3)$, $(-1, 1, -3)$, and $(0, 0, 0)$ must lie in the plane, $(1, 0, 3)$ and $(-1, 1, -3)$ must satisfy the equation of a plane through the origin $ax+by+cz = 0$.

We substitute the two nonzero points into $ax + by + cz = 0$ to create a homogenous system:

$$\begin{array}{rcl} a \quad + 3c &=& 0 \\ -a + b - 3c &=& 0 \end{array} \Rightarrow \left[\begin{array}{rrr|c} 1 & 0 & 3 & 0 \\ -1 & 1 & -3 & 0 \end{array}\right] \xrightarrow{R_2+R_1} \left[\begin{array}{rrr|c} 1 & 0 & 3 & 0 \\ 0 & 1 & 0 & 0 \end{array}\right]$$

So, we have $b = 0$ and $a = -3c$.

Letting $c = -1 \Rightarrow a = 3$ and $b = 0$ yields $3x + 0y - z = 3x - z = 0$ is the general equation.

Q: Does the free variable c imply infinitely many planes contain these three points?
A: Hint: We can divide the general solution $-3cx + cz = 0$ by $-c \neq 0 \Rightarrow 3x - z = 0$.

18. To show $\mathbf{u}$, $\mathbf{v}$, and $\mathbf{w}$ are in span($\mathbf{u}$, $\mathbf{v}$, $\mathbf{w}$), we must show that $\mathbf{u}$, $\mathbf{v}$, and $\mathbf{w}$ can be written as a linear combination of $\mathbf{u}$, $\mathbf{v}$, and $\mathbf{w}$.

Q: Can we simply let $\mathbf{u} = \mathbf{u}$, $\mathbf{v} = \mathbf{v}$, and $\mathbf{w} = \mathbf{w}$?
A: Yes and no. Technically, we have to write the linear combination using all the vectors:

$$\mathbf{u} = \mathbf{u} + 0\mathbf{v} + 0\mathbf{w},\ \mathbf{v} = 0\mathbf{u} + \mathbf{v} + 0\mathbf{w},\text{ and } \mathbf{w} = 0\mathbf{u} + 0\mathbf{v} + \mathbf{w}.$$

In the future, we will only write vectors with nonzero scalars in linear combinations.

19. To show $\mathbf{u}$, $\mathbf{v}$, and $\mathbf{w}$ are in span($\mathbf{u}$, $\mathbf{u}+\mathbf{v}$, $\mathbf{u}+\mathbf{v}+\mathbf{w}$), we must show that $\mathbf{u}$, $\mathbf{v}$, and $\mathbf{w}$ can be written as a linear combination of $\mathbf{u}$, $\mathbf{u}+\mathbf{v}$, $\mathbf{u}+\mathbf{v}+\mathbf{w}$.

Q: Can we simply let $\mathbf{u} = \mathbf{u}$, $\mathbf{v} = \mathbf{v}$, and $\mathbf{w} = \mathbf{w}$?
A: No. Why not? These vectors, except for $\mathbf{u}$, are not explicitly listed in the spanning set.

Instead, we need linear combinations of $\mathbf{u}$, $\mathbf{u}+\mathbf{v}$, $\mathbf{u}+\mathbf{v}+\mathbf{w}$ that yield $\mathbf{u}$, $\mathbf{v}$, and $\mathbf{w}$. So:

$$\mathbf{u} = \mathbf{u},\ \mathbf{v} = -\mathbf{u} + (\mathbf{u}+\mathbf{v}),\text{ and } \mathbf{w} = -(\mathbf{u}+\mathbf{v}) + (\mathbf{u}+\mathbf{v}+\mathbf{w}).$$

Note: We have now shown that we can use $\mathbf{u}$, $\mathbf{v}$, and $\mathbf{w}$. How?

20. Both span(S) and span(T) are contained in $\mathbb{R}^n$. That is, span(S) $\subseteq \mathbb{R}^n$, span(T) $\subseteq \mathbb{R}^n$.

(a) We need to show that any vector s in span(S) is also in span(T).
The idea is simple: Let the scalars for all the vectors $\mathbf{u}_{k+1}$ through $\mathbf{u}_m$ equal zero.
In symbols, we write:
$\mathbf{s} = s_1\mathbf{u}_1 + s_2\mathbf{u}_2 + \cdots + s_k\mathbf{u}_k = s_1\mathbf{u}_1 + s_2\mathbf{u}_2 + \cdots + s_k\mathbf{u}_k + 0\mathbf{u}_{k+1} + \cdots + 0\mathbf{u}_m.$
Therefore, any vector s in span(S) is in also span(T) as we were to show.

(b) We are told to *deduce* that if $\mathbb{R}^n$ =span(S), then $\mathbb{R}^n$ =span(T).
The word *deduce* tells us to use the basic properties of sets in our proof.
The basic property of sets we will use is: if $V \subseteq W$ and $W \subseteq V$, then $V = W$.
In the statement of the problem, we were told that span(T) $\subseteq \mathbb{R}^n$.
If we can show $\mathbb{R}^n \subseteq$ span(T), then we can deduce $\mathbb{R}^n =$ span(T).
From (a), we have span(S) $\subseteq$ span(T). For (b), we suppose span(S) $= \mathbf{R}^n$.
Therefore, we have span(S) $= \mathbb{R}^n \subseteq$ span(T) and deduce $\mathbb{R}^n =$ span(T).

21. When proving something for n, first let $n = 1$ or 2 to look for the underlying pattern.
Assume that there are only two vectors $\mathbf{u}_1$ and $\mathbf{u}_2$ and two vectors $\mathbf{v}_1$ and $\mathbf{v}_2$.
We are told $\mathbf{w}$ is a linear combination of $\mathbf{u}_1$ and $\mathbf{u}_2$. So: $\mathbf{w} = w_1\mathbf{u}_1 + w_2\mathbf{u}_2$.
We are also told that both $\mathbf{u}_1$ and $\mathbf{u}_2$ are linear combinations of $\mathbf{v}_1$ and $\mathbf{v}_2$.
So, we have both: $\mathbf{u}_1 = v_{11}\mathbf{v}_1 + v_{12}\mathbf{v}_2$ and $\mathbf{u}_2 = v_{21}\mathbf{v}_1 + v_{22}\mathbf{v}_2$.
We need to show these assumptions imply $\mathbf{w}$ is a linear combination of $\mathbf{v}_1$ and $\mathbf{v}_2$. How?
Let $\mathbf{u}_1 = u_{1_1}\mathbf{v}_1 + u_{1_2}\mathbf{v}_2$ and $\mathbf{u}_2 = u_{2_1}\mathbf{v}_1 + u_{2_2}\mathbf{v}_2$ in $\mathbf{w} = w_1\mathbf{u}_1 + w_2\mathbf{u}_2$.
This substitution yields: $\mathbf{w} = w_1(u_{1_1}\mathbf{v}_1 + u_{1_2}\mathbf{v}_2) + w_2(u_{2_1}\mathbf{v}_1 + u_{2_2}\mathbf{v}_2)$.
It is now obvious that $\mathbf{w}$ is a linear combination of $\mathbf{v}_1$ and $\mathbf{v}_2$. Why?
Observe that this reasoning holds for any n and proceed to the proof.

(a) Let $\mathbf{w} = w_1\mathbf{u}_1 + w_1\mathbf{u}_2 + \cdots + w_1\mathbf{u}_k$,
and assume that each $\mathbf{u}_i$ is a linear combination of vectors $\mathbf{v}_1, \mathbf{v}_2, \ldots, \mathbf{v}_m$.
Then each $\mathbf{u}_i = u_{i_1}\mathbf{v}_1 + u_{i_2}\mathbf{v}_2 + \cdots + u_{i_m}\mathbf{v}_m$, and

$$\begin{aligned}
\mathbf{w} &= w_1\mathbf{u}_1 + w_2\mathbf{u}_2 + \cdots + w_k\mathbf{u}_k \\
&= w_1\left(u_{1_1}\mathbf{v}_1 + u_{1_2}\mathbf{v}_2 + \cdots + u_{1_m}\mathbf{v}_m\right) + w_2\left(u_{2_1}\mathbf{v}_1 + u_{2_2}\mathbf{v}_2 + \cdots + u_{2_m}\mathbf{v}_m\right) + \cdots \\
&\qquad \cdots + w_k\left(u_{k_1}\mathbf{v}_1 + u_{k_2}\mathbf{v}_2 + \cdots + u_{k_m}\mathbf{v}_m\right) \\
&= \left(w_1u_{1_1} + w_2u_{2_1} + \cdots + w_ku_{k_1}\right)\mathbf{v}_1 + \left(w_1u_{1_2} + w_2u_{2_2} + \cdots + w_ku_{k_2}\right)\mathbf{v}_2 + \cdots \\
&\qquad \cdots + \left(w_1u_{1_m} + w_2u_{2_m} + \cdots + w_ku_{k_m}\right)\mathbf{v}_m \\
&= w_1'\mathbf{v}_1 + w_2'\mathbf{v}_2 + \cdots + w_m'\mathbf{v}_m.
\end{aligned}$$

So, any vector $\mathbf{w} \in \text{span}\,(\mathbf{u}_1, \mathbf{u}_2, \ldots, \mathbf{u}_k)$ is also in $\text{span}\,(\mathbf{v}_1, \mathbf{v}_2, \ldots, \mathbf{v}_m)$,
and $\text{span}\,(\mathbf{u}_1, \mathbf{u}_2, \ldots, \mathbf{u}_k) \subseteq \text{span}\,(\mathbf{v}_1, \mathbf{v}_2, \ldots, \mathbf{v}_m)$
.

(b) Suppose that in addition to (a), each $\mathbf{v}_j$ is a linear combination of $\mathbf{u}_1, \mathbf{u}_2, \ldots, \mathbf{u}_k$.
Let $\mathbf{w}$ be an arbitrary vector in $\text{span}\,(\mathbf{v}_1, \mathbf{v}_2, \ldots, \mathbf{v}_m)$.
Then $\mathbf{w} = w_1'\mathbf{v}_1 + w_2'\mathbf{v}_2 + \cdots + w_m'\mathbf{v}_m$, but each $\mathbf{v}_j = v_{j_1}\mathbf{u}_1 + v_{j_2}\mathbf{u}_2 + \cdots + v_{j_k}\mathbf{u}_k$, so

$$\begin{aligned}
\mathbf{w} &= w_1'\left(v_{1_1}\mathbf{u}_1 + v_{1_2}\mathbf{u}_2 + \cdots + v_{1_k}\mathbf{u}_k\right) + w_2'\left(v_{2_1}\mathbf{u}_1 + v_{2_2}\mathbf{u}_2 + \cdots + v_{2_k}\mathbf{u}_k\right) + \cdots \\
&\qquad \cdots + w_m'\left(v_{m_1}\mathbf{u}_1 + v_{m_2}\mathbf{u}_2 + \cdots + v_{m_k}\mathbf{u}_k\right) \\
&= \left(w_1'v_{1_1} + w_2'v_{2_1} + \cdots + w_m'v_{m_1}\right)\mathbf{u}_1 + \left(w_1'v_{1_2} + w_2'v_{2_2} + \cdots + w_m'v_{m_2}\right)\mathbf{u}_2 + \cdots \\
&\qquad \cdots + \left(w_1'v_{1_k} + w_2'v_{2_k} + \cdots + w_m'v_{m_k}\right)\mathbf{u}_k \\
&= w_1\mathbf{u}_1 + w_1\mathbf{u}_2 + \cdots + w_k\mathbf{u}_k.
\end{aligned}$$

So, any vector $\mathbf{w} \in \text{span}\,(\mathbf{v}_1, \mathbf{v}_2, \ldots, \mathbf{v}_m)$ is also in $\text{span}\,(\mathbf{u}_1, \mathbf{u}_2, \ldots, \mathbf{u}_k)$,
and $\text{span}\,(\mathbf{v}_1, \mathbf{v}_2, \ldots, \mathbf{v}_m) \subseteq \text{span}\,(\mathbf{u}_1, \mathbf{u}_2, \ldots, \mathbf{u}_k)$.
But we already had $\text{span}\,(\mathbf{u}_1, \mathbf{u}_2, \ldots, \mathbf{u}_k) \subseteq \text{span}\,(\mathbf{v}_1, \mathbf{v}_2, \ldots, \mathbf{v}_m)$,
so $\text{span}\,(\mathbf{u}_1, \mathbf{u}_2, \ldots, \mathbf{u}_k) = \text{span}\,(\mathbf{v}_1, \mathbf{v}_2, \ldots, \mathbf{v}_m)$.

(c) Need only show $\mathbf{e}_1, \mathbf{e}_2, \mathbf{e}_3$ are linear combinations of $\mathbf{v}_1 = \begin{bmatrix} 1 \\ 0 \\ 0 \end{bmatrix}$, $\mathbf{v}_2 = \begin{bmatrix} 1 \\ 1 \\ 0 \end{bmatrix}$, $\mathbf{v}_3 = \begin{bmatrix} 1 \\ 1 \\ 1 \end{bmatrix}$.

That's obvious since $\mathbf{e}_1 = \mathbf{v}_1$, $\mathbf{e}_2 = \mathbf{v}_2 - \mathbf{v}_1$, and $\mathbf{e}_3 = \mathbf{v}_3 - \mathbf{v}_2$. Why is that enough?
Because then we have $\mathbb{R}^3 = \text{span}(\mathbf{e}_1,\ \mathbf{e}_2,\ \mathbf{e}_3) = \text{span}(\mathbf{v}_1,\ \mathbf{v}_2,\ \mathbf{v}_3)$.

22. The vectors $\mathbf{v}_1 = \begin{bmatrix} 2 \\ -1 \\ 3 \end{bmatrix}$ and $\mathbf{v}_2 = \begin{bmatrix} 1 \\ 4 \\ 4 \end{bmatrix}$ are linearly independent.

This can be determined by inspection because they are not scalar multiples of each other.

23. Since there is no obvious dependence relation here, we follow Example 2.23.

Find scalars c_1, c_2, and c_3 such that: $c_1 \begin{bmatrix} 1 \\ 1 \\ 1 \end{bmatrix} + c_2 \begin{bmatrix} 1 \\ 2 \\ 3 \end{bmatrix} + c_3 \begin{bmatrix} 1 \\ -1 \\ 2 \end{bmatrix} = \begin{bmatrix} 0 \\ 0 \\ 0 \end{bmatrix}$.

Form the linear system, its associated augmented matrix, and row reduce to solve:

$$\begin{array}{rcrcrcl} c_1 & + & c_2 & + & c_3 & = & 0 \\ c_1 & + & 2c_2 & - & c_3 & = & 0 \\ c_1 & + & 3c_2 & + & 2c_3 & = & 0 \end{array} \quad \Rightarrow \quad \left[\begin{array}{ccc|c} 1 & 1 & 1 & 0 \\ 1 & 2 & -1 & 0 \\ 1 & 3 & 2 & 0 \end{array}\right] \longrightarrow \left[\begin{array}{ccc|c} 1 & 0 & 0 & 0 \\ 0 & 1 & 0 & 0 \\ 0 & 0 & 1 & 0 \end{array}\right]$$

Since $c_1 = c_2 = c_3 = 0$ is the unique solution, the vectors are linearly independent.

24. Since there is no obvious dependence relation here, we follow Example 2.23.

Find scalars c_1, c_2, and c_3 such that: $c_1 \begin{bmatrix} 2 \\ 2 \\ 1 \end{bmatrix} + c_2 \begin{bmatrix} 3 \\ 1 \\ 2 \end{bmatrix} + c_3 \begin{bmatrix} 1 \\ -5 \\ 2 \end{bmatrix} = \begin{bmatrix} 0 \\ 0 \\ 0 \end{bmatrix}$.

Form the linear system, its associated augmented matrix, and row reduce to solve:

$$\begin{array}{rcrcrcl} 2c_1 & + & 3c_2 & + & c_3 & = & 0 \\ 2c_1 & + & c_2 & - & 5c_3 & = & 0 \\ c_1 & + & 2c_2 & + & 2c_3 & = & 0 \end{array} \quad \Rightarrow \quad \left[\begin{array}{ccc|c} 2 & 3 & 1 & 0 \\ 2 & 1 & -5 & 0 \\ 1 & 2 & 2 & 0 \end{array}\right] \longrightarrow \left[\begin{array}{ccc|c} 1 & 0 & -4 & 0 \\ 0 & 1 & 3 & 0 \\ 0 & 0 & 0 & 0 \end{array}\right]$$

Since $c_1 = 4c_3$ and $c_2 = -3c_3$ is a solution, the vectors are linearly dependent.

One dependence relationship is: $4 \begin{bmatrix} 2 \\ 2 \\ 1 \end{bmatrix} - 3 \begin{bmatrix} 3 \\ 1 \\ 2 \end{bmatrix} + \begin{bmatrix} 1 \\ -5 \\ 2 \end{bmatrix} = \begin{bmatrix} 0 \\ 0 \\ 0 \end{bmatrix}$. Are there others?

25. The vectors $\mathbf{v}_1 = \begin{bmatrix} 0 \\ 1 \\ 2 \end{bmatrix}$, $\mathbf{v}_2 = \begin{bmatrix} 2 \\ 1 \\ 3 \end{bmatrix}$ and $\mathbf{v}_3 = \begin{bmatrix} 2 \\ 0 \\ 1 \end{bmatrix}$ are linearly dependent.

This can be determined by inspection because $\mathbf{v}_1 - \mathbf{v}_2 + \mathbf{v}_3 = \mathbf{0}$.

26. The vectors $\mathbf{v}_1 = \begin{bmatrix} -2 \\ 3 \\ 7 \end{bmatrix}$, $\mathbf{v}_2 = \begin{bmatrix} 4 \\ -1 \\ 5 \end{bmatrix}$, $\mathbf{v}_3 = \begin{bmatrix} 3 \\ 1 \\ 3 \end{bmatrix}$, and $\mathbf{v}_4 = \begin{bmatrix} 5 \\ 0 \\ 2 \end{bmatrix}$ are linearly dependent.

This can be determined by inspection because $[A|\mathbf{0}]$ obviously has a nontrivial solution. Why?
Because there are only 3 equations (corresponding to the number of rows),
but there are 4 variables (corresponding to the number of vectors which then become columns).
See Section 2.2, Theorem 2.6 for more detail.

Furthermore, this follows immediately from Theorem 2.8. Why?
This is a set of 4 vectors in $\mathbb{R}^3$. Why does Theorem 2.8 apply? Hint: $4 > 3$.

To find a specific dependence relationship, we follow Example 2.23.

Find scalars c_1, c_2, and c_3 such that: $c_1 \begin{bmatrix} -2 \\ 3 \\ 7 \end{bmatrix} + c_2 \begin{bmatrix} 4 \\ -1 \\ 5 \end{bmatrix} + c_3 \begin{bmatrix} 3 \\ 1 \\ 3 \end{bmatrix} + c_4 \begin{bmatrix} 5 \\ 0 \\ 2 \end{bmatrix} = \begin{bmatrix} 0 \\ 0 \\ 0 \end{bmatrix}$.

Form the linear system, its associated augmented matrix, and row reduce to solve:

$$\begin{aligned} -2c_1 + 4c_2 + 3c_3 + 5c_4 &= 0 \\ 3c_1 - c_2 + c_3 &= 0 \\ 7c_1 + 5c_2 + 3c_3 + 2c_4 &= 0 \end{aligned} \Rightarrow \left[\begin{array}{cccc|c} -2 & 4 & 3 & 5 & 0 \\ 3 & -1 & 1 & 0 & 0 \\ 7 & 5 & 3 & 2 & 0 \end{array}\right] \longrightarrow \left[\begin{array}{cccc|c} 37 & 0 & 0 & -13 & 0 \\ 0 & 37 & 0 & 6 & 0 \\ 0 & 0 & 37 & 45 & 0 \end{array}\right]$$

Since $c_1 = \frac{13}{37}c_4$, $c_2 = -\frac{6}{37}c_4$, $c_3 = -\frac{45}{37}c_4$ is a solution, the vectors are linearly dependent.

One dependence relationship is: $13 \begin{bmatrix} -2 \\ 3 \\ 7 \end{bmatrix} - 6 \begin{bmatrix} 4 \\ -1 \\ 5 \end{bmatrix} - 45 \begin{bmatrix} 3 \\ 1 \\ 3 \end{bmatrix} + 37 \begin{bmatrix} 5 \\ 0 \\ 2 \end{bmatrix} = \begin{bmatrix} 0 \\ 0 \\ 0 \end{bmatrix}$.

27. The vectors $\mathbf{v}_1 = \begin{bmatrix} 3 \\ 4 \\ 5 \end{bmatrix}$, $\mathbf{v}_2 = \begin{bmatrix} 6 \\ 7 \\ 8 \end{bmatrix}$, $\mathbf{v}_3 = \begin{bmatrix} 0 \\ 0 \\ 0 \end{bmatrix}$ are linearly dependent.

This can be determined by inspection because $\mathbf{v}_3$ is the zero vector. Why is that enough?
Because $0\mathbf{v}_1 + 0\mathbf{v}_2 + \mathbf{v}_3 = \mathbf{0}$.
Any set of vectors containing the zero vector is linearly dependent. Why?

28. Since there is no obvious dependence relation here, we follow Example 2.23.

Find scalars c_1, c_2, and c_3 such that: $c_1 \begin{bmatrix} -1 \\ 1 \\ 2 \\ 1 \end{bmatrix} + c_2 \begin{bmatrix} 3 \\ 2 \\ 2 \\ 4 \end{bmatrix} + c_3 \begin{bmatrix} 2 \\ 3 \\ 1 \\ -1 \end{bmatrix} = \begin{bmatrix} 0 \\ 0 \\ 0 \\ 0 \end{bmatrix}$.

Form the linear system, its associated augmented matrix, and row reduce to solve:

$$\begin{aligned} -c_1 + 3c_2 + 2c_3 &= 0 \\ c_1 + 2c_2 + 3c_3 &= 0 \\ 2c_1 + c_2 + c_3 &= 0 \\ c_1 + 4c_2 - c_3 &= 0 \end{aligned} \Rightarrow \left[\begin{array}{ccc|c} -1 & 3 & 2 & 0 \\ 1 & 2 & 3 & 0 \\ 2 & 2 & 1 & 0 \\ 1 & 4 & -1 & 0 \end{array}\right] \longrightarrow \left[\begin{array}{ccc|c} 1 & 0 & 0 & 0 \\ 0 & 1 & 0 & 0 \\ 0 & 0 & 1 & 0 \\ 0 & 0 & 0 & 0 \end{array}\right]$$

Since $c_1 = c_2 = c_3 = 0$ is the unique solution, the vectors are linearly independent.

29. Since there is no obvious dependence relation here, we follow Example 2.23.

Find scalars c_1, c_2, c_3, c_4 such that: $c_1 \begin{bmatrix} 1 \\ -1 \\ 1 \\ 0 \end{bmatrix} + c_2 \begin{bmatrix} -1 \\ 1 \\ 0 \\ 1 \end{bmatrix} + c_3 \begin{bmatrix} 1 \\ 0 \\ 1 \\ -1 \end{bmatrix} + c_4 \begin{bmatrix} 0 \\ 1 \\ -1 \\ 1 \end{bmatrix} = \begin{bmatrix} 0 \\ 0 \\ 0 \\ 0 \end{bmatrix}$.

Form the linear system, its associated augmented matrix, and row reduce to solve:

$$\begin{array}{rcrcrcrcl} c_1 & - & c_2 & + & c_3 & & & = & 0 \\ -c_1 & + & c_2 & & & + & c_4 & = & 0 \\ c_1 & & & + & c_3 & - & c_4 & = & 0 \\ & & c_2 & - & c_3 & + & c_4 & = & 0 \end{array} \Rightarrow \left[\begin{array}{rrrr|r} 1 & -1 & 1 & 0 & 0 \\ -1 & 1 & 0 & 1 & 0 \\ 1 & 0 & 1 & -1 & 0 \\ 0 & 1 & -1 & 1 & 0 \end{array}\right] \longrightarrow \left[\begin{array}{rrrr|r} 1 & 0 & 0 & 0 & 0 \\ 0 & 1 & 0 & 0 & 0 \\ 0 & 0 & 1 & 0 & 0 \\ 0 & 0 & 0 & 1 & 0 \end{array}\right]$$

Since $c_1 = c_2 = c_3 = c_4 = 0$ is the unique solution, the vectors are linearly independent.

30. The vectors $\mathbf{v}_1 = \begin{bmatrix} 0 \\ 0 \\ 0 \\ 1 \end{bmatrix}$, $\mathbf{v}_2 = \begin{bmatrix} 0 \\ 0 \\ 2 \\ 1 \end{bmatrix}$, $\mathbf{v}_3 = \begin{bmatrix} 0 \\ 3 \\ 2 \\ 1 \end{bmatrix}$, and $\mathbf{v}_4 = \begin{bmatrix} 4 \\ 3 \\ 2 \\ 1 \end{bmatrix}$ are linearly independent.

This can be determined by inspection. How?
To create a 0 in the first component, the coefficient of $\mathbf{v}_4$ must be 0. Why?
Given that, a 0 in the second component forces the coefficient of $\mathbf{v}_3$ to be 0.
Given those two facts, a 0 in the third component forces the coefficient of $\mathbf{v}_2$ to be 0.
And finally, given all that, a 0 in the fourth component forces the coefficient of $\mathbf{v}_1$ to be 0.

To verify this argument, we follow Example 2.23.

Find scalars c_1, c_2, c_3, and c_4 such that: $c_1 \begin{bmatrix} 0 \\ 0 \\ 0 \\ 1 \end{bmatrix} + c_2 \begin{bmatrix} 0 \\ 0 \\ 2 \\ 1 \end{bmatrix} + c_3 \begin{bmatrix} 0 \\ 3 \\ 2 \\ 1 \end{bmatrix} + c_4 \begin{bmatrix} 4 \\ 3 \\ 2 \\ 1 \end{bmatrix} = \begin{bmatrix} 0 \\ 0 \\ 0 \\ 0 \end{bmatrix}$.

Form the linear system, its associated augmented matrix, and row reduce to solve:

$$\begin{array}{rcrcrcrcl} & & & & & & c_4 & = & 0 \\ & & & & 2c_3 & + & c_4 & = & 0 \\ & & 3c_2 & + & 2c_3 & + & c_4 & = & 0 \\ 4c_1 & + & 3c_2 & + & 2c_3 & + & c_4 & = & 0 \end{array} \Rightarrow \left[\begin{array}{rrrr|r} 0 & 0 & 0 & 4 & 0 \\ 0 & 0 & 3 & 3 & 0 \\ 0 & 2 & 2 & 2 & 0 \\ 1 & 1 & 1 & 1 & 0 \end{array}\right] \longrightarrow \left[\begin{array}{rrrr|r} 1 & 0 & 0 & 0 & 0 \\ 0 & 1 & 0 & 0 & 0 \\ 0 & 0 & 1 & 0 & 0 \\ 0 & 0 & 0 & 1 & 0 \end{array}\right]$$

Since $c_1 = c_2 = c_3 = c_4 = 0$ is the unique solution, the vectors are linearly independent.

31. The vectors $\mathbf{v}_1 = \begin{bmatrix} 3 \\ -1 \\ 1 \\ -1 \end{bmatrix}$, $\mathbf{v}_2 = \begin{bmatrix} -1 \\ 3 \\ 1 \\ -1 \end{bmatrix}$, $\mathbf{v}_3 = \begin{bmatrix} 1 \\ 1 \\ 3 \\ 1 \end{bmatrix}$, and $\mathbf{v}_4 = \begin{bmatrix} -1 \\ -1 \\ 1 \\ 3 \end{bmatrix}$ are linearly dependent.

This can be determined by inspection because $\mathbf{v}_1 + \mathbf{v}_2 - \mathbf{v}_3 + \mathbf{v}_4 = \mathbf{0}$.

32. Exercises 32 through 41 provide a check on our solutions to Exercises 22 through 31. How? In these exercises the directions tell us to follow Example 2.25 and apply Theorem 2.7:

We construct a matrix with these vectors as its rows and proceed to reduce it to echelon form. Each time a row changes, we denote the new row by adding a prime symbol:

$$A = \begin{bmatrix} \mathbf{v}_1 \\ \mathbf{v}_2 \end{bmatrix} = \begin{bmatrix} 2 & -1 & 3 \\ 1 & 4 & 4 \end{bmatrix} \xrightarrow[R_2'=R_1]{R_1'=R_2} \begin{bmatrix} 1 & 4 & 4 \\ 2 & -1 & 3 \end{bmatrix} \xrightarrow{R_2''=R_2'-2R_1'} \begin{bmatrix} 1 & 4 & 4 \\ 0 & -9 & -5 \end{bmatrix}$$

We can stop. Why? We have put A into row echelon form. How can we tell?
Since the rank of a matrix is the number of nonzero rows in its row echelon form, $\text{rank}(A) = 2$.
What do we conclude? We conclude $\mathbf{v}_1$ and $\mathbf{v}_2$ are linearly independent. How?

Theorem 2.7 states that $\mathbf{v}_1, \mathbf{v}_2, \ldots, \mathbf{v}_m$ are linearly dependent if and only if $\text{rank}(A) < m$.
But that implies the following: If $\text{rank}(A) \geq m$, $\mathbf{v}_1, \mathbf{v}_2, \ldots, \mathbf{v}_m$ are linearly independent.

In this case, therefore, we argue as follows:
Since $\text{rank}(A) \geq 2$, Theorem 2.7 implies $\mathbf{v}_1$ and $\mathbf{v}_2$ are linearly independent.

Does the agree with the solution we found in Exercise 22? It should.
Which method was easier for this Exercise? Why?

33. Exercises 32 through 41 provide a check on our solutions to Exercises 22 through 31. How? In these exercises the directions tell us to follow Example 2.25 and apply Theorem 2.7:

We construct a matrix with these vectors as its rows and proceed to reduce it to echelon form. Each time a row changes, we denote the new row by adding a prime symbol:

$$A = \begin{bmatrix} \mathbf{v}_1 \\ \mathbf{v}_2 \\ \mathbf{v}_3 \end{bmatrix} = \begin{bmatrix} 1 & 1 & 1 \\ 1 & 2 & 3 \\ 1 & -1 & 2 \end{bmatrix} \xrightarrow[R_3'=R_3-R_1]{R_2'=R_2-R_1} \begin{bmatrix} 1 & 1 & 1 \\ 0 & 1 & 2 \\ 0 & -2 & 1 \end{bmatrix} \xrightarrow{R_3''=R_3'+2R_2'} \begin{bmatrix} 1 & 1 & 1 \\ 0 & 1 & 2 \\ 0 & 0 & 5 \end{bmatrix}$$

We can stop. Why? We have put A into row echelon form. How can we tell?
Since the rank of a matrix is the number of nonzero rows in its row echelon form, $\text{rank}(A) = 3$.
What do we conclude? We conclude $\mathbf{v}_1$, $\mathbf{v}_2$, and $\mathbf{v}_3$ are linearly independent. How?

Theorem 2.7 states that $\mathbf{v}_1, \mathbf{v}_2, \ldots, \mathbf{v}_m$ are linearly dependent if and only if $\text{rank}(A) < m$.
But that implies the following: If $\text{rank}(A) \geq m$, $\mathbf{v}_1, \mathbf{v}_2, \ldots, \mathbf{v}_m$ are linearly independent.

In this case, therefore, we argue as follows:
Since $\text{rank}(A) \geq 3$, Theorem 2.7 implies $\mathbf{v}_1$, $\mathbf{v}_2$, and $\mathbf{v}_3$ are linearly independent.

Does the agree with the solution we found in Exercise 23? It should.
Which method was easier for this Exercise? Why?

34. Exercises 32 through 41 provide a check on our solutions to Exercises 22 through 31. How? In these exercises the directions tell us to follow Example 2.25 and apply Theorem 2.7:

We construct a matrix with these vectors as its rows and proceed to reduce it to echelon form. Each time a row changes, we denote the new row by adding a prime symbol:

$$A = \begin{bmatrix} \mathbf{v}_1 \\ \mathbf{v}_2 \\ \mathbf{v}_3 \end{bmatrix} = \begin{bmatrix} 2 & 2 & 1 \\ 3 & 1 & 2 \\ 1 & -5 & 2 \end{bmatrix} \xrightarrow[R_3'=R_2]{\substack{R_1'=R_3 \\ R_2'=R_1}} \begin{bmatrix} 1 & -5 & 2 \\ 2 & 2 & 1 \\ 3 & 1 & 2 \end{bmatrix} \xrightarrow[R_3''=R_3'-3R_1']{R_2''=R_2'-2R_1'} \begin{bmatrix} 1 & -5 & 2 \\ 0 & 12 & -3 \\ 0 & 16 & -4 \end{bmatrix} \xrightarrow{R_3'''=R_3''-\frac{4}{3}R_2''} \begin{bmatrix} 1 & -5 & 2 \\ 0 & 12 & -3 \\ 0 & 0 & 0 \end{bmatrix}$$

We can stop. Why? We have created a zero row. What does that tell us?
Since the rank of a matrix is the number of nonzero rows in its row echelon form, $\text{rank}(A) = 2$.
What do we conclude? We conclude $\mathbf{v}_1$, $\mathbf{v}_2$, and $\mathbf{v}_3$ are linearly dependent. How?

Theorem 2.7 states that $\mathbf{v}_1, \mathbf{v}_2, \ldots, \mathbf{v}_m$ are linearly dependent if and only if $\text{rank}(A) < m$.
So, since $\text{rank}(A) = 2 < 3$, Theorem 2.7 implies $\mathbf{v}_1$, $\mathbf{v}_2$, and $\mathbf{v}_3$ are linearly dependent.

Furthermore, from the row reduction above, we see that: $\mathbf{0} = R_3''' = R_3'' - \frac{4}{3}R_2''$.

Multiplying both sides by 3 implies: $\mathbf{0} = 3R_3'' - 4R_2''$. Substituting, we have:

$\mathbf{0} = 3(R_3' - 3R_1') - 4(R_2' - 2R_1') = 3(R_2 - 3R_3) - 4(R_1 - 2R_3) = -4R_1 + 3R_2 - R_3$.

Multiplying both sides by -1 yields this dependence relation among the original vectors:

$$4\begin{bmatrix} 2 \\ 2 \\ 1 \end{bmatrix} - 3\begin{bmatrix} 3 \\ 1 \\ 2 \end{bmatrix} + \begin{bmatrix} 1 \\ -5 \\ 2 \end{bmatrix} = \begin{bmatrix} 0 \\ 0 \\ 0 \end{bmatrix}.$$ Compare this result to Exercise 24. Does it agree?

35. Exercises 32 through 41 provide a check on our solutions to Exercises 22 through 31. How? In these exercises the directions tell us to follow Example 2.25 and apply Theorem 2.7:

We construct a matrix with these vectors as its rows and proceed to reduce it to echelon form. Each time a row changes, we denote the new row by adding a prime symbol:

$$A = \begin{bmatrix} \mathbf{v}_1 \\ \mathbf{v}_2 \\ \mathbf{v}_3 \end{bmatrix} = \begin{bmatrix} 0 & 1 & 2 \\ 2 & 1 & 3 \\ 2 & 0 & 1 \end{bmatrix} \xrightarrow[R_2'=R_1]{R_1'=R_2} \begin{bmatrix} 2 & 1 & 3 \\ 0 & 1 & 2 \\ 2 & 0 & 1 \end{bmatrix} \xrightarrow{R_3'=R_3-R_1'} \begin{bmatrix} 2 & 1 & 3 \\ 0 & 1 & 2 \\ 0 & -1 & -2 \end{bmatrix} \xrightarrow{R_3''=R_3'+R_2'} \begin{bmatrix} 2 & 1 & 3 \\ 0 & 1 & 2 \\ 0 & 0 & 0 \end{bmatrix}$$

We can stop. Why? We have created a zero row. What does that tell us?
Since the rank of a matrix is the number of nonzero rows in its row echelon form, $\text{rank}(A) = 2$.
What do we conclude? We conclude $\mathbf{v}_1$, $\mathbf{v}_2$, and $\mathbf{v}_3$ are linearly dependent. How?

Theorem 2.7 states that $\mathbf{v}_1, \mathbf{v}_2, \ldots, \mathbf{v}_m$ are linearly dependent if and only if $\text{rank}(A) < m$.
So, since $\text{rank}(A) = 2 < 3$, Theorem 2.7 implies $\mathbf{v}_1$, $\mathbf{v}_2$, and $\mathbf{v}_3$ are linearly dependent.

From the row reduction, we see: $\mathbf{0} = R_3'' = R_3' + R_2' = (R_3' - R_1') + R_2' = R_3 - R_2 + R_1$.

This equation yields a dependence relation among the original vectors:

$$\begin{bmatrix} 0 \\ 1 \\ 2 \end{bmatrix} - \begin{bmatrix} 2 \\ 1 \\ 3 \end{bmatrix} + \begin{bmatrix} 2 \\ 0 \\ 1 \end{bmatrix} = \begin{bmatrix} 0 \\ 0 \\ 0 \end{bmatrix}.$$ Compare this result to Exercise 25. Does it agree?

36. Exercises 32 through 41 provide a check on our solutions to Exercises 22 through 31. How? In these exercises the directions tell us to follow Example 2.25 and apply Theorem 2.7:

We construct a matrix with these vectors as its rows and proceed to reduce it to echelon form. Each time a row changes, we denote the new row by adding a prime symbol:

$$A = \begin{bmatrix} \mathbf{v}_1 \\ \mathbf{v}_2 \\ \mathbf{v}_3 \\ \mathbf{v}_4 \end{bmatrix} = \begin{bmatrix} -2 & 3 & 7 \\ 4 & -1 & 5 \\ 3 & 1 & 3 \\ 5 & 0 & 2 \end{bmatrix} \xrightarrow{\substack{R_2'=R_2+2R_1 \\ R_3'=R_3+\frac{3}{2}R_1 \\ R_4'=R_4+\frac{5}{2}R_1}} \begin{bmatrix} -2 & 3 & 7 \\ 0 & 5 & 19 \\ 0 & 11/2 & 27/2 \\ 0 & 15/2 & 39/2 \end{bmatrix}$$

$$\xrightarrow{\substack{R_3''=R_3'-\frac{11}{10}R_2' \\ R_4''=R_4'-\frac{3}{2}R_2'}} \begin{bmatrix} -2 & 3 & 7 \\ 0 & 5 & 19 \\ 0 & 0 & -37/5 \\ 0 & 0 & -9 \end{bmatrix} \xrightarrow{R_4'''=R_4''-\frac{45}{37}R_3''} \begin{bmatrix} -2 & 3 & 7 \\ 0 & 5 & 19 \\ 0 & 0 & -37/5 \\ 0 & 0 & 0 \end{bmatrix}$$

We can stop. Why? We have created a zero row. What does that tell us?
Since the rank of a matrix is the number of nonzero rows in its row echelon form, $\text{rank}(A) = 3$.
What do we conclude? We conclude $\mathbf{v}_1$, $\mathbf{v}_2$, $\mathbf{v}_3$, and $\mathbf{v}_4$ are linearly dependent.

Theorem 2.7 states that $\mathbf{v}_1, \mathbf{v}_2, \ldots, \mathbf{v}_m$ are linearly dependent if and only if $\text{rank}(A) < m$.
So, since $\text{rank}(A) = 3 < 4$, Theorem 2.7 implies $\mathbf{v}_1$, $\mathbf{v}_2$, $\mathbf{v}_3$, and $\mathbf{v}_3$ are linearly dependent.

Furthermore, from the row reduction above, we see that: $\mathbf{0} = R_4''' = R_4'' - \frac{45}{37}R_3''$.
Multiplying both sides by 37 implies: $\mathbf{0} = 37R_4'' - 45R_3''$. Substituting, we have:
$\mathbf{0} = 37(R_4' - \frac{3}{2}R_2') - 45(R_3' - \frac{11}{10}R_2')$. Combining like terms yields: $\mathbf{0} = 37R_4' - 45R_3' - 6R_3'$.
Substituting again yields: $37(R_4 + \frac{5}{2}R_1) - 45(R_3 + \frac{3}{2}R_1) - 6(R_2 + 2R_1)$.
Combining like terms one last time yields: $\mathbf{0} = 37R_4 - 45R_3 - 6R_2 + 13R_1$.

This equation yields a dependence relation among the original vectors:

$$13\begin{bmatrix} -2 \\ 3 \\ 7 \end{bmatrix} - 6\begin{bmatrix} 4 \\ -1 \\ 5 \end{bmatrix} - 45\begin{bmatrix} 3 \\ 1 \\ 3 \end{bmatrix} + 37\begin{bmatrix} 5 \\ 0 \\ 2 \end{bmatrix} = \begin{bmatrix} 0 \\ 0 \\ 0 \end{bmatrix}.$$ Does this agree with Exercise 26?

Finally, linear dependence follows immediately from Theorem 2.8. Why?
This is a set of 4 vectors in $\mathbb{R}^3$. Why does Theorem 2.8 apply? Hint: $4 > 3$.

37. Exercises 32 through 41 provide a check on our solutions to Exercises 22 through 31. How? In these exercises the directions tell us to follow Example 2.25 and apply Theorem 2.7:

We construct a matrix with these vectors as its rows and proceed to reduce it to echelon form. Each time a row changes, we denote the new row by adding a prime symbol:

$$A = \begin{bmatrix} \mathbf{v}_1 \\ \mathbf{v}_2 \\ \mathbf{v}_3 \end{bmatrix} = \begin{bmatrix} 3 & 4 & 5 \\ 6 & 7 & 8 \\ 0 & 0 & 0 \end{bmatrix}$$

We can stop. Why? The matrix A has a zero row. What does that tell us?
Since the rank of a matrix is the number of nonzero rows in its row echelon form, $\text{rank}(A) \leq 2$.
What do we conclude? We conclude $\mathbf{v}_1$, $\mathbf{v}_2$, and $\mathbf{v}_3$ are linearly dependent. How?

Theorem 2.7 states that $\mathbf{v}_1, \mathbf{v}_2, \ldots, \mathbf{v}_m$ are linearly dependent if and only if $\text{rank}(A) < m$.
So, since $\text{rank}(A) \leq 2 < 3$, Theorem 2.7 implies $\mathbf{v}_1$, $\mathbf{v}_2$, and $\mathbf{v}_3$ are linearly dependent.

Furthermore, we have the obvious dependence relation among the original vectors:

$$0\begin{bmatrix} 3 \\ 4 \\ 5 \end{bmatrix} + 0\begin{bmatrix} 6 \\ 7 \\ 8 \end{bmatrix} + \begin{bmatrix} 0 \\ 0 \\ 0 \end{bmatrix} = \begin{bmatrix} 0 \\ 0 \\ 0 \end{bmatrix}.$$ Compare this result to Exercise 37. Does it agree?

This can be determined by inspection because $\mathbf{v}_3$ is the zero vector. Why is that enough?
Any set of vectors containing the zero vector is linearly dependent. Why?

38. Exercises 32 through 41 provide a check on our solutions to Exercises 22 through 31. How? In these exercises the directions tell us to follow Example 2.25 and apply Theorem 2.7:

We construct a matrix with these vectors as its rows and proceed to reduce it to echelon form. Each time a row changes, we denote the new row by adding a prime symbol:

$$A = \begin{bmatrix} \mathbf{v}_1 \\ \mathbf{v}_2 \\ \mathbf{v}_3 \end{bmatrix} = \begin{bmatrix} -1 & 1 & 2 & 1 \\ 3 & 2 & 2 & 4 \\ 2 & 3 & 1 & -1 \end{bmatrix} \xrightarrow[R_3'=R_3+2R_1]{R_2'=R_2+3R_1} \begin{bmatrix} -1 & 1 & 2 & 1 \\ 0 & 5 & 8 & 7 \\ 0 & 5 & 5 & 1 \end{bmatrix} \xrightarrow{R_3''=R_3'-R_2'} \begin{bmatrix} -1 & 1 & 2 & 1 \\ 0 & 5 & 8 & 7 \\ 0 & 0 & -3 & -6 \end{bmatrix}$$

We can stop. Why? We have put A into row echelon form. How can we tell?
Since the rank of a matrix is the number of nonzero rows in its row echelon form, $\text{rank}(A) = 3$.
What do we conclude? We conclude $\mathbf{v}_1$, $\mathbf{v}_2$, and $\mathbf{v}_3$ are linearly independent. How?

Theorem 2.7 states that $\mathbf{v}_1, \mathbf{v}_2, \ldots, \mathbf{v}_m$ are linearly dependent if and only if $\text{rank}(A) < m$.
But that implies the following: If $\text{rank}(A) \geq m$, $\mathbf{v}_1, \mathbf{v}_2, \ldots, \mathbf{v}_m$ are linearly independent.

In this case, therefore, we argue as follows:
Since $\text{rank}(A) \geq 3$, Theorem 2.7 implies $\mathbf{v}_1$, $\mathbf{v}_2$, and $\mathbf{v}_3$ are linearly independent.

Does the agree with the solution we found in Exercise 28? It should.
Which method was easier for this Exercise? Why?

39. Exercises 32 through 41 provide a check on our solutions to Exercises 22 through 31. How? In these exercises the directions tell us to follow Example 2.25 and apply Theorem 2.7:

We construct a matrix with these vectors as its rows and proceed to reduce it to echelon form. Each time a row changes, we denote the new row by adding a prime symbol:

$$A = \begin{bmatrix} \mathbf{v}_1 \\ \mathbf{v}_2 \\ \mathbf{v}_3 \\ \mathbf{v}_4 \end{bmatrix} = \begin{bmatrix} 1 & -1 & 1 & 0 \\ -1 & 1 & 0 & 1 \\ 1 & 0 & 1 & -1 \\ 0 & 1 & -1 & 1 \end{bmatrix} \xrightarrow[R_3'=R_3-R_1]{R_2'=R_2+R_1} \begin{bmatrix} 1 & -1 & 1 & 0 \\ 0 & 0 & 1 & 1 \\ 0 & 1 & 0 & -1 \\ 0 & 1 & -1 & 1 \end{bmatrix} \xrightarrow{R_4'=R_4-R_3'} \begin{bmatrix} 1 & -1 & 1 & 0 \\ 0 & 0 & 1 & 1 \\ 0 & 1 & 0 & -1 \\ 0 & 0 & -1 & 2 \end{bmatrix}$$

$$\xrightarrow[R_3''=R_2']{R_2''=R_3'} \begin{bmatrix} 1 & -1 & 1 & 0 \\ 0 & 1 & 0 & -1 \\ 0 & 0 & 1 & 1 \\ 0 & 0 & -1 & 2 \end{bmatrix} \xrightarrow{R_4''=R_4'+R_3'} \begin{bmatrix} 1 & -1 & 1 & 0 \\ 0 & 1 & 0 & -1 \\ 0 & 0 & 1 & 1 \\ 0 & 0 & 0 & 3 \end{bmatrix}$$

We can stop. Why? We have put A into row echelon form. How can we tell?
Since the rank of a matrix is the number of nonzero rows in its row echelon form, $\text{rank}(A) = 4$.
What do we conclude? We conclude $\mathbf{v}_1$, $\mathbf{v}_2$, $\mathbf{v}_3$, and $\mathbf{v}_4$ are linearly independent.

Theorem 2.7 states that $\mathbf{v}_1, \mathbf{v}_2, \ldots, \mathbf{v}_m$ are linearly dependent if and only if $\text{rank}(A) < m$.
But that implies the following: If $\text{rank}(A) \geq m$, $\mathbf{v}_1, \mathbf{v}_2, \ldots, \mathbf{v}_m$ are linearly independent.

In this case, therefore, we argue as follows:
Since $\text{rank}(A) \geq 4$, Theorem 2.7 implies $\mathbf{v}_1$, $\mathbf{v}_2$, $\mathbf{v}_3$, and $\mathbf{v}_4$ are linearly independent.

Does the agree with the solution we found in Exercise 29? It should.
Which method was easier for this Exercise? Why?

40. Exercises 32 through 41 provide a check on our solutions to Exercises 22 through 31. How? In these exercises the directions tell us to follow Example 2.25 and apply Theorem 2.7:

We construct a matrix with these vectors as its rows and proceed to reduce it to echelon form. Each time a row changes, we denote the new row by adding a prime symbol:

$$A = \begin{bmatrix} \mathbf{v}_1 \\ \mathbf{v}_2 \\ \mathbf{v}_3 \\ \mathbf{v}_4 \end{bmatrix} = \begin{bmatrix} 0 & 0 & 0 & 1 \\ 0 & 0 & 2 & 1 \\ 0 & 3 & 2 & 1 \\ 4 & 3 & 2 & 1 \end{bmatrix} \xrightarrow{\substack{R_1'=R_4 \\ R_2'=R_3 \\ R_3'=R_2 \\ R_4'=R_1}} \begin{bmatrix} 4 & 3 & 2 & 1 \\ 0 & 3 & 2 & 1 \\ 0 & 0 & 2 & 1 \\ 0 & 0 & 0 & 1 \end{bmatrix}$$

We can stop. Why? We have put A into row echelon form. How can we tell?
Since the rank of a matrix is the number of nonzero rows in its row echelon form, $\text{rank}(A) = 4$.
What do we conclude? We conclude $\mathbf{v}_1$, $\mathbf{v}_2$, $\mathbf{v}_3$, and $\mathbf{v}_4$ are linearly independent.

Theorem 2.7 states that $\mathbf{v}_1, \mathbf{v}_2, \ldots, \mathbf{v}_m$ are linearly dependent if and only if $\text{rank}(A) < m$.
But that implies the following: If $\text{rank}(A) \geq m$, $\mathbf{v}_1, \mathbf{v}_2, \ldots, \mathbf{v}_m$ are linearly independent.

In this case, therefore, we argue as follows:
Since $\text{rank}(A) \geq 4$, Theorem 2.7 implies $\mathbf{v}_1$, $\mathbf{v}_2$, $\mathbf{v}_3$, and $\mathbf{v}_4$ are linearly independent.

Does the agree with the solution we found in Exercise 30? It should.
Which method was easier for this Exercise? Why?

41. Exercises 32 through 41 provide a check on our solutions to Exercises 22 through 31. How? In these exercises the directions tell us to follow Example 2.25 and apply Theorem 2.7:

We construct a matrix with these vectors as its rows and proceed to reduce it to echelon form. Each time a row changes, we denote the new row by adding a prime symbol:

$$A = \begin{bmatrix} \mathbf{v}_1 \\ \mathbf{v}_2 \\ \mathbf{v}_3 \\ \mathbf{v}_4 \end{bmatrix} = \begin{bmatrix} 3 & -1 & 1 & -1 \\ -1 & 3 & 1 & -1 \\ 1 & 1 & 3 & 1 \\ -1 & -1 & 1 & 3 \end{bmatrix} \xrightarrow[R'_3=R_1]{R'_1=R_3} \begin{bmatrix} 1 & 1 & 3 & 1 \\ -1 & 3 & 1 & -1 \\ 3 & -1 & 1 & -1 \\ -1 & -1 & 1 & 3 \end{bmatrix} \xrightarrow[R'_4=R_4+R'_1]{\substack{R'_2=R_2+R'_1 \\ R''_3=R'_3-3R'_1}} \begin{bmatrix} 1 & 1 & 3 & 1 \\ 0 & 4 & 4 & 0 \\ 0 & -4 & -8 & -4 \\ 0 & 0 & 4 & 4 \end{bmatrix}$$

$$\xrightarrow{R'''_3=R''_3+R'_2} \begin{bmatrix} 1 & 1 & 3 & 1 \\ 0 & 4 & 4 & 0 \\ 0 & 0 & -4 & -4 \\ 0 & 0 & 4 & 4 \end{bmatrix} \xrightarrow{R''_4=R'_4+R'''_3} \begin{bmatrix} 1 & 1 & 3 & 1 \\ 0 & 4 & 4 & 0 \\ 0 & 0 & -4 & -4 \\ 0 & 0 & 0 & 0 \end{bmatrix}$$

We can stop. Why? We have created a zero row. What does that tell us?
Since the rank of a matrix is the number of nonzero rows in its row echelon form, $\text{rank}(A) = 3$.
What do we conclude? We conclude $\mathbf{v}_1$, $\mathbf{v}_2$, $\mathbf{v}_3$, and $\mathbf{v}_4$ are linearly dependent.

Theorem 2.7 states that $\mathbf{v}_1, \mathbf{v}_2, \ldots, \mathbf{v}_m$ are linearly dependent if and only if $\text{rank}(A) < m$.
So, since $\text{rank}(A) = 3 < 4$, Theorem 2.7 implies $\mathbf{v}_1$, $\mathbf{v}_2$, $\mathbf{v}_3$, and $\mathbf{v}_3$ are linearly dependent.

From the row reduction above, we see: $\mathbf{0} = R''_4 = R'_4 + R'''_3$. Substituting, we have:

$$\mathbf{0} = (R_4 + R'_1) + (R''_3 + R'_2) = (R_4 + R'_1) + (R'_3 - 3R'_1) + (R_2 + R'_1)$$
$$= R_4 + R'_3 + R_2 - R'_1 = R_4 + R_1 + R_2 - R_3.$$

This equation yields a dependence relation among the original vectors:

$$\begin{bmatrix} 3 \\ -1 \\ 1 \\ -1 \end{bmatrix} + \begin{bmatrix} -1 \\ 3 \\ 1 \\ -1 \end{bmatrix} - \begin{bmatrix} 1 \\ 1 \\ 3 \\ 1 \end{bmatrix} + \begin{bmatrix} -1 \\ -1 \\ 1 \\ 3 \end{bmatrix} = \begin{bmatrix} 0 \\ 0 \\ 0 \\ 0 \end{bmatrix}.$$ Does this agree with Exercise 31?

42. We will use the theorems of Sections 2.2 and 2.3 to support the conclusions below.

(a) Given $A = [\mathbf{v}_1\ \mathbf{v}_2 \ldots \mathbf{v}_n]$, we will show that $\text{rank}(A) = n$.
Theorem 2.6 of Section 2.3 implies:
Vectors $\mathbf{v}_1, \mathbf{v}_2, \ldots, \mathbf{v}_n$ are linearly independent if and only if the only solution of $[A \mid \mathbf{0}] = [\mathbf{v}_1\ \mathbf{v}_2 \ldots \mathbf{v}_n \mid \mathbf{0}]$ is the trivial solution.
Therefore, the number of free variables in the associated system is 0.
So, Theorem 2.2 of Section 2.2 (The Rank Theorem) implies:
number of free variables $= 0 = n - \text{rank}(A) \Rightarrow \text{rank}(A) = n$, as claimed.

(b) Given $A = \begin{bmatrix} \mathbf{v}_1 \\ \mathbf{v}_2 \\ \vdots \\ \mathbf{v}_n \end{bmatrix}$ we will show that $\text{rank}(A) = n$.

Theorem 2.7 of Section 2.3 implies:
Vectors $\mathbf{v}_1, \mathbf{v}_2, \ldots, \mathbf{v}_n$ are linearly independent if and only if

the rank of $A = \begin{bmatrix} \mathbf{v}_1 \\ \mathbf{v}_2 \\ \vdots \\ \mathbf{v}_n \end{bmatrix}$ is greater than or equal to n. That is, $\text{rank}(A) \geq n$.

But A has n rows therefore we have $n \leq \text{rank}(A) \leq n \Rightarrow \text{rank}(A) = n$, as claimed.

43. We apply the definition of linear independence and Examples 2.23 and 2.25 to prove our claims.

(a) We will show that $\mathbf{u}+\mathbf{v}$, $\mathbf{v}+\mathbf{w}$, and $\mathbf{u}+\mathbf{w}$ are linearly independent.
Given $c_1(\mathbf{u}+\mathbf{v}) + c_2(\mathbf{v}+\mathbf{w}) + c_3(\mathbf{u}+\mathbf{w}) = 0$, we will show $c_1 = c_2 = c_3 = 0$.
Multiplying and gathering like terms yields: $(c_1+c_3)\mathbf{u} + (c_1+c_2)\mathbf{v} + (c_2+c_3)\mathbf{w} = 0$.
Since $\mathbf{u}$, $\mathbf{v}$, and $\mathbf{w}$ are linearly independent, $c_1 + c_3 = c_1 + c_2 = c_2 + c_3 = 0$.
We create the matrix of coefficients A and row reduce to determine its rank:

$$\begin{array}{rcrcrcl} c_1 & + & & & c_3 & = & 0 \\ c_1 & + & c_2 & & & = & 0 \\ & & c_2 & + & c_3 & = & 0 \end{array} \Rightarrow \begin{bmatrix} 1 & 0 & 1 \\ 1 & 1 & 0 \\ 0 & 1 & 1 \end{bmatrix} \longrightarrow \begin{bmatrix} 1 & 0 & 1 \\ 0 & 1 & -1 \\ 0 & 0 & 2 \end{bmatrix}$$

Since $\text{rank}(A) = 3$ the only solution is the trivial one, so $c_1 = c_2 = c_3 = 0$.

(b) We will show that $\mathbf{u}-\mathbf{v}$, $\mathbf{v}-\mathbf{w}$, and $\mathbf{u}-\mathbf{w}$ are linearly dependent.
Given $c_1(\mathbf{u}-\mathbf{v}) + c_2(\mathbf{v}-\mathbf{w}) + c_3(\mathbf{u}-\mathbf{w}) = 0$, we will show $c_1 = c_2 = -c_3$.
Multiplying and gathering like terms yields: $(c_1+c_3)\mathbf{u} + (-c_1+c_2)\mathbf{v} + (-c_2-c_3)\mathbf{w} = 0$.
Since $\mathbf{u}$, $\mathbf{v}$, and $\mathbf{w}$ are linearly independent, $c_1 + c_3 = -c_1 + c_2 = -c_2 - c_3 = 0$.
We form the augmented matrix and row reduce to solve:

$$\begin{array}{rcrcrcl} c_1 & + & & & c_3 & = & 0 \\ -c_1 & + & c_2 & & & = & 0 \\ & & -c_2 & - & c_3 & = & 0 \end{array} \Rightarrow \left[\begin{array}{rrr|r} 1 & 0 & 1 & 0 \\ -1 & 1 & 0 & 0 \\ 0 & -1 & -1 & 0 \end{array}\right] \longrightarrow \left[\begin{array}{rrr|r} 1 & 0 & 1 & 0 \\ 0 & 1 & 1 & 0 \\ 0 & 0 & 0 & 0 \end{array}\right]$$

This clearly has the solution $c_1 = c_2 = -c_3$ as we were to show.

44. We will consider the case when one of the vectors the zero vector, $\mathbf{0}$, separately.

Case 1 Given $\mathbf{v}_1 = \mathbf{0}$. We will show $\mathbf{v}_1$ and $\mathbf{v}_2$ are linearly dependent and multiples.
They are linearly dependent since $\mathbf{v}_1 + 0\mathbf{v}_2 = \mathbf{0} + \mathbf{0} = \mathbf{0}$.
They are multiples since $\mathbf{v}_1 = 0\mathbf{v}_2 = \mathbf{0}$.

Case 2 Assume both $\mathbf{v}_1 \neq \mathbf{0}$ and $\mathbf{v}_2 \neq \mathbf{0}$.
To show two vectors are linearly dependent if and only if they are multiples,
we must prove two separate conditions:
(a) if two vectors are multiples of each other, then they are linearly dependent.
(b) if two vectors are linearly dependent, then they are multiples of each other.

(a) Assume $\mathbf{v}_1 = c\mathbf{v}_2$. We will show they are linearly dependent.
Since $\mathbf{v}_1 = c\mathbf{v}_2$ implies $\mathbf{v}_1 - c\mathbf{v}_2 = \mathbf{0}$, $\mathbf{v}_1$ and $\mathbf{v}_2$ are linearly dependent.

(b) Assume $\mathbf{v}_1$ and $\mathbf{v}_2$ are linearly dependent. We will show $\mathbf{v}_1 = c\mathbf{v}_2$.
Since $\mathbf{v}_1 \neq \mathbf{0}$ and $\mathbf{v}_2 \neq \mathbf{0}$ are linearly dependent,
we have $c_1 \neq 0$ and $c_2 \neq 0$ such that: $c_1\mathbf{v}_1 + c_2\mathbf{v}_2 = \mathbf{0}$.
So, $\mathbf{v}_1 = -\frac{c_2}{c_1}\mathbf{v}_2 = c\mathbf{v}_2$, as we were to show.

45. We will follow the proof of Theorem 2.8 and use the assumption that $n < m$.

PROOF: Let A be the $m \times n$ matrix with vectors $\mathbf{v}_1, \mathbf{v}_2, \ldots, \mathbf{v}_m$ in $\mathbb{R}^n$ as its rows.

By Theorem 2.7, $\mathbf{v}_1, \mathbf{v}_2, \ldots, \mathbf{v}_m$ are linearly dependent if and only if $\text{rank}(A) < m$.
By definition, $\text{rank}(A)$ = number of nonzero rows its its row echelon form. But, the definition of row echelon form implies that $\text{rank}(A)$ is always $\leq$ number of columns in $A = n$. Why?

So, we have $\text{rank}(A) \leq n < m$, as required to show $\mathbf{v}_1, \mathbf{v}_2, \ldots, \mathbf{v}_m$ are linearly dependent.

46. Idea: Given any linear combination of a subset that sums to the zero vector,
extend it to the entire set by including missing vectors with a coefficient of 0.

PROOF: Let $A = \{\mathbf{v}_1, \mathbf{v}_2, \ldots, \mathbf{v}_n\}$ be a set of linearly independent vectors.
By definition, $c_1\mathbf{v}_1 + c_2\mathbf{v}_2 + \cdots + c_n\mathbf{v}_n = \mathbf{0}$ if and only if $c_i = 0$ for all i.

Given any subset of $B = \{\ldots, \mathbf{v}_j, \ldots, \mathbf{v}_k, \ldots\}$, we must show:
$$\cdots + c_j\mathbf{v}_j + \cdots + c_k\mathbf{v}_k + \cdots = \mathbf{0} \text{ if and only if } c_i = 0 \text{ for all } i.$$

Extend $\cdots + c_j\mathbf{v}_j + \cdots + c_k\mathbf{v}_k + \cdots = \mathbf{0}$ to all of A
by letting $c_i = 0$ for any vector in set A that is not in subset B.

Then we have: $c_1\mathbf{v}_1 + c_2\mathbf{v}_2 + \cdots + c_n\mathbf{v}_n = \mathbf{0}$
which is possible if and only if $c_i = 0$ for all i, as we were to show.

47. As suggested in the hint, we will use the result of Exercise 21(b).

PROOF:

Since $S' = \{\mathbf{v}_1, \ldots, \mathbf{v}_k\} \subseteq S = \{\mathbf{v}_1, \ldots, \mathbf{v}_k, \mathbf{v}\}$, Exercise 21(b) implies $\text{span}(S') \subseteq \text{span}(S)$. So, we need only show $\text{span}(S) \subseteq \text{span}(S')$ to prove $\text{span}(S) = \text{span}(S')$.

Since $\mathbf{v}$ is a linear combination of $\mathbf{v}_1, \ldots, \mathbf{v}_k$, Exercise 21(b) implies $\text{span}(S) \subseteq \text{span}(S')$. Therefore, $\text{span}(S) = \text{span}(S')$.

48. We need to show $B = \{\mathbf{v}, \mathbf{v}_2, \ldots, \mathbf{v}_k\}$ is a set of linearly independent vectors. That is, $b_1\mathbf{v} + b_2\mathbf{v}_2 + \cdots + b_k\mathbf{v}_k = \mathbf{0}$ if and only if $b_i = 0$ for all i.

The key to the proof below is the fact that $c_1 \neq 0$.

PROOF: We are given $A = \{\mathbf{v}_1, \mathbf{v}_2, \ldots, \mathbf{v}_k\}$ is a set of linearly independent vectors. That is, $a_1\mathbf{v}_1 + a_2\mathbf{v}_2 + \cdots + a_k\mathbf{v}_k = \mathbf{0}$ if and only if $a_i = 0$ for all i. Furthermore, we are told $\mathbf{v} = c_1\mathbf{v}_1 + c_2\mathbf{v}_2 + \cdots + c_k\mathbf{v}_k$ with $c_1 \neq 0$.

When we substitute this expression for $\mathbf{v}$ in $b_1\mathbf{v} + b_2\mathbf{v}_2 + \cdots + b_k\mathbf{v}_k = \mathbf{0}$, we have:

$$b_1(c_1\mathbf{v}_1 + c_2\mathbf{v}_2 + \cdots + c_k\mathbf{v}_k) + b_2\mathbf{v}_2 + \cdots + b_k\mathbf{v}_k = \mathbf{0}.$$

Now distribute and combine like terms: $b_1c_1\mathbf{v}_1 + (b_1c_2 + b_2)\mathbf{v}_2 + \cdots + (b_1c_k + b_k)\mathbf{v}_k = 0$.

Since all the coefficients of $\mathbf{v}_i$ must be 0, we get the following equations:

$$b_1c_1 = 0 \qquad b_1c_2 + b_2 = 0 \qquad \ldots \qquad b_1c_k + b_k = 0$$

Since $c_1 \neq 0$ implies $b_1 = 0$, we have $b_i = 0$ for all i, as we were to show.

2.4 Applications

1. Let x_1, x_2, and x_3 be the number of bacteria of strains I, II, and III, respectively. Then from the consumption of A, B, and C, we get the following system:

$$\begin{aligned} x_1 + 2x_2 \quad &= 400 \\ 2x_1 + x_2 + x_3 &= 600 \\ x_1 + x_2 + 2x_2 &= 600 \end{aligned} \Rightarrow \left[\begin{array}{ccc|c} 1 & 2 & 0 & 400 \\ 2 & 1 & 1 & 600 \\ 1 & 1 & 2 & 600 \end{array}\right] \longrightarrow \left[\begin{array}{ccc|c} 1 & 0 & 0 & 160 \\ 0 & 1 & 0 & 120 \\ 0 & 0 & 1 & 160 \end{array}\right].$$

So, 160, 120, and 160 bacteria of strains I, II, and III respectively can coexist.

2. Let x_1, x_2, and x_3 be the number of bacteria of strains I, II, and III, respectively. Then from the consumption of A, B, and C, we get the system:

$$\begin{aligned} x_1 + 2x_2 \quad &= 400 \\ 2x_1 + x_2 + 3x_3 &= 500 \\ x_1 + x_2 + x_2 &= 600 \end{aligned} \Rightarrow \left[\begin{array}{ccc|c} 1 & 2 & 0 & 400 \\ 2 & 1 & 3 & 500 \\ 1 & 1 & 1 & 600 \end{array}\right] \longrightarrow \left[\begin{array}{ccc|c} 1 & 0 & 2 & 0 \\ 0 & 1 & -1 & 0 \\ 0 & 0 & 0 & 1 \end{array}\right].$$

This is an inconsistent system, so the bacteria cannot coexist.

3. Let x_1, x_2, and x_3 be the number of small, medium, and large arrangements. Then from the consumption of flowers in each arrangement we get:

$$\begin{aligned} x_1 + 2x_2 + 4x_3 &= 24 \\ 3x_1 + 4x_2 + 8x_3 &= 50 \\ 3x_1 + 6x_2 + 6x_2 &= 48 \end{aligned} \Rightarrow \left[\begin{array}{ccc|c} 1 & 2 & 4 & 24 \\ 3 & 4 & 8 & 50 \\ 3 & 6 & 6 & 48 \end{array}\right] \longrightarrow \left[\begin{array}{ccc|c} 1 & 0 & 0 & 2 \\ 0 & 1 & 0 & 3 \\ 0 & 0 & 1 & 4 \end{array}\right].$$

So, 2 small, 3 medium, and 4 large arrangements were sold that day.

4. (a) Let $x_1 =$ number of nickels, $x_2 =$ number of dimes, $x_3 =$ number of quarters.
So, we get $x_1 + x_2 + x_3 = 20$, $2x_1 - x_2 = 0$, $5x_1 + 10x_2 + 25x_2 = 300$

$$\left[\begin{array}{ccc|c} 1 & 1 & 1 & 20 \\ 2 & -1 & 0 & 0 \\ 5 & 10 & 25 & 300 \end{array}\right] \longrightarrow \left[\begin{array}{ccc|c} 1 & 0 & 0 & 4 \\ 0 & 1 & 0 & 8 \\ 0 & 0 & 1 & 8 \end{array}\right] \Rightarrow \text{There are 4 nickels, 8 dimes, and 8 quarters.}$$

(b) Similar to (a), but the constraint $2x_1 - x_2 = 0$ does not apply.

$$\left[\begin{array}{ccc|c} 1 & 1 & 1 & 20 \\ 5 & 10 & 25 & 300 \end{array}\right] \longrightarrow \left[\begin{array}{ccc|c} 1 & 0 & -3 & -20 \\ 0 & 1 & 4 & 40 \end{array}\right] \Rightarrow x_1 = -20 + 3t,\ x_2 = 40 - 4t,\ x_3 = t.$$

Summarizing, we have: $t \geq \frac{20}{3} \Rightarrow t \geq 7$ and $0 \leq t \leq 10$.
So with $7 \leq t \leq 10$, the possible combinations of 20 nickels, dimes, and quarters is $[x_1, x_2, x_3] \in \{[1, 12, 7], [4, 8, 8], [7, 4, 9], [10, 0, 10]\}$.

5. Let x_1, x_2, and x_3 be the number of house, special, and gourmet blends. Then from the consumption of beans in each blend we get

$$\begin{aligned} 300x_1 + 200x_2 + 100x_3 &= 30{,}000 \\ 200x_2 + 200x_3 &= 15{,}000 \\ 200x_1 + 100x_2 + 200x_3 &= 25{,}000 \end{aligned}$$

Form augmented matrix and reduce it: $\left[\begin{array}{ccc|c} 300 & 200 & 100 & 30{,}000 \\ 0 & 200 & 200 & 15{,}000 \\ 200 & 100 & 200 & 25{,}000 \end{array}\right] \longrightarrow \left[\begin{array}{ccc|c} 1 & 0 & 0 & 65 \\ 0 & 1 & 0 & 30 \\ 0 & 0 & 1 & 45 \end{array}\right].$

The merchant should make 65 house blend, 30 special blend, and 45 gourmet blend.

6. We let x_1, x_2, and x_3 be the number of house, special, and gourmet blends respectively. Then from the consumption of beans in each blends we get

$$\begin{aligned} 300x_1 + 200x_2 + 100x_3 &= 30,000 \\ 50x_1 + 200x_2 + 350x_3 &= 15,000 \\ 150x_1 + 100x_2 + 50x_3 &= 15,000 \end{aligned}$$

Form matrix and reduce it: $\left[\begin{array}{ccc|c} 300 & 200 & 100 & 30,000 \\ 50 & 200 & 350 & 15,000 \\ 150 & 100 & 50 & 15,000 \end{array}\right] \longrightarrow \left[\begin{array}{ccc|c} 1 & 0 & -1 & 60 \\ 0 & 1 & 2 & 60 \\ 0 & 0 & 0 & 0 \end{array}\right] \Rightarrow$

$x_1 = 60 + t$, $x_2 = 60 - 2t$, $x_3 = t$ with $t \geq -60$, $t \leq 30$, $t \geq 0 \Rightarrow 0 \leq t \leq 30$.
But we also need to maximize the profit $P \Rightarrow$
$\frac{1}{2}x_1 + \frac{3}{2}x_2 + 2x_3 = P \Rightarrow \frac{1}{2}(60 + t) + \frac{3}{2}(60 - 2t) + 2t = P \Rightarrow 120 - \frac{1}{2}t = P$.
Since $0 \leq t \leq 30$, the profit P is maximized if $t = 0$, in which case $x_1 = x_2 = 60$ and $x_3 = t$.
Therefore the merchant should make 60 house and special blends, and no gourmet blends.

The maximum profit is $120.

7. Let x, y, z, and w be the number of FeS_2, O_2, Fe_2O_3, and SO_2 molecules respectively. Then, compare the number of iron, sulfur, and oxygen atoms in reactants and products:

$$\begin{array}{lrcl} \text{Iron}: & x &=& 2z \\ \text{Sulfur}: & 2x &=& w \\ \text{Oxygen}: & 2y &=& 3z + 2w \end{array} \Rightarrow \left[\begin{array}{cccc|c} 1 & 0 & -2 & 0 & 0 \\ 2 & 0 & 0 & -1 & 0 \\ 0 & 2 & -3 & -2 & 0 \end{array}\right] \longrightarrow \left[\begin{array}{cccc|c} 1 & 0 & 0 & -\frac{1}{2} & 0 \\ 0 & 1 & 0 & -\frac{11}{8} & 0 \\ 0 & 0 & 1 & -\frac{1}{4} & 0 \end{array}\right].$$

Thus $z = \frac{1}{4}w$, $y = \frac{11}{8}w$, and $x = \frac{1}{2}w$.
The smallest positive value of w that will produce integer values for all four variables is the least common denominator of $\frac{1}{2}, \frac{11}{8}, \frac{1}{2} \Rightarrow w = 8$, $x = 4$, $y = 11$, and $z = 2$.
Therefore, the balanced chemical equation is $4FeS_2 + 11O_2 \longrightarrow 2Fe_2O_3 + 8SO_2$.

8. $6CO_2 + 6H_2O \longrightarrow C_6H_{12}O_6 + 6O_2$

9. $2C_4H_{10} + 13O_2 \longrightarrow 8CO_2 + 10H_2O$

10. $2C_7H_6O_2 + 15O_2 \longrightarrow 6H_2O + 14CO_2$

11. $2C_5H_{11}OH + 15O_2 \longrightarrow 12H_2O + 10CO_2$

12. $12HClO_4 + P_4O_{10} \longrightarrow 4H_3PO_4 + 6Cl_2O_7$

13. $Na_2CO_3 + 4C + N_2 \longrightarrow 2NaCN + 3CO$

14. $2C_2H_2Cl_4 + Ca(OH)_2 \longrightarrow 2C_2HCl_3 + CaCl_2 + 2H_2O$

15. (a) By applying the conservation of flow rule to each node we obtain the system of equations

$$f_1 + f_2 = 20 \qquad f_2 - f_3 = -10 \qquad f_1 + f_3 = 30$$

We form the augmented matrix and perform Gauss-Jordan elimination to get

$$\left[\begin{array}{ccc|c} 1 & 1 & 0 & 20 \\ 0 & 1 & -1 & -10 \\ 1 & 0 & 1 & 30 \end{array}\right] \longrightarrow \left[\begin{array}{ccc|c} 1 & 0 & 1 & 30 \\ 0 & 1 & -1 & -10 \\ 0 & 0 & 0 & 0 \end{array}\right]$$

Letting $f_3 = t$, the possible flows are $f_1 = 30 - t$, $f_2 = t - 10$, $f_3 = t$.

(b) In this case $f_2 = 5$, but $f_2 = t - 10$ so $t = 15$. Then the other flows are $f_1 = f_3 = 15$.

(c) Each flow must be nonnegative $\Rightarrow t \leq 30,\ t \geq 10,\ t \geq 0$
Thus $10 \leq t \leq 30$, so $0 \leq f_1 \leq 20$, $0 \leq f_2 \leq 20$, $10 \leq f_3 \leq 30$.

(d) A negative flow, if allowed, would indicate a transport in the opposite direction.
For example, a negative flow into a node is the same as a positive flow out of a node.
So, if $f_2 < 0$, then the arrow on f_2 could be changed and the flow taken as positive.

16. (a) By applying the conservation of flow rule to each node we obtain the system of equations

$$f_1 + f_2 = 20 \qquad f_1 + f_3 = 25 \qquad f_2 + f_4 = 25 \qquad f_3 + f_4 = 30$$

We form the augmented matrix and perform Gauss-Jordan elimination to get

$$\left[\begin{array}{cccc|c} 1 & 1 & 0 & 0 & 20 \\ 1 & 0 & 1 & 0 & 25 \\ 0 & 1 & 0 & 1 & 25 \\ 0 & 0 & 1 & 1 & 30 \end{array}\right] \longrightarrow \left[\begin{array}{cccc|c} 1 & 0 & 0 & -1 & -5 \\ 0 & 1 & 0 & 1 & 25 \\ 0 & 0 & 1 & 1 & 30 \\ 0 & 0 & 0 & 0 & 0 \end{array}\right]$$

Letting $f_4 = t$, the possible flows are $f_1 = -5 + t$, $f_2 = 25 - t$, $f_3 = 30 - t$, $f_4 = t$.

(b) If $f_4 = t = 10$, then average flows on other streets will be $f_1 = 5$, $f_2 = 15$, and $f_3 = 20$.

(c) Each flow must be nonnegative so the set of solutions gives the four constraints
$t \geq 5,\ t \leq 25,\ t \leq 30,\ t \geq 0$. To satisfy all four constraints t must satisfy $5 \leq t \leq 25$.
Combining the constraint on t with the four solutions,
we see that $0 \leq f_1 \leq 20$, $0 \leq f_2 \leq 20$, $5 \leq f_3 \leq 25$, $5 \leq f_4 \leq 25$.

(d) Reversing all directions would have no effect on the solution since this would be equivalent to multiplying each row of the augmented matrix by -1. But this is an elementary row operation so the new matrix would have the same reduced row echelon form, and the solutions will be the same.

17. (a) By applying the conservation of flow rule to each node we obtain the system of equations

$$f_1 + f_2 = 100 \qquad f_2 + f_3 = f_4 + 150 \qquad f_4 + f_5 = 150 \qquad f_1 + 200 = f_3 + f_5$$

We rearrange the equations, and perform Gauss-Jordan elimination

$$\left[\begin{array}{ccccc|c} 1 & 1 & 0 & 0 & 0 & 100 \\ 0 & 1 & 1 & -1 & 0 & 150 \\ 0 & 0 & 0 & 1 & 1 & 150 \\ 1 & 0 & -1 & 0 & -1 & -200 \end{array}\right] \longrightarrow \left[\begin{array}{ccccc|c} 1 & 0 & -1 & 0 & -1 & -200 \\ 0 & 1 & 1 & 0 & 1 & 300 \\ 0 & 0 & 0 & 1 & 1 & 150 \\ 0 & 0 & 0 & 0 & 0 & 0 \end{array}\right]$$

We parameterize the solution by letting $f_5 = t$, and $f_3 = s$, so the possible flows are $f_1 = -200 + s + t$, $f_2 = 300 - s - t$, $f_3 = s$, $f_4 = 150 - t$, $f_5 = t$.

(b) If DC is closed then $f_5 = t = 0$, so the flow through DB must be $200 \le f_3 \le 300$ $\frac{liters}{day}$.

(c) If DB were closed, then f_5 must carry away at least 200 $\frac{liters}{day}$.
But node C has a maximum outflow of 150 $\frac{liters}{day}$,
so, it will not be able to handle the inflow from f_5.
Thus DB cannot be closed.
From the solution in (a), with DB closed $f_3 = s = 0$, and the solution becomes $f_1 = -200 + t$, $f_2 = 300 - t$, $f_3 = 0$, $f_4 = 150 - t$, $f_5 = t$.
But each flow must be positive, so we get the following constraints on t:
$t \ge 200$, $t \le 300$, $t \le 150$, $t \ge 0$.
But this gives rise to a contradiction since it demands $t \le 150$ and $t \ge 200$, which is impossible, and again we see that DB cannot be closed.

(d) Each flow must be nonnegative, so the solutions give the following constraints:
$s \ge 200 - t$, $s \le 300 - t$, $s \ge 0$, $t \le 150$, $t \ge 0$.
From these we see that $0 \le t \le 150$ and $50 \le s \le 300$.
Combining the constraints on s and t with the five solutions,
we see that $0 \le f_1 \le 100$, $0 \le f_2 \le 100$, $50 \le f_3 \le 300$, $0 \le f_4 \le 150$, $0 \le f_5 \le 150$.

18. (a) By applying the conservation of flow rule to each node we obtain:

$$f_3 + 200 = f_1 + 100 \qquad f_1 + 150 = f_2 + f_4 \qquad f_2 + f_5 = 300$$
$$f_6 + 100 = f_3 + 200 \qquad f_4 + f_7 = f_6 + 100 \qquad f_5 + f_7 = 250$$

We rearrange the equations, and perform Gauss-Jordan elimination

$$\left[\begin{array}{ccccccc|c} 1 & 0 & -1 & 0 & 0 & 0 & 0 & 100 \\ 1 & -1 & 0 & -1 & 0 & 0 & 0 & -150 \\ 0 & 1 & 0 & 0 & 1 & 0 & 0 & 300 \\ 0 & 0 & 1 & 0 & 0 & -1 & 0 & -100 \\ 0 & 0 & 0 & 1 & 0 & -1 & 1 & 100 \\ 0 & 0 & 0 & 0 & 1 & 0 & 1 & 250 \end{array}\right] \longrightarrow \left[\begin{array}{ccccccc|c} 1 & 0 & 0 & 0 & 0 & -1 & 0 & 0 \\ 0 & 1 & 0 & 0 & 0 & 0 & -1 & 50 \\ 0 & 0 & 1 & 0 & 0 & -1 & 0 & -100 \\ 0 & 0 & 0 & 1 & 0 & -1 & 1 & 100 \\ 0 & 0 & 0 & 0 & 1 & 0 & 1 & 250 \\ 0 & 0 & 0 & 0 & 0 & 0 & 0 & 0 \end{array}\right]$$

We parameterize the solution by letting $f_7 = t$, and $f_6 = s$, so the possible flows are $f_1 = s,\ f_2 = 50 + t,\ f_3 = -100 + s,\ f_4 = 100 + s - t,\ f_5 = 250 - t,\ f_6 = s,\ f_7 = t.$

(b) From the solution in (a), we see that it is not possible for $f_1 = 100$ and $f_6 = 150$. From the diagram, if $f_1 = 100$ we see that node A demands that $f_3 = 0$. Then, node D demands $f_6 = 100$, so, it is impossible for $f_1 = 100$ and $f_6 = 150$.

(c) If $f_4 = 0$ then $100 + s - t = 0 \Rightarrow t = 100 + s$, so the possible flows are $f_1 = s,\ f_2 = 150 + s,\ f_3 = -100 + s,\ f_4 = 0,\ f_5 = 150 - s,\ f_6 = s,\ f_7 = 100 + s.$ If we demand that the flows be positive, then we see that $100 \le s \le 150$, and the flows will be restricted to $100 \le f_1 = 150$, $250 \le f_2 = 300$, $0 \le f_3 \le 50$, $0 \le f_4 = 5$, $50 \le f_5 \le 100$, $100 \le f_6 \le 150$, $200 \le f_7 \le 250$.

19. Applying the current law to node A gives the equation $I_1 + I_3 = I_2$ or $I_1 - I_2 + I_3 = 0$. Applying the voltage law to the top circuit (circuit $ABCA$) gives $-I_2 - I_1 + 8 = 0$. Similarly, for the circuit $ABDA$ we obtain $-I_2 + 13 - 4I_3 = 0$.
We get the following system:

$$\begin{aligned} I_1 - I_2 + I_3 &= 0 \\ I_1 + I_2 &= 8 \\ I_2 + 4I_3 &= 13 \end{aligned} \Rightarrow \left[\begin{array}{ccc|c} 1 & -1 & 1 & 0 \\ 1 & 1 & 0 & 8 \\ 0 & 1 & 4 & 13 \end{array}\right] \longrightarrow \left[\begin{array}{ccc|c} 1 & 0 & 0 & 3 \\ 0 & 1 & 0 & 5 \\ 0 & 0 & 1 & 2 \end{array}\right] \Rightarrow$$

$I_1 = 3$ amps, $I_2 = 5$ amps, and $I_3 = 2$ amps.

20. Applying the current and voltage laws, we have:
$$\begin{aligned} I_1 - I_2 + I_3 &= 0 \\ I_1 + 2I_2 + 2I_3 &= 5 \\ 0I_1 + 2I_2 + 4I_3 &= 8 \end{aligned} \Rightarrow$$

$$\left[\begin{array}{ccc|c} 1 & -1 & 1 & 0 \\ 1 & 2 & 0 & 5 \\ 0 & 2 & 4 & 8 \end{array}\right] \longrightarrow \left[\begin{array}{ccc|c} 1 & 0 & 0 & 1 \\ 0 & 1 & 0 & 2 \\ 0 & 0 & 1 & 1 \end{array}\right] \Rightarrow I_1 = 1 \text{ amp},\ I_2 = 2 \text{ amps, and } I_3 = 1 \text{ amp.}$$

21. (a) Applying the current and voltage laws to the circuit gives the system:

$$\begin{array}{ll} \text{Node B}: & I = I_1 + I_4 \\ \text{Node C}: & I_1 = I_1 + I_4 \\ \text{Node D}: & I_2 + I_5 = I \\ \text{Node E}: & I_3 + I_4 = I_5 \end{array} \Rightarrow \begin{array}{lrcl} \text{Circuit ABEDA}: & -2I_4 - I_5 + 14 & = & 0 \\ \text{Circuit BCEB}: & -I_1 - I_3 + 2I_4 & = & 0 \\ \text{Circuit CDEC}: & -2I_2 + I_5 + I_3 & = & 0 \end{array} \Rightarrow$$

Gauss-Jordan elimination gives

$$\left[\begin{array}{rrrrrr|r} 1 & -1 & 0 & 0 & -1 & 0 & 0 \\ 0 & 1 & -1 & -1 & 0 & 0 & 0 \\ 1 & 0 & -1 & 0 & 0 & -1 & 0 \\ 0 & 0 & 0 & 1 & 1 & -1 & 0 \\ 0 & 0 & 0 & 0 & 2 & 1 & 14 \\ 0 & 1 & 0 & 1 & -2 & 0 & 0 \\ 0 & 0 & 2 & -1 & 0 & -1 & 0 \end{array}\right] \longrightarrow \left[\begin{array}{cccccc|c} 1 & 0 & 0 & 0 & 0 & 0 & 10 \\ 0 & 1 & 0 & 0 & 0 & 0 & 6 \\ 0 & 0 & 1 & 0 & 0 & 0 & 4 \\ 0 & 0 & 0 & 1 & 0 & 0 & 2 \\ 0 & 0 & 0 & 0 & 1 & 0 & 4 \\ 0 & 0 & 0 & 0 & 0 & 1 & 6 \\ 0 & 0 & 0 & 0 & 0 & 0 & 0 \end{array}\right]$$

So, the currents are $I = 10$, $I_1 = 6$, $I_2 = 4$, $I_3 = 2$, $I_4 = 4$, and $I_5 = 6$ amps.

(b) From Ohm's Law, the effective resistance is found to be $R_{eff} = \frac{V}{I} = \frac{14}{10} = \frac{7}{5}$ ohms.

(c) In this case we force $I_3 = 0$, and assign r to the resistance in branch BC.
We then get the following system of equations:

$$\begin{array}{ll} \text{Node B}: & I = I_1 + I_4 \\ \text{Node C}: & I_1 = I_2 \\ \text{Node D}: & I_2 + I_5 = I \\ \text{Node E}: & I_4 = I_5 \end{array} \Rightarrow \begin{array}{lrcl} \text{Circuit ABEDA}: & -2I_4 - I_5 + 14 & = & 0 \\ \text{Circuit BCEB}: & -rI_1 + 2I_4 & = & 0 \\ \text{Circuit CDEC}: & -2I_2 + I_5 & = & 0 \end{array} \Rightarrow$$

Partial row reduction gives

$$\left[\begin{array}{rrrrrr|r} 1 & -1 & 0 & 0 & -1 & 0 & 0 \\ 0 & 1 & -1 & 0 & 0 & 0 & 0 \\ 1 & 0 & -1 & 0 & 0 & -1 & 0 \\ 0 & 0 & 0 & 0 & 1 & -1 & 0 \\ 0 & 0 & 0 & 0 & 2 & 1 & 14 \\ 0 & r & 0 & 0 & -2 & 0 & 0 \\ 0 & 0 & 2 & 0 & 0 & -1 & 0 \end{array}\right] \longrightarrow \left[\begin{array}{rrrrrr|r} 1 & -1 & 0 & 0 & -1 & 0 & 0 \\ 0 & 1 & -1 & 0 & 0 & 0 & 0 \\ 0 & 0 & 1 & 0 & 0 & -1/2 & 0 \\ 0 & 0 & 0 & 0 & 1 & -1 & 0 \\ 0 & 0 & 0 & 0 & 0 & 1 & 14/3 \\ 0 & r & 0 & 0 & -2 & 0 & 0 \\ 0 & 0 & 0 & 0 & 0 & 0 & 0 \end{array}\right]$$

So, substitution $\Rightarrow I_4 = I_5 = \frac{14}{3}$ and $I_1 = I_2 = \frac{7}{3} \Rightarrow \frac{7}{3}r - 2\frac{14}{3} = 0 \Rightarrow r = 4$.
So, if we let the resistance in $BC = r = 4$, then the current in $CE = 0$.

22. (a) Applying the voltage law to the first diagram of Figure 2.23 gives us: $IR_1 + IR_2 = E$. But from Ohm's Law, $E = IR_{\text{eff}}$, so the previous equation becomes $IR_1 + IR_2 = IR_{\text{eff}} \Rightarrow R_{\text{eff}} = R_1 + R_2$.

(b) Applying the current and voltage laws to the second circuit of Figure 2.23 gives the system of equations $I = I_1 + I_2$, $-I_1R_1 + I_2R_2 = 0$, $-I_1R_1 + E = 0$. Gauss-Jordan elimination gives

$$\left[\begin{array}{ccc|c} 1 & -1 & -1 & 0 \\ 0 & -R_1 & R_2 & 0 \\ 0 & -R_1 & 0 & -E \end{array}\right] \longrightarrow \left[\begin{array}{ccc|c} 1 & 0 & 0 & \frac{R_1+R_2}{R_1}\frac{E}{R_2} \\ 0 & 1 & 0 & \frac{1}{R_1}E \\ 0 & 0 & 1 & \frac{E}{R_2} \end{array}\right] \Rightarrow I = \frac{R_1+R_2}{R_1}\frac{E}{R_2},\ I_1 = \frac{E}{R_1},\ I_2 = \frac{E}{R_2}.$$

But $E = IR_{\text{eff}}$, which we plug into the expression for the I current to get:
$I = \frac{R_1+R_2}{R_1}\frac{IR_{\text{eff}}}{R_2} \Rightarrow R_{\text{eff}} = \frac{R_1R_2}{R_1+R_2} = \frac{1}{\frac{1}{R_1}+\frac{1}{R_2}}$.

23. (a) Over $\mathbb{Z}_2$, we need to solve $x_1\mathbf{a} + x_2\mathbf{b} + \cdots + x_5\mathbf{e} = \mathbf{t} + \mathbf{s}$:

$$\mathbf{s} = \begin{bmatrix} 0 \\ 0 \\ 0 \\ 0 \\ 0 \end{bmatrix}, \mathbf{t} = \begin{bmatrix} 0 \\ 1 \\ 0 \\ 1 \\ 0 \end{bmatrix} \Rightarrow \left[\begin{array}{ccccc|c} 1 & 1 & 0 & 0 & 0 & 0 \\ 1 & 1 & 1 & 0 & 0 & 1 \\ 0 & 1 & 1 & 1 & 0 & 0 \\ 0 & 0 & 1 & 1 & 1 & 1 \\ 0 & 0 & 0 & 1 & 1 & 0 \end{array}\right] \longrightarrow \left[\begin{array}{ccccc|c} 1 & 0 & 0 & 0 & 1 & 1 \\ 0 & 1 & 0 & 0 & 1 & 1 \\ 0 & 0 & 1 & 0 & 0 & 1 \\ 0 & 0 & 0 & 1 & 1 & 0 \\ 0 & 0 & 0 & 0 & 0 & 0 \end{array}\right]$$

Thus, x_5 is a free variable; hence there are exactly two solutions.
Solving for the other variables, we get: $x_1 = 1 + x_5$, $x_2 = 1 + x_5$, $x_3 = 1$, $x_4 = x_5 \Rightarrow$

$$x_5 = 0 \text{ and } x_5 = 1 \Rightarrow \text{the solutions } \begin{bmatrix} x_1 \\ x_2 \\ x_3 \\ x_4 \\ x_5 \end{bmatrix} = \begin{bmatrix} 1 \\ 1 \\ 1 \\ 0 \\ 0 \end{bmatrix} \text{ and } \begin{bmatrix} x_1 \\ x_2 \\ x_3 \\ x_4 \\ x_5 \end{bmatrix} = \begin{bmatrix} 0 \\ 0 \\ 1 \\ 1 \\ 1 \end{bmatrix}.$$

So, push switches 1, 2, and 3 or switches 3, 4, and 5.

(b) In this case, $\mathbf{t} = \mathbf{e}_2$ over $\mathbb{Z}_2 \Rightarrow$:

$$\left[\begin{array}{ccccc|c} 1 & 1 & 0 & 0 & 0 & 0 \\ 1 & 1 & 1 & 0 & 0 & 1 \\ 0 & 1 & 1 & 1 & 0 & 0 \\ 0 & 0 & 1 & 1 & 1 & 0 \\ 0 & 0 & 0 & 1 & 1 & 0 \end{array}\right] \longrightarrow \left[\begin{array}{ccccc|c} 1 & 0 & 0 & 0 & 1 & 0 \\ 0 & 1 & 0 & 0 & 1 & 0 \\ 0 & 0 & 1 & 0 & 0 & 0 \\ 0 & 0 & 0 & 1 & 1 & 0 \\ 0 & 0 & 0 & 0 & 0 & 1 \end{array}\right]$$

This shows that there is no solution in this case.
That is, it is impossible to start with all lights off and turn only the second light on.

24. (a) With $\mathbf{s} = \mathbf{e}_4$, $\mathbf{t} = \mathbf{e}_2 + \mathbf{e}_4$ over $\mathbb{Z}_2 \Rightarrow$:

$$\left[\begin{array}{ccccc|c} 1&1&0&0&0&0\\ 1&1&1&0&0&1\\ 0&1&1&1&0&0\\ 0&0&1&1&1&0\\ 0&0&0&1&1&0 \end{array}\right] \longrightarrow \left[\begin{array}{ccccc|c} 1&0&0&0&1&0\\ 0&1&0&0&-1&0\\ 0&0&1&0&0&0\\ 0&0&0&1&1&0\\ 0&0&0&0&0&1 \end{array}\right]$$

This shows that there is no solution in this case; that is, it is impossible to start with only the fourth light on and end up with only the second and fourth lights on.

(b) In this case $\mathbf{t} = \mathbf{e}_2$ over $\mathbf{Z}_2 \Rightarrow$:

$$\left[\begin{array}{ccccc|c} 1&1&0&0&0&0\\ 1&1&1&0&0&1\\ 0&1&1&1&0&0\\ 0&0&1&1&1&1\\ 0&0&0&1&1&0 \end{array}\right] \longrightarrow \left[\begin{array}{ccccc|c} 1&0&0&0&1&1\\ 0&1&0&0&1&1\\ 0&0&1&0&0&1\\ 0&0&0&1&1&0\\ 0&0&0&0&0&0 \end{array}\right] \Rightarrow \begin{bmatrix} x_1\\x_2\\x_3\\x_4\\x_5 \end{bmatrix} = \begin{bmatrix}1\\1\\1\\0\\0\end{bmatrix}, \begin{bmatrix}0\\0\\1\\1\\1\end{bmatrix}.$$

25. The possible configurations are

$$\begin{bmatrix} x_1\\x_2\\x_3\\x_4\\x_5 \end{bmatrix} \in \left\{ \begin{array}{c} \begin{bmatrix}0\\0\\0\\0\\0\end{bmatrix}, \begin{bmatrix}1\\1\\0\\0\\0\end{bmatrix}, \begin{bmatrix}1\\1\\1\\0\\0\end{bmatrix}, \begin{bmatrix}0\\1\\1\\1\\0\end{bmatrix}, \begin{bmatrix}0\\0\\1\\1\\1\end{bmatrix}, \begin{bmatrix}0\\0\\0\\1\\1\end{bmatrix}, \begin{bmatrix}0\\0\\1\\0\\0\end{bmatrix}, \begin{bmatrix}0\\1\\0\\1\\0\end{bmatrix}, \begin{bmatrix}0\\1\\1\\0\\1\end{bmatrix}, \begin{bmatrix}0\\1\\0\\0\\1\end{bmatrix} \\ \begin{bmatrix}1\\0\\1\\1\\0\end{bmatrix}, \begin{bmatrix}1\\0\\0\\0\\1\end{bmatrix}, \begin{bmatrix}1\\0\\1\\0\\1\end{bmatrix}, \begin{bmatrix}1\\0\\0\\1\\0\end{bmatrix}, \begin{bmatrix}1\\1\\0\\1\\0\end{bmatrix}, \begin{bmatrix}1\\1\\1\\0\\1\end{bmatrix}, \begin{bmatrix}1\\1\\0\\0\\1\end{bmatrix}, \begin{bmatrix}1\\1\\1\\1\\0\end{bmatrix}, \begin{bmatrix}1\\1\\0\\1\\1\end{bmatrix}, \begin{bmatrix}1\\1\\1\\1\\1\end{bmatrix} \end{array} \right\}$$

26. (a) Over $\mathbb{Z}_3$, we need to solve $x_1\mathbf{a} + x_2\mathbf{b} + x_3\mathbf{c} = \mathbf{t}$.

$$\mathbf{t} = \begin{bmatrix}0\\2\\1\end{bmatrix} \Rightarrow \left[\begin{array}{ccc|c} 1&1&0&0\\ 1&1&1&2\\ 0&1&1&1 \end{array}\right] \longrightarrow \left[\begin{array}{ccc|c} 1&0&0&1\\ 0&1&0&2\\ 0&0&1&2 \end{array}\right] \Rightarrow x_1 = 1,\ x_2 = 2, \text{ and } x_3 = 2.$$

In other words, we must push switch A once, and the other two switches twice each.

(b) $$\mathbf{t} = \begin{bmatrix}1\\0\\1\end{bmatrix} \Rightarrow \left[\begin{array}{ccc|c} 1&1&0&1\\ 1&1&1&0\\ 0&1&1&1 \end{array}\right] \longrightarrow \left[\begin{array}{ccc|c} 1&0&0&2\\ 0&1&0&2\\ 0&0&1&2 \end{array}\right] \Rightarrow x_1 = 2,\ x_2 = 2,\ x_3 = 2.$$

In other words, we must push each of the switches twice.

(c) Let x, y, and z be the final states of the three lights in the system:

$$\left[\begin{array}{ccc|c} 1&1&0&x\\ 1&1&1&y\\ 0&1&1&z \end{array}\right] \longrightarrow \left[\begin{array}{ccc|c} 1&0&0&y+2z\\ 0&1&0&x+2y+z\\ 0&0&1&2x+y \end{array}\right] \Rightarrow \begin{array}{c}\text{if the lights}\\ \text{are in the final configuration}\\ x,\ y, \text{ and } z,\end{array}$$

we can push switch A $y + 2z$ times, switch B $x + 2y + z$ times, and switch C $2x + y$ times (in $\mathbb{Z}_3$) in order to reach that state.

27. In this situation the switches correspond to the vectors

$$\mathbf{a} = \begin{bmatrix} 1 \\ 1 \\ 0 \\ 0 \\ 0 \end{bmatrix}, \mathbf{b} = \begin{bmatrix} 1 \\ 1 \\ 1 \\ 0 \\ 0 \end{bmatrix}, \mathbf{c} = \begin{bmatrix} 0 \\ 1 \\ 1 \\ 1 \\ 0 \end{bmatrix}, \mathbf{d} = \begin{bmatrix} 0 \\ 0 \\ 1 \\ 1 \\ 1 \end{bmatrix}, \mathbf{e} = \begin{bmatrix} 0 \\ 0 \\ 0 \\ 1 \\ 1 \end{bmatrix}$$

In $\mathbb{Z}_3^5$, with $\mathbf{s} = \mathbf{0}$, we need to solve $x_1\mathbf{a} + x_2\mathbf{b} + \cdots + x_5\mathbf{e} = \mathbf{t}$.

$$\mathbf{t} = \begin{bmatrix} 2 \\ 1 \\ 2 \\ 1 \\ 2 \end{bmatrix} \Rightarrow \left[\begin{array}{ccccc|c} 1 & 1 & 0 & 0 & 0 & 2 \\ 1 & 1 & 1 & 0 & 0 & 1 \\ 0 & 1 & 1 & 1 & 0 & 2 \\ 0 & 0 & 1 & 1 & 1 & 1 \\ 0 & 0 & 0 & 1 & 1 & 2 \end{array}\right] \longrightarrow \left[\begin{array}{ccccc|c} 1 & 0 & 0 & 0 & 1 & 1 \\ 0 & 1 & 0 & 0 & 2 & 1 \\ 0 & 0 & 1 & 0 & 0 & 2 \\ 0 & 0 & 0 & 1 & 1 & 2 \\ 0 & 0 & 0 & 0 & 0 & 0 \end{array}\right]$$

So, x_5 is free and there are exactly three solutions ($x_5 = 0, 1, 2$).
Solving for the other variables in terms of x_5 (over $\mathbb{Z}_3$), we get:

$$\begin{bmatrix} x_1 \\ x_2 \\ x_3 \\ x_4 \\ x_5 \end{bmatrix} = \begin{bmatrix} 1 \\ 1 \\ 2 \\ 2 \\ 0 \end{bmatrix}, \begin{bmatrix} x_1 \\ x_2 \\ x_3 \\ x_4 \\ x_5 \end{bmatrix} = \begin{bmatrix} 0 \\ 2 \\ 2 \\ 1 \\ 1 \end{bmatrix}, \begin{bmatrix} x_1 \\ x_2 \\ x_3 \\ x_4 \\ x_5 \end{bmatrix} = \begin{bmatrix} 2 \\ 0 \\ 2 \\ 0 \\ 2 \end{bmatrix}.$$

28. The possible configurations are $\begin{bmatrix} 1 & 1 & 0 & 0 & 0 \\ 1 & 1 & 1 & 0 & 0 \\ 0 & 1 & 1 & 1 & 0 \\ 0 & 0 & 1 & 1 & 1 \\ 0 & 0 & 0 & 1 & 1 \end{bmatrix} \begin{bmatrix} s_1 \\ s_2 \\ s_3 \\ s_4 \\ s_5 \end{bmatrix} = \begin{bmatrix} s_1 + s_2 \\ s_1 + s_2 + s_3 \\ s_2 + s_3 + s_4 \\ s_3 + s_4 + s_5 \\ s_4 + s_5 \end{bmatrix}$,
where $s_i \in \{0, 1, 2\}$ is the number of times that switch i is thrown.

29. (a) The matrix representing actions of touching the squares is $S = \begin{bmatrix} \mathbf{a} & \mathbf{b} & \mathbf{c} & \mathbf{d} & \mathbf{e} & \mathbf{f} & \mathbf{g} & \mathbf{h} & \mathbf{i} \end{bmatrix}$.
Over $\mathbb{Z}_2$, we need to solve $\mathbf{s} + x_1\mathbf{a} + x_2\mathbf{b} + \cdots + x_5\mathbf{e} = \mathbf{t}$ or $x_1\mathbf{a} + x_2\mathbf{b} + \cdots + x_9\mathbf{i} = \mathbf{t} - \mathbf{s} = \mathbf{s}$.

$$\mathbf{s} = \begin{bmatrix} 0 \\ 1 \\ 1 \\ 1 \\ 0 \\ 1 \\ 1 \\ 1 \\ 0 \end{bmatrix}, \mathbf{t} = \begin{bmatrix} 0 \\ 0 \\ 0 \\ 0 \\ 0 \\ 0 \\ 0 \\ 0 \\ 0 \end{bmatrix} \Rightarrow \left[\begin{array}{ccccccccc|c} 1 & 1 & 0 & 1 & 0 & 0 & 0 & 0 & 0 & 0 \\ 1 & 1 & 1 & 0 & 1 & 0 & 0 & 0 & 0 & 1 \\ 0 & 1 & 1 & 0 & 0 & 1 & 0 & 0 & 0 & 1 \\ 1 & 0 & 0 & 1 & 1 & 0 & 1 & 0 & 0 & 1 \\ 1 & 0 & 1 & 0 & 1 & 0 & 1 & 0 & 1 & 0 \\ 0 & 0 & 1 & 0 & 1 & 1 & 0 & 0 & 1 & 1 \\ 0 & 0 & 0 & 1 & 0 & 0 & 1 & 1 & 0 & 1 \\ 0 & 0 & 0 & 0 & 1 & 0 & 1 & 1 & 1 & 1 \\ 0 & 0 & 0 & 0 & 0 & 1 & 0 & 1 & 1 & 0 \end{array}\right] \longrightarrow \left[\begin{array}{ccccccccc|c} 1 & 0 & 0 & 0 & 0 & 0 & 0 & 0 & 0 & 0 \\ 0 & 1 & 0 & 0 & 0 & 0 & 0 & 0 & 0 & 0 \\ 0 & 0 & 1 & 0 & 0 & 0 & 0 & 0 & 0 & 1 \\ 0 & 0 & 0 & 1 & 0 & 0 & 0 & 0 & 0 & 0 \\ 0 & 0 & 0 & 0 & 1 & 0 & 0 & 0 & 0 & 0 \\ 0 & 0 & 0 & 0 & 0 & 1 & 0 & 0 & 0 & 0 \\ 0 & 0 & 0 & 0 & 0 & 0 & 1 & 0 & 0 & 1 \\ 0 & 0 & 0 & 0 & 0 & 0 & 0 & 1 & 0 & 0 \\ 0 & 0 & 0 & 0 & 0 & 0 & 0 & 0 & 1 & 0 \end{array}\right]$$

showing that touching the third and seventh squares will turn all nine squares black.

(b) Since $S \longrightarrow I_9$, we can always find a solution to the system of equations $x_1\mathbf{a} + x_2\mathbf{b} + \cdots + x_9\mathbf{i} = \mathbf{s}$ in part (a).

30. Over $\mathbb{Z}_3$, we need to go from $\mathbf{s}$ to $\mathbf{t}$.

$$\mathbf{s} = \begin{bmatrix} 2 \\ 2 \\ 0 \\ 1 \\ 1 \\ 0 \\ 2 \\ 0 \\ 1 \end{bmatrix}, \mathbf{t} = \begin{bmatrix} 2 \\ 2 \\ 2 \\ 2 \\ 2 \\ 2 \\ 2 \\ 2 \\ 2 \end{bmatrix} \Rightarrow \left[\begin{array}{ccccccccc|c} 1&1&0&1&0&0&0&0&0&0 \\ 1&1&1&0&1&0&0&0&0&0 \\ 0&1&1&0&0&1&0&0&0&2 \\ 1&0&0&1&1&0&1&0&0&1 \\ 1&0&1&0&1&0&1&0&1&1 \\ 0&0&1&0&1&1&0&0&1&2 \\ 0&0&0&1&0&0&1&1&0&0 \\ 0&0&0&0&1&0&1&1&1&2 \\ 0&0&0&0&0&1&0&1&1&1 \end{array}\right] \longrightarrow \left[\begin{array}{ccccccccc|c} 1&0&0&0&0&0&0&0&0&0 \\ 0&1&0&0&0&0&0&0&0&1 \\ 0&0&1&0&0&0&0&0&0&0 \\ 0&0&0&1&0&0&0&0&0&2 \\ 0&0&0&0&1&0&0&0&0&2 \\ 0&0&0&0&0&1&0&0&0&1 \\ 0&0&0&0&0&0&1&0&0&0 \\ 0&0&0&0&0&0&0&1&0&1 \\ 0&0&0&0&0&0&0&0&1&2 \end{array}\right]$$

showing that a solution exists.

31. Let Grace's and Hans' ages be g and h. Then we have $g = 3h$ and $g + 5 = 2(h+5) \Leftrightarrow g = 2(h+5) - 5$, so $3h = 2h + 5 \Leftrightarrow h = 5$ and $g = 15$. So, Hans is 5 and Grace is 15.

32. Let the ages be a, b, and c. Then we have $a + b + c = 60$, $a - b = b - c$, $a + (a - b) = 3c$. We can rewrite these as

$$\begin{aligned} a+b+c &= 60 \\ a-2b+c &= 0 \\ 2a-b-3c &= 0 \end{aligned} \quad \text{So, row reduction} \Rightarrow \left[\begin{array}{ccc|c} 1&1&1&60 \\ 1&-2&1&0 \\ 2&-1&-3&0 \end{array}\right] \longrightarrow \left[\begin{array}{ccc|c} 1&0&0&28 \\ 0&1&0&20 \\ 0&0&1&12 \end{array}\right].$$

Thus, Annie is 28, Bert is 20, and Chris is 12.

33. Let the areas be a and b. We have the following equations:

$$\begin{aligned} a+b &= 1800 \\ \frac{2}{3}a + \frac{1}{2}b &= 1100 \end{aligned}$$

We solve these to find that $a = 1200$ square yards and $b = 600$ square yards.

34. Let x_1, x_2, x_3 be the number of bundles of the first, second, and third types of corn. Then from the number of bundles in each given measure we get the following system:

$$\begin{aligned} 3x_1 + 2x_2 + x_3 &= 39 \\ 2x_1 + 3x_2 + x_3 &= 34 \\ x_1 + 2x_2 + 3x_2 &= 26 \end{aligned}$$

We form the augmented matrix and row reduce it into reduced row echelon form:

$$\left[\begin{array}{ccc|c} 3&2&1&39 \\ 2&3&1&34 \\ 1&2&3&26 \end{array}\right] \longrightarrow \left[\begin{array}{ccc|c} 1&0&0&\frac{37}{4} \\ 0&1&0&\frac{17}{4} \\ 0&0&1&\frac{11}{4} \end{array}\right]$$

Therefore there are 9.25 measures of corn in a bundle of the first type, 4.25 measures of corn in a bundle of the second type, and 2.75 measures of corn in a bundle of the third type.

35. (a) From the addition table we get: $a+c=2$, $a+d=4$, $b+c=3$, $b+d=5$.

$$\left[\begin{array}{cccc|c} 1&0&1&0&2\\ 1&0&0&1&4\\ 0&1&1&0&3\\ 0&1&0&1&5 \end{array}\right] \longrightarrow \left[\begin{array}{cccc|c} 1&0&0&1&4\\ 0&1&0&1&5\\ 0&0&1&-1&-2\\ 0&0&0&0&0 \end{array}\right]$$

So we see that d is a free variable. Solving for the other variables in terms of d we obtain $a=4-d$, $b=5-d$, and $c=-2+d$. Hence there are an infinite number of solutions.

(b) We have: $a+c=3$, $a+d=4$, $b+c=6$, and $b+d=5$

$$\left[\begin{array}{cccc|c} 1&0&1&0&3\\ 1&0&0&1&4\\ 0&1&1&0&6\\ 0&1&0&1&5 \end{array}\right] \longrightarrow \left[\begin{array}{cccc|c} 1&0&0&1&0\\ 0&1&0&1&0\\ 0&0&1&-1&0\\ 0&0&0&0&1 \end{array}\right]$$

So we see that this is an inconsistent system.

36. From the addition table we get the system of equations $a+c=w$, $a+d=y$, $b+c=x$, and $b+d=z$. We form the augmented matrix and perform Gaussian elimination to get

$$\left[\begin{array}{cccc|c} 1&0&1&0&w\\ 1&0&0&1&y\\ 0&1&1&0&x\\ 0&1&0&1&z \end{array}\right] \longrightarrow \left[\begin{array}{cccc|c} 1&0&1&0&w\\ 0&1&1&0&x\\ 0&0&-1&1&y-w\\ 0&0&0&0&z-x-y+w \end{array}\right]$$

These solutions $\Rightarrow$ to get a valid addition table we require $w-x-y+z=0$.

37. (a) From the addition table we get the following system of equations:

$$\begin{aligned} a+d&=3 \qquad & b+d&=2 \qquad & c+d&=1\\ a+e&=5 & b+e&=4 & c+e&=3\\ a+f&=4 & b+f&=3 & c+f&=1 \end{aligned}$$

We form the augmented matrix and perform Gauss-Jordan elimination to get

$$\left[\begin{array}{cccccc|c} 1&0&0&1&0&0&3\\ 1&0&0&0&1&0&5\\ 1&0&0&0&0&1&4\\ 0&1&0&1&0&0&2\\ 0&1&0&0&1&0&4\\ 0&1&0&0&0&1&3\\ 0&0&1&1&0&0&1\\ 0&0&1&0&1&0&3\\ 0&0&1&0&0&1&1 \end{array}\right] \longrightarrow \left[\begin{array}{cccccc|c} 1&0&0&0&0&1&0\\ 0&1&0&0&0&1&0\\ 0&0&1&0&0&1&0\\ 0&0&0&1&0&-1&0\\ 0&0&0&0&1&-1&0\\ 0&0&0&0&0&0&1\\ 0&0&0&0&0&0&0\\ 0&0&0&0&0&0&0\\ 0&0&0&0&0&0&0 \end{array}\right]$$

So, we see that this is an inconsistent system.

(b) From the addition table we get the following system of equations:

$$\begin{aligned} a+d&=1 \qquad & b+d&=2 \qquad & c+d&=3\\ a+e&=3 & b+e&=4 & c+e&=5\\ a+f&=4 & b+f&=5 & c+f&=6 \end{aligned}$$

We form the augmented matrix and perform Gauss-Jordan elimination to get

$$\left[\begin{array}{cccccc|c} 1&0&0&1&0&0&1\\ 1&0&0&0&1&0&3\\ 1&0&0&0&0&1&4\\ 0&1&0&1&0&0&2\\ 0&1&0&0&1&0&4\\ 0&1&0&0&0&1&5\\ 0&0&1&1&0&0&3\\ 0&0&1&0&1&0&5\\ 0&0&1&0&0&1&6 \end{array}\right] \longrightarrow \left[\begin{array}{cccccc|c} 1&0&0&0&0&1&4\\ 0&1&0&0&0&1&5\\ 0&0&1&0&0&1&6\\ 0&0&0&1&0&-1&-3\\ 0&0&0&0&1&-1&-1\\ 0&0&0&0&0&0&0\\ 0&0&0&0&0&0&0\\ 0&0&0&0&0&0&0\\ 0&0&0&0&0&0&0 \end{array}\right]$$

We see that f is a free variable, and there are an infinite number of solutions of the form $a=4-f$, $b=5-f$, $c=6-f$, $d=-3+f$, $e=-1+f$.

38. We generalize the 3×3 addition table to

$+$	a	b	c
d	m	n	o
e	p	q	r
f	s	t	u

From the addition table we get the system of equations

$$\begin{array}{lll} a+d=m & b+d=n & c+d=o \\ a+e=p & b+e=q & c+e=r \\ a+f=s & b+f=t & c+f=u \end{array}$$

We form the augmented matrix and perform Gaussian elimination to get

$$\left[\begin{array}{cccccc|c} 1&0&0&1&0&0&m \\ 1&0&0&0&1&0&p \\ 1&0&0&0&0&1&s \\ 0&1&0&1&0&0&n \\ 0&1&0&0&1&0&q \\ 0&1&0&0&0&1&t \\ 0&0&1&1&0&0&o \\ 0&0&1&0&1&0&r \\ 0&0&1&0&0&1&u \end{array}\right] \longrightarrow \left[\begin{array}{cccccc|c} 1&0&0&1&0&0&m \\ 0&1&0&1&0&0&n \\ 0&0&1&1&0&0&o \\ 0&0&0&1&-1&0&m-p \\ 0&0&0&0&1&-1&n+p-m-t \\ 0&0&0&0&0&0&m-n-p+q \\ 0&0&0&0&0&0&n-o-q+r \\ 0&0&0&0&0&0&p-q-s+t \\ 0&0&0&0&0&0&q-r-t+u \end{array}\right]$$

For a valid table we need: $m+q=n+p$, $n+r=o+q$, $p+t=q+s$, $q+u=r+t$. That is, the sum of the two diagonal entries must equal the sum of the two off-diagonal entries.

39. (a) We know that the three points $(0,1)$, $(-1,4)$, and $(2,1)$ must satisfy the equation $y=ax^2+bx+c$. Substitution $\Rightarrow c=1$, $a-b+c=4$, and $4a+2b+c=1$. So:

$$\left[\begin{array}{ccc|c} 0&0&1&1 \\ 1&-1&1&4 \\ 4&2&1&1 \end{array}\right] \longrightarrow \left[\begin{array}{ccc|c} 1&0&0&1 \\ 0&1&0&-2 \\ 0&0&1&1 \end{array}\right]$$

Thus $a=1$, $b=-2$,and $c=1$, and the equation of the parabola is $y=x^2-2x+1$.

(b) We know that the three points $(-3,1)$, $(-2,2)$, and $(-1,5)$ must satisfy the equation $y=ax^2+bx+c$. Substitution $\Rightarrow 9a-3b+c=1$, $4a-2b+c=2$, and $a-b+c=5$. So:

$$\left[\begin{array}{ccc|c} 9&-3&1&1 \\ 4&-2&1&2 \\ 1&1&1&5 \end{array}\right] \longrightarrow \left[\begin{array}{ccc|c} 1&0&0&1 \\ 0&1&0&6 \\ 0&0&1&10 \end{array}\right]$$

Thus $a=1$, $b=6$,and $c=10$, and the equation of the parabola is $y=x^2+6x+10$.

40. (a) We know that the three points $(0,1)$, $(-1,4)$, and $(2,1)$ must satisfy the equation $x^2+y^2+ax+by+c=0$. Plugging in these points, we get

$$\begin{aligned} 1+b+c &= 0 \\ 1+16-a+4b+c &= 0 \\ 4+1+2a+b+c &= 0 \end{aligned}$$

We form the augmented matrix and perform Gauss-Jordan elimination to get

$$\left[\begin{array}{ccc|c} 0 & 1 & 1 & -1 \\ 1 & -4 & -1 & 17 \\ 2 & 1 & 1 & -5 \end{array}\right] \longrightarrow \left[\begin{array}{ccc|c} 1 & 0 & 0 & -2 \\ 0 & 1 & 0 & -6 \\ 0 & 0 & 1 & 5 \end{array}\right]$$

Thus $a=-2$, $b=-6$, $c=5$, and the equation of the circle is $x^2+y^2-2x-6y+5=0$. By completing the square and simplifying we get the equation $(x-1)^2+(y-3)^2=5$. Thus the center of this circle is at $(1,3)$ and the radius is $r=\sqrt{5}$.

(b) We know that the three points $(-3,1)$, $(-2,2)$, and $(-1,5)$ must satisfy the equation $x^2+y^2+ax+by+c=0$. By plugging in these points we get the system of equations

$$\begin{aligned} 9+1-3a+b+c &= 0 \\ 4+4-2a+2b+c &= 0 \\ 1+25-a+5b+c &= 0 \end{aligned}$$

We form the augmented matrix and perform Gauss-Jordan elimination to get

$$\left[\begin{array}{ccc|c} 3 & -1 & -1 & 10 \\ 2 & -2 & -1 & 8 \\ 1 & -5 & -1 & 26 \end{array}\right] \longrightarrow \left[\begin{array}{ccc|c} 1 & 0 & 0 & 12 \\ 0 & 1 & 0 & -10 \\ 0 & 0 & 1 & 36 \end{array}\right]$$

$\Rightarrow a=12$, $b=-10$,and $c=36 \Rightarrow$ the equation is $x^2+y^2+12x-10y+36=0$.
By completing the square and simplifying we get the equation $(x+6)^2+(y-5)^2=25$.
The center of this circle is at $(-6,5)$ and the radius is $r=5$.

41. We have: $\frac{3x+1}{x^2+2x-3}=\frac{A}{x-1}+\frac{B}{x+3} \Leftrightarrow (x+3)\,A+(x-1)\,B=3x+1 \Leftrightarrow$
$x\,(A+B)+(3A-B)=3x+1$.
Equating the coefficients of x and constants we get:

$A+B=3,\ 3A-B=1 \Rightarrow \left[\begin{array}{cc|c} 1 & 1 & 3 \\ 3 & -1 & 1 \end{array}\right] \longrightarrow \left[\begin{array}{cc|c} 1 & 0 & 1 \\ 0 & 1 & 2 \end{array}\right] \Rightarrow A=1,\ B=2 \Rightarrow$

The partial fraction decomposition is $\frac{3x+1}{x^2+2x-3}=\frac{1}{x-1}+\frac{2}{x+3}$.

42. We have: $\frac{x^2-3x+3}{x^3+2x^2+x} = \frac{A}{x} + \frac{B}{x+1} + \frac{C}{(x+1)^2} \Leftrightarrow$

$$\left(x^2+2x+1\right)A + \left(x^2+x\right)B + (x)\,C = x^2-3x+3 \Leftrightarrow$$
$$x^2\left(A+B\right) + x\left(2A+B+C\right) + (A) = x^2-3x+3.$$

Equating coefficients, we get $A+B=1$, $2A+B+C=-3$, $A=3$:

$$\left[\begin{array}{ccc|c} 1 & 1 & 0 & 1 \\ 2 & 1 & 1 & -3 \\ 1 & 0 & 0 & 3 \end{array}\right] \longrightarrow \left[\begin{array}{ccc|c} 1 & 0 & 0 & 3 \\ 0 & 1 & 0 & -2 \\ 0 & 0 & 1 & -7 \end{array}\right] \Rightarrow A=3,\ B=-2,\ C=-7 \Rightarrow$$

The partial fraction decomposition is $\frac{x^2-3x+3}{x^3+2x^2+x} = \frac{3}{x} + \frac{-2}{x+1} + \frac{-7}{(x+1)^2}$.

43. We have: $\frac{x-1}{(x+1)(x^2+1)(x^2+4)} = \frac{A}{x+1} + \frac{Bx+C}{x^2+1} + \frac{Dx+E}{x^2+4} \Leftrightarrow$

$$\left(x^4+5x^2+4\right)A + \left(x^3+4x+x^2+4\right)\left(Bx+C\right) + \left(x^3+x+x^2+1\right)\left(Dx+E\right) = x-1 \Leftrightarrow$$
$$x^4\left(A+B+D\right) + x^3\left(B+C+D+E\right) + x^2\left(5A+4B+C+D+E\right)$$
$$+x\left(4C+4B+E+D\right) + \left(4A+4C+E\right) = x-1$$

Equating coefficients, we get

$$\begin{aligned} A+B+D &= 0 \\ B+C+D+E &= 0 \\ 5A+4B+C+D+E &= 0 \\ 4B+4C+D+E &= 1 \\ 4A+4C+E &= -1 \end{aligned}$$

From these, we form the augmented matrix and perform Gauss-Jordan elimination

$$\left[\begin{array}{cccccc} 1 & 1 & 0 & 1 & 0 & 0 \\ 0 & 1 & 1 & 1 & 1 & 0 \\ 5 & 4 & 1 & 1 & 1 & 0 \\ 0 & 4 & 4 & 1 & 1 & 1 \\ 4 & 0 & 4 & 0 & 1 & -1 \end{array}\right] \longrightarrow \left[\begin{array}{ccccc|c} 1 & 0 & 0 & 0 & 0 & -\frac{1}{5} \\ 0 & 1 & 0 & 0 & 0 & \frac{1}{3} \\ 0 & 0 & 1 & 0 & 0 & 0 \\ 0 & 0 & 0 & 1 & 0 & -\frac{2}{15} \\ 0 & 0 & 0 & 0 & 1 & -\frac{1}{5} \end{array}\right]$$

So, $A=-\frac{1}{5}$, $B=\frac{1}{3}$, $C=0$, $D=-\frac{2}{15}$, and $E=-\frac{1}{5}$, and we have

$$\frac{x-1}{(x+1)\left(x^2+1\right)\left(x^2+4\right)} = -\frac{\frac{1}{5}}{x+1} + \frac{\frac{1}{3}x}{x^2+1} - \frac{\frac{1}{15}\left(2x+3\right)}{x^2+4}.$$

44. We have the equation

$$\frac{x^3+x+1}{x\,(x-1)\,(x^2+x+1)\,(x^2+1)^3}=\frac{A}{x}+\frac{B}{x-1}+\frac{Cx+D}{x^2+x+1}+\frac{Ex+F}{x^2+1}+\frac{Gx+H}{(x^2+1)^2}+\frac{Ix+J}{(x^2+1)^3}$$

We multiply both sides of the above equation by $x\,(x-1)\left(x^2+x+1\right)\left(x^2+1\right)^3$, simplify, and equate the coefficients of x^n to get the system:

$$\begin{array}{rr}
A+B+C+E=0 & -3A+3B-3C+3D-2E+F-G+H+J=0\\
B-C+D+F=0 & A+4B+C-3D-2F-H=1\\
3A+4B+3C-D+2E+G=0 & -3A+B-C+D-E-G-I=0\\
-A+3B-3C+3D-E+2F+H=0 & B-D-F-H-J=1\\
3A+6B+3C-3D+E-F+G+I=0 & -A=1
\end{array}$$

From these we form the augmented matrix and perform Gauss-Jordan elimination to get

$$\left[\begin{array}{cccccccccc|c}
1&1&1&0&1&0&0&0&0&0&0\\
0&1&-1&1&0&1&0&0&0&0&0\\
3&4&3&-1&2&0&1&0&0&0&0\\
-1&3&-3&3&-1&2&0&1&0&0&0\\
3&6&3&-3&1&-1&1&0&1&0&0\\
-3&3&-3&3&-2&1&-1&1&0&1&0\\
1&4&1&-3&0&-2&0&-1&0&0&1\\
-3&1&-1&1&-1&0&-1&0&-1&0&0\\
0&1&0&-1&0&-1&0&-1&0&-1&1\\
-1&0&0&0&0&0&0&0&0&0&1
\end{array}\right]\longrightarrow\left[\begin{array}{cccccccccc|c}
1&0&0&0&0&0&0&0&0&0&-1\\
0&1&0&0&0&0&0&0&0&0&\frac{1}{8}\\
0&0&1&0&0&0&0&0&0&0&-1\\
0&0&0&1&0&0&0&0&0&0&0\\
0&0&0&0&1&0&0&0&0&0&\frac{15}{8}\\
0&0&0&0&0&1&0&0&0&0&-\frac{9}{8}\\
0&0&0&0&0&0&1&0&0&0&\frac{7}{4}\\
0&0&0&0&0&0&0&1&0&0&-\frac{1}{4}\\
0&0&0&0&0&0&0&0&1&0&\frac{1}{2}\\
0&0&0&0&0&0&0&0&0&1&\frac{1}{2}
\end{array}\right]$$

We find that the partial fraction decomposition is

$$\frac{x^3+x+1}{x\,(x-1)\,(x^2+x+1)\,(x^2+1)^3}=-\frac{1}{x}+\frac{\frac{1}{8}}{x-1}-\frac{x}{x^2+x+1}+\frac{\frac{3}{8}\,(5x-3)}{x^2+1}+\frac{\frac{1}{4}\,(7x-1)}{(x^2+1)^2}+\frac{\frac{1}{2}\,(x+1)}{(x^2+1)^3}$$

45. Assume $1+2+\cdots+n=an^2+bn+c$, and let $n=0,\ 1,\ 2$ to get

$$\begin{aligned}
c&=0\\
a+b+c&=1\\
4a+2b+c&=3
\end{aligned}$$

From these we form the augmented matrix and perform Gauss-Jordan elimination to get

$$\left[\begin{array}{ccc|c}0&0&1&0\\1&1&1&1\\4&2&1&3\end{array}\right]\longrightarrow\left[\begin{array}{ccc|c}1&0&0&\frac{1}{2}\\0&1&0&\frac{1}{2}\\0&0&1&0\end{array}\right]$$

Thus $a=\frac{1}{2}$, $b=\frac{1}{2}$, $c=0$, and we find that $1+2+\cdots+n=\frac{1}{2}n^2+\frac{1}{2}n=\frac{1}{2}n\,(n+1)$.

46. Assume $1^2 + 2^2 + \cdots + n^2 = an^3 + bn^2 + cn + d$, and let $n = 0, 1, 2, 3$ to get

$$\begin{aligned} d &= 0 \\ a + b + c + d &= 1 \\ 8a + 4b + 2c + d &= 5 \\ 27a + 9b + 3c + d &= 14 \end{aligned}$$

From these we form the augmented matrix and perform Gauss-Jordan elimination to get

$$\left[\begin{array}{cccc|c} 0 & 0 & 0 & 1 & 0 \\ 1 & 1 & 1 & 1 & 1 \\ 8 & 4 & 2 & 1 & 5 \\ 27 & 9 & 3 & 1 & 14 \end{array}\right] \longrightarrow \left[\begin{array}{cccc|c} 1 & 0 & 0 & 0 & \frac{1}{3} \\ 0 & 1 & 0 & 0 & \frac{1}{2} \\ 0 & 0 & 1 & 0 & \frac{1}{6} \\ 0 & 0 & 0 & 1 & 0 \end{array}\right]$$

Thus $a = \frac{1}{3}$, $b = \frac{1}{2}$, $c = \frac{1}{6}$, $d = 0$, and we find that

$$1^2 + 2^2 + \cdots + n^2 = \frac{1}{3}n^3 + \frac{1}{2}n^2 + \frac{1}{6}n = \frac{n(n+1)(2n+1)}{6}.$$

47. Assume $1^3 + 2^3 + \cdots + n^3 = an^4 + bn^3 + cn^2 + dn + e$, and let $n = 0, 1, 2, 3, 4$ to get

$$\begin{aligned} e &= 0 \\ a + b + c + d + e &= 1 \\ 16a + 8b + 4c + 2d + e &= 9 \\ 81a + 27b + 9c + 3d + e &= 36 \\ 256a + 64b + 16c + 4d + e &= 100 \end{aligned}$$

From these we form the augmented matrix and perform Gauss-Jordan elimination to get

$$\left[\begin{array}{ccccc|c} 0 & 0 & 0 & 0 & 1 & 0 \\ 1 & 1 & 1 & 1 & 1 & 1 \\ 16 & 8 & 4 & 2 & 1 & 9 \\ 81 & 27 & 9 & 3 & 1 & 36 \\ 256 & 64 & 16 & 4 & 1 & 100 \end{array}\right] \longrightarrow \left[\begin{array}{ccccc|c} 1 & 0 & 0 & 0 & 0 & \frac{1}{4} \\ 0 & 1 & 0 & 0 & 0 & \frac{1}{2} \\ 0 & 0 & 1 & 0 & 0 & \frac{1}{4} \\ 0 & 0 & 0 & 1 & 0 & 0 \\ 0 & 0 & 0 & 0 & 1 & 0 \end{array}\right]$$

Thus $a = \frac{1}{4}$, $b = \frac{1}{2}$, $c = \frac{1}{4}$, $d = 0$, $e = 0$, and we find that

$$1^3 + 2^3 + \cdots + n^3 = \frac{1}{4}n^4 + \frac{1}{2}n^3 + \frac{1}{4}n^2 = \frac{1}{4}n^2(n+1)^2 = \left(\frac{n(n+1)}{2}\right)^2.$$

2.5 Iterative Methods for Solving Linear Systems

1. Begin by solving the first equation for x_1 and the second equation for x_2 to obtain

$$x_1 = \frac{6}{7} + \frac{1}{7}x_2$$
$$x_2 = \frac{4}{5} + \frac{1}{5}x_1$$

Using the initial vector $[x_1, x_2] = [0, 0]$, we get a sequence of approximations:

n	0	1	2	3	4	5
x_1	0	0.857	0.971	0.996	0.999	1.000
x_2	0	0.800	0.971	0.994	0.999	0.999

The exact solution to this system is $[x_1, x_2] = [1, 1]$.

2.

n	0	1	2	3	4	5	6	7	8
x_1	0	2.5	3.0	1.75	1.5	2.125	2.25	1.9375	1.875
x_2	0	−1.0	1.5	2.0	0.75	0.5	1.125	1.25	0.9375

n	9	10	11	12	13	14	15	16
x_1	2.031	2.063	1.984	1.969	2.008	2.016	1.996	1.992
x_2	0.8750	1.031	1.063	0.984	0.969	1.008	1.016	0.996

n	17	18	19	20	21	22
x_1	2.002	2.004	1.999	1.998	2.001	2.000
x_2	0.9922	1.002	1.003	0.999	1.001	1.001

The exact solution is $[x_1, x_2] = [2, 1]$.

3.

n	0	1	2	3	4	5
x_1	0	0.2222	0.2540	0.2610	0.2620	0.2623
x_2	0	0.2857	0.3492	0.3583	0.3603	0.3606

The exact solution is $x = \frac{16}{61}$, $y = \frac{22}{61}$.

4.

n	0	1	2	3	4	5
x_1	0	0.8500	1.005	1.003	1.001	1.000
x_2	0	−1.300	−1.035	−0.9980	−0.9993	−1.000
x_3	0	1.800	2.015	2.004	2.000	2.000

The exact solution is $[x_1, x_2, x_3] = [1, -1, 2]$.

5.

n	0	1	2	3	4	5	6	7	8
x_1	0	0.3333	0.2500	0.3055	0.2916	0.3009	0.2986	0.3001	0.2997
x_2	0	0.2500	0.08337	0.1250	0.09722	0.1042	0.09957	0.1008	0.09997
x_3	0	0.3333	0.2500	0.3055	0.2916	0.3009	0.2986	0.3001	0.2997

The exact solution is $[x_1, x_2, x_3] = [0.3, 0.1, 0.3]$

6.

n	0	1	2	3	4	5
x_1	0	0.3333	0.3333	0.4074	0.4198	0.4403
x_2	0	0	0.2222	0.2593	0.3210	0.3333
x_3	0	0.3333	0.4444	0.5556	0.5803	0.6132
x_4	0	0.3333	0.4444	0.4815	0.5185	0.5268

n	6	7	8	9	10	11
x_1	0.4444	0.4504	0.4516	0.4533	0.4537	0.4542
x_2	0.3512	0.3548	0.3600	0.3611	0.3626	0.3629
x_3	0.6200	0.6296	0.6316	0.6344	0.6350	0.6358
x_4	0.5377	0.5400	0.5432	0.5439	0.5448	0.5450

The exact solution is $[x_1, x_2, x_3, x_4] = \left[\frac{5}{11}, \frac{4}{11}, \frac{7}{11}, \frac{6}{11}\right]$.

7.

n	0	1	2	3	4
x_1	0	0.8571	0.9959	0.9999	1.000
x_2	0	0.9714	0.9992	0.9999	1.000

The Gauss-Seidel method takes 4 steps while the Jacobi method takes 5.

8.

n	0	1	2	3	4	5	6	7	8	9	10	11	12
x_1	0	2.5	1.75	2.125	1.938	2.031	1.984	2.008	1.996	2.002	1.999	2.001	2.000
x_2	0	1.5	0.75	1.125	0.938	1.031	0.984	1.008	0.996	1.002	0.999	1.001	1.000

The Gauss-Seidel method takes 12 steps while the Jacobi method takes 22.

9.

n	0	1	2	3	4
x_1	0	0.2222	0.2610	0.2623	0.2623
x_2	0	0.3492	0.3603	0.3607	0.3607

The Gauss-Seidel method takes 4 steps while the Jacobi method takes 5.

10.

n	0	1	2	3	4
x_1	0	0.85	1.011	1.000	1.000
x_2	0	−1.215	−0.998	−1.000	−1.000
x_3	0	2.007	2.001	2.000	2.000

The Gauss-Seidel method takes 4 steps while the Jacobi method takes 5.

11.

n	0	1	2	3	4	5
x_1	0	0.3333	0.2778	0.2963	0.2994	0.2999
x_2	0	0.1667	0.1111	0.1019	0.1003	0.1001
x_3	0	0.2778	0.2963	0.2994	0.2999	0.2999

The Gauss-Seidel method takes 5 steps while the Jacobi method takes 8.

12.

n	0	1	2	3	4	5	6	7
x_1	0	0.3333	0.3704	0.4156	0.4426	0.4511	0.4535	0.4543
x_2	0	0.1111	0.2469	0.3279	0.3532	0.3606	0.3628	0.3634
x_3	0	0.3704	0.5679	0.6168	0.6307	0.6347	0.6359	0.6362
x_4	0	0.4568	0.5226	0.5389	0.5436	0.5449	0.5453	0.5454

The Gauss-Seidel method takes 7 steps while the Jacobi method takes 11.

13.

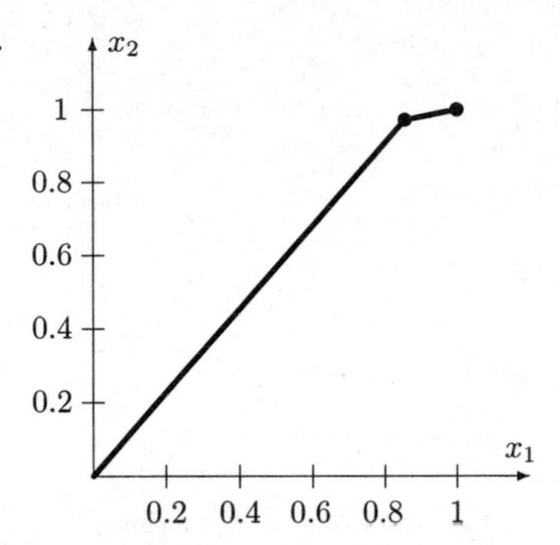

14.

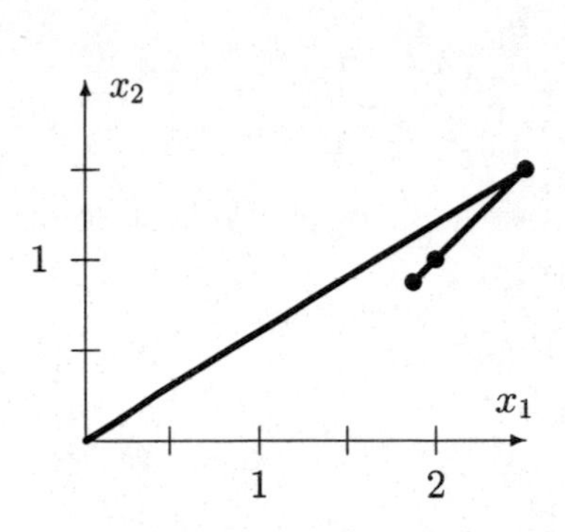

15. Applying the Gauss-Seidel method to $x_1 - 2x_2 = 3$, $3x_1 + 2x_2 = 1$ gives:

n	0	1	2	3	4
x_1	0	3	−5	19	−53
x_2	0	−4	8	−28	80

which evidently diverges. If, however, we swap the two equations to get $3x_1 + 2x_2 = 1$, $x_1 - 2x_2 = 3$ and use the Gauss-Seidel method on this system we get the table:

n	0	1	2	3	4	5	6	7	8	9
x_1	0	0.333	1.222	0.926	1.025	0.992	1.003	0.999	1.000	1.000
x_2	0	−1.334	−0.889	−1.037	−0.988	−1.004	−0.999	−1.000	−1.000	−1.000

Thus the solution to the system of equations is approximately $[x_1, x_2] = [1.000, -1.000]$. The exact solution is $[1, -1]$.

16. Applying the Gauss-Seidel method to the system of equations

$$\begin{aligned} x_1 - 4x_2 + 2x_3 &= 2 \\ 2x_2 + 4x_3 &= 1 \\ 6x_1 - x_2 - 2x_3 &= 1 \end{aligned}$$

gives the following table of iterations:

n	0	1	2	3	4
x_1	0	2	−6.5	−8	203.5
x_2	0	0.5	−10	30.5	80
x_3	0	5.250	−15	−39.75	570

which evidently diverges. If, however, we rearrange the equations to make them diagonally dominant we get the following system of equations:

$$\begin{aligned} 6x_1 - x_2 - 2x_3 &= 1 \\ x_1 - 4x_2 + 2x_3 &= 2 \\ 2x_2 + 4x_3 &= 1 \end{aligned}$$

Using the Gauss-Seidel method on this system we get the following table of iterations:

n	0	1	2	3	4	5	6	7
x_1	0	0.167	0.250	0.250	0.250	0.250	0.250	0.250
x_2	0	−0.458	−0.198	−0.263	−0.246	−0.251	−0.250	−0.250
x_3	0	0.479	0.349	0.382	0.373	0.376	0.375	0.375

Thus the solution to the system of equations is approximately $[x_1, x_2, x_3] = [0.250, -0.250, 0.375]$. The exact solution is $\left[\frac{1}{4}, -\frac{1}{4}, \frac{3}{8}\right]$.

17.

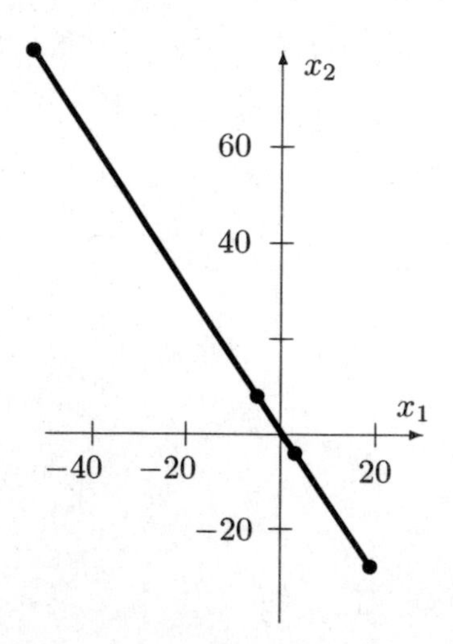

18. Applying the Gauss-Seidel method to the system of equations

$$\begin{aligned} -4x_1 + 5x_2 &= 14 \\ x_1 - 3x_2 &= -7 \end{aligned}$$

gives the following table of iterations:

n	0	1	2	3	4	5	6	7	8	9
x_1	0	−3.5	−2.041	−1.434	−1.181	−1.075	−1.031	−1.013	−1.006	−1.002
x_2	0	1.167	1.653	1.855	1.940	1.975	1.990	1.996	1.998	1.999

Thus the solution to the system of equations is approximately $[x_1, x_2] = [-1.01, 1.99]$. The exact solution is $[-1, 2]$.

19. Applying the Gauss-Seidel method to the system of equations

$$\begin{aligned} 5x_1 - 2x_2 + 3x_3 &= -8 \\ x_1 + 4x_2 - 4x_3 &= 102 \\ -2x_1 - 2x_2 + 4x_3 &= -90 \end{aligned}$$

gives the following table of iterations:

n	0	1	2	3	4	5	6
x_1	0	−1.60	14.97	8.55	10.74	9.84	10.12
x_2	0	25.9	11.41	14.05	11.62	11.72	11.25
x_3	0	−10.35	−9.31	−11.20	−11.32	−11.72	−11.82

n	7	8	9	10	11	12	13
x_1	9.99	10.02	10.00	10.01	10.00	10.00	10.00
x_2	11.18	11.08	11.05	11.02	11.01	11.00	11.00
x_3	−11.91	−11.95	−11.98	−11.99	−12.00	−12.00	−12.00

Thus the solution to the system of equations is approximately $[x_1, x_2, x_3] = [10.00, 11.00, -12.00]$. The exact solution is $[10, 11, -12]$.

20. Continuing the iterations of Exercise 18 to achieve a solution accurate to within 0.001 gives the following table:

n	0	1	⋯	9	10	12	13
x_1	0	−3.5	⋯	−1.002	−1.001	−1.000	−1.000
x_2	0	1.167	⋯	1.999	2.000	2.000	2.000

21. Continuing iterations of Exercise 19 to achieve a solution accurate to within 0.001:

n	0	1	⋯	13	14	15	16	17
x_1	0	−1.60	⋯	10.001	10.000	10.000	10.000	10.000
x_2	0	25.9	⋯	11.004	11.002	11.001	11.000	11.000
x_3	0	−10.35	⋯	−11.998	−11.999	−12.000	−12.000	−12.000

22. Let the equilibrium temperatures of the interior points be t_1, t_2, and t_3 as shown. Thus, by the temperature-averaging property, we have

$$t_1 = \tfrac{1}{4}(0 + 40 + 40 + t_2)$$
$$t_2 = \tfrac{1}{4}(0 + t_1 + t_3 + 5)$$
$$t_3 = \tfrac{1}{4}(t_2 + 40 + 40 + 5)$$

The Gauss-Seidel method gives the following with initial approximation $t_1 = t_2 = t_3 = 0$:

n	0	1	2	3	4	5	6	7	8
t_1	0	20.000	21.563	23.086	23.277	23.300	23.303	23.304	23.304
t_2	0	6.250	12.344	13.106	13.201	13.213	13.214	13.215	13.215
t_3	0	22.813	24.336	24.527	24.550	24.553	24.554	24.554	24.554

Thus the equilibrium temperatures at the interior points are $t_1 = 23.304$, $t_2 = 13.314$, and $t_3 = 24.544$ (to an accuracy of 0.001).

23. From the temperature-averaging property we get the system of four equations

$$\begin{aligned} t_1 &= \frac{1}{4}(t_2 + t_3) \\ t_2 &= \frac{1}{4}(t_1 + t_4) \\ t_3 &= \frac{1}{4}(t_1 + t_4 + 200) \\ t_4 &= \frac{1}{4}(t_2 + t_3 + 200) \end{aligned}$$

upon which we apply the Gauss-Seidel method. With $t_1 = t_2 = t_3 = t_4 = 0$ we get:

n	0	1	2	3	4	5	6	7	8	9	10
t_1	0	0	12.5	21.876	24.220	24.806	24.952	24.988	24.998	25.000	25.000
t_2	0	0	18.75	23.438	24.610	24.904	24.976	24.994	25.000	25.000	25.000
t_3	0	50	68.75	73.438	74.610	74.904	74.976	74.994	75.000	75.000	75.000
t_4	0	62.5	71.876	74.220	74.806	74.952	74.988	74.998	75.000	75.000	75.000

Thus the equilibrium temperatures at the interior points are $t_1 = 25.000$, $t_2 = 25.000$, $t_4 = 75.000$ (to an accuracy of 0.001).

24. As in the previous exercises we approach the problem using the Gauss-Seidel method. The equations are

$$\begin{aligned} t_1 &= \frac{1}{4}(t_2 + t_3) \\ t_2 &= \frac{1}{4}(t_1 + t_4 + 40) \\ t_3 &= \frac{1}{4}(t_1 + t_4 + 80) \\ t_4 &= \frac{1}{4}(t_2 + t_3 + 200) \end{aligned}$$

which gives the following table of iterations with an initial approximation $t_1 = t_2 = t_3 = 0$:

n	0	1	2	3	4	5	6	7	8	9	10
t_1	0	0	7.5	15.625	17.656	18.164	18.292	18.324	18.332	18.334	18.334
t_2	0	10	26.25	30.312	31.328	31.582	31.646	31.662	31.666	31.668	31.668
t_3	0	20	36.25	40.312	41.328	41.582	41.646	41.662	41.666	41.668	41.668
t_4	0	57.5	65.625	67.656	68.164	68.292	68.324	68.332	68.334	68.334	68.334

Thus the equilibrium temperatures at the interior points are $t_1 = 18.333$, $t_2 = 31.666$, $t_3 = 41.666$, and $t_4 = 68.333$ (to an accuracy of 0.001).

25. Here the equations are

$$\begin{aligned}
t_1 &= \frac{1}{4}(t_2 + 80) \\
t_2 &= \frac{1}{4}(t_1 + t_3 + t_4) \\
t_3 &= \frac{1}{4}(t_2 + t_5 + 80) \\
t_4 &= \frac{1}{4}(t_2 + t_5 + 5) \\
t_5 &= \frac{1}{4}(t_3 + t_4 + t_6 + 5) \\
t_6 &= \frac{1}{4}(t_5 + 85)
\end{aligned}$$

Following the same procedure as in the previous exercises, we get the following table:

n	0	1	2	3	4	5	6
t_1	0	20	21.250	22.813	23.330	23.660	23.773
t_2	0	5	11.250	13.321	14.639	15.093	15.237
t_3	0	21.25	24.609	26.988	27.731	27.963	28.035
t_4	0	2.5	5.859	8.238	8.981	9.213	9.285
t_5	0	7.188	14.629	16.283	16.758	16.904	16.949
t_6	0	23.047	24.907	25.321	25.440	25.476	25.487

n	7	8	9	10	11	12
t_1	23.809	23.821	23.824	23.825	23.826	23.826
t_2	15.282	15.297	15.301	15.302	15.304	15.304
t_3	28.058	28.065	28.067	28.068	28.069	28.069
t_4	9.308	9.315	9.317	9.318	9.319	9.319
t_5	16.963	16.968	16.969	16.970	16.970	16.970
t_6	25.491	25.492	25.492	25.493	25.493	25.493

So the equilibrium temperatures at the interior points are found to be about $t_1 = 23.826$, $t_2 = 15.304$,
$t_3 = 28.069$, $t_4 = 9.319$, $t_5 = 16.970$, and $t_6 = 25.493$.

26. Here the equations are

$$t_1 = \tfrac{1}{4}(t_2 + t_5) \qquad t_9 = \tfrac{1}{4}(t_5 + t_{10} + t_{13} + 40)$$
$$t_2 = \tfrac{1}{4}(t_1 + t_3 + t_6) \qquad t_{10} = \tfrac{1}{4}(t_6 + t_9 + t_{11} + t_{14})$$
$$t_3 = \tfrac{1}{4}(t_2 + t_4 + t_7 + 20) \qquad t_{11} = \tfrac{1}{4}(t_7 + t_{10} + t_{12} + t_{15})$$
$$t_4 = \tfrac{1}{4}(t_3 + t_8 + 40) \qquad t_{12} = \tfrac{1}{4}(t_8 + t_{11} + t_{16} + 100)$$
$$t_5 = \tfrac{1}{4}(t_1 + t_6 + t_9) \qquad t_{13} = \tfrac{1}{4}(t_9 + t_{14} + 80)$$
$$t_6 = \tfrac{1}{4}(t_2 + t_5 + t_7 + t_{10}) \qquad t_{14} = \tfrac{1}{4}(t_{10} + t_{13} + t_{15} + 40)$$
$$t_7 = \tfrac{1}{4}(t_3 + t_6 + t_8 + t_{11}) \qquad t_{15} = \tfrac{1}{4}(t_{11} + t_{14} + t_{16} + 100)$$
$$t_8 = \tfrac{1}{4}(t_4 + t_7 + t_{12} + 20) \qquad t_{16} = \tfrac{1}{4}(t_{12} + t_{15} + 200)$$

Following the same procedure as in the previous exercises, we get the following tables:

n	0	1	2	3	4	5	6	7	8	9	10
t_1	0	0	0	0.938	1.934	2.8529	3.996	5.194	6.143	6.817	7.275
t_2	0	0	1.25	2.813	4.444	6.6605	9.035	10.925	12.270	13.186	13.795
t_3	0	5	8.438	10.450	13.425	16.637	19.082	20.781	21.924	22.682	23.180
t_4	0	11.25	14.141	16.753	19.533	21.544	22.893	23.782	24.366	24.748	24.999
t_5	0	0	2.5	4.922	6.968	9.324	11.742	13.646	14.996	15.912	16.522
t_6	0	0	1.875	5.391	10.364	15.506	19.422	22.157	24.002	25.225	26.031
t_7	0	1.25	4.844	12.501	20.356	25.747	29.307	31.644	33.175	34.176	34.832
t_8	0	8.125	16.563	24.708	29.538	32.489	34.346	35.539	36.311	36.814	37.143
t_9	0	10	16.875	20.547	24.077	27.467	29.966	31.683	32.831	33.590	34.089
t_{10}	0	2.5	8.9845	17.544	25.683	31.162	34.749	37.094	38.627	39.630	40.286
t_{11}	0	0.938	17.598	32.929	41.308	46.235	49.286	51.230	52.485	53.300	53.832
t_{12}	0	27.266	49.575	58.264	62.662	65.183	66.727	67.704	68.333	68.741	69.007
t_{13}	0	22.5	28.281	31.797	34.859	36.959	38.335	39.233	39.819	40.203	40.453
t_{14}	0	16.25	26.641	35.359	40.368	43.372	45.247	46.445	47.220	47.724	48.052
t_{15}	0	29.297	52.095	60.927	65.369	67.904	69.452	70.430	71.060	71.468	71.734
t_{16}	0	64.141	75.418	79.798	82.008	83.272	84.045	84.534	84.848	85.052	85.186

n	21	22	23	24	25	26	27	28	29	30
t_1	8.156	8.159	8.161	8.162	8.163	8.163	8.164	8.164	8.164	8.164
t_2	14.954	14.957	14.960	14.962	14.963	14.963	14.964	14.964	14.964	14.964
t_3	24.119	24.123	24.125	24.127	24.127	24.127	24.127	24.127	24.127	24.127
t_4	25.469	25.471	25.472	25.473	25.473	25.473	25.473	25.473	25.473	25.473
t_5	17.681	17.685	17.687	17.689	17.690	17.691	17.691	17.691	17.691	17.691
t_6	27.551	27.555	27.559	27.561	27.563	27.564	27.564	27.564	27.564	27.564
t_7	36.063	36.067	36.070	36.072	36.073	36.073	36.073	36.073	36.073	36.073
t_8	37.759	37.762	37.763	37.763	37.764	37.764	37.764	37.764	37.764	37.764
t_9	35.028	35.031	35.033	35.034	35.037	35.037	35.037	35.037	35.037	35.037
t_{10}	41.517	41.521	41.523	41.526	41.527	41.527	41.527	41.527	41.527	41.527
t_{11}	54.829	54.832	54.835	54.836	54.836	54.836	54.836	54.836	54.836	54.836
t_{12}	69.506	69.508	69.509	69.509	69.509	69.509	69.509	69.509	69.509	69.509
t_{13}	40.923	40.925	40.925	40.927	40.927	40.927	40.927	40.927	40.927	40.927
t_{14}	48.668	48.669	48.671	48.673	48.673	48.673	48.673	48.673	48.673	48.673
t_{15}	72.233	72.234	72.236	72.236	72.236	72.236	72.236	72.236	72.236	72.236
t_{16}	85.435	85.436	85.436	85.436	85.436	85.436	85.436	85.436	85.436	85.436

Column 30 gives the equilibrium temperatures to an accuracy of 0.001.

27. (a) Let x_1 correspond to the left end of the paper and x_2 to the right end, and let n be the number of folds. Then the first six values of $[x_1, x_2]$ are

n	0	1	2	3	4	5	6
x_1	0	0	$\frac{1}{4}$	$\frac{1}{4}$	$\frac{5}{16}$	$\frac{5}{16}$	$\frac{21}{64}$
x_2	1	$\frac{1}{2}$	$\frac{1}{2}$	$\frac{3}{8}$	$\frac{3}{8}$	$\frac{11}{32}$	$\frac{11}{32}$

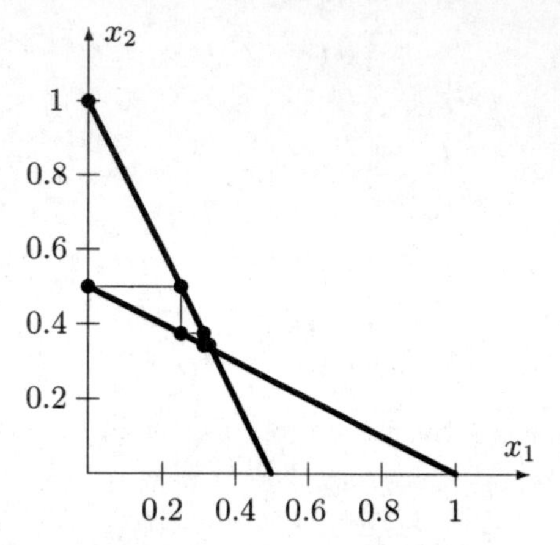

(b) Two linear equations that determine the new values of the endpoints at each iteration are $x_2 = -\frac{1}{2}x_1 + \frac{1}{2}$ and $x_1 = -\frac{1}{2}x_2 + \frac{1}{2}$. These two lines are plotted in part (a).

(c) Switching to decimal representation, we continue applying the Gauss-Seidel method to approximate the point of convergence, giving rise to the sequence of endpoints:

n	0	1	2	3	4	5	6	7	8	9	10	11	12
x_1	0	0	$\frac{1}{4}$	$\frac{1}{4}$	$\frac{5}{16}$	$\frac{5}{16}$	$\frac{21}{64}$	0.328	0.332	0.332	0.333	0.333	0.333
x_2	1	$\frac{1}{2}$	$\frac{1}{2}$	$\frac{3}{8}$	$\frac{3}{8}$	$\frac{11}{32}$	$\frac{11}{32}$	0.336	0.336	0.334	0.334	0.333	0.333

(d) We have the system of equations $x_2 = -\frac{1}{2}x_1 + \frac{1}{2}$, $x_1 = -\frac{1}{2}x_2 + \frac{1}{2} \Rightarrow$

$$\left[\begin{array}{cc|c} \frac{1}{2} & 1 & \frac{1}{2} \\ 1 & \frac{1}{2} & \frac{1}{2} \end{array}\right] \longrightarrow \left[\begin{array}{cc|c} 1 & 0 & \frac{1}{3} \\ 0 & 1 & \frac{1}{3} \end{array}\right]$$

Hence the exact solution to the system of equations is $[x_1, x_2] = \left[\frac{1}{3}, \frac{1}{3}\right]$. The ends of the paper converge at $\frac{1}{3}$.

28. The key is the ant always goes halfway to one of the original endpoints, 0 or 1.

(a) Let x_1 record the positions of the left-hand endpoints of the line segments and x_2 their right-hand endpoints at the end of each walk. Then the first six values of $[x_1, x_2]$ are

n	0	1	2	3	4	5	6
x_1	0	$\frac{1}{4}$	$\frac{1}{4}$	$\frac{5}{16}$	$\frac{5}{16}$	$\frac{21}{64}$	$\frac{21}{64}$
x_2	$\frac{1}{2}$	$\frac{1}{2}$	$\frac{5}{8}$	$\frac{5}{8}$	$\frac{21}{32}$	$\frac{21}{32}$	$\frac{85}{128}$

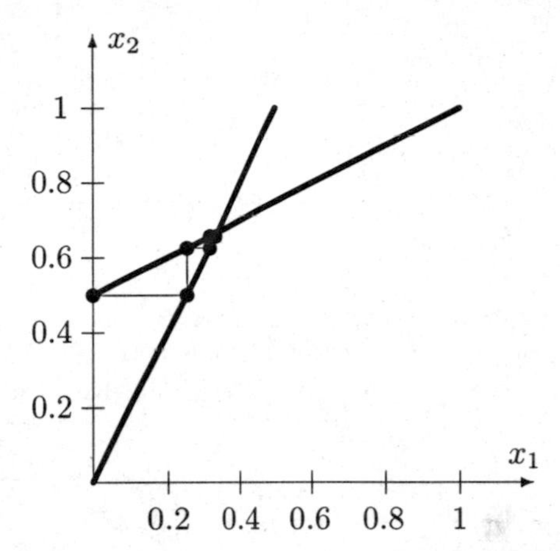

(b) Two linear equations that determine the new values of the endpoints at each iteration are $x_2 = \frac{1}{2}x_1 + \frac{1}{2}$ and $x_1 = \frac{1}{2}x_2$. These two lines are plotted in part (a).

(c) Switching to decimal representation, we continue applying the Gauss-Seidel method to approximate the point of convergence, giving rise to the sequence of endpoints presented in the following table:

n	0	1	2	3	4	5	6	7	8	9	10	11
x_1	0	$\frac{1}{4}$	$\frac{1}{4}$	$\frac{5}{16}$	$\frac{5}{16}$	$\frac{21}{64}$	$\frac{21}{64}$	0.332	0.332	0.333	0.333	0.333
x_2	$\frac{1}{2}$	$\frac{1}{2}$	$\frac{5}{8}$	$\frac{5}{8}$	$\frac{21}{32}$	$\frac{21}{32}$	$\frac{85}{128}$	0.664	0.666	0.666	0.666	0.666

(d) We have the system of equations $x_2 = \frac{1}{2}x_1 + \frac{1}{2}$, $x_1 = \frac{1}{2}x_2 \Rightarrow$

$$\left[\begin{array}{cc|c} -\frac{1}{2} & 1 & \frac{1}{2} \\ 1 & -\frac{1}{2} & 0 \end{array}\right] \longrightarrow \left[\begin{array}{cc|c} 1 & 0 & \frac{1}{3} \\ 0 & 1 & \frac{2}{3} \end{array}\right]$$

Hence the exact solution to the system of equations is $[x_1, x_2] = \left[\frac{1}{3}, \frac{2}{3}\right]$.
So, the ant eventually oscillates between $\frac{1}{3}$ and $\frac{2}{3}$.

Chapter 2 Review

1. We will explain and give counter examples to justify our answers below.

   (a) ***False***. In Section 2.1, see the definition of an *inconsistent* system.
   A useful counter example is parallel lines in $\mathbb{R}^2$, like $x + y = 0$ and $x + y = 1$.

   (b) ***True***. In Section 2.2, see remarks prior to the definition of an *homogenous* system.
   Q: Why does a homogenous system guarantee the associated lines intersect?
   A: Because all the associated lines pass through the origin.

   (c) ***False***. In Section 2.2, see Theorem 2.6 (system must be homogenous).
   When there are fewer conditions than variables,
   we can solve a homogenous system.

   (d) ***False***. In Section 2.2, see remarks under *Homogenous Systems* on p78.
   When a system has more equations than variables,
   it can either have a unique solution, infinitely many solutions, or no solution.

   (e) ***True***. In Section 2.3, see Theorem 2.4 ($[A\,|\,\mathbf{b}]$ is consistent $\Leftrightarrow \mathbf{b} = \sum c_i\mathbf{a}_i$).
   Q: How might we state an informal proof of this theorem?
   A: The fact that $\mathbf{b} = \sum c_i\mathbf{a}_i$ says $\mathbf{x} = [c_i]$ is a solution of $[A\,|\,\mathbf{b}]$.

   (f) ***False***. We need an additional condition to make this true. Which one?
   Q: If $\mathbf{u} \neq \mathbf{0}$ and $\mathbf{v}$ are linearly dependent, what is $\text{span}(\mathbf{u}, \mathbf{v})$?
   A: Then $\text{span}(\mathbf{u}, \mathbf{v}) = \text{span}(\mathbf{u}) = c\mathbf{u}$, a line through the origin.
   Q: If $\mathbf{u}$ and $\mathbf{v}$ are linearly independent, what is $\text{span}(\mathbf{u}, \mathbf{v})$?
   A: Then $\text{span}(\mathbf{u}, \mathbf{v}) = c\mathbf{u} + d\mathbf{v}$, a plane a line through the origin.

   (g) ***True***. Show this by proving the *contrapositive* (See Example 9 of Appendix A).
   Q: What is the *contrapositive* of this statement?
   A: If $\mathbf{u}$ and $\mathbf{v}$ are parallel, then they are linearly dependent.
   Q: How might we prove this statement is true?
   A: If $\mathbf{u}$ and $\mathbf{v}$ are parallel, then $\mathbf{v} = c\mathbf{u} \Rightarrow -c\mathbf{u} + \mathbf{v} = 0$.
   Q: Why does the fact that $\mathbf{u}$ and $\mathbf{v}$ are parallel imply $\mathbf{v} = c\mathbf{u}$?
   A: Vectors are defined to be *parallel* if they are scalar multiples of each other.
   Q: Why does $-c\mathbf{u} + \mathbf{v} = 0$ imply $\mathbf{u}$ and $\mathbf{v}$ are linearly dependent?
   A: Two vectors are linearly dependent if $c_1\mathbf{v}_1 + c_2\mathbf{v}_2 = 0$ (one $c_i \neq 0$).

   (h) ***True***. Why? A *closed* path means there has been no displacement.
   Q: Since there is no displacement, what do know about the sum of the vectors?
   A: Since there is no displacement, we have $\sum \mathbf{v}_i = \mathbf{0}$.
   Q: Why does $\sum \mathbf{v}_i = \mathbf{0}$ imply $\mathbf{v}_i$ are linearly dependent?
   A: Vectors are linearly dependent if $\sum c_i\mathbf{v}_i = \mathbf{0}$ (at least one $c_i \neq 0$).

   (i) ***False***. This pairwise condition is much *weaker* than linear independence.
   Consider this counter example: $\mathbf{u} = [1, 0, 0]$, $\mathbf{v} = [0, 1, 0]$, and $\mathbf{w} = [1, 1, 0]$.
   Geometrically, consider 3 lines in the same plane none of which are parallel.

   (j) ***True***. In Section 2.3, see Thm 2.8 (m vectors in $\mathbb{R}^n$ are linearly dependent if $m > n$).
   Q: What is one way of stating Theorem 2.8 informally in our own words?
   A: When there are more vectors than entries, we can solve $\sum c_i\mathbf{v}_i = \mathbf{0}$.

2. In Section 2.2, *rank* is defined to be the number of nonzero rows in row echelon form. So, we row reduce A using Gaussian Elimination to determine the number of nonzero rows.

$$\begin{bmatrix} 1 & -2 & 0 & 3 & 2 \\ 3 & -1 & 1 & 3 & 4 \\ 3 & 4 & 2 & -3 & 2 \\ 0 & -5 & -1 & 6 & 2 \end{bmatrix} \xrightarrow{R_2 \leftrightarrow R_4} \begin{bmatrix} 1 & -2 & 0 & 3 & 2 \\ 0 & -5 & -1 & 6 & 2 \\ 3 & 4 & 2 & -3 & 2 \\ 3 & -1 & 1 & 3 & 4 \end{bmatrix} \xrightarrow[R_4-3R_1+R_2]{R_3-3R_1+2R_2} \begin{bmatrix} 1 & -2 & 0 & 3 & 2 \\ 0 & -5 & -1 & 6 & 2 \\ 0 & 0 & 0 & 0 & 0 \\ 0 & 0 & 0 & 0 & 0 \end{bmatrix}$$

Since the reduced echelon form of A has 2 nonzero rows, the rank of A is 2.

3. As in Example 2.12 of Section 2.2, we form the augmented matrix and row reduce to solve.

$$\left[\begin{array}{ccc|c} 1 & 1 & -2 & 4 \\ 1 & 3 & -1 & 7 \\ 2 & 1 & -5 & 7 \end{array}\right] \xrightarrow[R_3-2R_1]{R_2-R_1} \left[\begin{array}{ccc|c} 1 & 1 & -2 & 4 \\ 0 & 2 & 1 & 3 \\ 0 & -1 & -1 & -1 \end{array}\right] \xrightarrow{-2R_3} \left[\begin{array}{ccc|c} 1 & 1 & -2 & 4 \\ 0 & 2 & 1 & 3 \\ 0 & 2 & 2 & 2 \end{array}\right] \xrightarrow{R_3-R_2} \left[\begin{array}{ccc|c} 1 & 1 & -2 & 4 \\ 0 & 2 & 1 & 3 \\ 0 & 0 & 1 & -1 \end{array}\right]$$

$$\xrightarrow[R_2+R_3]{R_1-2R_3} \left[\begin{array}{ccc|c} 1 & 1 & 0 & 2 \\ 0 & 2 & 0 & 4 \\ 0 & 0 & 1 & -1 \end{array}\right] \xrightarrow{\frac{1}{2}R_2} \left[\begin{array}{ccc|c} 1 & 1 & 0 & 2 \\ 0 & 1 & 0 & 2 \\ 0 & 0 & 1 & -1 \end{array}\right] \xrightarrow{R_1-R_2} \left[\begin{array}{ccc|c} 1 & 0 & 0 & 0 \\ 0 & 1 & 0 & 2 \\ 0 & 0 & 1 & -1 \end{array}\right] \Rightarrow \begin{bmatrix} x_1 \\ x_2 \\ x_3 \end{bmatrix} = \begin{bmatrix} 0 \\ 2 \\ -1 \end{bmatrix}.$$

4. As in Example 2.11 of Section 2.2, we form the augmented matrix and row reduce to solve.

$$\left[\begin{array}{cccc|c} 3 & 8 & -18 & 1 & 35 \\ 1 & 2 & -4 & 0 & 11 \\ 1 & 3 & -7 & 1 & 10 \end{array}\right] \longrightarrow \left[\begin{array}{cccc|c} 1 & 0 & 2 & 0 & 5 \\ 0 & 1 & -3 & 0 & 3 \\ 0 & 0 & 0 & 1 & -4 \end{array}\right]$$

So, $z = -4$, $x - 3y = 3$ and $w + 2y = 5$. Setting $y = t$ yields $x = 3 + 3t$ and $w = 5 - 2t$.

Therefore, the solution is $\begin{bmatrix} w \\ x \\ y \\ z \end{bmatrix} = \begin{bmatrix} 5 \\ 3 \\ 0 \\ -4 \end{bmatrix} + t \begin{bmatrix} -2 \\ 3 \\ 1 \\ 0 \end{bmatrix}.$

5. As in Example 2.16 of Section 2.2, we form the augmented matrix and row reduce over $\mathbb{Z}_7$. Since we are using modular arithmetic, we need only add and multiply. Why?

$$\left[\begin{array}{cc|c} 2 & 3 & 4 \\ 1 & 2 & 3 \end{array}\right] \xrightarrow{R_2+3R_1} \left[\begin{array}{cc|c} 2 & 3 & 4 \\ 0 & 4 & 1 \end{array}\right] \xrightarrow{2R_2} \left[\begin{array}{cc|c} 2 & 3 & 4 \\ 0 & 1 & 2 \end{array}\right] \xrightarrow{R_1+4R_2} \left[\begin{array}{cc|c} 2 & 0 & 5 \\ 0 & 1 & 2 \end{array}\right] \xrightarrow{4R_1} \left[\begin{array}{cc|c} 1 & 0 & 6 \\ 0 & 1 & 2 \end{array}\right]$$

So, the solution is $\begin{bmatrix} x \\ y \end{bmatrix} = \begin{bmatrix} 6 \\ 2 \end{bmatrix}$. We can check this: $\begin{array}{rcl} 2(6) + 3(2) & = 18 = & 4 \\ 6 + 2(2) & = 10 = & 3 \end{array}$ in $\mathbb{Z}_7$.

6. As in Example 2.16 of Section 2.2, we form the augmented matrix and row reduce over $\mathbb{Z}_5$. Since we are using modular arithmetic, we need only add and multiply. Why?

$$\left[\begin{array}{cc|c} 3 & 2 & 1 \\ 1 & 4 & 2 \end{array}\right] \xrightarrow{R_2+3R_1} \left[\begin{array}{cc|c} 3 & 2 & 1 \\ 0 & 0 & 0 \end{array}\right]$$

We now find all the solutions by setting $x = 0, 1, 2, 3, 4$ and solving for y.

When $x = 0$, $0 + 4y = 2 \Rightarrow y = 3$, so the solution is $\begin{bmatrix} x \\ y \end{bmatrix} = \begin{bmatrix} 0 \\ 3 \end{bmatrix}$.

When $x = 1$, $1 + 4y = 2 \Rightarrow y = 4$, so the solution is $\begin{bmatrix} x \\ y \end{bmatrix} = \begin{bmatrix} 1 \\ 4 \end{bmatrix}$.

When $x = 2$, $2 + 4y = 2 \Rightarrow y = 0$, so the solution is $\begin{bmatrix} x \\ y \end{bmatrix} = \begin{bmatrix} 2 \\ 0 \end{bmatrix}$.

When $x = 3$, $3 + 4y = 2 \Rightarrow y = 1$, so the solution is $\begin{bmatrix} x \\ y \end{bmatrix} = \begin{bmatrix} 3 \\ 1 \end{bmatrix}$.

When $x = 4$, $4 + 4y = 2 \Rightarrow y = 2$, so the solution is $\begin{bmatrix} x \\ y \end{bmatrix} = \begin{bmatrix} 4 \\ 2 \end{bmatrix}$.

Q: Each possible value for y occurs exactly one time. Is this what we should expect?

7. As in Exercise 40 of Section 2.2, we row reduce to find the restrictions on k.
Note: The system has no solution when A has a zero row with corresponding constant $\neq 0$.

$$\left[\begin{array}{cc|c} k & 2 & 1 \\ 1 & 2k & 1 \end{array}\right] \xrightarrow{R_1 \leftrightarrow R_1} \left[\begin{array}{cc|c} 1 & 2k & 1 \\ k & 2 & 1 \end{array}\right] \xrightarrow{R_2 - kR_1} \left[\begin{array}{cc|c} 1 & 2k & 1 \\ 0 & 2-2k^2 & 1-k \end{array}\right] \xrightarrow{-R_2} \left[\begin{array}{cc|c} 1 & 2k & 1 \\ 0 & 2(k-1)(k+1) & k-1 \end{array}\right]$$

So, the only value of k that creates a zero row with corresponding constant $\neq 0$ is $k = -1$.
That is, the only value of k for which this system is inconsistent is $k = -1$.

8. As in Example 2.14 of Section 2.2, we form the augmented matrix and row reduce to solve.

$$\left[\begin{array}{ccc|c} 1 & 2 & 3 & 4 \\ 5 & 6 & 7 & 8 \end{array}\right] \xrightarrow{R_2-5R_1} \left[\begin{array}{ccc|c} 1 & 2 & 3 & 4 \\ 0 & -4 & -8 & -12 \end{array}\right] \xrightarrow{-\frac{1}{4}R_2} \left[\begin{array}{ccc|c} 1 & 2 & 3 & 4 \\ 0 & 1 & 2 & 3 \end{array}\right] \xrightarrow{R_1-2R_2} \left[\begin{array}{ccc|c} 1 & 0 & -1 & -2 \\ 0 & 1 & 2 & 3 \end{array}\right]$$

So, $x - z = -2$, $y + 2z = 3$. Setting $z = t$ yields $x = -2 + t$ and $y = 3 - 2t$.

We get the line with parametric equation: $\begin{bmatrix} x \\ y \\ z \end{bmatrix} = \begin{bmatrix} -2 \\ 3 \\ 0 \end{bmatrix} + t \begin{bmatrix} 1 \\ -2 \\ 1 \end{bmatrix}$.

9. As in Example 2.15 of Section 2.2, we need to determine the point of intersection.
We want to find an $\mathbf{x} = [x, y, z]$ that satisfies both equations simultaneously.
That is, we want $\mathbf{x} = \mathbf{p} + s\mathbf{u} = \mathbf{q} + t\mathbf{v}$ or $s\mathbf{u} - t\mathbf{v} = \mathbf{q} - \mathbf{p}$.

Substituting the given $\mathbf{p}$, $\mathbf{q}$, $\mathbf{u}$, and $\mathbf{v}$ into $s\mathbf{u} - t\mathbf{v} = \mathbf{q} - \mathbf{p}$, we obtain the equations:

$$s\begin{bmatrix} 1 \\ -1 \\ 2 \end{bmatrix} - t\begin{bmatrix} -1 \\ 1 \\ 1 \end{bmatrix} = \begin{bmatrix} 5 \\ -2 \\ -4 \end{bmatrix} - \begin{bmatrix} 1 \\ 2 \\ 3 \end{bmatrix} \Rightarrow \begin{array}{rcrcr} s & + & t & = & 4 \\ -s & - & t & = & -4 \\ 2s & - & t & = & -7 \end{array}$$

We form the augmented matrix and row reduce to find values for s and t.

$$\left[\begin{array}{rr|r} 1 & 2 & 4 \\ -1 & -1 & -4 \\ 2 & -1 & -7 \end{array}\right] \longrightarrow \left[\begin{array}{rr|r} 1 & 0 & -1 \\ 0 & 0 & 0 \\ 0 & 1 & 5 \end{array}\right] \Rightarrow \text{So, } s = -1 \text{ and } t = 5.$$

Therefore, the point of intersection is: $\begin{bmatrix} x \\ y \\ z \end{bmatrix} = \begin{bmatrix} 1 \\ 2 \\ 3 \end{bmatrix} - 1\begin{bmatrix} 1 \\ -1 \\ 2 \end{bmatrix} = \begin{bmatrix} 0 \\ 3 \\ 1 \end{bmatrix}$.

Check that substituting $t = 5$ into the other equation gives the same point.

10. As in Example 2.18 of Section 2.3, we want to find scalars x and y such that:

$$x\begin{bmatrix} 1 \\ 1 \\ 3 \end{bmatrix} + y\begin{bmatrix} 1 \\ 2 \\ -2 \end{bmatrix} = \begin{bmatrix} 3 \\ 5 \\ -1 \end{bmatrix} \Rightarrow \begin{array}{rcrcr} x & + & y & = & 3 \\ x & + & 2y & = & 5 \\ 3x & - & 2y & = & -1 \end{array}$$

We form the augmented matrix and row reduce to find values for x and y.

$$\left[\begin{array}{rr|r} 1 & 1 & 3 \\ 1 & 2 & 5 \\ 3 & -2 & -1 \end{array}\right] \longrightarrow \left[\begin{array}{rr|r} 1 & 0 & 1 \\ 0 & 1 & 2 \\ 0 & 0 & 0 \end{array}\right] \Rightarrow \text{So } x = 1 \text{ and } y = 2.$$

Since $\begin{bmatrix} 3 \\ 5 \\ -1 \end{bmatrix} = \begin{bmatrix} 1 \\ 1 \\ 3 \end{bmatrix} + 2\begin{bmatrix} 1 \\ 2 \\ -2 \end{bmatrix}$, we conclude $\begin{bmatrix} 3 \\ 5 \\ -1 \end{bmatrix}$ is in the span of $\begin{bmatrix} 1 \\ 1 \\ 3 \end{bmatrix}$ and $\begin{bmatrix} 1 \\ 2 \\ -2 \end{bmatrix}$.

11. As in Example 2.21 of Section 2.3, the equation of the plane we are looking for is:

$$\begin{bmatrix} x \\ y \\ z \end{bmatrix} = s\begin{bmatrix} 1 \\ 1 \\ 1 \end{bmatrix} + t\begin{bmatrix} 3 \\ 2 \\ 1 \end{bmatrix} \Rightarrow \begin{array}{rcl} s + 3t &=& x \\ s + 2t &=& y \\ s + t &=& z \end{array}$$

We form the augmented matrix and row reduce to find conditions for x, y, and z.

$$\left[\begin{array}{cc|c} 1 & 3 & x \\ 1 & 2 & y \\ 1 & 1 & z \end{array}\right] \xrightarrow[R_3-R_1]{R_2-R_1} \left[\begin{array}{cc|c} 1 & 3 & x \\ 0 & -1 & -x+y \\ 0 & -2 & -x+z \end{array}\right] \xrightarrow{-R_2} \left[\begin{array}{cc|c} 1 & 3 & x \\ 0 & 1 & x-y \\ 0 & -2 & -x+z \end{array}\right] \xrightarrow{R_3+2R_2} \left[\begin{array}{cc|c} 1 & 3 & x \\ 0 & 1 & x-y \\ 0 & 0 & x-2y+z \end{array}\right]$$

By assumption the system is consistent so $x - 2y + z = 0$, the equation of the plane we sought.

Q: How can we verify that both these vectors lie in the plane?
A: By computing $1 - 2(1) + 1 = 0$ and $3 - 2(2) + 1 = 0$.

Q: What is the cross product of the given vectors?
A: $[-1, 2, -1]$. Should this agree with the normal of our plane? Does it?

12. As in Example 2.23 of Section 2.3, we want to find scalars c_1, c_2, and c_3 such that:

$$c_1\begin{bmatrix} 2 \\ 1 \\ -3 \end{bmatrix} + c_2\begin{bmatrix} 1 \\ -1 \\ -2 \end{bmatrix} + c_3\begin{bmatrix} 3 \\ 9 \\ -2 \end{bmatrix} = \begin{bmatrix} 0 \\ 0 \\ 0 \end{bmatrix}$$

We form the augmented matrix and row reduce to find values for c_1, c_2, and c_3.

$$\left[\begin{array}{ccc|c} 2 & 1 & 3 & 0 \\ 1 & -1 & 9 & 0 \\ -3 & -2 & -2 & 0 \end{array}\right] \longrightarrow \left[\begin{array}{ccc|c} 1 & 0 & 4 & 0 \\ 0 & 1 & -5 & 0 \\ 0 & 0 & 0 & 0 \end{array}\right]$$

Since $c_1 = -4c_3$, $c_2 = 5c_3$ is a solution, the vectors are linearly dependent.

Setting $c_3 = -1$ yields the dependence relation: $4\begin{bmatrix} 2 \\ 1 \\ -3 \end{bmatrix} - 5\begin{bmatrix} 1 \\ -1 \\ -2 \end{bmatrix} - \begin{bmatrix} 3 \\ 9 \\ -2 \end{bmatrix} = \begin{bmatrix} 0 \\ 0 \\ 0 \end{bmatrix}$.

13. We use Exercise 21 of Section 2.3 to determine whether or not $\mathbb{R}^3 = \text{span}(\mathbf{u}, \mathbf{v}, \mathbf{w})$.

(a) We need only show $\mathbf{e}_1$, $\mathbf{e}_2$, $\mathbf{e}_3$ are linear combinations of $\mathbf{u}$, $\mathbf{v}$, and $\mathbf{w}$.
Why is that enough? Because then we have $\mathbb{R}^3 = \text{span}(\mathbf{e}_1, \mathbf{e}_2, \mathbf{e}_3) = \text{span}(\mathbf{u}, \mathbf{v}, \mathbf{w})$.
We begin with $\mathbf{e}_1$. We want to find scalars x, y, and z such that:

$$x\begin{bmatrix}1\\1\\0\end{bmatrix} + y\begin{bmatrix}1\\0\\1\end{bmatrix} + z\begin{bmatrix}0\\1\\1\end{bmatrix} = \mathbf{e}_1 = \begin{bmatrix}1\\0\\0\end{bmatrix}$$

We form the augmented matrix and row reduce to find values for x, y, and z.

$$\left[\begin{array}{ccc|c}1&1&0&1\\1&0&1&0\\0&1&1&0\end{array}\right] \longrightarrow \left[\begin{array}{ccc|c}1&0&0&\frac{1}{2}\\0&1&0&\frac{1}{2}\\0&0&1&-\frac{1}{2}\end{array}\right] \Rightarrow \mathbf{e}_1 = \tfrac{1}{2}\mathbf{u} + \tfrac{1}{2}\mathbf{v} - \tfrac{1}{2}\mathbf{w}.$$

Likewise for $\mathbf{e}_2$, we form the augmented matrix and row reduce.

$$\left[\begin{array}{ccc|c}1&1&0&0\\1&0&1&1\\0&1&1&0\end{array}\right] \longrightarrow \left[\begin{array}{ccc|c}1&0&0&\frac{1}{2}\\0&1&0&-\frac{1}{2}\\0&0&1&\frac{1}{2}\end{array}\right] \Rightarrow \mathbf{e}_1 = \tfrac{1}{2}\mathbf{u} - \tfrac{1}{2}\mathbf{v} + \tfrac{1}{2}\mathbf{w}.$$

Finally for $\mathbf{e}_3$, we form the augmented matrix and row reduce.

$$\left[\begin{array}{ccc|c}1&1&0&0\\1&0&1&0\\0&1&1&1\end{array}\right] \longrightarrow \left[\begin{array}{ccc|c}1&0&0&-\frac{1}{2}\\0&1&0&\frac{1}{2}\\0&0&1&\frac{1}{2}\end{array}\right] \Rightarrow \mathbf{e}_1 = -\tfrac{1}{2}\mathbf{u} + \tfrac{1}{2}\mathbf{v} + \tfrac{1}{2}\mathbf{w}.$$

What do we conclude? $\mathbb{R}^3 = \text{span}(\mathbf{e}_1, \mathbf{e}_2, \mathbf{e}_3) = \text{span}(\mathbf{u}, \mathbf{v}, \mathbf{w})$.
We could also have used our intuition to solve this problem. How?

$$\begin{bmatrix}1\\1\\0\end{bmatrix} + \begin{bmatrix}1\\0\\1\end{bmatrix} + \begin{bmatrix}0\\1\\1\end{bmatrix} = \begin{bmatrix}2\\2\\2\end{bmatrix}. \text{ So } \tfrac{1}{2}\begin{bmatrix}2\\2\\2\end{bmatrix} = \begin{bmatrix}1\\1\\1\end{bmatrix} \Rightarrow \begin{bmatrix}1\\1\\1\end{bmatrix} - \begin{bmatrix}1\\1\\0\end{bmatrix} = \begin{bmatrix}0\\0\\1\end{bmatrix} = \mathbf{e}_3 \ldots$$

(b) These are clearly linearly dependent because $\mathbf{w} = \mathbf{u} + \mathbf{v}$.
Therefore, $\mathbb{R}^3$ is not equal to $\text{span}(\mathbf{u}, \mathbf{v}, \mathbf{w})$.
In order for a set to span $\mathbb{R}^3$ is must have 3 linearly independent vectors.

14. By Exercise 13, if $\mathbf{a}_1$, $\mathbf{a}_2$, and $\mathbf{a}_3$ are linearly independent, then $A \longrightarrow I_3$.

Note I_3 is the 3×3 identity matrix. That is $I_3 = \begin{bmatrix}1&0&0\\0&1&0\\0&0&1\end{bmatrix}$.

Furthermore, it is obvious that conditions (a), (b), and (c) are equivalent.
$A \longrightarrow I_3 \Rightarrow \text{rank}(A) = 3 \Rightarrow$ the system $[A \mid \mathbf{b}]$ has a unique solution.
Finally, if $[A \mid \mathbf{b}]$ has a unique solution, then $A \longrightarrow I_3$.

15. Since $\mathbf{a}_1$, $\mathbf{a}_2$, and $\mathbf{a}_3$ are linearly dependent, $\text{rank}(A) \leq 2 < 3$.
Since $\mathbf{a}_1$, $\mathbf{a}_2$, and $\mathbf{a}_3$ are not zero, $0 < \text{rank}(A)$.
Combining these two conditions, we see that $\text{rank}(A)$ must be 1 or 2.

16. By Theorem 2.8 of Section 2.3, the rank of any 5×3 matrix can be at most 3. Why?
Since the rows are vectors in $\mathbb{R}^3$, any set of 4 or more of them are linearly dependent.
What does that tell us? When we row reduce the matrix, we will create at least 2 zero rows.

There is no minimum rank. Why? Any matrix whose entries are all 0 has a rank of 0.

17. As in Exercise 43 of Section 2.3, we apply the definition of linear independence.

We will show that $\mathbf{u} + \mathbf{v}$ and $\mathbf{u} - \mathbf{v}$ are linearly independent.
Given $c_1(\mathbf{u} + \mathbf{v}) + c_2(\mathbf{u} - \mathbf{v}) = 0$, we will show $c_1 = c_2 = 0$.

Multiplying and gathering like terms yields: $(c_1 + c_2)\mathbf{u} + (c_1 - c_2)\mathbf{v} = 0$.
Since $\mathbf{u}$ and $\mathbf{v}$ are linearly independent, $c_1 + c_2 = c_1 - c_2 = 0 \Rightarrow c_1 = c_2 = 0$.

Also we could create the matrix of coefficients A and row reduce to determine its rank:

$$\begin{array}{rcl} c_1 + c_2 &=& 0 \\ c_1 - c_2 &=& 0 \end{array} \Rightarrow \begin{bmatrix} 1 & 1 \\ 1 & -1 \end{bmatrix} \longrightarrow \begin{bmatrix} 1 & 0 \\ 0 & 1 \end{bmatrix}$$

Since $\text{rank}(A) = 2$ the only solution is the trivial one, so $c_1 = c_2 = 0$.

18. We need only show that $\mathbf{v}$ can be written as a linear combination of $\mathbf{u}$ and $\mathbf{u} + \mathbf{v}$.
But this is obvious since $\mathbf{v} = -1(\mathbf{u}) + 1(\mathbf{u} + \mathbf{v})$.

Why is this enough?

What do we conclude?

We conclude that $\text{span}(\mathbf{u}, \mathbf{v}) = \text{span}(\mathbf{u}, \mathbf{u} + \mathbf{v})$.

19. In order for $[A \mid \mathbf{b}]$ to be consistent, $\text{rank}(A)$ must equal $\text{rank}([A \mid \mathbf{b}])$. Why?
A system has no solution when A has a zero row with corresponding constant $\neq 0$.

Note that $\text{rank}([A \mid \mathbf{b}])$ cannot be less than $\text{rank}(A)$. Why not?

20. By the definition in Section 2.2, A and B are row equivalent if $A \longrightarrow B$.
In this case, we note that $A \longrightarrow I_3$ and $B \longrightarrow I_3$. What does that imply?
Since A and B are row equivalent to a common matrix (in this case, I_3),
we conclude that they are row equivalent to each other.

Why?

The steps we take when we row reduce are completely reversible.

So, we have $A \longrightarrow I_3 \longrightarrow B$.

Chapter 3

Matrices

3.1 Matrix Operations

1. Following Examples 3.1 through 3.5, we have:

$$A+2D=\begin{bmatrix}3&0\\-1&5\end{bmatrix}+2\begin{bmatrix}0&-3\\-2&1\end{bmatrix}=\begin{bmatrix}3+2(0)&0+2(-3)\\-1+2(-2)&5+2(1)\end{bmatrix}=\begin{bmatrix}3&-6\\-5&7\end{bmatrix}.$$

2. Following Examples 3.1 through 3.5, we have:

$$3D-2A=3\begin{bmatrix}0&-3\\-2&1\end{bmatrix}-2\begin{bmatrix}3&0\\-1&5\end{bmatrix}=\begin{bmatrix}3(0)-2(3)&3(-3)-2(0)\\3(-2)-2(-1)&3(1)-2(5)\end{bmatrix}=\begin{bmatrix}-6&-9\\-4&-7\end{bmatrix}.$$

3. $B-C$ is not possible. Why not? B is a 2×3 matrix and C is a 3×2 matrix.
   We can only add and subtract matrices of the same size.

4. Applying the definition the transpose to C^T, the transpose of matrix C, we have:

$$\text{Since } C=\begin{bmatrix}1&2\\3&4\\5&6\end{bmatrix},\ C^T=\begin{bmatrix}1&3&5\\2&4&6\end{bmatrix},\ B-C^T=\begin{bmatrix}4&-2&1\\0&2&3\end{bmatrix}-\begin{bmatrix}1&3&5\\2&4&6\end{bmatrix}=\begin{bmatrix}3&-5&-4\\-2&-2&-3\end{bmatrix}.$$

5. By the definition of the matrix product, $C=AB$, and Example 3.6, we have: $AB=$

$$\begin{bmatrix}3&0\\-1&5\end{bmatrix}\begin{bmatrix}4&-2&1\\0&2&3\end{bmatrix}=\begin{bmatrix}3(4)+0(0)&3(-2)+0(2)&3(1)+0(3)\\(-1)(4)+5(0)&3(-2)+(-1)(2)&3(1)+(-1)(3)\end{bmatrix}=\begin{bmatrix}12&-6&3\\-4&12&14\end{bmatrix}.$$

6. From the remarks following the definition of matrix multiplication, we see BD is not possible.
   Why not? B is a 2×3 matrix and D is a 2×2 matrix. What does that tell us?
   The number of columns in $B=3\neq 2=$ the number of rows in D.
   For matrix multiplication, the number of columns in B has to equal the number of rows in D.

   In the future, when checking whether or not matrix multiplication is possible, we will write:
   B is $[2\times 3]$ and D is $[2\times 2]$, so BD is $[2\times 3][2\times 2]$ which is not possible
   because the *inner* numbers do not match.

   Q: Do they outer numbers have to match? If not, what do they tell us?
   A: No, they do not have to match. The *outer* numbers tell us the size, $[r\times c]$, of the result.

7. We begin by applying the definition of matrix multiplication to see if BC is possible:
   Since B is $[2\times 3]$ and C is $[3\times 2]$, BC, $[2\times 3][3\times 2]$, is possible. Why?
   Because the *inner* numbers match. What does that tell us?
   The number of columns in $B=3=$ the number of rows in C.

   Furthermore, since BC is $[2\times 3][3\times 2]$, BC will be a 2×2 matrix.
   Since D is also a 2×2 matrix, we can add them together. That is, $D+BC$ is possible.

$$\text{First, } BC=\begin{bmatrix}4&-2&1\\0&2&3\end{bmatrix}\begin{bmatrix}1&2\\3&4\\5&6\end{bmatrix}=\begin{bmatrix}4(1)-2(3)+1(5)&4(2)-2(4)+1(6)\\0(1)+2(3)+3(5)&0(2)+2(4)+3(6)\end{bmatrix}=\begin{bmatrix}3&6\\21&26\end{bmatrix}.$$

$$\text{So, } D+BC=\begin{bmatrix}0&-3\\-2&1\end{bmatrix}+\begin{bmatrix}3&6\\21&26\end{bmatrix}=\begin{bmatrix}3&3\\19&27\end{bmatrix}.$$

8. We should note that B^TB is *always* possible. Why?
the number of columns of B^T = the number of rows of B (by the definition of B^T).

Since $B = \begin{bmatrix} 4 & -2 & 1 \\ 0 & 2 & 3 \end{bmatrix}$ and $B^T = \begin{bmatrix} 4 & 0 \\ -2 & 2 \\ 1 & 3 \end{bmatrix}$,

$$B^TB = \begin{bmatrix} 4 & 0 \\ -2 & 2 \\ 1 & 3 \end{bmatrix} \begin{bmatrix} 4 & -2 & 1 \\ 0 & 2 & 3 \end{bmatrix} = \begin{bmatrix} 4(4)+0(0) & 4(-2)+0(2) & 4(1)+0(3) \\ (-2)(4)+2(0) & (-2)(-2)+2(2) & (-2)(1)+2(3) \\ 1(4)+3(0) & 1(-2)+3(2) & 1(1)+3(3) \end{bmatrix}$$

$$= \begin{bmatrix} 16 & -8 & 4 \\ -8 & 8 & 4 \\ 4 & 4 & 10 \end{bmatrix}.$$ Is BB^T always possible as well? Why or why not?

9. Before we begin, we should determine if AF and $E(AF)$ are possible.

Since A is $[2 \times 2]$ and F is $[2 \times 1]$, AF, $[2 \times 2][2 \times 1]$, is possible. Why?
Because the *inner* numbers match. What does that tell us?
The number of columns in $A = 2 =$ the number of rows in F.
Furthermore, since AF is $[2 \times 2][2 \times 1]$, AF will be a 2×1 matrix.

Since E is $[1 \times 2]$ and AF is $[2 \times 1]$, $E(AF)$, $[1 \times 2][2 \times 1]$, is possible. Why?
Because the *inner* numbers match. What does that tell us?
The number of columns in $E = 1 =$ the number of rows in AF.
Furthermore, since $E(AF)$ is $[1 \times 2][2 \times 1]$, $E(AF)$ will be a 1×1 matrix.

First, $AF = \begin{bmatrix} 3 & 0 \\ -1 & 5 \end{bmatrix} \begin{bmatrix} -1 \\ 2 \end{bmatrix} = \begin{bmatrix} 3(-1)+0(2) \\ (-1)(-1)+5(2) \end{bmatrix} = \begin{bmatrix} -3 \\ 11 \end{bmatrix}$.

So $E(AF) = [4 \;\; 2] \begin{bmatrix} -3 \\ 11 \end{bmatrix} = [4(-3)+2(11)] = [10]$.

10. Before we begin, we should determine if DF and $F(DF)$ are possible.

Since D is $[2 \times 2]$ and F is $[2 \times 1]$, DF, $[2 \times 2][2 \times 1]$, is possible. Why?
Because the *inner* numbers match. What does that tell us?
The number of columns in $D = 2 =$ the number of rows in F.
Furthermore, since DF is $[2 \times 2][2 \times 1]$, DF will be a 2×1 matrix.

Since F is $[2 \times 1]$ and DF is $[2 \times 1]$, $F(DF)$, $[2 \times 1][2 \times 1]$, is not possible. Why?
Because the *inner* numbers do not match. What does that tell us?
The number of columns in $F = 1 \neq 2 =$ the number of rows in DF.

So, $F(DF)$ is not possible.

11. Since FE, $[2 \times 1][1 \times 2]$, is possible and yields a 2×2 matrix, we have:

$$FE = \begin{bmatrix} -1 \\ 2 \end{bmatrix} [4\ 2] = \begin{bmatrix} -1(4) & (-1)(2) \\ 2(4) & 2(2) \end{bmatrix} = \begin{bmatrix} -4 & -2 \\ 8 & 4 \end{bmatrix}.$$

Q: Does FE equal EF?
A: No. In fact, note that EF, $[1 \times 2][2 \times 1]$, is possible and yields a 1×1 matrix.
This is a good example of the general fact that matrix multiplication does not commute.

12. Since EF, $[1 \times 2][2 \times 1]$, is possible and yields a 1×1 matrix, we have:

$$EF = [4\ 2] \begin{bmatrix} -1 \\ 2 \end{bmatrix} = [4(-1) + 2(2)] = [0].$$

Since Exercises 13 through 16 use the skills above, we simply present the answers below.

13. $B^T C^T - (CB)^T = \begin{bmatrix} 0 & 0 & 0 \\ 0 & 0 & 0 \\ 0 & 0 & 0 \end{bmatrix}.$

14. $DA - AD = \begin{bmatrix} 3 & -6 \\ 3 & -3 \end{bmatrix}.$

15. $A^3 = \begin{bmatrix} 27 & 0 \\ -49 & 125 \end{bmatrix}.$

16. $(I_2 - D)^2 = \begin{bmatrix} 7 & 3 \\ 2 & 6 \end{bmatrix}.$

17. We should use this exercise to increase our understanding of matrix multiplication.
To help us find an example, we should first take a look at the general pattern.

$$A^2 = AA = \begin{bmatrix} a & b \\ c & d \end{bmatrix} \begin{bmatrix} a & b \\ c & d \end{bmatrix} = \begin{bmatrix} a^2 + bc & ab + bd \\ ac + cd & bc + d^2 \end{bmatrix} = \begin{bmatrix} 0 & 0 \\ 0 & 0 \end{bmatrix}.$$ What do we have?

We have $ac + cd = 0 \Rightarrow ac = -cd \Rightarrow a = -d$, provided $c \neq 0$. Let $a = 1$.
Then $d = -1 \Rightarrow 1 + bc = 0 \Rightarrow bc = -1$, so $b = 1$ and $c = -1$ should work.

Check: $A^2 = AA = \begin{bmatrix} 1 & 1 \\ -1 & -1 \end{bmatrix} \begin{bmatrix} 1 & 1 \\ -1 & -1 \end{bmatrix} = \begin{bmatrix} 1-1 & 1-1 \\ -1+1 & -1+1 \end{bmatrix} = \begin{bmatrix} 0 & 0 \\ 0 & 0 \end{bmatrix}.$

So, $A = \begin{bmatrix} 1 & 1 \\ -1 & -1 \end{bmatrix}$ works.

Q: What other choices for a, b, c, and d work? Is there a pattern we can see?
A: Any matrix all of whose entries have the same absolute value where $\mathbf{A}_1 = -\mathbf{A}_2$ will work.

18. We should use this exercise to increase our understanding of matrix multiplication.
We begin by noting any essential patterns or features of A: $\mathbf{a}_1 = 2\mathbf{a}_2$.
Now multiply by simple matrices to see how to use this pattern.

$AB = \begin{bmatrix} 2 & 1 \\ 6 & 3 \end{bmatrix} \begin{bmatrix} 1 & 0 \\ 0 & 0 \end{bmatrix} = \begin{bmatrix} 2 & 0 \\ 6 & 0 \end{bmatrix}$. So, $\mathbf{b}_1 = 1\mathbf{a}_1 + 0\mathbf{a}_2 = \mathbf{a}_1$ and $\mathbf{b}_2 = 0\mathbf{a}_1 + 0\mathbf{a}_2 = \mathbf{0}$.

$AD = \begin{bmatrix} 2 & 1 \\ 6 & 3 \end{bmatrix} \begin{bmatrix} 0 & 0 \\ 1 & 0 \end{bmatrix} = \begin{bmatrix} 1 & 0 \\ 3 & 0 \end{bmatrix}$. So, $\mathbf{d}_1 = 0\mathbf{a}_1 + 1\mathbf{a}_2 = \mathbf{a}_2$ and $\mathbf{d}_2 = 0\mathbf{a}_1 + 0\mathbf{a}_2 = \mathbf{0}$.

What does this suggest? $AC = \begin{bmatrix} 2 & 1 \\ 6 & 3 \end{bmatrix} \begin{bmatrix} 0 & 0 \\ 2 & 0 \end{bmatrix} = \begin{bmatrix} 2 & 0 \\ 6 & 0 \end{bmatrix}$. So, $AB = AC$ but $B \neq C$.

19. The number of units of each product shipped to each warehouse is given by $A = \begin{bmatrix} 200 & 75 \\ 150 & 100 \\ 100 & 125 \end{bmatrix}$.

The cost of shipping one unit of each product is given by $B = \begin{bmatrix} 1.50 & 1.00 & 2.00 \\ 1.75 & 1.50 & 1.00 \end{bmatrix}$

(where b_{ij} is the cost of shipping a unit of product j by $i = 1$ truck, $i = 2$ train).

Compare the cost of shipping the products to each of the warehouses:

$$BA = \begin{bmatrix} 1.50 & 1.00 & 2.00 \\ 1.75 & 1.50 & 1.00 \end{bmatrix} \begin{bmatrix} 200 & 75 \\ 150 & 100 \\ 100 & 125 \end{bmatrix} = \begin{bmatrix} 650.00 & 462.50 \\ 675.00 & 406.25 \end{bmatrix} \Rightarrow$$

It is cheaper to ship the products to warehouse 1 by truck, but to warehouse 2 by train.

20. The costs of distributing one unit of product is given by $C = \begin{bmatrix} 0.75 & 0.75 & 0.75 \\ 1.00 & 1.00 & 1.00 \end{bmatrix}$

(where c_{ij} is the cost of shipping one unit of product j from warehouse i).

The cost of distribution is then given by:

$$CA^T = \begin{bmatrix} 0.75 & 0.75 & 0.75 \\ 1.00 & 1.00 & 1.00 \end{bmatrix} \begin{bmatrix} 200 & 75 \\ 150 & 100 \\ 100 & 125 \end{bmatrix} = \begin{bmatrix} 337.50 & 225.00 \\ 450.00 & 300.00 \end{bmatrix} \Rightarrow$$

It costs \$337.50 to ship all the products from warehouse 1, \$300.00 from warehouse 2.

21. $\begin{bmatrix} 1 & -2 & 3 \\ 2 & 1 & -5 \end{bmatrix} \begin{bmatrix} x_1 \\ x_2 \\ x_3 \end{bmatrix} = \begin{bmatrix} 0 \\ 4 \end{bmatrix}$.

22. $\begin{bmatrix} -1 & 0 & 2 \\ 1 & -1 & 0 \\ 0 & 1 & 1 \end{bmatrix} \begin{bmatrix} x_1 \\ x_2 \\ x_3 \end{bmatrix} = \begin{bmatrix} 1 \\ -2 \\ -1 \end{bmatrix}$.

23. $A\mathbf{b}_1 = 2\begin{bmatrix} 1 \\ -3 \\ 2 \end{bmatrix} + \begin{bmatrix} 0 \\ 1 \\ 0 \end{bmatrix} - \begin{bmatrix} -2 \\ 1 \\ -1 \end{bmatrix} = \begin{bmatrix} 4 \\ -6 \\ 5 \end{bmatrix}$, $A\mathbf{b}_2 = 3\begin{bmatrix} 1 \\ -3 \\ 2 \end{bmatrix} - \begin{bmatrix} 0 \\ 1 \\ 0 \end{bmatrix} + 6\begin{bmatrix} -2 \\ 1 \\ -1 \end{bmatrix} = \begin{bmatrix} -9 \\ -4 \\ 0 \end{bmatrix}$,

and $A\mathbf{b}_3 = \begin{bmatrix} 0 \\ 1 \\ 0 \end{bmatrix} + 4\begin{bmatrix} -2 \\ 1 \\ -1 \end{bmatrix} = \begin{bmatrix} -8 \\ 5 \\ -4 \end{bmatrix}$. Therefore, $AB = \begin{bmatrix} 4 & -9 & -8 \\ -6 & -4 & 5 \\ 5 & 0 & -4 \end{bmatrix}$.

24. $\mathbf{A}_1B = \begin{bmatrix} 2 & 3 & 0 \end{bmatrix} - 2\begin{bmatrix} -1 & 6 & 4 \end{bmatrix} = \begin{bmatrix} 4 & -9 & -8 \end{bmatrix}$,

$\mathbf{A}_2B = -3\begin{bmatrix} 2 & 3 & 0 \end{bmatrix} + \begin{bmatrix} 1 & -1 & 1 \end{bmatrix} + \begin{bmatrix} -1 & 6 & 4 \end{bmatrix} = \begin{bmatrix} -6 & -4 & 5 \end{bmatrix}$, and

$\mathbf{A}_3B = 2\begin{bmatrix} 2 & 3 & 0 \end{bmatrix} - \begin{bmatrix} -1 & 6 & 4 \end{bmatrix} = \begin{bmatrix} 5 & 0 & -4 \end{bmatrix}$. Thus, $AB = \begin{bmatrix} 4 & -9 & -8 \\ -6 & -4 & 5 \\ 5 & 0 & -4 \end{bmatrix}$.

25. The outer product expansion of AB is

$$\mathbf{a}_1\mathbf{B}_1 + \mathbf{a}_2\mathbf{B}_2 + \mathbf{a}_3\mathbf{B}_3 = \begin{bmatrix} 2 & 3 & 0 \\ -6 & -9 & 0 \\ 4 & 6 & 0 \end{bmatrix} + \begin{bmatrix} 0 & 0 & 0 \\ 1 & -1 & 1 \\ 0 & 0 & 0 \end{bmatrix} + \begin{bmatrix} 2 & -12 & -8 \\ -1 & 6 & 4 \\ 1 & -6 & -4 \end{bmatrix}$$

$$= \begin{bmatrix} 4 & -9 & -8 \\ -6 & -4 & 5 \\ 5 & 0 & -4 \end{bmatrix}.$$

26. $B\mathbf{a}_1 = 1\begin{bmatrix} 2 \\ 1 \\ -1 \end{bmatrix} - 3\begin{bmatrix} 3 \\ -1 \\ 6 \end{bmatrix} + 2\begin{bmatrix} 0 \\ 1 \\ 4 \end{bmatrix} = \begin{bmatrix} -7 \\ 6 \\ -11 \end{bmatrix}$, $B\mathbf{a}_2 = \begin{bmatrix} 3 \\ -1 \\ 6 \end{bmatrix} = \begin{bmatrix} 3 \\ -1 \\ 6 \end{bmatrix}$, and

$B\mathbf{a}_3 = -2\begin{bmatrix} 2 \\ 1 \\ -1 \end{bmatrix} + \begin{bmatrix} 3 \\ -1 \\ 6 \end{bmatrix} - \begin{bmatrix} 0 \\ 1 \\ 4 \end{bmatrix} = \begin{bmatrix} -1 \\ -4 \\ 4 \end{bmatrix}$. So, $BA = \begin{bmatrix} -7 & 3 & -1 \\ 6 & -1 & -4 \\ -11 & 6 & 4 \end{bmatrix}$.

27. $\mathbf{B}_1 A = 2\begin{bmatrix} 1 & 0 & -2 \end{bmatrix} + 3\begin{bmatrix} -3 & 1 & 1 \end{bmatrix} = \begin{bmatrix} -7 & 3 & -1 \end{bmatrix}$,

$\mathbf{B}_2 A = 1\begin{bmatrix} 1 & 0 & -2 \end{bmatrix} - \begin{bmatrix} -3 & 1 & 1 \end{bmatrix} + \begin{bmatrix} 2 & 0 & -1 \end{bmatrix} = \begin{bmatrix} 6 & -1 & -4 \end{bmatrix}$, and

$\mathbf{B}_3 A = -\begin{bmatrix} 1 & 0 & -2 \end{bmatrix} + 6\begin{bmatrix} -3 & 1 & 1 \end{bmatrix} + 4\begin{bmatrix} 2 & 0 & -1 \end{bmatrix} = \begin{bmatrix} -11 & 6 & 4 \end{bmatrix}$.

Therefore, $BA = \begin{bmatrix} -7 & 3 & -1 \\ 6 & -1 & -4 \\ -11 & 6 & 4 \end{bmatrix}$.

28. The outer product expansion of BA is

$$\mathbf{b}_1\mathbf{A}_1 + \mathbf{b}_2\mathbf{A}_2 + \mathbf{b}_3\mathbf{A}_3 = \begin{bmatrix} 2 & 0 & -4 \\ 1 & 0 & -2 \\ -1 & 0 & 2 \end{bmatrix} + \begin{bmatrix} -9 & 3 & 3 \\ 3 & -1 & -1 \\ -18 & 6 & 6 \end{bmatrix} + \begin{bmatrix} 0 & 0 & 0 \\ 2 & 0 & -1 \\ 8 & 0 & -4 \end{bmatrix}$$

$$= \begin{bmatrix} -7 & 3 & -1 \\ 6 & -1 & -4 \\ -11 & 6 & 4 \end{bmatrix}.$$

29. Assume that the columns of $B = \begin{bmatrix} \mathbf{b}_1 & \mathbf{b}_2 & \cdots & \mathbf{b}_n \end{bmatrix}$ are linearly dependent $\Rightarrow$
There exists a solution to $x_1\mathbf{b}_1 + x_2\mathbf{b}_2 + \cdots + x_n\mathbf{b}_n = \mathbf{0}$ (at least one $x_i \neq 0$).
Now consider the partition of AB in terms of column vectors:
$AB = A\begin{bmatrix} \mathbf{b}_1 & \mathbf{b}_2 & \cdots & \mathbf{b}_n \end{bmatrix} = \begin{bmatrix} A\mathbf{b}_1 & A\mathbf{b}_2 & \cdots & A\mathbf{b}_n \end{bmatrix}$.
But then $A(x_1\mathbf{b}_1 + x_2\mathbf{b}_2 + \cdots + x_n\mathbf{b}_n) = x_1(A\mathbf{b}_1) + x_2(A\mathbf{b}_2) + \cdots + x_n(A\mathbf{b}_n) = 0$,
showing that the columns of AB are linearly dependent.

30. Let $A = \begin{bmatrix} \mathbf{a}_1 \\ \mathbf{a}_2 \\ \vdots \\ \mathbf{a}_n \end{bmatrix}$ and $AB = \begin{bmatrix} \mathbf{a}_1 B \\ \mathbf{a}_2 B \\ \vdots \\ \mathbf{a}_n B \end{bmatrix}$. Then assume there exist x_i not all zero such that:

$x_1\mathbf{a}_1 + x_2\mathbf{a}_2 + \cdots + x_n\mathbf{a}_n = \mathbf{0}$, that is, the rows of A are linearly dependent.

So, we have: $(x_1\mathbf{a}_1 + x_2\mathbf{a}_2 + \cdots + x_n\mathbf{a}_n)B = x_1(\mathbf{a}_1 B) + x_2(\mathbf{a}_2 B) + \cdots + x_n(\mathbf{a}_n B) = 0 \Rightarrow$
The rows of AB are linearly dependent.

31. For matrices A, B we have the block structure $A = \begin{bmatrix} A_{11} & A_{12} \\ A_{21} & A_{22} \end{bmatrix}$ and $B = \begin{bmatrix} B_{11} & B_{12} \\ B_{21} & B_{22} \end{bmatrix} \Rightarrow$

$$AB = \begin{bmatrix} A_{11} & A_{12} \\ A_{21} & A_{22} \end{bmatrix}\begin{bmatrix} B_{11} & B_{12} \\ B_{21} & B_{22} \end{bmatrix} = \begin{bmatrix} A_{11}B_{11} + A_{12}B_{21} & A_{11}B_{12} + A_{12}B_{22} \\ A_{21}B_{11} + A_{22}B_{21} & A_{21}B_{12} + A_{22}B_{22} \end{bmatrix}$$

$$= \begin{bmatrix} \begin{bmatrix} 1 & -1 \\ 0 & 1 \end{bmatrix}\begin{bmatrix} 2 & 3 \\ -1 & 1 \end{bmatrix} + \begin{bmatrix} 0 & 0 \\ 0 & 0 \end{bmatrix}\begin{bmatrix} 0 & 0 \\ 0 & 0 \end{bmatrix} & \begin{bmatrix} 1 & -1 \\ 0 & 1 \end{bmatrix}\begin{bmatrix} 0 \\ 0 \end{bmatrix} + \begin{bmatrix} 0 & 0 \\ 0 & 0 \end{bmatrix}\begin{bmatrix} 1 \\ 1 \end{bmatrix} \\ \begin{bmatrix} 0 & 0 \end{bmatrix}\begin{bmatrix} 2 & 3 \\ -1 & 1 \end{bmatrix} + \begin{bmatrix} 2 & 3 \end{bmatrix}\begin{bmatrix} 0 & 0 \\ 0 & 0 \end{bmatrix} & \begin{bmatrix} 0 & 0 \end{bmatrix}\begin{bmatrix} 0 \\ 0 \end{bmatrix} + \begin{bmatrix} 2 & 3 \end{bmatrix}\begin{bmatrix} 1 \\ 1 \end{bmatrix} \end{bmatrix}$$

$$= \begin{bmatrix} 3 & 2 & 0 \\ -1 & 1 & 0 \\ 0 & 0 & 5 \end{bmatrix}.$$

32. $AB = \begin{bmatrix} 1 & 7 & 7 \\ -2 & 7 & 7 \end{bmatrix}$.

33. $AB = \begin{bmatrix} 1 & 2 & 2 & 0 \\ 3 & 4 & 5 & 3 \\ 1 & 0 & 1 & 2 \\ 0 & 1 & 0 & -1 \end{bmatrix}$.

34. $AB = \begin{bmatrix} 2 & 3 & 4 & 0 \\ 2 & 3 & 6 & -1 \\ 3 & 3 & 4 & -2 \\ 4 & 4 & 4 & -4 \end{bmatrix}$.

35. (a) Computing the powers of matrix A as required, we have:

$$A^2 = \begin{bmatrix} -1 & 1 \\ -1 & 0 \end{bmatrix}, A^3 = \begin{bmatrix} -1 & 0 \\ 0 & -1 \end{bmatrix}, A^4 = \begin{bmatrix} 0 & -1 \\ 1 & -1 \end{bmatrix},$$

$$A^5 = \begin{bmatrix} 1 & -1 \\ 1 & 0 \end{bmatrix}, A^6 = \begin{bmatrix} 1 & 0 \\ 0 & 1 \end{bmatrix}, A^7 = \begin{bmatrix} 0 & 1 \\ -1 & 1 \end{bmatrix} = A!$$

(b) From our work in (a), we see that $A^1 = A^7 = A^{1 \cdot 6 + 1}$.

So the powers of A that actually create *distinct* matrices act like $\mathbb{Z}^6$.

See Section 1.4, Examples 1.32 through 1.35.

So, to determine A^{2001}, we should first determine the value of 2001 in $\mathbb{Z}^6$.

How? Divide 2001 by 6 and look at the remainder: $2001 = 333 \cdot 6 + 3 = 3$ in $\mathbb{Z}^6$.

Therefore $A^{2001} = A^{333 \cdot 6 + 3} = A^3 = \begin{bmatrix} -1 & 0 \\ 0 & -1 \end{bmatrix}$.

36. As in Exercise 35, we compute the powers of B and look for patterns:

Since $B = \begin{bmatrix} \frac{1}{\sqrt{2}} & -\frac{1}{\sqrt{2}} \\ \frac{1}{\sqrt{2}} & \frac{1}{\sqrt{2}} \end{bmatrix}$ and $B^2 = \begin{bmatrix} 0 & -1 \\ 1 & 0 \end{bmatrix}$, $B^4 = (B^2)^2 = \begin{bmatrix} -1 & 0 \\ 0 & 1 \end{bmatrix}$.

So $B^8 = (B^4)^2 = \begin{bmatrix} 1 & 0 \\ 0 & 1 \end{bmatrix} = I \Rightarrow B^9 = B^8 \cdot B = I \cdot B = B$.

Since $2001 = 250 \cdot 8 + 1 = 1$ in $\mathbb{Z}^8$, so $B^{2001} = B^{250 \cdot 8+1} = B^1 = B$.

37. Given $A = \begin{bmatrix} 1 & 1 \\ 0 & 1 \end{bmatrix}$, we will prove $A^n = \begin{bmatrix} 1 & n \\ 0 & 1 \end{bmatrix}$ for $n \geq 1$ using *induction.*

See Appendix B for discussion and examples of *Mathematical Induction.*

1: $A^1 = \begin{bmatrix} 1 & 1 \\ 0 & 1 \end{bmatrix}$. This is obvious, so there is nothing to show.

n: $A^n = \begin{bmatrix} 1 & n \\ 0 & 1 \end{bmatrix}$. This is the induction hypothesis.

$n+1$: $A^{n+1} = \begin{bmatrix} 1 & n+1 \\ 0 & 1 \end{bmatrix}$. This is the statement we must prove using the induction hypothesis.

$$A^{n+1} = A^1 A^n \overset{\text{by induction}}{=} \begin{bmatrix} 1 & 1 \\ 0 & 1 \end{bmatrix} \begin{bmatrix} 1 & n \\ 0 & 1 \end{bmatrix} \overset{\text{by matrix multiplication}}{=} \begin{bmatrix} 1 & n+1 \\ 0 & 1 \end{bmatrix}$$

We have proven (by induction) that $A^n = \begin{bmatrix} 1 & n \\ 0 & 1 \end{bmatrix}$ for $n \geq 1$.

38. We will prove (b) using *induction.* See Appendix B for discussion and examples. We make use of the following trigonometric identities in our proof below:
$\cos\theta\cos n\theta - \sin\theta\sin n\theta = \cos(n+1)\theta$ and $\sin\theta\cos n\theta + \cos\theta\sin n\theta = \sin(n+1)\theta$.

(a) $A^2 = \begin{bmatrix} \cos\theta & -\sin\theta \\ \sin\theta & \cos\theta \end{bmatrix}\begin{bmatrix} \cos\theta & -\sin\theta \\ \sin\theta & \cos\theta \end{bmatrix} = \begin{bmatrix} \cos^2\theta - \sin^2\theta & -2\cos\theta\sin\theta \\ 2\cos\theta\sin\theta & \cos^2\theta - \sin^2\theta \end{bmatrix}$.

But $\cos^2\theta - \sin^2\theta = \cos 2\theta$, and $2\cos\theta\sin\theta = \sin 2\theta$, so $A^2 = \begin{bmatrix} \cos 2\theta & -\sin 2\theta \\ \sin 2\theta & \cos 2\theta \end{bmatrix}$.

(b) We will show that $A^n = \begin{bmatrix} \cos n\theta & -\sin n\theta \\ \sin n\theta & \cos n\theta \end{bmatrix}$ for $n \geq 1$ by induction.

1: $A^1 = \begin{bmatrix} \cos\theta & -\sin\theta \\ \sin\theta & \cos\theta \end{bmatrix}$. This is obvious, so there is nothing to show.

n: $A^n = \begin{bmatrix} \cos n\theta & -\sin n\theta \\ \sin n\theta & \cos n\theta \end{bmatrix}$. This is the induction hypothesis.

$n+1$: $A^{n+1} = \begin{bmatrix} \cos(n+1)\theta & -\sin(n+1)\theta \\ \sin(n+1)\theta & \cos(n+1)\theta \end{bmatrix}$.
This is the statement we must prove using the induction hypothesis.

$$A^{n+1} = A^1A^n \overset{\text{by induction}}{=} \begin{bmatrix} \cos\theta & -\sin\theta \\ \sin\theta & \cos\theta \end{bmatrix}\begin{bmatrix} \cos n\theta & -\sin n\theta \\ \sin n\theta & \cos n\theta \end{bmatrix}$$

$$\overset{\text{by matrix multiplication}}{=} \begin{bmatrix} \cos\theta\cos n\theta - \sin\theta\sin n\theta & -(\cos\theta\sin n\theta + \sin\theta\cos n\theta) \\ \sin\theta\cos n\theta + \cos\theta\sin n\theta & \cos\theta\cos n\theta - \sin\theta\sin n\theta \end{bmatrix}$$

$$\overset{\text{by trigonometric identities}}{=} \begin{bmatrix} \cos(n+1)\theta & -\sin(n+1)\theta \\ \sin(n+1)\theta & \cos(n+1)\theta \end{bmatrix}$$

We have shown (by induction) that $A^n = \begin{bmatrix} \cos n\theta & -\sin n\theta \\ \sin n\theta & \cos n\theta \end{bmatrix}$ for $n \geq 1$.

39. (a) $A = \begin{bmatrix} 1 & -1 & 1 & -1 \\ -1 & 1 & -1 & 1 \\ 1 & -1 & 1 & -1 \\ -1 & 1 & -1 & 1 \end{bmatrix}$. (b) $A = \begin{bmatrix} 0 & 1 & 2 & 3 \\ -1 & 0 & 1 & 2 \\ -2 & -1 & 0 & 1 \\ -3 & -2 & -1 & 0 \end{bmatrix}$.

(c) $A = \begin{bmatrix} 0 & 0 & 0 & 0 \\ 1 & 1 & 1 & 1 \\ 2 & 4 & 8 & 16 \\ 3 & 9 & 27 & 81 \end{bmatrix}$. (d) $A = \begin{bmatrix} \frac{1}{2}\sqrt{2} & 1 & \frac{1}{2}\sqrt{2} & 0 \\ 1 & \frac{1}{2}\sqrt{2} & 0 & -\frac{1}{2}\sqrt{2} \\ \frac{1}{2}\sqrt{2} & 0 & -\frac{1}{2}\sqrt{2} & -1 \\ 0 & -\frac{1}{2}\sqrt{2} & -1 & -\frac{1}{2}\sqrt{2} \end{bmatrix}$.

40. (a) $A = \begin{bmatrix} 2 & 3 & 4 & 5 & 6 & 7 \\ 0 & 4 & 5 & 6 & 7 & 8 \\ 0 & 0 & 6 & 7 & 8 & 9 \\ 0 & 0 & 0 & 8 & 9 & 10 \\ 0 & 0 & 0 & 0 & 10 & 11 \\ 0 & 0 & 0 & 0 & 0 & 12 \end{bmatrix}$. (b) $A = \begin{bmatrix} 1 & 1 & 0 & 0 & 0 & 0 \\ 1 & 1 & 1 & 0 & 0 & 0 \\ 0 & 1 & 1 & 1 & 0 & 0 \\ 0 & 0 & 1 & 1 & 1 & 0 \\ 0 & 0 & 0 & 1 & 1 & 1 \\ 0 & 0 & 0 & 0 & 1 & 1 \end{bmatrix}$. (c) $A = \begin{bmatrix} 0 & 0 & 0 & 0 & 0 & 1 \\ 0 & 0 & 0 & 0 & 1 & 1 \\ 0 & 0 & 0 & 1 & 1 & 1 \\ 0 & 0 & 1 & 1 & 1 & 1 \\ 0 & 1 & 1 & 1 & 1 & 1 \\ 1 & 1 & 1 & 1 & 1 & 1 \end{bmatrix}$.

41. Let A be an $m \times n$ matrix, and $\mathbf{e}_i$ a $1 \times m$ standard unit vector.
If $a_1, a_2, \ldots, a_m$ are the rows of A then the product $\mathbf{e}_i A$ can be written
$\mathbf{e}_i A = 0 \cdot a_1 + 0 \cdot a_2 + \cdots + 1 \cdot a_i + \cdots + 0 \cdot a_m = a_i$ which is the ith row of A.
We could also prove this by direct calculation:

$$\mathbf{e}_i A = \begin{bmatrix} 0 & \cdots & 1 & \cdots & 0 \end{bmatrix} \begin{bmatrix} a_{11} & a_{12} & \cdots & a_{1n} \\ \vdots & \vdots & \vdots & \vdots \\ a_{i1} & a_{i2} & \cdots & a_{in} \\ \vdots & \vdots & \vdots & \vdots \\ a_{m1} & a_{m2} & \cdots & a_{mn} \end{bmatrix} = \begin{bmatrix} a_{i1} & a_{i2} & \cdots & a_{in} \end{bmatrix}$$

since the 1 in $\mathbf{e}_i$ is the ith entry.

3.2 Matrix Algebra

1. Following remarks prior to Example 3.16, the key assumption is matrices are the same size. Then add, subtract, and multiply (by scalars only) as in *normal* algebra.

$$X - 2A + 3B = 0 \Rightarrow X = 2A - 3B = \begin{bmatrix} 5 & 4 \\ 3 & 5 \end{bmatrix}.$$

2. Following remarks prior to Example 3.16, the key assumption is matrices are the same size. Then add, subtract, and multiply (by scalars only) as in *normal* algebra.

$$2X = A - B \Rightarrow X = \tfrac{1}{2}(A - B) = \begin{bmatrix} 1 & 1 \\ 1 & \frac{3}{2} \end{bmatrix}.$$

3. $X = \frac{2}{3}(A + 2B) = \begin{bmatrix} -\frac{2}{3} & \frac{4}{3} \\ \frac{10}{3} & 4 \end{bmatrix}.$

4. $X = 5A - 2B = \begin{bmatrix} 7 & 10 \\ 13 & 18 \end{bmatrix}.$

5. As in Example 3.16, we want to find scalars c_1 and c_2 such that $c_1A_1 + c_2A_2 = B$.

$$c_1 \begin{bmatrix} 1 & 2 \\ -1 & 1 \end{bmatrix} + c_2 \begin{bmatrix} 0 & 1 \\ 2 & 1 \end{bmatrix} = \begin{bmatrix} 2 & 5 \\ 0 & 3 \end{bmatrix}$$

The left-hand side of this equation can be rewritten as $\begin{bmatrix} c_1 & 2c_1 + c_2 \\ -c_1 + 2c_2 & c_1 + c_2 \end{bmatrix}$

Comparing entries and the definition of matrix equality yields
$$\begin{aligned} c_1 \phantom{{}+2c_2} &= 2 \\ 2c_1 + c_2 &= 5 \\ -c_1 + 2c_2 &= 0 \\ c_1 + c_2 &= 3 \end{aligned}$$

Gauss-Jordan elimination easily gives $\left[\begin{array}{cc|c} 1 & 0 & 2 \\ 2 & 1 & 5 \\ -1 & 2 & 0 \\ 1 & 1 & 3 \end{array}\right] \longrightarrow \left[\begin{array}{cc|c} 1 & 0 & 2 \\ 0 & 1 & 1 \\ 0 & 0 & 0 \\ 0 & 0 & 0 \end{array}\right]$

So, $c_1 = 2$ and $c_2 = 1$. Thus, $2A_1 + A_2 = B$, which can be easily checked.

Having walked through the process, we note this pattern in our augmented matrix: the first column is the entries of A_1, the second column is the entries of A_2, and the third column, the augmented column, is the entries of B. Make use of this pattern!

6. As in Example 3.16, we form the augmented matrix and row reduce to solve.
As in Exercise 5, the first column is the entries of A_1, the second column is the entries of A_2, the third column is the entries of A_3, and the augmented column is the entries of B.

$$\left[\begin{array}{ccc|c} 1 & 0 & 1 & 2 \\ 0 & -1 & 1 & 3 \\ 0 & 1 & 0 & -4 \\ 1 & 0 & 1 & 2 \end{array}\right] \longrightarrow \left[\begin{array}{ccc|c} 1 & 0 & 0 & 3 \\ 0 & 1 & 0 & -4 \\ 0 & 0 & 1 & -1 \\ 0 & 0 & 0 & 0 \end{array}\right] \Rightarrow \quad B = 3\begin{bmatrix} 1 & 0 \\ 0 & 1 \end{bmatrix} - 4\begin{bmatrix} 0 & -1 \\ 1 & 0 \end{bmatrix} - \begin{bmatrix} 1 & 1 \\ 0 & 1 \end{bmatrix}.$$

7. As in Example 3.16, we form the augmented matrix and row reduce to solve.
As in Exercise 5, the first column is the entries of A_1, the second column is the entries of A_2, the third column is the entries of A_3, and the augmented column is the entries of B.

$$\left[\begin{array}{ccc|c} 1 & -1 & 1 & 3 \\ 0 & 2 & 1 & 1 \\ -1 & 0 & 1 & 1 \\ 0 & 0 & 0 & 0 \\ 1 & 1 & 0 & 1 \\ 0 & 0 & 0 & 0 \end{array}\right] \longrightarrow \left[\begin{array}{ccc|c} 1 & 0 & 0 & 0 \\ 0 & 1 & 0 & 0 \\ 0 & 0 & 1 & 0 \\ 0 & 0 & 0 & 1 \\ 0 & 0 & 0 & 0 \\ 0 & 0 & 0 & 0 \end{array}\right]$$

Since there is no solution, B is *not* a linear combination of A_1, A_2, and A_3.

Q: Why is it obvious that this system has no solution?
A: After row reduction, we have a row of zeroes with a corresponding constant term $\neq 0$.

8. As in Example 3.16, we form the augmented matrix and row reduce to solve.As in Exercise 5, the first column is the entries of A_1, the second column is the entries of A_2, the third column is the entries of A_3, the fourth column is the entries of A_4, and the augmented column is the entries of B.

$$\left[\begin{array}{cccc|c} 1 & 0 & -1 & 1 & 2 \\ 0 & 1 & 0 & -1 & -2 \\ 0 & 1 & -1 & 1 & 3 \\ 0 & 0 & 0 & 0 & 0 \\ 1 & 0 & 1 & -1 & 0 \\ 0 & 1 & 0 & -1 & -2 \\ 0 & 0 & 0 & 0 & 0 \\ 0 & 0 & 0 & 0 & 0 \\ 1 & 0 & -1 & 1 & 2 \end{array}\right] \longrightarrow \left[\begin{array}{cccc|c} 1 & 0 & 0 & 0 & 1 \\ 0 & 1 & 0 & 0 & 2 \\ 0 & 0 & 1 & 0 & 3 \\ 0 & 0 & 0 & 1 & 4 \\ 0 & 0 & 0 & 0 & 0 \\ 0 & 0 & 0 & 0 & 0 \\ 0 & 0 & 0 & 0 & 0 \\ 0 & 0 & 0 & 0 & 0 \\ 0 & 0 & 0 & 0 & 0 \end{array}\right] \Rightarrow$$

$$\begin{bmatrix} 2 & -2 & 3 \\ 0 & 0 & -2 \\ 0 & 0 & 2 \end{bmatrix} = \begin{bmatrix} 1 & 0 & 0 \\ 0 & 1 & 0 \\ 0 & 0 & 1 \end{bmatrix} + 2\begin{bmatrix} 0 & 1 & 1 \\ 0 & 0 & 1 \\ 0 & 0 & 0 \end{bmatrix} + 3\begin{bmatrix} -1 & 0 & -1 \\ 0 & 1 & 0 \\ 0 & 0 & -1 \end{bmatrix} + 4\begin{bmatrix} 1 & -1 & 1 \\ 0 & -1 & -1 \\ 0 & 0 & 1 \end{bmatrix}.$$

9. As in Example 3.17, we write out a general linear combination of A_1 and A_2.
As in Exercise 5, the first column is the entries of A_1, the second column is the entries of A_2, but the augmented column is now the variables of w, x, y, and z.

$$\left[\begin{array}{cc|c} 1 & 0 & w \\ 2 & 1 & x \\ -1 & 2 & y \\ 1 & 1 & z \end{array}\right] \longrightarrow \left[\begin{array}{cc|c} 1 & 0 & w \\ 0 & 1 & x-2w \\ 0 & 0 & y+5w-2x \\ 0 & 0 & z+w-x \end{array}\right]$$

Two restrictions come from the last two rows: $y + 5w - 2x = 0$ and $z + w - x = 0$.

So, the span of A_1 and A_2 consists of $\begin{bmatrix} w & x \\ y & z \end{bmatrix}$ for which $y = 2x - 5w$ and $z = x - w$.

That is, $\text{span}(A_1, A_2) = \left\{ \begin{bmatrix} w & x \\ 2x-5w & x-w \end{bmatrix} \right\}$.

Q: Why must $y + 5w - 2x = 0$ and $z + w - x = 0$?
A: After row reduction, a row of zeroes must have a corresponding constant term of zero.

Q: Is matrix B from Exercise 5 in $\text{span}(A_1, A_2)$? Should it be?
A: Yes, as well it should be since B is a linear combination of A_1 and A_2.

Q: How can we tell that B is in the $\text{span}(A_1, A_2)$?
A: By noting that $2(5) - 5(2) = 0$ and $5 - 2 = 3$.

10. Following Example 3.17, we create an augmented matrix and row reduce to find restrictions. As in Exercise 6, the first column is the entries of A_1, the second column is the entries of A_2, the third column is the entries of A_3, but the augmented column is now w, x, y, and z.

$$\left[\begin{array}{ccc|c} 1 & 0 & 1 & w \\ 0 & -1 & 1 & x \\ 0 & 1 & 0 & y \\ 1 & 0 & 1 & z \end{array}\right] \longrightarrow \left[\begin{array}{ccc|c} 1 & 0 & 0 & w-x-y \\ 0 & 1 & 0 & y \\ 0 & 0 & 1 & x+y \\ 0 & 0 & 0 & z-w \end{array}\right]$$

So the only restriction we get is $z - w = 0 \Rightarrow \text{span}(A_1, A_2, A_3) = \left\{ \begin{bmatrix} w & x \\ y & w \end{bmatrix} \right\}$.

11. Following Example 3.17, we create an augmented matrix and row reduce to find restrictions. As in Exercise 7, the first column is the entries of A_1, the second column is the entries of A_2, the third column is the entries of A_3, but the augmented column is a, b, c, d, e, and f.

$$\left[\begin{array}{rrr|c} 1 & -1 & 1 & a \\ 0 & 2 & 1 & b \\ -1 & 0 & 1 & c \\ 0 & 0 & 0 & d \\ 1 & 1 & 0 & e \\ 0 & 0 & 0 & f \end{array}\right] \longrightarrow \left[\begin{array}{rrr|c} 1 & 1 & 0 & e \\ 0 & 1 & 1 & c+e \\ 0 & 0 & -1 & b-2c-2e \\ 0 & 0 & 0 & a+3b-4c-5e \\ 0 & 0 & 0 & d \\ 0 & 0 & 0 & f \end{array}\right]$$

The restrictions are $a+3b-4c-5e=0$ and $d=f=0$.

So, $\operatorname{span}(A_1, A_2, A_3) = \left\{ \begin{bmatrix} -3b+4c+5e & b & c \\ 0 & e & 0 \end{bmatrix} \right\}$.

Q: Why did we not have to continue to reduced row echelon form?
A: The first 3 rows are obviously linearly independent, but restrictions come from zero rows.

Q: Is matrix B from Exercise 7 in $\operatorname{span}(A_1, A_2, A_3)$? Should it be?
A: No. It shouldn't be since B is *not* a linear combination of A_1, A_2, and A_3.

Q: How can we tell that B is *not* in the $\operatorname{span}(A_1, A_2)$?
A: By noting that $3+3(1)-4(c)-5(1) \neq 0$.

12. Following Example 3.17, we create an augmented matrix and row reduce to find restrictions. As in Exercise 8, the first column is the entries of A_1, the second column is the entries of A_2, the third column is the entries of A_3, the fourth column is the entries of A_4, but now the augmented column is the variables a, b, c, d, e, f, g, h, and i.

$$\left[\begin{array}{rrrr|c} 1 & 0 & -1 & 1 & a \\ 0 & 1 & 0 & -1 & b \\ 0 & 1 & -1 & 1 & c \\ 0 & 0 & 0 & 0 & d \\ 1 & 0 & 1 & -1 & e \\ 0 & 1 & 0 & -1 & f \\ 0 & 0 & 0 & 0 & g \\ 0 & 0 & 0 & 0 & h \\ 1 & 0 & -1 & 1 & i \end{array}\right] \longrightarrow \left[\begin{array}{rrrr|c} 1 & 0 & -1 & 1 & a \\ 0 & 1 & -1 & 1 & c \\ 0 & 0 & 1 & -2 & b-c \\ 0 & 0 & 0 & -6 & -a-2b+2c+e \\ 0 & 0 & 0 & 0 & -b+f \\ 0 & 0 & 0 & 0 & -a+i \\ 0 & 0 & 0 & 0 & d \\ 0 & 0 & 0 & 0 & g \\ 0 & 0 & 0 & 0 & h \end{array}\right]$$

The restrictions are $-b+f=0$, $-a+i=0$, $d=g=h=0$.

So, $\operatorname{span}(A_1, A_2, A_3, A_4) = \left\{ \begin{bmatrix} a & b & c \\ 0 & e & b \\ 0 & 0 & a \end{bmatrix} \right\}$.

13. Following Example 3.18, we create an augmented matrix and row reduce to solve.
As in Exercise 8, the first column is the entries of A_1, the second column is the entries of A_2, but now the augmented column is all zeroes.

$$\left[\begin{array}{cc|c} 1 & 4 & 0 \\ 2 & 3 & 0 \\ 3 & 2 & 0 \\ 4 & 1 & 0 \end{array}\right] \longrightarrow \left[\begin{array}{cc|c} 1 & 0 & 0 \\ 0 & 1 & 0 \\ 0 & 0 & 0 \\ 0 & 0 & 0 \end{array}\right]$$

Clearly, the only solution is $c_1 = c_2 = 0$. What do we conclude?
We conclude that A_1 and A_2 are linearly independent.

14. Following Example 3.18, we create an augmented matrix and row reduce to solve.
As in Exercise 8, the first column is the entries of A_1, the second column is the entries of A_2, the third column is the entries of A_3, but now the augmented column is all zeroes.

$$\left[\begin{array}{ccc|c} 1 & 2 & 1 & 0 \\ 1 & 1 & 2 & 0 \\ 1 & -1 & 4 & 0 \\ 1 & 0 & 3 & 0 \end{array}\right] \longrightarrow \left[\begin{array}{ccc|c} 1 & 0 & 3 & 0 \\ 0 & 1 & -1 & 0 \\ 0 & 0 & 0 & 0 \\ 0 & 0 & 0 & 0 \end{array}\right]$$

So, $c_1 = -3c_3$, $c_2 = c_3$ is a solution with at least one $c_i \neq 0$. What does that tell us?
That tells us that A_1, A_2, and A_3 are linearly dependent.
In particular, if we let $c_3 = -1$, we have the following dependence relation:

$$3\begin{bmatrix} 1 & 1 \\ 1 & 1 \end{bmatrix} - \begin{bmatrix} 2 & 1 \\ -1 & 0 \end{bmatrix} - \begin{bmatrix} 1 & 2 \\ 4 & 3 \end{bmatrix} = \begin{bmatrix} 0 & 0 \\ 0 & 0 \end{bmatrix}$$

15. Following Example 3.18, we create an augmented matrix and row reduce to solve.
As in Exercise 8, the first column is the entries of A_1, the second column is the entries of A_2, the third column is the entries of A_3, the fourth column is the entries of A_4, but now the augmented column is all zeroes.

$$\left[\begin{array}{cccc|c} 0 & 1 & -2 & -1 & 0 \\ 1 & 0 & -1 & -3 & 0 \\ 5 & 2 & 0 & 1 & 0 \\ 2 & 3 & 1 & 9 & 0 \\ -1 & 1 & 0 & 4 & 0 \\ 0 & 1 & 2 & 5 & 0 \end{array}\right] \longrightarrow \left[\begin{array}{cccc|c} 1 & 0 & 0 & 0 & 0 \\ 0 & 1 & 0 & 0 & 0 \\ 0 & 0 & 1 & 0 & 0 \\ 0 & 0 & 0 & 1 & 0 \\ 0 & 0 & 0 & 0 & 0 \\ 0 & 0 & 0 & 0 & 0 \end{array}\right]$$

Clearly, the only solution is $c_1 = c_2 = c_3 = c_4 = 0$. What do we conclude?
We conclude that A_1, A_2, A_3, and A_4 are linearly independent.

16. Following Example 3.18, we create an augmented matrix and row reduce to solve.
As in Exercise 8, the first column is the entries of A_1, the second column is the entries of A_2, the third column is the entries of A_3, the fourth column is the entries of A_4, but now the augmented column is all zeroes.

$$\left[\begin{array}{cccc|c} 1 & 2 & 1 & -1 & 0 \\ -1 & 1 & 2 & 1 & 0 \\ 0 & 0 & 0 & 0 & 0 \\ 0 & 0 & 0 & 0 & 0 \\ 2 & 3 & 1 & -1 & 0 \\ 0 & 0 & 0 & 0 & 0 \\ 2 & 4 & 3 & 0 & 0 \\ 6 & 9 & 5 & -4 & 0 \end{array}\right] \longrightarrow \left[\begin{array}{cccc|c} 1 & 0 & 0 & 0 & 0 \\ 0 & 1 & 0 & 0 & 0 \\ 0 & 0 & 1 & 0 & 0 \\ 0 & 0 & 0 & 1 & 0 \\ 0 & 0 & 0 & 0 & 0 \\ 0 & 0 & 0 & 0 & 0 \\ 0 & 0 & 0 & 0 & 0 \\ 0 & 0 & 0 & 0 & 0 \end{array}\right]$$

Clearly, the only solution is $c_1 = c_2 = c_3 = c_4 = 0$. What do we conclude?
We conclude that A_1, A_2, A_3, and A_4 are linearly independent.

17. Let A, B, and C be matrices of the same size $(m \times n)$ and let c and d be scalars.

(a) $$A + B = \begin{bmatrix} a_{11} & \cdots & a_{1n} \\ \vdots & \ddots & \vdots \\ a_{m1} & \cdots & a_{mn} \end{bmatrix} + \begin{bmatrix} b_{11} & \cdots & b_{1n} \\ \vdots & \ddots & \vdots \\ b_{m1} & \cdots & b_{mn} \end{bmatrix} = \begin{bmatrix} a_{11} + b_{11} & \cdots & a_{1n} + b_{1n} \\ \vdots & \ddots & \vdots \\ a_{m1} + b_{m1} & \cdots & a_{mn} + b_{mn} \end{bmatrix}.$$

But, a_{ij} and b_{ij} are scalars, which commute $\Rightarrow$

$$= \begin{bmatrix} b_{11} + a_{11} & \cdots & b_{1n} + a_{1n} \\ \vdots & \ddots & \vdots \\ b_{m1} + a_{m1} & \cdots & b_{mn} + a_{mn} \end{bmatrix} = B + A.$$

(b) $$\begin{aligned} (A + B) + C &= \left(\begin{bmatrix} a_{11} & \cdots & a_{1n} \\ \vdots & \ddots & \vdots \\ a_{m1} & \cdots & a_{mn} \end{bmatrix} + \begin{bmatrix} b_{11} & \cdots & b_{1n} \\ \vdots & \ddots & \vdots \\ b_{m1} & \cdots & b_{mn} \end{bmatrix} \right) + \begin{bmatrix} c_{11} & \cdots & c_{1n} \\ \vdots & \ddots & \vdots \\ c_{m1} & \cdots & c_{mn} \end{bmatrix} \\ &= \begin{bmatrix} b_{11} + a_{11} & \cdots & b_{1n} + a_{1n} \\ \vdots & \ddots & \vdots \\ b_{m1} + a_{m1} & \cdots & b_{mn} + a_{mn} \end{bmatrix} + \begin{bmatrix} c_{11} & \cdots & c_{1n} \\ \vdots & \ddots & \vdots \\ c_{m1} & \cdots & c_{mn} \end{bmatrix} \\ &= \begin{bmatrix} a_{11} + b_{11} + c_{11} & \cdots & a_{1n} + b_{1n} + c_{1n} \\ \vdots & \ddots & \vdots \\ a_{m1} + b_{m1} + c_{m1} & \cdots & a_{mn} + b_{mn} + c_{mn} \end{bmatrix} \\ &= \begin{bmatrix} a_{11} & \cdots & a_{1n} \\ \vdots & \ddots & \vdots \\ a_{m1} & \cdots & a_{mn} \end{bmatrix} + \left(\begin{bmatrix} b_{11} & \cdots & b_{1n} \\ \vdots & \ddots & \vdots \\ b_{m1} & \cdots & b_{mn} \end{bmatrix} + \begin{bmatrix} c_{11} & \cdots & c_{1n} \\ \vdots & \ddots & \vdots \\ c_{m1} & \cdots & c_{mn} \end{bmatrix} \right) \\ &= A + (B + C). \end{aligned}$$

(c) $$A + 0 = \begin{bmatrix} a_{11} & \cdots & a_{1n} \\ \vdots & \ddots & \vdots \\ a_{m1} & \cdots & a_{mn} \end{bmatrix} + \begin{bmatrix} 0 & \cdots & 0 \\ \vdots & \ddots & \vdots \\ 0 & \cdots & 0 \end{bmatrix} = \begin{bmatrix} a_{11} + 0 & \cdots & a_{1n} + 0 \\ \vdots & \ddots & \vdots \\ a_{m1} + 0 & \cdots & a_{mn} + 0 \end{bmatrix} = A.$$

(d) $$\begin{aligned} A + (-A) &= \begin{bmatrix} a_{11} & \cdots & a_{1n} \\ \vdots & \ddots & \vdots \\ a_{m1} & \cdots & a_{mn} \end{bmatrix} + \left(- \begin{bmatrix} a_{11} & \cdots & a_{1n} \\ \vdots & \ddots & \vdots \\ a_{m1} & \cdots & a_{mn} \end{bmatrix} \right) \\ &= \begin{bmatrix} a_{11} & \cdots & a_{1n} \\ \vdots & \ddots & \vdots \\ a_{m1} & \cdots & a_{mn} \end{bmatrix} + \left(\begin{bmatrix} -a_{11} & \cdots & -a_{1n} \\ \vdots & \ddots & \vdots \\ -a_{m1} & \cdots & -a_{mn} \end{bmatrix} \right) \\ &= \begin{bmatrix} a_{11} - a_{11} & \cdots & a_{1n} - a_{1n} \\ \vdots & \ddots & \vdots \\ a_{m1} - a_{m1} & \cdots & a_{mn} - a_{mn} \end{bmatrix} = \begin{bmatrix} 0 & \cdots & 0 \\ \vdots & \ddots & \vdots \\ 0 & \cdots & 0 \end{bmatrix} = O. \end{aligned}$$

18. (e)
$$c(A+B) = c\left(\begin{bmatrix} a_{11} & \cdots & a_{1n} \\ \vdots & \ddots & \vdots \\ a_{m1} & \cdots & a_{mn} \end{bmatrix} + \begin{bmatrix} b_{11} & \cdots & b_{1n} \\ \vdots & \ddots & \vdots \\ b_{m1} & \cdots & b_{mn} \end{bmatrix}\right)$$
$$= c\left(\begin{bmatrix} a_{11}+b_{11} & \cdots & a_{1n}+b_{1n} \\ \vdots & \ddots & \vdots \\ a_{m1}+b_{m1} & \cdots & a_{mn}+b_{mn} \end{bmatrix}\right)$$
$$= \begin{bmatrix} c(a_{11}+b_{11}) & \cdots & c(a_{1n}+b_{1n}) \\ \vdots & \ddots & \vdots \\ c(a_{m1}+b_{m1}) & \cdots & c(a_{mn}+b_{mn}) \end{bmatrix}$$
$$= \begin{bmatrix} ca_{11}+cb_{11} & \cdots & ca_{1n}+cb_{1n} \\ \vdots & \ddots & \vdots \\ ca_{m1}+cb_{m1} & \cdots & ca_{mn}+cb_{mn} \end{bmatrix}$$
$$= \begin{bmatrix} ca_{11} & \cdots & ca_{1n} \\ \vdots & \ddots & \vdots \\ ca_{m1} & \cdots & ca_{mn} \end{bmatrix} + \begin{bmatrix} cb_{11} & \cdots & cb_{1n} \\ \vdots & \ddots & \vdots \\ cb_{m1} & \cdots & cb_{mn} \end{bmatrix} = cA + cB.$$

(f)
$$(c+d)A = (c+d)\begin{bmatrix} a_{11} & \cdots & a_{1n} \\ \vdots & \ddots & \vdots \\ a_{m1} & \cdots & a_{mn} \end{bmatrix} = \begin{bmatrix} (c+d)a_{11} & \cdots & (c+d)a_{1n} \\ \vdots & \ddots & \vdots \\ (c+d)a_{m1} & \cdots & (c+d)a_{mn} \end{bmatrix}$$
$$= \begin{bmatrix} ca_{11}+da_{11} & \cdots & ca_{1n}+da_{1n} \\ \vdots & \ddots & \vdots \\ ca_{m1}+da_{m1} & \cdots & ca_{mn}+da_{mn} \end{bmatrix}$$
$$= \begin{bmatrix} ca_{11} & \cdots & ca_{1n} \\ \vdots & \ddots & \vdots \\ ca_{m1} & \cdots & ca_{mn} \end{bmatrix} + \begin{bmatrix} da_{11} & \cdots & da_{1n} \\ \vdots & \ddots & \vdots \\ da_{m1} & \cdots & da_{mn} \end{bmatrix}$$
$$= cA + dA.$$

(g)
$$c(dA) = c\begin{bmatrix} da_{11} & \cdots & da_{1n} \\ \vdots & \ddots & \vdots \\ da_{m1} & \cdots & da_{mn} \end{bmatrix} = \begin{bmatrix} cda_{11} & \cdots & cda_{1n} \\ \vdots & \ddots & \vdots \\ cda_{m1} & \cdots & cda_{mn} \end{bmatrix} = (cd)\begin{bmatrix} a_{11} & \cdots & a_{1n} \\ \vdots & \ddots & \vdots \\ a_{m1} & \cdots & a_{mn} \end{bmatrix} = (cd)A.$$

(h)
$$1A = 1\begin{bmatrix} a_{11} & \cdots & a_{1n} \\ \vdots & \ddots & \vdots \\ a_{m1} & \cdots & a_{mn} \end{bmatrix} = \begin{bmatrix} 1a_{11} & \cdots & 1a_{1n} \\ \vdots & \ddots & \vdots \\ 1a_{m1} & \cdots & 1a_{mn} \end{bmatrix} = \begin{bmatrix} a_{11} & \cdots & a_{1n} \\ \vdots & \ddots & \vdots \\ a_{m1} & \cdots & a_{mn} \end{bmatrix} = A.$$

19. Let A, B, and C be matrices of appropriate dimensions. Then

$$(A+B)C = \left(\begin{bmatrix} a_{11} & \cdots & a_{1n} \\ \vdots & \ddots & \vdots \\ a_{m1} & \cdots & a_{mn} \end{bmatrix} + \begin{bmatrix} b_{11} & \cdots & b_{1n} \\ \vdots & \ddots & \vdots \\ b_{m1} & \cdots & b_{mn} \end{bmatrix}\right)\begin{bmatrix} c_{11} & \cdots & c_{1n} \\ \vdots & \ddots & \vdots \\ c_{m1} & \cdots & c_{mn} \end{bmatrix}$$

$$= \begin{bmatrix} a_{11}+b_{11} & \cdots & a_{1n}+b_{1n} \\ \vdots & \ddots & \vdots \\ a_{m1}+b_{m1} & \cdots & a_{mn}+b_{mn} \end{bmatrix}\begin{bmatrix} c_{11} & \cdots & c_{1n} \\ \vdots & \ddots & \vdots \\ c_{m1} & \cdots & c_{mn} \end{bmatrix}$$

$$= \begin{bmatrix} (a_{11}+b_{11})c_{11}+\cdots+(a_{m1}+b_{m1})c_{1n} & \cdots & (a_{1n}+b_{1n})c_{11}+\cdots+(a_{mn}+b_{mn})c_{1n} \\ \vdots & \ddots & \vdots \\ (a_{11}+b_{11})c_{m1}+\cdots+(a_{m1}+b_{m1})c_{mn} & \cdots & (a_{1n}+b_{1n})c_{m1}+\cdots+(a_{mn}+b_{mn})c_{mn} \end{bmatrix}$$

$$= \begin{bmatrix} a_{11}c_{11}+b_{11}c_{11}+\cdots+a_{m1}c_{1n}+b_{m1}c_{1n} & \cdots & a_{1n}c_{11}+b_{1n}c_{11}+\cdots+a_{mn}c_{1n}+b_{mn}c_{1n} \\ \vdots & \ddots & \vdots \\ a_{11}c_{m1}+b_{11}c_{m1}+\cdots+a_{m1}c_{mn}+b_{m1}c_{mn} & \cdots & a_{1n}c_{m1}+b_{1n}c_{m1}+\cdots+a_{mn}c_{mn}+b_{mn}c_{mn} \end{bmatrix}$$

$$= \begin{bmatrix} \begin{matrix}(a_{11}c_{11}+\cdots+a_{m1}c_{1n})+ \\ +(b_{11}c_{11}+\cdots+b_{m1}c_{1n})\end{matrix} & \cdots & \begin{matrix}(a_{1n}c_{11}+\cdots+a_{mn}c_{1n})+ \\ +(b_{1n}c_{11}+\cdots+b_{mn}c_{1n})\end{matrix} \\ \vdots & \ddots & \vdots \\ \begin{matrix}(a_{11}c_{m1}+\cdots+a_{m1}c_{mn})+ \\ +(b_{11}c_{m1}+\cdots+b_{m1}c_{mn})\end{matrix} & \cdots & \\ \begin{matrix}(a_{1n}c_{m1}+\ldots++a_{mn}c_{mn})+ \\ +(b_{1n}c_{m1}+\cdots+b_{mn}c_{mn})\end{matrix} & & \end{bmatrix}$$

$$= \begin{bmatrix} (a_{11}c_{11}+\cdots+a_{m1}c_{1n}) & \cdots & (a_{1n}c_{11}+\cdots+a_{mn}c_{1n}) \\ \vdots & \ddots & \vdots \\ (a_{11}c_{m1}+\cdots+a_{m1}c_{mn}) & \cdots & (a_{1n}c_{m1}+\cdots++a_{mn}c_{mn}) \end{bmatrix}$$

$$+ \begin{bmatrix} (b_{11}c_{11}+\cdots+b_{m1}c_{1n}) & \cdots & (b_{1n}c_{11}+\cdots b_{mn}c_{1n}) \\ \vdots & \ddots & \vdots \\ (b_{11}c_{m1}+\cdots+b_{m1}c_{mn}) & \cdots & (b_{1n}c_{m1}+\cdots b_{mn}c_{mn}) \end{bmatrix}$$

$$= AC + BC.$$

20. Let A, and B be matrices of appropriate dimensions, and let k be a scalar. Then

$$\begin{aligned}
k(AB) &= k\left(\begin{bmatrix} a_{11}b_{11}+\cdots+a_{1n}b_{m1} & \cdots & a_{11}b_{1n}+\cdots+a_{1n}b_{mn} \\ \vdots & \ddots & \vdots \\ a_{m1}b_{11}+\cdots+a_{mn}b_{m1} & \cdots & a_{m1}b_{1n}+\cdots+a_{mn}b_{mn} \end{bmatrix}\right) \\
&= \begin{bmatrix} k(a_{11}b_{11}+\cdots+a_{1n}b_{m1}) & \cdots & k(a_{11}b_{1n}+\cdots+a_{1n}b_{mn}) \\ \vdots & \ddots & \vdots \\ k(a_{m1}b_{11}+\cdots+a_{mn}b_{m1}) & \cdots & k(a_{m1}b_{1n}+\cdots+a_{mn}b_{mn}) \end{bmatrix} \\
&= \begin{bmatrix} (ka_{11})b_{11}+\cdots+(ka_{1n})b_{m1} & \cdots & (ka_{11})b_{1n}+\cdots+(ka_{1n})b_{mn} \\ \vdots & \ddots & \vdots \\ (ka_{m1})b_{11}+\cdots+(ka_{mn})b_{m1} & \cdots & (ka_{m1})b_{1n}+\cdots+(ka_{mn})b_{mn} \end{bmatrix} \\
&= \begin{bmatrix} ka_{11} & \cdots & ka_{1n} \\ \vdots & \ddots & \vdots \\ ka_{m1} & \cdots & ka_{mn} \end{bmatrix}\begin{bmatrix} b_{11} & \cdots & b_{1n} \\ \vdots & \ddots & \vdots \\ b_{m1} & \cdots & b_{mn} \end{bmatrix} \\
&= (kA)B.
\end{aligned}$$

$$\begin{aligned}
&= \begin{bmatrix} a_{11}(kb_{11})+\cdots+a_{1n}(kb_{m1}) & \cdots & a_{11}(kb_{1n})+\cdots+a_{1n}(kb_{mn}) \\ \vdots & \ddots & \vdots \\ a_{m1}(kb_{11})+\cdots+a_{mn}(kb_{m1}) & \cdots & a_{m1}(kb_{1n})+\cdots+a_{mn}(kb_{mn}) \end{bmatrix} \\
&= \begin{bmatrix} a_{11} & \cdots & a_{1n} \\ \vdots & \ddots & \vdots \\ a_{m1} & \cdots & a_{mn} \end{bmatrix}\begin{bmatrix} kb_{11} & \cdots & kb_{1n} \\ \vdots & \ddots & \vdots \\ kb_{m1} & \cdots & kb_{mn} \end{bmatrix} \\
&= A(kB).
\end{aligned}$$

21. To prove $I_m A = A$, note $I_m = \begin{bmatrix} \mathbf{e}_1 \\ \mathbf{e}_2 \\ \vdots \\ \mathbf{e}_m \end{bmatrix} \Rightarrow I_m A = \begin{bmatrix} \mathbf{e}_1 A_1 \\ \mathbf{e}_2 A_2 \\ \vdots \\ \mathbf{e}_m A_m \end{bmatrix} = \begin{bmatrix} A_1 \\ A_2 \\ \vdots \\ A_m \end{bmatrix} = A.$

22. We need to show $AB = BA$ if and only if $(A-B)(A+B) = A^2 + B^2$.

Note: $(A-B)(A+B) \overset{\text{left distributivity}}{=} (A-B)A + (A-B)B \overset{\text{right distributivity}}{=} A^2 - BA + AB - B^2.$

If $AB = BA$, then $-BA + AB = O$,

so $(A-B)(A+B) = A^2 - BA + AB - B^2 \overset{\text{because } -BA+AB=O}{=} A^2 - B^2.$

If $(A-B)(A+B) = A^2 - BA + AB - B^2 = A^2 - B^2$,
then $-BA + AB = O$ so $AB = BA$.

23. We will compute AB and BA, then equate entries to find the conditions on a, b, c, and d.

$$AB = \begin{bmatrix} 1 & 1 \\ 0 & 1 \end{bmatrix} \begin{bmatrix} a & b \\ c & d \end{bmatrix} = \begin{bmatrix} a+c & b+d \\ c & d \end{bmatrix} = \begin{bmatrix} a & a+b \\ c & c+d \end{bmatrix} = \begin{bmatrix} a & b \\ c & d \end{bmatrix} \begin{bmatrix} 1 & 1 \\ 0 & 1 \end{bmatrix} = BA$$

Equating entries gives us the following four equations (conditions on a, b, c, and d):
$a+c=a$, $b+d=a+b$, $c=c$, and $d=c+d \Rightarrow$ The conditions are $a=d$ and $c=0$.

24. We will compute AB and BA, then equate entries to find the conditions on a, b, c, and d.

$$\begin{bmatrix} 1 & -1 \\ -1 & 1 \end{bmatrix} \begin{bmatrix} a & b \\ c & d \end{bmatrix} = \begin{bmatrix} a-c & b-d \\ -a+c & -b+d \end{bmatrix} = \begin{bmatrix} a-b & -a+b \\ c-d & -c+d \end{bmatrix} = \begin{bmatrix} a & b \\ c & d \end{bmatrix} \begin{bmatrix} 1 & -1 \\ -1 & 1 \end{bmatrix}$$

Equating entries gives us the following four equations (conditions on a, b, c, and d):
$a-c=a-b$, $b-d=-a+b$, $-a+c=c-d$, and $-b+d=-c+d$.
So, the conditions on a, b, c, and d are $a=d$ and $b=c$.

25. We will compute AB and BA, then equate entries to find the conditions on a, b, c, and d.

$$AB = \begin{bmatrix} 1 & 2 \\ 3 & 4 \end{bmatrix} \begin{bmatrix} a & b \\ c & d \end{bmatrix} = \begin{bmatrix} a+2c & b+2d \\ 3a+4c & 3b+4d \end{bmatrix} = \begin{bmatrix} a+3b & 2a+4b \\ c+3d & 2c+4d \end{bmatrix} = \begin{bmatrix} a & b \\ c & d \end{bmatrix} \begin{bmatrix} 1 & 2 \\ 3 & 4 \end{bmatrix} = BA$$

Equating entries gives us the following four equations (conditions on a, b, c, and d):
$a+2c=a+3b$, $b+2d=2a+4b$, $3a+4c=c+3d$, and $3b+4d=2c+4d$.
So, the conditions on a, b, c, and d are $3b=2c$ and $a=d-c$.

26. We will compute A_1B and BA_1, then equate entries to find conditions on a, b, c, and d.
We will then repeat the process for A_4B and BA_4, then combine conditions for our answer.

$$A_1B = \begin{bmatrix} 1 & 0 \\ 0 & 0 \end{bmatrix} \begin{bmatrix} a & b \\ c & d \end{bmatrix} = \begin{bmatrix} a & b \\ 0 & 0 \end{bmatrix} = \begin{bmatrix} a & 0 \\ c & 0 \end{bmatrix} = \begin{bmatrix} a & b \\ c & d \end{bmatrix} \begin{bmatrix} 1 & 0 \\ 0 & 0 \end{bmatrix} = BA_1$$

Equating entries gives us the following four equations (conditions on a, b, c, and d):
$a=a$, $b=0$, $c=0$, and $0=0 \Rightarrow$ The A_1 conditions are $b=0$ and $c=0$.

Repeating the process for A_4B and BA_4 yields:

$$A_4B = \begin{bmatrix} 0 & 0 \\ 0 & 1 \end{bmatrix} \begin{bmatrix} a & b \\ c & d \end{bmatrix} = \begin{bmatrix} 0 & 0 \\ c & d \end{bmatrix} = \begin{bmatrix} 0 & b \\ 0 & d \end{bmatrix} = \begin{bmatrix} a & b \\ c & d \end{bmatrix} \begin{bmatrix} 0 & 0 \\ 0 & 1 \end{bmatrix} = BA_4$$

Equating entries gives us the following four equations (conditions on a, b, c, and d):
$0=0$, $b=0$, $c=0$, and $d=d \Rightarrow$ The A_4 conditions are $b=0$ and $c=0$.

Combining the conditions for A_1 and A_4 (in this case they are the same) gives us:
The required conditions so that B will commute with A_1 and A_4 are $b=c=0$.

Q: Let $M = aA_1 + dA_4$. Does the B we found above commute with M?
A: Yes, since $BM = B(aA_1+dA_4) = aBA_1+dBA_4 = aA_1B+dA_4B = (aA_1+dA_4)B = MB$.

27. We should use linear combinations and our work in Exercise 26 to answer this question. How?

Consider $A_1 = \begin{bmatrix} 1 & 0 \\ 0 & 0 \end{bmatrix}$, $A_2 = \begin{bmatrix} 0 & 1 \\ 0 & 0 \end{bmatrix}$, $A_3 = \begin{bmatrix} 0 & 0 \\ 1 & 0 \end{bmatrix}$, and $A_4 = \begin{bmatrix} 0 & 0 \\ 0 & 1 \end{bmatrix}$. Then we have:

$$M = \begin{bmatrix} a & b \\ c & d \end{bmatrix} = aA_1 + bA_1 + cA_3 + dA_4 = a\begin{bmatrix} 1 & 0 \\ 0 & 0 \end{bmatrix} + b\begin{bmatrix} 0 & 1 \\ 0 & 0 \end{bmatrix} + c\begin{bmatrix} 0 & 0 \\ 1 & 0 \end{bmatrix} + d\begin{bmatrix} 0 & 0 \\ 0 & 1 \end{bmatrix}.$$

What does this tell us? Our matrix B need only commute with A_1, A_2, A_3, and A_4.

We will compute A_2B and BA_2, then equate entries to find conditions on a, b, c, and d. Repeat the process for A_3B and BA_3, then combine with our answer from Exercise 26.

$$A_2B = \begin{bmatrix} 0 & 1 \\ 0 & 0 \end{bmatrix}\begin{bmatrix} a & b \\ c & d \end{bmatrix} = \begin{bmatrix} c & d \\ 0 & 0 \end{bmatrix} = \begin{bmatrix} 0 & a \\ 0 & c \end{bmatrix} = \begin{bmatrix} a & b \\ c & d \end{bmatrix}\begin{bmatrix} 0 & 1 \\ 0 & 0 \end{bmatrix} = BA_2$$

Equating entries gives us the following four equations (conditions on a, b, c, and d):
$c = 0$, $d = a$, $0 = 0$, and $c = 0 \Rightarrow$ The A_2 conditions are $a = d$ and $c = 0$.

Repeating the process for A_3B and BA_3 yields:

$$A_3B = \begin{bmatrix} 0 & 0 \\ 1 & 0 \end{bmatrix}\begin{bmatrix} a & b \\ c & d \end{bmatrix} = \begin{bmatrix} 0 & 0 \\ a & b \end{bmatrix} = \begin{bmatrix} b & 0 \\ d & 0 \end{bmatrix} = \begin{bmatrix} a & b \\ c & d \end{bmatrix}\begin{bmatrix} 0 & 0 \\ 1 & 0 \end{bmatrix} = BA_3$$

Equating entries gives us the following four equations (conditions on a, b, c, and d):
$0 = b$, $0 = 0$, $a = d$, and $b = 0 \Rightarrow$ The A_3 conditions are $a = d$ and $b = 0$.

Combining the conditions for A_1, A_2, A_3, and A_4 gives us:
The required conditions so that B will commute with any 2×2 matrix are $a = d$ and $b = c = 0$.

28. Let A be an $m \times n$ matrix and B be an $a \times b$ matrix.

Since AB is $[m \times n][a \times b]$, if AB is defined we must have $n = a$.
Since BA is $[a \times b][m \times n]$, if BA is defined we must have $b = m$.

Therefore, when AB and BA are defined, B must in fact be an $n \times m$ matrix

This implies AB is $[m \times n][n \times m]$, so AB is an $m \times m$ matrix.
Likewise, BA is $[n \times m][m \times n]$, so BA is an $n \times n$ matrix.

29. Let A and B be two upper triangular $n \times n$ matrices.

Then, $A = \begin{bmatrix} a_{11} & a_{12} & a_{13} & \cdots & a_{1n} \\ 0 & a_{22} & a_{23} & \cdots & a_{2n} \\ 0 & 0 & a_{33} & \cdots & a_{3n} \\ \vdots & \vdots & \vdots & \ddots & \vdots \\ 0 & 0 & \cdots & 0 & a_{nn} \end{bmatrix}$ and $B = \begin{bmatrix} b_{11} & b_{12} & b_{13} & \cdots & b_{1n} \\ 0 & b_{22} & b_{23} & \cdots & b_{2n} \\ 0 & 0 & b_{33} & \cdots & b_{3n} \\ \vdots & \vdots & \vdots & \ddots & \vdots \\ 0 & 0 & \cdots & 0 & b_{nn} \end{bmatrix} \Rightarrow$

$$AB = \begin{bmatrix} a_{11} & a_{12} & a_{13} & \cdots & a_{1n} \\ 0 & a_{22} & a_{23} & \cdots & a_{2n} \\ 0 & 0 & a_{33} & \cdots & a_{3n} \\ \vdots & \vdots & \vdots & \ddots & \vdots \\ 0 & 0 & \cdots & 0 & a_{nn} \end{bmatrix} \begin{bmatrix} b_{11} & b_{12} & b_{13} & \cdots & b_{1n} \\ 0 & b_{22} & b_{23} & \cdots & b_{2n} \\ 0 & 0 & b_{33} & \cdots & b_{3n} \\ \vdots & \vdots & \vdots & \ddots & \vdots \\ 0 & 0 & \cdots & 0 & b_{nn} \end{bmatrix}$$

$$= \begin{bmatrix} a_{11}b_{11} & a_{11}b_{12} + a_{12}b_{22} & a_{11}b_{13} + a_{12}b_{23} + a_{13}b_{33} & \cdots & a_{11}b_{1n} + a_{12}b_{2n} + \cdots + a_{nn}b_{nn} \\ 0 & a_{22}b_{22} & a_{22}b_{23} + a_{23}b_{33} & \cdots & a_{22}b_{2n} + a_{23}b_{3n} + \cdots + a_{2n}b_{nn} \\ 0 & 0 & a_{33}b_{33} & \cdots & a_{33}b_{3n} + \cdots + a_{3n}b_{nn} \\ \vdots & \vdots & \vdots & \ddots & \vdots \\ 0 & 0 & 0 & 0 & a_{nn}b_{nn} \end{bmatrix}.$$

So we see that AB is also an upper triangular matrix.

30. Let A, B be matrices whose sizes permit the indicated operations and let k be a scalar. Denote the ith row of a matrix X by $\text{row}_i\,(X)$ and its jth column by $\text{col}_j\,(X)$.

Theorem 3.4(a): $\left[\left(A^T\right)^T\right]_{ij} = \left[A^T\right]_{ji} = [A]_{ij}$. So, i, j arbitrary $\Rightarrow \left(A^T\right)^T = A$.

Theorem 3.4(b): $\left[(A+B)^T\right]_{ij} = [A+B]_{ji} = [A]_{ji} + [B]_{ji} = \left[A^T\right]_{ij} + \left[B^T\right]_{ij}$.
i, j arbitrary $\Rightarrow (A+B)^T = A^T + B^T$.

Theorem 3.4(c): $\left[(kA)^T\right]_{ij} = [kA]_{ji} = k\,[A]_{ji} = k\left[A^T\right]_{ij}$.
i, j arbitrary $\Rightarrow (kA)^T = k\left(A^T\right)$.

31. We need to show $(A^r)^T = (A^T)^r$. We will prove this using *induction*.
See Appendix B for discussion and examples of *Mathematical Induction.*

1: $(A^1)^T = (A^T)^1$. This is obvious, so there is nothing to show.

r: $(A^r)^T = (A^T)^r$. This is the induction hypothesis.

$r+1$: $(A^{r+1})^T = (A^T)^{r+1}$. This is the statement we must prove using the induction hypothesis.

$$(A^{r+1})^T = (A \cdot A^r)^T \overset{\substack{\text{by Thm 3.4d} \\ (AB)^T = B^TA^T}}{=} (A^r)^T A^T \overset{\substack{\text{by induction} \\ (A^r)^T = (A^T)^r}}{=} (A^T)^r A^T = (A^T)^{r+1}.$$

Induction can seem a little bit like *magic* at first glance.
Pay close attention to how the induction hypothesis is used in the proof.

32. We need to show $(A_1 + A_2 + \cdots + A_n)^T = A_1^T + A_2^T + \cdots + A_n^T$ for $n \geq 1$.
We will prove this using *induction*.
See Appendix B for discussion and examples of *Mathematical Induction.*

1: $(A_1)^T = A_1^T$. This is obvious, so there is nothing to show.

k: $(A_1 + A_2 + \cdots + A_k)^T = A_1^T + A_2^T + \cdots + A_k^T$.
This is the induction hypothesis, so there is nothing to show.

$k+1$: $(A_1 + A_2 + \cdots + A_k + A_{k+1})^T = A_1^T + A_2^T + \cdots + A_k^T + A_{k+1}^T$
This is the statement we must prove using the induction hypothesis.

$$\begin{aligned}(A_1 + A_2 + \cdots + A_k + A_{k+1})^T &= ((A_1 + A_2 + \cdots + A_k) + A_{k+1})^T \\ &= (A_1 + A_2 + \cdots + A_k)^T + A_{k+1}^T \quad [\textit{by Thm } 3.4b] \\ &= A_1^T + A_2^T + \cdots + A_k^T + A_{k+1}^T \quad [\textit{by induction}]\end{aligned}$$

Thus $(A_1 + A_2 + \cdots + A_k + A_{k+1})^T = A_1^T + A_2^T + \cdots + A_k^T + A_{k+1}^T$

We have shown the pattern holds for $k+1$. What does that mean?
We have proven (by induction) that $(A_1 + A_2 + \cdots + A_n)^T = A_1^T + A_2^T + \cdots + A_n^T$.

33. We need to show $(A_1 A_2 \cdots A_n)^T = A_n^T \cdots A_2^T A_1^T$ for $n \geq 1$.
We will prove this using *induction*.
See Appendix B for discussion and examples of *Mathematical Induction.*

1: $(A_1)^T = A_1^T$. This is obvious, so there is nothing to show.

k: $(A_1 A_2 \cdots A_k)^T = A_k^T \cdots A_2^T A_1^T$.
This is the induction hypothesis, so there is nothing to show.

$k+1$: $(A_1 + A_2 \cdots A_k A_{k+1})^T = A_{k+1}^T A_k^T \cdots A_2^T A_1^T$
This is the statement we must prove using the induction hypothesis.

$$\begin{aligned}(A_1 A_2 \cdots A_k A_{k+1})^T &= ((A_1 A_2 \cdots A_k)\, A_{k+1})^T = A_{k+1}^T (A_1 A_2 \cdots A_k)^T \quad [\textit{by Thm } 3.4d] \\ &= A_{k+1}^T \left(A_k^T \cdots A_2^T A_1^T\right) \quad [\textit{by induction}]\end{aligned}$$

We have shown the pattern holds for $k+1$. What does that mean?
We have proven (by induction) that $(A_1 A_2 \cdots A_n)^T = A_n^T \cdots A_2^T A_1^T$.

34. We check that $\left(AA^T\right)^T = \left(A^T\right)^T A^T = AA^T$ and $\left(A^T A\right)^T = A^T \left(A^T\right)^T = A^T A \Rightarrow$ AA^T and $A^T A$ equal their own transposes and so, by definition, are symmetric.

35. Let A and B be symmetric $n \times n$ matrices and let k be a scalar.

(a) To prove $A + B$ is symmetric, we need to show $(A+B)^T = A + B$.

$$(A+B)^T \overset{\substack{\text{by} \\ \text{Thm 3.4b}}}{=} A^T + B^T \overset{\substack{A \text{ and } B \text{ are} \\ \text{symmetric}}}{=} A + B.$$

(b) To prove kA is symmetric, we need to show $(kA)^T = kA$.

$$(kA)^T \overset{\substack{\text{by} \\ \text{Thm 3.4c}}}{=} kA^T \overset{\substack{A \text{ is} \\ \text{symmetric}}}{=} kA.$$

36. (a) Let A be the $n \times n$ matrix with entries $a_{ij} = 1$ if either $i = 1$ or $j = 1$,, 0 otherwise, and let B be the $n \times n$ matrix with entries $b_{ij} = 1$ if $i + j = n + 1$, 0 otherwise.

So, $A = \begin{bmatrix} 1 & 1 & \cdots & 1 & 1 \\ 1 & 0 & \cdots & 0 & 0 \\ \vdots & \vdots & \ddots & \vdots & \vdots \\ 1 & 0 & \cdots & 0 & 0 \\ 1 & 0 & \cdots & 0 & 0 \end{bmatrix}$ and $B = \begin{bmatrix} 0 & 0 & \cdots & 0 & 1 \\ 0 & 0 & \cdots & 1 & 0 \\ \vdots & \vdots & \ddots & \vdots & \vdots \\ 0 & 1 & \cdots & 0 & 0 \\ 1 & 0 & \cdots & 0 & 0 \end{bmatrix} \Rightarrow AB = \begin{bmatrix} 1 & 1 & \cdots & 1 & 1 \\ 0 & 0 & \cdots & 0 & 1 \\ \vdots & \vdots & \ddots & \vdots & \vdots \\ 0 & 0 & \cdots & 0 & 1 \\ 0 & 0 & \cdots & 0 & 1 \end{bmatrix}$.

Clearly, A and B are symmetric, but AB is not symmetric.

When $n = 2$, for instance, we have $A = \begin{bmatrix} 1 & 1 \\ 1 & 0 \end{bmatrix}$, $B = \begin{bmatrix} 0 & 1 \\ 1 & 0 \end{bmatrix}$, and $AB = \begin{bmatrix} 1 & 1 \\ 0 & 1 \end{bmatrix}$.

(b) Since this is an *if and only if* statement, we have two claims to prove.

if: If A, B, and AB are symmetric, then $AB = BA$.

$$AB \overset{\substack{AB \text{ is} \\ \text{symmetric}}}{=} (AB)^T \overset{\substack{\text{by} \\ \text{Thm 3.4d}}}{=} B^T A^T \overset{\substack{A \text{ and } B \text{ are} \\ \text{symmetric}}}{=} BA$$

only if: If A, B are symmetric and $AB = BA$, then AB is symmetric, that is $(AB)^T = AB$.

$$(AB)^T \overset{\substack{\text{by} \\ \text{Thm 3.4d}}}{=} B^T A^T \overset{\substack{A \text{ and } B \text{ are} \\ \text{symmetric}}}{=} BA \overset{\substack{\text{by the given} \\ AB = BA}}{=} AB$$

37. For each matrix, we will simply check to see if $A^T = -A$ is satisfied.

(a) Since $A^T = \begin{bmatrix} 1 & -2 \\ 2 & 3 \end{bmatrix} \neq -\begin{bmatrix} 1 & 2 \\ -2 & 3 \end{bmatrix} = -A$, A is *not* skew-symmetric.

(b) Since $A^T = \begin{bmatrix} 0 & 1 \\ -1 & 0 \end{bmatrix} = -\begin{bmatrix} 0 & -1 \\ 1 & 0 \end{bmatrix} = -A$, A is skew-symmetric.

(c) Since $A^T = \begin{bmatrix} 0 & -3 & 1 \\ 3 & 0 & -2 \\ -1 & 2 & 0 \end{bmatrix} = -\begin{bmatrix} 0 & 3 & -1 \\ -3 & 0 & 2 \\ 1 & -2 & 0 \end{bmatrix} = -A$, A is skew-symmetric.

(d) Since $A^T = \begin{bmatrix} 0 & -1 & 2 \\ 1 & 0 & 5 \\ 2 & 5 & 0 \end{bmatrix} \neq -\begin{bmatrix} 0 & 1 & 2 \\ -1 & 0 & 5 \\ 2 & 5 & 0 \end{bmatrix} = -A$, A is *not* skew-symmetric.

38. A square matrix is called skew-symmetric if $A^T = -A \Leftrightarrow \left[A^T\right]_{ij} = [-A]_{ij} \Leftrightarrow [A]_{ji} = -[A]_{ij}$. Thus, the components must satisfy $a_{ji} = -a_{ij}$.

39. If A is skew-symmetric ($A^T = -A$), then the diagonal entries must be zero ($a_{ii} = 0$).

$$A^T = -A \quad \Rightarrow \quad \left[A^T\right]_{ij} = [-A]_{ij} \quad \Rightarrow \quad [A]_{ji} = -[A]_{ij}$$

So $a_{ji} = -a_{ij} \quad \Rightarrow \quad a_{ii} = -a_{ii} \quad \Rightarrow \quad 2a_{ii} = 0 \quad \Rightarrow \quad a_{ii} = 0$

40. If A and B are skew-symmetric, then so is $A + B$, that is $(A + B)^T = -(A + B)$.

$$(A+B)^T \overset{\substack{\text{by} \\ \text{Thm 3.4b}}}{=} A^T + B^T \overset{\substack{A \text{ and } B \text{ are} \\ \text{skew-symmetric}}}{=} (-A) + (-B) \overset{\substack{\text{by} \\ \text{Thm 3.2}}}{=} -(A+B)$$

41. Let A and B be skew-symmetric 2×2 matrices, so $A = \begin{bmatrix} 0 & a \\ -a & 0 \end{bmatrix}$ and $B = \begin{bmatrix} 0 & b \\ -b & 0 \end{bmatrix}$.

Demanding that AB be skew-symmetric gives us the equation:

$$(AB)^T = -AB \Leftrightarrow \left(\begin{bmatrix} 0 & a \\ -a & 0 \end{bmatrix} \begin{bmatrix} 0 & b \\ -b & 0 \end{bmatrix} \right)^T = - \begin{bmatrix} 0 & a \\ -a & 0 \end{bmatrix} \begin{bmatrix} 0 & b \\ -b & 0 \end{bmatrix} \Leftrightarrow$$

$$\begin{bmatrix} -ab & 0 \\ 0 & -ab \end{bmatrix}^T = \begin{bmatrix} ab & 0 \\ 0 & ab \end{bmatrix} \Leftrightarrow \begin{bmatrix} -ab & 0 \\ 0 & -ab \end{bmatrix} = \begin{bmatrix} ab & 0 \\ 0 & ab \end{bmatrix}.$$

So, $-ab = ab \Leftrightarrow ab = 0$. Letting $O =$ the zero matrix, we get:
AB will be skew-symmetric provided either $A = O$ or $B = O$ (or both).

42. If A is a square matrix, then $A - A^T$ is skew-symmetric, that is $(A - A^T)^T = -(A - A^T)$.

$$(A - A^T)^T \overset{\text{by Thm 3.4b}}{=} A^T - (A^T)^T \overset{\text{by Thm 3.4a}}{=} A^T - A \overset{\text{by Thm 3.2}}{=} -(A - A^T)$$

43. We will prove this claim in (a) and demonstrate it with an example in (b).

(a) If A is $n \times n$, then $A = B + C$, where B is symmetric and C is skew-symmetric.

$$S \overset{\text{symmetric by Thm 3.5a}}{=} A + A^T \text{ and } S' \overset{\text{skew-symmetric by Exercise 42}}{=} A - A^T$$

Now simply note $A = \frac{1}{2}(A + A^T) + \frac{1}{2}(A - A^T) = \frac{1}{2}S + \frac{1}{2}S'$

Q: When S is symmetric and S' is skew-symmetric, are cS and cS' also?
A: Yes, since $(cS)^T = cS^T = cS$ and $(cS')^T = c(S')^T = -cS'$.

(b)
$$A = \begin{bmatrix} 1 & 2 & 3 \\ 4 & 5 & 6 \\ 7 & 8 & 9 \end{bmatrix} = \frac{1}{2}\left(\begin{bmatrix} 1 & 2 & 3 \\ 4 & 5 & 6 \\ 7 & 8 & 9 \end{bmatrix} + \begin{bmatrix} 1 & 4 & 7 \\ 2 & 5 & 8 \\ 3 & 6 & 9 \end{bmatrix} \right) + \frac{1}{2}\left(\begin{bmatrix} 1 & 2 & 3 \\ 4 & 5 & 6 \\ 7 & 8 & 9 \end{bmatrix} - \begin{bmatrix} 1 & 4 & 7 \\ 2 & 5 & 8 \\ 3 & 6 & 9 \end{bmatrix} \right)$$
$$= \begin{bmatrix} 1 & 3 & 5 \\ 3 & 5 & 7 \\ 5 & 7 & 9 \end{bmatrix} + \begin{bmatrix} 0 & -1 & -2 \\ 1 & 0 & -1 \\ 2 & 1 & 0 \end{bmatrix}.$$

44. Let A and B be $n \times n$ matrices, and let k be a scalar. Then

(i) $$\operatorname{tr}(A + B) = (a_{11} + b_{11}) + (a_{22} + b_{22}) + \cdots + (a_{nn} + b_{nn})$$
$$= (a_{11} + a_{22} + \cdots + a_{nn}) + (b_{11} + b_{22} + \cdots + b_{nn}) = tr\,(A) + tr\,(B).$$

(ii) $\operatorname{tr}(kA) = ka_{11} + ka_{22} + \cdots + ka_{nn} = k\,(a_{11} + a_{22} + \cdots + a_{nn}) = k\operatorname{tr}(A)$.

45. Let A and B be $n \times n$ matrices. Then

$$\begin{aligned} \operatorname{tr}(AB) &= (a_{11}b_{11} + a_{12}b_{21} + \cdots + a_{1n}b_{n1}) + (a_{21}b_{12} + a_{22}b_{22} + \cdots + a_{2n}b_{n2}) + \\ &\qquad \cdots + (a_{n1}b_{1n} + a_{n2}b_{2n} + \cdots + a_{nn}b_{nn}) \\ &= (b_{11}a_{11} + b_{12}a_{21} + \cdots + b_{1n}a_{n1}) + (b_{21}a_{12} + b_{22}a_{22} + \cdots + b_{2n}a_{n2}) + \\ &\qquad \cdots + (b_{n1}a_{1n} + b_{n2}a_{2n} + \cdots + b_{nn}a_{nn}) \\ &= \operatorname{tr}(BA) \end{aligned}$$

46. Let A be an $n \times n$ matrix. Then
$\operatorname{tr}(AA^T) = (a_{11}^2 + a_{12}^2 + \cdots + a_{1n}^2) + (a_{21}^2 + a_{22}^2 + \cdots + a_{2n}^2) + \cdots + (a_{n1}^2 + a_{n2}^2 + \cdots + a_{nn}^2)$,
that is, the sum of the squares of the entries of A.

47. If A, B are 2×2, we will show $\operatorname{tr}(AB - BA) \neq \operatorname{tr}(I_2)$ which will imply $AB - BA \neq I_2$.

$$\operatorname{tr}(AB - BA) \overset{\substack{\text{by Exercises} \\ \text{44 and 45}}}{=} \operatorname{tr}(AB) - \operatorname{tr}(BA) = 0 \neq 2 = \operatorname{tr}(I_2) \Rightarrow AB - BA \neq I_2$$

Q: We have shown $AB - BA$ must equal a matrix of trace zero. Is this true for $n \times n$ matrices?
A: Yes, since $\operatorname{tr}(AB - BA) = 0$ for any square matrix.

3.3 The Inverse of a Matrix

1. As is Example 3.24, we begin by computing the determinant of A, $\det A$. Why?
Since if $\det A = 0$, then A is not invertible.

Since $A = \begin{bmatrix} a & b \\ c & d \end{bmatrix} = \begin{bmatrix} 4 & 7 \\ 1 & 2 \end{bmatrix}$, $\det A = ad - bc = 4(2) - 1(7) = 1$, A is invertible.

So, by Theorem 3.8, $A^{-1} = \frac{1}{ad-bc}\begin{bmatrix} d & -b \\ -c & a \end{bmatrix} = \frac{1}{\det A}\begin{bmatrix} d & -b \\ -c & a \end{bmatrix} = \frac{1}{1}\begin{bmatrix} 2 & -7 \\ -1 & 4 \end{bmatrix} = \begin{bmatrix} 2 & -7 \\ -1 & 4 \end{bmatrix}$.

Q: How can we check our answer for A^{-1}?
A: By applying the definition. That is, by checking that $AA^{-1} = I$. Let's do that:

$$AA^{-1} = \begin{bmatrix} 4 & 7 \\ 1 & 2 \end{bmatrix}\begin{bmatrix} 2 & -7 \\ -1 & 4 \end{bmatrix} = \begin{bmatrix} 4(2) + 7(-1) & 1(2) + 2(-1) \\ 4(-7) + 7(4) & 1(-7) + 2(4) \end{bmatrix} = \begin{bmatrix} 1 & 0 \\ 0 & 1 \end{bmatrix} = I.$$

Q: Should $A^{-1}A = I$, too?
A: Yes. We should check that as well. On the calculator, these are quick and simple to do.

2. As is Example 3.24, we begin by computing the determinant of A, $\det A$. Why?

Since $A = \begin{bmatrix} 4 & -2 \\ 2 & 0 \end{bmatrix}$, $\det A = 4(0) - (-2)(2) = 4$, A is invertible.

So, by Theorem 3.8, $A^{-1} = \frac{1}{4}\begin{bmatrix} 0 & 2 \\ -2 & 4 \end{bmatrix} = \begin{bmatrix} 0 & \frac{1}{2} \\ -\frac{1}{2} & 1 \end{bmatrix}$.

Check: $AA^{-1} = \begin{bmatrix} 4 & -2 \\ 2 & 0 \end{bmatrix}\begin{bmatrix} 0 & \frac{1}{2} \\ -\frac{1}{2} & 1 \end{bmatrix} = \begin{bmatrix} 1 & 0 \\ 0 & 1 \end{bmatrix} = I$ or $A^{-1}A = \begin{bmatrix} 0 & \frac{1}{2} \\ -\frac{1}{2} & 1 \end{bmatrix}\begin{bmatrix} 4 & -2 \\ 2 & 0 \end{bmatrix} = I$.

3. As is Example 3.24, we begin by computing the determinant of A, $\det A$. Why?

Since $A = \begin{bmatrix} 3 & 4 \\ 6 & 8 \end{bmatrix}$, $\det A = 3(8) - 4(6) = 0$, A is not invertible.

4. Since $\det A = 0(0) - (-1)(1) = 1$, A is invertible. So, by Theorem 3.8, we have:

$$A^{-1} = \frac{1}{1}\begin{bmatrix} 0 & 1 \\ -1 & 0 \end{bmatrix} = \begin{bmatrix} 0 & 1 \\ -1 & 0 \end{bmatrix}.$$

Check: $AA^{-1} = \begin{bmatrix} 0 & -1 \\ 1 & 0 \end{bmatrix}\begin{bmatrix} 0 & 1 \\ -1 & 0 \end{bmatrix} = \begin{bmatrix} 1 & 0 \\ 0 & 1 \end{bmatrix} = I$ or $A^{-1}A = \begin{bmatrix} 0 & 1 \\ -1 & 0 \end{bmatrix}\begin{bmatrix} 0 & -1 \\ 1 & 0 \end{bmatrix} = I$.

5. Since $\det A = \frac{3}{4}(\frac{2}{3}) - \frac{3}{5}(\frac{5}{6}) = 0$, A is not invertible.

6. Since $\det A = (\frac{\sqrt{2}}{2})(\sqrt{2}) - (-\sqrt{2})(2\sqrt{2}) = 5$, A is invertible. So, by Theorem 3.8:

$$A^{-1} = \frac{1}{5}\begin{bmatrix} \sqrt{2} & \sqrt{2} \\ -2\sqrt{2} & \frac{\sqrt{2}}{2} \end{bmatrix} = \begin{bmatrix} \frac{\sqrt{2}}{5} & \frac{\sqrt{2}}{5} \\ -\frac{2\sqrt{2}}{5} & \frac{\sqrt{2}}{10} \end{bmatrix}.$$

$$\text{Check: } AA^{-1} = \begin{bmatrix} \frac{\sqrt{2}}{2} & -\sqrt{2} \\ 2\sqrt{2} & \sqrt{2} \end{bmatrix} = \begin{bmatrix} \frac{\sqrt{2}}{5} & \frac{\sqrt{2}}{5} \\ -\frac{2\sqrt{2}}{5} & \frac{\sqrt{2}}{10} \end{bmatrix} = \begin{bmatrix} \frac{2}{10} + \frac{8}{10} & -\frac{2}{5} + \frac{2}{5} \\ -\frac{8}{10} + \frac{8}{10} & \frac{8}{10} + \frac{2}{10} \end{bmatrix} = \begin{bmatrix} 1 & 0 \\ 0 & 1 \end{bmatrix} = I.$$

7. Since $\det A = (-1.5)(2.4) - (-4.2)(0.5) = -1.5$, A is invertible. So, by Theorem 3.8:

$$A^{-1} = -\frac{1}{1.5}\begin{bmatrix} 2.4 & 4.2 \\ -0.5 & -1.5 \end{bmatrix} = \begin{bmatrix} -1.6 & -2.8 \\ 0.\overline{3} & 1 \end{bmatrix}$$

$$\text{Check: } AA^{-1} = \begin{bmatrix} -1.5 & -4.2 \\ 0.5 & 2.4 \end{bmatrix}\begin{bmatrix} -1.6 & -2.8 \\ 0.\overline{3} & 1 \end{bmatrix} = \begin{bmatrix} 1 & 0 \\ 0 & 1 \end{bmatrix} = I.$$

8. Since $\det A = (2.54)(0.8) - (8.128)(0.25) = 0$, A is not invertible.

9. Since $\det A = (a)(a) - (-b)(b) = a^2 + b^2$, provided a and b are not both zero, A is invertible.

$$A^{-1} = \frac{1}{a^2+b^2}\begin{bmatrix} a & b \\ -b & a \end{bmatrix} = \begin{bmatrix} \frac{a}{a^2+b^2} & \frac{b}{a^2+b^2} \\ -\frac{b}{a^2+b^2} & \frac{a}{a^2+b^2} \end{bmatrix} \quad (a \text{ and } b \text{ not both zero})$$

$$\text{Check: } AA^{-1} = \begin{bmatrix} a & -b \\ b & a \end{bmatrix}\begin{bmatrix} \frac{a}{a^2+b^2} & \frac{b}{a^2+b^2} \\ -\frac{b}{a^2+b^2} & \frac{a}{a^2+b^2} \end{bmatrix} = \begin{bmatrix} \frac{a^2+b^2}{a^2+b^2} & \frac{ab-ba}{a^2+b^2} \\ \frac{ba-ab}{a^2+b^2} & \frac{b^2+a^2}{a^2+b^2} \end{bmatrix} = \begin{bmatrix} 1 & 0 \\ 0 & 1 \end{bmatrix} = I.$$

10. Since $\det A = \frac{1}{a} \cdot \frac{1}{d} - \frac{1}{b} \cdot \frac{1}{c} = \frac{bc-ad}{abcd}$, when a, b, c, d not zero, $bc \neq ad$, A is invertible.

$$A^{-1} = \frac{abcd}{bc-ad}\begin{bmatrix} \frac{1}{d} & -\frac{1}{b} \\ -\frac{1}{c} & \frac{1}{a} \end{bmatrix} = \begin{bmatrix} \frac{abc}{bc-ad} & \frac{acd}{ad-bc} \\ \frac{abd}{ad-bc} & \frac{bcd}{bc-ad} \end{bmatrix} \quad (a, b, c, d \text{ all not zero and } bc \neq ad)$$

$$\text{Check: } AA^{-1} = \begin{bmatrix} \frac{1}{a} & \frac{1}{b} \\ \frac{1}{c} & \frac{1}{d} \end{bmatrix}\begin{bmatrix} \frac{abc}{bc-ad} & \frac{acd}{ad-bc} \\ \frac{abd}{ad-bc} & \frac{bcd}{bc-ad} \end{bmatrix} = \begin{bmatrix} \frac{bc-ad}{bc-ad} & \frac{-cd+cd}{ad-bc} \\ \frac{ab-ab}{ad-bc} & \frac{-ad+bc}{bc-ad} \end{bmatrix} = \begin{bmatrix} 1 & 0 \\ 0 & 1 \end{bmatrix} = I.$$

11. As in Example 3.25, we use the inverse of the coefficient matrix A to solve the system. That is, the reasoning we will employ here is: $A\mathbf{x} = \mathbf{b} \Rightarrow \mathbf{x} = A^{-1}\mathbf{b}$.

Since $A = \begin{bmatrix} 2 & 1 \\ 5 & 3 \end{bmatrix}$, we have $A^{-1} = \frac{1}{1}\begin{bmatrix} 3 & -1 \\ -5 & 2 \end{bmatrix} = \begin{bmatrix} 3 & -1 \\ -5 & 2 \end{bmatrix}$.

Therefore, since $\mathbf{b} = \begin{bmatrix} -1 \\ 2 \end{bmatrix}$, we have $\mathbf{x} = A^{-1}\mathbf{b} = \begin{bmatrix} 3 & -1 \\ -5 & 2 \end{bmatrix}\begin{bmatrix} -1 \\ 2 \end{bmatrix} = \begin{bmatrix} -5 \\ 9 \end{bmatrix}$.

Q: How can we check our answer for $\mathbf{x}$?
A: By applying the condition. That is, by checking that $A\mathbf{x} = \mathbf{b}$. Let's do that:

$$\text{Check: } A\mathbf{x} = \begin{bmatrix} 2 & 1 \\ 5 & 3 \end{bmatrix}\begin{bmatrix} -5 \\ 9 \end{bmatrix} = \begin{bmatrix} -1 \\ 2 \end{bmatrix} = \mathbf{b}.$$

12. As in Example 3.25, we use the inverse of the coefficient matrix A to solve the system.

Since $A = \begin{bmatrix} 1 & -1 \\ 2 & 1 \end{bmatrix}$, we have $A^{-1} = \frac{1}{3}\begin{bmatrix} 1 & 1 \\ -2 & 1 \end{bmatrix} = \begin{bmatrix} \frac{1}{3} & \frac{1}{3} \\ -\frac{2}{3} & \frac{1}{3} \end{bmatrix}$.

Therefore, since $\mathbf{b} = \begin{bmatrix} 1 \\ 2 \end{bmatrix}$, we have $\mathbf{x} = A^{-1}\mathbf{b} = \begin{bmatrix} \frac{1}{3} & \frac{1}{3} \\ -\frac{2}{3} & \frac{1}{3} \end{bmatrix}\begin{bmatrix} 1 \\ 2 \end{bmatrix} = \begin{bmatrix} 1 \\ 0 \end{bmatrix}$.

Check: $A\mathbf{x} = \begin{bmatrix} 1 & -1 \\ 2 & 1 \end{bmatrix}\begin{bmatrix} 1 \\ 0 \end{bmatrix} = \begin{bmatrix} 1 \\ 2 \end{bmatrix} = \mathbf{b}$.

13. As in Example 3.25, we use the inverse of the coefficient matrix A to solve the systems.

Since $A = \begin{bmatrix} 1 & 2 \\ 2 & 6 \end{bmatrix}$, we have$A^{-1} = \frac{1}{2}\begin{bmatrix} 6 & -2 \\ -2 & 1 \end{bmatrix} = \begin{bmatrix} 3 & -1 \\ -1 & \frac{1}{2} \end{bmatrix}$.

Also, note that we are given $\mathbf{b}_1 = \begin{bmatrix} 3 \\ 5 \end{bmatrix}$, $\mathbf{b}_2 = \begin{bmatrix} -1 \\ 2 \end{bmatrix}$, and $\mathbf{b}_3 = \begin{bmatrix} 2 \\ 0 \end{bmatrix}$.

(a) Since $A\mathbf{x}_i = \mathbf{b}_i$, the solution in each case is $\mathbf{x}_i = A^{-1}\mathbf{b}_i$. So we have:

$$\mathbf{x}_1 = A^{-1}\mathbf{b}_1 = \begin{bmatrix} 3 & -1 \\ -1 & \frac{1}{2} \end{bmatrix}\begin{bmatrix} 3 \\ 5 \end{bmatrix} = \begin{bmatrix} 4 \\ -\frac{1}{2} \end{bmatrix}$$

$$\mathbf{x}_2 = A^{-1}\mathbf{b}_2 = \begin{bmatrix} 3 & -1 \\ -1 & \frac{1}{2} \end{bmatrix}\begin{bmatrix} -1 \\ 2 \end{bmatrix} = \begin{bmatrix} -5 \\ 2 \end{bmatrix}$$

$$\mathbf{x}_3 = A^{-1}\mathbf{b}_3 = \begin{bmatrix} 3 & -1 \\ -1 & \frac{1}{2} \end{bmatrix}\begin{bmatrix} 2 \\ 0 \end{bmatrix} = \begin{bmatrix} 6 \\ -2 \end{bmatrix}$$

(b) We form the augmented matrix $\left[\, A \mid \mathbf{b}_1 \;\; \mathbf{b}_2 \;\; \mathbf{b}_3 \,\right]$ and row reduce to solve.

$$\left[\begin{array}{cc|ccc} 1 & 2 & 3 & -1 & 2 \\ 2 & 6 & 5 & 2 & 0 \end{array}\right] \longrightarrow \left[\begin{array}{cc|ccc} 1 & 0 & 4 & -5 & 6 \\ 0 & 1 & -\frac{1}{2} & 2 & -2 \end{array}\right].$$

(c) The A^{-1} method requires 7 multiplications, while row reduction requires only 6.

14. To prove X is the inverse of A, all we have to show is $AX = I$.
Theorem 3.9b. asserts $(cA)^{-1} = \frac{1}{c}A^{-1}$, so all we need to show is $(cA)\left(\frac{1}{c}A^{-1}\right) = I$.

$$(cA)\left(\tfrac{1}{c}A^{-1}\right) = \left(\tfrac{1}{c}c\right)\left(AA^{-1}\right) = AA^{-1} = I$$

15. To prove X is the inverse of A, all we have to show is $AX = I$.
Theorem 3.9d. asserts $(A^T)^{-1} = (A^{-1})^T$, so all we need to show is $(A^T)(A^{-1})^T = I$.

$$(A^T)(A^{-1})^T \overset{\substack{\text{by Thm 3.4d} \\ B^TA^T=(AB)^T}}{=} (A^{-1}A)^T = I^T \overset{\substack{\text{This is obvious.} \\ \text{Why?}}}{=} I$$

16. To prove X is the inverse of A, all we have to show is $AX = I$.
We claim $I_n^{-1} = I_n$, so all we need to show is $I_n \cdot I_n = I_n$, but that is obvious.

Q: But did we also prove that I_n is invertible?
A: Yes. How? By proving it has an inverse, namely it itself, I_n.

17. In (a) we will give an example, and in (b) we will prove $(AB)^{-1} = A^{-1}B^{-1} \Leftrightarrow AB = BA$.

(a) We choose A, B, then compute $(AB)^{-1}$, $A^{-1}B^{-1}$ to show $(AB)^{-1} \neq A^{-1}B^{-1}$.
Let $A = \begin{bmatrix} 1 & 0 \\ 2 & 1 \end{bmatrix}$ and $B = \begin{bmatrix} 1 & 0 \\ -2 & 1 \end{bmatrix}$. Then

$$(AB)^{-1} = \left(\begin{bmatrix} 1 & 0 \\ 2 & 1 \end{bmatrix}\begin{bmatrix} 1 & 0 \\ 1 & -1 \end{bmatrix}\right)^{-1} = \left(\begin{bmatrix} 1 & 0 \\ 3 & -1 \end{bmatrix}\right)^{-1} = \begin{bmatrix} 1 & 0 \\ 3 & -1 \end{bmatrix},$$

$$A^{-1}B^{-1} = \begin{bmatrix} 1 & 0 \\ 2 & 1 \end{bmatrix}^{-1}\begin{bmatrix} 1 & 0 \\ 1 & -1 \end{bmatrix}^{-1} = \begin{bmatrix} 1 & 0 \\ -2 & 1 \end{bmatrix}\begin{bmatrix} 1 & 0 \\ 1 & -1 \end{bmatrix} = \begin{bmatrix} 1 & 0 \\ -1 & -1 \end{bmatrix}.$$

So, clearly $(AB)^{-1} \neq A^{-1}B^{-1}$.

(b) Since this is an *if and only if* statement, we have two claims to prove.

if: If $(AB)^{-1} = A^{-1}B^{-1}$, then $AB = BA$. (Note $(AB)^{-1} = A^{-1}B^{-1}$ is *given*).

$$AB \overset{\substack{\text{by}\\ \text{Thm 3.9a}}}{=} ((AB)^{-1})^{-1} \overset{\substack{\text{by}\\ \text{given}}}{=} (A^{-1}B^{-1})^{-1} \overset{\substack{\text{by}\\ \text{Thm 3.9c}}}{=} (B^{-1})^{-1}(A^{-1})^{-1} \overset{\substack{\text{by}\\ \text{Thm 3.9a}}}{=} BA$$

only if: If $AB = BA$, then $(AB)^{-1} = A^{-1}B^{-1}$.

$$(AB)^{-1} \overset{\substack{\text{by given}\\ AB=BA}}{=} (BA)^{-1} \overset{\substack{\text{by}\\ \text{Thm 3.9c}}}{=} A^{-1}B^{-1}$$

18. We prove $A_1A_2\cdots A_n$ is invertible and $(A_1A_2\cdots A_n)^{-1} = A_n^{-1}\cdots A_2^{-1}A_1^{-1}$ by induction.

1: $(A_1)^{-1} = A_1^{-1}$. This is obvious, so there is nothing to show.

k: $(A_1A_2\cdots A_k)^{-1} = A_k^{-1}\cdots A_2^{-1}A_1^{-1}$. This is the induction hypothesis.

$k+1$: $(A_1A_2\cdots A_kA_{k+1})^{-1} = A_{k+1}^{-1}A_k^{-1}\cdots A_2^{-1}A_1^{-1}$.
This is the statement we must prove using the induction hypothesis.

$$(A_1A_2\cdots A_kA_{k+1})^{-1} \overset{\substack{\text{by}\\ \text{associativity}}}{=} ((A_1A_2\cdots A_k)A_{k+1})^{-1} \overset{\substack{\text{by}\\ \text{Thm 3.9c}}}{=} A_{k+1}^{-1}(A_1A_2\cdots A_k)^{-1}$$

$$\overset{\substack{\text{by}\\ \text{induction}}}{=} A_{k+1}^{-1}A_k^{-1}\cdots A_2^{-1}A_1^{-1}$$

We have shown the pattern holds for $k+1$. What does that mean?

We have proven (by induction) that $(A_1A_2\cdots A_n)^{-1} = A_n^{-1}\cdots A_2^{-1}A_1^{-1}$.

19. Let $A = \begin{bmatrix} 1 & 0 \\ 0 & 1 \end{bmatrix}$ and $B = \begin{bmatrix} 1 & 2 \\ 0 & 1 \end{bmatrix}$. Then

$$(A+B)^{-1} = \left(\begin{bmatrix} 1 & 0 \\ 0 & 1 \end{bmatrix} + \begin{bmatrix} 1 & 2 \\ 0 & 1 \end{bmatrix}\right)^{-1} = \begin{bmatrix} 2 & 2 \\ 0 & 2 \end{bmatrix}^{-1} = \begin{bmatrix} \frac{1}{2} & -\frac{1}{2} \\ 0 & \frac{1}{2} \end{bmatrix}$$

$$A^{-1} + B^{-1} = \begin{bmatrix} 1 & 0 \\ 0 & 1 \end{bmatrix}^{-1} + \begin{bmatrix} 1 & 2 \\ 0 & 1 \end{bmatrix}^{-1} = \begin{bmatrix} 1 & 0 \\ 0 & 1 \end{bmatrix} + \begin{bmatrix} 1 & -2 \\ 0 & 1 \end{bmatrix} = \begin{bmatrix} 2 & -2 \\ 0 & 2 \end{bmatrix}$$

So $(A+B)^{-1} \neq A^{-1} + B^{-1}$.

20. $XA^2 = A^{-1} \Rightarrow (XA^2)(A^2)^{-1} = A^{-1}(A^2)^{-1} \Rightarrow X\left((A^2)(A^2)^{-1}\right) = A^{-1}(A^2)^{-1}$
So $X = A^{-1}(A^{-1})^2 = (A^{-1})^3 = A^{-3}$.

21. $AXB = (BA)^2 \Rightarrow A^{-1}(AXB)B^{-1} = A^{-1}(BA)^2B^{-1} \Rightarrow (A^{-1}A)X(BB^{-1}) = A^{-1}(BA)^2B^{-1}$
So $X = A^{-1}(BA)^2B^{-1}$.

22. $(A^{-1}X)^{-1} = A(B^{-2}A)^{-1} \Rightarrow X^{-1}(A^{-1})^{-1} = A(A^{-1}(B^{-2})^{-1}) \Rightarrow X^{-1}A = B^2$
So $X = (B^2A^{-1})^{-1} = (A^{-1})^{-1}(B^2)^{-1} = AB^{-2}$.

23. $ABXA^{-1}B^{-1} = I + A \Rightarrow X = B^{-1}A^{-1}(I+A)BA = B^{-1}A^{-1}BA + B^{-1}A^{-1}ABA$
So $X = (AB)^{-1}BA + A$.

24. $E = \begin{bmatrix} 0 & 0 & 1 \\ 0 & 1 & 0 \\ 1 & 0 & 0 \end{bmatrix}$. **25.** $E = \begin{bmatrix} 0 & 0 & 1 \\ 0 & 1 & 0 \\ 1 & 0 & 0 \end{bmatrix}$. **26.** $E = \begin{bmatrix} 1 & 0 & 0 \\ 0 & 1 & 0 \\ 1 & 0 & 1 \end{bmatrix}$.

27. $E = \begin{bmatrix} 1 & 0 & 0 \\ 0 & 1 & 0 \\ -1 & 0 & 1 \end{bmatrix}$. **28.** $E = \begin{bmatrix} 1 & 0 & 0 \\ 0 & 1 & -2 \\ 0 & 0 & 1 \end{bmatrix}$. **29.** $E = \begin{bmatrix} 1 & 0 & 0 \\ 0 & 1 & 2 \\ 0 & 0 & 1 \end{bmatrix}$.

30. The row operations $R_2 - 2R_1 - 2R_3$, $R_3 + R_1$, will turn matrix A into matrix D. So,

$$E = \begin{bmatrix} 1 & 0 & 0 \\ -2 & 1 & -2 \\ 1 & 0 & 1 \end{bmatrix} \text{ satisfies } EA = \begin{bmatrix} 1 & 0 & 0 \\ -2 & 1 & -2 \\ 1 & 0 & 1 \end{bmatrix}\begin{bmatrix} 1 & 2 & -1 \\ 1 & 1 & 1 \\ 1 & -1 & 0 \end{bmatrix} = \begin{bmatrix} 1 & 2 & -1 \\ -3 & -1 & 3 \\ 2 & 1 & -1 \end{bmatrix} = D.$$

However, E is not an elementary matrix since it incorporates two elementary row operations.

31. $\begin{bmatrix} 3 & 0 \\ 0 & 1 \end{bmatrix}^{-1} = \begin{bmatrix} \frac{1}{3} & 0 \\ 0 & 1 \end{bmatrix}$. **32.** $\begin{bmatrix} 1 & 2 \\ 0 & 1 \end{bmatrix}^{-1} = \begin{bmatrix} 1 & -2 \\ 0 & 1 \end{bmatrix}$.

33. $\begin{bmatrix} 0 & 1 \\ 1 & 0 \end{bmatrix}^{-1} = \begin{bmatrix} 0 & 1 \\ 1 & 0 \end{bmatrix}$. **34.** $\begin{bmatrix} 1 & 0 \\ -\frac{1}{2} & 1 \end{bmatrix}^{-1} = \begin{bmatrix} 1 & 0 \\ \frac{1}{2} & 1 \end{bmatrix}$.

35. $\begin{bmatrix} 1 & 0 & 0 \\ 0 & 1 & -2 \\ 0 & 0 & 1 \end{bmatrix}^{-1} = \begin{bmatrix} 1 & 0 & 0 \\ 0 & 1 & 2 \\ 0 & 0 & 1 \end{bmatrix}$.

36. $\begin{bmatrix} 0 & 0 & 1 \\ 0 & 1 & 0 \\ 1 & 0 & 0 \end{bmatrix}^{-1} = \begin{bmatrix} 0 & 0 & 1 \\ 0 & 1 & 0 \\ 1 & 0 & 0 \end{bmatrix}$.

37. $\begin{bmatrix} 1 & 0 & 0 \\ 0 & c & 0 \\ 0 & 0 & 1 \end{bmatrix}^{-1} = \begin{bmatrix} 1 & 0 & 0 \\ 0 & \frac{1}{c} & 0 \\ 0 & 0 & 1 \end{bmatrix}$.

38. $\begin{bmatrix} 1 & 0 & 0 \\ 0 & 1 & c \\ 0 & 0 & 1 \end{bmatrix}^{-1} = \begin{bmatrix} 1 & 0 & 0 \\ 0 & 1 & -c \\ 0 & 0 & 1 \end{bmatrix}$.

39. As in Example 3.29, we attempt to express A as a product of elementary matrices. Why? To compute A^{-1} and to *factor* A, that is to see how A is created by elementary row operations.

As we row reduce A, we use the elementary operations involved to create E_i.

$$A = \begin{bmatrix} 1 & 0 \\ -1 & -2 \end{bmatrix} \xrightarrow{R_2+R_1} \begin{bmatrix} 1 & 0 \\ 0 & -2 \end{bmatrix} \xrightarrow{-\frac{1}{2}R_2} \begin{bmatrix} 1 & 0 \\ 0 & 1 \end{bmatrix}$$

We recreate those steps on the identity matrix I to create E_1 and E_2.
This calculation can be done mentally, but we write it here to demonstrate the process.

Since $I = \begin{bmatrix} 1 & 0 \\ 0 & 1 \end{bmatrix} \xrightarrow{R_2+R_1} \begin{bmatrix} 1 & 0 \\ 1 & 1 \end{bmatrix}$, we get $E_1 = \begin{bmatrix} 1 & 0 \\ 1 & 1 \end{bmatrix}$.

Since $I = \begin{bmatrix} 1 & 0 \\ 0 & 1 \end{bmatrix} \xrightarrow{-\frac{1}{2}R_2} \begin{bmatrix} 1 & 0 \\ 0 & -\frac{1}{2} \end{bmatrix}$, we get $E_2 = \begin{bmatrix} 1 & 0 \\ 0 & -\frac{1}{2} \end{bmatrix}$.

Since $(E_2E_1)A = I$, by definition $A^{-1} = E_2E_1$.
So, Theorem 3.9a implies $A = (A^{-1})^{-1} = (E_2E_1)^{-1} = E_1^{-1}E_2^{-1}$.

But E_1^{-1} and E_2^{-1} can be written down without doing any calculation. How?
Since E_1 was created by $R_2 + R_1$, we create E_1^{-1} by $R_2 - R_1$.
Likewise, since E_2 was created by $-\frac{1}{2}R_2$, we create E_2^{-1} by $-2R_2$.

Therefore, $E_1^{-1} = \begin{bmatrix} 1 & 0 \\ -1 & 1 \end{bmatrix}$ and $E_2^{-1} = \begin{bmatrix} 1 & 0 \\ 0 & -2 \end{bmatrix}$.

We could verify this by performing the indication operation on the identity matrix I.
Or we could simply verify the claim that $A = E_1^{-1}E_2^{-1}$ directly.

Check: $A = \begin{bmatrix} 1 & 0 \\ -1 & -2 \end{bmatrix} = \begin{bmatrix} 1 & 0 \\ -1 & 1 \end{bmatrix}\begin{bmatrix} 1 & 0 \\ 0 & -2 \end{bmatrix} = E_1^{-1}E_2^{-1}$

This answer is not unique. For example, we could have row reduced A as follows.

$$A = \begin{bmatrix} 1 & 0 \\ -1 & -2 \end{bmatrix} \xrightarrow{-\frac{1}{2}R_2} \begin{bmatrix} 1 & 0 \\ \frac{1}{2} & 1 \end{bmatrix} \xrightarrow{R_2-\frac{1}{2}R_1} \begin{bmatrix} 1 & 0 \\ 0 & 1 \end{bmatrix}$$

So $E_1 = \begin{bmatrix} 1 & 0 \\ 0 & -\frac{1}{2} \end{bmatrix}$, $E_2 = \begin{bmatrix} 1 & 0 \\ -\frac{1}{2} & 1 \end{bmatrix}$ and $E_1^{-1} = \begin{bmatrix} 1 & 0 \\ 0 & -2 \end{bmatrix}$, $E_2^{-1} = \begin{bmatrix} 1 & 0 \\ \frac{1}{2} & 1 \end{bmatrix}$.

Verify that $A = E_1^{-1}E_2^{-1}$ in this case as well.

40. As in Example 3.29, we attempt to express A as a product of elementary matrices. Why? To compute A^{-1} and to *factor* A, that is to see how A is created by elementary row operations.

As we row reduce A, we use the elementary operations involved to create E_i and E_i^{-1}.

$$A=\begin{bmatrix}2&4\\1&1\end{bmatrix}\xrightarrow{R_1\leftrightarrow R_2}\begin{bmatrix}1&1\\2&4\end{bmatrix}\xrightarrow{R_2-2R_1}\begin{bmatrix}1&1\\0&2\end{bmatrix}\xrightarrow{\frac{1}{2}R_2}\begin{bmatrix}1&1\\0&1\end{bmatrix}\xrightarrow{R_1-R_2}\begin{bmatrix}1&0\\0&1\end{bmatrix}$$

So $E_1=\begin{bmatrix}0&1\\1&0\end{bmatrix}$, $E_2=\begin{bmatrix}1&0\\-2&1\end{bmatrix}$, $E_3=\begin{bmatrix}1&0\\0&\frac{1}{2}\end{bmatrix}$, and $E_4=\begin{bmatrix}1&-1\\0&1\end{bmatrix}$.

And $E_1^{-1}=\begin{bmatrix}0&1\\1&0\end{bmatrix}$, $E_2^{-1}=\begin{bmatrix}1&0\\2&1\end{bmatrix}$, $E_3^{-1}=\begin{bmatrix}1&0\\0&2\end{bmatrix}$, and $E_4^{-1}=\begin{bmatrix}1&1\\0&1\end{bmatrix}$.

Since $E_4E_3E_2E_1A=I$, we should have $A=E_1^{-1}E_2^{-1}E_3^{-1}E_4^{-1}$.

$$\text{Check: } A=\begin{bmatrix}2&4\\1&1\end{bmatrix}=\begin{bmatrix}0&1\\1&0\end{bmatrix}\begin{bmatrix}1&0\\2&1\end{bmatrix}\begin{bmatrix}1&0\\0&2\end{bmatrix}\begin{bmatrix}1&1\\0&1\end{bmatrix}=E_1^{-1}E_2^{-1}E_3^{-1}E_4^{-1}$$

This answer is not unique. For example, we could have row reduced A as follows.

$$A=\begin{bmatrix}2&4\\1&1\end{bmatrix}\xrightarrow{\frac{1}{2}R_1}\begin{bmatrix}1&2\\1&1\end{bmatrix}\xrightarrow{R_2-R_1}\begin{bmatrix}1&2\\0&-1\end{bmatrix}\xrightarrow{-R_2}\begin{bmatrix}1&2\\0&1\end{bmatrix}\xrightarrow{R_1-2R_2}\begin{bmatrix}1&0\\0&1\end{bmatrix}$$

So $E_1=\begin{bmatrix}\frac{1}{2}&0\\0&1\end{bmatrix}$, $E_2=\begin{bmatrix}1&0\\-1&1\end{bmatrix}$, $E_3=\begin{bmatrix}1&0\\0&-1\end{bmatrix}$, and $E_4=\begin{bmatrix}1&-2\\0&1\end{bmatrix}$.

And $E_1^{-1}=\begin{bmatrix}2&0\\0&1\end{bmatrix}$, $E_2^{-1}=\begin{bmatrix}1&0\\1&1\end{bmatrix}$, $E_3^{-1}=\begin{bmatrix}1&0\\0&-1\end{bmatrix}$, and $E_4^{-1}=\begin{bmatrix}1&2\\0&1\end{bmatrix}$.

Verify that $A=E_1^{-1}E_2^{-1}E_3^{-1}E_4^{-1}$ in this case as well.

41. Suppose $AB=I$. Then, consider: $B\mathbf{x}=\mathbf{0}$. Left-multiplying by $A\Rightarrow AB\mathbf{x}=A\mathbf{0}$.
This implies that $I\mathbf{x}=\mathbf{x}=\mathbf{0}$. Thus $B\mathbf{x}=\mathbf{0}$ has the unique solution $\mathbf{x}=\mathbf{0}\Rightarrow$
The equivalence of (c) and (a) in the Fundamental Theorem (3.12) $\Rightarrow B$ is invertible.
If we now right-multiply both sides of $AB=I$ by B^{-1}, we obtain
$ABB^{-1}=IB^{-1}\Leftrightarrow AI=B^{-1}\Leftrightarrow A=B^{-1}\Leftrightarrow A^{-1}=B$.
Thus the inverse of A exists and $A^{-1}=B$.

42. (a) Let A be invertible and $AB=0$. Now left-multiply both sides of $AB=0$ by $A^{-1}\Rightarrow$
$A^{-1}AB=A^{-1}0\Leftrightarrow IB=0\Leftrightarrow B=0$.

(b) $A=\begin{bmatrix}1&2\\2&4\end{bmatrix}$, $B=\begin{bmatrix}2&-2\\-1&1\end{bmatrix}\Rightarrow AB=\begin{bmatrix}1&2\\2&4\end{bmatrix}\begin{bmatrix}2&-2\\-1&1\end{bmatrix}=\begin{bmatrix}0&0\\0&0\end{bmatrix}$, but $B\neq 0$.

43. (a) Suppose that A is invertible (thus A^{-1} exists) and $BA=CA$.
If we now right-multiply both sides of $BA=CA$ by A^{-1}, we obtain
$BAA^{-1}=CAA^{-1}\Leftrightarrow BI=CI\Leftrightarrow B=C$.

(b) Let $A=\begin{bmatrix}1&2\\2&4\end{bmatrix}$, $B=\begin{bmatrix}1&1\\1&1\end{bmatrix}$, and $C=\begin{bmatrix}3&0\\1&1\end{bmatrix}$. Then

$$BA=\begin{bmatrix}1&1\\1&1\end{bmatrix}\begin{bmatrix}1&2\\2&4\end{bmatrix}=\begin{bmatrix}3&6\\3&6\end{bmatrix}=\begin{bmatrix}3&0\\1&1\end{bmatrix}\begin{bmatrix}1&2\\2&4\end{bmatrix}=CA.$$ So $BA=CA$, but $B\neq C$.

44. (a) Let $A = \begin{bmatrix} 1 & 0 \\ 0 & 1 \end{bmatrix}$, $B = \begin{bmatrix} 1 & 0 \\ 1 & 0 \end{bmatrix}$, and $C = \begin{bmatrix} 1 & 1 \\ 0 & 0 \end{bmatrix}$. Then

$$A^2 = \begin{bmatrix} 1 & 0 \\ 0 & 1 \end{bmatrix}\begin{bmatrix} 1 & 0 \\ 0 & 1 \end{bmatrix} = \begin{bmatrix} 1 & 0 \\ 0 & 1 \end{bmatrix} = A,$$

$$B^2 = \begin{bmatrix} 1 & 0 \\ 1 & 0 \end{bmatrix}\begin{bmatrix} 1 & 0 \\ 1 & 0 \end{bmatrix} = \begin{bmatrix} 1 & 0 \\ 1 & 0 \end{bmatrix} = B, \text{ and}$$

$$C^2 = \begin{bmatrix} 1 & 1 \\ 0 & 0 \end{bmatrix}\begin{bmatrix} 1 & 1 \\ 0 & 0 \end{bmatrix} = \begin{bmatrix} 1 & 1 \\ 0 & 0 \end{bmatrix} = C.$$

So A, B, and C are idempotent 2×2 matrices.

(b) Let A be an invertible idempotent $n \times n$ matrix. Since A is idempotent we have $A^2 = A$. But, A is invertible, so if we now right-multiply both sides of $AA = A$ by A^{-1}, we get:

$$AAA^{-1} = AA^{-1} \Leftrightarrow AI = I \Leftrightarrow A = I.$$

Thus the only $n \times n$ invertible idempotent matrix is the identity matrix.

45. To prove X is the inverse of A, all we have to show is $AX = I$.
We claim $A^{-1} = 2I - A$, so all we need to show is $A(2I - A) = I$.

$$A^2 - 2A + I = O \overset{\text{given}}{\Rightarrow} 2A - A^2 = I \overset{\text{factor}}{\Rightarrow} A(2I - A) = I$$

46. Let A be an invertible symmetric matrix. Since A is invertible we have $AA^{-1} = I$.
If we now take the transpose of both sides of $AA^{-1} = I$, we obtain:

$$\left(AA^{-1}\right)^T = I^T \Rightarrow \left(A^{-1}\right)^T A^T = I \Rightarrow \left(A^{-1}\right)^T A = I \Rightarrow \left(A^{-1}\right)^T AA^{-1} = IA^{-1} \Rightarrow$$
$$\left(A^{-1}\right)^T I = A^{-1} \Rightarrow \left(A^{-1}\right)^T = A^{-1}.$$

Thus A^{-1} is equal to its own transpose and so, by definition, is symmetric.
We conclude that the inverse of a symmetric matrix is symmetric.

47. Let A and B be square matrices and assume that AB is invertible.

Then $AB\,(AB)^{-1} = A\left(B\,(AB)^{-1}\right) = I$, showing that A is invertible with $A^{-1} = B\,(AB)^{-1}$.

Therefore $B\,(AB)^{-1} A = I$, showing that B is invertible with $B^{-1} = (AB)^{-1} A$.

48. As in Example 3.30, we adjoin the identity matrix to A then row reduce $[\,A\,|\,I\,]$ to $[\,I\,|\,A^{-1}\,]$.

$$\left[\begin{array}{cc|cc} 1 & 3 & 1 & 0 \\ 2 & 5 & 0 & 1 \end{array}\right] \xrightarrow{R_2 - 2R_1} \left[\begin{array}{cc|cc} 1 & 3 & 1 & 0 \\ 0 & -1 & -2 & 1 \end{array}\right] \xrightarrow{-R_2} \left[\begin{array}{cc|cc} 1 & 3 & 1 & 0 \\ 0 & 1 & 2 & -1 \end{array}\right] \xrightarrow{R_1 - 3R_2} \left[\begin{array}{cc|cc} 1 & 0 & -5 & 3 \\ 0 & 1 & 2 & -1 \end{array}\right]$$

Therefore $A^{-1} = \begin{bmatrix} -5 & 3 \\ 2 & -1 \end{bmatrix}$.

Q: How does $[\,A\,|\,I\,] \longrightarrow [\,I\,|\,A^{-1}\,]$ relate to our understanding that $A^{-1} = E_3E_2E_1$?
A: Multiplication by the E_i performs the associated operations on the identity matrix.

49. As in Example 3.30, we adjoin the identity matrix to A then row reduce $[\,A\,|\,I\,]$ to $[\,I\,|\,A^{-1}\,]$.

$$[\,A\,|\,I\,] = \left[\begin{array}{cc|cc} -2 & 4 & 1 & 0 \\ 3 & -1 & 0 & 1 \end{array}\right] \longrightarrow \left[\begin{array}{cc|cc} 1 & 0 & \frac{1}{10} & \frac{2}{5} \\ 0 & 1 & \frac{3}{10} & \frac{1}{5} \end{array}\right] = [\,I\,|\,A^{-1}\,] \Rightarrow A^{-1} = \begin{bmatrix} \frac{1}{10} & \frac{2}{5} \\ \frac{3}{10} & \frac{1}{5} \end{bmatrix}$$

50. As in Example 3.30, we adjoin the identity matrix to A then row reduce $[\,A\,|\,I\,]$ to $[\,I\,|\,A^{-1}\,]$.

$$[\,A\,|\,I\,] = \left[\begin{array}{rr|rr} 6 & -4 & 1 & 0 \\ -3 & 2 & 0 & 1 \end{array}\right] \longrightarrow \left[\begin{array}{rr|rr} 6 & -4 & 1 & 0 \\ 0 & 0 & -\frac{1}{2} & 1 \end{array}\right] \neq [\,I\,|\,A^{-1}\,] \Rightarrow$$

Since the left matrix cannot be reduced to I (why?), we conclude A^{-1} does not exist.

51. As in Example 3.30, we adjoin the identity matrix to A then row reduce $[\,A\,|\,I\,]$ to $[\,I\,|\,A^{-1}\,]$.

$$[\,A\,|\,I\,] = \left[\begin{array}{rr|rr} 1 & a & 1 & 0 \\ -a & 1 & 0 & 1 \end{array}\right] \longrightarrow \left[\begin{array}{rr|rr} 1 & 0 & \frac{1}{a^2+1} & -\frac{a}{a^2+1} \\ 0 & 1 & \frac{a}{a^2+1} & \frac{1}{a^2+1} \end{array}\right] = [\,I\,|\,A^{-1}\,] \Rightarrow A^{-1} = \begin{bmatrix} \frac{1}{a^2+1} & -\frac{a}{a^2+1} \\ \frac{a}{a^2+1} & \frac{1}{a^2+1} \end{bmatrix}$$

Q: What is the restriction on a?
A: Since $a^2+1=0 \Rightarrow a^2=-1$, there is no restriction on a. Why not?

Q: Does this agree with the formula for A^{-1} given in Theorem 3.8, that is $A^{-1} = \frac{1}{ad-bc}\begin{bmatrix} d & -b \\ -c & a \end{bmatrix}$?

52. As in Example 3.30, we adjoin the identity matrix to A then row reduce $[\,A\,|\,I\,]$ to $[\,I\,|\,A^{-1}\,]$.

$$[\,A\,|\,I\,] = \left[\begin{array}{rrr|rrr} 2 & 3 & 0 & 1 & 0 & 0 \\ 1 & -2 & -1 & 0 & 1 & 0 \\ 2 & 0 & -1 & 0 & 0 & 1 \end{array}\right] \longrightarrow \left[\begin{array}{rrr|rrr} 1 & 0 & 0 & 2 & 3 & -3 \\ 0 & 1 & 0 & -1 & -2 & 2 \\ 0 & 0 & 1 & 4 & 6 & -7 \end{array}\right] \Rightarrow A^{-1} = \begin{bmatrix} 2 & 3 & -3 \\ -1 & -2 & 2 \\ 4 & 6 & -7 \end{bmatrix}$$

53. As in Example 3.30, we adjoin the identity matrix to A then row reduce $[\,A\,|\,I\,]$ to $[\,I\,|\,A^{-1}\,]$.

$$[\,A\,|\,I\,] = \left[\begin{array}{rrr|rrr} 1 & -1 & 2 & 1 & 0 & 0 \\ 3 & 1 & 2 & 0 & 1 & 0 \\ 2 & 3 & -1 & 0 & 0 & 1 \end{array}\right] \longrightarrow \left[\begin{array}{rrr|rrr} 1 & -1 & 2 & 1 & 0 & 0 \\ 0 & 4 & -4 & -3 & 1 & 0 \\ 0 & 5 & -5 & -2 & 0 & 1 \end{array}\right] \neq [\,I\,|\,A^{-1}\,] \Rightarrow$$

Since the left matrix cannot be reduced to I (why?), we conclude A^{-1} does not exist.

54. As in Example 3.30, we adjoin the identity matrix to A then row reduce $[\,A\,|\,I\,]$ to $[\,I\,|\,A^{-1}\,]$.

$$[\,A\,|\,I\,] = \left[\begin{array}{rrr|rrr} 1 & 1 & 0 & 1 & 0 & 0 \\ 1 & 0 & 1 & 0 & 1 & 0 \\ 0 & 1 & 1 & 0 & 0 & 1 \end{array}\right] \longrightarrow \left[\begin{array}{rrr|rrr} 1 & 0 & 0 & \frac{1}{2} & \frac{1}{2} & -\frac{1}{2} \\ 0 & 1 & 0 & \frac{1}{2} & -\frac{1}{2} & \frac{1}{2} \\ 0 & 0 & 1 & -\frac{1}{2} & \frac{1}{2} & \frac{1}{2} \end{array}\right] = [\,I\,|\,A^{-1}\,]$$

55. As in Example 3.30, we adjoin the identity matrix to A then row reduce $[\,A\,|\,I\,]$ to $[\,I\,|\,A^{-1}\,]$.

$$[\,A\,|\,I\,] = \left[\begin{array}{rrr|rrr} a & 0 & 0 & 1 & 0 & 0 \\ 1 & a & 0 & 0 & 1 & 0 \\ 0 & 1 & a & 0 & 0 & 1 \end{array}\right] \longrightarrow \left[\begin{array}{rrr|rrr} 1 & 0 & 0 & \frac{1}{a} & 0 & 0 \\ 0 & 1 & 0 & -\frac{1}{a^2} & \frac{1}{a} & 0 \\ 0 & 0 & 1 & \frac{1}{a^3} & -\frac{1}{a^2} & \frac{1}{a} \end{array}\right] = [\,I\,|\,A^{-1}\,]$$

Q: The entries of A^{-1} imply $a \neq 0$. Why is that obvious in the original matrix A?

A: If $a=0$ in A, we have $A = \begin{bmatrix} 0 & 0 & 0 \\ 1 & 0 & 0 \\ 0 & 1 & 0 \end{bmatrix}$, which is obviously not invertible. Why?

56. As in Example 3.30, we adjoin the identity matrix to A then row reduce $[\,A\,|\,I\,]$ to $[\,I\,|\,A^{-1}\,]$.

$$[\,A\,|\,I\,] = \left[\begin{array}{ccc|ccc} 0 & a & 0 & 1 & 0 & 0 \\ b & 0 & c & 0 & 1 & 0 \\ 0 & d & 0 & 0 & 0 & 1 \end{array}\right] \longrightarrow \left[\begin{array}{ccc|ccc} 0 & a & 0 & 1 & 0 & 0 \\ b & 0 & c & 0 & 1 & 0 \\ 0 & 0 & 0 & -\frac{d}{a} & 0 & 1 \end{array}\right] \neq [\,I\,|\,A^{-1}\,] \Rightarrow$$

Since the left matrix cannot be reduced to I (why?), we conclude A^{-1} does not exist.

57. $A^{-1} = \begin{bmatrix} -11 & -2 & 5 & -4 \\ 4 & 1 & -2 & 2 \\ 5 & 1 & -2 & 2 \\ 9 & 2 & -4 & 3 \end{bmatrix}$.

58. $A^{-1} = \begin{bmatrix} \frac{\sqrt{2}}{2} & 0 & -2 & 0 \\ 2\sqrt{2} & \frac{\sqrt{2}}{2} & -8 & 0 \\ 0 & 0 & 1 & 0 \\ 0 & 0 & -3 & 1 \end{bmatrix}$.

59. As in Example 3.30, we adjoin the identity matrix to A then row reduce $[\,A\,|\,I\,]$ to $[\,I\,|\,A^{-1}\,]$.

$$[\,A\,|\,I\,] = \left[\begin{array}{cccc|cccc} 1 & 0 & 0 & 0 & 1 & 0 & 0 & 0 \\ 0 & 1 & 0 & 0 & 0 & 1 & 0 & 0 \\ 0 & 0 & 1 & 0 & 0 & 0 & 1 & 0 \\ a & b & c & d & 0 & 0 & 0 & 1 \end{array}\right] \longrightarrow \left[\begin{array}{cccc|cccc} 1 & 0 & 0 & 0 & 1 & 0 & 0 & 0 \\ 0 & 1 & 0 & 0 & 0 & 1 & 0 & 0 \\ 0 & 0 & 1 & 0 & 0 & 0 & 1 & 0 \\ 0 & 0 & 0 & 1 & -\frac{a}{d} & -\frac{b}{d} & -\frac{c}{d} & \frac{1}{d} \end{array}\right] = [\,I\,|\,A^{-1}\,]$$

Q: The entries of A^{-1} imply $d \neq 0$. Why is that obvious in the original matrix A?

A: If $d = 0$ in A, we have $A = \begin{bmatrix} 1 & 0 & 0 & 0 \\ 0 & 1 & 0 & 0 \\ 0 & 0 & 1 & 0 \\ a & b & c & 0 \end{bmatrix}$, which is obviously not invertible. Why?

60. As in Example 3.32, we adjoin the identity matrix to A then row reduce $[\,A\,|\,I\,]$ to $[\,I\,|\,A^{-1}\,]$. Recall that we need only use addition and multiplication in $\mathbb{Z}_2$.

$$[\,A\,|\,I\,] = \left[\begin{array}{cc|cc} 0 & 1 & 1 & 0 \\ 1 & 1 & 0 & 1 \end{array}\right] \longrightarrow \left[\begin{array}{cc|cc} 1 & 0 & 1 & 1 \\ 0 & 1 & 1 & 0 \end{array}\right] = [\,I\,|\,A^{-1}\,] \Rightarrow A^{-1} = \begin{bmatrix} 1 & 1 \\ 1 & 0 \end{bmatrix}$$

Check by multiplication (in $\mathbb{Z}_2$): $AA^{-1} = \begin{bmatrix} 0 & 1 \\ 1 & 1 \end{bmatrix}\begin{bmatrix} 1 & 1 \\ 1 & 0 \end{bmatrix} = \begin{bmatrix} 1 & 0 \\ 0 & 1 \end{bmatrix} = I$

61. As in Example 3.32, we adjoin the identity matrix to A then row reduce $[\,A\,|\,I\,]$ to $[\,I\,|\,A^{-1}\,]$. Recall that we need only use addition and multiplication in $\mathbb{Z}_5$.

$$[\,A\,|\,I\,] = \left[\begin{array}{cc|cc} 4 & 2 & 1 & 0 \\ 3 & 4 & 0 & 1 \end{array}\right] \longrightarrow \left[\begin{array}{cc|cc} 1 & 0 & 1 & 0 \\ 0 & 0 & 3 & 1 \end{array}\right] \neq [\,I\,|\,A^{-1}\,] \Rightarrow$$

Since the left matrix cannot be reduced to I (why?), we conclude A^{-1} does not exist.

Q: How else could we have determined A is not invertible?
A: By noting that $\det A = 4 \cdot 4 - 2 \cdot 3 = 10 = 0$ in $\mathbb{Z}_5$. Why?

62. As in Example 3.32, we adjoin the identity matrix to A then row reduce $\left[\,A\,|\,I\,\right]$ to $\left[\,I\,|\,A^{-1}\,\right]$. Recall that we need only use addition and multiplication in $\mathbb{Z}_3$.

$$\left[\,A\,|\,I\,\right] = \left[\begin{array}{ccc|ccc} 2 & 1 & 0 & 1 & 0 & 0 \\ 0 & 1 & 2 & 0 & 1 & 0 \\ 0 & 2 & 1 & 0 & 0 & 1 \end{array}\right] \longrightarrow \left[\begin{array}{ccc|ccc} 1 & 0 & 0 & 0 & 1 & 1 \\ 0 & 1 & 0 & 1 & 1 & 1 \\ 0 & 0 & 1 & 1 & 1 & 2 \end{array}\right] = \left[\,I\,|\,A^{-1}\,\right]$$

Q: Can we verify that $AA^{-1} = I$ in $\mathbb{Z}_3$ on the calculator?
A: If $AA^{-1} = I$ entries on the diagonal should be multiples of 3 plus 1 (like 4, 7, 10). Entries off the diagonal should simply be multiples of 3. Why?

63. As in Example 3.32, we adjoin the identity matrix to A then row reduce $\left[\,A\,|\,I\,\right]$ to $\left[\,I\,|\,A^{-1}\,\right]$. Recall that we need only use addition and multiplication in $\mathbb{Z}_7$.

$$\left[\,A\,|\,I\,\right] = \left[\begin{array}{ccc|ccc} 1 & 5 & 0 & 1 & 0 & 0 \\ 1 & 2 & 4 & 0 & 1 & 0 \\ 3 & 6 & 1 & 0 & 0 & 1 \end{array}\right] \longrightarrow \left[\begin{array}{ccc|ccc} 1 & 0 & 0 & 4 & 6 & 4 \\ 0 & 1 & 0 & 5 & 3 & 2 \\ 0 & 0 & 1 & 0 & 6 & 5 \end{array}\right] = \left[\,I\,|\,A^{-1}\,\right]$$

Q: How many 3×3 matrices are there in $\mathbb{Z}_7$?
A: There are $40,453,607$ 3×3 matrices in $\mathbb{Z}_7$. Why? Because $7^9 = 40,453,607$.

64. $$\begin{bmatrix} A & B \\ O & D \end{bmatrix}\begin{bmatrix} A & B \\ O & D \end{bmatrix}^{-1} = \begin{bmatrix} A & B \\ O & D \end{bmatrix}\begin{bmatrix} A^{-1} & -A^{-1}BD^{-1} \\ O & D^{-1} \end{bmatrix} = \begin{bmatrix} AA^{-1}+BO & -AA^{-1}BD^{-1}+BD^{-1} \\ OA^{-1}+DO & -OA^{-1}BD^{-1}+DD^{-1} \end{bmatrix}$$
$$= \begin{bmatrix} I & -BD^{-1}+BD^{-1} \\ O & I \end{bmatrix} = \begin{bmatrix} I & O \\ O & I \end{bmatrix}.$$

65. $$\begin{bmatrix} O & B \\ C & I \end{bmatrix}\begin{bmatrix} O & B \\ C & I \end{bmatrix}^{-1} = \begin{bmatrix} O & B \\ C & I \end{bmatrix}\begin{bmatrix} -(BC)^{-1} & (BC)^{-1}B \\ C(BC)^{-1} & I-C(BC)^{-1}B \end{bmatrix}$$
$$= \begin{bmatrix} BC(BC)^{-1} & B-BC(BC)^{-1}B \\ -C(BC)^{-1}+C(BC)^{-1} & C(BC)^{-1}B+I-C(BC)^{-1}B \end{bmatrix}$$
$$= \begin{bmatrix} I & B-B \\ O & I \end{bmatrix} = \begin{bmatrix} I & O \\ O & I \end{bmatrix}.$$

66. $$\begin{bmatrix} I & B \\ C & I \end{bmatrix}\begin{bmatrix} I & B \\ C & I \end{bmatrix}^{-1} = \begin{bmatrix} I & B \\ C & I \end{bmatrix}\begin{bmatrix} (I-BC)^{-1} & -(I-BC)^{-1}B \\ -C(I-BC)^{-1} & I+C(I-BC)^{-1}B \end{bmatrix}$$
$$= \begin{bmatrix} (I-BC)^{-1}-BC(I-BC)^{-1} & -(I-BC)^{-1}B+BI+BC(I-BC)^{-1}B \\ C(I-BC)^{-1}-C(I-BC)^{-1} & -C(I-BC)^{-1}B+I+C(I-BC)^{-1}B \end{bmatrix}$$
$$= \begin{bmatrix} (I-BC)(I-BC)^{-1} & (-I+BC)(I-BC)^{-1}B-B \\ O & I \end{bmatrix}$$
$$= \begin{bmatrix} I & -B+B \\ O & I \end{bmatrix} = \begin{bmatrix} I & O \\ O & I \end{bmatrix}.$$

67. $$\begin{bmatrix} O & B \\ C & D \end{bmatrix}\begin{bmatrix} O & B \\ C & D \end{bmatrix}^{-1} = \begin{bmatrix} O & B \\ C & D \end{bmatrix}\begin{bmatrix} -(BD^{-1}C)^{-1} & (BD^{-1}C)^{-1}BD^{-1} \\ D^{-1}C(BD^{-1}C)^{-1} & D^{-1}-D^{-1}C(BD^{-1}C)^{-1}BD^{-1} \end{bmatrix}$$
$$= \begin{bmatrix} BD^{-1}C(BD^{-1}C)^{-1} & BD^{-1}-BD^{-1}C(BD^{-1}C)^{-1}BD^{-1} \\ -C(BD^{-1}C)^{-1}+DD^{-1}C(BD^{-1}C)^{-1} & \begin{matrix} C(BD^{-1}C)^{-1}BD^{-1}+DD^{-1}+ \\ \left(-DD^{-1}C(BD^{-1}C)^{-1}BD^{-1}\right) \end{matrix} \end{bmatrix}$$
$$= \begin{bmatrix} I & O \\ O & I \end{bmatrix}.$$

68. $$\begin{bmatrix} A & B \\ C & D \end{bmatrix}\begin{bmatrix} A & B \\ C & D \end{bmatrix}^{-1} = \begin{bmatrix} A & B \\ C & D \end{bmatrix}\begin{bmatrix} P & Q \\ R & S \end{bmatrix} = \begin{bmatrix} AP+BR & AQ+BS \\ CP+DR & CQ+DS \end{bmatrix}$$
$$= \begin{bmatrix} AP+B\left(-D^{-1}CP\right) & A\left(-PBD^{-1}\right)+B\left(D^{-1}+D^{-1}CPBD^{-1}\right) \\ CP+D\left(-D^{-1}CP\right) & C\left(-PBD^{-1}\right)+D\left(D^{-1}+D^{-1}CPBD^{-1}\right) \end{bmatrix}$$
$$= \begin{bmatrix} \left(A-BD^{-1}C\right)P & -\left(A-BD^{-1}C\right)PBD^{-1}+BD^{-1} \\ CP-CP & -CPBD^{-1}+I+CPBD^{-1} \end{bmatrix}$$
$$= \begin{bmatrix} P^{-1}P & -P^{-1}PBD^{-1}+BD^{-1} \\ O & I \end{bmatrix} = \begin{bmatrix} I & -BD^{-1}+BD^{-1} \\ O & I \end{bmatrix} = \begin{bmatrix} I & O \\ O & I \end{bmatrix}.$$

69. We partition the matrix $\begin{bmatrix} 1 & 0 & 0 & 0 \\ 0 & 1 & 0 & 0 \\ 2 & 3 & 1 & 0 \\ 1 & 2 & 0 & 1 \end{bmatrix}$ into the form $\begin{bmatrix} I_2 & B \\ C & I_2 \end{bmatrix}$ as in Exercise 66, where $B = O$.

Then $(I-BC)^{-1} = I$, $(I-BC)^{-1}B = OB = O$, $-C(I-BC)^{-1} = -CI = -C$, and

$I + C(I-BC)^{-1}B = I$, so $\begin{bmatrix} 1 & 0 & 0 & 0 \\ 0 & 1 & 0 & 0 \\ 2 & 3 & 1 & 0 \\ 1 & 2 & 0 & 1 \end{bmatrix}^{-1} = \begin{bmatrix} 1 & 0 & 0 & 0 \\ 0 & 1 & 0 & 0 \\ -2 & -3 & 1 & 0 \\ -1 & -2 & 0 & 1 \end{bmatrix}$.

70. We partition the matrix $\begin{bmatrix} \sqrt{2} & 0 & 2\sqrt{2} & 0 \\ -4\sqrt{2} & \sqrt{2} & 0 & 0 \\ 0 & 0 & 1 & 0 \\ 0 & 0 & 3 & 1 \end{bmatrix}$ into the form $\begin{bmatrix} A & B \\ O & D \end{bmatrix}$ as in Exercise 64.

Then $A^{-1} = \begin{bmatrix} \sqrt{2} & 0 \\ -4\sqrt{2} & \sqrt{2} \end{bmatrix}^{-1} = \begin{bmatrix} \frac{\sqrt{2}}{2} & 0 \\ 2\sqrt{2} & \frac{\sqrt{2}}{2} \end{bmatrix}$,

$-A^{-1}BD^{-1} = -\begin{bmatrix} \frac{\sqrt{2}}{2} & 0 \\ 2\sqrt{2} & \frac{\sqrt{2}}{2} \end{bmatrix}\begin{bmatrix} 2\sqrt{2} & 0 \\ 0 & 0 \end{bmatrix}\begin{bmatrix} 1 & 0 \\ 3 & 1 \end{bmatrix}^{-1} = \begin{bmatrix} -2 & 0 \\ -8 & 0 \end{bmatrix}$, and

$D^{-1} = \begin{bmatrix} 1 & 0 \\ 3 & 1 \end{bmatrix}^{-1} = \begin{bmatrix} 1 & 0 \\ -3 & 1 \end{bmatrix}$, so $\begin{bmatrix} \sqrt{2} & 0 & 2\sqrt{2} & 0 \\ -4\sqrt{2} & \sqrt{2} & 0 & 0 \\ 0 & 0 & 1 & 0 \\ 0 & 0 & 3 & 1 \end{bmatrix}^{-1} = \begin{bmatrix} \frac{\sqrt{2}}{2} & 0 & -2 & 0 \\ 2\sqrt{2} & \frac{\sqrt{2}}{2} & -8 & 0 \\ 0 & 0 & 1 & 0 \\ 0 & 0 & -3 & 1 \end{bmatrix}$.

71. We partition the matrix $\begin{bmatrix} 0 & 0 & 1 & 1 \\ 0 & 0 & 1 & 0 \\ 0 & -1 & 1 & 0 \\ 1 & 1 & 0 & 1 \end{bmatrix}$ into the form $\begin{bmatrix} O & B \\ C & I \end{bmatrix}$ as in Exercise 65. Then

$$-(BC)^{-1} = -\left(\begin{bmatrix} 1 & 1 \\ 1 & 0 \end{bmatrix}\begin{bmatrix} 0 & -1 \\ 1 & 1 \end{bmatrix}\right)^{-1} = -\left(\begin{bmatrix} 1 & 0 \\ 0 & -1 \end{bmatrix}\right)^{-1} = \begin{bmatrix} -1 & 0 \\ 0 & 1 \end{bmatrix},$$

$$(BC)^{-1}B = \begin{bmatrix} 1 & 0 \\ 0 & -1 \end{bmatrix}\begin{bmatrix} 1 & 1 \\ 1 & 0 \end{bmatrix} = \begin{bmatrix} 1 & 1 \\ -1 & 0 \end{bmatrix},$$

$$C(BC)^{-1} = \begin{bmatrix} 0 & -1 \\ 1 & 1 \end{bmatrix}\begin{bmatrix} 1 & 0 \\ 0 & -1 \end{bmatrix} = \begin{bmatrix} 0 & 1 \\ 1 & -1 \end{bmatrix}, \text{ and}$$

$$I - C(BC)^{-1}B = I - \begin{bmatrix} 0 & 1 \\ 1 & -1 \end{bmatrix}\begin{bmatrix} 1 & 1 \\ 1 & 0 \end{bmatrix} = \begin{bmatrix} 0 & 0 \\ 0 & 0 \end{bmatrix}.$$

Thus, $\begin{bmatrix} 0 & 0 & 1 & 1 \\ 0 & 0 & 1 & 0 \\ 0 & -1 & 1 & 0 \\ 1 & 1 & 0 & 1 \end{bmatrix}^{-1} = \begin{bmatrix} -1 & 0 & 1 & 1 \\ 0 & 1 & -1 & 0 \\ 0 & 1 & 0 & 0 \\ 1 & -1 & 0 & 0 \end{bmatrix}.$

72. We partition the matrix $\begin{bmatrix} 0 & 1 & 1 \\ 1 & 3 & 1 \\ -1 & 5 & 2 \end{bmatrix}$ into the form $\begin{bmatrix} O & B \\ C & D \end{bmatrix}$ as in Exercise 67. Then

$$-(BD^{-1}C)^{-1} = -\left(\begin{bmatrix} 1 & 1 \end{bmatrix}\begin{bmatrix} 3 & 1 \\ 5 & 2 \end{bmatrix}^{-1}\begin{bmatrix} 1 \\ -1 \end{bmatrix}\right)^{-1} = -([-5])^{-1} = \left[\tfrac{1}{5}\right],$$

$$(BD^{-1}C)^{-1}BD^{-1} = -\tfrac{1}{5}\begin{bmatrix} 1 & 1 \end{bmatrix}\begin{bmatrix} 3 & 1 \\ 5 & 2 \end{bmatrix}^{-1} = \begin{bmatrix} \tfrac{3}{5} & -\tfrac{2}{5} \end{bmatrix},$$

$$D^{-1}C(BD^{-1}C)^{-1} = \begin{bmatrix} 3 & 1 \\ 5 & 2 \end{bmatrix}^{-1}\begin{bmatrix} 1 \\ -1 \end{bmatrix}\left(-\tfrac{1}{5}\right) = \begin{bmatrix} -\tfrac{3}{5} \\ \tfrac{8}{5} \end{bmatrix}, \text{ and}$$

$$D^{-1} - D^{-1}C(BD^{-1}C)^{-1}BD^{-1} = \begin{bmatrix} 3 & 1 \\ 5 & 2 \end{bmatrix}^{-1} - \begin{bmatrix} -\tfrac{3}{5} \\ \tfrac{8}{5} \end{bmatrix}\begin{bmatrix} 1 & 1 \end{bmatrix}\begin{bmatrix} 3 & 1 \\ 5 & 2 \end{bmatrix}^{-1} = \begin{bmatrix} \tfrac{1}{5} & \tfrac{1}{5} \\ -\tfrac{1}{5} & -\tfrac{1}{5} \end{bmatrix}.$$

Thus, $\begin{bmatrix} 0 & 1 & 1 \\ 1 & 3 & 1 \\ -1 & 5 & 2 \end{bmatrix}^{-1} = \begin{bmatrix} \tfrac{1}{5} & \tfrac{3}{5} & -\tfrac{2}{5} \\ -\tfrac{3}{5} & \tfrac{1}{5} & \tfrac{1}{5} \\ \tfrac{8}{5} & -\tfrac{1}{5} & -\tfrac{1}{5} \end{bmatrix}.$

3.4 The LU Factorization

1. Since $A = LU$, where L is unit lower triangular and U is upper triangular, we have:

$$A\mathbf{x} = \mathbf{b} \overset{A=LU}{\Rightarrow} (LU)\mathbf{x} = \mathbf{b} \overset{(AB)C=A(BC)}{\Rightarrow} L(U\mathbf{x}) = \mathbf{b} \overset{U\mathbf{x}=\mathbf{y}}{\Rightarrow} L\mathbf{y} = \mathbf{b}$$

So we solve the system by the two-step method outlined after Theorem 3.15:

1) Solve $L\mathbf{y} = \mathbf{b}$ by *forward substitution* (see Exercises 25 and 26 in Section 2.1) and

2) Solve $U\mathbf{x} = \mathbf{y}$ by *back substitution* (see Example 2.5 in Section 2.1)

Since $A = \begin{bmatrix} -2 & 1 \\ 2 & 5 \end{bmatrix} = \begin{bmatrix} 1 & 0 \\ -1 & 1 \end{bmatrix}\begin{bmatrix} -2 & 1 \\ 0 & 6 \end{bmatrix} = LU$ and $\mathbf{b} = \begin{bmatrix} 5 \\ 1 \end{bmatrix}$, $L\mathbf{y} = \mathbf{b} \Rightarrow$

$$\begin{bmatrix} 1 & 0 \\ -1 & 1 \end{bmatrix}\begin{bmatrix} y_1 \\ y_2 \end{bmatrix} = \begin{bmatrix} 5 \\ 1 \end{bmatrix} \Rightarrow \begin{aligned} y_1 \quad &= 5 \\ -y_1 + y_2 &= 1 \end{aligned} \Rightarrow \begin{aligned} y_1 &= 5 \\ y_2 &= y_1 + 1 \end{aligned} \qquad \text{So, } \mathbf{y} = \begin{bmatrix} 5 \\ 6 \end{bmatrix}.$$

Likewise, since $U = \begin{bmatrix} -2 & 1 \\ 0 & 6 \end{bmatrix}$ and $\mathbf{y} = \begin{bmatrix} 5 \\ 6 \end{bmatrix}$, $U\mathbf{x} = \mathbf{y} \Rightarrow$

$$\begin{bmatrix} -2 & 1 \\ 0 & 6 \end{bmatrix}\begin{bmatrix} x_1 \\ x_2 \end{bmatrix} = \begin{bmatrix} 5 \\ 6 \end{bmatrix} \Rightarrow \begin{aligned} -2x_1 + x_2 &= 5 \\ 6x_2 &= 6 \end{aligned} \Rightarrow \begin{aligned} x_1 &= \tfrac{1}{2}x_2 - \tfrac{5}{2} \\ x_2 &= 1 \end{aligned} \qquad \text{So, } \mathbf{x} = \begin{bmatrix} -2 \\ 1 \end{bmatrix}.$$

Check: $A\mathbf{x} = \begin{bmatrix} -2 & 1 \\ 2 & 5 \end{bmatrix}\begin{bmatrix} -2 \\ 1 \end{bmatrix} = \begin{bmatrix} 5 \\ 1 \end{bmatrix} = \mathbf{b}$.

2. We solve the system by the two-step method outlined after Theorem 3.15:

$$A\mathbf{x} = \mathbf{b} \overset{A=LU}{\Rightarrow} (LU)\mathbf{x} = \mathbf{b} \overset{U\mathbf{x}=\mathbf{y}}{\Rightarrow} L\mathbf{y} = \mathbf{b}$$

Since $A = \begin{bmatrix} 4 & -2 \\ 2 & 3 \end{bmatrix} = \begin{bmatrix} 1 & 0 \\ \frac{1}{2} & 1 \end{bmatrix}\begin{bmatrix} 4 & -2 \\ 0 & 4 \end{bmatrix} = LU$ and $\mathbf{b} = \begin{bmatrix} 0 \\ 8 \end{bmatrix}$, $L\mathbf{y} = \mathbf{b} \Rightarrow$

$$\begin{bmatrix} 1 & 0 \\ \frac{1}{2} & 1 \end{bmatrix}\begin{bmatrix} y_1 \\ y_2 \end{bmatrix} = \begin{bmatrix} 0 \\ 8 \end{bmatrix} \Rightarrow \begin{aligned} y_1 \quad &= 0 \\ \tfrac{1}{2}y_1 + y_2 &= 8 \end{aligned} \Rightarrow \begin{aligned} y_1 &= 0 \\ y_2 &= -\tfrac{1}{2}y_1 + 8 \end{aligned} \qquad \text{So, } \mathbf{y} = \begin{bmatrix} 0 \\ 8 \end{bmatrix}.$$

Likewise, since $U = \begin{bmatrix} 4 & -2 \\ 0 & 4 \end{bmatrix}$ and $\mathbf{y} = \begin{bmatrix} 0 \\ 8 \end{bmatrix}$, $U\mathbf{x} = \mathbf{y} \Rightarrow$

$$\begin{bmatrix} 4 & -2 \\ 0 & 4 \end{bmatrix}\begin{bmatrix} x_1 \\ x_2 \end{bmatrix} = \begin{bmatrix} 0 \\ 8 \end{bmatrix} \Rightarrow \begin{aligned} 4x_1 - 2x_2 &= 0 \\ 4x_2 &= 8 \end{aligned} \Rightarrow \begin{aligned} x_1 &= \tfrac{1}{2}x_2 \\ x_2 &= 2 \end{aligned} \qquad \text{So, } \mathbf{x} = \begin{bmatrix} 1 \\ 2 \end{bmatrix}.$$

Check: $A\mathbf{x} = \begin{bmatrix} 4 & -2 \\ 2 & 3 \end{bmatrix}\begin{bmatrix} 1 \\ 2 \end{bmatrix} = \begin{bmatrix} 0 \\ 8 \end{bmatrix} = \mathbf{b}$.

3. Since $A = LU$, where L is unit lower triangular and U is upper triangular, we have:

$$A\mathbf{x} = \mathbf{b} \overset{A=LU}{\Rightarrow} (LU)\mathbf{x} = \mathbf{b} \overset{(AB)C=A(BC)}{\Rightarrow} L(U\mathbf{x}) = \mathbf{b} \overset{U\mathbf{x}=\mathbf{y}}{\Rightarrow} L\mathbf{y} = \mathbf{b}$$

So we solve the system by the two-step method outlined after Theorem 3.15:
1) Solve $L\mathbf{y} = \mathbf{b}$ by *forward substitution* (see Exercises 25 and 26 in Section 2.1) and
2) Solve $U\mathbf{x} = \mathbf{y}$ by *back substitution* (see Example 2.5 in Section 2.1)

$$A = \begin{bmatrix} 2 & 1 & -2 \\ 2 & 3 & -4 \\ 4 & -3 & 0 \end{bmatrix} = \begin{bmatrix} 1 & 0 & 0 \\ -1 & 1 & 0 \\ 2 & -\frac{5}{4} & 1 \end{bmatrix} \begin{bmatrix} 2 & 1 & -2 \\ 0 & 4 & -6 \\ 0 & 0 & -\frac{7}{2} \end{bmatrix} = LU \text{ and } \mathbf{b} = \begin{bmatrix} -3 \\ 1 \\ 0 \end{bmatrix}, \text{ so } L\mathbf{y} = \mathbf{b} \Rightarrow$$

$$\begin{bmatrix} 1 & 0 & 0 \\ -1 & 1 & 0 \\ 2 & -\frac{5}{4} & 1 \end{bmatrix} \begin{bmatrix} y_1 \\ y_2 \\ y_3 \end{bmatrix} = \begin{bmatrix} -3 \\ 1 \\ 0 \end{bmatrix} \Rightarrow \begin{matrix} y_1 & = -3 \\ -y_1 + y_2 & = 1 \\ 2y_1 - \frac{5}{4}y_2 + y_3 & = 0 \end{matrix} \Rightarrow \begin{matrix} y_1 = -3 \\ y_2 = y_1 + 1 \\ y_3 = -2y_1 + \frac{5}{4}y_2 \end{matrix} \Rightarrow$$

$$\begin{aligned} y_1 &= -3 \\ y_2 &= -3 + 1 = -2 \\ y_3 &= -2(-3) + \tfrac{5}{4}(-2) = \tfrac{7}{2} \end{aligned} \quad \text{So, } \mathbf{y} = \begin{bmatrix} -3 \\ -2 \\ \frac{7}{2} \end{bmatrix}$$

$$\text{Likewise, } U = \begin{bmatrix} 2 & 1 & -2 \\ 0 & 4 & -6 \\ 0 & 0 & -\frac{7}{2} \end{bmatrix} \text{ and } \mathbf{y} = \begin{bmatrix} -3 \\ -2 \\ \frac{7}{2} \end{bmatrix}, \text{ so } U\mathbf{x} = \mathbf{y} \Rightarrow$$

$$\begin{bmatrix} 2 & 1 & -2 \\ 0 & 4 & -6 \\ 0 & 0 & -\frac{7}{2} \end{bmatrix} \begin{bmatrix} x_1 \\ x_2 \\ x_3 \end{bmatrix} = \begin{bmatrix} -3 \\ -2 \\ \frac{7}{2} \end{bmatrix} \Rightarrow \begin{matrix} 2x_1 + x_2 - 2x_3 = -3 \\ 4x_2 - 6x_3 = -2 \\ -\frac{7}{2}x_3 = \frac{7}{2} \end{matrix} \Rightarrow \begin{matrix} x_1 = -\frac{1}{2}x_2 + x_3 - \frac{3}{2} \\ x_2 = \frac{3}{2}x_3 - \frac{1}{2} \\ x_3 = -1 \end{matrix} \Rightarrow$$

$$\begin{aligned} x_1 &= -\tfrac{1}{2}(-2) - 1 - \tfrac{3}{2} = -\tfrac{3}{2} \\ x_2 &= \tfrac{3}{2}(-1) - \tfrac{1}{2} = -2 \\ x_3 &= -1 \end{aligned} \Rightarrow \quad \text{So, } \mathbf{x} = \begin{bmatrix} -\frac{3}{2} \\ -2 \\ -1 \end{bmatrix}$$

$$\text{Check: } A\mathbf{x} = \begin{bmatrix} 2 & 1 & -2 \\ 2 & 3 & -4 \\ 4 & -3 & 0 \end{bmatrix} \begin{bmatrix} -\frac{3}{2} \\ -2 \\ -1 \end{bmatrix} = \begin{bmatrix} -3 \\ 1 \\ 0 \end{bmatrix} = \mathbf{b}.$$

4. We solve the system by the two-step method outlined after Theorem 3.15:

$$A\mathbf{x} = \mathbf{b} \overset{A=LU}{\Rightarrow} (LU)\mathbf{x} = \mathbf{b} \overset{U\mathbf{x}=\mathbf{y}}{\Rightarrow} L\mathbf{y} = \mathbf{b}$$

$$L\mathbf{y} = \mathbf{b} \Rightarrow \begin{bmatrix} 1 & 0 & 0 \\ \frac{3}{2} & 1 & 0 \\ -\frac{1}{2} & 0 & 1 \end{bmatrix} \begin{bmatrix} y_1 \\ y_2 \\ y_3 \end{bmatrix} = \begin{bmatrix} 2 \\ 0 \\ -5 \end{bmatrix} \Rightarrow \mathbf{y} = \begin{bmatrix} y_1 \\ y_2 \\ y_3 \end{bmatrix} = \begin{bmatrix} 2 \\ -3 \\ -4 \end{bmatrix} \Rightarrow$$

$$U\mathbf{x} = \mathbf{y} \Rightarrow \begin{bmatrix} 2 & -4 & 0 \\ 0 & 5 & 4 \\ 0 & 0 & 2 \end{bmatrix} \begin{bmatrix} x_1 \\ x_2 \\ x_3 \end{bmatrix} = \begin{bmatrix} 2 \\ -3 \\ -4 \end{bmatrix} \Rightarrow \mathbf{x} = \begin{bmatrix} x_1 \\ x_2 \\ x_3 \end{bmatrix} = \begin{bmatrix} 3 \\ 1 \\ -2 \end{bmatrix}$$

$$\text{Check: } A\mathbf{x} = \begin{bmatrix} 2 & -4 & 0 \\ 3 & -1 & 4 \\ -1 & 2 & 2 \end{bmatrix} \begin{bmatrix} 3 \\ 1 \\ -2 \end{bmatrix} = \begin{bmatrix} 2 \\ 0 \\ -5 \end{bmatrix} = \mathbf{b}$$

5. We solve the system by the two-step method outlined after Theorem 3.15:

$$A\mathbf{x} = \mathbf{b} \overset{A=LU}{\Rightarrow} (LU)\mathbf{x} = \mathbf{b} \overset{U\mathbf{x}=\mathbf{y}}{\Rightarrow} L\mathbf{y} = \mathbf{b}$$

$$L\mathbf{y} = \mathbf{b} \Rightarrow \begin{bmatrix} 1 & 0 & 0 & 0 \\ 3 & 1 & 0 & 0 \\ 4 & 0 & 1 & 0 \\ 2 & -1 & 5 & 1 \end{bmatrix} \begin{bmatrix} y_1 \\ y_2 \\ y_3 \\ y_4 \end{bmatrix} = \begin{bmatrix} 1 \\ 2 \\ 2 \\ 1 \end{bmatrix} \Rightarrow \mathbf{y} = \begin{bmatrix} y_1 \\ y_2 \\ y_3 \\ y_4 \end{bmatrix} = \begin{bmatrix} 1 \\ -1 \\ -2 \\ 8 \end{bmatrix} \Rightarrow$$

$$U\mathbf{x} = \mathbf{y} \Rightarrow \begin{bmatrix} 2 & -1 & 0 & 0 \\ 0 & -1 & 5 & -3 \\ 0 & 0 & 1 & 0 \\ 0 & 0 & 0 & 4 \end{bmatrix} \begin{bmatrix} x_1 \\ x_2 \\ x_3 \\ x_4 \end{bmatrix} = \begin{bmatrix} 1 \\ -1 \\ -2 \\ 8 \end{bmatrix} \Rightarrow \mathbf{x} = \begin{bmatrix} x_1 \\ x_2 \\ x_3 \\ x_4 \end{bmatrix} = \begin{bmatrix} -7 \\ -15 \\ -2 \\ 2 \end{bmatrix}$$

$$\text{Check: } A\mathbf{x} = \begin{bmatrix} 2 & -1 & 0 & 0 \\ 6 & -4 & 5 & -3 \\ 8 & -4 & 1 & 0 \\ 4 & -1 & 0 & 7 \end{bmatrix} \begin{bmatrix} -7 \\ -15 \\ -2 \\ 2 \end{bmatrix} = \begin{bmatrix} 1 \\ 2 \\ 2 \\ 1 \end{bmatrix} = \mathbf{b}$$

6. We solve the system by the two-step method outlined after Theorem 1:

$$A\mathbf{x} = \mathbf{b} \overset{A=LU}{\Rightarrow} (LU)\mathbf{x} = \mathbf{b} \overset{U\mathbf{x}=\mathbf{y}}{\Rightarrow} L\mathbf{y} = \mathbf{b}$$

$$L\mathbf{y} = \mathbf{b} \Rightarrow \begin{bmatrix} 1 & 0 & 0 & 0 \\ -2 & 1 & 0 & 0 \\ 3 & -2 & 1 & 0 \\ -5 & 4 & -2 & 1 \end{bmatrix} \begin{bmatrix} y_1 \\ y_2 \\ y_3 \\ y_4 \end{bmatrix} = \begin{bmatrix} 1 \\ -3 \\ -1 \\ 0 \end{bmatrix} \Rightarrow \mathbf{y} = \begin{bmatrix} y_1 \\ y_2 \\ y_3 \\ y_4 \end{bmatrix} = \begin{bmatrix} 1 \\ -1 \\ -6 \\ -3 \end{bmatrix} \Rightarrow$$

$$U\mathbf{x} = \mathbf{y} \Rightarrow \begin{bmatrix} 1 & 4 & 3 & 0 \\ 0 & 3 & 5 & 2 \\ 0 & 0 & -2 & 0 \\ 0 & 0 & 0 & 1 \end{bmatrix} \begin{bmatrix} x_1 \\ x_2 \\ x_3 \\ x_4 \end{bmatrix} = \begin{bmatrix} 1 \\ -1 \\ -6 \\ -3 \end{bmatrix} \Rightarrow \mathbf{x} = \begin{bmatrix} x_1 \\ x_2 \\ x_3 \\ x_4 \end{bmatrix} = \begin{bmatrix} \frac{16}{3} \\ -\frac{10}{3} \\ 3 \\ -3 \end{bmatrix}$$

$$\text{Check: } A\mathbf{x} = \begin{bmatrix} 1 & 4 & 3 & 0 \\ -2 & -5 & -1 & 2 \\ 3 & 6 & -3 & -4 \\ -5 & -8 & 9 & 9 \end{bmatrix} \begin{bmatrix} \frac{16}{3} \\ -\frac{10}{3} \\ 3 \\ -3 \end{bmatrix} = \begin{bmatrix} 1 \\ -3 \\ -1 \\ 0 \end{bmatrix} = \mathbf{b}$$

7. Following the *multiplier* method of Example 3.35, we find the LU factorization of A:

$$A = \begin{bmatrix} 1 & 2 \\ -3 & -1 \end{bmatrix} \xrightarrow{R_2' = R_2 + 3R_1} \begin{bmatrix} 1 & 2 \\ 0 & 5 \end{bmatrix} = U \Rightarrow l_{21} = -3 \Rightarrow L = \begin{bmatrix} 1 & 0 \\ -3 & 1 \end{bmatrix}$$

$$\text{Check: } LU = \begin{bmatrix} 1 & 0 \\ -3 & 1 \end{bmatrix} \begin{bmatrix} 1 & 2 \\ 0 & 5 \end{bmatrix} = \begin{bmatrix} 1 & 2 \\ -3 & -1 \end{bmatrix} = A$$

8. Following the *multiplier* method of Example 3.35, we find the LU factorization of A:

$$A = \begin{bmatrix} 2 & -4 \\ 3 & 1 \end{bmatrix} \xrightarrow{R_2' = R_2 - \frac{3}{2}R_1} \begin{bmatrix} 2 & -4 \\ 0 & 7 \end{bmatrix} = U \Rightarrow l_{21} = \tfrac{3}{2} \Rightarrow L = \begin{bmatrix} 1 & 0 \\ \frac{3}{2} & 1 \end{bmatrix}$$

$$\text{Check: } LU = \begin{bmatrix} 1 & 0 \\ \frac{3}{2} & 1 \end{bmatrix} \begin{bmatrix} 2 & -4 \\ 0 & 7 \end{bmatrix} = \begin{bmatrix} 2 & -4 \\ 3 & 1 \end{bmatrix} = A$$

9. Following the *multiplier* method of Example 3.35, we find the LU factorization of A:

$$A = \begin{bmatrix} 1 & 2 & 3 \\ 4 & 5 & 6 \\ 8 & 7 & 9 \end{bmatrix} \xrightarrow[R_3' = R_2 - 8R_1]{R_2' = R_2 - 4R_1} \begin{bmatrix} 1 & 2 & 3 \\ 0 & -3 & -6 \\ 0 & -9 & -15 \end{bmatrix} \xrightarrow{R_3'' = R_3' - 3R_2'} \begin{bmatrix} 1 & 2 & 3 \\ 0 & -3 & -6 \\ 0 & 0 & 3 \end{bmatrix} = U \Rightarrow$$

$$\begin{matrix} l_{21} = 4 & \\ l_{31} = 8 & l_{32} = 3 \end{matrix} \Rightarrow L = \begin{bmatrix} 1 & 0 & 0 \\ 4 & 1 & 0 \\ 8 & 3 & 1 \end{bmatrix}$$

$$\text{Check: } LU = \begin{bmatrix} 1 & 0 & 0 \\ 4 & 1 & 0 \\ 8 & 3 & 1 \end{bmatrix} \begin{bmatrix} 1 & 2 & 3 \\ 0 & -3 & -6 \\ 0 & 0 & 3 \end{bmatrix} = \begin{bmatrix} 1 & 2 & 3 \\ 4 & 5 & 6 \\ 8 & 7 & 9 \end{bmatrix} = A$$

10. Following the *multiplier* method of Example 3.35, we find the LU factorization of A:

$$A = \begin{bmatrix} 2 & 2 & -1 \\ 4 & 0 & 4 \\ 3 & 4 & 4 \end{bmatrix} \xrightarrow[R_3'=R_3-\frac{3}{2}R_1]{R_2'=R_2-2R_1} \begin{bmatrix} 2 & 2 & -1 \\ 0 & -4 & 6 \\ 0 & 1 & \frac{11}{2} \end{bmatrix} \xrightarrow{R_3''=R_3'+\frac{1}{4}R_2'} \begin{bmatrix} 2 & 2 & -1 \\ 0 & -4 & 6 \\ 0 & 0 & 7 \end{bmatrix} = U \Rightarrow$$

$$\begin{array}{ll} l_{21} = 2 & \\ l_{31} = \frac{3}{2} & l_{32} = -\frac{1}{4} \end{array} \Rightarrow L = \begin{bmatrix} 1 & 0 & 0 \\ 2 & 1 & 0 \\ \frac{3}{2} & -\frac{1}{4} & 1 \end{bmatrix}$$

$$\text{Check: } LU = \begin{bmatrix} 1 & 0 & 0 \\ 2 & 1 & 0 \\ \frac{3}{2} & -\frac{1}{4} & 1 \end{bmatrix} \begin{bmatrix} 2 & 2 & -1 \\ 0 & -4 & 6 \\ 0 & 0 & 7 \end{bmatrix} = \begin{bmatrix} 2 & 2 & -1 \\ 4 & 0 & 4 \\ 3 & 4 & 4 \end{bmatrix}$$

11. Following the *multiplier* method of Example 3.35, we find the LU factorization of A:

$$A = \begin{bmatrix} 1 & 2 & 3 & -1 \\ 2 & 6 & 3 & 0 \\ 0 & 6 & -6 & 7 \\ -1 & -2 & -9 & 0 \end{bmatrix} \xrightarrow[R_4'=R_4+R_1]{R_2'=R_2-2R_1} \begin{bmatrix} 1 & 2 & 3 & -1 \\ 0 & 4 & -3 & 2 \\ 0 & 6 & -6 & 7 \\ 0 & 0 & -6 & -1 \end{bmatrix} \xrightarrow{R_3'=R_3-\frac{3}{2}R_2'} \begin{bmatrix} 1 & 2 & 3 & -1 \\ 0 & 4 & -3 & 2 \\ 0 & 0 & -\frac{3}{2} & 4 \\ 0 & 0 & -6 & -1 \end{bmatrix}$$

$$\xrightarrow{R_4''=R_4'-4R_3'} \begin{bmatrix} 1 & 2 & 3 & -1 \\ 0 & 4 & -3 & 2 \\ 0 & 0 & -\frac{3}{2} & 4 \\ 0 & 0 & 0 & -17 \end{bmatrix} = U \Rightarrow \begin{array}{lll} l_{21} = 2 & & \\ l_{31} = 0 & l_{32} = \frac{3}{2} & \\ l_{41} = -1 & l_{42} = 0 & l_{43} = 4 \end{array} \Rightarrow L = \begin{bmatrix} 1 & 0 & 0 & 0 \\ 2 & 1 & 0 & 0 \\ 0 & \frac{3}{2} & 1 & 0 \\ -1 & 0 & 4 & 1 \end{bmatrix}$$

$$\text{Check: } LU = \begin{bmatrix} 1 & 0 & 0 & 0 \\ 2 & 1 & 0 & 0 \\ 0 & \frac{3}{2} & 1 & 0 \\ -1 & 0 & 4 & 1 \end{bmatrix} \begin{bmatrix} 1 & 2 & 3 & -1 \\ 0 & 4 & -3 & 2 \\ 0 & 0 & -\frac{3}{2} & 4 \\ 0 & 0 & 0 & -17 \end{bmatrix} = \begin{bmatrix} 1 & 2 & 3 & -1 \\ 2 & 6 & 3 & 0 \\ 0 & 6 & -6 & 7 \\ -1 & -2 & -9 & 0 \end{bmatrix}$$

12. Following the *multiplier* method of Example 3.35, we find the LU factorization of A:

$$A = \begin{bmatrix} 2 & 2 & 2 & 1 \\ -2 & 4 & -1 & 2 \\ 4 & 4 & 7 & 3 \\ 6 & 9 & 5 & 8 \end{bmatrix} \xrightarrow[\substack{R_3'=R_3-2R_1 \\ R_4'=R_4-3R_1}]{R_2'=R_2+R_1} \begin{bmatrix} 2 & 2 & 2 & 1 \\ 0 & 6 & 1 & 3 \\ 0 & 0 & 3 & 1 \\ 0 & 3 & -1 & 5 \end{bmatrix} \xrightarrow{R_4''=R_4'-\frac{1}{2}R_2'} \begin{bmatrix} 2 & 2 & 2 & 1 \\ 0 & 6 & 1 & 3 \\ 0 & 0 & 3 & 1 \\ 0 & 0 & -\frac{3}{2} & \frac{7}{2} \end{bmatrix}$$

$$\xrightarrow{R_4'''=R_4''+\frac{1}{2}R_3'} \begin{bmatrix} 2 & 2 & 2 & 1 \\ 0 & 6 & 1 & 3 \\ 0 & 0 & 3 & 1 \\ 0 & 0 & 0 & 4 \end{bmatrix} = U \Rightarrow \begin{array}{lll} l_{21} = -1 & & \\ l_{31} = 2 & l_{32} = 0 & \\ l_{41} = 3 & l_{42} = \frac{1}{2} & l_{43} = -\frac{1}{2} \end{array} \Rightarrow L = \begin{bmatrix} 1 & 0 & 0 & 0 \\ -1 & 1 & 0 & 0 \\ 2 & 0 & 1 & 0 \\ 3 & \frac{1}{2} & -\frac{1}{2} & 1 \end{bmatrix}$$

$$\text{Check: } LU = \begin{bmatrix} 1 & 0 & 0 & 0 \\ -1 & 1 & 0 & 0 \\ 2 & 0 & 1 & 0 \\ 3 & \frac{1}{2} & -\frac{1}{2} & 1 \end{bmatrix} \begin{bmatrix} 2 & 2 & 2 & 1 \\ 0 & 6 & 1 & 3 \\ 0 & 0 & 3 & 1 \\ 0 & 0 & 0 & 4 \end{bmatrix} = \begin{bmatrix} 2 & 2 & 2 & 1 \\ -2 & 4 & -1 & 2 \\ 4 & 4 & 7 & 3 \\ 6 & 9 & 5 & 8 \end{bmatrix}$$

13. By adapting the *multiplier* method of Example 3.35, we find the LU factorization of A. However, in this case, we simply note that A is already upper triangular so $L = I_3$.

$$\text{Check: } LU = \begin{bmatrix} 1 & 0 & 0 \\ 0 & 1 & 0 \\ 0 & 0 & 1 \end{bmatrix} \begin{bmatrix} 1 & 0 & 1 & -2 \\ 0 & 3 & 3 & 1 \\ 0 & 0 & 0 & 5 \end{bmatrix} = \begin{bmatrix} 1 & 0 & 1 & -2 \\ 0 & 3 & 3 & 1 \\ 0 & 0 & 0 & 5 \end{bmatrix}$$

Q: Why is $L = I_3$, the 3×3 identity matrix instead of I_4, the 4×4 identity matrix?
A: Because A has 3 rows. Considering only size, we have $A = LU = [3 \times 3][3 \times 4] = [3 \times 4]$.

14. By adapting the *multiplier* method of Example 3.35, we find the LU factorization of A:

$$\begin{bmatrix} 1 & 2 & 0 & -1 & 1 \\ -2 & -7 & 3 & 8 & -2 \\ 1 & 1 & 3 & 5 & 2 \\ 0 & 3 & -3 & -6 & 0 \end{bmatrix} \xrightarrow[R_3'=R_3-R_1]{R_2'=R_2+2R_1} \begin{bmatrix} 1 & 2 & 0 & -1 & 1 \\ 0 & -3 & 3 & 6 & 0 \\ 0 & -1 & 3 & 6 & 1 \\ 0 & 3 & -3 & -6 & 0 \end{bmatrix} \xrightarrow[R_4''=R_4+R_2']{R_3''=R_3'-\frac{1}{3}R_2'} \begin{bmatrix} 1 & 2 & 0 & -1 & 1 \\ 0 & -3 & 3 & 6 & 0 \\ 0 & 0 & 2 & 4 & 1 \\ 0 & 0 & 0 & 0 & 0 \end{bmatrix} = U$$

$$\begin{aligned} & l_{21} = -2 \\ \Rightarrow\ & l_{31} = 1 \quad l_{32} = \tfrac{1}{3} \\ & l_{41} = 0 \quad l_{42} = -1 \quad l_{43} = 0 \end{aligned} \qquad \Rightarrow L = \begin{bmatrix} 1 & 0 & 0 & 0 \\ -2 & 1 & 0 & 0 \\ 1 & \frac{1}{3} & 1 & 0 \\ 0 & -1 & 0 & 1 \end{bmatrix}$$

$$\text{Check: } LU = \begin{bmatrix} 1 & 0 & 0 & 0 \\ -2 & 1 & 0 & 0 \\ 1 & \frac{1}{3} & 1 & 0 \\ 0 & -1 & 0 & 1 \end{bmatrix} \begin{bmatrix} 1 & 2 & 0 & -1 & 1 \\ 0 & -3 & 3 & 6 & 0 \\ 0 & 0 & 2 & 4 & 1 \\ 0 & 0 & 0 & 0 & 0 \end{bmatrix} = \begin{bmatrix} 1 & 2 & 0 & -1 & 1 \\ -2 & -7 & 3 & 8 & -2 \\ 1 & 1 & 3 & 5 & 2 \\ 0 & 3 & -3 & -6 & 0 \end{bmatrix}$$

15. Since $A = LU \Rightarrow A^{-1} = U^{-1}L^{-1}$, we use the LU factorization of A to find A^{-1}:

$$\text{Since } A = \begin{bmatrix} -2 & 1 \\ 2 & 5 \end{bmatrix} = \begin{bmatrix} 1 & 0 \\ -1 & 1 \end{bmatrix} \begin{bmatrix} -2 & 1 \\ 0 & 6 \end{bmatrix} = LU, \text{ we have:}$$

$$L^{-1} = \frac{1}{1} \begin{bmatrix} 1 & 0 \\ 1 & 1 \end{bmatrix} = \begin{bmatrix} 1 & 0 \\ 1 & 1 \end{bmatrix} \text{ and } U^{-1} = -\frac{1}{12} \begin{bmatrix} 6 & -1 \\ 0 & -2 \end{bmatrix} = \begin{bmatrix} -\frac{1}{2} & \frac{1}{12} \\ 0 & \frac{1}{6} \end{bmatrix}.$$

$$\text{Therefore, } A^{-1} = U^{-1}L^{-1} = \begin{bmatrix} -\frac{1}{2} & \frac{1}{12} \\ 0 & \frac{1}{6} \end{bmatrix} \begin{bmatrix} 1 & 0 \\ 1 & 1 \end{bmatrix} = \begin{bmatrix} -\frac{5}{12} & \frac{1}{12} \\ \frac{1}{6} & \frac{1}{6} \end{bmatrix}$$

$$\text{Check: } AA^{-1} = \begin{bmatrix} -2 & 1 \\ 2 & 5 \end{bmatrix} \begin{bmatrix} -\frac{5}{12} & \frac{1}{12} \\ \frac{1}{6} & \frac{1}{6} \end{bmatrix} = \begin{bmatrix} 1 & 0 \\ 0 & 1 \end{bmatrix} = I.$$

Note: To compute L^{-1} and U^{-1}, we used the formula: $A^{-1} = \frac{1}{ad-bc} \begin{bmatrix} d & -b \\ -c & a \end{bmatrix}$.

16. Since $A = LU \Rightarrow A^{-1} = U^{-1}L^{-1}$, we use the LU factorization of A to find A^{-1}:

$$\text{Recall, } A = LU = \begin{bmatrix} 2 & -4 & 0 \\ 3 & -1 & 4 \\ -1 & 2 & 2 \end{bmatrix} = \begin{bmatrix} 1 & 0 & 0 \\ \frac{3}{2} & 1 & 0 \\ -\frac{1}{2} & 0 & 1 \end{bmatrix} \begin{bmatrix} 2 & -4 & 0 \\ 0 & 5 & 4 \\ 0 & 0 & 2 \end{bmatrix} = LU.$$

To compute L^{-1}, we adjoin the identity matrix to L then row reduce $\left[\, L \,|\, I \,\right]$ into $\left[\, I \,|\, L^{-1} \,\right]$.

$$\left[\, L \,|\, I \,\right] = \left[\begin{array}{ccc|ccc} 1 & 0 & 0 & 1 & 0 & 0 \\ \frac{3}{2} & 1 & 0 & 0 & 1 & 0 \\ -\frac{1}{2} & 0 & 0 & 0 & 0 & 1 \end{array}\right] \longrightarrow \left[\begin{array}{ccc|ccc} 1 & 0 & 0 & 1 & 0 & 0 \\ 0 & 1 & 0 & -\frac{3}{2} & 1 & 0 \\ 0 & 0 & 1 & \frac{1}{2} & 0 & 0 \end{array}\right] \Rightarrow L^{-1} = \begin{bmatrix} 1 & 0 & 0 \\ -\frac{3}{2} & 1 & 0 \\ \frac{1}{2} & 0 & 0 \end{bmatrix}$$

To compute U^{-1}, we adjoin the identity matrix to U then row reduce $\left[\, U \,|\, I \,\right]$ into $\left[\, I \,|\, U^{-1} \,\right]$.

$$\left[\, U \,|\, I \,\right] = \left[\begin{array}{ccc|ccc} 2 & -4 & 0 & 1 & 0 & 0 \\ 0 & 5 & 4 & 0 & 1 & 0 \\ 0 & 0 & 2 & 0 & 0 & 1 \end{array}\right] \longrightarrow \left[\begin{array}{ccc|ccc} 1 & 0 & 0 & \frac{1}{2} & \frac{2}{5} & -\frac{4}{5} \\ 0 & 1 & 0 & 0 & \frac{1}{5} & -\frac{2}{5} \\ 0 & 0 & 1 & 0 & 0 & \frac{1}{2} \end{array}\right] \Rightarrow U^{-1} = \begin{bmatrix} \frac{1}{2} & \frac{2}{5} & -\frac{4}{5} \\ 0 & \frac{1}{5} & -\frac{2}{5} \\ 0 & 0 & \frac{1}{2} \end{bmatrix}$$

$$\text{Therefore, } A^{-1} = U^{-1}L^{-1} = \begin{bmatrix} 1 & 0 & 0 \\ -\frac{3}{2} & 1 & 0 \\ \frac{1}{2} & 0 & 0 \end{bmatrix} \begin{bmatrix} \frac{1}{2} & \frac{2}{5} & -\frac{4}{5} \\ 0 & \frac{1}{5} & -\frac{2}{5} \\ 0 & 0 & \frac{1}{2} \end{bmatrix} = \begin{bmatrix} -\frac{1}{2} & \frac{2}{5} & -\frac{4}{5} \\ -\frac{1}{2} & \frac{1}{5} & -\frac{2}{5} \\ \frac{1}{4} & 0 & \frac{1}{2} \end{bmatrix}$$

$$\text{Check: } AA^{-1} = \begin{bmatrix} 2 & -4 & 0 \\ 3 & -1 & 4 \\ -1 & 2 & 2 \end{bmatrix} \begin{bmatrix} -\frac{1}{2} & \frac{2}{5} & -\frac{4}{5} \\ -\frac{1}{2} & \frac{1}{5} & -\frac{2}{5} \\ \frac{1}{4} & 0 & \frac{1}{2} \end{bmatrix} = \begin{bmatrix} 1 & 0 & 0 \\ 0 & 1 & 0 \\ 0 & 0 & 1 \end{bmatrix} = I.$$

17. We compute A^{-1} one column at a time using the method outlined below:

$$(LU)\mathbf{x}_1 = \mathbf{e}_1 \overset{U\mathbf{x}_1 = \mathbf{y}_1}{\Rightarrow} L\mathbf{y}_1 = \mathbf{e}_1 \quad \text{and} \quad (LU)\mathbf{x}_2 = \mathbf{e}_2 \overset{U\mathbf{x}_2 = \mathbf{y}_2}{\Rightarrow} L\mathbf{y}_2 = \mathbf{e}_2$$

We begin by computing column 1 of A^{-1}:

$$L\mathbf{y}_1 = \mathbf{e}_1 \Rightarrow \begin{bmatrix} 1 & 0 \\ -1 & 1 \end{bmatrix} \begin{bmatrix} y_{11} \\ y_{21} \end{bmatrix} = \begin{bmatrix} 1 \\ 0 \end{bmatrix} \Rightarrow \mathbf{y}_1 = \begin{bmatrix} 1 \\ 1 \end{bmatrix} \Rightarrow$$

$$U\mathbf{x}_1 = \mathbf{y}_1 \Rightarrow \begin{bmatrix} -2 & 1 \\ 0 & 6 \end{bmatrix} \begin{bmatrix} x_{11} \\ x_{21} \end{bmatrix} = \begin{bmatrix} 1 \\ 1 \end{bmatrix} \Rightarrow \mathbf{x}_1 = \begin{bmatrix} -\frac{5}{12} \\ \frac{1}{6} \end{bmatrix}$$

We repeat this process to compute column 2 of A^{-1}:

$$L\mathbf{y}_2 = \mathbf{e}_2 \Rightarrow \begin{bmatrix} 1 & 0 \\ -1 & 1 \end{bmatrix} \begin{bmatrix} y_{12} \\ y_{22} \end{bmatrix} = \begin{bmatrix} 0 \\ 1 \end{bmatrix} \Rightarrow \mathbf{y}_2 = \begin{bmatrix} 0 \\ 1 \end{bmatrix} \Rightarrow$$

$$U\mathbf{x}_2 = \mathbf{y}_2 \Rightarrow \begin{bmatrix} -2 & 1 \\ 0 & 6 \end{bmatrix} \begin{bmatrix} x_{12} \\ x_{22} \end{bmatrix} = \begin{bmatrix} 0 \\ 1 \end{bmatrix} \Rightarrow \mathbf{x}_2 = \begin{bmatrix} \frac{1}{12} \\ \frac{1}{6} \end{bmatrix}$$

Therefore, $A^{-1} = \left[\, \mathbf{x}_1 \;\; \mathbf{x}_2 \,\right] = \begin{bmatrix} -\frac{5}{12} & \frac{1}{12} \\ \frac{1}{6} & \frac{1}{6} \end{bmatrix}$ exactly as we found in Exercise 15.

18. We compute A^{-1} one column at a time using the method outlined below:

$$(LU)\mathbf{x}_i = \mathbf{e}_i \overset{U\mathbf{x}_i = \mathbf{y}_i}{\Rightarrow} L\mathbf{y}_i = \mathbf{e}_i$$

We begin by computing column 1 of A^{-1}:

$$L\mathbf{y}_1 = \mathbf{e}_1 \Rightarrow \begin{bmatrix} 1 & 0 & 0 \\ \frac{3}{2} & 1 & 0 \\ -\frac{1}{2} & 0 & 1 \end{bmatrix} \begin{bmatrix} y_{11} \\ y_{21} \\ y_{31} \end{bmatrix} = \begin{bmatrix} 1 \\ 0 \\ 0 \end{bmatrix} \Rightarrow \mathbf{y}_1 = \begin{bmatrix} 1 \\ -\frac{3}{2} \\ \frac{1}{2} \end{bmatrix} \Rightarrow$$

$$U\mathbf{x}_1 = \mathbf{y}_1 \Rightarrow \begin{bmatrix} 2 & -4 & 0 \\ 0 & 5 & 4 \\ 0 & 0 & 2 \end{bmatrix} \begin{bmatrix} x_{11} \\ x_{21} \\ x_{31} \end{bmatrix} = \begin{bmatrix} 1 \\ -\frac{3}{2} \\ \frac{1}{2} \end{bmatrix} \Rightarrow \mathbf{x}_1 = \begin{bmatrix} -\frac{1}{2} \\ -\frac{1}{2} \\ \frac{1}{4} \end{bmatrix} \Rightarrow$$

We repeat this process to compute column 2 of A^{-1}:

$$L\mathbf{y}_2 = \mathbf{e}_2 \Rightarrow \begin{bmatrix} 1 & 0 & 0 \\ \frac{3}{2} & 1 & 0 \\ -\frac{1}{2} & 0 & 1 \end{bmatrix} \begin{bmatrix} y_{12} \\ y_{22} \\ y_{32} \end{bmatrix} = \begin{bmatrix} 0 \\ 1 \\ 0 \end{bmatrix} \Rightarrow \mathbf{y}_2 = \begin{bmatrix} 0 \\ 1 \\ 0 \end{bmatrix} \Rightarrow$$

$$U\mathbf{x}_2 = \mathbf{y}_2 \Rightarrow \begin{bmatrix} 2 & -4 & 0 \\ 0 & 5 & 4 \\ 0 & 0 & 2 \end{bmatrix} \begin{bmatrix} x_{12} \\ x_{22} \\ x_{32} \end{bmatrix} = \begin{bmatrix} 0 \\ 1 \\ 0 \end{bmatrix} \Rightarrow \mathbf{x}_2 = \begin{bmatrix} \frac{2}{5} \\ \frac{1}{5} \\ 0 \end{bmatrix} \Rightarrow$$

We repeat one last time to compute column 3 of A^{-1}:

$$L\mathbf{y}_3 = \mathbf{e}_3 \Rightarrow \begin{bmatrix} 1 & 0 & 0 \\ \frac{3}{2} & 1 & 0 \\ -\frac{1}{2} & 0 & 1 \end{bmatrix} \begin{bmatrix} y_{13} \\ y_{23} \\ y_{33} \end{bmatrix} = \begin{bmatrix} 0 \\ 0 \\ 1 \end{bmatrix} \Rightarrow \mathbf{y}_3 = \begin{bmatrix} 0 \\ 0 \\ 1 \end{bmatrix} \Rightarrow$$

$$U\mathbf{x}_3 = \mathbf{y}_3 \Rightarrow \begin{bmatrix} 2 & -4 & 0 \\ 0 & 5 & 4 \\ 0 & 0 & 2 \end{bmatrix} \begin{bmatrix} x_{13} \\ x_{23} \\ x_{33} \end{bmatrix} = \begin{bmatrix} 0 \\ 0 \\ 1 \end{bmatrix} \Rightarrow \mathbf{x}_3 = \begin{bmatrix} -\frac{4}{5} \\ -\frac{2}{5} \\ \frac{1}{2} \end{bmatrix} \Rightarrow$$

Therefore, $A^{-1} = \begin{bmatrix} \mathbf{x}_1 & \mathbf{x}_2 & \mathbf{x}_3 \end{bmatrix} = \begin{bmatrix} -\frac{1}{2} & \frac{2}{5} & -\frac{4}{5} \\ -\frac{1}{2} & \frac{1}{5} & -\frac{2}{5} \\ \frac{1}{4} & 0 & \frac{1}{2} \end{bmatrix}$ exactly as we found in Exercise 16.

19. Since we need $R_1 \to R_2$, $R_2 \to R_3$, and $R_3 \to R_1$ one possibility is:

$$P = \begin{bmatrix} 0 & 0 & 1 \\ 1 & 0 & 0 \\ 0 & 1 & 0 \end{bmatrix} = E_{13}E_{12} = \begin{bmatrix} 0 & 0 & 1 \\ 0 & 1 & 0 \\ 1 & 0 & 0 \end{bmatrix} \begin{bmatrix} 0 & 1 & 0 \\ 1 & 0 & 0 \\ 0 & 0 & 1 \end{bmatrix}.$$

Q: Is this factorization of P or could we have found another one?
A: No, it is not unique. For instance, we could have chosen $P = E_{12}E_{23}$.

20. $P = E_{12}E_{14}E_{13}$, that is: $\begin{bmatrix} 0&0&0&1\\ 0&0&1&0\\ 0&1&0&0\\ 1&0&0&0 \end{bmatrix} = \begin{bmatrix} 0&1&0&0\\ 1&0&0&0\\ 0&0&1&0\\ 0&0&0&1 \end{bmatrix} \begin{bmatrix} 0&0&0&1\\ 0&1&0&0\\ 0&0&1&0\\ 1&0&0&0 \end{bmatrix} \begin{bmatrix} 0&0&1&0\\ 0&1&0&0\\ 1&0&0&0\\ 0&0&0&1 \end{bmatrix}$.

21. We find one possible factorization by tracing the row interchanges: $P = E_{13}E_{23}E_{34}$.

22. One possibility is: $P = E_{12}E_{45}E_{24}E_{34}$.

23. To find the P^TLU factorization of A, we begin by permuting the rows of A:

Since $A = \begin{bmatrix} 0&1&4\\ -1&2&1\\ 1&3&3 \end{bmatrix}$, we have $PA = \begin{bmatrix} 0&1&0\\ 1&0&0\\ 0&0&1 \end{bmatrix} \begin{bmatrix} 0&1&4\\ -1&2&1\\ 1&3&3 \end{bmatrix} = \begin{bmatrix} -1&2&1\\ 0&1&4\\ 1&3&3 \end{bmatrix}$.

Now we follow the *multiplier* method of Example 3.35 to find the LU factorization of PA:

$$PA = \begin{bmatrix} -1&2&1\\ 0&1&4\\ 1&3&3 \end{bmatrix} \xrightarrow{R_3'=R_3+R_1} \begin{bmatrix} -1&2&1\\ 0&1&4\\ 0&5&4 \end{bmatrix} \xrightarrow{R_3''=R_3'-5R_2} \begin{bmatrix} -1&2&1\\ 0&1&4\\ 0&0&-16 \end{bmatrix} = U \Rightarrow$$

$$\begin{matrix} l_{21} = 0 \\ l_{31} = -1 \quad 0 \quad l_{32} = 5 \end{matrix} \Rightarrow L = \begin{bmatrix} 1&0&0\\ 0&1&0\\ -1&5&1 \end{bmatrix}$$

Check: $LU = \begin{bmatrix} 1&0&0\\ 0&1&0\\ -1&5&1 \end{bmatrix} \begin{bmatrix} -1&2&1\\ 0&1&4\\ 0&0&-16 \end{bmatrix} = \begin{bmatrix} -1&2&1\\ 0&1&4\\ 1&3&3 \end{bmatrix} = PA$

Now $PA = LU \Rightarrow A = P^{-1}LU$, but $P^{-1} = P^T$, so we have $A = P^TLU$.

Check: $P^TLU = \begin{bmatrix} 0&1&0\\ 1&0&0\\ 0&0&1 \end{bmatrix} \begin{bmatrix} 1&0&0\\ 0&1&0\\ -1&5&1 \end{bmatrix} \begin{bmatrix} -1&2&1\\ 0&1&4\\ 0&0&-16 \end{bmatrix} = \begin{bmatrix} 0&1&4\\ -1&2&1\\ 1&3&3 \end{bmatrix} = A$

24. To find the P^TLU factorization of A, we begin by permuting the rows of A:

Since $A = \begin{bmatrix} 0 & 0 & 1 & 2 \\ -1 & 1 & 3 & 2 \\ 0 & 2 & 1 & 1 \\ 1 & 1 & -1 & 0 \end{bmatrix}$, we have $PA = \begin{bmatrix} 0 & 0 & 0 & 1 \\ 0 & 0 & 1 & 0 \\ 1 & 0 & 0 & 0 \\ 0 & 1 & 0 & 0 \end{bmatrix} \begin{bmatrix} 0 & 0 & 1 & 2 \\ -1 & 1 & 3 & 2 \\ 0 & 2 & 1 & 1 \\ 1 & 1 & -1 & 0 \end{bmatrix} = \begin{bmatrix} 1 & 1 & -1 & 0 \\ 0 & 2 & 1 & 1 \\ 0 & 0 & 1 & 2 \\ -1 & 1 & 3 & 2 \end{bmatrix}$.

Now we follow the *multiplier* method of Example 3.35 to find the LU factorization of PA:

$$PA = \begin{bmatrix} 1 & 1 & -1 & 0 \\ 0 & 2 & 1 & 1 \\ 0 & 0 & 1 & 2 \\ -1 & 1 & 3 & 2 \end{bmatrix} \xrightarrow{R_4' = R_4 + R_1} \begin{bmatrix} 1 & 1 & -1 & 0 \\ 0 & 2 & 1 & 1 \\ 0 & 0 & 1 & 2 \\ 0 & 2 & 2 & 2 \end{bmatrix} \xrightarrow{R_4'' = R_4' - R_2} \begin{bmatrix} 1 & 1 & -1 & 0 \\ 0 & 2 & 1 & 1 \\ 0 & 0 & 1 & 2 \\ 0 & 0 & 1 & 1 \end{bmatrix}$$

$$\xrightarrow{R_4''' = R_4'' - R_3} \begin{bmatrix} 1 & 1 & -1 & 0 \\ 0 & 2 & 1 & 1 \\ 0 & 0 & 1 & 2 \\ 0 & 0 & 0 & -1 \end{bmatrix} = U \Rightarrow \begin{matrix} l_{21} = 0 & & \\ l_{31} = 0 & l_{32} = 0 & \\ l_{41} = -1 & l_{42} = 1 & l_{43} = 1 \end{matrix} \Rightarrow L = \begin{bmatrix} 1 & 0 & 0 & 0 \\ 0 & 1 & 0 & 0 \\ 0 & 0 & 1 & 0 \\ -1 & 1 & 1 & 1 \end{bmatrix}$$

$$\text{Check: } LU = \begin{bmatrix} 1 & 0 & 0 & 0 \\ 0 & 1 & 0 & 0 \\ 0 & 0 & 1 & 0 \\ -1 & 1 & 1 & 1 \end{bmatrix} \begin{bmatrix} 1 & 1 & -1 & 0 \\ 0 & 2 & 1 & 1 \\ 0 & 0 & 1 & 2 \\ 0 & 0 & 0 & -1 \end{bmatrix} = \begin{bmatrix} 1 & 1 & -1 & 0 \\ 0 & 2 & 1 & 1 \\ 0 & 0 & 1 & 2 \\ -1 & 1 & 3 & 2 \end{bmatrix} = PA$$

Now $PA = LU \Rightarrow A = P^{-1}LU$, but $P^{-1} = P^T$, so we have $A = P^TLU$.

$$\text{Check: } P^TLU = \begin{bmatrix} 0 & 0 & 1 & 0 \\ 0 & 0 & 0 & 1 \\ 0 & 1 & 0 & 0 \\ 1 & 0 & 0 & 0 \end{bmatrix} \begin{bmatrix} 1 & 0 & 0 & 0 \\ 0 & 1 & 0 & 0 \\ 0 & 0 & 1 & 0 \\ -1 & 1 & 1 & 1 \end{bmatrix} \begin{bmatrix} 1 & 1 & -1 & 0 \\ 0 & 2 & 1 & 1 \\ 0 & 0 & 1 & 2 \\ 0 & 0 & 0 & -1 \end{bmatrix} = \begin{bmatrix} 0 & 0 & 1 & 2 \\ -1 & 1 & 3 & 2 \\ 0 & 2 & 1 & 1 \\ 1 & 1 & -1 & 0 \end{bmatrix} = A$$

25. To find the $P^T LU$ factorization of A, we begin by permuting the rows of A:

$$\text{Since } A = \begin{bmatrix} 0 & -1 & 1 & 3 \\ -1 & 1 & 1 & 2 \\ 0 & 1 & -1 & 1 \\ 0 & 0 & 1 & 1 \end{bmatrix}, \text{ we have } PA = \begin{bmatrix} 0 & 1 & 0 & 0 \\ 1 & 0 & 0 & 0 \\ 0 & 0 & 0 & 1 \\ 0 & 0 & 1 & 0 \end{bmatrix} \begin{bmatrix} 0 & -1 & 1 & 3 \\ -1 & 1 & 1 & 2 \\ 0 & 1 & -1 & 1 \\ 0 & 0 & 1 & 1 \end{bmatrix} = \begin{bmatrix} -1 & 1 & 1 & 2 \\ 0 & -1 & 1 & 3 \\ 0 & 0 & 1 & 1 \\ 0 & 1 & -1 & 1 \end{bmatrix}$$

Now we follow the *multiplier* method of Example 3.35 to find the LU factorization of PA:

$$PA = \begin{bmatrix} -1 & 1 & 1 & 2 \\ 0 & -1 & 1 & 3 \\ 0 & 0 & 1 & 1 \\ 0 & 1 & -1 & 1 \end{bmatrix} \xrightarrow{R_4' = R_4 + R_2} \begin{bmatrix} -1 & 1 & 1 & 2 \\ 0 & -1 & 1 & 3 \\ 0 & 0 & 1 & 1 \\ 0 & 0 & 0 & 4 \end{bmatrix} = U \Rightarrow$$

$$\begin{matrix} l_{21} = 0 & & \\ l_{31} = 0 & l_{32} = 0 & \\ l_{41} = 0 & l_{42} = -1 & l_{43} = 0 \end{matrix} \quad \Rightarrow L = \begin{bmatrix} 1 & 0 & 0 & 0 \\ 0 & 1 & 0 & 0 \\ 0 & 0 & 1 & 0 \\ 0 & -1 & 0 & 1 \end{bmatrix}$$

$$\text{Check: } LU = \begin{bmatrix} 1 & 0 & 0 & 0 \\ 0 & 1 & 0 & 0 \\ 0 & 0 & 1 & 0 \\ 0 & -1 & 0 & 1 \end{bmatrix} \begin{bmatrix} -1 & 1 & 1 & 2 \\ 0 & -1 & 1 & 3 \\ 0 & 0 & 1 & 1 \\ 0 & 0 & 0 & 4 \end{bmatrix} = \begin{bmatrix} -1 & 1 & 1 & 2 \\ 0 & -1 & 1 & 3 \\ 0 & 0 & 1 & 1 \\ 0 & 1 & -1 & 1 \end{bmatrix} = PA$$

Now $PA = LU \Rightarrow A = P^{-1}LU$, but $P^{-1} = P^T$, so we have $A = P^T LU$.

$$\text{Check: } P^T LU = \begin{bmatrix} 0 & 1 & 0 & 0 \\ 1 & 0 & 0 & 0 \\ 0 & 0 & 0 & 1 \\ 0 & 0 & 1 & 0 \end{bmatrix} \begin{bmatrix} 1 & 0 & 0 & 0 \\ 0 & 1 & 0 & 0 \\ 0 & 0 & 1 & 0 \\ 0 & -1 & 0 & 1 \end{bmatrix} \begin{bmatrix} -1 & 1 & 1 & 2 \\ 0 & -1 & 1 & 3 \\ 0 & 0 & 1 & 1 \\ 0 & 0 & 0 & 4 \end{bmatrix} = \begin{bmatrix} 0 & -1 & 1 & 3 \\ -1 & 1 & 1 & 2 \\ 0 & 1 & -1 & 1 \\ 0 & 0 & 1 & 1 \end{bmatrix} = A$$

26. As we construct the permutation matrix P, we will count the row interchange choices.

For the first row interchange, there are n choices.
That is, we can interchange row 1 with any of the n rows of P.

For the second row interchange, there are $n - 1$ choices.
That is, we can interchange row 2 with any of the remaining $n - 1$ rows of P.

So, for the first two row interchanges we have $n(n - 1)$ choices.
Continuing in this manner, we see there are $n(n - 1) \cdots 2 \cdot 1 = n!$ choices.

That is, there are $n!$ permutation matrices of size $n \times n$.

27. We solve the system $LU\mathbf{x} = P\mathbf{b}$ by the two-step method outlined after Theorem 3.15:

$$PA\mathbf{x} = P\mathbf{b} \overset{PA=LU}{\Rightarrow} (LU)\mathbf{x} = P\mathbf{b} \overset{U\mathbf{x}=\mathbf{y}}{\Rightarrow} L\mathbf{y} = \mathbf{b}$$

Since $P = \begin{bmatrix} 0 & 1 & 0 \\ 1 & 0 & 0 \\ 0 & 0 & 1 \end{bmatrix}$ and $\mathbf{b} = \begin{bmatrix} 1 \\ 1 \\ 5 \end{bmatrix}$, we have: $P\mathbf{b} = \begin{bmatrix} 0 & 1 & 0 \\ 1 & 0 & 0 \\ 0 & 0 & 1 \end{bmatrix} \begin{bmatrix} 1 \\ 1 \\ 5 \end{bmatrix} = \begin{bmatrix} 1 \\ 1 \\ 5 \end{bmatrix}$.

$$L\mathbf{y} = P\mathbf{b} \Rightarrow \begin{bmatrix} 1 & 0 & 0 \\ 0 & 1 & 0 \\ \frac{1}{2} & -\frac{1}{2} & 1 \end{bmatrix} \begin{bmatrix} y_1 \\ y_2 \\ y_3 \end{bmatrix} = \begin{bmatrix} 1 \\ 1 \\ 5 \end{bmatrix} \Rightarrow \mathbf{y} = \begin{bmatrix} y_1 \\ y_2 \\ y_3 \end{bmatrix} = \begin{bmatrix} 1 \\ 1 \\ 5 \end{bmatrix} \Rightarrow$$

$$U\mathbf{x} = \mathbf{y} \Rightarrow \begin{bmatrix} 2 & 3 & 2 \\ 0 & 1 & -1 \\ 0 & 0 & -\frac{5}{2} \end{bmatrix} \begin{bmatrix} x_1 \\ x_2 \\ x_3 \end{bmatrix} = \begin{bmatrix} 1 \\ 1 \\ 5 \end{bmatrix} \Rightarrow \mathbf{x} = \begin{bmatrix} x_1 \\ x_2 \\ x_3 \end{bmatrix} = \begin{bmatrix} 4 \\ -1 \\ -2 \end{bmatrix} \Rightarrow$$

$$\text{Check: } A\mathbf{x} = \begin{bmatrix} 0 & 1 & -1 \\ 2 & 3 & 2 \\ 1 & 1 & -1 \end{bmatrix} \begin{bmatrix} 4 \\ -1 \\ -2 \end{bmatrix} = \begin{bmatrix} 1 \\ 1 \\ 5 \end{bmatrix} = \mathbf{b}$$

28. We solve the system $LU\mathbf{x} = P\mathbf{b}$ by the two-step method outlined after Theorem 3.15:

$$PA\mathbf{x} = P\mathbf{b} \overset{PA=LU}{\Rightarrow} (LU)\mathbf{x} = P\mathbf{b} \overset{U\mathbf{x}=\mathbf{y}}{\Rightarrow} L\mathbf{y} = \mathbf{b}$$

Since $P = \begin{bmatrix} 0 & 1 & 0 \\ 0 & 0 & 1 \\ 1 & 0 & 0 \end{bmatrix}$ and $\mathbf{b} = \begin{bmatrix} 16 \\ -4 \\ 4 \end{bmatrix}$, we have: $P\mathbf{b} = \begin{bmatrix} 0 & 1 & 0 \\ 0 & 0 & 1 \\ 1 & 0 & 0 \end{bmatrix} \begin{bmatrix} 16 \\ -4 \\ 4 \end{bmatrix} = \begin{bmatrix} -4 \\ 4 \\ 16 \end{bmatrix}$.

$$L\mathbf{y} = P\mathbf{b} \Rightarrow \begin{bmatrix} 1 & 0 & 0 \\ 1 & 1 & 0 \\ 2 & -1 & 1 \end{bmatrix} \begin{bmatrix} y_1 \\ y_2 \\ y_3 \end{bmatrix} = \begin{bmatrix} -4 \\ 4 \\ 16 \end{bmatrix} \Rightarrow \mathbf{y} = \begin{bmatrix} y_1 \\ y_2 \\ y_3 \end{bmatrix} = \begin{bmatrix} -4 \\ 8 \\ 32 \end{bmatrix} \Rightarrow$$

$$U\mathbf{x} = \mathbf{y} \Rightarrow \begin{bmatrix} 4 & 1 & 2 \\ 0 & -1 & 1 \\ 0 & 0 & 2 \end{bmatrix} \begin{bmatrix} x_1 \\ x_2 \\ x_3 \end{bmatrix} = \begin{bmatrix} -4 \\ 8 \\ 32 \end{bmatrix} \Rightarrow \mathbf{x} = \begin{bmatrix} x_1 \\ x_2 \\ x_3 \end{bmatrix} = \begin{bmatrix} -11 \\ 8 \\ 16 \end{bmatrix} \Rightarrow$$

$$\text{Check: } A\mathbf{x} = \begin{bmatrix} 8 & 3 & 5 \\ 4 & 1 & 2 \\ 4 & 0 & 3 \end{bmatrix} \begin{bmatrix} -11 \\ 8 \\ 16 \end{bmatrix} = \begin{bmatrix} 16 \\ -4 \\ 4 \end{bmatrix} = \mathbf{b}$$

29. Let L and L' be unit lower diagonal matrices and $A = LL'$. Then:

$$a_{ij} = \sum_{k=1}^{n} l_{ik}l'_{ki} = \sum_{k=1}^{j-1} l_{ik}l'_{ki} + l_{ij}l'_{ji} + \sum_{k=j+1}^{n} l_{ik}l'_{ki} = \begin{cases} 0+1+0=1 & i=j \\ *+0+0=* & i>j \\ 0+0+0=0 & i<j \end{cases} \Rightarrow$$

A is unit lower diagonal.

30. Since $L \longrightarrow I$, the Fundamental Theorem of Invertible Matrices $\Rightarrow$ L is invertible.

$L \longrightarrow I$ because the 1s occurs in different rows for each column
and those 1s can be used to create zeroes below.

To see $B = L^{-1}$ is unit lower triangular note we only use rows above the current row
when reducing L to $I \Rightarrow b_{ij} = 0$ when $j > i \Rightarrow B = L^{-1}$ is lower triangular.
Furthermore, the 1s on the diagonals of $L \Rightarrow B = L^{-1}$ must also have 1s on the diagonal.

31. We begin by factoring $U = DU_1$, where D is diagonal and U_1 is *unit* upper triangular.
To factor $U = DU_1$, we adapt the *multiplier* method of Example 3.35.

In this case, however, our goal is to create 1s on the diagonals. So:

$$\begin{bmatrix} -2 & 1 \\ 0 & 6 \end{bmatrix} \overset{R_1' = -\frac{1}{2}R_1}{\longrightarrow} \begin{bmatrix} 1 & -\frac{1}{2} \\ 0 & 6 \end{bmatrix} \overset{R_2' = \frac{1}{6}R_2}{\longrightarrow} \begin{bmatrix} 1 & -\frac{1}{2} \\ 0 & 1 \end{bmatrix} = U_1 \text{ unit upper triangular} \Rightarrow$$

$$\begin{matrix} l_{11} = -2 \\ l_{22} = 6 \end{matrix} \Rightarrow D = \begin{bmatrix} -2 & 0 \\ 0 & 6 \end{bmatrix}$$

$$\text{Check: } DU_1 = \begin{bmatrix} -2 & 0 \\ 0 & 6 \end{bmatrix} \begin{bmatrix} 1 & -\frac{1}{2} \\ 0 & 1 \end{bmatrix} = \begin{bmatrix} -2 & 1 \\ 0 & 6 \end{bmatrix} = U$$

$$\text{Since } A = LU = \begin{bmatrix} 1 & 0 \\ -1 & 1 \end{bmatrix} \begin{bmatrix} -2 & 1 \\ 0 & 6 \end{bmatrix},\ A = LDU_1 = \begin{bmatrix} 1 & 0 \\ -1 & 1 \end{bmatrix} \begin{bmatrix} -2 & 0 \\ 0 & 6 \end{bmatrix} \begin{bmatrix} 1 & -\frac{1}{2} \\ 0 & 1 \end{bmatrix}.$$

Q: Why is $l_{11} = -2$ in D? $l_{22} = 6$?

A: -2 is the multiplicative inverse of $-\frac{1}{2}$, so $(-2)(-\frac{1}{2}) = 1$ ...

32. We begin by factoring $U = DU_1$, where D is diagonal and U_1 is *unit* upper triangular.
To factor $U = DU_1$, we adapt the *multiplier* method of Example 3.35.

In this case, however, our goal is to create 1s on the diagonals. So:

$$\begin{bmatrix} 2 & -4 & 0 \\ 0 & 5 & 4 \\ 0 & 0 & 2 \end{bmatrix} \overset{R_1' = \frac{1}{2}R_1}{\longrightarrow} \begin{bmatrix} 1 & -2 & 0 \\ 0 & 5 & 4 \\ 0 & 0 & 2 \end{bmatrix} \overset{R_2' = \frac{1}{5}R_2}{\longrightarrow} \begin{bmatrix} 1 & -2 & 0 \\ 0 & 1 & \frac{4}{5} \\ 0 & 0 & 2 \end{bmatrix} \overset{R_3' = \frac{1}{2}R_3}{\longrightarrow} \begin{bmatrix} 1 & -2 & 0 \\ 0 & 1 & \frac{4}{5} \\ 0 & 0 & 1 \end{bmatrix} = U_1 \Rightarrow$$

$$\begin{matrix} l_{11} = 2 \\ l_{22} = 5 \\ l_{33} = 2 \end{matrix} \Rightarrow D = \begin{bmatrix} 2 & 0 & 0 \\ 0 & 5 & 0 \\ 0 & 0 & 2 \end{bmatrix}$$

$$\text{Check: } DU_1 = \begin{bmatrix} 2 & 0 & 0 \\ 0 & 5 & 0 \\ 0 & 0 & 2 \end{bmatrix} \begin{bmatrix} 1 & -2 & 0 \\ 0 & 1 & \frac{4}{5} \\ 0 & 0 & 1 \end{bmatrix} = \begin{bmatrix} 2 & -4 & 0 \\ 0 & 5 & 4 \\ 0 & 0 & 2 \end{bmatrix} = U$$

$$A = LU = \begin{bmatrix} 1 & 0 & 0 \\ \frac{3}{2} & 1 & 0 \\ -\frac{1}{2} & 0 & 1 \end{bmatrix} \begin{bmatrix} 2 & -4 & 0 \\ 0 & 5 & 4 \\ 0 & 0 & 2 \end{bmatrix} \Rightarrow A = LDU_1 = \begin{bmatrix} 1 & 0 & 0 \\ \frac{3}{2} & 1 & 0 \\ -\frac{1}{2} & 0 & 1 \end{bmatrix} \begin{bmatrix} 2 & 0 & 0 \\ 0 & 5 & 0 \\ 0 & 0 & 2 \end{bmatrix} \begin{bmatrix} 1 & -2 & 0 \\ 0 & 1 & \frac{4}{5} \\ 0 & 0 & 1 \end{bmatrix}.$$

33. We need to show if $A = LDU$ is symmetric and invertible, then $U = L^T$.

The key to a successful proof is strong application of the given conditions. For example:

Q: Since $A = LDU$ is symmetric, what do we know about A?
A: We know $LDU = A = A^T = (LDU)^T = U^T D^T L^T = U^T D L^T$. Does that help?

Q: Since $A = LDU$ is invertible, what do we know about A?
A: Theorem 3.16 $\Rightarrow$ If $A = L_1 U_1$ and $A = L_2 U_2$, then $L_1 = L_2$ and $U_1 = U_2$.

Q: How can we put these two facts together to create a proof?
A: We use the fact that A is symmetric to create $A = L_1 U_1$ and $A = L_2 U_2$.
Then we use the fact that A is invertible to equate factors and see how $U = L^T$ results.

PROOF: $A = L(DU)$ and $A = U^T(DL^T)$ (because A is symmetric)
$\Rightarrow DU = DL^T$ (because A is invertible so $U_1 = U_2$)
$\Rightarrow U = L^T$ (because D is invertible so $D^{-1}DM = M$)

There are several underlying facts in this proof which we examine in detail below:

Q: If U is unit upper triangular, is U^T unit lower triangular?
A: Yes, since in U if $i > j \Rightarrow a_{ij} = 0 = a_{ji}$ if $j < i$ in U^T.
Also, it is obvious that $a_{ii} = 1$ in both U and U^T.

Q: Where did we use this fact in our proof?
A: When claiming that $A = U^T(DL^T)$ is an LU factorization of A.
In order for that to be true, U^T must be unit lower triangular. Why?

Q: If L is lower triangular, is L^T upper triangular?
A: Yes, since in L if $i < j \Rightarrow a_{ij} = 0 = a_{ji}$ if $j > i$ in L^T.

Q: If D is diagonal and U is upper triangular is DU upper triangular?
A: Yes. We proved this for lower triangular matrices in Exercise 29.

Q: Where did we use these two facts in our proof?
A: When claiming that $A = U^T(DL^T)$ is an LU factorization of A.
In order for that to be true, DL^T must be upper triangular. Why?

Q: When $A = LDU$ is invertible, is D invertible?
A: Yes. Since $A^{-1}A = (A^{-1}LD)U = I \Rightarrow U$ is invertible and
$AA^{-1} = L(DUA^{-1}) = I \Rightarrow L$ is invertible. So we have:
$D = L^{-1}AU^{-1} \Rightarrow D(UA^{-1}L) = (L^{-1}AU^{-1})(UA^{-1}L) = I \Rightarrow D$ is invertible.
In fact, if $A_1 A_2 \cdots A_n$ is invertible, then each A_i is invertible.

Q: Where did we use this fact in our proof?
A: When claiming that $DU = DL^T \Rightarrow U = L^T$ since $U = D^{-1}DU = D^{-1}DL^T = L^T$.

34. We need to show if $A = L_1 D_1 L_1^T = L_2 D_2 L_2^T$, then $L_1 = L_2$ and $D_1 = D_2$.

PROOF: $A = (L_1 D_1) L_1^T$, $A = (L_2 D_2) L_2^T$	(because A is symmetric)
$\Rightarrow L_1 D_1 = L_2 D_2$	(because A is invertible so $L' = L''$)
$\Rightarrow L_2^{-1} L_1 = D_2 D_1^{-1}$	(because L_2 and D_1 are invertible)
$\Rightarrow L_2^{-1} L_1 = D_2 D_1^{-1} = I$	(because both products are *unit* diagonal)
$\Rightarrow L_1 = L_2$ and $D_1 = D_2$	(because L_2 and D_1 are invertible)

3.5 Subspaces, Basis, Dimension, and Rank

1. As in Example 3.38, substituting the condition $x = 0$ into $\begin{bmatrix} x \\ y \end{bmatrix}$ yields $\begin{bmatrix} 0 \\ y \end{bmatrix} = y \begin{bmatrix} 0 \\ 1 \end{bmatrix}$.

Since y is arbitrary, $S = \text{span}\left(\begin{bmatrix} 0 \\ 1 \end{bmatrix}\right)$. So, S is a subspace of $\mathbb{R}^2$ by Theorem 3.19.

Q: Geometrically speaking, what is $x = 0$?
A: The equation $x = 0$ is a line through the origin. This should come as no surprise. Why? Because a line through the origin is a subspace in $\mathbb{R}^2, \mathbb{R}^3, \ldots, \mathbb{R}^n$.

2. We should suspect that set S defined by $x \geq 0$ and $y \geq 0$ is not a subspace. Why? Because multiplication by scalars (e.g., negative numbers) remove values from the set S.

How do we prove that S is *not* a subspace?
We provide a counterexample to show that one of the required properties does *not* hold.

So, note that $\begin{bmatrix} 1 \\ 0 \end{bmatrix}$ is in S, but $-1\begin{bmatrix} 1 \\ 0 \end{bmatrix} = \begin{bmatrix} -1 \\ 0 \end{bmatrix}$ is not.

Property (3) ($\mathbf{u}$ in S implies $c\mathbf{u}$ in S) fails so S is not a subspace of $\mathbb{R}^2$.

3. As in Example 3.38, substituting the condition $y = 2x$ into $\begin{bmatrix} x \\ y \end{bmatrix}$ yields $\begin{bmatrix} x \\ 2x \end{bmatrix} = x \begin{bmatrix} 1 \\ 2 \end{bmatrix}$.

Since x is arbitrary, $S = \text{span}\left(\begin{bmatrix} 1 \\ 2 \end{bmatrix}\right)$. So, S is a subspace of $\mathbb{R}^2$ by Theorem 3.19.

Q: Geometrically speaking, what is $y = 2x$?
A: Since $y = 2x \Rightarrow 2x - y = 0$, this is obviously a line through the origin.

4. We should suspect that set S defined by $xy \geq 0$ is not a subspace. Why? Because, in general, when an inequality is a condition it is hard to have closure.

How do we prove that S is *not* a subspace?
We provide a counterexample to show that one of the required properties does *not* hold.

Note $\begin{bmatrix} 1 \\ 0 \end{bmatrix}$ and $\begin{bmatrix} 0 \\ -1 \end{bmatrix}$ are in S, but $\begin{bmatrix} 1 \\ 0 \end{bmatrix} + \begin{bmatrix} 0 \\ -1 \end{bmatrix} = \begin{bmatrix} 1 \\ -1 \end{bmatrix}$ is not.

Property (2) ($\mathbf{u}$, $\mathbf{v}$ in S implies $\mathbf{u} + \mathbf{v}$ in S) fails so S is not a subspace of $\mathbb{R}^2$.

5. As in Example 3.38, substituting the condition $x = y = z$ into $\begin{bmatrix} x \\ y \\ z \end{bmatrix}$ yields $\begin{bmatrix} x \\ x \\ x \end{bmatrix} = x \begin{bmatrix} 1 \\ 1 \\ 1 \end{bmatrix}$.

Since x is arbitrary, $S = \text{span}\left(\begin{bmatrix} 1 \\ 1 \\ 1 \end{bmatrix}\right)$. So, S is a subspace of $\mathbb{R}^3$ by Theorem 3.19.

Q: Geometrically speaking, what is $x = y = z$?
A: Since $x = y = z \Rightarrow x - y = 0$ and $x - z = 0$, this is obviously a line through the origin. Why? Because $x - y = 0$ and $x - z = 0$ are planes through the origin that intersect in that line.

Note: a plane through the origin is a subspace in $\mathbb{R}^3, \mathbb{R}^4, \cdots, \mathbb{R}^n$.

6. As in Example 3.38, substituting $z = 2x$ and $y = 0$ into $\begin{bmatrix} x \\ y \\ z \end{bmatrix}$ yields $\begin{bmatrix} x \\ 0 \\ 2x \end{bmatrix} = x\begin{bmatrix} 1 \\ 0 \\ 2 \end{bmatrix}$.

Since x is arbitrary, $S = \text{span}\left(\begin{bmatrix} 1 \\ 0 \\ 2 \end{bmatrix}\right)$. So, S is a subspace of $\mathbb{R}^3$ by Theorem 3.19.

Q: Geometrically speaking, what is $z = 2x$ and $y = 0$?
A: Since $2x - z = 0$ and $y = 0$ are planes through the origin,
their intersection is a line through the origin.

7. We should recognize right away that set S defined by $x - y + z = 1$ is *not* a subspace. Why?
Because $x - y + z = 1$ is a plane that does *not* pass through the origin. So we conclude:

Property (1) ($\mathbf{0}$ is in S) fails so S is not a subspace of $\mathbb{R}^3$.

Q: How can we tell that $x - y + z = 1$? is a plane that does *not* pass through the origin?
A: The equation of a plane through the origin must be of the form $ax + by + cz = 0$. Why?
Because the zero vector must satisfy the equation. So, we must have: $a0 + b0 + c0 = 0$.

8. We should suspect that set S defined by $|x - y| = |y - z|$ is *not* a subspace. Why?
Because, in general, when an absolute value is a condition, it is difficult to have closure.

How do we prove that S is *not* a subspace?
We provide a counterexample to show that one of the required properties does *not* hold.

Note $\begin{bmatrix} 1 \\ 0 \\ 1 \end{bmatrix}$ and $\begin{bmatrix} 0 \\ 1 \\ 2 \end{bmatrix}$ are in S, but $\begin{bmatrix} 1 \\ 0 \\ 1 \end{bmatrix} + \begin{bmatrix} 0 \\ 1 \\ 2 \end{bmatrix} = \begin{bmatrix} 1 \\ 1 \\ 3 \end{bmatrix}$ is not.

Property (2) ($\mathbf{u}$, $\mathbf{v}$ in S implies $\mathbf{u} + \mathbf{v}$ in S) fails so S is not a subspace of $\mathbb{R}^3$.

9. Every line ℓ through the origin in $\mathbb{R}^3$ is a subspace of $\mathbb{R}^3$
because if ℓ has equation $\mathbf{x} = \mathbf{0} + t\mathbf{d}$ then $\ell = \text{span}(\mathbf{d})$.
Therefore, once again, Theorem 3.19 implies that ℓ is a subspace of $\mathbb{R}^3$.

We might also have stated our proof this way:

ℓ is a line $\Rightarrow$	ℓ has equation $\mathbf{x} = \mathbf{p} + t\mathbf{d}$
ℓ passes through the origin $\Rightarrow$	$\mathbf{x} = \mathbf{0} + t\mathbf{d} = t\mathbf{d}$
The definition of span $\Rightarrow$	$\ell = \text{span}(\mathbf{d})$
So, Theorem 3.19 $\Rightarrow$	ℓ is a subspace of $\mathbb{R}^3$

Q: Did we use the fact that ℓ passes through the origin?
A: Yes, in Step 2. Otherwise $\mathbf{0}$ will not be in ℓ as required.

Q: Did we use the fact that ℓ is a line in $\mathbb{R}^3$?
A: No. The parametric equation is the same in $\mathbb{R}^n$, so our conclusion is true in $\mathbb{R}^n$.

10. Let S be the set of points in $\mathbb{R}^2$ that are on the x-axis or the y-axis (or both).
Since $\mathbf{u}$, $\mathbf{v}$ in S does *not* imply $\mathbf{u} + \mathbf{v}$ is in S, S is not a subspace of $\mathbb{R}^2$.
For example, $\mathbf{u} = (1, 0)$, $\mathbf{v} = (0, 1)$ are in S, but $\mathbf{u} + \mathbf{v} = (1, 1)$ is not.

11. As in Example 3.41, we will determine whether $\mathbf{b}$ is in col(A) and $\mathbf{w}$ is in row(A).

col(A): To say $\mathbf{b}$ is in col(A) means $\mathbf{b}$ is a linear combination of the columns of A. So:

$\mathbf{b} = \begin{bmatrix} 3 \\ 2 \end{bmatrix}$ is in the column space of $A = \begin{bmatrix} 1 & 0 & -1 \\ 1 & 1 & 1 \end{bmatrix}$ if the system $A\mathbf{x} = \mathbf{b}$ is consistent.

We row reduce the augmented matrix: $\left[\begin{array}{ccc|c} 1 & 0 & -1 & 3 \\ 1 & 1 & 1 & 2 \end{array}\right] \longrightarrow \left[\begin{array}{ccc|c} 1 & 0 & -1 & 3 \\ 0 & 1 & 2 & -1 \end{array}\right] \Rightarrow$

The system is consistent. So, $\mathbf{b} \in$ col(A). In particular, $\mathbf{b} = 3\mathbf{a}_1 - \mathbf{a}_2$.

row(A): To say $\mathbf{w}$ is in row(A) means $\mathbf{w}$ is a linear combination of the rows of A. So:

$\mathbf{w} = \begin{bmatrix} -1 & 1 & 1 \end{bmatrix} \in$ row(A) if $\left[\begin{array}{c} A \\ \hline \mathbf{w} \end{array}\right] \longrightarrow \left[\begin{array}{c} A \\ \hline \mathbf{0} \end{array}\right]$

by elementary row operations *excluding* row interchanges involving the *last* row.

So, we have $\left[\begin{array}{c} A \\ \hline \mathbf{w} \end{array}\right] = \left[\begin{array}{ccc} 1 & 0 & -1 \\ 1 & 1 & 1 \\ \hline -1 & 1 & 1 \end{array}\right] \xrightarrow[R_3+R_1]{R_2-R_1} \left[\begin{array}{ccc} 1 & 0 & -1 \\ 0 & 1 & 2 \\ \hline 0 & 1 & 0 \end{array}\right] \xrightarrow{R_3-R_2} \left[\begin{array}{ccc} 1 & 0 & -1 \\ 0 & 1 & 2 \\ \hline 0 & 0 & -2 \end{array}\right] \Rightarrow$

We cannot make the last row all zeroes $\Rightarrow \mathbf{w} \notin$ row(A).

12. As in Example 3.41, we will determine whether $\mathbf{b}$ is in col(A) and $\mathbf{w}$ is in row(A).

col(A): To say $\mathbf{b}$ is in col(A) means $\mathbf{b}$ is a linear combination of the columns of A. So:

$\mathbf{b} = \begin{bmatrix} 1 \\ 1 \\ 0 \end{bmatrix}$ is in the column space of $A = \begin{bmatrix} 1 & 1 & -3 \\ 0 & 2 & 1 \\ 1 & -1 & -4 \end{bmatrix}$ if $A\mathbf{x} = \mathbf{b}$ is consistent.

We row reduce the augmented matrix: $\left[\begin{array}{ccc|c} 1 & 1 & -3 & 1 \\ 0 & 2 & 1 & 1 \\ 1 & -1 & -4 & 0 \end{array}\right] \longrightarrow \left[\begin{array}{ccc|c} 1 & 0 & -\frac{7}{2} & \frac{1}{2} \\ 0 & 1 & \frac{1}{2} & \frac{1}{2} \\ 0 & 0 & 0 & 0 \end{array}\right] \Rightarrow$

The system is consistent. So, $\mathbf{b} \in$ col(A). In particular, $\mathbf{b} = \frac{1}{2}\mathbf{a}_1 + \frac{1}{2}\mathbf{a}_2(A)$.

row(A): To say $\mathbf{w}$ is in row(A) means $\mathbf{w}$ is a linear combination of the rows of A. So:

$\mathbf{w} = \begin{bmatrix} 2 & 4 & -5 \end{bmatrix} \in$ row(A) if $\left[\begin{array}{c} A \\ \hline \mathbf{w} \end{array}\right] \longrightarrow \left[\begin{array}{c} A \\ \hline \mathbf{0} \end{array}\right]$

by elementary row operations *excluding* row interchanges involving the *last* row.

So, $\left[\begin{array}{c} A \\ \hline \mathbf{w} \end{array}\right] = \left[\begin{array}{ccc} 1 & 1 & -3 \\ 0 & 2 & 1 \\ 1 & -1 & -4 \\ \hline 2 & 4 & -5 \end{array}\right] \xrightarrow[R_4-2R_1]{R_3-R_1} \left[\begin{array}{ccc} 1 & 1 & -3 \\ 0 & 2 & 1 \\ 0 & -2 & -1 \\ \hline 0 & 2 & 1 \end{array}\right] \xrightarrow[R_4-R_2]{R_3+R_2} \left[\begin{array}{ccc} 1 & 1 & -3 \\ 0 & 2 & 1 \\ 0 & 0 & 0 \\ \hline 0 & 0 & 0 \end{array}\right] \Rightarrow$

So, $\mathbf{w}$ is a linear combination of the rows of $A \Rightarrow \mathbf{w} \in$ row (A).

13. As in the remarks following Example 3.41, we determine whether $\mathbf{w}$ is in row(A) using col(A^T).

To say $\mathbf{w}$ is in row(A) means $\mathbf{w}^T$ is a linear combination of the *columns* of A^T. So:

We row reduce to check: $\left[\, A^T \mid \mathbf{w}^T \,\right] = \left[\begin{array}{cc|c} 1 & 1 & -1 \\ 0 & 1 & 1 \\ -1 & 1 & 1 \end{array}\right] \longrightarrow \left[\begin{array}{cc|c} 1 & 0 & 0 \\ 0 & 1 & 0 \\ 0 & 0 & 1 \end{array}\right] \Rightarrow$

Since the last row is $\left[\begin{array}{cc|c} 0 & 0 & 1 \end{array}\right]$, this system is inconsistent. So, $\mathbf{w} \notin$ row(A).

14. As in the remarks following Example 3.41, we determine whether $\mathbf{w}$ is in row(A) using col(A^T).

To say $\mathbf{w}$ is in row(A) means $\mathbf{w}^T$ is a linear combination of the *columns* of A^T. So:

We row reduce to check: $\left[\, A^T \mid \mathbf{w}^T \,\right] = \left[\begin{array}{ccc|c} 1 & 0 & 1 & 2 \\ 1 & 2 & -1 & 4 \\ -3 & 1 & -4 & -5 \end{array}\right] \longrightarrow \left[\begin{array}{ccc|c} 1 & 0 & 1 & 2 \\ 0 & 1 & -1 & 1 \\ 0 & 0 & 0 & 0 \end{array}\right] \Rightarrow$

The system is consistent. So, $\mathbf{w} \in$ row(A). In particular, $\mathbf{w} = 2\mathbf{A}_1 + \mathbf{A}_2$.

15. Since $A\mathbf{v} = \mathbf{0}$ implies $\mathbf{v}$ is in null(A), we simply multiply $A\mathbf{v}$ to check:

$$A\mathbf{v} = \begin{bmatrix} 1 & 0 & -1 \\ 1 & 1 & 1 \end{bmatrix} \begin{bmatrix} -1 \\ 3 \\ -1 \end{bmatrix} = \begin{bmatrix} 0 \\ 1 \end{bmatrix} \neq \mathbf{0} \Rightarrow \mathbf{v} \notin \text{null}(A).$$

16. Since $A\mathbf{v} = \mathbf{0}$ implies $\mathbf{v}$ is in null(A), we simply multiply $A\mathbf{v}$ to check:

$$A\mathbf{v} = \begin{bmatrix} 1 & 1 & -3 \\ 0 & 2 & 1 \\ 1 & -1 & -4 \end{bmatrix} \begin{bmatrix} 7 \\ -1 \\ 2 \end{bmatrix} = \begin{bmatrix} 0 \\ 0 \\ 0 \end{bmatrix} = \mathbf{0} \Rightarrow \mathbf{v} \in \text{null}(A).$$

17. We find bases for row(A), col(A), and null(A) as in Examples 3.45, 3.47, and 3.48 respectively.

row(A): A basis for row(A) must span the rows of A and be linearly independent.
Given $A \longrightarrow R$, Theorem 3.20 asserts that the rows of R span the rows of A. Why?
Because the rows of A are linear combinations of the rows of R (and vice-versa).
Finally, we simply observe that the nonzero rows of R are linearly independent.

Since $A = \begin{bmatrix} 1 & 0 & -1 \\ 1 & 1 & 1 \end{bmatrix} \longrightarrow \begin{bmatrix} 1 & 0 & -1 \\ 0 & 1 & 2 \end{bmatrix} = R$,

we conclude that $\{ \begin{bmatrix} 1 & 0 & -1 \end{bmatrix}, \begin{bmatrix} 0 & 1 & 2 \end{bmatrix} \}$ is a basis for row(A).

We should also note that provided $A \longrightarrow R$ uses no row interchanges,
the corresponding rows in A are also linearly independent.
Whence, it is obvious that those rows form a basis for row(A).

col(A): A basis for col(A) must span the columns of A and be linearly independent.
When $A \longrightarrow R$, the columns with leading 1s in R are linearly independent.
As shown in Example 3.47, the corresponding columns in A are also linearly independent.
Whence, it is obvious that those columns form a basis for col(A).

Since $A = \begin{bmatrix} 1 & 0 & -1 \\ 1 & 1 & 1 \end{bmatrix} \longrightarrow \begin{bmatrix} 1 & 0 & -1 \\ 0 & 1 & 2 \end{bmatrix} = R$,

we conclude that $\left\{ \begin{bmatrix} 1 \\ 1 \end{bmatrix}, \begin{bmatrix} 0 \\ 1 \end{bmatrix} \right\}$ is a basis for col(A).

null(A): Since $A\mathbf{v} = \mathbf{0}$ implies $\mathbf{v}$ is in null(A), we solve $[\, A \,|\, \mathbf{0} \,] \longrightarrow [\, R \,|\, \mathbf{0} \,]$ to find the conditions:

$$[\, R \,|\, \mathbf{0} \,] = \left[\begin{array}{ccc|c} 1 & 0 & -1 & 0 \\ 0 & 1 & 2 & 0 \end{array}\right] \Rightarrow \begin{array}{l} x_1 - x_3 = 0 \\ x_2 + 2x_3 = 0 \\ x_3 \textit{ free} \end{array} \Rightarrow \begin{array}{rcr} x_1 & = & 1s \\ x_2 & = & -2s \\ x_3 & = & 1s \end{array}$$

Since t is arbitrary, null(A) = span $\left(\begin{bmatrix} 1 \\ -2 \\ 1 \end{bmatrix} \right)$. So, $\left\{ \begin{bmatrix} 1 \\ -2 \\ 1 \end{bmatrix} \right\}$ is a basis for null(A).

18. We find bases for row(A), col(A), and null(A) as in Examples 3.45, 3.47, and 3.48 respectively.

row(A): A basis for row(A) must span the rows of A and be linearly independent.
Given $A \longrightarrow U$, Theorem 3.20 asserts that the rows of U span the rows of A. Why?
Because the rows of A are linear combinations of the rows of U (and vice-versa).
Finally, we simply observe that the nonzero rows of U are linearly independent.

$$\text{Since } A = \begin{bmatrix} 1 & 1 & -3 \\ 0 & 2 & 1 \\ 1 & -1 & -4 \end{bmatrix} \xrightarrow{R_3 - R_1 + R_2} \begin{bmatrix} 1 & 1 & -3 \\ 0 & 2 & 1 \\ 0 & 0 & 0 \end{bmatrix} = U,$$

we conclude that $\{[\,1\ 1\ -3\,], [\,0\ 2\ 1\,]\}$ is a basis for row(A).

Q: In $A \longrightarrow U$, why is it sufficient to reduce A only to row echelon form U?

A: As the remark following Example 3.46 explains and then demonstrates by example, the nonzero rows of U are linearly independent. That is all that is required. Why?

We should also note that provided $A \longrightarrow U$ uses no row interchanges,
the corresponding rows in A are also linearly independent.
Whence, it is obvious that those rows form a basis for row(A).

col(A): A basis for col(A) must span the columns of A and be linearly independent.
When $A \longrightarrow U$, the columns with leading entries in U are linearly independent.
As in Example 3.47, the corresponding columns in A are also linearly independent.
Whence, it is obvious that those columns form a basis for col(A).

$$\text{Since } A = \begin{bmatrix} 1 & 1 & -3 \\ 0 & 2 & 1 \\ 1 & -1 & -4 \end{bmatrix} \longrightarrow \begin{bmatrix} 1 & 1 & -3 \\ 0 & 2 & 1 \\ 0 & 0 & 0 \end{bmatrix} = U,$$

$$\text{we conclude that } \left\{ \begin{bmatrix} 1 \\ 0 \\ 1 \end{bmatrix}, \begin{bmatrix} 1 \\ 2 \\ -1 \end{bmatrix} \right\} \text{ is a basis for col}(A).$$

null(A): Since $A\mathbf{v} = \mathbf{0}$ implies $\mathbf{v}$ is in null(A), we solve $[\,A\,|\,\mathbf{0}\,] \longrightarrow [\,U'\,|\,\mathbf{0}\,]$ to find the conditions.
We row reduce U one more step to U' make it easier to find the conditions:

$$A = \begin{bmatrix} 1 & 1 & -3 \\ 0 & 2 & 1 \\ 1 & -1 & -4 \end{bmatrix} \xrightarrow{R_3 - R_1 + R_2} \begin{bmatrix} 1 & 1 & -3 \\ 0 & 2 & 1 \\ 0 & 0 & 0 \end{bmatrix} \xrightarrow{R_1 + 3R_2} \begin{bmatrix} 1 & 7 & 0 \\ 0 & 2 & 1 \\ 0 & 0 & 0 \end{bmatrix} = U'$$

$$[\,U'\,|\,\mathbf{0}\,] = \left[\begin{array}{ccc|c} 1 & 7 & 0 & 0 \\ 0 & 2 & 1 & 0 \\ 0 & 0 & 0 & 0 \end{array}\right] \Rightarrow \begin{array}{c} x_1 + 7x_2 = 0 \\ 2x_2 + x_3 = 0 \\ x_2 \ free \end{array} \Rightarrow \begin{array}{rcl} x_1 & = & -7s \\ x_2 & = & s \\ x_3 & = & -2s \end{array}$$

$$\text{Since } s \text{ is arbitrary, null}(A) = \text{span}\left(\begin{bmatrix} -7 \\ 1 \\ -2 \end{bmatrix} \right). \text{ So, } \left\{ \begin{bmatrix} -7 \\ 1 \\ -2 \end{bmatrix} \right\} \text{ is a basis for null}(A).$$

19. We find bases for row(A), col(A), and null(A) as in Examples 3.45, 3.47, and 3.48 respectively.

row(A): A basis for row(A) must span the rows of A and be linearly independent.
Given $A \longrightarrow R$, Theorem 3.20 asserts that the rows of R span the rows of A. Why?
Because the rows of A are linear combinations of the rows of R (and vice-versa).
Finally, we simply observe that the nonzero rows of R are linearly independent.

$$\text{Since } A = \begin{bmatrix} 1 & 1 & 0 & 1 \\ 0 & 1 & -1 & 1 \\ 0 & 1 & -1 & -1 \end{bmatrix} \longrightarrow \begin{bmatrix} 1 & 0 & 1 & 0 \\ 0 & 1 & -1 & 0 \\ 0 & 0 & 0 & 1 \end{bmatrix} = R,$$

we conclude that $\{[\,1\ 0\ 1\ 0\,], [\,0\ 1\ -1\ 0\,], [\,0\ 0\ 0\ 1\,]\}$ is a basis for row(A).
We should also note that provided $A \longrightarrow R$ uses no row interchanges,
the corresponding rows in A are also linearly independent.
Whence, it is obvious that those rows form a basis for row(A).

col(A): A basis for col(A) must span the columns of A and be linearly independent.
When $A \longrightarrow R$, the columns with leading 1s in R are linearly independent.
As shown in Example 3.47, the corresponding columns in A are also linearly independent.
Whence, it is obvious that those columns form a basis for col(A).

$$\text{Since } A = \begin{bmatrix} 1 & 1 & 0 & 1 \\ 0 & 1 & -1 & 1 \\ 0 & 1 & -1 & -1 \end{bmatrix} \longrightarrow \begin{bmatrix} 1 & 0 & 1 & 0 \\ 0 & 1 & -1 & 0 \\ 0 & 0 & 0 & 1 \end{bmatrix} = R,$$

$$\text{we conclude that } \left\{ \begin{bmatrix} 1 \\ 0 \\ 0 \end{bmatrix}, \begin{bmatrix} 1 \\ 1 \\ 1 \end{bmatrix}, \begin{bmatrix} 1 \\ 1 \\ -1 \end{bmatrix} \right\} \text{ is a basis for col}(A).$$

null(A): Since $A\mathbf{v} = \mathbf{0}$ implies $\mathbf{v}$ is in null(A), we solve $[\,A\,|\,\mathbf{0}\,] \longrightarrow [\,R\,|\,\mathbf{0}\,]$ to find the conditions.

$$[\,R\,|\,\mathbf{0}\,] = \left[\begin{array}{cccc|c} 1 & 0 & 1 & 0 & 0 \\ 0 & 1 & -1 & 0 & 0 \\ 0 & 0 & 0 & 1 & 0 \end{array}\right] \Rightarrow \begin{array}{rcl} x_1 + x_3 & = & 0 \\ x_2 - x_3 & = & 0 \\ x_3\ free & & \\ x_4 & = & 0 \end{array} \Rightarrow \begin{array}{rcr} x_1 & = & -s \\ x_2 & = & s \\ x_3 & = & s \\ x_4 & = & 0 \end{array}$$

$$\text{Since } s \text{ is arbitrary, null}(A) = \text{span}\left(\begin{bmatrix} -1 \\ 1 \\ 1 \\ 0 \end{bmatrix} \right). \text{ So, } \left\{ \begin{bmatrix} -1 \\ 1 \\ 1 \\ 0 \end{bmatrix} \right\} \text{ is a basis for null}(A).$$

20. We find bases for row(A), col(A), and null(A) as in Examples 3.45, 3.47, and 3.48 respectively.

row(A): A basis for row(A) must span the rows of A and be linearly independent.
The linearly independent rows (which are simply the nonzero rows) of U do just that.

$$\text{Since } A = \begin{bmatrix} 2 & -4 & 0 & 2 & 1 \\ -1 & 2 & 1 & 2 & 3 \\ 1 & -2 & 1 & 4 & 4 \end{bmatrix} \longrightarrow \begin{bmatrix} 2 & -4 & 0 & 2 & 1 \\ 0 & 0 & 2 & 6 & 7 \\ 0 & 0 & 0 & 0 & 0 \end{bmatrix} = U$$

we conclude that $\{\begin{bmatrix} 2 & -4 & 0 & 2 & 1 \end{bmatrix}, \begin{bmatrix} 0 & 0 & 2 & 6 & 7 \end{bmatrix}\}$ is a basis for row(A).

Q: In $A \longrightarrow U$, why is it sufficient to reduce A only to row echelon form U?

A: As the remark following Example 3.46 explains and then demonstrates by example, the nonzero rows of U are linearly independent. That is all that is required. Why?

We should also note that provided $A \longrightarrow U$ uses no row interchanges,
the corresponding rows in A are also linearly independent.
Whence, it is obvious that those rows form a basis for row(A).

col(A): A basis for col(A) must span the columns of A and be linearly independent.
When $A \longrightarrow U$, the columns with leading entries in U are linearly independent.
As shown in Example 3.47, the corresponding columns in A are also linearly independent.
Whence, it is obvious that those columns form a basis for col(A).

$$\text{Since } A = \begin{bmatrix} 2 & -4 & 0 & 2 & 1 \\ -1 & 2 & 1 & 2 & 3 \\ 1 & -2 & 1 & 4 & 4 \end{bmatrix} \longrightarrow \begin{bmatrix} 2 & -4 & 0 & 2 & 1 \\ 0 & 0 & 2 & 6 & 7 \\ 0 & 0 & 0 & 0 & 0 \end{bmatrix} = U$$

we conclude that $\left\{ \begin{bmatrix} 2 \\ -1 \\ 1 \end{bmatrix}, \begin{bmatrix} 0 \\ 1 \\ 1 \end{bmatrix} \right\}$ is a basis for col(A).

null(A): Since $A\mathbf{v} = \mathbf{0}$ implies $\mathbf{v}$ is in null(A), we solve $[\, A \,|\, \mathbf{0} \,] \longrightarrow [\, U \,|\, \mathbf{0} \,]$ to find the conditions.

$$[\, U \,|\, \mathbf{0} \,] = \left[\begin{array}{ccccc|c} 2 & -4 & 0 & 2 & 1 & 0 \\ 0 & 0 & 2 & 6 & 7 & 0 \\ 0 & 0 & 0 & 0 & 0 & 0 \end{array}\right] \Rightarrow \begin{aligned} x_1 &= -t - s + 2w \\ x_2 &= w \\ x_3 &= -3s - 7t \\ x_4 &= s \\ x_5 &= 2t \end{aligned}$$

Therefore, $\left\{ \begin{bmatrix} 2 \\ 1 \\ 0 \\ 0 \\ 0 \end{bmatrix}, \begin{bmatrix} -1 \\ 0 \\ -3 \\ 1 \\ 0 \end{bmatrix}, \begin{bmatrix} -1 \\ 0 \\ -7 \\ 0 \\ 2 \end{bmatrix} \right\}$ is a basis for null(A).

21. We find bases for row(A) and col(A) following Examples 3.45 and 3.47 respectively.

row(A): A basis for col(A) must span the columns of A and be linearly independent.
Clearly, the linearly independent *columns* of A^T do just that.
When $A^T \longrightarrow R$, the columns with leading 1s in R are linearly independent.
As in Example 3.47, the corresponding columns in A^T are also linearly independent.
Whence, it is obvious that the *transposes* of those columns form a basis for row(A).

$$\text{Since } A^T = \begin{bmatrix} 1 & 1 \\ 0 & 1 \\ -1 & 1 \end{bmatrix} \longrightarrow \begin{bmatrix} 1 & 0 \\ 0 & 1 \\ 0 & 0 \end{bmatrix} = R$$

we conclude that $\{\begin{bmatrix} 1 & 0 & -1 \end{bmatrix}, \begin{bmatrix} 1 & 1 & 1 \end{bmatrix}\}$ is a basis for row(A).

col(A): A basis for col(A) must span the columns of A and be linearly independent.
When $A^T \longrightarrow R$, the linearly independent *rows* (the nonzero rows) of R do just that.
Whence, it is obvious that the *transposes* of those rows form a basis for col(A).

$$\text{Since } A^T = \begin{bmatrix} 1 & 1 \\ 0 & 1 \\ -1 & 1 \end{bmatrix} \longrightarrow \begin{bmatrix} 1 & 0 \\ 0 & 1 \\ 0 & 0 \end{bmatrix} = R$$

we conclude that $\left\{\begin{bmatrix} 1 \\ 0 \end{bmatrix}, \begin{bmatrix} 0 \\ 1 \end{bmatrix}\right\}$ is a basis for col(A).

We should also note that provided $A^T \longrightarrow R$ uses no row interchanges,
the corresponding rows in A^T are also linearly independent.
Whence, it is obvious that the *transposes* of those rows form a basis for col(A).

22. We find bases for row(A) and col(A) following Examples 3.45 and 3.47 respectively.

row(A): A basis for col(A) must span the columns of A and be linearly independent.
Clearly, the linearly independent *columns* of A^T do just that.
When $A^T \longrightarrow R$, the columns with leading 1s in R are linearly independent.
As in Example 3.47, the corresponding columns in A^T are also linearly independent.
Whence, it is obvious that the *transposes* of those columns form a basis for row(A).

$$\text{Since } A^T = \begin{bmatrix} 1 & 0 & 1 \\ 1 & 2 & -1 \\ -3 & 1 & -4 \end{bmatrix} \longrightarrow \begin{bmatrix} 1 & 0 & 1 \\ 0 & 1 & -1 \\ 0 & 0 & 0 \end{bmatrix} = R,$$

we conclude that $\{\begin{bmatrix} 1 & 1 & -3 \end{bmatrix}, \begin{bmatrix} 0 & 2 & 1 \end{bmatrix}\}$ is a basis for row(A).

col(A): A basis for col(A) must span the columns of A and be linearly independent.
When $A^T \longrightarrow R$, the linearly independent *rows* (the nonzero rows) of R do just that.
Whence, it is obvious that the *transposes* of those rows form a basis for col(A).

$$\text{Since } A^T = \begin{bmatrix} 1 & 0 & 1 \\ 1 & 2 & -1 \\ -3 & 1 & -4 \end{bmatrix} \longrightarrow \begin{bmatrix} 1 & 0 & 1 \\ 0 & 1 & -1 \\ 0 & 0 & 0 \end{bmatrix} = R,$$

we conclude that $\left\{\begin{bmatrix} 1 \\ 0 \\ 1 \end{bmatrix}, \begin{bmatrix} 0 \\ 1 \\ -1 \end{bmatrix}\right\}$ is a basis for col(A).

23. We find bases for row(A) and col(A) following Examples 3.45 and 3.47 respectively.

row(A): A basis for col(A) must span the columns of A and be linearly independent.
Clearly, the linearly independent *columns* of A^T do just that.
When $A^T \longrightarrow R$, the columns with leading 1s in R are linearly independent.
As in Example 3.47, the corresponding columns in A^T are also linearly independent.
Whence, it is obvious that the *transposes* of those columns form a basis for row(A).

$$\text{Since } A^T = \begin{bmatrix} 1 & 0 & 0 \\ 1 & 1 & 1 \\ 0 & -1 & -1 \\ 1 & 1 & -1 \end{bmatrix} \longrightarrow \begin{bmatrix} 1 & 0 & 0 \\ 0 & 1 & 0 \\ 0 & 0 & 1 \\ 0 & 0 & 0 \end{bmatrix} = R,$$

we conclude that $\{\begin{bmatrix} 1 & 1 & 0 & 1 \end{bmatrix}, \begin{bmatrix} 0 & 1 & -1 & 1 \end{bmatrix}, \begin{bmatrix} 0 & 1 & -1 & -1 \end{bmatrix}\}$ is a basis for row(A).

col(A): A basis for col(A) must span the columns of A and be linearly independent.
When $A^T \longrightarrow R$, the linearly independent *rows* (the nonzero rows) of R do just that.
Whence, it is obvious that the *transposes* of those rows form a basis for col(A).

$$\text{Since } A^T = \begin{bmatrix} 1 & 0 & 0 \\ 1 & 1 & 1 \\ 0 & -1 & -1 \\ 1 & 1 & -1 \end{bmatrix} \longrightarrow \begin{bmatrix} 1 & 0 & 0 \\ 0 & 1 & 0 \\ 0 & 0 & 1 \\ 0 & 0 & 0 \end{bmatrix} = R,$$

we conclude that $\left\{ \begin{bmatrix} 1 \\ 0 \\ 0 \end{bmatrix}, \begin{bmatrix} 0 \\ 1 \\ 0 \end{bmatrix}, \begin{bmatrix} 0 \\ 0 \\ 1 \end{bmatrix} \right\}$ is a basis for col(A).

We should also note that provided $A^T \longrightarrow R$ uses no row interchanges,
the corresponding rows in A^T are also linearly independent.
Whence, it is obvious that the *transposes* of those rows form a basis for col(A).

24. We find bases for row(A) and col(A) following Examples 3.45 and 3.47 respectively.

row(A): A basis for col(A) must span the columns of A and be linearly independent.
Clearly, the linearly independent *columns* of A^T do just that.
When $A^T \longrightarrow R$, the columns with leading 1s in R are linearly independent.
As in Example 3.47, the corresponding columns in A^T are also linearly independent.
Whence, it is obvious that the *transposes* of those columns form a basis for row(A).

$$\text{Since } A^T = \begin{bmatrix} 2 & -1 & 1 \\ -4 & 2 & -2 \\ 0 & 1 & 1 \\ 2 & 2 & 4 \\ 1 & 3 & 4 \end{bmatrix} \longrightarrow \begin{bmatrix} 1 & 0 & 1 \\ 0 & 1 & 1 \\ 0 & 0 & 0 \\ 0 & 0 & 0 \\ 0 & 0 & 0 \end{bmatrix} = R,$$

we conclude that $\left\{ \begin{bmatrix} 2 & -4 & 0 & 2 & 1 \end{bmatrix}, \begin{bmatrix} -1 & 2 & 1 & 2 & 3 \end{bmatrix} \right\}$ is a basis for row(A).

col(A): A basis for col(A) must span the columns of A and be linearly independent.
When $A^T \longrightarrow R$, the linearly independent *rows* (the nonzero rows) of R do just that.
Whence, it is obvious that the *transposes* of those rows form a basis for col(A).

$$\text{Since } A^T = \begin{bmatrix} 2 & -1 & 1 \\ -4 & 2 & -2 \\ 0 & 1 & 1 \\ 2 & 2 & 4 \\ 1 & 3 & 4 \end{bmatrix} \longrightarrow \begin{bmatrix} 1 & 0 & 1 \\ 0 & 1 & 1 \\ 0 & 0 & 0 \\ 0 & 0 & 0 \\ 0 & 0 & 0 \end{bmatrix} = R,$$

we conclude that $\left\{ \begin{bmatrix} 1 \\ 0 \\ 1 \end{bmatrix}, \begin{bmatrix} 0 \\ 1 \\ 1 \end{bmatrix} \right\}$ is a basis for col(A).

25. Our answers to Exercises 17 and 21 appear different because we used different methods. Let's compare the methods and answers for row(A) and col(A) from each of the exercises.

row(A): In Exercise 17, we found the basis for row(A) as follows:
Given $A \longrightarrow R$, the linearly independent (nonzero) rows of R span the rows of A. Thus:

$$\text{Since } A = \begin{bmatrix} 1 & 0 & -1 \\ 1 & 1 & 1 \end{bmatrix} \longrightarrow \begin{bmatrix} 1 & 0 & -1 \\ 0 & 1 & 2 \end{bmatrix} = R,$$

we conclude that $\{[\,1\ 0\ -1\,], [\,0\ 1\ 2\,]\}$ is a basis for row(A).

In Exercise 21, on the other hand, we found the basis for row(A) as follows:
When $A^T \longrightarrow R$, the *transposes* of the *columns* in A^T corresponding to the columns with leading 1s in R form a basis for row(A). Thus:

$$\text{Since } A^T = \begin{bmatrix} 1 & 1 \\ 0 & 1 \\ -1 & 1 \end{bmatrix} \longrightarrow \begin{bmatrix} 1 & 0 \\ 0 & 1 \\ 0 & 0 \end{bmatrix} = R$$

we conclude that $\{[\,1\ 0\ -1\,], [\,1\ 1\ 1\,]\}$ is a basis for row(A).

So, first we used rows of R as our basis, then we used rows of A.
Do the rows of A corresponding to the nonzero rows of R form a basis?
Yes. Since $A \longrightarrow R$ uses no row interchanges, those rows are linearly independent.
We prove this explicitly by showing the spans of these two sets are equal.
By Exercise 21 of Section 2.3, we need only observe:

$$-[\,1\ 0\ -1\,] + [\,1\ 1\ 1\,] = [\,0\ 1\ 2\,].$$

Why is this enough? The basis vectors in each set are linear combinations of each other.

col(A): In Exercise 17, we found the basis for col(A) as follows:
When $A \longrightarrow R$, the columns in A corresponding to the columns with leading 1s in R form a basis for col(A). Thus:

$$\text{Since } A = \begin{bmatrix} 1 & 0 & -1 \\ 1 & 1 & 1 \end{bmatrix} \longrightarrow \begin{bmatrix} 1 & 0 & -1 \\ 0 & 1 & 2 \end{bmatrix} = R,$$

we conclude that $\left\{ \begin{bmatrix} 1 \\ 1 \end{bmatrix}, \begin{bmatrix} 0 \\ 1 \end{bmatrix} \right\}$ is a basis for col(A).

In Exercise 21, on the other hand, we found the basis for col(A) as follows:
Given $A^T \longrightarrow R$, the *transposes* of the linearly independent (nonzero) *rows* of R span the columns of A. Thus:

$$\text{Since } A^T = \begin{bmatrix} 1 & 1 \\ 0 & 1 \\ -1 & 1 \end{bmatrix} \longrightarrow \begin{bmatrix} 1 & 0 \\ 0 & 1 \\ 0 & 0 \end{bmatrix} = R$$

we conclude that $\left\{ \begin{bmatrix} 1 \\ 0 \end{bmatrix}, \begin{bmatrix} 0 \\ 1 \end{bmatrix} \right\}$ is a basis for col(A).

So, first we used columns of A as our basis, then we used *transposes* of the rows of R.
Notice, however, those rows correspond to the columns of A found in Exercise 17.

For example, $[\,1\ 0\,]$ of R corresponds to $\begin{bmatrix} 1 \\ 1 \end{bmatrix}$ of A.

Explicitly, it is obvious that the span of both these sets is $\mathbb{R}^2$. Why?
Since both sets contain two vectors and the dimension of $\mathbb{R}^2$ is obviously 2.

26. Our answers to Exercises 18 and 22 appear different because we used different methods. Let's compare the methods and answers for row(A) and col(A) from each of the exercises.

row(A): In Exercise 18, we found the basis for row(A) as follows:
Given $A \longrightarrow U$, the linearly independent (nonzero) rows of U span the rows of A. Thus:

$$\text{Since } A = \begin{bmatrix} 1 & 1 & -3 \\ 0 & 2 & 1 \\ 1 & -1 & -4 \end{bmatrix} \xrightarrow{R_3 - R_1 + R_2} \begin{bmatrix} 1 & 1 & -3 \\ 0 & 2 & 1 \\ 0 & 0 & 0 \end{bmatrix} = U,$$

we conclude that $\{\begin{bmatrix} 1 & 1 & -3 \end{bmatrix}, \begin{bmatrix} 0 & 2 & 1 \end{bmatrix}\}$ is a basis for row(A).

In Exercise 22, on the other hand, we found the basis for row(A) as follows:
When $A^T \longrightarrow R$, the *transposes* of the *columns* in A^T corresponding to the columns with leading 1s in R form a basis for row(A). Thus:

$$\text{Since } A^T = \begin{bmatrix} 1 & 0 & 1 \\ 1 & 2 & -1 \\ -3 & 1 & -4 \end{bmatrix} \longrightarrow \begin{bmatrix} 1 & 0 & 1 \\ 0 & 1 & -1 \\ 0 & 0 & 0 \end{bmatrix} = R,$$

we conclude that $\{\begin{bmatrix} 1 & 1 & -3 \end{bmatrix}, \begin{bmatrix} 0 & 2 & 1 \end{bmatrix}\}$ is a basis for row(A).

So, first we used rows of U as our basis, then we used rows of A.
So, if the methods are so different, why are our answers identical?
Because $A \longrightarrow U$ did not change the two linearly independent rows of A.

col(A): In Exercise 18, we found the basis for col(A) as follows:
When $A \longrightarrow R$, the columns in A corresponding to the columns with leading 1s in R form a basis for col(A). Thus:

$$A = \begin{bmatrix} 1 & 1 & -3 \\ 0 & 2 & 1 \\ 1 & -1 & -4 \end{bmatrix} \longrightarrow \begin{bmatrix} 1 & 1 & -3 \\ 0 & 2 & 1 \\ 0 & 0 & 0 \end{bmatrix} = U \Rightarrow \left\{ \begin{bmatrix} 1 \\ 0 \\ 1 \end{bmatrix}, \begin{bmatrix} 1 \\ 2 \\ -1 \end{bmatrix} \right\} \text{ is a basis for col}(A).$$

In Exercise 22, on the other hand, we found the basis for col(A) as follows:
Given $A^T \longrightarrow R$, the *transposes* of the linearly independent (nonzero) *rows* of R span the columns of A. Thus:

$$A^T = \begin{bmatrix} 1 & 0 & 1 \\ 1 & 2 & -1 \\ -3 & 1 & -4 \end{bmatrix} \longrightarrow \begin{bmatrix} 1 & 0 & 1 \\ 0 & 1 & -1 \\ 0 & 0 & 0 \end{bmatrix} = R \Rightarrow \left\{ \begin{bmatrix} 1 \\ 0 \\ 1 \end{bmatrix}, \begin{bmatrix} 0 \\ 1 \\ -1 \end{bmatrix} \right\} \text{ is a basis for col}(A).$$

So, first we used columns of A as our basis, then we used *transposes* of the rows of R.
Notice, however, those rows correspond to the columns of A found in Exercise 17.

For example, $\begin{bmatrix} 0 & 1 & -1 \end{bmatrix}$ of R corresponds to $\begin{bmatrix} 1 \\ 2 \\ -1 \end{bmatrix}$ of A.

We can also show explicitly that the spans of these two sets are equal.
By Exercise 21 of Section 2.3, we need only observe:

$$\begin{bmatrix} 1 \\ 0 \\ 1 \end{bmatrix} + 2 \begin{bmatrix} 0 \\ 1 \\ -1 \end{bmatrix} = \begin{bmatrix} 1 \\ 2 \\ -1 \end{bmatrix}.$$

Why is this enough? The basis vectors in each set are linear combinations of each other.

27. As in Example 3.46, given $S = \{\mathbf{u}, \mathbf{v}, \mathbf{w}\}$ we form matrix B and row reduce:

$$\text{Since } B = \begin{bmatrix} \mathbf{u}^T \\ \mathbf{v}^T \\ \mathbf{w}^T \end{bmatrix} = \begin{bmatrix} 1 & -1 & 0 \\ -1 & 0 & 1 \\ 0 & 1 & -1 \end{bmatrix} \xrightarrow{R_3+R_1+R_2} \begin{bmatrix} 1 & -1 & 0 \\ -1 & 0 & 1 \\ 0 & 0 & 0 \end{bmatrix},$$

$$\text{we conclude that } \{\mathbf{u}, \mathbf{v}\} = \left\{ \begin{bmatrix} 1 \\ -1 \\ 0 \end{bmatrix}, \begin{bmatrix} -1 \\ 0 \\ 1 \end{bmatrix} \right\} \text{ is a basis for span}(S).$$

Note: We rearrange and transpose the vectors of S to simplify the row reduction.
As noted in the remark following Example 3.46, we need only reduce to row echelon form.
Then, we find the basis by identifying linearly independent vectors in the original set.

Q: Does $\{\mathbf{u}, \mathbf{w}\}$ also form a basis for S? What about $\{\mathbf{v}, \mathbf{w}\}$?
A: Yes, since no 2 vectors are multiples of each other. Why is that enough?

28. Before we begin, we should note that at most 3 of these vectors can be linearly independent.
Furthermore, we should let this insight inform our formation of matrix B. How?
As in Example 3.46, given $S = \{\mathbf{x}, \mathbf{u}, \mathbf{v}, \mathbf{w}\}$ we form matrix B and row reduce:

$$\text{Since } B = \begin{bmatrix} \mathbf{u}^T \\ \mathbf{w}^T \\ \mathbf{v}^T \end{bmatrix} = \begin{bmatrix} 1 & 2 & 0 \\ 0 & 1 & 1 \\ 2 & 1 & 2 \end{bmatrix} \xrightarrow{R_3-2R_1+3R_2} \begin{bmatrix} 1 & 2 & 0 \\ 0 & 1 & 1 \\ 0 & 0 & 5 \end{bmatrix} = U,$$

$$\text{we conclude that } \left\{ \begin{bmatrix} 1 \\ 2 \\ 0 \end{bmatrix}, \begin{bmatrix} 0 \\ 1 \\ 1 \end{bmatrix}, \begin{bmatrix} 0 \\ 0 \\ 5 \end{bmatrix} \right\} \text{ is a basis for span}(S).$$

Q: Could we also conclude $\{\mathbf{u}, \mathbf{v}, \mathbf{w}\}$ is a basis for S after only one step?
A: Yes, since the rows of B corresponding to the linearly independent (nonzero) rows of U are linearly independent. This is only true because we performed no row interchanges.

29. As in Example 3.46, given $S = \{\mathbf{u}, \mathbf{v}, \mathbf{w}\}$ we form matrix B and row reduce:

$$\text{Since } B = \begin{bmatrix} \mathbf{v} \\ \mathbf{u} \\ \mathbf{w} \end{bmatrix} = \begin{bmatrix} 1 & -1 & 0 \\ 2 & -3 & 1 \\ 4 & -4 & 1 \end{bmatrix} \xrightarrow[R_3-4R_1]{R_2-2R_1} \begin{bmatrix} 1 & -1 & 0 \\ 0 & -1 & 1 \\ 0 & 0 & 1 \end{bmatrix} = U,$$

$$\text{we conclude that } \left\{ \begin{bmatrix} 1 \\ -1 \\ 0 \end{bmatrix}, \begin{bmatrix} 0 \\ -1 \\ 1 \end{bmatrix}, \begin{bmatrix} 0 \\ 0 \\ 1 \end{bmatrix} \right\} \text{ is a basis for span}(S).$$

Furthermore, since we have 3 linearly independent vectors in $\mathbb{R}^3$,

$$\text{we can also conclude that } \left\{ \begin{bmatrix} 1 \\ 0 \\ 0 \end{bmatrix}, \begin{bmatrix} 0 \\ 1 \\ 0 \end{bmatrix}, \begin{bmatrix} 0 \\ 0 \\ 1 \end{bmatrix} \right\} \text{ is a basis for span}(S).$$

Q: How could we have reached this conclusion using row reduction?
A: By continuing to reduce to B to row reduced echelon form. So, $B \longrightarrow I$.

30. As in Example 3.46, given $S = \{\mathbf{u}, \mathbf{v}, \mathbf{w}\}$ we form matrix B and row reduce:

$$B = \begin{bmatrix} \mathbf{v} \\ \mathbf{u} \\ \mathbf{w} \end{bmatrix} = \begin{bmatrix} 3 & 1 & -1 & 0 \\ 0 & 1 & -2 & 1 \\ 2 & 1 & 5 & 1 \end{bmatrix} \xrightarrow{-3R_3} \begin{bmatrix} 3 & 1 & -1 & 0 \\ 0 & 1 & -2 & 1 \\ -6 & -3 & -15 & -3 \end{bmatrix} \xrightarrow{R_3+2R_1+R_2} \begin{bmatrix} 3 & 1 & -1 & 0 \\ 0 & 1 & -2 & 1 \\ 0 & 0 & -19 & -2 \end{bmatrix} = U$$

So, $\{\mathbf{u}, \mathbf{v}, \mathbf{w}\} = \{[\,3\ \ 1\ \ -1\ \ 0\,], [\,0\ \ 1\ \ -2\ \ 1\,], [\,0\ \ 0\ \ -19\ \ -2\,]\}$ is a basis for span(S).

31. As in Example 3.46, given $S = \{\mathbf{u}, \mathbf{v}, \mathbf{w}\}$ we form matrix B and row reduce:

$$\text{Since } B = \begin{bmatrix} \mathbf{v} \\ \mathbf{u} \\ \mathbf{w} \end{bmatrix} = \begin{bmatrix} 1 & -1 & 0 \\ 2 & -3 & 1 \\ 4 & -4 & 1 \end{bmatrix} \xrightarrow[R_3-4R_1]{R_2-2R_1} \begin{bmatrix} 1 & -1 & 0 \\ 0 & -1 & 1 \\ 0 & 0 & 1 \end{bmatrix} = U,$$

$$\text{we conclude that } \{\mathbf{u}, \mathbf{v}, \mathbf{w}\} = \left\{ \begin{bmatrix} 2 \\ -3 \\ 1 \end{bmatrix}, \begin{bmatrix} 1 \\ -1 \\ 0 \end{bmatrix}, \begin{bmatrix} 4 \\ -4 \\ 1 \end{bmatrix} \right\} \text{ is a basis for span}(S).$$

Q: Why can we conclude $\{\mathbf{u}, \mathbf{v}, \mathbf{w}\}$ is a basis for S after only one step?

A: Because the rows of B corresponding to the linearly independent (nonzero) rows of U are linearly independent. This is only true because we performed no row interchanges.

32. As in Example 3.46, given $S = \{\mathbf{u}, \mathbf{v}, \mathbf{w}\}$ we form matrix B and row reduce:

$$B = \begin{bmatrix} \mathbf{v} \\ \mathbf{u} \\ \mathbf{w} \end{bmatrix} = \begin{bmatrix} 3 & 1 & -1 & 0 \\ 0 & 1 & -2 & 1 \\ 2 & 1 & 5 & 1 \end{bmatrix} \xrightarrow{-3R_3} \begin{bmatrix} 3 & 1 & -1 & 0 \\ 0 & 1 & -2 & 1 \\ -6 & -3 & -15 & -3 \end{bmatrix} \xrightarrow{R_3+2R_1+R_2} \begin{bmatrix} 3 & 1 & -1 & 0 \\ 0 & 1 & -2 & 1 \\ 0 & 0 & -19 & -2 \end{bmatrix} = U$$

So, $\{\mathbf{u}, \mathbf{v}, \mathbf{w}\} = \{[\,0\ \ 1\ \ -2\ \ 1\,], [\,3\ \ 1\ \ -1\ \ 0\,], [\,2\ \ 1\ \ 5\ \ 1\,]\}$ is a basis for span(S).

Q: Why can we conclude $\{\mathbf{u}, \mathbf{v}, \mathbf{w}\}$ is a basis for S after only two steps?

A: Because the rows of B corresponding to the linearly independent (nonzero) rows of U are linearly independent. This is only true because we performed no row interchanges.

33. Let R be a matrix in echelon form. Then row(R) = span(the rows of R) by definition.
Nonzero rows of R are linearly independent (first entries are in different columns) $\Rightarrow$
Nonzero rows of R form a basis for row(R), by definition.

34. Recall, col(A) = span(the columns of A) by definition.
Furthermore, we are given that the columns of A are linearly independent $\Rightarrow$
The columns of A form a basis for col(A), by definition.

35. We use our work from Exercise 17 to determine rank(A) and nullity(A) below.

rank(A) = number of nonzero rows in R = number of vectors in a basis for row(A) or col(A)

$$\text{Since } A = \begin{bmatrix} 1 & 0 & -1 \\ 1 & 1 & 1 \end{bmatrix} \longrightarrow \begin{bmatrix} 1 & 0 & -1 \\ 0 & 1 & 2 \end{bmatrix} = R,$$

and $\{[\,1\ \ 0\ \ -1\,], [\,0\ \ 1\ \ 2\,]\}$ is a basis for row(A), we have rank(A) = 2.

nullity(A) = n − rank(A) = number of nonzero vectors in a basis for null(A)

$$\text{From Exercise 17, } \left\{ \begin{bmatrix} 1 \\ -2 \\ 1 \end{bmatrix} \right\} \text{ is a basis for null}(A), \text{ so nullity}(A) = 3 - 2 = 1.$$

36. We use our work from Exercise 18 to determine rank(A) and nullity(A) below.

rank(A) = number of nonzero rows in U = number of vectors in a basis for col(A) or row(A)

Since $A = \begin{bmatrix} 1 & 1 & -3 \\ 0 & 2 & 1 \\ 1 & -1 & -4 \end{bmatrix} \longrightarrow \begin{bmatrix} 1 & 1 & -3 \\ 0 & 2 & 1 \\ 0 & 0 & 0 \end{bmatrix} = U$,

and $\left\{ \begin{bmatrix} 1 \\ 0 \\ 1 \end{bmatrix}, \begin{bmatrix} 1 \\ 2 \\ -1 \end{bmatrix} \right\}$ is a basis for col(A), we have rank(A) = 2.

nullity(A) = n − rank(A) = number of nonzero vectors in a basis for null(A)

From Exercise 18, $\left\{ \begin{bmatrix} -7 \\ 1 \\ -2 \end{bmatrix} \right\}$ is a basis for null(A), so nullity(A) = 3 − 2 = 1.

37. We use Exercise 19 to determine rank(A) and The Rank Theorem to determine nullity(A).

rank(A) = number of vectors in a basis for col(A) or row(A)

Since $\left\{ \begin{bmatrix} 1 \\ 0 \\ 0 \end{bmatrix}, \begin{bmatrix} 1 \\ 1 \\ 1 \end{bmatrix}, \begin{bmatrix} 1 \\ 1 \\ -1 \end{bmatrix} \right\}$ is a basis for col(A), we have rank(A) = 3.

nullity(A) = n − rank(A) (by Theorem 3.26, The Rank Theorem)
Since $n = 4$ (because A is 3×4) and rank(A) = 3, nullity(A) = 4 − 3 = 1.
The basis for null(A) we found in Exercise 20 should have 3 vectors in it. Why?

38. We use Exercise 20 to determine rank(A) and The Rank Theorem to determine nullity(A).

rank(A) = number of vectors in a basis for col(A) or row(A)

Since $\left\{ \begin{bmatrix} 2 \\ -1 \\ 1 \end{bmatrix}, \begin{bmatrix} 0 \\ 1 \\ 1 \end{bmatrix} \right\}$ is a basis for col(A), we have rank(A) = 2.

nullity(A) = n − rank(A) (by Theorem 3.26, The Rank Theorem)
Since $n = 5$ (because A is 3×5) and rank(A) = 2, nullity(A) = 5 − 2 = 3.
The basis for null(A) we found in Exercise 20 should have 3 vectors in it. Why?

39. If nullity$(A) > 0$, then the columns of A are linearly dependent.
Though we could prove this using theorems, it is instructive to prove it directly.

If nullity$(A) > 0$, then there exists a vector $\mathbf{x} \neq \mathbf{0}$ such that $A\mathbf{x} = \mathbf{0}$.

$$\text{Let } A = \begin{bmatrix} \mathbf{a}_1 & \mathbf{a}_2 & \cdots & \mathbf{a}_n \end{bmatrix} \text{ and } \mathbf{x}^T = \begin{bmatrix} c_1 & c_2 & \cdots & c_n \end{bmatrix}.$$

Since $\mathbf{x} \neq \mathbf{0}$ at least one $c_i \neq 0$. Then $A\mathbf{x} = \sum c_i\mathbf{a}_i = \mathbf{0}$ where at least one $c_i \neq 0$.
Therefore, the columns of A are linearly dependent.

So all we have to show is: If A is a 3×5 matrix, then nullity$(A) > 0$.

$$\text{nullity}(A) \overset{\substack{\text{Rank}\\\text{Thm}}}{=} n - \text{rank}(A) \overset{\substack{A\text{ is}\\3\times5}}{=} 5 - \text{rank}(A) \overset{\text{rank}(A)\leq 3}{\geq} 5 - 3 = 2 > 0$$

Q: If A is 3×5, why is it obvious that rank$(A) \leq 3$?
A: Recall, that the number of vectors in a basis for row(A) = dim(row(A)).
Now note the rows contain a basis for row(A), so dim(row(A)) $\leq$ number of rows.
So, rank(A) = dim(row(A)) $\leq$ the number of rows = 3.

40. If the number of rows > the number of columns, then the rows are linearly dependent.
Why? Rank must satisfy the following two conditions simultaneously:
1) rank(A) = dim(row(A)) $\leq$ the number of rows
2) rank(A) = dim(col(A)) $\leq$ the number of columns
Therefore, rank must be less than or equal to the smaller of these two numbers.

So, dim(row(A)) = rank(A) $\leq$ the number of columns < the number of rows.
That is, dim(row(A)) = the number of vectors in a basis for row(A) < the number of rows.
Recall, we can find a basis for row(A) from among the rows of A.
Therefore, the fact that dim(row(A)) < the number of rows implies the following:
There exists at least one row that is a linear combination of the remaining rows.
That is, the rows of A are linearly dependent.

If A is a 4×2 matrix, then the rows are linearly dependent.
This is now obvious because the number of rows $= 4 > 2 =$ the number of columns.

Q: When we compare this result to Theorem 2.8 of Section 2.3, what do we notice?

Q: If the number of columns > the number of rows, are the columns are linearly dependent?

41. Rank must satisfy the following two conditions simultaneously:
1) rank(A) = dim(row(A)) $\leq$ the number of rows
2) rank(A) = dim(col(A)) $\leq$ the number of columns
Therefore, rank must be less than or equal to the smaller of these two numbers.

Since A is 3×5, rank(A) can equal 0, 1, 2, or 3.
Therefore, since $n = 5$ and nullity$(A) = n -$ rank(A), we have:

$$\text{nullity}(A) = 5 - 3 = 2,\ 5 - 2 = 3,\ 5 - 1 = 4, \text{ or } 5 - 0 = 5$$

42. Since A is 4×2, rank(A) can equal 0, 1, or 2.
Therefore, since $n = 2$ and nullity$(A) = n -$ rank(A), we have

$$\text{nullity}(A) = 2 - 2 = 0,\ 2 - 1 = 1, \text{ or } 2 - 0 = 2.$$

43. $A = \begin{bmatrix} 1 & 2 & a \\ -2 & 4a & 2 \\ a & -2 & 1 \end{bmatrix} \xrightarrow[R_3 - aR_1]{R_2+2R_1} \begin{bmatrix} 1 & 2 & a \\ 0 & a+1 & \frac{a+1}{2} \\ 0 & 0 & \frac{(a+1)(a-2)}{2} \end{bmatrix}.$

If $a = -1$, then $A \longrightarrow \begin{bmatrix} 1 & 2 & -1 \\ 0 & 0 & 0 \\ 0 & 0 & 0 \end{bmatrix} \Rightarrow \text{rank}(A) = 1.$

If $a = 2$, then $A \longrightarrow \begin{bmatrix} 1 & 2 & 2 \\ 0 & 3 & \frac{3}{2} \\ 0 & 0 & 0 \end{bmatrix} \Rightarrow \text{rank}(A) = 2.$ Otherwise, $\text{rank}(A) = 3$.

44. $A = \begin{bmatrix} a & 2 & -1 \\ 3 & 3 & -2 \\ -2 & -1 & a \end{bmatrix} \xrightarrow{R_1 \leftrightarrow R_3} \begin{bmatrix} -2 & -1 & a \\ 3 & 3 & -2 \\ a & 2 & -1 \end{bmatrix} \xrightarrow{R_1+R_2} \begin{bmatrix} 1 & 2 & a-2 \\ 3 & 3 & -2 \\ a & 2 & -1 \end{bmatrix} \longrightarrow$

$\begin{bmatrix} 1 & 2 & a-2 \\ 0 & 1 & a-\frac{4}{3} \\ 0 & 2-2a & -(a-1)^2 \end{bmatrix} \longrightarrow \begin{bmatrix} 1 & 2 & a-2 \\ 0 & 1 & a-\frac{4}{3} \\ 0 & 0 & (a-1)(a-\frac{5}{3}) \end{bmatrix} \Rightarrow$

If $a = 1, \frac{5}{3}$, then $A \longrightarrow \begin{bmatrix} 1 & * & * \\ 0 & 1 & * \\ 0 & 0 & 0 \end{bmatrix} \Rightarrow \text{rank}(A) = 2.$ Otherwise, $\text{rank}(A) = 3$.

45. As in Example 3.52, $\{\mathbf{u}, \mathbf{v}, \mathbf{w}\}$ form a basis for $\mathbb{R}^3 \Leftrightarrow$ When $A = \begin{bmatrix} \mathbf{u}^T \\ \mathbf{v}^T \\ \mathbf{w}^T \end{bmatrix}$, $\text{rank}(A) = 3$.

$$A = \begin{bmatrix} \mathbf{u}^T \\ \mathbf{v}^T \\ \mathbf{w}^T \end{bmatrix} = \begin{bmatrix} 1 & 1 & 0 \\ 1 & 0 & 1 \\ 0 & 1 & 1 \end{bmatrix} \xrightarrow{R_2 - R_1} \begin{bmatrix} 1 & 1 & 0 \\ 0 & -1 & 1 \\ 0 & 1 & 1 \end{bmatrix} \xrightarrow{R_3 + R_2} \begin{bmatrix} 1 & 1 & 0 \\ 0 & -1 & 1 \\ 0 & 0 & 2 \end{bmatrix}$$

Since $\text{rank}(A) = 3$, $\{\mathbf{u}, \mathbf{v}, \mathbf{w}\}$ form a basis for $\mathbb{R}^3$.

46. As in Example 3.52, $\{\mathbf{u}, \mathbf{v}, \mathbf{w}\}$ form a basis for $\mathbb{R}^3 \Leftrightarrow$ When $A = \begin{bmatrix} \mathbf{u}^T \\ \mathbf{v}^T \\ \mathbf{w}^T \end{bmatrix}$, $\text{rank}(A) = 3$.

$$A = \begin{bmatrix} \mathbf{u}^T \\ \mathbf{v}^T \\ \mathbf{w}^T \end{bmatrix} = \begin{bmatrix} 1 & -1 & 1 \\ -1 & 5 & -3 \\ 3 & 1 & 1 \end{bmatrix} \xrightarrow[R_3 - 3R_1]{R_2 + R_1} \begin{bmatrix} 1 & -1 & 1 \\ 0 & 4 & -2 \\ 0 & 4 & -2 \end{bmatrix}$$

Since $\text{rank}(A) < 3$, $\{\mathbf{u}, \mathbf{v}, \mathbf{w}\}$ does *not* form a basis for $\mathbb{R}^3$.

47. As in Example 3.52, $\{\mathbf{x}, \mathbf{u}, \mathbf{v}, \mathbf{w}\}$ form a basis for $\mathbb{R}^4 \Leftrightarrow$ When $A = \begin{bmatrix} \mathbf{x}^T \\ \mathbf{u}^T \\ \mathbf{v}^T \\ \mathbf{w}^T \end{bmatrix}$, $\text{rank}(A) = 4$.

$$A = \begin{bmatrix} \mathbf{x}^T \\ \mathbf{u}^T \\ \mathbf{v}^T \\ \mathbf{w}^T \end{bmatrix} = \begin{bmatrix} 1 & 1 & 1 & 0 \\ 1 & 1 & 0 & 1 \\ 1 & 0 & 1 & 1 \\ 0 & 1 & 1 & 1 \end{bmatrix} \xrightarrow[R_3 - R_1]{R_2 - R_1} \begin{bmatrix} 1 & 1 & 1 & 0 \\ 0 & 0 & -1 & 1 \\ 0 & -1 & 0 & 1 \\ 0 & 1 & 1 & 1 \end{bmatrix} \xrightarrow{R_3 + R_2 + R_4} \begin{bmatrix} 1 & 1 & 1 & 0 \\ 0 & 0 & -1 & 1 \\ 0 & 0 & 0 & 3 \\ 0 & 1 & 1 & 1 \end{bmatrix}$$

Since $\text{rank}(A) = 4$, $\{\mathbf{x}, \mathbf{u}, \mathbf{v}, \mathbf{w}\}$ form a basis for $\mathbb{R}^4$.

48. As in Example 3.52, $\{\mathbf{x}, \mathbf{u}, \mathbf{v}, \mathbf{w}\}$ form a basis for $\mathbb{R}^4 \Leftrightarrow$ When $A = \begin{bmatrix} \mathbf{x}^T \\ \mathbf{u}^T \\ \mathbf{v}^T \\ \mathbf{w}^T \end{bmatrix}$, $\text{rank}(A) = 4$.

$$A = \begin{bmatrix} \mathbf{x}^T \\ \mathbf{v}^T \\ \mathbf{w}^T \end{bmatrix} = \begin{bmatrix} 1 & 0 & 0 & -1 \\ -1 & 1 & 0 & 0 \\ 0 & 0 & -1 & 1 \\ 0 & -1 & 1 & 0 \end{bmatrix} \xrightarrow{R_2 \leftrightarrow R_4} \begin{bmatrix} 1 & 0 & 0 & -1 \\ 0 & -1 & 1 & 0 \\ 0 & 0 & -1 & 1 \\ -1 & 1 & 0 & 0 \end{bmatrix} \xrightarrow{R_4 + R_1 + R_2} \begin{bmatrix} 1 & 0 & 0 & -1 \\ 0 & -1 & 1 & 0 \\ 0 & 0 & -1 & 1 \\ 0 & 0 & 1 & -1 \end{bmatrix}$$

Since $\text{rank}(A) = 4$, $\{\mathbf{x}, \mathbf{v}, \mathbf{w}\}$ form a basis for $\mathbb{R}^4$.

49. We find the coordinate vector $[\mathbf{w}]_{\mathcal{B}}$ by finding c_1 and c_1 such that $\mathbf{w} = c_1\mathbf{b}_1 + c_1\mathbf{b}_2$.
As in Example 2.18 of Section 2.3, we form the matrix $A = \left[\, \mathbf{b}_1 \;\; \mathbf{b}_2 \,|\, \mathbf{w} \,\right]$ and row reduce.

$$A = \left[\, \mathbf{b}_1 \;\; \mathbf{b}_2 \,|\, \mathbf{w} \,\right] = \left[\begin{array}{cc|c} 1 & 1 & 1 \\ 2 & 0 & 6 \\ 0 & -1 & 2 \end{array}\right] \longrightarrow \left[\begin{array}{cc|c} 1 & 0 & 3 \\ 0 & 1 & -2 \\ 0 & 0 & 0 \end{array}\right]$$

Since $\mathbf{w} = 3\mathbf{b}_1 - 2\mathbf{b}_2$, we have the coordinate vector $[\mathbf{w}]_{\mathcal{B}} = \begin{bmatrix} 3 \\ -2 \end{bmatrix}$.

50. We find the coordinate vector $[\mathbf{w}]_{\mathcal{B}}$ by finding c_1 and c_1 such that $\mathbf{w} = c_1\mathbf{b}_1 + c_1\mathbf{b}_2$.
As in Example 2.18 of Section 2.3, we form the matrix $A = \left[\, \mathbf{b}_1 \;\; \mathbf{b}_2 \,|\, \mathbf{w} \,\right]$ and row reduce.

$$A = \left[\, \mathbf{b}_1 \;\; \mathbf{b}_2 \,|\, \mathbf{w} \,\right] = \left[\begin{array}{cc|c} 3 & 5 & 1 \\ 1 & 1 & 3 \\ 4 & 6 & 4 \end{array}\right] \longrightarrow \left[\begin{array}{cc|c} 1 & 0 & 7 \\ 0 & 1 & -4 \\ 0 & 0 & 0 \end{array}\right]$$

Since $\mathbf{w} = 7\mathbf{b}_1 - 4\mathbf{b}_2$, we have the coordinate vector $[\mathbf{w}]_{\mathcal{B}} = \begin{bmatrix} 7 \\ -4 \end{bmatrix}$.

51. We row reduce over $\mathbb{Z}_2$ to find $\text{rank}(A)$, then $\text{nullity}(A) = n - \text{rank}(A)$.

$$A = \begin{bmatrix} 1 & 1 & 0 \\ 0 & 1 & 1 \\ 1 & 0 & 1 \end{bmatrix} \xrightarrow{R_3 + R_1} \begin{bmatrix} 1 & 1 & 0 \\ 0 & 1 & 1 \\ 0 & 1 & 1 \end{bmatrix}$$

Since $\text{rank}(A) = 2$, we have $\text{nullity}(A) = 3 - 2 = 1$.

52. We row reduce over $\mathbb{Z}_3$ to find rank(A), then nullity$(A) = n - \text{rank}(A)$.

$$A = \begin{bmatrix} 1 & 1 & 2 \\ 2 & 1 & 2 \\ 2 & 0 & 0 \end{bmatrix} \xrightarrow{R_3+R_1} \begin{bmatrix} 1 & 1 & 2 \\ 0 & 2 & 1 \\ 0 & 1 & 2 \end{bmatrix}$$

Since rank$(A) = 3$, we have nullity$(A) = 3 - 3 = 0$.

53. We row reduce over $\mathbb{Z}_5$ to find rank(A), then nullity$(A) = n - \text{rank}(A)$.

$$A = \begin{bmatrix} 1 & 3 & 1 & 4 \\ 2 & 3 & 0 & 1 \\ 1 & 0 & 4 & 0 \end{bmatrix} \xrightarrow[R_3+4R_1]{R_2+3R_1} \begin{bmatrix} 1 & 3 & 1 & 4 \\ 0 & 2 & 4 & 2 \\ 0 & 2 & 3 & 1 \end{bmatrix}$$

Since rank$(A) = 3$, we have nullity$(A) = 4 - 3 = 1$.

54. We row reduce over $\mathbb{Z}_7$ to find rank(A), then nullity$(A) = n - \text{rank}(A)$.

$$A = \begin{bmatrix} 2 & 4 & 0 & 0 & 1 \\ 6 & 3 & 5 & 1 & 0 \\ 1 & 0 & 2 & 2 & 5 \\ 1 & 1 & 1 & 1 & 1 \end{bmatrix} \xrightarrow{R_2+R_4} \begin{bmatrix} 2 & 4 & 0 & 0 & 1 \\ 0 & 4 & 6 & 2 & 1 \\ 0 & 3 & 0 & 3 & 5 \\ 1 & 1 & 1 & 1 & 1 \end{bmatrix} \xrightarrow[R_4+3R_1+2R_2]{R_3+R_2} \begin{bmatrix} 2 & 4 & 0 & 0 & 1 \\ 0 & 4 & 6 & 2 & 1 \\ 0 & 0 & 6 & 5 & 6 \\ 0 & 0 & 6 & 5 & 6 \end{bmatrix}$$

Since rank$(A) = 3$, we have nullity$(A) = 5 - 3 = 2$.

55. We need to show if $\mathbf{v}$ is in row(A) and $\mathbf{x}$ is in null(A), then $\mathbf{v} \cdot \mathbf{x} = 0$.

First, we will show if $\mathbf{x}$ is in null(A) and $\mathbf{A}_i$ is the ith row of A, then $\mathbf{A}_i \cdot \mathbf{x} = 0$.

Let $A = \begin{bmatrix} \mathbf{A}_1 \\ \mathbf{A}_2 \\ \vdots \\ \mathbf{A}_m \end{bmatrix}$ and $\mathbf{x} = \begin{bmatrix} x_1 \\ x_2 \\ \vdots \\ x_n \end{bmatrix}$. Then we have: If $\mathbf{x}$ is in null(A), then $A\mathbf{x} = \mathbf{0}$.

So, $A\mathbf{x} = \begin{bmatrix} a_{11}x_1 + a_{12}x_2 + \cdots + a_{1n}x_n \\ a_{21}x_1 + a_{22}x_2 + \cdots + a_{2n}x_n \\ \vdots \\ a_{m1}x_1 + a_{m2}x_2 + \cdots + a_{mn}x_n \end{bmatrix} = \begin{bmatrix} \mathbf{A}_1 \cdot \mathbf{x} \\ \mathbf{A}_2 \cdot \mathbf{x} \\ \vdots \\ \mathbf{A}_m \cdot \mathbf{x} \end{bmatrix} = \begin{bmatrix} 0 \\ 0 \\ \vdots \\ 0 \end{bmatrix}$.

Therefore, if $\mathbf{x}$ is in null(A), then $\mathbf{A}_i \cdot \mathbf{x} = 0$.

If $\mathbf{v}$ is in row(A), then $\mathbf{v}$ is a linear combination of the rows of A so $\mathbf{v} = \sum c_i A_i$.
So, we have: $\mathbf{v} \cdot \mathbf{x} = (\sum c_i A_i) \cdot \mathbf{x} = \sum c_i (A_i \cdot \mathbf{x}) = \sum c_i 0 = 0$.

Q: What is the idea behind the proof of this exercise?
A: If a vector is orthogonal to a set of vectors,
it is orthogonal to all linear combinations of those vectors.

56. If A and B both have rank n, then they are both invertible, by the Fundamental Theorem. Therefore, by Theorem 3.9c, AB is invertible, which means that it, too, has rank n.

57. We will prove part (a) using the idea we suggested by Exercise 29 of Section 3.1.

(a) Let $\{\mathbf{A}\mathbf{b}_k\}$ be a basis for col(AB) formed from the columns of AB.
Then rank(AB) = number of vectors in $\{\mathbf{A}\mathbf{b}_k\}$ = number of vectors in $\{\mathbf{b}_k\}$.
First we show: If $\{\mathbf{A}\mathbf{b}_k\}$ is linearly independent, then $\{\mathbf{b}_k\}$ is linearly independent.
That is, if $\sum c_k\mathbf{b}_k = \mathbf{0}$, we need to show all the $c_k = 0$.
$\sum c_k\mathbf{b}_k = \mathbf{0} \Rightarrow A(\sum c_k\mathbf{b}_k) = \mathbf{0} \Rightarrow \sum c_k(A\mathbf{b}_k) = \mathbf{0}$
Since $\{\mathbf{A}\mathbf{b}_k\}$ is linearly independent by assumption, all the $c_k = 0$ as required.
Now, since $\{\mathbf{b}_k\}$ is a linearly independent subspace of the columns of B,
the number of vectors in $\{\mathbf{b}_k\} \leq$ rank(B).
Therefore, rank(AB) = number of vectors in $\{\mathbf{b}_k\} \leq$ rank(B).

(b) Let $A = \begin{bmatrix} 1 & 1 \\ 0 & 0 \end{bmatrix}$ and $B = \begin{bmatrix} 1 & 0 \\ 1 & 1 \end{bmatrix}$. Then B has rank 2, but $AB = \begin{bmatrix} 2 & 1 \\ 0 & 0 \end{bmatrix}$ has rank 1.

58. We will prove part (a) using the idea we suggested by Exercise 57 above.

(a) Let $\{\mathbf{A}_k\mathbf{B}\}$ be a basis for row(AB) formed from the rows of AB.
Then rank(AB) = number of vectors in $\{\mathbf{A}_k\mathbf{B}\}$ = number of vectors in $\{\mathbf{A}_k\}$.
First we show: If $\{\mathbf{A}_k\mathbf{B}\}$ is linearly independent, then $\{\mathbf{A}_k\}$ is linearly independent.
That is, if $\sum c_k\mathbf{b}_k = \mathbf{0}$, we need to show all the $c_k = 0$.
$\sum c_k\mathbf{A}_k = \mathbf{0} \Rightarrow (\sum c_k\mathbf{A}_k)B = \mathbf{0} \Rightarrow \sum c_k(\mathbf{A}_kB) = \mathbf{0}$
Since $\{\mathbf{A}_kB\}$ is linearly independent by assumption, all the $c_k = 0$ as required.
Now, since $\{\mathbf{A}_k\}$ is a linearly independent subset of the rows of A,
the number of vectors in $\{\mathbf{A}_k\} \leq$ rank(A).
Therefore, rank(AB) = number of vectors in $\{\mathbf{A}_k\} \leq$ rank(A).

(b) Let $A = \begin{bmatrix} 1 & 1 \\ 0 & 1 \end{bmatrix}$ and $B = \begin{bmatrix} 1 & 0 \\ 1 & 0 \end{bmatrix}$.
Then A has rank 2, but $AB = \begin{bmatrix} 1 & 1 \\ 0 & 1 \end{bmatrix}\begin{bmatrix} 1 & 0 \\ 1 & 0 \end{bmatrix} = \begin{bmatrix} 2 & 0 \\ 1 & 0 \end{bmatrix}$ has rank 1.

59. We prove this Exercise using the results of Exercise 57 and Exercise 58.

(a) By Exercise 53, it always true that rank(UA) $\leq$ rank(A).
So, it suffices to show rank(A) $\leq$ rank(UA) to conclude rank(UA) = rank(A).
But rank(A) = rank($U^{-1}(UA)$) $\leq$ rank(UA) as required.

(b) By Exercise 58, it always true that rank(AV) $\leq$ rank(A).
So, it suffices to show rank(A) $\leq$ rank(AV) to conclude rank(AV) = rank(A).
But rank(A) = rank($(AV)V^{-1}$) $\leq$ rank(AV) as required.

60. Since this is an if and only if statement, there are two statements to prove.

if: If $\text{rank}(A) = 1$ then $\text{col}(A) = \text{span}(\mathbf{u})$, so $A = \mathbf{u}\mathbf{v}^T$ where $\mathbf{a}_i = c_i\mathbf{u}$ are the columns of A and $\mathbf{v}^T = \begin{bmatrix} c_1 & \cdots & c_n \end{bmatrix}$.
If $\text{rank}(A) = 1$, a basis for $\text{col}(A)$ has only one vector. That is, $\text{col}(A) = \text{span}(\mathbf{u})$.
Furthermore, every column of A must be a multiple of that vector.
That is $\mathbf{a}_i = c_i\mathbf{u}$ for every column of A and $A = \begin{bmatrix} c_1\mathbf{u} & \cdots & c_n\mathbf{u} \end{bmatrix}$.
So, if we let $\mathbf{v}^T = \begin{bmatrix} c_1 & \cdots & c_n \end{bmatrix}$, then $A = \mathbf{u}\mathbf{v}^T$ as required.

only if: If $A = \mathbf{u}\mathbf{v}^T$ then $\text{col}(A) = \text{span}(\mathbf{u})$, so $\text{rank}(A) = 1$.
Why? Because $\text{rank}(A) =$ the number of vectors in a basis for $\text{col}(A)$.
Let $\mathbf{v}^T = \begin{bmatrix} c_1 & \cdots & c_n \end{bmatrix}$, then $A = \mathbf{u}\mathbf{v}^T = \begin{bmatrix} c_1\mathbf{u} & \cdots & c_n\mathbf{u} \end{bmatrix}$.
So, $\mathbf{a}_i = c_i\mathbf{u}$ for every column of $A \Rightarrow \text{col}(A) = \text{span}(\mathbf{u}) \Rightarrow \text{rank}(A) = 1$.

61. We will show $A = \sum \mathbf{u}_i\mathbf{v}_i^T$, where $\text{rank}(\mathbf{u}_i\mathbf{v}_i^T) = 1$ by Exercise 60.

Since $\text{rank}(A) = r$, a basis for $\text{col}(A)$ has r vectors, $\{\mathbf{u}_i\}$, in $\mathbb{R}^m$. So:

$$A = \begin{bmatrix} c_{11}\mathbf{u}_1 & & c_{n1}\mathbf{u}_1 \\ + & & + \\ \vdots & \cdots & \vdots \\ + & & + \\ c_{1r}\mathbf{u}_r & & c_{nr}\mathbf{u}_r \end{bmatrix} = \sum \begin{bmatrix} c_{1i}\mathbf{u}_i & \cdots & c_{ni}\mathbf{u}_i \end{bmatrix}$$

Now if we let $\mathbf{v}_i^T = \begin{bmatrix} c_{1i} & \cdots & c_{ni} \end{bmatrix}$, then $A = \sum \mathbf{u}_i\mathbf{v}_i^T$ as required.

62. Let $\mathcal{A} = \{\mathbf{u}_1, \cdots, \mathbf{u}_r\}$ be a basis for $\text{col}(A)$ and $\mathcal{B} = \{\mathbf{v}_1, \cdots, \mathbf{v}_s\}$ be a basis for $\text{col}(B)$.
Note: the number of vectors in $\mathcal{A} = \text{rank}(A)$ and the number of vectors in $\mathcal{B} = \text{rank}(B)$.

To prove $\text{rank}(A + B) \leq \text{rank}(A) + \text{rank}(B)$, we will show $\text{col}(A + B) \subseteq \text{span}(\mathcal{A} \cup \mathcal{B})$.

Let $\mathbf{x}$ be in $\text{col}(A + B)$, $\mathbf{a}_i$ be a column of A, and $\mathbf{b}_i$ be a column of B, then:

$$\mathbf{x} = \sum_{i=1}^{n} c_i(\mathbf{a}_i + \mathbf{b}_i) = \sum_{i=1}^{n} c_i(\sum_{j=1}^{r} \alpha_{ij}\mathbf{u}_j + \sum_{k=1}^{s} \beta_{ik}\mathbf{v}_k)$$

So, $\mathbf{x}$ is a linear combination of $\mathcal{A} = \{\mathbf{u}_1, \cdots, \mathbf{u}_r\}$ and $\mathcal{B} = \{\mathbf{v}_1, \cdots, \mathbf{v}_s\}$.

Therefore, $\mathcal{A} \cup \mathcal{B}$ spans $\text{col}(A + B)$. That is, $\text{col}(A + B) \subseteq \text{span}(\mathcal{A} \cup \mathcal{B})$, as required.
That implies a basis for $\text{col}(A + B) \subseteq \mathcal{A} \cup \mathcal{B}$.

So, $\text{rank}(A + B) =$ the number of vectors in a basis for $\text{col}(A + B) \leq$
the number of vectors in $\mathcal{A}$ + the number of vectors in $\mathcal{B} = \text{rank}(A) + \text{rank}(B)$.

Q: So, what have we shown?
A: We have proven that a basis for $\text{col}(A + B)$ can be found among the columns of A and B.

63. If $A^2 = O$, then $\text{col}(A) \subseteq \text{null}(A)$. That is, if $\mathbf{x} = \sum c_i\mathbf{a}_i$, then $A\mathbf{x} = \mathbf{0}$.

Recall that $\mathbf{a}_i = A\mathbf{e}_i$.

So, $A\mathbf{x} = A(\sum c_i\mathbf{a}_i) = \sum c_i(A\mathbf{a}_i) = \sum c_i(A(A\mathbf{e}_i)) = \sum c_i(A^2\mathbf{e}_i) = \sum c_i(O\mathbf{e}_i) = \mathbf{0}$.

Therefore, since $\text{col}(A) \subseteq \text{null}(A)$, $\text{rank}(A) = \dim(\text{col}(A)) \leq \dim(\text{null}(A)) = \text{nullity}(A)$.

So, $\text{rank}(A) + \text{rank}(A) \leq \text{rank}(A) + \text{nullity}(A) = n \Rightarrow 2\,\text{rank}(A) \leq n \Rightarrow \text{rank}(A) \leq \frac{n}{2}$.

64. We will show $\mathbf{x}^T A\mathbf{x} = 0$ and use that prove $I + A$ is invertible.

(a) Since $\mathbf{x}^T A\mathbf{x}$ is 1×1, $\mathbf{x}^T A\mathbf{x} = (\mathbf{x}^T A\mathbf{x})^T$. We use this to show $\mathbf{x}^T A\mathbf{x} = 0$.
Recall, A is skew-symmetric means $A^T = -A$.

So, $\mathbf{x}^T A\mathbf{x} = (\mathbf{x}^T A\mathbf{x})^T = \mathbf{x}^T A^T (\mathbf{x}^T)^T \overset{A^T=-A}{=} \mathbf{x}^T(-A)\mathbf{x} = -\mathbf{x}^T A\mathbf{x}$.
So, $\mathbf{x}^T A\mathbf{x} + \mathbf{x}^T A\mathbf{x} = 2\mathbf{x}^T A\mathbf{x} = 0 \Rightarrow \mathbf{x}^T A\mathbf{x} = 0$ as required.

(b) To show $I + A$ is invertible, we will show $\text{null}(I + A) = \mathbf{0}$.
That is, if $(I + A)\mathbf{x} = \mathbf{0}$, then $\mathbf{x} = \mathbf{0}$.
So, if $(I + A)\mathbf{x} = \mathbf{0}$ then $\mathbf{x}^T(I + A)\mathbf{x} = \mathbf{x}^T\mathbf{0} = \mathbf{x} \cdot \mathbf{0} = 0$. So we have:

$\mathbf{x}^T I\mathbf{x} + \mathbf{x}^T A\mathbf{x} = 0 \overset{\mathbf{x}^T A\mathbf{x}=0}{\Rightarrow} \mathbf{x}^T\mathbf{x} = 0 \overset{\mathbf{x}^T\mathbf{x}=\mathbf{x}\cdot\mathbf{x}}{\Rightarrow} \mathbf{x} \cdot \mathbf{x} = 0 \Rightarrow \mathbf{x} = \mathbf{0}$ as required.
Since $\text{null}(I + A) = \mathbf{0}$, $I + A$ is invertible.

3.6 Introduction to Linear Transformations

1. Since T is the linear transformation corresponding to matrix A, $T(\mathbf{x}) = A\mathbf{x}$. So:

$$T(\mathbf{u}) = \begin{bmatrix} 2 & -1 \\ 3 & 4 \end{bmatrix}\begin{bmatrix} 1 \\ 2 \end{bmatrix} = \begin{bmatrix} 0 \\ 11 \end{bmatrix}, T(\mathbf{v}) = \begin{bmatrix} 2 & -1 \\ 3 & 4 \end{bmatrix}\begin{bmatrix} 3 \\ -2 \end{bmatrix} = \begin{bmatrix} 8 \\ 1 \end{bmatrix}.$$

2. Since T is the linear transformation corresponding to matrix A, $T(\mathbf{x}) = A\mathbf{x}$. So:

$$T(\mathbf{u}) = \begin{bmatrix} 4 & 0 & -1 \\ -2 & 1 & 3 \end{bmatrix}\begin{bmatrix} 1 \\ -1 \\ 2 \end{bmatrix} = \begin{bmatrix} 2 \\ 3 \end{bmatrix}, T(\mathbf{v}) = \begin{bmatrix} 4 & 0 & -1 \\ -2 & 1 & 3 \end{bmatrix}\begin{bmatrix} 0 \\ 5 \\ -1 \end{bmatrix} = \begin{bmatrix} 1 \\ 2 \end{bmatrix}.$$

3. We prove T is a linear transformation by showing that $T(c_1\mathbf{v}_1 + c_2\mathbf{v}_2) = c_1T(\mathbf{v}_1) + c_2T(\mathbf{v}_2)$.

Let $T\begin{bmatrix} \mathrm{x} \\ y \end{bmatrix} = \begin{bmatrix} x+y \\ x-y \end{bmatrix}$, $\mathbf{v}_1 = \begin{bmatrix} x_1 \\ y_1 \end{bmatrix}$ and $\mathbf{v_2} = \begin{bmatrix} x_2 \\ y_2 \end{bmatrix}$. Then:

$$\begin{aligned}
T(c_1\mathbf{v}_1 + c_2\mathbf{v}_2) &= T\left(c_1\begin{bmatrix} x_1 \\ y_1 \end{bmatrix} + c_2\begin{bmatrix} x_2 \\ y_2 \end{bmatrix}\right) = T\left(\begin{bmatrix} c_1x_1 + c_2x_2 \\ c_1y_1 + c_2y_2 \end{bmatrix}\right) \\
&= \begin{bmatrix} c_1x_1 + c_2x_2 + c_1y_1 + c_2y_2 \\ c_1x_1 + c_2x_2 - c_1y_1 - c_2y_2 \end{bmatrix} = \begin{bmatrix} c_1x_1 + c_1y_1 \\ c_1x_1 - c_1y_1 \end{bmatrix} + \begin{bmatrix} c_2x_2 + c_2y_2 \\ c_2x_2 - c_2y_2 \end{bmatrix} \\
&= c_1\begin{bmatrix} x_1 + y_1 \\ x_1 - y_1 \end{bmatrix} + c_2\begin{bmatrix} x_2 + y_2 \\ x_2 - y_2 \end{bmatrix} = c_1T\begin{bmatrix} x_1 \\ y_1 \end{bmatrix} + c_2T\begin{bmatrix} x_2 \\ y_2 \end{bmatrix} \\
&= c_1T(\mathbf{v}_1) + c_2T(\mathbf{v}_2).
\end{aligned}$$

Therefore, we conclude that T is a linear transformation.

4. We prove T is a linear transformation by showing that $T(c_1\mathbf{v}_1 + c_2\mathbf{v}_2) = c_1T(\mathbf{v}_1) + c_2T(\mathbf{v}_2)$.

Let $T\begin{bmatrix} x \\ y \end{bmatrix} = \begin{bmatrix} -y \\ x+2y \\ 3x-4y \end{bmatrix}$, $\mathbf{v}_1 = \begin{bmatrix} x_1 \\ y_1 \end{bmatrix}$ and $\mathbf{v}_2 = \begin{bmatrix} x_2 \\ y_2 \end{bmatrix}$. Then

$$\begin{aligned}
T(c_1\mathbf{v}_1 + c_2\mathbf{v}_2) &= T\left(c_1\begin{bmatrix} x_1 \\ y_1 \end{bmatrix} + c_2\begin{bmatrix} x_2 \\ y_2 \end{bmatrix}\right) = T\left(\begin{bmatrix} c_1x_1 + c_2x_2 \\ c_1y_1 + c_2y_2 \end{bmatrix}\right) \\
&= \begin{bmatrix} -c_1y_1 - c_2y_2 \\ c_1x_1 + c_2x_2 + 2c_1y_1 + 2c_2y_2 \\ 3c_1x_1 + 3c_2x_2 - 4c_1y_1 - 4c_2y_2 \end{bmatrix} = \begin{bmatrix} -c_1y_1 \\ c_1x_1 + 2c_1y_1 \\ 3c_1x_1 - 4c_1y_1 \end{bmatrix} + \begin{bmatrix} -c_2y_2 \\ c_2x_2 + 2c_2y_2 \\ 3c_2x_2 - 4c_2y_2 \end{bmatrix} \\
&= c_1\begin{bmatrix} -y_1 \\ x_1 + 2y_1 \\ 3x_1 - 4y_1 \end{bmatrix} + c_2\begin{bmatrix} -y_2 \\ x_2 + 2y_2 \\ 3x_2 - 4y_2 \end{bmatrix} = c_1T\begin{bmatrix} x_1 \\ y_1 \end{bmatrix} + c_2T\begin{bmatrix} x_2 \\ y_2 \end{bmatrix} \\
&= c_1T(\mathbf{v}_1) + c_2T(\mathbf{v_2}).
\end{aligned}$$

So, this is indeed a linear transformation.

5. We prove T is a linear transformation by showing that $T(c_1\mathbf{v}_1 + c_2\mathbf{v}_2) = c_1T(\mathbf{v}_1) + c_2T(\mathbf{v}_2)$.

Let $T\begin{bmatrix} x \\ y \\ z \end{bmatrix} = \begin{bmatrix} x - y + z \\ 2x + y - 3z \end{bmatrix}$, $\mathbf{v}_1 = \begin{bmatrix} x_1 \\ y_1 \\ z_1 \end{bmatrix}$ and $\mathbf{v}_2 = \begin{bmatrix} x_2 \\ y_2 \\ z_2 \end{bmatrix}$. Then

$$\begin{aligned}
T\left(c_1\mathbf{v}_1 + c_2\mathbf{v}_2\right) &= T\left(c_1\begin{bmatrix} x_1 \\ y_1 \\ z_1 \end{bmatrix} + c_2\begin{bmatrix} x_2 \\ y_2 \\ z_2 \end{bmatrix}\right) = T\left(\begin{bmatrix} c_1x_1 + c_2x_2 \\ c_1y_1 + c_2y_2 \\ c_1z_1 + c_2z_2 \end{bmatrix}\right) \\
&= \begin{bmatrix} c_1x_1 + c_2x_2 - c_1y_1 - c_2y_2 + c_1z_1 + c_2z_2 \\ 2c_1x_1 + 2c_2x_2 + c_1y_1 + c_2y_2 - 3c_1z_1 - 3c_2z_2 \end{bmatrix} \\
&= \begin{bmatrix} c_1x_1 - c_1y_1 + c_1z_1 \\ 2c_1x_1 + c_1y_1 - 3c_1z_1 \end{bmatrix} + \begin{bmatrix} c_2x_2 - c_2y_2 + c_2z_2 \\ 2c_2x_2 + c_2y_2 - 3c_2z_2 \end{bmatrix} \\
&= c_1\begin{bmatrix} x_1 - y_1 + z_1 \\ 2x_1 + y_1 - 3z_1 \end{bmatrix} + c_2\begin{bmatrix} x_2 - y_2 + z_2 \\ 2x_2 + y_2 - 3z_2 \end{bmatrix} = c_1T\begin{bmatrix} x_1 \\ y_1 \\ z_1 \end{bmatrix} + c_2T\begin{bmatrix} x_1 \\ y_1 \\ z_1 \end{bmatrix} \\
&= c_1T\left(\mathbf{v}_1\right) + c_2T\left(\mathbf{v}_2\right).
\end{aligned}$$

6. We prove T is a linear transformation by showing that $T(c_1\mathbf{v}_1 + c_2\mathbf{v}_2) = c_1T(\mathbf{v}_1) + c_2T(\mathbf{v}_2)$.

Let $T\begin{bmatrix} x \\ y \\ z \end{bmatrix} = \begin{bmatrix} x + z \\ y + z \\ x + y \end{bmatrix}$, $\mathbf{v}_1 = \begin{bmatrix} x_1 \\ y_1 \\ z_1 \end{bmatrix}$ and $\mathbf{v}_2 = \begin{bmatrix} x_2 \\ y_2 \\ z_2 \end{bmatrix}$. Then

$$\begin{aligned}
T\left(c_1\mathbf{v}_1 + c_2\mathbf{v}_2\right) &= T\left(c_1\begin{bmatrix} x_1 \\ y_1 \\ z_1 \end{bmatrix} + c_2\begin{bmatrix} x_2 \\ y_2 \\ z_2 \end{bmatrix}\right) = T\left(\begin{bmatrix} c_1x_1 + c_2x_2 \\ c_1y_1 + c_2y_2 \\ c_1z_1 + c_2z_2 \end{bmatrix}\right) \\
&= \begin{bmatrix} c_1x_1 + c_2x_2 + c_1z_1 + c_2z_2 \\ c_1y_1 + c_2y_2 + c_1z_1 + c_2z_2 \\ c_1x_1 + c_2x_2 + c_1y_1 + c_2y_2 \end{bmatrix} = \begin{bmatrix} c_1x_1 + c_1z_1 \\ c_1y_1 + c_1z_1 \\ c_1x_1 + c_1y_1 \end{bmatrix} + \begin{bmatrix} c_2x_2 + c_2z_2 \\ c_2y_2 + c_2z_2 \\ c_2x_2 + c_2y_2 \end{bmatrix} \\
&= c_1\begin{bmatrix} x_1 + z_1 \\ y_1 + z_1 \\ x_1 + y_1 \end{bmatrix} + c_2\begin{bmatrix} x_2 + z_2 \\ y_2 + z_2 \\ x_2 + y_2 \end{bmatrix} = c_1T\begin{bmatrix} x_1 \\ y_1 \\ z_1 \end{bmatrix} + c_2T\begin{bmatrix} x_1 \\ y_1 \\ z_1 \end{bmatrix} \\
&= c_1T\left(\mathbf{v}_1\right) + c_2T\left(\mathbf{v}_2\right).
\end{aligned}$$

7. We prove T is *not* a linear transformation by showing that $T(c\mathbf{v}) \neq cT(\mathbf{v})$ (property (2) fails).

Let $T\begin{bmatrix} x \\ y \end{bmatrix} = \begin{bmatrix} y \\ x^2 \end{bmatrix}$ and $\mathbf{v} = \begin{bmatrix} x \\ y \end{bmatrix}$. Then

$$T(c\mathbf{v}) = T\left(c\begin{bmatrix} x \\ y \end{bmatrix}\right) = T\begin{bmatrix} cx \\ cy \end{bmatrix} = \begin{bmatrix} cy \\ c^2x^2 \end{bmatrix} \neq \begin{bmatrix} cy \\ cx^2 \end{bmatrix} = c\begin{bmatrix} y \\ cx^2 \end{bmatrix} = cT\begin{bmatrix} x \\ y \end{bmatrix} = cT(\mathbf{v})$$

Since $T(c\mathbf{v}) \neq cT(\mathbf{v})$ (property (2) fails), T is *not* a linear transformation.

Q: Is $S\begin{bmatrix} x \\ y \end{bmatrix} = \begin{bmatrix} x^2 \\ y \end{bmatrix}$ linear?

A: No, by a very similar argument to the one given above.
We should suspect both T and S are *not* linear because x^2 is *not* linear.

8. We prove T is *not* a linear transformation by showing that $T(\mathbf{u}+\mathbf{v}) \neq T(\mathbf{u}) + T(\mathbf{v})$.

Let $T\begin{bmatrix} x \\ y \end{bmatrix} = \begin{bmatrix} |x| \\ |y| \end{bmatrix}$, $\mathbf{v}_1 = \begin{bmatrix} 1 \\ 0 \end{bmatrix}$ and $\mathbf{v}_2 = \begin{bmatrix} -1 \\ 0 \end{bmatrix}$. Then

$$T(\mathbf{v}_1 + \mathbf{v}_2) = T\left(\begin{bmatrix} 1 \\ 0 \end{bmatrix} + \begin{bmatrix} -1 \\ 0 \end{bmatrix}\right) = T\left(\begin{bmatrix} 0 \\ 0 \end{bmatrix}\right) = \begin{bmatrix} 0 \\ 0 \end{bmatrix}$$

while

$$T(\mathbf{v}_1) + T(\mathbf{v}_2) = T\begin{bmatrix} 1 \\ 0 \end{bmatrix} + T\begin{bmatrix} -1 \\ 0 \end{bmatrix} = \begin{bmatrix} |1| \\ 0 \end{bmatrix} + \begin{bmatrix} |-1| \\ 0 \end{bmatrix} = \begin{bmatrix} 2 \\ 0 \end{bmatrix}.$$

Since $T(\mathbf{u}+\mathbf{v}) \neq T(\mathbf{u}) + T(\mathbf{v})$, T is *not* a linear transformation.

Q: Should we have anticipated that T would not be linear?

A: Probably. Since the graph of $|x|$ is not a line.

9. We prove T is *not* a linear transformation by showing that $T(c\mathbf{v}) \neq cT(\mathbf{v})$ (property (2) fails).

Let $T\begin{bmatrix} x \\ y \end{bmatrix} = \begin{bmatrix} xy \\ x+y \end{bmatrix}$ and $\mathbf{v} = \begin{bmatrix} x \\ y \end{bmatrix}$. Then

$$T(c\mathbf{v}) = T\begin{bmatrix} cx \\ cy \end{bmatrix} = \begin{bmatrix} cxcy \\ cx+cy \end{bmatrix} = c\begin{bmatrix} cxy \\ x+y \end{bmatrix} \neq c\begin{bmatrix} xy \\ x+y \end{bmatrix} = cT\begin{bmatrix} x \\ y \end{bmatrix} = cT(\mathbf{v})$$

Since $T(c\mathbf{v}) \neq cT(\mathbf{v})$ (property (2) fails), T is *not* a linear transformation.

Q: Is there any reason to suspect that T is not linear before completing the proof?

A: Yes, by a very similar argument to the one given above.
We should suspect T is *not* linear because xy is *not* linear.

10. We prove T is *not* a linear transformation by showing that $T(\mathbf{u}+\mathbf{v}) \neq T(\mathbf{v})+T(\mathbf{v})$.

Let $T\begin{bmatrix} x \\ y \end{bmatrix} = \begin{bmatrix} x+1 \\ y-1 \end{bmatrix}$, $\mathbf{v}_1 = \begin{bmatrix} x_1 \\ y_1 \end{bmatrix}$ and $\mathbf{v}_2 = \begin{bmatrix} x_2 \\ y_2 \end{bmatrix}$. Then

$$T(\mathbf{v}_1 + \mathbf{v}_2) = T\left(\begin{bmatrix} x_1 \\ y_1 \end{bmatrix} + \begin{bmatrix} x_2 \\ y_2 \end{bmatrix}\right) = T\left(\begin{bmatrix} x_1 + x_2 \\ y_1 + y_2 \end{bmatrix}\right) = \begin{bmatrix} x_1 + x_2 + 1 \\ y_1 + y_2 - 1 \end{bmatrix}$$

while

$$T(\mathbf{v}_1) + T(\mathbf{v}_2) = T\begin{bmatrix} x_1 \\ y_1 \end{bmatrix} + T\begin{bmatrix} x_2 \\ y_2 \end{bmatrix} = \begin{bmatrix} x_1 + 1 \\ y_1 - 1 \end{bmatrix} + \begin{bmatrix} x_2 + 1 \\ y_2 - 1 \end{bmatrix} = \begin{bmatrix} x_1 + x_2 + 2 \\ y_1 + y_2 - 2 \end{bmatrix}.$$

Since $T(\mathbf{u}+\mathbf{v}) \neq T(\mathbf{u})+T(\mathbf{v})$, T is *not* a linear transformation.

Q: Is $S\begin{bmatrix} x \\ y \end{bmatrix} = \begin{bmatrix} x+1 \\ y+1 \end{bmatrix}$ linear?

A: No. In fact, adding constants to components is never a linear transformation.
Why not? What goes wrong? That constant gets added twice when computing $T(\mathbf{v})+T(\mathbf{v})$
But only once when computing $T(\mathbf{u}+\mathbf{v})$.

11. As on p.212, we confirm T is a linear transformation by finding A such that $T(\mathbf{v}) = A\mathbf{v}$.

We have $T\begin{bmatrix} x \\ y \end{bmatrix} = \begin{bmatrix} x+y \\ x-y \end{bmatrix} = \begin{bmatrix} 1 \\ 1 \end{bmatrix} x + \begin{bmatrix} 1 \\ -1 \end{bmatrix} y = \begin{bmatrix} 1 & 1 \\ 1 & -1 \end{bmatrix}\begin{bmatrix} x \\ y \end{bmatrix}$.

So $T = T_A$ where $A = \begin{bmatrix} 1 & 1 \\ 1 & -1 \end{bmatrix}$.

12. As on p.212, we confirm T is a linear transformation by finding A such that $T(\mathbf{v}) = A\mathbf{v}$.

We have $T\begin{bmatrix} x \\ y \end{bmatrix} = \begin{bmatrix} -y \\ x+2y \\ 3x-4y \end{bmatrix} = \begin{bmatrix} 0 \\ 1 \\ 3 \end{bmatrix} x + \begin{bmatrix} -1 \\ 2 \\ -4 \end{bmatrix} y = \begin{bmatrix} 0 & -1 \\ 1 & 2 \\ 3 & -4 \end{bmatrix}\begin{bmatrix} x \\ y \end{bmatrix}$.

So $T = T_A$ where $A = \begin{bmatrix} 0 & -1 \\ 1 & 2 \\ 3 & -4 \end{bmatrix}$.

13. As on p.212, we confirm T is a linear transformation by finding A such that $T(\mathbf{v}) = A\mathbf{v}$.

We have $T\begin{bmatrix} x \\ y \\ z \end{bmatrix} = \begin{bmatrix} x-y+z \\ 2x+y-3z \end{bmatrix} = \begin{bmatrix} 1 \\ 2 \end{bmatrix} x + \begin{bmatrix} -1 \\ 1 \end{bmatrix} y + \begin{bmatrix} 1 \\ -3 \end{bmatrix} z = \begin{bmatrix} 1 & -1 & 1 \\ 2 & 1 & -3 \end{bmatrix}\begin{bmatrix} x \\ y \\ z \end{bmatrix}$.

So $T = T_A$ where $A = \begin{bmatrix} 1 & -1 & 1 \\ 2 & 1 & -3 \end{bmatrix}$.

14. As on p.212, we confirm T is a linear transformation by finding A such that $T(\mathbf{v}) = A\mathbf{v}$.

We have $T\begin{bmatrix} x \\ y \\ z \end{bmatrix} = \begin{bmatrix} x+z \\ y+z \\ x+y \end{bmatrix} = \begin{bmatrix} 1 \\ 0 \\ 1 \end{bmatrix} x + \begin{bmatrix} 0 \\ 1 \\ 1 \end{bmatrix} y + \begin{bmatrix} 1 \\ 1 \\ 0 \end{bmatrix} z = \begin{bmatrix} 1 & 0 & 1 \\ 0 & 1 & 1 \\ 1 & 1 & 0 \end{bmatrix} \begin{bmatrix} x \\ y \\ z \end{bmatrix}$.

So $T = T_A$ where $A = \begin{bmatrix} 1 & 0 & 1 \\ 0 & 1 & 1 \\ 1 & 1 & 0 \end{bmatrix}$.

15. As in Example 3.56, we confirm T is a linear transformation by finding A such that $T(\mathbf{v}) = A\mathbf{v}$.

We have $F\begin{bmatrix} x \\ y \end{bmatrix} = \begin{bmatrix} -x \\ y \end{bmatrix} = \begin{bmatrix} -1 \\ 0 \end{bmatrix} x + \begin{bmatrix} 0 \\ 1 \end{bmatrix} y = \begin{bmatrix} -1 & 0 \\ 0 & 1 \end{bmatrix} \begin{bmatrix} x \\ y \end{bmatrix}$.

So we identify $F = \begin{bmatrix} -1 & 0 \\ 0 & 1 \end{bmatrix}$ as the matrix performing the desired transformation.

16. As in Example 3.57, we confirm T is a linear transformation by finding A such that $T(\mathbf{v}) = A\mathbf{v}$.

We have $R\begin{bmatrix} x \\ y \end{bmatrix} = \begin{bmatrix} \frac{1}{\sqrt{2}}(x-y) \\ \frac{1}{\sqrt{2}}(x+y) \end{bmatrix} = \begin{bmatrix} \frac{1}{\sqrt{2}} \\ \frac{1}{\sqrt{2}} \end{bmatrix} x + \begin{bmatrix} -\frac{1}{\sqrt{2}} \\ \frac{1}{\sqrt{2}} \end{bmatrix} y = \begin{bmatrix} \frac{1}{\sqrt{2}} & -\frac{1}{\sqrt{2}} \\ \frac{1}{\sqrt{2}} & \frac{1}{\sqrt{2}} \end{bmatrix} \begin{bmatrix} x \\ y \end{bmatrix}$.

So, $R = \begin{bmatrix} \frac{1}{\sqrt{2}} & -\frac{1}{\sqrt{2}} \\ \frac{1}{\sqrt{2}} & \frac{1}{\sqrt{2}} \end{bmatrix}$ is the matrix performing the desired transformation.

17. As in Example 3.57, we confirm T is a linear transformation by finding A such that $T(\mathbf{v}) = A\mathbf{v}$.

We have $D\begin{bmatrix} x \\ y \end{bmatrix} = \begin{bmatrix} 2x \\ 3y \end{bmatrix} = \begin{bmatrix} 2 \\ 0 \end{bmatrix} x + \begin{bmatrix} 0 \\ 3 \end{bmatrix} y = \begin{bmatrix} 2 & 0 \\ 0 & 3 \end{bmatrix} \begin{bmatrix} x \\ y \end{bmatrix}$.

So, $D = \begin{bmatrix} 2 & 0 \\ 0 & 3 \end{bmatrix}$ is the matrix performing the desired transformation.

18. As in Example 3.59, we confirm T is a linear transformation by finding A such that $T(\mathbf{v}) = A\mathbf{v}$.

Since $x = y \Rightarrow x - y = 0$, the direction vector for the line $y = x$ is $\mathbf{d} = \begin{bmatrix} d_1 \\ d_2 \end{bmatrix} = \begin{bmatrix} 1 \\ 1 \end{bmatrix}$.

So, from the formula given for A at the end of Example 3.59, we have:

$A = \frac{1}{d_1^2+d_2^2}\begin{bmatrix} d_1^2 & d_1 d_2 \\ d_1 d_2 & d_2^2 \end{bmatrix}$. So, in this case: $A = \frac{1}{1+1}\begin{bmatrix} 1 & 1 \\ 1 & 1 \end{bmatrix} = \begin{bmatrix} \frac{1}{2} & \frac{1}{2} \\ \frac{1}{2} & \frac{1}{2} \end{bmatrix}$.

19. Let $A_1 = \begin{bmatrix} k & 0 \\ 0 & 1 \end{bmatrix}$, $A_2 = \begin{bmatrix} 1 & 0 \\ 0 & k \end{bmatrix}$, $B = \begin{bmatrix} 0 & 1 \\ 1 & 0 \end{bmatrix}$, $C_1 = \begin{bmatrix} 1 & k \\ 0 & 1 \end{bmatrix}$, and $C_2 = \begin{bmatrix} 1 & 0 \\ k & 1 \end{bmatrix}$.

$A_1\begin{bmatrix} x \\ y \end{bmatrix} = \begin{bmatrix} k & 0 \\ 0 & 1 \end{bmatrix}\begin{bmatrix} x \\ y \end{bmatrix} = \begin{bmatrix} kx \\ y \end{bmatrix}$: A_1 stretches vectors horizontally by a factor of k.

$A_2\begin{bmatrix} x \\ y \end{bmatrix} = \begin{bmatrix} 1 & 0 \\ 0 & k \end{bmatrix}\begin{bmatrix} x \\ y \end{bmatrix} = \begin{bmatrix} x \\ ky \end{bmatrix}$: A_2 stretches vectors vertically by a factor of k.

$B\begin{bmatrix} x \\ y \end{bmatrix} = \begin{bmatrix} 0 & 1 \\ 1 & 0 \end{bmatrix}\begin{bmatrix} y \\ x \end{bmatrix} = \begin{bmatrix} x \\ ky \end{bmatrix}$: B reflects vectors in the line $y = x$.

$C_1\begin{bmatrix} x \\ y \end{bmatrix} = \begin{bmatrix} 1 & k \\ 0 & 1 \end{bmatrix}\begin{bmatrix} x \\ y \end{bmatrix} = \begin{bmatrix} x + ky \\ y \end{bmatrix}$: C_1 extends vectors horizontally by ky.

$C_2\begin{bmatrix} x \\ y \end{bmatrix} = \begin{bmatrix} 1 & 0 \\ k & 1 \end{bmatrix}\begin{bmatrix} x \\ y \end{bmatrix} = \begin{bmatrix} x \\ y + kx \end{bmatrix}$: C_1 extends vectors vertically by kx.

The effects of these transformations are illustrated below (with $\mathbf{a} = \begin{bmatrix} 3 \\ 1 \end{bmatrix}$ and $k = 2$).

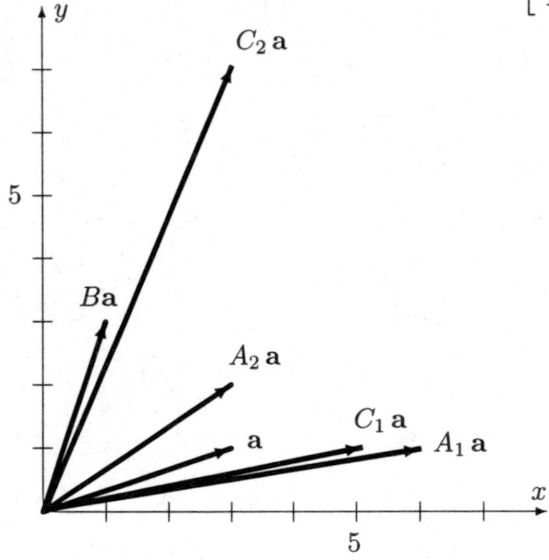

20. Using the formula from Example 3.58, we compute the matrix for a rotation of $120°$.

$$R_{120°}(\mathbf{e_1}) = \begin{bmatrix} \cos 120° \\ \sin 120° \end{bmatrix} = \begin{bmatrix} -\frac{1}{2} \\ \frac{\sqrt{3}}{2} \end{bmatrix} \text{ and } R_{120°}(\mathbf{e_2}) = \begin{bmatrix} -\sin 120° \\ \cos 120° \end{bmatrix} = \begin{bmatrix} -\frac{\sqrt{3}}{2} \\ -\frac{1}{2} \end{bmatrix},$$

So by Theorem 3.31, we have: $A = \begin{bmatrix} R_{120°}(\mathbf{e_1}) & R_{120°}(\mathbf{e_2}) \end{bmatrix} = \begin{bmatrix} -\frac{1}{2} & -\frac{\sqrt{3}}{2} \\ \frac{\sqrt{3}}{2} & -\frac{1}{2} \end{bmatrix}$.

21. Using the formula from Example 3.58, we compute the matrix for a rotation of $-30° = 330°$.

$$R_{330°}(\mathbf{e_1}) = \begin{bmatrix} \cos 330° \\ \sin 330° \end{bmatrix} = \begin{bmatrix} \frac{\sqrt{3}}{2} \\ -\frac{1}{2} \end{bmatrix} \text{ and } R_{330°}(\mathbf{e_2}) = \begin{bmatrix} -\sin 330° \\ \cos 330° \end{bmatrix} = \begin{bmatrix} \frac{1}{2} \\ \frac{\sqrt{3}}{2} \end{bmatrix},$$

So by Theorem 3.31, we have: $A = \begin{bmatrix} R_{330°}(\mathbf{e_1}) & R_{330°}(\mathbf{e_2}) \end{bmatrix} = \begin{bmatrix} \frac{\sqrt{3}}{2} & \frac{1}{2} \\ -\frac{1}{2} & \frac{\sqrt{3}}{2} \end{bmatrix}$.

22. To find the standard matrix of P_ℓ we apply Theorem 3.31. We let $\mathbf{d} = \begin{bmatrix} 1 \\ 2 \end{bmatrix}$, so

$$P_\ell(\mathbf{e}_1) = \left(\tfrac{\mathbf{d}\cdot\mathbf{e}_1}{\mathbf{d}\cdot\mathbf{d}}\right)\mathbf{d} = \tfrac{1}{1+4}\begin{bmatrix} 1 \\ 2 \end{bmatrix} = \begin{bmatrix} \frac{1}{5} \\ \frac{2}{5} \end{bmatrix} \text{ and } P_\ell(\mathbf{e}_2) = \left(\tfrac{\mathbf{d}\cdot\mathbf{e}_2}{\mathbf{d}\cdot\mathbf{d}}\right)\mathbf{d} = \tfrac{2}{1+4}\begin{bmatrix} 1 \\ 2 \end{bmatrix} = \begin{bmatrix} \frac{2}{5} \\ \frac{4}{5} \end{bmatrix}.$$

Thus the standard matrix of this projection is $A = \begin{bmatrix} P_\ell(\mathbf{e_1}) & P_\ell(\mathbf{e_2}) \end{bmatrix} = \begin{bmatrix} \frac{1}{5} & \frac{2}{5} \\ \frac{2}{5} & \frac{4}{5} \end{bmatrix}$.

23. We let $\mathbf{d} = \begin{bmatrix} 1 \\ -1 \end{bmatrix}$. Then $P_\ell(\mathbf{e_1}) = \left(\frac{\mathbf{d}\cdot\mathbf{e_1}}{\mathbf{d}\cdot\mathbf{d}}\right)\mathbf{d} = \frac{1}{1+1}\begin{bmatrix} 1 \\ -1 \end{bmatrix} = \begin{bmatrix} \frac{1}{2} \\ -\frac{1}{2} \end{bmatrix}$ and

$$P_\ell(\mathbf{e_2}) = \left(\tfrac{\mathbf{d}\cdot\mathbf{e_2}}{\mathbf{d}\cdot\mathbf{d}}\right)\mathbf{d} = \tfrac{-1}{1+1}\begin{bmatrix} 1 \\ -1 \end{bmatrix} = \begin{bmatrix} -\frac{1}{2} \\ \frac{1}{2} \end{bmatrix}$$

Thus the standard matrix of this projection is $A = \begin{bmatrix} P_\ell(\mathbf{e_1}) & P_\ell(\mathbf{e_2}) \end{bmatrix} = \begin{bmatrix} \frac{1}{2} & -\frac{1}{2} \\ -\frac{1}{2} & \frac{1}{2} \end{bmatrix}$.

24. Reflection in the line $y = x$ is given by $R\begin{bmatrix} x \\ y \end{bmatrix} = \begin{bmatrix} y \\ x \end{bmatrix}$. $R(\mathbf{e_1}) = \begin{bmatrix} 0 \\ 1 \end{bmatrix}$ and $R(\mathbf{e_2}) = \begin{bmatrix} 1 \\ 0 \end{bmatrix}$.

Thus the standard matrix of this projection is $A = \begin{bmatrix} R(\mathbf{e_1}) & R(\mathbf{e_2}) \end{bmatrix} = \begin{bmatrix} 0 & 1 \\ 1 & 0 \end{bmatrix}$.

25. Reflection in the line $y = -x$ is given by $R\begin{bmatrix} x \\ y \end{bmatrix} = \begin{bmatrix} -y \\ -x \end{bmatrix}$.

$R(\mathbf{e_1}) = \begin{bmatrix} 0 \\ -1 \end{bmatrix}$ and $R(\mathbf{e_2}) = \begin{bmatrix} -1 \\ 0 \end{bmatrix}$.

Thus the standard matrix of this projection is $A = \begin{bmatrix} R(\mathbf{e_1}) & R(\mathbf{e_2}) \end{bmatrix} = \begin{bmatrix} 0 & -1 \\ -1 & 0 \end{bmatrix}$.

26. (a)

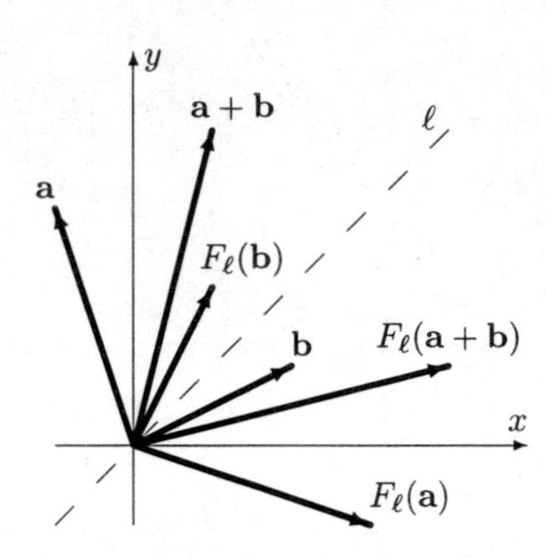

Using the fact that the diagonals of a parallelogram bisect one another we see that $F_\ell(\mathbf{x}) + \mathbf{x} = 2P_\ell(\mathbf{x})$ which rearranges to give $F_\ell(\mathbf{x}) = 2P_\ell(\mathbf{x}) - \mathbf{x}$. Now, we use the definition the projection matrix.

(b) Example 3.59 gave $P_\ell(\mathbf{x}) = \frac{1}{d_1^2+d_2^2}\begin{bmatrix} d_1^2 & d_1d_2 \\ d_1d_2 & d_2^2 \end{bmatrix}$, while $\mathbf{x}$ is the identity matrix. So:

$$\begin{aligned} F_\ell(\mathbf{x}) = 2P_\ell(\mathbf{x}) - x &= \tfrac{2}{d_1^2+d_2^2}\begin{bmatrix} d_1^2 & d_1d_2 \\ d_1d_2 & d_2^2 \end{bmatrix} - \begin{bmatrix} 1 & 0 \\ 0 & 1 \end{bmatrix} \\ &= \tfrac{2}{d_1^2+d_2^2}\begin{bmatrix} d_1^2 & d_1d_2 \\ d_1d_2 & d_2^2 \end{bmatrix} - \tfrac{1}{d_1^2+d_2^2}\begin{bmatrix} d_1^2+d_2^2 & 0 \\ 0 & d_1^2+d_2^2 \end{bmatrix} \\ &= \tfrac{1}{d_1^2+d_2^2}\begin{bmatrix} d_1^2-d_2^2 & 2d_1d_2 \\ 2d_1d_2 & -d_1^2+d_2^2 \end{bmatrix}. \end{aligned}$$

(c) Let the angle between ℓ and the positive x-axis be θ. Then $d_1 = \cos\theta$ and $d_2 = \sin\theta$. Plugging these values into the previous expression for $F_\ell(\mathbf{x})$ gives

$$\begin{aligned} F_\ell(\mathbf{x}) &= \frac{1}{\cos^2\theta+\sin^2\theta}\begin{bmatrix} \cos^2\theta-\sin^2\theta & 2\cos\theta\sin\theta \\ 2\cos\theta\sin\theta & -\cos^2\theta+\sin^2\theta \end{bmatrix} \\ &= \begin{bmatrix} 2\cos^2\theta-1 & \sin 2\theta \\ \sin 2\theta & -\left(2\cos^2\theta-1\right) \end{bmatrix} = \begin{bmatrix} \cos 2\theta & \sin 2\theta \\ \sin 2\theta & -\cos 2\theta \end{bmatrix}. \end{aligned}$$

27. We want to reflect about the line $y = 2x$ which lies in the direction of vector $\mathbf{d} = \begin{bmatrix} 1 \\ 2 \end{bmatrix}$

so that $F_\ell(\mathbf{x}) = \frac{1}{1+4}\begin{bmatrix} 1-4 & 2(2) \\ 2(2) & -1+4 \end{bmatrix} = \begin{bmatrix} -\frac{3}{5} & \frac{4}{5} \\ \frac{4}{5} & \frac{3}{5} \end{bmatrix}$.

Then $F_\ell(\mathbf{e_1}) = \begin{bmatrix} -\frac{3}{5} & \frac{4}{5} \\ \frac{4}{5} & \frac{3}{5} \end{bmatrix}\begin{bmatrix} 1 \\ 0 \end{bmatrix} = \begin{bmatrix} -\frac{3}{5} \\ \frac{4}{5} \end{bmatrix}$ and $F_\ell(\mathbf{e_2}) = \begin{bmatrix} -\frac{3}{5} & \frac{4}{5} \\ \frac{4}{5} & \frac{3}{5} \end{bmatrix}\begin{bmatrix} 0 \\ 1 \end{bmatrix} = \begin{bmatrix} \frac{4}{5} \\ \frac{3}{5} \end{bmatrix}$

and the standard matrix of this transformation is $A = \left[\, F_\ell(\mathbf{e_1}) \;\; F_\ell(\mathbf{e_2}) \,\right] = \begin{bmatrix} -\frac{3}{5} & \frac{4}{5} \\ \frac{4}{5} & \frac{3}{5} \end{bmatrix}$.

28. We want to reflect about the line $y = \sqrt{3}x$ which lies in the direction of vector $\mathbf{d} = \begin{bmatrix} 1 \\ \sqrt{3} \end{bmatrix}$

so that $F_\ell(\mathbf{x}) = \frac{1}{1+3}\begin{bmatrix} 1-3 & 2(\sqrt{3}) \\ 2(\sqrt{3}) & -1+3 \end{bmatrix} = \begin{bmatrix} -\frac{1}{2} & \frac{\sqrt{3}}{2} \\ \frac{\sqrt{3}}{2} & \frac{1}{2} \end{bmatrix}$.

Then $F_\ell(\mathbf{e}_1) = \begin{bmatrix} -\frac{1}{2} & \frac{1}{2}\sqrt{3} \\ \frac{1}{2}\sqrt{3} & \frac{1}{2} \end{bmatrix}\begin{bmatrix} 1 \\ 0 \end{bmatrix} = \begin{bmatrix} -\frac{1}{2} \\ \frac{1}{2}\sqrt{3} \end{bmatrix}$ and $F_\ell(\mathbf{e}_2) = \begin{bmatrix} -\frac{1}{2} & \frac{1}{2}\sqrt{3} \\ \frac{1}{2}\sqrt{3} & \frac{1}{2} \end{bmatrix}\begin{bmatrix} 0 \\ 1 \end{bmatrix} = \begin{bmatrix} \frac{1}{2}\sqrt{3} \\ \frac{1}{2} \end{bmatrix}$

and the standard matrix of this transformation is $A = \begin{bmatrix} F_\ell(\mathbf{e}_1) & F_\ell(\mathbf{e}_2) \end{bmatrix} = \begin{bmatrix} -\frac{1}{2} & \frac{\sqrt{3}}{2} \\ \frac{\sqrt{3}}{2} & \frac{1}{2} \end{bmatrix}$.

29. We have the two linear transformations defined by $T\begin{bmatrix} x_1 \\ x_2 \end{bmatrix} = \begin{bmatrix} x_1 \\ 2x_1 - x_2 \\ 3x_1 + 4x_2 \end{bmatrix}$ and

$S\begin{bmatrix} y_1 \\ y_2 \\ y_3 \end{bmatrix} = \begin{bmatrix} 2y_1 + y_3 \\ 3y_2 - y_3 \\ y_1 - y_2 \\ y_1 + y_2 + y_3 \end{bmatrix}$. To find $(S \circ T)\begin{bmatrix} x_1 \\ x_2 \end{bmatrix}$ directly, we calculate

$$(S \circ T)\begin{bmatrix} x_1 \\ x_2 \end{bmatrix} = S\begin{bmatrix} x_1 \\ 2x_1 - x_2 \\ 3x_1 + 4x_2 \end{bmatrix} = \begin{bmatrix} 2(x_1) + (3x_1 + 4x_2) \\ 3(2x_1 - x_2) - (3x_1 + 4x_2) \\ x_1 - (2x_1 - x_2) \\ x_1 + (2x_1 - x_2) + (3x_1 + 4x_2) \end{bmatrix}$$

$$= \begin{bmatrix} 5x_1 + 4x_2 \\ 3x_1 - 7x_2 \\ -x_1 + x_2 \\ 6x_1 + 3x_2 \end{bmatrix} = \begin{bmatrix} 5 & 4 \\ 3 & -7 \\ -1 & 1 \\ 6 & 3 \end{bmatrix}\begin{bmatrix} x_1 \\ x_2 \end{bmatrix}.$$

Finally, we identify the standard matrix of $(S \circ T)$ as $A = \begin{bmatrix} 5 & 4 \\ 3 & -7 \\ -1 & 1 \\ 6 & 3 \end{bmatrix}$.

30. (a) $(S \circ T)\begin{bmatrix} x_1 \\ x_2 \end{bmatrix} = S\left(T\begin{bmatrix} x_1 \\ x_2 \end{bmatrix}\right) = S\begin{bmatrix} x_1 - x_2 \\ x_1 + x_2 \end{bmatrix} = \begin{bmatrix} 2x_1 - 2x_2 \\ -x_1 - x_2 \end{bmatrix} = \begin{bmatrix} 2 & -2 \\ -1 & -1 \end{bmatrix}\begin{bmatrix} x_1 \\ x_2 \end{bmatrix}$.

So, by direct substitution we identify the matrix of $S \circ T$ as $\begin{bmatrix} 2 & -2 \\ -1 & -1 \end{bmatrix}$.

(b) We see that the standard matrices are $[S] = \begin{bmatrix} 2 & 0 \\ 0 & -1 \end{bmatrix}$ and $[T] = \begin{bmatrix} 1 & -1 \\ 1 & 1 \end{bmatrix}$.

So, Theorem 3.32 gives $[S \circ T] = [S][T] = \begin{bmatrix} 2 & 0 \\ 0 & -1 \end{bmatrix}\begin{bmatrix} 1 & -1 \\ 1 & 1 \end{bmatrix} = \begin{bmatrix} 2 & -2 \\ -1 & -1 \end{bmatrix}$

which is the same result as obtained by direct substitution.

31. (a) $$(S\circ T)\begin{bmatrix}x_1\\x_2\end{bmatrix}=S\begin{bmatrix}x_1+2x_2\\-3x_1+x_2\end{bmatrix}=\begin{bmatrix}(x_1+2x_2)+3(-3x_1+x_2)\\(x_1+2x_2)-(-3x_1+x_2)\end{bmatrix}$$
$$=\begin{bmatrix}-8x_1+5x_2\\4x_1+x_2\end{bmatrix}=\begin{bmatrix}-8&5\\4&1\end{bmatrix}\begin{bmatrix}x_1\\x_2\end{bmatrix}$$
So by direct substitution we identify the matrix of $S\circ T$ as $\begin{bmatrix}-8&5\\4&1\end{bmatrix}$.

(b) We see that the standard matrices are $[S]=\begin{bmatrix}1&3\\1&-1\end{bmatrix}$ and $[T]=\begin{bmatrix}1&2\\-3&1\end{bmatrix}$.

So Theorem 3.32 gives $[S\circ T]=[S]\,[T]=\begin{bmatrix}1&3\\1&-1\end{bmatrix}\begin{bmatrix}1&2\\-3&1\end{bmatrix}=\begin{bmatrix}-8&5\\4&1\end{bmatrix}$

which is the same result as obtained by direct substitution.

32. (a) $$(S\circ T)\begin{bmatrix}x_1\\x_2\end{bmatrix}=S\begin{bmatrix}x_2\\-x_1\end{bmatrix}=\begin{bmatrix}x_2-3x_1\\2x_2-x_1\\x_2+x_1\end{bmatrix}=\begin{bmatrix}-3&1\\-1&2\\1&1\end{bmatrix}\begin{bmatrix}x_1\\x_2\end{bmatrix}.$$
So by direct substitution we identify the matrix of $S\circ T$ as $\begin{bmatrix}3&1\\-1&2\\1&1\end{bmatrix}$.

(b) We see that the standard matrices are $[S]=\begin{bmatrix}1&3\\2&1\\1&-1\end{bmatrix}$ and $[T]=\begin{bmatrix}0&1\\-1&0\end{bmatrix}$.

So Theorem 3.32 gives $[S\circ T]=[S]\,[T]=\begin{bmatrix}1&3\\2&1\\1&-1\end{bmatrix}\begin{bmatrix}0&1\\-1&0\end{bmatrix}=\begin{bmatrix}-3&1\\-1&2\\1&1\end{bmatrix}$

which is the same result as obtained by direct substitution.

33. (a) $$(S\circ T)\begin{bmatrix}x_1\\x_2\\x_3\end{bmatrix}=S\begin{bmatrix}x_1+x_2-x_3\\2x_1-x_2+x_3\end{bmatrix}=\begin{bmatrix}4(x_1+x_2-x_3)-2(2x_1-x_2+x_3)\\-(x_1+x_2-x_3)+(2x_1-x_2+x_3)\end{bmatrix}$$
$$=\begin{bmatrix}6x_2-6x_3\\x_1-2x_2+2x_3\end{bmatrix}=\begin{bmatrix}0&6&-6\\1&-2&2\end{bmatrix}\begin{bmatrix}x_1\\x_2\\x_3\end{bmatrix}.$$
So by direct substitution we identify the matrix of $S\circ T$ as $\begin{bmatrix}0&6&-6\\1&-2&2\end{bmatrix}$.

(b) We see that the standard matrices are $[S]=\begin{bmatrix}4&-2\\-1&1\end{bmatrix}$ and $[T]=\begin{bmatrix}1&1&-1\\2&-1&1\end{bmatrix}$. So Theorem 3.32 gives
$$[S\circ T]=[S]\,[T]=\begin{bmatrix}4&-2\\-1&1\end{bmatrix}\begin{bmatrix}1&1&-1\\2&-1&1\end{bmatrix}=\begin{bmatrix}0&6&-6\\1&-2&2\end{bmatrix}$$
which is the same result as obtained by direct substitution.

34. (a) $$(S\circ T)\begin{bmatrix}x_1\\x_2\\x_3\end{bmatrix} = S\begin{bmatrix}x_1+2x_2\\2x_2-x_3\end{bmatrix} = \begin{bmatrix}x_1+2x_2-(2x_2-x_3)\\x_1+2x_2+2x_2-x_3\\-(x_1+2x_2)+2x_2-x_3\end{bmatrix}$$
$$= \begin{bmatrix}x_1+x_3\\x_1+4x_2-x_3\\-x_1-x_3\end{bmatrix} = \begin{bmatrix}1&0&1\\1&4&-1\\-1&0&-1\end{bmatrix}\begin{bmatrix}x_1\\x_2\\x_3\end{bmatrix}$$

So by direct substitution we identify the matrix of $S\circ T$ as $\begin{bmatrix}1&0&1\\1&4&-1\\-1&0&-1\end{bmatrix}$.

(b) We see the standard matrices are $[S] = \begin{bmatrix}1&-1\\1&1\\-1&1\end{bmatrix}$ and $[T] = \begin{bmatrix}1&2&0\\0&2&-1\end{bmatrix}$.

So, $[S\circ T] = [S]\,[T] = \begin{bmatrix}1&-1\\1&1\\-1&1\end{bmatrix}\begin{bmatrix}1&2&0\\0&2&-1\end{bmatrix} = \begin{bmatrix}1&0&1\\1&4&-1\\-1&0&-1\end{bmatrix}$

which is the same result as obtained by direct substitution.

35. (a) $$(S\circ T)\begin{bmatrix}x_1\\x_2\\x_3\end{bmatrix} = S\begin{bmatrix}x_1+x_2\\x_2+x_3\\x_1+x_3\end{bmatrix} = \begin{bmatrix}x_1+x_2-(x_2+x_3)\\x_2+x_3-(x_1+x_3)\\-(x_1+x_2)+x_1+x_3\end{bmatrix}$$
$$= \begin{bmatrix}x_1-x_3\\-x_1+x_2\\-x_2+x_3\end{bmatrix} = \begin{bmatrix}1&0&-1\\-1&1&0\\0&-1&1\end{bmatrix}\begin{bmatrix}x_1\\x_2\\x_3\end{bmatrix}$$

So by direct substitution we identify the matrix of $S\circ T$ as $\begin{bmatrix}1&0&-1\\-1&1&0\\0&-1&1\end{bmatrix}$.

(b) We see that the standard matrices are $[S] = \begin{bmatrix}1&-1&0\\0&1&-1\\-1&0&1\end{bmatrix}$ and $[T] = \begin{bmatrix}1&1&0\\0&1&1\\1&0&1\end{bmatrix}$.

So Theorem 3.32 gives $[S\circ T] = [S]\,[T] = \begin{bmatrix}1&-1&0\\0&1&-1\\-1&0&1\end{bmatrix}\begin{bmatrix}1&1&0\\0&1&1\\1&0&1\end{bmatrix} = \begin{bmatrix}1&0&-1\\-1&1&0\\0&-1&1\end{bmatrix}$

which is the same result as we obtained by direct substitution.

36. A counterclockwise rotation through 60° is given by $T = \begin{bmatrix}\cos 60^\circ & -\sin 60^\circ\\ \sin 60^\circ & \cos 60^\circ\end{bmatrix}$.

A reflection in the line $y = x$ with $\mathbf{d} = \begin{bmatrix}1\\1\end{bmatrix}$ is given by $S = \frac{1}{2}\begin{bmatrix}0&2\\2&0\end{bmatrix} = \begin{bmatrix}0&1\\1&0\end{bmatrix}$.

Then, by Theorem 3.32, the composite transformation is given by

$$[S\circ T] = [S]\,[T] = \begin{bmatrix}0&1\\1&0\end{bmatrix}\begin{bmatrix}\cos 60^\circ & -\sin 60^\circ\\ \sin 60^\circ & \cos 60^\circ\end{bmatrix} = \begin{bmatrix}\frac{\sqrt{3}}{2} & \frac{1}{2}\\ \frac{1}{2} & -\frac{\sqrt{3}}{2}\end{bmatrix}.$$

37. A reflection in the y-axis is given by $T = \begin{bmatrix} -1 & 0 \\ 0 & 1 \end{bmatrix}$,

while a clockwise rotation through 30° is given by $S = \begin{bmatrix} \cos(-30°) & -\sin(-30°) \\ \sin(-30°) & \cos(-30°) \end{bmatrix}$.

Then, by Theorem 3.32, the composite transformation is given by

$$[S \circ T] = [S]\,[T] = \begin{bmatrix} \cos 30° & \sin 30° \\ -\sin 30° & \cos 30° \end{bmatrix} \begin{bmatrix} -1 & 0 \\ 0 & 1 \end{bmatrix} = \begin{bmatrix} -\cos 30° & \sin 30° \\ \sin 30° & \cos 30° \end{bmatrix} = \begin{bmatrix} -\frac{\sqrt{3}}{2} & \frac{1}{2} \\ \frac{1}{2} & \frac{\sqrt{3}}{2} \end{bmatrix}.$$

38. A $-45°$ rotation is given by $T = \begin{bmatrix} \cos(-45°) & -\sin(-45°) \\ \sin(-45°) & \cos(-45°) \end{bmatrix} = \begin{bmatrix} \cos 45° & \sin 45° \\ -\sin 45° & \cos 45° \end{bmatrix}$.

Then, by Theorem 3.32, the composite transformation is given by

$$[T \circ S \circ T] = [T]\,[S]\,[T] = \begin{bmatrix} \frac{1}{\sqrt{2}} & \frac{1}{\sqrt{2}} \\ -\frac{1}{\sqrt{2}} & \frac{1}{\sqrt{2}} \end{bmatrix} \begin{bmatrix} 0 & 0 \\ 0 & 1 \end{bmatrix} \begin{bmatrix} \frac{1}{\sqrt{2}} & \frac{1}{\sqrt{2}} \\ -\frac{1}{\sqrt{2}} & \frac{1}{\sqrt{2}} \end{bmatrix} = \begin{bmatrix} -\frac{1}{2} & \frac{1}{2} \\ -\frac{1}{2} & \frac{1}{2} \end{bmatrix}$$

39. A reflection in the line $y = x$ with $\mathbf{d} = \begin{bmatrix} 1 \\ 1 \end{bmatrix}$ is given by $T = \frac{1}{2}\begin{bmatrix} 0 & 2 \\ 2 & 0 \end{bmatrix} = \begin{bmatrix} 0 & 1 \\ 1 & 0 \end{bmatrix}$.

A counterclockwise rotation through 30° is given by $S = \begin{bmatrix} \cos 30° & -\sin 30° \\ \sin 30° & \cos 30° \end{bmatrix} = \begin{bmatrix} \frac{\sqrt{3}}{2} & -\frac{1}{2} \\ \frac{1}{2} & \frac{\sqrt{3}}{2} \end{bmatrix}$.

A reflection in the line $y = -x$ with $\mathbf{d} = \begin{bmatrix} 1 \\ -1 \end{bmatrix}$ is given by $U = \frac{1}{2}\begin{bmatrix} 0 & -2 \\ -2 & 0 \end{bmatrix} = \begin{bmatrix} 0 & -1 \\ -1 & 0 \end{bmatrix}$.

Then, by Theorem 3.32, the composite transformation is given by

$$[U \circ S \circ T] = [U]\,[S]\,[T] = \begin{bmatrix} 0 & -1 \\ -1 & 0 \end{bmatrix} \begin{bmatrix} \frac{\sqrt{3}}{2} & -\frac{1}{2} \\ \frac{1}{2} & \frac{\sqrt{3}}{2} \end{bmatrix} \begin{bmatrix} 0 & 1 \\ 1 & 0 \end{bmatrix} = \begin{bmatrix} -\frac{\sqrt{3}}{2} & -\frac{1}{2} \\ \frac{1}{2} & -\frac{\sqrt{3}}{2} \end{bmatrix}.$$

40. Let R_θ denote a rotation through the angle θ. Then

$$\begin{aligned} R_\alpha \circ R_\beta &= \begin{bmatrix} \cos\alpha & -\sin\alpha \\ \sin\alpha & \cos\alpha \end{bmatrix} \begin{bmatrix} \cos\beta & -\sin\beta \\ \sin\beta & \cos\beta \end{bmatrix} \\ &= \begin{bmatrix} \cos\alpha\cos\beta - \sin\alpha\sin\beta & -\cos\alpha\sin\beta - \sin\alpha\cos\beta \\ \sin\alpha\cos\beta + \cos\alpha\sin\beta & \cos\alpha\cos\beta - \sin\alpha\sin\beta \end{bmatrix} \\ &= \begin{bmatrix} \cos(\alpha+\beta) & -\sin(\alpha+\beta) \\ \sin(\alpha+\beta) & \cos(\alpha+\beta) \end{bmatrix} = R_{\alpha+\beta}. \end{aligned}$$

41. Let θ be the angle between:

Line ℓ with direction vector $\mathbf{l} = \begin{bmatrix} \cos\alpha \\ \sin\alpha \end{bmatrix}$ and line m with the direction vector $\mathbf{m} = \begin{bmatrix} \cos\beta \\ \sin\beta \end{bmatrix}$

Now, we need to consider two cases: 1) $\alpha > \beta$ and 2) $\alpha < \beta$.

$\alpha > \beta$: If $\alpha > \beta$ then $\theta = \alpha - \beta$ and

$$\begin{aligned} F_m \circ F_\ell &= \begin{bmatrix} \cos 2\beta & \sin 2\beta \\ \sin 2\beta & -\cos 2\beta \end{bmatrix} \begin{bmatrix} \cos 2\alpha & \sin 2\alpha \\ \sin 2\alpha & -\cos 2\alpha \end{bmatrix} \\ &= \begin{bmatrix} \cos 2\beta \cos 2\alpha + \sin 2\beta \sin 2\alpha & \cos 2\beta \sin 2\alpha - \sin 2\beta \cos 2\alpha \\ \sin 2\beta \cos 2\alpha - \cos 2\beta \sin 2\alpha & \cos 2\beta \cos 2\alpha + \sin 2\beta \sin 2\alpha \end{bmatrix} \\ &= \begin{bmatrix} \cos(-2\beta + 2\alpha) & \sin(-2\beta + 2\alpha) \\ -\sin(-2\beta + 2\alpha) & \cos(-2\beta + 2\alpha) \end{bmatrix} = \begin{bmatrix} \cos 2\theta & \sin 2\theta \\ -\sin 2\theta & \cos 2\theta \end{bmatrix} = R_{-2\theta}. \end{aligned}$$

$\alpha < \beta$: If $\alpha < \beta$ then $\theta = \beta - \alpha$.
Making a similar calculation to the above, we get $F_m \circ F_\ell = R_{+2\theta}$.

42. (a) Let P be a projection. Then the standard matrix of P is $A = \frac{1}{d_1^2+d_2^2}\begin{bmatrix} d_1^2 & d_1d_2 \\ d_1d_2 & d_2^2 \end{bmatrix}$.

Then

$$\begin{aligned}
P \circ P &= \frac{1}{d_1^2+d_2^2}\begin{bmatrix} d_1^2 & d_1d_2 \\ d_1d_2 & d_2^2 \end{bmatrix}\frac{1}{d_1^2+d_2^2}\begin{bmatrix} d_1^2 & d_1d_2 \\ d_1d_2 & d_2^2 \end{bmatrix} \\
&= \frac{1}{\left(d_1^2+d_2^2\right)^2}\begin{bmatrix} d_1^4+d_1^2d_2^2 & d_1^3d_2+d_1d_2^3 \\ d_1^3d_2+d_1d_2^3 & d_1^2d_2^2+d_2^4 \end{bmatrix} \\
&= \frac{1}{\left(d_1^2+d_2^2\right)^2}\begin{bmatrix} d_1^2\left(d_1^2+d_2^2\right) & d_1d_2\left(d_1^2+d_2^2\right) \\ d_1d_2\left(d_1^2+d_2^2\right) & \left(d_1^2+d_2^2\right)d_2^2 \end{bmatrix} = \frac{1}{d_1^2+d_2^2}\begin{bmatrix} d_1^2 & d_1d_2 \\ d_1d_2 & d_2^2 \end{bmatrix} = P.
\end{aligned}$$

Note: We proved this directly in Exercise 64(a) of Chapter 1, Section 1.2:

$$P \circ P = \text{proj}_{\mathbf{u}}\left(\text{proj}_{\mathbf{u}}(\mathbf{v})\right) = \text{proj}_{\mathbf{u}}\left(\frac{\mathbf{u}\cdot\mathbf{v}}{\mathbf{u}\cdot\mathbf{u}}\mathbf{u}\right) = \left(\frac{\mathbf{u}\cdot\mathbf{v}}{\mathbf{u}\cdot\mathbf{u}}\right)\text{proj}_{\mathbf{u}}(\mathbf{u}) = \left(\frac{\mathbf{u}\cdot\mathbf{v}}{\mathbf{u}\cdot\mathbf{u}}\right)\mathbf{u} = \text{proj}_{\mathbf{u}}(\mathbf{v})$$

(b) We will show P is not invertible by proving that its matrix A is not invertible. We will show $P^2 = P \Rightarrow A^2 = A$ with $A \neq I$, so A is not invertible.

If A is a projection matrix then $A \neq I$.
If $a_{12} = d_1d_2 = 0$, then $d_1 = 0$ or $d_2 = 0$ (or both).
Therefore, $a_{11} = d_1^2 = 0$ or $a_{22} = d_2^2 = 0$, so $A \neq I$.

If $A^2 = A$ and $A \neq I$, then A is not invertible.
We will prove A is not invertible by showing that $\text{null}(A) \neq \mathbf{0}$.
That is, there exists a vector $\mathbf{x} \neq \mathbf{0}$ such that $A\mathbf{x} = \mathbf{0}$.

First note: $A^2 = A \Rightarrow A^2 - A = O \Rightarrow A(A-I) = O$.
Since $A \neq I$, we have $A - I \neq O$, so we let $\mathbf{x} = \mathbf{a}_k - \mathbf{e}_k = (A-I)\mathbf{e}_k \neq \mathbf{0}$.

$$A(A-I) = O \Rightarrow A(A-I)\mathbf{e}_k = O\mathbf{e}_k = \mathbf{0} \Rightarrow \overset{\mathbf{x}=(A-I)\mathbf{e}_k}{\Rightarrow} A\mathbf{x} = \mathbf{0}$$

Therefore, since A is not invertible, P is not invertible.

43. Let ℓ, m, and n are three lines through the origin with angles from the x-axis given by θ, α, and β respectively. Then

$$\begin{aligned}
F_n \circ F_m \circ F_\ell &= \begin{bmatrix} \cos 2\theta & \sin 2\theta \\ \sin 2\theta & -\cos 2\theta \end{bmatrix}\begin{bmatrix} \cos 2\alpha & \sin 2\alpha \\ \sin 2\alpha & -\cos 2\alpha \end{bmatrix}\begin{bmatrix} \cos 2\beta & \sin 2\beta \\ \sin 2\beta & -\cos 2\beta \end{bmatrix} \\
&= \begin{bmatrix} \cos(2\theta - 2\alpha + 2\beta) & \sin(2\theta - 2\alpha + 2\beta) \\ \sin(2\theta - 2\alpha + 2\beta) & -\cos(2\theta - 2\alpha + 2\beta) \end{bmatrix} \\
&= \begin{bmatrix} \cos(2(\theta - \alpha + \beta)) & \sin(2(\theta - \alpha + \beta)) \\ \sin(2(\theta - \alpha + \beta)) & -\cos(2(\theta - \alpha + \beta)) \end{bmatrix}.
\end{aligned}$$

So the reflection about the three lines ℓ, m, and n is the same as one reflection about the line which makes an angle with the x-axis of $\theta - \alpha + \beta$.

44. Before we begin our proof, it is important to note that $T(\mathbf{v})$ is a vector.

ℓ is a line $\Rightarrow$	ℓ has equation $\mathbf{x} = \mathbf{p} + t\mathbf{d}$
T is linear $\Rightarrow$	$T(\mathbf{x}) = T(\mathbf{p} + t\mathbf{d}) = T(\mathbf{p}) + tT(\mathbf{d})$
ℓ' has equation $T(\mathbf{x}) = T(\mathbf{p}) + tT(\mathbf{d})$	ℓ' is a line

45. By Exercise 44, we know that T maps ℓ, $\mathbf{p} + t\mathbf{d}$, and m, $\mathbf{q} + s\mathbf{d}$, onto ℓ', $T(\mathbf{p} + t\mathbf{d}) = T(\mathbf{p}) + tT(\mathbf{d})$ and m', $T(\mathbf{q} + s\mathbf{d}) = T(\mathbf{q}) + sT(\mathbf{d})$.

If $T(\mathbf{d}) \neq \mathbf{0}$ and $T(\mathbf{p}) \neq T(\mathbf{q})$, then ℓ' and m' are distinct parallel lines.

If $T(\mathbf{d}) \neq \mathbf{0}$ and $T(\mathbf{p}) = T(\mathbf{q})$, then ℓ' and m' are the same line.

If $T(\mathbf{d}) = \mathbf{0}$ and $T(\mathbf{p}) \neq T(\mathbf{q})$, then ℓ' and m' are distinct points $T(\mathbf{p})$ and $T(\mathbf{q})$.

If $T(\mathbf{d}) = \mathbf{0}$ and $T(\mathbf{p}) = T(\mathbf{q})$, then ℓ' and m' are a single point $T(\mathbf{p}) = T(\mathbf{q})$.

46. $T\begin{bmatrix}-1\\1\end{bmatrix} = \begin{bmatrix}1 & 1\\1 & -1\end{bmatrix}\begin{bmatrix}-1\\1\end{bmatrix} = \begin{bmatrix}0\\-2\end{bmatrix}$,

$T\begin{bmatrix}1\\1\end{bmatrix} = \begin{bmatrix}2\\0\end{bmatrix}$,

$T\begin{bmatrix}1\\-1\end{bmatrix} = \begin{bmatrix}0\\2\end{bmatrix}$,

$T\begin{bmatrix}-1\\-1\end{bmatrix} = \begin{bmatrix}-2\\0\end{bmatrix}$.

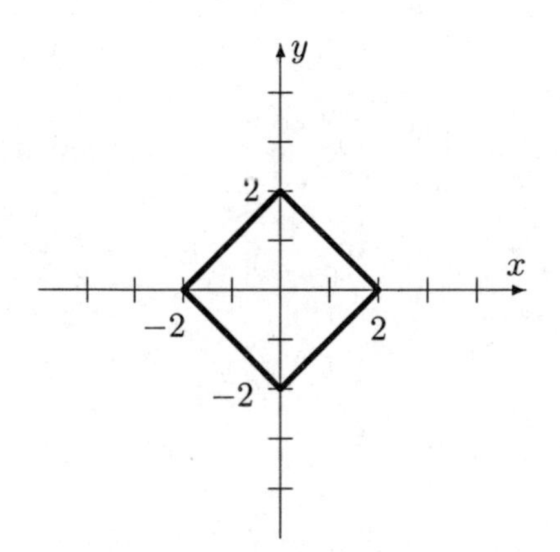

47. $D\begin{bmatrix}-1\\1\end{bmatrix} = \begin{bmatrix}2 & 0\\0 & 3\end{bmatrix}\begin{bmatrix}-1\\1\end{bmatrix} = \begin{bmatrix}-2\\3\end{bmatrix}$,

$D\begin{bmatrix}1\\1\end{bmatrix} = \begin{bmatrix}2\\3\end{bmatrix}$,

$D\begin{bmatrix}1\\-1\end{bmatrix} = \begin{bmatrix}2\\-3\end{bmatrix}$,

$D\begin{bmatrix}-1\\-1\end{bmatrix} = \begin{bmatrix}-2\\-3\end{bmatrix}$.

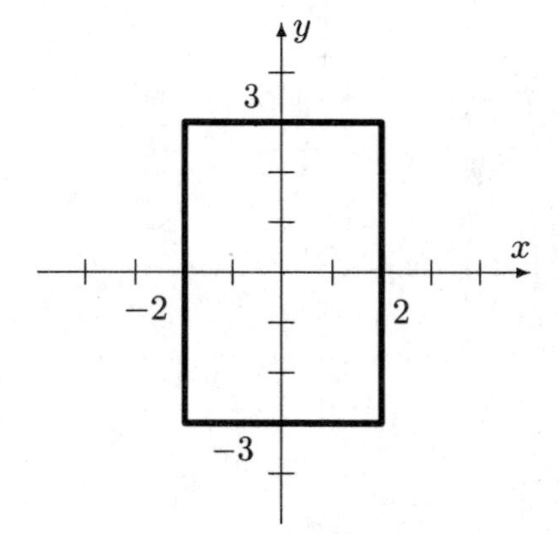

48. $P\begin{bmatrix}-1\\1\end{bmatrix}=\begin{bmatrix}\frac{1}{2}&\frac{1}{2}\\\frac{1}{2}&\frac{1}{2}\end{bmatrix}\begin{bmatrix}-1\\1\end{bmatrix}=\begin{bmatrix}0\\0\end{bmatrix}$,

$P\begin{bmatrix}1\\1\end{bmatrix}=\begin{bmatrix}1\\1\end{bmatrix}$, $P\begin{bmatrix}1\\-1\end{bmatrix}=\begin{bmatrix}0\\0\end{bmatrix}$,

$P\begin{bmatrix}-1\\-1\end{bmatrix}=\begin{bmatrix}-1\\-1\end{bmatrix}$.

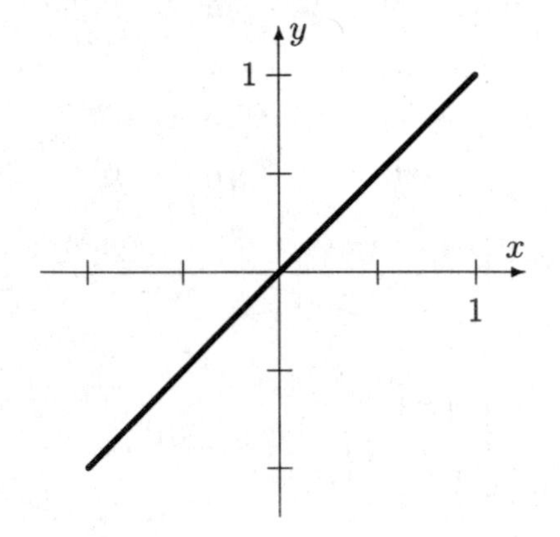

49. $P\begin{bmatrix}-1\\1\end{bmatrix}=\begin{bmatrix}\frac{1}{5}&\frac{2}{5}\\\frac{2}{5}&\frac{4}{5}\end{bmatrix}\begin{bmatrix}-1\\1\end{bmatrix}=\begin{bmatrix}\frac{1}{5}\\\frac{2}{5}\end{bmatrix}$,

$P\begin{bmatrix}1\\1\end{bmatrix}=\begin{bmatrix}\frac{3}{5}\\\frac{6}{5}\end{bmatrix}$,

$P\begin{bmatrix}1\\-1\end{bmatrix}=\begin{bmatrix}-\frac{1}{5}\\-\frac{2}{5}\end{bmatrix}$, $P\begin{bmatrix}-1\\-1\end{bmatrix}=\begin{bmatrix}-\frac{3}{5}\\-\frac{6}{5}\end{bmatrix}$.

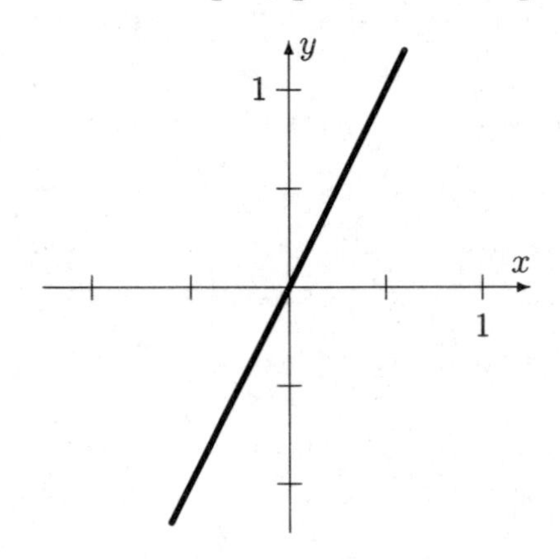

50. $T\begin{bmatrix}-1\\1\end{bmatrix}=\begin{bmatrix}1&2\\-3&1\end{bmatrix}\begin{bmatrix}-1\\1\end{bmatrix}=\begin{bmatrix}1\\4\end{bmatrix}$,

$T\begin{bmatrix}1\\1\end{bmatrix}=\begin{bmatrix}3\\-2\end{bmatrix}$,

$T\begin{bmatrix}1\\-1\end{bmatrix}=\begin{bmatrix}-1\\-4\end{bmatrix}$,

$T\begin{bmatrix}-1\\-1\end{bmatrix}=\begin{bmatrix}-3\\2\end{bmatrix}$

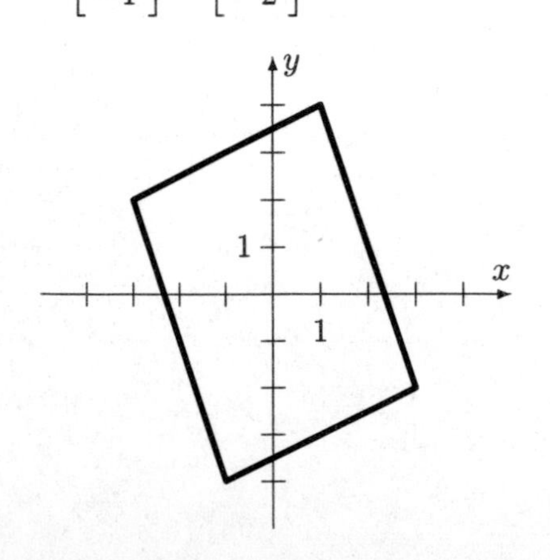

51. $T\begin{bmatrix}-1\\1\end{bmatrix}=\begin{bmatrix}\frac{1}{2}\sqrt{3}+\frac{1}{2}\\-\frac{1}{2}+\frac{1}{2}\sqrt{3}\end{bmatrix}$,

$T\begin{bmatrix}1\\1\end{bmatrix}=\begin{bmatrix}\frac{1}{2}-\frac{1}{2}\sqrt{3}\\\frac{1}{2}\sqrt{3}+\frac{1}{2}\end{bmatrix}$,

$T\begin{bmatrix}1\\-1\end{bmatrix}=\begin{bmatrix}-\frac{1}{2}-\frac{1}{2}\sqrt{3}\\\frac{1}{2}-\frac{1}{2}\sqrt{3}\end{bmatrix}$,

$T\begin{bmatrix}-1\\-1\end{bmatrix}=\begin{bmatrix}-\frac{1}{2}+\frac{1}{2}\sqrt{3}\\-\frac{1}{2}-\frac{1}{2}\sqrt{3}\end{bmatrix}$.

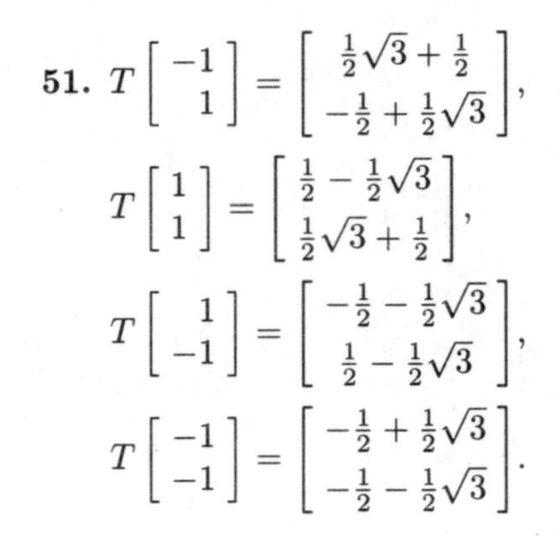

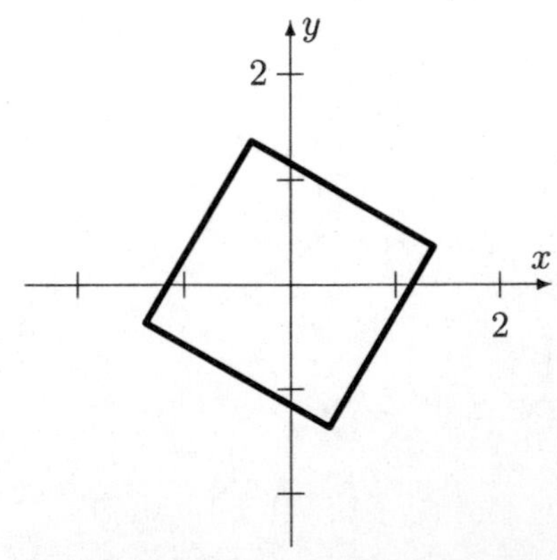

52. To complete the of proof in Example 5, we need to show: $P_\ell(c\mathbf{v}) = cP_\ell(\mathbf{v})$.

$$P_\ell(c\mathbf{v}) = \left(\frac{\mathbf{d}\cdot(c\mathbf{v})}{\mathbf{d}\cdot\mathbf{d}}\right)\mathbf{d} \overset{\substack{\mathbf{d}\cdot(c\mathbf{v})\\=c(\mathbf{d}\cdot\mathbf{v})}}{=} \left(\frac{c(\mathbf{d}\cdot\mathbf{v})}{\mathbf{d}\cdot\mathbf{d}}\right)\mathbf{d} = c\left(\left(\frac{\mathbf{d}\cdot\mathbf{v}}{\mathbf{d}\cdot\mathbf{d}}\right)\mathbf{d}\right) = cP_\ell(\mathbf{v})$$

Since $P_\ell(\mathbf{u}+\mathbf{v}) = P_\ell(\mathbf{u}) + P_\ell(\mathbf{v})$ and $P_\ell(c\mathbf{v}) = cP_\ell(\mathbf{v})$, P_ℓ is a linear transformation.

53. Since this is an if and only if statement, there are two statements to prove.

if: If T is linear, then $T(c_1\mathbf{v}_1 + c_2\mathbf{v}_2) = c_1T(\mathbf{v}_1) + c_2T(\mathbf{v}_2)$.

$$T(c_1\mathbf{v}_1 + c_2\mathbf{v}_2) \overset{\substack{T(\mathbf{u}+\mathbf{v})\\=T(\mathbf{u})+T(\mathbf{v})}}{=} T(c_1\mathbf{v}_1) + T(c_2\mathbf{v}_2) \overset{\substack{T(c\mathbf{v})\\=cT(\mathbf{v})}}{=} c_1T(\mathbf{v}_1) + c_2T(\mathbf{v}_2)$$

only if: If $T(c_1\mathbf{v}_1 + c_2\mathbf{v}_2) = c_1T(\mathbf{v}_1) + c_2T(\mathbf{v}_2)$, then T is linear.

$c_1 = c_2 = 1 \Rightarrow T(\mathbf{v}_1 + \mathbf{v}_2) = T(\mathbf{v}_1) + T(\mathbf{v}_2)$

$c_2 = 0 \Rightarrow T(c_1\mathbf{v}_1) = c_1T(\mathbf{v}_1)$

Since $T(\mathbf{u}+\mathbf{v}) = T(\mathbf{u}) + T(\mathbf{v})$ and $T(c\mathbf{v}) = cT(\mathbf{v})$, T is linear.

54. Let range(T) = the range of T and col($[T]$) = the column space of $[T]$.
We will prove range(T) = col($[T]$) by showing they both equal span($T(\mathbf{e}_i)$).

First, we show that range(T) = span($T(\mathbf{e}_i)$).

In range(T) $\Leftrightarrow \mathbf{x} = T(\mathbf{v}) = T(\sum c_i\mathbf{e}_i) \overset{\substack{T\text{ is}\\\text{linear}}}{=} \sum c_iT(\mathbf{e}_i) \Leftrightarrow$ in span($T(\mathbf{e}_i)$)

Therefore, range(T) = span($T(\mathbf{e}_i)$) as we were to show.

Second, we show that col($[T]$) = span($T(\mathbf{e}_i)$).

Note: By Theorem 2, $[T] = \begin{bmatrix} T(\mathbf{e}_1) & T(\mathbf{e}_2) & \dots & T(\mathbf{e}_n) \end{bmatrix}$.

In col($[T]$) $\Leftrightarrow \mathbf{x} = \sum c_iT(\mathbf{e}_i) \Leftrightarrow$ in span($T(\mathbf{e}_i)$)

Therefore, col($[T]$) = span($T(\mathbf{e}_i)$) as we were to show.

Since range(T) = span($T(\mathbf{e}_i)$) and col($[T]$) = span($T(\mathbf{e}_i)$), range(T) = col($[T]$).

55. The Fundamental Theorem implies every invertible matrix is a product of elementary matrices.
The matrices in Exercise 19 represent the 3 types of elementary 2×2 matrices.
So, T_A must be a composition of the three types of transformation represented in Exercise 19.

3.7 Applications

1. $\mathbf{x}_1 = P\mathbf{x}_0 = \begin{bmatrix} 0.5 & 0.3 \\ 0.5 & 0.7 \end{bmatrix}\begin{bmatrix} 0.5 \\ 0.5 \end{bmatrix} = \begin{bmatrix} 0.4 \\ 0.6 \end{bmatrix}$, $\mathbf{x}_2 = P\mathbf{x}_1 = \begin{bmatrix} 0.5 & 0.3 \\ 0.5 & 0.7 \end{bmatrix}\begin{bmatrix} 0.4 \\ 0.6 \end{bmatrix} = \begin{bmatrix} 0.38 \\ 0.62 \end{bmatrix}$.

2. Since there are 2 steps involved, we calculate: $P^2 = \begin{bmatrix} 0.5 & 0.3 \\ 0.5 & 0.7 \end{bmatrix}^2 = \begin{bmatrix} 0.40 & 0.36 \\ 0.60 & 0.64 \end{bmatrix}$.

So, the probability of the state 1 population being in state 2 is $[P^2]_{21} = 0.60$.

3. Since there are 2 steps involved, we calculate: $P^2 = \begin{bmatrix} 0.5 & 0.3 \\ 0.5 & 0.7 \end{bmatrix}^2 = \begin{bmatrix} 0.40 & 0.36 \\ 0.60 & 0.64 \end{bmatrix}$.

So, the probability of the state 2 population being in state 2 is $[P^2]_{22} = 0.64$.

4. The steady-state vector has the property that $P\mathbf{x} = \mathbf{x}$.

So, we form $\left[\, I - P \mid \mathbf{0} \,\right]$: $\left[\begin{array}{cc|c} 1-0.5 & -0.3 & 0 \\ -0.5 & 1-0.7 & 0 \end{array}\right] \longrightarrow \left[\begin{array}{cc|c} 1 & -0.6 & 0 \\ 0 & 0 & 0 \end{array}\right]$.

Thus, if $\mathbf{x} = \begin{bmatrix} x_1 \\ x_2 \end{bmatrix}$ is our steady state vector, then x_2 is free and $x_1 = \frac{3}{5}t$, $x_2 = t$.

If $\mathbf{x}$ is a probability vector, then $1 = x_1 + x_2 = \frac{3}{5}t + t = \frac{8}{5}t \Rightarrow t = \frac{5}{8} \Rightarrow \mathbf{x} = \begin{bmatrix} \frac{3}{8} \\ \frac{5}{8} \end{bmatrix}$.

5. $\mathbf{x}_1 = P\mathbf{x}_0 = \begin{bmatrix} \frac{1}{2} & \frac{1}{3} & \frac{1}{3} \\ 0 & \frac{1}{3} & \frac{2}{3} \\ \frac{1}{2} & \frac{1}{3} & 0 \end{bmatrix}\begin{bmatrix} 120 \\ 180 \\ 90 \end{bmatrix} = \begin{bmatrix} 150 \\ 120 \\ 120 \end{bmatrix}$, $\mathbf{x}_2 = P\mathbf{x}_1 = \begin{bmatrix} \frac{1}{2} & \frac{1}{3} & \frac{1}{3} \\ 0 & \frac{1}{3} & \frac{2}{3} \\ \frac{1}{2} & \frac{1}{3} & 0 \end{bmatrix}\begin{bmatrix} 150 \\ 120 \\ 120 \end{bmatrix} = \begin{bmatrix} 155 \\ 120 \\ 115 \end{bmatrix}$.

6. Since there are 2 steps involved, we calculate: $P^2 = \begin{bmatrix} \frac{1}{2} & \frac{1}{3} & \frac{1}{3} \\ 0 & \frac{1}{3} & \frac{2}{3} \\ \frac{1}{2} & \frac{1}{3} & 0 \end{bmatrix}^2 = \begin{bmatrix} \frac{5}{12} & \frac{7}{18} & \frac{7}{18} \\ \frac{1}{3} & \frac{1}{3} & \frac{2}{9} \\ \frac{1}{4} & \frac{5}{18} & \frac{7}{18} \end{bmatrix}$.

So, the probability of the state 1 population being in state 1 is $[P^2]_{11} = \frac{5}{12}$.

7. Since there are 2 steps involved, we calculate: $P^2 = \begin{bmatrix} \frac{1}{2} & \frac{1}{3} & \frac{1}{3} \\ 0 & \frac{1}{3} & \frac{2}{3} \\ \frac{1}{2} & \frac{1}{3} & 0 \end{bmatrix}^2 = \begin{bmatrix} \frac{5}{12} & \frac{7}{18} & \frac{7}{18} \\ \frac{1}{3} & \frac{1}{3} & \frac{2}{9} \\ \frac{1}{4} & \frac{5}{18} & \frac{7}{18} \end{bmatrix}$.

So, the probability of the state 3 population being in state 2 is $[P^2]_{32} = \frac{5}{18}$.

8. The steady-state vector has the property that $P\mathbf{x} = \mathbf{x}$.

So, we form $\left[\, I - P \mid \mathbf{0} \,\right]$: $\left[\begin{array}{ccc|c} 1-\frac{1}{2} & -\frac{1}{3} & -\frac{1}{3} & 0 \\ 0 & 1-\frac{1}{3} & -\frac{2}{3} & 0 \\ -\frac{1}{2} & -\frac{1}{3} & 1 & 0 \end{array}\right] \longrightarrow \left[\begin{array}{ccc|c} 1 & 0 & -\frac{4}{3} & 0 \\ 0 & 1 & -1 & 0 \\ 0 & 0 & 0 & 0 \end{array}\right]$.

Hence, if $\mathbf{x} \begin{bmatrix} x_1 \\ x_2 \\ x_3 \end{bmatrix}$, then $x_3 = t$ is free and $x_1 = \frac{4}{3}t$, $x_2 = t$.

Since $\mathbf{x}$ must be a probability vector, we need $1 = x_1 + x_2 + x_3 = \frac{10}{3}t$. Thus, $t = \frac{3}{10}$.

So, we see that our steady state vector is $\mathbf{x} = \begin{bmatrix} x_1 \\ x_2 \\ x_3 \end{bmatrix} = \begin{bmatrix} \frac{2}{5} \\ \frac{3}{10} \\ \frac{3}{10} \end{bmatrix}$.

9. Let A be a dry day and B a wet one. We are given the following probabilities:
Tomorrow will be wet is 0.662 if today is wet and 0.250 if today is dry.
Tomorrow will be dry is 0.750 if today is dry and 0.338 if today is wet.

(a) The transition matrix for this Markov chain is $P = \begin{bmatrix} 0.750 & 0.338 \\ 0.250 & 0.662 \end{bmatrix}$ with $\mathbf{x}_i = \begin{bmatrix} x_1 \\ x_2 \end{bmatrix}$, where x_1, x_2 denote the probability that tomorrow will be dry or wet respectively.

(b) We are told that Monday is a dry day; we can describe this state as $\mathbf{x}_0 = \begin{bmatrix} 1 \\ 0 \end{bmatrix}$.

The probability that Wednesday will be wet is then given by

$$\mathbf{x}_2 = P\,(P\mathbf{x}_0) = \begin{bmatrix} 0.750 & 0.338 \\ 0.250 & 0.662 \end{bmatrix} \begin{bmatrix} 0.750 & 0.338 \\ 0.250 & 0.662 \end{bmatrix} \begin{bmatrix} 1 \\ 0 \end{bmatrix} = \begin{bmatrix} 0.647 \\ 0.353 \end{bmatrix}.$$

So, given that Monday is dry, the probability that Wednesday will be wet is 0.353.

(c) To find the distribution of wet and dry days in the long run, we need $P\mathbf{x} = \mathbf{x}$.

So, we find $\left[\begin{array}{cc|c} 1-0.750 & -0.338 & 0 \\ -0.250 & 1-0.662 & 0 \end{array}\right] \longrightarrow \left[\begin{array}{cc|c} 1 & -1.352 & 0 \\ 0 & 0 & 0 \end{array}\right]$.

We parametrize the solution by letting $x_2 = t$, so $x_1 = 1.352t$.

However, a day will either be wet or dry so we impose the constraint:

$1 = x_1 + x_2 = 1.352t + t \Rightarrow t = \frac{1}{2.352} \approx 0.425$.

Thus, in the long run roughly 57.5% of days will be dry and 42.5% will be wet.

10. Let states A, B, and C denote a tall, medium-height, and short person respectively. A tall parent has a tall, medium, or short child with probabilities 0.6, 0.2, and 0.2 respectively. A medium parent has a tall, medium, or short child with probabilities 0.1, 0.7, and 0.2. A short parent has a tall, medium, or short child with probabilities 0.2, 0.4, and 0.4.

(a) The transition matrix for this Markov chain is $P = \begin{bmatrix} 0.6 & 0.1 & 0.2 \\ 0.2 & 0.7 & 0.4 \\ 0.2 & 0.2 & 0.4 \end{bmatrix}$.

(b) The probability that a short person will have a tall grandchild is given by $\left(P^2\right)_{13} = 0.24$.

(c) We are told that the current distribution of heights is $\mathbf{x} = \begin{bmatrix} 0.2 \\ 0.5 \\ 0.3 \end{bmatrix}$.

We wish to know the distribution after three generations. So, we find:

$\mathbf{x}_3 = P^3\mathbf{x} = \begin{bmatrix} 0.6 & 0.1 & 0.2 \\ 0.2 & 0.7 & 0.4 \\ 0.2 & 0.2 & 0.4 \end{bmatrix}^3 \begin{bmatrix} 0.2 \\ 0.5 \\ 0.3 \end{bmatrix} = \begin{bmatrix} 0.2457 \\ 0.5039 \\ 0.2504 \end{bmatrix}$. So, after three generations, roughly 24.6% of the population is tall, 50.4% medium, and 25.0% is short.

(d) The data are not changing over time, so for $\mathbf{x} = \begin{bmatrix} x_1 \\ x_2 \\ x_3 \end{bmatrix}$ we have $P\mathbf{x} = \mathbf{x}$.

We solve this system of equations by forming, and solving, the augmented matrix

$$\left[\, I - P \mid \mathbf{0} \,\right]: \left[\begin{array}{ccc|c} 1-0.6 & -0.1 & -0.2 & 0 \\ -0.2 & 1-0.7 & -0.4 & 0 \\ -0.2 & -0.2 & 1-0.4 & 0 \end{array}\right] \longrightarrow \left[\begin{array}{ccc|c} 1 & 0 & -1 & 0 \\ 0 & 1 & -2 & 0 \\ 0 & 0 & 0 & 0 \end{array}\right].$$

So, we see x_3 is a free variable and the parametric solution is $x_1 = t$, $x_2 = 2t$, $x_3 = t$. But x_i denotes the percent population of a certain height group, so the state components must be normalized by $1 = x_1 + x_2 + x_3 = 4t \Rightarrow t = \frac{1}{4}$. So after a long time, 25% of the population is tall, 50% is medium, and 25% is short.

11. Let states A, B, and C denote a good, fair, or poor pine nut crop.
If one year's crop is good, the following crop will be good, fair, or poor with probabilities 0.08, 0.07, and 0.85, respectively.
If one year's crop is fair then the following year's crop will be good, fair, or poor with probabilities 0.09, 0.11, and 0.80, respectively.
If a crop is good then the following year's crop will be good, fair, or poor with probabilities 0.11, 0.05, and 0.84, respectively.

(a) The transition matrix for this Markov chain is $P = \begin{bmatrix} 0.08 & 0.09 & 0.11 \\ 0.07 & 0.11 & 0.05 \\ 0.85 & 0.80 & 0.84 \end{bmatrix}$.

(b) We are told that the 1940 crop is good, so $\mathbf{x}_0 = \begin{bmatrix} 1 \\ 0 \\ 0 \end{bmatrix}$.

The probabilities of a good crop in the following five years are as follows:

$$\begin{aligned} 1941: &\quad P_{11} &= 0.0800 \\ 1942: &\quad (P)^2_{11} &= 0.1062 \\ 1943: &\quad (P)^3_{11} &= 0.1057 \\ 1944: &\quad (P)^4_{11} &= 0.1057 \\ 1945: &\quad (P)^5_{11} &= 0.1057 \end{aligned}$$

(c) To find the proportion of the crops in the long run, we need $P\mathbf{x} = \mathbf{x}$.
So, we have $\left[\, I - P \mid \mathbf{0} \,\right]$:

$$\left[\begin{array}{ccc|c} 1-0.08 & -0.09 & -0.11 & 0 \\ -0.07 & 1-0.11 & -0.05 & 0 \\ -0.85 & -0.80 & 1-0.84 & 0 \end{array}\right] \longrightarrow \left[\begin{array}{ccc|c} 1 & 0 & -\frac{1024}{8125} & 0 \\ 0 & 1 & -\frac{537}{8125} & 0 \\ 0 & 0 & 0 & 0 \end{array}\right]$$

So, x_3 is free and the parametric solution is $x_1 = 0.1260t$, $x_2 = 0.0661t$, $x_3 = t$.
But x_i denotes a probability, so the state components must be normalized by
$1 = x_1 + x_2 + x_3 = 1.1921t \Rightarrow t = 0.839$.
Thus, after a long time 10.6% of the crops are good, 5.5% are fair, and 83.9% are poor.

12. There are four states (positions) that the robots can be in (labeled 1 through 4).
If a robot is at position 1, it has a $\frac{1}{2}$ probability of going to either state 2 or state 3.
If a robot is at position 2, it has a $\frac{1}{3}$ probability each of going to states 1, 3, or 4.
If a robot is at position 3, it has a $\frac{1}{4}$ probability of going to state 1 or 2 and
a $\frac{2}{4} = \frac{1}{2}$ probability of going to state 4.
If a robot is at position 4, it has a $\frac{1}{3}$ probability each of going to state 2 and
a $\frac{2}{3}$ probability of going to state 3.

(a) The transition matrix for this Markov chain is $P = \begin{bmatrix} 0 & \frac{1}{3} & \frac{1}{4} & 0 \\ \frac{1}{2} & 0 & \frac{1}{4} & \frac{1}{3} \\ \frac{1}{2} & \frac{1}{3} & 0 & \frac{2}{3} \\ 0 & \frac{1}{3} & \frac{1}{2} & 0 \end{bmatrix}$.

(b) The steady-state distribution of robots is described by the state $\mathbf{x}$ such that $P\mathbf{x} = \mathbf{x}$.
We form and solve the augmented matrix $\left[\, I - P \mid \mathbf{0} \,\right]$:

$$\left[\begin{array}{cccc|c} 1 & -\frac{1}{3} & -\frac{1}{4} & 0 & 0 \\ -\frac{1}{2} & 1 & -\frac{1}{4} & -\frac{1}{3} & 0 \\ -\frac{1}{2} & -\frac{1}{3} & 1 & -\frac{2}{3} & 0 \\ 0 & -\frac{1}{3} & -\frac{1}{2} & 1 & 0 \end{array}\right] \longrightarrow \left[\begin{array}{cccc|c} 1 & 0 & 0 & -\frac{2}{3} & 0 \\ 0 & 1 & 0 & -1 & 0 \\ 0 & 0 & 1 & -\frac{4}{3} & 0 \\ 0 & 0 & 0 & 0 & 0 \end{array}\right]$$

So, x_4 is free and the parametric solution is $x_1 = 2t$, $x_2 = 3t$, $x_3 = 4t$, $x_4 = 3t$.
But this is the probability of a robot being at any of the junctions,
so the state components must be normalized by $1 = x_1 + x_2 + x_3 + x_4 = 12t \Rightarrow t = \frac{1}{12}$.
So after a long time, the probability of the robot being
at junction 1, 2, 3, or 4 is $\frac{1}{6}$, $\frac{1}{4}$, $\frac{1}{3}$, and $\frac{1}{4}$ respectively.
We are told that there are 15 robots at each junction to start with (60 total),
so in the steady-state distribution there will be 10 at 1, 15 at 2, 20 at 3, and 15 at 4.

13. Let $P = \left[\, \mathbf{p}_1 \ \cdots \ \mathbf{p}_n \,\right]$ be a stochastic matrix (so the elements of each $\mathbf{p}_i$ sum to 1), and let $\mathbf{j}$ be a row vector consisting entirely of 1s.

Then $\mathbf{j}P = \left[\, \mathbf{j}\cdot\mathbf{p}_1 \ \cdots \ \mathbf{j}\cdot\mathbf{p}_n \,\right] = \left[\, \sum(\mathbf{p}_1)_i \ \cdots \ \sum(\mathbf{p}_n)_i \,\right] = \left[\, 1 \ \cdots \ 1 \,\right] = \mathbf{j}$.

Conversely, suppose that $\mathbf{j}P = \mathbf{j}$.

Then we have $\mathbf{j}\cdot\mathbf{p}_i = 1$ for each i, so $\sum_j (\mathbf{p}_i)_j = 1$ for all i, showing P is stochastic.

14. (a) Let $P = \begin{bmatrix} x_1 & y_1 \\ x_2 & y_2 \end{bmatrix}$ and $Q = \begin{bmatrix} w_1 & z_1 \\ w_2 & z_2 \end{bmatrix}$ be stochastic matrices.

Then, consider $PQ = \begin{bmatrix} x_1 & y_1 \\ x_2 & y_2 \end{bmatrix} \begin{bmatrix} w_1 & z_1 \\ w_2 & z_2 \end{bmatrix} = \begin{bmatrix} x_1w_1 + y_1w_2 & x_1z_1 + y_1z_2 \\ x_2w_1 + y_2w_2 & x_2z_1 + y_2z_2 \end{bmatrix}$.

Summing on the first column gives:

$x_1w_1 + y_1w_2 + x_2w_1 + y_2w_2 = w_1(x_1 + x_2) + w_2(y_1 + y_2) = w_1 + w_2 = 1$,

where we've used the stochastic property $x_1 + x_2 = y_1 + y_2 \Rightarrow w_1 + w_2 = 1$.

Summing on the second column gives:

$x_1z_1 + y_1z_2 + x_2z_1 + y_2z_2 = z_1(x_1 + x_2) + z_2(y_1 + y_2) = z_1 + z_2 = 1$.

Thus the columns of PQ add to 1,

and we see the product of two 2×2 stochastic matrices is stochastic.

(b) Let P and Q be $n \times n$ stochastic matrices.

To show PQ is stochastic, we need to show each column of PQ sums to 1:

$\sum_{i=1}^{n} [PQ]_{ij} = \sum_{i=1}^{n} \sum_{k=1}^{n} P_{ik}Q_{kj} = \sum_{k=1}^{n} (Q_{kj} \sum_{i=1}^{n} P_{ik}) = \sum_{k=1}^{n} Q_{kj} = 1 \Rightarrow$

PQ is stochastic.

15. From Exercise 9 we have the following:

If Monday is a dry day, the expected number of days until a wet day can be determined by first deleting the second row and column of the transition matrix

$P = \begin{bmatrix} 0.750 & 0.338 \\ 0.250 & 0.662 \end{bmatrix}$ giving $Q = [0.750]$.

So, the expected number of days until the next wet day is given by:

$(I - Q)^{-1} = (1 - 0.750)^{-1} = 4.0$.

Thus, we expect four days until the next wet day.

16. From Exercise 10 we have the following:

If we are interested in the number of generations it will take for a short person to have a tall descendant we delete the row and column of the transition matrix

$P = \begin{bmatrix} 0.6 & 0.1 & 0.2 \\ 0.2 & 0.7 & 0.4 \\ 0.2 & 0.2 & 0.4 \end{bmatrix}$ corresponding to tall people (the first column and row). So,

$Q = \begin{bmatrix} 0.7 & 0.4 \\ 0.2 & 0.4 \end{bmatrix}$. Now, we sum on the last column of $(I - Q)^{-1} = \begin{bmatrix} 6.0 & 4.0 \\ 2.0 & 3.0 \end{bmatrix} \Rightarrow$

The expected number of generations before a short person has a tall descendant is 7.

17. From Exercise 11 we have the following:

If the pine nut crop is fair one year the expected number of years until a good crop occurs is given by deleting the first row and column of the transition matrix

$P = \begin{bmatrix} 0.08 & 0.09 & 0.11 \\ 0.07 & 0.11 & 0.05 \\ 0.85 & 0.80 & 0.84 \end{bmatrix}$ giving $Q = \begin{bmatrix} 0.11 & 0.05 \\ 0.80 & 0.84 \end{bmatrix}$.

Summing on the first (fair) column of $(I - Q)^{-1} = \begin{bmatrix} 1.562 & 0.488 \\ 7.812 & 8.691 \end{bmatrix} = 9.374$.

So, it will be roughly 9 or 10 years before a good crop occurs.

18. From Exercise 12, we have the following:
If a robot starts at a junction other than 4, the expected number of moves until the robot reaches junction 4 is found deleting the fourth row and column of P:

$$P = \begin{bmatrix} 0 & \frac{1}{3} & \frac{1}{4} & 0 \\ \frac{1}{2} & 0 & \frac{1}{4} & \frac{1}{3} \\ \frac{1}{2} & \frac{1}{3} & 0 & \frac{2}{3} \\ 0 & \frac{1}{3} & \frac{1}{2} & 0 \end{bmatrix} \text{ giving } Q = \begin{bmatrix} 0 & \frac{1}{3} & \frac{1}{4} \\ \frac{1}{2} & 0 & \frac{1}{4} \\ \frac{1}{2} & \frac{1}{3} & 0 \end{bmatrix}.$$

Then the vector $\mathbf{j}(I-Q)^{-1}$ gives the number of moves for every possible starting state.

$$\mathbf{j}(I-Q)^{-1} = \begin{bmatrix} 1 & 1 & 1 \end{bmatrix} \begin{bmatrix} 1 & -\frac{1}{3} & -\frac{1}{4} \\ -\frac{1}{2} & 1 & -\frac{1}{4} \\ -\frac{1}{2} & -\frac{1}{3} & 1 \end{bmatrix}^{-1} = \begin{bmatrix} 1 & 1 & 1 \end{bmatrix} \begin{bmatrix} \frac{22}{13} & \frac{10}{13} & \frac{8}{13} \\ \frac{15}{13} & \frac{21}{13} & \frac{9}{13} \\ \frac{6}{13} & \frac{12}{13} & \frac{20}{13} \end{bmatrix} = \begin{bmatrix} \frac{53}{13} & \frac{43}{13} & \frac{37}{13} \end{bmatrix}.$$

So, on average, we expect that a robot starting from junction 1, 2, or 3 will reach junction 4 in $\frac{53}{13} \approx 4.08$, $\frac{43}{13} \approx 3.31$, and $\frac{37}{13} \approx 2.85$ moves respectively.

19. A population with three age classes has a Leslie matrix $L = \begin{bmatrix} 1 & 1 & 3 \\ 0.7 & 0 & 0 \\ 0 & 0.5 & 0 \end{bmatrix}$.

The initial population vector is $\mathbf{x}_0 = \begin{bmatrix} 100 \\ 100 \\ 100 \end{bmatrix}$, so

$$\mathbf{x}_1 = L\mathbf{x}_0 = \begin{bmatrix} 1 & 1 & 3 \\ 0.7 & 0 & 0 \\ 0 & 0.5 & 0 \end{bmatrix} \begin{bmatrix} 100 \\ 100 \\ 100 \end{bmatrix} = \begin{bmatrix} 500 \\ 70.0 \\ 50.0 \end{bmatrix} \text{ and}$$

$$\mathbf{x}_2 = L\mathbf{x}_1 = \begin{bmatrix} 1 & 1 & 3 \\ 0.7 & 0 & 0 \\ 0 & 0.5 & 0 \end{bmatrix} \begin{bmatrix} 500 \\ 70.0 \\ 50.0 \end{bmatrix} = \begin{bmatrix} 720.0 \\ 350.0 \\ 35.0 \end{bmatrix}.$$

Alternatively,

$$\mathbf{x}_2 = L\mathbf{x}_1 = LL\mathbf{x}_1 = \begin{bmatrix} 1 & 1 & 3 \\ 0.7 & 0 & 0 \\ 0 & 0.5 & 0 \end{bmatrix} \begin{bmatrix} 1 & 1 & 3 \\ 0.7 & 0 & 0 \\ 0 & 0.5 & 0 \end{bmatrix} \begin{bmatrix} 100 \\ 100 \\ 100 \end{bmatrix} = \begin{bmatrix} 720.0 \\ 350.0 \\ 35.0 \end{bmatrix} \text{ and}$$

$$\mathbf{x}_3 = L\mathbf{x}_2 = \begin{bmatrix} 1 & 1 & 3 \\ 0.7 & 0 & 0 \\ 0 & 0.5 & 0 \end{bmatrix} \begin{bmatrix} 720.0 \\ 350.0 \\ 35.0 \end{bmatrix} = \begin{bmatrix} 1175.0 \\ 504.0 \\ 175.0 \end{bmatrix}.$$

20. A population with four age classes has a Leslie matrix $L = \begin{bmatrix} 0 & 1 & 2 & 5 \\ 0.5 & 0 & 0 & 0 \\ 0 & 0.7 & 0 & 0 \\ 0 & 0 & 0.3 & 0 \end{bmatrix}$.

So, $\mathbf{x}_0 = \begin{bmatrix} 10 \\ 10 \\ 10 \\ 10 \end{bmatrix} \Rightarrow \mathbf{x}_1 = L\mathbf{x}_0 = \begin{bmatrix} 0 & 1 & 2 & 5 \\ 0.5 & 0 & 0 & 0 \\ 0 & 0.7 & 0 & 0 \\ 0 & 0 & 0.3 & 0 \end{bmatrix} = \begin{bmatrix} 80 \\ 5.0 \\ 7.0 \\ 3.0 \end{bmatrix}$,

$$\mathbf{x}_2 = L\mathbf{x}_1 = \begin{bmatrix} 0 & 1 & 2 & 5 \\ 0.5 & 0 & 0 & 0 \\ 0 & 0.7 & 0 & 0 \\ 0 & 0 & 0.3 & 0 \end{bmatrix}\begin{bmatrix} 80 \\ 5.0 \\ 7.0 \\ 3.0 \end{bmatrix} = \begin{bmatrix} 34.0 \\ 40.0 \\ 3.5 \\ 2.1 \end{bmatrix}, \mathbf{x}_3 = \begin{bmatrix} 0 & 1 & 2 & 5 \\ 0.5 & 0 & 0 & 0 \\ 0 & 0.7 & 0 & 0 \\ 0 & 0 & 0.3 & 0 \end{bmatrix}\begin{bmatrix} 34.0 \\ 40.0 \\ 3.5 \\ 2.1 \end{bmatrix} = \begin{bmatrix} 57.5 \\ 17.0 \\ 28.0 \\ 1.05 \end{bmatrix}.$$

21. A certain species with two year-long age classes is described by two possible Leslie matrices $L_1 = \begin{bmatrix} 0 & 5 \\ 0.8 & 0 \end{bmatrix}$ and $L_2 = \begin{bmatrix} 4 & 1 \\ 0.8 & 0 \end{bmatrix}$.

(a) Starting with $\mathbf{x}_0 = \begin{bmatrix} 10 \\ 10 \end{bmatrix}$ the following table gives $\mathbf{x}_1, \ldots, \mathbf{x}_{10}$ for both cases.

	$\mathbf{x}_0$	$\mathbf{x}_1$	$\mathbf{x}_2$	$\mathbf{x}_3$	$\mathbf{x}_4$	$\mathbf{x}_5$
L_1	$\begin{bmatrix} 10 \\ 10 \end{bmatrix}$	$\begin{bmatrix} 50 \\ 8.0 \end{bmatrix}$	$\begin{bmatrix} 40.0 \\ 40.0 \end{bmatrix}$	$\begin{bmatrix} 200.0 \\ 32.0 \end{bmatrix}$	$\begin{bmatrix} 160.0 \\ 160.0 \end{bmatrix}$	$\begin{bmatrix} 800.0 \\ 128.0 \end{bmatrix}$
L_2	$\begin{bmatrix} 10 \\ 10 \end{bmatrix}$	$\begin{bmatrix} 50 \\ 8.0 \end{bmatrix}$	$\begin{bmatrix} 208.0 \\ 40.0 \end{bmatrix}$	$\begin{bmatrix} 872.0 \\ 166.4 \end{bmatrix}$	$\begin{bmatrix} 3654.4 \\ 697.6 \end{bmatrix}$	$\begin{bmatrix} 15,315.2 \\ 2923.52 \end{bmatrix}$

	$\mathbf{x}_6$	$\mathbf{x}_7$	$\mathbf{x}_8$	$\mathbf{x}_9$	$\mathbf{x}_{10}$
L_1	$\begin{bmatrix} 640.0 \\ 640.0 \end{bmatrix}$	$\begin{bmatrix} 3200.0 \\ 512.0 \end{bmatrix}$	$\begin{bmatrix} 2560.0 \\ 2560.0 \end{bmatrix}$	$\begin{bmatrix} 12,800.0 \\ 2048.0 \end{bmatrix}$	$\begin{bmatrix} 10,240.0 \\ 10,240.0 \end{bmatrix}$
L_2	$\begin{bmatrix} 64,184.32 \\ 12,252.16 \end{bmatrix}$	$\begin{bmatrix} 268,989.44 \\ 51,347.456 \end{bmatrix}$	$\begin{bmatrix} 1.127 \times 10^6 \\ 215,191.552 \end{bmatrix}$	$\begin{bmatrix} 4.724 \times 10^6 \\ 901,844.17 \end{bmatrix}$	$\begin{bmatrix} 1.979 \times 10^7 \\ 3.779 \times 10^6 \end{bmatrix}$

(b)

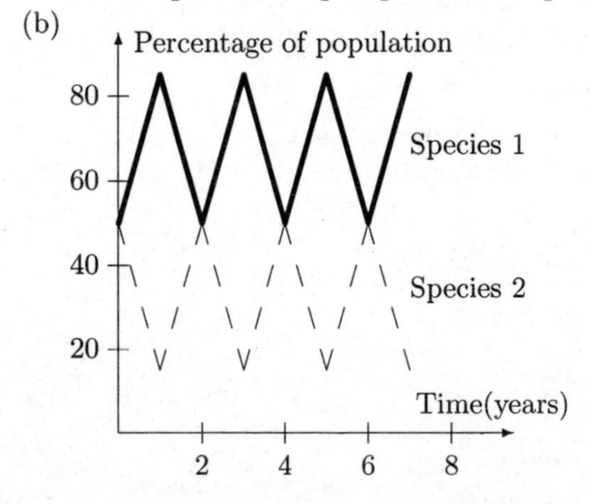

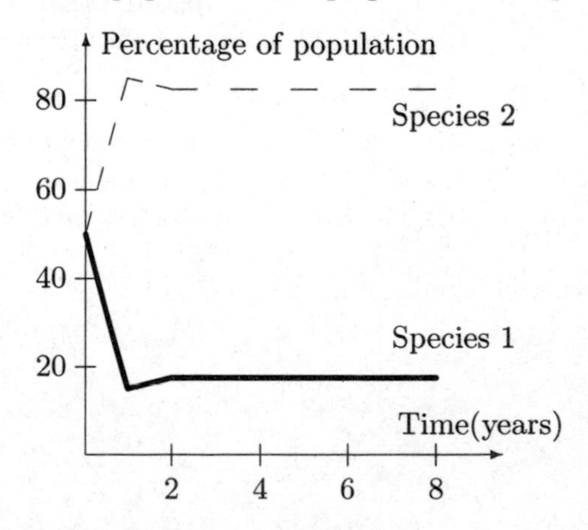

The graphs suggest that the first Leslie matrix does not have a steady state, while the second Leslie matrix quickly approaches its steady-state value.

22. Let the Leslie matrix for the VW Beetle be $L = \begin{bmatrix} 0 & 0 & 20 \\ 0.1 & 0 & 0 \\ 0 & 0.5 & 0 \end{bmatrix}$.

To determine the behavior, we calculate $\mathbf{x}_1$ to $\mathbf{x}_3$ for $\mathbf{x}_0 = \begin{bmatrix} 10 \\ 10 \\ 10 \end{bmatrix}$:

$$\begin{matrix} & \mathbf{x}_0 & \mathbf{x}_1 & \mathbf{x}_2 & \mathbf{x}_3 \\ L & \begin{bmatrix} 10 \\ 10 \\ 10 \end{bmatrix} & \begin{bmatrix} 200 \\ 1 \\ 5 \end{bmatrix} & \begin{bmatrix} 100 \\ 20 \\ 0.5 \end{bmatrix} & \begin{bmatrix} 10 \\ 10 \\ 10 \end{bmatrix} \end{matrix}$$

At this point we see that this population has a cyclic nature.

23. Let the Leslie matrix for the VW Beetle be $L = \begin{bmatrix} 0 & 0 & 20 \\ s & 0 & 0 \\ 0 & 0.5 & 0 \end{bmatrix}$.

To understand the effect of varying the survival probability s of the young beetles, we consider the evolution of the initial state vector over several years.

If the initial state is $\mathbf{x}_0 = \begin{bmatrix} 10 \\ 10 \\ 10 \end{bmatrix}$, then

$$\mathbf{x}_1 = L\mathbf{x}_0 = \begin{bmatrix} 0 & 0 & 20 \\ s & 0 & 0 \\ 0 & 0.5 & 0 \end{bmatrix} \begin{bmatrix} 10 \\ 10 \\ 10 \end{bmatrix} = \begin{bmatrix} 200 \\ 10s \\ 5.0 \end{bmatrix}, \mathbf{x}_2 = \begin{bmatrix} 0 & 0 & 20 \\ s & 0 & 0 \\ 0 & 0.5 & 0 \end{bmatrix} \begin{bmatrix} 200 \\ 10s \\ 5.0 \end{bmatrix} = \begin{bmatrix} 100.0 \\ 200s \\ 5.0s \end{bmatrix},$$

$$\mathbf{x}_3 = \begin{bmatrix} 0 & 0 & 20 \\ s & 0 & 0 \\ 0 & 0.5 & 0 \end{bmatrix} \begin{bmatrix} 100.0 \\ 200s \\ 5.0s \end{bmatrix} = \begin{bmatrix} 100.0s \\ 100.0s \\ 100.0s \end{bmatrix}.$$

At this point it is apparent that after three years of evolution,
the relative abundance of each class is equal to the initial distribution.

As well, we see the following patterns:

If $s < 0.1$, then the overall population declines after three years.
If $s = 0.1$, the overall population stays the same after three years.
If $s > 0.1$, the overall population increases after three years.

24. We split up the female caribou into seven age classes of duration two years each. The Leslie matrix for this system is

$$L = \begin{bmatrix} 0 & 0.4 & 1.8 & 1.8 & 1.8 & 1.6 & 0.6 \\ 0.3 & 0 & 0 & 0 & 0 & 0 & 0 \\ 0 & 0.7 & 0 & 0 & 0 & 0 & 0 \\ 0 & 0 & 0.9 & 0 & 0 & 0 & 0 \\ 0 & 0 & 0 & 0.9 & 0 & 0 & 0 \\ 0 & 0 & 0 & 0 & 0.9 & 0 & 0 \\ 0 & 0 & 0 & 0 & 0 & 0.6 & 0 \end{bmatrix}$$

$$1990\ \mathbf{x}_0 = \begin{bmatrix} 10 \\ 2 \\ 8 \\ 5 \\ 12 \\ 0 \\ 1 \end{bmatrix} \Rightarrow 1991,\ 1992 \text{ are } \mathbf{x}_1 = L\mathbf{x}_0 = \begin{bmatrix} 46.4 \\ 3.0 \\ 1.4 \\ 7.2 \\ 4.5 \\ 10.8 \\ 0 \end{bmatrix} \text{ and } \mathbf{x}_2 = L^2\mathbf{x}_0 = \begin{bmatrix} 42.06 \\ 13.92 \\ 2.1 \\ 1.26 \\ 6.48 \\ 4.05 \\ 6.48 \end{bmatrix}.$$

To project the population for the years 2000 and 2010 we note that $\mathbf{x}_n = L^n\mathbf{x}_0$.

$$\text{Thus } \mathbf{x}_{10} = L^{10}\mathbf{x}_0 = \begin{bmatrix} 69.12 \\ 18.61 \\ 12.24 \\ 10.59 \\ 8.29 \\ 6.51 \\ 3.57 \end{bmatrix} \text{ and } \mathbf{x}_{20} = \begin{bmatrix} 167.80 \\ 46.08 \\ 29.54 \\ 24.35 \\ 20.06 \\ 16.48 \\ 9.05 \end{bmatrix}.$$

Looking at the results it appears that the caribou population is increasing with time. This may not be accurate since the model assumed constant birth and survival rates.

25. $A = \begin{bmatrix} 0 & 1 & 0 & 1 \\ 1 & 0 & 1 & 0 \\ 0 & 1 & 0 & 1 \\ 1 & 0 & 1 & 0 \end{bmatrix}.$

26. $A = \begin{bmatrix} 1 & 0 & 0 & 1 \\ 0 & 1 & 1 & 0 \\ 0 & 1 & 0 & 0 \\ 1 & 0 & 0 & 0 \end{bmatrix}.$

27. $A = \begin{bmatrix} 0 & 1 & 1 & 1 & 1 \\ 1 & 0 & 1 & 0 & 0 \\ 1 & 1 & 0 & 1 & 0 \\ 1 & 0 & 1 & 0 & 1 \\ 1 & 0 & 0 & 1 & 0 \end{bmatrix}.$

28. $A = \begin{bmatrix} 0 & 0 & 0 & 1 & 1 \\ 0 & 0 & 0 & 1 & 0 \\ 0 & 0 & 0 & 1 & 0 \\ 1 & 1 & 1 & 0 & 0 \\ 1 & 0 & 0 & 0 & 0 \end{bmatrix}.$

29.

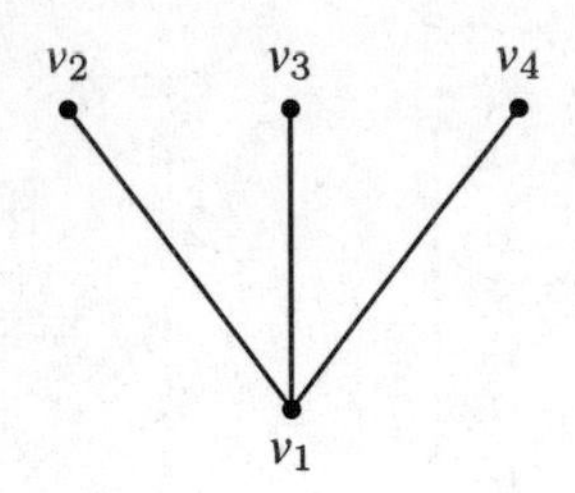

30.

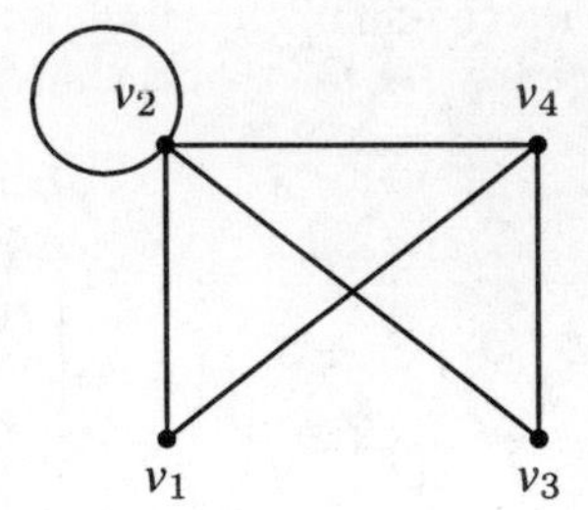

31.

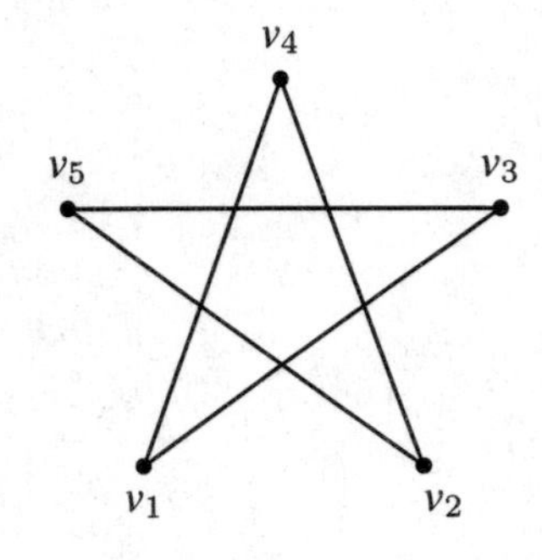

32.

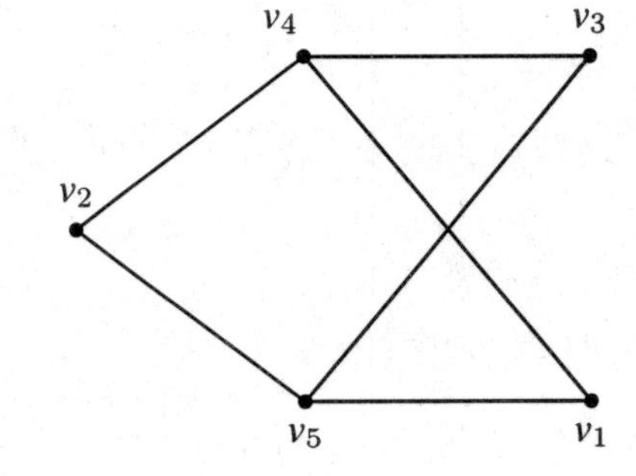

33. $A = \begin{bmatrix} 0 & 1 & 1 & 0 \\ 0 & 0 & 0 & 0 \\ 0 & 1 & 0 & 1 \\ 1 & 0 & 0 & 0 \end{bmatrix}.$

34. $A = \begin{bmatrix} 0 & 0 & 1 & 1 \\ 0 & 0 & 0 & 1 \\ 1 & 1 & 0 & 0 \\ 0 & 1 & 1 & 0 \end{bmatrix}.$

35. $A = \begin{bmatrix} 0 & 1 & 0 & 1 & 0 \\ 1 & 0 & 0 & 1 & 0 \\ 1 & 1 & 0 & 0 & 0 \\ 1 & 0 & 0 & 0 & 1 \\ 1 & 0 & 0 & 0 & 0 \end{bmatrix}.$

36. $A = \begin{bmatrix} 0 & 1 & 0 & 1 & 1 \\ 0 & 0 & 0 & 0 & 1 \\ 0 & 1 & 0 & 1 & 0 \\ 0 & 0 & 0 & 0 & 1 \\ 0 & 0 & 1 & 0 & 0 \end{bmatrix}.$

37.

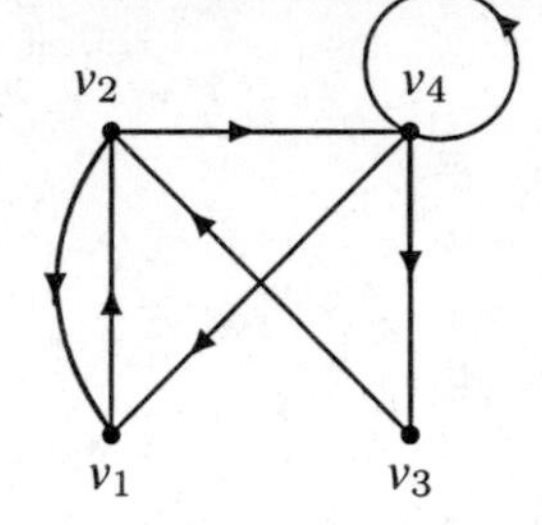

38.

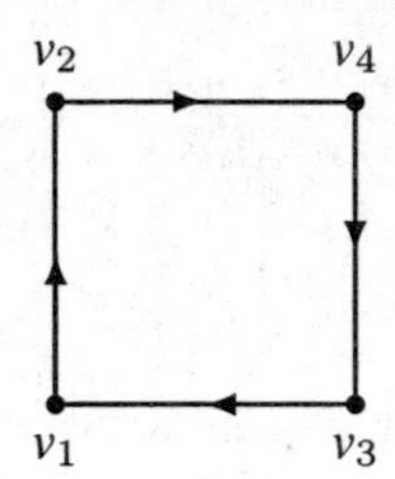

39.

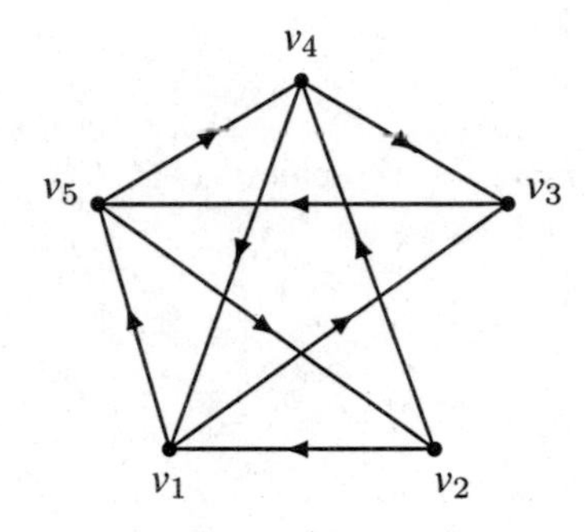

40.

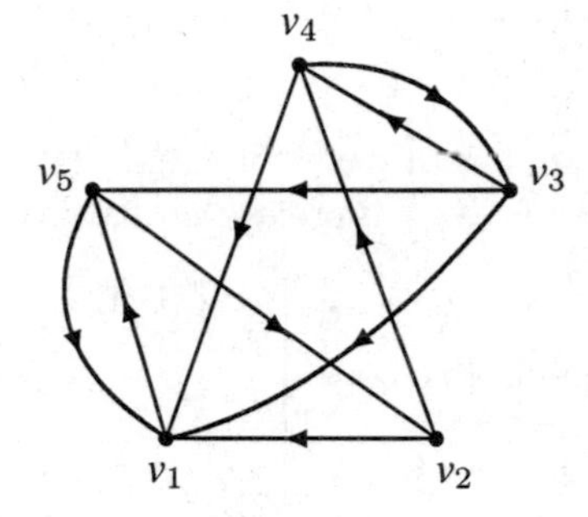

41. In Exercise 30, $A = \begin{bmatrix} 0 & 1 & 0 & 1 \\ 1 & 1 & 1 & 1 \\ 0 & 1 & 0 & 1 \\ 1 & 1 & 1 & 0 \end{bmatrix} \Rightarrow A^2 = \begin{bmatrix} 2 & 2 & 2 & 1 \\ 2 & 4 & 2 & 3 \\ 2 & 2 & 2 & 1 \\ 1 & 3 & 1 & 3 \end{bmatrix}$.

The number of paths of length 2 between v_1 and v_2 is given by the element $\left[A^2\right]_{12}$. So, we see that there are two paths of length 2 between vertices v_1 and v_2.

42. In Exercise 32, $A = \begin{bmatrix} 0 & 0 & 0 & 1 & 1 \\ 0 & 0 & 0 & 1 & 1 \\ 0 & 0 & 0 & 1 & 1 \\ 1 & 1 & 1 & 0 & 0 \\ 1 & 1 & 1 & 0 & 0 \end{bmatrix} \Rightarrow A^2 = \begin{bmatrix} 2 & 2 & 2 & 0 & 0 \\ 2 & 2 & 2 & 0 & 0 \\ 2 & 2 & 2 & 0 & 0 \\ 0 & 0 & 0 & 3 & 3 \\ 0 & 0 & 0 & 3 & 3 \end{bmatrix}$.

The number of paths of length 2 between v_1 and v_2 is given by the element $\left[A^2\right]_{12}$. So, we see that there are two paths of length 2 between vertices v_1 and v_2.

43. In Exercise 30, $A = \begin{bmatrix} 0 & 1 & 0 & 1 \\ 1 & 1 & 1 & 1 \\ 0 & 1 & 0 & 1 \\ 1 & 1 & 1 & 0 \end{bmatrix} \Rightarrow A^3 = \begin{bmatrix} 3 & 7 & 3 & 6 \\ 7 & 11 & 7 & 8 \\ 3 & 7 & 3 & 6 \\ 6 & 8 & 6 & 5 \end{bmatrix}$.

The number of paths of length 3 between v_1 and v_3 is given by the element $\left[A^3\right]_{13}$. So, we see that there are three paths of length 3 between vertices v_1 and v_3.

44. In Exercise 32, $A = \begin{bmatrix} 0 & 0 & 0 & 1 & 1 \\ 0 & 0 & 0 & 1 & 1 \\ 0 & 0 & 0 & 1 & 1 \\ 1 & 1 & 1 & 0 & 0 \\ 1 & 1 & 1 & 0 & 0 \end{bmatrix} \Rightarrow A^4 = \begin{bmatrix} 12 & 12 & 12 & 0 & 0 \\ 12 & 12 & 12 & 0 & 0 \\ 12 & 12 & 12 & 0 & 0 \\ 0 & 0 & 0 & 18 & 18 \\ 0 & 0 & 0 & 18 & 18 \end{bmatrix}$.

The number of paths of length 4 between v_2 and itself is given by the element $[A^4]_{22}$.
So, we see that there are twelve paths of length 4 between vertex v_2 and itself.

45. In Exercise 37, $A = \begin{bmatrix} 0 & 1 & 0 & 0 \\ 1 & 0 & 0 & 1 \\ 0 & 1 & 0 & 0 \\ 1 & 0 & 1 & 1 \end{bmatrix} \Rightarrow A^2 = \begin{bmatrix} 1 & 0 & 0 & 1 \\ 1 & 1 & 1 & 1 \\ 1 & 0 & 0 & 1 \\ 1 & 2 & 1 & 1 \end{bmatrix}$.

The number of paths of length 2 between v_1 and v_3 is given by the element $[A^2]_{13}$.
So, we see that there are no paths of length 2 between vertex v_1 and v_3.

46. In Exercise 37, $A = \begin{bmatrix} 0 & 1 & 0 & 0 \\ 1 & 0 & 0 & 1 \\ 0 & 1 & 0 & 0 \\ 1 & 0 & 1 & 1 \end{bmatrix} \Rightarrow A^3 = \begin{bmatrix} 1 & 1 & 1 & 1 \\ 2 & 2 & 1 & 2 \\ 1 & 1 & 1 & 1 \\ 3 & 2 & 1 & 3 \end{bmatrix}$.

The number of paths of length 3 between v_4 and v_1 is given by the element $[A^3]_{41}$.
So, we see that there are three paths of length 3 between vertex v_4 and v_1.

47. In Exercise 40, $A = \begin{bmatrix} 0 & 1 & 0 & 0 & 1 \\ 0 & 0 & 0 & 1 & 0 \\ 1 & 0 & 0 & 1 & 1 \\ 1 & 0 & 1 & 0 & 0 \\ 1 & 1 & 0 & 0 & 0 \end{bmatrix} \Rightarrow A^3 = \begin{bmatrix} 1 & 1 & 1 & 1 & 1 \\ 1 & 1 & 0 & 1 & 2 \\ 2 & 3 & 0 & 3 & 3 \\ 3 & 3 & 1 & 1 & 1 \\ 2 & 1 & 1 & 1 & 0 \end{bmatrix}$.

The number of paths of length 3 from vertex v_4 to vertex v_1 is given by the element $[A^3]_{41}$.
So, we see that there are three paths of length 3 from vertex v_4 to vertex v_1.

48. In Exercise 40, $A = \begin{bmatrix} 0 & 1 & 0 & 0 & 1 \\ 0 & 0 & 0 & 1 & 0 \\ 1 & 0 & 0 & 1 & 1 \\ 1 & 0 & 1 & 0 & 0 \\ 1 & 1 & 0 & 0 & 0 \end{bmatrix} \Rightarrow A^4 = \begin{bmatrix} 3 & 2 & 1 & 2 & 2 \\ 3 & 3 & 1 & 1 & 1 \\ 6 & 5 & 3 & 3 & 2 \\ 3 & 4 & 1 & 4 & 4 \\ 2 & 2 & 1 & 2 & 3 \end{bmatrix}$.

The number of paths of length 4 from vertex v_1 to vertex v_4 is given by the element $[A^4]_{14}$.
So, we see that there are two paths of length 4 from vertex v_1 to vertex v_4.

49. Let A be the adjacency matrix of a graph G.

(a) Assume that row i of A contains only zeros.
This implies that there are no edges joining vertex i to any of the other vertices.
We can conclude that this is a disconnected graph.

(b) Assume that column j of A contains only zeros.
This implies that there are no edges joining vertex j to any of the other vertices.
We can conclude that this is a disconnected graph.

50. Let A be the adjacency matrix of the digraph D.

(a) Assume that row i of A^2 is all zeros.
This implies that there is no path of length 2 *from* vertex i *to* any of the other vertices.

(b) Assume that column j of A^2 is all zeros.
This implies that there is no path of length 2 *to* vertex j *from* any of the other vertices.

51. Since $A = \begin{bmatrix} 0&0&0&0&1&0 \\ 1&0&1&0&1&1 \\ 1&0&0&1&1&0 \\ 1&1&0&0&1&0 \\ 0&0&0&0&0&1 \\ 1&0&1&1&0&0 \end{bmatrix}$ and $\mathbf{j} = \begin{bmatrix} 1 \\ 1 \\ 1 \\ 1 \\ 1 \\ 1 \end{bmatrix}$, we have: $A\mathbf{j} = \begin{bmatrix} 0&0&0&0&1&0 \\ 1&0&1&0&1&1 \\ 1&0&0&1&1&0 \\ 1&1&0&0&1&0 \\ 0&0&0&0&0&1 \\ 1&0&1&1&0&0 \end{bmatrix} \begin{bmatrix} 1 \\ 1 \\ 1 \\ 1 \\ 1 \\ 1 \end{bmatrix} = \begin{bmatrix} 1 \\ 4 \\ 3 \\ 3 \\ 1 \\ 3 \end{bmatrix}$.

The number of wins that each player had is given by $A\mathbf{j} = \begin{bmatrix} 1 & 4 & 3 & 3 & 1 & 3 \end{bmatrix}^T$,
so the ranking is: first, P_2; second: P_3, P_4, P_6 (tie); third: P_1, P_5 (tie).

If we use the notion of combined wins and indirect wins
the ranking will be determined by $\left(A + A^2\right)\mathbf{j} = \begin{bmatrix} 2 & 12 & 8 & 9 & 4 & 10 \end{bmatrix}^T$.
So, the ranking is: first: P_2; second: P_6; third: P_4.

52. The food web in this ecosystem consists of seven elements where vertices 1–7 correspond to rodent, plant, insect, bird, fish, fox, and bear respectively.

$$A = \begin{bmatrix} 0&1&0&0&0&0&0 \\ 0&0&0&0&0&0&0 \\ 0&1&0&0&0&0&0 \\ 0&1&1&0&1&0&0 \\ 0&1&1&0&0&0&0 \\ 1&0&0&1&0&0&0 \\ 1&0&0&0&1&1&0 \end{bmatrix} \Rightarrow \text{deleting second row and column } A' = \begin{bmatrix} 0&0&0&0&0&0 \\ 0&0&0&0&0&0 \\ 0&1&0&1&0&0 \\ 0&1&0&0&0&0 \\ 1&0&1&0&0&0 \\ 1&0&0&1&1&0 \end{bmatrix}.$$

(a) It is apparent that the bear and bird species have the most direct sources of food (3). We can see this by summing each row of A.

(b) It is apparent that the plant species is a direct source of food for the most other species. We can see this by summing each column of A. (This is analogous to the tennis tournament problem, where being a source of food to another species is like losing a tennis match.)

(c) Recall that $\left[A^2\right]_{ij}$ gives the number of paths of length 2 from vertex i to vertex j. Thus, an indirect source of food is given by $\left[A^2\right]_{ij}$, and the number of indirect food sources for each species is given by $A^2\mathbf{j} = \begin{bmatrix} 0 & 0 & 0 & 3 & 1 & 4 & 5 \end{bmatrix}^T$. So, we see that the bear has the most indirect food sources.

(d) If the plant population were to suddenly become extinct,we could model it with A'(above). Then, the species with the most direct and indirect sources is $A'\mathbf{j} = \begin{bmatrix} 0 & 0 & 2 & 1 & 2 & 3 \end{bmatrix}^T$. The bear has the most direct sources of food; it is unaffected by the killing of the plants. The species which is the source of food for the most other species is $(A')^T\mathbf{j} = \begin{bmatrix} 2 & 2 & 1 & 2 & 1 & 0 \end{bmatrix}^T$. Their number of predators is unchanged because the plants did not prey on any of them. Finally, the number of indirect sources of food each species has is $(A')^2\mathbf{j} = \begin{bmatrix} 0 & 0 & 1 & 0 & 2 & 3 \end{bmatrix}^T$ and combined direct and indirect sources of food are $\left(A' + (A')^2\right)\mathbf{j} = \begin{bmatrix} 0 & 0 & 3 & 1 & 4 & 6 \end{bmatrix}^T$.
Comparing these results to the original ecosystem we see:
the bird species has lost the most combined (direct and indirect) sources of food,
the bear, fox, and fish species have each lost two combined sources of food, and
the rodent and insect species have lost one combined source of food (the least affected).

(e) The long-term effect of pollution is to wipe out the entire ecosystem. How is this reflected in the matrix calculations? Look at the plant life.

53. The email list for this system defines a digraph with 5 vertices.
We let vertices 1–5 correspond to Annie, Bert, Carla, Daria, and Ehaz respectively.

(a) The adjacency matrix for this digraph is $A = \begin{bmatrix} 0&0&1&0&1 \\ 0&0&1&1&0 \\ 0&0&0&0&1 \\ 1&0&1&0&0 \\ 0&1&0&0&0 \end{bmatrix}$.

(b) The component of the adjacency matrix $\left[A^k\right]_{ij}$ has a value of 1 if there is a path of length k starting at i and ending at j.
Thus, if Bert hears a rumor we need to calculate $A'_m = \sum_{k=1}^{m} A^k$ (until each column in row 2 has a nonzero value).
So, we add all paths of increasing length until each person has heard the message.
Calculating the powers of A' gives:

$$A'_1 = A^1 = A = \begin{bmatrix} 0&0&1&0&1 \\ 0&0&1&1&0 \\ 0&0&0&0&1 \\ 1&0&1&0&0 \\ 0&1&0&0&0 \end{bmatrix} \text{ and } A'_2 = A^1 + A^2 = \begin{bmatrix} 0&1&1&0&2 \\ 1&0&2&1&1 \\ 0&1&0&0&1 \\ 1&0&2&0&2 \\ 0&1&1&1&0 \end{bmatrix}.$$

So, it takes two steps until everyone (other than Bert) has heard the rumor.
From A'_2 we see that by this time Carla has heard the message twice, while everyone else has heard it once.

(c) If Annie hears a rumor we need to calculate A'_m:
$A'_m = \sum_{k=1}^{m} A^k$ (until each column in row 1 has a nonzero value).
So, we add all paths of increasing length until each person has heard the message.

Continuing the previous calculation of the powers of A' gives $A'_3 = \begin{bmatrix} 0&2&2&1&2 \\ 1&1&3&1&3 \\ 0&1&1&1&1 \\ 1&2&2&0&3 \\ 1&1&2&1&1 \end{bmatrix}$.

So, it takes three steps until everyone (other than Annie) has heard the rumor.

(d) If A is the adjacency matrix of a digraph with n vertices, we can tell if vertex i is connected to vertex j by a path (of some length) by calculating $A'_n = \sum_{k=1}^{n} A^k$ (recall n is the number of vertices).
If there is an off-diagonal component with a value of zero, then we know that these two vertices are not connected.

54. Let A be the adjacency matrix of a graph G with m vertices.

(a) We wish to prove that for $n \geq 1$, the (i, j) entry of A^n is equal to the number of n-paths between the vertices i and j.
We prove this by induction.
Case $n = 1$: for if A_{ij} is either 1 or 0,
corresponding to one or no paths (of length 1) between vertices i and j.
Assume $\left[A^k\right]_{ij}$ is equal to the number of paths of length k between vertices i and j.
Finally, $\left[A^{k+1}\right]_{ij} = \sum_{l=1}^{m} \left[A^k\right]_{il} [A]_{lj}$. But $[A]_{lj} \in \{0, 1\}$ so that $\left[A^{k+1}\right]_{ij} = \sum_{l=1}^{m} \left[A^k\right]_{il}$ where there is a path from vertex l to vertex j.
Thus, $\left[A^{k+1}\right]_{ij}$ is a sum on all the paths from vertex i to vertex j, as desired.

(b) If G is a digraph the statement above has to be modified to:
If A is the adjacency matrix of digraph G, then the (i, j) entry of A^n is equal to the number of n-paths *from* vertex i *to* vertex j.
The above proof would not have to be altered.

55. The key thing to notice is the following:
$a_{ij}a_{jk} \neq 0 \Leftrightarrow$ there is an edge from v_i to v_k and v_j to v_k at the same time.
So, $(AA^T)_{ij} = \sum a_{ij}a_{jk}$ = number of vertices simultaneously adjacent to v_i and v_j.

56. $U = \{v_1\}$ and $V = \{v_2, v_3, v_4\}$.

57. $U = \{v_1, v_2, v_3\}$ and $V = \{v_4, v_5\}$.

58. This is *not* a bipartite graph.

59. $U = \{v_1, v_2, v_4\}$ and $V = \{v_3, v_5, v_6\}$.

60. (a) Let G be a bipartite graph.

Then its vertices can be subdivided into two sets U and V such that each edge has one endpoint in U and the other endpoint in V.
If U has n elements and we relabel the vertices of U as vertices 1, 2, ..., n and if V has m vertices which we relabel as $n+1, n+2, \ldots, n+m$ then the adjacency matrix can be partitioned as $A = \begin{bmatrix} O & B \\ B^T & O \end{bmatrix}$,
where B is a matrix representing the edges between vertices in U and V.

Conversely, if the adjacency matrix of G can be written as $A = \begin{bmatrix} O & B \\ B^T & O \end{bmatrix}$,
then by corresponding the columns in the top left zero matrix partition with vertices in U and the columns of B with vertices in V, we can write this graph as a bipartite graph.

(b) Let G be a bipartite graph.

Then using (a) we can write the adjacency matrix for this graph as $A = \begin{bmatrix} O & B \\ B^T & O \end{bmatrix}$.

Recall that the (i, j) entry of A^k represents the number of paths between vertex i and j.

But consider paths of odd lengths: $A^{2k+1} = \begin{bmatrix} O & (BB^T)^k B \\ (B^T B)^k B^T & O \end{bmatrix}$.

A circuit of odd length will have a nonzero entry in the (i, i) entry (that is, the path starts and ends at vertex i).
However, we see that A^{2k+1} has all diagonal elements equal to zero, so there are no closed circuits of odd length.

61. We have the following set of vectors in $\mathbb{Z}_2^2$ and their encoded vectors as:

$[0,0] \longrightarrow [0,0,0,0]$, $[0,1] \longrightarrow [0,1,0,1]$, $[1,0] \longrightarrow [1,0,1,0]$, and $[1,1] \longrightarrow [1,1,1,1]$.

This is not an error-correcting code.

For example, if we transmit $[0,0,0,0]$ and an error occurs in the last component, then $[0,0,0,1]$ is received and an error is detected since this is not a legal code vector. However, the receiver cannot correct the error since $[0,0,0,1]$ would also be the result of an error in the second component if $[0,1,0,1]$ had been transmitted.

62. If we encode the binary digits 0 and 1 with a 5-times-repetition code the vectors are:

$0 \longrightarrow \begin{bmatrix} 0, & 0, & 0, & 0, & 0 \end{bmatrix}$ and $1 \longrightarrow \begin{bmatrix} 1, & 1, & 1, & 1, & 1 \end{bmatrix}$.

The original vector could be some permutation of

$\begin{bmatrix} 1, & 1, & 1, & 1, & 0 \end{bmatrix}$, $\begin{bmatrix} 1, & 1, & 1, & 0, & 0 \end{bmatrix}$, $\begin{bmatrix} 1, & 1, & 0, & 0, & 0 \end{bmatrix}$, $\begin{bmatrix} 1, & 0, & 0, & 0, & 0 \end{bmatrix}$, or $\begin{bmatrix} 0, & 0, & 0, & 0, & 0 \end{bmatrix}$.

However, none of the first four types is valid.

So, we are left with the original vector $\begin{bmatrix} 0, & 0, & 0, & 0, & 0 \end{bmatrix} \Rightarrow$

This code can correct double errors.

63. $\mathbf{c} = \begin{bmatrix} 1 \\ 1 \\ 0 \\ 0 \\ 0 \\ 1 \\ 1 \end{bmatrix}$ **64.** $\mathbf{c} = \begin{bmatrix} 0 \\ 1 \\ 1 \\ 1 \\ 0 \\ 0 \\ 1 \end{bmatrix}$ **65.** $\mathbf{c} = \begin{bmatrix} 1 \\ 1 \\ 1 \\ 1 \\ 1 \\ 1 \\ 1 \end{bmatrix}$

66. The standard parity check matrix for the $(7,4)$ Hamming code is $P = \begin{bmatrix} 1\,1\,0\,1\,1\,0\,0 \\ 1\,0\,1\,1\,0\,1\,0 \\ 0\,1\,1\,1\,0\,0\,1 \end{bmatrix}$.

If $\mathbf{c}$ is a Hamming-encoded vector, then $P\mathbf{c} = \mathbf{0}$ implies that the vector is correct.

For this exercise, $\mathbf{c}' = [0,1,0,0,1,0,1]^T$, so $P\mathbf{c}' = \begin{bmatrix} 1\,1\,0\,1\,1\,0\,0 \\ 1\,0\,1\,1\,0\,1\,0 \\ 0\,1\,1\,1\,0\,0\,1 \end{bmatrix} \begin{bmatrix} 0 \\ 1 \\ 0 \\ 0 \\ 1 \\ 0 \\ 1 \end{bmatrix} = \begin{bmatrix} 0 \\ 0 \\ 0 \end{bmatrix}$

which satisfies the parity-check criterion.

Therefore no error has occurred and we have received the correct vector.
Also, the first four components of the Hamming code are the original message vector.
So, in this case, we decode $\mathbf{c}$ to get the original $\mathbf{x} = [0,1,0,0]^T$.

67. We compute $P\mathbf{c}' = \begin{bmatrix} 1\,1\,0\,1\,1\,0\,0 \\ 1\,0\,1\,1\,0\,1\,0 \\ 0\,1\,1\,1\,0\,0\,1 \end{bmatrix} \begin{bmatrix} 1 \\ 1 \\ 0 \\ 0 \\ 1 \\ 1 \\ 0 \end{bmatrix} = \begin{bmatrix} 1 \\ 0 \\ 1 \end{bmatrix}$ which we recognize as column 2 of P.

The error is in the third component of $\mathbf{c}'$. Changing it, we recover the correct code vector $\mathbf{c}$.
Thus, the original message was $\mathbf{x} = [1,0,0,0]^T$.

68. We compute $P\mathbf{c}' = \begin{bmatrix} 1\,1\,0\,1\,1\,0\,0 \\ 1\,0\,1\,1\,0\,1\,0 \\ 0\,1\,1\,1\,0\,0\,1 \end{bmatrix} \begin{bmatrix} 0 \\ 0 \\ 1 \\ 1 \\ 1 \\ 1 \\ 0 \end{bmatrix} = \begin{bmatrix} 0 \\ 1 \\ 0 \end{bmatrix}$ which we recognize as column 6 of P.

Therefore, the error is in the sixth component of $\mathbf{c}'$ and, by changing it,
we recover the correct code vector $\mathbf{c} = [0,0,1,1,1,0,0]^T$.
Note, however, that the original message is encoded in the first four columns.
Thus, the original message was $\mathbf{x} = [0,0,1,1]^T$.

69. The parity check code in Example 1.31 is a code $\mathbb{Z}_2^6 \longrightarrow \mathbb{Z}_2^7$.

(a) For some arbitrary vector $\mathbf{a} = [a_1, a_2, a_3, a_4, a_5, a_6]^T \in \mathbb{Z}_2^6$
the encoded vector has a 7th component a_7 such that $a_1 + a_2 + a_3 + a_4 + a_5 + a_6 + a_7 = 0$.
So, if we take $P = [1, 1, 1, 1, 1, 1, 1]$, then $P\,[\mathbf{a}{,}a_7] = 0$,
where $[\mathbf{a}{,}a_7] = [a_1, a_2, a_3, a_4, a_5, a_6, a_7]^T$.
Thus P is $\begin{bmatrix} 1 & 1 & 1 & 1 & 1 & 1 & 1 \end{bmatrix}$.

(b) A standard generator matrix for this code is $G = \begin{bmatrix} 1 & 0 & 0 & 0 & 0 & 0 \\ 0 & 1 & 0 & 0 & 0 & 0 \\ 0 & 0 & 1 & 0 & 0 & 0 \\ 0 & 0 & 0 & 1 & 0 & 0 \\ 0 & 0 & 0 & 0 & 1 & 0 \\ 0 & 0 & 0 & 0 & 0 & 1 \\ 1 & 1 & 1 & 1 & 1 & 1 \end{bmatrix}$.

This is not an error-correcting code because the columns of P are not distinct.

70. Define a code $\mathbb{Z}_2^2 \longrightarrow \mathbb{Z}_2^5$ using the standard generator matrix $G = \begin{bmatrix} 1 & 0 \\ 0 & 1 \\ 1 & 0 \\ 0 & 1 \\ 1 & 1 \end{bmatrix}$.

(a) The four code words are $\begin{bmatrix} 1 \\ 1 \\ 1 \\ 1 \\ 0 \end{bmatrix}, \begin{bmatrix} 1 \\ 0 \\ 1 \\ 0 \\ 1 \end{bmatrix}, \begin{bmatrix} 0 \\ 1 \\ 0 \\ 1 \\ 1 \end{bmatrix}$, and $\begin{bmatrix} 0 \\ 0 \\ 0 \\ 0 \\ 0 \end{bmatrix}$.

(b) The associated parity check matrix for this code is $P = \begin{bmatrix} 1 & 0 & 1 & 0 & 0 \\ 0 & 1 & 0 & 1 & 0 \\ 1 & 1 & 0 & 0 & 1 \end{bmatrix}$.

This is an error-correcting code since the columns are distinct.

71. Define a code $\mathbb{Z}_2^3 \longrightarrow \mathbb{Z}_2^6$ using the standard generator matrix $G = \begin{bmatrix} 1&0&0 \\ 0&1&0 \\ 0&0&1 \\ 1&0&0 \\ 1&1&0 \\ 1&1&1 \end{bmatrix}$.

(a) The eight code words are $\begin{bmatrix}1\\1\\1\\1\\0\\1\end{bmatrix}, \begin{bmatrix}1\\1\\0\\1\\0\\0\end{bmatrix}, \begin{bmatrix}1\\0\\1\\1\\1\\0\end{bmatrix}, \begin{bmatrix}0\\1\\1\\0\\1\\0\end{bmatrix}, \begin{bmatrix}1\\0\\0\\1\\1\\1\end{bmatrix}, \begin{bmatrix}0\\1\\0\\0\\1\\1\end{bmatrix}, \begin{bmatrix}0\\0\\1\\0\\0\\1\end{bmatrix}$, and $\begin{bmatrix}0\\0\\0\\0\\0\\0\end{bmatrix}$.

(b) The standard parity check matrix for this code is $P = \begin{bmatrix} 1&0&0&1&0&0 \\ 1&1&0&0&1&0 \\ 1&1&1&0&0&1 \end{bmatrix}$.

This is not an error-correcting code since columns three and six are the same.

72. The code in Example 3.69 has the generating matrix $G = \begin{bmatrix}1\\1\\1\end{bmatrix}$.

So, we identify $A = \begin{bmatrix}1\\1\end{bmatrix}$ and $P = \begin{bmatrix}1&1&0\\1&0&1\end{bmatrix}$.

From P we identify that $k = 1$ and $n - k = 2$, so that $n = 3$.

Thus, this is a $(3, 1)$ Hamming code.

73. We wish to construct the standard parity check and generator matrices for the $(15, 11)$ code. Thus, we have $n = 15$, $k = 11$, so that there are $n - k = 4$ parity check equations. The parity check equations give rise to the last four columns of P and by Theorem 3.34, the 15 columns of P need to be nonzero and distinct. Thus, the columns of P consist of $2^4 - 1 = 15$ unique vectors of $\mathbb{Z}_2^{n-k} = \mathbb{Z}_2^4$.

One such candidate for P is $P = \begin{bmatrix} 1&1&1&1&0&1&1&1&0&0&0&1&0&0&0 \\ 1&1&1&0&1&1&0&0&1&1&0&0&1&0&0 \\ 1&1&0&1&1&0&1&0&1&0&1&0&0&1&0 \\ 1&0&1&1&1&0&0&1&0&1&1&0&0&0&1 \end{bmatrix}$.

Then, by Theorem 3.34, we identify $A = \begin{bmatrix} 1&1&1&1&0&1&1&1&0&0&0 \\ 1&1&1&0&1&1&0&0&1&1&0 \\ 1&1&0&1&1&0&1&0&1&0&1 \\ 1&0&1&1&1&0&0&1&0&1&1 \end{bmatrix}$.

So, the generator matrix for the Hamming $(15, 11)$ code is $G = \begin{bmatrix} 1&0&0&0&0&0&0&0&0&0&0 \\ 0&1&0&0&0&0&0&0&0&0&0 \\ 0&0&1&0&0&0&0&0&0&0&0 \\ 0&0&0&1&0&0&0&0&0&0&0 \\ 0&0&0&0&1&0&0&0&0&0&0 \\ 0&0&0&0&0&1&0&0&0&0&0 \\ 0&0&0&0&0&0&1&0&0&0&0 \\ 0&0&0&0&0&0&0&1&0&0&0 \\ 0&0&0&0&0&0&0&0&1&0&0 \\ 0&0&0&0&0&0&0&0&0&1&0 \\ 0&0&0&0&0&0&0&0&0&0&1 \\ 1&1&1&1&0&1&1&1&0&0&0 \\ 1&1&1&0&1&1&0&0&1&1&0 \\ 1&1&0&1&1&0&1&0&1&0&1 \\ 1&0&1&1&1&0&0&1&0&1&1 \end{bmatrix}$.

74. Let $G = \begin{bmatrix} I_k \\ A \end{bmatrix}$ be a standard generator matrix and

$P = \begin{bmatrix} B \; I_{n-k} \end{bmatrix}$ be a standard parity check matrix for the same binary code.
Note, that the dimensions of A are $(n-k) \times k$ and the dimensions of B are $(n-k) \times k$.

Let $\mathbf{x}$ be in $\mathbb{Z}_2^n$, then $PG\mathbf{x} = \begin{bmatrix} B \; I_{n-k} \end{bmatrix} \begin{bmatrix} I_k \\ A \end{bmatrix} \mathbf{x} = (BI_k + I_{n-k}A)\,\mathbf{x} = (B + A)\,\mathbf{x}$.

Now, if $A = B$ then $PG\mathbf{x} = (A + A)\,\mathbf{x} = 2A\mathbf{x}$, but in $\mathbb{Z}_2$, $2 = 0$.
So, $PG\mathbf{x} = 0A\mathbf{x} = 0$. Therefore, if $A = B$ then $PG\mathbf{x} = \mathbf{0}$ for all vectors $\mathbf{x}$ in $\mathbb{Z}_2^k$.

75. Let $G = \begin{bmatrix} I_k \\ A \end{bmatrix}$ be a standard generator matrix and

$P = \begin{bmatrix} A \; I_{n-k} \end{bmatrix}$ be a standard parity check matrix for the same binary code.

Let $\mathbf{x}$ be a message vector in $\mathbb{Z}_2^k$ and let the corresponding code vector be $c = G\mathbf{x}$.

Then $P\mathbf{c} = \mathbf{0}$.

Suppose there has been an error in the ith component, resulting in the vector $\mathbf{c}'$.

It follows that $\mathbf{c}' = \mathbf{c} + \mathbf{e}_i$.

Also, assume that two columns of P are the same.
Then $\mathbf{p}_i = \mathbf{p}_j$. We now compute $P\mathbf{c}' = P\,(\mathbf{c} + \mathbf{e}_i) = P\mathbf{c} + P\mathbf{e}_i = \mathbf{0} + \mathbf{p}_i = \mathbf{p}_i$.

Normally, this would pinpoint the error to be in the ith component.
However, since $\mathbf{p}_i = \mathbf{p}_j$ we also get $P\mathbf{c}' = \mathbf{p}_i = \mathbf{p}_j$,
so that the error could have occurred in the jth column as well.

Chapter 3 Review

1. We will explain and give counter examples to justify our answers below.

(a) ***True***. See Exercise 28 in Section 3.2.
Let A be an $m \times n$ matrix, then A^T is an $n \times m$ matrix.
Since AA^T is $[m \times n][n \times m]$, AA^T is an $m \times m$ matrix.
Since A^TA is $[n \times m][m \times n]$, A^TA is an $n \times n$ matrix.
So not only are AA^T and A^TA defined, they are square matrices.

(b) ***False***. See Theorem 3.27 of Section 3.5.
Theorem 3.27 of Section 3.5 implies $AB = O \Rightarrow B = O$ if and only if A is invertible.
Since vectors are matrices, when $AB = O \Rightarrow B = O$, what do we know about null(A)?
Since $B = O$ means null(A) $= \mathbf{0}$, Theorem 3.27 of Section 3.5 implies A is invertible.
When A^{-1} exists, we have: $AB = O \Rightarrow (A^{-1}A)B = A^{-1}O = O \Rightarrow B = O$.
As a counterexample, consider: $\begin{bmatrix} 1 & 0 \\ 0 & 0 \end{bmatrix}\begin{bmatrix} 0 & 0 \\ 0 & 1 \end{bmatrix} = \begin{bmatrix} 0 & 0 \\ 0 & 0 \end{bmatrix}$.

(c) ***False***. See Exercises 20 through 23 in Section 3.3.
What is the source of the problem? Matrix multiplication does *not* commute.
Instead, we should solve for X using A^{-1}.
If $XA = B$ then $X(AA^{-1}) = BA^{-1}$, so $X = BA^{-1}$ (not always equal to $A^{-1}B$).

$$\text{When } X = \begin{bmatrix} 1 & 0 \\ -2 & 1 \end{bmatrix},\ A = \begin{bmatrix} 1 & 0 \\ 1 & -1 \end{bmatrix},\ \text{and } B = \begin{bmatrix} 1 & 0 \\ -1 & -1 \end{bmatrix},$$

$$\text{then } A^{-1}B = \begin{bmatrix} 1 & 0 \\ 2 & 1 \end{bmatrix} \neq \begin{bmatrix} 1 & 0 \\ -2 & 1 \end{bmatrix} = BA^{-1}.$$

So, $XA = B \Rightarrow X = A^{-1}B$ if and only if $A^{-1}B = BA^{-1}$.
Does $XA = AX$ if and only if $A^{-1}B = BA^{-1}$? Why or why not?

(d) ***True***. See Theorem 3.11 in Section 3.3.
What is the idea? E performs exactly 1 operation and E^{-1} undoes it.
For example, if E performs kR_i, then E^{-1} performs $\frac{1}{k}R_i$.
If E performs $R_i + R_j$, then E^{-1} performs $R_i - R_j$.
Let $E = \begin{bmatrix} 1 & 1 \\ 0 & 1 \end{bmatrix}$, then $E^{-1} = \begin{bmatrix} 1 & -1 \\ 0 & 1 \end{bmatrix}$. Confirm this by showing $EE^{-1} = I$.

(e) ***True***. See Section 3.3. Prove this by considering each type of E separately.
Since E is obtained from I, we need only consider the entries that differ from I.

If E performs kR_i, then $[E]_{ii} = k = [E^T]_{ii}$, so $E^T = E$ performs kR_i.
If E performs $R_i + kR_j$, then $[E]_{ij} = k = [E^T]_{ji}$.
So, E^T is an elementary matrix that performs $R_j + kR_i$.
If E performs $R_i \leftrightarrow R_j$, then $[E]_{ii} = [E]_{jj} = 0$ and $[E]_{ij} = [E]_{ji} = 1$.
So E^T performs $R_i \leftrightarrow R_j$, too since $[E^T]_{ii} = [E]_{ii}$ and $[E^T]_{ji} = [E]_{ij}$.

So, when E performs kR_i or $R_i \leftrightarrow R_j$, then $E^T = E$.
Furthermore, when E performs $R_i \leftrightarrow R_j$, $E^T = E^{-1} = E$.

1. We explain and give counter examples to justify our answers below (continued).

   (f) ***False***. See Section 3.3.

   An elementary matrix can only perform *one* elementary row operation on I.
   But E_2E_1 performs *two* elementary row operations on I.
   When $E_2 = \begin{bmatrix} 0 & 1 \\ 1 & 0 \end{bmatrix}$ and $E_1 = \begin{bmatrix} 2 & 0 \\ 0 & 1 \end{bmatrix}$, then $E_2E_1 = \begin{bmatrix} 0 & 1 \\ 2 & 0 \end{bmatrix}$.

   (g) ***True***. See Theorem 3.21 in Section 3.5.

   We should be able to easily recreate the proof in the text.
   Is the following enough? If $\mathbf{u}$, $\mathbf{v}$ are in null(A), then
   $A(c\mathbf{u} + d\mathbf{v}) = c(A\mathbf{u}) + d(A\mathbf{v}) = \mathbf{0} + \mathbf{0} = \mathbf{0}$ is in null(A).

   (h) ***False***. See the discussions of Exercises 1 through 6 in Section 3.5.

   What is the problem? Every subspace must contain the zero vector.
   What condition do we need to add make the statement true?
   Every plane *that passes through the origin* is a subspace.
   Is the dimension of the plane actually two?
   Yes. A basis for the subspace is the two direction vectors in the parametric form.
   For that form, $\mathbf{x} = \mathbf{p} + s\mathbf{u} + t\mathbf{v}$, and further discussion see Section 1.3.

   (i) ***True***. See the definition of a linear transformation in Section 3.6.

   We verify that $T(\mathbf{x}) = -\mathbf{x}$ satisfies the two necessary conditions.

   $T(\mathbf{u} + \mathbf{v}) = -(\mathbf{u} + \mathbf{v}) = (-\mathbf{u}) + (-\mathbf{v}) = T(\mathbf{u}) + T(\mathbf{v})$

   $T(c\mathbf{u}) = -(c\mathbf{u}) = c(-\mathbf{u}) = cT(\mathbf{u})$

   Theorems 3.30 and 3.31 of Section 3.6 imply an A with $A\mathbf{x} = -\mathbf{x}$ is also enough.
   Note that $A = \begin{bmatrix} T(\mathbf{e}_1) & T(\mathbf{e}_2) \end{bmatrix} = \begin{bmatrix} -1 & 0 \\ 0 & -1 \end{bmatrix}$ does just that.
   Finally, note that $A = -I$. Why does that make sense?

   (j) ***False***. See Theorems 3.30 and 3.31 of Section 3.6.

   The matrix A must be 5×4 not 4×5.

2. As in Section 3.1, we apply the definition of matrix multiplication to see if A^2B is possible: Since A is $[2 \times 2]$ and B is $[2 \times 3]$, A^2B, $[2 \times 2][2 \times 2][2 \times 3]$, is possible. Furthermore, A^2B will be a 2×3 matrix.

$$A^2 = \begin{bmatrix} 1 & 2 \\ 3 & 5 \end{bmatrix}\begin{bmatrix} 1 & 2 \\ 3 & 5 \end{bmatrix} = \begin{bmatrix} 7 & 12 \\ 18 & 31 \end{bmatrix}. \text{ So } A^2B = \begin{bmatrix} 7 & 12 \\ 18 & 31 \end{bmatrix}\begin{bmatrix} 2 & 0 & -1 \\ 3 & -3 & 4 \end{bmatrix} = \begin{bmatrix} 50 & -36 & 41 \\ 129 & -93 & 106 \end{bmatrix}.$$

3. Since B^2 is $[2 \times 3][2 \times 3]$, B^2 is not possible.

Q: If A is not a square matrix, is A^2 ever defined?
A: No, because $[m \times n][m \times n]$ is not possible unless $m = n$.

4. Since $B^TA^{-1}B$ is $[3 \times 2][2 \times 2][2 \times 3]$, $B^TA^{-1}B$ will be 3×3 matrix.

Furthermore, by Theorem 3.8 of Section 3.3, $A^{-1} = \frac{1}{-1}\begin{bmatrix} 5 & -2 \\ -3 & 1 \end{bmatrix} = \begin{bmatrix} -5 & 2 \\ 3 & -1 \end{bmatrix}$.

$$B^TA^{-1} = \begin{bmatrix} 2 & 3 \\ 0 & -3 \\ -1 & 4 \end{bmatrix}\begin{bmatrix} -5 & 2 \\ 3 & -1 \end{bmatrix} = \begin{bmatrix} -1 & 1 \\ -9 & 3 \\ 17 & -6 \end{bmatrix}.$$

$$\text{So } B^TA^{-1}B = \begin{bmatrix} -1 & 1 \\ -9 & 3 \\ 17 & -6 \end{bmatrix}\begin{bmatrix} 2 & 0 & -1 \\ 3 & -3 & 4 \end{bmatrix} = \begin{bmatrix} 1 & -3 & 5 \\ -9 & -9 & 21 \\ 16 & 18 & -41 \end{bmatrix}.$$

5. Since BB^T is $[2 \times 3][3 \times 2]$, BB^T is $[2 \times 2]$.

$$BB^T = \begin{bmatrix} 2 & 0 & -1 \\ 3 & -3 & 4 \end{bmatrix}\begin{bmatrix} 2 & 3 \\ 0 & -3 \\ -1 & 4 \end{bmatrix} = \begin{bmatrix} 5 & 2 \\ 2 & 34 \end{bmatrix}. \text{ So, } (BB^T)^{-1} = \frac{1}{166}\begin{bmatrix} 34 & -2 \\ -2 & 5 \end{bmatrix} = \begin{bmatrix} \frac{17}{83} & -\frac{1}{83} \\ -\frac{1}{83} & \frac{5}{166} \end{bmatrix}.$$

6. Since B^TB is $[3 \times 2][2 \times 3]$, BB^T is $[3 \times 3]$.

$$B^TB = \begin{bmatrix} 2 & 3 \\ 0 & -3 \\ -1 & 4 \end{bmatrix}\begin{bmatrix} 2 & 0 & -1 \\ 3 & -3 & 4 \end{bmatrix} = \begin{bmatrix} 13 & -9 & 10 \\ -9 & 9 & -12 \\ 10 & -12 & 17 \end{bmatrix}.$$

As in Example 3.31 of Section 3.3, we try to row reduce $[\, B^TB \,|\, I \,]$ into $[\, I \,|\, (B^TB)^{-1} \,]$.

$$[\, B^TB \,|\, I \,] = \left[\begin{array}{ccc|ccc} 13 & -9 & 10 & 1 & 0 & 0 \\ -9 & 9 & -12 & 0 & 1 & 0 \\ 10 & -12 & 17 & 0 & 0 & 1 \end{array}\right] \xrightarrow{R_3+\frac{1}{2}R_1+\frac{11}{6}R_2} \left[\begin{array}{ccc|ccc} 13 & -9 & 10 & 1 & 0 & 0 \\ -9 & 9 & -12 & 0 & 1 & 0 \\ 0 & 0 & 0 & \frac{1}{2} & \frac{11}{6} & 1 \end{array}\right]$$

Since the last row is all zeroes, $\text{rank}(BB^T) = 2 < 3$. So, $(BB^T)^{-1}$ does not exist.

7. As in Example 3.11 of Section 3.1, we compute the outer product expansion of AA^T.

$$\mathbf{a}_1A_1^T = \begin{bmatrix} 1 \\ 3 \end{bmatrix}\begin{bmatrix} 1 & 3 \end{bmatrix} = \begin{bmatrix} 1 & 3 \\ 3 & 9 \end{bmatrix}. \text{ and } \mathbf{a}_2A_2^T = \begin{bmatrix} 2 \\ 5 \end{bmatrix}\begin{bmatrix} 2 & 5 \end{bmatrix} = \begin{bmatrix} 4 & 10 \\ 10 & 25 \end{bmatrix}.$$

$$\text{So } AA^T = \begin{bmatrix} 1 & 3 \\ 3 & 9 \end{bmatrix} + \begin{bmatrix} 4 & 10 \\ 10 & 25 \end{bmatrix} = \begin{bmatrix} 5 & 13 \\ 13 & 34 \end{bmatrix}.$$

Note: The outer product expansion of AA^T is always defined.

8. By Theorem 3.9 of Section 3.3, we know that $(A^{-1})^{-1} = A$.

So, by Theorem 3.8 of Section 3.3, $A = (A^{-1})^{-1} = \frac{2}{1}\begin{bmatrix} 4 & 1 \\ \frac{3}{2} & \frac{1}{2} \end{bmatrix} = \begin{bmatrix} 8 & 2 \\ 3 & 1 \end{bmatrix}$.

Check: $AA^{-1} = \begin{bmatrix} 8 & 2 \\ 3 & 1 \end{bmatrix}\begin{bmatrix} \frac{1}{2} & -1 \\ -\frac{3}{2} & 4 \end{bmatrix} = \begin{bmatrix} 1 & 0 \\ 0 & 1 \end{bmatrix}$.

9. As in Exercises 22 and 52 in Section 3.3, since $AX = B$ we have $X = A^{-1}B$.

So, by Theorem 3.8 of Section 3.3, $A = (A^{-1})^{-1} = \frac{2}{1}\begin{bmatrix} 4 & 1 \\ \frac{3}{2} & \frac{1}{2} \end{bmatrix} = \begin{bmatrix} 8 & 2 \\ 3 & 1 \end{bmatrix}$.

As in Example 3.31 of Section 3.3, we try to row reduce $[\,A\,|\,I\,]$ into $[\,I\,|\,A^{-1}\,]$.

$$[\,A\,|\,I\,] = \left[\begin{array}{ccc|ccc} 1 & 0 & -1 & 1 & 0 & 0 \\ 2 & 3 & -1 & 0 & 1 & 0 \\ 0 & 1 & 1 & 0 & 0 & 1 \end{array}\right] \longrightarrow \left[\begin{array}{ccc|ccc} 1 & 0 & 0 & 2 & -\frac{1}{2} & \frac{3}{2} \\ 0 & 1 & 0 & -1 & \frac{1}{2} & -\frac{1}{2} \\ 0 & 0 & 1 & 1 & -\frac{1}{2} & \frac{3}{2} \end{array}\right] = [\,I\,|\,A^{-1}\,]$$

$$\text{So } X = A^{-1}B = \begin{bmatrix} 2 & -\frac{1}{2} & \frac{3}{2} \\ -1 & \frac{1}{2} & -\frac{1}{2} \\ 1 & -\frac{1}{2} & \frac{3}{2} \end{bmatrix}\begin{bmatrix} -1 & -3 \\ 5 & 0 \\ 3 & -2 \end{bmatrix} = \begin{bmatrix} 0 & -9 \\ 2 & 4 \\ 1 & -6 \end{bmatrix}.$$

$$\text{Check: } AX = \begin{bmatrix} 1 & 0 & -1 \\ 2 & 3 & -1 \\ 0 & 1 & 1 \end{bmatrix}\begin{bmatrix} 0 & -9 \\ 2 & 4 \\ 1 & -6 \end{bmatrix} = \begin{bmatrix} -1 & -3 \\ 5 & 0 \\ 3 & -2 \end{bmatrix}.$$

We could also have solved the problem directly as follows:

$$[\,A\,|\,AX\,] = \left[\begin{array}{ccc|cc} 1 & 0 & -1 & -1 & -3 \\ 2 & 3 & -1 & 5 & 0 \\ 0 & 1 & 1 & 3 & -2 \end{array}\right] \longrightarrow \left[\begin{array}{ccc|cc} 1 & 0 & 0 & 0 & -9 \\ 0 & 1 & 0 & 2 & 4 \\ 0 & 0 & 1 & 1 & -6 \end{array}\right] = [\,I\,|\,X\,]$$

10. Since $\det A = 6 - 8 = -2 \neq 0$, A is invertible.
So, by Theorem 3.12 of Section 3.3, A can be written as the product of elementary matrices.

As in Example 3.29 of Section 3.3, we express A as a product of elementary matrices.
To compute A^{-1} and to *factor* A, that is to see how A is created by elementary row operations.

As we row reduce A, we use the elementary operations involved to create E_i.

$$A = \begin{bmatrix} 1 & 2 \\ 4 & 6 \end{bmatrix} \xrightarrow{R_2 - 4R_1} \begin{bmatrix} 1 & 2 \\ 0 & -2 \end{bmatrix} \xrightarrow{-\frac{1}{2}R_2} \begin{bmatrix} 1 & 2 \\ 0 & 1 \end{bmatrix} \xrightarrow{R_1 - 2R_2} \begin{bmatrix} 1 & 0 \\ 0 & 1 \end{bmatrix}$$

We recreate those steps on the identity matrix I to create E_1, E_2, and E_3.
This calculation can be done mentally, but we write it here to demonstrate the process.

Since $I = \begin{bmatrix} 1 & 0 \\ 0 & 1 \end{bmatrix} \xrightarrow{R_2 - 4R_1} \begin{bmatrix} 1 & 0 \\ -4 & 1 \end{bmatrix}$, we get $E_1 = \begin{bmatrix} 1 & 0 \\ -4 & 1 \end{bmatrix}$.

Since $I = \begin{bmatrix} 1 & 0 \\ 0 & 1 \end{bmatrix} \xrightarrow{-\frac{1}{2}R_2} \begin{bmatrix} 1 & 0 \\ 0 & -\frac{1}{2} \end{bmatrix}$, we get $E_2 = \begin{bmatrix} 1 & 0 \\ 0 & -\frac{1}{2} \end{bmatrix}$.

Since $I = \begin{bmatrix} 1 & 0 \\ 0 & 1 \end{bmatrix} \xrightarrow{R_1 - 2R_1} \begin{bmatrix} 1 & -2 \\ 0 & 1 \end{bmatrix}$, we get $E_3 = \begin{bmatrix} 1 & -2 \\ 0 & 1 \end{bmatrix}$.

Since $(E_3E_2E_1)A = I$, by definition $A^{-1} = E_3E_2E_1$.
So, Theorem 3.9 of Section 3.3 implies $A = (A^{-1})^{-1} = (E_3E_2E_1)^{-1} = E_1^{-1}E_2^{-1}E_3^{-1}$.

But E_1^{-1}, E_2^{-1}, and E_3^{-1} can be written down without doing any calculation. How?
Since E_1 was created by $R_2 - 4R_1$, we create E_1^{-1} by $R_2 + 4R_1$.

Likewise, since E_2 was created by $-\frac{1}{2}R_2$, we create E_2^{-1} by $-2R_2$.

Finally, since E_3 was created by $R_1 - 2R_2$, we create E_3^{-1} by $R_1 + 2R_1$.

Therefore, $E_1^{-1} = \begin{bmatrix} 1 & 0 \\ -1 & 1 \end{bmatrix}$, $E_2^{-1} = \begin{bmatrix} 1 & 0 \\ 0 & -2 \end{bmatrix}$, and $E_3^{-1} = \begin{bmatrix} 1 & 2 \\ 0 & 1 \end{bmatrix}$.

We could verify this by performing the indication operation on the identity matrix I.
Or we could simply verify the claim that $A = E_1^{-1}E_2^{-1}E_3^{-1}$ directly.

Check: $A = \begin{bmatrix} 1 & 2 \\ 4 & 6 \end{bmatrix} = \begin{bmatrix} 1 & 0 \\ 4 & 1 \end{bmatrix}\begin{bmatrix} 1 & 0 \\ 0 & -2 \end{bmatrix}\begin{bmatrix} 1 & 2 \\ 0 & 1 \end{bmatrix} = E_1^{-1}E_2^{-1}E_3^{-1}$

This answer is not unique. For example, we could have row reduced A as follows.

$$A = \begin{bmatrix} 1 & 2 \\ 4 & 6 \end{bmatrix} \xrightarrow{-\frac{1}{2}R_2} \begin{bmatrix} 1 & 2 \\ -2 & -3 \end{bmatrix} \xrightarrow{R_2 + 2R_1} \begin{bmatrix} 1 & 2 \\ 0 & 1 \end{bmatrix} \xrightarrow{R_1 - 2R_1} \begin{bmatrix} 1 & 0 \\ 0 & 1 \end{bmatrix}$$

Find the E_i and verify that $A = E_1^{-1}E_2^{-1}E_3^{-1}$ in this case as well.

11. As in Exercise 45 of Section 3.3, to prove X is the inverse of B we show $BX = I$. We claim $(I - A)^{-1} = I + A + A^2$, so we show $(I - A)(I + A + A^2) = I$.

$$(I - A)(I + A + A^2) = I + A + A^2 - A - A^2 - A^3 = I - A^3 \overset{\substack{A^3=O \\ \text{given}}}{=} I$$

12. By the *multiplier* method of Example 3.35 in Section 3.4, we find the LU factorization of A:

$$A = \begin{bmatrix} 1 & 1 & 1 \\ 3 & 1 & 1 \\ 2 & -1 & 1 \end{bmatrix} \xrightarrow[R_3'=R_3-2R_1]{R_2'=R_2-3R_1} \begin{bmatrix} 1 & 1 & 1 \\ 0 & -2 & -2 \\ 0 & -3 & -1 \end{bmatrix} \xrightarrow{R_3''=R_3'-\frac{3}{2}R_2'} \begin{bmatrix} 1 & 1 & 1 \\ 0 & -2 & -2 \\ 0 & 0 & 2 \end{bmatrix} = U \Rightarrow$$

$$\begin{aligned} l_{21} &= 3 \\ l_{31} &= 2 \quad l_{32} = \tfrac{3}{2} \end{aligned} \Rightarrow L = \begin{bmatrix} 1 & 0 & 0 \\ 3 & 1 & 0 \\ 2 & \frac{3}{2} & 1 \end{bmatrix}$$

$$\text{Check: } LU = \begin{bmatrix} 1 & 0 & 0 \\ 3 & 1 & 0 \\ 2 & \frac{3}{2} & 1 \end{bmatrix} \begin{bmatrix} 1 & 1 & 1 \\ 0 & -2 & -2 \\ 0 & 0 & 2 \end{bmatrix} = \begin{bmatrix} 1 & 1 & 1 \\ 3 & 1 & 1 \\ 2 & -1 & 1 \end{bmatrix}$$

13. Find bases for row(A), col(A), and null(A) by Examples 3.45, 3.47, and 3.48 in Section 3.5.

row(A): A basis for row(A) must span the rows of A and be linearly independent.
The linearly independent rows (which are simply the nonzero rows) of U do just that.

$$\text{Since } A = \begin{bmatrix} 2 & -4 & 5 & 8 & 5 \\ 1 & -2 & 2 & 3 & 1 \\ 4 & -8 & 3 & 2 & 6 \end{bmatrix} \longrightarrow \begin{bmatrix} 2 & -4 & 5 & 8 & 5 \\ 0 & 0 & 1 & 2 & 3 \\ 0 & 0 & 0 & 0 & 1 \end{bmatrix} = U$$

we conclude $\{ [\,2 \;\; -4 \;\; 5 \;\; 8 \;\; 5\,], [\,0 \;\; 0 \;\; 1 \;\; 2 \;\; 3\,], [\,0 \;\; 0 \;\; 0 \;\; 0 \;\; 1\,] \}$ is a basis for row(A).

Q: In $A \longrightarrow U$, why is sufficient to reduce A only to row echelon form U?
A: As the remark after Example 3.46 in Section 3.5 explains and demonstrates, the nonzero rows of U are linearly independent. That is all that is required. Why?

We should also note that provided $A \longrightarrow U$ uses no row interchanges,
the corresponding rows in A are also linearly independent.
Whence, it is obvious that those rows form a basis for row(A).

col(A): A basis for col(A) must span the columns of A and be linearly independent.
When $A \longrightarrow U$, the columns with leading entries in U are linearly independent.
As in Example 3.47, the corresponding columns in A are also linearly independent.
Whence, it is obvious that those columns form a basis for col(A).

$$\text{Since } A = \begin{bmatrix} 2 & -4 & 5 & 8 & 5 \\ 1 & -2 & 2 & 3 & 1 \\ 4 & -8 & 3 & 2 & 6 \end{bmatrix} \longrightarrow \begin{bmatrix} 2 & -4 & 5 & 8 & 5 \\ 0 & 0 & 1 & 2 & 3 \\ 0 & 0 & 0 & 0 & 1 \end{bmatrix} = U$$

$$\text{we conclude that } \left\{ \begin{bmatrix} 2 \\ 1 \\ 4 \end{bmatrix}, \begin{bmatrix} 5 \\ 2 \\ 3 \end{bmatrix}, \begin{bmatrix} 5 \\ 1 \\ 6 \end{bmatrix} \right\} \text{ is a basis for col}(A).$$

null(A): Since $A\mathbf{v} = \mathbf{0}$ implies $\mathbf{v}$ is in null(A), we solve $[\,A\,|\,\mathbf{0}\,] \longrightarrow [\,R\,|\,\mathbf{0}\,]$ to find the conditions.

$$\text{Note } A = \begin{bmatrix} 2 & -4 & 5 & 8 & 5 \\ 1 & -2 & 2 & 3 & 1 \\ 4 & -8 & 3 & 2 & 6 \end{bmatrix} \longrightarrow \begin{bmatrix} 2 & -4 & 5 & 8 & 5 \\ 0 & 0 & 1 & 2 & 3 \\ 0 & 0 & 0 & 0 & 1 \end{bmatrix} \longrightarrow \begin{bmatrix} 1 & -2 & 0 & -1 & 0 \\ 0 & 0 & 1 & 2 & 0 \\ 0 & 0 & 0 & 0 & 1 \end{bmatrix} = R$$

$$[\,R\,|\,\mathbf{0}\,] = \left[\begin{array}{ccccc|c} 1 & -2 & 0 & -1 & 0 & 0 \\ 0 & 0 & 1 & 2 & 0 & 0 \\ 0 & 0 & 0 & 0 & 1 & 0 \end{array}\right] \Rightarrow \begin{array}{rcl} x_1 & = & 2s + 1t \\ x_2 & = & s \\ x_3 & = & -2t \\ x_4 & = & t \\ x_5 & = & 0 \end{array}$$

$$\text{Therefore, } \left\{ \begin{bmatrix} 2 \\ 1 \\ 0 \\ 0 \\ 0 \end{bmatrix}, \begin{bmatrix} 1 \\ 0 \\ -2 \\ 1 \\ 0 \end{bmatrix} \right\} \text{ is a basis for null}(A).$$

14. Given $A \longrightarrow B$, we compare row(A) to row(B) and col(A) to col(B) separately below.

row: By Theorem 3.20 in Section 3.5, if $A \longrightarrow B$ then row(A) = row(B).
Observe that the rows in A are linearly combinations the rows in B and vice versa.
See proof of Theorem 3.20 in Section 3.5 for details.

col: Since $A \longrightarrow B$ does *not* use linear combinations of columns, we suspect col$(A) \neq$ col(B).
Furthermore, recall a basis for col(A) is taken directly from A (see Exercise 13 above).
We prove $A \longrightarrow B$ does *not* imply col(A) = col(B) with the following counterexample.

Let $A = \begin{bmatrix} 1 & 0 & 0 \\ 0 & 1 & 0 \\ 1 & 0 & 0 \end{bmatrix}$ and $B = \begin{bmatrix} 1 & 0 & 0 \\ 0 & 1 & 0 \\ 0 & 0 & 0 \end{bmatrix}$. Then $A \overset{R_3 - R_1}{\longrightarrow} B$.

Furthermore, col(A) = span $\left(\begin{bmatrix} 1 \\ 0 \\ 1 \end{bmatrix}, \begin{bmatrix} 0 \\ 1 \\ 0 \end{bmatrix} \right)$ and col(B) = span $\left(\begin{bmatrix} 1 \\ 0 \\ 0 \end{bmatrix}, \begin{bmatrix} 0 \\ 1 \\ 0 \end{bmatrix} \right)$.

Then $\begin{bmatrix} 1 \\ 0 \\ 1 \end{bmatrix}$ is in col(A), but $\begin{bmatrix} 1 \\ 0 \\ 1 \end{bmatrix}$ is *not* in col(B), so col$(A) \neq$ col(B).

Q: If $A \longrightarrow B$ and $\mathbf{A}_i \neq \mathbf{0} \longrightarrow \mathbf{B}_i = \mathbf{0}$, does col$(A)$ = col(B)?
A: No, this is the generalization of the situation in the example above.
What is the problem? The ith component of the basis vectors for col(B) are all zero, but ith component of the basis vectors for col(A) are not all zero.

Consider $A = \begin{bmatrix} 1 & 0 & 0 \\ 1 & 0 & 0 \\ 0 & 0 & 0 \end{bmatrix}$ and $B = \begin{bmatrix} 1 & 0 & 0 \\ 0 & 0 & 0 \\ 0 & 0 & 0 \end{bmatrix}$. Does col$(A)$ = col(B)?

15. We consider when A *is* invertible and when A is *not* invertible separately below.

A^{-1}: By Theorem 3.9 in Section 3.3, if A is invertible then so is A^T.
Therefore, by Theorem 3.27 of Section 3.5, null(A) = null$(A^T) = \mathbf{0}$.
This conclusion is not stated directly but following immediately from c. or g. Why?
From c.: If $A\mathbf{x} = \mathbf{0}$ has only the trivial solution, then $\mathbf{x} = \mathbf{0}$ = null(A).
From g.: If nullity$(A) = 0$ then $\mathbf{0}$ is a basis for null(A) which implies null$(A) = \mathbf{0}$.

no A^{-1}: Since the column space is affected by row reduction, we suspect null$(A) \neq$ null(A^T).
We prove null$(A) \neq$ null(A^T) with the following counterexample.

Let $A = \begin{bmatrix} 1 & 1 & 1 \\ 0 & 0 & 0 \\ 0 & 0 & 0 \end{bmatrix}$ then $A^T = \begin{bmatrix} 1 & 0 & 0 \\ 1 & 0 & 0 \\ 1 & 0 & 0 \end{bmatrix}$. Note $A^T \longrightarrow \begin{bmatrix} 1 & 0 & 0 \\ 0 & 0 & 0 \\ 0 & 0 & 0 \end{bmatrix}$.

Then, null(A) = span $\left(\begin{bmatrix} -1 \\ 1 \\ 0 \end{bmatrix}, \begin{bmatrix} -1 \\ 0 \\ 1 \end{bmatrix} \right)$ and null(A^T) = span $\left(\begin{bmatrix} 0 \\ 1 \\ 0 \end{bmatrix}, \begin{bmatrix} 0 \\ 0 \\ 1 \end{bmatrix} \right)$.

Then $\begin{bmatrix} -1 \\ 1 \\ 0 \end{bmatrix}$ is in null(A), but $\begin{bmatrix} -1 \\ 1 \\ 0 \end{bmatrix}$ is *not* in null(A^T), so null$(A) \neq$ null(A^T).

16. If $\sum \mathbf{A}_i = \mathbf{0}$, then the rows of A are linearly dependent.
Therefore, by Theorem 3.27 in Section 3.5, A is not invertible.

Note this is the contrapositive of a. implies g. which states:
If A is invertible, then the rows of A are linearly independent.

Q: This also gives us a nontrivial solution to $A^T\mathbf{x} = \mathbf{0}$. Which one?
A: Let $\mathbf{x}$ be the vector each of whose components is equal to 1. For example:

$$\begin{bmatrix} 1 & -2 & 1 \\ 0 & 2 & 0 \\ -1 & 0 & -1 \end{bmatrix} \begin{bmatrix} 1 \\ 1 \\ 1 \end{bmatrix} = \begin{bmatrix} 0 \\ 0 \\ 0 \end{bmatrix}$$

Q: Does $A^T\mathbf{x} = \mathbf{0}$ having a nontrivial solution imply A is not invertible?
A: Yes. If $A^T\mathbf{x} = \mathbf{0}$ has a nontrivial solution, then A^T is not invertible.
That implies A is not invertible.

17. We consider A^TA and AA^T separately below.

A^TA: Since A has n linearly independent columns, $\text{rank}(A) = n$.
By Theorem 3.28 of Section 3.5, $\text{rank}(A^TA) = \text{rank}(A) = n$, so A^TA is invertible.

AA^T: Since m may be greater than n, we should suspect that AA^T is not necessarily invertible.
We prove AA^T is not necessarily invertible with the following counterexample.

Let $A = \begin{bmatrix} 1 & 0 \\ 0 & 1 \\ 1 & 0 \end{bmatrix}$ then $A^TA = \begin{bmatrix} 1 & 0 & 1 \\ 0 & 1 & 0 \end{bmatrix} \begin{bmatrix} 1 & 0 \\ 0 & 1 \\ 1 & 0 \end{bmatrix} = \begin{bmatrix} 2 & 0 \\ 0 & 1 \end{bmatrix}$ is obviously invertible.

On the other hand, $AA^T = \begin{bmatrix} 1 & 0 \\ 0 & 1 \\ 1 & 0 \end{bmatrix} \begin{bmatrix} 1 & 0 & 1 \\ 0 & 1 & 0 \end{bmatrix} = \begin{bmatrix} 1 & 0 & 1 \\ 0 & 1 & 0 \\ 1 & 0 & 1 \end{bmatrix}$ reduces to $\begin{bmatrix} 1 & 0 & 1 \\ 0 & 1 & 0 \\ 0 & 0 & 0 \end{bmatrix}$.

Since $\text{rank}(AA^T) = 2 < 3$, AA^T is *not* invertible.

Q: In this case, we have $\text{rank}(AA^T) \leq \text{rank}(A)$. Is that always true?
A: Yes, since we saw in Exercise 57 of Section 3.5 that $\text{rank}(AB) \leq \text{rank}(A)$.

Q: In this case, we saw $\text{rank}(AA^T) = \text{rank}(A^TA)$. Is that always true?
A: Hint: Consider the proof of Theorem 3.28 in Section 3.5.

18. We find the matrix $[T]$ and the formula for the linear transformation T separately below.

matrix: By Theorem 3.31 of Section 3.6, $[T] = \begin{bmatrix} T(\mathbf{e}_1) & T(\mathbf{e}_2) \end{bmatrix}$. Therefore, we have:

$$T\begin{bmatrix} 1 \\ 0 \end{bmatrix} = T\left(\tfrac{1}{2}\begin{bmatrix} 1 \\ 1 \end{bmatrix} + \tfrac{1}{2}\begin{bmatrix} 1 \\ -1 \end{bmatrix}\right) = \tfrac{1}{2}\begin{bmatrix} 2 \\ 3 \end{bmatrix} + \tfrac{1}{2}\begin{bmatrix} 0 \\ 5 \end{bmatrix} = \begin{bmatrix} 1 \\ 4 \end{bmatrix}$$

$$T\begin{bmatrix} 0 \\ 1 \end{bmatrix} = T\left(\tfrac{1}{2}\begin{bmatrix} 1 \\ 1 \end{bmatrix} - \tfrac{1}{2}\begin{bmatrix} 1 \\ -1 \end{bmatrix}\right) = \tfrac{1}{2}\begin{bmatrix} 2 \\ 3 \end{bmatrix} - \tfrac{1}{2}\begin{bmatrix} 0 \\ 5 \end{bmatrix} = \begin{bmatrix} 1 \\ -1 \end{bmatrix}$$

So, the matrix of the linear transformation is $[T] = \begin{bmatrix} T(\mathbf{e}_1) & T(\mathbf{e}_2) \end{bmatrix} = \begin{bmatrix} 1 & 1 \\ 4 & -1 \end{bmatrix}$.

formula: Since $T(\mathbf{v}) = [T]\mathbf{v}$, the formula for T is $\begin{bmatrix} 1 & 1 \\ 4 & -1 \end{bmatrix} \begin{bmatrix} x \\ y \end{bmatrix} = \begin{bmatrix} x + y \\ 4x - y \end{bmatrix}$.

We could have found the formula directly by using $\mathbf{v}$ instead of $\mathbf{e}_1$ and $\mathbf{e}_2$.

19. We find rotation $[R]$, projection $[P]$. Then by Theorem 3.32 of Section 3.6, $[P \circ R] = [P][R]$.

$[R]$: From Example 3.58 in Section 3.6, the rotation $[R]$ has matrix: $[R] = \begin{bmatrix} \cos\theta & -\sin\theta \\ \sin\theta & \cos\theta \end{bmatrix}$.

So, a rotation of 45° is given by $[R] = \begin{bmatrix} \cos 45° & -\sin 45° \\ \sin 45° & \cos 45° \end{bmatrix} = \begin{bmatrix} \frac{\sqrt{2}}{2} & -\frac{\sqrt{2}}{2} \\ \frac{\sqrt{2}}{2} & \frac{\sqrt{2}}{2} \end{bmatrix}$.

$[P]$: From Example 3.59 in Section 3.6, $[P]$ through ℓ with $\mathbf{d}$ is: $[P] = \frac{1}{d_1^2+d_1^2} \begin{bmatrix} d_1^2 & d_1 d_2 \\ d_1 d_2 & d_2^2 \end{bmatrix}$.

So, for $y = -2x$ with $\mathbf{d} = \begin{bmatrix} 1 \\ -2 \end{bmatrix}$ we have: $[P] = \frac{1}{5} \begin{bmatrix} 1 & -2 \\ -2 & 4 \end{bmatrix} = \begin{bmatrix} \frac{1}{5} & -\frac{2}{5} \\ -\frac{2}{5} & \frac{4}{5} \end{bmatrix}$.

So $[P][R] = \begin{bmatrix} \frac{1}{5} & -\frac{2}{5} \\ -\frac{2}{5} & \frac{4}{5} \end{bmatrix} \begin{bmatrix} \frac{\sqrt{2}}{2} & -\frac{\sqrt{2}}{2} \\ \frac{\sqrt{2}}{2} & \frac{\sqrt{2}}{2} \end{bmatrix} = \begin{bmatrix} \frac{\sqrt{2}-2\sqrt{2}}{10} & \frac{-\sqrt{2}-2\sqrt{2}}{10} \\ \frac{-2\sqrt{2}+4\sqrt{2}}{10} & \frac{2\sqrt{2}+4\sqrt{2}}{10} \end{bmatrix} = \begin{bmatrix} \frac{-\sqrt{2}}{10} & \frac{-3\sqrt{2}}{10} \\ \frac{\sqrt{2}}{5} & \frac{3\sqrt{2}}{5} \end{bmatrix}$.

20. To prove $\mathbf{v}$ and $T(\mathbf{v})$ are linearly independent, we need to show the following:
If $c\mathbf{v} + dT(\mathbf{v}) = \mathbf{0}$, then $c = d = 0$.

$$c\mathbf{v} + dT(\mathbf{v}) = \mathbf{0} \overset{\substack{T \text{ is} \\ \text{linear}}}{\Rightarrow} T(c\mathbf{v} + dT(\mathbf{v})) = T(\mathbf{0}) = \mathbf{0} \overset{\substack{T \text{ is} \\ \text{linear}}}{\Rightarrow}$$

$$cT(\mathbf{v}) + dT^2(\mathbf{v}) = \mathbf{0} \overset{\substack{T^2(\mathbf{v})=\mathbf{0} \\ \text{given}}}{\Rightarrow} cT(\mathbf{v}) = \mathbf{0} \overset{\substack{T(\mathbf{v})\neq\mathbf{0} \\ \text{given}}}{\Rightarrow} c = 0$$

So: $$c\mathbf{v} + dT(\mathbf{v}) = \mathbf{0} \overset{\substack{c=0 \\ \text{above}}}{\Rightarrow} dT(\mathbf{v}) = \mathbf{0} \overset{\substack{T(\mathbf{v})\neq\mathbf{0} \\ \text{given}}}{\Rightarrow} d = 0, \quad \text{as we were to show.}$$

Note: The key to this proof was using the givens: $T(\mathbf{v}) \neq \mathbf{0}$ and $T^2(\mathbf{v}) = \mathbf{0}$.

Chapter 4

Eigenvalues and Eigenvectors

4.1 Introduction to Eigenvalues and Eigenvectors

1. If $A\mathbf{x} = \lambda\mathbf{x}$, then $\mathbf{x}$ is an eigenvector of A corresponding to λ.

So, as in Example 4.1, since $A\mathbf{v} = \begin{bmatrix} 0 & 3 \\ 3 & 0 \end{bmatrix}\begin{bmatrix} 1 \\ 1 \end{bmatrix} = \begin{bmatrix} 3 \\ 3 \end{bmatrix} = 3\begin{bmatrix} 1 \\ 1 \end{bmatrix} = 3\mathbf{v}$,

we see $\mathbf{v}$ is an eigenvector of A corresponding to (the eigenvalue) 3.

2. If $A\mathbf{x} = \lambda\mathbf{x}$, then $\mathbf{x}$ is an eigenvector of A corresponding to λ.

So, as in Example 4.1, since $A\mathbf{v} = \begin{bmatrix} 1 & 2 \\ 2 & 1 \end{bmatrix}\begin{bmatrix} 3 \\ -3 \end{bmatrix} = \begin{bmatrix} -3 \\ 3 \end{bmatrix} = -1\begin{bmatrix} 3 \\ -3 \end{bmatrix} = -1\mathbf{v}$,

we see $\mathbf{v}$ is an eigenvector of A corresponding to (the eigenvalue) -1.

3. We compute $A\mathbf{v} = \begin{bmatrix} -1 & 1 \\ 6 & 0 \end{bmatrix}\begin{bmatrix} 1 \\ -2 \end{bmatrix} = \begin{bmatrix} -3 \\ 6 \end{bmatrix} = -3\begin{bmatrix} 1 \\ -2 \end{bmatrix} = -3\mathbf{v}$,

we see $\mathbf{v}$ is an eigenvector of A corresponding to (the eigenvalue) -3.

4. We compute $A\mathbf{v} = \begin{bmatrix} 4 & -2 \\ 5 & -7 \end{bmatrix}\begin{bmatrix} 4 \\ 2 \end{bmatrix} = \begin{bmatrix} 12 \\ 6 \end{bmatrix} = 3\begin{bmatrix} 4 \\ 2 \end{bmatrix} = 3\mathbf{v}$,

so $\mathbf{v}$ is an eigenvector of A corresponding to the eigenvalue 3.

5. We compute $A\mathbf{v} = \begin{bmatrix} 3 & 0 & 0 \\ 0 & 1 & -2 \\ 1 & 0 & 1 \end{bmatrix}\begin{bmatrix} 2 \\ -1 \\ 1 \end{bmatrix} = \begin{bmatrix} 6 \\ -3 \\ 3 \end{bmatrix} = 3\begin{bmatrix} 2 \\ -1 \\ 1 \end{bmatrix} = 3\mathbf{v}$,

so $\mathbf{v}$ is an eigenvector of A corresponding to the eigenvalue 3.

6. We compute $A\mathbf{v} = \begin{bmatrix} 0 & 1 & -1 \\ 1 & 1 & 1 \\ 1 & 2 & 0 \end{bmatrix}\begin{bmatrix} 2 \\ -1 \\ -1 \end{bmatrix} = \begin{bmatrix} 0 \\ 0 \\ 0 \end{bmatrix} = 0\mathbf{v}$,

so $\mathbf{v}$ is an eigenvector of A corresponding to the eigenvalue 0.

7. As in Example 4.2, we show $\text{null}(A - 3I) \neq \mathbf{0}$ then compute $\text{null}(A - 3I)$ to find $\mathbf{x}$.

Since $A\mathbf{x} = 3\mathbf{x}$ implies $(A - 3I)\mathbf{x} = \mathbf{0}$, we have:

$$A - 3I = \begin{bmatrix} 2 & 2 \\ 2 & -1 \end{bmatrix} - \begin{bmatrix} 3 & 0 \\ 0 & 3 \end{bmatrix} = \begin{bmatrix} -1 & 2 \\ 2 & -4 \end{bmatrix}$$

Since the columns of $A - 3I$ are clearly linearly dependent (because $\mathbf{a}_2 = -2\mathbf{a}_1$), the Fundamental Theorem of Invertible Matrices implies that $\text{null}(A - 3I) \neq \mathbf{0}$. That is $A\mathbf{x} = 3\mathbf{x}$ has a nontrivial solution, so 3 is an eigenvalue of A.

Since $A\mathbf{x} = 3\mathbf{x}$ implies $(A - 3I)\mathbf{x} = \mathbf{0}$, we now compute $\text{null}(A - 3I)$.

$$[A - 3I \mid \mathbf{0}] = \left[\begin{array}{cc|c} -1 & 2 & 0 \\ 2 & -4 & 0 \end{array}\right] \longrightarrow \left[\begin{array}{cc|c} 1 & -2 & 0 \\ 0 & 0 & 0 \end{array}\right]$$

So, if $\mathbf{x} = \begin{bmatrix} x_1 \\ x_2 \end{bmatrix}$ is an eigenvector corresponding to the eigenvalue 3, then $x_1 = 2x_2$.

These eigenvectors are of the form $\mathbf{x} = \begin{bmatrix} 2x_2 \\ x_2 \end{bmatrix}$. That is nonzero multiples of $\mathbf{x} = \begin{bmatrix} 2 \\ 1 \end{bmatrix}$.

Q: What does this tell us about $\text{null}(A - 3I)$? What about E_3?

A: The above shows $\text{null}(A - 3I) = \text{span}\left(\begin{bmatrix} 2 \\ 1 \end{bmatrix}\right) = E_3$, the *eigenspace* of 3.

8. As in Example 4.2, we show $\text{null}(A + 2I) \neq \mathbf{0}$ then compute $\text{null}(A + 2I)$ to find $\mathbf{x}$.

Since $A\mathbf{x} = -2\mathbf{x}$ implies $(A + 2I)\mathbf{x} = \mathbf{0}$, we have:

$$A + 2I = \begin{bmatrix} 2 & 2 \\ 2 & -1 \end{bmatrix} + \begin{bmatrix} 2 & 0 \\ 0 & 2 \end{bmatrix} = \begin{bmatrix} 4 & 2 \\ 2 & 1 \end{bmatrix}$$

Since the columns of $A + 2I$ are clearly linearly dependent (because $\mathbf{a}_2 = 2\mathbf{a}_1$), the Fundamental Theorem of Invertible Matrices implies that $\text{null}(A + 2I) \neq \mathbf{0}$. That is $A\mathbf{x} = -2\mathbf{x}$ has a nontrivial solution, so -2 is an eigenvalue of A.

Since $A\mathbf{x} = -2\mathbf{x}$ implies $(A + 2I)\mathbf{x} = \mathbf{0}$, we now compute $\text{null}(A + 2I)$.

$$[A + 2I \mid \mathbf{0}] = \left[\begin{array}{cc|c} 4 & 2 & 0 \\ 2 & 1 & 0 \end{array}\right] \longrightarrow \left[\begin{array}{cc|c} 2 & 1 & 0 \\ 0 & 0 & 0 \end{array}\right]$$

So, if $\mathbf{x} = \begin{bmatrix} x_1 \\ x_2 \end{bmatrix}$ is an eigenvector corresponding to the eigenvalue -2, then $x_2 = -2x_1$.

These eigenvectors are of the form $\mathbf{x} = \begin{bmatrix} x_1 \\ -2x_1 \end{bmatrix}$, nonzero multiples of $\mathbf{x} = \begin{bmatrix} 1 \\ -2 \end{bmatrix}$.

Q: What does this tell us about $\text{null}(A + 2I)$? What about E_{-2}?

A: The above shows $\text{null}(A + 2I) = \text{span}\left(\begin{bmatrix} 1 \\ -2 \end{bmatrix}\right) = E_{-2}$, the *eigenspace* of -2.

9. As in Example 4.2, we show null$(A - I) \neq \mathbf{0}$ then compute null$(A - I)$ to find $\mathbf{x}$.

Since $A\mathbf{x} = \mathbf{x}$ implies $(A - I)\mathbf{x} = \mathbf{0}$, we have:

$$A - I = \begin{bmatrix} 0 & 4 \\ -1 & 5 \end{bmatrix} - \begin{bmatrix} 1 & 0 \\ 0 & 1 \end{bmatrix} = \begin{bmatrix} -1 & 4 \\ -1 & 4 \end{bmatrix}$$

Since the columns of $A - I$ are clearly linearly dependent (because $\mathbf{a}_2 = -4\mathbf{a}_1$), the Fundamental Theorem of Invertible Matrices implies that null$(A - I) \neq \mathbf{0}$. That is $A\mathbf{x} = \mathbf{x}$ has a nontrivial solution, so 1 is an eigenvalue of A.

Since $A\mathbf{x} = \mathbf{x}$ implies $(A - I)\mathbf{x} = \mathbf{0}$, we now compute null$(A - I)$.

$$[A - I \mid \mathbf{0}] = \left[\begin{array}{cc|c} -1 & 4 & 0 \\ -1 & 4 & 0 \end{array}\right] \longrightarrow \left[\begin{array}{cc|c} -1 & 4 & 0 \\ 0 & 0 & 0 \end{array}\right]$$

So, if $\mathbf{x} = \begin{bmatrix} x_1 \\ x_2 \end{bmatrix}$ is an eigenvector corresponding to the eigenvalue 1, then $x_1 = 4x_2$.

These eigenvectors are of the form $\mathbf{x} = \begin{bmatrix} 4x_2 \\ x_2 \end{bmatrix}$. That is nonzero multiples of $\mathbf{x} = \begin{bmatrix} 4 \\ 1 \end{bmatrix}$.

Q: What does this tell us about null$(A - I)$? What about E_1?

A: The above shows null$(A - I) = \text{span}\left(\begin{bmatrix} 4 \\ 1 \end{bmatrix}\right) = E_1$, the *eigenspace* of 1.

10. As in Example 4.2, we show null$(A - 4I) \neq \mathbf{0}$ then compute null$(A - 4I)$ to find $\mathbf{x}$.

Since $A\mathbf{x} = 4\mathbf{x}$ implies $(A - 4I)\mathbf{x} = \mathbf{0}$, we have:

$$A - 4I = \begin{bmatrix} 0 & 4 \\ -1 & 5 \end{bmatrix} - \begin{bmatrix} 4 & 0 \\ 0 & 4 \end{bmatrix} = \begin{bmatrix} 1 & -1 \\ 0 & 0 \end{bmatrix}$$

Since the columns of $A - 4I$ are clearly linearly dependent (because $\mathbf{a}_2 = -\mathbf{a}_1$), the Fundamental Theorem of Invertible Matrices implies that null$(A - 4I) \neq \mathbf{0}$. That is $A\mathbf{x} = 4\mathbf{x}$ has a nontrivial solution, so 4 is an eigenvalue of A.

Since $A\mathbf{x} = 4\mathbf{x}$ implies $(A - 4I)\mathbf{x} = \mathbf{0}$, we now compute null$(A - 4I)$.

$$[A - 4I \mid \mathbf{0}] = \left[\begin{array}{cc|c} 1 & -1 & 0 \\ 0 & 0 & 0 \end{array}\right]$$

So, if $\mathbf{x} = \begin{bmatrix} x_1 \\ x_2 \end{bmatrix}$ is an eigenvector corresponding to the eigenvalue 4, then $x_2 = x_1$.

These eigenvectors are of the form $\mathbf{x} = \begin{bmatrix} x_1 \\ x_1 \end{bmatrix}$. That is nonzero multiples of $\mathbf{x} = \begin{bmatrix} 1 \\ 1 \end{bmatrix}$.

Q: What does this tell us about null$(A - 4I)$? What about E_4?

A: The above shows null$(A - 4I) = \text{span}\left(\begin{bmatrix} 1 \\ 1 \end{bmatrix}\right) = E_4$, the *eigenspace* of 4.

11. As in Example 4.2, we show $\text{null}(A+I) \neq \mathbf{0}$ then compute $\text{null}(A+I)$ to find $\mathbf{x}$.

Since $A\mathbf{x} = -1\mathbf{x}$ implies $(A+I)\mathbf{x} = \mathbf{0}$, we have:

$$A+I = \begin{bmatrix} 1 & 0 & 2 \\ -1 & 1 & 1 \\ 2 & 0 & 1 \end{bmatrix} + \begin{bmatrix} 1 & 0 & 0 \\ 0 & 1 & 0 \\ 0 & 0 & 1 \end{bmatrix} = \begin{bmatrix} 2 & 0 & 2 \\ -1 & 2 & 1 \\ 2 & 0 & 2 \end{bmatrix}$$

Since the columns of $A+I$ are clearly linearly dependent (because $\mathbf{a}_3 = \mathbf{a}_1 + \mathbf{a}_2$), the Fundamental Theorem of Invertible Matrices implies that $\text{null}(A+I) \neq \mathbf{0}$. That is $A\mathbf{x} = -1\mathbf{x}$ has a nontrivial solution, so -1 is an eigenvalue of A.

Since $A\mathbf{x} = -1\mathbf{x}$ implies $(A+I)\mathbf{x} = \mathbf{0}$, we now compute $\text{null}(A+I)$.

$$[A+I \mid \mathbf{0}] = \left[\begin{array}{ccc|c} 2 & 0 & 2 & 0 \\ -1 & 2 & 1 & 0 \\ 2 & 0 & 2 & 0 \end{array}\right] \longrightarrow \begin{bmatrix} 1 & 0 & 1 \\ 0 & 1 & 1 \\ 0 & 0 & 0 \end{bmatrix}$$

If $\mathbf{x} = \begin{bmatrix} x_1 \\ x_2 \\ x_3 \end{bmatrix}$ is an eigenvector corresponding to the eigenvalue -1, then $x_1 = x_2 = -x_3$.

These eigenvectors are of the form $\mathbf{x} = \begin{bmatrix} -x_3 \\ -x_3 \\ x_3 \end{bmatrix}$, nonzero multiples of $\mathbf{x} = \begin{bmatrix} -1 \\ -1 \\ 1 \end{bmatrix}$.

Q: What does this tell us about $\text{null}(A+I)$? What about E_{-1}?

A: The above shows $\text{null}(A+I) = \text{span}\left(\begin{bmatrix} -1 \\ -1 \\ 1 \end{bmatrix}\right) = E_{-1}$, the *eigenspace* of -1.

12. As in Example 4.2, we show $\text{null}(A-2I) \neq \mathbf{0}$ then compute $\text{null}(A-2I)$ to find $\mathbf{x}$.

Since $A\mathbf{x} = 2\mathbf{x}$ implies $(A-2I)\mathbf{x} = \mathbf{0}$, we have:

$$A-2I = \begin{bmatrix} 3 & 1 & -1 \\ 1 & 1 & 1 \\ 4 & 2 & 0 \end{bmatrix} - \begin{bmatrix} 2 & 0 & 0 \\ 0 & 2 & 0 \\ 0 & 0 & 2 \end{bmatrix} = \begin{bmatrix} 1 & 1 & -1 \\ 1 & -1 & 1 \\ 4 & 2 & -2 \end{bmatrix}$$

Since the columns of $A-2I$ are clearly linearly dependent (because $\mathbf{a}_3 = -\mathbf{a}_2$), the Fundamental Theorem of Invertible Matrices implies that $\text{null}(A-2I) \neq \mathbf{0}$. That is $A\mathbf{x} = 2\mathbf{x}$ has a nontrivial solution, so 2 is an eigenvalue of A.

Since $A\mathbf{x} = 2\mathbf{x}$ implies $(A-2I)\mathbf{x} = \mathbf{0}$, we now compute $\text{null}(A-2I)$.

$$[A-2I \mid \mathbf{0}] = \left[\begin{array}{ccc|c} 1 & 1 & -1 & 0 \\ 1 & -1 & 1 & 0 \\ 4 & 2 & -2 & 0 \end{array}\right] \longrightarrow \begin{bmatrix} 1 & 0 & 0 \\ 0 & 1 & -1 \\ 0 & 0 & 0 \end{bmatrix}$$

If $\mathbf{x} = \begin{bmatrix} x_1 \\ x_2 \\ x_3 \end{bmatrix}$ is an eigenvector corresponding to the eigenvalue 2, then $x_1 = 0$, $x_3 = x_2$.

These eigenvectors are of the form $\mathbf{x} = \begin{bmatrix} 0 \\ x_2 \\ x_3 \end{bmatrix}$, nonzero multiples of $\mathbf{x} = \begin{bmatrix} 0 \\ 1 \\ 1 \end{bmatrix}$.

13. Since A reflects F in the y-axis, the only vectors that F maps parallel to themselves are: vectors parallel to the x-axis (i.e. multiples of $\begin{bmatrix} 1 \\ 0 \end{bmatrix}$), which are reversed (eigenvalue -1), and vectors parallel to the y-axis (multiples of $\begin{bmatrix} 0 \\ 1 \end{bmatrix}$), which are sent to themselves (eigenvalue 1).

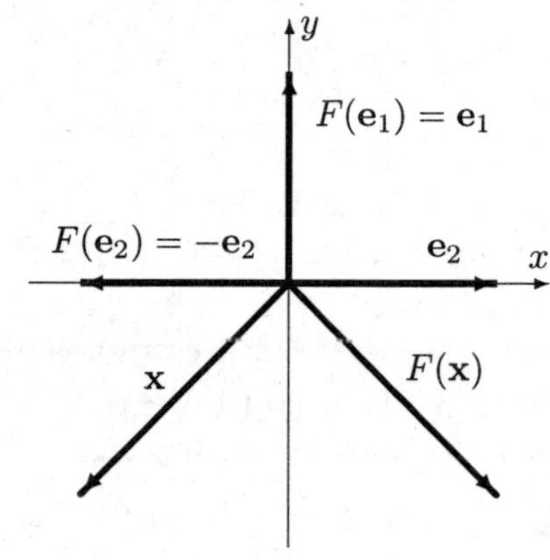

So, $\lambda = \pm 1$ are the eigenvalues of A with eigenspaces

$E_{-1} = \operatorname{span}\left(\begin{bmatrix} 1 \\ 0 \end{bmatrix}\right)$, and

$E_1 = \operatorname{span}\left(\begin{bmatrix} 0 \\ 1 \end{bmatrix}\right)$.

14. The only vectors that F maps parallel to themselves are: vectors perpendicular to this $y = x$, $t\begin{bmatrix} 1 \\ -1 \end{bmatrix}$, which are reversed (eigenvalue -1), and vectors parallel to $y = x$, $t\begin{bmatrix} 1 \\ 1 \end{bmatrix}$, which are sent to themselves (eigenvalue 1).

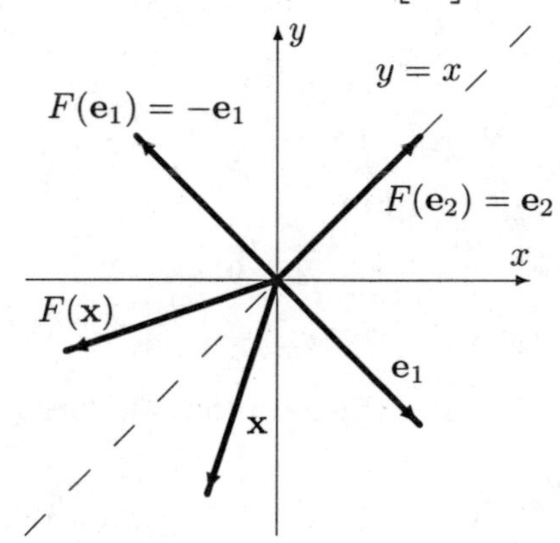

So, $\lambda = \pm 1$ are the eigenvalues of A with eigenspaces

$E_{-1} = \operatorname{span}\left(\begin{bmatrix} 1 \\ -1 \end{bmatrix}\right)$, and

$E_1 = \operatorname{span}\left(\begin{bmatrix} 1 \\ 1 \end{bmatrix}\right)$.

15. From the remarks prior to Example 4.4, we have the following key insight:
$\mathbf{x}$ is an eigenvector of A if and only if A transforms $\mathbf{x}$ to a parallel vector.
Why? Because then $A\mathbf{x}$ and $\mathbf{x}$ are multiples of each other. That is, $A\mathbf{x} = \lambda\mathbf{x}$.
Recall that $E_\lambda = \text{null}(A - \lambda I) = \{\text{eigenvectors of } \lambda\} \cup \{\text{the zero vector, } \mathbf{0}\}$.
We have to add the zero vector because eigenvectors are nonzero by definition.

Since $A\mathbf{x} = \begin{bmatrix} 1 & 0 \\ 0 & 0 \end{bmatrix}\begin{bmatrix} x \\ y \end{bmatrix} = \begin{bmatrix} x \\ 0 \end{bmatrix}$, A is the matrix of projection P onto the x-axis.

Consider vectors $\mathbf{v}$ parallel to the x-axis, parallel to the y-axis, and not parallel to either axis.

x-axis: If $\mathbf{v}$ is parallel to the x-axis, P transforms $\mathbf{v}$ to itself. That is, $P(\mathbf{v}) = \mathbf{v}$.
So, all nonzero vectors parallel to the x-axis are eigenvectors of A corresponding to 1.

y-axis: If $\mathbf{v}$ is parallel to the y-axis, P transforms $\mathbf{v}$ to $\mathbf{0}$. That is, $P(\mathbf{v}) = \mathbf{0}$.
So, all nonzero vectors parallel to the y-axis are eigenvectors of A corresponding to 0.

neither: If $\mathbf{v}$ is not parallel to either axis, P transforms $\mathbf{v}$ to a nonparallel vector.
So, all nonzero vectors not parallel to either axis are not eigenvectors of A.

So $E_1 = \text{span}(x\text{-axis}) = \text{span}\left(\begin{bmatrix} 1 \\ 0 \end{bmatrix}\right)$ and $E_0 = \text{span}(y\text{-axis}) = \text{span}\left(\begin{bmatrix} 0 \\ 1 \end{bmatrix}\right)$.

Q: Given that the x-axis is a line, how might we generalize this result?
A: Hint: Consider vectors parallel, perpendicular and neither to the given line.

16. From the remarks prior to Example 4.4, we have the following key insight:
$\mathbf{x}$ is an eigenvector of A if and only if A transforms $\mathbf{x}$ to a parallel vector.
Why? Because then $A\mathbf{x}$ and $\mathbf{x}$ are multiples of each other. That is, $A\mathbf{x} = \lambda\mathbf{x}$.
Recall that $E_\lambda = \text{null}(A - \lambda I) = \{\text{eigenvectors of } \lambda\} \cup \{\text{the zero vector, } \mathbf{0}\}$.
We have to add the zero vector because eigenvectors are nonzero by definition.

From Example 3.59 in Section 3.6, we have: $P_\ell(\mathbf{x}) = \frac{1}{d_1^2+d_2^2}\begin{bmatrix} d_1^2 & d_1d_2 \\ d_1d_2 & d_2^2 \end{bmatrix}$.

Since $A\mathbf{x} = \frac{1}{25}\begin{bmatrix} 16 & 12 \\ 12 & 9 \end{bmatrix}\begin{bmatrix} x \\ y \end{bmatrix} = \frac{1}{25}\begin{bmatrix} 16x + 12y \\ 12x + 9y \end{bmatrix} = \frac{4}{25}x\begin{bmatrix} 4 \\ 3 \end{bmatrix} + \frac{3}{25}y\begin{bmatrix} 4 \\ 3 \end{bmatrix}$,

A is the matrix of projection P_ℓ onto line ℓ with direction vector $\mathbf{d} = \begin{bmatrix} 4 \\ 3 \end{bmatrix}$.

Consider vectors $\mathbf{v}$ parallel to ℓ, perpendicular to ℓ, and neither to direction vector $\mathbf{d}$.

parallel: If $\mathbf{v}$ is parallel to ℓ, P_ℓ transforms $\mathbf{v}$ to itself. That is, $P_\ell(\mathbf{v}) = \mathbf{v}$.
So, all nonzero vectors parallel to ℓ are eigenvectors of A corresponding to 1.

perpendicular: If $\mathbf{v}$ is perpendicular to ℓ, P_ℓ transforms $\mathbf{v}$ to $\mathbf{0}$. That is, $P_\ell(\mathbf{v}) = \mathbf{0}$.
So, all nonzero vectors perpendicular to ℓ are eigenvectors of A corresponding to 0.

neither: If $\mathbf{v}$ is neither parallel nor perpendicular, P_ℓ transforms $\mathbf{v}$ to a nonparallel vector.
So, all vectors not parallel or perpendicular to ℓ are not eigenvectors of A.

$E_1 = \text{span(parallel to } \mathbf{d}) = \text{span}\left(\begin{bmatrix} 4 \\ 3 \end{bmatrix}\right)$

$E_0 = \text{span(perpendicular to } \mathbf{d}) = \text{span}\left(\begin{bmatrix} 3 \\ -4 \end{bmatrix}\right)$.

17. From the remarks prior to Example 4.4, we have the following key insight:
$\mathbf{x}$ is an eigenvector of A if and only if A transforms $\mathbf{x}$ to a parallel vector.
Why? Because then $A\mathbf{x}$ and $\mathbf{x}$ are multiples of each other. That is, $A\mathbf{x} = \lambda\mathbf{x}$.
Recall that $E_\lambda = \text{null}(A - \lambda I) = \{\text{eigenvectors of } \lambda\} \cup \{\text{the zero vector, } \mathbf{0}\}$.
We have to add the zero vector because eigenvectors are nonzero by definition.

Since $A\mathbf{x} = \begin{bmatrix} 2 & 0 \\ 0 & 3 \end{bmatrix}\begin{bmatrix} x \\ y \end{bmatrix} = \begin{bmatrix} 2x \\ 3y \end{bmatrix}$, A is the matrix of stretching S.

Consider vectors $\mathbf{v}$ parallel to the x-axis, parallel to the y-axis, and not parallel to either axis.

x-axis: If $\mathbf{v}$ is parallel to the x-axis, S transforms $\mathbf{v}$ to twice itself. That is, $S(\mathbf{v}) = 2\mathbf{v}$.
So, all nonzero vectors parallel to the x-axis are eigenvectors of A corresponding to 2.

y-axis: If $\mathbf{v}$ is parallel to the y-axis, S transforms $\mathbf{v}$ to thrice itself. That is, $S(\mathbf{v}) = 3\mathbf{v}$.
So, all nonzero vectors parallel to the y-axis are eigenvectors of A corresponding to 3.

neither: If $\mathbf{v}$ is not parallel to either axis, S transforms $\mathbf{v}$ to a nonparallel vector.
So, all vectors not parallel to either axis are not eigenvectors of A.

So $E_2 = \text{span}(x\text{-axis}) = \text{span}\left(\begin{bmatrix} 1 \\ 0 \end{bmatrix}\right)$ and $E_3 = \text{span}(y\text{-axis}) = \text{span}\left(\begin{bmatrix} 0 \\ 1 \end{bmatrix}\right)$.

Q: Following this exact same process, how might we generalize this result?

A: If $A = \begin{bmatrix} a & 0 \\ 0 & d \end{bmatrix}$, then $E_a = \text{span}(x\text{-axis})$ and $E_d = \text{span}(y\text{-axis})$.

18. From the remarks prior to Example 4.4, we have the following key insight:
$\mathbf{x}$ is an eigenvector of A if and only if A transforms $\mathbf{x}$ to a parallel vector.
Why? Because then $A\mathbf{x}$ and $\mathbf{x}$ are multiples of each other. That is, $A\mathbf{x} = \lambda\mathbf{x}$.
Recall that $E_\lambda = \text{null}(A - \lambda I) = \{\text{eigenvectors of } \lambda\} \cup \{\text{the zero vector, } \mathbf{0}\}$.
We have to add the zero vector because eigenvectors are nonzero by definition.

From Example 3.58 in Section 3.6, we have: $R_\theta = \begin{bmatrix} \cos\theta & \sin\theta \\ \sin\theta & \cos\theta \end{bmatrix}$.

$$A\mathbf{x} = \begin{bmatrix} \cos\theta & -\sin\theta \\ \sin\theta & \cos\theta \end{bmatrix}\begin{bmatrix} x \\ y \end{bmatrix} = \begin{bmatrix} 0 & -1 \\ 1 & 0 \end{bmatrix}\begin{bmatrix} x \\ y \end{bmatrix} = \begin{bmatrix} -y \\ x \end{bmatrix}, \text{ so } A \text{ is the matrix of rotation } R_{90^\circ}.$$

Consider the zero vector $\mathbf{0}$ and all nonzero vectors $\mathbf{v}$.

$\mathbf{v} = \mathbf{0}$: Since a rotation leaves the zero vector fixed, $R_{90^\circ}(\mathbf{0}) = \mathbf{0}$.
However, the zero vector is not an eigenvector of A corresponding to 0.
Why not? Because the zero vector is *zero* and eigenvectors must be nonzero by definition.

$\mathbf{v} \neq \mathbf{0}$: A rotation transforms any nonzero vector to a nonparallel vector.
So, all nonzero vectors are not eigenvectors of A when A is the matrix of any rotation.

Q: Is it still true that $E_0 = \text{span}(\mathbf{0}) = \mathbf{0}$?

A: Yes, because $E_0 = \{\text{eigenvectors of } \lambda\} \cup \{\text{the zero vector, } \mathbf{0}\} = \{\mathbf{0}\}$.

19. From the remarks prior to Example 4.4, we have the following key insight:
$\mathbf{x}$ is an eigenvector of A if and only if A transforms $\mathbf{x}$ to a parallel vector.
So, lines that do *not* bend at the unit circle represent eigenvectors.
The extension beyond the circle tells us if the vector has been stretched.

Since the lines do not bend on the x-axis and the y-axis, we consider
vectors $\mathbf{v}$ parallel to the x-axis, parallel to the y-axis, and not parallel to either axis.

x-axis: On the x-axis, the lines do not bend and extend precisely one unit beyond it. So:
If $\mathbf{v}$ is parallel to the x-axis, S transforms $\mathbf{v}$ to itself. That is, $S(\mathbf{v}) = \mathbf{v}$.
So, all nonzero vectors parallel to the x-axis are eigenvectors of A corresponding to 1.

y-axis: On the y-axis, the lines do not bend and extend precisely 2 units beyond it. So:
If $\mathbf{v}$ is parallel to the y-axis, S transforms $\mathbf{v}$ to twice itself. That is, $S(\mathbf{v}) = 2\mathbf{v}$.
So, all nonzero vectors parallel to the y-axis are eigenvectors of A corresponding to 2.

neither: Off the x-axis and y-axis, the lines *do* bend at the unit circle. So:
If $\mathbf{v}$ is not parallel to either axis, S transforms $\mathbf{v}$ to a nonparallel vector.
So, all vectors not parallel to either axis are not eigenvectors of A.

So $E_1 = \text{span}(x\text{-axis}) = \text{span}\left(\begin{bmatrix}1\\0\end{bmatrix}\right)$ and $E_2 = \text{span}(y\text{-axis}) = \text{span}\left(\begin{bmatrix}0\\1\end{bmatrix}\right)$.

20. From the remarks prior to Example 4.4, we have the following key insight:
$\mathbf{x}$ is an eigenvector of A if and only if A transforms $\mathbf{x}$ to a parallel vector.
So, lines that do *not* bend at the unit circle represent eigenvectors.
The extension beyond the circle tells us if the vector has been stretched.

Since the lines do not bend on the line ℓ $y = x$ with direction vector $\mathbf{d} = \begin{bmatrix}1\\1\end{bmatrix}$,

So, we consider vectors $\mathbf{v}$ parallel to $\mathbf{d}$ and vectors not parallel to $\mathbf{d}$.

parallel: On the line $y = x$, the lines do not bend and extend precisely 3 units beyond it. So:
If $\mathbf{v}$ is parallel to the $\mathbf{d}$, S transforms $\mathbf{v}$ to thrice itself. That is, $S(\mathbf{v}) = 3\mathbf{v}$.
So, all nonzero vectors parallel to $\mathbf{d}$ are eigenvectors of A corresponding to 3.

not parallel: Off the line $y = x$, the lines *do* bend at the unit circle. So:
If $\mathbf{v}$ is not parallel to $\mathbf{d}$, S transforms $\mathbf{v}$ to a nonparallel vector.
So, all vectors not parallel to $\mathbf{d}$ are not eigenvectors of A.

So $E_3 = \text{span}(\mathbf{d}) = \text{span}\left(\begin{bmatrix}1\\1\end{bmatrix}\right)$.

21. From the remarks prior to Example 4.4, we have the following key insight:
$\mathbf{x}$ is an eigenvector of A if and only if A transforms $\mathbf{x}$ to a parallel vector.
So, lines that do *not* bend at the unit circle represent eigenvectors.
The extension beyond the circle tells us if the vector has been stretched.

Since the lines do not bend on the line ℓ $y = -x$ with direction vector $\mathbf{d} = \begin{bmatrix} 1 \\ 1 \end{bmatrix}$,

So, we consider vectors $\mathbf{v}$ parallel to $\mathbf{d}$, perpendicular to $\mathbf{d}$, and neither.

parallel: On the line $y = x$, the lines do not bend and extend precisely 2 units beyond it. So:
If $\mathbf{v}$ is parallel to the $\mathbf{d}$, S transforms $\mathbf{v}$ to thrice itself. That is, $S(\mathbf{v}) = 2\mathbf{v}$.
So, all nonzero vectors parallel to $\mathbf{d}$ are eigenvectors of A corresponding to 2.

perp: On the line $y = -x$, the lines extend precisely 0 units beyond the unit circle. So:
If $\mathbf{v}$ is perpendicular to the $\mathbf{d}$, S transforms $\mathbf{v}$ to $\mathbf{0}$. That is, $S(\mathbf{v}) = 0\mathbf{v}$.
So, all nonzero vectors perpendicular to $\mathbf{d}$ are eigenvectors of A corresponding to 0.

neither: Off the lines $y = x$ and $y = -x$, the lines *do* bend at the unit circle. So:
If $\mathbf{v}$ is not parallel or perpendicular to $\mathbf{d}$, S transforms $\mathbf{v}$ to a nonparallel vector.
So, all vectors not parallel or perpendicular to $\mathbf{d}$ are not eigenvectors of A.

So $E_2 = \text{span}(\mathbf{d}) = \text{span}\left(\begin{bmatrix} 1 \\ 1 \end{bmatrix}\right)$ and $E_0 = \text{span}\left(\begin{bmatrix} 1 \\ -1 \end{bmatrix}\right)$.

22. From the remarks prior to Example 4.4, we have the following key insight:
$\mathbf{x}$ is an eigenvector of A if and only if A transforms $\mathbf{x}$ to a parallel vector.
So, lines that do *not* bend at the unit circle represent eigenvectors.
Here, however, all lines bend at the unit circle, so we conclude there are no eigenvectors.

Q: What types of transformations have no eigenvectors?
A: Rotations. See Exercise 18. Is the graph in Exercise 22 suggestive of a rotation?

23. As in Example 4.5, we find all solutions λ of the equation $\det(A - \lambda I) = 0$.

$$\det(A - \lambda I) = \det\begin{bmatrix} 4-\lambda & -1 \\ 2 & 1-\lambda \end{bmatrix} = \lambda^2 - 5\lambda + 6$$

Since $\lambda^2 - 5\lambda + 6 = (\lambda - 2)(\lambda - 3) = 0$, the solutions are $\lambda = 2$ and $\lambda = 3$.

$$\lambda = 2\text{:}\ A - 2I = \begin{bmatrix} 4-2 & -1 \\ 2 & 1-2 \end{bmatrix} = \begin{bmatrix} 2 & -1 \\ 2 & -1 \end{bmatrix} \longrightarrow \begin{bmatrix} 2 & -1 \\ 0 & 0 \end{bmatrix}$$

So, if $\mathbf{x} = \begin{bmatrix} x_1 \\ x_2 \end{bmatrix}$ is an eigenvector corresponding to the eigenvalue 2, then $x_2 = 2x_1$.

These eigenvectors are of the form $\mathbf{x} = \begin{bmatrix} x_1 \\ 2x_1 \end{bmatrix}$. That is nonzero multiples of $\begin{bmatrix} 1 \\ 2 \end{bmatrix}$.

So, $E_2 = \text{null}(A - 2I) = \text{span}\left(\begin{bmatrix} 1 \\ 2 \end{bmatrix}\right)$.

$$\lambda = 3\text{:}\ A - 3I = \begin{bmatrix} 4-3 & -1 \\ 2 & 1-3 \end{bmatrix} \longrightarrow \begin{bmatrix} 1 & -1 \\ 0 & 0 \end{bmatrix}$$

So, if $\mathbf{x} = \begin{bmatrix} x_1 \\ x_2 \end{bmatrix}$ is an eigenvector corresponding to the eigenvalue 3, then $x_2 = x_1$.

These eigenvectors are of the form $\mathbf{x} = \begin{bmatrix} x_1 \\ x_1 \end{bmatrix}$. That is nonzero multiples of $\begin{bmatrix} 1 \\ 1 \end{bmatrix}$.

So, $E_3 = \text{null}(A - 3I) = \text{span}\left(\begin{bmatrix} 1 \\ 1 \end{bmatrix}\right)$.

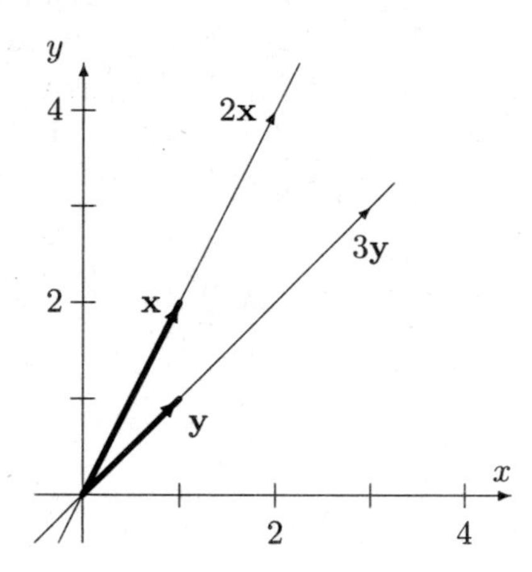

24. As in Example 4.5, we find all solutions λ of the equation $\det(A - \lambda I) = 0$.

$$\det(A - \lambda I) = \det \begin{bmatrix} 2-\lambda & 4 \\ 6 & -\lambda \end{bmatrix} = \lambda^2 - 2\lambda - 24$$

Since $\lambda^2 - 2\lambda - 24 = (\lambda + 4)(\lambda - 6) = 0$, the solutions are $\lambda = -4$ and $\lambda = 6$.

$$\lambda = -4\text{:}\quad A + 4I = \begin{bmatrix} 2+4 & 4 \\ 6 & 0+4 \end{bmatrix} = \begin{bmatrix} 6 & 4 \\ 6 & 4 \end{bmatrix} \longrightarrow \begin{bmatrix} 3 & 2 \\ 0 & 0 \end{bmatrix}$$

So, if $\mathbf{x} = \begin{bmatrix} x_1 \\ x_2 \end{bmatrix}$ is an eigenvector corresponding to -4, then $3x_1 = -2x_2 = 6t$.

These eigenvectors are of the form $\mathbf{x} = \begin{bmatrix} 2t \\ -3t \end{bmatrix}$. That is nonzero multiples of $\begin{bmatrix} 2 \\ -3 \end{bmatrix}$.

So, $E_{-4} - \text{null}(A + 4I) - \text{span}\left(\begin{bmatrix} 2 \\ -3 \end{bmatrix}\right)$.

$$\lambda = 6\text{:}\quad A - 6I = \begin{bmatrix} 2-6 & 4 \\ 6 & 0-6 \end{bmatrix} = \begin{bmatrix} -4 & 4 \\ 6 & -6 \end{bmatrix} \longrightarrow \begin{bmatrix} 1 & -1 \\ 0 & 0 \end{bmatrix}$$

So, if $\mathbf{x} = \begin{bmatrix} x_1 \\ x_2 \end{bmatrix}$ is an eigenvector corresponding to the eigenvalue 6, then $x_2 = x_1$.

These eigenvectors are of the form $\mathbf{x} = \begin{bmatrix} x_1 \\ x_1 \end{bmatrix}$. That is nonzero multiples of $\begin{bmatrix} 1 \\ 1 \end{bmatrix}$.

So, $E_6 = \text{null}(A - 6I) = \text{span}\left(\begin{bmatrix} 1 \\ 1 \end{bmatrix}\right)$.

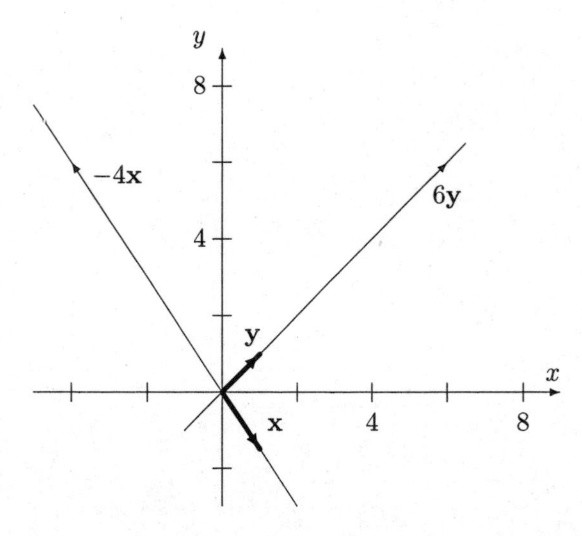

25. As in Example 4.5, we find all solutions λ of the equation $\det(A - \lambda I) = 0$.

$$\det(A - \lambda I) = \det \begin{bmatrix} 2-\lambda & 5 \\ 0 & 2-\lambda \end{bmatrix} = (2-\lambda)^2$$

Since $(2-\lambda)^2 = (\lambda - 2)(\lambda - 2) = 0$, the solution is $\lambda = 2$.

$$\lambda = 2\text{:}\ A - 2I = \begin{bmatrix} 2-2 & 5 \\ 0 & 2-2 \end{bmatrix} = \begin{bmatrix} 0 & 5 \\ 0 & 0 \end{bmatrix} \longrightarrow \begin{bmatrix} 0 & 1 \\ 0 & 0 \end{bmatrix}$$

So, if $\mathbf{x} = \begin{bmatrix} x_1 \\ x_2 \end{bmatrix}$ is an eigenvector corresponding to 2, then $x_1 = t$, $x_2 = 0$.

These eigenvectors are of the form $\mathbf{x} = \begin{bmatrix} t \\ 0 \end{bmatrix}$. That is nonzero multiples of $\begin{bmatrix} 1 \\ 0 \end{bmatrix}$.

So, $E_2 = \text{null}(A - 2I) = \text{span}\left(\begin{bmatrix} 1 \\ 0 \end{bmatrix}\right)$.

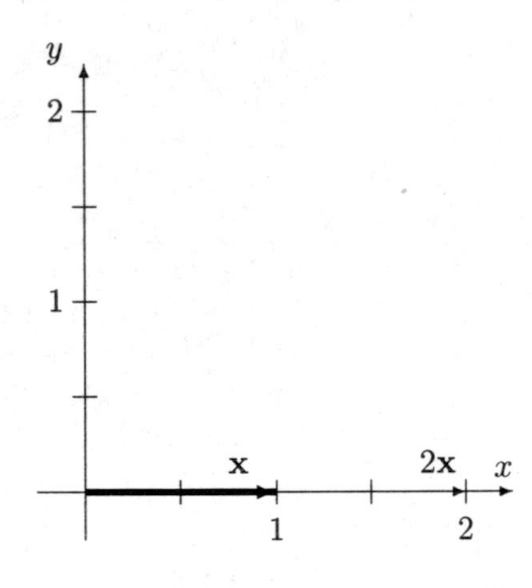

26. As in Example 4.5, we find all solutions λ of the equation $\det(A - \lambda I) = 0$.

$$\det(A - \lambda I) = \det \begin{bmatrix} 1-\lambda & 2 \\ -2 & 3-\lambda \end{bmatrix} = \lambda^2 - 4\lambda + 7$$

Since $\lambda^2 - 4\lambda + 7 = 0$ has no real solution, A has no real eigenvalues.

27. As in Example 4.7, we find all solutions λ of the equation $\det(A - \lambda I) = 0$.

$$\det(A - \lambda I) = \det \begin{bmatrix} 1-\lambda & 1 \\ -1 & 1-\lambda \end{bmatrix} = \lambda^2 - 2\lambda + 2$$

Since $\lambda^2 - 2\lambda + 2 = 0$, the solutions are $\lambda = 1+i, 1-i$.

$$1+i\colon\ A-(1+i)I = \begin{bmatrix} 1-(1+i) & 1 \\ -1 & 1-(1+i) \end{bmatrix} = \begin{bmatrix} -i & 1 \\ -1 & -i \end{bmatrix} \longrightarrow \begin{bmatrix} 1 & i \\ 0 & 0 \end{bmatrix}$$

So, if $\mathbf{x} = \begin{bmatrix} x_1 \\ x_2 \end{bmatrix}$ is an eigenvector corresponding to $1+i$, then $x_1 = -ix_2 = -it$.

These eigenvectors are of the form $\mathbf{x} = \begin{bmatrix} -it \\ t \end{bmatrix}$. That is nonzero multiples of $\begin{bmatrix} -i \\ 1 \end{bmatrix}$.

So, $E_{1+i} = \text{null}(A-(1+i)I) = \text{span}\left(\begin{bmatrix} -i \\ 1 \end{bmatrix}\right)$.

$$1-i\colon\ A-(1-i)I = \begin{bmatrix} 1-(1-i) & 1 \\ -1 & 1-(1-i) \end{bmatrix} = \begin{bmatrix} i & 1 \\ -1 & i \end{bmatrix} \longrightarrow \begin{bmatrix} 1 & -i \\ 0 & 0 \end{bmatrix}$$

So, if $\mathbf{x} = \begin{bmatrix} x_1 \\ x_2 \end{bmatrix}$ is an eigenvector corresponding to $1-i$, then $x_1 = ix_2 = it$.

These eigenvectors are of the form $\mathbf{x} = \begin{bmatrix} it \\ t \end{bmatrix}$. That is nonzero multiples of $\begin{bmatrix} i \\ 1 \end{bmatrix}$.

So, $E_{1-i} = \text{null}(A-(1-i)I) = \text{span}\left(\begin{bmatrix} i \\ 1 \end{bmatrix}\right)$.

28. As in Example 4.7, we find all solutions λ of the equation $\det(A - \lambda I) = 0$.

$$\det(A - \lambda I) = \det \begin{bmatrix} 2-\lambda & -3 \\ 1 & 0-\lambda \end{bmatrix} = \lambda^2 - 2\lambda + 3$$

Since $\lambda^2 - 2\lambda + 3 = 0$, the solutions are $\lambda = 1 + i\sqrt{2}, 1 - i\sqrt{2}$.

$1 + i\sqrt{2}$: $\begin{bmatrix} 2-(1+i\sqrt{2}) & -3 \\ 1 & 0-(1+i\sqrt{2}) \end{bmatrix} = \begin{bmatrix} 1-i\sqrt{2} & -3 \\ 1 & -1-i\sqrt{2} \end{bmatrix} \longrightarrow \begin{bmatrix} 1 & -1-i\sqrt{2} \\ 0 & 0 \end{bmatrix}$

So, if $\mathbf{x} = \begin{bmatrix} x_1 \\ x_2 \end{bmatrix}$ is an eigenvector corresponding to $1 + i\sqrt{2}$, then $x_1 = (1+i\sqrt{2})x_2$.

These eigenvectors are nonzero multiples of $\begin{bmatrix} 1+i\sqrt{2} \\ 1 \end{bmatrix}$.

So, $E_{1+i\sqrt{2}} = \text{null}(A - (1+i\sqrt{2})I) = \text{span}\left(\begin{bmatrix} 1+i\sqrt{2} \\ 1 \end{bmatrix}\right)$.

$1 - i\sqrt{2}$: $\begin{bmatrix} 2-(1-i\sqrt{2}) & -3 \\ 1 & 0-(1-i\sqrt{2}) \end{bmatrix} = \begin{bmatrix} 1+i\sqrt{2} & -3 \\ 1 & -1+i\sqrt{2}) \end{bmatrix} \longrightarrow \begin{bmatrix} 1 & -1+i\sqrt{2} \\ 0 & 0 \end{bmatrix}$

So, if $\mathbf{x} = \begin{bmatrix} x_1 \\ x_2 \end{bmatrix}$ is an eigenvector corresponding to $1 - i\sqrt{2}$, then $x_1 = (1-i\sqrt{2})x_2$.

These eigenvectors are nonzero multiples of $\begin{bmatrix} 1-i\sqrt{2} \\ 1 \end{bmatrix}$.

So, $E_{1-i\sqrt{2}} = \text{null}(A - (1-i\sqrt{2})I) = \text{span}\left(\begin{bmatrix} 1-i\sqrt{2} \\ 1 \end{bmatrix}\right)$.

29. As in Example 4.7, we find all solutions λ of the equation $\det(A - \lambda I) = 0$.

$$\det(A - \lambda I) = \det \begin{bmatrix} 1-\lambda & i \\ i & 1-\lambda \end{bmatrix} = \lambda^2 - 2\lambda + 2$$

Since $\lambda^2 - 2\lambda + 2 = 0$, the solutions are $\lambda = 1 + i, 1 - i$.

$1 + i$: $A - (1+i)I = \begin{bmatrix} 1-(1+i) & i \\ i & 1-(1+i) \end{bmatrix} = \begin{bmatrix} -i & i \\ i & -i \end{bmatrix} \longrightarrow \begin{bmatrix} 1 & -1 \\ 0 & 0 \end{bmatrix}$

So, if $\mathbf{x} = \begin{bmatrix} x_1 \\ x_2 \end{bmatrix}$ is an eigenvector corresponding to $1 + i$, then $x_1 = x_2 = t$.

These eigenvectors are of the form $\mathbf{x} = \begin{bmatrix} t \\ t \end{bmatrix}$. That is nonzero multiples of $\begin{bmatrix} 1 \\ 1 \end{bmatrix}$.

So, $E_{1+i} = \text{null}(A - (1+i)I) = \text{span}\left(\begin{bmatrix} 1 \\ 1 \end{bmatrix}\right)$.

$1 - i$: $A - (1-i)I = \begin{bmatrix} 1-(1-i) & i \\ i & 1-(1-i) \end{bmatrix} = \begin{bmatrix} i & i \\ i & i \end{bmatrix} \longrightarrow \begin{bmatrix} 1 & 1 \\ 0 & 0 \end{bmatrix}$

So, if $\mathbf{x} = \begin{bmatrix} x_1 \\ x_2 \end{bmatrix}$ is an eigenvector corresponding to $1 - i$, then $x_1 = -x_2 = t$.

These eigenvectors are of the form $\mathbf{x} = \begin{bmatrix} t \\ -t \end{bmatrix}$. That is nonzero multiples of $\begin{bmatrix} 1 \\ -1 \end{bmatrix}$.

So, $E_{1-i} = \text{null}(A - (1-i)I) = \text{span}\left(\begin{bmatrix} 1 \\ -1 \end{bmatrix}\right)$.

30. As in Example 4.7, we find all solutions λ of the equation $\det(A - \lambda I) = 0$.

$$\det(A - \lambda I) = \det \begin{bmatrix} 0-\lambda & 1+i \\ 1-i & 1-\lambda \end{bmatrix} = \lambda^2 - \lambda - 2$$

Since $\lambda^2 - \lambda - 2 = 0$, the solutions are $\lambda = -1, 2$.

$\lambda = -1$: $\begin{bmatrix} 0+1 & 1+i \\ 1-i & 1+1 \end{bmatrix} = \begin{bmatrix} 1 & 1+i \\ 1-i & 2 \end{bmatrix} \longrightarrow \begin{bmatrix} 1 & 1+i \\ 0 & 0 \end{bmatrix}$

So, if $\mathbf{x} = \begin{bmatrix} x_1 \\ x_2 \end{bmatrix}$ is an eigenvector corresponding to -1, then $x_1 = -(1+i)x_2$.

These eigenvectors are nonzero multiples of $\begin{bmatrix} -1-i \\ 1 \end{bmatrix}$.

So, $E_{-1} = \text{null}(A + I) = \text{span}\left(\begin{bmatrix} -1-i \\ 1 \end{bmatrix}\right)$.

$\lambda = 2$: $\begin{bmatrix} 0-2 & 1+i \\ 1-i & 1-2 \end{bmatrix} = \begin{bmatrix} -2 & 1+i \\ 1-i & -1 \end{bmatrix} \longrightarrow \begin{bmatrix} -2 & 1+i \\ 0 & 0 \end{bmatrix}$

So, if $\mathbf{x} = \begin{bmatrix} x_1 \\ x_2 \end{bmatrix}$ is an eigenvector corresponding to 2, then $2x_1 = (1+i)x_2 = 2(1+i)t$.

These eigenvectors are nonzero multiples of $\begin{bmatrix} 1+i \\ 2 \end{bmatrix}$.

So, $E_2 = \text{null}(A - 2I) = \text{span}\left(\begin{bmatrix} 1+i \\ 2 \end{bmatrix}\right)$.

31. As in Example 4.6, we find all solutions λ of the equation $\det(A - \lambda I) = 0$.

$$\det(A - \lambda I) = \det \begin{bmatrix} 1-\lambda & 0 \\ 1 & 2-\lambda \end{bmatrix} = \lambda^2 - 3\lambda + 2$$

Since $\lambda^2 - 3\lambda + 2 = (\lambda - 1)(\lambda - 2) = 0$ in $\mathbb{Z}_3$, the eigenvalues of A are $\lambda = 1, 2$.

32. As in Example 4.6, we find all solutions λ of the equation $\det(A - \lambda I) = 0$.

$$\det(A - \lambda I) = \det \begin{bmatrix} 2-\lambda & 1 \\ 1 & 2-\lambda \end{bmatrix} = \lambda^2 - 4\lambda + 3$$

Since $\lambda^2 - 4\lambda + 2 = \lambda^2 + 2\lambda = \lambda(\lambda + 2) = 0$ in $\mathbb{Z}_3$, the eigenvalues of A are $\lambda = 0, 1$.

33. As in Example 4.6, we find all solutions λ of the equation $\det(A - \lambda I) = 0$.

$$\det(A - \lambda I) = \det \begin{bmatrix} 3-\lambda & 1 \\ 4 & 0-\lambda \end{bmatrix} = \lambda^2 - 3\lambda - 4$$

Since $\lambda^2 - 3\lambda - 4 = \lambda^2 + 2\lambda + 1 = (\lambda + 1)^2 = 0$ in $\mathbb{Z}_5$, the eigenvalues of A are $\lambda = 4$.

34. As in Example 4.6, we find all solutions λ of the equation $\det(A - \lambda I) = 0$.

$$\det(A - \lambda I) = \det \begin{bmatrix} 1-\lambda & 4 \\ 4 & 0-\lambda \end{bmatrix} = \lambda^2 - \lambda - 16$$

Since $\lambda^2 - \lambda - 16 = \lambda^2 + 4\lambda + 4 = (\lambda + 2)^2 = 0$ in $\mathbb{Z}_5$, the eigenvalues of A are $\lambda = 3$.

35. (a) To find the eigenvalues of $A = \begin{bmatrix} a & b \\ c & d \end{bmatrix}$, we solve $\det(A - \lambda I) = 0 \Leftrightarrow$

$\det \begin{bmatrix} a-\lambda & b \\ c & d-\lambda \end{bmatrix} = \lambda^2 - (a+d)\lambda + (ad - bc) = \lambda^2 - \operatorname{tr}(A)\lambda + \det A = 0.$

(b) Using the quadratic formula, the solutions to the equation in part (a) are

$$\lambda = \frac{(a+d) \pm \sqrt{(a+d)^2 - 4(ad-bc)}}{2} = \frac{a+d \pm \sqrt{a^2+d^2+2ad-4ad+4bc}}{2}$$
$$= \frac{1}{2}\left(a+d \pm \sqrt{(a-d)^2+4bc}\right).$$

(c) Let $\lambda_1 = \frac{1}{2}(a+d) + \sqrt{(a-d)^2+4bc}))$ and $\lambda_2 = \frac{1}{2}(a+d) - \sqrt{(a-d)^2+4bc}))$.
So, $\lambda_1 + \lambda_2 = \frac{1}{2}(a+d) + \frac{1}{2}(a+d) = a+d = \operatorname{tr}(A)$.
Also, $\lambda_1\lambda_2 = \frac{1}{4}[(a+d)^2 - ((a-d)^2+4bc)] = \frac{1}{4}[4ad - 4bc] = ad - bc = \det A$.

36. (a) If A is to have two distinct real eigenvalues, the discriminant of the equation in Exercise 35(b) must be positive. That is, $(a-d)^2 + 4bc > 0 \Leftrightarrow (a-d)^2 > -4bc$.
If $a = d$, then neither b nor c can equal zero if this inequality is to hold.
If $a \neq d$, whenever b and c have the same sign this inequality holds.

(b) If A is to have one real eigenvalue, the discriminant must be zero.
That is, $(a-d)^2 + 4bc = 0 \Leftrightarrow (a-d)^2 = -4bc$.
If $a = d$, then neither b or c must be zero if this inequality is to hold.

(c) If A is to have no real eigenvalue, the discriminant must be negative.
That is, $(a-d)^2 + 4bc < 0 \Leftrightarrow (a-d)^2 < -4bc$.
If $a = d$, then neither b nor c can equal zero if this inequality is to hold.
If $a \neq d$, then b and c must have opposite signs if this inequality is to hold.

37. As in Example 4.5, we find all solutions λ of the equation $\det(A - \lambda I) = 0$.

$$\det(A - \lambda I) = \det \begin{bmatrix} a-\lambda & b \\ 0 & d-\lambda \end{bmatrix} = \lambda^2 - (a+d)\lambda + ad$$

Since $\lambda^2 - (a+d)\lambda + ad = (\lambda - a)(\lambda - d) = 0$, the solutions are $\lambda = a, d$.

$$\lambda = a: \; A - aI = \begin{bmatrix} a-a & b \\ 0 & a-d \end{bmatrix} = \begin{bmatrix} 0 & b \\ 0 & a-d \end{bmatrix} \longrightarrow \begin{bmatrix} 0 & 1 \\ 0 & 0 \end{bmatrix}$$

If $\mathbf{x} = \begin{bmatrix} x_1 \\ x_2 \end{bmatrix}$ is an eigenvector corresponding to a, $x_1 = t$, $x_2 = 0$.

These eigenvectors are of the form $\mathbf{x} = \begin{bmatrix} t \\ 0 \end{bmatrix}$, nonzero multiples of $\begin{bmatrix} 1 \\ 0 \end{bmatrix}$.

So, $E_a = \text{null}(A - aI) = \text{span}\left(\begin{bmatrix} 1 \\ 0 \end{bmatrix}\right)$.

$$\lambda = d: \; A - dI = \begin{bmatrix} a-d & b \\ 0 & d-d \end{bmatrix} = \begin{bmatrix} a-d & b \\ 0 & 0 \end{bmatrix} \longrightarrow \begin{bmatrix} a-d & b \\ 0 & 0 \end{bmatrix}$$

If $\mathbf{x} = \begin{bmatrix} x_1 \\ x_2 \end{bmatrix}$ is an eigenvector corresponding to d, $(d-a)x_1 = bx_2 = (d-a)bt$.

These eigenvectors are of the form $\mathbf{x} = \begin{bmatrix} bt \\ (d-a)t \end{bmatrix}$, nonzero multiples of $\begin{bmatrix} b \\ d-a \end{bmatrix}$.

So, $E_b = \text{null}(A - bI) = \text{span}\left(\begin{bmatrix} b \\ d-a \end{bmatrix}\right)$.

38. As in Example 4.5, we find all solutions λ of the equation $\det(A - \lambda I) = 0$.

$$\det(A - \lambda I) = \det \begin{bmatrix} a-\lambda & b \\ -b & a-\lambda \end{bmatrix} = \lambda^2 - 2a\lambda + a^2 + b^2$$

Since $\lambda^2 - 2a\lambda + a^2 + b^2 = 0$ implies $\lambda = \frac{2a \pm \sqrt{4a^2 - 4(a^2+b^2)}}{2}$, we have $\lambda = a + bi, a - bi$.

$$a + bi: \; A - (a+bi)I = \begin{bmatrix} a-(a+bi) & b \\ -b & a-(a+bi) \end{bmatrix} = \begin{bmatrix} -bi & b \\ -b & -bi \end{bmatrix} \longrightarrow \begin{bmatrix} -i & 1 \\ 0 & 0 \end{bmatrix}$$

If $\mathbf{x} = \begin{bmatrix} x_1 \\ x_2 \end{bmatrix}$ is an eigenvector corresponding to a, $x_2 = ix_1$.

These eigenvectors are of the form $\mathbf{x} = \begin{bmatrix} t \\ it \end{bmatrix}$, nonzero multiples of $\begin{bmatrix} 1 \\ i \end{bmatrix}$.

So, $E_{a+bi} = \text{null}(A - (a+bi)I) = \text{span}\left(\begin{bmatrix} 1 \\ i \end{bmatrix}\right)$.

$$a - bi: \; A - (a-bi)I = \begin{bmatrix} a-(a-bi) & b \\ -b & a-(a-bi) \end{bmatrix} = \begin{bmatrix} bi & b \\ -b & bi \end{bmatrix} \longrightarrow \begin{bmatrix} i & 1 \\ 0 & 0 \end{bmatrix}$$

If $\mathbf{x} = \begin{bmatrix} x_1 \\ x_2 \end{bmatrix}$ is an eigenvector corresponding to a, $x_2 = -ix_1$.

These eigenvectors are of the form $\mathbf{x} = \begin{bmatrix} t \\ -it \end{bmatrix}$, nonzero multiples of $\begin{bmatrix} 1 \\ -i \end{bmatrix}$.

So, $E_{a-bi} = \text{null}(A - (a-bi)I) = \text{span}\left(\begin{bmatrix} 1 \\ -i \end{bmatrix}\right)$.

4.2 Determinants

1. As in Example 4.8, we compute det A by expanding along the first *row* and the first *column*.

$$\textit{row}: \quad \begin{vmatrix} 1 & 0 & 3 \\ 5 & 1 & 1 \\ 0 & 1 & 2 \end{vmatrix} = 1\begin{vmatrix} 1 & 1 \\ 1 & 2 \end{vmatrix} - 0\begin{vmatrix} 5 & 1 \\ 0 & 2 \end{vmatrix} + 3\begin{vmatrix} 5 & 1 \\ 0 & 1 \end{vmatrix} = 1(1) + 3(5) = 16$$

$$\textit{column}: \quad \begin{vmatrix} 1 & 0 & 3 \\ 5 & 1 & 1 \\ 0 & 1 & 2 \end{vmatrix} = 1\begin{vmatrix} 1 & 1 \\ 1 & 2 \end{vmatrix} - 5\begin{vmatrix} 0 & 3 \\ 1 & 2 \end{vmatrix} + 0\begin{vmatrix} 0 & 3 \\ 1 & 1 \end{vmatrix} = 1(1) - 5(-3) = 16$$

2. As in Example 4.8, we compute det A by expanding along the first *row* and the first *column*.

$$\textit{row}: \quad \begin{vmatrix} 0 & 1 & -1 \\ 2 & 3 & -2 \\ -1 & 3 & 0 \end{vmatrix} = -1\begin{vmatrix} 2 & -2 \\ -1 & 0 \end{vmatrix} - 1\begin{vmatrix} 2 & 3 \\ -1 & 3 \end{vmatrix} = -1(-2) - 1(9) = -7$$

$$\textit{column}: \quad \begin{vmatrix} 0 & 1 & -1 \\ 2 & 3 & -2 \\ -1 & 3 & 0 \end{vmatrix} = -2\begin{vmatrix} 1 & -1 \\ 3 & 0 \end{vmatrix} - 1\begin{vmatrix} 1 & -1 \\ 3 & -2 \end{vmatrix} = -2(3) - 1(1) = -7$$

3. As in Example 4.8, we compute det A by expanding along the first *row* and the first *column*.

$$\textit{row}: \quad \begin{vmatrix} 1 & -1 & 0 \\ -1 & 0 & 1 \\ 0 & 1 & -1 \end{vmatrix} = 1\begin{vmatrix} 0 & 1 \\ 1 & -1 \end{vmatrix} - (-1)\begin{vmatrix} -1 & 1 \\ 0 & -1 \end{vmatrix} = 1(-1) + 1(1) = 0$$

$$\textit{column}: \quad \begin{vmatrix} 1 & -1 & 0 \\ -1 & 0 & 1 \\ 0 & 1 & -1 \end{vmatrix} = 1\begin{vmatrix} 0 & 1 \\ 1 & -1 \end{vmatrix} - (-1)\begin{vmatrix} -1 & 0 \\ 1 & -1 \end{vmatrix} = 1(-1) + 1(1) = 0$$

4. As in Example 4.8, we compute det A by expanding along the first *row* and the first *column*.

$$\textit{row}: \quad \begin{vmatrix} 1 & 1 & 0 \\ 1 & 0 & 1 \\ 0 & 1 & 1 \end{vmatrix} = 1\begin{vmatrix} 0 & 1 \\ 1 & 1 \end{vmatrix} - 1\begin{vmatrix} 1 & 1 \\ 0 & 1 \end{vmatrix} = 1(-1) - 1(1) = -2$$

$$\textit{column}: \quad \begin{vmatrix} 1 & 1 & 0 \\ 1 & 0 & 1 \\ 0 & 1 & 1 \end{vmatrix} = 1\begin{vmatrix} 0 & 1 \\ 1 & 1 \end{vmatrix} - 1\begin{vmatrix} 1 & 0 \\ 1 & 1 \end{vmatrix} = 1(-1) - 1(1) = -2$$

5. As in Example 4.8, we compute det A by expanding along the first *row* and the first *column*.

$$\textit{row}: \quad \begin{vmatrix} 1 & 2 & 3 \\ 2 & 3 & 1 \\ 3 & 1 & 2 \end{vmatrix} = 1\begin{vmatrix} 3 & 1 \\ 1 & 2 \end{vmatrix} - 2\begin{vmatrix} 2 & 1 \\ 3 & 2 \end{vmatrix} + 3\begin{vmatrix} 2 & 3 \\ 3 & 1 \end{vmatrix} = 1(5) - 2(1) + 3(-7) = -18$$

$$\textit{column}: \quad \begin{vmatrix} 1 & 2 & 3 \\ 2 & 3 & 1 \\ 3 & 1 & 2 \end{vmatrix} = 1\begin{vmatrix} 3 & 1 \\ 1 & 2 \end{vmatrix} - 2\begin{vmatrix} 2 & 3 \\ 1 & 2 \end{vmatrix} + 3\begin{vmatrix} 2 & 3 \\ 3 & 1 \end{vmatrix} = 1(5) - 2(1) + 3(-7) = -18$$

6. As in Example 4.8, we compute $\det A$ by expanding along the first *row* and the first *column*.

$$\textit{row}: \quad \begin{vmatrix} 1 & 2 & 3 \\ 4 & 5 & 6 \\ 7 & 8 & 9 \end{vmatrix} = 1\begin{vmatrix} 5 & 6 \\ 8 & 9 \end{vmatrix} - 2\begin{vmatrix} 4 & 6 \\ 7 & 9 \end{vmatrix} + 3\begin{vmatrix} 4 & 5 \\ 7 & 8 \end{vmatrix} = 3(-1) - 2(-6) + 3(-3) = 0$$

$$\textit{column}: \quad \begin{vmatrix} 1 & 2 & 3 \\ 4 & 5 & 6 \\ 7 & 8 & 9 \end{vmatrix} = 1\begin{vmatrix} 5 & 6 \\ 8 & 9 \end{vmatrix} - 4\begin{vmatrix} 2 & 3 \\ 8 & 9 \end{vmatrix} + 7\begin{vmatrix} 2 & 3 \\ 5 & 6 \end{vmatrix} = 3(-1) - 4(-6) + 7(-3) = 0$$

7. As in Example 4.10, we choose a row or column that minimizes the number of calculations. Since $\mathbf{A}_3 = \begin{bmatrix} 3 & 0 & 0 \end{bmatrix}$ contains two zeroes, $\det A = a_{31}C_{31} = a_{31}(-1)^{3+1}\det A_{31} = 3\det A_{31}$.

$$\textit{row } 3: \quad \begin{vmatrix} 5 & 2 & 2 \\ -1 & 1 & 2 \\ 3 & 0 & 0 \end{vmatrix} = 3\begin{vmatrix} 2 & 2 \\ 1 & 2 \end{vmatrix} = 3(2) = 6$$

Q: What should we look for when choosing a row or column to expand along?
A: A row or column with the maximum number of zeroes. Why?
The maximum number of zeroes minimizes the number of cofactors we have to compute.

8. As in Example 4.10, we choose a row or column that minimizes the number of calculations. Since $\mathbf{A}_2 = \begin{bmatrix} 2 & 0 & 1 \end{bmatrix}$ contains one zero, $\det A = -2\det A_{21} - 1\det A_{23}$.

$$\textit{row } 2: \quad \begin{vmatrix} 1 & 1 & -1 \\ 2 & 0 & 1 \\ 3 & -2 & 1 \end{vmatrix} = -2\begin{vmatrix} 1 & -1 \\ -2 & 1 \end{vmatrix} - 1\begin{vmatrix} 1 & 1 \\ 3 & -2 \end{vmatrix} = -2(-1) - 1(-5) = 7$$

Q: Why is the coefficient of $\det A_{21}$ equal to -2 instead of 2?
A: Because the cofactor $C_{21} = (-1)^{2+1}\det A_{21} = -\det A_{21}$.

Q: Why is the coefficient of $\det A_{23}$ equal to -1 instead of 1?

9. As in Example 4.10, we choose a row or column that minimizes the number of calculations. Since $\mathbf{A}_3 = \begin{bmatrix} 1 & -1 & 0 \end{bmatrix}$ contains one zero, $\det A = 1 \det A_{31} - (-1) \det A_{32}$.

$$\textit{row } 3: \quad \begin{vmatrix} -4 & 1 & 3 \\ 2 & -2 & 4 \\ 1 & -1 & 0 \end{vmatrix} = 1 \begin{vmatrix} 1 & 3 \\ -2 & 4 \end{vmatrix} - (-1) \begin{vmatrix} -4 & 3 \\ 2 & 4 \end{vmatrix} = 1(10) + 1(-22) = -12$$

Q: Since both row 3 and column 3 contain one zero, what makes row 3 a better choice?
A: Because the nonzero entries of row 3 are 1 and -1, but in column 3 they are 3 and 4.

10. As in Example 4.10, we choose a row or column that minimizes the number of calculations. Since $\mathbf{a}_1 = \begin{bmatrix} \cos\theta \\ 0 \\ 0 \end{bmatrix}$ contains two zeroes, $\det A = \cos\theta \det A_{11}$.

$$\textit{col } 1: \quad \begin{vmatrix} \cos\theta & \sin\theta & \tan\theta \\ 0 & \cos\theta & -\sin\theta \\ 0 & \sin\theta & \cos\theta \end{vmatrix} = \cos\theta \begin{vmatrix} \cos\theta & -\sin\theta \\ \sin\theta & \cos\theta \end{vmatrix} = \cos\theta(\cos^2\theta + \sin^2\theta) = \cos\theta$$

11. As in Example 4.10, we choose a row or column that minimizes the number of calculations. Since $\mathbf{A}_1 = \begin{bmatrix} a & b & 0 \end{bmatrix}$ contains one zero, $\det A = a \det A_{11} - b \det A_{12}$.

$$\textit{row 1:} \quad \begin{vmatrix} a & b & 0 \\ 0 & a & b \\ a & 0 & b \end{vmatrix} = a \begin{vmatrix} a & b \\ 0 & b \end{vmatrix} - b \begin{vmatrix} 0 & b \\ a & b \end{vmatrix} = a(ab) - b(-ab) = ab(a+b)$$

Q: Since $\det A = ab(a+b)$, if $a = -b$ what do we know about A?
A: If $a = -b$, then A is *not* invertible (because $\det A = 0$).

Q: Since $\det A = ab(a+b)$, if $a = 0$ what do we know about A?

Q: What are the conditions on a and b that will guarantee A is invertible?
A: Both a and b are nonzero and $a \neq -b$. Why is this enough?

12. As in Example 4.10, we choose a row or column that minimizes the number of calculations. Since $\mathbf{A}_1 = \begin{bmatrix} 0 & a & 0 \end{bmatrix}$ contains one zero, $\det A = -a \det A_{12}$.

$$\textit{row 1:} \quad \begin{vmatrix} 0 & a & 0 \\ b & c & d \\ 0 & e & 0 \end{vmatrix} = -a \begin{vmatrix} b & d \\ 0 & 0 \end{vmatrix} = -a(0) = 0$$

Q: Since $\det A = 0$, what do we know about A?
A: We know that A is *not* invertible.

Q: Why is it obvious that A is *not* invertible even *before* we compute $\det A$?
A: Since row 3 is a multiple of row 1, these rows are linearly dependent. Why is this enough?

Q: Is it equally obvious that $\det A = 0$ even *before* computing it?
A: Yes, because once we know A is *not* invertible, we know $\det A$ must equal zero.

13. Since $\mathbf{A}_3 = \begin{bmatrix} 0 & 1 & 0 & 0 \end{bmatrix}$ contains three zeroes, $\det A = -1 \det A_{32}$.
If we let $B = A_{32}$ and expand along $\mathbf{B}_1$, $\det A = -\det B = -(1 \det B_{11} + 3 \det B_{13})$.

$$\textit{row 3:} \quad \begin{vmatrix} 1 & -1 & 0 & 3 \\ 2 & 5 & 2 & 6 \\ 0 & 1 & 0 & 0 \\ 1 & 4 & 2 & 1 \end{vmatrix} = - \begin{vmatrix} 1 & 0 & 3 \\ 2 & 2 & 6 \\ 1 & 2 & 1 \end{vmatrix} = - \left(\begin{vmatrix} 2 & 6 \\ 2 & 1 \end{vmatrix} + 3 \begin{vmatrix} 2 & 2 \\ 1 & 2 \end{vmatrix} \right) = -(-10 + 3 \cdot 2) = 4.$$

Q: What does $\det A = 4$ tell us about the rows of A?
A: The rows of A are linearly independent (because $\det A \neq 0$. Why is that enough?).

Q: What does $\det A = 4$ tell us about the columns of A?
Q: What does $\det B \neq 0$ tell us about the rows and columns of B?

14. Since $\mathbf{a}_2 = \begin{bmatrix} 0 \\ 0 \\ -1 \\ 0 \end{bmatrix}$ contains three zeroes, $\det A = -1(-1)\det A_{32} = \det A_{32}$.

If we let $B = A_{32}$ and expand along $\mathbf{B}_1$, $\det A = \det B = 2\det B_{11} - 3\det B_{12} + (-1)\det B_{13}$.

$$\textit{col } 2\text{:} \quad \begin{vmatrix} 2 & 0 & 3 & -1 \\ 1 & 0 & 2 & 2 \\ 0 & -1 & 1 & 4 \\ 2 & 0 & 1 & -3 \end{vmatrix} = -(-1)\begin{vmatrix} 2 & 3 & -1 \\ 1 & 2 & 2 \\ 2 & 1 & -3 \end{vmatrix} = 2\begin{vmatrix} 2 & 2 \\ 1 & -3 \end{vmatrix} - 3\begin{vmatrix} 1 & 2 \\ 2 & -3 \end{vmatrix} - \begin{vmatrix} 1 & 2 \\ 2 & 1 \end{vmatrix} = 8.$$

Q: What does $\det A = 8$ tell us about the rows and columns of A and B?

15. Since $\mathbf{A}_1 = \begin{bmatrix} 0 & 0 & 0 & a \end{bmatrix}$ contains three zeroes, $\det A = -a\det A_{14}$.
If we let $B = A_{14}$ and expand along $\mathbf{B}_1$, $\det A = -a\det B = -a(b\det D_{13})$.

$$\textit{row } 1\text{:} \quad \begin{vmatrix} 0 & 0 & 0 & a \\ 0 & 0 & b & c \\ 0 & d & e & f \\ g & h & i & j \end{vmatrix} = -a\begin{vmatrix} 0 & 0 & b \\ 0 & d & e \\ g & h & i \end{vmatrix} = -ab\begin{vmatrix} 0 & d \\ g & h \end{vmatrix} = abdg.$$

Q: Since $\det A = abdg$, if a, b, d, g are nonzero then A is invertible. Why is this obvious?
A: Since if a, b, d, g are nonzero then A is *anti*-lower triangular. Why is that enough?

16. Following the method of Example 4.9, we have:

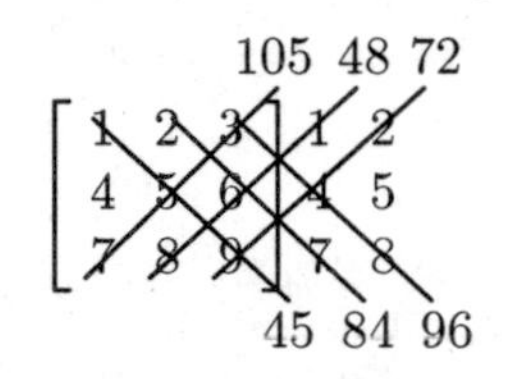

Adding the three products at the bottom and subtracting the three products at the top gives $\det A = 45 + 84 + 96 - 105 - 48 - 72 = 0$.

17. Following the method of Example 4.9, we have:

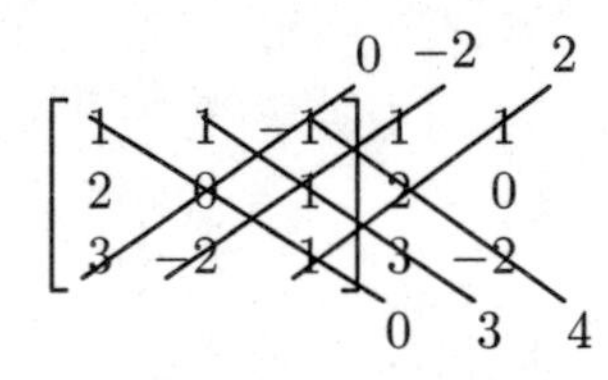

Adding the three products at the bottom and subtracting the three products at the top gives $\det A = 0 + 3 + 4 - 0 - (-2) - 2 = 7$.

18. Following the method of Example 4.9, we have:

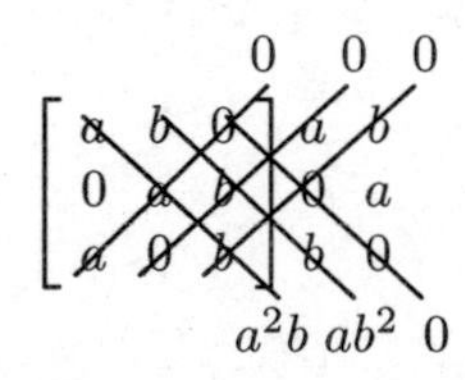

Adding the three products at the bottom and subtracting the three products at the top gives $\det A = a^2b + ab^2 + 0 - 0 = ab(a + b)$.

19. Suppose that $A = \begin{bmatrix} a_{11} & a_{12} & a_{13} \\ a_{21} & a_{22} & a_{23} \\ a_{31} & a_{32} & a_{33} \end{bmatrix}$. Following the method of Example 4.9, we have:

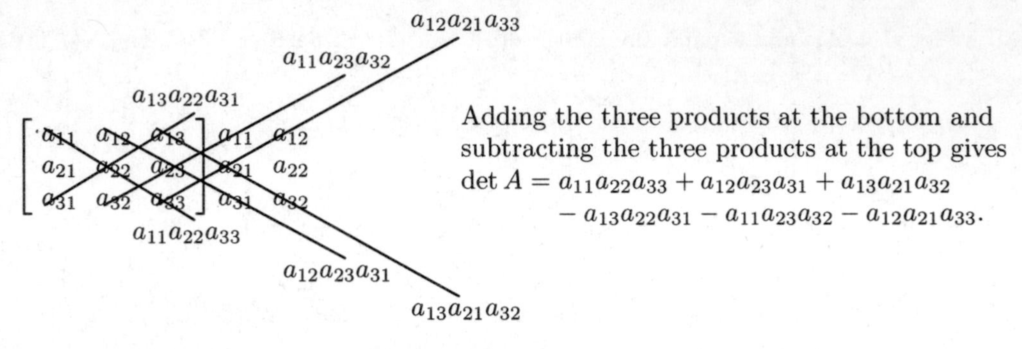

Adding the three products at the bottom and subtracting the three products at the top gives $\det A = a_{11}a_{22}a_{33} + a_{12}a_{23}a_{31} + a_{13}a_{21}a_{32} - a_{13}a_{22}a_{31} - a_{11}a_{23}a_{32} - a_{12}a_{21}a_{33}$.

Now note, as above, expanding definition (1) gives:

$$\begin{aligned}\det A &= a_{11}\begin{vmatrix} a_{22} & a_{23} \\ a_{32} & a_{33} \end{vmatrix} - a_{12}\begin{vmatrix} a_{21} & a_{23} \\ a_{31} & a_{33} \end{vmatrix} + a_{13}\begin{vmatrix} a_{21} & a_{22} \\ a_{31} & a_{32} \end{vmatrix} \\ &= a_{11}a_{22}a_{33} - a_{11}a_{23}a_{32} - a_{12}a_{21}a_{33} + a_{12}a_{23}a_{31} + a_{13}a_{21}a_{32} - a_{13}a_{22}a_{31}.\end{aligned}$$

20. For a 2×2 matrix $A = \begin{bmatrix} a_{11} & a_{12} \\ a_{21} & a_{22} \end{bmatrix}$.

We have $C_{11} = (-1)^{1+1} \det [a_{22}] = a_{22}$ and $C_{12} = (-1)^{1+2} \det [a_{21}] = -a_{21}$.

Thus, substituting $n = 2$ in definition (4) gives:

$$\det A = \sum_{j=1}^{2} a_{1j}C_{1j} = a_{11}C_{11} + a_{12}C_{12} = a_{11}(a_{22}) + a_{12}(-a_{21}) = a_{11}a_{22} - a_{12}a_{21}.$$

21. We use induction to prove Theorem 4.2: if A is triangular then $\det A = a_{11}a_{22}\cdots a_{nn}$.
See Appendix B for discussion and examples of *Mathematical Induction.*
Since Theorem 4.10 asserts $\det A = \det A^T$, we can assume A is upper triangular.

1: If A is 1×1, then $A = [a_{11}]$ so $\det A = \det([a_{11}]) = a_{11}$.
This is obvious, so there is nothing to show.

n: If A is $n \times n$ and upper triangular, then $\det A = a_{11}a_{22}\cdots a_{nn}$.
This is the induction hypothesis.

$n+1$: If A is $(n+1) \times (n+1)$ and upper triangular, then $\det A = a_{11}a_{22}\cdots a_{nn}a_{n+1n+1}$.
This is the statement we must prove using the induction hypothesis.

$$\det A \overset{\substack{A\text{ is upper}\\ \text{triangular}}}{=} a_{n+1n+1}\det A_{n+1n+1} \overset{\substack{\text{by}\\ \text{induction}}}{=} a_{n+1n+1}(a_{11}a_{22}\cdots a_{nn}) = a_{11}a_{22}\cdots a_{n+1n+1}$$

We have proven (by induction) that if A is $n \times n$ and triangular, then $\det A = a_{11}a_{22}\cdots a_{nn}$.

Q: Why does the fact that $\det A = \det A^T$ allow us to assume A is upper triangular?
A: Because if A is lower triangular, then A^T is upper triangular.
So, $\det A = \det A^T = a_{11}a_{22}\cdots a_{nn}$ because $[A]_{ii} = [A^T]_{ii}$.
That is, the diagonal entries of A and A^T are equal.

Q: Why does A being upper triangular imply $\det A = a_{n+1n+1}\det A_{n+1n+1}$?
A: Because if A is upper triangular, then row $n+1 = \mathbf{A}_{n+1} = \begin{bmatrix} 0 & 0 & \cdots & a_{n+1n+1} \end{bmatrix}$.
So when we expand along this row we have: $\det A = a_{n+1n+1}\det A_{n+1n+1}$.

Q: Why is the coefficient of $\det A_{n+1n+1}$ equal to a_{n+1n+1} instead of $-a_{n+1n+1}$?
A: The cofactor $C_{n+1n+1} = (-1)^{2n+2}\det A_{n+1n+1} = \det A_{n+1n+1}$ because $2n+2$ is even.

Q: Why do we get to apply the induction hypothesis to the matrix A_{n+1n+1}?
A: Since A_{n+1n+1} is created by removing the $n+1$ row and the $n+1$ column of A,
A_{n+1n+1} is $n \times n$ and upper triangular. Why is this obvious?
Since A is $(n+1) \times (n+1)$, A_{n+1n+1} (created by removing a row and column) is $n \times n$.
Since A is upper triangular, $[A]_{ij} = 0 = [A_{n+1n+1}]_{ij}$ for $i > j$ when $i, j \leq n$.
That is, A_{n+1n+1} is upper triangular.

All these details support the proof given above. We should carefully investigate them all.
In general, we should be critical of our reasoning and actively seek out oversights.

22. As in Example 4.13, we use Theorem 4.3 to track adjustments to $\det A$ required $A \longrightarrow U$.

$$\det A = \begin{vmatrix} 1 & 0 & 3 \\ 5 & 1 & 1 \\ 0 & 1 & 2 \end{vmatrix} \overset{R_2-5R_1}{=} \begin{vmatrix} 1 & 0 & 3 \\ 0 & 1 & -14 \\ 0 & 1 & 2 \end{vmatrix} \overset{R_3-R_2}{=} \begin{vmatrix} 1 & 0 & 3 \\ 0 & 1 & -14 \\ 0 & 0 & 16 \end{vmatrix} = 1 \cdot 1 \cdot 16 = 16 = \det U$$

Q: Even though row operations were used in $A \longrightarrow U$, we still have $\det A = \det U$. Why?
A: Both row operations were of the form $R_i + kRj$. So what does Theorem 4.3 say?
Part f. asserts a row operation like $R_i + kRj$ does not change the value of the determinant.

23. As in Example 4.13, we use Theorem 4.3 to track adjustments to $\det A$ required $A \longrightarrow U$.

$$\det A = \begin{vmatrix} -4 & 1 & 3 \\ 2 & -2 & 4 \\ 1 & -1 & 0 \end{vmatrix} \overset{\substack{R_1 \leftrightarrow R_3 \\ R_2 \leftrightarrow R_3}}{=} \begin{vmatrix} 1 & -1 & 0 \\ -4 & 1 & 3 \\ 2 & -2 & 4 \end{vmatrix} \overset{\substack{R_2+4R_1 \\ R_3-2R_1}}{=} \begin{vmatrix} 1 & -1 & 0 \\ 0 & -3 & 3 \\ 0 & 0 & 4 \end{vmatrix} = -12 = \det U$$

Q: Though row interchanges were used in $A \longrightarrow U$, we still have $\det A = \det U$. Why?
A: There were *two* row interchanges. So what does Theorem 4.3 say?
Part b. asserts a row interchange changes the sign of the determinant, $\det U = -\det A$.
If there are two interchanges the sign is changed twice, back to what it was originally.
That is, $\det U = -(-\det A) = \det A$. Does this hold for even numbers of row interchanges?

24. As in Example 4.13, we use Theorem 4.3 to track adjustments to $\det A$ required $A \longrightarrow U$.
Note: We use column operations below to gain experience in doing so.

$$\begin{vmatrix} 1 & -1 & 0 & 3 \\ 2 & 5 & 2 & 6 \\ 0 & 1 & 0 & 0 \\ 1 & 4 & 2 & 1 \end{vmatrix} \overset{\substack{C_1-C_3 \\ C_2-2C_3 \\ C_4-3C_3}}{=} \begin{vmatrix} 1 & -1 & 0 & 3 \\ 0 & 1 & 2 & 0 \\ 0 & 1 & 0 & 0 \\ -1 & 0 & 2 & -5 \end{vmatrix} \overset{\substack{C_1 \leftrightarrow C_4 \\ C_2 \leftrightarrow C_3}}{=} \begin{vmatrix} 3 & 0 & -1 & 1 \\ 0 & 2 & 1 & 0 \\ 0 & 0 & 1 & 0 \\ -5 & 2 & 0 & -1 \end{vmatrix} \overset{\substack{C_1-5C_4 \\ C_2+2C_4}}{=} \begin{vmatrix} -2 & 2 & -1 & 1 \\ 0 & 2 & 1 & 0 \\ 0 & 0 & 1 & 0 \\ 0 & 0 & 0 & -1 \end{vmatrix} = 4$$

Q: Are column operations allowed when solving a system with matrices?
A: No. Why not? Because the columns represent the coefficients of independent variables.
So, column operations change the solution to the system. For example:

$$\left[\begin{array}{ccc|c} 1 & -1 & -1 & 2 \\ 3 & -3 & 2 & 16 \\ 2 & -1 & 1 & 9 \end{array}\right] \longrightarrow \left[\begin{array}{ccc|c} 1 & 0 & 0 & 3 \\ 0 & 1 & 0 & -1 \\ 0 & 0 & 1 & 2 \end{array}\right], \text{ but}$$

$$\left[\begin{array}{ccc|c} 1 & -1 & -1 & 2 \\ 3 & -3 & 2 & 16 \\ 2 & -1 & 1 & 9 \end{array}\right] \xrightarrow{C_2+C_1} \left[\begin{array}{ccc|c} 1 & 0 & -1 & 2 \\ 3 & 0 & 2 & 16 \\ 2 & 1 & 1 & 9 \end{array}\right] \longrightarrow \left[\begin{array}{ccc|c} 1 & 0 & 0 & 4 \\ 0 & 1 & 0 & -1 \\ 0 & 0 & 1 & 2 \end{array}\right]$$

Q: So, how can we remember column operations are allowed when computing $\det A$?
A: The key idea here is the assertion of Theorem 4.10: $\det A = \det A^T$.
This implies that row and column operations are both allowed when computing $\det A$.
This is not trivial or obvious. For further discussion, see the proof of Theorem 4.3.

25. As in Example 4.13, we use Theorem 4.3 to track adjustments to $\det A$ required $A \longrightarrow U$. We use a combination of row and column operations below to gain experience in doing so.

$$\begin{vmatrix} 2 & 0 & 3 & -1 \\ 1 & 0 & 2 & 2 \\ 0 & -1 & 1 & 4 \\ 2 & 0 & 1 & -3 \end{vmatrix} \overset{\substack{C_3 \leftrightarrow C_4 \\ C_3 \leftrightarrow C_2 \\ C_1 \leftrightarrow C_2}}{=} - \begin{vmatrix} -1 & 2 & 0 & 3 \\ 2 & 1 & 0 & 2 \\ 4 & 0 & -1 & 1 \\ -3 & 2 & 0 & 1 \end{vmatrix} \overset{\substack{C_2+2C_1 \\ C_4+3C_1+13C_3}}{=} - \begin{vmatrix} -1 & 0 & 0 & 0 \\ 2 & 5 & 0 & 8 \\ 4 & 8 & -1 & 0 \\ -3 & -4 & 0 & -8 \end{vmatrix} \overset{\substack{-R_1 \\ R_2+R_4}}{=} \begin{vmatrix} 1 & 0 & 0 & 0 \\ 2 & 1 & 0 & 0 \\ 4 & 8 & -1 & 0 \\ 3 & -4 & 0 & -8 \end{vmatrix} = 8$$

Q: Strictly speaking, we did not reduce A to row echelon form. Why is this still sufficient?
A: The resulting matrix L is lower triangular, so $\det L$ is the product of the diagonal entries.

Q: Why did we have to introduce a negative sign after step one?
A: Because we performed 3 column interchanges. Is this true for any odd number?

Q: Why did we remove the negative sign after step three?
A: Since we multiplied row 1 by -1, we had to multiply the determinant by -1.
Which part of Theorem 4.3 tells us we have to do this multiplication? Part d.

26. Since $\mathbf{A}_3 = \begin{bmatrix} 2 & 2 & 2 \end{bmatrix} = 2\begin{bmatrix} 1 & 1 & 1 \end{bmatrix} = 2\mathbf{A}_1$, we have $\det A = 0$.

Q: Does Theorem 4.3 imply if $\mathbf{A}_i = k\mathbf{A}_j$ then $\det A = 0$?
A: Yes. How? Part f. asserts if $A \overset{R_i - kR_j}{\longrightarrow} B$, then $\det A = \det B$.
But $\mathbf{B}_i$ (row i of matrix B) is zero, so part c. says $\det B = \det A = 0$.

Q: Does the same hold true for columns? That is, does $\mathbf{a}_i = k\mathbf{a}_j$ imply $\det A = 0$?
A: Yes. Since $\det A = \det A^T$ this conclusion holds for both rows and columns.

Q: How might we word this conclusion as part g. (so to speak) of Theorem 4.3?

Q: These two statements are specific cases of what Theorem from this section?
A: Theorem 4.6 which asserts A is invertible if and only if $\det A \neq 0$.

Q: So what does Theorem 4.6 imply if A is *not* invertible?
A: Theorem 4.6 implies A is not invertible if and only if $\det A = 0$.

Q: How is this a generalization of the statements we have just proven?
A: We have shown if one row of A is a multiple of another row, then $\det A = 0$.
But we know that if one row of A is a multiple of another row, then A is not invertible.
Theorem 4.6 implies if there is *any* dependence relation among the rows then $\det A = 0$.
Why? Because the existence of any such dependence relation implies A is not invertible.

Q: Does Theorem 4.6 imply $\det A = 0$ if there is a dependence relation among the columns?
A: Yes. Since $\det A = \det A^T$ this conclusion holds for both rows and columns.

27. Since A is triangular, we have $\det A = a_{11}a_{22}a_{33} = (3)(-2)(4) = -24$.

Q: Which Theorem from this Section did we have employ in reaching this conclusion?
A: Theorem 4.2.

Q: How does the proof of this Theorem in Exercise 21 suggest solving this problem directly?
A: By expanding along row 3. So:

$$\text{row 3: } \begin{vmatrix} 3 & 1 & 0 \\ 0 & -2 & 5 \\ 0 & 0 & 4 \end{vmatrix} = 4\begin{vmatrix} 3 & 1 \\ 0 & -2 \end{vmatrix} = 4(3)(-2) = -24$$

Note: We should apply our proofs to specific examples to see if they make sense and work. Also: A_{33} is 2×2 and upper triangular as it should be according to proof in Exercise 21.

Q: Since $A \longrightarrow I$ (obviously), why is $\det A = -24 \neq 1 = \det I$?
A: For $A \longrightarrow I$ we have to multiply row 1 by $\frac{1}{3}$, row 2 by $-\frac{1}{2}$, and row 3 by $\frac{1}{4}$. What does that tell us? We have to do the same thing to $\det A$. Therefore $\det I = \left(\frac{1}{3}\right)\left(-\frac{1}{2}\right)\left(\frac{1}{4}\right)\det A = -\frac{1}{24}(-24) = 1$.

28. Since $A \overset{R_1 \leftrightarrow R_3}{\longrightarrow} B$, $\det A = -\det B = -(3)(5)(1) = -15$ because B is triangular.

Q: Why did we have to introduce a negative sign?
A: Because we performed 1 row interchange. Is this true for any odd number? Yes.

Q: If we call this type of matrix *anti*-triangular, is the following statement true:
If A is anti-triangular, then $\pm \det A$ equals the product of the entries on the anti-diagonal.
Hint: What are the only type of operations required to make A triangular?

Q: If A is $n \times n$ and *anti*-triangular, is the following statement true:
$(-1)^n \det A$ equals the product of the entries on the anti-diagonal.
If n is even, do we have to make an even number of row interchanges to make A triangular?

29. Since $\mathbf{a}_3 = \begin{bmatrix} -4 \\ -2 \\ 2 \end{bmatrix} = -2\begin{bmatrix} 2 \\ 1 \\ -1 \end{bmatrix} = -2\mathbf{a}_1$, we have $\det A = 0$.

Q: Does Theorem 4.3 imply if $\mathbf{a}_i = k\mathbf{a}_j$ then $\det A = 0$?
A: Yes. Thm 4.6 implies if there is *any* dependence relation among the columns, $\det A = 0$. For a full discussion see questions and answers following Exercise 26 above.

30. Since $\mathbf{A}_3 = \begin{bmatrix} 1 & 6 & 4 \end{bmatrix} = \begin{bmatrix} 1 & 2 & 3 \end{bmatrix} + \begin{bmatrix} 0 & 4 & 1 \end{bmatrix} = \mathbf{A}_1 + \mathbf{A}_2$, $\det A = 0$.

Q: Does Theorem 4.3 imply if $\mathbf{A}_i = c_j\mathbf{A}_j + c_k\mathbf{A}_k$ then $\det A = 0$?
A: Yes. Thm 4.6 implies if there is *any* dependence relation among the rows then $\det A = 0$. For a full discussion see questions and answers following Exercise 26 above.

31. Since $\mathbf{a}_3 = \begin{bmatrix} 3 \\ -2 \\ 1 \end{bmatrix} = \begin{bmatrix} 4 \\ -2 \\ 5 \end{bmatrix} - \begin{bmatrix} 1 \\ 0 \\ 4 \end{bmatrix} = \mathbf{a}_1 - \mathbf{a}_2$, $\det A = 0$.

Q: Does Theorem 4.3 imply if $\mathbf{a}_i = c_j\mathbf{a}_j + c_k\mathbf{a}_k$ then $\det A = 0$?
A: Yes. Thm 4.6 implies if there is *any* dependence relation among the columns, $\det A = 0$. For a full discussion see questions and answers following Exercise 26 above.

32. Since $A \overset{R_2 \leftrightarrow R_3}{\longrightarrow} B$, $\det A = -\det B = -(1)(1)(1)(1) = -1$ because B is triangular.

Q: Let E_{ij} be the elementary matrix interchanging row i and row j. Is $\det E_{ij} = -1$?
A: Yes. This is precisely the assertion of Theorem 4.4, part a.

Q: Can we generalize this result?
A: Hint: *Some* of the off diagonal entries do not affect our conclusion here.

33. Since $A \overset{\substack{R_1 \leftrightarrow R_2 \\ R_3 \leftrightarrow R_4}}{\longrightarrow} B$, $\det A = \det B = (-3)(2)(1)(4) = -24$ because B is triangular.

Q: In $A \longrightarrow B$ we used row interchanges, but $\det A = \det B$. Why?
A: Because we performed 2 row interchanges. Is this true for any even number? Yes.

34. Since $\mathbf{A}_1 + \mathbf{A}_2 = \mathbf{A}_3 + \mathbf{A}_4$, we have $\det A = 0$.

Q: What Theorem from this section supports our conclusion above that $\det A = 0$?
A: Theorem 4.6. It implies if there is *any* dependence relation among the rows then $\det A = 0$. For a full discussion see questions and answers following Exercise 26 above.

35. Let A be the matrix given at the beginning of Exercises 35 – 40 with $\det A = 4$.
Let B be the matrix given in this Exercise which is derived from A. So:
Since $A \overset{2R_1}{\longrightarrow} B$, $\det B = 2\det A = 2(4) = 8$.

Q: What Theorem from this section supports our conclusion above that $\det B = 2\det A$?
A: Theorem 4.3 part d. which asserts if $A \overset{kR_i}{\longrightarrow} B$, then $\det B = k\det A$.

36. Let A be the matrix given at the beginning of Exercises 35 – 40 with $\det A = 4$.
Let D be the matrix given in this Exercise which is derived from A in 3 steps.
$A \overset{3C_1}{\longrightarrow} B \overset{-C_2}{\longrightarrow} C \overset{2C_3}{\longrightarrow} D$, so $\det D = 2\det C = 2(-\det B) = 2(-(3\det A)) = -6\det A = -24$.

Q: How might we generalize this result?
A: If $A \overset{bC_i}{\longrightarrow} B \overset{cC_j}{\longrightarrow} C \overset{dC_k}{\longrightarrow} D$, $\det D = bcd(\det A)$.

Q: Does this result hold for rows? If $A \overset{bR_i}{\longrightarrow} B \overset{cR_j}{\longrightarrow} C \overset{dR_k}{\longrightarrow} D$, does $\det D = bcd(\det A)$?
A: Yes. Since $\det A = \det A^T$ this result holds for both rows and columns.

37. Let A be the matrix given at the beginning of Exercises 35 – 40 with $\det A = 4$.
Let B be the matrix given in this Exercise which is derived from A in 1 step.
Since $A \xrightarrow{R_1 \leftrightarrow R_2} B$, we conclude $\det B = -\det A = -4$.

Q: How might we generalize this result?
A: If $A \xrightarrow{R_i \leftrightarrow R_j} B$, $\det B = -\det A$. This is the assertion of Theorem 4.3, part b.

Q: Does this result hold for columns? If $A \xrightarrow{C_i \leftrightarrow C_j} B$, does $\det B = -\det A$?
A: Yes. Since $\det A = \det A^T$ this result holds for both rows and columns.

38. Let A be the matrix given at the beginning of Exercises 35 – 40 with $\det A = 4$.
Let B be the matrix given in this Exercise which is derived from A in 1 step.
Since $A \xrightarrow{R_1 + R_3} B$, we conclude $\det B = \det A = 4$.

Q: How might we generalize this result?
A: If $A \xrightarrow{R_i + kR_j} B$, $\det B = \det A$. This is the assertion of Theorem 4.3, part f.

Q: Does this result hold for columns? If $A \xrightarrow{C_i + kC_j} B$, does $\det B = \det A$?
A: Yes. Since $\det A = \det A^T$ this result holds for both rows and columns.

39. Let A be the matrix given at the beginning of Exercises 35 – 40 with $\det A = 4$.
Let C be the matrix given in this Exercise which is derived from A in 2 steps.
Since $A \xrightarrow{C_1 \leftrightarrow C_3} B \xrightarrow{2C_1} C$, we have $\det C = 2\det B = 2(-\det A) = 2(-4) = -8$.

Q: How might we generalize this result?
A: If $A \xrightarrow{C_i \leftrightarrow C_j} B \xrightarrow{cC_k} C$, $\det C = -c \det A$. This is precisely Theorem 4.3, parts b. and d.

Q: Does this result hold for rows? If $A \xrightarrow{R_i \leftrightarrow R_j} B \xrightarrow{cR_k} C$, does $\det C = -c \det A$?
A: Yes. Since $\det A = \det A^T$ this result holds for both rows and columns.

40. Let A be the matrix given at the beginning of Exercises 35 – 40 with $\det A = 4$.
Let C be the matrix given in this Exercise which is derived from A in 2 steps.
Since $A \xrightarrow{2R_2} B \xrightarrow{R_2 - 3R_3} C$, we have $\det C = \det B = 2(\det A) = 2(4) = 8$.

Q: How might we generalize this result?
A: If $A \xrightarrow{bR_i} B \xrightarrow{R_j - cR_k} C$, $\det C = b \det A$. This is precisely Theorem 4.3, parts d. and f.

Q: Does this result hold for columns? If $A \xrightarrow{bC_i} B \xrightarrow{C_j - cC_k} C$, does $\det C = b \det A$?
A: Yes. Since $\det A = \det A^T$ this result holds for both rows and columns.

41. We will first prove Theorem 4.3 part a. for rows and then for columns.

row: If A has a zero row, then $\det A = 0$. That is, if $\mathbf{A}_i = \mathbf{0}$, then $\det A = 0$.

By Theorem 4.1, $\det A = \sum_{j=1}^{n} a_{ij}C_{ij}$, where $\mathbf{A}_i = \begin{bmatrix} a_{i1} & a_{i2} & \cdots & a_{in} \end{bmatrix}$.

If $\mathbf{A}_i = \mathbf{0} = \begin{bmatrix} 0 & 0 & \cdots & 0 \end{bmatrix}$, then $a_{ij} = 0$ for all j. So, $\det A = \sum_{j=1}^{n} 0(C_{ij}) = 0$.

col: If A has a zero column, then $\det A = 0$. That is, if $\mathbf{a}_i = \mathbf{0}$, then $\det A = 0$.

By Theorem 4.1, $\det A = \sum_{i=1}^{n} a_{ij}C_{ij}$, where $\mathbf{a}_j = \begin{bmatrix} a_{1j} \\ a_{2j} \\ \vdots \\ a_{nj} \end{bmatrix}$.

If $\mathbf{a}_j = \mathbf{0} = \begin{bmatrix} 0 \\ 0 \\ \vdots \\ 0 \end{bmatrix}$, then $a_{ij} = 0$ for all i. So, $\det A = \sum_{i=1}^{n} 0(C_{ij}) = 0$.

Note: Our proof is trivial because the Laplace Expansion does all the work.

Q: How might we state our proof and conclusions about rows in words?

A: When we expand along a zero row, the coefficient of every cofactor is zero so $\det A = 0$.
This illustrates an important point: symbols are essential to an elegant proof,
but stating conclusions (loosely) in language is essential to solid understanding.

Q: Can we create a similar statement for our proof and conclusions about columns?

42. We will first prove Theorem 4.3(f) for rows and then for columns.

row: If $A \overset{R_i+kR_j}{\longrightarrow} C$, then $\det C = \det A$.

We will prove this using part e. of Theorem 4.3. That is, we will define B such that A, B, and C are identical except that $\mathbf{C}_i = \mathbf{A}_i + \mathbf{B}_i$.
Under these conditions on A, B, and C part e. asserts $\det C = \det A + \det B$.
We will then show $\det B = 0$, so we can conclude $\det C = \det A$.

Since $A \overset{R_i+kR_j}{\longrightarrow} C$ implies $C = \begin{bmatrix} \mathbf{A}_1 \\ \mathbf{A}_2 \\ \vdots \\ \mathbf{A}_i + k\mathbf{A}_j \\ \vdots \\ \mathbf{A}_n \end{bmatrix}$, we let $B = \begin{bmatrix} \mathbf{A}_1 \\ \mathbf{A}_2 \\ \vdots \\ k\mathbf{A}_j \\ \vdots \\ \mathbf{A}_n \end{bmatrix}$.

Then $\mathbf{A}_r = \mathbf{B}_r = \mathbf{C}_r$ when $r \neq i$ and row i of C is $\mathbf{C}_i = \mathbf{A}_i + k\mathbf{A}_j = \mathbf{A}_i + \mathbf{B}_i$ as required.
So part e. of Theorem 4.3. asserts $\det C = \det A + \det B$.

Next we will show that $\det B = 0$. (Note: row i of B, $\mathbf{B}_i = k\mathbf{A}_j$).
We will prove this using parts c. and d. of Theorem 4.3. That is, we will define D such that D has two identical rows so part c. will imply $\det D = 0$ and such that $D \overset{kR_i}{\longrightarrow} B$ so part d. will imply $\det B = k \det D = k(0) = 0$.

Since $B = \begin{bmatrix} \mathbf{A}_1 \\ \mathbf{A}_2 \\ \vdots \\ k\mathbf{A}_j \\ \vdots \\ \mathbf{A}_n \end{bmatrix}$, we let $D = \begin{bmatrix} \mathbf{A}_1 \\ \mathbf{A}_2 \\ \vdots \\ \mathbf{A}_j \\ \vdots \\ \mathbf{A}_n \end{bmatrix}$ so $\mathbf{D}_i = \mathbf{D}_j = \mathbf{A}_j$ and $\mathbf{B}_i = k\mathbf{D}_i$ as required.

So parts c. and d. assert $\det B \overset{\mathbf{B}_i = k\mathbf{D}_i}{=} k \det D \overset{\mathbf{D}_i = \mathbf{D}_j}{=} k(0) = 0$.

Therefore, we conclude $\det C \overset{\mathbf{C}_i = \mathbf{A}_i + \mathbf{B}_i}{=} \det A + \det B \overset{\det B = 0}{=} \det A$ as required.

col: If $A \overset{C_i+kC_j}{\longrightarrow} C$, then $\det C = \det A$.

Since $A \overset{C_i+kC_j}{\longrightarrow} C$ implies $C = \begin{bmatrix} \mathbf{a}_1 & \mathbf{a}_2 & \cdots & \mathbf{a}_i + k\mathbf{a}_j & \cdots & \mathbf{a}_n \end{bmatrix}$,
we let $B = \begin{bmatrix} \mathbf{a}_1 & \mathbf{a}_2 & \cdots & k\mathbf{a}_j & \cdots & \mathbf{a}_n \end{bmatrix}$.
Then $\mathbf{a}_r = \mathbf{b}_r = \mathbf{c}_r$ when $r \neq i$ and column i of C is $\mathbf{c}_i = \mathbf{a}_i + k\mathbf{a}_j = \mathbf{a}_i + \mathbf{b}_i$ as required.
So part e. of Theorem 4.3. asserts $\det C = \det A + \det B$.

Next we will show that $\det B = 0$. (Note: column i of B, $\mathbf{a}_i = k\mathbf{a}_j$).
Since $B = \begin{bmatrix} \mathbf{a}_1 & \mathbf{a}_2 & \cdots & k\mathbf{a}_j & \cdots & \mathbf{a}_n \end{bmatrix}$, we let $D = \begin{bmatrix} \mathbf{a}_1 & \mathbf{a}_2 & \cdots & \mathbf{a}_j & \cdots & \mathbf{a}_n \end{bmatrix}$.
So $\mathbf{d}_i = \mathbf{d}_j = \mathbf{a}_j$ and $\mathbf{b}_i = k\mathbf{d}_i$ as required.

So parts c. and d. assert $\det B \overset{\mathbf{b}_i = k\mathbf{d}_i}{=} k \det D \overset{\mathbf{d}_i = \mathbf{d}_j}{=} k(0) = 0$.

Therefore, we conclude $\det C \overset{\mathbf{c}_i = \mathbf{a}_i + \mathbf{b}_i}{=} \det A + \det B \overset{\det B = 0}{=} \det A$ as required.

43. We will apply Theorems 4.3 and 4.4 to prove Lemma 4.5: $\det(EB) = (\det E)(\det B)$. We consider each type of elementary matrix separately.

$R_i \leftrightarrow R_j$: Given $I \overset{R_i \leftrightarrow R_j}{\longrightarrow} E$, we need to show $\det(EB) = (\det E)(\det B)$.
Since $B \overset{R_i \leftrightarrow R_j}{\longrightarrow} EB$, Theorem 4.3(b) implies $\det(EB) = -\det B$.
Since $I \overset{R_i \leftrightarrow R_j}{\longrightarrow} E$, Theorem 4.4(a) implies $\det E = -\det I = -1$.
So, $\det(EB) = -\det B = (-1)(\det B) = (\det E)(\det B)$.

Briefly in symbols: $\det(EB) \overset{\substack{R_i \leftrightarrow R_j \\ \text{Thm 4.3(b)}}}{=} -\det B \overset{\substack{\det E=-1 \\ \text{Thm 4.4(a)}}}{=} (\det E)(\det B)$

kR_i: Given $I \overset{kR_i}{\longrightarrow} E$, we need to show $\det(EB) = (\det E)(\det B)$.
Since $B \overset{kR_i}{\longrightarrow} EB$, Theorem 4.3(d) implies $\det(EB) = k\det B$.
Since $I \overset{kR_i}{\longrightarrow} E$, Theorem 4.4(b) implies $\det E = k\det I = k$.
So, $\det(EB) = k\det B = (k)(\det B) = (\det E)(\det B)$.

Briefly in symbols: $\det(EB) \overset{\substack{R_i \to kR_i \\ \text{Thm 4.3(d)}}}{=} k\det B \overset{\substack{\det E=k \\ \text{Thm 4.4(b)}}}{=} (\det E)(\det B)$

$R_i + kR_j$: Given $I \overset{R_i + kR_j}{\longrightarrow} E$, we need to show $\det(EB) = (\det E)(\det B)$.
Since $B \overset{R_i + kR_j}{\longrightarrow} EB$, Theorem 4.3(f) implies $\det(EB) = \det B$.
Since $I \overset{R_i + kR_j}{\longrightarrow} E$, Theorem 4.4(c) implies $\det E = \det I = 1$.
So, $\det(EB) = 1\det B = (1)(\det B) = (\det E)(\det B)$.

Briefly in symbols: $\det(EB) \overset{\substack{R_i \to R_i + kR_j \\ \text{Thm 4.3(f)}}}{=} \det B \overset{\substack{\det E=1 \\ \text{Thm 4.4(c)}}}{=} (\det E)(\det B)$

Q: How might we describe our result for $R_i \leftrightarrow R_j$ more completely in words?
A: If E results from interchanging two rows of I, then EB is derived from B by interchanging two rows of B. So by Theorem 4.3(b), $\det(EB) = -\det B$. But by Theorem 4.4(a), $\det E = -1$, so $\det(EB) = (\det E)(\det B)$ as required. Repeat this exercise for kR_i and $R_i + kR_j$.

44. To prove Theorem 4.7, we need to show if $B = kA$, then $\det B = k^n \det A$.

We will begin by proving the following slightly more general result by induction:
If the first m rows of A have been multiplied by k to create B then $\det B = k^m \det A$.

1: If the first row of A has been multiplied by k to create B, then $\det B = k \det A$.
But if $A \xrightarrow{kR_1} B$, Theorem 4.3(d) implies $\det B = k \det A$.

$$\text{That is, if } B = \begin{bmatrix} k\mathbf{A}_1 \\ \mathbf{A}_2 \\ \vdots \\ \mathbf{A}_n \end{bmatrix}, \text{ then } \det(B) \overset{\substack{R_1 \to kR_1 \\ \text{Thm 4.3(d)}}}{=} k \det A$$

r: If the first r rows of A have been multiplied by k to create B, then $\det B = k^r \det A$.

$$\text{That is, if } B = \begin{bmatrix} k\mathbf{A}_1 \\ k\mathbf{A}_2 \\ \vdots \\ k\mathbf{A}_r \\ \mathbf{A}_{r+1} \\ \vdots \\ \mathbf{A}_n \end{bmatrix}, \text{ then } \det B = k^r \det A$$

This is the induction hypothesis so there is nothing to show.

$r+1$: If the first $r+1$ rows of A have been multiplied by k to create B, then $\det B = k^{r+1} \det A$.
This is the key step we must show using the induction hypothesis.

$$\text{That is, if } B = \begin{bmatrix} k\mathbf{A}_1 \\ k\mathbf{A}_2 \\ \vdots \\ k\mathbf{A}_r \\ k\mathbf{A}_{r+1} \\ \mathbf{A}_{r+2} \\ \vdots \\ \mathbf{A}_n \end{bmatrix}, \text{ we must show } \det B = k^{r+1} \det A. \text{ So let } C = \begin{bmatrix} k\mathbf{A}_1 \\ k\mathbf{A}_2 \\ \vdots \\ k\mathbf{A}_r \\ \mathbf{A}_{r+1} \\ \mathbf{A}_{r+2} \\ \vdots \\ \mathbf{A}_n \end{bmatrix}.$$

$$\text{Then } \det B \overset{\substack{R_{r+1} \to kR_{r+1} \\ \text{Thm 4.3(f)}}}{=} k \det C \overset{\substack{\text{by} \\ \text{induction}}}{=} k(k^r \det A) = k^{r+1} \det A$$

So if the first m rows of A have been multiplied by k to create B, then $\det B = k^m \det A$.

$$\text{Therefore, if } B = kA = \begin{bmatrix} k\mathbf{A}_1 \\ k\mathbf{A}_2 \\ \vdots \\ k\mathbf{A}_n \end{bmatrix}, \text{ then } \det A = k^n \det A \text{ as we were to show.}$$

45. Theorem 4.6 asserts A is invertible if and only if $\det A \neq 0$.
So, the contrapositive of Theorem 4.6 is: A is *not* invertible if and only if $\det A = 0$.
So we solve $\det A = 0$ to find the values of k we need to exclude in order for A to be invertible.

$$\det A = \begin{vmatrix} k & -k & 3 \\ 0 & k+1 & 1 \\ k & -8 & k-1 \end{vmatrix} \overset{R_3-R_1}{=} \begin{vmatrix} k & -k & 3 \\ 0 & k+1 & 1 \\ 0 & k-8 & k-4 \end{vmatrix} \overset{\substack{\text{along} \\ C_1}}{=} k \begin{vmatrix} k+1 & 1 \\ k-8 & k-4 \end{vmatrix}$$

$$= k\left(k^2 - 4k + 4\right) = k\,(k-2)^2 = 0 \Rightarrow k = 0, 2.$$

So, the values of k we need to exclude in order for A to be invertible are $k = 0, 2$.
That is, A is invertible if and only if $k \neq 0, 2$.

46. Since Theorem 4.6 asserts A is invertible if and only if $\det A \neq 0$,
we solve $\det A = 0$ to find the values of k we need to exclude in order for A to be invertible.

$$\det A = \begin{vmatrix} k & k & 0 \\ k^2 & 2 & k \\ 0 & k & k \end{vmatrix} \overset{C_2-C_1-C_3}{=} \begin{vmatrix} k & 0 & 0 \\ k^2 & 2-k^2-k & k \\ 0 & 0 & k \end{vmatrix} \overset{\substack{\text{along} \\ R_1}}{=} k \begin{vmatrix} 2-k^2-k & k \\ 0 & k \end{vmatrix}$$

$$= k(2-k^2-k)k = -k^2(k+2)(k-1) = 0 \Rightarrow k = -2, 0, 1.$$

So, the values of k we need to exclude in order for A to be invertible are $k = -2, 0, 1$.
That is, A is invertible if and only if $k \neq -2, 0, 1$.

47. We use Theorem 4.8, $\det(AB) = (\det A)(\det B)$, and the given values to compute $\det(AB)$.

$$\det(AB) \overset{\text{Thm 4.8}}{=} (\det A)(\det B) \overset{\text{givens}}{=} (3)(-2) = -6$$

Q: How might we state the conclusion $\det(AB) = (\det A)(\det B)$ in words?
A: The determinant of the product equals the product of the determinant.

48. We use Theorem 4.8 and $\det A = 3$ to compute $\det(B^{-1}A)$.

$$\det(A^2) = \det(A \cdot A) \overset{\text{Thm 4.8}}{=} (\det A)(\det A) = (\det A)^2 \overset{\text{givens}}{=} (3)^2 = 9$$

Q: How might we state the conclusion $\det(A^2) = (\det A)^2$ in words?
A: The determinant of the square equals the square of the determinant.

49. We use Theorems 4.8, 4.9 and $\det A = 3$, $\det B = -2$ to compute $\det(B^{-1}A)$.

$$\det(B^{-1}A) \overset{\text{Thm 4.8}}{=} (\det(B^{-1}))(\det A) \overset{\text{Thm 4.9}}{=} \left(\tfrac{1}{\det B}\right)(\det A) \overset{\text{givens}}{=} (-\tfrac{1}{2})(3) = -\tfrac{3}{2}$$

50. We use Theorem 4.7 with $k = 2$ and $\det A = 3$, $\det B = -2$ to compute $\det(2A)$.

$$\det(2A) \overset{\substack{k=2 \\ \text{Thm 4.7}}}{=} 2^n \det A \overset{\text{givens}}{=} 2^n(3) = 3 \cdot 2^n$$

51. We use Theorem 4.7 with $k = 3$, Theorem 4.10, and the givens to compute $\det(3B^T)$.

$$\det(3B^T) \overset{\substack{k=3 \\ \text{Thm 4.7}}}{=} 3^n \det(B^T) \overset{\text{Thm 4.10}}{=} 3^n \det B \overset{\text{givens}}{=} 3^n(-2) = -2 \cdot 3^n$$

52. We use Theorems 4.8, 4.10 and $\det A = 3$, $\det B = -2$ to compute $\det(AA^T)$.

$$\det(AA^T) \overset{\text{Thm 4.8}}{=} (\det A)(\det A^T) \overset{\text{Thm 4.10}}{=} (\det A)(\det A) \overset{\text{givens}}{=} (3)(3) = 9$$

53. We use Theorem 4.8 to prove $\det(AB) = \det(BA)$.

$$\det(AB) \overset{\text{Thm 4.8}}{=} (\det A)(\det B) \overset{\substack{\det M \text{ is a} \\ \text{scalar}}}{=} (\det B)(\det A) \overset{\text{Thm 4.8}}{=} \det(BA)$$

Q: What is the key insight to take away from this exercise?
A: That $\det A$ is a scalar (not a matrix), so it commutes.

54. We use Exercise 53, associativity, and $MM^{-1} = I$ to prove $\det(B^{-1}AB) = \det A$.

$$\det(B^{-1}AB) \overset{\text{assoc}}{=} \det(B^{-1}(AB)) \overset{\text{Ex 53}}{=} \det((AB)B^{-1}) \overset{\text{assoc}}{=} \det(A(BB^{-1})) \overset{MM^{-1}=I}{=} \det A$$

Q: What is the key insight to take away from this exercise?
A: That it is important to justify each step and pay attention to the details.

Q: Could we have proven this equality directly as well? How?
A: Use Thms 4.8, 4.10, and $\left(\frac{1}{k}\right) k = 1$ to prove $\det(B^{-1}AB) = \det A$.

$$\det(B^{-1}AB) \overset{\text{Thm 4.8}}{=} (\det(B^{-1}))(\det A)(\det B) \overset{\text{Thm 4.10}}{=} \left(\frac{1}{\det B}\right)(\det A)(\det B) \overset{(1/k)k=1}{=} \det A$$

Q: What detail do we need to prove to complete the proof just given?
A: We need to show $\det(ABC) = (\det A)(\det B)(\det C)$.

Q: What are the strengths and weaknesses of the two proofs given?

Q: Could we construct a proof based on the elementary matrices?
A: See Section 3.3 where B and B^{-1} are constructed from elementary matrices.

55. We use the fact that $A^2 = A \overset{\text{Ex 47}}{\Rightarrow} (\det A)^2 = \det A$ to find all possible values of $\det A$.

$$(\det A)^2 = \det A \Rightarrow (\det A)^2 - \det A = 0 \Rightarrow \det A(\det A - 1) = 0 \Rightarrow \det A = 0, 1$$

Q: If we let $x = \det A$, what does the above calculation look like?
A: With $x = \det A$: $x^2 = x \Rightarrow x^2 - x = 0 \Rightarrow x(x-1) = 0 \Rightarrow x = 0, 1$.
This clarifies the algebra and helps us remember that $\det A$ is a scalar.

56. First we use induction on Theorem 4.8 to show that $\det(A^m) = (\det A)^m$.
Then we use that fact to find all possible values of $\det A$ when $A^m = O$.

1: $\det(A^1) = (\det A)^1$
This is obvious, so there is nothing to show.

k: $\det(A^k) = (\det A)^k$
This is the induction hypothesis, so there is nothing to show.

$k+1$: $\det(A^{k+1}) = (\det A)^{k+1}$
This is the statement we must prove using the induction on Theorem 4.8.

$$\det(A^{k+1}) = \det(AA^k) \overset{\text{Thm 4.8}}{=} \det A \det(A^k) \overset{\text{induction}}{=} \det A(\det A)^k = (\det A)^{k+1}$$

Therefore, $\det(A^m) = (\det A)^m$ as we were to show.

Since $(\det A)^m = \det(A^m) \overset{\text{given}}{=} \det O \overset{\text{Thm 4.3(a)}}{=} 0$, the only possible value for $\det A$ is zero.

57. We solve the system using Theorem 4.11 (Cramer's Rule). So we have:

$$\det A = \begin{vmatrix} 1 & 1 \\ 1 & -1 \end{vmatrix} = -2,\ \det(A_1(\mathbf{b})) = \begin{vmatrix} 1 & 1 \\ 2 & -1 \end{vmatrix} = -3,\ \det(A_2(\mathbf{b})) = \begin{vmatrix} 1 & 1 \\ 1 & 2 \end{vmatrix} = 1.$$

By Cramer's Rule, $x = \dfrac{\det(A_1(\mathbf{b}))}{\det A} = \dfrac{-3}{-2} = \dfrac{3}{2}$ and $y = \dfrac{\det(A_2(\mathbf{b}))}{\det A} = \dfrac{1}{-2} = -\dfrac{1}{2}$.

58. We solve the system using Theorem 4.11 (Cramer's Rule). So we have:

$$\det A = \begin{vmatrix} 2 & -1 \\ 1 & 3 \end{vmatrix} = 7,\ \det(A_1(\mathbf{b})) = \begin{vmatrix} 5 & -1 \\ -1 & 3 \end{vmatrix} = 14,\ \det(A_2(\mathbf{b})) = \begin{vmatrix} 2 & 5 \\ 1 & -1 \end{vmatrix} = -1.$$

By Cramer's Rule, $x = \dfrac{\det(A_1(\mathbf{b}))}{\det A} = \dfrac{14}{7} = 2$ and $y = \dfrac{\det(A_2(\mathbf{b}))}{\det A} = \dfrac{-7}{7} = 1$.

59. We solve the system using Theorem 4.11 (Cramer's Rule). So we have:

$$\det A = \begin{vmatrix} 2 & 1 & 3 \\ 0 & 1 & 1 \\ 0 & 0 & 1 \end{vmatrix} = 2,\ \det(A_1(\mathbf{b})) = \begin{vmatrix} 1 & 1 & 3 \\ 1 & 1 & 1 \\ 1 & 0 & 1 \end{vmatrix} = 1 - 0 + 3(-1) = -2,$$

$$\det(A_2(\mathbf{b})) = \begin{vmatrix} 2 & 1 & 3 \\ 0 & 1 & 1 \\ 0 & 1 & 1 \end{vmatrix} = 0, \text{ and } \det(A_3(\mathbf{b})) = \begin{vmatrix} 2 & 1 & 1 \\ 0 & 1 & 1 \\ 0 & 0 & 1 \end{vmatrix} = 2. \text{ By Cramer's Rule,}$$

$$x = \frac{\det(A_1(\mathbf{b}))}{\det A} = \frac{-2}{2} = -1,\ y = \frac{\det(A_2(\mathbf{b}))}{\det A} = 0, \text{ and } z = \frac{\det(A_3(\mathbf{b}))}{\det A} = \frac{2}{2} = 1.$$

60. We compute $\det A = \begin{vmatrix} 1 & 1 & -1 \\ 1 & 1 & 1 \\ 1 & -1 & 0 \end{vmatrix} = 4$, $\det(A_1(\mathbf{b})) = \begin{vmatrix} 1 & 1 & -1 \\ 2 & 1 & 1 \\ 3 & -1 & 0 \end{vmatrix} = 9$,

$$\det(A_2(\mathbf{b})) = \begin{vmatrix} 1 & 1 & -1 \\ 1 & 2 & 1 \\ 1 & 3 & 0 \end{vmatrix} = -3, \text{ and } \det(A_3(\mathbf{b})) = \begin{vmatrix} 1 & 1 & 1 \\ 1 & 1 & 2 \\ 1 & -1 & 3 \end{vmatrix} = 2.$$

By Cramer's Rule, $x = \dfrac{\det(A_1(\mathbf{b}))}{\det A} = \dfrac{9}{4}$, $y = \dfrac{\det(A_2(\mathbf{b}))}{\det A} = -\dfrac{3}{4}$, $z = \dfrac{\det(A_3(\mathbf{b}))}{\det A} = \dfrac{1}{2}$.

61. We compute $\det A = \begin{vmatrix} 1 & 1 \\ 1 & -1 \end{vmatrix} = -2$ and the four cofactors $C_{11} = +(-1) = -1$,

$C_{12} = -1$, $C_{21} = -1$, and $C_{22} = 1$. The adjoint is the transpose of the matrix of cofactors:

$$\operatorname{adj} A = \begin{bmatrix} -1 & -1 \\ -1 & 1 \end{bmatrix}^T = \begin{bmatrix} -1 & -1 \\ -1 & 1 \end{bmatrix}. \text{ Then } A^{-1} = \frac{1}{\det A}\operatorname{adj} A = -\frac{1}{2}\begin{bmatrix} -1 & -1 \\ -1 & 1 \end{bmatrix} = \begin{bmatrix} \frac{1}{2} & \frac{1}{2} \\ \frac{1}{2} & -\frac{1}{2} \end{bmatrix}.$$

Q: Is adj A a matrix or a scalar? What about $\det A$?
A: By definition, adj A is the matrix of cofactors. So adj A is a matrix.
On the other hand, $\det A$ is a scalar. The names have a similar form. Be careful!

62. $\det A = \begin{vmatrix} 2 & -1 \\ 1 & 3 \end{vmatrix} = 7$, $C_{11} = 3$, $C_{12} = -1$, $C_{21} = -(-1) = 1$, and $C_{22} = 2$.

$$\text{So adj } A = \begin{bmatrix} 3 & -1 \\ 1 & 2 \end{bmatrix}^T = \begin{bmatrix} 3 & 1 \\ -1 & 2 \end{bmatrix} \text{ and } A^{-1} = \frac{1}{\det A}\operatorname{adj} A = \frac{1}{7}\begin{bmatrix} 3 & 1 \\ -1 & 2 \end{bmatrix} = \begin{bmatrix} \frac{3}{7} & \frac{1}{7} \\ -\frac{1}{7} & \frac{2}{7} \end{bmatrix}.$$

63. $\det A = \begin{vmatrix} 2 & 1 & 3 \\ 0 & 1 & 1 \\ 0 & 0 & 1 \end{vmatrix} = 2$, $C_{11} = \begin{vmatrix} 1 & 1 \\ 0 & 1 \end{vmatrix} = 1$, $C_{12} = -\begin{vmatrix} 0 & 1 \\ 0 & 1 \end{vmatrix} = 0$

$C_{13} = 0$, $C_{21} = -1$, $C_{22} = 2$, $C_{23} = 0$, $C_{31} = -2$, $C_{32} = -2$, and $C_{33} = 2$.

Thus $\text{adj } A = \begin{bmatrix} 1 & -1 & -2 \\ 0 & 2 & -2 \\ 0 & 0 & 2 \end{bmatrix}$ and $A^{-1} = \dfrac{1}{\det A}\text{adj } A = \dfrac{1}{2}\begin{bmatrix} 1 & -1 & -2 \\ 0 & 2 & -2 \\ 0 & 0 & 2 \end{bmatrix} = \begin{bmatrix} \frac{1}{2} & -\frac{1}{2} & -1 \\ 0 & 1 & -1 \\ 0 & 0 & 1 \end{bmatrix}$.

64. $\det A = \begin{vmatrix} 1 & 1 & -1 \\ 1 & 1 & 1 \\ 1 & -1 & 0 \end{vmatrix} = 4$ and $\text{adj } A = \begin{bmatrix} C_{11} & C_{12} & C_{13} \\ C_{21} & C_{22} & C_{23} \\ C_{31} & C_{32} & C_{33} \end{bmatrix}^T = \begin{bmatrix} 1 & 1 & -2 \\ 1 & 1 & 2 \\ 2 & -2 & 0 \end{bmatrix}^T = \begin{bmatrix} 1 & 1 & 2 \\ 1 & 1 & -2 \\ -2 & 2 & 0 \end{bmatrix}$.

Thus, $A^{-1} = \dfrac{1}{\det A}\text{adj } A = \dfrac{1}{4}\begin{bmatrix} 1 & 1 & 2 \\ 1 & 1 & -2 \\ -2 & 2 & 0 \end{bmatrix} = \begin{bmatrix} \frac{1}{4} & \frac{1}{4} & \frac{1}{2} \\ \frac{1}{4} & \frac{1}{4} & -\frac{1}{2} \\ -\frac{1}{2} & \frac{1}{2} & 0 \end{bmatrix}$.

65. We must show adj A is invertible, $(\text{adj } A)^{-1} = \frac{1}{\det A}A$, and $\text{adj }(A^{-1}) = \frac{1}{\det A}A$.

adj A: We will first show if A is invertible, then $B = kA$ ($k \neq 0$) is invertible.
We will then use this fact to prove that $\text{adj } A = (\det A)A^{-1}$ is invertible.
Since A is invertible implies $\det A \neq 0$, if $B = kA$, then $\det B = k^n \det A \neq 0$.
So, Theorem 4.6 implies that B is invertible.
Now by Theorem 4.12, $A^{-1} = \frac{1}{\det A}\text{adj } A$, so $\text{adj } A = (\det A)A^{-1}$.
Since A^{-1} is invertible, $\det A \neq 0$. Therefore, $\text{adj } A = (\det A)A^{-1}$ is invertible.
Q: How might we convey the above argument (loosely) in words?
A: Since adj A is a (nonzero) scalar multiple of an invertible matrix, adj A is invertible.

$(\text{adj } A)^{-1}$: As in Section 3.3, we prove X is the inverse of adj A by showing $(\text{adj } A)X = I$.
We will show $(\text{adj } A)(\frac{1}{\det A}A) = I$ which will imply $(\text{adj } A)^{-1} = \frac{1}{\det A}A$.
$(\text{adj } A)(\frac{1}{\det A}A) = ((\det A)A^{-1})(\frac{1}{\det A}A) = (\det A\frac{1}{\det A})(A^{-1}A) = A^{-1}A = I$

$\text{adj }(A^{-1})$: Note that Theorem 4.12 asserts $A^{-1} = \frac{1}{\det A}\text{adj } A$, so $\text{adj } A = (\det A)A^{-1}$.
So $(A^{-1})^{-1} = A = \frac{1}{\det(A^{-1})}\text{adj }(A^{-1})$, so $\text{adj }(A^{-1}) = (\det(A^{-1}))A$.
Now recall that $\det(A^{-1}) = \frac{1}{\det A}$, so $\text{adj }(A^{-1}) = (\det(A^{-1}))A = \frac{1}{\det A}A$ as claimed.

66. We use Theorems 4.7, 4.9, and $\frac{k^n}{k} = k^{n-1}$ to prove $\det(\text{adj } A) = (\det A)^{n-1}$.

$$\det(\text{adj } A) \stackrel{\text{Ex 65}}{=} \det((\det A)A^{-1}) \stackrel{\text{Thm 4.7}}{=} (\det A)^n(\det(A^{-1})) \stackrel{\text{Thm 4.9}}{=} (\det A)^n \frac{1}{\det A}$$

$$\stackrel{k^n/k=k^{n-1}}{=} (\det A)^{n-1}$$

Q: If we let $k = \det A$ how does the above argument look?
A: This is given below. Is this easier to follow?

$$\det(\text{adj } A) \stackrel{\text{Ex 67}}{=} \det(kA^{-1}) \stackrel{\text{Thm 4.7}}{=} k^n(\det(A^{-1})) \stackrel{\text{Thm 4.9}}{=} k^n \frac{1}{k} = k^{n-1}$$

Note: This exercise reminds us that adj A is a matrix and $\det A$ is a scalar.

67. We use induction to show it requires n adjacent interchanges to move row s to row $s-n$. Then we prove it requires $2(s-r)-1$ adjacent interchanges to interchange rows r and s.

1: Since $R_{s-1} \leftrightarrow R_s$ moves row s to row $s-1$, only 1 adjacent interchange is required.

k: It requires k adjacent interchanges to move row s to row $s-k$.
This is the induction hypothesis, so there is nothing to show.

$k+1$: It requires $k+1$ adjacent interchanges to move row s to row $s-(k+1)$.
This is the step we must prove using the induction hypothesis.
By induction, to move row s to row $s-k$ requires k adjacent interchanges.
But to move row $s-k$ to row $s-(k+1)$ requires only 1 additional adjacent interchange.
So, it requires $k+1$ adjacent interchanges to move row s to row $s-(k+1)$.

Now let $r = s-n$, then it takes $n = s-r$ adjacent interchanges to move row s to row r.
Since this process is symmetric, it also $s-r$ adjacent interchanges to move row r to row s.
However, the $R_r \leftrightarrow R_s$ adjacent exchange involves both row r and and row s simultaneously.
Therefore, we must subtract 1 in order not to double-count this shared interchange.
So, the total number of adjacent interchanges required is $(s-r)+(s-r)-1 = 2(s-r)-1$.

Note: This argument explains the presence of the -1 in the formula.

Q: What is the basic insight underlying the justification of this conclusion?
A: To move a row one row up or down requires one adjacent interchange.

68. What is it that we are trying to show? Lemma 4.13 shows we can expand along column 1. Our goal is to generalize that result to show we can expand along column j.

Similar to Exercise 67, we will move column j of A to column 1 of A to create matrix B.

We will then show $\det A = (-1)^{j-1} \det B = \sum_{i=1}^{n} a_{ij}C_{ij}$ as claimed by Laplace.

First we show: if B is obtained from A by n row interchanges, then $\det B = (-1)^n \det A$.

1: If B is obtained from A by 1 row interchange, then $\det B = (-1)^1 \det A$.
This is precisely the statement of Lemma 4.14 so there is nothing to show.

k: If B is obtained from A by k row interchanges, then $\det B = (-1)^k \det A$.
This is the induction hypothesis, so there is nothing to show.

$k+1$: If B is obtained from A by $k+1$ row interchanges, then $\det B = (-1)^{k+1} \det A$.
This is the step we must prove using the induction hypothesis.
Let C be obtained from A by any k of the $k+1$ row interchanges used to create B.
Let E be the elementary matrix that performs the 1 remaining row interchange.
That is, if $C \xrightarrow{R_i \leftrightarrow R_j} B$, then $I \xrightarrow{R_i \leftrightarrow R_j} E$. So $B = EC$.

$$\det B = \det(EC) \overset{\substack{\text{Thm}\\4.8}}{=} (\det E)(\det C) \overset{\substack{\text{Thm}\\4.4}}{=} (-1)(\det C) \overset{\substack{\text{by}\\\text{induc}}}{=} (-1)((-1)^k \det A) = (-1)^{k+1} \det A$$

So given $A = \begin{bmatrix} \mathbf{a}_1 & \mathbf{a}_2 & \cdots & \mathbf{a}_j & \cdots & \mathbf{a}_n \end{bmatrix}$, we have $B = \begin{bmatrix} \mathbf{a}_j & \mathbf{a}_1 & \mathbf{a}_2 & \cdots & \mathbf{a}_n \end{bmatrix}$.

Let C_{ij}^B be the cofactors of B, then Lemma 4.13 asserts: $\det B = \sum_{i=1}^{n} b_{i1}C_{i1}^B$.

This is simply the expansion along column 1 for B which was established by Lemma 4.13.

Since $\mathbf{b}_1 = \mathbf{a}_j$ implies $B_{i1} = A_{ij}$, we have $C_{i1}^B = (-1)^{i+1} \det B_{i1} = (-1)^{i+1} \det A_{ij}$.

Therefore, $\det B = \sum_{i=1}^{n} b_{i1}C_{i1}^B = \sum_{i=1}^{n} b_{i1}((-1)^{i+1} \det B_{i1}) = \sum_{i=1}^{n} b_{i1}((-1)^{i+1} \det A_{ij})$.

Also $\mathbf{b}_1 = \mathbf{a}_j$ implies $b_{i1} = a_{ij}$, so $\det B = \sum_{i=1}^{n} b_{i1}((-1)^{i+1} \det A_{ij}) = \sum_{i=1}^{n} a_{ij}((-1)^{i+1} \det A_{ij})$.

Note that $(-1)^{i+1} \det A_{ij}$ is *almost* the (i,j)-cofactor of A, C_{ij}, $(-1)^{i+j} \det A_{ij}$.
They differ by a factor $(-1)^{j-1}$. That is, the (i,j)-cofactor of A, $C_{ij} = (-1)^{j-1}((-1)^{i+1} \det A_{ij})$.
This is the key insight for the final step of this proof.

Since B is obtained from A by $j-1$ interchanges, $\det B = (-1)^{j-1} \det A$.
However, multiplying both sides of this equation by $(-1)^{j-1}$, we see $\det A = (-1)^{j-1} \det B$.

$$\det A = (-1)^{j-1} \det B = (-1)^{j-1} \sum_{i=1}^{n} a_{ij}((-1)^{i+1} \det A_{ij}) = \sum_{i=1}^{n} a_{ij}((-1)^{j-1}(-1)^{i+1} \det A_{ij})$$

$$\overset{\substack{\text{key}\\\text{insight}}}{=} \sum_{i=1}^{n} a_{ij}((-1)^{i+j} \det A_{ij}) \overset{\substack{\text{def of}\\C_{ij}}}{=} \sum_{i=1}^{n} a_{ij}C_{ij} \text{ as claimed by Laplace.}$$

69. The result for *block* triangular matrices is neither trivial nor obvious.
It requires careful application of the definitions and use of all the given conditions.

As suggested in the hint, we proceed by induction on the number of rows of P.
Note that A, P, and S *must* be square but O and Q may or may not be square.
However, because A is square O *must* have the same number of columns as P.
Also, because A is square, Q *must* have the same number of rows as P.

1: If P has 1 row, we must show $\det A = (\det P)(\det S)$.

If $P = [p_{11}]$ and A is $m \times m$, then $\det P = p_{11}$ and $A = \begin{bmatrix} p_{11} & Q \\ O & S \end{bmatrix}$.

Compute $\det A$ by expansion along column 1: $\det A = \sum_{i=1}^{m} a_{i1}((-1)^{i+1} \det A_{i1})$.

Recall O is the matrix whose entries are all zero. So, $a_{11} = p_{11}$ and $a_{i1} = 0$ for $i > 1$.

So, $\det A = p_{11}((-1)^{1+1} \det A_{11}) + \sum_{i=2}^{m} 0((-1)^{i+1} \det A_{i1}) = p_{11} \det A_{11}$.

Recall, A_{11} is the submatrix of A obtained by deleting row 1 and column 1.
Since P is 1×1, O has 1 column and Q has 1 row. Therefore, $A_{11} = S$.

So, $\det A = p_{11} \det A_{11} \overset{A_{11}=S}{=} p_{11}(\det S) \overset{p_{11}=\det P}{=} (\det P)(\det S)$ as we were to show.

k: If P has k rows, then $\det A = (\det P)(\det S)$.
This is the induction hypothesis, so there is nothing to show.

$k+1$: If P has $k+1$ rows, then $\det A = (\det P)(\det S)$.
This is the step we must prove using the induction hypothesis.

As in step 1, we compute $\det A$ using column 1: $\det A = \sum_{i=1}^{m} a_{i1}((-1)^{i+1} \det A_{i1})$.

Similar to step 1, $a_{i1} = p_{i1}$ for $i \le k+1$ and $a_{i1} = 0$ for $i > k+1$.

So, $\det A = \sum_{i=1}^{k+1} p_{i1}((-1)^{i+1} \det A_{i1}) + \sum_{i=k+2}^{m} 0((-1)^{i+1} \det A_{i1}) = \sum_{i=1}^{k+1} p_{i1}((-1)^{i+1} \det A_{i1})$.

Recall, A_{i1} is the submatrix of A obtained from by deleting row i and column 1.
So, for $i \le k+1$ this process also removes row i and column 1 of P.
Since Q has $k+1$ rows, for $i \le k+1$ this process does not alter S.

Since $A_{ij} = \begin{bmatrix} P_{i1} & Q_i \\ O_1 & S \end{bmatrix}$, where P_{i1} has k rows, the induction hypothesis applies.

By induction, therefore, we have $\det A_{ij} = (\det P_{i1})(\det S)$ for $i \le k+1$.

So, $\det A = \sum_{i=1}^{k+1} p_{i1}((-1)^{i+1} \det A_{i1}) = \sum_{i=1}^{k+1} p_{i1}((-1)^{i+1} (\det P_{i1})(\det S))$.

Since every term in the sum is multiplied by $\det S$, we can factor it out.

So, $\det A = \sum_{i=1}^{k+1} p_{i1}((-1)^{i+1} (\det P_{i1})(\det S)) = (\det S) \sum_{i=1}^{k+1} p_{i1}((-1)^{i+1} \det P_{i1})$.

Recall, by expansion along column 1, $\det P = \sum_{i=1}^{k+1} p_{i1}((-1)^{i+1} \det P_{i1})$.

So, $\det A = (\det S) \left(\sum_{i=1}^{k+1} p_{i1}((-1)^{i+1} \det P_{i1}) \right) = (\det S)(\det P) = (\det P)(\det S)$.

70. We show that *block* form can be used to compute determinants with certain restrictions.

(a) Below is an example of P, Q, R, and S all square such that $\det A \neq (\det P)(\det S) - (\det Q)(\det R)$.

Let $P = \begin{bmatrix} 1 & 0 \\ 0 & 1 \end{bmatrix}$, $S = \begin{bmatrix} 1 & 0 \\ 0 & 1 \end{bmatrix}$, $Q = \begin{bmatrix} 1 & 0 \\ 1 & 0 \end{bmatrix}$, $R = \begin{bmatrix} 0 & 1 \\ 0 & 1 \end{bmatrix}$.

Therefore, $A = \begin{bmatrix} P & Q \\ R & S \end{bmatrix} = \begin{bmatrix} 1 & 0 & 1 & 0 \\ 0 & 1 & 1 & 0 \\ 0 & 1 & 1 & 0 \\ 0 & 1 & 0 & 1 \end{bmatrix}$.

So, $\det A = \begin{vmatrix} 1 & 0 & 1 & 0 \\ 0 & 1 & 1 & 0 \\ 0 & 1 & 1 & 0 \\ 0 & 1 & 0 & 1 \end{vmatrix} = 0 \neq 1 = (1)(1) - (0)(0) = (\det P)(\det S) - (\det Q)(\det R)$.

(b) Given the proof of Exercise 69, it is clear the result holds for *lower* block form as well.

That is, if A, P, and S are square and $A = \begin{bmatrix} P & O \\ Q & S \end{bmatrix}$, then $\det A = (\det P)(\det S)$.

So, since $B = \begin{bmatrix} P^{-1} & O \\ -RP^{-1} & I \end{bmatrix}$, Exercise 69 implies $\det B = (\det P^{-1})(\det I) = \det P^{-1}$.

Since P is invertible $\det B = \det P^{-1} \neq 0$. So, B is invertible and $A = B^{-1}BA$.

Furthermore, $\det B^{-1} \overset{\text{Thm 4.9}}{=} \frac{1}{\det B} = \frac{1}{\det(P^{-1})} \overset{\text{Thm 4.9}}{=} \frac{1}{1/\det P} = \det P$.

Also since $B = \begin{bmatrix} P^{-1} & O \\ -RP^{-1} & I \end{bmatrix}$ and $A = \begin{bmatrix} P & Q \\ R & S \end{bmatrix}$, we have the following:

$$BA = \begin{bmatrix} P^{-1} & O \\ -RP^{-1} & I \end{bmatrix}\begin{bmatrix} P & Q \\ R & S \end{bmatrix} = \begin{bmatrix} I & P^{-1}Q \\ -RP^{-1}P + R & -RP^{-1}Q + S \end{bmatrix} = \begin{bmatrix} I & P^{-1}Q \\ O & S - RP^{-1}Q \end{bmatrix}.$$

So Exercise 69 implies $\det(BA) = (\det I)(\det(S - RP^{-1}Q)) = \det(S - RP^{-1}Q)$.

Now, $\det A = \det(B^{-1}(BA)) \overset{\text{Thm 4.8}}{=} (\det(B^{-1}))(\det(BA)) = (\det P)(\det(S - RP^{-1}Q))$.

(c) We use Theorem 4.8 to prove $\det A = \det(PS - RQ)$ provided $PR = RP$. So:

$\det A = (\det P)(\det(S - RP^{-1}Q)) \overset{\text{Thm 4.8}}{=} \det(P(S - RP^{-1}Q))$

$= \det(PS - (PR)P^{-1}Q) \overset{PR=RP}{=} \det(PS - (RP)P^{-1}Q) \overset{\text{assoc}}{=} \det(PS - R(PP^{-1})Q)$

$\overset{P \text{ invertible}}{=} \det(PS - RQ)$ as we were to show.

Exploration: Geometric Applications of Determinants

1. (a) $\mathbf{u}\times\mathbf{v}=\det\begin{bmatrix}\mathbf{e}_1 & u_1 & v_1\\ \mathbf{e}_2 & u_2 & v_2\\ \mathbf{e}_3 & u_3 & v_3\end{bmatrix}=\begin{vmatrix}\mathbf{e}_1 & 0 & 3\\ \mathbf{e}_2 & 1 & -1\\ \mathbf{e}_3 & 1 & 2\end{vmatrix}$

$$=\mathbf{e}_1(2-(-1))-\mathbf{e}_2(0-3)+\mathbf{e}_3(0-3)=3\mathbf{e}_1+3\mathbf{e}_2-3\mathbf{e}_3 \text{ or } \begin{bmatrix}3\\3\\-3\end{bmatrix}.$$

(b) $\mathbf{u}\times\mathbf{v}=\det\begin{bmatrix}\mathbf{e}_1 & u_1 & v_1\\ \mathbf{e}_2 & u_2 & v_2\\ \mathbf{e}_3 & u_3 & v_3\end{bmatrix}=\begin{vmatrix}\mathbf{e}_1 & 3 & 0\\ \mathbf{e}_2 & -1 & 1\\ \mathbf{e}_3 & 2 & 1\end{vmatrix}$

$$=\mathbf{e}_1(-1-2)-\mathbf{e}_2(3-0)+\mathbf{e}_3(3-0)=\begin{bmatrix}-3\\3\\3\end{bmatrix}.$$

(c) $\mathbf{u}\times\mathbf{v}=\begin{vmatrix}\mathbf{e}_1 & -1 & 2\\ \mathbf{e}_2 & 2 & -4\\ \mathbf{e}_3 & 3 & -6\end{vmatrix}=O.$

(d) $\mathbf{u}\times\mathbf{v}=\begin{vmatrix}\mathbf{e}_1 & 1 & 1\\ \mathbf{e}_2 & 1 & 2\\ \mathbf{e}_3 & 1 & 3\end{vmatrix}=\begin{bmatrix}1\\-2\\1\end{bmatrix}.$

2. LHS $=\mathbf{u}\cdot(\mathbf{v}\times\mathbf{w})=\begin{bmatrix}u_1\\u_2\\u_3\end{bmatrix}\cdot\det\begin{bmatrix}\mathbf{e}_1 & v_1 & w_1\\ \mathbf{e}_2 & v_2 & w_2\\ \mathbf{e}_3 & v_3 & w_3\end{bmatrix}=\begin{bmatrix}u_1\\u_2\\u_3\end{bmatrix}\cdot\begin{bmatrix}v_2w_3-v_3w_2\\ v_3w_1-v_1w_3\\ v_1w_2-v_2w_1\end{bmatrix}$

$$=u_1v_2w_3-u_1v_3w_2+u_2v_3w_1-u_2v_1w_3+u_3v_1w_2-u_3v_2w_1.$$

RHS $=\det\begin{bmatrix}u_1 & v_1 & w_1\\ u_2 & v_2 & w_2\\ u_3 & v_3 & w_3\end{bmatrix}=u_1\begin{vmatrix}v_2 & w_2\\ v_3 & w_3\end{vmatrix}-u_2\begin{vmatrix}v_1 & w_1\\ v_3 & w_3\end{vmatrix}+u_3\begin{vmatrix}v_1 & w_1\\ v_2 & w_2\end{vmatrix}$

$$=u_1v_2w_3-u_1v_3w_2+u_2v_3w_1-u_2v_1w_3+u_3v_1w_2-u_3v_2w_1.$$

3. (a) $\mathbf{v}\times\mathbf{u} = \det\begin{bmatrix} \mathbf{e}_1 & v_1 & u_1 \\ \mathbf{e}_2 & v_2 & u_2 \\ \mathbf{e}_3 & v_3 & u_3 \end{bmatrix} \overset{C_2\leftrightarrow C_3}{=} -\det\begin{bmatrix} \mathbf{e}_1 & u_1 & v_1 \\ \mathbf{e}_2 & u_2 & v_2 \\ \mathbf{e}_3 & u_3 & v_3 \end{bmatrix} = -(\mathbf{u}\times\mathbf{v})$.

(b) $\mathbf{u}\times\mathbf{0} = \begin{vmatrix} \mathbf{e}_1 & u_1 & 0 \\ \mathbf{e}_2 & u_2 & 0 \\ \mathbf{e}_3 & u_3 & 0 \end{vmatrix} = \mathbf{0}$. (c) $\mathbf{u}\times\mathbf{u} = \begin{vmatrix} \mathbf{e}_1 & u_1 & u_1 \\ \mathbf{e}_2 & u_2 & u_2 \\ \mathbf{e}_3 & u_3 & u_3 \end{vmatrix} = \mathbf{0}$ (col 2 = col = 3).

(d) $\mathbf{u}\times k\mathbf{v} = \det\begin{bmatrix} \mathbf{e}_1 & u_1 & kv_1 \\ \mathbf{e}_2 & u_2 & kv_2 \\ \mathbf{e}_3 & u_3 & kv_3 \end{bmatrix} \overset{C_3/k}{=} k\det\begin{bmatrix} \mathbf{e}_1 & u_1 & v_1 \\ \mathbf{e}_2 & u_2 & v_2 \\ \mathbf{e}_3 & u_3 & v_3 \end{bmatrix} = k(\mathbf{u}\times\mathbf{v})$.

(e) $\mathbf{u}\times(\mathbf{v}+\mathbf{w}) = \det\begin{bmatrix} \mathbf{e}_1 & u_1 & v_1+w_1 \\ \mathbf{e}_2 & u_2 & v_2+w_2 \\ \mathbf{e}_3 & u_3 & v_3+w_3 \end{bmatrix} = \det\begin{bmatrix} \mathbf{e}_1 & u_1 & v_1 \\ \mathbf{e}_2 & u_2 & v_2 \\ \mathbf{e}_3 & u_3 & v_3 \end{bmatrix} + \det\begin{bmatrix} \mathbf{e}_1 & u_1 & w_1 \\ \mathbf{e}_2 & u_2 & w_2 \\ \mathbf{e}_3 & u_3 & w_3 \end{bmatrix}$.

$= \mathbf{u}\times\mathbf{v} + \mathbf{u}\times\mathbf{w}$

(f) By Problem 2, $\mathbf{u}\cdot(\mathbf{u}\times\mathbf{v}) = \det\begin{bmatrix} u_1 & u_1 & v_1 \\ u_2 & u_2 & v_2 \\ u_3 & u_3 & v_3 \end{bmatrix} = 0$ since columns 1 and 2 are identical.

Similarly, $\mathbf{v}\cdot(\mathbf{u}\times\mathbf{v}) = \det\begin{bmatrix} v_1 & u_1 & v_1 \\ v_2 & u_2 & v_2 \\ v_3 & u_3 & v_3 \end{bmatrix} = 0$.

(g) By Problem 2, $\mathbf{u}\cdot(\mathbf{v}\times\mathbf{w}) = \det\begin{bmatrix} u_1 & v_1 & w_1 \\ u_2 & v_2 & w_2 \\ u_3 & v_3 & w_3 \end{bmatrix}$.

But $(\mathbf{u}\times\mathbf{v})\cdot\mathbf{w} = \mathbf{w}\cdot(\mathbf{u}\times\mathbf{v}) = \det\begin{bmatrix} w_1 & u_1 & v_1 \\ w_2 & u_2 & v_2 \\ w_3 & u_3 & v_3 \end{bmatrix} = \det\begin{bmatrix} u_1 & v_1 & w_1 \\ u_2 & v_2 & w_2 \\ u_3 & v_3 & w_3 \end{bmatrix}$.

4. Since $\mathbf{u} = \begin{bmatrix} u_1 \\ u_2 \\ 0 \end{bmatrix}$ and $\mathbf{v} = \begin{bmatrix} v_1 \\ v_2 \\ 0 \end{bmatrix}$, we have:

$$\mathcal{A} = \|\mathbf{u}\times\mathbf{v}\| = \left\|\det\begin{bmatrix} \mathbf{e}_1 & u_1 & v_1 \\ \mathbf{e}_2 & u_2 & v_2 \\ \mathbf{e}_3 & 0 & 0 \end{bmatrix}\right\| = \left\|\mathbf{0} - \mathbf{0} + \mathbf{e}_3\det\begin{bmatrix} u_1 & v_1 \\ u_2 & v_2 \end{bmatrix}\right\| = \left|\det\begin{bmatrix} u_1 & v_1 \\ u_2 & v_2 \end{bmatrix}\right|.$$

5. The area of the large rectangle is $(a+c)(b+d)$, the area of each of the large triangles is $\frac{1}{2}ab$, $\frac{1}{2}cd$, and of each of the small rectangles is bc.

Thus, $\mathcal{A} = (a+c)(b+d) - 2\left(\frac{1}{2}ab\right) - 2\left(\frac{1}{2}cd\right) - 2bc = ad - bc = \det\begin{bmatrix} a & b \\ c & d \end{bmatrix}$.

We must add an absolute value sign in case the two vectors are interchanged,

in which case $ad < bc$. So $\mathcal{A} = \left|\det\begin{bmatrix} a & b \\ c & d \end{bmatrix}\right|$.

6. (a) $\mathcal{A} = \|\mathbf{u} \times \mathbf{v}\| = \left|\det \begin{bmatrix} u_1 & v_1 \\ u_2 & v_2 \end{bmatrix}\right| = \left|\det \begin{bmatrix} 2 & -1 \\ 3 & 4 \end{bmatrix}\right| = |11| = 11.$

(b) $\mathcal{A} = \|\mathbf{u} \times \mathbf{v}\| = \left|\det \begin{bmatrix} u_1 & v_1 \\ u_2 & v_2 \end{bmatrix}\right| = \left|\det \begin{bmatrix} 3 & 5 \\ 4 & 5 \end{bmatrix}\right| = |-5| = 5.$

7. Recall, the dot product of two vectors is $\mathbf{a} \cdot \mathbf{b} = \|\mathbf{a}\| \, \|\mathbf{b}\| \cos\theta$, θ the angle between $\mathbf{a}$ and $\mathbf{b}$.

In this case, the cross product $\mathbf{v} \times \mathbf{w}$ is perpendicular to both $\mathbf{v}$ and $\mathbf{w}$,
so $\mathbf{u} \cdot (\mathbf{v} \times \mathbf{w}) = (\|\mathbf{u}\| \cos\theta) \, \|\mathbf{v} \times \mathbf{w}\|$.

But $\|\mathbf{v} \times \mathbf{w}\|$ is the area of the parallelogram lying in the xy-plane determined by $\mathbf{v}$ and $\mathbf{w}$, and $\|\mathbf{u}\| \cos\theta$ is the height h of the parallelepiped.

So the volume of the parallelepiped is:
$\mathcal{V} = \|\mathbf{v} \times \mathbf{w}\| \, h = (\|\mathbf{u}\| \cos\theta) \, \|\mathbf{v} \times \mathbf{w}\| = |\mathbf{u} \cdot (\mathbf{v} \times \mathbf{w})| = \left|\det \begin{bmatrix} \mathbf{u} & \mathbf{v} & \mathbf{w} \end{bmatrix}\right|$ as required.

8. As in Problem 7, the height of the tetrahedron is $h = \|\mathbf{u}\| \cos\theta$.
Also, the area of the triangular base is half the area of the parallelogram formed by $\mathbf{v}$ and $\mathbf{w}$.
So, using the formula in the hint,

$$\mathcal{V} = \tfrac{1}{3}\,(\text{area of the base})\,(\text{height}) = \tfrac{1}{3}\left(\tfrac{1}{2}\,\|\mathbf{v} \times \mathbf{w}\|\right)(\|\mathbf{u}\| \cos\theta) = \tfrac{1}{6}\,\|\mathbf{u}\|\,\|\mathbf{v} \times \mathbf{w}\| \cos\theta = \tfrac{1}{6}\,|\mathbf{u} \cdot (\mathbf{v} \times \mathbf{w})|\,.$$

9. By Problem 4, the area of the parallelogram determined by $A\mathbf{u}$ and $A\mathbf{v}$ is

$$\left|\det \begin{bmatrix} A\mathbf{u} & A\mathbf{v} \end{bmatrix}\right| = \left|\det \left(A \begin{bmatrix} \mathbf{u} & \mathbf{v} \end{bmatrix}\right)\right| = \left|(\det A)\left(\det \begin{bmatrix} \mathbf{u} & \mathbf{v} \end{bmatrix}\right)\right| = |\det A|\,(\text{area of } P)\,.$$

10. By Problem 7, the volume of the parallelepiped determined by $A\mathbf{u}$, $A\mathbf{v}$, and $A\mathbf{w}$ is

$$\left|\det \begin{bmatrix} A\mathbf{u} & A\mathbf{v} & A\mathbf{w} \end{bmatrix}\right| = \left|\det \left(A \begin{bmatrix} \mathbf{u} & \mathbf{v} & \mathbf{w} \end{bmatrix}\right)\right| = \left|(\det A)\left(\det \begin{bmatrix} \mathbf{u} & \mathbf{v} & \mathbf{w} \end{bmatrix}\right)\right| = |\det A|\,(\text{volume of } P)\,.$$

11. (a) The equation of the line through $(2, 3)$ and $(-1, 0)$ is $\begin{vmatrix} x & y & 1 \\ 2 & 3 & 1 \\ -1 & 0 & 1 \end{vmatrix} = 3x - 3y + 3 = 0.$

(b) The equation of the line through $(1, 2)$ and $(4, 3)$ is $\begin{vmatrix} x & y & 1 \\ 1 & 2 & 1 \\ 4 & 3 & 1 \end{vmatrix} = -x + 2y - 2 = 0.$

12. Suppose that (x_1, y_1), (x_2, y_2) and (x_3, y_3) lie on the line $ax + by + c = 0$. Then

$$\begin{aligned} ax_1 + by_1 + c &= 0 \\ ax_2 + by_2 + c &= 0 \\ ax_3 + by_3 + c &= 0 \end{aligned}$$

We can write this system as $XA = O$, where $X = \begin{bmatrix} x_1 & y_1 & 1 \\ x_2 & y_2 & 1 \\ x_3 & y_3 & 1 \end{bmatrix}$ and $A = \begin{bmatrix} a \\ b \\ c \end{bmatrix}$.

Now assume that $|X| \neq 0$, that is, X^{-1} exists. We left-multiply by X^{-1} to get $X^{-1}XA = O \Leftrightarrow A = O$, which is plainly false. Thus, $|X| = 0$.

13. There is a unique plane passing through these points, $ax+by+cz+d=0$.
Since the three given points are on this plane, their coordinates satisfy the equation:

$$\begin{aligned} ax_1+by_1+cz_1+d &= 0 \\ ax_2+by_2+cz_2+d &= 0 \\ ax_3+by_3+cz_3+d &= 0 \end{aligned}$$

This system has a nontrivial solution,

so the coefficient matrix $X=\begin{bmatrix} x & y & z & 1 \\ x_1 & y_1 & z_1 & 1 \\ x_2 & y_2 & z_2 & 1 \\ x_3 & y_3 & z_3 & 1 \end{bmatrix}$ cannot be invertible.

Thus, $|X|=0$.

If the three points are collinear,

then the vectors $\begin{bmatrix} x_2-x_1 \\ y_2-y_1 \\ z_2-z_1 \end{bmatrix}$ and $\begin{bmatrix} x_3-x_1 \\ y_3-y_1 \\ z_3-z_1 \end{bmatrix}$ are multiples of one another.

That is, $\begin{bmatrix} x_3-x_1 \\ y_3-y_1 \\ z_3-z_1 \end{bmatrix}=k\begin{bmatrix} x_2-x_1 \\ y_2-y_1 \\ z_2-z_1 \end{bmatrix}$, $k\neq 0$.

So we can row-reduce the 4×4 matrix as follows:

$$\begin{bmatrix} x & y & z & 1 \\ x_1 & y_1 & z_1 & 1 \\ x_2 & y_2 & z_2 & 1 \\ x_3 & y_3 & z_3 & 1 \end{bmatrix} \xrightarrow[R_4-R_2]{R_3-R_2} \begin{bmatrix} x & y & z & 1 \\ x_1 & y_1 & z_1 & 1 \\ x_2-x_1 & y_2-y_1 & z_2-z_1 & 0 \\ x_3-x_1 & y_3-y_1 & z_3-z_1 & 0 \end{bmatrix} \xrightarrow{R_4-kR_3} \begin{bmatrix} x & y & z & 1 \\ x_1 & y_1 & z_1 & 1 \\ x_2-x_1 & y_2-y_1 & z_2-z_1 & 0 \\ 0 & 0 & 0 & 0 \end{bmatrix}.$$

In this case $|X|$ is trivially 0 and calculating the determinant gives us no information.

14. Suppose that the four points lie on the plane $ax+by+cz+d=0$. Then:

$$\begin{aligned} ax_1+by_1+cz_1+d &= 0 \\ ax_2+by_2+cz_2+d &= 0 \\ ax_3+by_3+cz_3+d &= 0 \\ ax_4+by_4+cz_4+d &= 0 \end{aligned}$$

We can write this system as $XA=O$ where $X=\begin{bmatrix} x_1 & y_1 & z_1 & 1 \\ x_2 & y_2 & z_2 & 1 \\ x_3 & y_3 & z_3 & 1 \\ x_4 & y_4 & z_4 & 1 \end{bmatrix}$ and $A=\begin{bmatrix} a \\ b \\ c \\ d \end{bmatrix}$.

Now assume that $|X|\neq 0$, that is, X^{-1} exists.
We left-multiply by X^{-1} to get $X^{-1}XA=O \Leftrightarrow A=O$, which is plainly false.

Thus, $|X|=0$.

15. The three equations are

$$\begin{aligned} a-b+c &= 10 \\ a &= 5 \\ a+3b+9c &= 2 \end{aligned}$$

which we can write as $PX = Q$ where $P = \begin{bmatrix} 1 & -1 & 1 \\ 1 & 0 & 0 \\ 1 & 3 & 9 \end{bmatrix}$, $X = \begin{bmatrix} a \\ b \\ c \end{bmatrix}$, and $Q \begin{bmatrix} 10 \\ 5 \\ 2 \end{bmatrix}$.

Since $|P| = 12 \neq 0$, P is invertible and the system has the unique solution $X = P^{-1}Q$.

Solving the system, we add three times the first equation to the third equation to get:
$4a + 12c = 32 \Rightarrow 12c = 32 - 4(5) \Rightarrow c = 1$, so $5 - b + 1 = 10 \Rightarrow b = -4$
from the first equation.

Thus, the equation of the parabola is $y = 5 - 4x + x^2$.

16. In this case, we have:

$$\begin{aligned} a-b+c &= -3 \\ a+b+c &= -1 \\ a+3b+9c &= 1 \end{aligned} \Rightarrow |P| = \begin{vmatrix} 1 & -1 & 1 \\ 1 & 1 & 1 \\ 1 & 3 & 9 \end{vmatrix} = 16 \neq 0.$$

Solving the system, we obtain $a = -2$, $b = 1$, and $c = 0$.
So, the parabola is degenerate: a straight line with equation $y = -2 + x$.

17. If all three points are on the parabola $y = a + bx + cx^2$, then the system is:

$$\begin{aligned} a + ba_1 + ca_1^2 &= b_1 \\ a + ba_2 + ca_2^2 &= b_2 \\ a + ba_3 + ca_3^2 &= b_3 \end{aligned} \Rightarrow \text{the coefficient matrix } P = \begin{bmatrix} 1 & a_1 & a_1^2 \\ 1 & a_2 & a_2^2 \\ 1 & a_3 & a_3^2 \end{bmatrix},$$

and since the a_i are distinct:

$$|P| \overset{\substack{R_2-R_1 \\ R_3-R_1}}{=} \begin{vmatrix} 1 & a_1 & a_1^2 \\ 0 & a_2-a_1 & a_2^2-a_1^2 \\ 0 & a_3-a_1 & a_3^2-a_1^2 \end{vmatrix} \overset{\substack{R_2/(a_2-a_1) \\ R_3/(a_3-a_1)}}{=} (a_2-a_1)(a_3-a_1) \begin{vmatrix} 1 & a_1 & a_1^2 \\ 0 & 1 & a_1+a_2 \\ 0 & 1 & a_1+a_3 \end{vmatrix}$$

$$\overset{R_3-R_2}{=} (a_2-a_1)(a_3-a_1) \begin{vmatrix} 1 & a_1 & a_1^2 \\ 0 & 1 & a_1+a_2 \\ 0 & 0 & a_3-a_2 \end{vmatrix} = (a_2-a_1)(a_3-a_1)(a_3-a_2) \neq 0.$$

18. At the third and sixth lines below, we expand along the first column

$$\begin{vmatrix} 1 & a_1 & a_1^2 & a_1^3 \\ 1 & a_2 & a_2^2 & a_2^3 \\ 1 & a_3 & a_3^2 & a_3^3 \\ 1 & a_4 & a_4^2 & a_4^3 \end{vmatrix} \overset{\substack{R_2-R_1\\R_3-R_1\\R_4-R_1}}{=} \begin{vmatrix} 1 & a_1 & a_1^2 & a_1^3 \\ 0 & a_2-a_1 & (a_2-a_1)(a_2+a_1) & (a_2-a_1)(a_1^2+a_1a_2+a_2^2) \\ 0 & a_3-a_1 & (a_3-a_1)(a_3+a_1) & (a_3-a_1)(a_1^2+a_1a_3+a_3^2) \\ 0 & a_4-a_1 & (a_4-a_1)(a_4+a_1) & (a_4-a_1)(a_1^2+a_1a_4+a_4^2) \end{vmatrix}$$

$$\overset{\substack{R_2/(a_2-a_1)\\R_3/(a_3-a_1)\\R_4/(a_4-a_1)}}{=} (a_2-a_1)(a_3-a_1)(a_4-a_1) \begin{vmatrix} 1 & a_1 & a_1^2 & a_1^3 \\ 0 & 1 & a_1+a_2 & a_1^2+a_1a_2+a_2^2 \\ 0 & 1 & a_1+a_3 & a_1^2+a_1a_3+a_3^2 \\ 0 & 1 & a_1+a_4 & a_1^2+a_1a_4+a_4^2 \end{vmatrix}$$

$$= (a_2-a_1)(a_3-a_1)(a_4-a_1) \begin{vmatrix} 1 & a_1+a_2 & a_1^2+a_1a_2+a_2^2 \\ 1 & a_1+a_3 & a_1^2+a_1a_3+a_3^2 \\ 1 & a_1+a_4 & a_1^2+a_1a_4+a_4^2 \end{vmatrix}$$

$$\overset{\substack{R_2-R_1\\R_3-R_1}}{=} (a_2-a_1)(a_3-a_1)(a_4-a_1) \begin{vmatrix} 1 & a_1+a_2 & a_1^2+a_1a_2+a_2^2 \\ 0 & a_3-a_2 & (a_3-a_2)(a_1+a_2+a_3) \\ 0 & a_4-a_2 & (a_4-a_2)(a_1+a_2+a_4) \end{vmatrix}$$

$$= (a_2-a_1)(a_3-a_1)(a_4-a_1)(a_3-a_2)(a_4-a_2) \begin{vmatrix} 1 & a_1+a_2+a_3 \\ 1 & a_1+a_2+a_4 \end{vmatrix}$$

$$\overset{R_2-R_1}{=} (a_2-a_1)(a_3-a_1)(a_4-a_1)(a_3-a_2)(a_4-a_2) \begin{vmatrix} 1 & a_1+a_2+a_3 \\ 0 & a_4-a_3 \end{vmatrix}$$

$$= (a_2-a_1)(a_3-a_1)(a_4-a_1)(a_3-a_2)(a_4-a_2)(a_4-a_3).$$

If we interpret the original matrix as the coefficient matrix of the system

$$\begin{aligned} a+ba_1+ca_1^2+da_1^3 &= b_1 \\ a+ba_2+ca_2^2+da_2^3 &= b_2 \\ a+ba_3+ca_3^2+da_3^3 &= b_3 \\ a+ba_4+ca_4^2+da_4^3 &= b_4 \end{aligned}$$

Then we have shown that if the a_i are distinct, then there is a unique solution to this system. So, a unique $y = a + bx + cx^2 + dx^3$ is satisfied by (a_i, b_i), $1 \le i \le 4$.

19. For the $n \times n$ determinant given, we proceed as we did in Problem 18.
At the ith step, we subtract row 1 from all subsequent rows,
then divide row j by $(a_j - a_i)$ for $2 \le j \le n$, then expand along the first column.
So at the ith step we obtain a factor of $\prod_{j=i+1}^{n} (a_j - a_i)$.
The first step is as follows:

$$D = \begin{vmatrix} 1 & a_1 & a_1^2 & \cdots & a_1^{n-1} \\ 1 & a_2 & a_2^2 & \cdots & a_2^{n-1} \\ 1 & a_3 & a_3^2 & \cdots & a_3^{n-1} \\ \vdots & \vdots & \vdots & \ddots & \vdots \\ 1 & a_n & a_n^2 & \cdots & a_n^{n-1} \end{vmatrix} = \prod_{k=2}^{n} (a_k - a_1) \begin{vmatrix} 1 & a_1 & a_1^2 & \cdots & a_1^{n-1} \\ 0 & 1 & a_1 + a_2 & \cdots & \dfrac{a_2^{n-1} - a_1^{n-1}}{a_2 - a_1} \\ 0 & 1 & a_1 + a_3 & \cdots & \dfrac{a_3^{n-1} - a_1^{n-1}}{a_3 - a_1} \\ \vdots & \vdots & \vdots & \ddots & \vdots \\ 0 & 1 & a_1 + a_n & \cdots & \dfrac{a_n^{n-1} - a_1^{n-1}}{a_n - a_1} \end{vmatrix}$$

$$= \prod_{k=2}^{n} (a_k - a_1) \begin{vmatrix} 1 & a_1 + a_2 & \cdots & \dfrac{a_2^{n-1} - a_1^{n-1}}{a_2 - a_1} \\ 1 & a_1 + a_3 & \cdots & \dfrac{a_3^{n-1} - a_1^{n-1}}{a_3 - a_1} \\ \vdots & \vdots & \ddots & \vdots \\ 1 & a_1 + a_n & \cdots & \dfrac{a_n^{n-1} - a_1^{n-1}}{a_n - a_1} \end{vmatrix}.$$

We see that after $n - 1$ iterations, we have

$$D = \left[\prod_{k=2}^{n} (a_k - a_1)\right] \left[\prod_{k=3}^{n} (a_k - a_2)\right] \cdots \left[\prod_{k=n-1}^{n} (a_k - a_{n-2})\right] \left[\prod_{k=n}^{n} (a_k - a_{n-1})\right]$$

$$= \prod_{1 \le i < j \le n} (a_j - a_i).$$

This means that for distinct $\{a_i\}$ and any $\{b_i\}$, $1 \le i \le n$,
there is a unique polynomial of degree $n - 1$ whose graph passes through the points (a_i, b_i).

4.3 Eigenvalues and Eigenvectors of $n \times n$ Matrices

1. We follow the procedure outlined before Example 1.

(a) The characteristic polynomial is $\det(A - \lambda I) = 0$, so we have:

$$\det(A - \lambda I) = \begin{vmatrix} 1-\lambda & 3 \\ -2 & 6-\lambda \end{vmatrix} = (1-\lambda)(6-\lambda) - 3(-2) = \lambda^2 - 7\lambda + 12 = (\lambda - 3)(\lambda - 4).$$

(b) The characteristic equation is $(\lambda - 3)(\lambda - 4) = 0$, which has solutions $\lambda_1 = 3$ and $\lambda_2 = 4$.

(c) To find the eigenvectors corresponding to λ_1, we find the null space of $A - 3I = \begin{bmatrix} -2 & 3 \\ -2 & 3 \end{bmatrix}$.

Row reduction produces $\left[\begin{array}{cc|c} -2 & 3 & 0 \\ -2 & 3 & 0 \end{array}\right] \longrightarrow \left[\begin{array}{cc|c} 1 & -\frac{3}{2} & 0 \\ 0 & 0 & 0 \end{array}\right]$.

Thus, $\mathbf{x} = \begin{bmatrix} x_1 \\ x_2 \end{bmatrix}$ is in the eigenspace E_3 if and only if $x_1 - \frac{3}{2}x_2 = 0 \Leftrightarrow x_2 = \frac{2}{3}x_1$.

Thus, $E_3 = \text{span}\left(\begin{bmatrix} 1 \\ \frac{2}{3} \end{bmatrix}\right)$.

Similarly, $A - 4I = \begin{bmatrix} -3 & 3 \\ -2 & 2 \end{bmatrix} \longrightarrow \begin{bmatrix} 1 & -1 \\ 0 & 0 \end{bmatrix}$, so $E_4 = \text{span}\left(\begin{bmatrix} 1 \\ 1 \end{bmatrix}\right)$.

(d) Each eigenvalue has algebraic and geometric multiplicity 1.

2. (a) $\det(A - \lambda I) = \begin{vmatrix} 2-\lambda & 1 \\ -1 & -\lambda \end{vmatrix} = \lambda^2 - 2\lambda + 1 = (\lambda - 1)^2$.

(b) $(\lambda - 1)^2 = 0 \Leftrightarrow \lambda_1 = \lambda_2 = 1$.

(c) $A - I = \begin{bmatrix} 1 & 1 \\ -1 & -1 \end{bmatrix} \longrightarrow \begin{bmatrix} 1 & 1 \\ 0 & 0 \end{bmatrix}$, so $E_1 = \text{span}\left(\begin{bmatrix} 1 \\ -1 \end{bmatrix}\right)$.

(d) The eigenvalue 1 has algebraic multiplicity 2 and geometric multiplicity 1.

3. (a) $\det(A - \lambda I) = \begin{vmatrix} 1-\lambda & 1 & 0 \\ 0 & -2-\lambda & 1 \\ 0 & 0 & 3-\lambda \end{vmatrix} = (1-\lambda)(-2-\lambda)(3-\lambda)$.

(b) $(1-\lambda)(-2-\lambda)(3-\lambda) = 0 \Leftrightarrow \lambda_1 = -2$, $\lambda_2 = 1$, or $\lambda_3 = 3$.

(c) $A + 2I = \begin{bmatrix} 3 & 1 & 0 \\ 0 & 0 & 1 \\ 0 & 0 & 5 \end{bmatrix} \longrightarrow \begin{bmatrix} 1 & \frac{1}{3} & 0 \\ 0 & 0 & 1 \\ 0 & 0 & 0 \end{bmatrix}$, so $E_{-2} = \text{span}\left(\begin{bmatrix} 1 \\ -3 \\ 0 \end{bmatrix}\right)$.

$A - I = \begin{bmatrix} 0 & 1 & 0 \\ 0 & -3 & 1 \\ 0 & 0 & 2 \end{bmatrix} \longrightarrow \begin{bmatrix} 0 & 1 & 0 \\ 0 & 0 & 1 \\ 0 & 0 & 0 \end{bmatrix}$, so $E_1 = \text{span}\left(\begin{bmatrix} 1 \\ 0 \\ 0 \end{bmatrix}\right)$.

$A - 3I = \begin{bmatrix} -2 & 1 & 0 \\ 0 & -5 & 1 \\ 0 & 0 & 0 \end{bmatrix} \longrightarrow \begin{bmatrix} 1 & 0 & -\frac{1}{10} \\ 0 & 1 & -\frac{1}{5} \\ 0 & 0 & 0 \end{bmatrix}$, so $E_3 = \text{span}\left(\begin{bmatrix} 1 \\ 2 \\ 10 \end{bmatrix}\right)$.

(d) Each eigenvalue has algebraic and geometric multiplicity 1.

4. (a) $$\det(A - \lambda I) = \begin{vmatrix} 1-\lambda & 0 & 1 \\ 0 & 1-\lambda & 1 \\ 1 & 1 & -\lambda \end{vmatrix} = -(1-\lambda) - (1-\lambda) - \lambda(1-\lambda)^2$$
$$= -2(\lambda - 1) - \lambda(\lambda - 1)^2 = (\lambda - 1)(\lambda - 2)(\lambda + 1).$$

(b) $(\lambda - 1)(\lambda - 2)(\lambda + 1) = 0 \Leftrightarrow \lambda_1 = -1,\ \lambda_2 = 1,\ \lambda_3 = 2.$

(c) $$A + I = \begin{bmatrix} 2 & 0 & 1 \\ 0 & 2 & 1 \\ 1 & 1 & 1 \end{bmatrix} \longrightarrow \begin{bmatrix} 1 & 0 & \frac{1}{2} \\ 0 & 1 & \frac{1}{2} \\ 0 & 0 & 0 \end{bmatrix}, \text{ so } E_{-1} = \text{span}\left(\begin{bmatrix} 1 \\ 1 \\ -2 \end{bmatrix}\right).$$

$$A - I = \begin{bmatrix} 0 & 0 & 1 \\ 0 & 0 & 1 \\ 1 & 1 & -1 \end{bmatrix} \longrightarrow \begin{bmatrix} 1 & 1 & 0 \\ 0 & 0 & 1 \\ 0 & 0 & 0 \end{bmatrix}, \text{ so } E_1 = \text{span}\left(\begin{bmatrix} 1 \\ -1 \\ 0 \end{bmatrix}\right).$$

$$A - 2I = \begin{bmatrix} -1 & 0 & 1 \\ 0 & -1 & 1 \\ 1 & 1 & -2 \end{bmatrix} \longrightarrow \begin{bmatrix} 1 & 0 & -1 \\ 0 & 1 & -1 \\ 0 & 0 & 0 \end{bmatrix}, \text{ so } E_2 = \text{span}\left(\begin{bmatrix} 1 \\ 1 \\ 1 \end{bmatrix}\right).$$

(d) Each eigenvalue has algebraic and geometric multiplicity 1.

5. (a) $$\det(A - \lambda I) = \begin{vmatrix} 1-\lambda & 2 & 0 \\ -1 & -1-\lambda & 1 \\ 0 & 1 & 1-\lambda \end{vmatrix} = -(1-\lambda) + (1-\lambda)\left[(1-\lambda)(-1-\lambda) - (-2)\right]$$
$$= \lambda - 1 + \left(1 - \lambda + \lambda^2 - \lambda^3\right) = \lambda^2 - \lambda^3 = \lambda^2(1 - \lambda).$$

(b) $\lambda^2(1 - \lambda) = 0 \Leftrightarrow \lambda_1 = \lambda_2 = 0,\ \lambda_3 = 1.$

(c) $$A = \begin{bmatrix} 1 & 2 & 0 \\ -1 & -1 & 1 \\ 0 & 1 & 1 \end{bmatrix} \longrightarrow \begin{bmatrix} 1 & 0 & -2 \\ 0 & 1 & 1 \\ 0 & 0 & 0 \end{bmatrix}, \text{ so } E_0 = \text{span}\left(\begin{bmatrix} 1 \\ -\frac{1}{2} \\ \frac{1}{2} \end{bmatrix}\right).$$

$$A - I = \begin{bmatrix} 0 & 2 & 0 \\ -1 & -2 & 1 \\ 0 & 1 & 0 \end{bmatrix} \longrightarrow \begin{bmatrix} 1 & 0 & -1 \\ 0 & 1 & 0 \\ 0 & 0 & 0 \end{bmatrix}, \text{ so } E_1 = \text{span}\left(\begin{bmatrix} 1 \\ 0 \\ 1 \end{bmatrix}\right).$$

(d) 0 has algebraic multiplicity 2 and geometric multiplicity 1,
while 1 has algebraic and geometric multiplicity 1.

6. (a) $\det(A-\lambda I) = \begin{vmatrix} 1-\lambda & 0 & 2 \\ 3 & -1-\lambda & 3 \\ 2 & 0 & 1-\lambda \end{vmatrix} = (-1-\lambda)\left[(1-\lambda)^2 - 4\right] = (-1-\lambda)^2(3-\lambda).$

(b) $(-1-\lambda)^2(3-\lambda) = 0 \Leftrightarrow \lambda_1 = \lambda_2 = -1,\ \lambda_3 = 3.$

(c) $A + I = \begin{bmatrix} 2 & 0 & 2 \\ 3 & 0 & 3 \\ 2 & 0 & 2 \end{bmatrix} \longrightarrow \begin{bmatrix} 1 & 0 & 1 \\ 0 & 0 & 0 \\ 0 & 0 & 0 \end{bmatrix}$, so $E_{-1} = \text{span}\left(\begin{bmatrix} 1 \\ 0 \\ -1 \end{bmatrix}, \begin{bmatrix} 0 \\ 1 \\ 0 \end{bmatrix}\right).$

$A - 3I = \begin{bmatrix} -2 & 0 & 2 \\ 3 & -4 & 3 \\ 2 & 0 & -2 \end{bmatrix} \longrightarrow \begin{bmatrix} 1 & 0 & -1 \\ 0 & 1 & -\frac{3}{2} \\ 0 & 0 & 0 \end{bmatrix}$, so $E_3 = \text{span}\left(\begin{bmatrix} 1 \\ \frac{3}{2} \\ 1 \end{bmatrix}\right).$

(d) -1 has algebraic and geometric multiplicity 2,
while 3 has algebraic and geometric multiplicity 1.

7. (a) $\det(A-\lambda I) = \begin{vmatrix} 4-\lambda & 0 & 1 \\ 2 & 3-\lambda & 2 \\ -1 & 0 & 2-\lambda \end{vmatrix} = (3-\lambda)\left[(4-\lambda)(2-\lambda) - (-1)\right]$

$= (3-\lambda)\left(\lambda^2 - 6\lambda + 9\right) = (3-\lambda)^3.$

(b) $(3-\lambda)^3 = 0 \Leftrightarrow \lambda_1 = \lambda_2 = \lambda_3 = 3.$

(c) $A - 3I = \begin{bmatrix} 1 & 0 & 1 \\ 2 & 0 & 2 \\ -1 & 0 & -1 \end{bmatrix} \longrightarrow \begin{bmatrix} 1 & 0 & 1 \\ 0 & 0 & 0 \\ 0 & 0 & 0 \end{bmatrix}$, so $E_3 = \text{span}\left(\begin{bmatrix} 1 \\ 0 \\ -1 \end{bmatrix}, \begin{bmatrix} 0 \\ 1 \\ 0 \end{bmatrix}\right).$

(d) 3 has algebraic multiplicity 3 and geometric multiplicity 2.

8. (a) $\det(A-\lambda I) = \begin{vmatrix} 1-\lambda & -1 & -1 \\ 0 & 2-\lambda & 0 \\ -1 & -1 & 1-\lambda \end{vmatrix} = (2-\lambda)\left[(1-\lambda)^2 - 1\right]$

$= (2-\lambda)\left(\lambda^2 - 2\lambda\right) = -\lambda(2-\lambda)^2.$

(b) $-\lambda(2-\lambda)^2 = 0 \Leftrightarrow \lambda_1 = 0,\ \lambda_2 = \lambda_3 = 2.$

(c) $A = \begin{bmatrix} 1 & -1 & -1 \\ 0 & 2 & 0 \\ -1 & -1 & 1 \end{bmatrix} \longrightarrow \begin{bmatrix} 1 & 0 & -1 \\ 0 & 1 & 0 \\ 0 & 0 & 0 \end{bmatrix}$, so $E_0 = \text{span}\left(\begin{bmatrix} 1 \\ 0 \\ 1 \end{bmatrix}\right).$

$A - 2I = \begin{bmatrix} -1 & -1 & -1 \\ 0 & 0 & 0 \\ -1 & -1 & -1 \end{bmatrix} \longrightarrow \begin{bmatrix} 1 & 1 & 1 \\ 0 & 0 & 0 \\ 0 & 0 & 0 \end{bmatrix}$, so $E_2 = \text{span}\left(\begin{bmatrix} 1 \\ -1 \\ 0 \end{bmatrix}, \begin{bmatrix} 1 \\ 0 \\ -1 \end{bmatrix}\right).$

(d) 0 has algebraic and geometric multiplicity 1,
while 2 has algebraic and geometric multiplicity 2.

9. (a) $$\det(A-\lambda I) = \begin{vmatrix} 3-\lambda & 1 & 0 & 0 \\ -1 & 1-\lambda & 0 & 0 \\ 0 & 0 & 1-\lambda & 4 \\ 0 & 0 & 1 & 1-\lambda \end{vmatrix} = \begin{vmatrix} 3-\lambda & 1 \\ -1 & 1-\lambda \end{vmatrix} \begin{vmatrix} 1-\lambda & 4 \\ 1 & 1-\lambda \end{vmatrix}$$
$$= [(3-\lambda)(1-\lambda)+1]\left[(\lambda-1)^2-4\right] = (\lambda-2)^2(\lambda-3)(\lambda+1).$$

(b) $(\lambda-2)^2(\lambda-3)(\lambda+1) = 0 \Leftrightarrow \lambda_1 = -1,\ \lambda_2 = \lambda_3 = 2,\ \lambda_4 = 3.$

(c) $$A+I = \begin{bmatrix} 4 & 1 & 0 & 0 \\ -1 & 2 & 0 & 0 \\ 0 & 0 & 2 & 4 \\ 0 & 0 & 1 & 2 \end{bmatrix} \longrightarrow \begin{bmatrix} 1 & 0 & 0 & 0 \\ 0 & 1 & 0 & 0 \\ 0 & 0 & 1 & 2 \\ 0 & 0 & 0 & 0 \end{bmatrix}, \text{ so } E_{-1} = \text{span}\left(\begin{bmatrix} 0 \\ 0 \\ 1 \\ -\frac{1}{2} \end{bmatrix}\right).$$

$$A-2I = \begin{bmatrix} 1 & 1 & 0 & 0 \\ -1 & -1 & 0 & 0 \\ 0 & 0 & -1 & 4 \\ 0 & 0 & 1 & -1 \end{bmatrix} \longrightarrow \begin{bmatrix} 1 & 1 & 0 & 0 \\ 0 & 0 & 1 & 0 \\ 0 & 0 & 0 & 1 \\ 0 & 0 & 0 & 0 \end{bmatrix}, \text{ so } E_2 = \text{span}\left(\begin{bmatrix} 1 \\ -1 \\ 0 \\ 0 \end{bmatrix}\right).$$

$$A-3I = \begin{bmatrix} 0 & 1 & 0 & 0 \\ -1 & -2 & 0 & 0 \\ 0 & 0 & -2 & 4 \\ 0 & 0 & 1 & -2 \end{bmatrix} \longrightarrow \begin{bmatrix} 1 & 0 & 0 & 0 \\ 0 & 1 & 0 & 0 \\ 0 & 0 & 1 & -2 \\ 0 & 0 & 0 & 0 \end{bmatrix}, \text{ so } E_3 = \text{span}\left(\begin{bmatrix} 0 \\ 0 \\ 1 \\ \frac{1}{2} \end{bmatrix}\right).$$

(d) -1 and 3 have algebraic and geometric multiplicity 1,
while 2 has algebraic multiplicity 2 and geometric multiplicity 1.

10. (a) $$\det(A-\lambda I) = \begin{vmatrix} 2-\lambda & 1 & 1 & 0 \\ 0 & 1-\lambda & 4 & 5 \\ 0 & 0 & 3-\lambda & 1 \\ 0 & 0 & 0 & 2-\lambda \end{vmatrix} = (1-\lambda)(2-\lambda)^2(3-\lambda).$$

(b) $(1-\lambda)(2-\lambda)^2(3-\lambda) = 0 \Leftrightarrow \lambda_1 = 1,\ \lambda_2 = \lambda_3 = 2,\ \lambda_4 = 3.$

(c) $$A-I = \begin{bmatrix} 1 & 1 & 1 & 0 \\ 0 & 0 & 4 & 5 \\ 0 & 0 & 2 & 1 \\ 0 & 0 & 0 & 1 \end{bmatrix} \longrightarrow \begin{bmatrix} 1 & 1 & 0 & 0 \\ 0 & 0 & 1 & 0 \\ 0 & 0 & 0 & 1 \\ 0 & 0 & 0 & 0 \end{bmatrix}, \text{ so } E_1 = \text{span}\left(\begin{bmatrix} 1 \\ -1 \\ 0 \\ 0 \end{bmatrix}\right).$$

$$A-2I = \begin{bmatrix} 0 & 1 & 1 & 0 \\ 0 & -1 & 4 & 5 \\ 0 & 0 & 1 & 1 \\ 0 & 0 & 0 & 0 \end{bmatrix} \longrightarrow \begin{bmatrix} 0 & 1 & 0 & -1 \\ 0 & 0 & 1 & 1 \\ 0 & 0 & 0 & 0 \\ 0 & 0 & 0 & 0 \end{bmatrix}, \text{ so } E_2 = \text{span}\left(\begin{bmatrix} 1 \\ 0 \\ 0 \\ 0 \end{bmatrix}, \begin{bmatrix} 0 \\ 1 \\ -1 \\ 1 \end{bmatrix}\right).$$

$$A-3I = \begin{bmatrix} -1 & 1 & 1 & 0 \\ 0 & -2 & 4 & 5 \\ 0 & 0 & 0 & 1 \\ 0 & 0 & 0 & -1 \end{bmatrix} \longrightarrow \begin{bmatrix} 1 & 0 & -3 & 0 \\ 0 & 1 & -2 & 0 \\ 0 & 0 & 0 & 1 \\ 0 & 0 & 0 & 0 \end{bmatrix}, \text{ so } E_3 = \text{span}\left(\begin{bmatrix} 1 \\ \frac{2}{3} \\ \frac{1}{3} \\ 0 \end{bmatrix}\right).$$

(d) 1 and 3 have algebraic and geometric multiplicity 1,
while 2 has algebraic and geometric multiplicity 2.

11. (a) $\det(A-\lambda I) = \begin{vmatrix} 1-\lambda & 0 & 0 & 0 \\ 0 & 1-\lambda & 0 & 0 \\ 1 & 1 & 3-\lambda & 0 \\ -2 & 1 & 2 & -1-\lambda \end{vmatrix} = (-1-\lambda)(1-\lambda)^2(3-\lambda).$

(b) $(-1-\lambda)(1-\lambda)^2(3-\lambda) = 0 \Leftrightarrow \lambda_1 = -1,\ \lambda_2 = \lambda_3 = 1,\ \lambda_4 = 3.$

(c) $A+I = \begin{bmatrix} 2 & 0 & 0 & 0 \\ 0 & 2 & 0 & 0 \\ 1 & 1 & 4 & 0 \\ -2 & 1 & 2 & 0 \end{bmatrix} \longrightarrow \begin{bmatrix} 1 & 0 & 0 & 0 \\ 0 & 1 & 0 & 0 \\ 0 & 0 & 1 & 0 \\ 0 & 0 & 0 & 0 \end{bmatrix}$, so $E_{-1} = \operatorname{span}\left(\begin{bmatrix} 0 \\ 0 \\ 0 \\ 1 \end{bmatrix}\right)$.

$A-I = \begin{bmatrix} 0 & 0 & 0 & 0 \\ 0 & 0 & 0 & 0 \\ 1 & 1 & 2 & 0 \\ -2 & 1 & 2 & -2 \end{bmatrix} \longrightarrow \begin{bmatrix} 1 & 0 & 0 & \frac{2}{3} \\ 0 & 1 & 2 & -\frac{2}{3} \\ 0 & 0 & 0 & 0 \\ 0 & 0 & 0 & 0 \end{bmatrix}$, so $E_1 = \operatorname{span}\left(\begin{bmatrix} 1 \\ -1 \\ 0 \\ -\frac{3}{2} \end{bmatrix}, \begin{bmatrix} 1 \\ 0 \\ -\frac{1}{2} \\ -\frac{3}{2} \end{bmatrix}\right)$.

$A-3I = \begin{bmatrix} -2 & 0 & 0 & 0 \\ 0 & -2 & 0 & 0 \\ 1 & 1 & 0 & 0 \\ -2 & 1 & 2 & -4 \end{bmatrix} \longrightarrow \begin{bmatrix} 1 & 0 & 0 & 0 \\ 0 & 1 & 0 & 0 \\ 0 & 0 & 1 & -2 \\ 0 & 0 & 0 & 0 \end{bmatrix}$, so $E_3 = \operatorname{span}\left(\begin{bmatrix} 0 \\ 0 \\ 1 \\ \frac{1}{2} \end{bmatrix}\right)$

(d) -1 and 3 have algebraic and geometric multiplicity 1, while 1 has algebraic and geometric multiplicity 2.

12. (a) $\det(A-\lambda I) = \begin{vmatrix} 4-\lambda & 0 & 1 & 0 \\ 0 & 4-\lambda & 1 & 1 \\ 0 & 0 & 1-\lambda & 2 \\ 0 & 0 & 3 & -\lambda \end{vmatrix} = (4-\lambda)^2[(1-\lambda)(-\lambda)-6]$

$= (4-\lambda)^2(\lambda-3)(\lambda+2).$

(b) $(4-\lambda)^2(\lambda-3)(\lambda+2) = 0 \Leftrightarrow \lambda_1 = -2,\ \lambda_2 = 3,\ \lambda_3 = \lambda_4 = 4.$

(c) $A+2I = \begin{bmatrix} 6 & 0 & 1 & 0 \\ 0 & 6 & 1 & 1 \\ 0 & 0 & 3 & 2 \\ 0 & 0 & 3 & 2 \end{bmatrix} \longrightarrow \begin{bmatrix} 1 & 0 & 0 & -\frac{1}{9} \\ 0 & 1 & 0 & \frac{1}{18} \\ 0 & 0 & 1 & \frac{2}{3} \\ 0 & 0 & 0 & 0 \end{bmatrix}$, so $E_{-2} = \operatorname{span}\left(\begin{bmatrix} 1 \\ -\frac{1}{2} \\ -6 \\ 9 \end{bmatrix}\right)$.

$A-3I = \begin{bmatrix} 1 & 0 & 1 & 0 \\ 0 & 1 & 1 & 1 \\ 0 & 0 & -2 & 2 \\ 0 & 0 & 3 & -3 \end{bmatrix} \longrightarrow \begin{bmatrix} 1 & 0 & 0 & 1 \\ 0 & 1 & 0 & 2 \\ 0 & 0 & 1 & -1 \\ 0 & 0 & 0 & 0 \end{bmatrix}$, so $E_3 = \operatorname{span}\left(\begin{bmatrix} 1 \\ 2 \\ -1 \\ -1 \end{bmatrix}\right)$.

$A-4I = \begin{bmatrix} 0 & 0 & 1 & 0 \\ 0 & 0 & 1 & 1 \\ 0 & 0 & -3 & 2 \\ 0 & 0 & 3 & -4 \end{bmatrix} \longrightarrow \begin{bmatrix} 0 & 0 & 1 & 0 \\ 0 & 0 & 0 & 1 \\ 0 & 0 & 0 & 0 \\ 0 & 0 & 0 & 0 \end{bmatrix}$, so $E_4 = \operatorname{span}\left(\begin{bmatrix} 0 \\ 0 \\ 1 \\ 0 \end{bmatrix}, \begin{bmatrix} 0 \\ 0 \\ 0 \\ 1 \end{bmatrix}\right)$.

(d) -2 and 3 have algebraic and geometric multiplicity 1, while 4 has algebraic and geometric multiplicity 2.

13. We need to show if $A\mathbf{x} = \lambda\mathbf{x}$, then $A^{-1}\mathbf{x} = \frac{1}{\lambda}\mathbf{x} = \lambda^{-1}\mathbf{x}$.

Since $A\mathbf{x} = \lambda\mathbf{x}$, we have $A^{-1}(A\mathbf{x}) = A^{-1}(\lambda\mathbf{x}) = \lambda(A^{-1}\mathbf{x})$.

So, $\lambda(A^{-1}\mathbf{x}) = (A^{-1}A)\mathbf{x} = \mathbf{x}$ which implies $A^{-1}\mathbf{x} = \frac{1}{\lambda}\mathbf{x} = \lambda^{-1}\mathbf{x}$ as required.

14. Using induction and the proofs of 4.18(a) and (b), we will prove Theorem 4.18(c): For any integer n if $A\mathbf{x} = \lambda\mathbf{x}$, then $A^n\mathbf{x} = \lambda^n\mathbf{x}$.

As suggested, we will also use the fourth *Remark* following Theorem 3.9 in Section 3.3: If A is invertible and n is a positive integer, then $A^{-n} = (A^{-1})^n = (A^n)^{-1}$.

Since (a) gives us the result for positive integers, we proceed by induction on $-n$.

1: If $A\mathbf{x} = \lambda\mathbf{x}$, then $A^{-1}\mathbf{x} = \lambda^{-1}\mathbf{x}$.
Since $\lambda^{-1} = \frac{1}{\lambda}$, this is the statement of Theorem 4.18(b). So there is nothing to show.

k: If $A\mathbf{x} = \lambda\mathbf{x}$, then $A^{-k}\mathbf{x} = \lambda^{-k}\mathbf{x}$.
This is the induction hypothesis.

$k+1$: If $A\mathbf{x} = \lambda\mathbf{x}$, then $A^{-(k+1)}\mathbf{x} = \lambda^{-(k+1)}\mathbf{x}$.
This is the statement we must prove using the induction hypothesis.

$$A^{-(k+1)}\mathbf{x} \overset{\substack{\text{by}\\ \text{Remark}}}{=} A^{-1}(A^{-k}\mathbf{x}) \overset{\substack{\text{by}\\ \text{induc}}}{=} A^{-1}(\lambda^{-k}\mathbf{x}) = \lambda^{-k}(A^{-1}\mathbf{x}) \overset{\substack{\text{by}\\ n=1}}{=} \lambda^{-k}(\lambda^{-1}\mathbf{x}) = \lambda^{-(k+1)}\mathbf{x}$$

Q: Why does the *Remark* imply that $A^{-(k+1)} = A^{-1}A^{-k}$?

A: The remark implies both $A^{-k} = (A^k)^{-1}$ and $A^{-(k+1)} = (A^{k+1})^{-1}$.
So we need only show $A^{-1}A^{-k} = (A^{k+1})^{-1}$. That is, $(A^{-1}A^{-k})(A^{k+1}) = I$.
That is obvious since: $(A^{-1}A^{-k})(A^{k+1}) = A^{-1}(A^{-k}A^k)A = A^{-1}A = I$.

Q: What does the *Remark* and our work above suggest about integer exponents of A?
A: They behave precisely as we would hope. That is, like the exponents of real variables.

15. Since $\mathbf{x} = 2\mathbf{v}_1 + 3\mathbf{v}_2$ we have $A^{10}\mathbf{x} = A^{10}(2\mathbf{v}_1 + 3\mathbf{v}_2) = 2A^{10}\mathbf{v}_1 + 3A^{10}\mathbf{v}_2$.
But by Theorem 4.4(a), $\mathbf{v}_1$ and $\mathbf{v}_2$ are eigenvectors of A^{10} with eigenvalues λ_1^{10} and λ_2^{10}.

So, $2\lambda_1^{10}\mathbf{v}_1 + 3\lambda_2^{10}\mathbf{v}_2 = 2\left(\frac{1}{2}\right)^{10}\mathbf{v}_1 + 3\,(2)^{10}\,\mathbf{v}_2 = \frac{1}{512}\mathbf{v}_1 + 3072\mathbf{v}_2 = \begin{bmatrix} 3072 + \frac{1}{512} \\ 3072 - \frac{1}{512} \end{bmatrix}$.

16. As in Exercise 15, $A^k\mathbf{x} = 2\left(\frac{1}{2}\right)^k\mathbf{v}_1 + 3\,(2)^k\,\mathbf{v}_2 = \begin{bmatrix} 3\cdot 2^k + 2^{1-k} \\ 3\cdot 2^k - 2^{1-k} \end{bmatrix}$.

As $k \to \infty$, the $\mathbf{v}_2$ term dominates and $A^k\mathbf{x} \approx 3\cdot 2^k\mathbf{v}_2$.

17. We must find $\mathbf{x}$ as a linear combination $a_1\mathbf{v}_1 + a_2\mathbf{v}_2 + a_3\mathbf{v}_3$ of the eigenvectors. So:

$$\begin{bmatrix} 2 \\ 1 \\ 2 \end{bmatrix} = a_1 \begin{bmatrix} 1 \\ 0 \\ 0 \end{bmatrix} + a_2 \begin{bmatrix} 1 \\ 1 \\ 0 \end{bmatrix} + a_3 \begin{bmatrix} 1 \\ 1 \\ 1 \end{bmatrix}$$

We must have $a_3 = 2$, so this reduces to

$$\begin{bmatrix} 0 \\ -1 \\ 0 \end{bmatrix} = a_1 \begin{bmatrix} 1 \\ 0 \\ 0 \end{bmatrix} + a_2 \begin{bmatrix} 1 \\ 1 \\ 0 \end{bmatrix}$$

which has solution $a_1 = 1$, $a_2 = -1$. Thus,

$$A^{20}\mathbf{x} = \lambda_1^{20}\mathbf{v}_1 - \lambda_2^{20}\mathbf{v}_2 + 2 \cdot \lambda_3^{20}\mathbf{v}_3 = \begin{bmatrix} -\frac{1}{3^{20}} - \frac{1}{3^{20}} + 2 \\ -\frac{1}{3^{20}} + 2 \\ 2 \end{bmatrix} = \begin{bmatrix} 2 \\ 2 - \frac{1}{3^{20}} \\ 2 \end{bmatrix}$$

18. As in Exercise 17, $A^k\mathbf{x} = \begin{bmatrix} 2 - \frac{2}{3^k} \\ 2 - \frac{1}{3^k} \\ 2 \end{bmatrix}$. As $k \to \infty$, $A^k\mathbf{x} \to \begin{bmatrix} 2 \\ 2 \\ 2 \end{bmatrix}$.

19. (a) The key observation is that $A^T - \lambda I = A^T - (\lambda I)^T = (A - \lambda I)^T$.
Thus, using Theorem 4.10, the characteristic polynomial of A^T is

$$\det\left(A^T - \lambda I\right) = \det\left(A - \lambda I\right)^T = \det\left(A - \lambda I\right)$$

But $\det(A - \lambda I)$ is the characteristic polynomial of A.
Therefore, A and A^T have the same eigenvalues as we were to show.

(b) $A = \begin{bmatrix} 1 & 0 \\ 1 & 2 \end{bmatrix}$ has eigenspaces $E_1 = \text{span}\left(\begin{bmatrix} 1 \\ -1 \end{bmatrix}\right)$ and $E_2 = \text{span}\left(\begin{bmatrix} 0 \\ 1 \end{bmatrix}\right)$,
while $A^T = \begin{bmatrix} 1 & 1 \\ 0 & 2 \end{bmatrix}$ has eigenspaces $E_1 = \text{span}\left(\begin{bmatrix} 1 \\ 0 \end{bmatrix}\right)$ and $E_2 = \text{span}\left(\begin{bmatrix} 1 \\ 1 \end{bmatrix}\right)$.

20. Given $A^n = O$, we need to show if $A\mathbf{x} = \lambda\mathbf{x}$ then $\lambda = 0$. That is:
1) 0 is an eigenvalue of A and 2) if λ is an eigenvalue of A, then $\lambda = 0$.

To prove assertion 1 we make the following observation:

Q: What is the contrapositive of Theorem 4.16?
A: A is *not* invertible if and only if 0 *is* an eigenvalue of A.

Q: What does the contrapositive of Theorem 4.16 imply?
A: $\det A = 0$ if and only if 0 *is* an eigenvalue of A. Why? Because the contrapositive of Theorem 4.6 in Section 4.2 implies $\det A = 0$ if and only if A is *not* invertible.

So, to prove 0 is an eigenvalue of A it suffices to show $\det A = 0$.
That is, if $A^n = O$, then $\det A = 0$. So, 0 is an eigenvalue of A.

Since $(\det A)^n \overset{\substack{\text{Thm 4.8}\\ \text{Sect 4.2}}}{=} \det(A^n) \overset{\substack{A^n=O\\ \text{given}}}{=} \det O = 0$, $\det A = 0$. So, 0 is an eigenvalue of A.

Next we show if λ is an eigenvalue of A, then $\lambda = 0$.

If $A\mathbf{x} = \lambda\mathbf{x}$, then Theorem 4.18(c) implies $A^n\mathbf{x} = \lambda^n\mathbf{x} = O\mathbf{x} = \mathbf{0}$.
Since $\mathbf{x}$ is an eigenvector, $\mathbf{x} \neq \mathbf{0}$. So $\lambda^n\mathbf{x} = \mathbf{0}$ implies $\lambda^n = 0$. Therefore, $\lambda = 0$ as claimed.

21. Suppose A is idempotent with eigenvector $\mathbf{x}$ corresponding to λ.
Then $\lambda\mathbf{x} = A\mathbf{x} = A^2\mathbf{x} = A(A\mathbf{x}) = A(\lambda\mathbf{x}) = \lambda(A\mathbf{x}) = \lambda(\lambda\mathbf{x}) = \lambda^2\mathbf{x}$.

So, we get $\lambda\mathbf{x} = \lambda^2\mathbf{x} \Rightarrow \lambda = \lambda^2$ (because $\mathbf{x} \neq \mathbf{0}$) $\Rightarrow \lambda^2 - \lambda = \lambda(\lambda - 1) = 0 \Rightarrow \lambda = 0$ or 1.

22. $A\mathbf{v} = \lambda\mathbf{v} \Rightarrow A\mathbf{v} - cI\mathbf{v} = \lambda\mathbf{v} - cI\mathbf{v} \Leftrightarrow (A - cI)\,\mathbf{v} = (\lambda - c)\,\mathbf{v}$.
So $\mathbf{v}$ is an eigenvector of $A - cI$ with corresponding eigenvalue $\lambda - c$.

23. (a) $\det(A - \lambda I) = \begin{vmatrix} 3-\lambda & 2 \\ 5 & 0-\lambda \end{vmatrix} = \lambda^2 - 3\lambda - 10 = (\lambda+2)(\lambda-5) = 0 \Leftrightarrow \lambda = -2$ or 5.

$A + 2I = \begin{bmatrix} 5 & 2 \\ 5 & 2 \end{bmatrix}$, so $E_{-2} = \text{span}\left(\begin{bmatrix} 1 \\ -\frac{5}{2} \end{bmatrix}\right)$, and $A - 5I = \begin{bmatrix} -2 & 2 \\ 5 & -5 \end{bmatrix}$.

So, $E_5 = \text{span}\left(\begin{bmatrix} 1 \\ 1 \end{bmatrix}\right)$.

(b) By Theorem 4.4(b), A^{-1} has eigenvalues $-\frac{1}{2}$ and $\frac{1}{5}$ with

$E_{-1/2} = \text{span}\left(\begin{bmatrix} 1 \\ -\frac{5}{2} \end{bmatrix}\right)$ and $E_{1/5} = \text{span}\left(\begin{bmatrix} 1 \\ 1 \end{bmatrix}\right)$.

By Exercise 22, $A - 2I$ has eigenvalues -4 and 3 with

$E_{-4} = \text{span}\left(\begin{bmatrix} 1 \\ -\frac{5}{2} \end{bmatrix}\right)$ and $E_3 = \text{span}\left(\begin{bmatrix} 1 \\ 1 \end{bmatrix}\right)$, and

$A + 2I$ has eigenvalues 0 and 7 with

$E_0 = \text{span}\left(\begin{bmatrix} 1 \\ -\frac{5}{2} \end{bmatrix}\right)$ and $E_7 = \text{span}\left(\begin{bmatrix} 1 \\ 1 \end{bmatrix}\right)$.

24. (a) Let $A = \begin{bmatrix} 0 & 1 \\ 1 & 0 \end{bmatrix}$ and $B = \begin{bmatrix} 1 & 1 \\ 0 & 1 \end{bmatrix}$. A has eigenvalues ± 1 and B has eigenvalue 1, but $A + B = \begin{bmatrix} 1 & 2 \\ 1 & 1 \end{bmatrix}$ has neither 0 nor 2 as an eigenvalue (in fact it has eigenvalues $1 \pm \sqrt{2}$.)

(b) If A and B are as above, then $AB = \begin{bmatrix} 0 & 1 \\ 1 & 1 \end{bmatrix}$, which has eigenvalues $\frac{1}{2} \pm \frac{1}{2}\sqrt{5}$.

(c) If λ and μ correspond to $\mathbf{x}$, then we have $\lambda \mathbf{x} = A\mathbf{x}$ and $\mu \mathbf{x} = B\mathbf{x}$.
Thus, $(\lambda + \mu)\,\mathbf{x} = (A + B)\,\mathbf{x}$, showing that $\lambda + \mu$ is an eigenvalue of $A + B$.
Now if either $\lambda = 0$ or $\mu = 0$, then one of A and B has determinant 0,
in which case AB has determinant 0 and thus has 0 as an eigenvalue.
So suppose that $\lambda \neq 0$, $\mu \neq 0$.
Then $\mathbf{x} = \lambda A^{-1}\mathbf{x} \Rightarrow \mu\left(\lambda A^{-1}\mathbf{x}\right) = B\mathbf{x} \Rightarrow A^{-1}(\lambda\mu\mathbf{x}) = B\mathbf{x} \Rightarrow \lambda\mu\mathbf{x} = (AB)\,\mathbf{x}$.
Note that $\lambda\mu$ is an eigenvalue of BA as well.

25. As noted in Theorem 4.17(d), $A \longrightarrow I$ if and only if A is invertible.

Q: If the conjecture of this exercise were true, what would that imply?
A: Since the only eigenvalue of I is 1, all invertible matrices would only have eigenvalue 1. This is clearly nonsense. However, they may be *related.*

Q: Let $\mathbf{x}$ be an eigenvector of A corresponding to eigenvalue λ.
If $A \xrightarrow{R_i \leftrightarrow R_j} B$, that is $B = E_{ij}A$, what goes wrong?
A: Since $B = E_{ij}A$, we have $B\mathbf{x} = E_{ij}(A\mathbf{x}) = \lambda(E_{ij}\mathbf{x})$.
So the components of $\mathbf{x}$ are interchanged and $\mathbf{x}$ fails to be an eigenvector for B.

Q: If $A \xrightarrow{kR_i} B$, what goes wrong?
Q: If $A \xrightarrow{R_i + kR_j} B$, what goes wrong?

Q: So, if $A \longrightarrow B$, we have seen their eigenvalues are not necessarily equal. However:
If $A \longrightarrow B$, is there a relationship among the eigenvalues and eigenvectors?
A: Hint: $2I$ has eigenvalue 2. Can this process be generalized? See Exercise 41.

26. By definition, the companion matrix of $p(x) = x^2 - 7x + 12$ is

$C(p) = \begin{bmatrix} -(-7) & -12 \\ 1 & 0 \end{bmatrix} = \begin{bmatrix} 7 & -12 \\ 1 & 0 \end{bmatrix}$, and the characteristic polynomial of $C(p)$ is

$$\det \begin{bmatrix} 7 - \lambda & -12 \\ 1 & -\lambda \end{bmatrix} = (7 - \lambda)(-\lambda) - (-12)(1) = -(7 - \lambda)\lambda + 12 = \lambda^2 - 7\lambda + 12.$$

27. The companion matrix of $p(x) = x^3 + 3x^2 - 4x + 12$ is $C(p) = \begin{bmatrix} -3 & 4 & -12 \\ 1 & 0 & 0 \\ 0 & 1 & 0 \end{bmatrix}$,

and the characteristic polynomial of $C(p)$ is

$$\det \begin{bmatrix} -3-\lambda & 4 & -12 \\ 1 & -\lambda & 0 \\ 0 & 1 & -\lambda \end{bmatrix} = -12 \begin{vmatrix} 1 & -\lambda \\ 0 & 1 \end{vmatrix} - \lambda \begin{vmatrix} -3-\lambda & 4 \\ 1 & -\lambda \end{vmatrix} = -12 - \lambda[\lambda(3+\lambda) - 4]$$

$$= -\lambda^3 - 3\lambda^2 + 4\lambda - 12.$$

28. (a) $C(p) = \begin{bmatrix} -a & -b \\ 1 & 0 \end{bmatrix}$.

So, the characteristic polynomial of $C(p)$ is $\begin{vmatrix} -a-\lambda & -b \\ 1 & -\lambda \end{vmatrix} = \lambda^2 + \lambda a + b$.

(b) Suppose that λ is an eigenvalue of $C(p)$, with eigenvector $\mathbf{x} = \begin{bmatrix} x_1 \\ x_2 \end{bmatrix}$.

Then $\lambda \begin{bmatrix} x_1 \\ x_2 \end{bmatrix} = C(p) \begin{bmatrix} x_1 \\ x_2 \end{bmatrix} \Leftrightarrow \begin{bmatrix} \lambda x_1 \\ \lambda x_2 \end{bmatrix} = \begin{bmatrix} -a & -b \\ 1 & 0 \end{bmatrix} \begin{bmatrix} x_1 \\ x_2 \end{bmatrix} = \begin{bmatrix} -ax_1 - bx_2 \\ x_1 \end{bmatrix} \Leftrightarrow$

$x_1 = \lambda x_2$, so $\begin{bmatrix} \lambda \\ 1 \end{bmatrix}$ is a corresponding eigenvector.

29. (a) $C(p) = \begin{bmatrix} -a & -b & -c \\ 1 & 0 & 0 \\ 0 & 1 & 0 \end{bmatrix}$, so the characteristic polynomial of $C(p)$ is

$$\begin{vmatrix} -a-\lambda & -b & -c \\ 1 & -\lambda & 0 \\ 0 & 1 & -\lambda \end{vmatrix} = -c \begin{vmatrix} 1 & -\lambda \\ 0 & 1 \end{vmatrix} - \lambda \begin{vmatrix} -a-\lambda & -b \\ 1 & -\lambda \end{vmatrix}$$

$$= -c - \lambda^3 - a\lambda^2 - b\lambda = -\left(\lambda^3 + a\lambda^2 + b\lambda + c\right).$$

(b) Suppose that λ is an eigenvalue of $C(p)$, with eigenvector $\mathbf{x} = \begin{bmatrix} x_1 \\ x_2 \\ x_3 \end{bmatrix}$. Then

$$\lambda \begin{bmatrix} x_1 \\ x_2 \\ x_3 \end{bmatrix} = C(p) \begin{bmatrix} x_1 \\ x_2 \\ x_3 \end{bmatrix} \Leftrightarrow \begin{bmatrix} \lambda x_1 \\ \lambda x_2 \\ \lambda x_3 \end{bmatrix} = \begin{bmatrix} -a & -b & -c \\ 1 & 0 & 0 \\ 0 & 1 & 0 \end{bmatrix} \begin{bmatrix} x_1 \\ x_2 \\ x_3 \end{bmatrix} = \begin{bmatrix} -ax_1 - bx_2 - cx_3 \\ x_1 \\ x_2 \end{bmatrix} \Leftrightarrow$$

$x_2 = \lambda x_3$ and $x_1 = \lambda x_2$, so $\begin{bmatrix} \lambda^2 \\ \lambda \\ 1 \end{bmatrix}$ is a corresponding eigenvector.

30. According to Exercise 28, the characteristic polynomial of the non-diagonal matrix

$C(p) = \begin{bmatrix} -a & -b \\ 1 & 0 \end{bmatrix}$ is $\lambda^2 + a\lambda + b$. So we require that

$\lambda^2 + a\lambda + b = (\lambda - 2)(\lambda - 5) \Leftrightarrow a = -7$ and $b = 10$.

Clearly, the matrix $\begin{bmatrix} 7 & -10 \\ 1 & 0 \end{bmatrix}$ is such a matrix.

31. According to Exercise 29, the characteristic polynomial of the non-diagonal matrix

$C(p) = \begin{bmatrix} -a & -b & -c \\ 1 & 0 & 0 \\ 0 & 1 & 0 \end{bmatrix}$ is $-\left(\lambda^3 + a\lambda^2 + b\lambda + c\right)$. So we require that

$-\left(\lambda^3 + a\lambda^2 + b\lambda + c\right) = -(\lambda + 2)(\lambda - 1)(\lambda - 3) = -\left(\lambda^3 - 2\lambda^2 - 5\lambda + 6\right) \Leftrightarrow$

$a = -2$, $b = -5$, and $c = 6$.

Clearly, the matrix $\begin{bmatrix} 2 & 5 & -6 \\ 1 & 0 & 0 \\ 0 & 1 & 0 \end{bmatrix}$ is such a matrix.

32. (a) We have shown in Exercise 28 that this is true for $n = 2$. So assume that it is true for $n = k$, that is, $\begin{vmatrix} -a_{k-1} - \lambda & -a_{k-2} & \cdots & -a_1 & -a_0 \\ 1 & -\lambda & \cdots & 0 & 0 \\ 0 & 1 & \ddots & 0 & 0 \\ 0 & 0 & \cdots & -\lambda & 0 \\ 0 & 0 & \cdots & 1 & -\lambda \end{vmatrix} = (-1)^k p(\lambda)$,

where $p(x) = x^k + a_{k-1}x^{k-1} + \cdots + a_1 x + a_0$.

Then, expanding along the last column, we calculate

$$\begin{vmatrix} -a_k - \lambda & -a_{k-1} & \cdots & -a_1 & -a_0 \\ 1 & -\lambda & \cdots & 0 & 0 \\ 0 & 1 & \ddots & 0 & 0 \\ 0 & 0 & \cdots & -\lambda & 0 \\ 0 & 0 & \cdots & 1 & -\lambda \end{vmatrix}$$

$$= (-1)^k (-a_0) \begin{vmatrix} 1 & -\lambda & \cdots & 0 \\ 0 & 1 & \ddots & 0 \\ 0 & 0 & \cdots & -\lambda \\ 0 & 0 & \cdots & 1 \end{vmatrix} + (-\lambda) \begin{vmatrix} -a_k - \lambda & -a_{k-1} & \cdots & -a_2 & -a_1 \\ 1 & -\lambda & \cdots & 0 & 0 \\ 0 & 1 & \ddots & 0 & 0 \\ 0 & 0 & \cdots & -\lambda & 0 \\ 0 & 0 & \cdots & 1 & -\lambda \end{vmatrix}.$$

But the second determinant can be evaluated using the induction hypothesis with each a_i replaced by a_{i+1}.

We can write the entire expression as

$(-1)^{k+1} a_0 - \lambda \left[(-1)^k \left(\lambda^k + a_k \lambda^{k-1} \cdots + a_2 \lambda + a_1 \right) \right]$

$= (-1)^{k+1} \left(\lambda^{k+1} + a_k \lambda^k + \cdots + a_2 \lambda^2 + a_1 \lambda + a_0 \right)$, completing the proof.

(b) Suppose that λ is an eigenvalue of $C(p)$

with eigenvector $\mathbf{x} = \begin{bmatrix} x_1 \\ x_2 \\ \vdots \\ x_{n-1} \\ x_n \end{bmatrix}$. Then $\lambda \begin{bmatrix} x_1 \\ x_2 \\ \vdots \\ x_{n-1} \\ x_n \end{bmatrix} = C(p) \begin{bmatrix} x_1 \\ x_2 \\ \vdots \\ x_{n-1} \\ x_n \end{bmatrix} \Leftrightarrow$

$$\begin{bmatrix} \lambda x_1 \\ \lambda x_2 \\ \vdots \\ \lambda x_{n-1} \\ \lambda x_n \end{bmatrix} = \begin{bmatrix} -a_{n-1} & -a_{n-2} & \cdots & -a_1 & -a_0 \\ 1 & 0 & \cdots & 0 & 0 \\ 0 & 1 & \cdots & 0 & 0 \\ \vdots & \vdots & \ddots & \vdots & \vdots \\ 0 & 0 & \cdots & 1 & 0 \end{bmatrix} \begin{bmatrix} x_1 \\ x_2 \\ \vdots \\ x_{n-1} \\ x_n \end{bmatrix}$$

$$= \begin{bmatrix} -a_{n-1}x_1 - a_{n-2}x_2 - \cdots - a_1 x_{n-1} - a_0 x_n \\ x_1 \\ x_2 \\ \vdots \\ x_{n-1} \end{bmatrix} \Leftrightarrow \mathbf{x}_\lambda = \begin{bmatrix} \lambda^{n-1} \\ \lambda^{n-2} \\ \vdots \\ \lambda \\ 1 \end{bmatrix}$$

that is $x_i = \lambda x_{i+1}$ for $1 \le i \le n-1$. So, $\mathbf{x}_\lambda$ is a corresponding eigenvector.

33. The characteristic polynomial $c_A(\lambda)$ of A is $\det(A - \lambda I) = \begin{vmatrix} 1-\lambda & -1 \\ 2 & 3-\lambda \end{vmatrix} = \lambda^2 - 4\lambda + 5$.

We verify that

$$\begin{aligned} A^2 - 4A + 5I &= \begin{bmatrix} 1 & -1 \\ 2 & 3 \end{bmatrix}^2 - 4\begin{bmatrix} 1 & -1 \\ 2 & 3 \end{bmatrix} + 5\begin{bmatrix} 1 & 0 \\ 0 & 1 \end{bmatrix} \\ &= \begin{bmatrix} -1 & -4 \\ 8 & 7 \end{bmatrix} - \begin{bmatrix} 4 & -4 \\ 8 & 12 \end{bmatrix} + \begin{bmatrix} 5 & 0 \\ 0 & 5 \end{bmatrix} = O. \end{aligned}$$

34. The characteristic polynomial $c_A(\lambda)$ of A is

$$\det(A - \lambda I) = \begin{vmatrix} 1-\lambda & 1 & 0 \\ -1 & -\lambda & 1 \\ -2 & 1 & -\lambda \end{vmatrix} = (1-\lambda)\begin{vmatrix} -\lambda & 1 \\ 1 & -\lambda \end{vmatrix} - \begin{vmatrix} -1 & 1 \\ -2 & -\lambda \end{vmatrix} = -\lambda^3 + \lambda^2 - 3.$$

We verify that

$$\begin{aligned} -A^3 + A^2 - 3I &= -\begin{bmatrix} 1 & 1 & 0 \\ -1 & 0 & 1 \\ -2 & 1 & 0 \end{bmatrix}^3 + \begin{bmatrix} 1 & 1 & 0 \\ -1 & 0 & 1 \\ -2 & 1 & 0 \end{bmatrix}^2 - 3\begin{bmatrix} 1 & 0 & 0 \\ 0 & 1 & 0 \\ 0 & 0 & 1 \end{bmatrix} \\ &= \begin{bmatrix} 3 & -1 & -1 \\ 3 & 3 & 0 \\ 3 & 2 & 2 \end{bmatrix} + \begin{bmatrix} 0 & 1 & 1 \\ -3 & 0 & 0 \\ -3 & -2 & 1 \end{bmatrix} + \begin{bmatrix} -3 & 0 & 0 \\ 0 & -3 & 0 \\ 0 & 0 & -3 \end{bmatrix} = O. \end{aligned}$$

35. In Exercise 33, $c_A(\lambda) = \lambda^2 - 4\lambda + 5$, so $a = -4$ and $b = 5$.

Thus, $A^2 = -aA - bI = 4A - 5I = 4\begin{bmatrix} 1 & -1 \\ 2 & 3 \end{bmatrix} - 5\begin{bmatrix} 1 & 0 \\ 0 & 1 \end{bmatrix} = \begin{bmatrix} -1 & -4 \\ 8 & 7 \end{bmatrix}$.

Similarly, $A^3 = (a^2 - b)A + abI = 11A - 20I = 11\begin{bmatrix} 1 & -1 \\ 2 & 3 \end{bmatrix} - 20\begin{bmatrix} 1 & 0 \\ 0 & 1 \end{bmatrix} = \begin{bmatrix} -9 & -11 \\ 22 & 13 \end{bmatrix}$.

The corresponding formula for A^4 is given by

$$\begin{aligned} A^4 &= AA^3 = A\left[(a^2 - b)A + abI\right] = (a^2 - b)A^2 + abA = (a^2 - b)\left[-aA - bI\right] + abA \\ &= (-a^3 + 2ab)A + (b^2 - a^2 b)I. \end{aligned}$$

So, $A^4 = 24A - 55I = 24\begin{bmatrix} 1 & -1 \\ 2 & 3 \end{bmatrix} - 55\begin{bmatrix} 1 & 0 \\ 0 & 1 \end{bmatrix} = \begin{bmatrix} -31 & -24 \\ 48 & 17 \end{bmatrix}$.

36. For convenience, we write $-A^3 + aA^2 + bA + cI = O$, where $a = 1$, $b = 0$, $c = -3$:

$$\begin{aligned} A^3 &= aA^2 + bA + cI = A^2 - 3I = \begin{bmatrix} 1 & 1 & 0 \\ -1 & 0 & 1 \\ -2 & 1 & 0 \end{bmatrix}^2 - 3\begin{bmatrix} 1 & 0 & 0 \\ 0 & 1 & 0 \\ 0 & 0 & 1 \end{bmatrix} = \begin{bmatrix} -3 & 1 & 1 \\ -3 & -3 & 0 \\ -3 & -2 & -2 \end{bmatrix}. \\ A^4 &= AA^3 = A\left(aA^2 + bA + cI\right) \\ &= a\left(aA^2 + bA + cI\right) + bA^2 + cA = \left(a^2 + b\right)A^2 + (ab + c)A + acI \\ &= A^2 - 3A - 3I = \begin{bmatrix} 1 & 1 & 0 \\ -1 & 0 & 1 \\ -2 & 1 & 0 \end{bmatrix}^2 - 3\begin{bmatrix} 1 & 1 & 0 \\ -1 & 0 & 1 \\ -2 & 1 & 0 \end{bmatrix} - 3\begin{bmatrix} 1 & 0 & 0 \\ 0 & 1 & 0 \\ 0 & 0 & 1 \end{bmatrix} = \begin{bmatrix} -6 & -2 & 1 \\ 0 & -3 & -3 \\ 3 & -5 & -2 \end{bmatrix}. \end{aligned}$$

37. $A^{-1} = -\frac{1}{b}A - \frac{a}{b}I = -\frac{1}{5}\begin{bmatrix} 1 & -1 \\ 2 & 3 \end{bmatrix} + \frac{4}{5}\begin{bmatrix} 1 & 0 \\ 0 & 1 \end{bmatrix} = \begin{bmatrix} \frac{3}{5} & \frac{1}{5} \\ -\frac{2}{5} & \frac{1}{5} \end{bmatrix}$. Also,

$$\begin{aligned} A^{-2} &= \left(A^{-1}\right)^2 = \left(-\frac{1}{b}A - \frac{a}{b}I\right)^2 = \frac{1}{b^2}A^2 + \frac{2a}{b^2}A + \frac{a^2}{b^2}I = \frac{1}{b^2}\left(-aA - bI\right) + \frac{2a}{b^2}A + \frac{a^2}{b^2}I \\ &= \frac{a}{b^2}A + \frac{a^2 - b}{b^2}I = -\frac{4}{25}\begin{bmatrix} 1 & -1 \\ 2 & 3 \end{bmatrix} + \frac{11}{25}\begin{bmatrix} 1 & 0 \\ 0 & 1 \end{bmatrix} = \begin{bmatrix} \frac{7}{25} & \frac{4}{25} \\ -\frac{8}{25} & -\frac{1}{25} \end{bmatrix}. \end{aligned}$$

38. $-A^3 + aA^2 + bA + cI = -A\left(A^2 - aA - bI\right) = -cI \Leftrightarrow$

$$A^{-1} = \frac{1}{c}\left(A^2 - aA - bI\right) = -\frac{1}{3}\left(\begin{bmatrix} 1 & 1 & 0 \\ -1 & 0 & 1 \\ -2 & 1 & 0 \end{bmatrix}^2 - \begin{bmatrix} 1 & 1 & 0 \\ -1 & 0 & 1 \\ -2 & 1 & 0 \end{bmatrix}\right) = \begin{bmatrix} \frac{1}{3} & 0 & -\frac{1}{3} \\ \frac{2}{3} & 0 & \frac{1}{3} \\ \frac{1}{3} & 1 & -\frac{1}{3} \end{bmatrix}.$$

$$\begin{aligned} A^{-2} &= \left(A^{-1}\right)^2 = \left[\frac{1}{c}\left(A^2 - aA - bI\right)\right]^2 = \frac{1}{c^2}\left[A^4 - 2aA^3 + \left(a^2 - 2b\right)A^2 + 2abA + b^2 I\right] \\ &= \frac{1}{c^2}\left\{\left[\left(a^2 + b\right)A^2 + (ab + c)A + acI\right] - 2a\left[aA^2 + bA + cI\right] + \left(a^2 - 2b\right)A^2 + 2abA + b^2 I\right\} \\ &= \frac{1}{c^2}\left[-bA^2 + (ab + c)A + \left(b^2 - ac\right)I\right] \\ &= \frac{1}{9}(-3A + 3I) = \frac{1}{3}\left(-\begin{bmatrix} 1 & 1 & 0 \\ -1 & 0 & 1 \\ -2 & 1 & 0 \end{bmatrix} + \begin{bmatrix} 1 & 0 & 0 \\ 0 & 1 & 0 \\ 0 & 0 & 1 \end{bmatrix}\right) = \begin{bmatrix} 0 & -\frac{1}{3} & 0 \\ \frac{1}{3} & \frac{1}{3} & -\frac{1}{3} \\ \frac{2}{3} & -\frac{1}{3} & \frac{1}{3} \end{bmatrix}. \end{aligned}$$

39. According to Exercise 69 in Section 4.2, $\det A = \det \begin{bmatrix} P & Q \\ O & S \end{bmatrix} = (\det P)(\det S)$.

We apply this result to $A - \lambda I$ instead of A below.

Note $c_A(\lambda) = \det(A - \lambda I)$, $c_P(\lambda) = \det(P - \lambda I)$, and $c_S(\lambda) = \det(S - \lambda I)$.

So, $c_A(\lambda) = \det(A - \lambda I) = \det(P - \lambda I)\det(S - \lambda I) = c_P(\lambda)c_S(\lambda)$.

Q: Why can we apply the result of Exercise 69 to $A - \lambda I$?

A: Because P becomes $P - \lambda I$ and S becomes $S - \lambda I$ so the proof holds.

That is, we have $A - \lambda I = \begin{bmatrix} P - \lambda I & Q \\ O & S - \lambda I \end{bmatrix}$.

Q: Neither O nor Q are affected. Why not?

A: Because P and S are square and I has all zero entries off the diagonal.

40. We prove $\det A = \lambda_1\lambda_2\cdots\lambda_n = \prod_{i=1}^{n}\lambda_i$ and $\text{tr}(A) = \lambda_1 + \lambda_2 + \cdots + \lambda_n = \sum_{i=1}^{n}\lambda_i$ separately.

$\det A$: As suggested, we equate the formulas for the characteristic polynomial.
That is, $\det(A - \lambda I) = (-1)^n(\lambda - \lambda_1)(\lambda - \lambda_2)\cdots(\lambda - \lambda_n) = (-1)^n \prod_{i=1}^{n}(\lambda - \lambda_i)$.

Q: Given $p(x) = \sum_{i=0}^{n} a_i x^i$, what value of x returns the constant term?

A: The value $x = 0$, since $p(0) = \sum_{i=0}^{n} a_i x^i = a_0 + \sum_{i=1}^{n} a_i 0^i = a_0 + 0 = a_0$.

We use this to show the constant term of $p(\lambda) = (-1)^n \prod_{i=1}^{n}(\lambda - \lambda_i)$ is $\prod_{i=1}^{n}\lambda_i$.

Letting $\lambda = 0$ in $\det(A - \lambda I) = p(\lambda) = (-1)^n \prod_{i=1}^{n}(\lambda - \lambda_i)$ yields:

$$\det(A - 0I) = p(0) = (-1)^n \prod_{i=1}^{n}(0 - \lambda_i) = (-1)^n \prod_{i=1}^{n}(-\lambda_i) = (-1)^{2n} \prod_{i=1}^{n}\lambda_i = \prod_{i=1}^{n}\lambda_i.$$

Therefore, $\det A = p(0) = \prod_{i=1}^{n}\lambda_i = \lambda_1\lambda_2\cdots\lambda_n$ as we were to show.

Since the conclusion for $\text{tr}(A)$ requires identifying the coefficient of λ^{n-1},
we need to understand how $\det(A - \lambda I)$ gives rise to λ^n and λ^{n-1} terms.
To get a sense of this we compute $\det(A - \lambda I)$ when A is 2×2 and 3×3.
In each case, we compute $\det(A - \lambda I)$ by expanding along the first column.
That is, $\det(A - \lambda I) = (a_{11} - \lambda)\det(A_{11} - \lambda I) + \sum_{i=2}^{n} a_{i1}(-1)^{i+1}\det(A_{i1} - \lambda I)$.

2×2: $$\det(A - \lambda I) = \begin{vmatrix} a_{11} - \lambda & a_{12} \\ a_{21} & a_{22} - \lambda \end{vmatrix}$$
$$= (a_{11} - \lambda)(a_{22} - \lambda) - a_{12}a_{21}$$
$$= (a_{11}a_{12} - a_{21}a_{22}) - (a_{11} + a_{22})\lambda + \lambda^2.$$
So, the coefficient of λ^n is $(-1)^n$ and of λ^{n-1} is $(-1)^{n-1}\text{tr}(A)$.

3×3: $$\det(A - \lambda I) = \begin{vmatrix} a_{11} - \lambda & a_{12} & a_{13} \\ a_{21} & a_{22} - \lambda & a_{23} \\ a_{31} & a_{32} & a_{33} - \lambda \end{vmatrix}$$
$$= (a_{11} - \lambda)\begin{vmatrix} a_{22} - \lambda & a_{23} \\ a_{32} & a_{33} - \lambda \end{vmatrix} - a_{21}\begin{vmatrix} a_{12} & a_{13} \\ a_{32} & a_{33} - \lambda \end{vmatrix} + a_{31}\begin{vmatrix} a_{12} & a_{13} \\ a_{22} - \lambda & a_{23} \end{vmatrix}$$
Since A_{21} and A_{31} only have 1 row with entries in which λ appears,
the λ^n and λ^{n-1} terms come entirely from $(a_{11} - \lambda)\det(A_{11} - \lambda I)$. So:

$$(a_{11} - \lambda)\det(A_{11} - \lambda I) = \prod_{i=1}^{3}(a_{ii} - \lambda) - a_{23}a_{32} = \{\lambda \text{ and constant terms}\} + \text{tr}(A)\lambda^2 - \lambda^3$$

These results tell us what we need to show using induction.

40. We continue to work on showing $\text{tr}(A) = \lambda_1 + \lambda_2 + \cdots + \lambda_n = \sum_{i=1}^{n} \lambda_i$.

In the characteristic polynomial $\det(A - \lambda I)$,

$$\det(A - \lambda I) = (a_{11} - \lambda)\det(A_{11} - \lambda I) + \sum_{i=2}^{n} a_{i1}(-1)^{i+1}\det(A_{i1} - \lambda I),$$

the λ^n and λ^{n-1} terms come entirely from $(a_{11} - \lambda)\det(A_{11} - \lambda I)$.
Furthermore, the coefficient of λ^n is $(-1)^n$ and the coefficient of λ^{n-1} is $(-1)^{n-1}\text{tr}(A)$.

1: If $\det(A - \lambda I) = a_{11} - \lambda$, then the coefficient of λ^1 is $(-1)^1$ and the coefficient of λ^0 is $a_{11} = \text{tr}(A)$ both of which come from $(a_{11} - \lambda)\det(A_{11} - \lambda I)$.

k: If A is $k \times k$, the assertion holds. This is the induction hypothesis.

$k+1$: If A is $(k+1) \times (k+1)$, we must show the assertion holds.

Recall $\det(A - \lambda I) = (a_{11} - \lambda)\det(A_{11} - \lambda I) + \sum_{i=2}^{k+1} a_{i1}(-1)^{i+1}\det(A_{i1} - \lambda I)$.

First we show $\sum_{i=2}^{k+1} a_{i1}(-1)^{i+1}\det(A_{i1} - \lambda I)$ gives rise only to λ^m terms where $m \le k-1$.
Recall $A_{i1} - \lambda I$ is the submatrix created by deleting row i and column 1 of $A - \lambda I$.
Since row i and column 1 contain a λ entry, $A_{i1} - \lambda I$ contains $k-1$ entries with λ.
So, $\sum_{i=2}^{k+1} a_{i1}(-1)^{i+1}\det(A_{i1} - \lambda I)$ gives rise only to λ^m terms where $m \le k-1$.
Therefore, the λ^{k+1} and λ^k terms come entirely from $(a_{11} - \lambda)\det(A_{11} - \lambda I)$.
Now since A_{11} is $k \times k$, we can apply the induction hypothesis to $\det(A_{11} - \lambda I)$.
So, its coefficient of λ^k is $(-1)^k$ and its coefficient of λ^{k-1} is $(-1)^{k-1}\text{tr}(A_{11})$. So:
$(a_{11}-\lambda)\det(A_{11}-\lambda I) = (a_{11}-\lambda)(\{\lambda^m \text{ terms, } m \le k-1\} + (-1)^{k-2}\text{tr}(A)\lambda^{k-1} + (-1)^k\lambda^k)$
Distributing a_{11} and λ implies the coefficient of λ^{k+1} is $(-1)^{k+1}$ and the coefficient of λ^k is $(-1)^k\text{tr}(A_{11}) + (-1)^k a_{11} = (-1)^k\text{tr}(A)$ exactly as required.

If $p(\lambda) = (-1)^n \prod_{i=1}^{n}(\lambda - \lambda_i)$, then the coefficient of λ^n is $(-1)^n$ and of λ^{n-1} is $(-1)^{n-1}\sum_{i=1}^{n} \lambda_i$.

1: $p(\lambda) = (-1)^1(\lambda - \lambda_1)$, so the assertion is obvious.

k: Assume the assertion holds. This is the induction hypothesis.

$k+1$: We must show the assertion holds in this case using the induction hyptothesis.

$$p(\lambda) = (-1)^{k+1}\prod_{i=1}^{k+1}(\lambda - \lambda_i) = (\lambda_{k+1} - \lambda)\left((-1)^k \prod_{i=1}^{k}(\lambda - \lambda_i)\right)$$

$$\overset{\text{by induc}}{=} (\lambda_{k+1} - \lambda)(\{\lambda^m \text{ terms, } m \le k-1\} + \left((-1)^{k-2}\sum_{i=1}^{k} \lambda_i\right)\lambda^{k-1} + (-1)^k\lambda^k)$$

Distributing λ_{k+1} and λ gives the desired result. Verify this. It is not trivial.

$\text{tr}(A)$: Equate the coefficient of λ^{n-1} from $\det(A - \lambda I)$, namely $(-1)^{n-1}\text{tr}(A)$ and the coefficient of λ^{n-1} from $p(\lambda)$, namely $(-1)^{n-1}\sum_{i=1}^{n} \lambda_i$.

This shows $\text{tr}(A) = \sum_{i=1}^{n} \lambda_i = \lambda_1 + \lambda_2 + \cdots + \lambda_n$ as required.

41. Given A has eigenvalues α_i and B has eigenvalues β_i,we need to show the following:

1) If $C = A + B$ has eigenvalues γ_i, then $\sum_{i=1}^{n} \gamma_i = \sum_{i=1}^{n} (\alpha_i + \beta_i)$ and

2) If $C = AB$ has eigenvalues γ_i, then $\prod_{i=1}^{n} \gamma_i = \prod_{i=1}^{n} (\alpha_i \beta_i)$.

Q: How does this differ from the statement of Exercise 24?

A: In Exercise 24, we show that if $C = A + B$ it does *not* follow that $\gamma_i = \alpha_i + \beta_i$. In this Exercise, we consider the sum (and product) of *all* the eigenvalues.

tr(C): If $C = A + B$ has eigenvalues γ_i, then $\sum_{i=1}^{n} \gamma_i = \sum_{i=1}^{n} (\alpha_i + \beta_i)$.

Since $[C]_{ii} = [A]_{ii} + [B]_{ii}$, we have $\text{tr}(C) = \text{tr}(A) + \text{tr}(B)$.

Furthermore, Exercise 40 asserts for any matrix A, $\text{tr}(A) = \sum_{i=1}^{n} \lambda_i$.

So, $\text{tr}(C) = \text{tr}(A) + \text{tr}(B)$ implies $\sum_{i=1}^{n} \gamma_i = \sum_{i=1}^{n} \alpha_i + \sum_{i=1}^{n} \beta_i = \sum_{i=1}^{n} (\alpha_i + \beta_i)$.

Q: How might we state $[C]_{ii} = [A]_{ii} + [B]_{ii}$ in words?

A: A diagonal entry of C equals the sum of the corresponding entries in A and B.

Q: How might we state $\sum_{i=1}^{n} \gamma_i = \sum_{i=1}^{n} (\alpha_i + \beta_i)$ in words?

A: The sum of the eigenvalues of C equals the sum of all the eigenvalues of both A and B.

det(C): If $C = AB$ has eigenvalues γ_i, then $\prod_{i=1}^{n} \gamma_i = \prod_{i=1}^{n} (\alpha_i \beta_i)$.

By Theorem 4.8 in Section 4.2, $C = AB$ implies $\det C = (\det A)(\det B)$.

Furthermore, Exercise 40 asserts for any matrix A, $\det(A) = \prod_{i=1}^{n} \lambda_i$.

So, $\det C = (\det A)(\det B)$ implies $\prod_{i=1}^{n} \gamma_i = \left(\prod_{i=1}^{n} \alpha_i \right) \left(\prod_{i=1}^{n} \beta_i \right) = \prod_{i=1}^{n} (\alpha_i \beta_i)$.

Q: How might we state $\prod_{i=1}^{n} \gamma_i = \prod_{i=1}^{n} (\alpha_i \beta_i)$ in words?

A: The product of the eigenvalues of C equals the product of all eigenvalues of A, B.

42. Given A has eigenvalues λ_i with corresponding eigenvectors $\mathbf{v}_i$, we must show:

If $\mathbf{x} = \sum_{i=1}^{n} c_i \mathbf{v}_i$, then $A^k \mathbf{x} = \sum_{i=1}^{n} c_i \lambda_i^k \mathbf{v}_i$.

But that is obvious since: $A^k \mathbf{x} = A^k \left(\sum_{i=1}^{n} c_i \mathbf{v}_i \right) = \sum_{i=1}^{n} c_i (A^k \mathbf{v}_i) \overset{\text{by Thm 4.18(c)}}{=} \sum_{i=1}^{n} c_i \lambda_i^k \mathbf{v}_i$

4.4 Similarity and Diagonalization

1. $\det(A - \lambda I) = \begin{vmatrix} 4-\lambda & 1 \\ 3 & 1-\lambda \end{vmatrix} = \lambda^2 - 5\lambda + 1$, while

$\det(B - \lambda I) = \begin{vmatrix} 1-\lambda & 0 \\ 0 & 1-\lambda \end{vmatrix} = \lambda^2 - 2\lambda + 1$. Thus, by Theorem 4.22(d), $A \nsim B$.

2. $\det(A - \lambda I) = \begin{vmatrix} 3-\lambda & -1 \\ -5 & 7-\lambda \end{vmatrix} = \lambda^2 - 10\lambda + 16$, while

$\det(B - \lambda I) = \begin{vmatrix} 2-\lambda & 1 \\ -4 & 6-\lambda \end{vmatrix} = \lambda^2 - 8\lambda + 16$. Thus, by Theorem 4.22(d), $A \nsim B$.

3. A has eigenvalues 2, 2, 4; B has eigenvalues 1, 4, and 4 $\Rightarrow A \nsim B$.

4. $\det(A - \lambda I) = \begin{vmatrix} 1-\lambda & 2 & 0 \\ 0 & 1-\lambda & -1 \\ 0 & -1 & 1-\lambda \end{vmatrix} = -\lambda^3 + 3\lambda^2 - 3\lambda$, while

$\det(B - \lambda I) = \begin{vmatrix} 2-\lambda & 1 & 1 \\ 0 & 1-\lambda & 0 \\ 2 & 0 & 1-\lambda \end{vmatrix} = -\lambda^3 + 4\lambda^2 - 3\lambda \Rightarrow A \nsim B$.

5. Since $\begin{bmatrix} 2 & -1 \\ -1 & 1 \end{bmatrix} \begin{bmatrix} 5 & -1 \\ 2 & 2 \end{bmatrix} \begin{bmatrix} 1 & 1 \\ 1 & 2 \end{bmatrix} = \begin{bmatrix} 4 & 0 \\ 0 & 3 \end{bmatrix}$ is of the form $P^{-1}AP = D$,

we see that the eigenvalues of A are $\lambda_1 = 4$ and $\lambda_2 = 3$.

So, E_4, E_3 have bases $\mathbf{p}_1 = \begin{bmatrix} 1 \\ 1 \end{bmatrix}$ and $\mathbf{p}_2 = \begin{bmatrix} 1 \\ 2 \end{bmatrix}$.

6. The eigenvalues of A are 2, 0, $-1 \Rightarrow E_2, E_0, E_{-1}$ have bases $\mathbf{p}_1 = \begin{bmatrix} 3 \\ 1 \\ 2 \end{bmatrix}, \begin{bmatrix} 1 \\ -1 \\ 0 \end{bmatrix}$, and $\begin{bmatrix} 0 \\ 1 \\ -1 \end{bmatrix}$.

7. The eigenvalues of A are 6 and $-2 \Rightarrow E_6, E_{-2}$ have bases $\begin{bmatrix} 3 \\ 2 \\ 3 \end{bmatrix}$ and $\left\{ \begin{bmatrix} 0 \\ 1 \\ -1 \end{bmatrix}, \begin{bmatrix} 1 \\ 0 \\ -1 \end{bmatrix} \right\}$.

8. $\det(A-\lambda I) = \begin{vmatrix} 5-\lambda & 2 \\ 2 & 5-\lambda \end{vmatrix} = \lambda^2 - 10\lambda + 21 = (\lambda-3)(\lambda-7)$,

so A has eigenvalues 3 and 7 and is thus diagonalizable, by Theorem 4.25.
We find bases for the eigenspaces:

$A - 3I = \begin{bmatrix} 2 & 2 \\ 2 & 2 \end{bmatrix} \longrightarrow \begin{bmatrix} 1 & 1 \\ 0 & 0 \end{bmatrix}$, so $\mathbf{p}_1 = \begin{bmatrix} 1 \\ -1 \end{bmatrix}$, and

$A - 7I = \begin{bmatrix} -2 & 2 \\ 2 & -2 \end{bmatrix} \longrightarrow \begin{bmatrix} 1 & -1 \\ 0 & 0 \end{bmatrix}$, so $\mathbf{p}_2 = \begin{bmatrix} 1 \\ 1 \end{bmatrix}$.

Thus $P = \begin{bmatrix} \mathbf{p}_1 & \mathbf{p}_2 \end{bmatrix} = \begin{bmatrix} 1 & 1 \\ -1 & 1 \end{bmatrix}$ diagonalizes A, and

$$P^{-1}AP = D \Leftrightarrow AP = PD \Leftrightarrow \begin{bmatrix} 5 & 2 \\ 2 & 5 \end{bmatrix}\begin{bmatrix} 1 & 1 \\ -1 & 1 \end{bmatrix} = \begin{bmatrix} 1 & 1 \\ -1 & 1 \end{bmatrix}\begin{bmatrix} 3 & 0 \\ 0 & 7 \end{bmatrix} = \begin{bmatrix} 3 & 7 \\ -3 & 7 \end{bmatrix}.$$

9. $\det(A-\lambda I) = \begin{vmatrix} -3-\lambda & 4 \\ -1 & 1-\lambda \end{vmatrix} = \lambda^2 + 2\lambda + 1 = (\lambda+1)^2$,

so A has eigenvalue -1 with algebraic multiplicity 2.
To find the corresponding eigenspace, we calculate

$A + I = \begin{bmatrix} -2 & 4 \\ -1 & 2 \end{bmatrix} \longrightarrow \begin{bmatrix} 1 & -2 \\ 0 & 0 \end{bmatrix}$, so the eigenvalue -1 has geometric multiplicity 1,

and thus A is not diagonalizable, by the Diagonalization Theorem.

10. A has eigenvalue 3 with algebraic multiplicity 3.

Furthermore $A - 3I = \begin{bmatrix} 0 & 1 & 0 \\ 0 & 0 & 1 \\ 0 & 0 & 0 \end{bmatrix}$, so this eigenvalue has geometric multiplicity 1,

and thus A is not diagonalizable, by the Diagonalization Theorem.

11. Expanding along the first row,

$$\det(A-\lambda I)=\begin{vmatrix}1-\lambda & 0 & 1\\ 0 & 1-\lambda & 1\\ 1 & 1 & -\lambda\end{vmatrix}=(1-\lambda)\left[(1-\lambda)(-\lambda)-1\right]-(1-\lambda)$$
$$=(1-\lambda)\left(\lambda^2-\lambda-2\right)=-(\lambda+1)(\lambda-1)(\lambda-2).$$

So, A has eigenvalues -1, 1, and 2, and Theorem 4.25 tells us that that A is diagonalizable. We find bases for the eigenspaces:

$$A+I=\begin{bmatrix}2&0&1\\0&2&1\\1&1&1\end{bmatrix}\longrightarrow\begin{bmatrix}1&0&\frac{1}{2}\\0&1&\frac{1}{2}\\0&0&0\end{bmatrix}, \text{ so } E_{-1} \text{ has basis } \mathbf{p}_1=\begin{bmatrix}1\\1\\-2\end{bmatrix}.$$

$$A-I=\begin{bmatrix}0&0&1\\0&0&1\\1&1&-1\end{bmatrix}\longrightarrow\begin{bmatrix}1&1&0\\0&0&1\\0&0&0\end{bmatrix}, \text{ so } E_1 \text{ has basis } \mathbf{p}_2=\begin{bmatrix}1\\-1\\0\end{bmatrix}.$$

$$A-2I=\begin{bmatrix}-1&0&1\\0&-1&1\\1&1&-2\end{bmatrix}\longrightarrow\begin{bmatrix}1&0&-1\\0&1&-1\\0&0&0\end{bmatrix}, \text{ so } E_2 \text{ has basis } \mathbf{p}_3=\begin{bmatrix}1\\1\\1\end{bmatrix}.$$

Thus, $P=\begin{bmatrix}1&1&1\\1&-1&1\\-2&0&1\end{bmatrix}$ and $D=\begin{bmatrix}-1&0&0\\0&1&0\\0&0&2\end{bmatrix}$ satisfy $P^{-1}AP=D$.

12. $\det(A-\lambda I)=\begin{vmatrix}1-\lambda & 0 & 0\\ 2 & 2-\lambda & 1\\ 3 & 0 & 1-\lambda\end{vmatrix}=-\lambda^3+4\lambda^2-5\lambda+2=-(\lambda-1)^2(\lambda-2).$

So, A has eigenvalues 1 and 2.

But $A-I=\begin{bmatrix}0&0&0\\2&1&1\\3&0&0\end{bmatrix}\longrightarrow\begin{bmatrix}1&0&0\\0&1&1\\0&0&0\end{bmatrix}$, so the eigenvalue 1 has geometric multiplicity 1.

So, A is not diagonalizable, by the Diagonalization Theorem.

13. $\det(A-\lambda I)=\begin{vmatrix}1-\lambda & 2 & 1\\ -1 & -\lambda & 1\\ 1 & 1 & -\lambda\end{vmatrix}=\lambda^2-\lambda^3=-\lambda^2(\lambda-1)$. Thus, A has eigenvalues 0 and 1.

But $A-0I=\begin{bmatrix}1&2&1\\-1&0&1\\1&1&0\end{bmatrix}\longrightarrow\begin{bmatrix}1&0&-1\\0&1&1\\0&0&0\end{bmatrix}$, so 0 has geometric multiplicity 1,

and thus A is not diagonalizable, by the Diagonalization Theorem.

14. A has eigenvalues 1, 2, and 3. Since the eigenvalue 3 has algebraic multiplicity 2,

we check $A - 3I = \begin{bmatrix} -1 & 0 & 0 & 2 \\ 0 & 0 & 2 & 1 \\ 0 & 0 & 0 & 0 \\ 0 & 0 & 0 & -2 \end{bmatrix} \longrightarrow \begin{bmatrix} 1 & 0 & 0 & 0 \\ 0 & 0 & 1 & 0 \\ 0 & 0 & 0 & 1 \\ 0 & 0 & 0 & 0 \end{bmatrix}$.

Thus 3 has geometric multiplicity 1, and A is thus not diagonalizable.

15. A has eigenvalues -2 and 2. We check

$$A + 2I = \begin{bmatrix} 4 & 0 & 0 & 4 \\ 0 & 4 & 0 & 0 \\ 0 & 0 & 0 & 0 \\ 0 & 0 & 0 & 0 \end{bmatrix} \longrightarrow \begin{bmatrix} 1 & 0 & 0 & 1 \\ 0 & 1 & 0 & 0 \\ 0 & 0 & 0 & 0 \\ 0 & 0 & 0 & 0 \end{bmatrix}, \text{ so } E_{-2} \text{ has basis } \left\{ \begin{bmatrix} 1 \\ 0 \\ 0 \\ -1 \end{bmatrix}, \begin{bmatrix} 0 \\ 0 \\ 1 \\ 0 \end{bmatrix} \right\}.$$

$$A - 2I = \begin{bmatrix} 0 & 0 & 0 & 4 \\ 0 & 0 & 0 & 0 \\ 0 & 0 & -4 & 0 \\ 0 & 0 & 0 & -4 \end{bmatrix} \longrightarrow \begin{bmatrix} 0 & 0 & 1 & 0 \\ 0 & 0 & 0 & 1 \\ 0 & 0 & 0 & 0 \\ 0 & 0 & 0 & 0 \end{bmatrix}, \text{ so } E_2 \text{ has basis } \left\{ \begin{bmatrix} 1 \\ 0 \\ 0 \\ 0 \end{bmatrix}, \begin{bmatrix} 0 \\ 1 \\ 0 \\ 0 \end{bmatrix} \right\}.$$

Thus $P = \begin{bmatrix} 1 & 0 & 1 & 0 \\ 0 & 0 & 0 & 1 \\ 0 & 1 & 0 & 0 \\ -1 & 0 & 0 & 0 \end{bmatrix}$ and $D = \begin{bmatrix} -2 & 0 & 0 & 0 \\ 0 & -2 & 0 & 0 \\ 0 & 0 & 2 & 0 \\ 0 & 0 & 0 & 2 \end{bmatrix}$ satisfy $P^{-1}AP = D$.

16. We diagonalize A, first finding its eigenvalues and eigenvectors:

$\det(A - \lambda I) = \begin{vmatrix} -4-\lambda & 6 \\ -3 & 5-\lambda \end{vmatrix} = \lambda^2 - \lambda - 2 = (\lambda+1)(\lambda-2)$, so A has eigenvalues -1, 2.

$A + I = \begin{bmatrix} -3 & 6 \\ -3 & 6 \end{bmatrix} \longrightarrow \begin{bmatrix} 1 & -2 \\ 0 & 0 \end{bmatrix}$, so E_{-1} has basis $\begin{bmatrix} 2 \\ 1 \end{bmatrix}$.

$A - 2I = \begin{bmatrix} -6 & 6 \\ -3 & 3 \end{bmatrix} \longrightarrow \begin{bmatrix} 1 & -1 \\ 0 & 0 \end{bmatrix}$, so E_2 has basis $\begin{bmatrix} 1 \\ 1 \end{bmatrix}$.

Therefore, $P^{-1}AP = D$ is satisfied if $P = \begin{bmatrix} 2 & 1 \\ 1 & 1 \end{bmatrix}$ and $D = \begin{bmatrix} -1 & 0 \\ 0 & 2 \end{bmatrix}$. So

$$\begin{bmatrix} -4 & 6 \\ -3 & 5 \end{bmatrix}^9 = (PDP^{-1})^9 = PD^9P^{-1} = \begin{bmatrix} 2 & 1 \\ 1 & 1 \end{bmatrix} \begin{bmatrix} -1 & 0 \\ 0 & 2 \end{bmatrix}^9 \begin{bmatrix} 2 & 1 \\ 1 & 1 \end{bmatrix}^{-1}$$
$$= \begin{bmatrix} 2 & 1 \\ 1 & 1 \end{bmatrix} \begin{bmatrix} (-1)^9 & 0 \\ 0 & 2^9 \end{bmatrix} \begin{bmatrix} 1 & -1 \\ -1 & 2 \end{bmatrix} = \begin{bmatrix} -514 & 1026 \\ -513 & 1025 \end{bmatrix}.$$

17. We diagonalize A, first finding its eigenvalues and eigenvectors:

$\det(A-\lambda I) = \begin{vmatrix} -1-\lambda & 6 \\ 1 & -\lambda \end{vmatrix} = \lambda^2+\lambda-6 = (\lambda+3)(\lambda-2)$, so A has eigenvalues -3 and 2.

$A+3I = \begin{bmatrix} 2 & 6 \\ 1 & 3 \end{bmatrix} \longrightarrow \begin{bmatrix} 1 & 3 \\ 0 & 0 \end{bmatrix}$, so E_{-3} has basis $\begin{bmatrix} 3 \\ -1 \end{bmatrix}$.

$A-2I = \begin{bmatrix} -3 & 6 \\ 1 & -2 \end{bmatrix} \longrightarrow \begin{bmatrix} 1 & -2 \\ 0 & 0 \end{bmatrix}$, so E_2 has basis $\begin{bmatrix} 2 \\ 1 \end{bmatrix}$.

Therefore, $P^{-1}AP = D$ is satisfied if $P = \begin{bmatrix} 3 & 2 \\ -1 & 1 \end{bmatrix}$ and $D = \begin{bmatrix} -3 & 0 \\ 0 & 2 \end{bmatrix}$. Therefore,

$$\begin{bmatrix} -1 & 6 \\ 1 & 0 \end{bmatrix}^{10} = \left(PDP^{-1}\right)^{10} - PD^{10}P^{-1} = \begin{bmatrix} 3 & 2 \\ -1 & 1 \end{bmatrix} \begin{bmatrix} -3 & 0 \\ 0 & 2 \end{bmatrix}^{10} \begin{bmatrix} 3 & 2 \\ -1 & 1 \end{bmatrix}^{-1}$$

$$= \begin{bmatrix} 3 & 2 \\ -1 & 1 \end{bmatrix} \begin{bmatrix} (-3)^{10} & 0 \\ 0 & 2^{10} \end{bmatrix} \begin{bmatrix} \frac{1}{5} & -\frac{2}{5} \\ \frac{1}{5} & \frac{3}{5} \end{bmatrix} = \begin{bmatrix} 35,839 & -69,630 \\ -11,605 & 24,234 \end{bmatrix}.$$

18. $\det(A-\lambda I) = \begin{vmatrix} 4-\lambda & -3 \\ -1 & 2-\lambda \end{vmatrix} = \lambda^2-6\lambda+5 = (\lambda-1)(\lambda-5)$.

$A-I = \begin{bmatrix} 3 & -3 \\ -1 & 1 \end{bmatrix} \Rightarrow E_1 = \text{span}\left(\begin{bmatrix} 1 \\ 1 \end{bmatrix}\right)$ and $A-5I = \begin{bmatrix} -1 & -3 \\ -1 & -3 \end{bmatrix} \Rightarrow E_5 = \text{span}\left(\begin{bmatrix} 3 \\ -1 \end{bmatrix}\right)$.
Therefore

$$\begin{bmatrix} 4 & -3 \\ -1 & 2 \end{bmatrix}^{-6} = \left(PDP^{-1}\right)^{-6} = PD^{-6}P^{-1} = \begin{bmatrix} 1 & 3 \\ 1 & -1 \end{bmatrix} \begin{bmatrix} 1^{-6} & 0 \\ 0 & 5^{-6} \end{bmatrix} \begin{bmatrix} 1 & 3 \\ 1 & -1 \end{bmatrix}^{-1}$$

$$= \begin{bmatrix} 1 & 3 \\ 1 & -1 \end{bmatrix} \begin{bmatrix} 1 & 0 \\ 0 & \frac{1}{15,625} \end{bmatrix} \begin{bmatrix} \frac{1}{4} & \frac{3}{4} \\ \frac{1}{4} & -\frac{1}{4} \end{bmatrix} = \frac{1}{15,625} \begin{bmatrix} 3907 & 11,718 \\ 3906 & 11,719 \end{bmatrix}.$$

19. $\det(A-\lambda I) = \begin{vmatrix} -\lambda & 3 \\ 1 & 2-\lambda \end{vmatrix} = \lambda^2-2\lambda-3 = (\lambda+1)(\lambda-3)$.

$A+I = \begin{bmatrix} 1 & 3 \\ 1 & 3 \end{bmatrix} \Rightarrow E_{-1} = \text{span}\left(\begin{bmatrix} 3 \\ -1 \end{bmatrix}\right)$ and $A-3I = \begin{bmatrix} -3 & 3 \\ 1 & -1 \end{bmatrix} \Rightarrow E_3 = \text{span}\left(\begin{bmatrix} 1 \\ 1 \end{bmatrix}\right)$.
Therefore

$$\begin{bmatrix} 0 & 3 \\ 1 & 2 \end{bmatrix}^{k} = \left(PDP^{-1}\right)^{k} = \begin{bmatrix} 3 & 1 \\ -1 & 1 \end{bmatrix} \begin{bmatrix} -1 & 0 \\ 0 & 3 \end{bmatrix}^{k} \begin{bmatrix} 3 & 1 \\ -1 & 1 \end{bmatrix}^{-1}$$

$$= \begin{bmatrix} 3 & 1 \\ -1 & 1 \end{bmatrix} \begin{bmatrix} (-1)^k & 0 \\ 0 & 3^k \end{bmatrix} \frac{1}{4} \begin{bmatrix} 1 & -1 \\ 1 & 3 \end{bmatrix} = \frac{1}{4} \begin{bmatrix} 3(-1)^k + 3^k & 3(-1)^{k+1} + 3^{k+1} \\ (-1)^{k+1} + 3^k & (-1)^k + 3^{k+1} \end{bmatrix}.$$

20. $\det(A-\lambda I) = \begin{vmatrix} 2-\lambda & 1 & 2 \\ 2 & 1-\lambda & 2 \\ 2 & 1 & 2-\lambda \end{vmatrix} = -\lambda^2(\lambda-5)$. $E_0 = \text{span}\left(\begin{bmatrix} 1 \\ -2 \\ 0 \end{bmatrix}, \begin{bmatrix} 1 \\ 0 \\ -1 \end{bmatrix}\right)$ and

$A - 5I = \begin{bmatrix} -3 & 1 & 2 \\ 2 & -4 & 2 \\ 2 & 1 & -3 \end{bmatrix} \longrightarrow \begin{bmatrix} 1 & 0 & -1 \\ 0 & 1 & -1 \\ 0 & 0 & 0 \end{bmatrix}$, so $E_5 = \text{span}\left(\begin{bmatrix} 1 \\ 1 \\ 1 \end{bmatrix}\right)$. Therefore

$$\begin{bmatrix} 2 & 1 & 2 \\ 2 & 1 & 2 \\ 2 & 1 & 2 \end{bmatrix}^8 = \begin{bmatrix} 1 & 1 & 1 \\ -2 & 0 & 1 \\ 0 & -1 & 1 \end{bmatrix} \begin{bmatrix} 0 & 0 & 0 \\ 0 & 0 & 0 \\ 0 & 0 & 5 \end{bmatrix}^8 \begin{bmatrix} 1 & 1 & 1 \\ -2 & 0 & 1 \\ 0 & -1 & 1 \end{bmatrix}^{-1}$$
$$= \begin{bmatrix} 0 & 0 & 5^8 \\ 0 & 0 & 5^8 \\ 0 & 0 & 5^8 \end{bmatrix} \frac{1}{5} \begin{bmatrix} 1 & -2 & 1 \\ 2 & 1 & -3 \\ 2 & 1 & 2 \end{bmatrix} = 5^7 \begin{bmatrix} 2 & 1 & 2 \\ 2 & 1 & 2 \\ 2 & 1 & 2 \end{bmatrix}.$$

21. Since $D = \begin{bmatrix} 1 & 0 & 0 \\ 0 & -1 & 0 \\ 0 & 0 & -1 \end{bmatrix}$, we have

$$A^{2002} = \left(PDP^{-1}\right)^{2002} = P \begin{bmatrix} 1 & 0 & 0 \\ 0 & -1 & 0 \\ 0 & 0 & -1 \end{bmatrix}^{2002} P^{-1} = PP^{-1} = I_3.$$

22. $\det(A-\lambda I) = \begin{vmatrix} 2-\lambda & 0 & 1 \\ 1 & 1-\lambda & 1 \\ 1 & 0 & 2-\lambda \end{vmatrix} = -\lambda^3 + 5\lambda^2 - 7\lambda + 3 = -(\lambda-1)^2(\lambda-3)$.

$E_1 = \text{span}\left(\begin{bmatrix} 1 \\ 0 \\ -1 \end{bmatrix}, \begin{bmatrix} 0 \\ 1 \\ 0 \end{bmatrix}\right)$ and $A - 3I = \begin{bmatrix} -1 & 0 & 1 \\ 1 & -2 & 1 \\ 1 & 0 & -1 \end{bmatrix} \longrightarrow \begin{bmatrix} 1 & 0 & -1 \\ 0 & 1 & -1 \\ 0 & 0 & 0 \end{bmatrix}$,

so $E_3 = \text{span}\left(\begin{bmatrix} 1 \\ 1 \\ 1 \end{bmatrix}\right)$. Therefore

$$\begin{bmatrix} 2 & 0 & 1 \\ 1 & 1 & 1 \\ 1 & 0 & 2 \end{bmatrix}^{-5} = \begin{bmatrix} 1 & 0 & 1 \\ 0 & 1 & 1 \\ -1 & 0 & 1 \end{bmatrix} \begin{bmatrix} 1 & 0 & 0 \\ 0 & 1 & 0 \\ 0 & 0 & 3 \end{bmatrix}^{-5} \begin{bmatrix} 1 & 0 & 1 \\ 0 & 1 & 1 \\ -1 & 0 & 1 \end{bmatrix}^{-1}$$
$$= \begin{bmatrix} 1 & 0 & 1 \\ 0 & 1 & 1 \\ -1 & 0 & 1 \end{bmatrix} \begin{bmatrix} 1 & 0 & 0 \\ 0 & 1 & 0 \\ 0 & 0 & \frac{1}{243} \end{bmatrix} \frac{1}{2} \begin{bmatrix} 1 & 0 & -1 \\ -1 & 2 & -1 \\ 1 & 0 & 1 \end{bmatrix} = \frac{1}{243} \begin{bmatrix} 122 & 0 & -121 \\ -121 & 243 & -121 \\ -121 & 0 & 122 \end{bmatrix}.$$

23. $\det(A-\lambda I) = \begin{vmatrix} 1-\lambda & 1 & 0 \\ 2 & -2-\lambda & 2 \\ 0 & 1 & 1-\lambda \end{vmatrix} = -\lambda^3 + 7\lambda - 6 = -(\lambda+3)(\lambda-1)(\lambda-2).$

$$A+3I = \begin{bmatrix} 4 & 1 & 0 \\ 2 & 1 & 2 \\ 0 & 1 & 4 \end{bmatrix} \longrightarrow \begin{bmatrix} 1 & 0 & -1 \\ 0 & 1 & 4 \\ 0 & 0 & 0 \end{bmatrix} \Rightarrow E_{-3} = \text{span}\left(\begin{bmatrix} 1 \\ -4 \\ 1 \end{bmatrix}\right).$$

$$A-I = \begin{bmatrix} 0 & 1 & 0 \\ 2 & -3 & 2 \\ 0 & 1 & 0 \end{bmatrix} \longrightarrow \begin{bmatrix} 1 & 0 & 1 \\ 0 & 1 & 0 \\ 0 & 0 & 0 \end{bmatrix} \Rightarrow E_1 = \text{span}\left(\begin{bmatrix} 1 \\ 0 \\ -1 \end{bmatrix}\right).$$

$$A-2I = \begin{bmatrix} -1 & 1 & 0 \\ 2 & -4 & 2 \\ 0 & 1 & -1 \end{bmatrix} \longrightarrow \begin{bmatrix} 1 & 0 & -1 \\ 0 & 1 & -1 \\ 0 & 0 & 0 \end{bmatrix}, \text{ so } E_2 = \text{span}\left(\begin{bmatrix} 1 \\ 1 \\ 1 \end{bmatrix}\right). \text{ Therefore}$$

$$\begin{bmatrix} 1 & 1 & 0 \\ 2 & -2 & 2 \\ 0 & 1 & 1 \end{bmatrix}^k = \begin{bmatrix} 1 & 1 & 1 \\ -4 & 0 & 1 \\ 1 & -1 & 1 \end{bmatrix} \begin{bmatrix} -3 & 0 & 0 \\ 0 & 1 & 0 \\ 0 & 0 & 2 \end{bmatrix}^k \begin{bmatrix} 1 & 1 & 1 \\ -4 & 0 & 1 \\ 1 & 1 & 1 \end{bmatrix}^{-1}$$

$$= \frac{1}{10}\begin{bmatrix} 1 & 1 & 1 \\ -4 & 0 & 1 \\ 1 & -1 & 1 \end{bmatrix} \begin{bmatrix} (-3)^k & 0 & 0 \\ 0 & 1 & 0 \\ 0 & 0 & 2^k \end{bmatrix} \begin{bmatrix} 1 & -2 & 1 \\ 5 & 0 & -5 \\ 4 & 2 & 4 \end{bmatrix}$$

$$= \frac{1}{10}\begin{bmatrix} 4(2)^k + (-3)^k + 5 & 2(2)^k - 2(-3)^k & 4(2)^k + (-3)^k - 5 \\ 4(2)^k - 4(-3)^k & 2(2)^k + 8(-3)^k & 4(2)^k - 4(-3)^k \\ 4(2)^k + (-3)^k - 5 & 2(2)^k - 2(-3)^k & 4(2)^k + (-3)^k + 5 \end{bmatrix}.$$

24. $\det(A-\lambda I) = \begin{vmatrix} 1-\lambda & 1 \\ 0 & k-\lambda \end{vmatrix} = k - \lambda - k\lambda + \lambda^2 = (\lambda-1)(\lambda-k).$

So A is certainly diagonalizable if $k \neq 1$.

$k=1$ then $A - I = \begin{bmatrix} 0 & 1 \\ 0 & 0 \end{bmatrix} \Rightarrow 1$ has geometric multiplicity $1 \Rightarrow A$ is not diagonalizable.

Thus, A is diagonalizable provided $k \neq 1$.

25. $\det(A-\lambda I) = \begin{vmatrix} 1-\lambda & k \\ 0 & 1-\lambda \end{vmatrix}$. The only eigenvalue is 1, and $A - I = \begin{bmatrix} 0 & k \\ 0 & 0 \end{bmatrix}$.

If $k=0$, then A is already diagonal;
otherwise 1 has geometric multiplicity 1 and thus A is not diagonalizable.
Thus, A is diagonalizable only if $k=0$.

26. $\det(A-\lambda I) = \begin{vmatrix} k-\lambda & 1 \\ 1 & -\lambda \end{vmatrix} = \lambda^2 - k\lambda - 1 = 0 \Leftrightarrow \lambda = \frac{k \pm \sqrt{k^2+4}}{2} \Rightarrow$

There are two distinct eigenvalues for all $k \Rightarrow A$ is diagonalizable for all k.

27. A has eigenvalue 1 with algebraic multiplicity 3. $A - I = \begin{bmatrix} 0 & 0 & k \\ 0 & 0 & 0 \\ 0 & 0 & 0 \end{bmatrix}$.

If $k = 0$ then A is already diagonal;
otherwise 1 has geometric multiplicity 2 and thus A is not diagonalizable.

Thus, A is diagonalizable only if $k = 0$.

28. A has eigenvalues 1 (with algebraic multiplicity 2) and 2 (with algebraic multiplicity 1).

$A - I = \begin{bmatrix} 0 & k & 0 \\ 0 & 1 & 0 \\ 0 & 0 & 0 \end{bmatrix} \longrightarrow \begin{bmatrix} 0 & 1 & 0 \\ 0 & 0 & 0 \\ 0 & 0 & 0 \end{bmatrix}$, so the eigenvalue 1 has geometric multiplicity 2

regardless of the value of k, and A is diagonalizable for all k.

29. $\det(A - \lambda I) = \begin{vmatrix} 1-\lambda & 1 & k \\ 1 & 1-\lambda & k \\ 1 & 1 & k-\lambda \end{vmatrix}$

$$= (1-\lambda)\left[(1-\lambda)(k-\lambda) - k\right] - \left[(k-\lambda) - k\right] + k\left[1 - (1-\lambda)\right]$$
$$= (1-\lambda)\left(\lambda^2 - k\lambda - \lambda\right) + \lambda + k\lambda = -\lambda^3 + 2\lambda^2 + k\lambda^2 = \lambda^2(k - \lambda + 2).$$

Now if $k = -2$, then A has only 0 as an eigenvalue (with algebraic multiplicity 3) and since

$A = \begin{bmatrix} 1 & 1 & -2 \\ 1 & 1 & -2 \\ 1 & 1 & -2 \end{bmatrix} \longrightarrow \begin{bmatrix} 1 & 1 & -2 \\ 0 & 0 & 0 \\ 0 & 0 & 0 \end{bmatrix}$, 0 has geometric multiplicity 2 and A is not diagonalizable.

If $k \neq -2$, then 0 has algebraic multiplicity 2 and $A = \begin{bmatrix} 1 & 1 & k \\ 1 & 1 & k \\ 1 & 1 & k \end{bmatrix} \longrightarrow \begin{bmatrix} 1 & 1 & k \\ 0 & 0 & 0 \\ 0 & 0 & 0 \end{bmatrix}$,

so 0 has geometric multiplicity 2 and A is diagonalizable.

Thus, A is diagonalizable provided $k \neq -2$.

30. We must show that similarity is transitive. That is, if $A \sim B$ and $B \sim C$, then $A \sim C$.
Since $M \sim N$ means $S^{-1}MS = N$ by the definition of similarity, it suffices to show:
If 1) $P^{-1}AP = B$ and 2) $Q^{-1}BQ = C$, then $R^{-1}AR = C$ where $R = PQ$.

This is obvious since letting $B = P^{-1}AP$ in equation 2) yields $Q^{-1}(P^{-1}AP)Q = C$.
Furthermore, since $Q^{-1}P^{-1} = (PQ)^{-1}$, we see in fact that $R = PQ$.

Q: How might we summarize this stronger result in words?
A: Not only are A and C similar, but the matrix relating them can be determined by computing the product of the matrix relating A to B and the matrix relating B to C.

31. $AP = PB \Rightarrow \det(AP) = \det(PB) \Rightarrow \det(A)\det(P) = \det(P)\det(B) \Rightarrow \det(A) = \det(B)$.
So A is invertible if and only if B is invertible.

32. Since we are proving Theorem 4.22(c), we need to show if $A \sim B$ then $\text{rank}(A) = \text{rank}(B)$. That is, if $P^{-1}AP = B$ or equivalently $A = PBP^{-1}$, then $\text{rank}(A) = \text{rank}(B)$.

The key insight is: if Q and R are invertible, then $\text{rank}(M) \overset{(1)}{=} \text{rank}(QM) \overset{(2)}{=} \text{rank}(MR)$. This result is not trivial. It was proven in Exercise 59 of Section 3.5.

Note, $P^{-1}AP = B$ implies $A = PBP^{-1}$, where P is invertible and so is P^{-1}. Therefore:

$$\text{rank}(A) \overset{\text{by def}}{=} \text{rank}(PBP^{-1}) \overset{\text{by assoc}}{=} \text{rank}(P(BP^{-1})) \overset{\text{by (1)}}{=} \text{rank}(BP^{-1}) \overset{\text{by (2)}}{=} \text{rank}(B)$$

33. Since we are proving Theorem 4.22(e), we need to prove the following two statements:
1) Given $A \sim B$, if α is an eigenvalue of A then α is an eigenvalue of B and
2) Given $A \sim B$, if β is an eigenvalue of B then β is an eigenvalue of A.

Since λ if an eigenvalue of M if and only if there exists an eigenvector $\mathbf{v} \neq \mathbf{0}$ such that $M\mathbf{v} = \lambda\mathbf{v}$, it suffices to show the following:

1) If $A\mathbf{x} = \alpha\mathbf{x}$, $\mathbf{x} \neq \mathbf{0}$, then $B(P^{-1}\mathbf{x}) = \alpha(P^{-1}\mathbf{x})$ and
2) If $B\mathbf{y} = \beta\mathbf{y}$, $\mathbf{y} \neq \mathbf{0}$, then $A(P\mathbf{y}) = \beta(P\mathbf{y})$.

Note, $P^{-1}AP = B$ implies $A = PBP^{-1}$, where P is invertible and so is P^{-1}. Therefore:

If $A\mathbf{x} = \alpha\mathbf{x}$, $\mathbf{x} \neq \mathbf{0}$, then $(PBP^{-1})\mathbf{x} = \alpha\mathbf{x}$.
So, left multiplying both sides by P^{-1} implies $B(P^{-1}\mathbf{x}) = P^{-1}(\alpha\mathbf{x}) = \alpha(P^{-1}\mathbf{x})$.

Likewise, if $B\mathbf{y} = \beta\mathbf{y}$, $\mathbf{y} \neq \mathbf{0}$, then $(P^{-1}AP)\mathbf{y} = \beta\mathbf{y}$.
So, left multiplying both sides by P implies $A(P\mathbf{y}) = P(\beta\mathbf{y}) = \beta(P\mathbf{y})$.

Q: Why do we *not* have to specify that $P^{-1}\mathbf{x}$ and $P\mathbf{y}$ are not the zero vector?
A: Because P is invertible and so P^{-1} is invertible, too. Therefore:
If $\mathbf{u} \neq 0$, then $P\mathbf{u} \neq 0$ and $P^{-1}\mathbf{u} \neq 0$. Why?

Q: How might we state our stronger result in words?
A: Not only is α an eigenvalue for B, one associated eigenvector of α is $P^{-1}\mathbf{x}$.
Not only is β an eigenvalue for A, one associated eigenvector of β is $P\mathbf{y}$.

34. Let $P = B^{-1}$. Then $(AB)\,P = (AB)\,B^{-1} = A$ and $P\,(BA) = B^{-1}\,(BA) = A$.
So $(AB)\,P = P\,(BA) \Leftrightarrow P^{-1}\,(AB)\,P = BA \Leftrightarrow AB \sim BA$.

35. We need to show if $A \sim B$, then $\text{tr}(A) = \text{tr}(B)$.
That is, if $P^{-1}AP = B$ or equivalently $A = PBP^{-1}$, then $\text{tr}(A) = \text{tr}(B)$.

The key insight is: $\text{tr}(MN) \overset{(1)}{=} \text{tr}(NM)$.
This result is not trivial. It was proven in Exercise 45 of Section 3.2.

$$\text{tr}(A) \overset{\text{by def}}{=} \text{tr}(PBP^{-1}) \overset{\text{by assoc}}{=} \text{tr}(P(BP^{-1})) \overset{\text{by (1)}}{=} \text{tr}((BP^{-1})P) \overset{\text{by assoc}}{=} \text{tr}(B(P^{-1}P)) = \text{tr}(B)$$

Alternate: We know that A and B have the same eigenvalues $\lambda_1, \lambda_2, \ldots, \lambda_n$.
So by Exercise 40 from Section 4.3, $\text{tr}\,(A) = \lambda_1 + \lambda_2 + \cdots + \lambda_n = \text{tr}\,(B)$.

36. Because A is triangular, we see that it has eigenvalues -1 and 3.

On the other hand, $\det(B - \lambda I) = \begin{vmatrix} 1-\lambda & 2 \\ 2 & 1-\lambda \end{vmatrix} = (1-\lambda)^2 - 4 = (\lambda+1)(\lambda-3)$.

So, the eigenvalues of B are also -1 and $3 \Rightarrow A \sim B \sim \begin{bmatrix} -1 & 0 \\ 0 & 3 \end{bmatrix}$.

For A, we have $E_{-1} = \text{span}\left(\begin{bmatrix} 1 \\ -4 \end{bmatrix}\right)$ and $E_3 = \text{span}\left(\begin{bmatrix} 1 \\ 0 \end{bmatrix}\right)$.

For B, $E_{-1} = \text{span}\left(\begin{bmatrix} 1 \\ -1 \end{bmatrix}\right)$ and $E_3 = \text{span}\left(\begin{bmatrix} 1 \\ 1 \end{bmatrix}\right)$.

So $Q^{-1}AQ = D = R^{-1}BR$, where $Q = \begin{bmatrix} 1 & 1 \\ -4 & 0 \end{bmatrix}$, $D = \begin{bmatrix} -1 & 0 \\ 0 & 3 \end{bmatrix}$, and $R = \begin{bmatrix} 1 & 1 \\ -1 & 1 \end{bmatrix}$.

Thus, $RQ^{-1}AQR^{-1} = (QR^{-1})^{-1} A (QR^{-1}) = B$. So we have

$P^{-1}AP = B$ where $P = QR^{-1} = \begin{bmatrix} 1 & 1 \\ -4 & 0 \end{bmatrix} \begin{bmatrix} \frac{1}{2} & -\frac{1}{2} \\ \frac{1}{2} & \frac{1}{2} \end{bmatrix} = \begin{bmatrix} 1 & 0 \\ -2 & 2 \end{bmatrix}$.

37. $\det(A - \lambda I) = \begin{vmatrix} 5-\lambda & -3 \\ 4 & -2-\lambda \end{vmatrix} = \lambda^2 - 3\lambda + 2 = (\lambda - 1)(\lambda - 2)$, so A has eigenvalues 1, 2.

$\det(B - \lambda I) = \begin{vmatrix} -1-\lambda & 1 \\ -6 & 4-\lambda \end{vmatrix} = \lambda^2 - 3\lambda + 2$, so B has eigenvalues 1, 2.

Therefore, A and B are similar to $\begin{bmatrix} 1 & 0 \\ 0 & 2 \end{bmatrix}$.

For A, we have $E_1 = \text{span}\left(\begin{bmatrix} 3 \\ 4 \end{bmatrix}\right)$ and $E_2 = \text{span}\left(\begin{bmatrix} 1 \\ 1 \end{bmatrix}\right)$.

For B, $E_1 = \text{span}\left(\begin{bmatrix} 1 \\ 2 \end{bmatrix}\right)$ and $E_2 = \text{span}\left(\begin{bmatrix} 1 \\ 3 \end{bmatrix}\right)$.

So $Q^{-1}AQ = D = R^{-1}BR$, where $Q = \begin{bmatrix} 3 & 1 \\ 4 & 1 \end{bmatrix}$, $D = \begin{bmatrix} 1 & 0 \\ 0 & 2 \end{bmatrix}$, and $R = \begin{bmatrix} 1 & 1 \\ 2 & 3 \end{bmatrix}$.

Thus, $RQ^{-1}AQR^{-1} = (QR^{-1})^{-1} A (QR^{-1}) = B \Rightarrow$

$P^{-1}AP = B$ with $P = QR^{-1} = \begin{bmatrix} 3 & 1 \\ 4 & 1 \end{bmatrix} \begin{bmatrix} 1 & 1 \\ 2 & 3 \end{bmatrix}^{-1} = \begin{bmatrix} 3 & 1 \\ 4 & 1 \end{bmatrix} \begin{bmatrix} 3 & -1 \\ -2 & 1 \end{bmatrix} = \begin{bmatrix} 7 & -2 \\ 10 & -3 \end{bmatrix}$.

38. A has eigenvalues -2, 1, and 2, while

$$\begin{aligned}\det(B-\lambda I) &= \begin{vmatrix} 3-\lambda & 2 & -5 \\ 1 & 2-\lambda & -1 \\ 2 & 2 & -4-\lambda \end{vmatrix} \overset{R_1-R_3}{=} \begin{vmatrix} 1-\lambda & 0 & -(1-\lambda) \\ 1 & 2-\lambda & -1 \\ 2 & 2 & -4-\lambda \end{vmatrix} \\ &= (1-\lambda)\left[(2-\lambda)(-4-\lambda)+2-[2-2(2-\lambda)]\right] = (1-\lambda)\left[(2-\lambda)(-4-\lambda+2)\right] \\ &= -(\lambda+2)(\lambda-1)(\lambda-2).\end{aligned}$$

So, B has the same eigenvalues as A. Thus both are similar to $\begin{bmatrix} -2 & 0 & 0 \\ 0 & 1 & 0 \\ 0 & 0 & 2 \end{bmatrix}$.

For A, $E_{-2} = \operatorname{span}\left(\begin{bmatrix} 1 \\ -4 \\ 0 \end{bmatrix}\right)$, $E_1 = \operatorname{span}\left(\begin{bmatrix} 1 \\ -1 \\ -3 \end{bmatrix}\right)$, and $E_2 = \operatorname{span}\left(\begin{bmatrix} 1 \\ 0 \\ 0 \end{bmatrix}\right)$.

For B, $E_{-2} = \left(\begin{bmatrix} 1 \\ 0 \\ 1 \end{bmatrix}\right)$, $E_1 = \operatorname{span}\left(\begin{bmatrix} 1 \\ -1 \\ 0 \end{bmatrix}\right)$, and $E_2 = \operatorname{span}\left(\begin{bmatrix} 1 \\ 2 \\ 1 \end{bmatrix}\right)$.

Thus, as in Exercises 36 and 37, we have

$$P^{-1}AP = B \text{ where } P = QR^{-1} = \begin{bmatrix} 1 & 1 & 1 \\ -4 & -1 & 0 \\ 0 & -3 & 0 \end{bmatrix}\begin{bmatrix} 1 & 1 & 1 \\ 0 & -1 & 2 \\ 1 & 0 & 1 \end{bmatrix}^{-1} = \begin{bmatrix} 1 & 0 & 0 \\ 1 & 2 & -5 \\ -3 & 0 & 3 \end{bmatrix}.$$

39. $\det(A-\lambda I) = \begin{vmatrix} 1-\lambda & 0 & 2 \\ 1 & -1-\lambda & 1 \\ 2 & 0 & 1-\lambda \end{vmatrix} = (-1-\lambda)\left[(1-\lambda)^2-4\right] = -(\lambda+1)^2(\lambda-3)$, while

$$\det(B-\lambda I) = \begin{vmatrix} -3-\lambda & -2 & 0 \\ 6 & 5-\lambda & 0 \\ 4 & 4 & -1-\lambda \end{vmatrix} = -(\lambda+1)\left[(-3-\lambda)(5-\lambda)+12\right] = -(\lambda+1)^2(\lambda-3).$$

So, A and B both have eigenvalues -1 and 3.

For A, $E_{-1} = \operatorname{span}\left(\begin{bmatrix} 1 \\ 0 \\ -1 \end{bmatrix}, \begin{bmatrix} 0 \\ 1 \\ 0 \end{bmatrix}\right)$ and $E_3 = \operatorname{span}\left(\begin{bmatrix} 2 \\ 1 \\ 2 \end{bmatrix}\right)$.

For B, $E_{-1} = \left(\begin{bmatrix} 1 \\ -1 \\ 0 \end{bmatrix}, \begin{bmatrix} 0 \\ 0 \\ 1 \end{bmatrix}\right)$ and $E_3 = \operatorname{span}\left(\begin{bmatrix} 1 \\ -3 \\ -2 \end{bmatrix}\right)$.

Thus, as in Exercises 36 and 37, we have

$$P^{-1}AP = B \text{ where } P = QR^{-1} = \begin{bmatrix} 1 & 0 & 2 \\ 0 & 1 & 1 \\ -1 & 0 & 2 \end{bmatrix}\begin{bmatrix} 1 & 0 & 1 \\ -1 & 0 & -3 \\ 0 & 1 & -2 \end{bmatrix}^{-1} = \begin{bmatrix} \frac{1}{2} & -\frac{1}{2} & 0 \\ -\frac{3}{2} & -\frac{3}{2} & 1 \\ -\frac{5}{2} & -\frac{3}{2} & 0 \end{bmatrix}.$$

40. $A \sim B \Rightarrow AP = PB \Rightarrow (AP)^T = (PB)^T \Rightarrow P^TA^T = B^TP^T$, by Theorem 3.4(d).
So we have $B^TP^T = P^TA^T \Rightarrow B^T \sim A^T \Rightarrow A^T \sim B^T$.

41. Given A is diagonalizable, we need to show A^T is diagonalizable.
It suffices to show if $P^{-1}AP = D$, then $P^T A^T (P^T)^{-1} = D$.

This is obvious since $D \overset{\substack{D \\ \text{diagonal}}}{=} D^T \overset{\substack{\text{by} \\ \text{def}}}{=} (P^{-1}AP)^T \overset{\substack{\text{Thm 3.9} \\ \text{Sect 3.3}}}{=} P^T A^T (P^T)^{-1}$

Q: Why is this sufficient?
A: Because P is invertible, so is P^T. Why? $\det(P^T) = \det P$.

Q: How might we summarize this stronger result in words?
A: Not only is A^T diagonalizable, its diagonal matrix is the same as that of A.
Furthermore, its matrix of diagonalization is the inverse tranpose of the one for A.

42. Given A is invertible and diagonalizable, we need to show A^{-1} is diagonalizable.
It suffices to show if $P^{-1}AP = D$, then $P^{-1}A^{-1}P = D^{-1}$.

This is obvious since $D^{-1} \overset{\substack{\text{by} \\ \text{def}}}{=} (P^{-1}AP)^{-1} \overset{\substack{\text{Thm 3.9} \\ \text{Sect 3.3}}}{=} P^{-1}A^{-1}P$.

Q: Why does D being diagonal imply that D^{-1} is diagonal?
A: If $D = \begin{bmatrix} \lambda_1 \mathbf{e}_1 & \lambda_2 \mathbf{e}_2 & \cdots & \lambda_n \mathbf{e}_n \end{bmatrix}$, then $D^{-1} = \begin{bmatrix} \frac{1}{\lambda_1}\mathbf{e}_1 & \frac{1}{\lambda_2}\mathbf{e}_2 & \cdots & \frac{1}{\lambda_n}\mathbf{e}_n \end{bmatrix}$. Prove this.

Hint: When D is 2×2, $D = \begin{bmatrix} \lambda_1 & 0 \\ 0 & \lambda_2 \end{bmatrix}$ and $D^{-1} = \begin{bmatrix} \frac{1}{\lambda_1} & 0 \\ 0 & \frac{1}{\lambda_2} \end{bmatrix}$.

Q: How do we know that that $\frac{1}{\lambda_i}$ is defined?
A: Because D is diagonal, D is also invertible. So, $\lambda_i \neq 0$. Why?

Q: How might we summarize this stronger result in words?
A: Not only is A^{-1} diagonalizable, its diagonal matrix is the inverse of that for A.
Furthermore, its matrix of diagonalization is the *same* as the one for A.

43. Suppose $P^{-1}AP = D$. Now each entry on the diagonal of D is λ (the only eigenvalue of A),
So $D = \lambda I$. Thus, $A = PDP^{-1} = P(\lambda I)P^{-1} = \lambda PP^{-1}I = \lambda I$.

44. Since this is an if and only if proposition, there are two statements we have to prove:
1) If A and B have the same eigenvectors, then $AB = BA$ and
2) If $AB = BA$, then A and B have the same eigenvectors.

We begin by noting two essential consequences of A and B having n distinct eigenvalues:
Theorem 4.25 implies both A and B are diagonalizable. Therefore,
Theorem 4.23 implies that the columns of P for both A and B are their eigenvectors.

Q: If D and E are diagonal matrices, does $DE = ED$?
A: Yes. Prove this. Hint: A diagonal matrix has the form $\begin{bmatrix} \lambda_1 \mathbf{e}_1 & \lambda_2 \mathbf{e}_2 & \cdots & \lambda_n \mathbf{e}_n \end{bmatrix}$.

1: If A and B have the same eigenvectors, then $AB = BA$.
If A and B have the *same* eigenvectors, then for the *same* matrix P
we have both $P^{-1}AP = D_A$ and $P^{-1}BP = D_B$. So:

$$AB = (PD_AP^{-1})(PD_BP^{-1}) = (PD_AD_BP^{-1}) \overset{\substack{\text{diag}\\ \text{commute}}}{=} (PD_BP^{-1})(PD_AP^{-1}) = BA$$

Q: Can we write P explicitly in terms of the eigenvectors of A and B?
A: Yes. Theorem 4.23 implies the columns of P are the eigenvectors of A and B.
Since {eigenvectors of A} $= \{\mathbf{p}_i\} =$ {eigenvectors of B}, $P = \begin{bmatrix} \mathbf{p}_1 & \mathbf{p}_2 & \cdots & \mathbf{p}_n \end{bmatrix}$.

2: If $AB = BA$, then A and B have the same eigenvectors.
If $\mathbf{p}_i$ is the *distinct* eigenvalue of A corresponding to α_i, then

$$A(B\mathbf{p}_i) \overset{\substack{AB=BA\\ \text{given}}}{=} B(A\mathbf{p}_i) = B(\alpha_i \mathbf{p}_i) = \alpha_i(B\mathbf{p}_i).$$

Therefore, $B\mathbf{p}_i$ is also an eigenvector of A corresponding to α_i.
Since the eigenvectors of A are *distinct*, $B\mathbf{p}_i$ must be a multiple of $\mathbf{p}_i$.
That is, $B\mathbf{p}_i = \beta_i \mathbf{p}_i$.
Therefore, $\mathbf{p}_i$ is also an eigenvector of B corresponding to β_i.
Since this argument is symmetric with respect to A and B,
A and B have the same eigenvectors as claimed.

Q: Since the eigenvectors of A are *distinct* and $A(B\mathbf{p}_i) = \alpha_i(B\mathbf{p}_i)$,
we claim above that $B\mathbf{p}_i$ must be a multiple of $\mathbf{p}_i$. Why? Prove this.
A: Since A has n distinct eigenvalues, Theorem 4.20 in Section 4.3 implies
the corresponding eigenvectors $\mathbf{p}_i$ are linearly independent and therefore span $\mathbb{R}^n$.

So, we can express $\mathbf{x}$ as a linear combination of the $\mathbf{p}_i$: $\mathbf{x} = \sum_{k=1}^{n} \beta_k \mathbf{p}_k$.

Now let $\mathbf{x}$ be *any* nonzero vector such that $A\mathbf{x} = \alpha_i \mathbf{x}$.

On the *right* hand side, we have $A\mathbf{x} = A\left(\sum_{k=1}^{n} \beta_k \mathbf{p}_k\right) = \sum_{k=1}^{n} \beta_k (A\mathbf{p}_k) = \sum_{k=1}^{n} \beta_k \alpha_k \mathbf{p}_k$.

On the *left* hand side, we have $\alpha_i \mathbf{x} = \alpha_i \left(\sum_{k=1}^{n} \beta_k \mathbf{p}_k\right) = \sum_{k=1}^{n} \beta_k \alpha_i \mathbf{p}_k$.

Setting these two sides equal yields: $\sum_{k=1}^{n} \beta_k \alpha_k \mathbf{p}_k = \sum_{k=1}^{n} \beta_k \alpha_i \mathbf{p}_k$.

It is important to note that α_i is the one *fixed* eigenvalue we are considering.
However, the α_k are changing with the index value in the summation. Finish the proof.

45. A and B are similar and thus have the same characteristic polynomial, by Theorem 4.22(d). It follows that the algebraic multiplicities of the eigenvalues of A and B are the same. Why?

46. Since λ is an eigenvalue of M if and only if there exists an eigenvector $\mathbf{v} \neq \mathbf{0}$ such that $M\mathbf{v} = \lambda\mathbf{v}$, as suggested in the hint it suffices to show the following:

If $A\mathbf{x} = \alpha\mathbf{x}$, $\mathbf{x} \neq \mathbf{0}$, then $B(P^{-1}\mathbf{x}) = \alpha(P^{-1}\mathbf{x})$.

Note, $P^{-1}AP = B$ implies $A = PBP^{-1}$, where P is invertible and so is P^{-1}. Therefore:

If $A\mathbf{x} = \alpha\mathbf{x}$, $\mathbf{x} \neq \mathbf{0}$, then $(PBP^{-1})\mathbf{x} = \alpha\mathbf{x}$.
So, left multiplying both sides by P^{-1} implies $B(P^{-1}\mathbf{x}) = P^{-1}(\alpha\mathbf{x}) = \alpha(P^{-1}\mathbf{x})$.

So, not only is α an eigenvalue for B, one associated eigenvector of α is $P^{-1}\mathbf{x}$.

For more details see Exercise 33.

47. We need to show if $A = P^{-1}DP$ is diagonalizable and $\lambda_i = 0$ or 1, then $A^2 = A$.

The key observation is: if $\lambda_i = 0$ or 1, then $\lambda_i^2 = \lambda_i$. Prove this.

So, since $\lambda_i^2 = \lambda_i$ and $D = \begin{bmatrix} \lambda_1\mathbf{e}_1 & \lambda_2\mathbf{e}_2 & \cdots & \lambda_n\mathbf{e}_n \end{bmatrix}$, $D^2 \overset{(1)}{=} D$. Prove this. So:

$$A^2 = (P^{-1}DP)^2 = (P^{-1}DP)(P^{-1}DP) = P^{-1}D^2P \overset{\text{by (1)}}{=} P^{-1}DP = A$$

48. We need to show if $A = P^{-1}DP$ is diagonalizable and $A^m = O$, then $A = O$.

The key observation is: if $\lambda_i^m = 0$, then $\lambda_i = 0$.
So, since $D = \begin{bmatrix} \lambda_1\mathbf{e}_1 & \lambda_2\mathbf{e}_2 & \cdots & \lambda_n\mathbf{e}_n \end{bmatrix}$, $D^m = O$ implies $D = O$.
Therefore, $A = P^{-1}DP = P^{-1}OP = O$.

49. We note that Lemma 4.26 is the key to (a) and Theorem 4.27(c) is the key to (b).

(a) Since $A\mathbf{v}_1 = \mathbf{v}_1$, $A\mathbf{v}_2 = \mathbf{v}_2$, $A\mathbf{v}_3 = \mathbf{v}_3$, implies $\mathbf{v}_1$, $\mathbf{v}_2$, $\mathbf{v}_3$ are in E_1,
it suffices to show that if $c_A(\lambda) = (1+\lambda)(1-\lambda)^2(2-\lambda)^3$, then $\dim(E_1) \leq 2$.
Q: Why is this sufficient?
A: If the dimension of a subspace is ≤ 2, any 3 vectors in it must be linearly dependent.
The key observation is: if $c_A(\lambda) = \prod_{i=1}^{m} (\lambda_i - \lambda)^{k_i}$, then $\dim E_{\lambda_i} \leq k_i$. That is:

The dimension of the eigenspace associated with an eigenvector is less than or equal to the exponent of the factor of its eigenvalue in the characteristic equation.

Since $c_A(\lambda) = (1+\lambda)(1-\lambda)^2(2-\lambda)^3$, $\dim(E_1) \leq 2$.
Therefore any three vectors in E_1 are linearly dependent. So:
If $A\mathbf{v}_1 = \mathbf{v}_1$, $A\mathbf{v}_2 = \mathbf{v}_2$, $A\mathbf{v}_3 = \mathbf{v}_3$, then $\mathbf{v}_1$, $\mathbf{v}_2$, $\mathbf{v}_3$ must be linearly dependent.

(b) Since A is diagonalizable and $c_A(\lambda) = (1+\lambda)(1-\lambda)^2(2-\lambda)^3$,
Theorem 4.27(c) implies $\dim(E_{-1}) = 1$, $\dim(E_1) = 2$, and $\dim(E_2) = 3$.

50. Recall Exercise 35 in Section 4.1. We repeat the relevant parts of its solution here.

(a) If the eigenvalues of A are distinct, then A is diagonalizable. So it suffices to show: If $(a-d)^2+4bc>0$, then the eigenvalues of A are distinct.

To find the eigenvalues of $A=\begin{bmatrix} a & b \\ c & d \end{bmatrix}$, we solve $\det(A-\lambda I)=0 \Leftrightarrow$

$\det\begin{bmatrix} a-\lambda & b \\ c & d-\lambda \end{bmatrix} = \lambda^2-(a+d)\lambda+(ad-bc)=\lambda^2-\operatorname{tr}(A)\lambda+\det A=0.$

Using the quadratic formula, the solutions to the equation above are

$$\begin{aligned}\lambda &= \frac{(a+d)\pm\sqrt{(a+d)^2-4(ad-bc)}}{2} = \frac{a+d\pm\sqrt{a^2+d^2+2ad-4ad+4bc}}{2} \\ &= \frac{1}{2}\left(a+d\pm\sqrt{(a-d)^2+4bc}\right).\end{aligned}$$

From this, the two key observations follow immediately:
1) If $(a-d)^2+4bc>0$, then A has two distinct real eigenvalues, and
2) If $(a-d)^2+4bc<0$, then A has no real eigenvalues.

So: If $(a-d)^2+4bc>0$, then A is diagonalizable.
If $(a-d)^2+4bc<0$, then A is *not* diagonalizable.

(b) Furthermore, if $(a-d)^2+4bc=0$, then A has one real eigenvalue.
Note this does not imply either that A is or is not diagonalizable.

For example: $O=\begin{bmatrix} 0 & 0 \\ 0 & 0 \end{bmatrix}$, is clearly diagonalizable since

$$O=P^{-1}DP=I^{-1}OI=\begin{bmatrix} 1 & 0 \\ 0 & 1 \end{bmatrix}\begin{bmatrix} 0 & 0 \\ 0 & 0 \end{bmatrix}\begin{bmatrix} 1 & 0 \\ 0 & 1 \end{bmatrix}$$

The one real eigenvalue of O is 0.

However, $A=\begin{bmatrix} 1 & -1 \\ 1 & -1 \end{bmatrix}$ satisfies $(a-d)^2+4bc=2^2+4(-1)=0$.

But $E_0=\operatorname{span}\left(\begin{bmatrix} 1 \\ 1 \end{bmatrix}\right)$.

Prove this. Also, prove any multiple of this matrix also satisfies the same condition. Since A does not have 2 linearly independent eigenvectors, A is not diagonalizable.

4.5 Iterative Methods for Computing Eigenvalues

1. (a) $\mathbf{x}_5 = \begin{bmatrix} 4443 \\ 11,109 \end{bmatrix}$, so we estimate a dominant eigenvector of A to be $\begin{bmatrix} 1 \\ \frac{11,109}{4443} \end{bmatrix} \approx \begin{bmatrix} 1 \\ 2.500 \end{bmatrix}$.

To find the dominant eigenvalue λ_1, we calculate

$$\mathbf{x}_6 = A\mathbf{x}_5 = \begin{bmatrix} 1 & 2 \\ 5 & 4 \end{bmatrix}\begin{bmatrix} 4443 \\ 11,109 \end{bmatrix} = \begin{bmatrix} 26,661 \\ 66,651 \end{bmatrix} \approx \lambda_1 \begin{bmatrix} 4443 \\ 11,109 \end{bmatrix} \Rightarrow \lambda_1 \approx \tfrac{26,661}{4443} \approx 6.001.$$

(b) $\det(A - \lambda I) = \begin{vmatrix} 1-\lambda & 2 \\ 5 & 4-\lambda \end{vmatrix} = \lambda^2 - 5\lambda - 6 = (\lambda+1)(\lambda-6)$.

So, 6 is the dominant eigenvalue.

2. (a) $\mathbf{x}_5 = \begin{bmatrix} 7811 \\ -3904 \end{bmatrix}$, so we estimate a dominant eigenvector of A to be $\begin{bmatrix} 1 \\ -\frac{3904}{7811} \end{bmatrix} \approx \begin{bmatrix} 1 \\ -0.500 \end{bmatrix}$.

To find the dominant eigenvalue λ_1, we calculate

$$\mathbf{x}_6 = A\mathbf{x}_5 = \begin{bmatrix} 7 & 4 \\ -3 & -1 \end{bmatrix}\begin{bmatrix} 7811 \\ -3904 \end{bmatrix} = \begin{bmatrix} 39,061 \\ -19,529 \end{bmatrix} \approx \lambda_1 \begin{bmatrix} 7811 \\ -3904 \end{bmatrix} \Rightarrow$$

$\lambda_1 \approx \frac{39,061}{7811} \approx 5.001.$

(b) $\det(A - \lambda I) = \begin{vmatrix} 7-\lambda & 4 \\ -3 & -1-\lambda \end{vmatrix} = \lambda^2 - 6\lambda + 5 = (\lambda-1)(\lambda-5)$.

So, 5 is the dominant eigenvalue.

3. (a) $\mathbf{x}_5 = \begin{bmatrix} 144 \\ 89 \end{bmatrix}$, so we estimate a dominant eigenvector of A to be $\begin{bmatrix} 1 \\ \frac{89}{144} \end{bmatrix} \approx \begin{bmatrix} 1 \\ 0.618 \end{bmatrix}$.

To find the dominant eigenvalue λ_1, we calculate

$$\mathbf{x}_6 = A\mathbf{x}_5 = \begin{bmatrix} 2 & 1 \\ 1 & 1 \end{bmatrix}\begin{bmatrix} 144 \\ 89 \end{bmatrix} = \begin{bmatrix} 377 \\ 233 \end{bmatrix} \Rightarrow \lambda_1 \approx \tfrac{377}{144} = 2.618.$$

(b) $\det(A - \lambda I) = \begin{vmatrix} 2-\lambda & 1 \\ 1 & 1-\lambda \end{vmatrix} = \lambda^2 - 3\lambda + 1 = 0 \Rightarrow \lambda = \frac{3\pm\sqrt{5}}{2}$.

So, $\frac{3+\sqrt{5}}{2}$ is the dominant eigenvalue.

4. (a) $\mathbf{x}_5 = \begin{bmatrix} 60.625 \\ 239.500 \end{bmatrix} \Rightarrow$ dominant eigenvector of $A \begin{bmatrix} 1 \\ \frac{239.500}{60.625} \end{bmatrix} \approx \begin{bmatrix} 1 \\ 3.951 \end{bmatrix}$.

To find the dominant eigenvalue λ_1, we calculate

$$\mathbf{x}_6 = A\mathbf{x}_5 = \begin{bmatrix} 1.5 & 0.5 \\ 2.0 & 3.0 \end{bmatrix}\begin{bmatrix} 60.625 \\ 239.500 \end{bmatrix} = \begin{bmatrix} 210.6875 \\ 839.75 \end{bmatrix} \Rightarrow \lambda_1 \approx \tfrac{210.6875}{60.625} \approx 3.475.$$

(b) $\det(A - \lambda I) = \begin{vmatrix} 1.5-\lambda & 0.5 \\ 2.0 & 3.0-\lambda \end{vmatrix} = \lambda^2 - 4.5\lambda + 3.5 = \frac{1}{2}(\lambda-1)(2\lambda-7)$,

so $\lambda = \frac{7}{2}$ is the dominant eigenvalue.

5. (a) $\mathbf{x}_5 = \begin{bmatrix} -3.667 \\ 11.001 \end{bmatrix} \Rightarrow m_5 = 11.001$ and $\mathbf{y}_5 = \begin{bmatrix} -\frac{3.667}{11.001} \\ 1 \end{bmatrix} = \begin{bmatrix} -0.333 \\ 1 \end{bmatrix}$.

(b) $A\mathbf{y}_5 = \begin{bmatrix} 2 & -3 \\ -3 & 10 \end{bmatrix}\begin{bmatrix} -0.333 \\ 1 \end{bmatrix} = \begin{bmatrix} -3.666 \\ 10.999 \end{bmatrix}$ while $\lambda_1\mathbf{y}_5 = 11.001 \begin{bmatrix} -0.333 \\ 1 \end{bmatrix} = \begin{bmatrix} -3.663 \\ 11.001 \end{bmatrix}$.

So we have indeed approximated an eigenvalue and an eigenvector of A.

6. (a) $\mathbf{x}_{10} = \begin{bmatrix} 5.530 \\ 1.470 \end{bmatrix} \Rightarrow m_{10} = 5.530$ and $\mathbf{y}_{10} = \begin{bmatrix} 1 \\ \frac{1.470}{5.530} \end{bmatrix} = \begin{bmatrix} 1 \\ 0.266 \end{bmatrix}$.

(b) $A\mathbf{y}_{10} = \begin{bmatrix} 5 & 2 \\ 2 & -2 \end{bmatrix} \begin{bmatrix} 1 \\ 0.266 \end{bmatrix} = \begin{bmatrix} 5.532 \\ 1.468 \end{bmatrix}$ while $\lambda_1 \mathbf{y}_{10} = 5.530 \begin{bmatrix} 1 \\ 0.266 \end{bmatrix} = \begin{bmatrix} 5.53 \\ 1.471 \end{bmatrix}$.

7. (a) $\mathbf{x}_8 = \begin{bmatrix} 10.000 \\ 0.001 \\ 10.000 \end{bmatrix} \Rightarrow m_8 = 10.000$ and $\mathbf{y}_8 = \begin{bmatrix} 1 \\ \frac{0.001}{10.000} \\ 1 \end{bmatrix} = \begin{bmatrix} 1 \\ 0.0001 \\ 1 \end{bmatrix}$.

(b) $A\mathbf{y}_8 = \begin{bmatrix} 4 & 0 & 6 \\ -1 & 3 & 1 \\ 6 & 0 & 4 \end{bmatrix} \begin{bmatrix} 1 \\ 0.0001 \\ 1 \end{bmatrix} = \begin{bmatrix} 10 \\ 0.0003 \\ 10 \end{bmatrix}$ while $\lambda_1 \mathbf{y}_8 = 10.000 \begin{bmatrix} 1 \\ 0.0001 \\ 1 \end{bmatrix} = \begin{bmatrix} 10 \\ 0.001 \\ 10 \end{bmatrix}$.

8. (a) $\mathbf{x}_{10} = \begin{bmatrix} 3.415 \\ 2.914 \\ -1.207 \end{bmatrix} \Rightarrow m_{10} = 3.415$ and $\mathbf{y}_{10} = \begin{bmatrix} 1 \\ \frac{2.914}{3.415} \\ -\frac{1.207}{3.415} \end{bmatrix} = \begin{bmatrix} 1 \\ 0.853 \\ -0.353 \end{bmatrix}$.

(b) $A\mathbf{y}_{10} = \begin{bmatrix} 1 & 2 & -2 \\ 1 & 1 & -3 \\ 0 & -1 & 1 \end{bmatrix} \begin{bmatrix} 1 \\ 0.853 \\ -0.353 \end{bmatrix} = \begin{bmatrix} 3.412 \\ 2.912 \\ -1.206 \end{bmatrix}$ while

$$\lambda_1 \mathbf{y}_{10} = 3.415 \begin{bmatrix} 1 \\ 0.853 \\ -0.353 \end{bmatrix} = \begin{bmatrix} 3.415 \\ 2.913 \\ -1.205 \end{bmatrix}.$$

9. With $A = \begin{bmatrix} 14 & 12 \\ 5 & 3 \end{bmatrix}$ and $\mathbf{x}_0 = \begin{bmatrix} 1 \\ 1 \end{bmatrix}$ as the initial vector, we get the values in the table.

k	0	1	2	3	4	5
$\mathbf{x}_k$	$\begin{bmatrix} 1 \\ 1 \end{bmatrix}$	$\begin{bmatrix} 26 \\ 8 \end{bmatrix}$	$\begin{bmatrix} 17.696 \\ 5.924 \end{bmatrix}$	$\begin{bmatrix} 18.017 \\ 6.004 \end{bmatrix}$	$\begin{bmatrix} 17.999 \\ 6.000 \end{bmatrix}$	$\begin{bmatrix} 18.000 \\ 6.000 \end{bmatrix}$
$\mathbf{y}_k$	$\begin{bmatrix} 1 \\ 1 \end{bmatrix}$	$\begin{bmatrix} 1 \\ 0.308 \end{bmatrix}$	$\begin{bmatrix} 1 \\ 0.335 \end{bmatrix}$	$\begin{bmatrix} 1 \\ 0.333 \end{bmatrix}$	$\begin{bmatrix} 1 \\ 0.333 \end{bmatrix}$	$\begin{bmatrix} 1 \\ 0.333 \end{bmatrix}$
m_k	1	26	17.696	18.017	17.999	18.000

We deduce that a dominant eigenvector of A is $\begin{bmatrix} 3 \\ 1 \end{bmatrix}$ with eigenvalue 18.

10. With $A = \begin{bmatrix} -6 & 4 \\ 8 & -2 \end{bmatrix}$ and $\mathbf{x}_0 = \begin{bmatrix} 1 \\ 0 \end{bmatrix}$ as the initial vector, we get the values in the table.

k	0	1	2	3	4	5	6
$\mathbf{x}_k$	$\begin{bmatrix} 1 \\ 0 \end{bmatrix}$	$\begin{bmatrix} -6 \\ 8 \end{bmatrix}$	$\begin{bmatrix} 8.5 \\ -8.0 \end{bmatrix}$	$\begin{bmatrix} -9.764 \\ 9.882 \end{bmatrix}$	$\begin{bmatrix} 9.928 \\ -9.904 \end{bmatrix}$	$\begin{bmatrix} -9.992 \\ 9.996 \end{bmatrix}$	$\begin{bmatrix} 10 \\ -10 \end{bmatrix}$
$\mathbf{y}_k$	$\begin{bmatrix} 1 \\ 0 \end{bmatrix}$	$\begin{bmatrix} -0.75 \\ 1 \end{bmatrix}$	$\begin{bmatrix} 1 \\ -0.941 \end{bmatrix}$	$\begin{bmatrix} -0.988 \\ 1 \end{bmatrix}$	$\begin{bmatrix} 1 \\ -0.998 \end{bmatrix}$	$\begin{bmatrix} -1 \\ 1 \end{bmatrix}$	$\begin{bmatrix} 1 \\ -1 \end{bmatrix}$
m_k	1	8	8.5	9.882	9.928	9.996	10

We deduce that a dominant eigenvector of A is $\begin{bmatrix} 1 \\ -1 \end{bmatrix}$.

Note that the corresponding eigenvalue is -10 and not 10,
since at each step the sign of each component of $\mathbf{y}_k$ is reversed.

11. With $A = \begin{bmatrix} 7 & 2 \\ 2 & 3 \end{bmatrix}$ and $\mathbf{x}_0 = \begin{bmatrix} 1 \\ 0 \end{bmatrix}$ as the initial vector, we get the values in the table.

k	0	1	2	3	4	5	6
$\mathbf{x}_k$	$\begin{bmatrix} 1 \\ 0 \end{bmatrix}$	$\begin{bmatrix} 7 \\ 2 \end{bmatrix}$	$\begin{bmatrix} 7.572 \\ 2.858 \end{bmatrix}$	$\begin{bmatrix} 7.754 \\ 3.131 \end{bmatrix}$	$\begin{bmatrix} 7.808 \\ 3.212 \end{bmatrix}$	$\begin{bmatrix} 7.822 \\ 3.233 \end{bmatrix}$	$\begin{bmatrix} 7.826 \\ 3.239 \end{bmatrix}$
$\mathbf{y}_k$	$\begin{bmatrix} 1 \\ 0 \end{bmatrix}$	$\begin{bmatrix} 1 \\ 0.286 \end{bmatrix}$	$\begin{bmatrix} 1 \\ 0.377 \end{bmatrix}$	$\begin{bmatrix} 1 \\ 0.404 \end{bmatrix}$	$\begin{bmatrix} 1 \\ 0.411 \end{bmatrix}$	$\begin{bmatrix} 1 \\ 0.413 \end{bmatrix}$	$\begin{bmatrix} 1 \\ 0.414 \end{bmatrix}$
m_k	1	7	7.572	7.754	7.808	7.822	7.826

We deduce that a dominant eigenvector of A is approximately $\begin{bmatrix} 1 \\ 0.414 \end{bmatrix}$ and

the corresponding eigenvalue is about 7.83.

(In fact, a dominant eigenvector is $\begin{bmatrix} 1 \\ \sqrt{2}-1 \end{bmatrix}$ and the corresponding eigenvalue is $5 + 2\sqrt{2}$.)

12. With $A = \begin{bmatrix} 3.5 & 1.5 \\ 1.5 & -0.5 \end{bmatrix}$ and $\mathbf{x}_0 = \begin{bmatrix} 1 \\ 0 \end{bmatrix}$ as the initial vector, we get the values in the table.

k	0	1	2	3	4	5	6
$\mathbf{x}_k$	$\begin{bmatrix} 1 \\ 0 \end{bmatrix}$	$\begin{bmatrix} 3.5 \\ 1.5 \end{bmatrix}$	$\begin{bmatrix} 4.144 \\ 1.286 \end{bmatrix}$	$\begin{bmatrix} 3.965 \\ 1.345 \end{bmatrix}$	$\begin{bmatrix} 4.009 \\ 1.331 \end{bmatrix}$	$\begin{bmatrix} 3.998 \\ 1.334 \end{bmatrix}$	$\begin{bmatrix} 4.001 \\ 1.333 \end{bmatrix}$
$\mathbf{y}_k$	$\begin{bmatrix} 1 \\ 0 \end{bmatrix}$	$\begin{bmatrix} 1 \\ 0.429 \end{bmatrix}$	$\begin{bmatrix} 1 \\ 0.310 \end{bmatrix}$	$\begin{bmatrix} 1 \\ 0.339 \end{bmatrix}$	$\begin{bmatrix} 1 \\ 0.332 \end{bmatrix}$	$\begin{bmatrix} 1 \\ 0.334 \end{bmatrix}$	$\begin{bmatrix} 1 \\ 0.333 \end{bmatrix}$
m_k	1	3.5	4.144	3.965	4.009	3.998	4.001

We deduce that a dominant eigenvector of A is $\begin{bmatrix} 3 \\ 1 \end{bmatrix}$ with eigenvalue 4.

13. With $A = \begin{bmatrix} 9 & 4 & 8 \\ 4 & 15 & -4 \\ 8 & -4 & 9 \end{bmatrix}$ and $\mathbf{x}_0 = \begin{bmatrix} 1 \\ 1 \\ 1 \end{bmatrix}$ as the initial vector, we get the values in the table.

k	0	1	2	3	4	5
$\mathbf{x}_k$	$\begin{bmatrix} 1 \\ 1 \\ 1 \end{bmatrix}$	$\begin{bmatrix} 21 \\ 15 \\ 13 \end{bmatrix}$	$\begin{bmatrix} 16.808 \\ 12.234 \\ 10.715 \end{bmatrix}$	$\begin{bmatrix} 17.008 \\ 12.372 \\ 10.821 \end{bmatrix}$	$\begin{bmatrix} 16.996 \\ 12.361 \\ 10.816 \end{bmatrix}$	$\begin{bmatrix} 16.996 \\ 12.361 \\ 10.816 \end{bmatrix}$
$\mathbf{y}_k$	$\begin{bmatrix} 1 \\ 1 \\ 1 \end{bmatrix}$	$\begin{bmatrix} 1 \\ 0.714 \\ 0.619 \end{bmatrix}$	$\begin{bmatrix} 1 \\ 0.728 \\ 0.637 \end{bmatrix}$	$\begin{bmatrix} 1 \\ 0.727 \\ 0.636 \end{bmatrix}$	$\begin{bmatrix} 1 \\ 0.727 \\ 0.636 \end{bmatrix}$	$\begin{bmatrix} 1 \\ 0.727 \\ 0.636 \end{bmatrix}$
m_k	1	21	16.808	17.008	16.996	16.996

We deduce that a dominant eigenvector of $A \approx \begin{bmatrix} 1 \\ 0.727 \\ 0.636 \end{bmatrix}$ with eigenvalue 17.

(In fact, this eigenvalue corresponds to the eigenspace spanned by $\begin{bmatrix} 1 \\ 2 \\ 0 \end{bmatrix}$ and $\begin{bmatrix} 0 \\ -2 \\ 1 \end{bmatrix}$.)

14. With $A = \begin{bmatrix} 3 & 1 & 0 \\ 1 & 3 & 1 \\ 0 & 1 & 3 \end{bmatrix}$ and $\mathbf{x}_0 = \begin{bmatrix} 1 \\ 1 \\ 1 \end{bmatrix}$ as the initial vector, we get the values in the table.

k	0	1	2	3	4	5	6
$\mathbf{x}_k$	$\begin{bmatrix} 1 \\ 1 \\ 1 \end{bmatrix}$	$\begin{bmatrix} 4 \\ 5 \\ 4 \end{bmatrix}$	$\begin{bmatrix} 3.4 \\ 4.6 \\ 3.4 \end{bmatrix}$	$\begin{bmatrix} 3.217 \\ 4.478 \\ 3.217 \end{bmatrix}$	$\begin{bmatrix} 3.154 \\ 4.436 \\ 3.154 \end{bmatrix}$	$\begin{bmatrix} 3.133 \\ 4.422 \\ 3.133 \end{bmatrix}$	$\begin{bmatrix} 3.127 \\ 4.418 \\ 3.127 \end{bmatrix}$
$\mathbf{y}_k$	$\begin{bmatrix} 1 \\ 1 \\ 1 \end{bmatrix}$	$\begin{bmatrix} 0.8 \\ 1 \\ 0.8 \end{bmatrix}$	$\begin{bmatrix} 0.739 \\ 1 \\ 0.739 \end{bmatrix}$	$\begin{bmatrix} 0.718 \\ 1 \\ 0.718 \end{bmatrix}$	$\begin{bmatrix} 0.711 \\ 1 \\ 0.711 \end{bmatrix}$	$\begin{bmatrix} 0.709 \\ 1 \\ 0.709 \end{bmatrix}$	$\begin{bmatrix} 0.708 \\ 1 \\ 0.708 \end{bmatrix}$
m_k	1	1	4.6	4.478	4.437	4.422	4.418

We deduce that a dominant eigenvector of $A \approx \begin{bmatrix} 0.708 \\ 1 \\ 0.708 \end{bmatrix}$ with eigenvalue ≈ 4.4.

(In fact, a dominant eigenvector is $\begin{bmatrix} \frac{\sqrt{2}}{2} \\ 1 \\ \frac{\sqrt{2}}{2} \end{bmatrix}$, and the corresponding eigenvalue is $3 + \sqrt{2}$.)

15. With $A = \begin{bmatrix} 4 & 1 & 3 \\ 0 & 2 & 0 \\ 1 & 1 & 2 \end{bmatrix}$ and $\mathbf{x}_0 = \begin{bmatrix} 1 \\ 1 \\ 1 \end{bmatrix}$ as the initial vector, we get the values in the table.

k	0	1	2	4	6	7	8
$\mathbf{x}_k$	$\begin{bmatrix} 1 \\ 1 \\ 1 \end{bmatrix}$	$\begin{bmatrix} 8 \\ 2 \\ 4 \end{bmatrix}$	$\begin{bmatrix} 5.75 \\ 0.5 \\ 2.25 \end{bmatrix}$	$\begin{bmatrix} 5.11 \\ 0.06 \\ 1.75 \end{bmatrix}$	$\begin{bmatrix} 5.02 \\ 0 \\ 1.68 \end{bmatrix}$	$\begin{bmatrix} 4.99 \\ 0 \\ 1.66 \end{bmatrix}$	$\begin{bmatrix} 4.99 \\ 0 \\ 1.66 \end{bmatrix}$
$\mathbf{y}_k$	$\begin{bmatrix} 1 \\ 1 \\ 1 \end{bmatrix}$	$\begin{bmatrix} 1 \\ 0.25 \\ 0.5 \end{bmatrix}$	$\begin{bmatrix} 1 \\ 0.09 \\ 0.39 \end{bmatrix}$	$\begin{bmatrix} 1 \\ 0.01 \\ 0.34 \end{bmatrix}$	$\begin{bmatrix} 1 \\ 0 \\ 0.33 \end{bmatrix}$	$\begin{bmatrix} 1 \\ 0 \\ 0.33 \end{bmatrix}$	$\begin{bmatrix} 1 \\ 0 \\ 0.33 \end{bmatrix}$
m_k	1	8	5.75	5.11	5.02	4.99	4.99

We deduce that a dominant eigenvector of A is $\begin{bmatrix} 3 \\ 0 \\ 1 \end{bmatrix}$ with eigenvalue 5.

16. With $A = \begin{bmatrix} 12 & 6 & -6 \\ 2 & 0 & -2 \\ -6 & 6 & 12 \end{bmatrix}$ and $\mathbf{x}_0 = \begin{bmatrix} 2 \\ 1 \\ 1 \end{bmatrix}$ as the initial vector, we get the values in the table.

k	0	1	2	4	6	7	8
$\mathbf{x}_k$	$\begin{bmatrix} 2 \\ 1 \\ 1 \end{bmatrix}$	$\begin{bmatrix} 24 \\ 2 \\ 6 \end{bmatrix}$	$\begin{bmatrix} 10.98 \\ 1.5 \\ -2.52 \end{bmatrix}$	$\begin{bmatrix} 16.38 \\ 3.12 \\ -11.7 \end{bmatrix}$	$\begin{bmatrix} 17.82 \\ 3.54 \\ -14.04 \end{bmatrix}$	$\begin{bmatrix} 17.94 \\ 3.58 \\ -14.28 \end{bmatrix}$	$\begin{bmatrix} 18.00 \\ 3.60 \\ -14.40 \end{bmatrix}$
$\mathbf{y}_k$	$\begin{bmatrix} 2 \\ 1 \\ 1 \end{bmatrix}$	$\begin{bmatrix} 1 \\ 0.08 \\ 0.25 \end{bmatrix}$	$\begin{bmatrix} 1 \\ 0.14 \\ -0.23 \end{bmatrix}$	$\begin{bmatrix} 1 \\ 0.19 \\ -0.71 \end{bmatrix}$	$\begin{bmatrix} 1 \\ 0.20 \\ -0.79 \end{bmatrix}$	$\begin{bmatrix} 1 \\ 0.20 \\ -0.80 \end{bmatrix}$	$\begin{bmatrix} 1 \\ 0.20 \\ -0.80 \end{bmatrix}$
m_k	1	24	10.98	16.38	17.82	17.94	18

We deduce that a dominant eigenvector of A is $\begin{bmatrix} 1 \\ 0.2 \\ -0.8 \end{bmatrix}$ with eigenvalue 18.

17. With $A = \begin{bmatrix} 7 & 2 \\ 2 & 3 \end{bmatrix}$ and $\mathbf{x}_0 = \begin{bmatrix} 1 \\ 0 \end{bmatrix}$ as the initial vector, we get the values in the table.

k	0	1	2	3
$\mathbf{x}_k$	$\begin{bmatrix} 1 \\ 0 \end{bmatrix}$	$\begin{bmatrix} 7 \\ 2 \end{bmatrix}$	$\begin{bmatrix} 7.572 \\ 2.858 \end{bmatrix}$	$\begin{bmatrix} 7.754 \\ 3.131 \end{bmatrix}$
$R(\mathbf{x}_k)$	7	7.755	7.823	7.828

18. With $A = \begin{bmatrix} 3.5 & 1.5 \\ 1.5 & -0.5 \end{bmatrix}$ and $\mathbf{x}_0 = \begin{bmatrix} 1 \\ 0 \end{bmatrix}$ as the initial vector, we get the values in the table.

k	0	1	2	3
$\mathbf{x}_k$	$\begin{bmatrix} 1 \\ 0 \end{bmatrix}$	$\begin{bmatrix} 3.5 \\ 1.5 \end{bmatrix}$	$\begin{bmatrix} 4.144 \\ 1.286 \end{bmatrix}$	$\begin{bmatrix} 3.965 \\ 1.345 \end{bmatrix}$
$R(\mathbf{x}_k)$	3.5	3.966	3.998	4.000

19. With $A = \begin{bmatrix} 9 & 4 & 8 \\ 4 & 15 & -4 \\ 8 & -4 & 9 \end{bmatrix}$ and $\mathbf{x}_0 = \begin{bmatrix} 1 \\ 1 \\ 1 \end{bmatrix}$ as the initial vector, we get the values in the table.

k	0	1	2
$\mathbf{x}_k$	$\begin{bmatrix} 1 \\ 1 \\ 1 \end{bmatrix}$	$\begin{bmatrix} 21 \\ 15 \\ 13 \end{bmatrix}$	$\begin{bmatrix} 16.808 \\ 12.234 \\ 10.715 \end{bmatrix}$
$R(\mathbf{x}_k)$	16.333	16.998	17.000

20. With $A = \begin{bmatrix} 3 & 1 & 0 \\ 1 & 3 & 1 \\ 0 & 1 & 3 \end{bmatrix}$ and $\mathbf{x}_0 = \begin{bmatrix} 1 \\ 1 \\ 1 \end{bmatrix}$ as the initial vector, we get the values in the table.

k	0	1	2	3
$\mathbf{x}_k$	$\begin{bmatrix} 1 \\ 1 \\ 1 \end{bmatrix}$	$\begin{bmatrix} 4 \\ 5 \\ 4 \end{bmatrix}$	$\begin{bmatrix} 3.4 \\ 4.6 \\ 3.4 \end{bmatrix}$	$\begin{bmatrix} 3.217 \\ 4.478 \\ 3.217 \end{bmatrix}$
$R(\mathbf{x}_k)$	4.333	4.404	4.413	4.414

21. With $A = \begin{bmatrix} 4 & 1 \\ 0 & 4 \end{bmatrix}$ and $\mathbf{x}_0 = \begin{bmatrix} 1 \\ 1 \end{bmatrix}$ as the initial vector, we get the values in the table.

k	0	1	2	3	4	5	6	7	8
$\mathbf{x}_k$	$\begin{bmatrix} 1 \\ 1 \end{bmatrix}$	$\begin{bmatrix} 5 \\ 4 \end{bmatrix}$	$\begin{bmatrix} 4.8 \\ 3.2 \end{bmatrix}$	$\begin{bmatrix} 4.67 \\ 2.68 \end{bmatrix}$	$\begin{bmatrix} 4.57 \\ 2.28 \end{bmatrix}$	$\begin{bmatrix} 4.5 \\ 2.0 \end{bmatrix}$	$\begin{bmatrix} 4.44 \\ 1.76 \end{bmatrix}$	$\begin{bmatrix} 4.4 \\ 1.6 \end{bmatrix}$	$\begin{bmatrix} 4.36 \\ 1.44 \end{bmatrix}$
$\mathbf{y}_k$	$\begin{bmatrix} 1 \\ 1 \end{bmatrix}$	$\begin{bmatrix} 1 \\ 0.8 \end{bmatrix}$	$\begin{bmatrix} 1 \\ 0.67 \end{bmatrix}$	$\begin{bmatrix} 1 \\ 0.57 \end{bmatrix}$	$\begin{bmatrix} 1 \\ 0.50 \end{bmatrix}$	$\begin{bmatrix} 1 \\ 0.44 \end{bmatrix}$	$\begin{bmatrix} 1 \\ 0.40 \end{bmatrix}$	$\begin{bmatrix} 1 \\ 0.36 \end{bmatrix}$	$\begin{bmatrix} 1 \\ 0.33 \end{bmatrix}$
m_k	1	5	4.8	4.67	4.57	4.5	4.44	4.4	4.36

A has the double eigenvalue 4 with corresponding eigenvector $\begin{bmatrix} 1 \\ 0 \end{bmatrix}$.

It seems that the power method is converging, but very slowly.

22. With $A = \begin{bmatrix} 3 & 1 \\ -1 & 1 \end{bmatrix}$ and $\mathbf{x}_0 = \begin{bmatrix} 1 \\ 1 \end{bmatrix}$ as the initial vector, we get the values in the table.

k	0	1	2	4	6	7	8
$\mathbf{x}_k$	$\begin{bmatrix} 1 \\ 1 \end{bmatrix}$	$\begin{bmatrix} 4 \\ 0 \end{bmatrix}$	$\begin{bmatrix} 3 \\ -1 \end{bmatrix}$	$\begin{bmatrix} 2.5 \\ -1.5 \end{bmatrix}$	$\begin{bmatrix} 2.33 \\ -1.67 \end{bmatrix}$	$\begin{bmatrix} 2.28 \\ -1.72 \end{bmatrix}$	$\begin{bmatrix} 2.25 \\ -1.75 \end{bmatrix}$
$\mathbf{y}_k$	$\begin{bmatrix} 1 \\ 1 \end{bmatrix}$	$\begin{bmatrix} 1 \\ 0 \end{bmatrix}$	$\begin{bmatrix} 1 \\ -0.33 \end{bmatrix}$	$\begin{bmatrix} 1 \\ -0.6 \end{bmatrix}$	$\begin{bmatrix} 1 \\ -0.72 \end{bmatrix}$	$\begin{bmatrix} 1 \\ -0.75 \end{bmatrix}$	$\begin{bmatrix} 1 \\ -0.78 \end{bmatrix}$
m_k	1	4	3	2.5	2.33	2.28	2.25

A has the double eigenvalue 2 with eigenvector $\begin{bmatrix} 1 \\ -1 \end{bmatrix}$.

If the power method is converging, it is doing so very slowly.

23. With $A = \begin{bmatrix} 4 & 0 & 1 \\ 0 & 4 & 0 \\ 0 & 0 & 1 \end{bmatrix}$ and $\mathbf{x}_0 = \begin{bmatrix} 1 \\ 1 \\ 1 \end{bmatrix}$ as the initial vector, we get the values in the table.

k	0	1	2	3	4	5	6
$\mathbf{x}_k$	$\begin{bmatrix} 1 \\ 1 \\ 1 \end{bmatrix}$	$\begin{bmatrix} 5 \\ 4 \\ 1 \end{bmatrix}$	$\begin{bmatrix} 4.2 \\ 3.2 \\ 0.2 \end{bmatrix}$	$\begin{bmatrix} 4.05 \\ 3.04 \\ 0.05 \end{bmatrix}$	$\begin{bmatrix} 4.01 \\ 3 \\ 0.01 \end{bmatrix}$	$\begin{bmatrix} 4 \\ 3 \\ 0 \end{bmatrix}$	$\begin{bmatrix} 4 \\ 3 \\ 0 \end{bmatrix}$
$\mathbf{y}_k$	$\begin{bmatrix} 1 \\ 1 \\ 1 \end{bmatrix}$	$\begin{bmatrix} 1 \\ 0.8 \\ 0.2 \end{bmatrix}$	$\begin{bmatrix} 1 \\ 0.76 \\ 0.05 \end{bmatrix}$	$\begin{bmatrix} 1 \\ 0.75 \\ 0.01 \end{bmatrix}$	$\begin{bmatrix} 1 \\ 0.75 \\ 0 \end{bmatrix}$	$\begin{bmatrix} 1 \\ 0.75 \\ 0 \end{bmatrix}$	$\begin{bmatrix} 1 \\ 0.75 \\ 0 \end{bmatrix}$
m_k	1	5	4.2	4.05	4.01	4	4

It seems that an eigenvector of A is $\begin{bmatrix} 4 \\ 3 \\ 0 \end{bmatrix}$ with eigenvalue 4.

In fact, the eigenspace of 4 is span $\left(\begin{bmatrix} 1 \\ 0 \\ 0 \end{bmatrix}, \begin{bmatrix} 0 \\ 1 \\ 0 \end{bmatrix} \right)$.

24. With $A = \begin{bmatrix} 0 & 0 & 0 \\ 0 & 5 & 1 \\ 0 & 0 & 5 \end{bmatrix}$ and $\mathbf{x}_0 = \begin{bmatrix} 1 \\ 1 \\ 1 \end{bmatrix}$ as the initial vector, we get:

k	0	1	2	4	6	7	8
$\mathbf{x}_k$	$\begin{bmatrix} 1 \\ 1 \\ 1 \end{bmatrix}$	$\begin{bmatrix} 0 \\ 6 \\ 5 \end{bmatrix}$	$\begin{bmatrix} 0 \\ 5.83 \\ 4.15 \end{bmatrix}$	$\begin{bmatrix} 0 \\ 5.62 \\ 3.1 \end{bmatrix}$	$\begin{bmatrix} 0 \\ 5.5 \\ 2.5 \end{bmatrix}$	$\begin{bmatrix} 0 \\ 5.45 \\ 2.25 \end{bmatrix}$	$\begin{bmatrix} 0 \\ 5.41 \\ 2.05 \end{bmatrix}$
$\mathbf{y}_k$	$\begin{bmatrix} 1 \\ 1 \\ 1 \end{bmatrix}$	$\begin{bmatrix} 0 \\ 1 \\ 0.83 \end{bmatrix}$	$\begin{bmatrix} 0 \\ 1 \\ 0.71 \end{bmatrix}$	$\begin{bmatrix} 0 \\ 1 \\ 0.55 \end{bmatrix}$	$\begin{bmatrix} 0 \\ 1 \\ 0.45 \end{bmatrix}$	$\begin{bmatrix} 0 \\ 1 \\ 0.41 \end{bmatrix}$	$\begin{bmatrix} 0 \\ 1 \\ 0.38 \end{bmatrix}$
m_k	1	6	5.83	5.62	5.5	5.45	5.41

A has the double eigenvalue 5 with corresponding eigenvector $\begin{bmatrix} 0 \\ 1 \\ 1 \end{bmatrix}$.

If the power method is converging, it is doing so very slowly.

25. With $A = \begin{bmatrix} -1 & 2 \\ -1 & 1 \end{bmatrix}$ and $\mathbf{x}_0 = \begin{bmatrix} 1 \\ 1 \end{bmatrix}$ $\Rightarrow$ A has no real eigenvalue:

k	0	1	2
$\mathbf{x}_k$	$\begin{bmatrix} 1 \\ 1 \end{bmatrix}$	$\begin{bmatrix} 1 \\ 0 \end{bmatrix}$	$\begin{bmatrix} -1 \\ -1 \end{bmatrix}$
$\mathbf{y}_k$	$\begin{bmatrix} 1 \\ 1 \end{bmatrix}$	$\begin{bmatrix} 1 \\ 0 \end{bmatrix}$	$\begin{bmatrix} 1 \\ 1 \end{bmatrix}$
m_k	1	1	-1

26. With $A = \begin{bmatrix} 2 & 1 \\ -2 & 5 \end{bmatrix}$ and $\mathbf{x}_0 = \begin{bmatrix} 1 \\ 1 \end{bmatrix} \Rightarrow \begin{bmatrix} 1 \\ 1 \end{bmatrix}$ with eigenvalue 3:

k	0	1
$\mathbf{x}_k$	$\begin{bmatrix} 1 \\ 1 \end{bmatrix}$	$\begin{bmatrix} 3 \\ 3 \end{bmatrix}$
$\mathbf{y}_k$	$\begin{bmatrix} 1 \\ 1 \end{bmatrix}$	$\begin{bmatrix} 1 \\ 1 \end{bmatrix}$
m_k	1	3

But A also has eigenvalue 4 with corresponding eigenvector $\begin{bmatrix} 1 \\ 2 \end{bmatrix}$.

We have found the non-dominant eigenvalue 3 because we took $\mathbf{x}_0$ to be its eigenvector $\begin{bmatrix} 1 \\ 1 \end{bmatrix}$.

27. With $A = \begin{bmatrix} -5 & 1 & 7 \\ 0 & 4 & 0 \\ 7 & 1 & -5 \end{bmatrix}$ and $\mathbf{x}_0 = \begin{bmatrix} 1 \\ 1 \\ 1 \end{bmatrix}$ as the initial vector, we get:

k	0	1	2	3	4	5	6
$\mathbf{x}_k$	$\begin{bmatrix} 1 \\ 1 \\ 1 \end{bmatrix}$	$\begin{bmatrix} 3 \\ 4 \\ 3 \end{bmatrix}$	$\begin{bmatrix} 2.5 \\ 4 \\ 2.5 \end{bmatrix}$	$\begin{bmatrix} 2.26 \\ 4 \\ 2.26 \end{bmatrix}$	$\begin{bmatrix} 2.14 \\ 4 \\ 2.14 \end{bmatrix}$	$\begin{bmatrix} 2.08 \\ 4 \\ 2.08 \end{bmatrix}$	$\begin{bmatrix} 2.04 \\ 4 \\ 2.04 \end{bmatrix}$
$\mathbf{y}_k$	$\begin{bmatrix} 1 \\ 1 \\ 1 \end{bmatrix}$	$\begin{bmatrix} 0.75 \\ 1 \\ 0.75 \end{bmatrix}$	$\begin{bmatrix} 0.63 \\ 1 \\ 0.63 \end{bmatrix}$	$\begin{bmatrix} 0.57 \\ 1 \\ 0.57 \end{bmatrix}$	$\begin{bmatrix} 0.54 \\ 1 \\ 0.54 \end{bmatrix}$	$\begin{bmatrix} 0.52 \\ 1 \\ 0.52 \end{bmatrix}$	$\begin{bmatrix} 0.51 \\ 1 \\ 0.51 \end{bmatrix}$
m_k	1	4	4	4	4	4	4

The power method is converging to $\begin{bmatrix} 1 \\ 2 \\ 1 \end{bmatrix}$ and 4,

which are indeed an eigenvector and eigenvalue of A.

But, in fact, the dominant eigenvalue of A is -12, with corresponding eigenvector $\begin{bmatrix} 1 \\ 0 \\ -1 \end{bmatrix}$.

We did not find these values because our initial vector $\mathbf{x}_0$ is orthogonal to $\begin{bmatrix} 1 \\ 0 \\ -1 \end{bmatrix}$.

28. With $A = \begin{bmatrix} 1 & -1 & 0 \\ 1 & 1 & 0 \\ 1 & -1 & 1 \end{bmatrix}$ and $\mathbf{x}_0 = \begin{bmatrix} 1 \\ 1 \\ 1 \end{bmatrix}$ as the initial vector, we get:

k	0	1	2	4	6	7	8
$\mathbf{x}_k$	$\begin{bmatrix} 1 \\ 1 \\ 1 \end{bmatrix}$	$\begin{bmatrix} 0 \\ 2 \\ 1 \end{bmatrix}$	$\begin{bmatrix} -1 \\ 1 \\ -0.5 \end{bmatrix}$	$\begin{bmatrix} 0.8 \\ 0.8 \\ 1.8 \end{bmatrix}$	$\begin{bmatrix} -0.88 \\ 0.88 \\ 0.12 \end{bmatrix}$	$\begin{bmatrix} 2 \\ 0 \\ 1.86 \end{bmatrix}$	$\begin{bmatrix} 1 \\ 1 \\ 1.93 \end{bmatrix}$
$\mathbf{y}_k$	$\begin{bmatrix} 1 \\ 1 \\ 1 \end{bmatrix}$	$\begin{bmatrix} 0 \\ 1 \\ 0.5 \end{bmatrix}$	$\begin{bmatrix} 1 \\ -1 \\ 0.5 \end{bmatrix}$	$\begin{bmatrix} 0.44 \\ 0.44 \\ 1 \end{bmatrix}$	$\begin{bmatrix} 1 \\ -1 \\ -0.14 \end{bmatrix}$	$\begin{bmatrix} 1 \\ 0 \\ 0.93 \end{bmatrix}$	$\begin{bmatrix} 0.52 \\ 0.52 \\ 1 \end{bmatrix}$
m_k	1	2	-1	1.8	-0.88	1.86	1.93

The problem here is that A has complex eigenvalues $1 \pm i$,
each of which has a larger absolute value than the real eigenvalue 1.

29. In Exercise 9 we found that $\lambda_1 = 18$.

To find λ_2, we apply the power method to $A - 18I = \begin{bmatrix} -4 & 12 \\ 5 & -15 \end{bmatrix}$ with $\mathbf{x}_0 = \begin{bmatrix} 1 \\ 1 \end{bmatrix}$:

k	0	1	2	3
$\mathbf{x}_k$	$\begin{bmatrix} 1 \\ 1 \end{bmatrix}$	$\begin{bmatrix} 8 \\ -10 \end{bmatrix}$	$\begin{bmatrix} 15.2 \\ -19.0 \end{bmatrix}$	$\begin{bmatrix} 15.2 \\ -19.0 \end{bmatrix}$
$\mathbf{y}_k$	$\begin{bmatrix} 1 \\ 1 \end{bmatrix}$	$\begin{bmatrix} -0.8 \\ 1 \end{bmatrix}$	$\begin{bmatrix} -0.8 \\ 1 \end{bmatrix}$	$\begin{bmatrix} -0.8 \\ 1 \end{bmatrix}$
m_k	1	-10	-19	-19

We deduce that $\lambda_2 - \lambda_1 = -19$, so $\lambda_2 = -1$ is the second eigenvalue of A.

30. $\lambda_1 = -10$, so we apply the power method to $A + 10I = \begin{bmatrix} 4 & 4 \\ 8 & 8 \end{bmatrix}$ with $\mathbf{x}_0 = \begin{bmatrix} 1 \\ 0 \end{bmatrix}$:

k	0	1	2	3
$\mathbf{x}_k$	$\begin{bmatrix} 1 \\ 0 \end{bmatrix}$	$\begin{bmatrix} 4 \\ 8 \end{bmatrix}$	$\begin{bmatrix} 6 \\ 12 \end{bmatrix}$	$\begin{bmatrix} 6 \\ 12 \end{bmatrix}$
$\mathbf{y}_k$	$\begin{bmatrix} 1 \\ 0 \end{bmatrix}$	$\begin{bmatrix} 0.5 \\ 1 \end{bmatrix}$	$\begin{bmatrix} 0.5 \\ 1 \end{bmatrix}$	$\begin{bmatrix} 0.5 \\ 1 \end{bmatrix}$
m_k	1	8	12	12

We see that $\lambda_2 - \lambda_1 = 12 \Rightarrow \lambda_2 = 2$.

31. $\lambda_1 = 17$, so we apply the power method to $A - 17I = \begin{bmatrix} -8 & 4 & 8 \\ 4 & -2 & -4 \\ 8 & -4 & -8 \end{bmatrix}$ with $\mathbf{x}_0 = \begin{bmatrix} 1 \\ 1 \\ 1 \end{bmatrix}$:

k	0	1	2	3
$\mathbf{x}_k$	$\begin{bmatrix} 1 \\ 1 \\ 1 \end{bmatrix}$	$\begin{bmatrix} 4 \\ -2 \\ -4 \end{bmatrix}$	$\begin{bmatrix} -18 \\ 9 \\ 18 \end{bmatrix}$	$\begin{bmatrix} -18 \\ 9 \\ 18 \end{bmatrix}$
$\mathbf{y}_k$	$\begin{bmatrix} 1 \\ 1 \\ 1 \end{bmatrix}$	$\begin{bmatrix} 1 \\ -0.5 \\ -1 \end{bmatrix}$	$\begin{bmatrix} 1 \\ -0.5 \\ -1 \end{bmatrix}$	$\begin{bmatrix} 1 \\ -0.5 \\ -1 \end{bmatrix}$
m_k	1	4	-18	-18

We deduce that $\lambda_2 - \lambda_1 = -18 \Rightarrow \lambda_2 = -1$.

32. $\lambda_1 \approx 4.4$, so we apply the power method to $A - 4.4I \approx \begin{bmatrix} -1.4 & 1 & 0 \\ 1 & -1.4 & 1 \\ 0 & 1 & -1.4 \end{bmatrix}$ and $\mathbf{x}_0 = \begin{bmatrix} 1 \\ 1 \\ 1 \end{bmatrix}$:

k	0	1	2	3	4
$\mathbf{x}_k$	$\begin{bmatrix} 1 \\ 1 \\ 1 \end{bmatrix}$	$\begin{bmatrix} -0.4 \\ 0.6 \\ -0.4 \end{bmatrix}$	$\begin{bmatrix} 1.94 \\ -2.74 \\ 1.94 \end{bmatrix}$	$\begin{bmatrix} 1.99 \\ -2.82 \\ 1.99 \end{bmatrix}$	$\begin{bmatrix} 1.99 \\ -2.82 \\ 1.99 \end{bmatrix}$
$\mathbf{y}_k$	$\begin{bmatrix} 1 \\ 1 \\ 1 \end{bmatrix}$	$\begin{bmatrix} -0.67 \\ 1 \\ -0.67 \end{bmatrix}$	$\begin{bmatrix} -0.71 \\ 1 \\ -0.71 \end{bmatrix}$	$\begin{bmatrix} -0.71 \\ 1 \\ -0.71 \end{bmatrix}$	$\begin{bmatrix} -0.71 \\ 1 \\ -0.71 \end{bmatrix}$
m_k	1	0.6	2.74	−2.82	−2.82

We see that $\lambda_2 - \lambda_1 \approx -2.82$, so $\lambda_2 \approx 1.58$. (In fact, $\lambda_2 = 3 - \sqrt{2}$.)

33. We solve $A\mathbf{x}_1 = \mathbf{y}_0$ using row reduction: $\left[\, A \mid \mathbf{y}_0 \,\right] = \left[\begin{array}{cc|c} 14 & 12 & 1 \\ 5 & 3 & 1 \end{array}\right] \longrightarrow \left[\begin{array}{cc|c} 1 & 0 & 0.5 \\ 0 & 1 & -0.5 \end{array}\right]$.

Thus, $\mathbf{x}_1 = \begin{bmatrix} 0.5 \\ -0.5 \end{bmatrix}$, so $\mathbf{y}_1 = \begin{bmatrix} 1 \\ -1 \end{bmatrix}$. Continuing in this way we calculate:

k	0	1	2	3	4	5
$\mathbf{x}_k$	$\begin{bmatrix} 1 \\ 1 \end{bmatrix}$	$\begin{bmatrix} 0.5 \\ -0.5 \end{bmatrix}$	$\begin{bmatrix} -0.833 \\ 1.056 \end{bmatrix}$	$\begin{bmatrix} 0.798 \\ -0.997 \end{bmatrix}$	$\begin{bmatrix} 0.8 \\ -1 \end{bmatrix}$	$\begin{bmatrix} 0.8 \\ -1 \end{bmatrix}$
$\mathbf{y}_k$	$\begin{bmatrix} 1 \\ 1 \end{bmatrix}$	$\begin{bmatrix} 1 \\ -1 \end{bmatrix}$	$\begin{bmatrix} -0.789 \\ 1 \end{bmatrix}$	$\begin{bmatrix} -0.800 \\ 1 \end{bmatrix}$	$\begin{bmatrix} -0.8 \\ 1 \end{bmatrix}$	$\begin{bmatrix} -0.8 \\ 1 \end{bmatrix}$
m_k	1	0.5	1.056	−0.997	−1	−1

We deduce that the eigenvalue of A smallest in magnitude is $\frac{1}{-1} = -1$.

34. We solve $A\mathbf{x}_1 = \mathbf{y}_0$ using row reduction: $\left[\, A \mid \mathbf{y}_0 \,\right] = \left[\begin{array}{cc|c} -6 & 4 & 1 \\ 8 & -2 & 0 \end{array}\right] \longrightarrow \left[\begin{array}{cc|c} 1 & 0 & 0.1 \\ 0 & 1 & 0.4 \end{array}\right]$.

Thus, $\mathbf{x}_1 = \begin{bmatrix} 0.1 \\ 0.4 \end{bmatrix}$, so $\mathbf{y}_1 = \begin{bmatrix} 0.25 \\ 1 \end{bmatrix}$. Continuing in this way we calculate:

k	0	1	2	3	4	5	6
$\mathbf{x}_k$	$\begin{bmatrix} 1 \\ 1 \end{bmatrix}$	$\begin{bmatrix} 0.1 \\ 0.4 \end{bmatrix}$	$\begin{bmatrix} 0.225 \\ 0.4 \end{bmatrix}$	$\begin{bmatrix} 0.256 \\ 0.525 \end{bmatrix}$	$\begin{bmatrix} 0.249 \\ 0.495 \end{bmatrix}$	$\begin{bmatrix} 0.250 \\ 0.501 \end{bmatrix}$	$\begin{bmatrix} 0.25 \\ 0.5 \end{bmatrix}$
$\mathbf{y}_k$	$\begin{bmatrix} 1 \\ 1 \end{bmatrix}$	$\begin{bmatrix} 0.25 \\ 1 \end{bmatrix}$	$\begin{bmatrix} 0.563 \\ 1 \end{bmatrix}$	$\begin{bmatrix} 0.488 \\ 1 \end{bmatrix}$	$\begin{bmatrix} 0.503 \\ 1 \end{bmatrix}$	$\begin{bmatrix} 0.499 \\ 1 \end{bmatrix}$	$\begin{bmatrix} 0.5 \\ 1 \end{bmatrix}$
m_k	1	0.4	0.4	0.525	0.495	0.501	0.5

We deduce that the eigenvalue of A that is smallest in magnitude is $\frac{1}{0.5} = 2$.

35. We solve $A\mathbf{x}_1 = \mathbf{y}_0$ using row reduction: $\left[\begin{array}{c|c} A & \mathbf{y}_0 \end{array}\right] = \left[\begin{array}{ccc|c} 4 & 0 & 6 & 1 \\ -1 & 3 & 1 & 1 \\ 6 & 0 & 4 & -1 \end{array}\right] \longrightarrow \left[\begin{array}{ccc|c} 1 & 0 & 0 & -0.5 \\ 0 & 1 & 0 & 0 \\ 0 & 0 & 1 & 0.5 \end{array}\right]$.

Thus, $\mathbf{x}_1 = \begin{bmatrix} -0.5 \\ 0 \\ 0.5 \end{bmatrix}$, so $\mathbf{y}_1 = \begin{bmatrix} 1 \\ 0 \\ -1 \end{bmatrix}$. Continuing in this way we calculate:

k	0	1	2	3	4	5
$\mathbf{x}_k$	$\begin{bmatrix} 1 \\ 1 \\ -1 \end{bmatrix}$	$\begin{bmatrix} -0.5 \\ 0 \\ 0.5 \end{bmatrix}$	$\begin{bmatrix} -0.5 \\ -0.333 \\ 0.5 \end{bmatrix}$	$\begin{bmatrix} -0.5 \\ -0.111 \\ 0.5 \end{bmatrix}$	$\begin{bmatrix} -0.5 \\ -0.259 \\ 0.5 \end{bmatrix}$	$\begin{bmatrix} -0.5 \\ -0.161 \\ 0.5 \end{bmatrix}$
$\mathbf{y}_k$	$\begin{bmatrix} 1 \\ 1 \\ -1 \end{bmatrix}$	$\begin{bmatrix} 1 \\ 0 \\ -1 \end{bmatrix}$	$\begin{bmatrix} 1 \\ 0.666 \\ -1 \end{bmatrix}$	$\begin{bmatrix} 1 \\ 0.222 \\ -1 \end{bmatrix}$	$\begin{bmatrix} 1 \\ 0.518 \\ -1 \end{bmatrix}$	$\begin{bmatrix} 1 \\ 0.322 \\ -1 \end{bmatrix}$
m_k	1	−0.5	−0.5	−0.5	−0.5	−0.5

We deduce that the eigenvalue of A that is smallest in magnitude is $\frac{1}{-0.5} = -2$.

36. We solve $A\mathbf{x}_1 = \mathbf{y}_0$ using row reduction: $\left[\begin{array}{c|c} A & \mathbf{y}_0 \end{array}\right] = \left[\begin{array}{ccc|c} 3 & 1 & 0 & 1 \\ 1 & 3 & 1 & 1 \\ 0 & 1 & 3 & 1 \end{array}\right] \longrightarrow \left[\begin{array}{ccc|c} 1 & 0 & 0 & 0.286 \\ 0 & 1 & 0 & 0.143 \\ 0 & 0 & 1 & 0.286 \end{array}\right]$.

Thus, $\mathbf{x}_1 = \begin{bmatrix} 0.286 \\ 0.143 \\ 0.286 \end{bmatrix}$, so $\mathbf{y}_1 = \begin{bmatrix} 1 \\ 0.5 \\ 1 \end{bmatrix}$. Continuing in this way we calculate:

k	0	1	2	3	4	5	6
$\mathbf{x}_k$	$\begin{bmatrix} 1 \\ 1 \\ 1 \end{bmatrix}$	$\begin{bmatrix} 0.286 \\ 0.143 \\ 0.286 \end{bmatrix}$	$\begin{bmatrix} 0.357 \\ -0.071 \\ 0.357 \end{bmatrix}$	$\begin{bmatrix} 0.181 \\ -0.187 \\ 0.181 \end{bmatrix}$	$\begin{bmatrix} -0.558 \\ 0.705 \\ -0.558 \end{bmatrix}$	$\begin{bmatrix} -0.482 \\ 0.655 \\ -0.482 \end{bmatrix}$	$\begin{bmatrix} -0.458 \\ 0.639 \\ -0.458 \end{bmatrix}$
$\mathbf{y}_k$	$\begin{bmatrix} 1 \\ 1 \\ 1 \end{bmatrix}$	$\begin{bmatrix} 1 \\ 0.5 \\ 1 \end{bmatrix}$	$\begin{bmatrix} 1 \\ -0.199 \\ 1 \end{bmatrix}$	$\begin{bmatrix} -0.968 \\ 1 \\ -0.968 \end{bmatrix}$	$\begin{bmatrix} -0.791 \\ 1 \\ -0.791 \end{bmatrix}$	$\begin{bmatrix} -0.736 \\ 1 \\ -0.736 \end{bmatrix}$	$\begin{bmatrix} -0.717 \\ 1 \\ -0.717 \end{bmatrix}$
m_k	1	0.286	0.357	−0.187	0.705	0.655	0.639

We deduce that the eigenvalue of A that is smallest in magnitude is approximately $\frac{1}{0.6} \approx 1.7$. (In fact, this eigenvalue is exactly $3 - \sqrt{2}$.)

37. Since $\alpha = 0$, this is the inverse power method. So, the solution to Exercise 33 applies.

38. Since $\alpha = 0$, there is no shift required. So, taking $\mathbf{y}_0 = \begin{bmatrix} 1 \\ 1 \end{bmatrix}$, we solve $A\mathbf{x}_1 = \mathbf{y}_0$ using row reduction:

$$\left[\, A \mid \mathbf{y}_0 \,\right] = \left[\begin{array}{cc|c} 3.5 & 1.5 & 1 \\ 1.5 & -0.5 & 1 \end{array}\right] \longrightarrow \left[\begin{array}{cc|c} 1 & 0 & 0.5 \\ 0 & 1 & -0.5 \end{array}\right].$$

Thus, $\mathbf{x}_1 = \begin{bmatrix} 0.5 \\ -0.5 \end{bmatrix}$, so $\mathbf{y}_1 = \begin{bmatrix} 1 \\ -1 \end{bmatrix}$. Continuing in this way we calculate:

k	0	1	2	3	4	5	6
$\mathbf{x}_k$	$\begin{bmatrix} 1 \\ 1 \end{bmatrix}$	$\begin{bmatrix} 0.5 \\ -0.5 \end{bmatrix}$	$\begin{bmatrix} -0.25 \\ 1.25 \end{bmatrix}$	$\begin{bmatrix} 0.35 \\ -0.95 \end{bmatrix}$	$\begin{bmatrix} 0.329 \\ -1.013 \end{bmatrix}$	$\begin{bmatrix} 0.334 \\ -0.997 \end{bmatrix}$	$\begin{bmatrix} 0.333 \\ -1.001 \end{bmatrix}$
$\mathbf{y}_k$	$\begin{bmatrix} 1 \\ 1 \end{bmatrix}$	$\begin{bmatrix} 1 \\ -1 \end{bmatrix}$	$\begin{bmatrix} -0.2 \\ 1 \end{bmatrix}$	$\begin{bmatrix} -0.368 \\ 1 \end{bmatrix}$	$\begin{bmatrix} -0.325 \\ 1 \end{bmatrix}$	$\begin{bmatrix} -0.335 \\ 1 \end{bmatrix}$	$\begin{bmatrix} -0.333 \\ 1 \end{bmatrix}$
m_k	1	0.5	1.25	−0.95	−1.013	−0.997	−1.001

We deduce that the eigenvector of A closest to 0 is $\frac{1}{-1} = -1$.

39. We first shift: $A - 5I = \begin{bmatrix} -1 & 0 & 6 \\ -1 & -2 & 1 \\ 6 & 0 & -1 \end{bmatrix}$. Now, taking $\mathbf{x}_0 = \mathbf{y}_0 = \begin{bmatrix} 1 \\ 1 \\ 1 \end{bmatrix}$,

we solve $\left[\, A - 5I \mid \mathbf{y}_0 \,\right] = \left[\begin{array}{ccc|c} -1 & 0 & 6 & 1 \\ -1 & -2 & 1 & 1 \\ 6 & 0 & -1 & 1 \end{array}\right] \longrightarrow \left[\begin{array}{ccc|c} 1 & 0 & 0 & 0.2 \\ 0 & 1 & 0 & -0.5 \\ 0 & 0 & 1 & 0.2 \end{array}\right].$

Thus, $\mathbf{x}_1 = \begin{bmatrix} 0.2 \\ -0.5 \\ 0.2 \end{bmatrix}$ and $\mathbf{y}_1 = \begin{bmatrix} -0.4 \\ 1 \\ -0.4 \end{bmatrix}$. Continuing, we get the values shown in the table.

k	0	1	2	3
$\mathbf{x}_k$	$\begin{bmatrix} 1 \\ 1 \\ 1 \end{bmatrix}$	$\begin{bmatrix} 0.2 \\ -0.5 \\ 0.2 \end{bmatrix}$	$\begin{bmatrix} -0.08 \\ -0.5 \\ -0.08 \end{bmatrix}$	$\begin{bmatrix} 0.032 \\ -0.5 \\ 0.032 \end{bmatrix}$
$\mathbf{y}_k$	$\begin{bmatrix} 1 \\ 1 \\ 1 \end{bmatrix}$	$\begin{bmatrix} -0.4 \\ 1 \\ -0.4 \end{bmatrix}$	$\begin{bmatrix} 0.16 \\ 1 \\ 0.16 \end{bmatrix}$	$\begin{bmatrix} -0.064 \\ 1 \\ -0.064 \end{bmatrix}$
m_k	1	−0.5	−0.5	−0.5

It appears that the eigenvalue of A closest to 5 is $5 + \frac{1}{-0.5} = 5 - 2 = 3$.

40. $A + 2I = \begin{bmatrix} 11 & 4 & 8 \\ 4 & 17 & -4 \\ 8 & -4 & 11 \end{bmatrix}$, so taking $\mathbf{x}_0 = \mathbf{y}_0 = \begin{bmatrix} 1 \\ 1 \\ 1 \end{bmatrix}$, we solve

$$\left[\, A + 2I \mid \mathbf{y}_0 \,\right] = \left[\begin{array}{ccc|c} 11 & 4 & 8 & 1 \\ 4 & 17 & -4 & 1 \\ 8 & -4 & 11 & 1 \end{array}\right] \longrightarrow \left[\begin{array}{ccc|c} 1 & 0 & 0 & -0.158 \\ 0 & 1 & 0 & 0.158 \\ 0 & 0 & 1 & 0.263 \end{array}\right].$$

Thus, $\mathbf{x}_1 = \begin{bmatrix} -0.158 \\ 0.158 \\ 0.263 \end{bmatrix}$ and $\mathbf{y}_1 = \begin{bmatrix} -0.601 \\ 0.601 \\ 1 \end{bmatrix}$. Continuing, we get the values in the table.

k	0	1	2	3	4	5
$\mathbf{x}_k$	$\begin{bmatrix} 1 \\ 1 \\ 1 \end{bmatrix}$	$\begin{bmatrix} -0.158 \\ 0.158 \\ 0.263 \end{bmatrix}$	$\begin{bmatrix} -0.832 \\ 0.432 \\ 0.853 \end{bmatrix}$	$\begin{bmatrix} -0.989 \\ 0.496 \\ 0.991 \end{bmatrix}$	$\begin{bmatrix} -0.999 \\ 0.5 \\ 0.999 \end{bmatrix}$	$\begin{bmatrix} -1 \\ 0.5 \\ 1 \end{bmatrix}$
$\mathbf{y}_k$	$\begin{bmatrix} 1 \\ 1 \\ 1 \end{bmatrix}$	$\begin{bmatrix} -0.601 \\ 0.601 \\ 1 \end{bmatrix}$	$\begin{bmatrix} -0.975 \\ 0.506 \\ 1 \end{bmatrix}$	$\begin{bmatrix} -0.998 \\ 0.501 \\ 1 \end{bmatrix}$	$\begin{bmatrix} -1 \\ 0.5 \\ 1 \end{bmatrix}$	$\begin{bmatrix} -1 \\ 0.5 \\ 1 \end{bmatrix}$
m_k	1	0.263	0.853	0.991	0.999	1

It appears that the eigenvalue of A closest to -2 is $-2 + \frac{1}{1} = -1$.

41. The companion matrix of $p(x) = x^2 + 2x - 2$ is $C(p) = \begin{bmatrix} -2 & 2 \\ 1 & 0 \end{bmatrix}$.

So, we apply the inverse power method with $\mathbf{x}_0 = \mathbf{y}_0 = \begin{bmatrix} 1 \\ 1 \end{bmatrix}$.

We first row-reduce $\left[\, C(p) \mid \mathbf{y}_0 \,\right] = \left[\begin{array}{cc|c} -2 & 2 & 1 \\ 1 & 0 & 1 \end{array}\right] \longrightarrow \left[\begin{array}{cc|c} 1 & 0 & 1 \\ 0 & 1 & 1.5 \end{array}\right]$.

Thus, $\mathbf{x}_1 = \begin{bmatrix} 1 \\ 1.5 \end{bmatrix}$. Continuing, we get:

k	0	1	2	3	4	5	6
$\mathbf{x}_k$	$\begin{bmatrix} 1 \\ 1 \end{bmatrix}$	$\begin{bmatrix} 1 \\ 1.5 \end{bmatrix}$	$\begin{bmatrix} 1 \\ 1.333 \end{bmatrix}$	$\begin{bmatrix} 1 \\ 1.375 \end{bmatrix}$	$\begin{bmatrix} 1 \\ 1.364 \end{bmatrix}$	$\begin{bmatrix} 1 \\ 1.367 \end{bmatrix}$	$\begin{bmatrix} 1 \\ 1.366 \end{bmatrix}$
$\mathbf{y}_k$	$\begin{bmatrix} 1 \\ 1 \end{bmatrix}$	$\begin{bmatrix} 0.666 \\ 1 \end{bmatrix}$	$\begin{bmatrix} 0.750 \\ 1 \end{bmatrix}$	$\begin{bmatrix} 0.727 \\ 1 \end{bmatrix}$	$\begin{bmatrix} 0.733 \\ 1 \end{bmatrix}$	$\begin{bmatrix} 0.732 \\ 1 \end{bmatrix}$	$\begin{bmatrix} 0.732 \\ 1 \end{bmatrix}$
m_k	1	1.5	1.333	1.375	1.364	1.367	1.366

Thus, we estimate that the root of $p(x)$ closest to $\alpha = 0$ is approximately $\frac{1}{1.366} \approx 0.732$.

42. $C(p) = \begin{bmatrix} 1 & 3 \\ 1 & 0 \end{bmatrix}$ and $\alpha = 2$.

So, we apply the inverse power method to $C(p) - 2I = \begin{bmatrix} -1 & 3 \\ 1 & -2 \end{bmatrix}$ with $\mathbf{x}_0 = \mathbf{y}_0 = \begin{bmatrix} 1 \\ 1 \end{bmatrix}$.

First row-reducing $\left[\, C(p) \mid \mathbf{y}_0 \,\right] = \left[\begin{array}{cc|c} -1 & 3 & 1 \\ 1 & -2 & 1 \end{array}\right] \longrightarrow \left[\begin{array}{cc|c} 1 & 0 & 5 \\ 0 & 1 & 2 \end{array}\right]$. Continuing, we get:

k	0	1	2	3	4
$\mathbf{x}_k$	$\begin{bmatrix} 1 \\ 1 \end{bmatrix}$	$\begin{bmatrix} 5 \\ 2 \end{bmatrix}$	$\begin{bmatrix} 3.2 \\ 1.4 \end{bmatrix}$	$\begin{bmatrix} 3.314 \\ 1.438 \end{bmatrix}$	$\begin{bmatrix} 3.302 \\ 1.434 \end{bmatrix}$
$\mathbf{y}_k$	$\begin{bmatrix} 1 \\ 1 \end{bmatrix}$	$\begin{bmatrix} 1 \\ 0.4 \end{bmatrix}$	$\begin{bmatrix} 1 \\ 0.438 \end{bmatrix}$	$\begin{bmatrix} 1 \\ 0.434 \end{bmatrix}$	$\begin{bmatrix} 1 \\ 0.434 \end{bmatrix}$
m_k	1	5	3.2	3.314	3.302

Thus, we estimate that the root of $p(x)$ closest to $\alpha = 2$ is approximately $2 + \frac{1}{3.302} \approx 2.303$.

43. $C(p) = \begin{bmatrix} 2 & 0 & 1 \\ 1 & 0 & 0 \\ 0 & 1 & 0 \end{bmatrix}$, so we apply the inverse power method with $\mathbf{x}_0 = \mathbf{y}_0 = \begin{bmatrix} 1 \\ 1 \\ 1 \end{bmatrix}$.

Row-reducing $\left[\, C(p) \mid \mathbf{y}_0 \,\right] = \left[\begin{array}{ccc|c} 2 & 0 & -1 & 1 \\ 1 & 0 & 0 & 1 \\ 0 & 1 & 0 & 1 \end{array}\right] \longrightarrow \left[\begin{array}{ccc|c} 1 & 0 & 0 & 1 \\ 0 & 1 & 0 & 1 \\ 0 & 0 & 1 & 1 \end{array}\right]$.

We see that $\mathbf{y}_0$ is an eigenvector, but this is undesirable since the root $x = 1$ is not necessarily the closest to 0.

So we start again with $\mathbf{x}_0 = \mathbf{y}_0 = \begin{bmatrix} 1 \\ 0 \\ 1 \end{bmatrix}$, obtaining the values:

k	0	1	2	4	6	8	9
$\mathbf{x}_k$	$\begin{bmatrix} 1 \\ 0 \\ 1 \end{bmatrix}$	$\begin{bmatrix} 0 \\ 1 \\ -1 \end{bmatrix}$	$\begin{bmatrix} 1 \\ -1 \\ 2 \end{bmatrix}$	$\begin{bmatrix} -0.666 \\ 1 \\ -1.665 \end{bmatrix}$	$\begin{bmatrix} -0.625 \\ 1 \\ -1.625 \end{bmatrix}$	$\begin{bmatrix} -0.619 \\ 1 \\ -1.619 \end{bmatrix}$	$\begin{bmatrix} -0.618 \\ 1 \\ -1.618 \end{bmatrix}$
$\mathbf{y}_k$	$\begin{bmatrix} 1 \\ 0 \\ 1 \end{bmatrix}$	$\begin{bmatrix} 0 \\ 1 \\ -1 \end{bmatrix}$	$\begin{bmatrix} 0.5 \\ -0.5 \\ 1 \end{bmatrix}$	$\begin{bmatrix} 0.4 \\ -0.6 \\ 1 \end{bmatrix}$	$\begin{bmatrix} 0.385 \\ -0.615 \\ 1 \end{bmatrix}$	$\begin{bmatrix} 0.382 \\ -0.618 \\ 1 \end{bmatrix}$	$\begin{bmatrix} 0.382 \\ -0.618 \\ 1 \end{bmatrix}$
m_k	1	1	2	−1.665	−1.625	−1.619	−1.618

Thus, we conclude that the root of $p(x)$ closest to 0 is $\frac{1}{-1.618} \approx -0.618$.

44. $C(p) = \begin{bmatrix} 5 & -1 & -1 \\ 1 & 0 & 0 \\ 0 & 1 & 0 \end{bmatrix}$, so we apply the inverse power method with $\mathbf{x}_0 = \mathbf{y}_0 = \begin{bmatrix} 1 \\ 1 \\ 1 \end{bmatrix}$.

Row-reducing $\left[\, C(p) - 5I \mid \mathbf{y}_0 \,\right] = \left[\begin{array}{ccc|c} 0 & -1 & -1 & 1 \\ 1 & -5 & 0 & 1 \\ 0 & 1 & -5 & 1 \end{array}\right] \longrightarrow \left[\begin{array}{ccc|c} 1 & 0 & 0 & -2.333 \\ 0 & 1 & 0 & -0.667 \\ 0 & 0 & 1 & -0.333 \end{array}\right]$.

So, $\mathbf{x}_1 = \begin{bmatrix} -2.333 \\ -0.667 \\ -0.333 \end{bmatrix}$. Continuing, we get:

k	0	1	2	3	4
$\mathbf{x}_k$	$\begin{bmatrix} 1 \\ 1 \\ 1 \end{bmatrix}$	$\begin{bmatrix} -2.333 \\ -0.667 \\ -0.333 \end{bmatrix}$	$\begin{bmatrix} -3.762 \\ -0.810 \\ -0.191 \end{bmatrix}$	$\begin{bmatrix} -3.91 \\ -0.825 \\ -0.175 \end{bmatrix}$	$\begin{bmatrix} -3.922 \\ -0.827 \\ -0.173 \end{bmatrix}$
$\mathbf{y}_k$	$\begin{bmatrix} 1 \\ 1 \\ 1 \end{bmatrix}$	$\begin{bmatrix} 1 \\ 0.286 \\ 0.143 \end{bmatrix}$	$\begin{bmatrix} 1 \\ 0.215 \\ 0.05 \end{bmatrix}$	$\begin{bmatrix} 1 \\ 0.211 \\ 0.04 \end{bmatrix}$	$\begin{bmatrix} 1 \\ 0.211 \\ 0.04 \end{bmatrix}$
m_k	1	−2.333	−3.762	−3.91	−3.922

Thus, we conclude that the root of $p(x)$ closest to 5 is $5 + \frac{1}{-3.922} \approx 4.745$.

45. To prove $\frac{1}{\lambda-\alpha}$ is an eigenvalue of $(A-\alpha I)^{-1}$ we must show:
There exists $\mathbf{x} \neq \mathbf{0}$ such that $(A-\alpha I)^{-1}\mathbf{x} = \frac{1}{\lambda-\alpha}\mathbf{x}$.

Since λ is an eigenvalue of A, there exists $\mathbf{x} \neq \mathbf{0}$ such that $A\mathbf{x} = \lambda\mathbf{x}$.
Subtracting $\alpha\mathbf{x}$ from both sides yields: $A\mathbf{x} - \alpha\mathbf{x} = \lambda\mathbf{x} - \alpha\mathbf{x}$. So, $(A-\alpha I)\mathbf{x} = (\lambda-\alpha)\mathbf{x}$.
Left multiplying both sides by $(A-\alpha I)^{-1}$ gives us: $\mathbf{x} = (A-\alpha I)^{-1}(\lambda-\alpha)\mathbf{x}$.
Dividing both sides by $\lambda-\alpha$ yields: $(A-\alpha I)^{-1}\mathbf{x} = \frac{1}{\lambda-\alpha}\mathbf{x}$ as required.

Q: Why does the fact that α is *not* an eigenvalue of A imply $A-\alpha I$ is invertible?
A: If $\det(A-\alpha I) = 0$, then α *is* an eigenvalue of A.

Q: What is the contrapositive of this statement?
A: If α is *not* an eigenvalue of A, then $\det(A-\alpha I) \neq 0$. What does that tell us?
If $\det(A-\alpha I) \neq 0$, then $A-\alpha I$ is invertible.

46. Given A has a dominant eigenvalue λ_1, we must show that $\dim(E_{\lambda_1}) = 1$.
It suffices to show λ_1 is distinct from all the other eigenvalues of A. That is:
If $\lambda_1 \neq \lambda_j$ for $j > 1$, then $\dim(E_{\lambda_1}) = 1$.

Since λ_1 is dominant, $|\lambda_1| > |\lambda_j|$ for all eigenvalues of A where $j > 1$.
Therefore, $\lambda_1 \neq \lambda_j$ for $j > 1$, so $\dim(E_{\lambda_1}) = 1$.

Q: Why is this sufficient?
A: If λ_i is distinct from all the other eigenvalues of A,
then the algebraic multiplicity of λ_i is 1.

Q: Why does an algebraic multiplicity for λ_k of 1 imply $\dim(E_{\lambda_k}) = 1$?
A: Since Lemma 4.26 of Section 4.4 asserts $\dim(E_{\lambda_1})$ (the geometric multiplicity)
is always less than or equal than the algebraic multiplicity, k_i.

Q: Why does λ_k being distinct from all the other eigenvalues of A imply
that the algebraic multiplicity of λ_k is 1?
A: Given characteristic polynomial $c_A(\lambda) = \prod_{i=1}^{n}(\lambda_i - \lambda) = \prod_{i=1}^{m}(\lambda_i - \lambda)^{k_i}$.

In the second product, the λ_i are distinct.
In this product, we see k_i is equal to the number of times λ_i occurs as a root of $c_A(\lambda)$.
So, k_i is the algebraic multiplicity (the number of times λ_i occurs as a root of $c_A(\lambda)$).
So, the algebraic multiplicity is the number of times λ_i occurs as a eigenvalue of A.
Therefore, if λ_i is distinct from all the other eigenvalues of A,
then the algebraic multiplicity of λ_i is 1.

47. We find the Gershgorin's Disk Theorem for rows and columns separately.

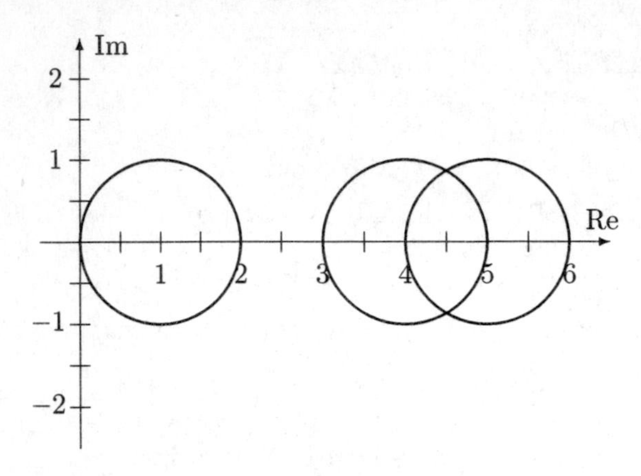

Gershgorin's Disk Theorem for rows:

$\begin{bmatrix} 1 & 1 & 0 \end{bmatrix} \Rightarrow c = a_{11} = 1,\ r = a_{12} + a_{13} = 1$

$\begin{bmatrix} \frac{1}{2} & 4 & \frac{1}{2} \end{bmatrix} \Rightarrow c = 4,\ r = \frac{1}{2} + \frac{1}{2} = 1$

$\begin{bmatrix} 1 & 0 & 5 \end{bmatrix} \Rightarrow c = 5,\ r = 1 + 0 = 1$

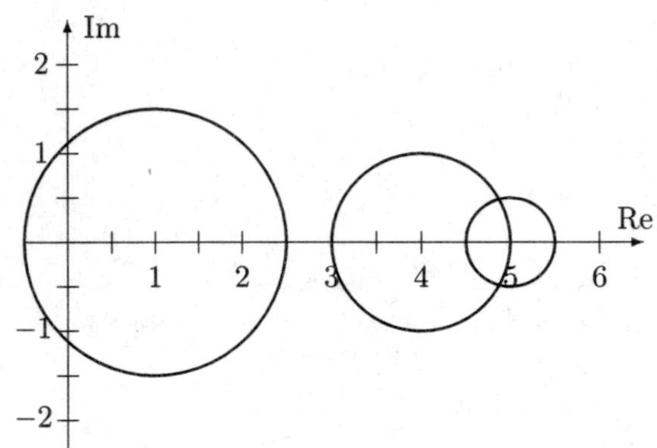

Gershgorin's Disk Theorem for columns:

$\begin{bmatrix} 1 \\ \frac{1}{2} \\ 1 \end{bmatrix} \Rightarrow c = a_{11} = 1,\ r = a_{21} + a_{31} = \frac{3}{2}$

$\mathbf{a}_2 \Rightarrow c = 4,\ r = 1 + 0 = 1$

$\mathbf{a}_3 \Rightarrow c = 5,\ r = 0 + \frac{1}{2} = \frac{1}{2}$

Q: The centers of the disks do not change for the rows and columns. Why?
A: Because they come from the diagonal entries.

Q: The radii of the disks *do* change for the rows and columns. Why?
A: Because they are created from the rows and columns respectively.

Q: What does the fact that the disk of radius 1 centered at 1 is disjoint tell us?
A: There is a real eigenvalue of A in the interval $[0, 2]$.

48. We find the Gershgorin's Disk Theorem for rows and columns separately.

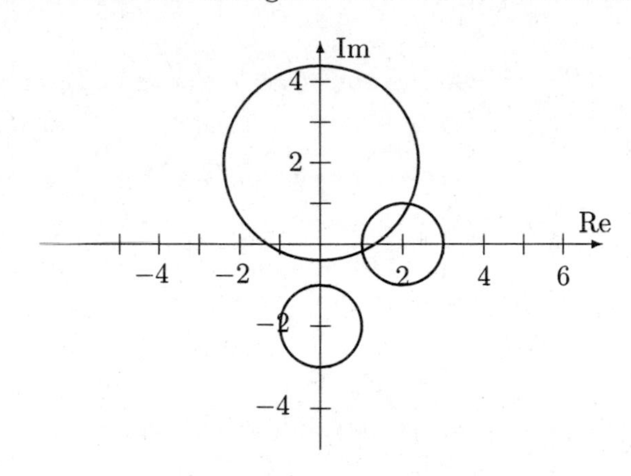

Gershgorin's Disk Theorem for rows:

$\begin{bmatrix} 2 & -i & 0 \end{bmatrix} \Rightarrow c = a_{11} = 2,\ r = |-i| = 1$

$\begin{bmatrix} 1 & 2i & 1+i \end{bmatrix} \Rightarrow c = 2i,\ r = 2 + \sqrt{2}$

$\begin{bmatrix} 0 & 1 & -2i \end{bmatrix} \Rightarrow c = -2i,\ r = 0 + 1 = 1$

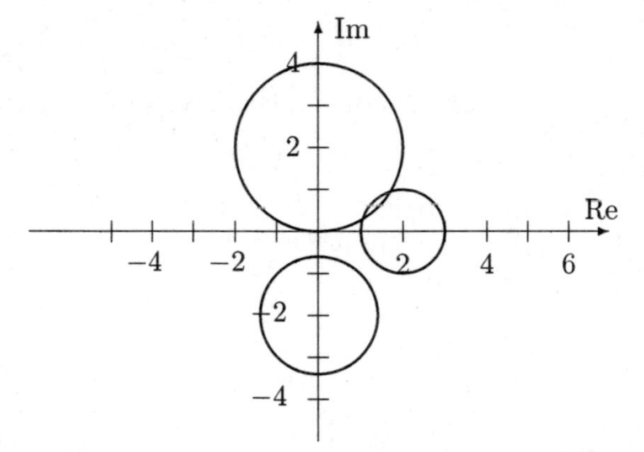

Gershgorin's Disk Theorem for columns:

$$\begin{bmatrix} 2 \\ 1 \\ 0 \end{bmatrix} \Rightarrow c = a_{11} = 2,\ r = 1 + 0 = 1$$

$\mathbf{a}_2 \Rightarrow c = 2i,\ r = 1 + 1 = 2$

$\mathbf{a}_3 \Rightarrow c = -2i,\ r = |1 + i| = \sqrt{2}$

Q: The centers of the disks do not change for the rows and columns. Why?
A: Because they come from the diagonal entries.

Q: The radii of the disks *do* change for the rows and columns. Why?
A: Because they are created from the rows and columns respectively.

Q: What does the fact that the disk of radius $\sqrt{2}$ centered at $-2i$ is disjoint tell us?
A: Not much. The characteristic polynomial of A does not have strictly *real* coefficients because some of the entries of A are complex numbers. So we cannot conclude the complex eigenvalues occur in conjugate pairs. Also, this disk does not intersect the *real* axis as it did in Exercise 47.

49. We find the Gershgorin's Disk Theorem for rows and columns separately.

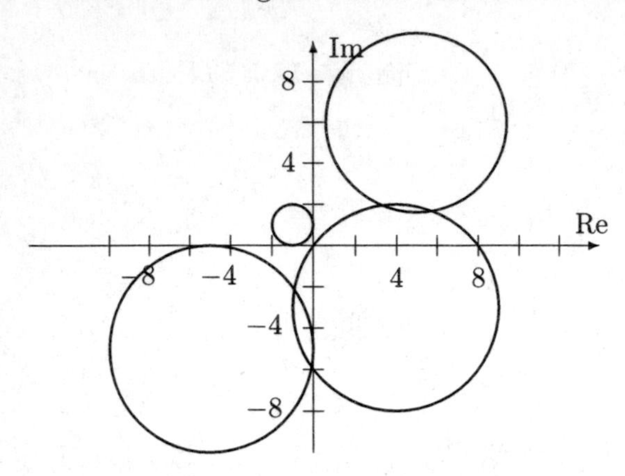

Gershgorin's Disk Theorem for rows:

$\mathbf{A}_1 \Rightarrow c = 4 - 3i,\ r = 5$

$\mathbf{A}_2 \Rightarrow c = -1 + i,\ r = 1$

$\mathbf{A}_3 \Rightarrow c = 5 + 6i,\ r = 3 + \sqrt{2}$

$\mathbf{A}_4 \Rightarrow c = -5 - 5i,\ r = 5$

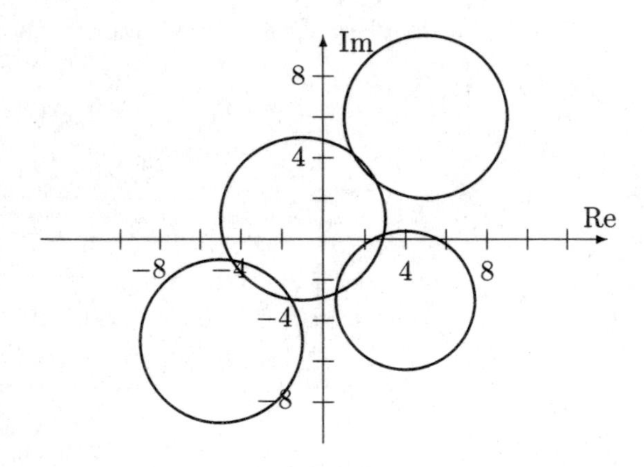

Gershgorin's Disk Theorem for columns:

$\mathbf{a}_1 \Rightarrow c = 4 - 3i,\ r = 2 + \sqrt{2}$

$\mathbf{a}_2 \Rightarrow c = -1 + i,\ r = 4$

$\mathbf{a}_3 \Rightarrow c = 5 + 6i,\ r = 4$

$\mathbf{a}_4 \Rightarrow c = -5 - 5i,\ r = 4$

50. We find the Gershgorin's Disk Theorem for rows and columns separately.

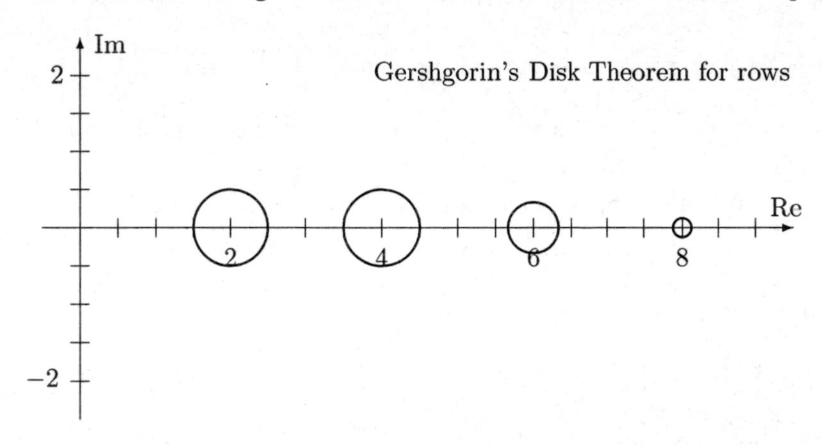

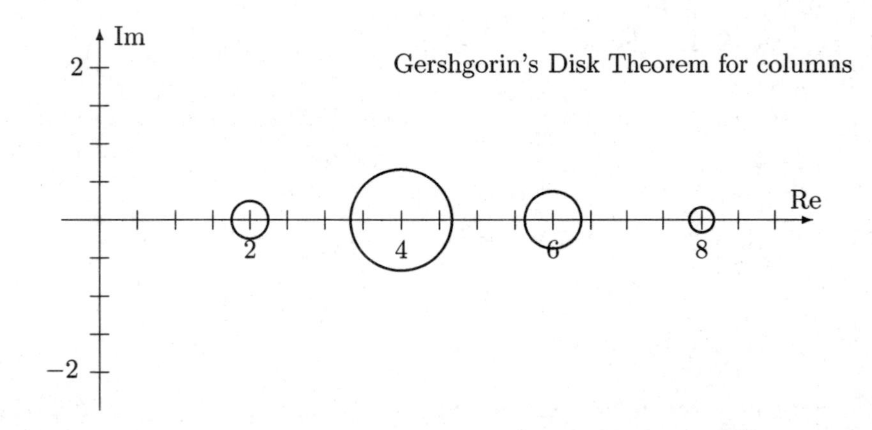

Q: What do all the disks being distinct tell us?
A: There is precisely one real eigenvalue in each of the corresponding intervals.

51. Given A *strictly diagonally dominant*, $|a_{ii}| > \sum_{j \neq i} |a_{ij}|$, we must show A is invertible.

We will use $|a_{ii}| > \sum_{j \neq i} |a_{ij}|$, to prove that all the eigenvalues of A are nonzero.

By Theorem 4.17(o) in Section 4.3, if 0 is not an eigenvalue of A then A is invertible.

Let λ be an eigenvalue of A. Then by the Gerchgorin's Disk Theorem, Theorem 4.29, we know λ is contained in a Gershgorin disk for some row i. So: $r_i = \sum_{j \neq i} |a_{ij}| \geq |a_{ii} - \lambda|$.

Since A is *strictly diagonally dominant*, we have: $|a_{ii}| > \sum_{j \neq i} |a_{ij}| \geq |a_{ii} - \lambda|$.

That is, $|a_{ii}| > |a_{ii} - \lambda|$. So, by the Triangle Inequality for complex numbers, we have $|a_{ii}| > |a_{ii} - \lambda| \geq |a_{ii}| - |\lambda|$. That is, $|a_{ii}| > |a_{ii}| - |\lambda|$.
Subtracting $|a_{ii}|$ from both sides and adding $|\lambda|$ to both sides, yields: $|\lambda| > 0$.

But $|\lambda| > 0$ implies $\lambda \neq 0$, as we were to show.
Therefore, all the eigenvalues of A are nonzero which implies A is invertible.

Q: How might we generalize this result?
A: Hint: If the dominant entry is *not* on the diagonal, can we permute the rows?

Q: What does the Triangle Inequality for complex numbers state?
A: It states $|z| + |w| \geq |z + w|$. See Appendix C, *Complex Numbers.*

Q: Why does $|z| + |w| \geq |z + w|$ imply $|a_{ii} - \lambda| \geq |a_{ii}| - |\lambda|$?
A: Hint: Let $z = a_{ii} - \lambda$ and $w = \lambda$.

52. We need to show $|\lambda| \leq \|A\|$, where $\|A\| = \max_{1 \leq i \leq n} \left(\sum_{j=1}^{n} |a_{ij}| \right)$.

The *key fact* is similar to the one used in Exercise 51 (See Appendix C, *Complex Numbers*): $|\lambda| - |a_{ii}| \leq |\lambda - a_{ii}|$ which implies $|\lambda| \leq |\lambda - a_{ii}| + |a_{ii}|$.

Let λ be an eigenvalue of A. Then by the Gerchgorin's Disk Theorem, Theorem 4.29, we know λ is contained in a Gershgorin disk for some row i. So: $|\lambda - a_{ii}| \leq \sum_{j \neq i} |a_{ij}| = r_i$.

By the *key fact* $|\lambda| \leq |\lambda - a_{ii}| + |a_{ii}|$. So: $|\lambda| \leq |\lambda - a_{ii}| + |a_{ii}| \leq \sum_{j \neq i} |a_{ij}| + |a_{ii}| \leq \sum_{j=1}^{n} |a_{ij}|$.

That is, $|\lambda| \leq \sum_{j=1}^{n} |a_{ij}|$ which implies $|\lambda| \leq \max_{1 \leq i \leq n} \left(\sum_{j=1}^{n} |a_{ij}| \right) = \|A\|$ as we were to show.

53. We need to show $|\lambda| \leq 1$, where λ is the eigenvalue of a *stochastic* matrix A.
We will show this using the result of Exercise 52 applied to A^T, after showing $\|A^T\| = 1$.

From Section 3.7, a *stochastic* matrix is a square matrix whose columns are probability vectors.

That is, the sum of all the entries in any column of A is 1: $\sum_{i=1}^{n} |a_{ij}| = 1$.

Therefore, the sum of all the entries in any row of A^T is 1: $\sum_{i=1}^{n} |a_{ij}| = 1$.

So by Exercise 52, we have $|\lambda| \leq \|A^T\| = 1$,

where $\|A^T\| = \max_{1 \leq j \leq n} \left(\sum_{i=1}^{n} |a_{ij}| \right) = 1$ and λ is an eigenvalue of A^T.

Recall from Exercise 19 of Section 4.3, that the eigenvalues of A^T and A are the same.

Therefore $|\lambda| \leq 1$, where λ is the eigenvalue of a *stochastic* matrix A as claimed.

54. We need to show that each eigenvalue of A is real and give the interval in which it lies.
Since all the entries of A are real, its characteristic polynomial has all real coefficients.
Therefore, as argued in Example 4.36, if the disks are disjoint the eigenvalues are real.

We will use the Gerchgorin disks from the rows and columns to find disks that are disjoint.

From row 1, we have one eigenvalue in the disk $|\lambda - 0| \leq 1$.
From column 2, we have one eigenvalue in the disk $|\lambda - 5| \leq 1$.

From column 3, we have one eigenvalue in the disk $|\lambda - 3| \leq \frac{3}{4}$.

From column 4, we have one eigenvalue in the disk $|\lambda - 7| \leq \frac{1}{2}$.

These disks are clearly disjoint, therefore each of these four eigenvalues is real.
Furthermore, there is exactly one eigenvalue of A in each of the following intervals:
$[-1, 1]$, $[4, 6]$, $[\frac{9}{4}, \frac{15}{4}]$, and $[\frac{13}{2}, \frac{15}{2}]$.

Converting to decimals and putting in number line order, we have:
$[-1, 1]$, $[2.25, 3.75]$, $[4, 6]$, and $[6.5, 7.5]$.

4.6 Applications and the Perron-Frobenius Theorem

1. $\begin{bmatrix} 0 & 1 \\ 1 & 0 \end{bmatrix}^2 = \begin{bmatrix} 1 & 0 \\ 0 & 1 \end{bmatrix}$, so any power of $\begin{bmatrix} 0 & 1 \\ 1 & 0 \end{bmatrix}$ contains zeros. Thus $\begin{bmatrix} 0 & 1 \\ 1 & 0 \end{bmatrix}$ is not regular.

2. $\begin{bmatrix} 1 & \frac{1}{2} \\ 0 & \frac{1}{2} \end{bmatrix}^n$ has 0 as its $(2,1)$ entry for all n, so $\begin{bmatrix} 1 & \frac{1}{2} \\ 0 & \frac{1}{2} \end{bmatrix}$ is not regular.

3. $\begin{bmatrix} \frac{1}{3} & 1 \\ \frac{2}{3} & 0 \end{bmatrix}^2 = \begin{bmatrix} \frac{7}{9} & \frac{1}{3} \\ \frac{2}{9} & \frac{2}{3} \end{bmatrix}$, which is positive, so $\begin{bmatrix} \frac{1}{3} & 1 \\ \frac{2}{3} & 0 \end{bmatrix}$ is regular.

4. $\begin{bmatrix} \frac{1}{2} & 0 & 1 \\ \frac{1}{2} & 0 & 0 \\ 0 & 1 & 0 \end{bmatrix}^4 = \begin{bmatrix} \frac{9}{16} & \frac{1}{4} & \frac{5}{8} \\ \frac{5}{16} & \frac{1}{4} & \frac{1}{8} \\ \frac{1}{8} & \frac{1}{2} & \frac{1}{4} \end{bmatrix}$, so $\begin{bmatrix} \frac{1}{2} & 0 & 1 \\ \frac{1}{2} & 0 & 0 \\ 0 & 1 & 0 \end{bmatrix}$ is regular.

5. $\begin{bmatrix} 0.1 & 0 & 0.5 \\ 0.5 & 1 & 0 \\ 0.4 & 0 & 0.5 \end{bmatrix}^n$ has 0s as its $(1,2)$ and $(3,2)$ entries for all n, so $\begin{bmatrix} 0.1 & 0 & 0.5 \\ 0.5 & 1 & 0 \\ 0.4 & 0 & 0.5 \end{bmatrix}$ is not regular.

6. The third row of $\begin{bmatrix} 0.1 & 1 & 0 \\ 0.5 & 0 & 1 \\ 0 & 0 & 0 \end{bmatrix}^n$ consists of zeros for all n, so $\begin{bmatrix} 0.1 & 1 & 0 \\ 0.5 & 0 & 1 \\ 0 & 0 & 0 \end{bmatrix}$ is not regular.

7. The transition matrix $P = \begin{bmatrix} \frac{1}{3} & \frac{1}{6} \\ \frac{2}{3} & \frac{5}{6} \end{bmatrix}$ has characteristic equation

$\det(P - \lambda I) = \begin{vmatrix} \frac{1}{3} - \lambda & \frac{1}{6} \\ \frac{2}{3} & \frac{5}{6} - \lambda \end{vmatrix} = \frac{1}{6}(\lambda - 1)(6\lambda - 1)$ so its eigenvalues are $1, \frac{1}{6}$.

The eigenspaces are $E_1 = \text{span}\left(\begin{bmatrix} 1 \\ 4 \end{bmatrix}\right)$ and $E_{1/6} = \text{span}\left(\begin{bmatrix} 1 \\ -1 \end{bmatrix}\right)$.

So, taking $Q = \begin{bmatrix} 1 & 1 \\ 4 & -1 \end{bmatrix}$, we know that $Q^{-1}PQ = \begin{bmatrix} 1 & 0 \\ 0 & \frac{1}{6} \end{bmatrix} = D$, and

$P^k = QD^kQ^{-1} = \begin{bmatrix} 1 & 1 \\ 4 & -1 \end{bmatrix} \begin{bmatrix} 1^k & 0 \\ 0 & \left(\frac{1}{6}\right)^k \end{bmatrix} \begin{bmatrix} 1 & 1 \\ 4 & -1 \end{bmatrix}^{-1}$.

As $k \to \infty$, $\left(\frac{1}{6}\right)^k \to 0$, so $D^k \to \begin{bmatrix} 1 & 0 \\ 0 & 0 \end{bmatrix}$, $P^k \to \begin{bmatrix} 1 & 1 \\ 4 & -1 \end{bmatrix} \begin{bmatrix} 1 & 0 \\ 0 & 0 \end{bmatrix} \begin{bmatrix} 1 & 1 \\ 4 & -1 \end{bmatrix}^{-1} = \begin{bmatrix} \frac{1}{5} & \frac{1}{5} \\ \frac{4}{5} & \frac{4}{5} \end{bmatrix} = L$.

8. We find the eigenspace of $P = \begin{bmatrix} \frac{1}{2} & \frac{1}{3} & \frac{1}{6} \\ \frac{1}{2} & \frac{1}{2} & \frac{1}{3} \\ 0 & \frac{1}{6} & \frac{1}{2} \end{bmatrix}$ corresponding to the eigenvalue 1 to be

$$E_1 = \text{span}\left(\begin{bmatrix} 7 \\ 9 \\ 3 \end{bmatrix}\right), \text{ so } L = \frac{1}{19}\begin{bmatrix} 7 & 7 & 7 \\ 9 & 9 & 9 \\ 3 & 3 & 3 \end{bmatrix}.$$

9. The eigenspace of $P = \begin{bmatrix} 0.2 & 0.3 & 0.4 \\ 0.6 & 0.1 & 0.4 \\ 0.2 & 0.6 & 0.2 \end{bmatrix}$ corresponding to the eigenvalue 1 is

$$E_1 = \text{span}\left(\begin{bmatrix} 24 \\ 28 \\ 27 \end{bmatrix}\right), \text{ so } L = \frac{1}{79}\begin{bmatrix} 24 & 24 & 24 \\ 28 & 28 & 28 \\ 27 & 27 & 27 \end{bmatrix}.$$

10. Suppose that a regular Markov chain has two steady state probability vectors $\mathbf{x}$ and $\mathbf{y}$.
Then for any $\mathbf{x}_0$, the sequence of iterates $\mathbf{x}_k$ approaches both $\mathbf{x}$ and $\mathbf{y}$.
Then we can make each of $|\mathbf{x}_k - \mathbf{x}|$ and $|\mathbf{x}_k - \mathbf{y}|$ arbitrarily small for sufficiently large k.
So, $|\mathbf{x} - \mathbf{y}| = |-(\mathbf{x}_k - \mathbf{x}) + (\mathbf{x}_k - \mathbf{y})|$ can also be made arbitrarily small — that is, $\mathbf{x} = \mathbf{y}$.

11. $\det(L - \lambda I) = \begin{vmatrix} -\lambda & 2 \\ 0.5 & -\lambda \end{vmatrix} = (\lambda - 1)(\lambda + 1)$, so the positive eigenvalue of L is 1.

$$[\, L - I \mid \mathbf{0} \,] = \left[\begin{array}{cc|c} -1 & 2 & 0 \\ 0.5 & -1 & 0 \end{array}\right] \longrightarrow \left[\begin{array}{cc|c} 1 & -2 & 0 \\ 0 & 0 & 0 \end{array}\right], \text{ so } E_1 = \text{span}\left(\begin{bmatrix} 2 \\ 1 \end{bmatrix}\right).$$

12. $\det(L - \lambda I) = \begin{vmatrix} 1-\lambda & 1.5 \\ 0.5 & -\lambda \end{vmatrix} = (\lambda + 0.5)(\lambda - 1.5) \Rightarrow$positive eigenvalue is 1.5.

$$[\, L - 1.5I \mid \mathbf{0} \,] = \left[\begin{array}{cc|c} -0.5 & 1.5 & 0 \\ 0.5 & -1.5 & 0 \end{array}\right] \longrightarrow \left[\begin{array}{cc|c} 1 & -3 & 0 \\ 0 & 0 & 0 \end{array}\right], \text{ so } E_1 = \text{span}\left(\begin{bmatrix} 3 \\ 1 \end{bmatrix}\right).$$

13. $\det(L - \lambda I) = \begin{vmatrix} -\lambda & 7 & 4 \\ 0.5 & -\lambda & 0 \\ 0 & 0.5 & -\lambda \end{vmatrix} = -\lambda^3 + 3.5\lambda + 1 = -\frac{1}{2}(\lambda - 2)(2\lambda^2 + 4\lambda + 1)$.

So, the positive eigenvalue of L is 2.

$$[\, L - 2I \mid \mathbf{0} \,] = \left[\begin{array}{ccc|c} -2 & 7 & 4 & 0 \\ 0.5 & -2 & 0 & 0 \\ 0 & 0.5 & -2 & 0 \end{array}\right] \longrightarrow \left[\begin{array}{ccc|c} 1 & 0 & -16 & 0 \\ 0 & 1 & -4 & 0 \\ 0 & 0 & 0 & 0 \end{array}\right], \text{ so } E_2 = \text{span}\left(\begin{bmatrix} 16 \\ 4 \\ 1 \end{bmatrix}\right).$$

14. $\det(L - \lambda I) = \begin{vmatrix} 1-\lambda & 5 & 3 \\ \frac{1}{3} & -\lambda & 0 \\ 0 & \frac{2}{3} & -\lambda \end{vmatrix} = -\lambda^3 + \lambda^2 + \frac{5}{3}\lambda + \frac{2}{3} = -\frac{1}{3}(\lambda - 2)(3\lambda^2 + 3\lambda + 1)$.

So, the positive eigenvalue of L is 2.

$$[\, L - 2I \mid \mathbf{0} \,] = \left[\begin{array}{ccc|c} -1 & 5 & 3 & 0 \\ \frac{1}{3} & -2 & 0 & 0 \\ 0 & \frac{2}{3} & -2 & 0 \end{array}\right] \longrightarrow \left[\begin{array}{ccc|c} 1 & 0 & -18 & 0 \\ 0 & 1 & -3 & 0 \\ 0 & 0 & 0 & 0 \end{array}\right], \text{ so } E_2 = \text{span}\left(\begin{bmatrix} 18 \\ 3 \\ 1 \end{bmatrix}\right).$$

15. If $\lambda_1 > 1$, then the population will increase without limit.
If $\lambda_1 < 1$, then the population will decline and eventually vanish.
If $\lambda_1 = 1$, then the population will be stable.

16. For $n = 2$, $L = \begin{bmatrix} b_1 & b_2 \\ s_1 & 0 \end{bmatrix}$ and $c_L(\lambda) = \begin{vmatrix} b_1 - \lambda & b_2 \\ s_1 & -\lambda \end{vmatrix} = \lambda^2 - \lambda b_1 - b_2 s_1 = (-1)^2\left(\lambda^2 - b_1\lambda - b_2 s_1\right)$.

For $n = k$, assume $c_L(\lambda) = (-1)^k\left(\lambda^k - b_1\lambda^{k-1} - b_2 s_1 \lambda^{k-2} - \cdots - b_k s_1 s_2 \cdots s_{k-1}\right)$.

Then, for $n = k+1$, expanding along the last column gives

$$\det(L - \lambda I) = \begin{vmatrix} b_1 - \lambda & b_2 & b_3 & \cdots & b_k & b_{k+1} \\ s_1 & -\lambda & 0 & \cdots & 0 & 0 \\ 0 & s_2 & -\lambda & \cdots & 0 & 0 \\ 0 & 0 & s_3 & \cdots & 0 & 0 \\ \vdots & \vdots & \vdots & \ddots & \vdots & \vdots \\ 0 & 0 & 0 & \cdots & s_k & -\lambda \end{vmatrix}$$

$$= (-1)^k b_{k+1} \begin{vmatrix} s_1 & -\lambda & 0 & \cdots & 0 \\ 0 & s_2 & -\lambda & \cdots & 0 \\ 0 & 0 & s_3 & \cdots & 0 \\ \vdots & \vdots & \vdots & \ddots & \vdots \\ 0 & 0 & 0 & \cdots & s_k \end{vmatrix} - \lambda \begin{vmatrix} b_1 - \lambda & b_2 & b_3 & \cdots & b_k \\ s_1 & -\lambda & 0 & \cdots & 0 \\ 0 & s_2 & -\lambda & \cdots & 0 \\ \vdots & \vdots & \vdots & \ddots & 0 \\ 0 & 0 & 0 & 0 & -\lambda \end{vmatrix}$$

$$= (-1)^k b_{k+1} s_1 s_2 \cdots s_k - \lambda\left[(-1)^k\left(\lambda^k - b_1\lambda^{k-1} - b_2 s_1 \lambda^{k-2} - \cdots - b_k s_1 s_2 \cdots s_{k-1}\right)\right]$$

$$= (-1)^{k+1}\left(\lambda^{k+1} - b_1\lambda^k - b_2 s_1 \lambda^{k-1} - \cdots - b_k s_1 s_2 \cdots s_{k-1}\lambda - b_{k+1} s_1 s_2 \cdots s_k\right)$$

by the induction hypothesis. So we have verified the formula by mathematical induction.

17. $P^{-1} = \begin{bmatrix} 1 & 0 & 0 & \cdots & 0 \\ 0 & \frac{1}{s_1} & 0 & \cdots & 0 \\ 0 & 0 & \frac{1}{s_1 s_2} & \cdots & 0 \\ \vdots & \vdots & \vdots & \ddots & \vdots \\ 0 & 0 & 0 & \cdots & \frac{1}{s_1 s_2 \cdots s_{n-1}} \end{bmatrix}$ and $L = \begin{bmatrix} b_1 & b_2 & b_3 & \cdots & b_{n-1} & b_n \\ s_1 & 0 & 0 & \cdots & 0 & 0 \\ 0 & s_2 & 0 & \cdots & 0 & 0 \\ 0 & 0 & s_3 & \cdots & 0 & 0 \\ \vdots & \vdots & \vdots & \ddots & \vdots & \vdots \\ 0 & 0 & 0 & \cdots & s_{n-1} & 0 \end{bmatrix}$.

So, $P^{-1}L = \begin{bmatrix} b_1 & b_2 & b_3 & \cdots & b_{n-1} & b_n \\ 1 & 0 & 0 & \cdots & 0 & 0 \\ 0 & \frac{1}{s_1} & 0 & \cdots & 0 & 0 \\ 0 & 0 & \frac{1}{s_1 s_2} & \cdots & 0 & 0 \\ \vdots & \vdots & \vdots & \ddots & \vdots & \vdots \\ 0 & 0 & 0 & \cdots & \frac{1}{s_1 s_2 \cdots s_{n-2}} & 0 \end{bmatrix}$ and

$$P^{-1}LP = \begin{bmatrix} b_1 & b_2 s_1 & b_3 s_1 s_2 & \cdots & b_{n-1} s_1 s_2 \cdots s_{n-2} & b_n s_1 s_2 \cdots s_{n-1} \\ 1 & 0 & 0 & \cdots & 0 & 0 \\ 0 & 1 & 0 & \cdots & 0 & 0 \\ 0 & 0 & 1 & \cdots & 0 & 0 \\ \vdots & \vdots & \vdots & \ddots & \vdots & \vdots \\ 0 & 0 & 0 & \cdots & 1 & 0 \end{bmatrix}.$$

This is the companion matrix of the polynomial:
$p(x) = x^n - b_1 x^{n-1} - b_2 s_1 x^{n-2} - \cdots - b_{n-1} s_1 s_2 \cdots s_{n-2} x - b_n s_1 s_2 \cdots s_{n-1}$.
So, by Exercise 32 in Section 4.3, the characteristic polynomial of $P^{-1}LP$ is $(-1)^n p(\lambda)$.

So, the characteristic polynomial of L is:

$$\begin{aligned} \det(L - \lambda I) &= \det\left(P^{-1}LP - \lambda I\right) = (-1)^n p(\lambda) \\ &= (-1)^n \left(\lambda^n - b_1 \lambda^{n-1} - b_2 s_1 \lambda^{n-2} - \cdots - b_n s_1 s_2 \cdots s_{n-1}\right). \end{aligned}$$

18. By Exercise 46 in Section 4.4, every eigenvector of $P^{-1}LP$ is of the form $P^{-1}\mathbf{x}$ for some eigenvector $\mathbf{x}$ of L.
By Exercise 32 in Section 4.3, an eigenvector of $P^{-1}LP$ for λ_1 is $\begin{bmatrix} \lambda_1^{n-1} & \lambda_1^{n-2} & \cdots & \lambda_1 & 1 \end{bmatrix}^T$.

Thus, an eigenvector of L corresponding to λ_1 is $P \begin{bmatrix} \lambda_1^{n-1} & \lambda_1^{n-2} & \lambda_1^{n-3} & \cdots & \lambda_1 & 1 \end{bmatrix}^T$
$= \begin{bmatrix} \lambda_1^{n-1} & s_1 \lambda_1^{n-2} & s_1 s_2 \lambda_1^{n-3} & \cdots & s_1 s_2 \cdots s_{n-2} \lambda_1 & s_1 s_2 \cdots s_{n-2} s_{n-1} \end{bmatrix}^T$.

Or, multiplying through by λ_1^{1-n}:

$\begin{bmatrix} 1 & s_1/\lambda_1 & s_1 s_2/\lambda_1^2 & \cdots & s_1 s_2 \cdots s_{n-2}/\lambda_1^{n-2} & s_1 s_2 \cdots s_{n-2} s_{n-1}/\lambda_1^{n-1} \end{bmatrix}^T$.

19. Using a CAS, we find the positive eigenvalue of $L = \begin{bmatrix} 1 & 1 & 3 \\ 0.7 & 0 & 0 \\ 0 & 0.5 & 0 \end{bmatrix}$ is ≈ 1.7456.

So, this is the steady state growth rate of the population with this Leslie matrix.
Now according to Exercise 18, a corresponding eigenvector is
$\begin{bmatrix} 1 & 0.7/1.7456 & (0.7 \cdot 0.5)/(1.7456)^2 \end{bmatrix}^T \approx \begin{bmatrix} 1 & 0.401 & 0.115 \end{bmatrix}^T$.
So, the age classes are in the ratio $1 : 0.401 : 0.115$.

20. The positive eigenvalue of $\begin{bmatrix} 0 & 1 & 2 & 5 \\ 0.5 & 0 & 0 & 0 \\ 0 & 0.7 & 0 & 0 \\ 0 & 0 & 0.3 & 0 \end{bmatrix}$ is ≈ 1.2023, the steady state growth rate.

A corresponding eigenvector is
$\begin{bmatrix} 1 & 0.5/1.2023 & (0.5 \cdot 0.7)/(1.2023)^2 & (0.5 \cdot 0.7 \cdot 0.3)/(1.2023)^3 \end{bmatrix}^T = \begin{bmatrix} 1 & 0.416 & 0.242 & 0.060 \end{bmatrix}^T$.
So, the age classes are in the ratio $1 : 0.416 : 0.242 : 0.060$.

21. The positive eigenvalue of L is approximately 1.0924, so this is the steady state growth rate.

A corresponding eigenvector is $\begin{bmatrix} 1 \\ \frac{0.3}{1.0924} \\ \frac{0.3 \cdot 0.7}{(1.0924)^2} \\ \frac{0.3 \cdot 0.7 \cdot 0.9}{(1.0924)^3} \\ \frac{0.3 \cdot 0.7 \cdot 0.9 \cdot 0.9}{(1.0924)^4} \\ \frac{0.3 \cdot 0.7 \cdot 0.9 \cdot 0.9 \cdot 0.9}{(1.0924)^5} \\ \frac{0.3 \cdot 0.7 \cdot 0.9 \cdot 0.9 \cdot 0.9 \cdot 0.6}{(1.0924)^6} \end{bmatrix} \approx \begin{bmatrix} 1 \\ 0.275 \\ 0.176 \\ 0.145 \\ 0.119 \\ 0.098 \\ 0.054 \end{bmatrix}$.

22. (a) $L = \begin{bmatrix} 0 & 0.02 & 0.70 & 1.53 & 1.67 & 1.65 & 1.56 & 1.45 & 1.22 & 0.91 & 0.70 & 0.22 & 0 \\ 0.91 & 0 & 0 & 0 & 0 & 0 & 0 & 0 & 0 & 0 & 0 & 0 & 0 \\ 0 & 0.88 & 0 & 0 & 0 & 0 & 0 & 0 & 0 & 0 & 0 & 0 & 0 \\ 0 & 0 & 0.85 & 0 & 0 & 0 & 0 & 0 & 0 & 0 & 0 & 0 & 0 \\ 0 & 0 & 0 & 0.80 & 0 & 0 & 0 & 0 & 0 & 0 & 0 & 0 & 0 \\ 0 & 0 & 0 & 0 & 0.74 & 0 & 0 & 0 & 0 & 0 & 0 & 0 & 0 \\ 0 & 0 & 0 & 0 & 0 & 0.67 & 0 & 0 & 0 & 0 & 0 & 0 & 0 \\ 0 & 0 & 0 & 0 & 0 & 0 & 0.59 & 0 & 0 & 0 & 0 & 0 & 0 \\ 0 & 0 & 0 & 0 & 0 & 0 & 0 & 0.49 & 0 & 0 & 0 & 0 & 0 \\ 0 & 0 & 0 & 0 & 0 & 0 & 0 & 0 & 0.38 & 0 & 0 & 0 & 0 \\ 0 & 0 & 0 & 0 & 0 & 0 & 0 & 0 & 0 & 0.27 & 0 & 0 & 0 \\ 0 & 0 & 0 & 0 & 0 & 0 & 0 & 0 & 0 & 0 & 0.17 & 0 & 0 \\ 0 & 0 & 0 & 0 & 0 & 0 & 0 & 0 & 0 & 0 & 0 & 0.15 & 0 \end{bmatrix}$

has positive eigenvalue ≈ 1.333 with corresponding eigenvector $\begin{bmatrix} 36.1 \\ 24.7 \\ 16.3 \\ 10.4 \\ 6.23 \\ 3.46 \\ 1.74 \\ 0.77 \\ 0.28 \\ 0.081 \\ 0.016 \\ 0.0021 \\ 0.00023 \end{bmatrix}$.

(b) In the long run, the percentage of seals in each age group is given by the entries in the eigenvector in part (a), and the growth rate will be about 1.333.

23. (a) Each term in r corresponds to the probability of a given female having a daughter while she is a member of a particular age class. Thus, the sum of the terms gives the average number of daughters born to a single female over her lifetime.

(b) We define $g(\lambda)$ as suggested. Then $g(\lambda) = 1 \Leftrightarrow$

$\frac{b_1}{\lambda} + \frac{b_2 s_1}{\lambda^2} + \cdots + \frac{b_n s_1 s_2 \cdots s_{n-1}}{\lambda^n} = 1 \Leftrightarrow \lambda^n - b_1\lambda^{n-1} - b_2 s_1 \lambda^{n-2} - \cdots - b_n s_1 s_2 \cdots s_{n-1} = 0.$

But from Exercise 16, this is true if and only if λ is a zero of the characteristic polynomial of the Leslie matrix, indicating that λ is an eigenvalue of L.

Now $r = b_1 + b_2 s_1 + \cdots + b_n s_1 s_2 \cdots s_{n-1} = 1 \Leftrightarrow \lambda = 1$, as desired.

(c) If $r < 1$, each female averages less than 1 daughter $\Rightarrow$ decreasing population.
If $r > 1$, each female averages more than 1 daughter $\Rightarrow$ increasing population.

24. If λ_1 is the unique positive eigenvalue, then $\lambda_1 \mathbf{x} = L\mathbf{x}$ for the corresponding eigenvector $\mathbf{x}$.

Sustainability requires $(1-h)\,L\mathbf{x} = \mathbf{x} \Leftrightarrow L\mathbf{x} = \frac{1}{1-h}\mathbf{x}$.

So, substituting, we get $\lambda_1 \mathbf{x} = \frac{1}{1-h}\mathbf{x} \Leftrightarrow \lambda_1 = \frac{1}{1-h} \Leftrightarrow h = 1 - 1/\lambda_1$.

25. (a) In Exercise 21 we found the positive eigenvalue of L to be $\lambda_1 \approx 1.0924$, so from Exercise 24, $h = 1 - 1/\lambda_1 \approx 1 - 1/1.0924 \approx 0.0846$.

(b) We reduce the initial population levels by a factor of 0.0846, so:
$\mathbf{x}_0' = (1 - 0.0846)\begin{bmatrix} 10 & 2 & 8 & 5 & 12 & 0 & 1 \end{bmatrix}^T = \begin{bmatrix} 9.15 & 1.831 & 7.323 & 4.577 & 10.985 & 0 & 0.915 \end{bmatrix}^T$.

$$\text{Then } \mathbf{x}_1' = L\mathbf{x}_0' = \begin{bmatrix} 0 & 0.4 & 1.8 & 1.8 & 1.8 & 1.6 & 0.6 \\ 0.3 & 0 & 0 & 0 & 0 & 0 & 0 \\ 0 & 0.7 & 0 & 0 & 0 & 0 & 0 \\ 0 & 0 & 0.9 & 0 & 0 & 0 & 0 \\ 0 & 0 & 0 & 0.9 & 0 & 0 & 0 \\ 0 & 0 & 0 & 0 & 0.9 & 0 & 0 \\ 0 & 0 & 0 & 0 & 0 & 0.6 & 0 \end{bmatrix}\begin{bmatrix} 9.15 \\ 1.831 \\ 7.323 \\ 4.577 \\ 10.985 \\ 0 \\ 0.915 \end{bmatrix} = \begin{bmatrix} 42.47 \\ 2.75 \\ 1.28 \\ 6.59 \\ 4.12 \\ 9.89 \\ 0 \end{bmatrix}.$$

In fact, the population increases substantially.

26. $h = 1 - 1/\lambda_1 = 1 - 1/1.333 \approx 0.25$ — about one-quarter of the population!

27. $\lambda_1 = r_1(\cos 0 + i \sin 0) = r_1$ since it is positive.
Let $\lambda = r(\cos\theta + i\sin\theta)$ be some other eigenvector with $\theta \neq 0$.
Then equating the two expressions for $g(\lambda)$, applying De Moivre's Theorem, and taking the real part of each side, we have:

$$\frac{b_1}{r_1} + \frac{b_2 s_1}{r_1^2} + \frac{b_3 s_1 s_2}{r_1^3} + \cdots + \frac{b_n s_1 s_2 \cdots s_{n-1}}{r_1^n}$$
$$= \frac{b_1}{r}\cos\theta + \frac{b_2 s_1}{r^2}\cos 2\theta + \frac{b_3 s_1 s_2}{r^3}\cos 3\theta + \cdots + \frac{b_n s_1 s_2 \cdots s_{n-1}}{r^n}\cos n\theta$$

But $\cos k\theta \leq 1$ for all k, so

$$\frac{b_1}{r_1} + \frac{b_2 s_1}{r_1^2} + \frac{b_3 s_1 s_2}{r_1^3} + \cdots + \frac{b_n s_1 s_2 \cdots s_{n-1}}{r_1^n} \leq \frac{b_1}{r} + \frac{b_2 s_1}{r^2} + \frac{b_3 s_1 s_2}{r^3} + \cdots + \frac{b_n s_1 s_2 \cdots s_{n-1}}{r^n}$$

showing that $r_1 \geq r$.

28. $\det(A - \lambda I) = \begin{vmatrix} 2-\lambda & 0 \\ 1 & 1-\lambda \end{vmatrix} = (\lambda - 2)(\lambda - 1)$, so $\lambda_1 = 2$ is the Perron root.

$$A - 2I = \begin{bmatrix} 0 & 0 \\ 1 & -1 \end{bmatrix} \longrightarrow \begin{bmatrix} 1 & -1 \\ 0 & 0 \end{bmatrix}.$$

So, the Perron eigenvector (whose entries must sum to 1, remember) is $\begin{bmatrix} \frac{1}{2} \\ \frac{1}{2} \end{bmatrix}$.

29. $\det(A - \lambda I) = \begin{vmatrix} 1-\lambda & 3 \\ 2 & -\lambda \end{vmatrix} = \lambda^2 - \lambda - 6 = (\lambda + 2)(\lambda - 3)$, so $\lambda_1 = 3$ is the Perron root.

$A - 3I = \begin{bmatrix} -2 & 3 \\ 2 & -3 \end{bmatrix} \longrightarrow \begin{bmatrix} 2 & -3 \\ 0 & 0 \end{bmatrix}$, so the Perron eigenvector is $\begin{bmatrix} \frac{3}{5} \\ \frac{2}{5} \end{bmatrix}$.

30. $\det(A-\lambda I) = \begin{vmatrix} -\lambda & 1 & 1 \\ 1 & -\lambda & 1 \\ 1 & 1 & -\lambda \end{vmatrix} = -(\lambda+1)^2(\lambda-2)$, so $\lambda_1 = 2$ is the Perron root.

$A - 2I = \begin{bmatrix} -2 & 1 & 1 \\ 1 & -2 & 1 \\ 1 & 1 & -2 \end{bmatrix} \longrightarrow \begin{bmatrix} 1 & 0 & -1 \\ 0 & 1 & -1 \\ 0 & 0 & 0 \end{bmatrix}$, so the Perron eigenvector is $\begin{bmatrix} \frac{1}{3} \\ \frac{1}{3} \\ \frac{1}{3} \end{bmatrix}$.

31. $\det(A-\lambda I) = \begin{vmatrix} 2-\lambda & 1 & 1 \\ 1 & 1-\lambda & 0 \\ 1 & 0 & 1-\lambda \end{vmatrix} = -\lambda(\lambda-1)(\lambda-3)$, so $\lambda_1 = 3$ is the Perron root.

$A - 3I = \begin{bmatrix} -1 & 1 & 1 \\ 1 & -2 & 0 \\ 1 & 0 & -2 \end{bmatrix} \longrightarrow \begin{bmatrix} 1 & 0 & -2 \\ 0 & 1 & -1 \\ 0 & 0 & 0 \end{bmatrix}$, so the Perron eigenvector is $\begin{bmatrix} \frac{1}{2} \\ \frac{1}{4} \\ \frac{1}{4} \end{bmatrix}$.

32. $(I+A)^{n-1} = \begin{bmatrix} 1 & 0 & 1 & 0 \\ 0 & 1 & 0 & 1 \\ 0 & 1 & 1 & 0 \\ 1 & 0 & 0 & 1 \end{bmatrix}^3 = \begin{bmatrix} 1 & 3 & 3 & 1 \\ 3 & 1 & 1 & 3 \\ 1 & 3 & 1 & 3 \\ 3 & 1 & 3 & 1 \end{bmatrix} > O$, so A is irreducible.

33. $(I+A)^{n-1} = \begin{bmatrix} 1 & 0 & 1 & 0 \\ 0 & 1 & 1 & 1 \\ 1 & 0 & 1 & 0 \\ 1 & 1 & 0 & 1 \end{bmatrix}^3 = \begin{bmatrix} 4 & 0 & 4 & 0 \\ 6 & 4 & 6 & 4 \\ 4 & 0 & 4 & 0 \\ 6 & 4 & 6 & 4 \end{bmatrix}$, so A is reducible.

In particular, the permutations $R_1 \leftrightarrow R_3$ and $R_2 \leftrightarrow R_4$ put A into the desired block form.

34. $(I+A)^{n-1} = \begin{bmatrix} 1 & 1 & 0 & 0 & 0 \\ 0 & 1 & 1 & 0 & 1 \\ 1 & 0 & 2 & 0 & 1 \\ 0 & 0 & 1 & 2 & 0 \\ 1 & 0 & 0 & 0 & 1 \end{bmatrix}^4 = \begin{bmatrix} 11 & 6 & 11 & 0 & 11 \\ 22 & 11 & 17 & 0 & 17 \\ 28 & 16 & 23 & 0 & 22 \\ 23 & 7 & 33 & 16 & 18 \\ 6 & 6 & 5 & 0 & 6 \end{bmatrix}$, so A is reducible.

In particular, the permutation $C_1 \leftrightarrow C_2$ puts A into the desired block form.

35. $(I+A)^{n-1} = \begin{bmatrix} 1 & 1 & 0 & 0 & 0 \\ 0 & 1 & 0 & 0 & 1 \\ 1 & 0 & 1 & 0 & 1 \\ 0 & 0 & 1 & 1 & 0 \\ 0 & 0 & 0 & 1 & 2 \end{bmatrix}^4 = \begin{bmatrix} 1 & 4 & 1 & 5 & 11 \\ 1 & 1 & 5 & 11 & 16 \\ 5 & 6 & 6 & 12 & 21 \\ 6 & 4 & 5 & 6 & 12 \\ 5 & 1 & 11 & 16 & 22 \end{bmatrix}$, so A is irreducible.

36. (a) Given a graph G, consider a graph $G^* = G$ excluding edges from a vertex to itself. Further, consider $G' = G^*$ plus edges from each vertex to itself.
G is connected $\Leftrightarrow$ G^* is connected $\Leftrightarrow$ G' is connected.
Now if A corresponds to graph G^*, then $I + A$ corresponds to G', and we know that A is irreducible if and only if $(I + A)^{n-1} > O$.
But this is the case if and only if G' is connected.
So A is irreducible if and only if G is connected.

(b) Since all Section 4.0 graphs are connected, they have irreducible adjacency matrices. The graph in Figure 1 is has a primitive adjacency matrix:

$$A^2 = \begin{bmatrix} 0 & 1 & 1 & 1 \\ 1 & 0 & 1 & 1 \\ 1 & 1 & 0 & 1 \\ 1 & 1 & 1 & 0 \end{bmatrix}^2 = \begin{bmatrix} 3 & 2 & 2 & 2 \\ 2 & 3 & 2 & 2 \\ 2 & 2 & 3 & 2 \\ 2 & 2 & 2 & 3 \end{bmatrix} > O.$$

So does the graph in Figure 2:

$$A^4 = \begin{bmatrix} 0&1&0&0&1&0&1&0&0&0 \\ 1&0&1&0&0&0&0&1&0&0 \\ 0&1&0&1&0&0&0&0&1&0 \\ 0&0&1&0&1&0&0&0&0&1 \\ 1&0&0&1&0&1&0&0&0&0 \\ 0&0&0&0&1&0&0&1&1&0 \\ 1&0&0&0&0&0&0&0&1&1 \\ 0&1&0&0&0&1&0&0&0&1 \\ 0&0&1&0&0&1&1&0&0&0 \\ 0&0&0&1&0&0&1&1&0&0 \end{bmatrix}^4 = \begin{bmatrix} 15&4&9&9&4&9&4&9&9&9 \\ 4&15&4&9&9&9&9&4&9&9 \\ 9&4&15&4&9&9&9&9&4&9 \\ 9&9&4&15&4&9&9&9&9&4 \\ 4&9&9&4&15&4&9&9&9&9 \\ 9&9&9&9&4&15&9&4&4&9 \\ 4&9&9&9&9&9&15&9&4&4 \\ 9&4&9&9&9&4&9&15&9&4 \\ 9&9&4&9&9&4&4&9&15&9 \\ 9&9&9&4&9&9&4&4&9&15 \end{bmatrix} > O.$$

The graph in Figure 3 also has a primitive adjacency matrix:

$$A^4 = \begin{bmatrix} 0&1&0&0&1 \\ 1&0&1&0&0 \\ 0&1&0&1&0 \\ 0&0&1&0&1 \\ 1&0&0&1&0 \end{bmatrix}^4 = \begin{bmatrix} 6&1&4&4&1 \\ 1&6&1&4&4 \\ 4&1&6&1&4 \\ 4&4&1&6&1 \\ 1&4&4&1&6 \end{bmatrix} > O. \text{ For Figure 4, } A = \begin{bmatrix} 0&0&0&1&1&1 \\ 0&0&0&1&1&1 \\ 0&0&0&1&1&1 \\ 1&1&1&0&0&0 \\ 1&1&1&0&0&0 \\ 1&1&1&0&0&0 \end{bmatrix}.$$

Since any power of A contains zeroes,
the graph in Figure 4 does not have a primitive adjacency matrix.

37. (a) Suppose G is bipartite and let v and w be vertices which are connected.
Then all paths between v and w have odd length,
whereas all paths between v and itself have even length.
So, any power of adjacency matrix A contains zeros, showing A is not primitive.

(b) We can write the adjacency matrix of G as $A = \begin{bmatrix} O & B \\ B^T & O \end{bmatrix}$ and partition the eigenvector $\mathbf{v}$ of λ as $\begin{bmatrix} \mathbf{v}_1 \\ \mathbf{v}_2 \end{bmatrix}$, where $\mathbf{v}_1$, $\mathbf{v}_2$ are column vectors and $\mathbf{v}_2$ has the same number of rows as B does columns.
Then $A\mathbf{v} = \lambda\mathbf{v} \Leftrightarrow \begin{bmatrix} B\mathbf{v}_2 \\ B^T\mathbf{v}_1 \end{bmatrix} = \begin{bmatrix} \lambda\mathbf{v}_1 \\ \lambda\mathbf{v}_2 \end{bmatrix}$, so $B\mathbf{v}_2 = \lambda\mathbf{v}_1$ and $B^T\mathbf{v}_1 = \lambda\mathbf{v}_2$.
Then $-\lambda \begin{bmatrix} \mathbf{v}_1 \\ -\mathbf{v}_2 \end{bmatrix} = \begin{bmatrix} -B\mathbf{v}_2 \\ B^T\mathbf{v}_1 \end{bmatrix} = A \begin{bmatrix} \mathbf{v}_1 \\ -\mathbf{v}_2 \end{bmatrix}$,
showing that $-\lambda$ is an eigenvalue of A with corresponding eigenvector $\begin{bmatrix} \mathbf{v}_1 \\ -\mathbf{v}_2 \end{bmatrix}$.

38. (a) Each column of A sums to k, since each vertex connects to k others.
Thus $A = kP$, where each of P's columns sum to 1.
By Theorem 4.30, 1 is an eigenvalue of P, so k is an eigenvalue of A.

(b) If A is primitive, then the matrix P defined in part (a) is regular.
So, Theorem 4.3(b) guarantees that the other eigenvalues of P are strictly less than 1.
Therefore, the other eigenvalues of $A = kP$ are strictly less than k.

39. This exploration and the subsequent explanation are left to the reader.

40. (a) The (i, j) entry of $|cA|$ is $|ca_{ij}| = |c|\,|a_{ij}|$, so $|cA| = |c|\,|A|$.

(b) The (i, j) entry of $|A + B|$ is $|a_{ij} + b_{ij}| \le |a_{ij}| + |b_{ij}|$, so $|A + B| \le |A| + |B|$.

(c) The ith entry of $|A\mathbf{x}|$ is $|a_{i1}\mathbf{x}_1 + \cdots + a_{in}\mathbf{x}_n| \le |a_{i1}\mathbf{x}_1| + \cdots + |a_{in}\mathbf{x}_n|$, so $|A\mathbf{x}| \le |A|\,|\mathbf{x}|$.

(d) The (i, j) entry of $|AB|$ is $|a_{i1}b_{1j} + \cdots + a_{in}b_{nj}| \le |a_{i1}b_{1j}| + \cdots + |a_{in}b_{nj}|$,
so $|AB| \le |A|\,|B|$.

(e) The (i, j) entry of AC is $a_{i1}c_{1j} + \cdots + a_{in}c_{nj} \ge b_{i1}d_{i1} + \cdots + b_{in}d_{nj}$, so $AC \ge BD$.
Similarly $b_{i1}d_{i1} + \cdots + b_{in}d_{nj} \ge 0$, so $BD \ge 0$.

41. $x_0 = 1$, $x_1 = 2$, $x_2 = 4$, $x_3 = 8$, $x_4 = 16$.

42. $a_1 = 128$, $a_2 = 64$, $a_3 = 32$, $a_4 = 16$.

43. $y_0 = 0$, $y_1 = 1$, $y_2 = 1$, $y_3 = 0$, $y_4 = -1$.

44. $b_0 = 1$, $b_1 = 1$, $b_2 = 3$, $b_3 = 7$, $b_4 = 17$.

45. We compute the eigenvalues of $A = \begin{bmatrix} 3 & 4 \\ 1 & 0 \end{bmatrix}$ to be -1 and 4.

So, we can immediately write $x_n = c_1(-1)^n + c_2 4^n$, where c_1 and c_2 are to be determined.
Using the initial conditions, we have:
$0 = x_0 = c_1(-1)^0 + c_2 4^0 = c_1 + c_2$ and $5 = x_1 = c_1(-1)^1 + c_2 4^1 = -c_1 + 4c_2$.
We solve the system to find that $c_1 = -1$ and $c_2 = 1$, so $x_n = -(-1)^n + 4^n$.

46. The recurrence is $x_n - 4x_{n-1} + 3x_{n-2} = 0$,
so, the characteristic equation is $x^2 - 4x - 3 = 0$ and the eigenvalues are 1 and 3.
Thus, $x_n = c_1(1)^n + c_2(3)^n$.
Using the initial conditions, we have:
$x_0 = 0 = c_1(1)^0 + c_2(3)^0 = c_1 + c_2$ and $x_1 = 1 = c_1(1)^1 + c_2(3)^1$.
We solve this system to find that $c_1 = -\frac{1}{2}$ and $c_2 = \frac{1}{2}$.
Thus, $x_n = -\frac{1}{2}(1)^n + \frac{1}{2}(3)^n$.

47. The recurrence is $y_n - 4y_{n-1} + 4y_{n-2} = 0$, so the only eigenvalue is 2.
Thus, by Theorem 4.38(b), $y_n = c_1(2)^n + c_2 n 2^n$.
We solve $y_1 = 1 = 2c_1 + 2c_2$ and $y_2 = 6 = 4c_1 + 8c_2$ to get $c_1 = -\frac{1}{2}$ and $c_2 = 1$.
Thus, $y_n = -\frac{1}{2}(2)^n + n2^n$.

48. The recurrence is $a_n - a_{n-1} + \frac{1}{4}a_{n-2} = 0$, so the eigenvalue is $\frac{1}{2}$.
So, $a_n = c_1\left(\frac{1}{2}\right)^n + c_2 n\left(\frac{1}{2}\right)^n$.
We solve $a_0 = 4 = c_1$ and $a_1 = 1 = \frac{1}{2}c_1 + \frac{1}{2}c_2$ to get $c_1 = 4$, $c_2 = -2$.
Thus, $a_n = 4\left(\frac{1}{2}\right)^n - 2n\left(\frac{1}{2}\right)^n$.

49. The recurrence is $b_n - 2b_{n-1} - 2b_{n-2} = 0$, so the eigenvalues are $1 \pm \sqrt{3}$. We solve:
$b_0 = 0 = c_1 + c_2$ and $b_1 = 1 = \left(1 - \sqrt{3}\right)c_1 + \left(1 + \sqrt{3}\right)c_2$ to get $c_1 = -\frac{\sqrt{3}}{6}$, $c_2 = \frac{\sqrt{3}}{6}$.
Thus, $b_n = -\frac{\sqrt{3}}{6}\left(1 - \sqrt{3}\right)^n + \frac{\sqrt{3}}{6}\left(1 + \sqrt{3}\right)^n$.

50. The recurrence is $y_n - y_{n-1} + y_{n-2} = 0$, so the eigenvalues are $\frac{1}{2} \pm \frac{\sqrt{3}}{2}i$. We solve:
$y_0 = 0 = c_1 + c_2$, $y_1 = 1 = c_1\left(\frac{1}{2} - \frac{\sqrt{3}}{2}i\right) + c_2\left(\frac{1}{2} + \frac{\sqrt{3}}{2}i\right)$ to get $c_1 = \frac{\sqrt{3}}{3}i$ and $c_2 = -\frac{\sqrt{3}}{3}i$.
Thus, $y_n = \frac{\sqrt{3}}{3}i\left(\frac{1}{2} - \frac{\sqrt{3}}{2}i\right)^n - \frac{\sqrt{3}}{3}i\left(\frac{1}{2} + \frac{\sqrt{3}}{2}i\right)^n$.
Using this formula, we calculate:
$y_0 = 0$, $y_1 = 1$, $y_2 = 1$, $y_3 = 0$, $y_4 = -1$.

51. Since A is diagonalizable, we have $P^{-1}AP = D = \begin{bmatrix} \lambda_1 & 0 \\ 0 & \lambda_2 \end{bmatrix}$.

Suppose $P = \begin{bmatrix} a & b \\ c & d \end{bmatrix}$, so $P^{-1} = \frac{1}{ad-bc}\begin{bmatrix} d & -b \\ -c & a \end{bmatrix}$ Then:

$A^k = PD^kP^{-1} = \frac{1}{ad-bc}\begin{bmatrix} a & b \\ c & d \end{bmatrix}\begin{bmatrix} \lambda_1^k & 0 \\ 0 & \lambda_2^k \end{bmatrix}\begin{bmatrix} d & -b \\ -c & a \end{bmatrix} = \frac{1}{ad-bc}\begin{bmatrix} ad\lambda_1^k - bc\lambda_2^k & -ab\lambda_1^k + ab\lambda_2^k \\ cd\lambda_1^k - cd\lambda_2^k & -bc\lambda_1^k + ad\lambda_2^k \end{bmatrix}$.

If the two given terms of the sequence are x_{j-1} and x_j, then for any n we have

$$\begin{aligned}
\mathbf{x}_n &= A^{n-j}\mathbf{x}_j = \frac{1}{ad-bc}\begin{bmatrix} ad\lambda_1^{n-j} - bc\lambda_2^{n-j} & -ab\lambda_1^{n-j} + ab\lambda_2^{n-j} \\ cd\lambda_1^{n-j} - cd\lambda_2^{n-j} & -bc\lambda_1^{n-j} + ad\lambda_2^{n-j} \end{bmatrix}\begin{bmatrix} x_j \\ x_{j-1} \end{bmatrix} \\
&= \frac{1}{ad-bc}\begin{bmatrix} \left(ad\lambda_1^{n-j} - bc\lambda_2^{n-j}\right)x_j + \left(-ab\lambda_1^{n-j} + ab\lambda_2^{n-j}\right)x_{j-1} \\ \left(cd\lambda_1^{n-j} - cd\lambda_2^{n-j}\right)x_j + \left(-bc\lambda_1^{n-j} + ad\lambda_2^{n-j}\right)x_{j-1} \end{bmatrix} \\
&= \frac{1}{ad-bc}\begin{bmatrix} \left[(adx_j - abx_{j-1})\lambda_1^{-j}\right]\lambda_1^n + \left[(-bcx_j + abx_{j-1})\lambda_2^{-j}\right]\lambda_2^n \\ \left[(cdx_j - bcx_{j-1})\lambda_1^{-j}\right]\lambda_1^n + \left[(-cdx_j + adx_{j-1})\lambda_2^{-j}\right]\lambda_2^n \end{bmatrix}.
\end{aligned}$$

Thus, for constants $c_1 = \frac{adx_j - abx_{j-1}}{ad-bc}\lambda_1^{-j}$ and $c_2 = -\frac{(bcx_j - abx_{j-1})}{ad-bc}\lambda_2^{-j}$,
we have $x_n = c_1\lambda_1^n + c_2\lambda_2^n$ as required.

52. The scalars c_1 and c_2 were found explicitly in Exercise 51.

53. (a) For $n = 1$, $A = \begin{bmatrix} 1 & 1 \\ 1 & 0 \end{bmatrix} = \begin{bmatrix} f_2 & f_1 \\ f_1 & f_0 \end{bmatrix}$. So, assume that $A^k = \begin{bmatrix} f_{k+1} & f_k \\ f_k & f_{k-1} \end{bmatrix}$. Then:

$$A^{k+1} = A^kA = \begin{bmatrix} f_{k+1} & f_k \\ f_k & f_{k-1} \end{bmatrix}\begin{bmatrix} 1 & 1 \\ 1 & 0 \end{bmatrix} = \begin{bmatrix} f_{k+1} + f_k & f_{k+1} \\ f_k + f_{k-1} & f_k \end{bmatrix} = \begin{bmatrix} f_{k+2} & f_{k+1} \\ f_{k+1} & f_k \end{bmatrix}.$$

(b) $\det A = -1$, so $\det(A^n) = (\det A)^n \Leftrightarrow \begin{vmatrix} f_{n+1} & f_n \\ f_n & f_{n-1} \end{vmatrix} = f_{n+1}f_{n-1} - f_n^2 = (-1)^n$.

(c) If we interpret the 5×13 "rectangle" as being composed of two "triangles," we see that, in fact, they are not triangles.
The "diagonal" of the "rectangle" is not straight because
the 8×3 triangular pieces are not similar to the entire triangles (in particular, $\frac{5}{13} \neq \frac{3}{8}$).
So there is empty space along the diagonal of the figure (1 square unit of space, in fact).

54. (a) $t_1 = 1$, $t_2 = 3$, $t_3 = 5$, $t_4 = 11$, $t_5 = 21$.
We can interpret t_0 as the number of ways to tile a 1×0 rectangle.
Since all 1×0 rectangles are tiled the same way, $t_0 = 1$.

(b) If we add one square to an $(n-1) \times 1$ rectangle,
we can complete the tiling in only one way, using a 1×1 tile.
If we add two squares to an $(n-2) \times 1$ rectangle,
we can complete the tiling in three ways:
using one or the other 1×2 tile or using two 1×1 tiles.
But the latter completion is identical to the $(n-1) \times 1$ case.
Thus, $t_n = t_{n-1} + 2t_{n-2}$.

(c) The eigenvalues are -1 and 2, so we solve
$t_1 = 1 = -c_1 + 2c_2$ and $t_2 = 3 = c_1 + 4c_2$ to get $c_1 = \frac{1}{3}$ and $c_2 = \frac{2}{3}$.
Thus, $t_n = \frac{1}{3}(-1)^n + \frac{2}{3}(2)^n$.
This formula agrees with the data in part (a), including our value for t_0.

55. (a) $d_1 = 1$, $d_2 = 2$, $d_3 = 3$, $d_4 = 5$, $d_5 = 8$. As in Exercise 54, $d_0 = 1$.

(b) If we lengthen an $(n-1) \times 2$ rectangle by 1, we can complete the tiling in only one way.
If we lengthen an $(n-2) \times 2$ rectangle by 2, there is only one way to complete the tiling that is distinct from the $(n-1) \times 2$ case. Thus, $d_n = d_{n-1} + d_{n-2}$.

(c) This is simply the Fibonacci sequence shifted by 1,
so we know that $d_n = \frac{1}{\sqrt{5}}\left(\frac{1+\sqrt{5}}{2}\right)^{n+1} - \frac{1}{\sqrt{5}}\left(\frac{1-\sqrt{5}}{2}\right)^{n+1}$.
This formula agrees with the data in part (a), including our value for d_0.

56. $$A - \frac{1+\sqrt{5}}{2}I = \begin{bmatrix} 1 - \frac{1+\sqrt{5}}{2} & 1 \\ 1 & -\frac{1+\sqrt{5}}{2} \end{bmatrix} \longrightarrow \begin{bmatrix} 1 & \frac{-1-\sqrt{5}}{2} \\ 0 & 0 \end{bmatrix} \text{ and}$$

$$A - \frac{1-\sqrt{5}}{2}I = \begin{bmatrix} 1 - \frac{1-\sqrt{5}}{2} & 1 \\ 1 & -\frac{1-\sqrt{5}}{2} \end{bmatrix} \longrightarrow \begin{bmatrix} 1 & \frac{-1+\sqrt{5}}{2} \\ 0 & 0 \end{bmatrix}.$$

So, the eigenvectors are $\mathbf{v}_1 = \begin{bmatrix} \frac{1+\sqrt{5}}{2} \\ 1 \end{bmatrix}$ and $\mathbf{v}_2 = \begin{bmatrix} \frac{1-\sqrt{5}}{2} \\ 1 \end{bmatrix}$.

$$\text{Thus, } \lim_{k\to\infty} \frac{\mathbf{x}_k}{\lambda_1^k} = \lim_{k\to\infty} \frac{\begin{bmatrix} \frac{1}{\sqrt{5}}\left(\frac{1+\sqrt{5}}{2}\right)^k - \frac{1}{\sqrt{5}}\left(\frac{1-\sqrt{5}}{2}\right)^k \\ \frac{1}{\sqrt{5}}\left(\frac{1+\sqrt{5}}{2}\right)^{k-1} - \frac{1}{\sqrt{5}}\left(\frac{1-\sqrt{5}}{2}\right)^{k-1} \end{bmatrix}}{\left(\frac{1+\sqrt{5}}{2}\right)^k} = \begin{bmatrix} \frac{\sqrt{5}}{5} \\ \frac{5-\sqrt{5}}{10} \end{bmatrix}.$$

So, since $\mathbf{v}_1 = \begin{bmatrix} \frac{1+\sqrt{5}}{2} \\ 1 \end{bmatrix}$, we check $c_1 = \frac{5-\sqrt{5}}{10}$ satisfies the equation $\frac{1+\sqrt{5}}{2}c_1 = \frac{1}{5}\sqrt{5}$.

57. The coefficient matrix is $\begin{bmatrix} 1 & 3 \\ 2 & 2 \end{bmatrix}$, which has eigenvalues $\lambda_1 = 4$ and $\lambda_2 = -1$

with corresponding eigenvectors $\mathbf{v}_1 = \begin{bmatrix} 1 \\ 1 \end{bmatrix}$ and $\mathbf{v}_2 = \begin{bmatrix} 3 \\ -2 \end{bmatrix}$.

Thus, $\mathbf{y}' = P^{-1}AP\mathbf{y} = D\mathbf{y}$ is the system $y_1' = 4y_1$, $y_2' = -y_2$ which has the general solution

$\mathbf{y} = \begin{bmatrix} C_1e^{4t} \\ C_2e^{-t} \end{bmatrix}$ and $\mathbf{x} = P\mathbf{y} = \begin{bmatrix} 1 & 3 \\ 1 & -2 \end{bmatrix}\begin{bmatrix} C_1e^{4t} \\ C_2e^{-t} \end{bmatrix} = \begin{bmatrix} C_1e^{4t} + 3C_2e^{-t} \\ C_1e^{4t} - 2C_2e^{-t} \end{bmatrix}$.

We substitute the initial conditions $x(0) = 0$ and $y(0) = 5$ to find that $C_1 = 3$ and $C_2 = -1$.

So, the solution to the initial value problem is $x = 3e^{4t} - 3e^{-t}$, $y = 3e^{4t} + 2e^{-t}$.

58. The coefficient matrix is $\begin{bmatrix} 2 & -1 \\ -1 & 2 \end{bmatrix}$, which has eigenvalues $\lambda_1 = 3$ and $\lambda_2 = 1$

with corresponding eigenvectors $\mathbf{v}_1 = \begin{bmatrix} 1 \\ -1 \end{bmatrix}$ and $\mathbf{v}_2 = \begin{bmatrix} 1 \\ 1 \end{bmatrix}$.

Thus, $\mathbf{y} = \begin{bmatrix} C_1e^{3t} \\ C_2e^{-t} \end{bmatrix}$ and $\mathbf{x} = P\mathbf{y} = \begin{bmatrix} C_1e^{3t} + C_2e^{t} \\ -C_1e^{3t} + C_2e^{t} \end{bmatrix}$.

We substitute $x(0) = 1$ and $y(0) = 1$ to find that $C_1 = 0$ and $C_2 = 1$.

So, the solution to the initial value problem is $x = e^t$, $y = e^t$.

59. The coefficient matrix is $\begin{bmatrix} 1 & 1 \\ 1 & -1 \end{bmatrix}$, which has eigenvalues $\lambda_1 = \sqrt{2}$ and $\lambda_2 = -\sqrt{2}$

with corresponding eigenvectors $\mathbf{v}_1 = \begin{bmatrix} 1 \\ -1+\sqrt{2} \end{bmatrix}$ and $\mathbf{v}_2 = \begin{bmatrix} 1 \\ -1-\sqrt{2} \end{bmatrix}$.

Thus, $\mathbf{y} = \begin{bmatrix} C_1e^{\sqrt{2}t} \\ C_2e^{-\sqrt{2}t} \end{bmatrix}$ and $\mathbf{x} = P\mathbf{y} = \begin{bmatrix} C_1e^{\sqrt{2}t} + C_2e^{-\sqrt{2}t} \\ (-1+\sqrt{2})\,C_1e^{\sqrt{2}t} + (-1-\sqrt{2})\,C_2e^{-\sqrt{2}t} \end{bmatrix}$.

We substitute $x_1(0) = 1$ and $x_2(0) = 0$ to find that $C_1 = \frac{2+\sqrt{2}}{4}$ and $C_2 = \frac{2-\sqrt{2}}{4}$.

So, the solution to the initial value problem is

$x_1 = \frac{2+\sqrt{2}}{4}e^{\sqrt{2}t} + \frac{2-\sqrt{2}}{4}e^{-\sqrt{2}t}$,

$x_2 = (-1+\sqrt{2})\left(\frac{2+\sqrt{2}}{4}\right)e^{\sqrt{2}t} + (-1-\sqrt{2})\left(\frac{2-\sqrt{2}}{4}\right)e^{-\sqrt{2}t} = \frac{\sqrt{2}}{4}e^{\sqrt{2}t} - \frac{\sqrt{2}}{4}e^{-\sqrt{2}t}$.

60. The coefficient matrix is $\begin{bmatrix} 1 & -1 \\ 1 & 1 \end{bmatrix}$, which has eigenvalues $\lambda_1 = 1 + i$ and $\lambda_2 = 1 - i$

with corresponding eigenvectors $\mathbf{v}_1 = \begin{bmatrix} 1 \\ -i \end{bmatrix}$ and $\mathbf{v}_2 = \begin{bmatrix} 1 \\ i \end{bmatrix}$.

Thus, $\mathbf{y} = \begin{bmatrix} C_1 e^{(1+i)t} \\ C_2 e^{(1-i)t} \end{bmatrix}$ and $\mathbf{x} = P\mathbf{y} = \begin{bmatrix} C_1 e^{(1+i)t} + C_2 e^{(1-i)t} \\ -iC_1 e^{(1+i)t} + iC_2 e^{(1-i)t} \end{bmatrix}$.

We substitute $y_1(0) = 1$ and $y_2(0) = 1$ to find that $C_1 = \frac{1}{2} + \frac{1}{2}i$ and $C_2 = \frac{1}{2} - \frac{1}{2}i$.

So, the solution to the initial value problem is

$$\begin{aligned} y_1 &= \left(\tfrac{1}{2} + \tfrac{1}{2}i\right) e^{(1+i)t} + \left(\tfrac{1}{2} - \tfrac{1}{2}i\right) e^{(1-i)t} \\ &= \left(\tfrac{1}{2} + \tfrac{1}{2}i\right) e^t (\cos t + i \sin t) + \left(\tfrac{1}{2} - \tfrac{1}{2}i\right) e^t (\cos(-t) + i \sin(-t)) = e^t (\cos t - \sin t), \\ y_2 &= -i\left(\tfrac{1}{2} + \tfrac{1}{2}i\right) e^{(1+i)t} + i\left(\tfrac{1}{2} - \tfrac{1}{2}i\right) e^{(1-i)t} \\ &= -i\left(\tfrac{1}{2} + \tfrac{1}{2}i\right) e^t (\cos t + i \sin t) + i\left(\tfrac{1}{2} - \tfrac{1}{2}i\right) e^t (\cos(-t) + i \sin(-t)) = e^t (\cos t + \sin t). \end{aligned}$$

61. The coefficient matrix $\begin{bmatrix} 0 & 1 & -1 \\ 1 & 0 & 1 \\ 1 & 1 & 0 \end{bmatrix}$ has eigenvalues $\lambda_1 = 1$, $\lambda_2 = 0$, $\lambda_3 = -1$

with corresponding eigenvectors $\mathbf{v}_1 = \begin{bmatrix} 0 \\ 1 \\ 1 \end{bmatrix}$, $\mathbf{v}_2 = \begin{bmatrix} 1 \\ -1 \\ -1 \end{bmatrix}$, and $\mathbf{v}_3 = \begin{bmatrix} 1 \\ -1 \\ 0 \end{bmatrix}$.

Thus, $\mathbf{y} = \begin{bmatrix} C_1 e^t \\ C_2 \\ C_3 e^{-t} \end{bmatrix}$ and $\mathbf{x} = P\mathbf{y} = \begin{bmatrix} C_2 + C_3 e^{-t} \\ C_1 e^t - C_2 - C_3 e^{-t} \\ C_1 e^t - C_2 \end{bmatrix}$.

We substitute $x(0) = 1$, $y(0) = 0$, and $z(0) = -1$ to find that $C_1 = 1$, $C_2 = 2$, and $C_3 = -1$.

So, the solution to the initial value problem is $x = 2 - e^{-t}$, $y = e^t - 2 + e^{-t}$, and $z = e^t - 2$.

62. The coefficient matrix is $\begin{bmatrix} 1 & 0 & 3 \\ 1 & -2 & 1 \\ 3 & 0 & 1 \end{bmatrix}$, which has eigenvalues $\lambda_1 = 4$ and $\lambda_2 = \lambda_3 = -2$,

with corresponding eigenvectors $\mathbf{v}_1 = \begin{bmatrix} 3 \\ 1 \\ 3 \end{bmatrix}$, $\mathbf{v}_2 = \begin{bmatrix} 1 \\ 0 \\ -1 \end{bmatrix}$, and $\mathbf{v}_3 = \begin{bmatrix} 0 \\ 1 \\ 0 \end{bmatrix}$.

Thus, $\mathbf{y} = \begin{bmatrix} C_1 e^{4t} \\ C_2 e^{-2t} \\ C_3 e^{-2t} \end{bmatrix}$ and $\mathbf{x} = P\mathbf{y} = \begin{bmatrix} 3C_1 e^{4t} + C_2 e^{-2t} \\ C_1 e^{4t} + C_3 e^{-2t} \\ 3C_1 e^{4t} - C_2 e^{-2t} \end{bmatrix}$.

We substitute $x(0) = 2$, $y(0) = 3$, and $z(0) = 4$ to find that $C_1 = 1$, $C_2 = -1$, and $C_3 = 2$.

So, the solution is $x = 3e^{4t} - e^{-2t}$, $y = e^{4t} + 2e^{-2t}$, and $z = 3e^{4t} + e^{-2t}$.

63. (a) The coefficient matrix is $\begin{bmatrix} 1.2 & -0.2 \\ -0.2 & 1.5 \end{bmatrix}$, which has eigenvalues $\lambda_1 = 1.6$ and $\lambda_2 = 1.1$ with corresponding eigenvectors $\mathbf{v}_1 = \begin{bmatrix} 1 \\ -2 \end{bmatrix}$ and $\mathbf{v}_2 = \begin{bmatrix} 2 \\ 1 \end{bmatrix}$.

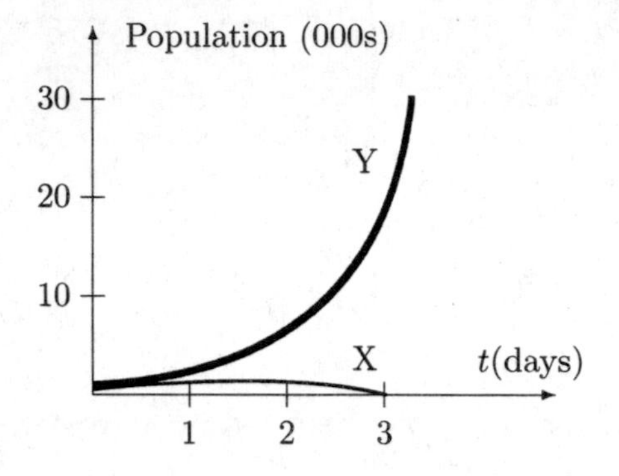

Thus, $\mathbf{y} = \begin{bmatrix} C_1e^{1.6t} \\ C_2e^{1.1t} \end{bmatrix}$

and

$\mathbf{x} = P\mathbf{y} = \begin{bmatrix} C_1e^{1.6t} + 2C_2e^{1.1t} \\ -2C_1e^{1.6t} + C_2e^{1.1t} \end{bmatrix}$.

We substitute $x(0) = 400$ and $y(0) = 500$ to find that $C_1 = -120$ and $C_2 = 260$.
So, the solution is $x = -120e^{1.6t} + 520e^{1.1t}$, $y = 240e^{1.6t} + 260e^{1.1t}$.
We see from the graph:
Bacteria X dies out after about 3 days, while bacteria Y increases indefinitely.

(b) Solving the general case, we find that $C_1 = \frac{1}{5}a - \frac{2}{5}b$ and $C_2 = \frac{2}{5}a + \frac{1}{5}b$.
Since $e^{1.6t}$ term is dominant, the fate of the populations depends on the positivity of C_1.

Thus, if $a = 2b$, then both strains of bacteria will thrive.
Otherwise, only one strain of bacteria will survive indefinitely.
$a > 2b \Rightarrow$, bacteria X survives, $a < 2b \Rightarrow$ bacteria Y survives.

64. The coefficient matrix is $\begin{bmatrix} -0.8 & 0.4 \\ 0.4 & -0.2 \end{bmatrix}$, which has eigenvalues $\lambda_1 = -1$ and $\lambda_2 = 0$

with corresponding eigenvectors $\mathbf{v}_1 = \begin{bmatrix} -2 \\ 1 \end{bmatrix}$ and $\mathbf{v}_2 = \begin{bmatrix} 1 \\ 2 \end{bmatrix}$.

Thus, $\mathbf{y} = \begin{bmatrix} C_1e^{-t} \\ C_2 \end{bmatrix}$ and $\mathbf{x} = P\mathbf{y} = \begin{bmatrix} -2C_1e^{-t} + C_2 \\ C_1e^{-t} + 2C_2 \end{bmatrix}$.

We substitute $x(0) = 15$ and $y(0) = 10$ to find that $C_1 = -4$ and $C_2 = 7$.

So, the solution is $x = 8e^{-t} + 7$, $y = -4e^{-t} + 14$.

So, the populations eventually stabilize at 7 of X and 14 of Y.

65. $\mathbf{x}' = A\mathbf{x} + \mathbf{b} = \begin{bmatrix} x+y-30 \\ -x+y-10 \end{bmatrix}$. We wish to have $\mathbf{x}' = A\mathbf{u}$ where $\mathbf{u} = \begin{bmatrix} u \\ v \end{bmatrix} = \mathbf{x} + \begin{bmatrix} a \\ b \end{bmatrix}$, so we solve $\begin{bmatrix} x+y-30 \\ -x+y-10 \end{bmatrix} = \begin{bmatrix} (x+a)+(y+b) \\ -(x+a)+(y+b) \end{bmatrix}$ to get $a = -10$, $b = -20$.

Our new initial conditions are $u(0) = x(0) - 10 = 10$ and $v(0) = y(0) - 20 = 10$.

A has eigenvalues $1+i$ and $1-i$ corresponding to eigenvectors $\begin{bmatrix} 1 \\ i \end{bmatrix}$ and $\begin{bmatrix} 1 \\ -i \end{bmatrix}$.

So, our solution has the form $\mathbf{u}(t) = C_1 e^{(1+i)t} \begin{bmatrix} 1 \\ i \end{bmatrix} + C_2 e^{(1-i)t} \begin{bmatrix} 1 \\ -i \end{bmatrix}$.

From $\mathbf{u}(0) = \begin{bmatrix} 10 \\ 10 \end{bmatrix}$ we get $C_1 = 5 - 5i$ and $C_2 = 5 + 5i$.

So applying Euler's Formula and expanding, we get:

$$\mathbf{u}(t) = (5-5i)\, e^{(1+i)t} \begin{bmatrix} 1 \\ i \end{bmatrix} + (5+5i)\, e^{(1-i)t} \begin{bmatrix} 1 \\ -i \end{bmatrix}$$

$$= (5-5i)\, e^{t} (\cos t + i \sin t) \begin{bmatrix} 1 \\ i \end{bmatrix} + (5+5i)\, e^{t} (\cos(-t) + i \sin(-t)) \begin{bmatrix} 1 \\ -i \end{bmatrix}$$

$$= \begin{bmatrix} 10e^t \cos t + 10e^t \sin t \\ 10e^t \cos t - 10e^t \sin t \end{bmatrix}.$$

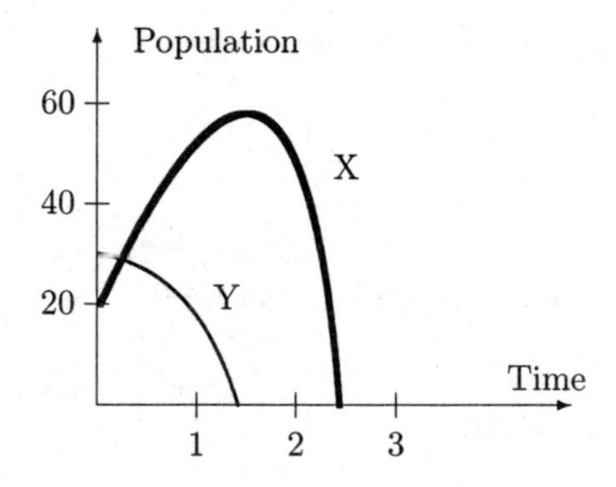

Rewriting in terms of the original variables, we get:

$$\mathbf{x}(t) = \mathbf{u}(t) - \begin{bmatrix} -10 \\ -20 \end{bmatrix}$$

$$= \begin{bmatrix} 10e^t \cos t + 10e^t \sin t + 10 \\ 10e^t \cos t - 10e^t \sin t + 20 \end{bmatrix}.$$

From the graph, we see that both populations die out.
Species Y after about 1.2 units of time and species X after about 2.4 units.

66. $\mathbf{x}' = A\mathbf{x} + \mathbf{b} = \begin{bmatrix} -x+y \\ -x-y+40 \end{bmatrix}$. Substituting $\mathbf{u} = \begin{bmatrix} u \\ v \end{bmatrix} = \mathbf{x} + \begin{bmatrix} -20 \\ -20 \end{bmatrix}$, we have $\mathbf{u}' = A\mathbf{u}$.

Our new initial conditions are $\mathbf{u}(0) = \mathbf{x}(0) + \begin{bmatrix} -20 \\ -20 \end{bmatrix} = \begin{bmatrix} -10 \\ 10 \end{bmatrix}$.

A has eigenvalues $-1+i$ and $-1-i$ corresponding to eigenvectors $\begin{bmatrix} -i \\ 1 \end{bmatrix}$ and $\begin{bmatrix} i \\ 1 \end{bmatrix}$.

So, our solution has the form $\mathbf{u}(t) = C_1 e^{(-1+i)t}\begin{bmatrix} -i \\ 1 \end{bmatrix} + C_2 e^{(-1-i)t}\begin{bmatrix} i \\ 1 \end{bmatrix}$.

From $\mathbf{u}(0) = \begin{bmatrix} -10 \\ 10 \end{bmatrix}$ we get $C_1 = 5 - 5i$ and $C_2 = 5 + 5i$.

So applying Euler's Formula and expanding, we get:

$$\begin{aligned}\mathbf{u}(t) &= (5-5i)\,e^{(-1+i)t}\begin{bmatrix} -i \\ 1 \end{bmatrix} + (5+5i)\,e^{(-1-i)t}\begin{bmatrix} i \\ 1 \end{bmatrix} \\ &= (5-5i)\,e^{-t}(\cos t + i\sin t)\begin{bmatrix} -i \\ 1 \end{bmatrix} + (5+5i)\,e^{-t}(\cos(-t) + i\sin(-t))\begin{bmatrix} i \\ 1 \end{bmatrix} \\ &= \begin{bmatrix} -10e^{-t}\cos t + 10e^{-t}\sin t \\ 10e^{-t}\cos t + 10e^{-t}\sin t \end{bmatrix}.\end{aligned}$$

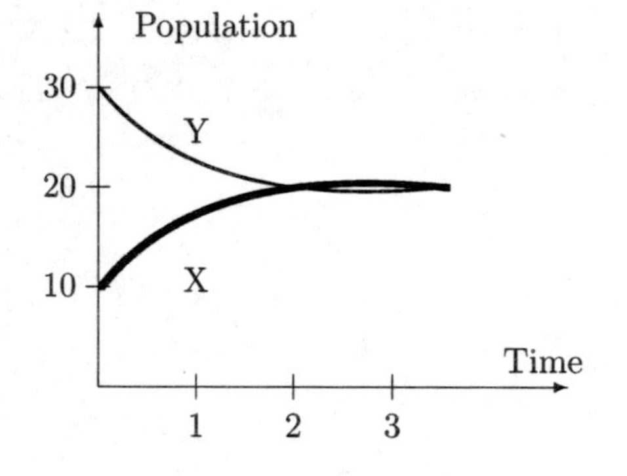

Rewriting in terms of the original variables, we get:

$$\begin{aligned}\mathbf{x}(t) &= \mathbf{u}(t) - \begin{bmatrix} -20 \\ -20 \end{bmatrix} \\ &= \begin{bmatrix} -10e^{-t}\cos t + 10e^{-t}\sin t + 20 \\ 10e^{-t}\cos t + 10e^{-t}\sin t + 20 \end{bmatrix}.\end{aligned}$$

From the graph, we see both populations eventually stabilize at 20.

67. (a) Making the change of variables $y = x'$ and $z = x$, we have $x'' + ax' + bx = 0 \Leftrightarrow y' + ay + bz = 0$ and $y' + az' + bz = 0 \Leftrightarrow y' = -ay - bz$ and $z' = y$.

(b) The coefficient matrix is $\begin{bmatrix} -a & -b \\ 1 & 0 \end{bmatrix}$,

whose characteristic polynomial is $\begin{vmatrix} -a-\lambda & -b \\ 1 & -\lambda \end{vmatrix} = \lambda^2 + \lambda a + b$.

68. We make the substitutions $y_1 = x^{(n-1)}$, $y_2 = x^{(n-2)}$, ..., $y_n = x$. Then the equation

$$x^{(n)} + a_{n-1}x^{(n-1)} + a_{n-2}x^{(n-2)} \cdots + a_2 x'' + a_1 x' + a_0 x = 0$$

can be written in n different ways:

$$\begin{aligned}
y_1' + a_{n-1}y_1 + a_{n-2}y_2 + \cdots + a_2 y_{n-2} + a_1 y_{n-1} + a_0 y_n &= 0 \\
y_1' + a_{n-1}y_2' + a_{n-2}y_2 + \cdots + a_2 y_{n-2} + a_1 y_{n-1} + a_0 y_n &= 0 \\
y_1' + a_{n-1}y_2' + a_{n-2}y_3' + \cdots + a_2 y_{n-2} + a_1 y_{n-1} + a_0 y_n &= 0 \\
&\vdots \\
y_1' + a_{n-1}y_2' + a_{n-2}y_3' + \cdots + a_2 y_{n-1}' + a_1 y_{n-1} + a_0 y_n &= 0 \\
y_1' + a_{n-1}y_2' + a_{n-2}y_3' + \cdots + a_2 y_{n-1}' + a_1 y_n' + a_0 y_n &= 0
\end{aligned}$$

which reduce to the n equations

$$\begin{aligned}
y_1' &= -a_{n-1}y_1 + a_{n-2}y_2 + \cdots + a_2 y_{n-2} - a_1 y_{n-1} - a_0 y_n \\
y_2' &= y_1 \\
y_3' &= y_2 \\
&\vdots \\
y_n' &= y_{n-1}
\end{aligned}$$

whose coefficient matrix is $\begin{bmatrix} -a_{n-1} & -a_{n-2} & \cdots & -a_1 & -a_0 \\ 1 & 0 & \cdots & 0 & 0 \\ 0 & 1 & \cdots & 0 & 0 \\ \vdots & \vdots & \ddots & \vdots & \vdots \\ 0 & 0 & \cdots & 1 & 0 \end{bmatrix}$, the companion matrix of $p(\lambda)$.

69. From Exercise 67, we know that the characteristic polynomial is $\lambda^2 - 5\lambda + 6$, so the eigenvalues are 2 and 3. The general solution is thus $x(t) = C_1 e^{2t} + C_2 e^{3t}$.

70. From Exercise 67, we know that the characteristic polynomial is $\lambda^2 + 4\lambda + 3$, so the eigenvalues are -3 and -1. The general solution is thus $x(t) = C_1 e^{-3t} + C_2 e^{-t}$.

71. We know that $A = \begin{bmatrix} 1 & 3 \\ 2 & 2 \end{bmatrix}$ has eigenvalues $\lambda_1 = 4$ and $\lambda_2 = -1$

with corresponding eigenvectors $\mathbf{v}_1 = \begin{bmatrix} 1 \\ 1 \end{bmatrix}$ and $\mathbf{v}_2 = \begin{bmatrix} 3 \\ -2 \end{bmatrix}$.

So by Theorem 4.41, the system $\mathbf{x}' = A\mathbf{x}$ has solution $\mathbf{x} = e^{At}\mathbf{x}(0)$.

In this case $\mathbf{x}(0) = \begin{bmatrix} 0 \\ 5 \end{bmatrix}$ and we calculate

$$e^{At} = P\left(e^t\right)^D P^{-1} = \begin{bmatrix} 1 & 3 \\ 1 & -2 \end{bmatrix} \begin{bmatrix} e^{4t} & 0 \\ 0 & e^{-t} \end{bmatrix} \begin{bmatrix} 1 & 3 \\ 1 & -2 \end{bmatrix}^{-1} = \begin{bmatrix} \frac{2}{5}e^{4t} + \frac{3}{5}e^{-t} & \frac{3}{5}e^{4t} - \frac{3}{5}e^{-t} \\ \frac{2}{5}e^{4t} - \frac{2}{5}e^{-t} & \frac{3}{5}e^{4t} + \frac{2}{5}e^{-t} \end{bmatrix}.$$

So, the solution is $\begin{bmatrix} \frac{2}{5}e^{4t} + \frac{3}{5}e^{-t} & \frac{3}{5}e^{4t} - \frac{3}{5}e^{-t} \\ \frac{2}{5}e^{4t} - \frac{2}{5}e^{-t} & \frac{3}{5}e^{4t} + \frac{2}{5}e^{-t} \end{bmatrix} \begin{bmatrix} 0 \\ 5 \end{bmatrix} = \begin{bmatrix} 3e^{-4t} - 3e^{-t} \\ 3e^{-t} + 2e^{-t} \end{bmatrix}$ as before.

72. $A = \begin{bmatrix} 2 & -1 \\ -1 & 2 \end{bmatrix}$, which has eigenvalues $\lambda_1 = 3$ and $\lambda_2 = 1$

with corresponding eigenvectors $\mathbf{v}_1 = \begin{bmatrix} 1 \\ -1 \end{bmatrix}$ and $\mathbf{v}_2 = \begin{bmatrix} 1 \\ 1 \end{bmatrix}$.

Thus, by Theorem 4.41, the solution is

$$\mathbf{x} = e^{At}\mathbf{x}(0) = \begin{bmatrix} 1 & 1 \\ -1 & 1 \end{bmatrix} \begin{bmatrix} e^{3t} & 0 \\ 0 & e^t \end{bmatrix} \begin{bmatrix} 1 & 1 \\ -1 & 1 \end{bmatrix}^{-1} \begin{bmatrix} 1 \\ 1 \end{bmatrix} = \begin{bmatrix} e^t \\ e^t \end{bmatrix}.$$

73. $A = \begin{bmatrix} 0 & 1 & -1 \\ 1 & 0 & 1 \\ 1 & 1 & 0 \end{bmatrix}$ has eigenvalues $\lambda_1 = 1$, $\lambda_2 = 0$, and $\lambda_3 = -1$

with corresponding eigenvectors $\mathbf{v}_1 = \begin{bmatrix} 0 \\ 1 \\ 1 \end{bmatrix}$, $\mathbf{v}_2 = \begin{bmatrix} 1 \\ -1 \\ -1 \end{bmatrix}$, and $\mathbf{v}_3 = \begin{bmatrix} 1 \\ -1 \\ 0 \end{bmatrix}$.

Thus, by Theorem 4.41, the solution is

$$\mathbf{x} = e^{At}\mathbf{x}(0) = \begin{bmatrix} 0 & 1 & 1 \\ 1 & -1 & -1 \\ 1 & -1 & 0 \end{bmatrix} \begin{bmatrix} e^t & 0 & 0 \\ 0 & 1 & 0 \\ 0 & 0 & e^{-t} \end{bmatrix} \begin{bmatrix} 0 & 1 & 1 \\ 1 & -1 & -1 \\ 1 & -1 & 0 \end{bmatrix}^{-1} \begin{bmatrix} 1 \\ 0 \\ -1 \end{bmatrix} = \begin{bmatrix} 2 - e^{-t} \\ e^t - 2 + e^{-t} \\ e^t - 2 \end{bmatrix}.$$

74. $A = \begin{bmatrix} 1 & 0 & 3 \\ 1 & -2 & 1 \\ 3 & 0 & 1 \end{bmatrix}$ has eigenvalues $\lambda_1 = 4$ and $\lambda_2 = \lambda_3 = -2$,

with corresponding eigenvectors $\mathbf{v}_1 = \begin{bmatrix} 3 \\ 1 \\ 3 \end{bmatrix}$, $\mathbf{v}_2 = \begin{bmatrix} 1 \\ 0 \\ -1 \end{bmatrix}$, and $\mathbf{v}_3 = \begin{bmatrix} 0 \\ 1 \\ 0 \end{bmatrix}$

Thus, by Theorem 4.41, the solution is

$$\mathbf{x} = e^{At}\mathbf{x}(0) = \begin{bmatrix} 3 & 1 & 0 \\ 1 & 0 & 1 \\ 3 & -1 & 0 \end{bmatrix} \begin{bmatrix} e^{4t} & 0 & 0 \\ 0 & e^{-2t} & 0 \\ 0 & 0 & e^{-2t} \end{bmatrix} \begin{bmatrix} 3 & 1 & 0 \\ 1 & 0 & 1 \\ 3 & -1 & 0 \end{bmatrix}^{-1} \begin{bmatrix} 2 \\ 3 \\ 4 \end{bmatrix} = \begin{bmatrix} 3e^{4t} - e^{-2t} \\ e^{4t} + 2e^{-2t} \\ 3e^{4t} + e^{-2t} \end{bmatrix}.$$

75. We follow the insights and methods outlined in Examples 4.48 through 4.51.

(a) Since $A = \begin{bmatrix} 2 & 1 \\ 0 & 3 \end{bmatrix}$, we have $\mathbf{x}_1 = A\mathbf{x}_0 = \begin{bmatrix} 3 \\ 3 \end{bmatrix}$, $\mathbf{x}_2 = A\mathbf{x}_1 = \begin{bmatrix} 9 \\ 9 \end{bmatrix}$, $\mathbf{x}_3 = A\mathbf{x}_2 = \begin{bmatrix} 27 \\ 27 \end{bmatrix}$.

(b) Since $A = \begin{bmatrix} 2 & 1 \\ 0 & 3 \end{bmatrix}$, we have $\mathbf{x}_1 = A\mathbf{x}_0 = \begin{bmatrix} 2 \\ 0 \end{bmatrix}$, $\mathbf{x}_2 = A\mathbf{x}_1 = \begin{bmatrix} 4 \\ 0 \end{bmatrix}$, $\mathbf{x}_3 = A\mathbf{x}_2 = \begin{bmatrix} 8 \\ 0 \end{bmatrix}$.

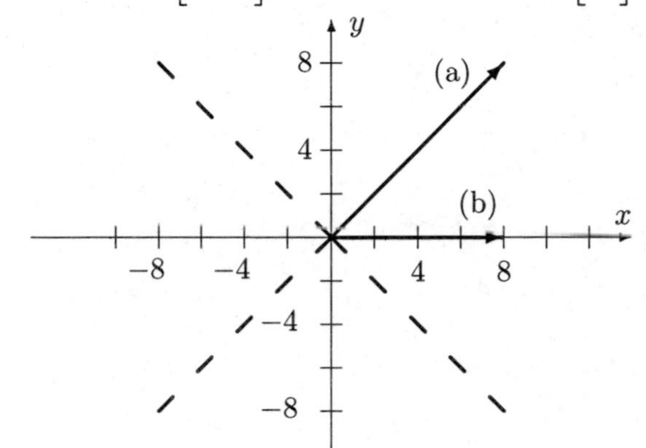

(c) Since $\lambda = 2, 3$, we see 0 is a repeller.

(d) Dashed lines are other trajectories.

76. We follow the insights and methods outlined in Examples 4.48 through 4.51.

(a) $\mathbf{x}_1 = A\mathbf{x}_0 = \begin{bmatrix} 0 \\ 0.5 \end{bmatrix}$, $\mathbf{x}_2 = A\mathbf{x}_1 = \begin{bmatrix} -0.25 \\ 0.25 \end{bmatrix}$, $\mathbf{x}_3 = A\mathbf{x}_2 = \begin{bmatrix} 0.500 \\ 0.125 \end{bmatrix}$.

(b) $\mathbf{x}_1 = A\mathbf{x}_0 = \begin{bmatrix} 0.5 \\ 0 \end{bmatrix}$, $\mathbf{x}_2 = A\mathbf{x}_1 = \begin{bmatrix} 0.25 \\ 0 \end{bmatrix}$, $\mathbf{x}_3 = A\mathbf{x}_2 = \begin{bmatrix} 0.125 \\ 0 \end{bmatrix}$.

(c) Since $\lambda = 0.5$, we see 0 is an attractor.

(d) Sketches are left to the reader.

77. We follow the insights and methods outlined in Examples 4.48 through 4.51.

(a) Since $A = \begin{bmatrix} 2 & -1 \\ -1 & 2 \end{bmatrix}$, we have $\mathbf{x}_1 = A\mathbf{x}_0 = \begin{bmatrix} 1 \\ 1 \end{bmatrix}$, $\mathbf{x}_2 = A\mathbf{x}_1 = \begin{bmatrix} 1 \\ 1 \end{bmatrix}$, $\mathbf{x}_3 = A\mathbf{x}_2 = \begin{bmatrix} 1 \\ 1 \end{bmatrix}$.

(b) $\mathbf{x}_1 = A\mathbf{x}_0 = \begin{bmatrix} 2 \\ -1 \end{bmatrix}$, $\mathbf{x}_2 = A\mathbf{x}_1 = \begin{bmatrix} 5 \\ -4 \end{bmatrix}$, $\mathbf{x}_3 = A\mathbf{x}_2 = \begin{bmatrix} 14 \\ -13 \end{bmatrix}$.

(c) Since $\lambda = 1, 3$, we see 0 is a saddle point.

(d) Sketches are left to the reader.

78. We follow the insights and methods outlined in Examples 4.48 through 4.51.

(a) Since $A = \begin{bmatrix} -4 & 2 \\ 1 & -3 \end{bmatrix}$, $\mathbf{x}_1 = A\mathbf{x}_0 = \begin{bmatrix} -2 \\ -2 \end{bmatrix}$, $\mathbf{x}_2 = A\mathbf{x}_1 = \begin{bmatrix} 4 \\ 4 \end{bmatrix}$, $\mathbf{x}_3 = A\mathbf{x}_2 = \begin{bmatrix} -8 \\ -8 \end{bmatrix}$.

(b) Since $A = \begin{bmatrix} -4 & 2 \\ 1 & -3 \end{bmatrix}$, $\mathbf{x}_1 = A\mathbf{x}_0 = \begin{bmatrix} -4 \\ 1 \end{bmatrix}$, $\mathbf{x}_2 = A\mathbf{x}_1 = \begin{bmatrix} 18 \\ -7 \end{bmatrix}$, $\mathbf{x}_3 = A\mathbf{x}_2 = \begin{bmatrix} -86 \\ 39 \end{bmatrix}$.

(c) Since $|\lambda| = 2, 5$, we see 0 is a repeller.

(d) Sketches are left to the reader.

79. We follow the insights and methods outlined in Examples 4.48 through 4.51.

(a) $\mathbf{x}_1 = A\mathbf{x}_0 = \begin{bmatrix} 0.5 \\ -1.0 \end{bmatrix}$, $\mathbf{x}_2 = A\mathbf{x}_1 = \begin{bmatrix} 1.75 \\ -0.50 \end{bmatrix}$, $\mathbf{x}_3 = A\mathbf{x}_2 = \begin{bmatrix} 3.125 \\ -1.750 \end{bmatrix}$.

(b) $\mathbf{x}_1 = A\mathbf{x}_0 = \begin{bmatrix} 1.5 \\ -1.0 \end{bmatrix}$, $\mathbf{x}_2 = A\mathbf{x}_1 = \begin{bmatrix} 3.25 \\ -1.50 \end{bmatrix}$, $\mathbf{x}_3 = A\mathbf{x}_2 = \begin{bmatrix} 6.375 \\ -3.250 \end{bmatrix}$.

(c) Since $|\lambda| = 0.5, 2$, we see 0 is a saddle point.

(d) Sketches are left to the reader.

80. We follow the insights and methods outlined in Examples 4.48 through 4.51.

(a) Since $A = \begin{bmatrix} 0.1 & 0.9 \\ 0.5 & 0.5 \end{bmatrix}$, we have $\mathbf{x}_1 = A\mathbf{x}_0 = \begin{bmatrix} 1 \\ 1 \end{bmatrix}$, $\mathbf{x}_2 = A\mathbf{x}_1 = \begin{bmatrix} 1 \\ 1 \end{bmatrix}$, $\mathbf{x}_3 = A\mathbf{x}_2 = \begin{bmatrix} 1 \\ 1 \end{bmatrix}$.

(b) $\mathbf{x}_1 = A\mathbf{x}_0 = \begin{bmatrix} 0.1 \\ 0.5 \end{bmatrix}$, $\mathbf{x}_2 = A\mathbf{x}_1 = \begin{bmatrix} 0.46 \\ 0.30 \end{bmatrix}$, $\mathbf{x}_3 = A\mathbf{x}_2 = \begin{bmatrix} 0.316 \\ 0.380 \end{bmatrix}$.

(c) Since $|\lambda| = 0.4, 1$, we see 0 is a saddle point.

(d) Sketches are left to the reader.

81. We follow the insights and methods outlined in Examples 4.48 through 4.51.

(a) $\mathbf{x}_1 = A\mathbf{x}_0 = \begin{bmatrix} 0.6 \\ 0.6 \end{bmatrix}$, $\mathbf{x}_2 = A\mathbf{x}_1 = \begin{bmatrix} 0.36 \\ 0.36 \end{bmatrix}$, $\mathbf{x}_3 = A\mathbf{x}_2 = \begin{bmatrix} 0.216 \\ 0.216 \end{bmatrix}$.

(b) $\mathbf{x}_1 = A\mathbf{x}_0 = \begin{bmatrix} 0.2 \\ -0.2 \end{bmatrix}$, $\mathbf{x}_2 = A\mathbf{x}_1 = \begin{bmatrix} -0.04 \\ -0.20 \end{bmatrix}$, $\mathbf{x}_3 = A\mathbf{x}_2 = \begin{bmatrix} -0.088 \\ -0.152 \end{bmatrix}$.

(c) Since $|\lambda| = 0.4, 0.6$, we see 0 is an attractor (every trajectory converges to 0).

(d) Sketches are left to the reader.

82. We follow the insights and methods outlined in Examples 4.48 through 4.51.

(a) $\mathbf{x}_1 = A\mathbf{x}_0 = \begin{bmatrix} -1.5 \\ 4.8 \end{bmatrix}$, $\mathbf{x}_2 = A\mathbf{x}_1 = \begin{bmatrix} -7.20 \\ 15.48 \end{bmatrix}$, $\mathbf{x}_3 = A\mathbf{x}_2 = \begin{bmatrix} -23.220 \\ 47.088 \end{bmatrix}$.

(b) $\mathbf{x}_1 = A\mathbf{x}_0 = \begin{bmatrix} 0.0 \\ 1.2 \end{bmatrix}$, $\mathbf{x}_2 = A\mathbf{x}_1 = \begin{bmatrix} -1.80 \\ 4.32 \end{bmatrix}$, $\mathbf{x}_3 = A\mathbf{x}_2 = \begin{bmatrix} -6.480 \\ 13.392 \end{bmatrix}$.

(c) Since $\lambda = 0.6, 3$, we see 0 is a saddle point.

(d) Sketches are left to the reader.

83. We follow the insights in the proof of Theorem 4.42 and the remarks following it.

Since $A = \begin{bmatrix} 1 & -1 \\ 1 & 1 \end{bmatrix}$ and $r = \sqrt{a^2+b^2}$, the scaling factor $r = \sqrt{1^2+1^2} = \sqrt{2}$.

So, we factor A as $A = \begin{bmatrix} r & 0 \\ 0 & r \end{bmatrix} \begin{bmatrix} \cos\theta & -\sin\theta \\ \sin\theta & \cos\theta \end{bmatrix} = \begin{bmatrix} \sqrt{2} & 0 \\ 0 & \sqrt{2} \end{bmatrix} \begin{bmatrix} \frac{1}{\sqrt{2}} & -\frac{1}{\sqrt{2}} \\ \frac{1}{\sqrt{2}} & \frac{1}{\sqrt{2}} \end{bmatrix}$.

Since $\cos\theta = \frac{1}{\sqrt{2}}$ and $\sin\theta = \frac{1}{\sqrt{2}}$, the angle of rotation $\theta = 45°$.

Finally, since $r = |\lambda| = \sqrt{2} > 1$, the origin is a spiral repeller.

Plots and sketches are left to the reader.

84. We follow the insights in the proof of Theorem 4.42 and the remarks following it.

Since $A = \begin{bmatrix} 0 & 0.5 \\ -0.5 & 0 \end{bmatrix}$ and $r = \sqrt{a^2+b^2}$, the scaling factor $r = \sqrt{0^2+(0.5)^2} = 0.5$.

So, we factor A as $A = \begin{bmatrix} r & 0 \\ 0 & r \end{bmatrix} \begin{bmatrix} \cos\theta & -\sin\theta \\ \sin\theta & \cos\theta \end{bmatrix} = \begin{bmatrix} 0.5 & 0 \\ 0 & 0.5 \end{bmatrix} \begin{bmatrix} 0 & 1 \\ -1 & 0 \end{bmatrix}$.

Since $\cos\theta = 0$ and $\sin\theta = -1$, the angle of rotation $\theta = 270°$.

Finally, since $r = |\lambda| = 0.5 < 1$, the origin is a spiral attractor.

Plots and sketches are left to the reader.

85. We follow the insights in the proof of Theorem 4.42 and the remarks following it.

Since $A = \begin{bmatrix} 1 & \sqrt{3} \\ -\sqrt{3} & 1 \end{bmatrix}$ and $r = \sqrt{a^2+b^2}$, the scaling factor $r = \sqrt{1^2+(\sqrt{3})^2} = 2$.

So, we factor A as $A = \begin{bmatrix} r & 0 \\ 0 & r \end{bmatrix} \begin{bmatrix} \cos\theta & -\sin\theta \\ \sin\theta & \cos\theta \end{bmatrix} = \begin{bmatrix} 2 & 0 \\ 0 & 2 \end{bmatrix} \begin{bmatrix} \frac{1}{2} & \frac{\sqrt{3}}{2} \\ -\frac{\sqrt{3}}{2} & \frac{1}{\sqrt{2}} \end{bmatrix}$.

Since $\cos\theta = \frac{1}{2}$ and $\sin\theta = -\frac{\sqrt{3}}{2}$, the angle of rotation $\theta = 300°$.

Finally, since $r = |\lambda| = 2 > 1$, the origin is a spiral repeller.

Plots and sketches are left to the reader.

86. We follow the insights in the proof of Theorem 4.42 and the remarks following it.

Since $A = \begin{bmatrix} -\frac{\sqrt{3}}{2} & -\frac{1}{2} \\ \frac{1}{2} & -\frac{\sqrt{3}}{2} \end{bmatrix}$ and $r = \sqrt{a^2+b^2}$, the scaling factor $r = \sqrt{\left(-\frac{\sqrt{3}}{2}\right)^2 + \left(\frac{1}{2}\right)^2} = 1$.

So, we factor A as $A = \begin{bmatrix} r & 0 \\ 0 & r \end{bmatrix} \begin{bmatrix} \cos\theta & -\sin\theta \\ \sin\theta & \cos\theta \end{bmatrix} = \begin{bmatrix} 1 & 0 \\ 0 & 1 \end{bmatrix} \begin{bmatrix} -\frac{\sqrt{3}}{2} & -\frac{1}{2} \\ \frac{1}{2} & -\frac{\sqrt{3}}{2} \end{bmatrix}$.

Since $\cos\theta = -\frac{\sqrt{3}}{2}$ and $\sin\theta = \frac{1}{2}$, the angle of rotation $\theta = 150°$.

Finally, since $r = |\lambda| = 1 = 1$, the origin is the orbital center.

Plots and sketches are left to the reader.

87. We follow the insights in the proof of Theorem 4.43 and the remarks following it.

Since $A = \begin{bmatrix} 0.1 & -0.2 \\ 0.1 & 0.3 \end{bmatrix}$ has eigenvalues $\lambda = 0.2 \pm 0.1i$ with eigenvector $\mathbf{x} = \begin{bmatrix} -1-i \\ 1+0i \end{bmatrix}$.

So $P = \begin{bmatrix} \text{Re}\,\mathbf{x} & \text{Im}\,\mathbf{x} \end{bmatrix} = \begin{bmatrix} -1 & -1 \\ 1 & 0 \end{bmatrix}$.

Therefore $C = P^{-1}AP = \begin{bmatrix} 0 & 1 \\ -1 & -1 \end{bmatrix} \begin{bmatrix} 0.1 & -0.2 \\ 0.1 & 0.3 \end{bmatrix} \begin{bmatrix} -1 & -1 \\ 1 & 0 \end{bmatrix} = \begin{bmatrix} 0.2 & -0.1 \\ 0.1 & 0.2 \end{bmatrix}$.

Since $|\lambda| = 0.2 < 1$, the origin is a spiral attractor. Plots and sketches are left to the reader.

88. We follow the insights in the proof of Theorem 4.43 and the remarks following it.

Since $A = \begin{bmatrix} 2 & 1 \\ -2 & 0 \end{bmatrix}$ has eigenvalues $\lambda = 1 \pm i$ with eigenvector $\mathbf{x} = \begin{bmatrix} -1-i \\ 2+0i \end{bmatrix}$.

So $P = \begin{bmatrix} \text{Re}\,\mathbf{x} & \text{Im}\,\mathbf{x} \end{bmatrix} = \begin{bmatrix} -1 & -1 \\ 2 & 0 \end{bmatrix}$.

Therefore $C = P^{-1}AP = \begin{bmatrix} 0 & \frac{1}{2} \\ -1 & -\frac{1}{2} \end{bmatrix} \begin{bmatrix} 2 & 1 \\ -2 & 0 \end{bmatrix} \begin{bmatrix} -1 & -1 \\ 2 & 0 \end{bmatrix} = \begin{bmatrix} 1 & 1 \\ -1 & 1 \end{bmatrix}$.

Since $|\lambda| = \sqrt{2} > 1$, the origin is a spiral repeller. Plots and sketches are left to the reader.

89. We follow the insights in the proof of Theorem 4.43 and the remarks following it.

Since $A = \begin{bmatrix} 1 & -1 \\ 1 & 0 \end{bmatrix}$ has eigenvalues $\lambda = \frac{1}{2} \pm \frac{\sqrt{3}}{2}i$ with eigenvector $\mathbf{x} = \begin{bmatrix} \frac{1}{2} - \frac{\sqrt{3}}{2}i \\ 1+0i \end{bmatrix}$.

So $P = \begin{bmatrix} \text{Re}\,\mathbf{x} & \text{Im}\,\mathbf{x} \end{bmatrix} = \begin{bmatrix} \frac{1}{2} & -\frac{\sqrt{3}}{2} \\ 1 & 0 \end{bmatrix}$.

Therefore $C = P^{-1}AP = \begin{bmatrix} 0 & 1 \\ -\frac{2}{\sqrt{3}} & \frac{1}{\sqrt{3}} \end{bmatrix} \begin{bmatrix} 1 & -1 \\ 1 & 0 \end{bmatrix} \begin{bmatrix} \frac{1}{2} & -\frac{\sqrt{3}}{2} \\ 1 & 0 \end{bmatrix} = \begin{bmatrix} \frac{1}{2} & -\frac{\sqrt{3}}{2} \\ \frac{\sqrt{3}}{2} & \frac{1}{2} \end{bmatrix}$.

Since $|\lambda| = 1$, the origin is an orbital center. Plots and sketches are left to the reader.

90. We follow the insights in the proof of Theorem 4.43 and the remarks following it.

Since $A = \begin{bmatrix} 0 & -1 \\ 1 & \sqrt{3} \end{bmatrix}$ has eigenvalues $\lambda = \frac{\sqrt{3}}{2} \pm \frac{1}{2}i$ with eigenvector $\mathbf{x} = \begin{bmatrix} -\frac{\sqrt{3}}{2} - \frac{1}{2}i \\ 1+0i \end{bmatrix}$.

So $P = \begin{bmatrix} \text{Re}\,\mathbf{x} & \text{Im}\,\mathbf{x} \end{bmatrix} = \begin{bmatrix} -\frac{\sqrt{3}}{2} & -\frac{1}{2} \\ 1 & 0 \end{bmatrix}$.

Therefore $C = P^{-1}AP = \begin{bmatrix} 0 & 1 \\ -2 & -\sqrt{3} \end{bmatrix} \begin{bmatrix} 0 & -1 \\ 1 & \sqrt{3} \end{bmatrix} \begin{bmatrix} -\frac{\sqrt{3}}{2} & -\frac{1}{2} \\ 1 & 0 \end{bmatrix} = \begin{bmatrix} \frac{\sqrt{3}}{2} & -\frac{1}{2} \\ \frac{1}{2} & \frac{\sqrt{3}}{2} \end{bmatrix}$.

Since $|\lambda| = 1$, the origin is an orbital center. Plots and sketches are left to the reader.

Chapter 4 Review

1. We will explain and give counter examples to justify our answers below.

   (a) ***False***. See Theorem 4.7 of Section 4.2.
   Not quite accurate. Theorem 4.7 of Section 4.2 states $\det(kA) = k^n A$.
   So letting $k = -1$, we see $\det(-A) = (-1)^n \det A$, where A is $n \times n$.

   (b) ***True***. See Theorem 4.8 in Section 4.2.
   This is precisely the statement of Theorem 4.8 in Section 4.2.

   (c) ***False***. See Theorem 4.3(b) in Section 4.2.
   Not necessarily. Theorem 4.3(b) states: $A \overset{R_i \leftrightarrow R_j}{\longrightarrow} B \Rightarrow \det B = -\det A$.
   So if B is obtained from A by one interchange of two columns, then $\det B = -\det A$.
   However, if we perform a second interchange, we get $\det B = -(-\det A) = \det A$.
   So, if $A \longrightarrow B$ by an *odd* number of column exchanges, then $\det B = -\det A$.
   But, if $A \longrightarrow B$ by an *even* number of column exchanges, then $\det B = \det A$.

   (d) ***False***. See Theorems 4.9 and 4.10 in Section 4.2.
   Theorem 4.9 asserts $\det(A^{-1}) = \frac{1}{\det A}$ while Theorem 4.10 asserts $\det A = \det(A^T)$.
   Therefore, $\det(A^{-1}) = \frac{1}{\det(A^T)} = [\det(A^T)]^{-1}$.

   (e) ***False***. Consider the counter example, $\begin{bmatrix} 0 & 1 \\ 0 & 0 \end{bmatrix}$.
   Q: Why is it obvious that 0 is the only eigenvalue of this matrix?
   Q: What is the general pattern underlying this specific counterexample?
   Q: If all the eigenvalues of a matrix A are 0, what can we conclude about $\det A$?

   (f) ***False***. See Exercise 7 in Section 4.3.
   A simple but constructive counter example is the matrix I.
   Since $I\mathbf{e}_i = \mathbf{e}_i$, all nonzero vectors correspond to the eigenvalue $\lambda = 1$.
   That is, $E_1 = \mathbb{R}^n$ for I, the identity matrix.
   So, there are n linearly independent vectors corresponding to the eigenvalue $\lambda = 1$.

   (g) ***True***. See Theorem 4.25 in Section 4.4.

   (h) ***False***. A simple counterexample is the identity matrix, I.
   Is every diagonal matrix diagonalizable? Why or why not?

   (i) ***False***. See Theorem 4.22 and Exercise 33 in Section 4.4.
   Theorem 4.22(e) states λ is an eigenvalue of A if and only if λ is an eigenvalue of B.
   So if A is similar to B, then they have the same eigen*values* not eigen*vectors*.
   In Exercise 33, we see how the eigenvectors are related given $A = PBP^{-1}$.
   If $A\mathbf{x} = \alpha\mathbf{x}$, $\mathbf{x} \neq \mathbf{0}$, then $(PBP^{-1})\mathbf{x} = \alpha\mathbf{x}$.
   So, left multiplying both sides by P^{-1} implies $B(P^{-1}\mathbf{x}) = P^{-1}(\alpha\mathbf{x}) = \alpha(P^{-1}\mathbf{x})$.
   That is, not only is α an eigenvalue for B, one associated eigenvector of α is $P^{-1}\mathbf{x}$.

   (j) ***False***. An instructive counter example is all invertible matrices.
   If A is invertible, then $A \longrightarrow I$.
   But $A \sim I$ implies the only eigenvalue of A is $\lambda = 1$. Why?
   We would be forced to conclude the only eigenvalue of any invertible matrix is 1.
   That conclusion is clearly nonsense.

2. See Examples 4.10 and 4.13 of Section 4.2.

(a) We proceed as in Example 4.10 of Section 4.2.

$$\begin{vmatrix} 5 & 7 \\ 9 & 11 \end{vmatrix} - 3\begin{vmatrix} 3 & 7 \\ 7 & 11 \end{vmatrix} + 5\begin{vmatrix} 3 & 5 \\ 7 & 9 \end{vmatrix} = (55-63) - 3(33-49) + 5(27-35) = -8+48-40 = 0$$

(b) We proceed as in Example 4.13 of Section 4.2.

$$\begin{bmatrix} 1 & 3 & 5 \\ 3 & 5 & 7 \\ 7 & 9 & 11 \end{bmatrix} \longrightarrow \begin{bmatrix} 1 & 3 & 5 \\ 0 & -4 & -8 \\ 0 & -12 & -24 \end{bmatrix} \longrightarrow \begin{bmatrix} 1 & 3 & 5 \\ 0 & -4 & -8 \\ 0 & 0 & 0 \end{bmatrix} \Rightarrow \det A = (1)(-4)(0) = 0$$

3. See Exercises 35 through 40 in Section 4.2.

Let A be the first matrix given at the beginning of this exercise with $\det A = 3$.
Let E be the second matrix given in this Exercise which is derived from A in 4 steps.

$$A \overset{R_1 \leftrightarrow R_2}{\longrightarrow} B \overset{3C_1}{\longrightarrow} C \overset{2C_2}{\longrightarrow} D \overset{C_2-4C_3}{\longrightarrow} E$$

So, $\det E = \det D = 2(\det C) = 2((3\det B)) = 2((3(-\det A))) = -6\det A = -18$.

We show the steps and matrices created below:

$$A = \begin{bmatrix} a & b & c \\ d & e & f \\ g & h & i \end{bmatrix} \overset{R_1 \leftrightarrow R_2}{\longrightarrow} \begin{bmatrix} d & e & f \\ a & b & c \\ g & h & i \end{bmatrix} \overset{3C_1}{\longrightarrow} \begin{bmatrix} 3d & e & f \\ 3a & b & c \\ 3g & h & i \end{bmatrix} \overset{2C_2}{\longrightarrow} \begin{bmatrix} 3d & 2e & f \\ 3a & 2b & c \\ 3g & 2h & i \end{bmatrix}$$

$$\overset{C_2-4C_3}{\longrightarrow} \begin{bmatrix} 3d & 2e-4f & f \\ 3a & 2b-4c & c \\ 3g & 2h-4i & i \end{bmatrix} = E$$

4. See Exercises 47 through 51 in Section 4.2.
We use Theorems 4.7, 4.8, 4.9, 4.10 and $\det A = 2$, $\det B = -\frac{1}{4}$ to compute $\det C$.

We use Theorem 4.8, $\det(AB) = (\det A)(\det B)$, and the given values to compute $\det(AB)$.

(a) We use Theorems 4.8, 4.9 and $\det A = 2$, $\det B = -\frac{1}{4}$ to compute $\det((AB)^{-1})$.

$$\det((AB)^{-1}) \overset{\text{Thm 4.9}}{=} \frac{1}{\det(AB)} \overset{\text{Thm 4.8}}{=} \frac{1}{\det A \det B} = \frac{1}{(2)(-1/4)} = -2$$

(b) We use Theorems 4.7, 4.8, 4.9, 4.10 and $\det A = 2$, $\det B = -\frac{1}{4}$ to compute $\det((A^2B)(3A^T))$.

$$\det((A^2B)(3A^T)) \overset{\text{Thm 4.8}}{=} (\det A)^2(\det B)(\det(3A^T)) \overset{\text{Thm 4.7}}{=} (\det A)^2(\det B)(3^4\det(A^T))$$
$$\overset{\text{Thm 4.10}}{=} (\det A)^2(\det B)(3^4\det A) = (3^4)(\det A)^3(\det B) = (81)(8)(-\tfrac{1}{4}) = -162$$

5. See Exercises 37 to 43 in Section 3.2 including the definition of skew-symmetric, $A^T = -A$.

Since $A^T = -A$, $\det A = \det(A^T) = \det(-A) = (-1)^n \det A$. So, $\det A = (-1)^n \det A$.
So when n is odd, $(-1)^n = -1$ which implies $\det A = -1\det A$.
Therefore when n is odd, $2\det A = 0$ which implies $\det A = 0$.

6. See Exercises 45 and 46 in Section 4.2.
To simplify the computation of the determinant, we create zeroes in the first column. Since this reduction is achieved by subtracting a multiple of one row from another, the determinant is unchanged.

$$|A| = \begin{vmatrix} 1 & -1 & 2 \\ 1 & 1 & k \\ 2 & 4 & k^2 \end{vmatrix} \overset{R_2-R_1}{=} \begin{vmatrix} 1 & -1 & 2 \\ 0 & 2 & k-2 \\ 2 & 4 & k^2 \end{vmatrix} \overset{R_3-2R_1}{=} \begin{vmatrix} 1 & -1 & 2 \\ 0 & 2 & k-2 \\ 0 & 6 & k^2-4 \end{vmatrix} = \begin{vmatrix} 2 & k-2 \\ 6 & k^2-4 \end{vmatrix}$$

$$= 2(k^2-4) - 6(k-2) = 2(k-2)(k+2) - 6(k-2) = 2(k-2)[(k+2)-6]$$

$$= 2(k-2)(k-4) = 0 \Rightarrow k = 2, 4.$$

So, the values of k that make the determinant zero are $k = 2, 4$.

7. See Exercises 1 through 5 and Example 4.1 in Section 4.1.
If $A\mathbf{x} = \lambda\mathbf{x}$, then $\mathbf{x}$ is an eigenvector of A corresponding to λ.

So, as in Example 4.1, since $A\mathbf{v} = \begin{bmatrix} 3 & 1 \\ 4 & 3 \end{bmatrix} \begin{bmatrix} 1 \\ 2 \end{bmatrix} = \begin{bmatrix} 5 \\ 10 \end{bmatrix} = 5 \begin{bmatrix} 1 \\ 2 \end{bmatrix} = 5\mathbf{v}$,

we see $\mathbf{v}$ is an eigenvector of A corresponding to (the eigenvalue) 5.

8. See Exercises 1 through 5 and Example 4.1 in Section 4.1.
If $A\mathbf{x} = \lambda\mathbf{x}$, then $\mathbf{x}$ is an eigenvector of A corresponding to λ.

So, as in Example 4.1, since $A\mathbf{v} = \begin{bmatrix} 13 & -60 & -45 \\ -5 & 18 & 15 \\ 10 & -40 & -32 \end{bmatrix} \begin{bmatrix} 3 \\ -1 \\ 2 \end{bmatrix} = \begin{bmatrix} 9 \\ -3 \\ 6 \end{bmatrix} = 3 \begin{bmatrix} 3 \\ -1 \\ 2 \end{bmatrix} = 3\mathbf{v}$,

we see $\mathbf{v}$ is an eigenvector of A corresponding to (the eigenvalue) 3.

9. See Examples 4.18, 4.19 and Exercises 1 through 12 in Section 4.3.
See the box outlining this procedure in Section 4.3 on p.289.

(a) See the definition of the characteristic polynomial given in Section 4.3 on p.289.
Recall the characteristic polynomial of A is defined as $c_A(\lambda) = \det(A - \lambda I)$.
We expand along the third row because that row contains two zeroes.

$$c_A(\lambda) = \det(A - \lambda I) = \begin{vmatrix} -5-\lambda & -6 & 3 \\ 3 & 4-\lambda & -3 \\ 0 & 0 & -2-\lambda \end{vmatrix}$$

$$= (-2-\lambda)\begin{vmatrix} -5-\lambda & -6 \\ 3 & 4-\lambda \end{vmatrix} = (-2-\lambda)(\lambda^2+\lambda-2) = -(\lambda+2)^2(\lambda-1)$$

(b) See the definition of eigenvalues given in Section 4.3 on p.289.
Recall the eigenvalues of A are the solutions of $\det(A - \lambda I) = 0$.
We expand along the third row because that row contains two zeroes.
$\det(A - \lambda I) = -(\lambda+2)^2(\lambda-1) = 0$ which implies the eigenvalues of A are $\lambda = -2, 1$.

(c) The eigenspace $E_\lambda = \text{null}(A - \lambda I)$, so $E_{-2} = \text{null}(A + 2I)$ and $E_1 = \text{null}(A - I)$.

$$\left[\, A+2I \,|\, \mathbf{0} \,\right] = \left[\begin{array}{ccc|c} -3 & -6 & 3 & 0 \\ 3 & 6 & -3 & 0 \\ 0 & 0 & 0 & 0 \end{array}\right] \xrightarrow{R_2+R_1} \left[\begin{array}{ccc|c} -3 & -6 & 3 & 0 \\ 0 & 0 & 0 & 0 \\ 0 & 0 & 0 & 0 \end{array}\right] \xrightarrow{\frac{1}{3}R_1} \left[\begin{array}{ccc|c} -1 & -2 & 1 & 0 \\ 0 & 0 & 0 & 0 \\ 0 & 0 & 0 & 0 \end{array}\right]$$

So $x_1 = -2x_2 + x_3$, from which it follows that $E_{-2} = \text{span}\left(\begin{bmatrix} 2 \\ -1 \\ 0 \end{bmatrix}, \begin{bmatrix} 1 \\ 0 \\ 1 \end{bmatrix}\right)$.

$$\left[\, A-I \,|\, \mathbf{0} \,\right] = \left[\begin{array}{ccc|c} -6 & -6 & 3 & 0 \\ 3 & 3 & -3 & 0 \\ 0 & 0 & -3 & 0 \end{array}\right] \xrightarrow{2R_3} \left[\begin{array}{ccc|c} -6 & -6 & 3 & 0 \\ 6 & 6 & -6 & 0 \\ 0 & 0 & -3 & 0 \end{array}\right] \xrightarrow{R_2+R_1} \left[\begin{array}{ccc|c} -6 & -6 & 3 & 0 \\ 0 & 0 & -3 & 0 \\ 0 & 0 & -3 & 0 \end{array}\right]$$

So $x_3 = 0$ which implies $x_1 = -x_2$, from which it follows that $E_1 = \text{span}\left(\begin{bmatrix} 1 \\ -1 \\ 0 \end{bmatrix}\right)$.

10. Since A is diagonalizable, $A \sim D$, where D is diagonal.
So, A and D have the same eigenvalues and $\det A = \det D$. Which theorem implies this?

Since D is a diagonal matrix,
$\det D =$ the product of its diagonal entries $=$ the product of its eigenvalues (why?).

Therefore, $\det A = \det D = (-2)(3)(4) = -24$.

11. This is a straightforward application of Theorem 4.19 on p.294 in Section 4.3.

Since $\begin{bmatrix} 3 \\ 7 \end{bmatrix} = 5\begin{bmatrix} 1 \\ 1 \end{bmatrix} - 2\begin{bmatrix} 1 \\ -1 \end{bmatrix}$, $A^{-5}\begin{bmatrix} 3 \\ 7 \end{bmatrix} = 5\left(\frac{1}{2}\right)^{-5}\begin{bmatrix} 1 \\ 1 \end{bmatrix} - 2(-1)^{-5}\begin{bmatrix} 1 \\ -1 \end{bmatrix} = \begin{bmatrix} 162 \\ 158 \end{bmatrix}$.

12. This is an application of Theorem 4.19 on p.294 in Section 4.3.
We will show $A^n \longrightarrow O$ by showing that $\|A^n\mathbf{x}\| \longrightarrow \mathbf{0}$ for all vectors $\mathbf{x}$.

Since $|\lambda_i| \le 1$ for all the eigenvalues of A, $|\lambda_i|^n \longrightarrow 0$ as $n \longrightarrow \infty$. Why?

Since $A^n\mathbf{x} = \sum c_i\lambda_i^n\mathbf{v}_i$, $\|A^n\mathbf{x}\| = \|\sum c_i\lambda_i^n\mathbf{v}_i\| \le \sum |c_i||\lambda_i|^n\|\mathbf{v}_i\| \longrightarrow \sum |c_i| \cdot 0 \cdot \|\mathbf{v}_i\| = \mathbf{0}$.

13. Since the characteristic polynomial of A is not equal to the characteristic polynomial of B, A is not similar to B ($A \nsim B$). That is:

Since $c_A(\lambda) = (4-\lambda)(1-\lambda) - 6 \neq (2-\lambda)(2-\lambda) - 6 = c_B(\lambda)$, we have $A \nsim B$.

14. See Exercises 36 through 39 in Section 4.4.

Since A and B have the same eigenvalues, A and B are similar.
The eigenvalues of both A and B are $2, 3$. Why is that obvious?

Clearly, $P = \begin{bmatrix} 0 & \frac{3}{2} \\ \frac{2}{3} & 0 \end{bmatrix}$. Verify that $P^{-1}AP = B$.

Q: Is there a simpler and more obvious choice for P?
A: Hint: Yes, consider a matrix whose only entries are zeroes or ones.

15. A and B are *not* similar. Why?
This is an example of a pair of matrices that share the characteristics of Theorem 4.22 in Section 4.4 but are still *not* similar.

See the remarks on p.300 in Section 4.4.

16. **NOTE**: The matrix A given in the text is in ***error***. It should be: $A = \begin{bmatrix} 2 & k \\ 1 & 0 \end{bmatrix}$.

Since $\det(A - \lambda I) = \begin{vmatrix} 2-\lambda & k \\ 1 & -\lambda \end{vmatrix} = \lambda^2 - 2\lambda - k$, the eigenvalues of A are $\lambda = 1 \pm \sqrt{1+k}$.

(a) Therefore, $\lambda = 3, -1$ if and only $k = 3$. Why?

(b) Furthermore, $\lambda = 1$ (with multiplicity 2) if and only $k = -1$. Why?

(c) Finally, A has no real eigenvalues if and only $k < -1$. Why?

17. If $\mathbf{x}$ is an eigenvector of A, $A^3\mathbf{x} = \lambda^3\mathbf{x}$. What theorem asserts this?

Since $A^3 = A$, we have $\lambda\mathbf{x} = A\mathbf{x} = A^3\mathbf{x} = \lambda^3\mathbf{x}$.

So, $\lambda\mathbf{x} = \lambda^3\mathbf{x}$. Since $\mathbf{x} \neq \mathbf{0}$, we have:
$\lambda = \lambda^3 \Rightarrow \lambda^3 - \lambda = 0 \Rightarrow \lambda(\lambda^2 - 1) = 0 \Rightarrow \lambda(\lambda-1)(\lambda+1) = 0 \Rightarrow \lambda = -1, 0, 1.$

18. This is an application of Theorem 4.3c and The Fundamental Theorem (Theorem 4.17).

Theorem 4.3c. says: if A has two identical rows then $\det A = 0$.
So, the contrapositive of Theorem 4.17 (NOT n implies NOT o) shows 0 must be an eigenvalue of A.

Q: What is the precise statement of Theorem 4.17 (NOT n implies NOT o) cited above?
A: If $\det A = 0$ (NOT n), then 0 must be an eigenvalue of A (NOT o).

19. Given $\mathbf{x}$ is an eigenvector and $B = A^2 - 5A + 2$, we compute $A^2\mathbf{x} - 5A\mathbf{x} + 2\mathbf{x}$ directly.

We get: $B\mathbf{x} = A^2\mathbf{x} - 5A\mathbf{x} + 2\mathbf{x} = 3^2\mathbf{x} - 5 \cdot 3\mathbf{x} + 2\mathbf{x} = -4\mathbf{x}$.

This shows that $\mathbf{x}$ is an eigenvector of B with eigenvalue $\lambda = -4$. Why?

20. See Exercise 46 in Section 4.4. The solution to Exercise 33 is also included below.

Since $\mathbf{x} \neq \mathbf{0}$ is an eigenvector of M if and only if there exists an eigenvalue λ such that $M\mathbf{v} = \lambda\mathbf{v}$, it suffices to show the following:

If $A\mathbf{x} = \alpha\mathbf{x}$, $\mathbf{x} \neq \mathbf{0}$, then $B(P^{-1}\mathbf{x}) = \alpha(P^{-1}\mathbf{x})$.

Note, $P^{-1}AP = B$ implies $A = PBP^{-1}$, where P is invertible and so is P^{-1}. Therefore:

If $A\mathbf{x} = \alpha\mathbf{x}$, $\mathbf{x} \neq \mathbf{0}$, then $(PBP^{-1})\mathbf{x} = \alpha\mathbf{x}$.
So, left multiplying both sides by P^{-1} implies $B(P^{-1}\mathbf{x}) = P^{-1}(\alpha\mathbf{x}) = \alpha(P^{-1}\mathbf{x})$.

So, not only is $P^{-1}\mathbf{x}$ an eigenvector of B, but its associated eigenvalue is α.

The additional details from the solution of Exercise 33 are included below:

Since we are proving Theorem 4.22(e), we need to prove the following two statements:
1) Given $A \sim B$, if α is an eigenvalue of A then α is an eigenvalue of B and
2) Given $A \sim B$, if β is an eigenvalue of B then β is an eigenvalue of A.

Since λ if an eigenvalue of M if and only if there exists an eigenvector $\mathbf{v} \neq \mathbf{0}$ such that $M\mathbf{v} = \lambda\mathbf{v}$, it suffices to show the following:

1) If $A\mathbf{x} = \alpha\mathbf{x}$, $\mathbf{x} \neq \mathbf{0}$, then $B(P^{-1}\mathbf{x}) = \alpha(P^{-1}\mathbf{x})$ and
2) If $B\mathbf{y} = \beta\mathbf{y}$, $\mathbf{y} \neq \mathbf{0}$, then $A(P\mathbf{y}) = \beta(P\mathbf{y})$.

Note, $P^{-1}AP = B$ implies $A = PBP^{-1}$, where P is invertible and so is P^{-1}. Therefore:

If $A\mathbf{x} = \alpha\mathbf{x}$, $\mathbf{x} \neq \mathbf{0}$, then $(PBP^{-1})\mathbf{x} = \alpha\mathbf{x}$.
So, left multiplying both sides by P^{-1} implies $B(P^{-1}\mathbf{x}) = P^{-1}(\alpha\mathbf{x}) = \alpha(P^{-1}\mathbf{x})$.

Likewise, if $B\mathbf{y} = \beta\mathbf{y}$, $\mathbf{y} \neq \mathbf{0}$, then $(P^{-1}AP)\mathbf{y} = \beta\mathbf{y}$.
So, left multiplying both sides by P implies $A(P\mathbf{y}) = P(\beta\mathbf{y}) = \beta(P\mathbf{y})$.

Q: Why do we *not* have to specify that $P^{-1}\mathbf{x}$ and $P\mathbf{y}$ are not the zero vector?
A: Because P is invertible and so P^{-1} is invertible, too. Therefore:
If $\mathbf{u} \neq 0$, then $P\mathbf{u} \neq 0$ and $P^{-1}\mathbf{u} \neq 0$. Why?

Q: How might we state our stronger result in words?
A: Not only is α an eigenvalue for B, one associated eigenvector of α is $P^{-1}\mathbf{x}$.
Not only is β an eigenvalue for A, one associated eigenvector of β is $P\mathbf{y}$.

Chapter 5

Orthogonality

5.1 Orthogonality in $\mathbb{R}^n$

1. $\mathbf{v}_1 \cdot \mathbf{v}_2 = (-3)(2) + 1(4) + 2(1) = 0$, $\mathbf{v}_2 \cdot \mathbf{v}_3 = 2(1) + 4(-1) + 1(2) = 0$, $\mathbf{v}_1 \cdot \mathbf{v}_3 = (-3) + 1(-1) + 2(2) = 0 \Rightarrow$ This set of vectors is orthogonal.

2. $\mathbf{v}_1 \cdot \mathbf{v}_2 = 4(-1) + 2(2) + (-5)(0) = 0$, $\mathbf{v}_2 \cdot \mathbf{v}_3 = (-1)(2) + 2(1) + 0(2) = 0$, $\mathbf{v}_1 \cdot \mathbf{v}_3 = 4(2) + 2(1) + (-5)(2) = 0 \Rightarrow$ This set of vectors is orthogonal.

3. $\mathbf{v}_1 \cdot \mathbf{v}_2 = 3(-1) + 1(2) + (-1)(1) \neq 0 \Rightarrow$ *not* orthogonal.

4. $\mathbf{v}_1 \cdot \mathbf{v}_3 = 5(3) + 3(1) + 1(-1) \neq 0 \Rightarrow$ *not* orthogonal.

5. $\mathbf{v}_1 \cdot \mathbf{v}_2 = \mathbf{v}_2 \cdot \mathbf{v}_3 = \mathbf{v}_1 \cdot \mathbf{v}_3 = 0 \Rightarrow$ This set of vectors is orthogonal.

6. $\mathbf{v}_2 \cdot \mathbf{v}_4 = 0(1) + (-1)(0) + 1(1) + 1(2) \neq 0 \Rightarrow$ *not* orthogonal.

7. $c_1 = \frac{4+6}{16+4} = \frac{1}{2}$, $c_2 = \frac{1-6}{1+4} = -1 \Rightarrow [\mathbf{w}]_{\mathcal{B}} = \begin{bmatrix} 1/2 \\ -1 \end{bmatrix}$. $\mathbf{w} = \frac{1}{2}\begin{bmatrix} 4 \\ -2 \end{bmatrix} + (-1)\begin{bmatrix} 1 \\ -2 \end{bmatrix} = \begin{bmatrix} 1 \\ -3 \end{bmatrix}$.

8. $\mathbf{v}_1 \cdot \mathbf{v}_2 = 0 \Rightarrow c_1 = \frac{1+3}{1+9} = \frac{2}{5}$, $c_2 = \frac{-6+2}{36+4} = -\frac{1}{10} \Rightarrow [\mathbf{w}]_{\mathcal{B}} = \begin{bmatrix} 2/5 \\ -1/10 \end{bmatrix}$.

9. $c_1 = \frac{1+0-1}{1+0+1} = 0$, $c_2 = \frac{1+2+1}{1+4+1} = \frac{2}{3}$, $c_3 = \frac{1-1+1}{1+1+1} = \frac{1}{3} \Rightarrow [\mathbf{w}]_{\mathcal{B}} = \begin{bmatrix} 0 \\ 2/3 \\ 1/3 \end{bmatrix}$.

10. $c_1 = \frac{1+2+3}{1+1+1} = 2$, $c_2 = \frac{1-2+0}{1+1+0} = -\frac{1}{2}$, $c_3 = \frac{1+2-6}{1+1+4} = -\frac{1}{2} \Rightarrow [\mathbf{w}]_{\mathcal{B}} = \begin{bmatrix} 2 \\ -1/2 \\ -1/2 \end{bmatrix}$.

11. $\|\mathbf{v}_1\| = \sqrt{(\frac{3}{5})^2 + (\frac{4}{5})^2} = 1$, $\|\mathbf{v}_2\| = \sqrt{(-\frac{4}{5})^2 + (\frac{3}{5})^2} = 1 \Rightarrow$ This set is orthonormal.

12. $\|\mathbf{v}_1\| = \sqrt{(\frac{1}{2})^2 + (\frac{1}{2})^2} = \frac{1}{\sqrt{2}}$, $\|\mathbf{v}_2\| = \sqrt{(\frac{1}{2})^2 + (-\frac{1}{2})^2} = \frac{1}{\sqrt{2}}$
$\Rightarrow$ We need to normalize both vectors.
So, multiply each of them by $\frac{1}{length} = \sqrt{2} \Rightarrow \left\{ \begin{bmatrix} \sqrt{2}/2 \\ \sqrt{2}/2 \end{bmatrix}, \begin{bmatrix} \sqrt{2}/2 \\ -\sqrt{2}/2 \end{bmatrix} \right\}$.

13. $\|\mathbf{v}_1\| = \sqrt{(\frac{1}{3})^2 + 2(\frac{2}{3})^2} = 1$, $\|\mathbf{v}_2\| = \sqrt{(\frac{2}{3})^2 + (-\frac{1}{3})^2} = \frac{\sqrt{5}}{3}$, $\|\mathbf{v}_3\| = \sqrt{(1^2 + 2^2 + (-\frac{5}{2})^2} = \frac{3\sqrt{5}}{2}$

$$\Rightarrow \left\{ \begin{bmatrix} 1/3 \\ 2/3 \\ 2/3 \end{bmatrix}, \frac{3}{\sqrt{5}} \begin{bmatrix} 2/3 \\ -1/3 \\ 0 \end{bmatrix} = \begin{bmatrix} 2/\sqrt{5} \\ -1/\sqrt{5} \\ 0 \end{bmatrix}, \frac{2}{3\sqrt{5}} \begin{bmatrix} 1 \\ 2 \\ -5/2 \end{bmatrix} = \begin{bmatrix} 2/3\sqrt{5} \\ 4/3\sqrt{5} \\ -5/3\sqrt{5} \end{bmatrix} \right\}.$$

14. $\|\mathbf{v}_1\| = 1, \|\mathbf{v}_2\| = \frac{2}{\sqrt{6}}$, $\|\mathbf{v}_3\| = \frac{1}{\sqrt{3}} \Rightarrow$

$$\left\{ \begin{bmatrix} 1/2 \\ 1/2 \\ -1/2 \\ 1/2 \end{bmatrix}, \frac{\sqrt{6}}{2} \begin{bmatrix} 0 \\ 1/3 \\ 2/3 \\ 1/3 \end{bmatrix} = \begin{bmatrix} 0 \\ \sqrt{6}/6 \\ \sqrt{6}/3 \\ \sqrt{6}/6 \end{bmatrix}, \sqrt{3} \begin{bmatrix} 1/2 \\ -1/6 \\ 1/6 \\ -1/6 \end{bmatrix} = \begin{bmatrix} \sqrt{3}/2 \\ -\sqrt{3}/6 \\ \sqrt{3}/6 \\ -\sqrt{3}/6 \end{bmatrix} \right\}.$$

15. $\|\mathbf{v}_1\| = \|\mathbf{v}_2\| = \|\mathbf{v}_3\| = \|\mathbf{v}_4\| = 1 \Rightarrow$ orthonormal.

16. $QQ^T = \begin{bmatrix} 0 & 1 \\ 1 & 0 \end{bmatrix}\begin{bmatrix} 0 & 1 \\ 1 & 0 \end{bmatrix} = \begin{bmatrix} 1 & 0 \\ 0 & 1 \end{bmatrix} \Rightarrow$ orthogonal, $Q^{-1} = Q^T = \begin{bmatrix} 0 & 1 \\ 1 & 0 \end{bmatrix}$.

17. $QQ^T = I \Rightarrow Q$ is orthogonal and $Q^{-1} = Q^T = \begin{bmatrix} 1/\sqrt{2} & -1/\sqrt{2} \\ 1/\sqrt{2} & 1/\sqrt{2} \end{bmatrix}$.

18. $\|\mathbf{q}_2\| = \sqrt{(\frac{1}{3})^2 + (-\frac{1}{3})^2 + 0^2} \neq 1 \Rightarrow Q$ is *not* orthogonal (columns must have length 1).

19. $QQ^T = I \Rightarrow Q$ is orthogonal and $Q^{-1} = Q^T = \begin{bmatrix} \cos\theta\sin\theta & \cos^2\theta & \sin\theta \\ -\cos\theta & \sin\theta & 0 \\ -\sin^2\theta & -\cos\theta\sin\theta & \cos\theta \end{bmatrix}$.

20. $QQ^T = I \Rightarrow Q$ is orthogonal and $Q^{-1} = Q^T = \begin{bmatrix} 1/2 & 1/2 & -1/2 & 1/2 \\ -1/2 & 1/2 & 1/2 & 1/2 \\ 1/2 & 1/2 & 1/2 & -1/2 \\ 1/2 & -1/2 & 1/2 & 1/2 \end{bmatrix}$.

21. $\mathbf{q}_1 \cdot \mathbf{q}_4 = 1(\frac{1}{\sqrt{6}}) + 0(\frac{1}{\sqrt{6}}) + 0(-\frac{1}{\sqrt{6}}) + 0(-\frac{1}{\sqrt{2}}) \neq 0 \Rightarrow$ not orthogonal (columns must be $\perp$).

22. By Theorem 5.6(c), we need only show that $Q^{-1}\mathbf{x} \cdot Q^{-1}\mathbf{y} = \mathbf{x} \cdot \mathbf{y}$, for every $\mathbf{x}$ and $\mathbf{y}$.
We will show this using Theorem 5.5: Q is orthogonal $\Leftrightarrow Q^{-1} = Q^T \Leftrightarrow QQ^T = I$.
$Q^{-1}\mathbf{x} \cdot Q^{-1}\mathbf{y} = Q^T\mathbf{x} \cdot Q^T\mathbf{y} = (Q^T\mathbf{x})^T Q^T\mathbf{y} = \mathbf{x}^T QQ^T\mathbf{y} = \mathbf{x}^T I\mathbf{y} = \mathbf{x} \cdot \mathbf{y}$.

23. We will show this using the fact that $\det I = 1, QQ^T = I$, and $\det Q = \det Q^T$.
$1 = \det I = \det(QQ^T) = \det Q \det Q^T = \det Q \det Q \Rightarrow \sqrt{(\det Q)^2} = \sqrt{1} \Rightarrow \det Q = \pm 1$.

24. By Theorem 5.5, we need only show $(Q_1Q_2)^{-1} = (Q_1Q_2)^T$, i.e. $(Q_1Q_2)^T Q_1Q_2 = I$.
$(Q_1Q_2)^T Q_1Q_2 = Q_2^T(Q_1^T Q_1)Q_2 = Q_2^T I Q_2 = Q_2^T Q_2 = I$.

25. Induction on Exercise 24 $\Rightarrow Q_1, Q_2, \ldots, Q_n$ orthogonal $\Rightarrow Q = Q_n \ldots Q_2Q_1$ orthogonal.
Let $P = P_n \ldots P_2P_1$, where P_k is an elementary matrix corresponding to a row interchange.
Then $P_kP_k^T = I$ and Theorem 5.5 $\Rightarrow P_k$ is orthogonal $\Rightarrow P = P_n \ldots P_2P_1$ is orthogonal.

26. Let $Q' = PQ$, be obtained by rearranging the rows of Q, so P is a permutation matrix.
Then by Exercise 25, the permutation matrix P is orthogonal. Therefore, Exercise 24 in which we proved Theorem 5.8(d) $\Rightarrow Q' = PQ$ is orthogonal because P and Q are orthogonal.

27. Let θ be the angle between $\mathbf{x}$ and $\mathbf{y}$ and ϕ be the angle between $Q\mathbf{x}$ and $Q\mathbf{y}$.
We need to show that $\theta = \phi$. We will use Theorem 6(b,c): $\|Q\mathbf{x}\| = \|\mathbf{x}\|, Q\mathbf{x} \cdot Q\mathbf{y} = \mathbf{x} \cdot \mathbf{y}$.
$0 \leq \theta, \phi \leq \pi \Rightarrow \cos\theta = \cos\phi \Rightarrow \theta = \phi$. So $\cos\theta = \frac{\mathbf{x}\cdot\mathbf{y}}{\|\mathbf{x}\|\|\mathbf{y}\|} = \frac{Q\mathbf{x}\cdot Q\mathbf{y}}{\|Q\mathbf{x}\|\|Q\mathbf{y}\|} = \cos\phi \Rightarrow \theta = \phi$.

28. (a) $QQ^T = I \Rightarrow \begin{bmatrix} a & c \\ b & d \end{bmatrix}\begin{bmatrix} a & b \\ c & d \end{bmatrix} = \begin{bmatrix} 1 & 0 \\ 0 & 1 \end{bmatrix} \Rightarrow a^2 + b^2 = 1 \Rightarrow \begin{bmatrix} a \\ b \end{bmatrix}$ is a unit vector.

Now all we have left to show is ($d = a$ and $c = -b$) or ($d = -a$ and $c = b$).
From the fact that $QQ^T = I$, we have these three equations: $a^2 + b^2 = 1 \Rightarrow b^2 = 1 - a^2$ (1), $c^2 + d^2 = 1 \Rightarrow c^2 = 1 - d^2$ (2), $ac + bd = 0 \Rightarrow a^2c^2 = b^2d^2$ (3).
Substituting (1), (2) into (3) $\Rightarrow a^2(1 - d^2) = (1 - a^2)d^2 \Rightarrow a^2 = d^2 \Rightarrow a = \pm d$.
When $d = a = 0 \Rightarrow b = c = \pm 1$, we have the following four possibilities:
$\begin{bmatrix} 0 & 1 \\ 1 & 0 \end{bmatrix}, \begin{bmatrix} 0 & -1 \\ 1 & 0 \end{bmatrix}, \begin{bmatrix} 0 & 1 \\ -1 & 0 \end{bmatrix}$, and $\begin{bmatrix} 0 & -1 \\ -1 & 0 \end{bmatrix}$
all of which have one of the two forms listed.
When $d = a \neq 0$, we have:
$\begin{bmatrix} a & c \\ b & a \end{bmatrix}\begin{bmatrix} a & b \\ c & a \end{bmatrix} = \begin{bmatrix} 1 & 0 \\ 0 & 1 \end{bmatrix} \Rightarrow ab + ac = 0 \Rightarrow ac = -ab \Rightarrow c = -b.$
So, we get $\begin{bmatrix} a & -b \\ b & a \end{bmatrix}$.
Likewise, when $d = -a \neq 0$, we have:
$\begin{bmatrix} a & c \\ b & -a \end{bmatrix}\begin{bmatrix} a & b \\ c & -a \end{bmatrix} = \begin{bmatrix} 1 & 0 \\ 0 & 1 \end{bmatrix} \Rightarrow ab - ac = 0 \Rightarrow ac = ab \Rightarrow c = b.$
So, we get $\begin{bmatrix} a & b \\ b & -a \end{bmatrix}$.

(b) $\begin{bmatrix} a \\ b \end{bmatrix}$ unit vector $\Rightarrow 0 \le |a|, |b| \le 1, a^2 + b^2 = 1 \Rightarrow$ There exists θ with $a = \cos\theta, b = \sin\theta$.

(c) Let $\mathbf{x} = \begin{bmatrix} r\cos\phi \\ r\sin\phi \end{bmatrix}, Q = \begin{bmatrix} \cos\theta & -\sin\theta \\ \sin\theta & \cos\theta \end{bmatrix}$, and $R = \begin{bmatrix} \cos\theta & \sin\theta \\ \sin\theta & -\cos\theta \end{bmatrix}$. Then we have:

$Q\mathbf{x} = \begin{bmatrix} r(\cos\theta\cos\phi - \sin\theta\sin\phi) \\ r(\sin\theta\cos\phi + \cos\theta\sin\phi) \end{bmatrix} = \begin{bmatrix} r\cos(\phi+\theta) \\ r\sin(\phi+\theta) \end{bmatrix} \Rightarrow Q$ represents a rotation,

$R\mathbf{x} = \begin{bmatrix} r(\cos\theta\cos\phi + \sin\theta\sin\phi) \\ r(\sin\theta\cos\phi - \cos\theta\sin\phi) \end{bmatrix} = \begin{bmatrix} r\cos(\phi-\theta) \\ -r\sin(\phi-\theta) \end{bmatrix} \Rightarrow R$ represents a reflection.

(d) We will show this using the fact that $\cos^2\theta + \sin^2\theta = 1$. So, we have:
$\det Q = \cos^2\theta + \sin^2\theta = 1$ and $\det R = -\cos^2\theta - \sin^2\theta = -(\cos^2\theta + \sin^2\theta) = -1$.

29. By Exercise 28(d), $\det Q = 1 \Rightarrow$ rotation. So, since $\cos\theta = \frac{1}{\sqrt{2}}$, $\sin\theta = \frac{1}{\sqrt{2}}$, $\theta = \frac{\pi}{4}$ or $45°$.

30. $\det Q = 1 \Rightarrow$ rotation. Furthermore, since $\cos\theta = -\frac{1}{2}$, $\sin\theta = -\frac{\sqrt{3}}{2}$, $\theta = \frac{4\pi}{3}$ or $240°$.

31. $\det Q = -1 \Rightarrow$ reflection. Line of reflection is $y = -\tan\theta\, x = -\frac{\sin\theta}{\cos\theta}\, x = -(\frac{\sqrt{3}/2}{-1/2})\, x = \sqrt{3}\, x$.

32. $\det Q = -1 \Rightarrow$ reflection. Line of reflection is $y = -\tan\theta\, x = -\frac{\sin\theta}{\cos\theta}\, x = -(\frac{-4/5}{-3/5})\, x = -\frac{4}{3}\, x$.

33. (a) $\Rightarrow AA^T = I,\ B^TB = I \Rightarrow A(A^T + B^T)B = AA^TB + AB^TB = IB + AI = A + B.$

(b) Note: B orthogonal and Theorem 5.8(b) $\Rightarrow \det B = \pm 1 \Rightarrow \det B \det B = 1$.
If $\det A + \det B = 0$ also, then $\det A = -\det B \Rightarrow \det A \det B = -\det B \det B = -1$.
Also, recall: $\det(AB) = \det A \det B$ and $\det(A^T + B^T) = \det(A+B)^T = \det(A+B)$.
In order to prove $A + B$ is not invertible, we need only show that $\det(A+B) = 0$.
$A+B = A(A^T+B^T)B \Rightarrow \det(A+B) = \det(A(A^T+B^T)B) = \det A \det B \det(A^T+B^T)$.
Now use the fact that A, B orthogonal and $\det A + \det B = 0 \Rightarrow \det A \det B = -1$.
So, $\det(A+B) = \det A \det B \det(A^T + B^T) = -\det(A^T + B^T) = -\det(A+B)$.
Therefore, $\det(A+B) = -\det(A+B) \Rightarrow 2\det(A+B) = 0 \Rightarrow \det(A+B) = 0 \Rightarrow$
$A + B$ is not invertible.

34. We will show $QQ^T = I \Rightarrow Q$ is orthogonal.

The fact that $\mathbf{x}$ is a unit vector $\Rightarrow x_1^2 + \ldots + x_n^2 = 1 \Rightarrow x_2^2 + \ldots + x_n^2 = 1 - x_1^2$.
That is, $\mathbf{y}^T\mathbf{y} = x_2^2 + \ldots + x_n^2 = 1 - x_1^2 \Rightarrow x_1^2 + \mathbf{y}^T\mathbf{y} = 1$.

So, $\left(\frac{1}{1-x_1}\right)\mathbf{y}^T\mathbf{y} = \frac{x_2^2+\ldots+x_n^2}{1-x_1} = \frac{1-x_1^2}{1-x_1} = 1 + x_1$. Also:

$$\mathbf{y}^T\left(I - \left(\tfrac{1}{1-x_1}\right)\mathbf{y}\mathbf{y}^T\right) = \mathbf{y}^T - \left(\tfrac{1}{1-x_1}\right)(\mathbf{y}^T\mathbf{y})\mathbf{y}^T = \mathbf{y}^T - \tfrac{1-x_1^2}{1-x_1}\mathbf{y}^T = \mathbf{y}^T - (1+x_1)\mathbf{y}^T = -x_1\mathbf{y}^T.$$

Similarly, $\left(I - \left(\frac{1}{1-x_1}\right)\mathbf{y}\mathbf{y}^T\right)\mathbf{y} = \mathbf{y} - \left(\frac{1}{1-x_1}\right)(\mathbf{y}(\mathbf{y}^T\mathbf{y}) = \mathbf{y} - \frac{1-x_1^2}{1-x_1}\mathbf{y} = \mathbf{y} - (1+x_1)\mathbf{y} = -x_1\mathbf{y}$.

Finally, $\left[I - \left(\frac{1}{1-x_1}\right)\mathbf{y}\mathbf{y}^T\right]\left[I - \left(\frac{1}{1-x_1}\right)\mathbf{y}\mathbf{y}^T\right] = I^2 - 2\left(\frac{1}{1-x_1}\right)\mathbf{y}\mathbf{y}^T + \left(\frac{1}{1-x_1}\right)^2\mathbf{y}(\mathbf{y}^T\mathbf{y})\mathbf{y}^T$
$= I - \frac{2-2x_1}{(1-x_1)^2}\mathbf{y}\mathbf{y}^T + \frac{1-x_1^2}{(1-x_1)^2}\mathbf{y}\mathbf{y}^T = I - \frac{1-2x_1+x_1^2}{(1-x_1)^2}\mathbf{y}\mathbf{y}^T = I - \mathbf{y}\mathbf{y}^T$.

$$\text{So, } QQ^T = \left[\begin{array}{c|c} x_1 & \mathbf{y}^T \\ \hline \mathbf{y} & I - \left(\frac{1}{1-x_1}\right)\mathbf{y}\mathbf{y}^T \end{array}\right]\left[\begin{array}{c|c} x_1 & \mathbf{y}^T \\ \hline \mathbf{y} & I - \left(\frac{1}{1-x_1}\right)\mathbf{y}\mathbf{y}^T \end{array}\right]$$

$$= \left[\begin{array}{c|c} x_1^2 + \mathbf{y}^T\mathbf{y} = 1 & x_1\mathbf{y}^T - x_1\mathbf{y}^T = 0 \\ \hline x_1\mathbf{y} - x_1\mathbf{y} = 0 & \mathbf{y}\mathbf{y}^T + (I - \mathbf{y}\mathbf{y}^T) = I \end{array}\right] = I.$$

35. If Q is orthogonal and upper triangular, we must show Q is diagonal.
That is, if Q is orthogonal and $q_{ij} = 0$ for $j > i$, we must show $q_{ij} = 0$ for $j < i$.

Note: Since Q is orthogonal, $Q^T Q = I$. That is, as remarked in the proof of Theorem 5.4:

$$\mathbf{q}_i \cdot \mathbf{q}_j = \sum_{k=1}^{n} q_{ik} q_{jk} \overset{\substack{Q \text{ is upper} \\ \text{triangular}}}{=} \sum_{k=1}^{j} q_{ik} q_{jk} = \begin{cases} 0 & \text{if } i \neq j \\ 1 & \text{if } i = j \end{cases}$$

We proceed to show $q_{ij} = 0$ for $j < i$ by induction.

1: $q_{i1} = 0$ for $1 < i$.

$$\mathbf{q}_i \cdot \mathbf{q}_1 = \sum_{k=1}^{n} q_{ik} q_{1k} \overset{\substack{Q \text{ is upper} \\ \text{triangular}}}{=} \sum_{k=1}^{1} q_{ik} q_{1k} = q_{i1} q_{11} = 0.$$

Therefore, since $q_{11} \neq 0$, we have $q_{i1} = 0$ for $1 < i$ as required.

j: $q_{ij} = 0$ for $j < i$.
This is the induction hypothesis.

$j+1$: $q_{i(j+1)} = 0$ for $j + 1 < i$.
This is the statement we must prove using the induction hypothesis.

$$\mathbf{q}_i \cdot \mathbf{q}_{j+1} = \sum_{k=1}^{n} q_{ik} q_{(j+1)k} \overset{\substack{Q \text{ is upper} \\ \text{triangular}}}{=} \sum_{k=1}^{j+1} q_{ik} q_{(j+1)k} \overset{\substack{\text{by} \\ \text{induction}}}{=} q_{i(j+1)} q_{(j+1)(j+1)} = 0.$$

Therefore, since $q_{(j+1)(j+1)} \neq 0$, we have $q_{i(j+1)} = 0$ for $j + 1 < i$ as required.

We have proven (by induction) that $q_{ij} = 0$ for $j < i$. That is, Q is diagonal.

Q: Why does Q being upper triangular imply $\sum_{k=1}^{n} q_{ik} q_{jk} \overset{\substack{Q \text{ is upper} \\ \text{triangular}}}{=} \sum_{k=1}^{j} q_{ik} q_{jk}$?

A: Because $q_{jk} = 0$ for all $k > j$ by definition of upper triangular.

Q: Why does the induction hypothesis imply $\sum_{k=1}^{j+1} q_{ik} q_{(j+1)k} \overset{\substack{\text{by} \\ \text{induction}}}{=} q_{i(j+1)} q_{(j+1)(j+1)} = 0$?

A: Because $q_{ik} = 0$ for all $k \leq j$ by the induction hypothesis.

36. If $n > m$, we need to show there exists an $\mathbf{x}$ such that $\|A\mathbf{x}\| \neq \|\mathbf{x}\|$.
If $n > m$, we will show there exists an $\mathbf{x} \neq \mathbf{0}$ such that $A\mathbf{x} = 0$.

By The Rank Theorem, that is Theorem 3.26 of Section 3.5, we have the following:
If A is an $m \times n$ matrix, then $\text{rank}(A) + \text{nullity}(A) = n$

Therefore, if $n > m$, we have $\text{nullity}(A) = n - \text{rank}(A) \geq n - m > 0$.

Since $\text{nullity}(A) \neq 0$, $\text{null}(A) \neq \mathbf{0}$. That is, there are nonzero vectors in $\text{null}(A)$.
That is, there exists an $\mathbf{x} \neq 0$ such that $A\mathbf{x} = 0$.

Clearly, $\|A\mathbf{x}\| = \|\mathbf{0}\| = 0 \neq \|\mathbf{x}\|$ because $\mathbf{x} \neq \mathbf{0}$.

Q: Why is it obvious that $\text{rank}(A)$ must be less than or equal to m?
A: Because $\text{rank}(A)$ must be less than or equal to the number of rows. Why?

Q: Where is the fact that $\text{rank}(A) \leq m$ used in the above proof?

Q: We could also argue directly that the columns of A are linearly dependent. Why?
A: Because they are a set of n vectors in $\mathbb{R}^m$ where $n > m$.
Use this fact to show that $\text{null}(A) \neq \mathbf{0}$ directly.

37. We are given $\mathcal{B} = \{\mathbf{v}_1, \mathbf{v}_2, \ldots, \mathbf{v}_n\}$ is an orthonormal basis for $\mathbb{R}^n$.

(a) We need to show $\mathbf{x} \cdot \mathbf{y} = (\mathbf{x} \cdot \mathbf{v}_1)(\mathbf{y} \cdot \mathbf{v}_1) + \cdots + (\mathbf{x} \cdot \mathbf{v}_n)(\mathbf{y} \cdot \mathbf{v}_n) = \sum_{k=1}^{n} (\mathbf{x} \cdot \mathbf{v}_k)(\mathbf{y} \cdot \mathbf{v}_k)$.

Since $\mathcal{B}$ is orthonormal, we have the following two key properties:

$$\mathbf{v}_i \cdot \mathbf{v}_j = \begin{cases} 0 & \text{if } i \neq j \quad \text{Property 1} \\ 1 & \text{if } i = j \quad \text{Property 2} \end{cases}$$

Since $\mathcal{B}$ is a basis, there exist α_k and β_k such that $\mathbf{x} = \sum_{k=1}^{n} \alpha_k \mathbf{v}_k$, $\mathbf{y} = \sum_{k=1}^{n} \beta_k \mathbf{v}_k$.

The key fact is $\mathbf{x} \cdot \mathbf{v}_i = \left(\sum_{k=1}^{n} \alpha_k \mathbf{v}_k\right) \cdot \mathbf{v}_i = \alpha_i$ and $\mathbf{y} \cdot \mathbf{v}_i = \left(\sum_{k=1}^{n} \beta_k \mathbf{v}_k\right) \cdot \mathbf{v}_i = \beta_i$.

$$\text{So, } \mathbf{x} \cdot \mathbf{y} = \left(\sum_{k=1}^{n} \alpha_k \mathbf{v}_k\right) \cdot \left(\sum_{k=1}^{n} \beta_k \mathbf{v}_k\right) \overset{\text{Prop 1}}{=} \left(\sum_{k=1}^{n} (\alpha_k \mathbf{v}_k) \cdot (\beta_k \mathbf{v}_k)\right) \overset{\text{Prop 2}}{=} \left(\sum_{k=1}^{n} \alpha_k \beta_k\right)$$

$$\overset{\text{key fact}}{=} \sum_{k=1}^{n} (\mathbf{x} \cdot \mathbf{v}_k)(\mathbf{y} \cdot \mathbf{v}_k) \text{ as we were to show.}$$

(b) What does our proof in part (a) suggest is the relationship between $\mathbf{x} \cdot \mathbf{y}$ and $[\mathbf{x}]_{\mathcal{B}} \cdot [\mathbf{y}]_{\mathcal{B}}$?

Since $[\mathbf{x}]_{\mathcal{B}} = [\alpha_k]$, $[\mathbf{y}]_{\mathcal{B}} = [\beta_k]$, as shown above $\mathbf{x} \cdot \mathbf{y} = \left(\sum_{k=1}^{n} \alpha_k \beta_k\right) = [\mathbf{x}]_{\mathcal{B}} \cdot [\mathbf{y}]_{\mathcal{B}}$.

That is, these two dot products are exactly the same.

Q: How might we summarize this finding in words?
A: The dot product is invariant under representation by orthonormal bases.

Q: The *key fact* cited above is implicit in what Theorem from this Section?
A: Theorem 5.3.

5.2 Orthogonal Complements and Projections

1. Since $\mathbf{w} = \begin{bmatrix} 1 \\ 2 \end{bmatrix}$ is a basis for W, for all $\mathbf{v} = \begin{bmatrix} x \\ y \end{bmatrix}$ in $W^{\perp}$ we have $\mathbf{v} \cdot \mathbf{w} = x + 2y = 0$.

So, $W^{\perp} = \left\{ \begin{bmatrix} x \\ y \end{bmatrix} : x + 2y = 0 \right\}$ which implies $x = -2y$. Therefore, $\begin{bmatrix} -2 \\ 1 \end{bmatrix}$ is a basis.

Note the following general pattern:
The lines that describe the necessary condition to be in W and $W^{\perp}$
are necessarily perpendicular.
So, in this case, $2x - y = 0$ and $x + 2y = 0$ are necessarily perpendicular.

2. $\mathbf{v} \cdot \mathbf{w} = 4x - 3y = 0 \Rightarrow W^{\perp} = \left\{ \begin{bmatrix} x \\ y \end{bmatrix} : 4x - 3y = 0 \right\} \Rightarrow 4x = 3y \Rightarrow \begin{bmatrix} 3 \\ 4 \end{bmatrix}$ is a basis.

Note the following general pattern:
The lines that describe the necessary condition to be in W and $W^{\perp}$
are necessarily perpendicular.
So, in this case, $3x + 4y = 0$ and $4x - 3y = 0$ are necessarily perpendicular.

3. W consists of $\begin{bmatrix} x \\ y \\ x + y \end{bmatrix} = x \begin{bmatrix} 1 \\ 0 \\ 1 \end{bmatrix} + y \begin{bmatrix} 0 \\ 1 \\ 1 \end{bmatrix} \Rightarrow \left\{ \begin{bmatrix} 1 \\ 0 \\ 1 \end{bmatrix}, \begin{bmatrix} 0 \\ 1 \\ 1 \end{bmatrix} \right\}$ is a basis for W.

So, for all $\mathbf{v} = \begin{bmatrix} x \\ y \\ z \end{bmatrix}$ in $W^{\perp}$ we have $\begin{array}{l} \mathbf{v} \cdot \mathbf{w_1} = x + z = 0 \\ \mathbf{v} \cdot \mathbf{w_2} = y + z = 0 \end{array}$ which imply $\begin{array}{l} x = y \\ z = -x \end{array}$.

Therefore, $W^{\perp} = \left\{ \begin{bmatrix} x \\ y \\ z \end{bmatrix} : x = t, y = t, z = -t \right\}$ which has basis $\begin{bmatrix} 1 \\ 1 \\ -1 \end{bmatrix}$.

Note the following general pattern:
If W is a plane with normal $\mathbf{n}$, then $W^{\perp}$ is a line that has direction vector $\mathbf{n}$.
Likewise, if W is a line that has direction vector $\mathbf{n}$, then $W^{\perp}$ is a plane with normal $\mathbf{n}$.

4. W consists of $\begin{bmatrix} x \\ 2x + 3z \\ z \end{bmatrix} = x \begin{bmatrix} 1 \\ 2 \\ 0 \end{bmatrix} + z \begin{bmatrix} 0 \\ 3 \\ 1 \end{bmatrix} \Rightarrow \left\{ \begin{bmatrix} 1 \\ 2 \\ 0 \end{bmatrix}, \begin{bmatrix} 0 \\ 3 \\ 1 \end{bmatrix} \right\}$ is a basis for W.

$\begin{array}{l} \mathbf{v} \cdot \mathbf{w_1} = x + 2y = 0 \\ \mathbf{v} \cdot \mathbf{w_2} = 3y + z = 0 \end{array} \Rightarrow W^{\perp} = \left\{ \begin{bmatrix} x \\ y \\ z \end{bmatrix} : x = -2t, y = t, z = -3t \right\}$ with basis $\begin{bmatrix} -2 \\ 1 \\ -3 \end{bmatrix}$.

Note the following general pattern:
If W is a plane with normal $\mathbf{n}$, then $W^{\perp}$ is a line that has direction vector $\mathbf{n}$.
Likewise, if W is a line that has direction vector $\mathbf{n}$, then $W^{\perp}$ is a plane with normal $\mathbf{n}$.

5. Since $\mathbf{w} = \begin{bmatrix} 1 \\ -1 \\ 3 \end{bmatrix}$ is a basis for W, for $\mathbf{v} = \begin{bmatrix} x \\ y \\ z \end{bmatrix}$ in $W^{\perp}$, $\mathbf{v} \cdot \mathbf{w} = x - y + 3z = 0$.

Therefore, $W^{\perp} = \left\{ \begin{bmatrix} x \\ y \\ z \end{bmatrix} : x - y + 3z = 0 \right\}$, which implies $y = x + 3z$.

So, $W^{\perp}$ consists of $\begin{bmatrix} x \\ x + 3z \\ z \end{bmatrix} = x \begin{bmatrix} 1 \\ 1 \\ 0 \end{bmatrix} + z \begin{bmatrix} 0 \\ 3 \\ 1 \end{bmatrix} \Rightarrow \left\{ \begin{bmatrix} 1 \\ 1 \\ 0 \end{bmatrix}, \begin{bmatrix} 0 \\ 3 \\ 1 \end{bmatrix} \right\}$ is a basis for $W^{\perp}$.

6. Since $\mathbf{w} = \begin{bmatrix} 2 \\ 2 \\ -1 \end{bmatrix}$ is a basis for W, for $\mathbf{v} = \begin{bmatrix} x \\ y \\ z \end{bmatrix}$ in $W^{\perp}$ $\mathbf{v} \cdot \mathbf{w} = 2x + 2y - z = 0 \Rightarrow$

$W^{\perp} = \left\{ \begin{bmatrix} x \\ y \\ z \end{bmatrix} : 2x + 2y - z = 0 \right\} \Rightarrow \left\{ \begin{bmatrix} 1 \\ 0 \\ 2 \end{bmatrix}, \begin{bmatrix} 0 \\ 1 \\ 2 \end{bmatrix} \right\}$ is a basis for $W^{\perp}$.

7. Since $A \longrightarrow \begin{bmatrix} 1 & 0 & 1 \\ 0 & 1 & -2 \\ 0 & 0 & 0 \\ 0 & 0 & 0 \end{bmatrix}$, $\{ \begin{bmatrix} 1 & 0 & 1 \end{bmatrix}, \begin{bmatrix} 0 & 1 & -2 \end{bmatrix} \}$ is a basis for row(A).

Since $\left[\begin{array}{ccc|c} 1 & 0 & 1 & 0 \\ 0 & 1 & -2 & 0 \\ 0 & 0 & 0 & 0 \\ 0 & 0 & 0 & 0 \end{array}\right] \Rightarrow \mathbf{x}$ in null$(A) = \begin{bmatrix} -x_3 \\ 2x_3 \\ x_3 \end{bmatrix}$, $\left\{ \begin{bmatrix} -1 \\ 2 \\ 1 \end{bmatrix} \right\}$ is a basis for null(A).

$\begin{bmatrix} 1 & 0 & 1 \end{bmatrix} \begin{bmatrix} -1 \\ 2 \\ 1 \end{bmatrix} = 0$, $\begin{bmatrix} 0 & 1 & -2 \end{bmatrix} \begin{bmatrix} -1 \\ 2 \\ 1 \end{bmatrix} = 0 \Rightarrow \mathbf{v}$ in row$(A) \perp \mathbf{x}$ in null(A).

8. Note: We can stop row reduction at any step to keep calculations simple.

$$A \longrightarrow \begin{bmatrix} 1 & 1 & -1 & 0 & 2 \\ 0 & 1 & 0 & 2 & 4 \\ 0 & 0 & 0 & 0 & 1 \\ 0 & 0 & 0 & 0 & 0 \end{bmatrix}, \{[1\ 1\ -1\ 0\ 2], [0\ 1\ 0\ 2\ 4]\} \text{ is a basis for row}(A).$$

$$\left[\begin{array}{ccccc|c} 1 & 1 & -1 & 0 & 2 & 0 \\ 0 & 1 & 0 & 2 & 4 & 0 \\ 0 & 0 & 0 & 0 & 1 & 0 \\ 0 & 0 & 0 & 0 & 0 & 0 \end{array}\right] \Rightarrow \mathbf{x} = \begin{bmatrix} x_1 \\ -2x_4 \\ x_1 - 2x_4 \\ x_4 \\ 0 \end{bmatrix} \Rightarrow \left\{ \begin{bmatrix} 1 \\ 0 \\ 1 \\ 0 \\ 0 \end{bmatrix}, \begin{bmatrix} 0 \\ -2 \\ -2 \\ 1 \\ 0 \end{bmatrix} \right\} \text{ is a basis for null}(A).$$

We now verify that the basis vectors for row(A) and null(A) are orthogonal:

$$\begin{bmatrix} 1 & 1 & -1 & 0 & 2 \end{bmatrix} \begin{bmatrix} 1 \\ 0 \\ 1 \\ 0 \\ 0 \end{bmatrix} = 1 - 1 = 0. \qquad \begin{bmatrix} 1 & 1 & -1 & 0 & 2 \end{bmatrix} \begin{bmatrix} 0 \\ -2 \\ -2 \\ 1 \\ 0 \end{bmatrix} = 2 + 2 = 0.$$

$$\begin{bmatrix} 0 & 1 & 0 & 2 & 4 \end{bmatrix} \begin{bmatrix} 1 \\ 0 \\ 1 \\ 0 \\ 0 \end{bmatrix} = 0. \qquad \begin{bmatrix} 0 & 1 & 0 & 2 & 4 \end{bmatrix} \begin{bmatrix} 0 \\ -2 \\ -2 \\ 1 \\ 0 \end{bmatrix} = -2 + 2 = 0.$$

9. Note: We can stop row reduction at any step to keep calculations simple.

$$A^T \longrightarrow \begin{bmatrix} 1 & 5 & 0 & -1 \\ -1 & 2 & 1 & -1 \\ 0 & 0 & 0 & 0 \end{bmatrix}, \left\{ \begin{bmatrix} 1 \\ 5 \\ 0 \\ -1 \end{bmatrix}, \begin{bmatrix} -1 \\ 2 \\ 1 \\ -1 \end{bmatrix} \right\} \text{ is a basis for row}(A^T) = \text{col}(A).$$

$$\left[\begin{array}{cccc|c} 1 & 5 & 0 & -1 & 0 \\ -1 & 2 & 1 & -1 & 0 \\ 0 & 0 & 0 & 0 & 0 \end{array}\right] \Rightarrow \mathbf{y} = \begin{bmatrix} y_4 - 5y_2 \\ y_2 \\ 2y_4 - 7y_2 \\ y_4 \end{bmatrix}, \left\{ \begin{bmatrix} 1 \\ 0 \\ 2 \\ 1 \end{bmatrix}, \begin{bmatrix} -5 \\ 1 \\ -7 \\ 0 \end{bmatrix} \right\} \text{ is a basis for null}(A^T).$$

We now verify that the basis vectors for col(A) and null(A^T) are orthogonal:

$$\begin{bmatrix} 1 & 5 & 0 & -1 \end{bmatrix} \begin{bmatrix} 1 \\ 0 \\ 2 \\ 1 \end{bmatrix} = 1 - 1 = 0. \qquad \begin{bmatrix} 1 & 5 & 0 & -1 \end{bmatrix} \begin{bmatrix} -5 \\ 1 \\ -7 \\ 0 \end{bmatrix} = -5 + 5 = 0.$$

$$\begin{bmatrix} -1 & 2 & 1 & -1 \end{bmatrix} \begin{bmatrix} 1 \\ 0 \\ 2 \\ 1 \end{bmatrix} = -1 + 2 - 1 = 0. \qquad \begin{bmatrix} -1 & 2 & 1 & -1 \end{bmatrix} \begin{bmatrix} -5 \\ 1 \\ -7 \\ 0 \end{bmatrix} = 5 + 2 - 7 = 0.$$

10. $A^T \longrightarrow \begin{bmatrix} 1 & 0 & 0 & 1 \\ 0 & 1 & 0 & 1 \\ 0 & 0 & -1 & 1 \\ 0 & 0 & 0 & 0 \\ 0 & 0 & 0 & 0 \end{bmatrix}$, $\left\{ \begin{bmatrix} 1 \\ 0 \\ 0 \\ 1 \end{bmatrix}, \begin{bmatrix} 0 \\ 1 \\ 0 \\ 1 \end{bmatrix}, \begin{bmatrix} 0 \\ 0 \\ -1 \\ 1 \end{bmatrix} \right\}$ is a basis for $\text{row}(A^T) = \text{col}(A)$.

$$\left[\begin{array}{cccc|c} 1 & 0 & 0 & 1 & 0 \\ 0 & 1 & 0 & 1 & 0 \\ 0 & 0 & -1 & 1 & 0 \\ 0 & 0 & 0 & 0 & 0 \\ 0 & 0 & 0 & 0 & 0 \end{array}\right] \Rightarrow \mathbf{y} = \begin{bmatrix} -y_4 \\ -y_4 \\ y_4 \\ y_4 \end{bmatrix}, \left\{ \begin{bmatrix} -1 \\ -1 \\ 1 \\ 1 \end{bmatrix} \right\} \text{ is a basis for null}(A^T).$$

We now verify that the basis vectors for $\text{col}(A)$ and $\text{null}(A^T)$ are orthogonal:

$$\begin{bmatrix} 1 & 0 & 0 & 1 \end{bmatrix} \begin{bmatrix} -1 \\ -1 \\ 1 \\ 1 \end{bmatrix} = -1 + 1 = 0. \qquad \begin{bmatrix} 0 & 1 & 0 & 1 \end{bmatrix} \begin{bmatrix} -1 \\ -1 \\ 1 \\ 1 \end{bmatrix} = -1 + 1 = 0.$$

$$\begin{bmatrix} 0 & 0 & -1 & 1 \end{bmatrix} \begin{bmatrix} -1 \\ -1 \\ 1 \\ 1 \end{bmatrix} = -1 + 1 = 0.$$

11. $\left[A \,|\, \mathbf{0} \right] = \left[\begin{array}{ccc|c} 2 & 1 & -2 & 0 \\ 4 & 0 & 1 & 0 \end{array}\right] \Rightarrow \mathbf{x} = \begin{bmatrix} x_1 \\ -10x_1 \\ -4x_1 \end{bmatrix} \Rightarrow \left\{ \begin{bmatrix} 1 \\ -10 \\ -4 \end{bmatrix} \right\}$ is a basis for $W^\perp$.

12. $\left[A \,|\, \mathbf{0} \right] \longrightarrow \left[\begin{array}{cccc|c} 1 & 0 & 1 & -1 & 0 \\ 0 & 1 & -2 & 1 & 0 \end{array}\right] \Rightarrow \begin{array}{c} x_1 + x_3 - x_4 = 0 \\ x_2 - 2x_3 + x_4 = 0 \end{array} \Rightarrow \begin{array}{c} x_1 = x_4 - x_3 \\ x_2 = -x_4 - 2x_3 \end{array} \Rightarrow$

Every $\mathbf{x}$ in $\text{null}(A)$ is of the form $\begin{bmatrix} x_4 - x_3 \\ -x_4 - 2x_3 \\ x_3 \\ x_4 \end{bmatrix} \Rightarrow \left\{ \begin{bmatrix} 1 \\ -1 \\ 0 \\ 1 \end{bmatrix}, \begin{bmatrix} -1 \\ -2 \\ 1 \\ 0 \end{bmatrix} \right\}$ is a basis for $W^\perp$.

13. $\left[\begin{array}{cccc|c} 2 & -1 & 6 & 3 & 0 \\ 0 & 3 & 0 & -1 & 0 \\ 0 & 0 & 0 & 0 & 0 \end{array}\right] \Rightarrow \begin{array}{c} 2x_1 - x_2 + 6x_3 + 3x_4 = 0 \\ 3x_2 - x_4 = 0 \end{array} \Rightarrow \begin{array}{c} x_1 = -4x_2 - 3x_3 \\ x_4 = 3x_2 \end{array} \Rightarrow$

Every $\mathbf{x}$ in $\text{null}(A) = \begin{bmatrix} -4x_2 - 3x_3 \\ x_2 \\ x_3 \\ 3x_2 \end{bmatrix} \Rightarrow \left\{ \begin{bmatrix} -4 \\ 1 \\ 0 \\ 3 \end{bmatrix}, \begin{bmatrix} -3 \\ 0 \\ 1 \\ 0 \end{bmatrix} \right\}$ is a basis for $W^\perp$.

14. $\left[\begin{array}{ccccc|c} 1 & 0 & -1 & -2 & 8 & 0 \\ 0 & 2 & 1 & 3 & -11 & 0 \\ 0 & 0 & 1 & 0 & -1 & 0 \end{array}\right] \Rightarrow \begin{array}{c} x_1 - x_3 - 2x_4 + 8x_5 = 0 \\ 2x_2 + x_3 + 3x_4 - 11x_5 = 0 \\ x_3 - x_5 = 0 \end{array} \Rightarrow \begin{array}{c} x_1 = -7x_3 + 2x_4 \\ x_2 = 5x_3 - \frac{3}{2}x_4 \\ x_5 = x_3 \end{array} \Rightarrow$

Every $\mathbf{x}$ in null$(A) = \begin{bmatrix} -7x_3 + 2x_4 \\ 5x_3 - \frac{3}{2}x_4 \\ x_3 \\ x_4 \\ x_5 \end{bmatrix} \Rightarrow \left\{ \begin{bmatrix} -7 \\ 5 \\ 1 \\ 0 \\ 1 \end{bmatrix}, \begin{bmatrix} 2 \\ -\frac{3}{2} \\ 0 \\ 1 \\ 0 \end{bmatrix} \right\}$ is a basis for $W^{\perp}$.

15. So, $\mathbf{u}_1 \cdot \mathbf{v} = 3$, $\mathbf{u}_1 \cdot \mathbf{u}_1 = 2$, and $\mathbf{u}_1 = \begin{bmatrix} 1 \\ 1 \end{bmatrix} \Rightarrow \text{proj}_W(\mathbf{v}) = \frac{3}{2}\begin{bmatrix} 1 \\ 1 \end{bmatrix} = \begin{bmatrix} 3/2 \\ 3/2 \end{bmatrix}$.

16. $\mathbf{u}_1 \cdot \mathbf{v} = 2$, $\mathbf{u}_2 \cdot \mathbf{v} = 2$, $\mathbf{u}_1 \cdot \mathbf{u}_1 = 3$, $\mathbf{u}_2 \cdot \mathbf{u}_2 = 2 \Rightarrow \text{proj}_W(\mathbf{v}) = \frac{2}{3}\begin{bmatrix} 1 \\ 1 \\ 1 \end{bmatrix} + \begin{bmatrix} 1 \\ -1 \\ 0 \end{bmatrix} = \begin{bmatrix} 5/3 \\ -1/3 \\ 2/3 \end{bmatrix}$.

17. $\text{proj}_W(\mathbf{v}) = \frac{1}{9}\begin{bmatrix} 2 \\ -2 \\ 1 \end{bmatrix} + \frac{13}{18}\begin{bmatrix} -1 \\ 1 \\ 4 \end{bmatrix} = \begin{bmatrix} -1/2 \\ 1/2 \\ 3 \end{bmatrix}$.

18. $\text{proj}_W(\mathbf{v}) = \frac{1}{2}\begin{bmatrix} 1 \\ 1 \\ 0 \\ 0 \end{bmatrix} - \frac{1}{2}\begin{bmatrix} 1 \\ -1 \\ -1 \\ 1 \end{bmatrix} + \frac{1}{2}\begin{bmatrix} 0 \\ 0 \\ 1 \\ 1 \end{bmatrix} = \begin{bmatrix} 0 \\ 1 \\ 1 \\ 0 \end{bmatrix}$.

19. $\text{proj}_W(\mathbf{v}) + (\mathbf{v} - \text{proj}_W(\mathbf{v})) = -\frac{4}{10}\begin{bmatrix} 1 \\ 3 \end{bmatrix} + \left(\begin{bmatrix} 2 \\ -2 \end{bmatrix} - \frac{-4}{10}\begin{bmatrix} 1 \\ 3 \end{bmatrix}\right) = \begin{bmatrix} -2/5 \\ -6/5 \end{bmatrix} + \begin{bmatrix} 12/5 \\ -4/5 \end{bmatrix}$.

20. $\text{proj}_W(\mathbf{v}) + (\mathbf{v} - \text{proj}_W(\mathbf{v})) = \frac{3}{6}\begin{bmatrix} 1 \\ 2 \\ 1 \end{bmatrix} + \left(\begin{bmatrix} 4 \\ -2 \\ 3 \end{bmatrix} - \frac{3}{6}\begin{bmatrix} 1 \\ 2 \\ 1 \end{bmatrix}\right) = \begin{bmatrix} 1/2 \\ 1 \\ 1/2 \end{bmatrix} + \begin{bmatrix} 7/2 \\ -3 \\ 5/2 \end{bmatrix}$.

21. $\left(\frac{3}{6}\begin{bmatrix} 1 \\ 2 \\ 1 \end{bmatrix} + \frac{9}{3}\begin{bmatrix} 1 \\ -1 \\ 1 \end{bmatrix}\right) + \left(\begin{bmatrix} 4 \\ -2 \\ 3 \end{bmatrix} - \left(\frac{3}{6}\begin{bmatrix} 1 \\ 2 \\ 1 \end{bmatrix} + \frac{9}{3}\begin{bmatrix} 1 \\ -1 \\ 1 \end{bmatrix}\right)\right) = \begin{bmatrix} 7/2 \\ -2 \\ 7/2 \end{bmatrix} + \begin{bmatrix} 1/2 \\ 0 \\ -1/2 \end{bmatrix}$.

22. $\left(\frac{6}{3}\begin{bmatrix} 1 \\ -1 \\ 1 \\ 0 \end{bmatrix} + \frac{9}{3}\begin{bmatrix} 0 \\ 1 \\ 1 \\ 1 \end{bmatrix}\right) + \left(\begin{bmatrix} 2 \\ 1 \\ 5 \\ 3 \end{bmatrix} - \left(\frac{6}{3}\begin{bmatrix} 1 \\ -1 \\ 1 \\ 0 \end{bmatrix} + \frac{9}{3}\begin{bmatrix} 0 \\ 1 \\ 1 \\ 1 \end{bmatrix}\right)\right) = \begin{bmatrix} 2 \\ 1 \\ 5 \\ 3 \end{bmatrix} + \begin{bmatrix} 0 \\ 0 \\ 0 \\ 0 \end{bmatrix}$.

23. Need to show if $\mathbf{w}$ in W *and* $W^\perp$, then $\mathbf{w} = \mathbf{0}$: $\mathbf{w}$ in W *and* $W^\perp \Rightarrow \mathbf{w}\cdot\mathbf{w} = 0 \Rightarrow \mathbf{w} = \mathbf{0}$.

24. If $\mathbf{v}$ is in $W^\perp$, then $\mathbf{v}\cdot\mathbf{w}_i = 0$ because $\mathbf{w}_i$ is in W.
If $\mathbf{v}\cdot\mathbf{w}_i = 0$, then $\mathbf{v}\cdot\mathbf{w} = \mathbf{v}\cdot(c_1\mathbf{w}_1 + \ldots + c_k\mathbf{w}_k) = c_1\mathbf{v}\cdot\mathbf{w}_1 + \ldots + c_k\mathbf{v}\cdot\mathbf{w}_k = 0 \Rightarrow \mathbf{v}$ in $W^\perp$.

25. *Orthogonality to* $\mathbf{w}$ *in subspace* W *does not guarantee orthogonality to* W. Example:

$$\text{Let } \mathbf{w}' = \begin{bmatrix} -1 \\ 1 \\ 0 \end{bmatrix} \text{ and } W = \text{span}\left(\mathbf{w} = \begin{bmatrix} 1 \\ 1 \\ 0 \end{bmatrix}, \mathbf{v}' = \begin{bmatrix} 0 \\ 1 \\ 1 \end{bmatrix}\right), \text{ then } \mathbf{w}'\cdot\mathbf{w} = 0, \text{ but } \mathbf{w}'\cdot\mathbf{v}' \neq 0.$$

26. Let $V = \text{span}(\mathbf{v}_{k+1} \ldots, \mathbf{v}_n)$. We need to show $V = W^\perp$.

However, Theorem 5.13 $\Rightarrow \dim V = \dim W^\perp \Rightarrow$ So, we only need to show $V \subseteq W^\perp$.

By Theorem 5.9d., all we have to show is for every $\mathbf{v}$ in V, $\mathbf{v}_i\cdot\mathbf{v} = 0$ for i from 1 to k.

Then $\mathbf{v}_i\cdot\mathbf{v} = \mathbf{v}_i\cdot(c_{k+1}\mathbf{v}_{k+1} + \ldots + c_n\mathbf{v}_n) = c_{k+1}\mathbf{v}_i\cdot\mathbf{v}_{k+1} + \ldots + c_n\mathbf{v}_i\cdot\mathbf{v}_n = 0$
because $\mathbf{v}_1, \ldots, \mathbf{v}_n$ is an orthogonal basis. Therefore, $\mathbf{v}$ in $W^\perp \Rightarrow V \subseteq W^\perp$.

Therefore, since $\dim V = \dim W^\perp$, $V = W^\perp$.

27. Since this is an if and only if statement, there are two statements to prove.
Let $\{\mathbf{u}_k\}$ be an orthogonal basis for W.

if: If $\mathbf{x} = \text{proj}_W(\mathbf{x})$ then $\mathbf{x}$ is in W.

If $\mathbf{x} = \text{proj}_W(\mathbf{x}) = \sum_{k=1}^{n}\left(\frac{\mathbf{u}_k\cdot\mathbf{x}}{\mathbf{u}_k\cdot\mathbf{u}_k}\right)\mathbf{u}_k \overset{\mathbf{u}_k\cdot\mathbf{u}_k=1}{=} \sum_{k=1}^{n}(\mathbf{u}_k\cdot\mathbf{x})\mathbf{u}_k$, then $\mathbf{x}$ is in $\text{span}(\mathbf{u}_k) = W$.

only if: If $\mathbf{x}$ is in W then $\mathbf{x} = \text{proj}_W(\mathbf{x})$.

If $\mathbf{x}$ is in W then $\mathbf{x} = \sum_{k=1}^{n}\alpha_k\mathbf{u}_k$, but $\mathbf{u}_i\cdot\mathbf{x} = \mathbf{u}_i\cdot\left(\sum_{k=1}^{n}\alpha_k\mathbf{u}_k\right) = \alpha_i$. Why?

Because $\{\mathbf{u}_k\}$ is an orthogonal basis for W.
For a basis to be orthonormal it must have two key properties:

$$\mathbf{u}_i\cdot\mathbf{u}_j = \begin{cases} 0 & \text{if } i \neq j \quad \text{Property 1} \\ 1 & \text{if } i = j \quad \text{Property 2} \end{cases}$$

Therefore, $\mathbf{x} = \sum_{k=1}^{n}\alpha_k\mathbf{u}_k = \sum_{k=1}^{n}(\mathbf{u}_k\cdot\mathbf{x})\mathbf{u}_k = \text{proj}_W(\mathbf{x})$.

28. Since this is an if and only if statement, there are two statements to prove. Let $\{\mathbf{u}_k\}$ be an orthogonal basis for W.

if: If $\text{proj}_W(\mathbf{x}) = \mathbf{0}$ then $\mathbf{x}$ is orthogonal to W.
That is, if $\text{proj}_W(\mathbf{x}) = \mathbf{0}$ then $\mathbf{x}$ is in $W^\perp$.
That is, $\mathbf{v} \cdot \mathbf{x} = \mathbf{0}$ for all $\mathbf{v}$ in W.
If $\sum_{k=1}^{n} (\mathbf{u}_k \cdot \mathbf{x})\mathbf{u}_k = \mathbf{0}$, then $\mathbf{u}_k \cdot \mathbf{x} = \mathbf{0}$ for all $\mathbf{u}_k$ in the basis for W.
Therefore, $\mathbf{v} \cdot \mathbf{x} = \mathbf{0}$ for all $\mathbf{v}$ in W as required. Why?
If $\mathbf{v}$ is in W then $\mathbf{v} = \sum_{k=1}^{n} \beta_k \mathbf{u}_k$, but $\mathbf{v} \cdot \mathbf{x} = \left(\sum_{k=1}^{n} \beta_k \mathbf{u}_k \right) \cdot \mathbf{x} = \sum_{k=1}^{n} \beta_k (\mathbf{u}_k \cdot \mathbf{x}) = \mathbf{0}$.

only if: If $\mathbf{x}$ is orthogonal to W, then $\text{proj}_W(\mathbf{x}) = \mathbf{0}$.
If $\mathbf{x}$ is orthogonal to W, then $\mathbf{v} \cdot \mathbf{x} = 0$ for all $\mathbf{v}$ in W.
So, if $\mathcal{B} = \{\mathbf{u}_1, \ldots, \mathbf{u}_k\}$ is the basis for W, then $\mathbf{u}_i \cdot \mathbf{x} = 0$ for all i.
So, $\text{proj}_W(\mathbf{x}) = \sum_{i=1}^{k} \left(\frac{\mathbf{u}_i \cdot \mathbf{x}}{\mathbf{u}_i \cdot \mathbf{u}_i} \right) \mathbf{u}_i = \sum_{i=1}^{k} \left(\frac{0}{\mathbf{u}_i \cdot \mathbf{u}_i} \right) \mathbf{u}_i = \mathbf{0}$ as required.

29. We will use our insights from Exercise 27 to prove this assertion. Note that $\text{proj}_W(\mathbf{x})$ is in W. Why?

Because $\text{proj}_W(\mathbf{x}) = \sum_{k=1}^{n} (\mathbf{u}_k \cdot \mathbf{x})\mathbf{u}_k$, so $\text{proj}_W(\mathbf{x})$ is in $\text{span}(\mathbf{u}_k) = W$.

By Exercise 27, if $\mathbf{v}$ is in W then $\text{proj}_W(\mathbf{v}) = \mathbf{v}$.
Since $\mathbf{v} = \text{proj}_W(\mathbf{x})$ is in W, $\text{proj}_W(\text{proj}_W(\mathbf{x})) = \text{proj}_W(\mathbf{x})$.

30. We are given $\mathcal{S} = \{\mathbf{v}_1, \ldots, \mathbf{v}_k\}$ is an orthonormal set in $\mathbb{R}^n$.
Let $W = \text{span}(\mathcal{S})$ and $\mathcal{T} = \{\mathbf{b}_1, \ldots, \mathbf{b}_j\}$ be an orthonormal basis for $W^{\perp}$.
By Theorem 5.13, $\dim W + \dim W^{\perp} = k + j = n$.
Therefore, $\mathcal{S} \cup \mathcal{T}$ is an orthonormal basis for $\mathbb{R}^n$. Prove this.

(a) We need to show $\|\mathbf{x}\|^2 = \mathbf{x} \cdot \mathbf{x} = |\mathbf{x} \cdot \mathbf{x}| \geq |\mathbf{x} \cdot \mathbf{v}_1|^2 + \cdots + |\mathbf{x} \cdot \mathbf{v}_k|^2 = \sum_{s=1}^{k} |\mathbf{x} \cdot \mathbf{v}_s|^2$.

Since $\mathcal{S}$ and $\mathcal{T}$ are orthonormal bases, we have the following two key properties:

$$\mathbf{v}_c \cdot \mathbf{v}_d \text{ and } \mathbf{b}_c \cdot \mathbf{b}_d = \begin{cases} 0 & \text{if } c \neq d \quad \text{Property 1} \\ 1 & \text{if } c = d \quad \text{Property 2} \end{cases}$$

Since $\mathcal{S} \cup \mathcal{T}$ is a basis for $\mathbb{R}^n$, there exist α_s, β_t such that $\mathbf{x} = \sum_{s=1}^{k} \alpha_s \mathbf{v}_s + \sum_{t=1}^{j} \beta_t \mathbf{b}_t$.

The key facts are:

$$\mathbf{x} \cdot \mathbf{v}_i = \left(\sum_{s=1}^{k} \alpha_s \mathbf{v}_s + \sum_{t=1}^{k} \beta_t \mathbf{b}_t\right) \cdot \mathbf{v}_i = \alpha_i \text{ and } \mathbf{x} \cdot \mathbf{b}_i = \left(\sum_{s=1}^{k} \alpha_s \mathbf{v}_s + \sum_{t=1}^{k} \beta_t \mathbf{b}_t\right) \cdot \mathbf{b}_i = \beta_i.$$

Why?

Because $\mathcal{T}$ is a basis for $W^{\perp}$. Therefore, $\mathbf{v}_s \cdot \mathbf{b}_t = 0$ for all s and for all t.

$$\text{So, } \|\mathbf{x}\|^2 = \mathbf{x} \cdot \mathbf{x} = \left(\sum_{s=1}^{k} \alpha_s \mathbf{v}_s + \sum_{t=1}^{j} \beta_t \mathbf{b}_t\right) \cdot \left(\sum_{s=1}^{k} \alpha_s \mathbf{v}_s + \sum_{t=1}^{j} \beta_t \mathbf{b}_t\right) \overset{\substack{\text{Prop} \\ \text{1 and 2}}}{=} \sum_{s=1}^{k} \alpha_s^2 + \sum_{t=1}^{j} \beta_t^2$$

$$\overset{\substack{\text{key} \\ \text{facts}}}{=} \sum_{s=1}^{k} |\mathbf{x} \cdot \mathbf{v}_s|^2 + \sum_{t=1}^{j} |\mathbf{x} \cdot \mathbf{b}_t|^2 \geq \sum_{s=1}^{k} |\mathbf{x} \cdot \mathbf{v}_s|^2 \text{ as we were to show.}$$

(b) Our proof in part (a) implies that equality holds if and only if $\mathbf{x}$ is in span(S).
Why?

Since $\mathbf{x}$ is in span(S) if and only if $\sum_{t=1}^{j} |\mathbf{x} \cdot \mathbf{b}_t|^2 = 0$.

Why?

Because $\mathcal{T} = \{\mathbf{b}_t\}$ is a basis for $W^{\perp}$, where $W = \text{span}(S)$.
Therefore, $\mathbf{x} \cdot \mathbf{b}_t = 0$ for all t.

Q: Do we actually have a process for creating the orthonormal basis $\mathcal{T}$?
A: No, but that is the goal of the next section, the Gram-Schmidt Process .

5.3 The Gram-Schmidt Process and the QR Factorization

1. Applying Gram-Schmidt $\Rightarrow \mathbf{v}_1 = \mathbf{x}_1$, $\mathbf{v}_2 = \begin{bmatrix}1\\2\end{bmatrix} - \frac{3}{2}\begin{bmatrix}1\\1\end{bmatrix} = \begin{bmatrix}-1/2\\1/2\end{bmatrix}$.

$\|\mathbf{v}_1\| = \sqrt{2}$, $\|\mathbf{v}_2\| = \frac{\sqrt{2}}{2} \Rightarrow \mathbf{q}_1 = \frac{1}{\sqrt{2}}\begin{bmatrix}1\\1\end{bmatrix} = \begin{bmatrix}1/\sqrt{2}\\1/\sqrt{2}\end{bmatrix}$, $\mathbf{q}_2 = \frac{2}{\sqrt{2}}\begin{bmatrix}-1/2\\1/2\end{bmatrix} = \begin{bmatrix}-1/\sqrt{2}\\1/\sqrt{2}\end{bmatrix}$.

2. Applying Gram-Schmidt $\Rightarrow \mathbf{v}_1 = \mathbf{x}_1$, $\mathbf{v}_2 = \begin{bmatrix}3\\1\end{bmatrix} - \frac{6}{18}\begin{bmatrix}3\\-3\end{bmatrix} = \begin{bmatrix}2\\2\end{bmatrix}$.

$\|\mathbf{v}_1\| = 3\sqrt{2}$, $\|\mathbf{v}_2\| = 2\sqrt{2} \Rightarrow \mathbf{q}_1 = \frac{1}{3\sqrt{2}}\begin{bmatrix}3\\-3\end{bmatrix} = \begin{bmatrix}1/\sqrt{2}\\-1/\sqrt{2}\end{bmatrix}$, $\mathbf{q}_2 = \frac{1}{2\sqrt{2}}\begin{bmatrix}2\\2\end{bmatrix} = \begin{bmatrix}1/\sqrt{2}\\1/\sqrt{2}\end{bmatrix}$.

3. G-S $\Rightarrow \mathbf{v}_2 = \begin{bmatrix}0\\3\\3\end{bmatrix} + 2\begin{bmatrix}1\\-1\\-1\end{bmatrix} = \begin{bmatrix}2\\1\\1\end{bmatrix}$, $\mathbf{v}_3 = \begin{bmatrix}3\\2\\4\end{bmatrix} + 1\begin{bmatrix}1\\-1\\-1\end{bmatrix} - 2\begin{bmatrix}2\\1\\1\end{bmatrix} = \begin{bmatrix}0\\-1\\1\end{bmatrix}$.

Therefore, $\|\mathbf{v}_1\| = \sqrt{3}$, $\|\mathbf{v}_2\| = \sqrt{6}$, $\|\mathbf{v}_3\| = \sqrt{2} \Rightarrow$

$$\mathbf{q}_1 = \begin{bmatrix}1/\sqrt{3}\\-1/\sqrt{3}\\-1/\sqrt{3}\end{bmatrix}, \mathbf{q}_2 = \begin{bmatrix}2/\sqrt{6}\\1/\sqrt{6}\\1/\sqrt{6}\end{bmatrix}, \mathbf{q}_3 = \begin{bmatrix}0\\-1/\sqrt{2}\\1/\sqrt{2}\end{bmatrix}.$$

4. $\mathbf{v}_2 = \begin{bmatrix}1\\1\\0\end{bmatrix} - \frac{2}{3}\begin{bmatrix}1\\1\\1\end{bmatrix} = \begin{bmatrix}1/3\\1/3\\-2/3\end{bmatrix}$, $\mathbf{v}_3 = \begin{bmatrix}1\\0\\0\end{bmatrix} - \frac{1}{3}\begin{bmatrix}1\\1\\1\end{bmatrix} - \frac{1}{2}\begin{bmatrix}1/3\\1/3\\-2/3\end{bmatrix} = \begin{bmatrix}1/2\\-1/2\\0\end{bmatrix}$.

Therefore, $\|\mathbf{v}_1\| = \sqrt{3}$, $\|\mathbf{v}_2\| = \frac{\sqrt{6}}{3}$, $\|\mathbf{v}_3\| = \frac{\sqrt{2}}{2} \Rightarrow$

$$\mathbf{q}_1 = \begin{bmatrix}1/\sqrt{3}\\1/\sqrt{3}\\1/\sqrt{3}\end{bmatrix}, \mathbf{q}_2 = \begin{bmatrix}1/\sqrt{6}\\1/\sqrt{6}\\-2/\sqrt{6}\end{bmatrix}, \mathbf{q}_3 = \begin{bmatrix}1/\sqrt{2}\\-1/\sqrt{2}\\0\end{bmatrix}.$$

5. $\mathbf{v}_2 = \begin{bmatrix}3\\4\\2\end{bmatrix} - \frac{7}{2}\begin{bmatrix}1\\1\\0\end{bmatrix} = \begin{bmatrix}-1/2\\1/2\\2\end{bmatrix} \Rightarrow$ Orthogonal basis $= \left\{\begin{bmatrix}1\\1\\0\end{bmatrix}, \begin{bmatrix}-1/2\\1/2\\2\end{bmatrix}\right\}$.

6. $\mathbf{v}_2 = \begin{bmatrix}3\\-1\\0\\4\end{bmatrix} - \frac{15}{10}\begin{bmatrix}2\\-1\\1\\2\end{bmatrix} = \begin{bmatrix}0\\1/2\\-3/2\\1\end{bmatrix} \Rightarrow$ Orthogonal basis $= \left\{\begin{bmatrix}2\\-1\\1\\2\end{bmatrix}, \begin{bmatrix}0\\1/2\\-3/2\\1\end{bmatrix}\right\}$.

7. $$\left(0\begin{bmatrix}1\\1\\0\end{bmatrix} + \frac{4}{9}\begin{bmatrix}-1/2\\1/2\\2\end{bmatrix}\right) + \left(\begin{bmatrix}4\\-4\\3\end{bmatrix} - \left(0\begin{bmatrix}1\\1\\0\end{bmatrix} + \frac{4}{9}\begin{bmatrix}-1/2\\1/2\\2\end{bmatrix}\right)\right) = \begin{bmatrix}-2/9\\2/9\\8/9\end{bmatrix} + \begin{bmatrix}38/9\\-38/9\\19/9\end{bmatrix}.$$

8. $\left(\frac{1}{5}\begin{bmatrix}2\\-1\\1\\2\end{bmatrix}+\frac{8}{7}\begin{bmatrix}0\\-1/2\\-3/2\\1\end{bmatrix}\right)+\left(\begin{bmatrix}1\\4\\0\\2\end{bmatrix}-\left(\frac{1}{5}\begin{bmatrix}2\\-1\\1\\2\end{bmatrix}+\frac{8}{7}\begin{bmatrix}0\\-1/2\\-3/2\\1\end{bmatrix}\right)\right)=$

$$\begin{bmatrix}2/5\\-27/35\\-53/35\\54/35\end{bmatrix}+\begin{bmatrix}3/5\\167/35\\53/35\\16/35\end{bmatrix}.$$

9. $\mathbf{v}_2=\begin{bmatrix}1\\0\\1\end{bmatrix}-\frac{1}{2}\begin{bmatrix}0\\1\\1\end{bmatrix}=\begin{bmatrix}1\\-1/2\\1/2\end{bmatrix}$, $\mathbf{v}_3=\begin{bmatrix}1\\1\\0\end{bmatrix}-\frac{1}{2}\begin{bmatrix}0\\1\\1\end{bmatrix}-\frac{1}{3}\begin{bmatrix}1\\-1/2\\1/2\end{bmatrix}=\begin{bmatrix}2/3\\2/3\\-2/3\end{bmatrix}$.

Therefore, $\left\{\begin{bmatrix}0\\1\\1\end{bmatrix},\begin{bmatrix}1\\-1/2\\1/2\end{bmatrix},\begin{bmatrix}2/3\\2/3\\-2/3\end{bmatrix}\right\}$ is an orthogonal basis for col(A).

10. $\mathbf{v}_2=\begin{bmatrix}1\\-1\\1\\5\end{bmatrix}-\begin{bmatrix}1\\1\\-1\\1\end{bmatrix}=\begin{bmatrix}0\\-2\\2\\4\end{bmatrix}$, $\mathbf{v}_3=\begin{bmatrix}1\\2\\0\\1\end{bmatrix}-\begin{bmatrix}1\\1\\-1\\1\end{bmatrix}-0\begin{bmatrix}0\\-2\\2\\4\end{bmatrix}=\begin{bmatrix}0\\1\\1\\0\end{bmatrix}$.

11. $\mathbf{v}_2=\begin{bmatrix}0\\1\\0\end{bmatrix}-\frac{1}{35}\begin{bmatrix}3\\1\\5\end{bmatrix}=\begin{bmatrix}-3/35\\34/35\\-1/7\end{bmatrix}$, $\mathbf{v}_3=\begin{bmatrix}0\\0\\1\end{bmatrix}-\frac{1}{7}\begin{bmatrix}3\\1\\5\end{bmatrix}+\frac{5}{34}\begin{bmatrix}-3/35\\34/35\\-1/7\end{bmatrix}=\begin{bmatrix}-15/34\\0\\9/34\end{bmatrix}$.

Therefore, $\left\{\begin{bmatrix}3\\1\\5\end{bmatrix},\begin{bmatrix}-3/35\\34/35\\-1/7\end{bmatrix},\begin{bmatrix}-15/34\\0\\9/34\end{bmatrix}\right\}$ is an orthogonal basis.

12. $\mathbf{v}_2=\begin{bmatrix}1\\0\\3\\2\end{bmatrix}-0\begin{bmatrix}2\\1\\0\\-1\end{bmatrix}=\begin{bmatrix}1\\0\\3\\2\end{bmatrix}$, $\mathbf{v}_3=\begin{bmatrix}0\\1\\0\\0\end{bmatrix}-\frac{1}{6}\begin{bmatrix}2\\1\\0\\-1\end{bmatrix}-0\begin{bmatrix}1\\0\\3\\2\end{bmatrix}=\begin{bmatrix}-1/3\\5/6\\0\\1/6\end{bmatrix}$,

$$\mathbf{v}_4=\begin{bmatrix}0\\0\\1\\0\end{bmatrix}-0\begin{bmatrix}2\\1\\0\\-1\end{bmatrix}-\frac{3}{14}\begin{bmatrix}1\\0\\3\\2\end{bmatrix}-0\begin{bmatrix}-1/3\\5/6\\0\\1/6\end{bmatrix}=\begin{bmatrix}-3/14\\0\\5/14\\-3/7\end{bmatrix}.$$

13. We need to choose an $\mathbf{x}_3$ and then use the Gram-Schmidt Process.

Since $\mathbf{q}_1 = \begin{bmatrix} \frac{1}{\sqrt{2}} \\ 0 \\ -\frac{1}{\sqrt{2}} \end{bmatrix}$ and $\mathbf{q}_2 = \begin{bmatrix} \frac{1}{\sqrt{3}} \\ \frac{1}{\sqrt{3}} \\ \frac{1}{\sqrt{3}} \end{bmatrix}$, we can let $\mathbf{x}_3 = \begin{bmatrix} 0 \\ 0 \\ 1 \end{bmatrix}$. Why?

Q: How can we see that this is a good choice for $\mathbf{x}_3$?
A: It may help to temporarily ignore the fact that the columns of Q have been normalized.

That is, consider $P = \begin{bmatrix} 1 & 1 & * \\ 0 & 1 & * \\ -1 & 1 & * \end{bmatrix}$ then let $\mathbf{p}_3 = \mathbf{x}_3$. So $P = \begin{bmatrix} 1 & 1 & 0 \\ 0 & 1 & 0 \\ -1 & 1 & 1 \end{bmatrix}$.

Then $P \longrightarrow I$. That is, the columns of P are linearly independent.

Applying the Gram-Schmidt Process, we have:

$$\mathbf{v}_3 = \mathbf{x}_3 - \left(\frac{\mathbf{q}_1 \cdot \mathbf{x}_3}{\mathbf{q}_1 \cdot \mathbf{q}_1}\right)\mathbf{q}_1 - \left(\frac{\mathbf{q}_2 \cdot \mathbf{x}_3}{\mathbf{q}_2 \cdot \mathbf{q}_2}\right)\mathbf{q}_2 = \begin{bmatrix} 0 \\ 0 \\ 1 \end{bmatrix} + \frac{1}{\sqrt{2}}\begin{bmatrix} \frac{1}{\sqrt{2}} \\ 0 \\ -\frac{1}{\sqrt{2}} \end{bmatrix} - \frac{1}{\sqrt{3}}\begin{bmatrix} \frac{1}{\sqrt{3}} \\ \frac{1}{\sqrt{3}} \\ \frac{1}{\sqrt{3}} \end{bmatrix} = \begin{bmatrix} -\frac{1}{2} \\ 1 \\ \frac{1}{2} \end{bmatrix}.$$

In order for Q to be an orthogonal matrix, its columns must be orthonormal.
Therefore, we must compute the length of $\mathbf{x}_3$ and divide each of the components by it.

Clearly, $\|\mathbf{x}_3\| = \sqrt{\left(\frac{1}{2}\right)^2 + 1^2 \left(-\frac{1}{2}\right)^2} = \frac{\sqrt{6}}{2}$.

Therefore, the completed matrix $Q = \begin{bmatrix} \frac{1}{\sqrt{2}} & \frac{1}{\sqrt{3}} & -\frac{1}{\sqrt{6}} \\ 0 & \frac{1}{\sqrt{3}} & \frac{2}{\sqrt{6}} \\ -\frac{1}{\sqrt{2}} & \frac{1}{\sqrt{3}} & \frac{1}{\sqrt{6}} \end{bmatrix}$.

Q: Why is Q an orthogonal matrix?
A: Because it has orthonormal columns.

Q: Therefore, what two conditions do the columns of Q satisfy?
A: For a set to be orthonormal it must have two key properties:

$$\mathbf{q}_i \cdot \mathbf{q}_j = \begin{cases} 0 \text{ if } i \neq j & \text{Property 1} \\ 1 \text{ if } i = j & \text{Property 2} \end{cases}$$

Verify this.

Q: Do the rows of Q also form an orthonormal set?
A: According to Theorem 5.7 of Section 5.1, they should. Verify this.
This verification also provides a quick check on our final result.

Q: If we have an orthogonal set can we divide by the lengths to create an orthonormal set?
A: Yes. Why?
If $\mathbf{v} \cdot c\mathbf{w} = \mathbf{0}$ $(c \neq 0)$, then $\mathbf{v} \cdot \mathbf{w} = \mathbf{0}$.
Why is this sufficient? Verify that this claim is true.

14. We need to choose an $\mathbf{x}_3$, $\mathbf{x}_4$ and then use the Gram-Schmidt Process.
We will also use our insight to simplify the computation.

As noted in Exercise 13, we can start by creating an orthogonal set of vectors.
Normalizing at the end simplifies all the calculations prior to that last step.

That is, consider $P = \begin{bmatrix} 1 & 2 & * & * \\ 1 & 1 & * & * \\ 1 & 0 & * & * \\ 1 & -3 & * & * \end{bmatrix}$. Now let $\mathbf{x}_3 = \begin{bmatrix} 0 \\ 0 \\ 0 \\ 1 \end{bmatrix}$, $\mathbf{x}_4 = \begin{bmatrix} 0 \\ 0 \\ 1 \\ 0 \end{bmatrix}$ so $P = \begin{bmatrix} 1 & 2 & 0 & 0 \\ 1 & 1 & 0 & 0 \\ 1 & 0 & 0 & 1 \\ 1 & -3 & 1 & 0 \end{bmatrix}$.

Q: How do we know that the columns of P are linearly independent?
A: Because $P \longrightarrow I$. Verify this.

Applying Gram-Schmidt to the first three columns of P yields:

$$\mathbf{v}_3 = \begin{bmatrix} 0 \\ 0 \\ 0 \\ 1 \end{bmatrix} - \frac{1}{4}\begin{bmatrix} 1 \\ 1 \\ 1 \\ 1 \end{bmatrix} + \frac{3}{14}\begin{bmatrix} 2 \\ 1 \\ 0 \\ -3 \end{bmatrix} = \frac{1}{28}\begin{bmatrix} 5 \\ -1 \\ -7 \\ 3 \end{bmatrix}. \text{ Let } \mathbf{p}_3 = \begin{bmatrix} 5 \\ -1 \\ -7 \\ 3 \end{bmatrix}, \text{ so } P = \begin{bmatrix} 1 & 2 & 5 & 0 \\ 1 & 1 & -1 & 0 \\ 1 & 0 & -7 & 1 \\ 1 & -3 & 3 & 0 \end{bmatrix}.$$

Q: Why can we ignore the scalar multiple when selecting $\mathbf{p}_3$?
A: If $\mathbf{v} \cdot c\mathbf{w} = \mathbf{0}$ $(c \neq 0)$, then $\mathbf{v} \cdot \mathbf{w} = \mathbf{0}$. Verify this.

Q: Where did we already use this principle in our selection process?
A: In our selection of P. We ignored the scalar factors that normalized the columns of Q.

Applying Gram-Schmidt to all the columns of P yields:

$$\mathbf{v}_4 = \begin{bmatrix} 0 \\ 0 \\ 1 \\ 0 \end{bmatrix} - \frac{1}{4}\begin{bmatrix} 1 \\ 1 \\ 1 \\ 1 \end{bmatrix} + \frac{1}{12}\begin{bmatrix} 5 \\ -1 \\ -7 \\ 3 \end{bmatrix} = \frac{1}{6}\begin{bmatrix} 1 \\ -2 \\ 1 \\ 0 \end{bmatrix}. \text{ Let } \mathbf{p}_4 = \begin{bmatrix} 1 \\ -2 \\ 1 \\ 0 \end{bmatrix}, \text{ so } P = \begin{bmatrix} 1 & 2 & 5 & 1 \\ 1 & 1 & -1 & -2 \\ 1 & 0 & -7 & 1 \\ 1 & -3 & 3 & 0 \end{bmatrix}.$$

Q: As noted above the last step is simply to normalize all the columns. How?
A: Calculate the lengths of each column and divide the components by it.
Note that the two given columns of Q were already normalized.

Therefore, the completed matrix $Q = \begin{bmatrix} \frac{1}{2} & \frac{2}{\sqrt{14}} & \frac{5}{\sqrt{84}} & \frac{1}{\sqrt{6}} \\ \frac{1}{2} & \frac{1}{\sqrt{14}} & -\frac{1}{\sqrt{84}} & -\frac{2}{\sqrt{6}} \\ \frac{1}{2} & 0 & -\frac{7}{\sqrt{84}} & \frac{1}{\sqrt{6}} \\ \frac{1}{2} & -\frac{3}{\sqrt{14}} & \frac{3}{\sqrt{84}} & 0 \end{bmatrix}$.

Q: Do the rows of Q also form an orthonormal set?
A: According to Theorem 5.7 of Section 5.1, they should. Verify this.
This verification also provides a quick check on our final result.

Q: What is another means of completing the orthogonal set?
A: As noted throughout the text, sometimes we can simply use our insight to *guess*.
This is a worthwhile exercise both in terms of saving time and developing our insight.

15. From 9, $\|\mathbf{v}_1\| = \sqrt{2}$, $\|\mathbf{v}_2\| = \frac{\sqrt{6}}{2}$, $\|\mathbf{v}_3\| = \frac{2\sqrt{3}}{3} \Rightarrow Q = \begin{bmatrix} 0 & 2/\sqrt{6} & 1/\sqrt{3} \\ 1/\sqrt{2} & -1/\sqrt{6} & 1/\sqrt{3} \\ 1/\sqrt{2} & 1/\sqrt{6} & -1/\sqrt{3} \end{bmatrix}$.

Therefore, $R = Q^T A = \begin{bmatrix} 0 & 1/\sqrt{2} & 1/\sqrt{2} \\ 2/\sqrt{6} & -1/\sqrt{6} & 1/\sqrt{6} \\ 1/\sqrt{3} & 1/\sqrt{3} & -1/\sqrt{3} \end{bmatrix} \begin{bmatrix} 0 & 1 & 1 \\ 1 & 0 & 1 \\ 1 & 1 & 0 \end{bmatrix} = \begin{bmatrix} \sqrt{2} & 1/\sqrt{2} & 1/\sqrt{2} \\ 0 & 3/\sqrt{6} & 1/\sqrt{6} \\ 0 & 0 & 2/\sqrt{3} \end{bmatrix}$.

Verify that $A = QR = \begin{bmatrix} 0 & 2/\sqrt{6} & 1/\sqrt{3} \\ 1/\sqrt{2} & -1/\sqrt{6} & 1/\sqrt{3} \\ 1/\sqrt{2} & 1/\sqrt{6} & -1/\sqrt{3} \end{bmatrix} \begin{bmatrix} \sqrt{2} & 1/\sqrt{2} & 1/\sqrt{2} \\ 0 & 3/\sqrt{6} & 1/\sqrt{6} \\ 0 & 0 & 2/\sqrt{3} \end{bmatrix} = \begin{bmatrix} 0 & 1 & 1 \\ 1 & 0 & 1 \\ 1 & 1 & 0 \end{bmatrix}$.

16. From 10, $\|\mathbf{v}_1\| = 2$, $\|\mathbf{v}_2\| = 2\sqrt{6}$, $\|\mathbf{v}_3\| = \sqrt{2} \Rightarrow Q = \begin{bmatrix} 1/2 & 0 & 0 \\ 1/2 & -1/\sqrt{6} & 1/\sqrt{2} \\ -1/2 & 1/\sqrt{6} & 1/\sqrt{2} \\ 1/2 & 2/\sqrt{6} & 0 \end{bmatrix}$.

Therefore, $R = Q^T A = \begin{bmatrix} 1/2 & 1/2 & -1/2 & 1/2 \\ 0 & -1/\sqrt{6} & 1/\sqrt{6} & 2/\sqrt{6} \\ 0 & 1/\sqrt{2} & 1/\sqrt{2} & 0 \end{bmatrix} \begin{bmatrix} 1 & 1 & 1 \\ 1 & -1 & 2 \\ -1 & 1 & 0 \\ 1 & 5 & 1 \end{bmatrix} = \begin{bmatrix} 2 & 2 & 2 \\ 0 & 2\sqrt{6} & 0 \\ 0 & 0 & \sqrt{2} \end{bmatrix}$.

17. $R = Q^T A = \begin{bmatrix} 2/3 & 1/3 & -2/3 \\ 1/3 & 2/3 & 2/3 \\ 2/3 & -2/3 & 1/3 \end{bmatrix} \begin{bmatrix} 2 & 8 & 2 \\ 1 & 7 & -1 \\ -2 & -2 & 1 \end{bmatrix} = \begin{bmatrix} 3 & 9 & 1/3 \\ 0 & 6 & 2/3 \\ 0 & 0 & 7/3 \end{bmatrix}$.

Verify that $A = QR = \begin{bmatrix} 2/3 & 1/3 & 2/3 \\ 1/3 & 2/3 & -2/3 \\ -2/3 & 2/3 & 1/3 \end{bmatrix} \begin{bmatrix} 3 & 9 & 1/3 \\ 0 & 6 & 2/3 \\ 0 & 0 & 7/3 \end{bmatrix} = \begin{bmatrix} 2 & 8 & 2 \\ 1 & 7 & -1 \\ -2 & -2 & 1 \end{bmatrix}$.

18. $R = Q^T A = \begin{bmatrix} 1/\sqrt{6} & 2/\sqrt{6} & -1/\sqrt{6} & 0 \\ 1/\sqrt{3} & 0 & 1/\sqrt{3} & 1/\sqrt{3} \end{bmatrix} \begin{bmatrix} 1 & 3 \\ 2 & 4 \\ -1 & -1 \\ 0 & 1 \end{bmatrix} = \begin{bmatrix} \sqrt{6} & 2\sqrt{6} \\ 0 & \sqrt{3} \end{bmatrix}$.

Verify that $A = QR = \begin{bmatrix} 1/\sqrt{6} & 1/\sqrt{3} \\ 2/\sqrt{6} & 0 \\ -1/\sqrt{6} & 1/\sqrt{3} \\ 0 & 1/\sqrt{3} \end{bmatrix} \begin{bmatrix} \sqrt{6} & 2\sqrt{6} \\ 0 & \sqrt{3} \end{bmatrix} = \begin{bmatrix} 1 & 3 \\ 2 & 4 \\ -1 & -1 \\ 0 & 1 \end{bmatrix}$.

19. Since A is orthogonal, simply let $A = Q \Rightarrow A = AI$, where I is obviously upper triangular.

20. A invertible $\Rightarrow A$ has linearly independent columns, so Theorem 5.16 $\Rightarrow A = QR$.
$A = QR$, R upper triangular with nonzero diagonal entries $\Rightarrow Q, R$ invertible $\Rightarrow A$ invertible.

Note: $Q^{-1} = Q^T$, $A = QR \Rightarrow Q^T A = R \Rightarrow R^{-1} Q^T A = R^{-1} R = I \Rightarrow A^{-1} = R^{-1} Q^T$.

21. $A^{-1} = R^{-1}Q^T = \begin{bmatrix} 1/\sqrt{2} & -1/\sqrt{6} & -1/2\sqrt{3} \\ 0 & 2/\sqrt{6} & -1/2\sqrt{3} \\ 0 & 0 & 3/2\sqrt{3} \end{bmatrix} \begin{bmatrix} 0 & 1/\sqrt{2} & 1/\sqrt{2} \\ 2/\sqrt{6} & -1/\sqrt{6} & 1/\sqrt{6} \\ 1/\sqrt{3} & 1/\sqrt{3} & -1/\sqrt{3} \end{bmatrix}$

$= \begin{bmatrix} -1/2 & 1/2 & 1/2 \\ 1/2 & -1/2 & 1/2 \\ 1/2 & 1/2 & -1/2 \end{bmatrix}$, so $AA^{-1} = \begin{bmatrix} 0 & 1 & 1 \\ 1 & 0 & 1 \\ 1 & 1 & 0 \end{bmatrix} \begin{bmatrix} -1/2 & 1/2 & 1/2 \\ 1/2 & -1/2 & 1/2 \\ 1/2 & 1/2 & -1/2 \end{bmatrix} = \begin{bmatrix} 1 & 0 & 0 \\ 0 & 1 & 0 \\ 0 & 0 & 1 \end{bmatrix}$.

22. We present a solution for Exercise 17 instead of Exercise 15 as cited in the text.

$A^{-1} = R^{-1}Q^T = \begin{bmatrix} 1/3 & -1/2 & 2/21 \\ 0 & 1/6 & -1/21 \\ 0 & 0 & 3/7 \end{bmatrix} \begin{bmatrix} 2/3 & 1/3 & -2/3 \\ 1/3 & 2/3 & 2/3 \\ 2/3 & -2/3 & 1/3 \end{bmatrix}$

$= \begin{bmatrix} 5/42 & -2/7 & -11/21 \\ 1/42 & 1/7 & 2/21 \\ 2/7 & -2/7 & 1/7 \end{bmatrix}$. Verify $AA^{-1} = I$.

23. Recall, Q orthogonal $\Rightarrow Q^T$ orthogonal $\Rightarrow \|Q^T\mathbf{x}\| = \|\mathbf{x}\|$.

We need to show $R\mathbf{x} = \mathbf{0} \Rightarrow \mathbf{x} = \mathbf{0}$, where $A = QR \Rightarrow R = Q^TA$.

So, $R\mathbf{x} = \mathbf{0} \Rightarrow \|R\mathbf{x}\| = 0 \Rightarrow \|(Q^TA)\mathbf{x}\| = \|Q^T(A\mathbf{x})\| = \|A\mathbf{x}\| = 0 \Rightarrow$
$\sum \mathbf{a}_i x_i = 0$ (since $A\mathbf{x}$ is just a linear combination of the columns of A) $\Rightarrow$
$x_i = 0$ because the columns of A, $\mathbf{a}_i$, are linearly independent $\Rightarrow$
$A\mathbf{x} = 0 \Rightarrow \mathbf{x} = 0$.

Therefore, R is invertible by property (c) of the Fundamental Theorem.

24. Recall, $\text{row}(A) = \text{row}(B) \Leftrightarrow$ there exists an invertible matrix M such that $A = MB$.

$A = QR \Rightarrow A^T = R^TQ^T$, R^T invertible $\Rightarrow \text{row}(A^T) = \text{row}(Q^T) \Rightarrow \text{col}(A) = \text{col}(Q)$.

Exploration: The Modified QR Factorization

Explorations are self-contained, so only solutions will be provided.

1. $Q = \begin{bmatrix} 1-2d_1^2 & -2d_1d_2 \\ -2d_1d_2 & 1-2d_2^2 \end{bmatrix} = \begin{bmatrix} 1 & 0 \\ 0 & 1 \end{bmatrix} - 2\begin{bmatrix} d_1^2 & d_1d_2 \\ d_1d_2 & d_2^2 \end{bmatrix} = I - 2\mathbf{u}\mathbf{u}^T.$

2. We will compute Q for (a) and start the process for (b).

(a) $\mathbf{u} = \begin{bmatrix} \frac{3}{5} \\ \frac{4}{5} \end{bmatrix} \Rightarrow \mathbf{u}\mathbf{u}^T = \begin{bmatrix} \frac{9}{25} & \frac{12}{25} \\ \frac{12}{25} & \frac{16}{25} \end{bmatrix} \Rightarrow Q = I - 2\mathbf{u}\mathbf{u}^T = \begin{bmatrix} 1 & 0 \\ 0 & 1 \end{bmatrix} - 2\begin{bmatrix} \frac{9}{25} & \frac{12}{25} \\ \frac{12}{25} & \frac{16}{25} \end{bmatrix} = \begin{bmatrix} \frac{7}{25} & -\frac{24}{25} \\ -\frac{24}{25} & -\frac{7}{25} \end{bmatrix}.$

(b) We start with $\mathbf{u} = \left(\frac{1}{\|\mathbf{x}-\mathbf{y}\|}\right)(\mathbf{x}-\mathbf{y}) = \frac{1}{2\sqrt{5}}\begin{bmatrix} 4 \\ -2 \end{bmatrix} = \begin{bmatrix} \frac{2}{\sqrt{5}} \\ -\frac{1}{\sqrt{5}} \end{bmatrix}.$

Also, note the definition of $\mathbf{u} \Rightarrow \|\mathbf{u}\| = 1$ since $\|\mathbf{u}\| = \left\|\frac{\mathbf{x}-\mathbf{y}}{\|\mathbf{x}-\mathbf{y}\|}\right\| = \frac{\|\mathbf{x}-\mathbf{y}\|}{\|\mathbf{x}-\mathbf{y}\|} = 1.$
It follows that $\mathbf{u}^T\mathbf{u} = \|\mathbf{u}\|^2 = 1^2 = 1.$

3. (a) $Q^T = (I - 2\mathbf{u}\mathbf{u}^T)^T = I - 2(\mathbf{u}^T)^T\mathbf{u}^T = I - 2\mathbf{u}\mathbf{u}^T.$

(b) Show $Q^T = Q^{-1}$, then apply Section 5.1, Theorem 5.5.

(c) Follows from (a) and (b) since $I = QQ^T = QQ = Q^2 = I.$
In particular, note that $\mathbf{u}^T\mathbf{u} = 1$ is key to this property.

$$QQ^T = Q^2 = (I - 2\mathbf{u}\mathbf{u}^T)(I - 2\mathbf{u}\mathbf{u}^T) = I^2 - 4\mathbf{u}\mathbf{u}^T - (2\mathbf{u}\mathbf{u}^T)(2\mathbf{u}\mathbf{u}^T)$$
$$= I - 4\mathbf{u}\mathbf{u}^T - 4\mathbf{u}(\mathbf{u}^T\mathbf{u})\mathbf{u}^T = I - 4\mathbf{u}\mathbf{u}^T - 4\mathbf{u}\mathbf{u}^T = I.$$

4. Again, we will use the fact that $\|\mathbf{u}\| = 1 \Rightarrow \mathbf{u}^T\mathbf{u} = \mathbf{u}\cdot\mathbf{u} = \|\mathbf{u}\|^2 = 1.$

Let $\mathbf{v} = c\mathbf{u}$, so: $Q\mathbf{v} = Q(c\mathbf{u}) = (I - 2\mathbf{u}\mathbf{u}^T)(c\mathbf{u}) = c(\mathbf{u} - 2\mathbf{u}(\mathbf{u}^T\mathbf{u})) = c(\mathbf{u} - 2\mathbf{u}) = -c\mathbf{u} = -\mathbf{v}.$

Let $\mathbf{v}\cdot\mathbf{u} = \mathbf{u}^T\mathbf{v} = 0$: $Q\mathbf{v} = (I - 2\mathbf{u}\mathbf{u}^T)\mathbf{v} = \mathbf{v} - 2\mathbf{u}(\mathbf{u}^T\mathbf{v}) = \mathbf{v}.$

5. Correction: Note that we have to normalize the given $\mathbf{u}$ to get:

$$\mathbf{u} = \begin{bmatrix} \frac{1}{\sqrt{6}} \\ -\frac{1}{\sqrt{6}} \\ \frac{2}{\sqrt{6}} \end{bmatrix} \Rightarrow \mathbf{u}\mathbf{u}^T = \begin{bmatrix} \frac{1}{6} & -\frac{1}{6} & \frac{1}{3} \\ -\frac{1}{6} & \frac{1}{6} & -\frac{1}{3} \\ \frac{1}{3} & -\frac{1}{3} & \frac{2}{3} \end{bmatrix} \Rightarrow Q = I - 2\mathbf{u}\mathbf{u}^T = \begin{bmatrix} \frac{2}{3} & \frac{1}{3} & -\frac{2}{3} \\ \frac{1}{3} & \frac{2}{3} & \frac{2}{3} \\ -\frac{2}{3} & \frac{2}{3} & -\frac{1}{3} \end{bmatrix}.$$

Verification that $Q^T = Q$ and $QQ^T = Q^2 = I$.

$$Q^T = \begin{bmatrix} \frac{2}{3} & \frac{1}{3} & -\frac{2}{3} \\ \frac{1}{3} & \frac{2}{3} & \frac{2}{3} \\ -\frac{2}{3} & \frac{2}{3} & -\frac{1}{3} \end{bmatrix} = Q. \qquad QQ^T = Q^2 = \begin{bmatrix} \frac{2}{3} & \frac{1}{3} & -\frac{2}{3} \\ \frac{1}{3} & \frac{2}{3} & \frac{2}{3} \\ -\frac{2}{3} & \frac{2}{3} & -\frac{1}{3} \end{bmatrix} \begin{bmatrix} \frac{2}{3} & \frac{1}{3} & -\frac{2}{3} \\ \frac{1}{3} & \frac{2}{3} & \frac{2}{3} \\ -\frac{2}{3} & \frac{2}{3} & -\frac{1}{3} \end{bmatrix} = I.$$

Verification that if $\mathbf{v} = c\mathbf{u}$, then $Q\mathbf{v} = -\mathbf{v}$.

$$Q\mathbf{v} = Q(c\mathbf{u}) = c\,Q\mathbf{u} = c \begin{bmatrix} \frac{2}{3} & \frac{1}{3} & -\frac{2}{3} \\ \frac{1}{3} & \frac{2}{3} & \frac{2}{3} \\ -\frac{2}{3} & \frac{2}{3} & -\frac{1}{3} \end{bmatrix} \begin{bmatrix} \frac{1}{\sqrt{6}} \\ -\frac{1}{\sqrt{6}} \\ \frac{2}{\sqrt{6}} \end{bmatrix} = c \begin{bmatrix} -\frac{1}{\sqrt{6}} \\ \frac{1}{\sqrt{6}} \\ -\frac{2}{\sqrt{6}} \end{bmatrix} = -c\mathbf{u} = -\mathbf{v}.$$

Verification that if $\mathbf{v} \cdot \mathbf{u} = 0$, then $Q\mathbf{v} = \mathbf{v}$.

Note, $\mathbf{v} \cdot \mathbf{u} = \mathbf{u}^T\mathbf{v} = 0 \Rightarrow \begin{bmatrix} \frac{1}{\sqrt{6}} & -\frac{1}{\sqrt{6}} & \frac{2}{\sqrt{6}} \end{bmatrix} \begin{bmatrix} x_1 \\ x_2 \\ x_3 \end{bmatrix} = \frac{1}{\sqrt{6}}x_1 - \frac{1}{\sqrt{6}}x_2 + \frac{2}{\sqrt{6}}x_3 = 0 \Rightarrow$

$x_1 - x_2 + 2x_3 = 0 \Rightarrow x_2 = x_1 + 2x_3 \Rightarrow$ If $\mathbf{v} \cdot \mathbf{u} = 0$ then $\mathbf{v} = s \begin{bmatrix} 1 \\ 1 \\ 0 \end{bmatrix} + t \begin{bmatrix} 0 \\ 2 \\ 1 \end{bmatrix}$.

So, we have:

$$Q\mathbf{v} = Q\left(s \begin{bmatrix} 1 \\ 1 \\ 0 \end{bmatrix} + t \begin{bmatrix} 0 \\ 2 \\ 1 \end{bmatrix}\right) = s\,Q\left(\begin{bmatrix} 1 \\ 1 \\ 0 \end{bmatrix}\right) + t\,Q\left(\begin{bmatrix} 0 \\ 2 \\ 1 \end{bmatrix}\right)$$

$$= s \begin{bmatrix} \frac{2}{3} & \frac{1}{3} & -\frac{2}{3} \\ \frac{1}{3} & \frac{2}{3} & \frac{2}{3} \\ -\frac{2}{3} & \frac{2}{3} & -\frac{1}{3} \end{bmatrix} \begin{bmatrix} 1 \\ 1 \\ 0 \end{bmatrix} + t \begin{bmatrix} \frac{2}{3} & \frac{1}{3} & -\frac{2}{3} \\ \frac{1}{3} & \frac{2}{3} & \frac{2}{3} \\ -\frac{2}{3} & \frac{2}{3} & -\frac{1}{3} \end{bmatrix} \begin{bmatrix} 0 \\ 2 \\ 1 \end{bmatrix} = s \begin{bmatrix} 1 \\ 1 \\ 0 \end{bmatrix} + t \begin{bmatrix} 0 \\ 2 \\ 1 \end{bmatrix} = \mathbf{v}.$$

6. Exercise 51 in Section 1.2 $\Rightarrow (\mathbf{u}+\mathbf{v})\cdot(\mathbf{u}-\mathbf{v}) = 0 \Leftrightarrow \|\mathbf{u}\| = \|\mathbf{v}\|$.

In this case, we have: $\|\mathbf{x}\| = \|\mathbf{y}\| \Rightarrow (\mathbf{x}+\mathbf{y})\cdot\mathbf{u} = (\mathbf{x}+\mathbf{y})\cdot\left(\frac{\mathbf{x}-\mathbf{y}}{\|\mathbf{x}-\mathbf{y}\|}\right) = \frac{1}{\|\mathbf{x}-\mathbf{y}\|}(\mathbf{x}+\mathbf{y})\cdot(\mathbf{x}-\mathbf{y}) = 0.$

So, Exercise 4 $\Rightarrow \begin{matrix} Q(\mathbf{x}-\mathbf{y}) = -(\mathbf{x}-\mathbf{y}) \\ Q(\mathbf{x}+\mathbf{y}) = \mathbf{x}+\mathbf{y} \end{matrix} \Rightarrow \begin{matrix} Q\mathbf{x}-Q\mathbf{y} = -\mathbf{x}+\mathbf{y} \\ Q\mathbf{x}+Q\mathbf{y} = \mathbf{x}+\mathbf{y} \end{matrix} \Rightarrow 2Q\mathbf{x} = 2\mathbf{y} \Rightarrow Q\mathbf{x} = \mathbf{y}.$

7. Note, $\mathbf{u} = \frac{1}{2\sqrt{3}}\begin{bmatrix} -2 \\ 2 \\ 2 \end{bmatrix} \Rightarrow \mathbf{u}\mathbf{u}^T = \begin{bmatrix} \frac{1}{3} & -\frac{1}{3} & -\frac{1}{3} \\ -\frac{1}{3} & \frac{1}{3} & \frac{1}{3} \\ -\frac{1}{3} & \frac{1}{3} & \frac{1}{3} \end{bmatrix} \Rightarrow Q = I - 2\mathbf{u}\mathbf{u}^T = \begin{bmatrix} \frac{1}{3} & \frac{2}{3} & \frac{2}{3} \\ \frac{2}{3} & \frac{1}{3} & -\frac{2}{3} \\ \frac{2}{3} & -\frac{2}{3} & \frac{1}{3} \end{bmatrix}.$

So, we verify: $Q\mathbf{x} = \begin{bmatrix} \frac{1}{3} & \frac{2}{3} & \frac{2}{3} \\ \frac{2}{3} & \frac{1}{3} & -\frac{2}{3} \\ \frac{2}{3} & -\frac{2}{3} & \frac{1}{3} \end{bmatrix}\begin{bmatrix} 1 \\ 2 \\ 2 \end{bmatrix} = \begin{bmatrix} 3 \\ 0 \\ 0 \end{bmatrix} = \mathbf{y}.$

8. Since $\|\mathbf{x}\| = \|\, \|\mathbf{x}\|\, \| = \|\mathbf{y}\|$, applying the results of Exercise 6, we have:
$Q_1A = \begin{bmatrix} Q\mathbf{a}_1 & Q\mathbf{a}_2 & \ldots & Q\mathbf{a}_n \end{bmatrix} = \begin{bmatrix} Q\mathbf{x} & Q\mathbf{a}_2 & \ldots & Q\mathbf{a}_n \end{bmatrix} = \begin{bmatrix} \mathbf{y} & Q\mathbf{a}_2 & \ldots & Q\mathbf{a}_n \end{bmatrix} = \begin{bmatrix} * & * \\ \mathbf{0} & A_1 \end{bmatrix}.$

9. Show $Q_2Q_2^T = I$, using the fact that $P_2P_2^T = I$.

Then: $Q_2Q_1A = \begin{bmatrix} 1 & \mathbf{0} \\ \mathbf{0} & P_2 \end{bmatrix}\begin{bmatrix} * & * \\ \mathbf{0} & A_1 \end{bmatrix} = \begin{bmatrix} * & * \\ \mathbf{0} & P_2A_1 \end{bmatrix} = \begin{bmatrix} * & * & * \\ 0 & * & * \\ \mathbf{0} & \mathbf{0} & A_2 \end{bmatrix}.$

10. Follows easily by induction.

11. Since $Q_{m-1}\cdots Q_2Q_1A = R$, take $Q^T = Q_{m-1}\cdots Q_2Q_1$, so we have:
$Q^TA = R \Rightarrow A = QR$, where $Q = (Q_{m-1}\cdots Q_2Q_1)^T = Q_1^TQ_2^T\cdots Q_{m-1}^T$ orthogonal.

12. We will work through (a) and give some hints for (b).

(a) $A = \begin{bmatrix} 3 & 9 & 1 \\ -4 & 3 & 2 \end{bmatrix} \Rightarrow \mathbf{x} = \begin{bmatrix} 3 \\ -4 \end{bmatrix}, \mathbf{y} = \begin{bmatrix} \sqrt{3^2+(-4)^2} = 5 \\ 0 \end{bmatrix} \Rightarrow \mathbf{u} = \begin{bmatrix} -\frac{1}{\sqrt{5}} \\ -\frac{2}{\sqrt{5}} \end{bmatrix} \Rightarrow$

$\mathbf{u}\mathbf{u}^T = \begin{bmatrix} \frac{1}{5} & \frac{2}{5} \\ \frac{2}{5} & \frac{4}{5} \end{bmatrix} \Rightarrow Q_1 = I - 2\mathbf{u}\mathbf{u}^T = \begin{bmatrix} \frac{3}{5} & -\frac{4}{5} \\ -\frac{4}{5} & -\frac{3}{5} \end{bmatrix} \Rightarrow Q_1A = \begin{bmatrix} 5 & 3 & -1 \\ 0 & -9 & -2 \end{bmatrix} \Rightarrow A = Q_1^TR.$

(b) Note $Q_1 = Q$ found in Exercise 7. Why?

Exploration: Approximating Eigenvalues with the QR Algorithm

1. The definition of similar in Section 4.4 states: $B = P^{-1}AP$, where P is invertible $\Rightarrow A \sim B$.
So, when Q is orthogonal this definition becomes $B = QAQ^T \Rightarrow A \sim B$.
Now in this case, we have $A_1 = RQ$, where $A = QR$.
So, $A = QR = QR(QQ^T) = Q(RQ)Q^T = QA_1Q^T \Rightarrow A \sim A_1$.

Since eigenvalues are the solutions of $\det(A - \lambda I) = 0$, we need only show $\det(A - \lambda I) = \det(A_1 - \lambda I)$, that is A and A_1 have the same characteristic polynomial.

$\det(A - \lambda I) = \det(QA_1Q^T - \lambda I) = \det(QA_1Q^T - \lambda QIQ^T) = \det Q \det(A_1 - \lambda I) \det Q^T$
$= \det Q \det Q^T \det(A_1 - \lambda I) = \det QQ^T \det(A_1 - \lambda I) = \det I \det(A_1 - \lambda I) = \det(A_1 - \lambda I)$.

2. Using Gram-Schmidt, we find:

$$A = QR = \begin{bmatrix} \frac{1}{\sqrt{2}} & \frac{1}{\sqrt{2}} \\ \frac{1}{\sqrt{2}} & -\frac{1}{\sqrt{2}} \end{bmatrix} \begin{bmatrix} \sqrt{2} & \frac{3\sqrt{2}}{2} \\ 0 & -\frac{3\sqrt{2}}{2} \end{bmatrix} \Rightarrow A_1 = RQ = \begin{bmatrix} \sqrt{2} & \frac{3\sqrt{2}}{2} \\ 0 & -\frac{3\sqrt{2}}{2} \end{bmatrix} \begin{bmatrix} \frac{1}{\sqrt{2}} & \frac{1}{\sqrt{2}} \\ \frac{1}{\sqrt{2}} & -\frac{1}{\sqrt{2}} \end{bmatrix} = \begin{bmatrix} \frac{5}{2} & -\frac{1}{2} \\ -\frac{3}{2} & \frac{3}{2} \end{bmatrix}.$$

$$\det(A - \lambda I) = \begin{vmatrix} 1-\lambda & 0 \\ 1 & 3-\lambda \end{vmatrix} = (1-\lambda)(3-\lambda) = 0 \Rightarrow \lambda = 1, 3.$$

$$\det(A_1 - \lambda I) = \begin{vmatrix} \frac{5}{2}-\lambda & -\frac{1}{2} \\ -\frac{3}{2} & \frac{3}{2}-\lambda \end{vmatrix} = (\tfrac{5}{2}-\lambda)(\tfrac{3}{2}-\lambda) - \tfrac{3}{4} = \lambda^2 - 4\lambda - 3 = 0 \Rightarrow \lambda = 1, 3.$$

3. We will proceed by induction.

$n = 1$: This case was proven in Exercise 1.

$n = k-1$: Assume $A_n \sim A$, for $n \leq k-1$. We need to show $A_k \sim A$.

We have $A_k = R_{k-1}Q_{k-1}$, where $A_{k-1} = Q_{k-1}R_{k-1}$.
So, $A_k = Q_{k-1}R_{k-1} = Q_{k-1}R_{k-1}(Q_{k-1}Q_{k-1}^T) = Q_{k-1}(R_{k-1}Q_{k-1})Q_{k-1}^T = Q_{k-1}A_kQ_{k-1}^T \Rightarrow$ $A_k \sim A_{k-1}$. Therefore, $A_k \sim A$, by the transitive property of similarity.

4. We pick it up from A_1 from Exercise 2 (now written in decimal form):

$$A_1 = Q_1R_1 \approx \begin{bmatrix} 0.86 & 0.51 \\ -0.51 & 0.86 \end{bmatrix} \begin{bmatrix} 2.92 & -1.20 \\ 0 & 1.04 \end{bmatrix} \Rightarrow A_2 = R_1Q_1 \approx \begin{bmatrix} 3.12 & 0.47 \\ -0.53 & 0.88 \end{bmatrix} \Rightarrow$$

$$A_2 = Q_2R_2 \approx \begin{bmatrix} -0.99 & 0.17 \\ 0.17 & 0.99 \end{bmatrix} \begin{bmatrix} -3.16 & -0.32 \\ 0 & 0.95 \end{bmatrix} \Rightarrow A_3 = R_2Q_2 \approx \begin{bmatrix} 3.06 & -0.84 \\ 0.16 & 0.93 \end{bmatrix} \Rightarrow$$

$$A_3 = Q_3R_3 \approx \begin{bmatrix} -1.00 & -0.05 \\ -0.05 & 1.0 \end{bmatrix} \begin{bmatrix} -3.06 & 0.79 \\ 0 & 0.97 \end{bmatrix} \Rightarrow A_4 = R_3Q_3 \approx \begin{bmatrix} 3.02 & 0.94 \\ -0.05 & 0.97 \end{bmatrix} \Rightarrow$$

$$A_4 = Q_4R_4 \approx \begin{bmatrix} -1.00 & 0.02 \\ 0.02 & 1.00 \end{bmatrix} \begin{bmatrix} -3.02 & -0.92 \\ 0 & 0.99 \end{bmatrix} \Rightarrow A_5 = R_4Q_4 \approx \begin{bmatrix} 3.00 & -0.98 \\ 0.02 & 0.99 \end{bmatrix}.$$

We notice the A_k are approaching an upper triangular matrix U.

5. It's clear that the diagonal entries of U will be the eigenvalues of A because the eigenvalues of U and A are the same.

6. We follow the same process as in Exercise 4.

(a) $A = QR \approx \begin{bmatrix} 0.71 & 0.71 \\ 0.71 & -0.71 \end{bmatrix} \begin{bmatrix} 2.83 & 2.83 \\ 0 & 1.41 \end{bmatrix} \Rightarrow A_1 = RQ \approx \begin{bmatrix} 4.02 & 0.00 \\ 1.00 & -1.00 \end{bmatrix} \Rightarrow$

$A_1 = Q_1R_1 \approx \begin{bmatrix} -0.97 & -0.24 \\ -0.24 & 0.97 \end{bmatrix} \begin{bmatrix} -4.14 & 0.24 \\ 0 & -0.97 \end{bmatrix} \Rightarrow A_2 = R_1Q_1 \approx \begin{bmatrix} 3.96 & 1.23 \\ 0.23 & -0.94 \end{bmatrix} \Rightarrow$

$A_2 = Q_2R_2 \approx \begin{bmatrix} -1.00 & -0.06 \\ -0.06 & 1.00 \end{bmatrix} \begin{bmatrix} -3.97 & -1.17 \\ 0 & -1.01 \end{bmatrix} \Rightarrow A_3 = R_2Q_2 \approx \begin{bmatrix} 4.04 & -0.93 \\ 0.06 & -1.01 \end{bmatrix} \Rightarrow$

$A_3 = Q_3R_3 \approx \begin{bmatrix} -1.00 & -0.01 \\ -0.01 & 1.00 \end{bmatrix} \begin{bmatrix} -4.04 & 0.94 \\ 0 & -1.00 \end{bmatrix} \Rightarrow A_4 = R_3Q_3 \approx \begin{bmatrix} 4.03 & 0.98 \\ 0.01 & -1.00 \end{bmatrix} \Rightarrow$

$A_4 = Q_4R_4 \approx \begin{bmatrix} -1.00 & 0.00 \\ 0.00 & 1.00 \end{bmatrix} \begin{bmatrix} -4.03 & -0.98 \\ 0 & -1.00 \end{bmatrix} \Rightarrow A_5 = R_4Q_4 \approx \begin{bmatrix} 4.03 & -0.98 \\ 0.00 & -1.00 \end{bmatrix} \Rightarrow$

The eigenvalues of A are approximately 4.03 and -1.00.

(b) Proceed as shown in (a).

(c) $A \approx \begin{bmatrix} 0.24 & -0.06 & -0.97 \\ 0.24 & 0.97 & 0 \\ -0.94 & 0.23 & -0.24 \end{bmatrix} \begin{bmatrix} 4.24 & 0.47 & -0.94 \\ 0 & 1.94 & 1.26 \\ 0 & 0 & 0.73 \end{bmatrix} \Rightarrow A_1 \approx \begin{bmatrix} 2.01 & -0.01 & -3.89 \\ -0.72 & 2.17 & -0.30 \\ -0.69 & 0.17 & -0.18 \end{bmatrix} \Rightarrow$

$A_1 \approx \begin{bmatrix} -0.90 & -0.33 & 0.30 \\ 0.32 & -0.94 & -0.07 \\ 0.31 & 0.03 & 0.95 \end{bmatrix} \begin{bmatrix} -2.24 & 0.76 & 3.33 \\ 0 & -2.04 & 1.55 \\ 0 & 0 & -1.32 \end{bmatrix} \Rightarrow A_2 \approx \begin{bmatrix} 3.29 & 0.12 & 2.44 \\ -0.17 & 1.96 & 1.62 \\ -0.41 & -0.04 & -1.25 \end{bmatrix} \Rightarrow$

$A_2 \approx \begin{bmatrix} -0.99 & -0.05 & 0.12 \\ 0.05 & -1.00 & 0.01 \\ 0.12 & 0.02 & 0.99 \end{bmatrix} \begin{bmatrix} -3.32 & -0.02 & -2.50 \\ 0 & -1.96 & -1.76 \\ 0 & 0 & -0.92 \end{bmatrix} \Rightarrow A_3 \approx \begin{bmatrix} 2.99 & 0.14 & -2.87 \\ -0.31 & 1.92 & -1.76 \\ -0.11 & 0.02 & -0.91 \end{bmatrix} \Rightarrow$

$A_3 \approx \begin{bmatrix} -0.99 & -0.10 & 0.04 \\ 0.10 & -0.99 & -0.01 \\ 0.04 & -0.01 & 1.00 \end{bmatrix} \begin{bmatrix} -3.01 & 0.06 & 2.64 \\ 0 & -1.92 & 2.06 \\ 0 & 0 & -0.99 \end{bmatrix} \Rightarrow A_4 \approx \begin{bmatrix} 3.09 & 0.22 & 2.52 \\ -0.11 & 1.88 & 2.08 \\ -0.04 & 0.01 & -0.99 \end{bmatrix} \Rightarrow$

$A_4 \approx \begin{bmatrix} -1.00 & -0.04 & 0.01 \\ 0.04 & -1.00 & -0.01 \\ 0.01 & -0.01 & 1.00 \end{bmatrix} \begin{bmatrix} -3.09 & -0.15 & -2.46 \\ 0 & -1.89 & -2.16 \\ 0 & 0 & -0.97 \end{bmatrix} \Rightarrow A_5 \approx \begin{bmatrix} 3.06 & 0.30 & -2.49 \\ -0.10 & 1.91 & -2.14 \\ -0.01 & 0.01 & -0.97 \end{bmatrix} \Rightarrow$

The eigenvalues of A are approximately 3.06, 1.91, and -0.97.

(d) Proceed as shown in (c).

7. We follow the same process as in Exercise 2.

$$A = QR = \begin{bmatrix} \frac{2}{\sqrt{5}} & -\frac{1}{\sqrt{5}} \\ -\frac{1}{\sqrt{5}} & -\frac{2}{\sqrt{5}} \end{bmatrix} \begin{bmatrix} \sqrt{5} & \frac{8}{\sqrt{5}} \\ 0 & \frac{1}{\sqrt{5}} \end{bmatrix} \Rightarrow A_1 = RQ = \begin{bmatrix} \sqrt{5} & \frac{8}{\sqrt{5}} \\ 0 & \frac{1}{\sqrt{5}} \end{bmatrix} \begin{bmatrix} \frac{2}{\sqrt{5}} & -\frac{1}{\sqrt{5}} \\ -\frac{1}{\sqrt{5}} & -\frac{2}{\sqrt{5}} \end{bmatrix} = \begin{bmatrix} \frac{2}{5} & -\frac{21}{5} \\ -\frac{1}{5} & -\frac{2}{5} \end{bmatrix}.$$

$$A_1 = Q_1R_1 = \begin{bmatrix} \frac{2}{\sqrt{5}} & -\frac{1}{\sqrt{5}} \\ -\frac{1}{\sqrt{5}} & -\frac{2}{\sqrt{5}} \end{bmatrix} \begin{bmatrix} \frac{1}{\sqrt{5}} & -\frac{8}{\sqrt{5}} \\ 0 & \sqrt{5} \end{bmatrix} \Rightarrow A_2 = R_1Q_2 = \begin{bmatrix} 2 & 3 \\ -1 & -2 \end{bmatrix} = A.$$

In this case, the QR algorithm returns to A because $Q_1 = Q^{-1}$ (and $R_1 = R^{-1}$).

In general, $A_1 = Q^{-1}AQ$ and $A_2 = Q_1^{-1}A_1Q_1$, so we have:
$A_2 = Q_1^{-1}A_1Q_1 = Q_1^{-1}(Q^{-1}AQ)Q_1 = (Q_1Q)^{-1}A(Q_1Q) = A.$

8. Setting $B = A + 0.9I = \begin{bmatrix} 2.9 & 3 \\ -1 & -1.1 \end{bmatrix}$, we have:

$$B = QR \approx \begin{bmatrix} -0.95 & 0.33 \\ 0.33 & 0.95 \end{bmatrix} \begin{bmatrix} -3.07 & -3.19 \\ 0 & -0.06 \end{bmatrix} \Rightarrow A_1 = RQ \approx \begin{bmatrix} 1.86 & -4.02 \\ -0.02 & -0.06 \end{bmatrix} \Rightarrow$$

$$A_1 = Q_1R_1 \approx \begin{bmatrix} -1.00 & 0.01 \\ 0.01 & 1.00 \end{bmatrix} \begin{bmatrix} -1.86 & 4.04 \\ 0 & -0.10 \end{bmatrix} \Rightarrow A_2 = R_1Q_1 \approx \begin{bmatrix} 1.90 & 4.02 \\ 0.00 & -0.10 \end{bmatrix} \Rightarrow$$

The eigenvalues of B are approximately 1.90 and $-0.10 \Rightarrow$
The eigenvalues of A are approximately $1.90 - 0.9 = 1$ and $-0.10 - 0.9 = -1$.

9. Note in this problem, unlike above, we take $A = A_1 = Q_1R_1$. We proceed by induction.

$n = 2$: $Q_1A_2 = Q_1(R_1Q_1) = (Q_1R_1)Q_1 = AQ_1$.

$n = k - 1$: Assume $Q_1Q_2 \cdots Q_{k-2}A_{k-1} = AQ_1Q_2 \cdots Q_{k-2}$, then we must show
$Q_1Q_2 \cdots Q_{k-1}A_k = AQ_1Q_2 \cdots Q_{k-1}$.

Note, $A_k = R_{k-1}Q_{k-1}$. We use this fact in the second step below.

$Q_1Q_2 \cdots Q_{k-1}A_k = Q_1Q_2 \cdots Q_{k-2}(Q_{k-1}A_k) = Q_1Q_2 \cdots Q_{k-2}(Q_{k-1}(R_{k-1}Q_{k-1}))$
$= Q_1Q_2 \cdots Q_{k-2}(Q_{k-1}R_{k-1})Q_{k-1} = (Q_1Q_2 \cdots Q_{k-2}A_{k-1})Q_{k-1}$
$= (AQ_1Q_2 \cdots Q_{k-2})Q_{k-1}$ (by the $k - 1$th assumption) $= AQ_1Q_2 \cdots Q_{k-1}$.

So, we have shown: $Q_1Q_2 \cdots Q_{k-1}A_k = AQ_1Q_2 \cdots Q_{k-1}$.

Next, we need to show $(Q_1Q_2 \cdots Q_k)(R_kR_{k-1} \cdots R_1) = A(Q_1Q_2 \cdots Q_{k-1})(R_{k-1} \cdots R_2R_1)$.
This follows immediately from the above by multiplying both sides by $R_{k-1} \cdots R_2R_1$.

We need only note that $A_kR_{k-1} = Q_kR_kR_{k-1}$.

$Q_1Q_2 \cdots Q_{k-1}A_k(R_{k-1} \cdots R_2R_1) = AQ_1Q_2 \cdots Q_{k-1}(R_{k-1} \cdots R_2R_1) \Rightarrow$
$(Q_1Q_2 \cdots Q_k)(R_kR_{k-1} \cdots R_1) = A(Q_1Q_2 \cdots Q_{k-1})(R_{k-1} \cdots R_2R_1)$.

Finally, we need to show $A^k = (Q_1Q_2 \cdots Q_k)(R_kR_{k-1} \cdots R_1)$.
We proceed by induction.

$n = 1$: $A = Q_1R_1$.

$n = k - 1$: Assume $A^{k-1} = (Q_1Q_2 \cdots Q_{k-1})(R_{k-1}R_{k-2} \cdots R_1)$, then we must show
$A^k = (Q_1Q_2 \cdots Q_k)(R_kR_{k-1} \cdots R_1)$.

Note, from the first equation we proved above, we have: $A_k = (Q_1Q_2 \cdots Q_{k-1})^{-1}AQ_1Q_2 \cdots Q_{k-1}$.
We will use this fact in step 3 below.

$(Q_1Q_2 \cdots Q_k)(R_kR_{k-1} \cdots R_1) = (Q_1Q_2 \cdots Q_{k-1})(Q_kR_k)(R_{k-1}R_{k-2} \cdots R_1)$
$= (Q_1Q_2 \cdots Q_{k-1})(A_k)(R_{k-1}R_{k-2} \cdots R_1)$
$= (Q_1Q_2 \cdots Q_{k-1})((Q_1Q_2 \cdots Q_{k-1})^{-1}AQ_1Q_2 \cdots Q_{k-1})(R_{k-1}R_{k-2} \cdots R_1)$
$= A(Q_1Q_2 \cdots Q_{k-1})(R_{k-1}R_{k-2} \cdots R_1) = AA^{k-1}$ (by the $k - 1$th assumption) $= A^k$.

5.4 Orthogonal Diagonalization of Symmetric Matrices

1. $\begin{vmatrix} 4-\lambda & 1 \\ 1 & 4-\lambda \end{vmatrix} = 0 \Rightarrow \lambda^2 - 8\lambda + 15 = (\lambda-5)(\lambda-3) = 0 \Rightarrow \lambda_1 = 5, \lambda_2 = 3 \Rightarrow D = \begin{bmatrix} 5 & 0 \\ 0 & 3 \end{bmatrix}$.

As a reminder, we show how to find $\mathbf{v}_1$ (make sure $\mathbf{v}_2$ is orthogonal to $\mathbf{v}_1$).

$(A-5I)\mathbf{v}_1 = \begin{bmatrix} -1 & 1 \\ 1 & -1 \end{bmatrix}\begin{bmatrix} x_1 \\ x_2 \end{bmatrix} = \begin{bmatrix} 0 \\ 0 \end{bmatrix} \Rightarrow x_1 = x_2 \Rightarrow \mathbf{v}_1 = \begin{bmatrix} 1 \\ 1 \end{bmatrix}$ is a basis.

So, $\mathbf{v}_1 = \begin{bmatrix} 1 \\ 1 \end{bmatrix}$, $\mathbf{v}_2 = \begin{bmatrix} 1 \\ -1 \end{bmatrix}$. Normalizing $\Rightarrow Q = \begin{bmatrix} 1/\sqrt{2} & 1/\sqrt{2} \\ 1/\sqrt{2} & -1/\sqrt{2} \end{bmatrix}$.

Verify $Q^TAQ = D = \begin{bmatrix} 1/\sqrt{2} & 1/\sqrt{2} \\ 1/\sqrt{2} & -1/\sqrt{2} \end{bmatrix}\begin{bmatrix} 4 & 1 \\ 1 & 4 \end{bmatrix}\begin{bmatrix} 1/\sqrt{2} & 1/\sqrt{2} \\ 1/\sqrt{2} & -1/\sqrt{2} \end{bmatrix} = \begin{bmatrix} 5 & 0 \\ 0 & 3 \end{bmatrix}$.

2. $\begin{vmatrix} -1-\lambda & 3 \\ 3 & -1-\lambda \end{vmatrix} = 0 \Rightarrow \lambda^2 + 2\lambda - 8 = 0 \Rightarrow \lambda_1 = -4, \lambda_2 = 2 \Rightarrow D = \begin{bmatrix} -4 & 0 \\ 0 & 2 \end{bmatrix}$.

Furthermore, $\mathbf{v}_1 = \begin{bmatrix} 1 \\ -1 \end{bmatrix}$, $\mathbf{v}_2 = \begin{bmatrix} 1 \\ 1 \end{bmatrix}$. Normalizing $\Rightarrow Q = \begin{bmatrix} 1/\sqrt{2} & 1/\sqrt{2} \\ -1/\sqrt{2} & 1/\sqrt{2} \end{bmatrix}$.

3. $\begin{vmatrix} 1-\lambda & \sqrt{2} \\ \sqrt{2} & 0-\lambda \end{vmatrix} = 0 \Rightarrow \lambda^2 - \lambda - 2 = (\lambda-2)(\lambda+1) = 0 \Rightarrow \lambda_1 = 2, \lambda_2 = -1 \Rightarrow D = \begin{bmatrix} 2 & 0 \\ 0 & -1 \end{bmatrix}$.

Furthermore, $\mathbf{v}_1 = \begin{bmatrix} \sqrt{2} \\ 1 \end{bmatrix}$, $\mathbf{v}_2 = \begin{bmatrix} 1 \\ -\sqrt{2} \end{bmatrix}$. Normalizing $\Rightarrow Q = \begin{bmatrix} 2/\sqrt{6} & 1/\sqrt{3} \\ 1/\sqrt{3} & -2/\sqrt{6} \end{bmatrix}$.

4. $\begin{vmatrix} 9-\lambda & -2 \\ -2 & 6-\lambda \end{vmatrix} = 0 \Rightarrow \lambda^2 - 15\lambda + 50 = 0 \Rightarrow \lambda_1 = 5, \lambda_2 = 10 \Rightarrow D = \begin{bmatrix} 5 & 0 \\ 0 & 10 \end{bmatrix}$.

Furthermore, $\mathbf{v}_1 = \begin{bmatrix} 1 \\ 2 \end{bmatrix}$, $\mathbf{v}_2 = \begin{bmatrix} -2 \\ 1 \end{bmatrix}$. Normalizing $\Rightarrow Q = \begin{bmatrix} 1/\sqrt{5} & -2/\sqrt{5} \\ 2/\sqrt{5} & 1/\sqrt{5} \end{bmatrix}$.

5. $\begin{vmatrix} 5-\lambda & 0 & 0 \\ 0 & 1-\lambda & 3 \\ 0 & 3 & 1-\lambda \end{vmatrix} = 0 \Rightarrow (5-\lambda)(\lambda^2 - 2\lambda - 8) = 0 \Rightarrow D = \begin{bmatrix} 5 & 0 & 0 \\ 0 & 4 & 0 \\ 0 & 0 & -2 \end{bmatrix}$.

As a reminder, in this case we show how to find $\mathbf{v}_1$ ($\mathbf{v}_2$ and $\mathbf{v}_3$ are extremely similar).

$(A-5I)\mathbf{v}_1 = \begin{bmatrix} 0 & 0 & 0 \\ 0 & -4 & 3 \\ 0 & 3 & -4 \end{bmatrix}\begin{bmatrix} x_1 \\ x_2 \\ x_3 \end{bmatrix} = \begin{bmatrix} 0 \\ 0 \\ 0 \end{bmatrix} \Rightarrow \begin{matrix} x_1 \textit{ is free} \\ x_2 = x_3 = 0 \end{matrix} \Rightarrow \mathbf{v}_1 = \begin{bmatrix} 1 \\ 0 \\ 0 \end{bmatrix}$ is a basis.

So, $\mathbf{v}_1 = \begin{bmatrix} 1 \\ 0 \\ 0 \end{bmatrix}$, $\mathbf{v}_2 = \begin{bmatrix} 0 \\ 1 \\ 1 \end{bmatrix}$, $\mathbf{v}_3 = \begin{bmatrix} 0 \\ -1 \\ 1 \end{bmatrix}$. Normalizing $\Rightarrow Q = \begin{bmatrix} 1 & 0 & 0 \\ 0 & 1/\sqrt{2} & -1/\sqrt{2} \\ 0 & 1/\sqrt{2} & 1/\sqrt{2} \end{bmatrix}$.

6. Since $A = \begin{bmatrix} 2 & 3 & 0 \\ 3 & 2 & 4 \\ 0 & 4 & 2 \end{bmatrix} \Rightarrow D = \begin{bmatrix} 2 & 0 & 0 \\ 0 & 7 & 0 \\ 0 & 0 & -3 \end{bmatrix}$.

So, $\mathbf{v}_1 = \begin{bmatrix} 4 \\ 0 \\ -3 \end{bmatrix}$, $\mathbf{v}_2 = \begin{bmatrix} 3 \\ 5 \\ 4 \end{bmatrix}$, $\mathbf{v}_3 = \begin{bmatrix} 3 \\ -5 \\ 4 \end{bmatrix}$. Normalizing $\Rightarrow Q = \begin{bmatrix} 4/5 & 3/5\sqrt{2} & 3/5\sqrt{2} \\ 0 & 1/\sqrt{2} & -1/\sqrt{2} \\ -3/5 & 4/5\sqrt{2} & 4/5\sqrt{2} \end{bmatrix}$.

7. $\begin{vmatrix} 1-\lambda & 0 & -1 \\ 0 & 1-\lambda & 0 \\ -1 & 0 & 1-\lambda \end{vmatrix} = 0 \Rightarrow (1-\lambda)(\lambda^2 - 2\lambda) = 0 \Rightarrow D = \begin{bmatrix} 2 & 0 & 0 \\ 0 & 1 & 0 \\ 0 & 0 & 0 \end{bmatrix}$.

So, $\mathbf{v}_1 = \begin{bmatrix} -1 \\ 0 \\ 1 \end{bmatrix}$, $\mathbf{v}_2 = \begin{bmatrix} 0 \\ 1 \\ 0 \end{bmatrix}$, $\mathbf{v}_3 = \begin{bmatrix} 1 \\ 0 \\ 1 \end{bmatrix}$. Normalizing $\Rightarrow Q = \begin{bmatrix} -1/\sqrt{2} & 0 & 1/\sqrt{2} \\ 0 & 1 & 0 \\ 1/\sqrt{2} & 0 & 1/\sqrt{2} \end{bmatrix}$.

8. $\begin{vmatrix} 1-\lambda & 2 & 2 \\ 0 & 1-\lambda & 2 \\ 2 & 2 & 1-\lambda \end{vmatrix} = 0 \Rightarrow (\lambda - 5)(\lambda + 1)^2 = 0 \Rightarrow D = \begin{bmatrix} 5 & 0 & 0 \\ 0 & -1 & 0 \\ 0 & 0 & -1 \end{bmatrix}$.

In this case, $(A - 5I)\mathbf{v} = \mathbf{0} \Rightarrow x_1 = \frac{1}{2}(x_2 + x_3)$. So, $x_2 = x_3 = 1 \Rightarrow \mathbf{v}_1 = \begin{bmatrix} 1 \\ 1 \\ 1 \end{bmatrix}$.

$(A - (-1)I)\mathbf{v} = \mathbf{0}$ (of multiplicity 2) $\Rightarrow \begin{matrix} x_1 = -(x_2 + x_3) \\ or\ x_2 = 1, x_3 = -1 \end{matrix} \Rightarrow \mathbf{v}_2 = \begin{bmatrix} -2 \\ 1 \\ 1 \end{bmatrix}$, $\mathbf{v}_3 = \begin{bmatrix} 0 \\ 1 \\ -1 \end{bmatrix}$.

These choices make the $\mathbf{v}_i$ orthogonal. Normalizing $\Rightarrow Q = \begin{bmatrix} 1/\sqrt{3} & -2/\sqrt{6} & 0 \\ 1/\sqrt{3} & 1/\sqrt{6} & 1/\sqrt{2} \\ 1/\sqrt{3} & 1/\sqrt{6} & -1/\sqrt{2} \end{bmatrix}$.

9. $\begin{vmatrix} 1-\lambda & 1 & 0 & 0 \\ 1 & 1-\lambda & 0 & 0 \\ 0 & 0 & 1-\lambda & 1 \\ 0 & 0 & 1 & 1-\lambda \end{vmatrix} = 0 \Rightarrow \lambda^2(\lambda - 2)^2 = 0 \Rightarrow D = \begin{bmatrix} 2 & 0 & 0 & 0 \\ 0 & 2 & 0 & 0 \\ 0 & 0 & 0 & 0 \\ 0 & 0 & 0 & 0 \end{bmatrix}$.

$\lambda_1 = 2$ of multiplicity 2 $\Rightarrow (A - 2I)\mathbf{v} = \mathbf{0} \Rightarrow \begin{matrix} x_2 = x_1 \\ x_4 = x_3 \end{matrix} \Rightarrow \mathbf{v}_1 = \begin{bmatrix} 1 \\ 1 \\ 0 \\ 0 \end{bmatrix}$, $\mathbf{v}_2 = \begin{bmatrix} 0 \\ 0 \\ 1 \\ 1 \end{bmatrix}$.

$\lambda_2 = 0$ of multiplicity 2 $\Rightarrow (A - 0I)\mathbf{v} = \mathbf{0} \Rightarrow \begin{matrix} x_2 = -x_1 \\ x_4 = -x_3 \end{matrix} \Rightarrow \mathbf{v}_3 = \begin{bmatrix} 1 \\ -1 \\ 0 \\ 0 \end{bmatrix}$, $\mathbf{v}_4 = \begin{bmatrix} 0 \\ 0 \\ 1 \\ -1 \end{bmatrix}$.

These choices make the $\mathbf{v}_i$ orthogonal. Normalizing $\Rightarrow Q = \begin{bmatrix} 1/\sqrt{2} & 0 & 1/\sqrt{2} & 0 \\ 1/\sqrt{2} & 0 & -1/\sqrt{2} & 0 \\ 0 & 1/\sqrt{2} & 0 & 1/\sqrt{2} \\ 0 & 1/\sqrt{2} & 0 & -1/\sqrt{2} \end{bmatrix}$.

10. $\begin{vmatrix} 2-\lambda & 0 & 0 & 1 \\ 0 & 1-\lambda & 0 & 0 \\ 0 & 0 & 1-\lambda & 0 \\ 1 & 0 & 0 & 2-\lambda \end{vmatrix} = 0 \Rightarrow (\lambda-3)(\lambda-1)^3 = 0 \Rightarrow D = \begin{bmatrix} 3 & 0 & 0 & 0 \\ 0 & 1 & 0 & 0 \\ 0 & 0 & 1 & 0 \\ 0 & 0 & 0 & 1 \end{bmatrix}.$

Multiplicity 1 $\Rightarrow (A - 3I)\mathbf{v} = \mathbf{0} \Rightarrow x_4 = x_1 \Rightarrow \mathbf{v}_1 = \begin{bmatrix} 1 \\ 0 \\ 0 \\ 1 \end{bmatrix}.$

Multiplicity 3 $\Rightarrow (A - 1I)\mathbf{v} = \mathbf{0} \Rightarrow \begin{matrix} x_4 = -x_1 \\ x_2,\ x_3\ free \end{matrix} \Rightarrow \mathbf{v}_2 = \begin{bmatrix} 1 \\ 0 \\ 0 \\ -1 \end{bmatrix},\ \mathbf{v}_3 = \begin{bmatrix} 0 \\ 1 \\ 0 \\ 0 \end{bmatrix},\ \mathbf{v}_4 = \begin{bmatrix} 0 \\ 0 \\ 1 \\ 0 \end{bmatrix}.$

These choices make the $\mathbf{v}_i$ orthogonal. Normalizing $\Rightarrow Q = \begin{bmatrix} 1/\sqrt{2} & 1/\sqrt{2} & 0 & 0 \\ 0 & 0 & 1 & 0 \\ 0 & 0 & 0 & 1 \\ 1/\sqrt{2} & -1/\sqrt{2} & 0 & 0 \end{bmatrix}.$

11. $\begin{vmatrix} a-\lambda & b \\ b & a-\lambda \end{vmatrix} = 0 \Rightarrow (a-\lambda)^2 - b^2 = 0 \Rightarrow \lambda = a \pm b, \Rightarrow D = \begin{bmatrix} a+b & 0 \\ 0 & a-b \end{bmatrix}.$

We show how to find $\mathbf{v}_1$ (make sure $\mathbf{v}_2$ is orthogonal to $\mathbf{v}_1$).

$(A - (a+b)I)\mathbf{v}_1 = \begin{bmatrix} -b & b \\ b & -b \end{bmatrix} \begin{bmatrix} x_1 \\ x_2 \end{bmatrix} = \begin{bmatrix} 0 \\ 0 \end{bmatrix} \Rightarrow x_1 = x_2 \Rightarrow \mathbf{v}_1 = \begin{bmatrix} 1 \\ 1 \end{bmatrix}$ is a basis.

So, $\mathbf{v}_1 = \begin{bmatrix} 1 \\ 1 \end{bmatrix},\ \mathbf{v}_2 = \begin{bmatrix} 1 \\ -1 \end{bmatrix}$. Normalizing $\Rightarrow Q = \begin{bmatrix} 1/\sqrt{2} & 1/\sqrt{2} \\ 1/\sqrt{2} & -1/\sqrt{2} \end{bmatrix}.$

Verify $Q^TAQ = D = \begin{bmatrix} 1/\sqrt{2} & 1/\sqrt{2} \\ 1/\sqrt{2} & -1/\sqrt{2} \end{bmatrix} \begin{bmatrix} a & b \\ b & a \end{bmatrix} \begin{bmatrix} 1/\sqrt{2} & 1/\sqrt{2} \\ 1/\sqrt{2} & -1/\sqrt{2} \end{bmatrix} = \begin{bmatrix} a+b & 0 \\ 0 & a-b \end{bmatrix}.$

12. $\begin{vmatrix} a-\lambda & 0 & b \\ 0 & a-\lambda & 0 \\ b & 0 & a-\lambda \end{vmatrix} = 0 \Rightarrow (\lambda - a)[(a-\lambda)^2 - b^2] = 0 \Rightarrow D = \begin{bmatrix} a+b & 0 & 0 \\ 0 & a & 0 \\ 0 & 0 & a-b \end{bmatrix}.$

As a reminder, in this case we show how to find $\mathbf{v}_1$ ($\mathbf{v}_2$ and $\mathbf{v}_3$ are extremely similar).

$(A - (a+b)I)\mathbf{v}_1 = \begin{bmatrix} -b & 0 & b \\ 0 & -b & 0 \\ b & 0 & -b \end{bmatrix} \begin{bmatrix} x_1 \\ x_2 \\ x_3 \end{bmatrix} = \begin{bmatrix} 0 \\ 0 \\ 0 \end{bmatrix} \Rightarrow \begin{matrix} x_3 = x_1 \\ x_2 = 0 \end{matrix} \Rightarrow \mathbf{v}_1 = \begin{bmatrix} 1 \\ 0 \\ 1 \end{bmatrix}.$

So, $\mathbf{v}_1 = \begin{bmatrix} 1 \\ 0 \\ 1 \end{bmatrix},\ \mathbf{v}_2 = \begin{bmatrix} 0 \\ 1 \\ 0 \end{bmatrix},\ \mathbf{v}_3 = \begin{bmatrix} 1 \\ 0 \\ -1 \end{bmatrix}$. Normalizing $\Rightarrow Q = \begin{bmatrix} 1/\sqrt{2} & 0 & 1/\sqrt{2} \\ 0 & 1 & 0 \\ 1/\sqrt{2} & 0 & -1/\sqrt{2} \end{bmatrix}.$

13. (a) A, B orthogonally diagonalizable $\Rightarrow A$, B symmetric $\Rightarrow A+B$ symmetric $\Rightarrow Q^T(A+B)Q = D$.

(b) A orthogonally diagonalizable $\Rightarrow A$ symmetric $\Rightarrow cA$ symmetric $\Rightarrow Q^T(cA)Q = D$.

(c) AA^T is symmetric $\Rightarrow$ if A symmetric, then $AA^T = AA = A^2$ symmetric.
So, A orthogonally diagonalizable $\Rightarrow A$ symmetric $\Rightarrow A^2$ symmetric $\Rightarrow Q^T(A^2)Q = D$.

14. A invertible $\Rightarrow Q^TAQ = D$ invertible $\Rightarrow Q^TA^{-1}Q = D^{-1}$.

15. By Exercise 36 in Section 3.2, A, B symmetric and $AB = BA \Rightarrow AB$ symmetric.
So, A, B symmetric and $AB = BA$ (given) $\Rightarrow AB$ symmetric $\Rightarrow Q^T(AB)Q = D$.

16. Note, D diagonal $\Rightarrow B = QDQ^T$ symmetric since $(QDQ^T)^T = (Q^T)^TD^TQ^T = QDQ^T$.

$$A \text{ symmetric with } \lambda_i \geq 0 \Leftrightarrow Q^TAQ = \begin{bmatrix} \sqrt{\lambda_1} & \cdots & 0 \\ \vdots & \ddots & \vdots \\ 0 & \cdots & \sqrt{\lambda_n} \end{bmatrix}^2 \Leftrightarrow Q^TAQ = D^2.$$

A symmetric with $\lambda_i \geq 0 \Rightarrow Q^TAQ = D^2 \Rightarrow A = QD^2Q^T = (QDQ^T)(QDQ^T) = B^2$.
$A = B^2 \Rightarrow Q^TAQ = Q^TB^2Q = (Q^TBQ)(Q^TBQ) = D^2 \Rightarrow A$ symmetric with $\lambda_i \geq 0$.

17. $A = \lambda_1\mathbf{q}_1\mathbf{q}_1{}^T + \lambda_2\mathbf{q}_2\mathbf{q}_2{}^T$

$$= 5\begin{bmatrix} 1/2 & 1/2 \\ 1/2 & 1/2 \end{bmatrix} + 3\begin{bmatrix} 1/2 & -1/2 \\ -1/2 & 1/2 \end{bmatrix} = \begin{bmatrix} 5/2 & 5/2 \\ 5/2 & 5/2 \end{bmatrix} + \begin{bmatrix} 3/2 & -3/2 \\ -3/2 & 3/2 \end{bmatrix}.$$

18. $$A = -4\begin{bmatrix} 1/2 & -1/2 \\ -1/2 & 1/2 \end{bmatrix} + 2\begin{bmatrix} 1/2 & 1/2 \\ 1/2 & 1/2 \end{bmatrix} = \begin{bmatrix} -1 & 3 \\ 3 & -1 \end{bmatrix}.$$

19. $$A = 5\begin{bmatrix} 1 & 0 & 0 \\ 0 & 0 & 0 \\ 0 & 0 & 0 \end{bmatrix} + 4\begin{bmatrix} 0 & 0 & 0 \\ 0 & 1/2 & 1/2 \\ 0 & 1/2 & 1/2 \end{bmatrix} - 2\begin{bmatrix} 0 & 0 & 0 \\ 0 & 1/2 & -1/2 \\ 0 & -1/2 & 1/2 \end{bmatrix}$$

$$= \begin{bmatrix} 5 & 0 & 0 \\ 0 & 0 & 0 \\ 0 & 0 & 0 \end{bmatrix} + \begin{bmatrix} 0 & 0 & 0 \\ 0 & 2 & 2 \\ 0 & 2 & 2 \end{bmatrix} + \begin{bmatrix} 0 & 0 & 0 \\ 0 & -1 & 1 \\ 0 & 1 & -1 \end{bmatrix}.$$

20. $$A = 5\begin{bmatrix} 1/3 & 1/3 & 1/3 \\ 1/3 & 1/3 & 1/3 \\ 1/3 & 1/3 & 1/3 \end{bmatrix} + (-1)\begin{bmatrix} 2/3 & -1/3 & -1/3 \\ -1/3 & 1/6 & 1/6 \\ -1/3 & 1/6 & 1/6 \end{bmatrix} + (-1)\begin{bmatrix} 0 & 0 & 0 \\ 0 & 1/2 & -1/2 \\ 0 & -1/2 & 1/2 \end{bmatrix}$$

$$= \begin{bmatrix} 5/3 & 5/3 & 5/3 \\ 5/3 & 5/3 & 5/3 \\ 5/3 & 5/3 & 5/3 \end{bmatrix} + \begin{bmatrix} -2/3 & 1/3 & 1/3 \\ 1/3 & -1/6 & -1/6 \\ 1/3 & -1/6 & -1/6 \end{bmatrix} + \begin{bmatrix} 0 & 0 & 0 \\ 0 & -1/2 & 1/2 \\ 0 & 1/2 & -1/2 \end{bmatrix}.$$

21. $$A = \lambda_1\mathbf{q}_1\mathbf{q}_1{}^T + \lambda_2\mathbf{q}_2\mathbf{q}_2{}^T = (-1)\begin{bmatrix} 1/2 & 1/2 \\ 1/2 & 1/2 \end{bmatrix} + 2\begin{bmatrix} 1/2 & -1/2 \\ -1/2 & 1/2 \end{bmatrix} = \begin{bmatrix} 1/2 & -3/2 \\ -3/2 & 1/2 \end{bmatrix}.$$

22. $$A = 3\begin{bmatrix} 1/5 & 2/5 \\ 2/5 & 4/5 \end{bmatrix} + (-3)\begin{bmatrix} 4/5 & -2/5 \\ -2/5 & 1/5 \end{bmatrix} = \begin{bmatrix} -9/5 & 12/5 \\ 12/5 & 9/5 \end{bmatrix}.$$

23. $A = 1\begin{bmatrix} 1/2 & 1/2 & 0 \\ 1/2 & 1/2 & 0 \\ 0 & 0 & 0 \end{bmatrix} + 2\begin{bmatrix} 1/3 & -1/3 & 1/3 \\ -1/3 & 1/3 & -1/3 \\ 1/3 & -1/3 & 1/3 \end{bmatrix} + 3\begin{bmatrix} 1/6 & -1/6 & -1/3 \\ -1/6 & 1/6 & 1/3 \\ -1/3 & 1/3 & 2/3 \end{bmatrix}$

$= \begin{bmatrix} 5/3 & -2/3 & -1/3 \\ -2/3 & 5/3 & 1/3 \\ -1/3 & 1/3 & 8/3 \end{bmatrix}.$

24. $A = 0A + (-4)\begin{bmatrix} 1/3 & -1/3 & -1/3 \\ -1/3 & 1/3 & 1/3 \\ -1/3 & 1/3 & 1/3 \end{bmatrix} + (-4)\begin{bmatrix} 2/7 & -1/7 & 3/7 \\ -1/7 & 1/14 & -3/14 \\ 3/7 & -3/14 & 9/14 \end{bmatrix}$

$= \begin{bmatrix} -52/21 & 40/21 & -8/21 \\ 40/21 & -34/21 & -10/21 \\ -8/21 & -10/21 & -82/21 \end{bmatrix}.$

25. $\text{proj}_W(\mathbf{v}) = (\frac{\mathbf{q}\cdot\mathbf{v}}{\mathbf{q}\cdot\mathbf{q}})\mathbf{q} = (\mathbf{q}^T\mathbf{v})\mathbf{q} = \mathbf{q}(\mathbf{q}^T\mathbf{v}) = (\mathbf{q}\mathbf{q}^T)\mathbf{v}.$

Note: $(\mathbf{q}^T\mathbf{v})\mathbf{q} = \mathbf{q}(\mathbf{q}^T\mathbf{v})$ because $\mathbf{q}^T\mathbf{v}$ is a scalar.

26. (a) This follows from induction on Exercise 25.

(b) $P^T = (\mathbf{q}_1\mathbf{q}_1{}^T + \ldots + \mathbf{q}_k\mathbf{q}_k{}^T)^T = \mathbf{q}_1{}^T\mathbf{q}_1 + \ldots + \mathbf{q}_k{}^T\mathbf{q}_k = \mathbf{q}_1\mathbf{q}_1{}^T + \ldots + \mathbf{q}_k\mathbf{q}_k{}^T = P.$
$P^2 = (\mathbf{q}_1\mathbf{q}_1{}^T)(\mathbf{q}_1\mathbf{q}_1{}^T) + \ldots + (\mathbf{q}_k\mathbf{q}_k{}^T)(\mathbf{q}_k\mathbf{q}_k{}^T)$ (because $\mathbf{q}_i \cdot \mathbf{q}_j = 0$ for $i \neq j$)
$= \mathbf{q}_1(\mathbf{q}_1{}^T\mathbf{q}_1)\mathbf{q}_1{}^T + \ldots + \mathbf{q}_k(\mathbf{q}_k{}^T\mathbf{q}_k)\mathbf{q}_k = \mathbf{q}_1\mathbf{q}_1{}^T + \ldots + \mathbf{q}_k\mathbf{q}_k{}^T = P.$

(c) $QQ^T = \begin{bmatrix} \mathbf{q}_1 & \ldots & \mathbf{q}_k \end{bmatrix}\begin{bmatrix} \mathbf{q}_1{}^T \\ \vdots \\ \mathbf{q}_k{}^T \end{bmatrix} = \mathbf{q}_1\mathbf{q}_1{}^T + \ldots + \mathbf{q}_k\mathbf{q}_k{}^T = P.$

So, $\{\mathbf{q}_1, \ldots, \mathbf{q}_k\}$ orthonormal basis $\Rightarrow$
$\text{rank}(P) = \text{rank}(QQ^T) = \text{rank}(Q) = k$ by Section 3.4, Theorem 3.28.

27. Let $\mathbf{v}_1$ be the normalized eigenvector for λ_1 then extend $\mathbf{v}_1$ to orthonormal basis $\{\mathbf{v}_1, \ldots, \mathbf{v}_n\} \Rightarrow$

$$Q_1{}^T A Q_1 = \begin{bmatrix} \mathbf{v}_1{}^T \\ \vdots \\ \mathbf{v}_n{}^T \end{bmatrix} A \begin{bmatrix} \mathbf{v}_1 & \ldots & \mathbf{v}_n \end{bmatrix} = \begin{bmatrix} \mathbf{v}_1{}^T \\ \vdots \\ \mathbf{v}_n{}^T \end{bmatrix} \begin{bmatrix} A\mathbf{v}_1 & \ldots & A\mathbf{v}_n \end{bmatrix}$$

$$= \begin{bmatrix} \mathbf{v}_1{}^T \\ \vdots \\ \mathbf{v}_n{}^T \end{bmatrix} \begin{bmatrix} \lambda_1\mathbf{v}_1 & \ldots & \lambda_n\mathbf{v}_n \end{bmatrix} = \left[\begin{array}{c|c} \lambda_1 & * \\ \hline \mathbf{0} & A_1 \end{array}\right] = T.$$

Note that T is upper triangular and that the result follows by induction.

28. We need to show if A is nilpotent, then there exists Q (orthogonal) and T (upper triangular with zeroes on the diagonal) such that $Q^T AQ = T$.

From Exercise 27, we have $Q^T AQ = T$, provided the eigenvalues of A are real.
It remains to show that the diagonal entries of T are zero.
It is obvious from the solution of Exercise 27,
that the diagonal entries of T are simply the eigenvalues of A.

So, it suffices to show that if A is nilpotent, then the only eigenvalue of A is zero.
That is, if $A^m = O$ and $A\mathbf{x} = \lambda\mathbf{x}$ then $\lambda = 0$.

By Theorem 4.18 in Section 4.3, if $A\mathbf{x} = \lambda\mathbf{x}$ then $A^m\mathbf{x} = \lambda^m\mathbf{x}$.
So if $A^m = O$ then $A^m\mathbf{x} = \lambda^m\mathbf{x} = \mathbf{0}$.
However, since $\mathbf{x}$ is an eigenvector we know $\mathbf{x} \neq \mathbf{0}$.
Therefore, $A^m\mathbf{x} = \lambda^m\mathbf{x} = \mathbf{0}$ implies $\lambda^m = 0$ so $\lambda = 0$ as required.

5.5 Applications

1. $C_1{}' = C_1 + C_2 \Rightarrow G' \longrightarrow \begin{bmatrix} 1 & 0 \\ 0 & 1 \\ 1 & 0 \end{bmatrix}$ (standard form, $\left[\frac{I}{A}\right]$) $\Rightarrow C' = C$ (R1 not required).

2. $\begin{matrix} C_1{}' = C_2 + C_3 \\ C_2{}' = C_1 + C_2 + C_3 \\ C_3{}' = C_2 \end{matrix} \Rightarrow G' \longrightarrow \begin{bmatrix} 1 & 0 & 0 \\ 0 & 1 & 0 \\ 0 & 0 & 1 \\ 1 & 0 & 1 \\ 1 & 1 & 0 \end{bmatrix} \Rightarrow C' = C$ (R1 not required).

3. $R_1 = 0 \Rightarrow$ R1 required $\Rightarrow C' \neq C$.

$$\begin{matrix} R_1{}' = R_4 \\ R_4{}' = R_1 \end{matrix} \Rightarrow G \longrightarrow \begin{bmatrix} 1 & 1 & 1 \\ 1 & 0 & 1 \\ 0 & 1 & 1 \\ 0 & 0 & 0 \end{bmatrix}, \text{ then } \begin{matrix} C_1{}' = C_1 + C_2 + C_3 \\ C_2{}' = C_2 + C_3 \\ C_3{}' = C_1 + C_3 \end{matrix} \Rightarrow G' \longrightarrow \begin{bmatrix} 1 & 0 & 0 \\ 0 & 1 & 0 \\ 0 & 0 & 1 \\ 0 & 0 & 0 \end{bmatrix}.$$

4. $R_1 = R_2 \Rightarrow$ R1 required $\Rightarrow C' \neq C$.

$$\begin{matrix} R_1{}' = R_3 \\ R_3{}' = R_1 \end{matrix} \Rightarrow G' \longrightarrow \begin{bmatrix} 1 & 0 \\ 1 & 1 \\ 1 & 1 \\ 0 & 0 \\ 1 & 0 \end{bmatrix}, \text{ then } \begin{matrix} C_1{}' = C_1 + C_2 \\ C_2{}' = C_2 \end{matrix} \Rightarrow G \longrightarrow \begin{bmatrix} 1 & 0 \\ 0 & 1 \\ 0 & 1 \\ 0 & 0 \\ 1 & 0 \end{bmatrix}.$$

5. Only one row $\Rightarrow$ C1 required $\Rightarrow C' \neq C$.

$$\begin{matrix} C_1{}' = C_1 \\ C_2{}' = C_1 + C_2 \\ C_3{}' = C_1 \end{matrix} \Rightarrow P' \longrightarrow \begin{bmatrix} 1 & 0 & 1 \end{bmatrix}.$$

6. $\begin{matrix} R_1{}' = R_1 + R_2 \\ R_2{}' = R_1 \end{matrix} \Rightarrow P' \longrightarrow \begin{bmatrix} 0 & 0 & 1 & 0 \\ 1 & 1 & 0 & 1 \end{bmatrix} \Rightarrow C' = C$ (C1 not required).

7. $C_3 = C_4 \Rightarrow$ C1 required $\Rightarrow C' \neq C$.

$$\begin{matrix} C_1{}' = C_4 \\ C_4{}' = C_1 \end{matrix} \Rightarrow P \longrightarrow \begin{bmatrix} 1 & 1 & 1 & 0 & 0 \\ 0 & 1 & 0 & 1 & 1 \\ 1 & 0 & 1 & 0 & 1 \end{bmatrix}, \text{ then } \begin{matrix} R_1{}' = R_1 \\ R_2{}' = R_1 + R_2 + R_3 \\ R_3{}' = R_1 + R_3 \end{matrix} \Rightarrow P' \longrightarrow \begin{bmatrix} 1 & 1 & 1 & 0 & 0 \\ 0 & 0 & 0 & 1 & 0 \\ 0 & 1 & 0 & 0 & 1 \end{bmatrix}.$$

8. $C_3 = 0 \Rightarrow$ C1 required $\Rightarrow C' \neq C$.

$$\begin{matrix} C_2{}' = C_3 \\ C_3{}' = C_2 \end{matrix} \Rightarrow P' \longrightarrow \begin{bmatrix} 0 & 0 & 1 & 1 \\ 1 & 0 & 0 & 1 \end{bmatrix}, \text{ then } \begin{matrix} R_1{}' = R_1 + R_2 \\ R_2{}' = R_2 \end{matrix} \Rightarrow P \longrightarrow \begin{bmatrix} 1 & 0 & 1 & 0 \\ 1 & 0 & 0 & 1 \end{bmatrix}.$$

9. $C = \left\{ \begin{bmatrix} 0 \\ 0 \\ 0 \end{bmatrix}, \begin{bmatrix} 0 \\ 1 \\ 0 \end{bmatrix} \right\} \Rightarrow A = \begin{bmatrix} 0 & 0 & 0 \\ 0 & 1 & 0 \end{bmatrix}$. Therefore, to find $null(A)$ we compute $A\mathbf{x} = 0$.

$$\Rightarrow \begin{bmatrix} 0 & 0 & 0 \\ 0 & 1 & 0 \end{bmatrix} \begin{bmatrix} x_1 \\ x_2 \\ x_3 \end{bmatrix} = \begin{bmatrix} 0 \\ 0 \\ 0 \end{bmatrix} \Rightarrow \begin{matrix} x_1,\ x_3 = 0 \text{ or } 1 \\ x_2 = 0 \end{matrix} \Rightarrow C^{\perp} = \left\{ \begin{bmatrix} 0 \\ 0 \\ 0 \end{bmatrix}, \begin{bmatrix} 0 \\ 0 \\ 1 \end{bmatrix}, \begin{bmatrix} 1 \\ 0 \\ 0 \end{bmatrix}, \begin{bmatrix} 1 \\ 0 \\ 1 \end{bmatrix} \right\}.$$

10. $\begin{bmatrix} 0 & 0 & 0 \\ 1 & 1 & 0 \\ 0 & 0 & 1 \\ 1 & 1 & 1 \end{bmatrix} \begin{bmatrix} x_1 \\ x_2 \\ x_3 \end{bmatrix} = \mathbf{0} \Rightarrow \begin{matrix} x_1 = x_2 = 0, 1 \\ x_3 = 0 \end{matrix} \Rightarrow C^{\perp} = \left\{ \begin{bmatrix} 0 \\ 0 \\ 0 \end{bmatrix}, \begin{bmatrix} 1 \\ 1 \\ 0 \end{bmatrix} \right\}.$

11. $\begin{bmatrix} 0 & 0 & 0 & 0 \\ 0 & 1 & 0 & 0 \\ 0 & 1 & 0 & 1 \\ 0 & 0 & 0 & 1 \end{bmatrix} \begin{bmatrix} x_1 \\ x_2 \\ x_3 \\ x_4 \end{bmatrix} = \mathbf{0} \Rightarrow \begin{matrix} x_1,\ x_3 = 0 \textit{ or } 1 \\ x_2 = x_4 = 0 \end{matrix} \Rightarrow C^{\perp} = \left\{ \begin{bmatrix} 0 \\ 0 \\ 0 \\ 0 \end{bmatrix}, \begin{bmatrix} 0 \\ 0 \\ 1 \\ 0 \end{bmatrix}, \begin{bmatrix} 1 \\ 0 \\ 0 \\ 0 \end{bmatrix}, \begin{bmatrix} 1 \\ 0 \\ 1 \\ 0 \end{bmatrix} \right\}.$

12. $A\mathbf{x} = 0 \Rightarrow \begin{matrix} x_4 = x_1 \\ x_5 = x_2 + x_3 \end{matrix} \Rightarrow C^{\perp} = \left\{ \begin{bmatrix} 0 \\ 0 \\ 0 \\ 0 \\ 0 \end{bmatrix}, \begin{bmatrix} 0 \\ 1 \\ 1 \\ 0 \\ 0 \end{bmatrix}, \begin{bmatrix} 0 \\ 1 \\ 0 \\ 0 \\ 1 \end{bmatrix}, \begin{bmatrix} 0 \\ 0 \\ 1 \\ 0 \\ 1 \end{bmatrix}, \begin{bmatrix} 1 \\ 0 \\ 0 \\ 1 \\ 0 \end{bmatrix}, \begin{bmatrix} 1 \\ 1 \\ 1 \\ 1 \\ 0 \end{bmatrix}, \begin{bmatrix} 1 \\ 1 \\ 0 \\ 1 \\ 1 \end{bmatrix}, \begin{bmatrix} 1 \\ 0 \\ 1 \\ 1 \\ 1 \end{bmatrix} \right\}.$

13. $P^{\perp} = G^T = \begin{bmatrix} 1 & 1 & 1 & 0 \\ 1 & 1 & 0 & 1 \end{bmatrix}$ (standard) $\Rightarrow A = \begin{bmatrix} 1 & 1 \\ 1 & 1 \end{bmatrix} \Rightarrow G^{\perp} = \left[\dfrac{I}{A} \right] = \begin{bmatrix} 1 & 0 \\ 0 & 1 \\ 1 & 1 \\ 1 & 1 \end{bmatrix}.$

14. $\begin{matrix} R_1' = R_1 \\ R_2' = R_1 + R_2 \end{matrix} \Rightarrow P^{\perp} \longrightarrow \begin{bmatrix} 1 & 0 & 1 & 1 & 0 \\ 1 & 1 & 1 & 0 & 1 \end{bmatrix} \Rightarrow A = \begin{bmatrix} 1 & 0 & 1 \\ 1 & 1 & 1 \end{bmatrix} \Rightarrow G^{\perp} = \begin{bmatrix} 1 & 0 & 0 \\ 0 & 1 & 0 \\ 0 & 0 & 1 \\ 1 & 0 & 1 \\ 1 & 1 & 1 \end{bmatrix}.$

15. $\begin{matrix} C_1' = C_1 + C_2 \\ C_2' = C_2 \end{matrix} \Rightarrow G^{\perp} \longrightarrow \begin{bmatrix} 1 & 0 \\ 0 & 1 \\ 1 & 0 \\ 1 & 1 \end{bmatrix} \Rightarrow A = \begin{bmatrix} 1 & 0 \\ 1 & 1 \end{bmatrix} \Rightarrow P^{\perp} = \begin{bmatrix} 1 & 0 & 1 & 0 \\ 1 & 1 & 0 & 1 \end{bmatrix}.$

16. $\begin{matrix} C_1' = C_1 + C_3 \\ C_2' = C_2 \\ C_3' = C_3 \end{matrix} \Rightarrow G^{\perp} \longrightarrow \begin{bmatrix} 1 & 0 & 0 \\ 0 & 1 & 0 \\ 0 & 0 & 1 \\ 1 & 0 & 0 \\ 1 & 1 & 1 \end{bmatrix} \Rightarrow A = \begin{bmatrix} 1 & 0 & 0 \\ 1 & 1 & 1 \end{bmatrix} \Rightarrow P^{\perp} = \begin{bmatrix} 1 & 0 & 0 & 1 & 0 \\ 1 & 1 & 1 & 0 & 1 \end{bmatrix}.$

17. $P^{\perp} = G^T = \begin{bmatrix} 1 & 0 & 0 & 0 & 1 & 1 & 0 \\ 0 & 1 & 0 & 0 & 1 & 0 & 1 \\ 0 & 0 & 1 & 0 & 0 & 1 & 1 \\ 0 & 0 & 0 & 1 & 1 & 1 & 1 \end{bmatrix}$ and $G^{\perp} = P^T = \begin{bmatrix} 1 & 1 & 0 \\ 1 & 0 & 1 \\ 0 & 1 & 1 \\ 1 & 1 & 1 \\ 1 & 0 & 0 \\ 0 & 1 & 0 \\ 0 & 0 & 1 \end{bmatrix}.$

18. (a) For E_3, clearly, $P = \left[\, A \,|\, I \,\right] = \left[\, 1 \;\, 1 \;\, 1 \,\right] \Rightarrow A = \left[\, 1 \;\, 1 \,\right] \Rightarrow G = \left[\frac{I}{A}\right] = \begin{bmatrix} 1 & 0 \\ 0 & 1 \\ 1 & 1 \end{bmatrix}$.

For Rep_3, clearly, $G' = \left[\frac{I}{A'}\right] = \begin{bmatrix} 1 \\ 1 \\ 1 \end{bmatrix} \Rightarrow A' = \begin{bmatrix} 1 \\ 1 \end{bmatrix} \Rightarrow P' = \left[\, I \,|\, A' \,\right] = \begin{bmatrix} 1 & 0 & 1 \\ 0 & 1 & 1 \end{bmatrix}$.

(b) Since $G' = P^T$ and $P' = G^T$ in (a), E_3 and Rep_3 are dual.

19. We will show $Rep_n \subseteq {E_n}^{\perp}$ and ${E_n}^{\perp} \subseteq Rep_n \Rightarrow Rep_n = {E_n}^{\perp}$.

$\mathbf{x}$ in $E_n \Rightarrow w(\mathbf{x}) = 0$ because vectors in E_n have even weight.
Therefore, $\mathbf{x} \cdot \mathbf{0} = 0$ and $\mathbf{x} \cdot \mathbf{1} = w(\mathbf{x}) = 0 \Rightarrow Rep_n \subseteq {E_n}^{\perp}$.

To show ${E_n}^{\perp} \subset Rep_n$ we will show that any vector $\mathbf{v}$ that is *not* in Rep_n is *not* in ${E_n}^{\perp}$.

$\mathbf{v}$ not in $Rep_n \Rightarrow \mathbf{v}$ is not all 0s, 1s $\Rightarrow$ there is an ith position $= 0$ and a jth position $= 1$.
$\mathbf{e}_{ij}$ with a 1 in the ith and j^{th} positions and 0s elsewhere has even weight $\Rightarrow \mathbf{e}_{ij}$ is in E_n.
However, $\mathbf{e}_{ij} \cdot \mathbf{v} = 1 \neq 0 \Rightarrow \mathbf{v}$ is not in ${E_n}^{\perp} \Rightarrow {E_n}^{\perp} \subseteq Rep_n \Rightarrow Rep_n = {E_n}^{\perp}$.

20. We need only show that $\mathbf{x}$ in $D^{\perp} \Rightarrow \mathbf{x}$ in $C^{\perp}$.

$\mathbf{x}$ in $D^{\perp} \Rightarrow \mathbf{d} \cdot \mathbf{x} = 0$ for all $\mathbf{d}$ in $D \Rightarrow \mathbf{c} \cdot \mathbf{x} = 0$ for all $\mathbf{c}$ in C because $C \subseteq D \Rightarrow \mathbf{x}$ in $C^{\perp}$.

21. We need only show that $(C^{\perp})^{\perp}$ and C have the same generator matrix G.

$G = \left[\frac{I}{A}\right] \Rightarrow P = \left[\, A \,|\, I \,\right] \Rightarrow G^{\perp} = P^T \Rightarrow (C^{\perp})^{\perp}$ has $(P^{\perp})^{\perp} = (G^{\perp})^T = (P^T)^T = P$

So, $P = \left[\, A \,|\, I \,\right] \Rightarrow G = \left[\frac{I}{A}\right] \Rightarrow (C^{\perp})^{\perp}$ has generator matrix $G \Rightarrow (C^{\perp})^{\perp} = C$.

22. We are given $n = 6$ and we need $G^{\perp} = P^T = G \Rightarrow n - k = k \Rightarrow k = 3$.

Therefore, C generated by $G = \begin{bmatrix} I_3 \\ I_3 \end{bmatrix}$ with $P = \left[\, I_3 \;\; I_3 \,\right]$
is a self dual code of length 6 because $G^{\perp} = P^T = G$.

23. $f(\mathbf{x}) = \mathbf{x}^T A \mathbf{x} = 2x^2 + 4y^2 + 6xy$.

24. $f(\mathbf{x}) = \mathbf{x}^T A \mathbf{x} = 5x_1^2 - x_2^2 + 2x_1x_2$.

25. $f(\mathbf{x}) = \mathbf{x}^T A \mathbf{x} = 3(1)^2 + 4(6)^2 - 4(1)(6) = 123$.

26. $f(\mathbf{x}) = \mathbf{x}^T A \mathbf{x} = 1x^2 + 2y^2 + 3z^2 + 0xy - 6xz + 2yz = x^2 + 2y^2 + 3z^2 - 6xz + 2yz$.

27. $f(\mathbf{x}) = \mathbf{x}^T A \mathbf{x} = 1(2)^2 + 2(-1)^2 + 3(1)^2 + 0(2)(-1) - 6(2)(1) + 2(-1)(1) = -5$.

28. $f(\mathbf{x}) = \mathbf{x}^T A \mathbf{x} = 2(1)^2 + 0(2)^2 + 1(3)^2 + 4(1)(2) + 0(1)(3) + 2(2)(3) = 31$.

29. $x_1^2 + 2x_2^2 + 6x_1x_2 \Rightarrow$ diagonal $= 1, 2$ and corners $= \frac{1}{2}6 = 3 \Rightarrow A = \begin{bmatrix} 1 & 3 \\ 3 & 2 \end{bmatrix}$.

30. $x_1x_2 \Rightarrow$ diagonal $= 0, 0$ and corners $= \frac{1}{2}1 = \frac{1}{2} \Rightarrow A = \begin{bmatrix} 0 & 1/2 \\ 1/2 & 0 \end{bmatrix}$.

31. $3x^2 - 3xy - y^2 \Rightarrow$ diagonal $= 3, -1$ and corners $= \frac{1}{2}(-3) = -\frac{3}{2} \Rightarrow A = \begin{bmatrix} 3 & -3/2 \\ -3/2 & -1 \end{bmatrix}$.

32. $\Rightarrow$ diagonal $= 1, 0, -1$ and off diagonal $= \frac{1}{2}8 = 4, \frac{1}{2}(-6) = -3 \Rightarrow A = \begin{bmatrix} 1 & 4 & 0 \\ 4 & 0 & -3 \\ 0 & -3 & -1 \end{bmatrix}$.

33. $\Rightarrow$ diagonal $= 5, -1, 2$ and off diagonal $= \frac{1}{2}2 = 1, \frac{1}{2}(-4) = -2, \frac{1}{2}4 = 2 \Rightarrow A = \begin{bmatrix} 5 & 1 & -2 \\ 1 & -1 & 2 \\ -2 & 2 & 2 \end{bmatrix}$.

34. $\Rightarrow$ diagonal $= 2, -3, 1$ and off diagonal $= \frac{1}{2}(-4) = -2 \Rightarrow A = \begin{bmatrix} 2 & 0 & -2 \\ 0 & -3 & 0 \\ -2 & 0 & 1 \end{bmatrix}$.

35. $A = \begin{bmatrix} 2 & -2 \\ -2 & 5 \end{bmatrix} \Rightarrow \begin{vmatrix} 2-\lambda & -2 \\ -2 & 5-\lambda \end{vmatrix} = \lambda^2 - 7\lambda + 6 = 0 \Rightarrow D = \begin{bmatrix} 1 & 0 \\ 0 & 6 \end{bmatrix}$.

Furthermore, $\mathbf{v}_1 = \begin{bmatrix} 2 \\ 1 \end{bmatrix}$, $\mathbf{v}_2 = \begin{bmatrix} 1 \\ -2 \end{bmatrix}$. Normalizing $\Rightarrow Q = \begin{bmatrix} 2/\sqrt{5} & 1/\sqrt{5} \\ 1/\sqrt{5} & -2/\sqrt{5} \end{bmatrix}$.

So, the new quadratic form is $f(\mathbf{y}) = \mathbf{y}^T D\mathbf{y} = \begin{bmatrix} y_1 & y_2 \end{bmatrix} \begin{bmatrix} 1 & 0 \\ 0 & 6 \end{bmatrix} \begin{bmatrix} y_1 \\ y_2 \end{bmatrix} = y_1^2 + 6y_2^2$.

36. $A = \begin{bmatrix} 1 & 4 \\ 4 & 1 \end{bmatrix} \Rightarrow \begin{vmatrix} 1-\lambda & 4 \\ 4 & 1-\lambda \end{vmatrix} = \lambda^2 - 2\lambda - 15 = 0 \Rightarrow D = \begin{bmatrix} 5 & 0 \\ 0 & -3 \end{bmatrix}$.

$\mathbf{v}_1 = \begin{bmatrix} 1 \\ 1 \end{bmatrix}$, $\mathbf{v}_2 = \begin{bmatrix} 1 \\ -1 \end{bmatrix} \Rightarrow Q = \begin{bmatrix} 1/\sqrt{2} & 1/\sqrt{2} \\ 1/\sqrt{2} & -1/\sqrt{2} \end{bmatrix} \Rightarrow f(\mathbf{y}) = \mathbf{y}^T D\mathbf{y} = 5y_1^2 - 3y_2^2$.

37. $A = \begin{bmatrix} 7 & 4 & 4 \\ 4 & 1 & -8 \\ 4 & -8 & 1 \end{bmatrix} \Rightarrow \begin{vmatrix} 7-\lambda & 4 & 4 \\ 4 & 1-\lambda & -8 \\ 4 & -8 & 1-\lambda \end{vmatrix} = (\lambda - 9)^2(\lambda + 9) = 0 \Rightarrow D = \begin{bmatrix} 9 & 0 & 0 \\ 0 & 9 & 0 \\ 0 & 0 & -9 \end{bmatrix}$.

$(A - 9I)\mathbf{v} = \mathbf{0}$ (of multiplicity 2) $\Rightarrow x_1 = 2(x_2 + x_3) \Rightarrow \mathbf{v}_1 = \begin{bmatrix} 2 \\ 0 \\ 1 \end{bmatrix}$, $\mathbf{v}_2 = \begin{bmatrix} 2 \\ 1 \\ 0 \end{bmatrix}$.

$(A - (-9)I)\mathbf{v} = \mathbf{0}$ (of multiplicity 1) $\Rightarrow x_1 = -\frac{1}{4}(x_2 + x_3) \Rightarrow \mathbf{v}_3 = \begin{bmatrix} -1 \\ 2 \\ 2 \end{bmatrix}$.

$\Rightarrow Q = \begin{bmatrix} 2/\sqrt{5} & 2/\sqrt{5} & -1/3 \\ 0 & 1/\sqrt{5} & 2/3 \\ 1/\sqrt{5} & 0 & 2/3 \end{bmatrix} \Rightarrow f(\mathbf{y}) = \mathbf{y}^T D\mathbf{y} = 9y_1^2 + 9y_2^2 - 9y_2^3$.

38. $A = \begin{bmatrix} 1 & -2 & 0 \\ -2 & 1 & 0 \\ 0 & 0 & 3 \end{bmatrix} \Rightarrow \begin{vmatrix} 1-\lambda & -2 & 0 \\ -2 & 1-\lambda & 0 \\ 0 & 0 & 3-\lambda \end{vmatrix} = (\lambda-3)^2(\lambda+1) = 0 \Rightarrow D = \begin{bmatrix} 3 & 0 & 0 \\ 0 & 3 & 0 \\ 0 & 0 & -1 \end{bmatrix}.$

$(A - \lambda I)\mathbf{v} = \mathbf{0} \Rightarrow \mathbf{v}_1 = \begin{bmatrix} 1 \\ -1 \\ 0 \end{bmatrix}, \mathbf{v}_2 = \begin{bmatrix} 0 \\ 0 \\ 1 \end{bmatrix}, \mathbf{v}_3 = \begin{bmatrix} 1 \\ 1 \\ 0 \end{bmatrix}.$

$\Rightarrow Q = \begin{bmatrix} 1/\sqrt{2} & 0 & 1/\sqrt{2} \\ -1/\sqrt{2} & 0 & 1/\sqrt{2} \\ 0 & 1 & 0 \end{bmatrix} \Rightarrow f(\mathbf{y}) = \mathbf{y}^T D\mathbf{y} = 3y_1^2 + 3y_2^2 - y_2^3.$

39. $A = \begin{bmatrix} 1 & -1 & 0 \\ -1 & 0 & 1 \\ 0 & 1 & 1 \end{bmatrix} \Rightarrow \begin{vmatrix} 1-\lambda & -1 & 0 \\ -1 & 0-\lambda & 1 \\ 0 & 1 & 1-\lambda \end{vmatrix} = (\lambda-2)(\lambda^2-1) = 0 \Rightarrow D = \begin{bmatrix} 2 & 0 & 0 \\ 0 & 1 & 0 \\ 0 & 0 & -1 \end{bmatrix}.$

$(A - \lambda I)\mathbf{v} = \mathbf{0} \Rightarrow \mathbf{v}_1 = \begin{bmatrix} 1 \\ -1 \\ -1 \end{bmatrix}, \mathbf{v}_2 = \begin{bmatrix} 1 \\ 0 \\ 1 \end{bmatrix}, \mathbf{v}_3 = \begin{bmatrix} 1 \\ 2 \\ -1 \end{bmatrix}.$

$\Rightarrow Q = \begin{bmatrix} 1/\sqrt{3} & 1/\sqrt{2} & 1/\sqrt{6} \\ -1/\sqrt{3} & 0 & 2/\sqrt{6} \\ -1/\sqrt{3} & 1/\sqrt{2} & -1/\sqrt{6} \end{bmatrix} \Rightarrow f(\mathbf{x}') = (\mathbf{x}')^T D(\mathbf{x}') = 2(x')^2 + (y')^2 - (z')^2.$

40. $A = \begin{bmatrix} 0 & 1 & 1 \\ 1 & 0 & 1 \\ 1 & 1 & 0 \end{bmatrix} \Rightarrow \begin{vmatrix} 0-\lambda & 1 & 1 \\ 1 & 0-\lambda & 1 \\ 1 & 1 & 0-\lambda \end{vmatrix} = (\lambda-2)(\lambda+1)^2 = 0 \Rightarrow D = \begin{bmatrix} 2 & 0 & 0 \\ 0 & -1 & 0 \\ 0 & 0 & -1 \end{bmatrix}.$

$(A - \lambda I)\mathbf{v} = \mathbf{0} \Rightarrow \mathbf{v}_1 = \begin{bmatrix} 1 \\ 1 \\ 1 \end{bmatrix}, \mathbf{v}_2 = \begin{bmatrix} -2 \\ 1 \\ 1 \end{bmatrix}, \mathbf{v}_3 = \begin{bmatrix} 0 \\ 1 \\ -1 \end{bmatrix}.$

$\Rightarrow Q = \begin{bmatrix} 1/\sqrt{3} & -2/\sqrt{6} & 0 \\ 1/\sqrt{3} & 1/\sqrt{6} & 1/\sqrt{2} \\ 1/\sqrt{3} & 1/\sqrt{6} & -1/\sqrt{2} \end{bmatrix} \Rightarrow f(\mathbf{x}') = (\mathbf{x}')^T D(\mathbf{x}') = 2(x')^2 - (y')^2 - (z')^2.$

41. $A = \begin{bmatrix} 1 & 0 \\ 0 & 2 \end{bmatrix} \Rightarrow \begin{vmatrix} 1-\lambda & 0 \\ 0 & 2-\lambda \end{vmatrix} = 0 \Rightarrow \lambda = 1, 2 > 0 \Rightarrow f$ is positive definite.

42. $A = \begin{bmatrix} 1 & -1 \\ -1 & 1 \end{bmatrix} \Rightarrow \begin{vmatrix} 1-\lambda & -1 \\ -1 & 1-\lambda \end{vmatrix} = 0 \Rightarrow \lambda = 0, 2 \geq 0 \Rightarrow f$ is positive semidefinite.

43. $A = \begin{bmatrix} -2 & 1 \\ 1 & -2 \end{bmatrix} \Rightarrow \begin{vmatrix} -2-\lambda & 1 \\ 1 & -2-\lambda \end{vmatrix} = 0 \Rightarrow \lambda = -3, -1 < 0 \Rightarrow f$ is negative definite.

44. $A = \begin{bmatrix} 1 & 2 \\ 2 & 1 \end{bmatrix} \Rightarrow \begin{vmatrix} 1-\lambda & 2 \\ 2 & 1-\lambda \end{vmatrix} = 0 \Rightarrow \lambda = -1, 3 \Rightarrow f$ is indefinite.

45. $A = \begin{bmatrix} 2 & 1 & 1 \\ 1 & 2 & 1 \\ 1 & 1 & 2 \end{bmatrix} \Rightarrow \begin{vmatrix} 2-\lambda & 1 & 1 \\ 1 & 2-\lambda & 1 \\ 1 & 1 & 2-\lambda \end{vmatrix} = 0 \Rightarrow \lambda = 1, 4 > 0 \Rightarrow f$ is positive definite.

46. $\Rightarrow \begin{vmatrix} 1-\lambda & 0 & 1 \\ 0 & 1-\lambda & 0 \\ 1 & 0 & 1-\lambda \end{vmatrix} = 0 \Rightarrow \lambda = 0, 1, 2 \geq 0 \Rightarrow f$ is positive semidefinite.

47. $\Rightarrow \begin{vmatrix} 1-\lambda & 2 & 0 \\ 2 & 1-\lambda & 0 \\ 0 & 0 & -1-\lambda \end{vmatrix} = 0 \Rightarrow \lambda = -1, 3 \Rightarrow f$ is indefinite.

48. $\Rightarrow \begin{vmatrix} -1-\lambda & -1 & -1 \\ -1 & -1-\lambda & -1 \\ -1 & -1 & -1-\lambda \end{vmatrix} = 0 \Rightarrow \lambda = -3, 0 \leq 0 \Rightarrow f$ is negative semidefinite.

49. For 5.24a., Theorem 5.23 $\Rightarrow f(\mathbf{x}) = \lambda_1 y_1^2 + \ldots + \lambda_n y_n^2 > 0 \Leftrightarrow \mathbf{x} \neq 0$ and $\lambda_i > 0$.
Likewise 5.24b. to 5.24e.

50. We need to show that $\det A > 0$ and $a > 0 \Leftrightarrow \lambda > 0$.

Note $\det A = ad - b^2 > 0 \Rightarrow ad > b^2 \Leftrightarrow$ both $a, d > 0$ or both $a, d < 0$.
So, when $a, d > 0$ or $a, d < 0$, then $a + d > 0 \Leftrightarrow a, d > 0$ and $a + d < 0 \Leftrightarrow a, d < 0$.

So, since $\det A = \lambda((a + d) - \lambda)$, $\det A > 0 \Leftrightarrow \lambda > 0$ and $a + d > \lambda$ or $\lambda < 0$ and $a + d < \lambda$.

Therefore, $\det A > 0 \Leftrightarrow \lambda > 0$ and $a > 0 \Rightarrow \det A > 0$ and $a > 0 \Leftrightarrow \lambda > 0$.

51. Note $\mathbf{x}^T A\mathbf{x} = \mathbf{x}^T B^T B\mathbf{x} = (B\mathbf{x})^T (B\mathbf{x}) = \|B\mathbf{x}\|^2 \geq 0$.

So, if $\mathbf{x}^T A\mathbf{x} = 0$, then $\|B\mathbf{x}\|^2 = 0 \Rightarrow \mathbf{x} = 0$ because B is invertible
$\Rightarrow \mathbf{x}^T A\mathbf{x} > 0$ for all $\mathbf{x} \neq 0 \Rightarrow A = B^T B$ is positive definite.

52. A symmetric, positive definite $\Rightarrow A = QDQ^T$, where $D = \begin{bmatrix} \lambda_1 & \ldots & 0 \\ \vdots & \ddots & \vdots \\ 0 & \ldots & \lambda_n \end{bmatrix}$, and $\lambda_i > 0$.

Then let $C = C^T = \begin{bmatrix} \sqrt{\lambda_1} & \ldots & 0 \\ \vdots & \ddots & \vdots \\ 0 & \ldots & \sqrt{\lambda_n} \end{bmatrix}$, so $D = C^T C$.

$\Rightarrow A = Q(C^T C)Q^T = (QC^T)CQ^T = (CQ^T)^T CQ^T$, where CQ^T is invertible.

53. (a) $\mathbf{x}^T(cA)\mathbf{x} = c(\mathbf{x}^T A\mathbf{x}) > 0$ because $c > 0$ and $\mathbf{x}^T A\mathbf{x} > 0$ when $\mathbf{x} \neq 0$.

(b) $\mathbf{x}^T A^2 \mathbf{x} = (\mathbf{x}^T A\mathbf{x})(\mathbf{x}^T A\mathbf{x}) = (\mathbf{x}^T A\mathbf{x})^2 > 0$.

(c) $\mathbf{x}^T(A + B)\mathbf{x} = (\mathbf{x}^T A\mathbf{x}) + (\mathbf{x}^T B\mathbf{x}) > 0$ because $\mathbf{x}^T A\mathbf{x} > 0$ and $\mathbf{x}^T B\mathbf{x} > 0$.

(d) Note, A is invertible by Exercise 52 since $A = (CQ^T)^T CQ^T$, where CQ^T is invertible.
Therefore, the eigenvalues of A^{-1} are $\frac{1}{\lambda_i} > 0 \Rightarrow A^{-1}$ is positive definite by Theorem 4a.

54. Q orthogonal $\Rightarrow QCQ^T$ positive definite because $\mathbf{x}^T(QCQ^T)\mathbf{x} = (Q\mathbf{x})^T C(Q^T\mathbf{x}) = \mathbf{x}^T C\mathbf{x}$.
From Exercise 52 and recalling that $C = C^T$, we have $A = QCCQ^T = (QCQ^T)(QCQ^T)$.

55. $\lambda = 0, 2 \Rightarrow$ max $= 2$, min $= 0$. So, $(A - 2I)\mathbf{v} = \mathbf{0} \Rightarrow \mathbf{v}_1 = \begin{bmatrix} 1 \\ -1 \end{bmatrix}$, $A\mathbf{v} = \mathbf{0} \Rightarrow \mathbf{v}_2 = \begin{bmatrix} 1 \\ 1 \end{bmatrix}$.

For $\|\mathbf{x}\| = 1$ max occurs at $\mathbf{x} = \pm \begin{bmatrix} 1/\sqrt{2} \\ -1/\sqrt{2} \end{bmatrix}$ and min occurs at $\mathbf{x} = \pm \begin{bmatrix} 1/\sqrt{2} \\ 1/\sqrt{2} \end{bmatrix}$.

56. $\lambda = -1, 3 \Rightarrow$ max $= 3$, min $=$ -1. So, $(A - \lambda I)\mathbf{v} = \mathbf{0} \Rightarrow \mathbf{v}_1 = \begin{bmatrix} 1 \\ 1 \end{bmatrix}$, $\mathbf{v}_2 = \begin{bmatrix} 1 \\ -1 \end{bmatrix}$.

For $\|\mathbf{x}\| = 1$ max occurs at $\mathbf{x} = \pm \begin{bmatrix} 1/\sqrt{2} \\ 1/\sqrt{2} \end{bmatrix}$ and min occurs at $\mathbf{x} = \pm \begin{bmatrix} 1/\sqrt{2} \\ -1/\sqrt{2} \end{bmatrix}$.

57. $\lambda = 1, 4 \Rightarrow$ max $= 4$, min $= 1$. So, $(A - \lambda I)\mathbf{v} = \mathbf{0} \Rightarrow \mathbf{v}_1 = \begin{bmatrix} 1 \\ 1 \\ 1 \end{bmatrix}$, $\mathbf{v}_2 = \begin{bmatrix} 1 \\ 0 \\ -1 \end{bmatrix}$, $\mathbf{v}_3 = \begin{bmatrix} -1 \\ 1 \\ 0 \end{bmatrix}$.

So, max occurs at $\mathbf{x} = \pm \begin{bmatrix} 1/\sqrt{3} \\ 1/\sqrt{3} \\ 1/\sqrt{3} \end{bmatrix}$ and min occurs at $\mathbf{x} = \pm \begin{bmatrix} 1/\sqrt{2} \\ 0 \\ -1/\sqrt{2} \end{bmatrix}$ or $\pm \begin{bmatrix} -1/\sqrt{2} \\ 1/\sqrt{2} \\ 0 \end{bmatrix}$.

58. $\lambda = 0, 1, 2 \Rightarrow$ max $= 2$, min $= 0$. So, $(A - \lambda I)\mathbf{v} = \mathbf{0} \Rightarrow \mathbf{v}_1 = \begin{bmatrix} 1 \\ 0 \\ 1 \end{bmatrix}$, $\mathbf{v}_2 = \begin{bmatrix} -1 \\ 0 \\ 1 \end{bmatrix}$.

So, max occurs at $\mathbf{x} = \pm \begin{bmatrix} 1/\sqrt{2} \\ 0 \\ 1/\sqrt{2} \end{bmatrix}$ and min occurs at $\mathbf{x} = \pm \begin{bmatrix} -1/\sqrt{2} \\ 0 \\ 1/\sqrt{2} \end{bmatrix}$.

59. $f(x) = \lambda_1 y_1^2 + \lambda_2 y_2^2 + \ldots + \lambda_n y_n^2 \geq \lambda_n(y_1^2 + y_2^2 + \ldots + y_n^2) = \lambda_n \|\mathbf{y}\| = \lambda_n$ because $\lambda_1 \geq \ldots \geq \lambda_n$.

60. Let $\mathbf{q}_n$ be an eigenvector for λ_n, then $f(\mathbf{q}_n) = \lambda_n \Rightarrow \lambda_n =$ min by Theorem 5.25a.

61. $x^2 + 5y^2 = 25 \Rightarrow \frac{x^2}{25} + \frac{y^2}{5} = 1 \Rightarrow$ ellipse.

62. $x^2 - y^2 - 4 = 0 \Rightarrow \frac{x^2}{4} - \frac{y^2}{4} = 1 \Rightarrow$ hyperbola.

63. $x^2 - y - 1 = 0 \Rightarrow y = x^2 - 1 \Rightarrow$ parabola.

64. $2x^2 + y^2 - 8 = 0 \Rightarrow \frac{x^2}{4} + \frac{y^2}{8} = 1 \Rightarrow$ ellipse.

65. $3x^2 = y^2 - 1 \Rightarrow \frac{y^2}{1} - \frac{x^2}{1/3} = 1 \Rightarrow$ hyperbola.

66. $x = -2y^2 \Rightarrow$ parabola.

67.

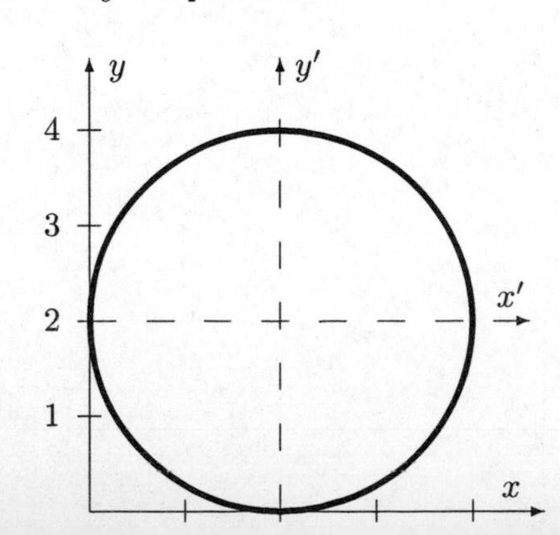

$x^2 + y^2 - 4x - 4y + 4 = 0 \Rightarrow$
$(x-2)^2 + (y-2)^2 = 4 \Rightarrow$
circle $(x')^2 + (y')^2 = 4$.

68.

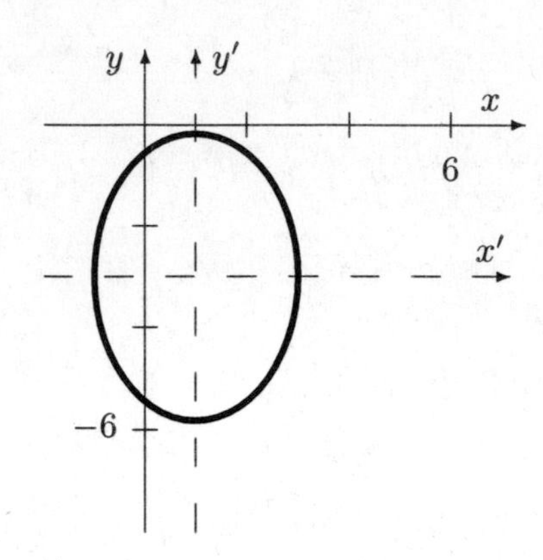

$4x^2 + 2y^2 - 8x + 12y + 6 = 0 \Rightarrow$
$\frac{(x-1)^2}{4} + \frac{(y+3)^2}{8} = 1 \Rightarrow$
ellipse $\frac{(x')^2}{4} + \frac{(y')^2}{8} = 1.$

69.

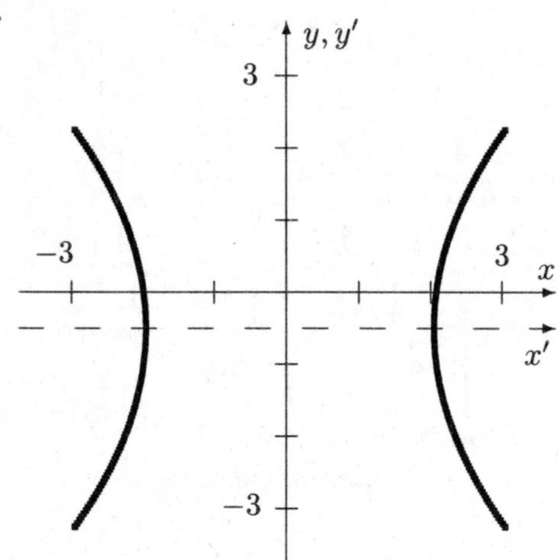

$9x^2 - 4y^2 - 4y = 37 \Rightarrow$
$\frac{(x-0)^2}{4} - \frac{(y+1/2)^2}{9} = 1 \Rightarrow$
hyperbola $\frac{(x')^2}{4} - \frac{(y')^2}{9} = 1.$

70.

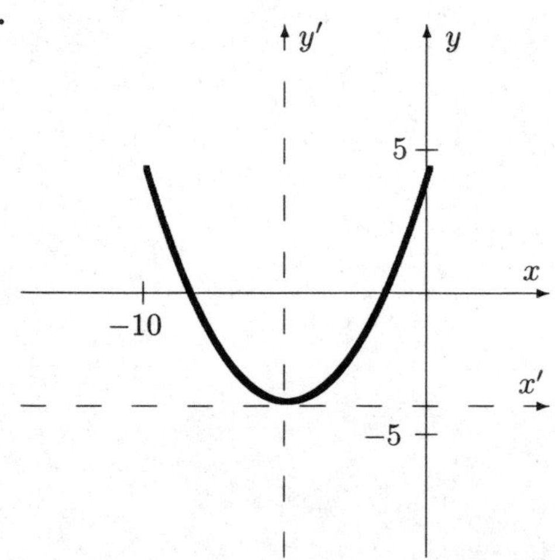

$x^2 + 10x - 3y = -13 \Rightarrow$
$y = \frac{1}{3}(x+5)^2 - 4 \Rightarrow$
parabola $y' = \frac{1}{3}(x')^2.$

71.

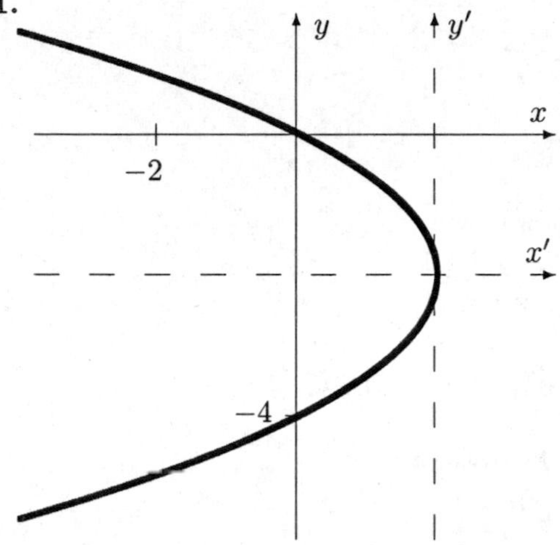

$2y^2 + 8y + 4x = 0 \Rightarrow$
$x = -\frac{1}{2}(y+2)^2 + 2 \Rightarrow$
parabola $x' = -\frac{1}{2}(y')^2$.

72.

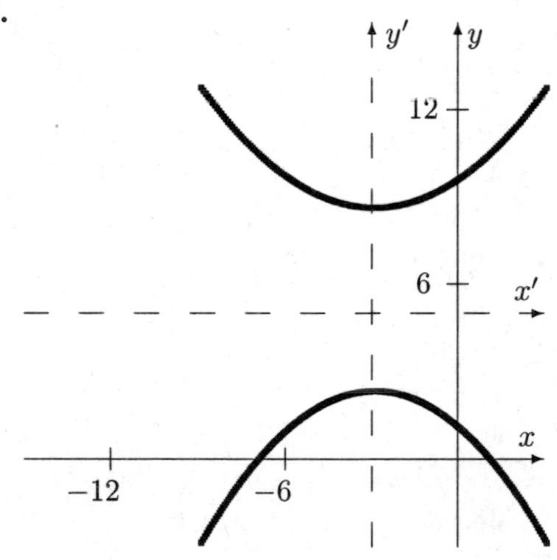

$2y^2 - 3x^2 - 18x - 20y + 11 = 0 \Rightarrow$
$\frac{(y-5)^2}{6} - \frac{(x+3)^2}{4} = 1 \Rightarrow$
hyperbola $\frac{(y')^2}{6} - \frac{(x')^2}{4} = 1$.

73. $x^2 + xy + y^2 = 6 \Rightarrow A = \begin{bmatrix} 1 & \frac{1}{2} \\ \frac{1}{2} & 1 \end{bmatrix} \Rightarrow \lambda = \frac{3}{2}, \frac{1}{2} \Rightarrow \frac{3}{2}(x')^2 + \frac{1}{2}(y')^2 = 6 \Rightarrow \frac{(x')^2}{4} + \frac{(y')^2}{12} = 1.$

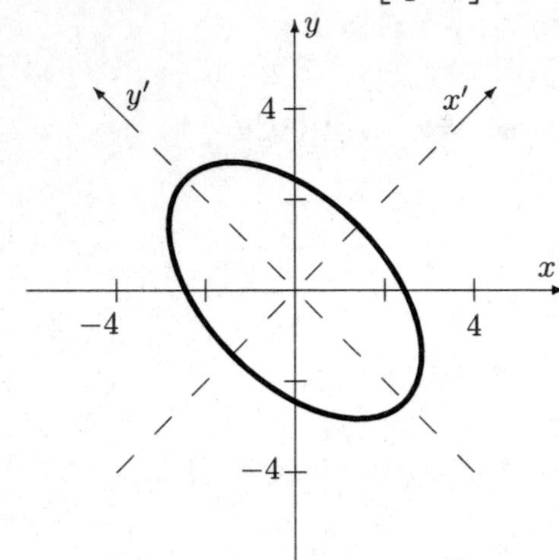

$Q = \begin{bmatrix} \frac{1}{\sqrt{2}} & -\frac{1}{\sqrt{2}} \\ \frac{1}{\sqrt{2}} & \frac{1}{\sqrt{2}} \end{bmatrix} \Rightarrow$

$\mathbf{e}_1 = \begin{bmatrix} 1 \\ 0 \end{bmatrix} \longrightarrow \begin{bmatrix} \frac{1}{\sqrt{2}} \\ \frac{1}{\sqrt{2}} \end{bmatrix}$ and

$\mathbf{e}_2 = \begin{bmatrix} 0 \\ 1 \end{bmatrix} \longrightarrow \begin{bmatrix} -\frac{1}{\sqrt{2}} \\ \frac{1}{\sqrt{2}} \end{bmatrix} \Rightarrow$

45° rotation

Ellipse

74. $4x^2 + 10xy + 4y^2 = 9 \Rightarrow A = \begin{bmatrix} 4 & 5 \\ 5 & 4 \end{bmatrix} \Rightarrow \lambda = 9, -1 \Rightarrow 9(x')^2 - (y')^2 = 9 \Rightarrow \frac{(x')^2}{1} - \frac{(y')^2}{9} = 1.$

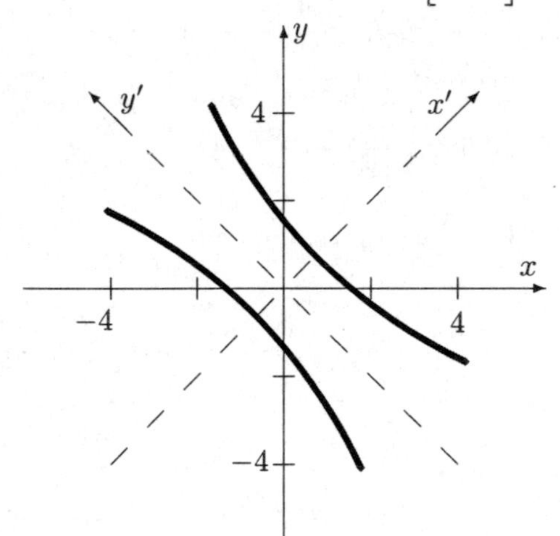

$Q = \begin{bmatrix} \frac{1}{\sqrt{2}} & -\frac{1}{\sqrt{2}} \\ \frac{1}{\sqrt{2}} & \frac{1}{\sqrt{2}} \end{bmatrix} \Rightarrow$

$\mathbf{e}_1 = \begin{bmatrix} 1 \\ 0 \end{bmatrix} \longrightarrow \begin{bmatrix} \frac{1}{\sqrt{2}} \\ \frac{1}{\sqrt{2}} \end{bmatrix}$ and

$\mathbf{e}_2 = \begin{bmatrix} 0 \\ 1 \end{bmatrix} \longrightarrow \begin{bmatrix} -\frac{1}{\sqrt{2}} \\ \frac{1}{\sqrt{2}} \end{bmatrix} \Rightarrow$

45° rotation

Hyperbola

75. $4x^2+6xy-4y^2=5 \Rightarrow A=\begin{bmatrix}4&3\\3&-4\end{bmatrix} \Rightarrow \lambda=5,\,-5 \Rightarrow 5(x')^2-5(y')^2=5 \Rightarrow (x')^2-(y')^2=1.$

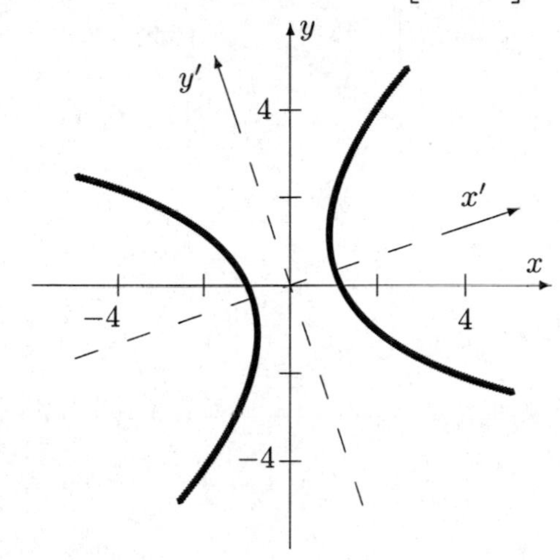

$Q=\begin{bmatrix}\frac{3}{\sqrt{10}}&-\frac{1}{\sqrt{10}}\\\frac{1}{\sqrt{10}}&\frac{3}{\sqrt{10}}\end{bmatrix} \Rightarrow$

$\mathbf{e}_1=\begin{bmatrix}1\\0\end{bmatrix} \longrightarrow \begin{bmatrix}\frac{3}{\sqrt{10}}\\\frac{1}{\sqrt{10}}\end{bmatrix}$ and

$\mathbf{e}_2=\begin{bmatrix}0\\1\end{bmatrix} \longrightarrow \begin{bmatrix}-\frac{1}{\sqrt{10}}\\\frac{3}{\sqrt{10}}\end{bmatrix} \Rightarrow$

$\approx 18.435^\circ$ rotation

Hyperbola

76. $3x^2-2xy+3y^2=8 \Rightarrow A=\begin{bmatrix}3&-1\\-1&3\end{bmatrix} \Rightarrow \lambda=4,\,2 \Rightarrow 4(x')^2+2(y')^2=8 \Rightarrow \frac{(x')^2}{2}+\frac{(y')^2}{4}=1.$

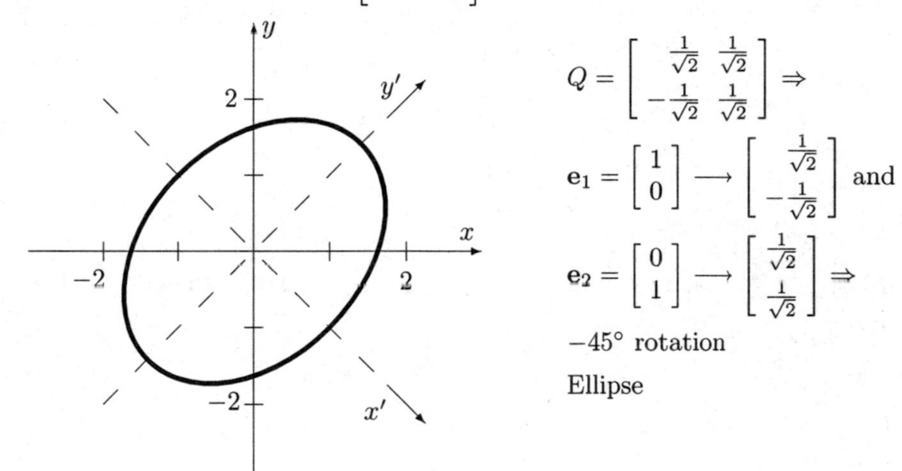

$Q=\begin{bmatrix}\frac{1}{\sqrt{2}}&\frac{1}{\sqrt{2}}\\-\frac{1}{\sqrt{2}}&\frac{1}{\sqrt{2}}\end{bmatrix} \Rightarrow$

$\mathbf{e}_1=\begin{bmatrix}1\\0\end{bmatrix} \longrightarrow \begin{bmatrix}\frac{1}{\sqrt{2}}\\-\frac{1}{\sqrt{2}}\end{bmatrix}$ and

$\mathbf{e}_2=\begin{bmatrix}0\\1\end{bmatrix} \longrightarrow \begin{bmatrix}\frac{1}{\sqrt{2}}\\\frac{1}{\sqrt{2}}\end{bmatrix} \Rightarrow$

-45° rotation

Ellipse

77. $3x^2 - 4xy + 3y^2 - 28\sqrt{2}\,x + 22\sqrt{2}\,y + 84 = 0 \Rightarrow$

$A = \begin{bmatrix} 3 & -2 \\ -2 & 3 \end{bmatrix}$, $B = \begin{bmatrix} -28\sqrt{2} & 22\sqrt{2} \end{bmatrix} \Rightarrow \lambda = 1, 5$ and $Q = \begin{bmatrix} \frac{1}{\sqrt{2}} & \frac{1}{\sqrt{2}} \\ -\frac{1}{\sqrt{2}} & \frac{1}{\sqrt{2}} \end{bmatrix} \Rightarrow$

$BQ = \begin{bmatrix} -6 & -50 \end{bmatrix} \Rightarrow (x')^2 - 6x' + 5(y')^2 - 50y' = -84 \Rightarrow \frac{(x'-3)^2}{50} + \frac{(y'-5)^2}{10} = 1 \Rightarrow$

$\frac{(x'')^2}{50} + \frac{(y'')^2}{10} = 1$, an ellipse.

78. $6x^2 - 4xy + 9y^2 - 20x - 10y - 5 = 0 \Rightarrow$

$A = \begin{bmatrix} 6 & -2 \\ -2 & 9 \end{bmatrix}$, $B = \begin{bmatrix} -20 & -10 \end{bmatrix} \Rightarrow \lambda = 5, 10$ and $Q = \begin{bmatrix} \frac{2}{\sqrt{5}} & -\frac{1}{\sqrt{5}} \\ \frac{1}{\sqrt{5}} & \frac{2}{\sqrt{5}} \end{bmatrix} \Rightarrow$

$BQ = \begin{bmatrix} -10\sqrt{5} & 0 \end{bmatrix} \Rightarrow 5(x')^2 + 10(y')^2 - 10\sqrt{5}\,x' = 5 \Rightarrow \frac{(x'-\sqrt{5})^2}{6} + \frac{(y')^2}{3} = 1 \Rightarrow$

$\frac{(x'')^2}{6} + \frac{(y'')^2}{3} = 1$, an ellipse.

79. $2xy + 2\sqrt{2}\,x - 1 = 0 \Rightarrow$

$A = \begin{bmatrix} 0 & 1 \\ 1 & 0 \end{bmatrix}$, $B = \begin{bmatrix} 2\sqrt{2} & 0 \end{bmatrix} \Rightarrow \lambda = 1, -1$ and $Q = \begin{bmatrix} \frac{1}{\sqrt{2}} & \frac{1}{\sqrt{2}} \\ \frac{1}{\sqrt{2}} & -\frac{1}{\sqrt{2}} \end{bmatrix} \Rightarrow$

$BQ = \begin{bmatrix} 2 & 2 \end{bmatrix} \Rightarrow (x')^2 + 2x' - (y')^2 + 2y' = 1 \Rightarrow (x' + 1)^2 - (y' - 1)^2 = 1 \Rightarrow$

$(x'')^2 - (y'')^2 = 1$, a hyperbola.

80. $x^2 - 2xy + y^2 + 4\sqrt{2}\,x - 4 = 0 \Rightarrow$

$A = \begin{bmatrix} 1 & -1 \\ -1 & 1 \end{bmatrix}$, $B = \begin{bmatrix} 4\sqrt{2} & 0 \end{bmatrix} \Rightarrow \lambda = 2, 0$ and $Q = \begin{bmatrix} \frac{1}{\sqrt{2}} & \frac{1}{\sqrt{2}} \\ -\frac{1}{\sqrt{2}} & \frac{1}{\sqrt{2}} \end{bmatrix} \Rightarrow$

$BQ = \begin{bmatrix} 4 & 4 \end{bmatrix} \Rightarrow 2(x')^2 + 2x' + 0(y')^2 + 4x' + 4y' = 4 \Rightarrow y' = -\frac{1}{2}(x' + 1)^2 + \frac{1}{2} \Rightarrow$

$y'' = -\frac{1}{2}(x'')^2$, a parabola.

81.

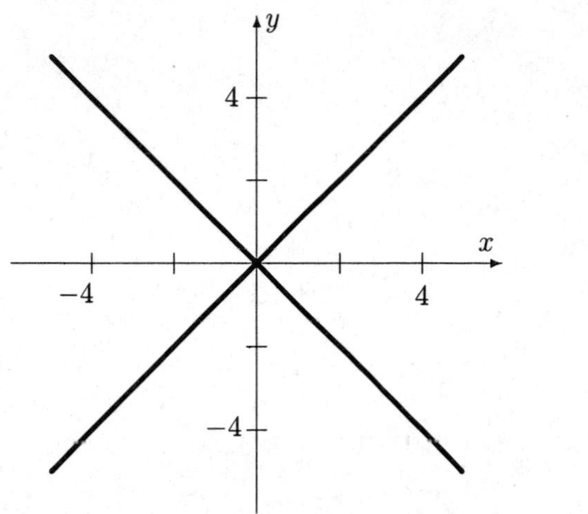

$x^2 - y^2 = 0 \Rightarrow$

$y = \pm x$

Degenerate

Two lines

82. $x^2 + 2y^2 + 2 = 0 \Rightarrow \frac{x^2}{2} + \frac{y^2}{1} = -1 \Rightarrow$ No solution $\Rightarrow$ Imaginary conic.

83.

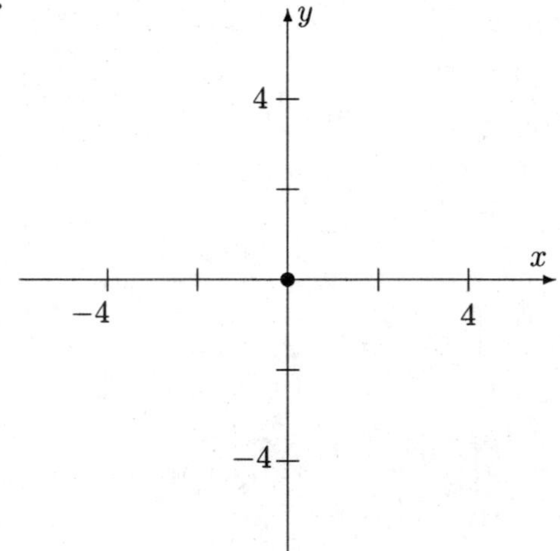

$3x^2 + y^2 = 0 \Rightarrow$

$y^2 = -3x^2 \Rightarrow$

$y = x = 0 \Rightarrow$

Degenerate

A point (0, 0)

84. $x^2 + 2xy + y^2 = 0 \Rightarrow A = \begin{bmatrix} 1 & 1 \\ 1 & 1 \end{bmatrix} \Rightarrow \lambda = 2, 0 \Rightarrow$

$2(x')^2 + 0(y')^2 = 0 \Rightarrow x' = 0$, degenerate, a line $\Rightarrow$

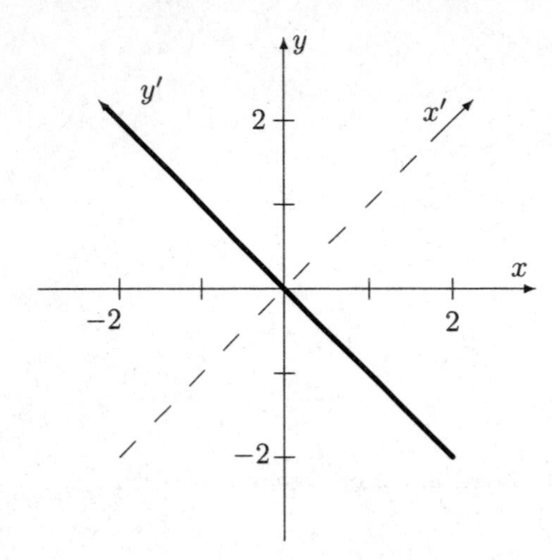

$Q = \begin{bmatrix} \frac{1}{\sqrt{2}} & \frac{1}{\sqrt{2}} \\ \frac{1}{\sqrt{2}} & -\frac{1}{\sqrt{2}} \end{bmatrix} \Rightarrow$

$\mathbf{e}_1 = \begin{bmatrix} 1 \\ 0 \end{bmatrix} \longrightarrow \begin{bmatrix} \frac{1}{\sqrt{2}} \\ \frac{1}{\sqrt{2}} \end{bmatrix}$ and

$\mathbf{e}_2 = \begin{bmatrix} 0 \\ 1 \end{bmatrix} \longrightarrow \begin{bmatrix} \frac{1}{\sqrt{2}} \\ -\frac{1}{\sqrt{2}} \end{bmatrix} \Rightarrow$

$45°$ rotation

85. $x^2 - 2xy + y^2 + 2\sqrt{2}\,x - 2\sqrt{2}\,y = 0 \Rightarrow$

$A = \begin{bmatrix} 1 & -1 \\ -1 & 1 \end{bmatrix}$, $B = \begin{bmatrix} 2\sqrt{2} & 2\sqrt{2} \end{bmatrix} \Rightarrow \lambda = 2,\ 0$ and $Q = \begin{bmatrix} \frac{1}{\sqrt{2}} & \frac{1}{\sqrt{2}} \\ -\frac{1}{\sqrt{2}} & \frac{1}{\sqrt{2}} \end{bmatrix} \Rightarrow$

$BQ = \begin{bmatrix} 4 & 0 \end{bmatrix} \Rightarrow 2(x')^2 + 2x' + 0(y')^2 + 4x' = 0 \Rightarrow x' = 0, -2 \Rightarrow$ degenerate, two lines $\Rightarrow$

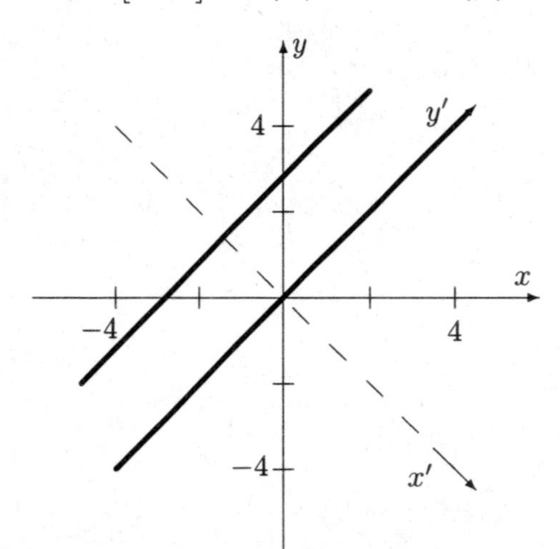

$Q = \begin{bmatrix} \frac{1}{\sqrt{2}} & \frac{1}{\sqrt{2}} \\ -\frac{1}{\sqrt{2}} & \frac{1}{\sqrt{2}} \end{bmatrix} \Rightarrow$

$\mathbf{e}_1 = \begin{bmatrix} 1 \\ 0 \end{bmatrix} \longrightarrow \begin{bmatrix} \frac{1}{\sqrt{2}} \\ -\frac{1}{\sqrt{2}} \end{bmatrix}$ and

$\mathbf{e}_2 = \begin{bmatrix} 0 \\ 1 \end{bmatrix} \longrightarrow \begin{bmatrix} \frac{1}{\sqrt{2}} \\ \frac{1}{\sqrt{2}} \end{bmatrix} \Rightarrow$

$-45°$ rotation

86. $2x^2 + 2xy + 2y^2 + 2\sqrt{2}\,x - 2\sqrt{2}\,y + 6 = 0 \Rightarrow$

$A = \begin{bmatrix} 2 & 1 \\ 1 & 2 \end{bmatrix}$, $B = \begin{bmatrix} 2\sqrt{2} & -2\sqrt{2} \end{bmatrix} \Rightarrow \lambda = 3,\ 1$ and $Q = \begin{bmatrix} \frac{1}{\sqrt{2}} & \frac{1}{\sqrt{2}} \\ \frac{1}{\sqrt{2}} & -\frac{1}{\sqrt{2}} \end{bmatrix} \Rightarrow$

$BQ = \begin{bmatrix} 0 & 4 \end{bmatrix} \Rightarrow 3(x')^2 + (y')^2 + 4y' = -6 \Rightarrow 3(x')^2 + (y'+2)^2 = -2 \Rightarrow$

No solution $\Rightarrow$ Imaginary conic.

87. Recall, $Q^TAQ = D \Rightarrow A = QDQ^T$, where $D = \begin{bmatrix} \lambda_1 & 0 \\ 0 & \lambda_2 \end{bmatrix}$.

So, $\det A = (\det Q)(\det D)(\det Q^T) = \det D = \lambda_1\lambda_2$.

Furthermore, The Principal Axes Theorem $\Rightarrow \mathbf{x}^TA\mathbf{x} = \mathbf{x}'^TD\mathbf{x}' = \lambda_1(x')^2 + \lambda_2(y')^2 = k$.

(a) We are given $k \neq 0$ and $\det A = \lambda_1\lambda_2 < 0 \Rightarrow$ The coefficients of $(x')^2$ and $(y')^2$ in $\mathbf{x}^TA\mathbf{x} = \lambda_1(x')^2 + \lambda_2(y')^2 = k$, that is $\frac{\lambda_1}{k}(x')^2 + \frac{\lambda_2}{k}(y')^2 = 1$, are opposite in sign $\Rightarrow$ The graph of $\mathbf{x}^TA\mathbf{x}$ is a hyperbola.
Note, we use $k \neq 0$ when we divide by k to put the equation into standard form.

(b) We are given $k \neq 0$ and $\det A = \lambda_1\lambda_2 > 0 \Rightarrow$ The coefficients of $(x')^2$ and $(y')^2$ in $\mathbf{x}^TA\mathbf{x} = \lambda_1(x')^2 + \lambda_2(y')^2 = k$, that is $\frac{\lambda_1}{k}(x')^2 + \frac{\lambda_2}{k}(y')^2 = 1$, have the same sign $\Rightarrow$
If $k < 0$, then the graph of $\mathbf{x}^TA\mathbf{x}$ is an imaginary conic.
If $k > 0$ and $\lambda_1 = \lambda_2$, then the graph of $\mathbf{x}^TA\mathbf{x}$ is a circle.
If $k > 0$ and $\lambda_1 \neq \lambda_2$, then the graph of $\mathbf{x}^TA\mathbf{x}$ is an ellipse.

(c) We are given $k \neq 0$ and $\det A = \lambda_1\lambda_2 = 0 \Rightarrow$ One of the coefficients of $(x')^2$ and $(y')^2$ in $\mathbf{x}^TA\mathbf{x} = \lambda_1(x')^2 + \lambda_2(y')^2 = k$, that is $\frac{\lambda_1}{k}(x')^2 + \frac{\lambda_2}{k}(y')^2 = 1$, is zero $\Rightarrow$
If $k < 0$, then the graph of $\mathbf{x}^TA\mathbf{x}$ is an imaginary conic.
If $k > 0$ and $\lambda_1 = 0$, then we have $y' = \pm\sqrt{\frac{k}{\lambda_2}}$, so the graph is two straight lines.
If $k > 0$ and $\lambda_2 = 0$, then we have $x' = \pm\sqrt{\frac{k}{\lambda_1}}$, so the graph is two straight lines.

(d) We are given $k = 0$ and $\det A = \lambda_1\lambda_2 \neq 0 \Rightarrow$ The coefficients of $(x')^2$ and $(y')^2$ in $\mathbf{x}^TA\mathbf{x} = \lambda_1(x')^2 + \lambda_2(y')^2 = 0$, are nonzero $\Rightarrow y' = \pm\sqrt{\frac{-\lambda_1}{\lambda_2}}\,x'$.
If λ_1 and λ_2 have the same sign, then $x' = y' = 0$, the graph is a single point (0,0).
If λ_1 and λ_2 have opposite signs, the graph is two straight lines.

(e) We are given $k = 0$ and $\det A = \lambda_1\lambda_2 = 0 \Rightarrow$ One of the coefficients of $(x')^2$ and $(y')^2$ in $\mathbf{x}^TA\mathbf{x} = \lambda_1(x')^2 + \lambda_2(y')^2 = 0$, is zero $\Rightarrow$
If $\lambda_1 = 0$, then $y' = 0$, so the graph of $\mathbf{x}^TA\mathbf{x}$ is a straight line.
If $\lambda_2 = 0$, then $x' = 0$, so the graph of $\mathbf{x}^TA\mathbf{x}$ is a straight line.

88. $4x^2+4y^2+4z^2+4xy+4xz+4yz=8 \Rightarrow$

$$A=\begin{bmatrix}4&2&2\\2&4&2\\2&2&4\end{bmatrix} \Rightarrow \lambda=2,\,8,\,2 \Rightarrow 2(x')^2+8(y')^2+2(z')^2=8 \Rightarrow \tfrac{(x')^2}{4}+\tfrac{(y')^2}{1}+\tfrac{(z')^2}{4}=1 \Rightarrow$$

The graph is an ellipsoid.

89. $x^2+y^2+z^2-4yz=1 \Rightarrow$

$$A=\begin{bmatrix}1&0&0\\0&1&-2\\0&-2&1\end{bmatrix} \Rightarrow \lambda=1,\,-1,\,3 \Rightarrow (x')^2-(y')^2+3(z')^2=1 \Rightarrow$$

The graph is a hyperboloid of one sheet.

90. $-x^2-y^2-z^2+4xy+4xz+4yz=12 \Rightarrow$

$$A=\begin{bmatrix}-1&2&2\\2&-1&2\\2&2&-1\end{bmatrix} \Rightarrow \lambda=-3,\,3,\,-3 \Rightarrow -3(x')^2+3(y')^2-3(z')^2=12 \Rightarrow$$

$\tfrac{(x')^2}{4}-\tfrac{(y')^2}{4}-\tfrac{(z')^2}{4}=-1 \Rightarrow$ The graph is a hyperboloid of two sheets.

91. $2xy+z=0 \Rightarrow$

$$A=\begin{bmatrix}0&1&0\\1&0&0\\0&0&0\end{bmatrix} \text{ and } B=\begin{bmatrix}0&0&1\end{bmatrix} \Rightarrow \lambda=1,\,-1,\,0 \Rightarrow Q=\begin{bmatrix}\frac{1}{\sqrt{2}}&\frac{1}{\sqrt{2}}&0\\\frac{1}{\sqrt{2}}&-\frac{1}{\sqrt{2}}&0\\0&0&1\end{bmatrix} \Rightarrow$$

$BQ=\begin{bmatrix}0&0&1\end{bmatrix} \Rightarrow z'=z$ and $(x')^2-(y')^2+0(z)^2+z=0 \Rightarrow$

$z=-(x')^2+(y')^2 \Rightarrow$ The graph is a hyperbolic paraboloid.

92. $16x^2+100y^2+9z^2-24xz-60x-80z=0 \Rightarrow$

$$A=\begin{bmatrix}16&0&-12\\0&100&0\\-12&0&9\end{bmatrix} \text{ and } B=\begin{bmatrix}-60&0&-80\end{bmatrix} \Rightarrow \lambda=25,\,0,\,100 \Rightarrow Q=\begin{bmatrix}\frac{4}{5}&\frac{3}{5}&0\\0&0&1\\-\frac{3}{5}&\frac{4}{5}&0\end{bmatrix} \Rightarrow$$

$BQ=\begin{bmatrix}0&-100&0\end{bmatrix} \Rightarrow 25(x')^2+0(y')^2+100(z)^2-100y'=0 \Rightarrow$

$y'=\tfrac{(x')^2}{4}+\tfrac{(z')^2}{1} \Rightarrow$ The graph is an elliptic paraboloid.

93. $x^2+y^2-2z^2+4xy-2xz+2yz-x+y+z=0 \Rightarrow$

$$A=\begin{bmatrix}1&2&-1\\2&1&1\\-1&1&-2\end{bmatrix} \text{ and } B=\begin{bmatrix}-1&1&1\end{bmatrix} \Rightarrow \lambda=0,\,3,\,-3 \Rightarrow Q=\begin{bmatrix}-\frac{1}{\sqrt{3}}&\frac{1}{\sqrt{2}}&\frac{1}{\sqrt{6}}\\\frac{1}{\sqrt{3}}&\frac{1}{\sqrt{2}}&-\frac{1}{\sqrt{6}}\\\frac{1}{\sqrt{3}}&0&\frac{2}{\sqrt{6}}\end{bmatrix} \Rightarrow$$

$BQ=\begin{bmatrix}\sqrt{3}&0&0\end{bmatrix} \Rightarrow 0(x')^2+3(y')^2-3(z)^2+\sqrt{3}\,x'=0 \Rightarrow$

$x'=-\sqrt{3}(y')^2+\sqrt{3}(z')^2 \Rightarrow$ The graph is a hyperbolic paraboloid.

94. $10x^2 + 25y^2 + 10z^2 - 40xz + 20\sqrt{2}\,x + 50y + 20\sqrt{2}\,z = 15 \Rightarrow$

$$A = \begin{bmatrix} 10 & 0 & -20 \\ 0 & 25 & 0 \\ -20 & 0 & 10 \end{bmatrix} \text{ and } B = \begin{bmatrix} 20\sqrt{2} & 50 & 20\sqrt{2} \end{bmatrix} \Rightarrow \lambda = 30,\ -10,\ 25 \Rightarrow$$

$$Q = \begin{bmatrix} \frac{1}{\sqrt{2}} & \frac{1}{\sqrt{2}} & 0 \\ 0 & 0 & 1 \\ -\frac{1}{\sqrt{2}} & \frac{1}{\sqrt{2}} & 0 \end{bmatrix} \Rightarrow BQ = \begin{bmatrix} 0 & 40 & 50 \end{bmatrix} \Rightarrow 30(x')^2 - 10(y')^2 + 25(z')^2 + 40y' + 50z' = 15 \Rightarrow$$

$(y'+2)^2 = \frac{(x')^2}{1/3} + \frac{(z'+1)^2}{2/5} \Rightarrow (y'')^2 = \frac{(x'')^2}{1/3} + \frac{(z'')^2}{2/5} \Rightarrow$ The graph is an elliptic cone.

Note that $x'' = x'$.

95. $11x^2 + 11y^2 + 14z^2 + 2xy + 8xz - 8yz - 12x + 12y + 12z = 6 \Rightarrow$

$$A = \begin{bmatrix} 11 & 1 & 4 \\ 1 & 11 & -4 \\ 4 & -4 & 14 \end{bmatrix} \text{ and } B = \begin{bmatrix} -12 & 12 & 12 \end{bmatrix} \Rightarrow \lambda = 18,\ 6,\ 12 \Rightarrow$$

$$Q = \begin{bmatrix} \frac{1}{\sqrt{6}} & -\frac{1}{\sqrt{3}} & \frac{1}{\sqrt{2}} \\ -\frac{1}{\sqrt{6}} & \frac{1}{\sqrt{3}} & \frac{1}{\sqrt{2}} \\ \frac{2}{\sqrt{6}} & \frac{1}{\sqrt{3}} & 0 \end{bmatrix} \Rightarrow BQ = \begin{bmatrix} 0 & 12\sqrt{3} & 0 \end{bmatrix} \Rightarrow 18(x')^2 + 6(y')^2 + 12(z')^2 + 12\sqrt{3}\,y' = 6 \Rightarrow$$

$18(x')^2 + 6(y' + \sqrt{3})^2 + 12(z')^2 = 24 \Rightarrow 3(x'')^2 + (y'')^2 + 2(z'')^2 = 4 \Rightarrow$

The graph is an ellipsoid.

Note that $x'' = x'$ and $z'' = z'$.

96. We will use the criteria of Exercise 87b to show $\mathbf{x}^T B\mathbf{x} = k$ defines an ellipse. Namely that if $k > 0$ and $\det B > 0$, then $\mathbf{x}^T B\mathbf{x} = k$ defines an ellipse.

Note: $C = \begin{bmatrix} a & -b \\ b & a \end{bmatrix}$, then $C^T C = I$. Why?

Recall, $\lambda = a - bi$, and $|\lambda| = a^2 + b^2 = 1$.

So: $C^T C = \begin{bmatrix} a & b \\ -b & a \end{bmatrix}\begin{bmatrix} a & -b \\ b & a \end{bmatrix} = \begin{bmatrix} a^2+b^2 & ab-ab \\ ab-ab & a^2+b^2 \end{bmatrix} = \begin{bmatrix} 1 & 0 \\ 0 & 1 \end{bmatrix} = I.$

So let $B = (PP^T)^{-1}$ and $C = \begin{bmatrix} a & -b \\ b & a \end{bmatrix}$.

Then $A = PCP^{-1}$ which implies $A^T = (PCP^{-1})^T = (P^{-1})^T C^T P^T \overset{(P^{-1})^T = (P^T)^{-1}}{=} (P^T)^{-1} C^T P^T$.

We will first show that $(A\mathbf{x})^T B(A\mathbf{x}) = \mathbf{x}^T B\mathbf{x} = k$.
Therefore, if $\mathbf{x}$ lies on the ellipse, then so does $A\mathbf{x}$.

$(A\mathbf{x})^T B(A\mathbf{x}) =$

$$\overset{B=(PP^T)^{-1}}{=} (A\mathbf{x})^T (PP^T)^{-1}(A\mathbf{x})$$

$$\overset{(PP^T)^{-1}=(P^T)^{-1}P^{-1}}{=} \mathbf{x}^T A^T (P^T)^{-1} P^{-1}(A\mathbf{x})$$

$$\overset{A^T=(P^T)^{-1}C^T P^T}{=} \mathbf{x}^T((P^T)^{-1}C^T P^T)(P^T)^{-1}P^{-1}(A\mathbf{x}) = \mathbf{x}^T((P^T)^{-1}C^T)(P^T(P^T)^{-1})P^{-1}(A\mathbf{x})$$

$$\overset{P^T(P^T)^{-1}=I}{=} \mathbf{x}^T((P^T)^{-1}C^T)P^{-1}(A\mathbf{x})$$

$$\overset{A=PCP^{-1}}{=} \mathbf{x}^T((P^T)^{-1}C^T)P^{-1}(PCP^{-1})\mathbf{x} = \mathbf{x}^T((P^T)^{-1}C^T)(P^{-1}P)(CP^{-1})\mathbf{x}$$

$$\overset{P^{-1}P=I}{=} \mathbf{x}^T((P^T)^{-1}C^T)(CP^{-1})\mathbf{x} = \mathbf{x}^T(P^T)^{-1}(C^T C)P^{-1}\mathbf{x}$$

$$\overset{C^T C=I}{=} \mathbf{x}^T(P^T)^{-1}P^{-1}\mathbf{x}$$

$$\overset{(P^T)^{-1}P^{-1}=(PP^T)^{-1}}{=} \mathbf{x}^T(PP^T)^{-1}\mathbf{x}$$

$$\overset{B=(PP^T)^{-1}}{=} \mathbf{x}^T B\mathbf{x} = k.$$

Next we need to show that $\det B > 0$. We begin by noting that since P is invertible, $\det P \neq 0$.
Recall: $\det A^{-1} = \frac{1}{\det A}$, $\det(AC) = \det A \det C$, and $\det A^T = \det A$.
Where are these facts used in the equation below?

Now $\det B = \det((PP^T)^{-1}) = \frac{1}{\det(PP^T)} = \frac{1}{\det P \det P^T} = \frac{1}{\det P \det P} = \left(\frac{1}{\det P}\right)^2 > 0.$

Therefore, when $k > 0$, $\mathbf{x}^T B\mathbf{x} = (A\mathbf{x})^T B(A\mathbf{x}) = k$ defines an ellipse.

Chapter 5 Review

1. We will explain and give counter examples to justify our answers below.

(a) ***True***. See definition on p.369 of Section 5.1.

It is important to note the following hierarchy of definitions:
ortho*normal* $\Rightarrow$ orthogonal $\Rightarrow$ linearly independent.
Why?
For a set of vectors $\{\mathbf{q}_i\}$ to be orthonormal it must satisfy two properties:

$$\mathbf{q}_i \cdot \mathbf{q}_j = \begin{cases} 0 & \text{if } i \neq j \quad \text{Property 1} \\ 1 & \text{if } i = j \quad \text{Property 2} \end{cases}$$

A set of vectors $\{\mathbf{q}_i\}$ that satisfies only Property 1, $\mathbf{q}_i \cdot \mathbf{q}_j = 0$ if $i \neq j$,
is only orthogonal, not orthonormal.
That is, we can make an orthogonal set orthonormal by normalizing all of the vectors.
What does that mean? We have to make the length of each of the vectors 1.
That is, an orthogonal set of *unit* vectors is an ortho*normal* set.

Q: Why is it obvious that an orthogonal set of vectors is linearly independent?

A: A set of vectors is linearly independent if $\sum\limits_{i=1}^{n} c_i \mathbf{v}_i = \mathbf{0}$ implies all the $c_i = 0$.
If that set of vectors is *orthogonal* with each $\mathbf{v}_j \neq \mathbf{0}$, then

$$\left(\sum_{i=1}^{n} c_i \mathbf{v}_i\right) \mathbf{v}_j = \sum_{i \neq j} c_i \mathbf{0} + c_j \mathbf{v}_j \cdot \mathbf{v}_j = c_j \|\mathbf{v}_j\|^2 = \mathbf{0} \text{ implies } c_j = 0.$$

So, both orthogonality and orthonormality imply linear independence.

(b) ***True***. See Theorem 5.17 in Section 5.3.

This is what the Gram-Schmidt Process of Theorem 5.17 in Section 5.3 asserts:
Given any basis for a subspace W, we can construct an orthogonal basis.

(c) ***True***. See definition on p.371 in Section 5.1.

A *square* matrix Q whose columns $\{\mathbf{q}_i\}$ form an orthonormal set.
It is important that Q be square to satisfy the definition.

(d) ***True***. See definition on p.371 in Section 5.1.

A *square* matrix Q whose columns $\{\mathbf{q}_i\}$ form an orthonormal set.
Since its columns are orthonormal, they are linearly independent.
Therefore, Q is invertible.

Q: Theorem 5.8(b) of Section 5.1 asserts if Q is orthogonal, then $|\det Q| = 1$.
Does this also imply if Q is orthogonal then Q is invertible? Why or why not?

A: Hint: See Theorem 4.6 in Section 4.2.

1. We will explain and give counter examples to justify our answers below. (continued)

(e) ***False***. See Theorem 5.8(b) in Section 5.1 and Theorem 4.2 in Section 4.2.

Theorem 5.8(b) of Section 5.1 asserts if Q is orthogonal, then $|\det Q| = 1$.
However, Theorem 4.2 of Section 4.2 asserts if A is triangular, then $\det A = a_{11}a_{22}\cdots a_{nn}$.
This tells us how to construct many counter examples. For instance, consider:

$$A = \begin{bmatrix} 1 & 1 \\ 0 & 1 \end{bmatrix}, B = \begin{bmatrix} 1 & 2 \\ 0 & 1 \end{bmatrix}, C = \begin{bmatrix} 1 & 3 \\ 0 & 1 \end{bmatrix}, \text{ and } D = \begin{bmatrix} 1 & 4 \\ 0 & 1 \end{bmatrix}.$$

Q: Why is it obvious that that none of A, B, C, or D are orthogonal?
A: Hint: Are the columns orthogonal? Do they have length 1?
Again, we should note what a strong condition orthogonality in matrices really is.

(f) ***True***. See Theorem 5.10 in Section 5.2.

Theorem 5.10 in Section 5.2 asserts that $(\text{row}(A))^{\perp} = \text{null}(A)$.
So if $(\text{row}(A))^{\perp} = \text{null}(A) = \mathbb{R}$, then $A\mathbf{x} = \mathbf{0}$ for every $\mathbf{x}$.

Q: Why does $A\mathbf{x} = \mathbf{0}$ for every $\mathbf{x}$ imply that $A = O$?
A: Hint: What does $A\mathbf{e}_i$ equal?

(g) ***False***. See Theorems 5.11 to 5.14 in Section 5.2 and Exercises 1 to 5 in Section 5.2.

As a counterexample, consider: if $W = \text{span}(\mathbf{e}_1)$, then $\text{proj}_W(\mathbf{e}_2) = \mathbf{0}$.
As another counterexample, see the solution to Exercise 3 of Section 5.2:
More generally, what is going on? $\mathbb{R}^n$ breaks into two pieces: W and $W^{\perp}$.
One of the underlying assertions of Theorems 5.11 to 5.14 is the following:

$$W^{\perp} = \{\mathbf{v} : \text{proj}_W(\mathbf{v}) = \mathbf{0}\}.$$

See Exercise 28 in Section 5.2.

Q: Given the above, if $\text{proj}_W(\mathbf{v}) = \mathbf{0} \Rightarrow \mathbf{v} = \mathbf{0}$, what is W?
A: Hint: If $\mathbf{0}$ is the only vector in $W^{\perp}$, what is $\dim W^{\perp}$?

(h) ***True***. See Theorem 5.5 in Section 5.1 and *symmetric* defined on p.149 in Section 3.1.

By Thm 5.5, if A is orthogonal then $A^{-1} = A^T$. If A is also symmetric, then $A^T = A$.
Therefore, if A is symmetric and orthogonal, then $A = A^T = A^{-1}$.
So, if A is symmetric and orthogonal, then $A^2 = AA = AA^T = AA^{-1} = I$.

(i) ***False***. See definition on p.397 in Section 5.4. Consider diagonal matrices.

If A is *orthogonally diagonalizable*, then there exists an orthogonal matrix Q and a diagonal matrix D such that $Q^TAQ = D$. Therefore, $A = QDQ^T$.
So if the diagonal matrix D is not invertible, neither is A.

(j) ***True***. It is left to the reader to research and prove this assertion.

2. See Example 5.1 in Section 5.1 and recall $\mathbf{u}$ is orthogonal to $\mathbf{v}$ means $\mathbf{u}\cdot\mathbf{v}=0$.

Let $\mathbf{u}=\begin{bmatrix}1\\2\\3\end{bmatrix}$, $\mathbf{v}=\begin{bmatrix}4\\1\\-2\end{bmatrix}$, and $\mathbf{w}=\begin{bmatrix}a\\b\\3\end{bmatrix}$.

We want $\mathbf{w}$ to be orthogonal to both $\mathbf{u}$ and $\mathbf{v}$. So, $\mathbf{u}\cdot\mathbf{w}=0$ and $\mathbf{v}\cdot\mathbf{w}=0$.

So $\mathbf{u}\cdot\mathbf{w}=\begin{bmatrix}1&2&3\end{bmatrix}\begin{bmatrix}a\\b\\3\end{bmatrix}=a+2b+9=0$ and $\mathbf{v}\cdot\mathbf{w}=\begin{bmatrix}4&1&-2\end{bmatrix}\begin{bmatrix}a\\b\\3\end{bmatrix}=4a+b-6=0.$

So we have: $\begin{array}{rcrcrcl} a&+&2b&+&9&=&0\\ 4a&+&b&-&6&=&0\end{array}$ or $\begin{array}{rcrcr} a&+&2b&=&-9\\ 4a&+&b&=&6\end{array}.$

We reduce the matrix of coefficients: $\left[\begin{array}{cc|c}1&2&-9\\4&1&6\end{array}\right]\longrightarrow\left[\begin{array}{cc|c}1&0&3\\0&1&-6\end{array}\right].$

Therefore, $a=3$ and $b=-6$.

Check: $\mathbf{u}\cdot\mathbf{w}=\begin{bmatrix}1&2&3\end{bmatrix}\begin{bmatrix}3\\-6\\3\end{bmatrix}=0$ and $\mathbf{v}\cdot\mathbf{w}=\begin{bmatrix}4&1&-2\end{bmatrix}\begin{bmatrix}3\\-6\\3\end{bmatrix}==0.$

Q: We find that a and b are uniquely determined. Why does that make sense?
A: The two dot products create two equations, one for each unknown.

3. See Example 5.4, Exercises 7 to 10, and Theorem 5.2 in Section 5.1.

As in Example 5.4, we apply the formula from Theorem 5.2, $c_i=\frac{\mathbf{v}\cdot\mathbf{u}_i}{\mathbf{u}_i\cdot\mathbf{u}_i}$.

Let $\mathbf{u}_1=\begin{bmatrix}1\\0\\1\end{bmatrix}$, $\mathbf{u}_2=\begin{bmatrix}1\\1\\-1\end{bmatrix}$, $\mathbf{u}_3=\begin{bmatrix}-1\\2\\1\end{bmatrix}$ and $\mathbf{v}=\begin{bmatrix}7\\-3\\2\end{bmatrix}$.

So $c_1=\frac{7+2}{1+1}=\frac{9}{2}$, $c_2=\frac{7-3-2}{1+1+1}=\frac{2}{3}$, and $c_3=\frac{-7-6+2}{1+4+1}=-\frac{11}{6}$.

Therefore, $[\mathbf{v}]_{\mathbf{B}}=\begin{bmatrix}\frac{9}{2}\\\frac{2}{3}\\-\frac{11}{6}\end{bmatrix}$.

Q: What is the basic definition of $[\mathbf{v}]_{\mathbf{B}}$?

A: If $\mathbf{B}=\{\mathbf{u}_1,\ldots,\mathbf{u}_n\}$ and $\mathbf{v}=\sum_{i=1}^{n}c_i\mathbf{u}_i$, then $[\mathbf{v}]_{\mathbf{B}}=\begin{bmatrix}c_1\\\vdots\\c_n\end{bmatrix}$.

Q: How might we put this basic definition of $[\mathbf{v}]_{\mathbf{B}}$ into words?
A: $[\mathbf{v}]_{\mathbf{B}}$ is the coefficients in the linear combination equal to $\mathbf{v}$.

Q: Given this basic definition, what is another method we might use to find $[\mathbf{v}]_{\mathbf{B}}$?
A: Simply row reduce $\left[\begin{array}{ccc|c}\mathbf{u}_1&\mathbf{u}_2&\mathbf{u}_3&\mathbf{v}\end{array}\right]\longrightarrow\left[\begin{array}{c|c}I&[\mathbf{v}]_{\mathbf{B}}\end{array}\right]$. Verify this.

See Examples 3.53 and 3.54 in Section 3.4 for further detail.

4. See Example 3.54 in Section 3.5 and Example 5.20 in Section 5.4.

As in Example 5.20 in Section 5.4, we find $\mathbf{v}_2$ given it is orthogonal to $\mathbf{v}_1$ and has length 1.

$$\mathbf{v}_1 \cdot \mathbf{v}_2 = \begin{bmatrix} \frac{3}{5} \\ \frac{4}{5} \end{bmatrix} \cdot \begin{bmatrix} x \\ y \end{bmatrix} = \frac{3}{5}x + \frac{4}{5}y = 0 \Rightarrow y = -\frac{3}{4}x$$

Also $x^2 + y^2 = 1 \Rightarrow x^2 + \left(-\frac{3}{4}x\right)^2 = 1 \Rightarrow x^2 + \frac{9}{16}x^2 = 1 \Rightarrow \frac{25}{16}x^2 = 1 \Rightarrow x = \pm\frac{4}{5}$.

So take $x = \frac{4}{5}$ then $y = -\frac{3}{4}x = -\frac{3}{4}\left(\frac{4}{5}\right) = -\frac{3}{5}$.

Now since $[\mathbf{v}]_{\mathbf{B}} = \begin{bmatrix} -3 \\ \frac{1}{2} \end{bmatrix}$, we have $\mathbf{v} = -3\mathbf{v}_1 + \frac{1}{2}\mathbf{v}_2 = -3\begin{bmatrix} \frac{3}{5} \\ \frac{4}{5} \end{bmatrix} + \frac{1}{2}\begin{bmatrix} \frac{4}{5} \\ -\frac{3}{5} \end{bmatrix} = \begin{bmatrix} -\frac{7}{5} \\ -\frac{27}{10} \end{bmatrix}$.

Q: What pattern do we notice between the components of $\mathbf{v}_1$ and $\mathbf{v}_2$?
A: The components have been switched and one has been made negative.

Q: Will this pattern always produce the missing vector in a orthonormal basis for $\mathbb{R}^2$?

Now take $x = -\frac{4}{5}$ then $y = -\frac{3}{4}x = -\frac{3}{4}\left(-\frac{4}{5}\right) = \frac{3}{5}$. (Does the suggested pattern hold?)

So since $[\mathbf{v}]_{\mathbf{B}} = \begin{bmatrix} -3 \\ \frac{1}{2} \end{bmatrix}$, we have $\mathbf{v} = -3\mathbf{v}_1 + \frac{1}{2}\mathbf{v}_2 = -3\begin{bmatrix} \frac{3}{5} \\ \frac{4}{5} \end{bmatrix} + \frac{1}{2}\begin{bmatrix} -\frac{4}{5} \\ \frac{3}{5} \end{bmatrix} = \begin{bmatrix} -\frac{11}{5} \\ -\frac{21}{10} \end{bmatrix}$.

5. As in Theorem 5.5 of Section 5.1, we will show Q is orthogonal by showing that $QQ^T = I$.

$$QQ^T = \begin{bmatrix} \frac{6}{7} & \frac{2}{7} & \frac{3}{7} \\ -\frac{1}{\sqrt{5}} & 0 & \frac{2}{\sqrt{5}} \\ \frac{4}{7\sqrt{5}} & -\frac{15}{7\sqrt{5}} & \frac{2}{7\sqrt{5}} \end{bmatrix} \begin{bmatrix} \frac{6}{7} & -\frac{1}{\sqrt{5}} & \frac{4}{7\sqrt{5}} \\ \frac{2}{7} & 0 & -\frac{15}{7\sqrt{5}} \\ \frac{3}{7} & \frac{2}{\sqrt{5}} & \frac{2}{7\sqrt{5}} \end{bmatrix} = \begin{bmatrix} 1 & 0 & 0 \\ 0 & 1 & 0 \\ 0 & 0 & 1 \end{bmatrix} = I$$

Q: Why is the entry in the first row and first column of the resulting matrix equal to 1?

A: Because $\frac{36+4+9}{49} = \frac{49}{49} = 1$. Verify all the entries in the product matrix.

6. We will use Theorem 5.5 of Section 5.1 ($QQ^T = I$) to find the possible values of a, b, c.

$$QQ^T = \begin{bmatrix} \frac{1}{2} & a \\ b & c \end{bmatrix} \begin{bmatrix} \frac{1}{2} & b \\ a & c \end{bmatrix} = \begin{bmatrix} 1 & 0 \\ 0 & 1 \end{bmatrix} = I$$

So we have the following three equations: $\frac{1}{4} + a^2 = 1$, $\frac{1}{2}b + ac = 0$, $b^2 + c^2 = 1$.

So: $\frac{1}{4} + a^2 = 1 \Rightarrow a^2 = \frac{3}{4} \Rightarrow a = \pm\frac{\sqrt{3}}{2}$

$\frac{1}{2}b \pm \frac{\sqrt{3}}{2}c = 0 \Rightarrow b = \pm\sqrt{3}c \Rightarrow b^2 = 3c^2$

$b^2 + c^2 = 1 \Rightarrow 3c^2 + c^2 = 1 \Rightarrow 4c^2 = 1 \Rightarrow c^2 = \frac{1}{4} \Rightarrow c = \pm\frac{1}{2} \Rightarrow$

So since $b = \pm\sqrt{3}c$, $b = \pm\sqrt{3}\left(\frac{1}{2}\right) = \pm\frac{\sqrt{3}}{2}$.

So we have: $\begin{bmatrix} \frac{1}{2} & \frac{\sqrt{3}}{2} \\ -\frac{\sqrt{3}}{2} & \frac{1}{2} \end{bmatrix}$, $\begin{bmatrix} \frac{1}{2} & -\frac{\sqrt{3}}{2} \\ \frac{\sqrt{3}}{2} & \frac{1}{2} \end{bmatrix}$, $\begin{bmatrix} \frac{1}{2} & \frac{\sqrt{3}}{2} \\ \frac{\sqrt{3}}{2} & -\frac{1}{2} \end{bmatrix}$.

Q: Are there any other orthogonal matrices we can create using these values for a, b, and c?

7. See the definition of orthonormal given in Section 5.1 on p.369.

Given Q is orthogonal and $\mathbf{v}_i \cdot \mathbf{v}_j = 0$ and $\mathbf{v}_i \cdot \mathbf{v}_i = 1$,
we need to show $Q\mathbf{v}_i \cdot Q\mathbf{v}_j = 0$ and $Q\mathbf{v}_i \cdot Q\mathbf{v}_i = 1$.

First: $Q\mathbf{v}_i \cdot Q\mathbf{v}_j = (Q\mathbf{v}_i)^T Q\mathbf{v}_j = \mathbf{v}_i^T (Q^T Q)\mathbf{v}_j \overset{Q \text{ is orthogonal}}{=} \mathbf{v}_i^T I \mathbf{v}_j = \mathbf{v}_i^T \mathbf{v}_j = \mathbf{v}_i \cdot \mathbf{v}_j = 0$.

Likewise: $Q\mathbf{v}_i \cdot Q\mathbf{v}_i = \mathbf{v}_i \cdot \mathbf{v}_i = 1$.

Q: What does the fact that $Q\mathbf{v}_i \cdot Q\mathbf{v}_i = \mathbf{v}_i \cdot \mathbf{v}_i = 1$ tell us?
A: Multiplication by an orthogonal matrix preserves length.

Q: What does the fact that $Q\mathbf{v}_i \cdot Q\mathbf{v}_j = \mathbf{v}_i \cdot \mathbf{v}_j = 0$ tell us?
A: Multiplication by an orthogonal matrix preserves orthogonality.

Q: Does multiplication by an orthogonal matrix preserve all angles at which vectors meet?

8. Since the angles are equal, $\theta^{-1} = \dfrac{\mathbf{x} \cdot \mathbf{y}}{\|\mathbf{x}\| \, \|\mathbf{y}\|} = \dfrac{Q\mathbf{x} \cdot Q\mathbf{y}}{\|Q\mathbf{x}\| \, \|Q\mathbf{y}\|}$. Is this enough?

9. See Exercises 1 through 6 and Example 5.8 in Section 5.2.

$\mathbf{v} \cdot \mathbf{w} = 5x + 2y = 0 \Rightarrow W^{\perp} = \left\{ \begin{bmatrix} x \\ y \end{bmatrix} : 5x + 2y = 0 \right\} \Rightarrow 5x = -2y \Rightarrow \begin{bmatrix} -2 \\ 5 \end{bmatrix}$ is a basis.

Note the following general pattern:
The lines that describe the necessary condition to be in W and $W^{\perp}$
are necessarily perpendicular.
So, in this case, $2x - 5y = 0$ and $5x + 2y = 0$ are necessarily perpendicular.

10. See Exercises 1 through 6 and Example 5.8 in Section 5.2.

Since $\mathbf{w} = \begin{bmatrix} 1 \\ 2 \\ -1 \end{bmatrix}$ is a basis for W, for $\mathbf{v} = \begin{bmatrix} x \\ y \\ z \end{bmatrix}$ in $W^{\perp}$ $\mathbf{v} \cdot \mathbf{w} = x + 2y - z = 0 \Rightarrow$

$W^{\perp} = \left\{ \begin{bmatrix} x \\ y \\ z \end{bmatrix} : x + 2y - z = 0 \right\} \Rightarrow \left\{ \begin{bmatrix} 1 \\ 0 \\ 1 \end{bmatrix}, \begin{bmatrix} 0 \\ 1 \\ 2 \end{bmatrix} \right\}$ is a basis for $W^{\perp}$.

11. We use the *cross product* to find a vector orthogonal to both of the given vectors. See *Exploration: The Cross Product* following Section 1.3.

Since $\mathbf{x} = \mathbf{u} \times \mathbf{v} = \begin{bmatrix} u_2v_3 - u_3v_2 \\ u_3v_1 - u_1v_3 \\ u_1v_2 - u_2v_1 \end{bmatrix} = \begin{bmatrix} (-1)(-3) - (4)(1) \\ (4)(0) - (1)(-3) \\ (1)(1) - (-1)(0) \end{bmatrix} = \begin{bmatrix} -1 \\ 3 \\ 1 \end{bmatrix}$.

Verify that $\mathbf{x} \cdot \mathbf{u} = \mathbf{x} \cdot \mathbf{v} = 0$. So $W^\perp = \left\{ \begin{bmatrix} -1 \\ 3 \\ 1 \end{bmatrix} \right\}$ is a basis for $W^\perp$.

12. See Example 5.10 in Section 5.2.

We need both $\mathbf{u} \cdot \mathbf{x} = \mathbf{v} \cdot \mathbf{x} = 0$.

We have $\mathbf{u} \cdot \mathbf{x} = r + s + t + u = 0$ and $\mathbf{v} \cdot \mathbf{x} = r - s + t + 2u = 0$.

Subtracting implies $u - 2s = 0 \Rightarrow u = 2s$. So $r = -s - t - u = -s - t - 2s = -3s - t$.

So $W^\perp = \left\{ \begin{bmatrix} -3 \\ 1 \\ 0 \\ 2 \end{bmatrix}, \begin{bmatrix} -1 \\ 0 \\ 1 \\ 0 \end{bmatrix} \right\}$ is a basis for $W^\perp$.

13. See Example 5.9 and Exercises 7 through 10 in Section 5.2.

$A \longrightarrow \begin{bmatrix} 1 & 0 & 2 & 3 & 4 \\ 0 & 1 & 0 & 2 & 1 \\ 0 & 0 & 0 & 0 & 0 \\ 0 & 0 & 0 & 0 & 0 \end{bmatrix}$, $\{[\,1\ 0\ 2\ 3\ 4\,], [\,0\ 1\ 0\ 2\ 1\,]\}$ is a basis for row(A).

$\left[\begin{array}{ccccc|c} 1 & 0 & 2 & 3 & 4 & 0 \\ 0 & 1 & 0 & 2 & 1 & 0 \\ 0 & 0 & 0 & 0 & 0 & 0 \\ 0 & 0 & 0 & 0 & 0 & 0 \end{array}\right] \Rightarrow \mathbf{x} = \begin{bmatrix} -2x_3 - 3x_4 - 4x_5 \\ -2x_4 - x_5 \\ x_3 \\ x_4 \\ x_5 \end{bmatrix} \Rightarrow \left\{ \begin{bmatrix} -2 \\ 0 \\ 1 \\ 0 \\ 0 \end{bmatrix}, \begin{bmatrix} -3 \\ -2 \\ 0 \\ 1 \\ 0 \end{bmatrix}, \begin{bmatrix} -4 \\ -1 \\ 0 \\ 0 \\ 1 \end{bmatrix} \right\}$ is a basis for null(A).

$A^T \longrightarrow \begin{bmatrix} 1 & 0 & 5 & 1 \\ 0 & 1 & 3 & -2 \\ 0 & 0 & 0 & 0 \\ 0 & 0 & 0 & 0 \\ 0 & 0 & 0 & 0 \end{bmatrix}$, $\left\{ \begin{bmatrix} 1 \\ -1 \\ 2 \\ 3 \end{bmatrix}, \begin{bmatrix} -1 \\ 2 \\ 1 \\ -5 \end{bmatrix} \right\}$ is a basis for row$(A^T) = $ col(A).

$\left[\begin{array}{cccc|c} 1 & 0 & 5 & 1 & 0 \\ 0 & 1 & 3 & -2 & 0 \\ 0 & 0 & 0 & 0 & 0 \\ 0 & 0 & 0 & 0 & 0 \\ 0 & 0 & 0 & 0 & 0 \end{array}\right] \Rightarrow \mathbf{y} = \begin{bmatrix} -5y_3 - y_4 \\ -3y_3 + 2y_4 \\ y_3 \\ y_4 \end{bmatrix}, \left\{ \begin{bmatrix} -5 \\ -3 \\ 1 \\ 0 \end{bmatrix}, \begin{bmatrix} -1 \\ 2 \\ 0 \\ 1 \end{bmatrix} \right\}$ is a basis for null(A^T).

14. See Example 5.11 and Exercises 19 through 22 in Section 5.2.
By Theorem 5.11 of Section 5.2, $\mathbf{v} = \mathbf{w} + \mathbf{w}^{\perp} = \text{proj}_W(\mathbf{v}) + (\mathbf{v} - \text{proj}_W(\mathbf{v}))$.

We begin by computing $\text{proj}_W(\mathbf{v})$.

$$\text{proj}_W(\mathbf{v}) = \left(\frac{\mathbf{u}_1 \cdot \mathbf{v}}{\mathbf{u}_1 \cdot \mathbf{u}_1}\right)\mathbf{u}_1 + \left(\frac{\mathbf{u}_2 \cdot \mathbf{v}}{\mathbf{u}_2 \cdot \mathbf{u}_2}\right)\mathbf{u}_2 + \left(\frac{\mathbf{u}_3 \cdot \mathbf{v}}{\mathbf{u}_3 \cdot \mathbf{u}_3}\right)\mathbf{u}_3$$

$$= \frac{1}{3}\begin{bmatrix} 0 \\ 1 \\ 1 \\ 1 \end{bmatrix} - \frac{2}{3}\begin{bmatrix} 1 \\ 0 \\ 1 \\ -1 \end{bmatrix} + \frac{7}{15}\begin{bmatrix} 3 \\ 1 \\ -2 \\ 1 \end{bmatrix} = \begin{bmatrix} \frac{11}{15} \\ \frac{4}{5} \\ -\frac{19}{15} \\ \frac{22}{15} \end{bmatrix}.$$

So $\mathbf{v} = \text{proj}_W(\mathbf{v}) + (\mathbf{v} - \text{proj}_W(\mathbf{v})) = \begin{bmatrix} \frac{11}{15} \\ \frac{4}{5} \\ -\frac{19}{15} \\ \frac{22}{15} \end{bmatrix} + \left(\begin{bmatrix} 1 \\ 0 \\ -1 \\ 2 \end{bmatrix} - \begin{bmatrix} \frac{11}{15} \\ \frac{4}{5} \\ -\frac{19}{15} \\ \frac{22}{15} \end{bmatrix}\right) = \begin{bmatrix} \frac{11}{15} \\ \frac{4}{5} \\ -\frac{19}{15} \\ \frac{22}{15} \end{bmatrix} + \begin{bmatrix} \frac{4}{15} \\ -\frac{4}{5} \\ \frac{4}{15} \\ \frac{8}{15} \end{bmatrix}.$

Q: How do we verify the above vectors are an orthogonal decomposition of $\mathbf{v}$?
A: Compute $\mathbf{w} \cdot \mathbf{w}^{\perp}$. What result should we get? Do we?

15. For (a), see Example 5.13 and Exercise 12 in Section 5.3.
For (b), see Example 5.15 and Exercises 15 and 16 in Section 5.3.

(a) We compute the orthogonal basis vectors $\mathbf{v}_1$, $\mathbf{v}_2$, and $\mathbf{v}_3$ in sequence:

$$\mathbf{v}_1 = \mathbf{x}_1 = \begin{bmatrix} 1 \\ 1 \\ 1 \\ 1 \end{bmatrix}, \ \mathbf{v}_2 = \mathbf{x}_2 - \left(\frac{\mathbf{v}_1 \cdot \mathbf{x}_2}{\mathbf{v}_1 \cdot \mathbf{v}_1}\right)\mathbf{v}_1 = \begin{bmatrix} 1 \\ 1 \\ 1 \\ 0 \end{bmatrix} - \frac{3}{4}\begin{bmatrix} 1 \\ 1 \\ 1 \\ 1 \end{bmatrix} = \begin{bmatrix} \frac{1}{4} \\ \frac{1}{4} \\ \frac{1}{4} \\ -\frac{3}{4} \end{bmatrix},$$

$$\mathbf{v}_3 = \mathbf{x}_3 - \left(\frac{\mathbf{v}_1 \cdot \mathbf{x}_3}{\mathbf{v}_1 \cdot \mathbf{v}_1}\right)\mathbf{v}_1 - \left(\frac{\mathbf{v}_2 \cdot \mathbf{x}_3}{\mathbf{v}_2 \cdot \mathbf{v}_2}\right)\mathbf{v}_2 = \begin{bmatrix} 0 \\ 1 \\ 1 \\ 1 \end{bmatrix} - \frac{3}{4}\begin{bmatrix} 1 \\ 1 \\ 1 \\ 1 \end{bmatrix} + \frac{1}{3}\begin{bmatrix} \frac{1}{4} \\ \frac{1}{4} \\ \frac{1}{4} \\ -\frac{3}{4} \end{bmatrix} = \begin{bmatrix} -\frac{2}{3} \\ \frac{1}{3} \\ \frac{1}{3} \\ 0 \end{bmatrix}.$$

(b) To find $A = QR$ we normalize the vectors from (a) to find Q and then set $R = Q^T A$:

Normalizing $\mathbf{v}_1$, $\mathbf{v}_2$, and $\mathbf{v}_3$ yields: $\mathbf{q}_1 = \dfrac{\mathbf{v}_1}{\|\mathbf{v}_1\|} = \frac{1}{2}\begin{bmatrix} 1 \\ 1 \\ 1 \\ 1 \end{bmatrix} = \begin{bmatrix} \frac{1}{2} \\ \frac{1}{2} \\ \frac{1}{2} \\ \frac{1}{2} \end{bmatrix}$,

$$\mathbf{q}_2 = \frac{\mathbf{v}_2}{\|\mathbf{v}_2\|} = \frac{2\sqrt{3}}{3}\begin{bmatrix} \frac{1}{4} \\ \frac{1}{4} \\ \frac{1}{4} \\ -\frac{3}{4} \end{bmatrix} = \begin{bmatrix} \frac{\sqrt{3}}{6} \\ \frac{\sqrt{3}}{6} \\ \frac{\sqrt{3}}{6} \\ -\frac{\sqrt{3}}{2} \end{bmatrix}, \ \mathbf{q}_3 = \frac{\mathbf{v}_3}{\|\mathbf{v}_3\|} = \frac{\sqrt{6}}{2}\begin{bmatrix} -\frac{2}{3} \\ \frac{1}{3} \\ \frac{1}{3} \\ 0 \end{bmatrix} = \begin{bmatrix} -\frac{\sqrt{6}}{3} \\ \frac{\sqrt{6}}{6} \\ \frac{\sqrt{6}}{6} \\ 0 \end{bmatrix}.$$

So $R = Q^T A = \begin{bmatrix} \frac{1}{2} & \frac{1}{2} & \frac{1}{2} & \frac{1}{2} \\ \frac{\sqrt{3}}{6} & \frac{\sqrt{3}}{6} & \frac{\sqrt{3}}{6} & -\frac{\sqrt{3}}{2} \\ -\frac{\sqrt{6}}{3} & \frac{\sqrt{6}}{6} & \frac{\sqrt{6}}{6} & 0 \end{bmatrix}\begin{bmatrix} 1 & 1 & 0 \\ 1 & 1 & 1 \\ 1 & 1 & 1 \\ 1 & 0 & 1 \end{bmatrix} = \begin{bmatrix} 2 & \frac{3}{2} & \frac{3}{2} \\ 0 & \frac{\sqrt{3}}{2} & -\frac{\sqrt{3}}{6} \\ 0 & 0 & \frac{\sqrt{6}}{3} \end{bmatrix}.$

Verify that $A = QR$.

That is: $\begin{bmatrix} 1 & 1 & 0 \\ 1 & 1 & 1 \\ 1 & 1 & 1 \\ 1 & 0 & 1 \end{bmatrix} = \begin{bmatrix} \frac{1}{2} & \frac{\sqrt{3}}{6} & -\frac{\sqrt{6}}{3} \\ \frac{1}{2} & \frac{\sqrt{3}}{6} & \frac{\sqrt{6}}{6} \\ \frac{1}{2} & \frac{\sqrt{3}}{6} & \frac{\sqrt{6}}{6} \\ \frac{1}{2} & -\frac{\sqrt{3}}{2} & 0 \end{bmatrix}\begin{bmatrix} 2 & \frac{3}{2} & \frac{3}{2} \\ 0 & \frac{\sqrt{3}}{2} & -\frac{\sqrt{3}}{6} \\ 0 & 0 & \frac{\sqrt{6}}{3} \end{bmatrix}.$

16. See Example 5.14 and Exercise 12 in Section 5.3.

Create a basis by adding two standard basis vectors, $\mathbf{x}_3 = \mathbf{e}_3$ and $\mathbf{x}_4 = \mathbf{e}_4$. Now we compute orthogonal basis vectors $\mathbf{v}_1$, $\mathbf{v}_2$, $\mathbf{v}_3$, and $\mathbf{v}_4$ in sequence:

$$\mathbf{v}_1 = \mathbf{x}_1 = \begin{bmatrix} 1 \\ 0 \\ 2 \\ 2 \end{bmatrix}, \mathbf{v}_2 = \mathbf{x}_2 - \left(\frac{\mathbf{v}_1 \cdot \mathbf{x}_2}{\mathbf{v}_1 \cdot \mathbf{v}_1}\right)\mathbf{v}_1 = \begin{bmatrix} 0 \\ 1 \\ 1 \\ -1 \end{bmatrix} - \frac{0}{9}\begin{bmatrix} 1 \\ 0 \\ 2 \\ 2 \end{bmatrix} = \begin{bmatrix} 0 \\ 1 \\ 1 \\ -1 \end{bmatrix}$$

$$\mathbf{v}_3 = \mathbf{x}_3 - \left(\frac{\mathbf{v}_1 \cdot \mathbf{x}_3}{\mathbf{v}_1 \cdot \mathbf{v}_1}\right)\mathbf{v}_1 - \left(\frac{\mathbf{v}_2 \cdot \mathbf{x}_3}{\mathbf{v}_2 \cdot \mathbf{v}_2}\right)\mathbf{v}_2 = \begin{bmatrix} 0 \\ 0 \\ 1 \\ 0 \end{bmatrix} - \frac{2}{9}\begin{bmatrix} 1 \\ 0 \\ 2 \\ 2 \end{bmatrix} - \frac{1}{3}\begin{bmatrix} 0 \\ 1 \\ 1 \\ -1 \end{bmatrix} = \begin{bmatrix} -\frac{2}{9} \\ -\frac{1}{3} \\ \frac{2}{9} \\ -\frac{1}{9} \end{bmatrix},$$

$$\mathbf{v}_4 = \mathbf{x}_4 - \left(\frac{\mathbf{v}_1 \cdot \mathbf{x}_4}{\mathbf{v}_1 \cdot \mathbf{v}_1}\right)\mathbf{v}_1 - \left(\frac{\mathbf{v}_2 \cdot \mathbf{x}_4}{\mathbf{v}_2 \cdot \mathbf{v}_2}\right)\mathbf{v}_2 - \left(\frac{\mathbf{v}_3 \cdot \mathbf{x}_4}{\mathbf{v}_3 \cdot \mathbf{v}_3}\right)\mathbf{v}_3$$

$$= \begin{bmatrix} 0 \\ 0 \\ 0 \\ 1 \end{bmatrix} - \frac{2}{9}\begin{bmatrix} 1 \\ 0 \\ 2 \\ 2 \end{bmatrix} + \frac{1}{3}\begin{bmatrix} 0 \\ 1 \\ 1 \\ -1 \end{bmatrix} + \frac{1}{2}\begin{bmatrix} -\frac{2}{9} \\ -\frac{1}{3} \\ \frac{2}{9} \\ -\frac{1}{9} \end{bmatrix} = \begin{bmatrix} -\frac{1}{3} \\ \frac{1}{6} \\ 0 \\ \frac{1}{6} \end{bmatrix}.$$

17. Similar to Exercise 16 above. However, we begin by finding a basis for W.

Since $x_1 + x_2 + x_3 + x_4 = 0 \Rightarrow x_4 = -x_1 - x_2 - x_3$,

$$W \text{ has basis } \mathcal{B} = \left\{ \begin{bmatrix} 1 \\ 0 \\ 0 \\ -1 \end{bmatrix}, \begin{bmatrix} 0 \\ 1 \\ 0 \\ -1 \end{bmatrix}, \begin{bmatrix} 0 \\ 0 \\ 1 \\ -1 \end{bmatrix} \right\}.$$ Why is this obvious?

Now we compute orthogonal basis vectors $\mathbf{v}_1$, $\mathbf{v}_2$, and $\mathbf{v}_3$ in sequence:

$$\mathbf{v}_1 = \mathbf{x}_1 = \begin{bmatrix} 1 \\ 0 \\ 0 \\ -1 \end{bmatrix}, \mathbf{v}_2 = \mathbf{x}_2 - \left(\frac{\mathbf{v}_1 \cdot \mathbf{x}_2}{\mathbf{v}_1 \cdot \mathbf{v}_1}\right)\mathbf{v}_1 = \begin{bmatrix} 0 \\ 1 \\ 0 \\ -1 \end{bmatrix} - \frac{1}{2}\begin{bmatrix} 1 \\ 0 \\ 0 \\ -1 \end{bmatrix} = \begin{bmatrix} -\frac{1}{2} \\ 1 \\ 0 \\ -\frac{1}{2} \end{bmatrix},$$

$$\mathbf{v}_3 = \mathbf{x}_3 - \left(\frac{\mathbf{v}_1 \cdot \mathbf{x}_3}{\mathbf{v}_1 \cdot \mathbf{v}_1}\right)\mathbf{v}_1 - \left(\frac{\mathbf{v}_2 \cdot \mathbf{x}_3}{\mathbf{v}_2 \cdot \mathbf{v}_2}\right)\mathbf{v}_2 = \begin{bmatrix} 0 \\ 0 \\ 1 \\ -1 \end{bmatrix} - \frac{1}{2}\begin{bmatrix} 1 \\ 0 \\ 0 \\ -1 \end{bmatrix} - \frac{1}{3}\begin{bmatrix} -\frac{1}{2} \\ 1 \\ 0 \\ -\frac{1}{2} \end{bmatrix} = \begin{bmatrix} -\frac{1}{3} \\ -\frac{1}{3} \\ 1 \\ -\frac{1}{3} \end{bmatrix}.$$

Q: Is it reasonable to think of W as a hyperplane in $\mathbb{R}^4$? Why or why not?
A: Consider $ax + by + cz = 0$ in $\mathbb{R}^3$. Is this familiar? If so, what is it?

18. For (a), see Example 5.14 and Exercises 1 through 10 in Section 5.4.
For (b), see Example 5.19 and Exercises 17 through 20 in Section 5.4.

(a) We begin by finding the eigenvalues of A.

$$\begin{vmatrix} 2-\lambda & 1 & -1 \\ 1 & 2-\lambda & 1 \\ -1 & 1 & 2-\lambda \end{vmatrix} = 0 \Rightarrow \lambda(\lambda-3)^2 = 0 \Rightarrow D = \begin{bmatrix} 3 & 0 & 0 \\ 0 & 3 & 0 \\ 0 & 0 & 0 \end{bmatrix}.$$

$$(A-3I)\mathbf{v} = \mathbf{0} \Rightarrow x_1 = x_2 - x_3. \text{ So, } \mathbf{v}_1 = \begin{bmatrix} 1 \\ 1 \\ 0 \end{bmatrix} \text{ and } \mathbf{v}_2 = \begin{bmatrix} -1 \\ 0 \\ 1 \end{bmatrix}.$$

$$(A-0I)\mathbf{v} = \mathbf{0} \Rightarrow \begin{array}{rcl} x_1 &=& x_3 \\ x_2 &=& -x_3 \end{array} \Rightarrow \mathbf{v}_3 = \begin{bmatrix} 1 \\ -1 \\ 1 \end{bmatrix}.$$

Now we compute orthogonal basis vectors $\mathbf{v}_1$, $\mathbf{v}_2$, and $\mathbf{v}_3$ in sequence:

$$\mathbf{v}_1 = \mathbf{x}_1 = \begin{bmatrix} 1 \\ 1 \\ 0 \end{bmatrix}, \ \mathbf{v}_2 = \mathbf{x}_2 - \left(\frac{\mathbf{v}_1 \cdot \mathbf{x}_2}{\mathbf{v}_1 \cdot \mathbf{v}_1}\right)\mathbf{v}_1 = \begin{bmatrix} -1 \\ 0 \\ 1 \end{bmatrix} + \tfrac{1}{2}\begin{bmatrix} 1 \\ 1 \\ 0 \end{bmatrix} = \begin{bmatrix} -\frac{1}{2} \\ \frac{1}{2} \\ 1 \end{bmatrix},$$

$$\mathbf{v}_3 = \mathbf{x}_3 - \left(\frac{\mathbf{v}_1 \cdot \mathbf{x}_3}{\mathbf{v}_1 \cdot \mathbf{v}_1}\right)\mathbf{v}_1 - \left(\frac{\mathbf{v}_2 \cdot \mathbf{x}_3}{\mathbf{v}_2 \cdot \mathbf{v}_2}\right)\mathbf{v}_2 = \begin{bmatrix} 1 \\ -1 \\ 1 \end{bmatrix} - \tfrac{0}{2}\begin{bmatrix} 1 \\ 1 \\ 0 \end{bmatrix} - \tfrac{0}{1.5}\begin{bmatrix} -\frac{1}{2} \\ \frac{1}{2} \\ 1 \end{bmatrix} = \begin{bmatrix} 1 \\ -1 \\ 1 \end{bmatrix}.$$

$$\text{Normalizing} \Rightarrow Q = \begin{bmatrix} \frac{\sqrt{2}}{2} & -\frac{\sqrt{6}}{6} & \frac{\sqrt{3}}{3} \\ \frac{\sqrt{2}}{2} & \frac{\sqrt{6}}{6} & -\frac{\sqrt{3}}{3} \\ 0 & \frac{\sqrt{6}}{3} & \frac{\sqrt{3}}{3} \end{bmatrix}. \text{ Verify that } Q^TAQ = D = \begin{bmatrix} 3 & 0 & 0 \\ 0 & 3 & 0 \\ 0 & 0 & 0 \end{bmatrix}.$$

(b) From (a) we have $\lambda_1 = 3$, $\lambda_2 = 3$, and $\lambda_3 = 0$ with:

$$\mathbf{q}_1 = \begin{bmatrix} \frac{\sqrt{2}}{2} \\ \frac{\sqrt{2}}{2} \\ 0 \end{bmatrix}, \ \mathbf{q}_2 = \begin{bmatrix} -\frac{\sqrt{6}}{6} \\ \frac{\sqrt{6}}{6} \\ \frac{\sqrt{6}}{3} \end{bmatrix}, \text{ and } \mathbf{q}_3 = \begin{bmatrix} \frac{\sqrt{3}}{3} \\ -\frac{\sqrt{3}}{3} \\ \frac{\sqrt{3}}{3} \end{bmatrix}.$$

$$\text{Therefore, } \mathbf{q}_1\mathbf{q}_1^T = \begin{bmatrix} \frac{\sqrt{2}}{2} \\ \frac{\sqrt{2}}{2} \\ 0 \end{bmatrix} \begin{bmatrix} \frac{\sqrt{2}}{2} & \frac{\sqrt{2}}{2} & 0 \end{bmatrix} = \begin{bmatrix} \frac{1}{2} & \frac{1}{2} & 0 \\ \frac{1}{2} & \frac{1}{2} & 0 \\ 0 & 0 & 0 \end{bmatrix}.$$

$$\text{Likewise, } \mathbf{q}_2\mathbf{q}_2^T = \begin{bmatrix} -\frac{\sqrt{6}}{6} \\ \frac{\sqrt{6}}{6} \\ \frac{\sqrt{6}}{3} \end{bmatrix} \begin{bmatrix} -\frac{\sqrt{6}}{6} & \frac{\sqrt{6}}{6} & \frac{\sqrt{6}}{3} \end{bmatrix} = \begin{bmatrix} \frac{1}{6} & -\frac{1}{6} & -\frac{1}{3} \\ -\frac{1}{6} & \frac{1}{6} & \frac{1}{3} \\ -\frac{1}{3} & \frac{1}{3} & \frac{2}{3} \end{bmatrix}.$$

We do not need to compute $\mathbf{q}_3\mathbf{q}_3^T$. Why not?

Verify that $A = 3\mathbf{q}_1\mathbf{q}_1^T + 3\mathbf{q}_2\mathbf{q}_2^T$.

19. See Example 5.20 and Exercises 23 and 24 in Section 5.4.

We are given $\mathbf{x}_1 = \begin{bmatrix} 1 \\ 1 \\ 0 \end{bmatrix}$, $\mathbf{x}_2 = \begin{bmatrix} 1 \\ 1 \\ 1 \end{bmatrix}$ (from E_1), and $\mathbf{x}_3 = \begin{bmatrix} 1 \\ -1 \\ 0 \end{bmatrix}$ (from E_{-2}).

Now we compute orthogonal basis vectors $\mathbf{v}_1$, $\mathbf{v}_2$, and $\mathbf{v}_3$ in sequence:

$$\mathbf{v}_1 = \mathbf{x}_1 = \begin{bmatrix} 1 \\ 1 \\ 0 \end{bmatrix}, \ \mathbf{v}_2 = \mathbf{x}_2 - \left(\frac{\mathbf{v}_1 \cdot \mathbf{x}_2}{\mathbf{v}_1 \cdot \mathbf{v}_1}\right)\mathbf{v}_1 = \begin{bmatrix} 1 \\ 1 \\ 1 \end{bmatrix} - \frac{2}{2}\begin{bmatrix} 1 \\ 1 \\ 0 \end{bmatrix} = \begin{bmatrix} 0 \\ 0 \\ 1 \end{bmatrix},$$

$$\mathbf{v}_3 = \mathbf{x}_3 - \left(\frac{\mathbf{v}_1 \cdot \mathbf{x}_3}{\mathbf{v}_1 \cdot \mathbf{v}_1}\right)\mathbf{v}_1 - \left(\frac{\mathbf{v}_2 \cdot \mathbf{x}_3}{\mathbf{v}_2 \cdot \mathbf{v}_2}\right)\mathbf{v}_2 = \begin{bmatrix} 1 \\ -1 \\ 0 \end{bmatrix} - \frac{0}{2}\begin{bmatrix} 1 \\ 1 \\ 0 \end{bmatrix} - \frac{0}{1}\begin{bmatrix} 0 \\ 0 \\ 1 \end{bmatrix} = \begin{bmatrix} 1 \\ -1 \\ 0 \end{bmatrix}.$$

Normalizing $\Rightarrow \mathbf{q}_1 = \begin{bmatrix} \frac{\sqrt{2}}{2} \\ \frac{\sqrt{2}}{2} \\ 0 \end{bmatrix}$, $\mathbf{q}_2 = \begin{bmatrix} 0 \\ 0 \\ 1 \end{bmatrix}$, and $\mathbf{q}_3 = \begin{bmatrix} \frac{\sqrt{2}}{2} \\ -\frac{\sqrt{2}}{2} \\ 0 \end{bmatrix}$.

Therefore, $\mathbf{q}_1\mathbf{q}_1^T = \begin{bmatrix} \frac{\sqrt{2}}{2} \\ \frac{\sqrt{2}}{2} \\ 0 \end{bmatrix} \begin{bmatrix} \frac{\sqrt{2}}{2} & \frac{\sqrt{2}}{2} & 0 \end{bmatrix} = \begin{bmatrix} \frac{1}{2} & \frac{1}{2} & 0 \\ \frac{1}{2} & \frac{1}{2} & 0 \\ 0 & 0 & 0 \end{bmatrix}$.

Likewise, $\mathbf{q}_2\mathbf{q}_2^T = \begin{bmatrix} 0 \\ 0 \\ 1 \end{bmatrix} \begin{bmatrix} 0 & 0 & 1 \end{bmatrix} = \begin{bmatrix} 0 & 0 & 0 \\ 0 & 0 & 0 \\ 0 & 0 & 1 \end{bmatrix}$.

Finally, $\mathbf{q}_3\mathbf{q}_3^T = \begin{bmatrix} \frac{\sqrt{2}}{2} \\ -\frac{\sqrt{2}}{2} \\ 0 \end{bmatrix} \begin{bmatrix} \frac{\sqrt{2}}{2} & -\frac{\sqrt{2}}{2} & 0 \end{bmatrix} = \begin{bmatrix} \frac{1}{2} & -\frac{1}{2} & 0 \\ -\frac{1}{2} & \frac{1}{2} & 0 \\ 0 & 0 & 0 \end{bmatrix}$.

Now let $A = 1\mathbf{q}_1\mathbf{q}_1^T + 1\mathbf{q}_2\mathbf{q}_2^T - 2\mathbf{q}_3\mathbf{q}_3^T$

$$= \begin{bmatrix} \frac{1}{2} & \frac{1}{2} & 0 \\ \frac{1}{2} & \frac{1}{2} & 0 \\ 0 & 0 & 0 \end{bmatrix} + \begin{bmatrix} 0 & 0 & 0 \\ 0 & 0 & 0 \\ 0 & 0 & 1 \end{bmatrix} - 2\begin{bmatrix} \frac{1}{2} & -\frac{1}{2} & 0 \\ -\frac{1}{2} & \frac{1}{2} & 0 \\ 0 & 0 & 0 \end{bmatrix} = \begin{bmatrix} -\frac{1}{2} & \frac{3}{2} & 0 \\ \frac{3}{2} & -\frac{1}{2} & 0 \\ 0 & 0 & 1 \end{bmatrix}.$$

Verify that the eigenvalues of A are 1 and -2 with the given eigenspaces E_1 and E_{-2}.

Q: How do we verify the eigenvalues and eigenspaces E_1 and E_{-2}?
A: Hint: Is $A\mathbf{x}_1 = \mathbf{x}_1$? Should it be? What about $A\mathbf{x}_2$ and $A\mathbf{x}_3$?

20. Since $\{\mathbf{v}_1, \mathbf{v}_2, \ldots, \mathbf{v}_n\}$ is an orthonormal basis, we have:

$$\mathbf{v}_i \cdot \mathbf{v}_j = \begin{cases} 0 & \text{if } i \neq j \text{ Property 1} \\ 1 & \text{if } i = j \text{ Property 2} \end{cases}$$

Since we are given $A = \sum c_i \mathbf{v}_i \mathbf{v}_i^T$, we have:

$A\mathbf{v}_j = \sum c_i(\mathbf{v}_i \mathbf{v}_i^T)\mathbf{v}_j = \sum c_i \mathbf{v}_i(\mathbf{v}_i \cdot \mathbf{v}_j) = c_j \mathbf{v}_j(\mathbf{v}_j \cdot \mathbf{v}_j) = c_j \mathbf{v}_j(1) = c_j \mathbf{v}_j.$

So, A has eigenvalues $c_1, c_2, \ldots, c_n$ with corresponding eigenvectors $\mathbf{v}_1, \mathbf{v}_2, \ldots, \mathbf{v}_n$.

Since A is a linear combination of symmetric matrices $(c_i \mathbf{v}_i \mathbf{v}_i^T)$, A is symmetric.

Q: Why is it obvious that $c_i \mathbf{v}_i \mathbf{v}_i^T$ is a symmetric matrix?
A: Hint: The (j, k) entry is $v_{ij} v_{ik}$. What is the (k, j) entry?

Q: Why is a linear combination of symmetric matrices clearly symmetric?

Chapter 6

Vector Spaces

6.1 Vector Spaces and Subspaces

1. Let $V = \left\{ \begin{bmatrix} x \\ x \end{bmatrix} \right\}$. We will show V satisfies all 10 axioms required of a vector space.

1. $\begin{bmatrix} x \\ x \end{bmatrix}, \begin{bmatrix} y \\ y \end{bmatrix} \in V \Rightarrow \begin{bmatrix} x+y \\ x+y \end{bmatrix} \in V.$

2. $\begin{bmatrix} x \\ x \end{bmatrix} + \begin{bmatrix} y \\ y \end{bmatrix} = \begin{bmatrix} x+y \\ x+y \end{bmatrix} = \begin{bmatrix} y \\ y \end{bmatrix} + \begin{bmatrix} x \\ x \end{bmatrix}.$

3. $\left(\begin{bmatrix} x \\ x \end{bmatrix} + \begin{bmatrix} y \\ y \end{bmatrix} \right) + \begin{bmatrix} z \\ z \end{bmatrix} = \begin{bmatrix} x+y+z \\ x+y+z \end{bmatrix} = \begin{bmatrix} x \\ x \end{bmatrix} + \left(\begin{bmatrix} y \\ y \end{bmatrix} + \begin{bmatrix} z \\ z \end{bmatrix} \right).$

4. $\begin{bmatrix} 0 \\ 0 \end{bmatrix} \in V$ and $\begin{bmatrix} x \\ x \end{bmatrix} + \begin{bmatrix} 0 \\ 0 \end{bmatrix} = \begin{bmatrix} x \\ x \end{bmatrix}.$

5. $\begin{bmatrix} x \\ x \end{bmatrix} \in V \Rightarrow \begin{bmatrix} -x \\ -x \end{bmatrix} \in V$ and $\begin{bmatrix} x \\ x \end{bmatrix} + \begin{bmatrix} -x \\ -x \end{bmatrix} = \begin{bmatrix} 0 \\ 0 \end{bmatrix}.$

6. $\begin{bmatrix} x \\ x \end{bmatrix} \in V \Rightarrow \begin{bmatrix} cx \\ cx \end{bmatrix} \in V.$

7. $c\left(\begin{bmatrix} x \\ x \end{bmatrix} + \begin{bmatrix} y \\ y \end{bmatrix} \right) = \begin{bmatrix} cx+cy \\ cx+cy \end{bmatrix} = c\begin{bmatrix} x \\ x \end{bmatrix} + c\begin{bmatrix} y \\ y \end{bmatrix}.$

8. $(c+d)\begin{bmatrix} x \\ x \end{bmatrix} = \begin{bmatrix} cx+dx \\ cx+dx \end{bmatrix} = c\begin{bmatrix} x \\ x \end{bmatrix} + d\begin{bmatrix} x \\ x \end{bmatrix}.$

9. $c\left(d\begin{bmatrix} x \\ x \end{bmatrix} \right) = \begin{bmatrix} cdx \\ cdx \end{bmatrix} = (cd)\begin{bmatrix} x \\ x \end{bmatrix}.$

10. $1\begin{bmatrix} x \\ x \end{bmatrix} = \begin{bmatrix} 1x \\ 1x \end{bmatrix} = \begin{bmatrix} x \\ x \end{bmatrix}.$

2. Since $V = \left\{ \begin{bmatrix} x \\ y \end{bmatrix} : x, y \geq 0 \right\}$ fails to satisfy Axioms 5 and 6, V is not a vector space.

5. $\begin{bmatrix} x \\ y \end{bmatrix} \in V, c < 0 \Rightarrow c \begin{bmatrix} x \\ y \end{bmatrix} = \begin{bmatrix} cx \\ cy \end{bmatrix} \notin V$ because $cx, cy \leq 0$.

6. $\begin{bmatrix} x \\ y \end{bmatrix} \in V \Rightarrow \begin{bmatrix} -x \\ -y \end{bmatrix} \notin V$ because $-x, -y \leq 0$.

3. Since $V = \left\{ \begin{bmatrix} x \\ y \end{bmatrix} : xy \geq 0 \right\}$ fails to satisfy axiom 1, V is not a vector space.

1. $x \neq y, \begin{bmatrix} x \\ y \end{bmatrix}, \begin{bmatrix} -y \\ -x \end{bmatrix} \in V \Rightarrow \begin{bmatrix} x \\ y \end{bmatrix} + \begin{bmatrix} -y \\ -x \end{bmatrix} = \begin{bmatrix} x - y \\ y - x \end{bmatrix} \notin V$ since $(x - y)(y - x) < 0$.

4. Since $V = \left\{ \begin{bmatrix} x \\ y \end{bmatrix} : x \geq y \right\}$ fails to satisfy axioms 5 and 6, V is not a vector space.

5. $\begin{bmatrix} x \\ -y \end{bmatrix} \in V$ then $\begin{bmatrix} x \\ -y \end{bmatrix} + \begin{bmatrix} -x \\ y \end{bmatrix} = \begin{bmatrix} 0 \\ 0 \end{bmatrix}$ but $\begin{bmatrix} -x \\ y \end{bmatrix} \notin V$ because $-x \leq y$.

6. $\begin{bmatrix} x \\ -y \end{bmatrix} \in V \Rightarrow (-1) \begin{bmatrix} x \\ -y \end{bmatrix} = \begin{bmatrix} -x \\ y \end{bmatrix} \notin V$ because $-x \leq y$.

5. $\begin{bmatrix} cx \\ y \end{bmatrix} + \begin{bmatrix} dx \\ y \end{bmatrix} \neq \begin{bmatrix} (c+d)x \\ y \end{bmatrix} \Rightarrow$ axiom 8 fails $\Rightarrow V$ is not a vector space.

6. $\begin{bmatrix} c(x_1 + x_1 + 1) \\ c(x_2 + x_2 + 1 \end{bmatrix} \neq \begin{bmatrix} cx_1 + cx_1 + 1 \\ cx_2 + cx_2 + 1 \end{bmatrix} \Rightarrow$ axiom 7 fails $\Rightarrow V$ is not a vector space.

7. All axioms apply $\Rightarrow V$ is a vector space.

1. $x > 0, y > 0 \Rightarrow xy > 0$ 2. $xy = yx$ 3. $(xy)z = x(yz)$
4. $x1 = x$ (so $1 = \mathbf{0}$!) 5. $x\frac{1}{x} = 1$ 6. $x > 0 x^c > 0$
7. $(xy)^c = x^c y^c$ 8. $x^{(c+d)} = x^c x^d$ 9. $(x^d)^c = x^{dc}$
10. $x^1 = x$

8. All axioms apply $\Rightarrow V$ is a vector space. We will show only the key axioms of 1 and 6 are satisfied.

1. $\frac{a}{b} + \frac{c}{d} = \frac{ad+bc}{bd}$ 6. $c\frac{a}{b} = \frac{ca}{b}$

9. All axioms apply $\Rightarrow V$ is a vector space. We will show only the key axioms of 1 and 6.

1. $\begin{bmatrix} a & b \\ 0 & c \end{bmatrix} + \begin{bmatrix} a' & b' \\ 0 & c' \end{bmatrix} = \begin{bmatrix} a + a' & b + b' \\ 0 & c + c' \end{bmatrix}$ 6. $d \begin{bmatrix} a & b \\ 0 & c \end{bmatrix} = \begin{bmatrix} da & db \\ 0 & dc \end{bmatrix}$

10. $\begin{bmatrix} 1 & 0 \\ 0 & 0 \end{bmatrix}, \begin{bmatrix} 0 & 0 \\ 0 & 1 \end{bmatrix} \in V \Rightarrow \begin{bmatrix} 1 & 0 \\ 0 & 0 \end{bmatrix} + \begin{bmatrix} 0 & 0 \\ 0 & 1 \end{bmatrix} = \begin{bmatrix} 1 & 0 \\ 0 & 1 \end{bmatrix} \notin V \Rightarrow V$ is not a vector space.

11. All axioms apply $\Rightarrow V$ is a vector space. We will show only the key axioms of 1 and 6.

1. $(A + B)^T = B^T + A^T = -B + (-A) = -(A + B)$ 6. $(cA)^T = cA^T = c(-A) = -(cA)$

12. Need only show axioms 8, 9 and 10 are satisfied.

8. $c(p(x)+q(x)) = c(a_0+a_1x+a_2x^2)+c(b_0+b_1x+b_2x^2) = cp(x)+cq(x)$.
9. $c(d\,p(x)) = c(da_0+da_1x+da_2x^2) = cda_0+cda_1x+cda_2x^2 = (cd)p(x)$.
10. $1\,p(x) = 1(a_0+a_1x+a_2x^2) = a_0+a_1x+a_2x^2 = p(x)$.

13. Need only show axioms 7, 8, 9 and 10 are satisfied.

7. $c(f+g)(x) = cf(x)+cg(x) = (cf)(x)+(cg)(x)$.
8. $(c+d)((f(x)) = cf(x)+df(x) = (cf)(x)+(df)(x)$.
9. $c((df)(x)) = c(d(f(x))) = (cd)f(x)$.
10. $1(f(x) = 1\,f(x) = f(x)$.

14. $c\begin{bmatrix} z \\ \overline{z} \end{bmatrix} = \begin{bmatrix} cz \\ c\overline{z} \end{bmatrix} \neq \begin{bmatrix} z \\ \overline{cz} \end{bmatrix} \Rightarrow$ axiom 6 fails $\Rightarrow V$ is not a complex vector space.

15. All axioms apply $\Rightarrow V$ is a complex vector space.

16. All axioms apply $\Rightarrow V$ is a complex vector space. We will show only 8 and 9.

8) $(\overline{c+d})z = (\overline{c}+\overline{d})z = \overline{c}z+\overline{d}z$. 9) $c(\overline{d}z) = \overline{c}\overline{d}z = \overline{cd}z$.

17. $c \in \mathbb{C}$ and $c \notin \mathbb{R} \Rightarrow c\mathbf{x} \notin \mathbb{R}^n \Rightarrow$ axiom 6 fails $\Rightarrow V$ is not a complex vector space.

18. All axioms apply $\Rightarrow V$ is a vector space. We will show only the key axioms of 1 and 6.

1. $w(\mathbf{x}+\mathbf{y}) = w(\mathbf{x})+w(\mathbf{y}) - 2\,\text{overlap} \Rightarrow w(\mathbf{x}+\mathbf{y})$ is even.
6. $w(\mathbf{x})$ even $\Rightarrow cw(\mathbf{x})$ is even.

19. Axioms 1, 4, and 6 fail $\Rightarrow V$ is not a vector space.

1. $w(\mathbf{x}+\mathbf{y}) = w(\mathbf{x})+w(\mathbf{y}) - 2\,\text{overlap} \Rightarrow w(\mathbf{x}+\mathbf{y})$ is even $\Rightarrow \mathbf{x}+\mathbf{y} \notin V$.
4. $\mathbf{0} \notin V$ because it has an even number (0) of 1s.
6. Given $\mathbf{x} \in V$, $0\mathbf{x} = \mathbf{0} \notin V$.

20. All axioms apply $\Rightarrow V$ is a vector space.

21. Addition, multiplication not well-defined (do not match) $\Rightarrow V$ is not a vector space.

22. Will show $\mathbf{u}+0\mathbf{u} = \mathbf{u} \Rightarrow 0\mathbf{u} = \mathbf{0}$: $\mathbf{u}+0\mathbf{u} = 1\mathbf{u}+0\mathbf{u} = (1+0)\mathbf{u} = \mathbf{u}$.

23. Will show $\mathbf{u}+(-1)\mathbf{u} = \mathbf{0} \Rightarrow (-1)\mathbf{u} = -\mathbf{u}$: $\mathbf{u}+(-1)\mathbf{u} = 1\mathbf{u}+(-1)\mathbf{u} = (1+(-1))\mathbf{u} = 0\mathbf{u} = \mathbf{0}$.

24. $\begin{bmatrix} a \\ 0 \\ a \end{bmatrix} + \begin{bmatrix} b \\ 0 \\ b \end{bmatrix} = \begin{bmatrix} a+b \\ 0 \\ a+b \end{bmatrix}, c\begin{bmatrix} a \\ 0 \\ a \end{bmatrix} = \begin{bmatrix} ca \\ 0 \\ ca \end{bmatrix} \Rightarrow W$ is a subspace.

25. $\begin{bmatrix} a \\ -a \\ 2a \end{bmatrix} + \begin{bmatrix} b \\ -b \\ 2b \end{bmatrix} = \begin{bmatrix} a+b \\ -(a+b) \\ 2(a+b) \end{bmatrix}, c\begin{bmatrix} a \\ -a \\ 2a \end{bmatrix} = \begin{bmatrix} ca \\ -ca \\ 2ca \end{bmatrix} \Rightarrow W$ is a subspace.

26. $\begin{bmatrix} a \\ b \\ a+b+1 \end{bmatrix} + \begin{bmatrix} a' \\ b' \\ a'+b'+1 \end{bmatrix} \neq \begin{bmatrix} a+a' \\ b+b' \\ (a+a')+(b+b')+1 \end{bmatrix} \Rightarrow W$ is not a subspace.

27. $\begin{bmatrix} a \\ b \\ |a| \end{bmatrix} + \begin{bmatrix} a' \\ b' \\ |a'| \end{bmatrix} \begin{bmatrix} a+a' \\ b+b' \\ |a|+|a'| \end{bmatrix} \neq \begin{bmatrix} a+a' \\ b+b' \\ |a+a'| \end{bmatrix} \Rightarrow W$ is not a subspace.

28. $\begin{bmatrix} a & b \\ b & 2a \end{bmatrix} + \begin{bmatrix} a' & b' \\ b' & 2a' \end{bmatrix} = \begin{bmatrix} a+a' & b+b' \\ b+b' & 2(a+a') \end{bmatrix}, c\begin{bmatrix} a & b \\ b & 2a \end{bmatrix} = \begin{bmatrix} ca & cb \\ cb & 2ca \end{bmatrix} \Rightarrow$ yes.

29. $a, b \neq 0, \begin{bmatrix} a & 0 \\ 0 & b \end{bmatrix}, \begin{bmatrix} -a & a \\ b & -b \end{bmatrix} \in W$ but $\begin{bmatrix} a & 0 \\ 0 & b \end{bmatrix} + \begin{bmatrix} -a & a \\ b & -b \end{bmatrix} = \begin{bmatrix} 0 & a \\ b & 0 \end{bmatrix} \notin W \Rightarrow$ no.

30. $a \neq 0, \begin{bmatrix} a & 0 \\ 0 & \frac{1}{a} \end{bmatrix}, \begin{bmatrix} -a & 0 \\ 0 & -\frac{1}{a} \end{bmatrix} \in W$ but $\begin{bmatrix} a & 0 \\ 0 & \frac{1}{a} \end{bmatrix} + \begin{bmatrix} -a & 0 \\ 0 & -\frac{1}{a} \end{bmatrix} = \begin{bmatrix} 0 & 0 \\ 0 & 0 \end{bmatrix} \notin W \Rightarrow$ no.

31. Note, $D \in W \Leftrightarrow D = \sum \lambda_i I$. We will show W satisfies axioms 1 and 6 $\Rightarrow W$ is a subspace.
1. $D + D' = \sum \lambda_i I + \sum \lambda_i' I = \sum(\lambda_i + \lambda_i')I \Rightarrow D + D' \in W$.
6. $cD = c\sum \lambda_i I \sum_c \lambda_i I \Rightarrow cD \in W$.

32. $(cA)^2 = c^2A^2 = c^2A \neq cA \Rightarrow cA \notin W \Rightarrow W$ is not a subspace.

33. We will show that W satisfies axioms 1 and 6 $\Rightarrow W$ is a subspace.
1. $(A + C)B = AB + CB = BA + BC = B(A + C) \Rightarrow A + C \in W$.
6. $(cA)B = c(AB) = c(BA) = B(cA) \Rightarrow cA \in W$.

34. We will show that W satisfies axioms 1 and 6 $\Rightarrow W$ is a subspace.
1. $\mathbf{v} + \mathbf{v}' = (bx + cx^2) + (b'x + c'x^2) = (b + b')x + (c + c')x^2 \Rightarrow \mathbf{v} + \mathbf{v}' \in W$.
6. $d\mathbf{v} = d(bx + cx^2) = dbx + dcx^2 \Rightarrow d\mathbf{v} \in W$.

35. We will show that W satisfies axioms 1 and 6 $\Rightarrow W$ is a subspace.
1. Let $\mathbf{v} = a + b + cx^2$, $\mathbf{v}' = a' + b' + c'x^2 \in W$, that is $a + b + c = 0$ and $a' + b' + c' = 0$.
   Then $(a + a') + (b + b') + (c + c') = 0 \Rightarrow \mathbf{v} + \mathbf{v}' \in W$.
6. Let $\mathbf{v} = a + b + cx^2 \in W$, that is $a + b + c = 0$.
   Then $c(a + b + c) = 0 \Rightarrow c\mathbf{v} \in W$.

36. $\mathbf{v} = 1 + x, \mathbf{w} = x^2 \in W \Rightarrow \mathbf{v} + \mathbf{w} = 1 + x + x^2 \notin W$ (because $abc = 1$) $\Rightarrow W$ is not a subspace.

37. $\mathbf{v} = x^3, \mathbf{w} = -x^3 \in W \Rightarrow \mathbf{v} + \mathbf{w} = x^3 + (-x^3) = 0 \notin W$ (because degree $= 0$) $\Rightarrow$ no.

38. We will show that W satisfies axioms 1 and 6 $\Rightarrow W$ is a subspace.
1. $(f + g)(-x) = f(-x) + g(-x) = f(x) + g(x) = (f + g)(x) \Rightarrow f + g \in W$.
6. $(cf)(-x) = cf(-x) = cf(x) = (cf)(x) \Rightarrow cf \in W$.

39. We will show that W satisfies axioms 1 and 6 $\Rightarrow W$ is a subspace.
1. $(f + g)(-x) = f(-x) + g(-x) = -(f(x) + g(x)) = -(f + g)(x) \Rightarrow f + g \in W$.
6. $(cf)(-x) = cf(-x) = -cf(x) = -(cf)(x) \Rightarrow cf \in W$.

40. $(f + g)(0) = f(0) + g(0) = 1 + 1 \neq 1 \Rightarrow f + g \notin W \Rightarrow W$ is not a subspace.

41. $(f + g)(x) = f(x) + g(x) = 0 + 0 = 0$ and $(cf)(x) = cf(x) = c0 = 0 \Rightarrow W$ is a subspace.

42. $\int(f+g) = \int f + \int g$ and $\int(cf) = c\int f \Rightarrow W$ is closed.

43. $f(x) = x \in W$ but $-f(x) = -x \notin W$ $(-f'(x) = -1) \Rightarrow W$ is not a subspace.

44. $(f+g)'' = f'' + g''$ and $(cf)'' = cf'' \Rightarrow W$ is a subspace.

45. $f(x) = \frac{1}{x} \in W$ but $-f(x) = -\frac{1}{x} \notin W$ (since $\lim(-f) = -\lim f = -\infty$) $\Rightarrow$ W is not a subspace.

46. $\mathbf{v}, \mathbf{v}' \in U \cap W \Rightarrow c\mathbf{v}, \mathbf{v}+\mathbf{v}' \in U \cap W$ since $c\mathbf{v}, \mathbf{v}+\mathbf{v}' \in U$ and $c\mathbf{v}, \mathbf{v}+\mathbf{v}' \in W$ because they are subspaces $\Rightarrow U \cap W$ is a subspace.
Note: $U \cap W$ is not empty because $0 \in U$ and $0 \in V \Rightarrow 0 \in U \cap W$.

47. Let $U = \left\{\begin{bmatrix} x \\ 0 \end{bmatrix}\right\}$ (the x-axis) and $W = \left\{\begin{bmatrix} 0 \\ y \end{bmatrix}\right\}$ (the y-axis).

$x, y \neq 0$, $\begin{bmatrix} x \\ 0 \end{bmatrix} \in U$, $\begin{bmatrix} 0 \\ y \end{bmatrix} \in W$, but $\begin{bmatrix} x \\ 0 \end{bmatrix} + \begin{bmatrix} 0 \\ y \end{bmatrix} = \begin{bmatrix} x \\ y \end{bmatrix} \notin U \cup W$ ($\begin{bmatrix} x \\ y \end{bmatrix}$ not on axes).

48. (a) $U + W = \left\{\begin{bmatrix} x \\ y \end{bmatrix}\right\}$, the x-y plane.

(b) We will show that $U+W$ satisfies axioms 1 and 6 $\Rightarrow U+W$ is a subspace.
1. $\mathbf{x}, \mathbf{x}' \in U+W \Rightarrow \mathbf{x}+\mathbf{x}' = (\mathbf{u}+\mathbf{v})+(\mathbf{u}'+\mathbf{v}') = (\mathbf{u}+\mathbf{u}')+(\mathbf{v}+\mathbf{v}') \Rightarrow \mathbf{x}+\mathbf{x}' \in U+W$.
6. $\mathbf{x} \in U+W \Rightarrow c\mathbf{x} = c(\mathbf{u}+\mathbf{v}) = c\mathbf{u}+c\mathbf{v} \Rightarrow c\mathbf{x} \in U+W$.
Also, note that $U+W$ is not empty since $0+0 = 0 \in U+W$.

49. We will show that $U \times W$ satisfies the axioms of a vector space.

1. $\mathbf{x}, \mathbf{x}' \in U \times W \Rightarrow \mathbf{x}+\mathbf{x}' = (\mathbf{u},\mathbf{v}) + (\mathbf{u}',\mathbf{v}') = (\mathbf{u}+\mathbf{u}', \mathbf{v}+\mathbf{v}') \Rightarrow \mathbf{x}+\mathbf{x}' \in U \times W$.
2. $\mathbf{x}, \mathbf{x}' \in U \times W \Rightarrow \mathbf{x}+\mathbf{x}' = (\mathbf{u}+\mathbf{u}', \mathbf{v}+\mathbf{v}') = (\mathbf{u}'+\mathbf{u}, \mathbf{v}'+\mathbf{v}) = \mathbf{x}'+\mathbf{x}$.
6. $\mathbf{x} \in U \times W \Rightarrow c\mathbf{x} = (c\mathbf{u}, c\mathbf{v}) \Rightarrow c\mathbf{x} \in U \times W$.

The remaining axioms are proved similarly. The details are left to the reader.

50. We will show that Δ satisfies axioms 1 and 6 $\Rightarrow \Delta$ is a subspace of $V \times V$.
1. $\mathbf{x}, \mathbf{x}' \in \Delta \Rightarrow \mathbf{x}+\mathbf{x}' = (\mathbf{w},\mathbf{w}) + (\mathbf{w}',\mathbf{w}') = (\mathbf{w}+\mathbf{w}', \mathbf{w}+\mathbf{w}') \Rightarrow \mathbf{x}+\mathbf{x}' \in \Delta$.
6. $\mathbf{x} \in \Delta \Rightarrow c\mathbf{x} = (c\mathbf{w}, c\mathbf{w}) \Rightarrow c\mathbf{x} \in \Delta$.

Finally, since $W \subseteq V$, Δ is a subspace of $V \times V$.

Q: In the proof above, where did we use the fact that W is a subspace of V?
A: When asserting (indirectly) that both $\mathbf{w}+\mathbf{w}'$ and $c\mathbf{w}$ are in W.

Q: Is Δ that same thing as $W \times W$? Why or why not?
A: Hint: Does $W \times W$ require that the components in the ordered pair be identical?

Q: Is $W \times W$ also a subspace of $V \times V$? Why or why not?

Q: Is $\Delta_V = \{(\mathbf{v}, \mathbf{v}) : \mathbf{v} \text{ is in } V\}$ a subspace of $V \times V$?
A: Can V be thought of as a subspace of itself?

51. $aA + bB = C \Rightarrow \begin{bmatrix} a+b & a-b \\ b-a & a \end{bmatrix} = \begin{bmatrix} 1 & 2 \\ 3 & 4 \end{bmatrix} \Rightarrow \begin{matrix} a+b=1 \\ b-a=3 \end{matrix} \Rightarrow a = -1 \neq 4 \Rightarrow$ no.

52. $\begin{bmatrix} a+b & a-b \\ b-a & a \end{bmatrix} = \begin{bmatrix} 3 & -5 \\ 4 & -1 \end{bmatrix} \Rightarrow a=-1, b=4 \Rightarrow -A+4B=C \Rightarrow C \in \text{span}(A,B).$

53. $a(1-2x)+b(x-x^2)+c(-2+3x+x^2)=3-5x-x^2 \Rightarrow \begin{array}{r} a-2c=3 \\ -2a+b+3c=-5 \\ -b+c=-1 \end{array} \Rightarrow$

$a=-2, b=-1, c=0 \Rightarrow s(x)=-2p(x)-q(x) \Rightarrow s(x) \in \text{span}(p(x),q(x),r(x)).$

54. $a(1-2x)+b(x-x^2)+c(-2+3x+x^2)=1+x+x^2 \Rightarrow \begin{array}{r} a-2c=1 \\ -2a+b+3c=1 \\ -b+c=1 \end{array} \Rightarrow 0=4 \Rightarrow$

This system has no solution $\Rightarrow s(x) \notin \text{span}(p(x),q(x),r(x))$.

55. $h(x)=1=\cos^2 x+\sin^2 x=f(x)+g(x) \Rightarrow h(x) \in \text{span}(f(x),g(x)).$

56. $h(x)=\cos 2x=\cos^2 x-\sin^2 x=-f(x)+g(x) \Rightarrow h(x) \in \text{span}(f(x),g(x)).$

57. $h(x)=\sin 2x=a\,f(x)+b\,g(x)=a\sin^2 x+b\cos^2 x$ for all $x \Rightarrow$

$\begin{array}{l} x=0 \Rightarrow a(0)^2+b(1)^2=0 \Rightarrow b=0 \\ x=\frac{\pi}{2} \Rightarrow a(1)^2+b(0)^2=0 \Rightarrow a=0 \end{array} \Rightarrow h(x)=\sin 2x$ would be the zero function $\Rightarrow$

This system has no solution $\Rightarrow h(x) \notin \text{span}(f(x),g(x))$.

To see the absurdity of $h(x)=\sin 2x=$ the zero function consider $\sin\left(2\,\frac{\pi}{4}\right)=1 \Rightarrow 1 \neq 0$.

58. $h(x)=\sin x=a\,f(x)+b\,g(x)=a\sin^2 x+b\cos^2 x$ for all $x \Rightarrow$

$\begin{array}{l} x=0 \Rightarrow a(0)^2+b(1)^2=0 \Rightarrow b=0 \\ x=\frac{\pi}{2} \Rightarrow a(1)^2+b(0)^2=1 \Rightarrow a=1 \end{array} \Rightarrow$

$h(x)=\sin x=\sin^2 x \Rightarrow$ This system has no solution $\Rightarrow h(x) \notin \text{span}(f(x),g(x))$.

To see the absurdity of $h(x)=\sin x=\sin^2 x$ consider $\sin\frac{\pi}{6}=\frac{1}{2} \Rightarrow \frac{1}{2}=(\frac{1}{2})^2 \neq \frac{1}{4}$.

59. Let $V_1=\begin{bmatrix} 1 & 1 \\ 0 & 1 \end{bmatrix}, V_2=\begin{bmatrix} 0 & 1 \\ 1 & 0 \end{bmatrix}, V_3=\begin{bmatrix} 1 & 0 \\ 1 & 1 \end{bmatrix}, V_4=\begin{bmatrix} 0 & -1 \\ 1 & 0 \end{bmatrix}$.

Then since $V_4=V_3-V_1$, $\text{span}(V_1,V_2,V_3,V_4)=\text{span}(V_1,V_2,V_3)$.

To show $\text{span}(V_1,V_2,V_3) \neq M_{22}$ we will show $E_{11} \notin \text{span}(V_1,V_2,V_3)$.

$aV_1+bV_2+cV_3=E_{11} \Rightarrow \begin{bmatrix} a+c=1 & a+b=0 \\ b+c=0 & a+c=0 \end{bmatrix} \Rightarrow 0=1 \Rightarrow$

This system has no solution $\Rightarrow E_{11} \notin \text{span}(V_1,V_2,V_3) \Rightarrow \text{span}(V_1,V_2,V_3) \neq M_{22}$.

60. Let $V_1 = \begin{bmatrix} 1 & 0 \\ 1 & 0 \end{bmatrix}, V_2 = \begin{bmatrix} 1 & 1 \\ 1 & 0 \end{bmatrix}, V_3 = \begin{bmatrix} 1 & 1 \\ 1 & 1 \end{bmatrix}, V_4 = \begin{bmatrix} 0 & -1 \\ 1 & 0 \end{bmatrix}$. So,

$$\begin{matrix} E_{11} = 2V_1 - V_2 - V_4 & E_{12} = V_2 - V_1 \\ E_{21} = V_2 + V_4 - V_1 & E_{22} = V_3 - V_2 \end{matrix} \Rightarrow \text{span}(V_i) = \text{span}(E_{11}, E_{12}, E_{21}, E_{22}) = M_{22}.$$

Explicitly, $A \in M_{22} \Rightarrow A = \begin{bmatrix} a & b \\ c & d \end{bmatrix} \Rightarrow$

$$\begin{aligned} A &= aE_{11} + bE_{12} + cE_{21} + dE_{22} \\ &= a(2V_1 - V_2 - V_4) + b(V_2 - V_1) + c(V_2 + V_4 - V_1) + d(V_3 - V_2) \Rightarrow \end{aligned}$$

$A \in \text{span}(V_1, V_2, V_3, V_4) \Rightarrow \text{span}(V_1, V_2, V_3, V_4) = \text{span}(E_{11}, E_{12}, E_{21}, E_{22}) = M_{22}$.

61. Let $p(x) = 1 + x$, $q(x) = x + x^2$, $r(x) = 1 + x^2$. Then note, $1 = a\,p(x) + b\,q(x) + c\,r(x) \Rightarrow$

$$\begin{matrix} a + c = 1 \\ a + b = 0 \\ b + c = 0 \end{matrix} \Rightarrow a = c = \tfrac{1}{2},\ b = -\tfrac{1}{2} \Rightarrow 1 = \tfrac{1}{2}\,p(x) - \tfrac{1}{2}\,q(x) + \tfrac{1}{2}\,r(x).$$

Likewise, $x = \frac{1}{2}\,p(x) + \frac{1}{2}\,q(x) - \frac{1}{2}\,r(x)$, $x^2 = -\frac{1}{2}\,p(x) + \frac{1}{2}\,q(x) + \frac{1}{2}\,r(x) \Rightarrow$

$\text{span}(p(x), q(x), r(x)) = \text{span}(1, x, x^2) = \mathscr{P}_2$. Explicitly, $s(x) \in \mathscr{P}_2 \Rightarrow$

$$\begin{aligned} s(x) &= a + bx + cx^2 \\ &= a(\tfrac{1}{2}\,p(x) - \tfrac{1}{2}\,q(x) + \tfrac{1}{2}\,r(x)) + b(\tfrac{1}{2}\,p(x) + \tfrac{1}{2}\,q(x) - \tfrac{1}{2}\,r(x)) + c(-\tfrac{1}{2}\,p(x) + \tfrac{1}{2}\,q(x) + \tfrac{1}{2}\,r(x)) \Rightarrow \end{aligned}$$

$s(x) \in \text{span}(p(x), q(x), r(x)) = \text{span}(1, x, x^2) = \mathscr{P}_2$.

62. Let $p(x) = 1 + x + 2x^2$, $q(x) = 2 + x + 2x^2$, $r(x) = -1 + x + 2x^2$.
Then note, $r(x) = 3\,p(x) - 2\,q(x) \Rightarrow \text{span}(p(x), q(x), r(x)) = \text{span}(p(x), q(x))$.

To show $\text{span}(p(x), q(x)) \neq \mathscr{P}_2$ we will show $s(x) = 1 \notin \text{span}(p(x), q(x))$.

$$1 = a\,p(x) + b\,q(x) \Rightarrow \begin{matrix} a + b = 1 \\ a + b = 0 \\ a + b = 0 \end{matrix} \Rightarrow 0 = 1 \Rightarrow$$

This system has no solution $\Rightarrow s(x) = 1 \notin \text{span}(p(x), q(x)) \Rightarrow \text{span}(p(x), q(x)) \neq \mathscr{P}_2$.

63. Let $\mathbf{0}$ and $\mathbf{0}'$ satisfy axiom 4. We will show $\mathbf{0}' = \mathbf{0}$. Axiom 4 $\Rightarrow \mathbf{0}' = \mathbf{0}' + \mathbf{0} = \mathbf{0}$.

64. Let $\mathbf{v}'$ and $\mathbf{v}''$ satisfy axiom 5 for $\mathbf{v}$. We will show $\mathbf{v}'' = \mathbf{v}'$.
Axiom 5 $\Rightarrow \mathbf{v}'' = \mathbf{v}'' + \mathbf{0} = \mathbf{v}'' + (\mathbf{v} + \mathbf{v}') = (\mathbf{v}'' + \mathbf{v}) + \mathbf{v}' = \mathbf{0} + \mathbf{v}' = \mathbf{v}'$.

6.2 Linear Independence, Basis, and Dimension

1. Let $V_1 = \begin{bmatrix} 1 & 1 \\ 0 & -1 \end{bmatrix}, V_2 = \begin{bmatrix} 1 & -1 \\ 1 & 0 \end{bmatrix}, V_3 = \begin{bmatrix} 1 & 0 \\ 3 & 2 \end{bmatrix}$.

Then, $aV_1 + bV_2 + cV_3 = \begin{bmatrix} 0 & 0 \\ 0 & 0 \end{bmatrix} \Rightarrow \begin{bmatrix} a+b+c=0 & a-b=0 \\ b+3c=0 & -a+2c=0 \end{bmatrix} \Rightarrow a=b=c=0 \Rightarrow$

V_1, V_2, V_3 linearly independent by definition.

2. Let $V_1 = \begin{bmatrix} 2 & -3 \\ 4 & 2 \end{bmatrix}, V_2 = \begin{bmatrix} 1 & -1 \\ 3 & 3 \end{bmatrix}, V_3 = \begin{bmatrix} -1 & 3 \\ 1 & 5 \end{bmatrix}$.

Then, $aV_1 + bV_2 + cV_3 = \begin{bmatrix} 0 & 0 \\ 0 & 0 \end{bmatrix} \Rightarrow \begin{bmatrix} 2a+b-c=0 & -3a-b+3c=0 \\ 4a+3b+c=0 & 2a+3b+5c=0 \end{bmatrix} \Rightarrow$

$a=-2$, $b=3$, $c=-1 \Rightarrow V_3 = -2V_1 + 3V_2$.

3. $aV_1 + bV_2 + cV_3 + dV_4 = \begin{bmatrix} 0 & 0 \\ 0 & 0 \end{bmatrix} \Rightarrow \begin{bmatrix} -a+3b-d=0 & a+2c=0 \\ -2a+b+-3c-d=0 & 2a+b+c+7d=0 \end{bmatrix} \Rightarrow$

$a=4$, $b=1$, $c=-2$, $d=-1 \Rightarrow V_4 = 4V_1 + V_2 - 2V_3$.

4. $aV_1 + bV_2 + cV_3 + dV_4 = \begin{bmatrix} 0 & 0 \\ 0 & 0 \end{bmatrix} \Rightarrow \begin{bmatrix} a+b+d=0 & a+c+d=0 \\ b+c+d=0 & a+b+c=0 \end{bmatrix} \Rightarrow$

$a=b=c=0 \Rightarrow V_1$, V_2, V_3 linearly independent.

5. Let $p(x)=x$, $q(x)=1+x$, then $a\,p(x)+b\,q(x)=a(x)+b(1+x)=b+(a+b)x=0 \Rightarrow$

$b=0 \;\; a+b=0 \;\Rightarrow a=b=0 \Rightarrow p(x)$, $q(x)$ linearly independent.

6. Let $p(x)=1+x$, $q(x)=1+x^2$, $r(x)=1-x+x^2$.

Then $a\,p(x)+b\,q(x)+c\,r(x)=(a+b+c)+(a-c)x+(b+c)x^2=0 \Rightarrow$

$a+b+c=0$, $a-c=0$, $b+c=0 \;\Rightarrow a=b=c=0 \Rightarrow$ linearly independent.

7. $a\,p(x)+b\,q(x)+c\,r(x)=(a+2b+3c)x+(-b+2c)x^2=0 \Rightarrow$

$0=0$, $a+2b+3c=0$, $-b+2c=0 \;\Rightarrow a=7$, $b=-2$, $c=-1 \Rightarrow r(x)=7p(x)-2q(x)$.

Since we had only 2 equations and 3 unknowns a was free so there are infinitely many solutions.

8. $a\,p(x)+b\,q(x)+c\,r(x)+d\,s(x)=(c+2d)+(2a+b)x+(-b-c)x^2+(c+d)x^3=0 \Rightarrow$

$\begin{matrix} c+2d=0 & 2a+b=0 \\ -b-c=0 & c+d=0 \end{matrix} \Rightarrow a=b=c=d=0 \Rightarrow p(x)$, $q(x)$, $r(x)$, $s(x)$ linearly independent.

9. $a\,p(x)+b\,q(x)+c\,r(x)+d\,s(x) = (a+c+3d)+(-2a+3b+2)x+(b+c)x^2+(-b+2c+3d)x^3 = 0 \Rightarrow$

$$\begin{matrix} a+c+3d=0 & -2a+3b+2=0 \\ b+c=0 & -b+2c+3d=0 \end{matrix} \Rightarrow a=b=c=d=0 \Rightarrow \text{linearly independent.}$$

10. $a + b\,\sin x + c\,\cos x = 0$ for all $x \Rightarrow$

$$\begin{matrix} x=0 \Rightarrow a+c=0 \\ x=\frac{\pi}{2} \Rightarrow a+b=0 \\ x=\pi \Rightarrow a-c=0 \end{matrix} \Rightarrow a=b=c=0 \Rightarrow \{1, \sin x, \cos x\} \text{ linearly independent.}$$

11. $1 = \sin^2 x + \cos^2 x \Rightarrow \{1, \sin^2 x, \cos^2 x\}$ linearly dependent.

12. $a\,e^x + b\,e^{-x} = 0$ for all $x \Rightarrow$

$$\begin{matrix} x=\ln 1 \Rightarrow a+b=0 \\ x=\ln 2 \Rightarrow 2a+\frac{1}{2}\,b=0 \end{matrix} \Rightarrow a=b=0 \Rightarrow \{e^x, e^{-x}\} \text{ linearly independent.}$$

13. $\ln(x^2) = -2\ln 2 \cdot 1 + 2 \cdot \ln(2x) \Rightarrow \{1, \ln(2x), \ln(x^2)\}$ linearly dependent.

14. $a\,\sin x + b\,\sin 2x + c\,\sin 3x = 0$ for all $x \Rightarrow$

$x=\frac{\pi}{2} \Rightarrow a-c=0,\; x=\frac{\pi}{6} \Rightarrow \frac{1}{2}a+\frac{\sqrt{3}}{2}b+c=0,\; x=\frac{\pi}{3} \Rightarrow \frac{\sqrt{3}}{2}a-\frac{\sqrt{3}}{2}b=0$

$\Rightarrow a=b=c=0 \Rightarrow \{\sin x, \sin 2x, \sin 3x\}$ linearly independent.

15. $W(x) \neq 0$ for some $x \Rightarrow f(x),\ g(x)$ linearly independent $\Leftrightarrow$
$f(x),\ g(x)$ linearly dependent $\Rightarrow W(x) = g'(x)f(x) - f'(x)g(x) = 0$ for all x.
This is what we will show.

$f(x), g(x)$ linearly dependent $\Rightarrow f(x)+ag(x) = 0,\ a \neq 0 \Rightarrow\Rightarrow f'(x)+a'g(x) = 0 \Rightarrow a = -\frac{f'(x)}{g'(x)}$.

Substituting this value into $f(x) + ag(x) = 0 \Rightarrow f(x) - \frac{f'(x)}{g'(x)}g(x) = 0 \Rightarrow$
$W(x) = g'(x)f(x) - f'(x)g(x) = 0$ for all x.

16. To prove linear independence, we need only one x such that $W(x) \neq 0$.
When we have already demonstrated linear dependence, we should show $W(x) = 0$ for all x.

So, for 10, 12, and 14, we need to find a value of x such that $W(x) \neq 0$.
For 11 and 13, we need to show $W(x) = 0$ for all x.

$$10.\ W(x) = \begin{vmatrix} 1 & \sin x & \cos x \\ 0 & \cos x & -\sin x \\ 0 & -\sin x & -\cos x \end{vmatrix}, \text{ so } W(0) = \begin{vmatrix} 1 & 0 & 1 \\ 0 & 1 & 0 \\ 0 & 0 & -1 \end{vmatrix} = -1.$$

$$11.\ W(x) = \begin{vmatrix} 1 & \sin^2 x & \cos^2 x \\ 0 & \sin(2x) & -\sin(2x) \\ 0 & 2\cos(2x) & -2\cos(2x) \end{vmatrix} = \begin{vmatrix} \sin(2x) & -\sin(2x) \\ 2\cos(2x) & -2\cos(2x) \end{vmatrix} = 0.$$

$$12.\ W(x) = \begin{vmatrix} e^x & e^{-x} \\ e^x & -e^{-x} \end{vmatrix} = e^x(-e^{-x}) - e^x(e^{-x}) = -2 \text{ (this is true for all } x\text{)}.$$

$$13.\ W(x) = \begin{vmatrix} 1 & \ln(2x) & \ln(x^2) \\ 0 & 1/x & 2/x \\ 0 & -1/x^2 & -2/x^2 \end{vmatrix} = \begin{vmatrix} 1/x & 2/x \\ -1/x^2 & -2/x^2 \end{vmatrix} = 0.$$

$$14.\ W(x) = \begin{vmatrix} \sin x & \sin(2x) & \sin(3x) \\ \cos x & 2\cos(2x) & 3\cos(3x) \\ -\sin x & -4\sin(2x) & -9\sin(3x) \end{vmatrix}, \text{ so } W(\tfrac{\pi}{6}) = \begin{vmatrix} 1/2 & \sqrt{3}/2 & 1 \\ \sqrt{3}/2 & 1 & 0 \\ -1/2 & -2\sqrt{3} & -9 \end{vmatrix} = -\tfrac{1}{4}.$$

17. (a) $a(\mathbf{u}+\mathbf{v}) + b(\mathbf{v}+\mathbf{w}) + c(\mathbf{u}+\mathbf{w}) = (a+c)\mathbf{u} + (a+b)\mathbf{v} + (b+c)\mathbf{w} = \mathbf{0} = 0 \Rightarrow$
$a+c=0,\ a+b=0,\ b+c=0 \Rightarrow a=b=c=0 \Rightarrow$
$\mathbf{u}+\mathbf{v},\ \mathbf{v}+\mathbf{w},\ \mathbf{u}+\mathbf{w}$ are linearly independent.

(b) $a(\mathbf{u}-\mathbf{v}) + b(\mathbf{v}-\mathbf{w}) + c(\mathbf{u}-\mathbf{w}) = (a+c)\mathbf{u} + (-a+b)\mathbf{v} + (-b-c)\mathbf{w} = \mathbf{0} = 0 \Rightarrow$
$a+c=0,\ -a+b=0,\ -b-c=0 \Rightarrow a=b=1,\ c=-1$ is a solution
$\Rightarrow \mathbf{u}-\mathbf{v},\ \mathbf{v}-\mathbf{w},\ \mathbf{u}-\mathbf{w}$ are linearly dependent.
Note, a was free so there are infinitely many solutions.
Eg.: Let $\mathbf{u}=1,\ \mathbf{v}=x,\ \mathbf{w}=x^2$, then $\mathbf{u}-\mathbf{v}=1-x,\ \mathbf{v}-\mathbf{w}=x-x^2,\ \mathbf{u}-\mathbf{w}=1-x^2$.
Since $1-x^2 = 1(1-x)+1(x-x^2)$, $\mathbf{u}-\mathbf{v},\ \mathbf{v}-\mathbf{w},\ \mathbf{u}-\mathbf{w}$ are clearly linearly dependent.

18. Number of linearly independent vectors in $\mathcal{B} \leq 3 < 4 = \dim M_{22} \Rightarrow \mathcal{B}$ is not a basis for M_{22}.
Note $\dim M_{22} = 4$ because $E_{11},\ E_{12},\ E_{21},\ E_{22}$ are a basis for it.

19. Let $V_1 = \begin{bmatrix} 1 & 0 \\ 0 & 1 \end{bmatrix}, V_2 = \begin{bmatrix} 0 & -1 \\ 1 & 0 \end{bmatrix}, V_3 = \begin{bmatrix} 1 & 1 \\ 1 & 1 \end{bmatrix}, V_4 = \begin{bmatrix} 1 & 1 \\ 1 & -1 \end{bmatrix}$. So,

$$aV_1 + bV_2 + cV_3 + dV_4 = \mathbf{0} \Rightarrow \begin{bmatrix} a+c+d=0 & -b+c+d=0 \\ b+c+d=0 & a+c-d=0 \end{bmatrix} \Rightarrow a=b=c=d=0 \Rightarrow$$

$V_1,\ V_2,\ V_3,\ V_4$ linearly independent $\Rightarrow$
The number of linearly independent vectors in $\mathcal{B} = 4 = \dim M_{22} \Rightarrow \mathcal{B}$ is a basis for M_{22}.

20. $V_4 = 2V_1 + V_2 - V_3 \Rightarrow$ number of linearly independent vectors in $\mathcal{B} \leq 3 < 4 = \dim M_{22} \Rightarrow$
$\mathcal{B}$ is not a basis for M_{22}.

21. The number of vectors in $\mathcal{B} = 5 > 4 = \dim M_{22} \Rightarrow \mathcal{B}$ is not a basis for M_{22}.

22. Let $p(x) = x$, $q(x) = 1 + x$, $r(x) = x - x^2$.

Then $a\,p(x) + b\,q(x) + c\,r(x) = b + (a + b + c)x + (-c)x^2 = 0 \Rightarrow$

$b = 0$, $a + b + c = 0$, $-c = 0 \Rightarrow a = b = c = 0 \Rightarrow$ linearly independent.

$\Rightarrow$ the number of linearly independent vectors in $\mathcal{B} = 3 = \dim \mathscr{P}_2 \Rightarrow \mathcal{B}$ is a basis for $\mathscr{P}_2$.

Note $\dim \mathscr{P}_2 = 3$ since $\{1, x, x^2\}$ is a basis for it.

23. $x - x^2 = -(1 - x) + (1 - x^2) \Rightarrow$ number of linearly independent vectors in $\mathcal{B} \leq 2 < \dim \mathscr{P}_2 \Rightarrow$ $\mathcal{B}$ is not a basis for $\mathscr{P}_2$.

24. The number of vectors in $\mathcal{B} = 2 < 3 = \dim \mathscr{P}_2 \Rightarrow \mathcal{B}$ is not a basis for $\mathscr{P}_2$.

25. The number of vectors in $\mathcal{B} = 4 > 3 = \dim \mathscr{P}_2 \Rightarrow \mathcal{B}$ is not a basis for $\mathscr{P}_2$.

26. $A = \begin{bmatrix} 1 & 2 \\ 3 & 4 \end{bmatrix} = a\,E_{12} + b\,E_{11} + c\,E_{22} + d\,E_{21} \Rightarrow \begin{bmatrix} a = 2 & b = 1 \\ c = 4 & d = 3 \end{bmatrix} \Rightarrow$

$$A = 2\,E_{11} + 1\,E_{11} + 4\,E_{22} + 3\,E_{21} \Rightarrow [A]_{\mathcal{B}} = \begin{bmatrix} 2 \\ 1 \\ 4 \\ 3 \end{bmatrix}.$$

27. Let $V_1 = \begin{bmatrix} 1 & 0 \\ 0 & 0 \end{bmatrix}$, $V_2 = \begin{bmatrix} 1 & 1 \\ 0 & 0 \end{bmatrix}$, $V_3 = \begin{bmatrix} 1 & 1 \\ 1 & 0 \end{bmatrix}$, $V_4 = \begin{bmatrix} 1 & 1 \\ 1 & 1 \end{bmatrix}$.

Then $A = \begin{bmatrix} 1 & 2 \\ 3 & 4 \end{bmatrix} = a\,V_1 + b\,V_2 + c\,V_3 + d\,V_4 \Rightarrow \begin{bmatrix} a + b + c + d = 1 & b + c + d = 2 \\ c + d = 3 & d = 4 \end{bmatrix} \Rightarrow$

$$A = -V_1 - V_2 - V_3 + 4\,V_4 \Rightarrow [A]_{\mathcal{B}} = \begin{bmatrix} -1 \\ -1 \\ -1 \\ 4 \end{bmatrix}.$$

28. $p(x) = 2 - x + 3x^2 = a(1 + x) + b(1 - x) + cx^2 = (a + b) + (a - b)x + cx^2 \Rightarrow$

$$a + b = 2,\ a - b = -1,\ c = 3 \Rightarrow a = \tfrac{1}{2},\ b = \tfrac{3}{2},\ c = 3 \Rightarrow [p(x)]_{\mathcal{B}} = \begin{bmatrix} 1/2 \\ 3/2 \\ 3 \end{bmatrix}.$$

29. $p(x) = 2 - x + 3x^2 = a(1) + b(1 + x) + c(-1 + x^2) = (a + b - c) + bx + cx^2 \Rightarrow$

$$a + b - c = 2,\ b = -1,\ c = 3 \Rightarrow a = 6,\ b = -1,\ c = 3 \Rightarrow [p(x)]_{\mathcal{B}} = \begin{bmatrix} 6 \\ -1 \\ 3 \end{bmatrix}.$$

30. We need to show $\operatorname{span}(\mathcal{B}) = \operatorname{span}(\mathbf{u}_i) = V$ and $\mathbf{u}_i$ are linearly independent.

Every vector in V can be written as a linear combination of the vectors in $\mathcal{B} \Rightarrow \operatorname{span}(\mathcal{B}) = V$.
$\mathbf{0}$ can be written as $\sum 0\mathbf{u}_i$ but the uniqueness of this representation $\Rightarrow$
$\sum c_i\mathbf{u}_i = \mathbf{0} \Rightarrow c_i = 0 \Rightarrow \mathbf{u}_i$ are linearly independent.

31. $[c_1\mathbf{u}_1 + \ldots + c_n\mathbf{u}_n]_{\mathcal{B}} = [c_1\mathbf{u}_1]_{\mathcal{B}} + \ldots + [c_n\mathbf{u}_n]_{\mathcal{B}} = c_1[\mathbf{u}_1]_{\mathcal{B}} + \ldots + c_n[\mathbf{u}_n]_{\mathcal{B}}$.

32. We only need to show $c_1\mathbf{u}_1 + \ldots + c_n\mathbf{u}_n = \mathbf{0} \Rightarrow c_i = 0$ for all i.
$c_1\mathbf{u}_1 + \ldots + c_n\mathbf{u}_n = \mathbf{0} \Rightarrow 0 = [c_1\mathbf{u}_1 + \ldots + c_n\mathbf{u}_n]_{\mathcal{B}}$ by definition.
Now, Exercise 31 $\Rightarrow 0 = c_1[\mathbf{u}_1 + \ldots + c_n\mathbf{u}_n]_{\mathcal{B}} = c_1[\mathbf{u}_1]_{\mathcal{B}} + \ldots + c_n[\mathbf{u}_n]_{\mathcal{B}} \Rightarrow$
$c_i = 0$ because $[\mathbf{u}_i]_{\mathcal{B}}$ are linearly independent in $\mathbb{R} \Rightarrow \mathbf{u}_i$ are linearly independent in V.

33. Let span$(S) = \mathbb{R}^n$ and let $\mathbf{v} \in V \Rightarrow \mathbf{x} = [\mathbf{v}]_{\mathcal{B}} = \sum c_i[\mathbf{u}_i]_{\mathcal{B}} = \sum [c_i\mathbf{u}_i]_{\mathcal{B}} \Rightarrow \mathbf{v} = \sum c_i\mathbf{u}_i$ because the representation of $\mathbf{v}$ with respect to basis $\mathcal{B}$ is unique.

Therefore, $V = \text{span}(\mathbf{u}_i)$.

Let $V = \text{span}(\mathbf{u}_i)$ and let $\mathbf{x} = \begin{bmatrix} 1 \\ c_2 \\ \vdots \\ c_n \end{bmatrix} \in \mathbb{R}^n \Rightarrow \sum c_i\mathbf{u}_i = \mathbf{v} \in V$ because $V = \text{span}(\mathbf{u}_i)$.

Therefore, $\mathbf{x} = [\mathbf{v}]_{\mathcal{B}} = \sum [c_i\mathbf{u}_i]_{\mathcal{B}} = \sum c_i[\mathbf{u}_i]_{\mathcal{B}} \Rightarrow \mathbf{x} \in \text{span}(S)$.

Therefore, span$(S) = \mathbb{R}^n$.

34. $p(0) = a(0)^2 + b(0) + c = 0 \Rightarrow c = 0$, a, b free $\Rightarrow \mathcal{B} = \{x, x^2\}$ is a basis $\Rightarrow \dim V = 2$.

35. $p(1) = a(1)^2 + b(1) + c = 0 \Rightarrow a = -(b+c)$, b, c free $\Rightarrow$
$\mathcal{B} = \{1 - x, 1 - x^2\}$ is a basis $\Rightarrow \dim V = 2$.

36. $xp'(x) = x(2ax + b) = p(x) = ax^2 + bx + c \Rightarrow a = c = 0$, b free $\Rightarrow$
$\mathcal{B} = \{x\}$ is a basis $\Rightarrow \dim V = 1$.

37. $\mathcal{B} = \{E_{11}, E_{12}, E_{22}\}$ is a basis $\Rightarrow \dim V = 3$.

38. Recall *skew* symmetric means $A^T = -A \Rightarrow \begin{bmatrix} a & c \\ b & d \end{bmatrix} = -\begin{bmatrix} a & b \\ c & d \end{bmatrix} \Rightarrow$

$a = d = 0$, $c = -b$, b free $\Rightarrow \mathcal{B} = \left\{\begin{bmatrix} 0 & 1 \\ -1 & 0 \end{bmatrix}\right\}$ is a basis $\Rightarrow \dim V = 1$.

39. $AB = BA \Rightarrow \begin{bmatrix} a & a+b \\ c & c+d \end{bmatrix} = \begin{bmatrix} a+c & b+d \\ c & d \end{bmatrix} \Rightarrow$

$c = 0$, $a = d$, a, b free $\Rightarrow \mathcal{B} = \left\{\begin{bmatrix} 1 & 0 \\ 0 & 1 \end{bmatrix}, \begin{bmatrix} 0 & 1 \\ 0 & 0 \end{bmatrix}\right\}$ is a basis $\Rightarrow \dim V = 2$.

40. $\mathcal{B} = \{E_{ij}\}$ is a basis for $M_{nn} \Rightarrow \dim M_{nn} = n^2$. Now, for symmetric matrices, we have: $A^T = A \Rightarrow a_{ij} = a_{ji}$ reduces the off-diagonal free variables by a factor of $\frac{1}{2}$.
So, since there are n diagonal free variables, we have $n^2 - \frac{1}{2}(n^2 - n) = \frac{n^2+n}{2}$.

41. We should begin by noting that we need to count the number of free variables.
We start with n^2 free variables from M_{nn}, then reduce that number by applying the conditions.

$A^T = -A \Rightarrow a_{ii} = 0$ which reduces the number of free variables by n to $n^2 - n$.
Then $A^T = -A \Rightarrow a_{ij} = -a_{ji}$ cuts the remaining number in half, so $\dim V = \frac{n^2-n}{2}$.

42. We need to prove *Grassmann's Identity*: $\dim(U+W) = \dim U + \dim W - \dim(U \cap W)$.

We will follow the hint: Let $\mathcal{B} = \{\mathbf{v}_1, \dots, \mathbf{v}_k\}$ be a basis for $U \cap W$.
Then extend $\mathcal{B}$ to a basis $\mathcal{C}$ of U and a basis $\mathcal{D}$ of W.
We will show $C \cup D$ is a basis for $U + W$.

Clearly, $\text{span}(C \cup D) = U + W$, so we need only show $C \cup D$ is linearly independent.
If $\dim U = m$ and $\dim W = n$, we have: $\mathcal{C} = \{\mathbf{u}_{k+1}, \dots, \mathbf{u}_m\} \cup \mathcal{B}$ and $\mathcal{D} = \{\mathbf{w}_{k+1}, \dots, \mathbf{w}_n\} \cup \mathcal{B}$.
Since $\mathcal{C}$ and $\mathcal{D}$ are bases, $\{\mathbf{u}_{k+1}, \dots, \mathbf{u}_m\} \cap \mathcal{B} = \emptyset$ and $\{\mathbf{w}_{k+1}, \dots, \mathbf{w}_n\} \cap \mathcal{B} = \emptyset$.

Therefore, $\mathcal{C} \cap \mathcal{D} = \mathcal{B}$. Why?
Clearly, $\mathcal{B} \subseteq \mathcal{C} \cap \mathcal{D}$, so we need only show $\mathcal{C} \cap \mathcal{D} \subseteq \mathcal{B}$.

If $\mathbf{v} \in \mathcal{C} \cap \mathcal{D}$, then $\mathbf{v} \in U \cap W$, therefore $\mathbf{v} = \sum_{i=1}^{k} c_1 \mathbf{u}_i$ because $\mathcal{B}$ is a basis for $U \cap W$.
But since $\mathbf{v} \in (\mathcal{C} \cap \mathcal{B}) \cup (\mathcal{D} \cap \mathcal{B})$, $\mathbf{v}$ must equal one of the $\mathbf{u}_i$.

So, $\mathcal{C} \cup \mathcal{D}$ is a basis for $U + W$. Therefore, $\dim(U + W) = \{\text{the number of vectors in } \mathcal{C} \cup \mathcal{D}\}$.

Since $\mathcal{C} \cap \mathcal{D} = \mathcal{B}$, the number of vectors in $\mathcal{C} \cup \mathcal{D}$ is
the number of vectors in $\mathcal{C} \cup \mathcal{D}$ minus the number of vectors in $\mathcal{B}$.
We subtract the number of vectors in $\mathcal{B}$ because they are counted twice in the intersection.

So, $\dim(U + W) = m + n - k = \dim U + \dim W - \dim(U \cap W)$.

Q: Why must $\{\mathbf{u}_{k+1}, \dots, \mathbf{u}_m\} \cap \mathcal{B} = \emptyset$ and $\{\mathbf{w}_{k+1}, \dots, \mathbf{w}_n\} \cap \mathcal{B} = \emptyset$?
A: If there were any vectors in common, $\mathcal{C}$ and $\mathcal{D}$ would not be linearly independent.

Q: Why is it possible to extend $\mathcal{B}$ to a basis $\mathcal{C}$ of U and a basis $\mathcal{D}$ of W?
A: By Thm 6.10e: Any linearly independent set in V can be extended to a basis for V.

43. We will find bases, then use those bases to compute the dimension.
Let $\dim U = m$ with basis $\mathcal{B} = \{\mathbf{u}_1, \dots, \mathbf{u}_m\}$ and $\dim V = n$ with basis $\mathcal{C} = \{\mathbf{v}_1, \dots, \mathbf{v}_n\}$.

(a) We will show that $\mathcal{D} = \{(\mathbf{u}_i, \mathbf{0}), (\mathbf{0}, \mathbf{v}_j)\}$ is a basis for $U \times V$.
This will imply that $\dim(U \times V) = \dim U + \dim V$.

spans: We will show that $\text{span}(\mathcal{D}) = U \times V$. Let $\mathbf{x} = (\mathbf{u}, \mathbf{v}) \in U \times V$ then:

$$\mathbf{x} = (\mathbf{u}, \mathbf{v}) = \left(\sum_{i=1}^{m} c_i \mathbf{u}_i, \sum_{j=1}^{n} d_j \mathbf{v}_j\right) = \sum_{i=1}^{m} c_i (\mathbf{u}_i, \mathbf{0}) + \sum_{j=1}^{n} d_j (\mathbf{0}, \mathbf{v}_j).$$

lin ind: We will show that $\mathcal{D}$ is linearly independent.
That is, we will show if $\mathbf{0} = (\mathbf{0}, \mathbf{0}) = \sum_{i=1}^{m} c_i (\mathbf{u}_i, \mathbf{0}) + \sum_{j=1}^{n} d_j (\mathbf{0}, \mathbf{v}_j)$, then $c_i = d_j = 0$.

$$\mathbf{0} = (\mathbf{0}, \mathbf{0}) = \sum_{i=1}^{m} c_i (\mathbf{u}_i, \mathbf{0}) + \sum_{j=1}^{n} d_j (\mathbf{0}, \mathbf{v}_j) = \left(\sum_{i=1}^{m} c_i \mathbf{u}_i = \mathbf{0}, \sum_{j=1}^{n} d_j \mathbf{v}_j = \mathbf{0}\right)$$

Since $\mathbf{u}_i$ and $\mathbf{v}_j$ are linearly independent, $\sum_{i=1}^{m} c_i \mathbf{u}_i = \mathbf{0}$, $\sum_{j=1}^{n} d_j \mathbf{v}_j = \mathbf{0}$
implies $c_i = d_j = 0$, as we were to show.

(b) Let $\dim W = k$ with basis $\mathcal{B} = \{\mathbf{w}_1, \dots, \mathbf{w}_k\}$. Then:
Similar to (a), $\mathcal{C} = \{(\mathbf{w}_i, \mathbf{w}_i)\}$ is a basis for Δ, so $\dim \Delta = \dim W$.

44. Clearly span$(x^0 = 1, x, \ldots, x^n, \ldots) = \mathscr{P}$. Furthermore, $\sum a_i x^i = 0 \Leftrightarrow a_i = 0$ for all i.

45. $a(1+x) + b(1+x+x^2) + c(1) = (a+b+c) + (a+b)x + bx^2 = 0 \Rightarrow$
$a+b+c = 0$, $a+b = 0$, $b = 0 \Rightarrow$
$a = b = c = 0 \Rightarrow \mathcal{B} = \{1, 1+x, 1+x+x^2\}$ linearly independent $\Rightarrow$
$\mathcal{B}$ is a basis for $\mathscr{P}_2$ because $\dim \mathscr{P}_2 = 3 =$ the number of vectors in $\mathcal{B}$.

46. Let $V_1 = \begin{bmatrix} 0 & 1 \\ 0 & 1 \end{bmatrix}, V_2 = \begin{bmatrix} 1 & 1 \\ 0 & 1 \end{bmatrix}$. Then, $a\,V_1 + b\,V_2 + +c\,E_{21} + d\,E_{22} = 0 \Rightarrow$

$$\begin{bmatrix} b = 0 & a+b = 0 \\ c = 0 & a+b+d = 0 \end{bmatrix} \Rightarrow a = b = c = d = 0 \Rightarrow$$

$\mathcal{B} = \{V_1, V_2, E_{21}, E_{22}\}$ linearly independent $\Rightarrow$

$\mathcal{B}$ is a basis for M_{22} because $\dim M_{22} = 4 =$ the number of vectors in $\mathcal{B}$.

47. Let $V_1 = \begin{bmatrix} 1 & 0 \\ 0 & 1 \end{bmatrix}, V_2 = \begin{bmatrix} 0 & 1 \\ 1 & 0 \end{bmatrix}, V_3 = \begin{bmatrix} 0 & -1 \\ 1 & 0 \end{bmatrix}$.

Then, $a\,V_1 + b\,V_2 + +c\,V_3 + d\,E_{11} = 0 \Rightarrow \begin{bmatrix} a+d = 0 & b-c = 0 \\ b+c = 0 & a = 0 \end{bmatrix} \Rightarrow$

$a = b = c = d = 0 \Rightarrow \mathcal{B} = \{V_1, V_2, V_3, E_{11}\}$ linearly independent $\Rightarrow$
$\mathcal{B}$ is a basis for M_{22} because $\dim M_{22} = 4 =$ the number of vectors in $\mathcal{B}$.

48. $V =$ vector space of symmetric $2 \times 2 = \left\{ \begin{bmatrix} a & b \\ b & d \end{bmatrix} \right\} \Rightarrow$

$\left\{ E_{11}, \begin{bmatrix} 0 & 1 \\ 1 & 0 \end{bmatrix}, E_{22} \right\}$ is a basis $\Rightarrow \dim V = 3$.

Let $V_1 = \begin{bmatrix} 1 & 0 \\ 0 & 1 \end{bmatrix}, V_2 = \begin{bmatrix} 0 & 1 \\ 1 & 0 \end{bmatrix}$. Then, $a\,V_1 + b\,V_2 + c\,E_{22} = 0 \Rightarrow \begin{bmatrix} a = 0 & b = 0 \\ b = 0 & a+c = 0 \end{bmatrix} \Rightarrow$

$a = b = c = 0 \Rightarrow \mathcal{B} = \{V_1, V_2, E_{22}\}$ linearly independent $\Rightarrow$
$\mathcal{B}$ is a basis for V because $\dim V = 3 =$ the number of vectors in $\mathcal{B}$.

49. We will show span$(1, 1+x, 2x) =$ span$(1, 1+x)$ and $\{1, 1+x\}$ linearly independent which will imply $\{1, 1+x\}$ is a basis for span$(1, 1+x, 2x)$.

$2x = -2(1) + 2(1+x) \Rightarrow$ span$(1, 1+x, 2x) =$ span$(1, 1+x)$.

$a(1) + b(1+x) = (a+b) + bx = 0 \Rightarrow a+b = 0,\ b = 0 \Rightarrow a = b = 0 \Rightarrow$
$\{1, 1+x\}$ linearly independent. Therefore, $\{1, 1+x\}$ is a basis for span$(1, 1+x, 2x)$.

50. We will show span$(1-2x, 2x-x^2, 1-x^2, 1+x^2) =$ span$(1-2x, 2x-x^2, 1+x^2)$ and $\{1-2x, 2x-x^2, 1+x^2\}$ linearly independent which will imply
$\{1-2x, 2x-x^2, 1+x^2\}$ is a basis for span$(1-2x, 2x-x^2, 1-x^2, 1+x^2)$.

$1-x^2 = 1(1-2x) + 1(2x-x^2) \Rightarrow$ span$(1-2x, 2x-x^2, 1-x^2, 1+x^2) =$ span$(1-2x, 2x-x^2, 1+x^2)$.

$a(1-2x) + b(2x-x^2) + c(1+x^2) = (a+c) + 2(-a+b)x + (-b+c)x^2 = 0 \Rightarrow$
$\Rightarrow a = b = c = 0 \Rightarrow \{1-2x, 2x-x^2, 1+x^2\}$ linearly independent.

Therefore, $\{1-2x, 2x-x^2, 1+x^2\}$ is a basis for span$(1-2x, 2x-x^2, 1-x^2, 1+x^2)$.

51. We will show $\text{span}(1-x, x-x^2, 1-x^2, 1-2x+x^2) = \text{span}(1-x, x-x^2)$ and $\{1-x, x-x^2\}$ linearly independent which will imply
$\{1-x, x-x^2\}$ is a basis for $\text{span}(1-x, x-x^2, 1-x^2, 1-2x+x^2)$.

$1-x^2 = 1(1-x) + 1(x-x^2)$ and $1-2x+x^2 = 1(1-x) - (x-x^2) \Rightarrow$
$\text{span}(1-x, x-x^2, 1-x^2, 1-2x+x^2) = \text{span}(1-x, x-x^2)$.

$a(1-x) + b(x-x^2) = a + (-a+b)x + (-b)x^2 = 0 \Rightarrow$
$\Rightarrow a = b = 0 \Rightarrow \{1-x, x-x^2\}$ linearly independent.

Therefore, $\{1-x, x-x^2\}$ is a basis for $\text{span}(1-x, x-x^2, 1-x^2, 1-2x+x^2)$.

52. Let $V_1 = \begin{bmatrix} 1 & 0 \\ 0 & 1 \end{bmatrix}, V_2 = \begin{bmatrix} 0 & 1 \\ 1 & 0 \end{bmatrix}, V_3 = \begin{bmatrix} -1 & 1 \\ 1 & -1 \end{bmatrix}, V_4 = \begin{bmatrix} 1 & -1 \\ -1 & 1 \end{bmatrix}$.

We will show $\text{span}(V_1, V_2, V_3, V_4) = \text{span}(V_1, V_2)$ and $\{V_1, V_2\}$ linearly independent which will imply $\{V_1, V_2\}$ is a basis for $\text{span}(V_1, V_2, V_3, V_4)$.

$V_3 = V_2 - V_1$ and $V_4 = V_1 - V_2 \Rightarrow \text{span}(V_1, V_2, V_3, V_4) = \text{span}(V_1, V_2)$.

$a\,V_1 + b\,V_2 = 0 \Rightarrow a = b = 0 \Rightarrow \{V_1, V_2\}$ linearly independent $\Rightarrow$
$\{V_1, V_2\}$ is a basis for $\text{span}(V_1, V_2, V_3, V_4)$.

53. We will show $\text{span}(\sin^2 x, \cos^2 x, \cos 2x) = \text{span}(\sin^2 x, \cos^2 x)$ and
$\{\sin^2 x, \cos^2 x\}$ linearly independent which will imply
$\{\sin^2 x, \cos^2 x\}$ is a basis for $\text{span}(\sin^2 x, \cos^2 x, \cos 2x)$.

$\cos 2x = \cos^2 x - \sin^2 x \Rightarrow \text{span}(\sin^2 x, \cos^2 x, \cos 2x) = \text{span}(\sin^2 x, \cos^2 x)$.

$a\sin^2 x + b\cos^2 x = 0 \Rightarrow x = 0 \Rightarrow b = 0$ and $x = \frac{\pi}{2} \Rightarrow a = 0 \Rightarrow a = b = 0 \Rightarrow$
$\{\sin^2 x, \cos^2 x\}$ linearly independent $\Rightarrow \{\sin^2 x, \cos^2 x\}$ is a basis for $\text{span}(\sin^2 x, \cos^2 x, \cos 2x)$.

54. Note, proving $\mathbf{v} \notin \text{span}(S) \Rightarrow \{\mathbf{v}\} \cup S$ linearly independent is the same thing as proving $\{\mathbf{v}\} \cup S$ linearly dependent $\Rightarrow \mathbf{v} \in \text{span}(S)$.

$\{\mathbf{v}\} \cup S$ linearly dependent $\Rightarrow c\mathbf{v} + \sum c_i \mathbf{v}_i = 0$ with $c \neq 0$ and at least one $c_i \neq 0 \Rightarrow$
$\mathbf{v} + \sum \frac{c_i}{c} \mathbf{v}_i = 0 \Rightarrow \mathbf{v} = -\sum c_i \mathbf{v}_i \Rightarrow \mathbf{v} \in \text{span}(S)$ by definition.

Therefore, $\{\mathbf{v}\} \cup S = \{\mathbf{v}_1, \ldots, \mathbf{v}_n, \mathbf{v}\}$ linearly independent as required.

55. Must show $\mathbf{v} \in V \Leftrightarrow$ there exist c_i such that $\mathbf{v} = c_1\mathbf{v}_1 + \ldots + c_{n-1}\mathbf{v}_{n-1}$.
Note this will prove that $V = \text{span}(S) = \text{span}(S')$.

$V = \text{span}(S) \Rightarrow \mathbf{v} = c_1\mathbf{v}_1 + \ldots + c_{n-1}\mathbf{v}_{n-1} + c_n\mathbf{v}_n$.
$\mathbf{v}_n \in \text{span}(S') \Rightarrow \mathbf{v}_n = d_1\mathbf{v}_1 + \ldots + d_{n-1}\mathbf{v}_{n-1}$. Therefore, substitution gives us:

$$\begin{aligned}\mathbf{v} &= c_1\mathbf{v}_1 + \ldots + c_{n-1}\mathbf{v}_{n-1} + c_n(d_n\mathbf{v}_1 + \ldots + d_{n-1}\mathbf{v}_{n-1}) \\ &= (c_1 + c_n d_1)\mathbf{v}_1 + \ldots + (c_{n-1} + c_n d_{n-1})\mathbf{v}_{n-1} \Rightarrow \mathbf{v} \in \text{span}(S') \text{ as required.}\end{aligned}$$

56. Let S be a spanning set for V, a vector space with $\dim V = n$.
We will show there exists a subset S' of S that is a basis for V.

Note, since S is a spanning set for V, Theorem 6.8b. $\Rightarrow S$ contains at least n vectors.
Furthermore, S must contain a subset S' of n linearly independent vectors
(otherwise S could not span V).

Order the vectors in S so that the first n vectors are linearly independent. Specifically, let $S = \{\mathbf{v}_1, \ldots, \mathbf{v}_n, \cdots\}$ and $S' = \{\mathbf{v}_1, \ldots, \mathbf{v}_n\}$ be a subset of n linearly independent vectors.

Since S' consists of exactly n vectors, Theorem 6.9 $\Rightarrow S'$ is a basis for V.

57. Note $c_i \neq 0 \Rightarrow$ we can divide by every c_i. We will show $V = \text{span}(c_1\mathbf{v}_1, \ldots, c_n\mathbf{v}_n)$ and $\{c_1\mathbf{v}_1, \ldots, c_n\mathbf{v}_n\}$ linearly independent which will imply $\{c_1\mathbf{v}_1, \ldots, c_n\mathbf{v}_n\}$ is a basis for V.

$\{\mathbf{v}_1, \ldots, \mathbf{v}_n\}$ is a basis for $V \Rightarrow \mathbf{v} = d_1\mathbf{v}_1 + \ldots + d_n\mathbf{v}_n = \frac{d_1}{c_1}\,c_1\mathbf{v}_1 + \ldots + \frac{d_n}{c_n}\,c_n\mathbf{v}_n \Rightarrow$
$V = \text{span}(c_1\mathbf{v}_1, \ldots, c_n\mathbf{v}_n)$.

$\mathbf{0} = d_1\,c_1\mathbf{v}_1 + \ldots + d_1\,c_n\mathbf{v}_n \Rightarrow d_ic_i = 0$ for all i because $\mathbf{v}_i$ are linearly independent $\Rightarrow$
$d_i = 0$ because c_i are all nonzero $\Rightarrow \{c_1\mathbf{v}_1, \ldots, c_n\mathbf{v}_n\}$ linearly independent.

Therefore, $\{c_1\mathbf{v}_1, \ldots, c_n\mathbf{v}_n\}$ is a basis for V.

58. By induction on number of vectors in the basis. Let $S_n^+ = \{\mathbf{v}_1, \mathbf{v}_1 + \mathbf{v}_2, \ldots, \mathbf{v}_1 + \ldots + \mathbf{v}_n\}$.
$n = 1$ Then $S_1 = \{\mathbf{v}_1\} = S_1^+$.
By the induction hypothesis, $\text{span}(S_{n-1}) = \text{span}(S_{n-1}^+)$.

We will show $S_n^+ = \{\mathbf{v}_1, \mathbf{v}_1 + \mathbf{v}_2, \ldots, \mathbf{v}_1 + \ldots + \mathbf{v}_n\}$ spans V and is linearly independent.

$\mathbf{v}_n = (\mathbf{v}_1 + \ldots + \mathbf{v}_n) - (\mathbf{v}_1 + \ldots + \mathbf{v}_{n-1}) \Rightarrow S_n^+$ spans V since $\text{span}(S_n^+) = \text{span}(S_{n-1}^+, \mathbf{v}_n)$.

Furthermore, since $\mathbf{v}_1 + \ldots + \mathbf{v}_n \notin \text{span}(S_{n-1}^+) = \text{span}(\mathbf{v}_1, \mathbf{v}_1 + \mathbf{v}_2, \ldots, \mathbf{v}_1 + \ldots + \mathbf{v}_{n-1}) \Rightarrow$
$S_n^+ = \{\mathbf{v}_1, \mathbf{v}_1 + \mathbf{v}_2, \ldots, \mathbf{v}_1 + \ldots + \mathbf{v}_n\}$ is linearly independent.

Therefore, $\{\mathbf{v}_1, \mathbf{v}_1 + \mathbf{v}_2, \ldots, \mathbf{v}_1 + \ldots + \mathbf{v}_n\}$ is a basis for V.

59. (a) $p_0(x) = \frac{(x-2)(x-3)}{(1-2)(1-3)} = \frac{1}{2}\,x^2 - \frac{5}{2}\,x + 3.$

$p_1(x) = \frac{(x-1)(x-3)}{(2-1)(2-3)} = -x^2 + 4x - 3.$

$p_2(x) = \frac{(x-1)(x-2)}{(3-1)(3-2)} = \frac{1}{2}x^2 - \frac{3}{2}x + 1.$

(b) $p_i(a_i) = \frac{(a_i-a_0)\ldots(a_i-a_{i-1})(a_i-a_{i+1})\ldots(a_i-a_n)}{(a_i-a_0)\ldots(a_i-a_{i-1})(a_i-a_{i+1})\ldots(a_i-a_n)} = 1.$

$p_i(a_j) = \frac{(a_j-a_0)\ldots(a_j-a_j)\ldots(a_j-a_n)}{(a_i-a_0)\ldots(a_i-a_j)\ldots(a_i-a_n)} = 0.$

60. (a) Given $c_0\,p_0(x) + \ldots + c_n\,p_n(x) = 0$ we want to show $c_i = 0$ for all i.
Exercise 59(b) $\Rightarrow 0 = c_0\,p_0(a_i) + \ldots + c_n\,p_n(a_i) = c_ip_i(a_i) = c_i$ for all $i \Rightarrow$
$\mathcal{B}$ is linearly independent.

(b) Since $\dim \mathscr{P}_n = n =$ number of linearly independent vectors in $\mathcal{B}$, $\mathcal{B}$ is a basis for $\mathscr{P}_n$.

61. (a) Exercise 59(b) $\Rightarrow q(a_i) = c_0\, p_0(a_i) + \ldots + c_n\, p_n(a_i) = c_i p_i(a_i) = c_i$ for all $i \Rightarrow$
$q(x) = q(a_0)\, p_0(a_i) + \ldots + q(a_n)\, p_n(a_i)$.

(b) Exercise 59(b) $\Rightarrow q(a_i) = c_0\, p_0(a_i) + \ldots + c_n\, p_n(a_i) = c_i \Rightarrow$
q passes through $(a_0, c_0), \ldots, (a_n, c_n)$. So, q is unique because $\mathcal{B}$ is a basis for $\mathscr{P}_n$.
Note that a_i must be distinct to avoid division by zero.

(c) (i) $a_0 = 1$, $a_1 = 2$, $a_3 = 3 \Rightarrow$ we can use $p_0(x)$, $p_1(x)$, and $p_2(x)$ from Exercise 59(a).

Therefore, (c) $\Rightarrow q(x) = 6\, p_0(x) - p_1(x) - 2\, p_2(x) = 3x^2 - 16x + 19$.

(ii) With $a_0 = -1$, $a_1 = 0$, and $a_3 = 3$, we find:

$$p_0(x) = \frac{(x-0)(x-3)}{(-1-0)(-1-3)} = \frac{1}{4}\, x^2 - \frac{3}{4}\, x.$$

$$p_1(x) = \frac{(x+1)(x-3)}{(0+1)(0-3)} = -\frac{1}{3}\, x^2 + \frac{2}{3}\, x + 1.$$

$$p_2(x) = \frac{(x+1)(x-0)}{(3+1)(3-0)} = \frac{1}{12}\, x^2 + \frac{1}{12}\, x.$$

Therefore, (c) $\Rightarrow q(x) = 10\, p_0(x) + 5\, p_1(x) + 2\, p_2(x) = x^2 - 4x + 5$.

62. Let $(a_0, 0), \ldots, (a_n, 0)$ be the zeroes of the polynomial which we will call $q(x)$.
Then Exercise 61(a) $\Rightarrow q(x) = q(a_0)\, p_0(x) + \ldots + q(a_n)\, p_n(x)$
but that is identically zero because $q(a_i) = 0$ for all $i \Rightarrow q$ is the zero polynomial.

63. Recall, a $n \times n$ matrix A is invertible $\Leftrightarrow$ its columns form a basis for col(I).
Therefore, a matrix in $M_{nn}(\mathbb{Z}_p)$ is invertible $\Leftrightarrow$ its columns form a basis for col$(I) = \mathbb{Z}_p^n$.

As suggested in the hint, we will count the number of ways to construct a basis for $\mathbb{Z}_p^n$.

Note, when constructing a basis, we can start with any vector except for $\mathbf{0}$.
Therefore, since there are p^n vectors in $\mathbb{Z}_p^n$, there are $p^n - 1$ ways to choose the first vector.
Call this vector $\mathbf{v}_1$.

For our second vector, we can choose any vector not in span$(\mathbf{v}_1) = \{a_1\mathbf{v}_1 : a_1 = 1, \ldots, p\}$.
Since there are p vectors in this span, therefore there are $p^n - p$ ways to choose $\mathbf{v}_2$.

Likewise, choose $\mathbf{v}_3 \notin \text{span}(\mathbf{v}_1, \mathbf{v}_2) = \{a_1\,\mathbf{v}_1 + a_2\,\mathbf{v}_2 : a_1, a_2 = 1, \ldots, p\}$.
Since there are p^2 vectors in this span, therefore there are $p^n - p^2$ ways to choose $\mathbf{v}_3$.

In general, therefore, there are $p^n - p^{i-1}$ ways to choose the ith vector in our basis.
Taking the product of these to find the total number of ways possible, we find:

the number of different bases for $\mathbb{Z}_p^n$ = the number of invertible matrices in $M_{nn}(\mathbb{Z}_p)$
$= (p^n - 1)(p^n - p)(p^n - p^2) \ldots (p^n - p^{n-1})$.

Exploration: Magic Squares

Since Explorations are self-contained, only solutions will be provided.

1. We want to show: $\text{wt}(M) = \frac{n(n^2+1)}{2}$.

A *classical magic square* contains each of entries $1, 2, \cdots, n^2$ exactly once.
Since $\text{wt}(M)$ is the common sum of any row or column we have the following:

$$\text{wt}(M) = \tfrac{1}{n}(1 + 2 + \cdots + n^2)$$

Using induction, we will prove $1 + 2 + \cdots + n^2 = \frac{n^2(n^2+1)}{2}$ for all $n \geq 1$.

1: $1 = \frac{1^2(1+1)}{2}$
This is obvious, so there is nothing to show.

k: $1 + 2 + \cdots + k^2 = \frac{k^2(k^2+1)}{2}$
This is the induction hypothesis, so there is nothing to show.

$k+1$: $1 + 2 + \cdots + k^2 + (k+1)^2 = \frac{(k+1)^2((k+1)^2+1)}{2}$
This is the statement we must prove using the induction hypothesis.

$$1 + 2 + \cdots + k^2 + (k+1)^2 \overset{\text{by induction}}{=} \frac{k^2(k^2+1)}{2} + (k^2+1) + (k^2+2) + \cdots + (k^2+2k+1).$$

By induction $1 + 2 + \cdots + (2k+1) = \frac{(2k+1)(2k+2)}{2} = (2k+1)(k+1)$.
See Appendix B for further discussion of *Mathematical Induction.*

We also have $2k+1$ copies of k^2. So:

$$\begin{aligned}1 + 2 + \cdots + k^2 + (k+1)^2 &= \frac{k^2(k^2+1)}{2} + (k^2+1) + (k^2+2) + \cdots + (k^2+2k+1)\\ &= \frac{k^2(k^2+1)}{2} + (2k+1)k^2 + (2k+1)(k+1)\\ &= \frac{k^4+4k^3+7k^2+6k+2}{2} = \frac{(k+1)^2((k+1)^2+1)}{2}\end{aligned}$$

We have proven (by induction) that

$$1 + 2 + \cdots + n^2 = \frac{n^2(n^2+1)}{2} \text{ for all } n \geq 1.$$

So we have $\text{wt}(M)$ is the the common sum of any row or column we have the following:

$$\text{wt}(M) = \tfrac{1}{n}(1 + 2 + \cdots + n^2) = \frac{1}{n}\left(\frac{n^2(n^2+1)}{2}\right) = \frac{n(n^2+1)}{2}.$$

Q: Why do we multiply the sum by the factor $\frac{1}{n}$?
A: Because each of the n columns of M sum to $\text{wt}(M)$.
So the sum of all the entries equals n times the $\text{wt}(M)$.
That is, we have *total* $= n \times \text{wt}(M)$, so $\text{wt}(M) = \frac{1}{n} \times$ *total.*

2. We want to construct a classical 3×3 magic square.
Since it is 3×3, we know the entries will be $1, 2, \ldots, 3^2 = 9$.
From Exercise 1, we know $\text{wt}(M) = \frac{3(3^2+1)}{2} = 15$.

Q: What clues about how to construct the magic square does $\text{wt}(M) = 15$ give us?
A: The numbers, 7, 8, and 9 must be in separate rows. Why?

Q: What other clues about how to construct the magic square can we deduce?
A: We should mid-range numbers in the central square. Why?
Because that number must sum to 15 with 4 separate pairs of other numbers.

With these ideas and some trial and error, we get:

$$M = \begin{bmatrix} 8 & 3 & 4 \\ 1 & 5 & 9 \\ 6 & 7 & 2 \end{bmatrix}.$$

To create another classical 3×3 magic square, we simply take the transpose of M:

$$M^T = \begin{bmatrix} 8 & 1 & 6 \\ 3 & 5 & 7 \\ 4 & 9 & 2 \end{bmatrix}.$$

Q: Is there another unrelated classical 3×3 magic square we can construct?

3. We want to construct a 3×3 magic square of $\text{wt}(N) = 1$ with all different entries.
From Exercise 2, we have M with $\text{wt}(M) = 15$, so we can let $N = \frac{1}{15}M$.

$$N = \tfrac{1}{15}M = \tfrac{1}{15}\begin{bmatrix} 8 & 3 & 4 \\ 1 & 5 & 9 \\ 6 & 7 & 2 \end{bmatrix} = \begin{bmatrix} \frac{8}{15} & \frac{3}{15} & \frac{4}{15} \\ \frac{1}{15} & \frac{5}{15} & \frac{9}{15} \\ \frac{6}{15} & \frac{7}{15} & \frac{2}{15} \end{bmatrix} = \begin{bmatrix} \frac{8}{15} & \frac{1}{5} & \frac{4}{15} \\ \frac{1}{15} & \frac{1}{3} & \frac{3}{5} \\ \frac{2}{5} & \frac{7}{15} & \frac{2}{15} \end{bmatrix}.$$

Given any weight w, we can construct the classical magic square M. Then let $N = \frac{w}{\text{wt}(M)}M$.

4. We need to prove Mag_3 is a subspace of M_{33} and Mag_3^0 is a subspace of Mag_3.

(a) Prove Mag_3 is a subspace of M_{33}.
Need show that Mag_3 is closed with respect to matrix addition and scalar multiplication.
Let M and N be in Mag_3.
Then the kth column of $P = M + N$ is $\mathbf{p}_k = \mathbf{m}_k + \mathbf{n}_k$.
So the sum of the entries of the kth column of P is $\text{wt}(M) + \text{wt}(N)$.
Likewise for the diagonals and rows of P. Therefore, $\text{wt}(P) = \text{wt}(M) + \text{wt}(N)$.
Since this sum is the same for all columns, rows, and diagonals of P, P is in Mag_3.
Furthermore, this same argument shows $Q = kM$ is also in Mag_3 with $\text{wt}(Q) = k{\cdot}\text{wt}(M)$.
Since Mag_3 is closed with respect to matrix addition and scalar multiplication,
Mag_3 is a subspace of M_{33}.

(b) Prove Mag_3^0 is a subspace of Mag_3.
Need to show Mag_3^0 is closed with respect to matrix addition and scalar multiplication.
Let M and N be in Mag_3^0.
Then the kth column of $P = M + N$ is $\mathbf{p}_k = \mathbf{m}_k + \mathbf{n}_k$.
So the sum of the entries of the kth column of P is $\text{wt}(M) + \text{wt}(N) = 0 + 0 = 0$.
Likewise for the diagonals and rows of P. Therefore, $\text{wt}(P) = 0$.
Since this sum is the same for all columns, rows, and diagonals P, P is in Mag_3^0.
This same argument shows $Q = kM$ is also in Mag_3^0 since $\text{wt}(Q) = k \cdot \text{wt}(M) = k \cdot 0 = 0$.
Since Mag_3^0 is closed with respect to matrix addition and scalar multiplication,
Mag_3^0 is a subspace of Mag_3.

5. We need to show if M is in Mag_3 with $\text{wt}(M) = w$ then $M = M_0 + kJ$,
where M_0 has weight zero and J is the matrix whose entries are all ones.

Following the hint, we will show that $M - kJ$ is in Mag_3^0.

Let $k = \frac{w}{3} = \frac{\text{wt}(M)}{3}$ and $M_0 = M - kJ = M - \frac{\text{wt}(M)}{3}J = M - \frac{w}{3}J$.

With this value of k, note that $\text{wt}(\frac{w}{3}J) = w$.

Why?

Because, for example, the first row of $\frac{w}{3}J$ is $\begin{bmatrix} \frac{w}{3} & \frac{w}{3} & \frac{w}{3} \end{bmatrix}$.

When we sum these three entries, we get: $\frac{w}{3} + \frac{w}{3} + \frac{w}{3} = w$.

So $\text{wt}(M_0) = \text{wt}(M) - \text{wt}\left(\frac{w}{3}J\right) = w - w = 0$.

6. The solution of the system described in the text is left to the reader.

7. Since there are two unknowns s and t, it is clear that $\dim Mag_3^0 = 2$. Why?

From the general description of M given in the text,
it is obvious that the following set is a basis for Mag_3^0:

$$\left\{ M_s = \begin{bmatrix} 1 & -1 & 0 \\ -1 & 0 & 1 \\ 0 & 1 & -1 \end{bmatrix}, M_t = \begin{bmatrix} 0 & -1 & 1 \\ 1 & 0 & -1 \\ -1 & 1 & 0 \end{bmatrix}, \right\}$$

Q: Think of these matrices as basis vectors for Mag_3^0.
Is it obvious that these matrices are linearly independent? Why or why not?
A: Consider $kM_s + jM_t = O$. What do k and j have to be?

Q: Define the dot product between these matrices as $M \cdot N = \sum m_{ij} \cdot n_{ij}$.
Using this definition, are these matrices orthogonal? Is this a good definition?

8. Note: The directions should say: find the dimension of Mag_3.

Since there are three unknowns r, s and t, it is clear that $\dim Mag_3 = 3$. Why?

From the general description of M given in the text,
it is obvious that the following set is a basis for Mag_3:

$$\left\{ M_r = \begin{bmatrix} 1 & 1 & 1 \\ 1 & 1 & 1 \\ 1 & 1 & 1 \end{bmatrix}, M_s = \begin{bmatrix} 1 & -1 & 0 \\ -1 & 0 & 1 \\ 0 & 1 & -1 \end{bmatrix}, M_t = \begin{bmatrix} 0 & -1 & 1 \\ 1 & 0 & -1 \\ -1 & 1 & 0 \end{bmatrix}, \right\}$$

Q: Think of these matrices as basis vectors for Mag_3.
Is it obvious that these matrices are linearly independent? Why or why not?
A: Consider $kM_r + jM_s + lM_t = O$. What do k, j, and l have to be?

Q: Define the dot product between these matrices as $M \cdot N = \sum m_{ij} \cdot n_{ij}$.
Using this definition, are these matrices orthogonal? Is this a good definition?

9. Using the hint we note the following: If we add the diagonals and the middle column,
we get row 1 and row 2 and central $(2,2)$ entry 3 times.
So, if we subtract row 1 and row 2 we are left with three times the central entry, c.

That is: $(\mathbf{d}_1 + \mathbf{d}_2 + \mathbf{c}_2) - (\mathbf{r}_1 + \mathbf{r}_2) = 3c$.

Also, since we subtracted two rows from two diagonals and a column we get the weight, w.

That is: $w = (\mathbf{d}_1 + \mathbf{d}_2 + \mathbf{c}_2) - (\mathbf{r}_1 + \mathbf{r}_2) = 3c$.

Therefore, the central entry $(2,2)$ must be $c = \frac{w}{3}$ as we were to show.

10. We leave the computation of the equation for the circle to the reader.

However, we note that all 8 matrices are created by:
switching top and bottom rows, left and right columns, and transposing in each case.

6.3 Change of Basis

1. (a) $\begin{bmatrix} 2 \\ 3 \end{bmatrix} = a_1 \begin{bmatrix} 1 \\ 0 \end{bmatrix} + a_2 \begin{bmatrix} 0 \\ 1 \end{bmatrix} \Rightarrow \begin{bmatrix} a_1 = 2 \\ a_2 = 3 \end{bmatrix} \Rightarrow [\mathbf{x}]_{\mathcal{B}} = \begin{bmatrix} 2 \\ 3 \end{bmatrix}.$

$\begin{bmatrix} 2 \\ 3 \end{bmatrix} = b_1 \begin{bmatrix} 1 \\ 1 \end{bmatrix} + b_2 \begin{bmatrix} 1 \\ -1 \end{bmatrix} \Rightarrow \begin{bmatrix} b_1 + b_2 = 2 \\ b_1 - b_2 = 3 \end{bmatrix} \Rightarrow \begin{matrix} b_1 = \frac{5}{2} \\ b_2 = -\frac{1}{2} \end{matrix} \Rightarrow [\mathbf{x}]_{\mathcal{C}} = \begin{bmatrix} \frac{5}{2} \\ -\frac{1}{2} \end{bmatrix}.$

(b) Let $\mathcal{E} = \left\{ \begin{bmatrix} 1 \\ 0 \end{bmatrix}, \begin{bmatrix} 0 \\ 1 \end{bmatrix} \right\}$. Then Theorem 6.13, which states $[\,C \mid B\,] \longrightarrow [\,I \mid P_{\mathcal{C}\leftarrow\mathcal{B}}\,]$, implies

$\left[\begin{array}{cc|cc} 1 & 1 & 1 & 0 \\ 1 & -1 & 0 & 1 \end{array}\right] \longrightarrow \left[\begin{array}{cc|cc} 1 & 0 & \frac{1}{2} & \frac{1}{2} \\ 0 & 1 & \frac{1}{2} & -\frac{1}{2} \end{array}\right] \Rightarrow P_{\mathcal{C}\leftarrow\mathcal{B}} = \begin{bmatrix} \frac{1}{2} & \frac{1}{2} \\ \frac{1}{2} & -\frac{1}{2} \end{bmatrix}.$

(c) $[\mathbf{x}]_{\mathcal{C}} = P_{\mathcal{C}\leftarrow\mathcal{B}}\,[\mathbf{x}]_{\mathcal{B}} = \begin{bmatrix} \frac{1}{2} & \frac{1}{2} \\ \frac{1}{2} & -\frac{1}{2} \end{bmatrix} \begin{bmatrix} 2 \\ 3 \end{bmatrix} = \begin{bmatrix} \frac{5}{2} \\ -\frac{1}{2} \end{bmatrix}.$

(d) Let $\mathcal{E} = \left\{ \begin{bmatrix} 1 \\ 0 \end{bmatrix}, \begin{bmatrix} 0 \\ 1 \end{bmatrix} \right\}$. Then Theorem 6.13, which states $[\,B \mid C\,] \longrightarrow [\,I \mid P_{\mathcal{B}\leftarrow\mathcal{C}}\,]$, implies

$\left[\begin{array}{cc|cc} 1 & 0 & 1 & 1 \\ 0 & 1 & 1 & -1 \end{array}\right] = \left[\begin{array}{cc|cc} 1 & 0 & 1 & 1 \\ 0 & 1 & 1 & -1 \end{array}\right] \Rightarrow P_{\mathcal{B}\leftarrow\mathcal{C}} = \begin{bmatrix} 1 & 1 \\ 1 & -1 \end{bmatrix}.$

(e) $[\mathbf{x}]_{\mathcal{B}} = P_{\mathcal{B}\leftarrow\mathcal{C}}\,[\mathbf{x}]_{\mathcal{C}} = \begin{bmatrix} 1 & 1 \\ 1 & -1 \end{bmatrix} \begin{bmatrix} \frac{5}{2} \\ -\frac{1}{2} \end{bmatrix} = \begin{bmatrix} 2 \\ 3 \end{bmatrix}.$

2. (a) $\begin{bmatrix} 4 \\ -1 \end{bmatrix} = a_1 \begin{bmatrix} 1 \\ 0 \end{bmatrix} + a_2 \begin{bmatrix} 1 \\ 1 \end{bmatrix} \Rightarrow \begin{bmatrix} a_1 + a_2 = 4 \\ a_2 = -1 \end{bmatrix} \Rightarrow \begin{matrix} a_1 = 5 \\ a_2 = -1 \end{matrix} \Rightarrow [\mathbf{x}]_{\mathcal{B}} = \begin{bmatrix} 5 \\ -1 \end{bmatrix}.$

$\begin{bmatrix} 4 \\ -1 \end{bmatrix} = b_1 \begin{bmatrix} 0 \\ 1 \end{bmatrix} + b_2 \begin{bmatrix} 2 \\ 3 \end{bmatrix} \Rightarrow \begin{matrix} b_1 = -7 \\ b_2 = 2 \end{matrix} \Rightarrow [\mathbf{x}]_{\mathcal{C}} = \begin{bmatrix} -7 \\ 2 \end{bmatrix}.$

(b) $\left[\begin{array}{cc|cc} 0 & 2 & 1 & 1 \\ 1 & 3 & 0 & 1 \end{array}\right] \longrightarrow \left[\begin{array}{cc|cc} 1 & 0 & -\frac{3}{2} & -\frac{1}{2} \\ 0 & 1 & \frac{1}{2} & \frac{1}{2} \end{array}\right] \Rightarrow P_{\mathcal{C}\leftarrow\mathcal{B}} = \begin{bmatrix} -\frac{3}{2} & -\frac{1}{2} \\ \frac{1}{2} & \frac{1}{2} \end{bmatrix}.$

(c) $[\mathbf{x}]_{\mathcal{C}} = P_{\mathcal{C}\leftarrow\mathcal{B}}\,[\mathbf{x}]_{\mathcal{B}} = \begin{bmatrix} -\frac{3}{2} & -\frac{1}{2} \\ \frac{1}{2} & \frac{1}{2} \end{bmatrix} \begin{bmatrix} 5 \\ -1 \end{bmatrix} = \begin{bmatrix} -7 \\ 2 \end{bmatrix}.$

(d) $\left[\begin{array}{cc|cc} 0 & 2 & 1 & 1 \\ 1 & 3 & 0 & 1 \end{array}\right] = \left[\begin{array}{cc|cc} 1 & 0 & -1 & -1 \\ 0 & 1 & 1 & 3 \end{array}\right] \Rightarrow P_{\mathcal{B}\leftarrow\mathcal{C}} = \begin{bmatrix} -1 & -1 \\ 1 & 3 \end{bmatrix}.$

(e) $[\mathbf{x}]_{\mathcal{B}} = P_{\mathcal{B}\leftarrow\mathcal{C}}\,[\mathbf{x}]_{\mathcal{C}} = \begin{bmatrix} -1 & -1 \\ 1 & 3 \end{bmatrix} \begin{bmatrix} -7 \\ 2 \end{bmatrix} = \begin{bmatrix} 5 \\ -1 \end{bmatrix}.$

3. (a) $\begin{bmatrix} 1 \\ 0 \\ -1 \end{bmatrix} = a_1 \begin{bmatrix} 1 \\ 0 \\ 0 \end{bmatrix} + a_2 \begin{bmatrix} 0 \\ 1 \\ 0 \end{bmatrix} + a_3 \begin{bmatrix} 0 \\ 0 \\ 1 \end{bmatrix} \Rightarrow \begin{matrix} a_1 = 1 \\ a_2 = 0 \\ a_3 = -1 \end{matrix} \Rightarrow [\mathbf{x}]_{\mathcal{B}} = \begin{bmatrix} 1 \\ 0 \\ -1 \end{bmatrix}.$

$\begin{bmatrix} 1 \\ 0 \\ -1 \end{bmatrix} = b_1 \begin{bmatrix} 1 \\ 1 \\ 1 \end{bmatrix} + b_2 \begin{bmatrix} 0 \\ 1 \\ 1 \end{bmatrix} + b_3 \begin{bmatrix} 0 \\ 0 \\ 1 \end{bmatrix} \Rightarrow \begin{matrix} b_1 = 1 \\ b_2 = -1 \\ b_3 = -1 \end{matrix} \Rightarrow [\mathbf{x}]_{\mathcal{C}} = \begin{bmatrix} 1 \\ -1 \\ -1 \end{bmatrix}.$

(b) $\left[\begin{array}{ccc|ccc} 1 & 0 & 0 & 1 & 0 & 0 \\ 1 & 1 & 0 & 0 & 1 & 0 \\ 1 & 1 & 1 & 0 & 0 & 1 \end{array}\right] \longrightarrow \left[\begin{array}{ccc|ccc} 1 & 0 & 0 & 1 & 0 & 0 \\ 0 & 1 & 0 & -1 & 1 & 0 \\ 0 & 0 & 1 & 0 & -1 & 1 \end{array}\right] \Rightarrow P_{\mathcal{C}\leftarrow\mathcal{B}} = \begin{bmatrix} 1 & 0 & 0 \\ -1 & 1 & 0 \\ 0 & -1 & 1 \end{bmatrix}.$

(c) $[\mathbf{x}]_{\mathcal{C}} = P_{\mathcal{C}\leftarrow\mathcal{B}}\,[\mathbf{x}]_{\mathcal{B}} = \begin{bmatrix} 1 & 0 & 0 \\ -1 & 1 & 0 \\ 0 & -1 & 1 \end{bmatrix} \begin{bmatrix} 1 \\ 0 \\ -1 \end{bmatrix} = \begin{bmatrix} 1 \\ -1 \\ -1 \end{bmatrix}.$

(d) $\left[\begin{array}{ccc|ccc} 1 & 0 & 0 & 1 & 0 & 0 \\ 0 & 1 & 0 & 1 & 1 & 0 \\ 0 & 0 & 1 & 1 & 1 & 1 \end{array}\right] \longrightarrow \left[\begin{array}{ccc|ccc} 1 & 0 & 0 & 1 & 0 & 0 \\ 0 & 1 & 0 & 1 & 1 & 0 \\ 0 & 0 & 1 & 1 & 1 & 1 \end{array}\right] \Rightarrow P_{\mathcal{B}\leftarrow\mathcal{C}} = \begin{bmatrix} 1 & 0 & 0 \\ 1 & 1 & 0 \\ 1 & 1 & 1 \end{bmatrix}.$

(e) $[\mathbf{x}]_{\mathcal{B}} = P_{\mathcal{B}\leftarrow\mathcal{C}}\,[\mathbf{x}]_{\mathcal{C}} = \begin{bmatrix} 1 & 0 & 0 \\ 1 & 1 & 0 \\ 1 & 1 & 1 \end{bmatrix} \begin{bmatrix} 1 \\ -1 \\ -1 \end{bmatrix} = \begin{bmatrix} 1 \\ 0 \\ -1 \end{bmatrix}.$

4. (a) $\begin{bmatrix} 3 \\ 1 \\ 5 \end{bmatrix} = a_1 \begin{bmatrix} 0 \\ 1 \\ 0 \end{bmatrix} + a_2 \begin{bmatrix} 0 \\ 0 \\ 1 \end{bmatrix} + a_3 \begin{bmatrix} 1 \\ 0 \\ 0 \end{bmatrix} \Rightarrow \begin{matrix} a_1 = 1 \\ a_2 = 5 \\ a_3 = 3 \end{matrix} \Rightarrow [\mathbf{x}]_{\mathcal{B}} = \begin{bmatrix} 1 \\ 5 \\ 3 \end{bmatrix}.$

$\begin{bmatrix} 3 \\ 1 \\ 5 \end{bmatrix} = b_1 \begin{bmatrix} 1 \\ 1 \\ 0 \end{bmatrix} + b_2 \begin{bmatrix} 0 \\ 1 \\ 1 \end{bmatrix} + b_3 \begin{bmatrix} 1 \\ 0 \\ 1 \end{bmatrix} \Rightarrow \begin{matrix} b_1 = -\frac{1}{2} \\ b_2 = \frac{3}{2} \\ b_3 = \frac{7}{2} \end{matrix} \Rightarrow [\mathbf{x}]_{\mathcal{C}} = \begin{bmatrix} -\frac{1}{2} \\ \frac{3}{2} \\ \frac{7}{2} \end{bmatrix}.$

(b) $\left[\begin{array}{ccc|ccc} 1 & 0 & 1 & 0 & 0 & 1 \\ 1 & 1 & 0 & 1 & 0 & 0 \\ 0 & 1 & 1 & 0 & 1 & 0 \end{array}\right] \longrightarrow \left[\begin{array}{ccc|ccc} 1 & 0 & 0 & \frac{1}{2} & -\frac{1}{2} & \frac{1}{2} \\ 0 & 1 & 0 & \frac{1}{2} & \frac{1}{2} & -\frac{1}{2} \\ 0 & 0 & 1 & -\frac{1}{2} & \frac{1}{2} & \frac{1}{2} \end{array}\right] \Rightarrow P_{\mathcal{C}\leftarrow\mathcal{B}} = \begin{bmatrix} \frac{1}{2} & -\frac{1}{2} & \frac{1}{2} \\ \frac{1}{2} & \frac{1}{2} & -\frac{1}{2} \\ -\frac{1}{2} & \frac{1}{2} & \frac{1}{2} \end{bmatrix}.$

(c) $[\mathbf{x}]_{\mathcal{C}} = P_{\mathcal{C}\leftarrow\mathcal{B}}\,[\mathbf{x}]_{\mathcal{B}} = \begin{bmatrix} \frac{1}{2} & -\frac{1}{2} & \frac{1}{2} \\ \frac{1}{2} & \frac{1}{2} & -\frac{1}{2} \\ -\frac{1}{2} & \frac{1}{2} & \frac{1}{2} \end{bmatrix} \begin{bmatrix} 1 \\ 5 \\ 3 \end{bmatrix} = \begin{bmatrix} -\frac{1}{2} \\ \frac{3}{2} \\ \frac{7}{2} \end{bmatrix}.$

(d) $\left[\begin{array}{ccc|ccc} 0 & 0 & 1 & 1 & 0 & 1 \\ 1 & 0 & 0 & 1 & 1 & 0 \\ 0 & 1 & 0 & 0 & 1 & 1 \end{array}\right] \longrightarrow \left[\begin{array}{ccc|ccc} 1 & 0 & 0 & 1 & 1 & 0 \\ 0 & 1 & 0 & 0 & 1 & 1 \\ 0 & 0 & 1 & 1 & 0 & 1 \end{array}\right] \Rightarrow P_{\mathcal{B}\leftarrow\mathcal{C}} = \begin{bmatrix} 1 & 1 & 0 \\ 0 & 1 & 1 \\ 1 & 0 & 1 \end{bmatrix}.$

(e) $[\mathbf{x}]_{\mathcal{B}} = P_{\mathcal{B}\leftarrow\mathcal{C}}\,[\mathbf{x}]_{\mathcal{C}} = \begin{bmatrix} 1 & 1 & 0 \\ 0 & 1 & 1 \\ 1 & 0 & 1 \end{bmatrix} \begin{bmatrix} -\frac{1}{2} \\ \frac{3}{2} \\ \frac{7}{2} \end{bmatrix} = \begin{bmatrix} 1 \\ 5 \\ 3 \end{bmatrix}.$

5. (a) $\begin{bmatrix} 2 \\ -1 \end{bmatrix} = a_1 \begin{bmatrix} 1 \\ 0 \end{bmatrix} + a_2 \begin{bmatrix} 0 \\ 1 \end{bmatrix} \Rightarrow \begin{bmatrix} a_1 = 2 \\ a_2 = -1 \end{bmatrix} \Rightarrow [p(x)]_{\mathcal{B}} = \begin{bmatrix} 2 \\ -1 \end{bmatrix}.$

$\begin{bmatrix} 2 \\ -1 \end{bmatrix} = b_1 \begin{bmatrix} 0 \\ 1 \end{bmatrix} + b_2 \begin{bmatrix} 1 \\ 1 \end{bmatrix} \Rightarrow \begin{bmatrix} b_2 = 2 \\ b_1 + b_2 = -1 \end{bmatrix} \Rightarrow \begin{matrix} b_1 = -3 \\ b_2 = 2 \end{matrix} \Rightarrow [p(x)]_{\mathcal{C}} = \begin{bmatrix} -3 \\ 2 \end{bmatrix}.$

(b) $\left[\begin{array}{cc|cc} 0 & 1 & 1 & 0 \\ 1 & 1 & 0 & 1 \end{array}\right] \longrightarrow \left[\begin{array}{cc|cc} 1 & 0 & -1 & 1 \\ 0 & 1 & 1 & 0 \end{array}\right] \Rightarrow P_{\mathcal{C}\leftarrow\mathcal{B}} = \begin{bmatrix} -1 & 1 \\ 1 & 0 \end{bmatrix}.$

(c) $[p(x)]_{\mathcal{C}} = P_{\mathcal{C}\leftarrow\mathcal{B}}\,[p(x)]_{\mathcal{B}} = \begin{bmatrix} -1 & 1 \\ 1 & 0 \end{bmatrix} \begin{bmatrix} 2 \\ -1 \end{bmatrix} = \begin{bmatrix} -3 \\ 2 \end{bmatrix}.$

(d) $\left[\begin{array}{cc|cc} 1 & 0 & 0 & 1 \\ 0 & 1 & 1 & 1 \end{array}\right] = \left[\begin{array}{cc|cc} 1 & 0 & 0 & 1 \\ 0 & 1 & 1 & 1 \end{array}\right] \Rightarrow P_{\mathcal{B}\leftarrow\mathcal{C}} = \begin{bmatrix} 0 & 1 \\ 1 & 1 \end{bmatrix}.$

(e) $[p(x)]_{\mathcal{B}} = P_{\mathcal{B}\leftarrow\mathcal{C}}\,[p(x)]_{\mathcal{C}} = \begin{bmatrix} 0 & 1 \\ 1 & 1 \end{bmatrix} \begin{bmatrix} -3 \\ 2 \end{bmatrix} = \begin{bmatrix} 2 \\ -1 \end{bmatrix}.$

6. (a) $\begin{bmatrix}1\\3\end{bmatrix} = a_1\begin{bmatrix}1\\1\end{bmatrix} + a_2\begin{bmatrix}1\\-1\end{bmatrix} \Rightarrow \begin{bmatrix}a_1 = 2\\ a_2 = -1\end{bmatrix} \Rightarrow [p(x)]_{\mathcal{B}} = \begin{bmatrix}2\\-1\end{bmatrix}.$

$\begin{bmatrix}1\\3\end{bmatrix} = b_1\begin{bmatrix}0\\2\end{bmatrix} + b_2\begin{bmatrix}4\\0\end{bmatrix} \Rightarrow \begin{bmatrix}4b_2 = 1\\ 2b_1 = 3\end{bmatrix} \Rightarrow \begin{matrix}b_1 = \frac{3}{2}\\ b_2 = \frac{1}{4}\end{matrix} \Rightarrow [p(x)]_{\mathcal{C}} = \begin{bmatrix}\frac{3}{2}\\ \frac{1}{4}\end{bmatrix}.$

(b) $\left[\begin{array}{cc|cc}0 & 4 & 1 & 1\\ 2 & 0 & 1 & -1\end{array}\right] \longrightarrow \left[\begin{array}{cc|cc}1 & 0 & \frac{1}{2} & -\frac{1}{2}\\ 0 & 1 & \frac{1}{4} & -\frac{1}{4}\end{array}\right] \Rightarrow P_{\mathcal{C}\leftarrow\mathcal{B}} = \begin{bmatrix}\frac{1}{2} & -\frac{1}{2}\\ \frac{1}{4} & -\frac{1}{4}\end{bmatrix}.$

(c) $[p(x)]_{\mathcal{C}} = P_{\mathcal{C}\leftarrow\mathcal{B}}\,[p(x)]_{\mathcal{B}} = \begin{bmatrix}\frac{1}{2} & -\frac{1}{2}\\ \frac{1}{4} & -\frac{1}{4}\end{bmatrix}\begin{bmatrix}2\\-1\end{bmatrix} = \begin{bmatrix}\frac{3}{2}\\ \frac{1}{4}\end{bmatrix}.$

(d) $\left[\begin{array}{cc|cc}1 & 1 & 0 & 4\\ 2 & 0 & 1 & -1\end{array}\right] \longrightarrow \left[\begin{array}{cc|cc}1 & 0 & 1 & 2\\ 0 & 1 & -1 & 2\end{array}\right] \Rightarrow P_{\mathcal{B}\leftarrow\mathcal{C}} = \begin{bmatrix}1 & 2\\ -1 & 2\end{bmatrix}.$

(e) $[p(x)]_{\mathcal{B}} = P_{\mathcal{B}\leftarrow\mathcal{C}}\,[p(x)]_{\mathcal{C}} = \begin{bmatrix}1 & 2\\ -1 & 2\end{bmatrix}\begin{bmatrix}\frac{3}{2}\\ \frac{1}{4}\end{bmatrix} = \begin{bmatrix}2\\-1\end{bmatrix}.$

7. (a) $\begin{bmatrix}1\\0\\1\end{bmatrix} = a_1\begin{bmatrix}1\\1\\1\end{bmatrix} + a_2\begin{bmatrix}0\\1\\1\end{bmatrix} + a_3\begin{bmatrix}0\\0\\1\end{bmatrix} \Rightarrow \begin{matrix}a_1 = 1\\ a_2 = -1\\ a_3 = 1\end{matrix} \Rightarrow [p(x)]_{\mathcal{B}} = \begin{bmatrix}1\\-1\\1\end{bmatrix}.$

$\begin{bmatrix}1\\0\\1\end{bmatrix} = b_1\begin{bmatrix}1\\0\\0\end{bmatrix} + b_2\begin{bmatrix}0\\1\\0\end{bmatrix} + b_3\begin{bmatrix}0\\0\\1\end{bmatrix} \Rightarrow \begin{matrix}b_1 = 1\\ b_2 = 0\\ b_3 = 1\end{matrix} \Rightarrow [p(x)]_{\mathcal{C}} = \begin{bmatrix}1\\0\\1\end{bmatrix}.$

(b) $\left[\begin{array}{ccc|ccc}1 & 0 & 0 & 1 & 0 & 0\\ 0 & 1 & 0 & 1 & 1 & 0\\ 0 & 0 & 1 & 1 & 1 & 1\end{array}\right] \longrightarrow \left[\begin{array}{ccc|ccc}1 & 0 & 0 & 1 & 0 & 0\\ 0 & 1 & 0 & 1 & 1 & 0\\ 0 & 0 & 1 & 1 & 1 & 1\end{array}\right] \Rightarrow P_{\mathcal{C}\leftarrow\mathcal{B}} = \begin{bmatrix}1 & 0 & 0\\ 1 & 1 & 0\\ 1 & 1 & 1\end{bmatrix}.$

(c) $[p(x)]_{\mathcal{C}} = P_{\mathcal{C}\leftarrow\mathcal{B}}\,[p(x)]_{\mathcal{B}} = \begin{bmatrix}1 & 0 & 0\\ 1 & 1 & 0\\ 1 & 1 & 1\end{bmatrix}\begin{bmatrix}1\\-1\\1\end{bmatrix} = \begin{bmatrix}1\\0\\1\end{bmatrix}.$

(d) $\left[\begin{array}{ccc|ccc}1 & 0 & 0 & 1 & 0 & 0\\ 1 & 1 & 0 & 0 & 1 & 0\\ 1 & 1 & 1 & 0 & 0 & 1\end{array}\right] \longrightarrow \left[\begin{array}{ccc|ccc}1 & 0 & 0 & 1 & 0 & 0\\ 0 & 1 & 0 & -1 & 1 & 0\\ 0 & 0 & 1 & 0 & -1 & 1\end{array}\right] \Rightarrow P_{\mathcal{B}\leftarrow\mathcal{C}} = \begin{bmatrix}1 & 0 & 0\\ -1 & 1 & 0\\ 0 & -1 & 1\end{bmatrix}.$

(e) $[p(x)]_{\mathcal{B}} = P_{\mathcal{B}\leftarrow\mathcal{C}}\,[p(x)]_{\mathcal{C}} = \begin{bmatrix}1 & 0 & 0\\ -1 & 1 & 0\\ 1 & -1 & 1\end{bmatrix}\begin{bmatrix}1\\0\\1\end{bmatrix} = \begin{bmatrix}1\\-1\\1\end{bmatrix}.$

8. (a) $\begin{bmatrix} 4 \\ -2 \\ -1 \end{bmatrix} = a_1 \begin{bmatrix} 0 \\ 1 \\ 0 \end{bmatrix} + a_2 \begin{bmatrix} 1 \\ 0 \\ 1 \end{bmatrix} + a_3 \begin{bmatrix} 0 \\ 1 \\ 1 \end{bmatrix} \Rightarrow \begin{matrix} a_1 = 3 \\ a_2 = 4 \\ a_3 = -5 \end{matrix} \Rightarrow [p(x)]_{\mathcal{B}} = \begin{bmatrix} 3 \\ 4 \\ -5 \end{bmatrix}.$

$\begin{bmatrix} 4 \\ -2 \\ -1 \end{bmatrix} = b_1 \begin{bmatrix} 1 \\ 0 \\ 0 \end{bmatrix} + b_2 \begin{bmatrix} 1 \\ 1 \\ 0 \end{bmatrix} + b_3 \begin{bmatrix} 0 \\ 0 \\ 1 \end{bmatrix} \Rightarrow \begin{matrix} b_1 = 6 \\ b_2 = -2 \\ b_3 = -1 \end{matrix} \Rightarrow [p(x)]_{\mathcal{C}} = \begin{bmatrix} 6 \\ -2 \\ -1 \end{bmatrix}.$

(b) $\left[\begin{array}{ccc|ccc} 1 & 1 & 0 & 0 & 1 & 0 \\ 0 & 1 & 0 & 1 & 0 & 1 \\ 0 & 0 & 1 & 0 & 1 & 1 \end{array}\right] \longrightarrow \left[\begin{array}{ccc|ccc} 1 & 0 & 0 & -1 & 1 & -1 \\ 0 & 1 & 0 & 1 & 0 & 1 \\ 0 & 0 & 1 & 0 & 1 & 1 \end{array}\right] \Rightarrow P_{\mathcal{C}\leftarrow\mathcal{B}} = \begin{bmatrix} -1 & 1 & -1 \\ 1 & 0 & 1 \\ 0 & 1 & 1 \end{bmatrix}.$

(c) $[p(x)]_{\mathcal{C}} = P_{\mathcal{C}\leftarrow\mathcal{B}}\,[p(x)]_{\mathcal{B}} = \begin{bmatrix} -1 & 1 & -1 \\ 1 & 0 & 1 \\ 0 & 1 & 1 \end{bmatrix} \begin{bmatrix} 3 \\ 4 \\ -5 \end{bmatrix} = \begin{bmatrix} 6 \\ -2 \\ -1 \end{bmatrix}.$

(d) $\left[\begin{array}{ccc|ccc} 0 & 1 & 0 & 1 & 1 & 0 \\ 1 & 0 & 1 & 0 & 1 & 0 \\ 0 & 1 & 1 & 0 & 0 & 1 \end{array}\right] \longrightarrow \left[\begin{array}{ccc|ccc} 1 & 0 & 0 & 1 & 2 & -1 \\ 0 & 1 & 0 & 1 & 1 & 0 \\ 0 & 0 & 1 & -1 & -1 & 1 \end{array}\right] \Rightarrow P_{\mathcal{B}\leftarrow\mathcal{C}} = \begin{bmatrix} 1 & 2 & -1 \\ 1 & 1 & 0 \\ -1 & -1 & 1 \end{bmatrix}.$

(e) $[p(x)]_{\mathcal{B}} = P_{\mathcal{B}\leftarrow\mathcal{C}}\,[p(x)]_{\mathcal{C}} = \begin{bmatrix} 1 & 2 & -1 \\ 1 & 1 & 0 \\ -1 & -1 & 1 \end{bmatrix} \begin{bmatrix} 6 \\ -2 \\ -1 \end{bmatrix} = \begin{bmatrix} 3 \\ 4 \\ -5 \end{bmatrix}.$

9. (a) $\begin{bmatrix} 4 \\ 2 \\ 0 \\ -1 \end{bmatrix} = a_1 \begin{bmatrix} 1 \\ 0 \\ 0 \\ 0 \end{bmatrix} + a_2 \begin{bmatrix} 0 \\ 1 \\ 0 \\ 0 \end{bmatrix} + a_3 \begin{bmatrix} 0 \\ 0 \\ 1 \\ 0 \end{bmatrix} + a_4 \begin{bmatrix} 0 \\ 0 \\ 0 \\ 1 \end{bmatrix} \Rightarrow \begin{matrix} a_1 = 4 \\ a_2 = 2 \\ a_3 = 0 \\ a_4 = -1 \end{matrix} \Rightarrow [A]_{\mathcal{B}} = \begin{bmatrix} 4 \\ 2 \\ 0 \\ -1 \end{bmatrix}.$

$\begin{bmatrix} 4 \\ 2 \\ 0 \\ -1 \end{bmatrix} = b_1 \begin{bmatrix} 1 \\ 2 \\ 0 \\ -1 \end{bmatrix} + b_2 \begin{bmatrix} 2 \\ 1 \\ 1 \\ 0 \end{bmatrix} + b_3 \begin{bmatrix} 1 \\ 1 \\ 0 \\ 1 \end{bmatrix} + b_4 \begin{bmatrix} 1 \\ 0 \\ 0 \\ 1 \end{bmatrix} \Rightarrow$

$\begin{matrix} b_1 = 5/2 \\ b_2 = 0 \\ b_3 = -3 \\ b_4 = 9/2 \end{matrix} \Rightarrow [A]_{\mathcal{C}} = \begin{bmatrix} 5/2 \\ 0 \\ -3 \\ 9/2 \end{bmatrix}.$

(b) $\left[\begin{array}{rrrr|rrrr} 1 & 2 & 1 & 1 & 1 & 0 & 0 & 0 \\ 2 & 1 & 1 & 0 & 0 & 1 & 0 & 0 \\ 0 & 1 & 0 & 0 & 0 & 0 & 1 & 0 \\ -1 & 0 & 1 & 1 & 0 & 0 & 0 & 1 \end{array}\right] \longrightarrow \left[\begin{array}{rrrr|rrrr} 1 & 0 & 0 & 0 & 1/2 & 0 & -1 & -1/2 \\ 0 & 1 & 0 & 0 & 0 & 0 & 1 & 0 \\ 0 & 0 & 1 & 0 & -1 & 1 & 1 & 1 \\ 0 & 0 & 0 & 1 & 3/2 & -1 & -2 & -1/2 \end{array}\right] \Rightarrow$

$P_{\mathcal{C}\leftarrow\mathcal{B}} = \begin{bmatrix} 1/2 & 0 & -1 & -1/2 \\ 0 & 0 & 1 & 0 \\ -1 & 1 & 1 & 1 \\ 3/2 & -1 & -2 & -1/2 \end{bmatrix}.$

(c) $[A]_{\mathcal{C}} = P_{\mathcal{C}\leftarrow\mathcal{B}}\,[A]_{\mathcal{B}} = \begin{bmatrix} 1/2 & 0 & -1 & -1/2 \\ 0 & 0 & 1 & 0 \\ -1 & 1 & 1 & 1 \\ 3/2 & -1 & -2 & -1/2 \end{bmatrix} \begin{bmatrix} 4 \\ 2 \\ 0 \\ -1 \end{bmatrix} = \begin{bmatrix} 5/2 \\ 0 \\ -3 \\ 9/2 \end{bmatrix}.$

(d) $\left[\begin{array}{rrrr|rrrr} 1 & 0 & 0 & 0 & 1 & 2 & 1 & 1 \\ 0 & 1 & 0 & 0 & 2 & 1 & 1 & 0 \\ 0 & 0 & 1 & 0 & 0 & 1 & 0 & 0 \\ 0 & 0 & 0 & 1 & -1 & 0 & 1 & 1 \end{array}\right] \longrightarrow \left[\begin{array}{rrrr|rrrr} 1 & 0 & 0 & 0 & 1 & 2 & 1 & 1 \\ 0 & 1 & 0 & 0 & 2 & 1 & 1 & 0 \\ 0 & 0 & 1 & 0 & 0 & 1 & 0 & 0 \\ 0 & 0 & 0 & 1 & -1 & 0 & 1 & 1 \end{array}\right] \Rightarrow P_{\mathcal{B}\leftarrow\mathcal{C}} = \begin{bmatrix} 1 & 2 & 1 & 1 \\ 2 & 1 & 1 & 0 \\ 0 & 1 & 0 & 0 \\ -1 & 0 & 1 & 1 \end{bmatrix}.$

(e) $[A]_{\mathcal{B}} = P_{\mathcal{B}\leftarrow\mathcal{C}}\,[A]_{\mathcal{C}} = \begin{bmatrix} 1 & 2 & 1 & 1 \\ 2 & 1 & 1 & 0 \\ 0 & 1 & 0 & 0 \\ -1 & 0 & 1 & 1 \end{bmatrix} \begin{bmatrix} 5/2 \\ 0 \\ -3 \\ 9/2 \end{bmatrix} = \begin{bmatrix} 4 \\ 2 \\ 0 \\ -1 \end{bmatrix}.$

10. (a) $\begin{bmatrix}1\\1\\1\\1\end{bmatrix} = a_1\begin{bmatrix}1\\0\\0\\1\end{bmatrix} + a_2\begin{bmatrix}0\\1\\1\\0\end{bmatrix} + a_3\begin{bmatrix}1\\1\\0\\0\end{bmatrix} + a_4\begin{bmatrix}1\\0\\1\\0\end{bmatrix} \Rightarrow \begin{array}{c} a_1 = 1\\ a_2 = 1\\ a_3 = 0\\ a_4 = 0\end{array} \Rightarrow [A]_{\mathcal{C}} = \begin{bmatrix}1\\1\\0\\0\end{bmatrix}.$

$\begin{bmatrix}1\\1\\1\\1\end{bmatrix} = b_1\begin{bmatrix}1\\1\\0\\1\end{bmatrix} + b_2\begin{bmatrix}1\\1\\1\\0\end{bmatrix} + b_3\begin{bmatrix}1\\0\\1\\1\end{bmatrix} + b_4\begin{bmatrix}0\\1\\1\\1\end{bmatrix} \Rightarrow \begin{array}{c} b_1 = \frac{1}{3}\\ b_2 = \frac{1}{3}\\ b_3 = \frac{1}{3}\\ b_4 = \frac{1}{3}\end{array} \Rightarrow [A]_{\mathcal{B}} = \begin{bmatrix}\frac{1}{3}\\\frac{1}{3}\\\frac{1}{3}\\\frac{1}{3}\end{bmatrix}.$

(b) $\left[\begin{array}{cccc|cccc} 1&1&1&0&1&0&1&1\\ 1&1&0&1&0&1&1&0\\ 0&1&1&1&0&1&0&1\\ 1&0&1&1&1&0&0&0 \end{array}\right] \longrightarrow \left[\begin{array}{cccc|cccc} 1&0&0&0&2/3&-1/3&2/3&-1/3\\ 0&1&0&0&-1/3&2/3&2/3&2/3\\ 0&0&1&0&2/3&-1/3&-1/3&2/3\\ 0&0&0&1&-1/3&2/3&-1/3&-1/3 \end{array}\right] \Rightarrow$

$P_{\mathcal{C}\leftarrow\mathcal{B}} = \begin{bmatrix} \frac{2}{3} & -\frac{1}{3} & \frac{2}{3} & -\frac{1}{3}\\ -\frac{1}{3} & \frac{2}{3} & \frac{2}{3} & \frac{2}{3}\\ \frac{2}{3} & -\frac{1}{3} & -\frac{1}{3} & \frac{2}{3}\\ -\frac{1}{3} & \frac{2}{3} & -\frac{1}{3} & -\frac{1}{3} \end{bmatrix}.$

(c) $[A]_{\mathcal{C}} = P_{\mathcal{C}\leftarrow\mathcal{B}}\,[A]_{\mathcal{B}} = \begin{bmatrix} \frac{2}{3} & -\frac{1}{3} & \frac{2}{3} & -\frac{1}{3}\\ -\frac{1}{3} & \frac{2}{3} & \frac{2}{3} & \frac{2}{3}\\ \frac{2}{3} & -\frac{1}{3} & -\frac{1}{3} & \frac{2}{3}\\ -\frac{1}{3} & \frac{2}{3} & -\frac{1}{3} & -\frac{1}{3} \end{bmatrix}\begin{bmatrix}1\\1\\0\\0\end{bmatrix} = \begin{bmatrix}\frac{1}{3}\\\frac{1}{3}\\\frac{1}{3}\\\frac{1}{3}\end{bmatrix}.$

(d) $\left[\begin{array}{cccc|cccc} 1&0&1&1&1&1&1&0\\ 0&1&1&0&1&1&0&1\\ 0&1&0&1&0&1&1&1\\ 1&0&0&0&1&0&1&1 \end{array}\right] \longrightarrow \left[\begin{array}{cccc|cccc} 1&0&0&0&1&0&1&1\\ 0&1&0&0&1/2&1/2&1/2&3/2\\ 0&0&1&0&1/2&1/2&-1/2&-1/2\\ 0&0&0&1&-1/2&1/2&1/2&-1/2 \end{array}\right] \Rightarrow$

$P_{\mathcal{B}\leftarrow\mathcal{C}} = \begin{bmatrix} 1&0&1&1\\ 1/2&1/2&1/2&3/2\\ 1/2&1/2&-1/2&-1/2\\ -1/2&1/2&1/2&-1/2 \end{bmatrix}.$

(e) $[A]_{\mathcal{B}} = P_{\mathcal{B}\leftarrow\mathcal{C}}\,[A]_{\mathcal{C}} = \begin{bmatrix} 1&0&1&1\\ 1/2&1/2&1/2&3/2\\ 1/2&1/2&-1/2&-1/2\\ -1/2&1/2&1/2&-1/2 \end{bmatrix}\begin{bmatrix}1/3\\1/3\\1/3\\1/3\end{bmatrix} = \begin{bmatrix}1\\1\\0\\0\end{bmatrix}.$

11. (a) To find $[f(x)]_{\mathcal{B}}$, we need to solve the following function equation for a and b.
$f(x) = 2\sin x - 3\cos x = a(\sin x + \cos x) + b(\cos x) = a\sin x + (a+b)\cos x \Rightarrow$
$a = -2$ and $a + b = -3 \Rightarrow b = -5 \Rightarrow [f(x)]_{\mathcal{B}} = \begin{bmatrix} 2 \\ -5 \end{bmatrix}$.

Using coordinate vectors, we would find $[f(x)]_{\mathcal{B}}$ as follows:

$$\begin{bmatrix} 2 \\ -3 \end{bmatrix} = a_1 \begin{bmatrix} 1 \\ 1 \end{bmatrix} + a_2 \begin{bmatrix} 0 \\ 1 \end{bmatrix} \Rightarrow \begin{bmatrix} a_1 = 2 \\ a_2 = -5 \end{bmatrix} \Rightarrow [f(x)]_{\mathcal{B}} == \begin{bmatrix} 2 \\ -5 \end{bmatrix}.$$

Likewise, to find $[f(x)]_{\mathcal{C}}$, we need to solve the following for a and b.

$f(x) = 2\sin x - 3\cos x = a(\sin x) + b(\cos x) \Rightarrow a = 2$ and $b = -3 \Rightarrow [f(x)]_{\mathcal{C}} = \begin{bmatrix} 2 \\ -3 \end{bmatrix}$.

Using coordinate vectors, we would find $[f(x)]_{\mathcal{C}}$ as follows:

$$\begin{bmatrix} 2 \\ -3 \end{bmatrix} = b_1 \begin{bmatrix} 1 \\ 0 \end{bmatrix} + b_2 \begin{bmatrix} 0 \\ 1 \end{bmatrix} \Rightarrow \begin{matrix} b_1 = 2 \\ b_2 = -3 \end{matrix} \Rightarrow [f(x)]_{\mathcal{C}} = \begin{bmatrix} 2 \\ -3 \end{bmatrix}.$$

(b) $\left[\begin{array}{cc|cc} 1 & 0 & 1 & 0 \\ 1 & & 0 & 1 \end{array}\right] \longrightarrow \left[\begin{array}{cc|cc} 1 & 0 & 1 & 0 \\ 0 & 1 & 1 & 1 \end{array}\right] \Rightarrow P_{\mathcal{C}\leftarrow\mathcal{B}} = \begin{bmatrix} 1 & 0 \\ 1 & 1 \end{bmatrix} \Rightarrow.$

(c) $[f(x)]_{\mathcal{C}} = P_{\mathcal{C}\leftarrow\mathcal{B}}\,[f(x)]_{\mathcal{B}} = \begin{bmatrix} 1 & 0 \\ 1 & 1 \end{bmatrix} \begin{bmatrix} 2 \\ -5 \end{bmatrix} = \begin{bmatrix} 2 \\ -3 \end{bmatrix}.$

(d) $\left[\begin{array}{cc|cc} 1 & 0 & 1 & 0 \\ 0 & 1 & 1 & 1 \end{array}\right] \longrightarrow \left[\begin{array}{cc|cc} 1 & 0 & 1 & 0 \\ 0 & 1 & -1 & 1 \end{array}\right] \Rightarrow P_{\mathcal{B}\leftarrow\mathcal{C}} = \begin{bmatrix} 1 & 0 \\ -1 & 1 \end{bmatrix} \Rightarrow.$

(e) $[f(x)]_{\mathcal{B}} = P_{\mathcal{B}\leftarrow\mathcal{C}}\,[f(x)]_{\mathcal{C}} = \begin{bmatrix} 1 & 0 \\ -1 & 1 \end{bmatrix} = \begin{bmatrix} 2 \\ -3 \end{bmatrix} = \begin{bmatrix} 2 \\ -5 \end{bmatrix}.$

12. (a) To find $[f(x)]_{\mathcal{B}}$, we need to solve the following function equation for a and b.
$f(x) = \sin x = a(\sin x + \cos x) + b(\cos x) = a \sin x + (a+b)\cos x \Rightarrow$

$$a = 1 \text{ and } a + b = 0 \Rightarrow b = -1 \Rightarrow [f(x)]_{\mathcal{B}} = \begin{bmatrix} 1 \\ -1 \end{bmatrix}.$$

Using coordinate vectors, we would find $[f(x)]_{\mathcal{B}}$ as follows:

$$\begin{bmatrix} 1 \\ 0 \end{bmatrix} = a_1 \begin{bmatrix} 1 \\ 1 \end{bmatrix} + a_2 \begin{bmatrix} 0 \\ 1 \end{bmatrix} \Rightarrow \begin{bmatrix} a_1 = 1 \\ a_2 = -1 \end{bmatrix} \Rightarrow [f(x)]_{\mathcal{B}} = \begin{bmatrix} 1 \\ -1 \end{bmatrix}.$$

To find $[f(x)]_{\mathcal{C}}$, we need to solve the following function equation for a and b.
$f(x) = \sin x = a(\cos x - \sin x) + b(\sin x + \cos x) = (b-a)\sin x + (a+b)\cos x \Rightarrow$

$$b - a = 1 \text{ and } a + b = 0 \Rightarrow a = -\tfrac{1}{2} \text{ and } b = \tfrac{1}{2} \Rightarrow [f(x)]_{\mathcal{C}} = \begin{bmatrix} -\frac{1}{2} \\ \frac{1}{2} \end{bmatrix}.$$

Using coordinate vectors, we would find $[f(x)]_{\mathcal{B}}$ as follows:

$$\begin{bmatrix} 1 \\ 0 \end{bmatrix} = b_1 \begin{bmatrix} 1 \\ -1 \end{bmatrix} + b_2 \begin{bmatrix} 1 \\ 1 \end{bmatrix} \Rightarrow \begin{matrix} b_1 = -\frac{1}{2} \\ b_2 = \frac{1}{2} \end{matrix} \Rightarrow [f(x)]_{\mathcal{C}} = \begin{bmatrix} -\frac{1}{2} \\ \frac{1}{2} \end{bmatrix}.$$

(b) $$\left[\begin{array}{cc|cc} 1 & 1 & 1 & 0 \\ -1 & 1 & 1 & 1 \end{array}\right] \longrightarrow \left[\begin{array}{cc|cc} 1 & 0 & 0 & 1/2 \\ 0 & 1 & 1 & 1/2 \end{array}\right] \Rightarrow P_{\mathcal{C} \leftarrow \mathcal{B}} = \begin{bmatrix} 0 & 1/2 \\ 1 & 1/2 \end{bmatrix}.$$

(c) $$[f(x)]_{\mathcal{C}} = P_{\mathcal{C} \leftarrow \mathcal{B}}\,[f(x)]_{\mathcal{B}} = \begin{bmatrix} 0 & 1/2 \\ 1 & 1/2 \end{bmatrix} \begin{bmatrix} -1 \\ 1 \end{bmatrix} = \begin{bmatrix} -\frac{1}{2} \\ \frac{1}{2} \end{bmatrix}.$$

(d) $$\left[\begin{array}{cc|cc} 1 & 0 & 1 & 0 \\ 0 & 1 & 1 & 1 \end{array}\right] \longrightarrow \left[\begin{array}{cc|cc} 1 & 0 & -1 & 1 \\ 0 & 1 & 2 & 0 \end{array}\right] \Rightarrow P_{\mathcal{B} \leftarrow \mathcal{C}} = \begin{bmatrix} -1 & 1 \\ 2 & 0 \end{bmatrix}.$$

(e) $$[f(x)]_{\mathcal{B}} = P_{\mathcal{B} \leftarrow \mathcal{C}}\,[f(x)]_{\mathcal{C}} = \begin{bmatrix} -1 & 1 \\ 2 & 0 \end{bmatrix} \begin{bmatrix} -\frac{1}{2} \\ \frac{1}{2} \end{bmatrix} = \begin{bmatrix} 1 \\ -1 \end{bmatrix}.$$

13. (a) Section 3.6, Example 3.58 $\Rightarrow P_{\mathcal{C} \leftarrow \mathcal{B}} = \begin{bmatrix} \cos 60^\circ & \sin 60^\circ \\ -\sin 60^\circ & \cos 60^\circ \end{bmatrix} = \begin{bmatrix} 1/2 & \sqrt{3}/2 \\ -\sqrt{3}/2 & 1/2 \end{bmatrix}.$

So, $[\mathbf{x}]_{\mathcal{C}} = P_{\mathcal{C} \leftarrow \mathcal{B}}[\mathbf{x}]_{\mathcal{B}} = \begin{bmatrix} 1/2 & \sqrt{3}/2 \\ -\sqrt{3}/2 & 1/2 \end{bmatrix} \begin{bmatrix} 3 \\ 2 \end{bmatrix} = \begin{bmatrix} (3+2\sqrt{3})/2 \\ (2-3\sqrt{3})/2 \end{bmatrix}.$

(b) We have: $[\mathbf{y}]_{\mathcal{C}} = P_{\mathcal{C} \leftarrow \mathcal{B}}[\mathbf{y}]_{\mathcal{B}} = \begin{bmatrix} 1/2 & -\sqrt{3}/2 \\ \sqrt{3}/2 & 1/2 \end{bmatrix} \begin{bmatrix} 4 \\ -4 \end{bmatrix} = \begin{bmatrix} 2+2\sqrt{3} \\ 2\sqrt{3}-2 \end{bmatrix}.$

14. (a) Section 3.6, Example 3.58 $\Rightarrow P_{\mathcal{C} \leftarrow \mathcal{B}} = \begin{bmatrix} \cos 135^\circ & -\sin 135^\circ \\ \sin 135^\circ & \cos 135^\circ \end{bmatrix} = \begin{bmatrix} -\sqrt{2}/2 & -\sqrt{2}/2 \\ \sqrt{2}/2 & -\sqrt{2}/2 \end{bmatrix}.$

So, $[\mathbf{x}]_{\mathcal{C}} = P_{\mathcal{C} \leftarrow \mathcal{B}}[\mathbf{x}]_{\mathcal{B}} = \begin{bmatrix} -\sqrt{2}/2 & -\sqrt{2}/2 \\ \sqrt{2}/2 & -\sqrt{2}/2 \end{bmatrix} \begin{bmatrix} 3 \\ 2 \end{bmatrix} = \begin{bmatrix} -5\sqrt{2}/2 \\ \sqrt{2}/2 \end{bmatrix}.$

(b) Likewise, $[\mathbf{y}]_{\mathcal{C}} = P_{\mathcal{C} \leftarrow \mathcal{B}}[\mathbf{y}]_{\mathcal{B}} = \begin{bmatrix} -\sqrt{2}/2 & -\sqrt{2}/2 \\ \sqrt{2}/2 & -\sqrt{2}/2 \end{bmatrix} \begin{bmatrix} 4 \\ -4 \end{bmatrix} = \begin{bmatrix} 0 \\ 4\sqrt{2} \end{bmatrix}.$

15. $P_{\mathcal{C} \leftarrow \mathcal{B}} = \begin{bmatrix} 1 & -1 \\ -1 & 2 \end{bmatrix} \Rightarrow [\mathbf{u}_1]_{\mathcal{C}} = \begin{bmatrix} 1 \\ -1 \end{bmatrix}, [\mathbf{u}_2]_{\mathcal{C}} = \begin{bmatrix} -1 \\ 2 \end{bmatrix} \Rightarrow \mathbf{u}_1 = \begin{bmatrix} 1 \\ 2 \end{bmatrix} - \begin{bmatrix} 2 \\ 3 \end{bmatrix} = \begin{bmatrix} -1 \\ -1 \end{bmatrix}.$

Likewise, $\mathbf{u}_2 = -\begin{bmatrix} 1 \\ 2 \end{bmatrix} + 2\begin{bmatrix} 2 \\ 3 \end{bmatrix} = \begin{bmatrix} 3 \\ 4 \end{bmatrix} \Rightarrow \mathcal{B} = \{\mathbf{u}_1, \mathbf{u}_2\} = \left\{ \begin{bmatrix} -1 \\ -1 \end{bmatrix}, \begin{bmatrix} 3 \\ 4 \end{bmatrix} \right\}.$

16. Let $\mathcal{C} = \{p_1(x), p_2(x), p_3(x)\}$. Then we have:

$$P_{\mathcal{C}\leftarrow\mathcal{B}} = \begin{bmatrix} 1 & 0 & 0 \\ 0 & 2 & 1 \\ -1 & 1 & 1 \end{bmatrix} \Rightarrow [x]_{\mathcal{B}} = \begin{bmatrix} 1 \\ 0 \\ -1 \end{bmatrix}, [1+x]_{\mathcal{B}} = \begin{bmatrix} 0 \\ 2 \\ 1 \end{bmatrix}, [1-x+x^2]_{\mathcal{B}} = \begin{bmatrix} 0 \\ 1 \\ 1 \end{bmatrix} \Rightarrow$$

$$\begin{array}{l} x = v_1(x) - v_3(x) \\ 1+x = 2v_2(x) + v_3(x) \\ 1 - x + x^2 = v_2(x) + v_3(x) \end{array} \Rightarrow \mathcal{C} = \left\{ \begin{array}{l} v_1(x) = 1 - 2x + 2x^2 \\ v_2(x) = 2x - x^2 \\ v_3(x) = 1 - 3x + 2x^2 \end{array} \right\}.$$

17. $a = 1 \Rightarrow \mathcal{B} = \{1, x-1, (x-1)^2 = 1 - 2x + x^2\}$ is the Taylor Polynomial basis.

Let $\mathcal{C} = \{1, x, x^2\} \Rightarrow [p(x)]_{\mathcal{C}} = \begin{bmatrix} 1 \\ 2 \\ -5 \end{bmatrix}$.

$$[\,B\,|\,C\,] \longrightarrow [\,I\,|\,P_{\mathcal{B}\leftarrow\mathcal{C}}\,] \Rightarrow \left[\begin{array}{ccc|ccc} 1 & -1 & 1 & 1 & 0 & 0 \\ 0 & 1 & -2 & 0 & 1 & 0 \\ 0 & 0 & 1 & 0 & 0 & 1 \end{array}\right] \longrightarrow \left[\begin{array}{ccc|ccc} 1 & 0 & 0 & 1 & 1 & 1 \\ 0 & 1 & 0 & 0 & 1 & 2 \\ 0 & 0 & 1 & 0 & 0 & 1 \end{array}\right] \Rightarrow$$

$$P_{\mathcal{B}\leftarrow\mathcal{C}} = \begin{bmatrix} 1 & 1 & 1 \\ 0 & 1 & 2 \\ 0 & 0 & 1 \end{bmatrix} \Rightarrow [p(x)]_{\mathcal{B}} = P_{\mathcal{B}\leftarrow\mathcal{C}}\,[p(x)]_{\mathcal{C}} = \begin{bmatrix} 1 & 1 & 1 \\ 0 & 1 & 2 \\ 0 & 0 & 1 \end{bmatrix} \begin{bmatrix} 1 \\ 2 \\ -5 \end{bmatrix} = \begin{bmatrix} -2 \\ -8 \\ -5 \end{bmatrix} \Rightarrow$$

$p(x) = -2(1) - 8(x-1) - 5(x-1)^2$.

18. $a = -2 \Rightarrow \mathcal{B} = \{1, x+2, (x+2)^2 = 4 + 4x + x^2\}$ is the Taylor Polynomial basis.

Let $\mathcal{C} = \{1, x, x^2\} \Rightarrow [p(x)]_{\mathcal{C}} = \begin{bmatrix} 1 \\ 2 \\ -5 \end{bmatrix}$.

$$[\,B\,|\,C\,] \longrightarrow [\,I\,|\,P_{\mathcal{B}\leftarrow\mathcal{C}}\,] \Rightarrow \left[\begin{array}{ccc|ccc} 1 & 2 & 4 & 1 & 0 & 0 \\ 0 & 1 & 4 & 0 & 1 & 0 \\ 0 & 0 & 1 & 0 & 0 & 1 \end{array}\right] \longrightarrow \left[\begin{array}{ccc|ccc} 1 & 0 & 0 & 1 & -2 & 4 \\ 0 & 1 & 0 & 0 & 1 & -4 \\ 0 & 0 & 1 & 0 & 0 & 1 \end{array}\right] \Rightarrow$$

$$P_{\mathcal{B}\leftarrow\mathcal{C}} = \begin{bmatrix} 1 & -2 & 4 \\ 0 & 1 & -4 \\ 0 & 0 & 1 \end{bmatrix} \Rightarrow [p(x)]_{\mathcal{B}} = P_{\mathcal{B}\leftarrow\mathcal{C}}\,[p(x)]_{\mathcal{C}} = \begin{bmatrix} 1 & -2 & 4 \\ 0 & 1 & -4 \\ 0 & 0 & 1 \end{bmatrix} \begin{bmatrix} 1 \\ 2 \\ -5 \end{bmatrix} = \begin{bmatrix} -23 \\ 22 \\ -5 \end{bmatrix} \Rightarrow$$

$p(x) = -23(1) + 22(x+2) - 5(x+2)^2$.

19. $a = -1 \Rightarrow \mathcal{B} = \{1, x+1, (x+1)^2 = 1 + 2x + x^2, (x+1)^3 = 1 + 3x + 3x^2 + x^3\}$ is the Taylor Polynomial basis. Let $\mathcal{C} = \{1, x, x^2, x^3\} \Rightarrow$

$$[1]_{\mathcal{C}} = \begin{bmatrix} 1 \\ 0 \\ 0 \\ 0 \end{bmatrix}, \ [x+1]_{\mathcal{C}} = \begin{bmatrix} 1 \\ 1 \\ 0 \\ 0 \end{bmatrix}, \ [(x+1)^2]_{\mathcal{C}} = \begin{bmatrix} 1 \\ 2 \\ 1 \\ 0 \end{bmatrix}, \ [(x+1)^3]_{\mathcal{C}} = \begin{bmatrix} 1 \\ 3 \\ 3 \\ 1 \end{bmatrix} \Rightarrow$$

$[\,B \,|\, C\,] \longrightarrow [\,I \,|\, P_{\mathcal{B}\leftarrow\mathcal{C}}\,] \Rightarrow$

$$\left[\begin{array}{cccc|cccc} 1 & 1 & 1 & 1 & 1 & 0 & 0 & 0 \\ 0 & 1 & 2 & 3 & 0 & 1 & 0 & 0 \\ 0 & 0 & 1 & 3 & 0 & 0 & 1 & 0 \\ 0 & 0 & 0 & 1 & 0 & 0 & 0 & 1 \end{array}\right] \longrightarrow \left[\begin{array}{cccc|cccc} 1 & 0 & 0 & 0 & 1 & -1 & 1 & -1 \\ 0 & 1 & 0 & 0 & 0 & 1 & -2 & 3 \\ 0 & 0 & 1 & 0 & 0 & 0 & 1 & -3 \\ 0 & 0 & 0 & 1 & 0 & 0 & 0 & 1 \end{array}\right] \Rightarrow$$

$$P_{\mathcal{B}\leftarrow\mathcal{C}} = \begin{bmatrix} 1 & -1 & 1 & 1 \\ 0 & 1 & -2 & 3 \\ 0 & 0 & 1 & -3 \\ 0 & 0 & 0 & 1 \end{bmatrix} \Rightarrow$$

$$[p(x)]_{\mathcal{B}} = P_{\mathcal{B}\leftarrow\mathcal{C}}\,[p(x)]_{\mathcal{C}} = \begin{bmatrix} 1 & -1 & 1 & -1 \\ 0 & 1 & -2 & 3 \\ 0 & 0 & 1 & -3 \\ 0 & 0 & 0 & 1 \end{bmatrix} \begin{bmatrix} 0 \\ 0 \\ 0 \\ 1 \end{bmatrix} = \begin{bmatrix} -1 \\ 3 \\ -3 \\ 1 \end{bmatrix} \Rightarrow$$

$p(x) = -(1) + 3(x+1) - 3(x+1)^2 + 1(x+1)^3.$

20. $a = \frac{1}{2} \Rightarrow \mathcal{B} = \{1, x - \frac{1}{2}, (x - \frac{1}{2})^2 = \frac{1}{4} - x + x^2, (x - \frac{1}{2})^3 = -\frac{1}{8} + \frac{3}{4}x - \frac{3}{2}x^2 + x^3\}$ is the Taylor Polynomial basis. Let $\mathcal{C} = \{1, x, x^2, x^3\} \Rightarrow$

$$[1]_{\mathcal{C}} = \begin{bmatrix} 1 \\ 0 \\ 0 \\ 0 \end{bmatrix}, \ [x - \tfrac{1}{2}]_{\mathcal{C}} = \begin{bmatrix} -1/2 \\ 1 \\ 0 \\ 0 \end{bmatrix}, \ [(x - \tfrac{1}{2})^2]_{\mathcal{C}} = \begin{bmatrix} 1/4 \\ -1 \\ 1 \\ 0 \end{bmatrix}, \ [(x - \tfrac{1}{2})^3]_{\mathcal{C}} = \begin{bmatrix} -1/8 \\ 3/4 \\ -3/2 \\ 1 \end{bmatrix} \Rightarrow$$

$[\,B \,|\, C\,] \longrightarrow [\,I \,|\, P_{\mathcal{B}\leftarrow\mathcal{C}}\,] \Rightarrow$

$$[\,I \,|\, P_{\mathcal{B}\leftarrow\mathcal{C}}\,] \Rightarrow \left[\begin{array}{cccc|cccc} 1 & -1/2 & 1/4 & -1/8 & 1 & 0 & 0 & 0 \\ 0 & 1 & -1 & 3/4 & 0 & 1 & 0 & 0 \\ 0 & 0 & 1 & -3/2 & 0 & 0 & 1 & 0 \\ 0 & 0 & 0 & 1 & 0 & 0 & 0 & 1 \end{array}\right] \longrightarrow \left[\begin{array}{cccc|cccc} 1 & 0 & 0 & 0 & 1 & 1/2 & 1/4 & 1/8 \\ 0 & 1 & 0 & 0 & 0 & 1 & 1 & 3/4 \\ 0 & 0 & 1 & 0 & 0 & 0 & 1 & 3/2 \\ 0 & 0 & 0 & 1 & 0 & 0 & 0 & 1 \end{array}\right] \Rightarrow$$

$$P_{\mathcal{B}\leftarrow\mathcal{C}} = \begin{bmatrix} 1 & 1/2 & 1/4 & 1/8 \\ 0 & 1 & 1 & 3/4 \\ 0 & 0 & 1 & 3/2 \\ 0 & 0 & 0 & 1 \end{bmatrix} \Rightarrow$$

$$[p(x)]_{\mathcal{B}} = P_{\mathcal{B}\leftarrow\mathcal{C}}\,[p(x)]_{\mathcal{C}} = \begin{bmatrix} 1 & 1/2 & 1/4 & 1/8 \\ 0 & 1 & 1 & 3/4 \\ 0 & 0 & 1 & 3/2 \\ 0 & 0 & 0 & 1 \end{bmatrix} \begin{bmatrix} 0 \\ 0 \\ 0 \\ 1 \end{bmatrix} = \begin{bmatrix} 1/8 \\ 3/4 \\ 3/2 \\ 1 \end{bmatrix} \Rightarrow$$

$p(x) = \frac{1}{8}(1) + \frac{3}{4}(x - \frac{1}{2}) + \frac{3}{2}(x - \frac{1}{2})^2 + (x - \frac{1}{2})^3.$

21. Need only show $[\mathbf{x}]_{\mathcal{D}} = P_{\mathcal{D}\leftarrow\mathcal{C}}\, P_{\mathcal{C}\leftarrow\mathcal{B}}\, [\mathbf{x}]_{\mathcal{B}}$ since this matrix with this property is unique.

By definition, $[\mathbf{x}]_{\mathcal{D}} = P_{\mathcal{D}\leftarrow\mathcal{C}}\, [\mathbf{x}]_{\mathcal{C}}$ (1) and $[\mathbf{x}]_{\mathcal{C}} = P_{\mathcal{C}\leftarrow\mathcal{B}}\, [\mathbf{x}]_{\mathcal{B}}$ (2).

Substituting (2) into (1) $\Rightarrow$ $[\mathbf{x}]_{\mathcal{D}} = P_{\mathcal{D}\leftarrow\mathcal{C}}\, P_{\mathcal{C}\leftarrow\mathcal{B}}\, [\mathbf{x}]_{\mathcal{B}}$ which is what we needed to show.

22. Section 6.2, Theorem 6.10c. $\Rightarrow$ only need show $\{\mathbf{u}_1, \ldots, \mathbf{u}_n\}$ linearly independent since $\dim V = n$.

Note, by definition $[\mathbf{u}_i]_{\mathcal{B}} = \mathbf{p}_i$, the ith column of P.

Recall, P invertible $\Rightarrow$ the columns of P are linearly independent $\Rightarrow$ $\{[\mathbf{u}_1]_{\mathcal{B}}, \ldots, [\mathbf{u}_n]_{\mathcal{B}}\} = \{\mathbf{p}_1, \ldots, \mathbf{p}_n\}$ are linearly independent.

Therefore, Section 6.2, Theorem 6.7 $\Rightarrow$ $\{\mathbf{u}_1, \ldots, \mathbf{u}_n\}$ are linearly independent $\Rightarrow$ $\{\mathbf{u}_1, \ldots, \mathbf{u}_n\}$is a basis for V.

Need only show $[\mathbf{v}]_{\mathcal{B}} = P\, [\mathbf{v}]_{\mathcal{C}}$for all $\mathbf{v}$ in Vsince this matrix with this property is unique.

By definition $[\mathbf{u}_i]_{\mathcal{C}} = \mathbf{e}_i$, the ith standard column vector of $\mathbb{R}^n$ $\Rightarrow$

$[\mathbf{u}_i]_{\mathcal{B}} = P\, [\mathbf{u}_i]_{\mathcal{C}} = P\, \mathbf{e}_i = \mathbf{p}_i$.

Now since $\{\mathbf{u}_1, \ldots, \mathbf{u}_n\}$ is a basis for V for any vector in V we can write $\mathbf{v} = \sum c_i\, \mathbf{u}_i$ $\Rightarrow$

$[\mathbf{v}]_{\mathcal{B}} = [\sum c_i\, \mathbf{u}_i]_{\mathcal{B}} = \sum c_i\, [\mathbf{u}_i]_{\mathcal{B}} = \sum c_i\, P[\mathbf{u}_i]_{\mathcal{C}} = P(\sum c_i\, [\mathbf{u}_i]_{\mathcal{C}}) = P([\sum c_i\, \mathbf{u}_i]_{\mathcal{C}}) = P([\mathbf{v}]_{\mathcal{C}}$.

6.4 Linear Transformations

1. $T\left(\begin{bmatrix} a & b \\ c & d \end{bmatrix} + \begin{bmatrix} a' & b' \\ c' & d' \end{bmatrix}\right) = T\begin{bmatrix} a+a' & b+b' \\ c+c' & d+d' \end{bmatrix} = \begin{bmatrix} a+a'+b+b' & 0 \\ 0 & c+c'+d+d' \end{bmatrix}$

$= \begin{bmatrix} a+b & 0 \\ 0 & c+d \end{bmatrix} + \begin{bmatrix} a'+b' & 0 \\ 0 & c'+d' \end{bmatrix} = T\begin{bmatrix} a & b \\ c & d \end{bmatrix} + T\begin{bmatrix} a' & b' \\ c' & d' \end{bmatrix}.$

$T\left(\alpha\begin{bmatrix} a & b \\ c & d \end{bmatrix}\right) = T\begin{bmatrix} \alpha a & \alpha b \\ \alpha c & \alpha d \end{bmatrix} = \begin{bmatrix} \alpha a + \alpha b & 0 \\ 0 & \alpha c + \alpha d \end{bmatrix} = \alpha\begin{bmatrix} a+b & 0 \\ 0 & c+d \end{bmatrix} = \alpha T\begin{bmatrix} a & b \\ c & d \end{bmatrix} \Rightarrow$

T is a linear transformation.

2. $T\begin{bmatrix} 0 & 0 \\ 0 & 0 \end{bmatrix} = \begin{bmatrix} 1 & 0 \\ 0 & 1 \end{bmatrix} \neq \begin{bmatrix} 0 & 0 \\ 0 & 0 \end{bmatrix} \Rightarrow T$ is *not* a linear transformation (it fails Theorem 6.14a).

Note this is an application of the principle $p \to q \leftrightarrow -q \to -p$.

3. $T(A+C) = (A+C)B = AB + CB = T(A) + T(C)$ and
$T(\alpha A) = (\alpha A)B = \alpha(AB) = \alpha T(A) \Rightarrow T$ is a linear transformation.

4. $T(A+C) = (A+C)B - B(A+C) = AB + CB - BA - BC = (AB - BA) + (CB - BC) =$
$T(A) + T(C)$ and $T(\alpha A) = (\alpha A)B - B(\alpha A) = \alpha(AB - BA) = \alpha T(A) \Rightarrow$
T is a linear transformation.

5. By Exercise 44, Section 3.2 $\text{tr}(A+C) = \text{tr}(A) + \text{tr}(C)$ and $\text{tr}(\alpha A) = \alpha\,\text{tr}(A) \Rightarrow$
$T = \text{tr}$ is a linear transformation.

6. No, T is not a linear transformation. Consider a counter example in M_{22}.

$T\left(\begin{bmatrix} 1 & 0 \\ 0 & 1 \end{bmatrix} + \begin{bmatrix} 1 & 0 \\ 0 & 2 \end{bmatrix}\right) = T\begin{bmatrix} 2 & 0 \\ 0 & 3 \end{bmatrix} = 6 \neq 1 + 2 = T\begin{bmatrix} 1 & 0 \\ 0 & 1 \end{bmatrix} + T\begin{bmatrix} 1 & 0 \\ 0 & 2 \end{bmatrix}.$

T's failure comes from the fact that $(a_{11} + b_{11})(a_{22} + b_{22}) \neq a_{11}b_{11} + a_{22}b_{22}$ in general.

7. Let A be a matrix with $\text{rank}(A) \neq 0$, then $T(A) = T(-A) = \text{rank}(A) \Rightarrow$ $T(-\mathbf{v}) \neq -T(\mathbf{v}) \Rightarrow T$ is *not* a linear transformation because it fails Theorem 6.14b. Note this is another application of the principle $p \to q \Leftrightarrow -q \to -p$. We specify $\text{rank}(A) \neq 0$ since zero is the only number such that $0 = -0$.

8. $T(0) = 1 + x + x^2 \neq \mathbf{0} \Rightarrow T(\mathbf{0}) \neq \mathbf{0} \Rightarrow T$ is *not* a linear transformation.

9. $T(\,(a + bx + cx^2) + (a' + b'x + c'x^2)\,) = T(\,(a + a') + (b + b')x + (c + c')x^2\,)$
$= (a+a')+(b+b')(x+1)+(c+c')(x+1)^2 = (a+b(x+1)+c(x+1)^2)+(a'+b'(x+1)+c'(x+1)^2)$
$= T((a + bx + cx^2) + T(a' + b'x + c'x^2)$ and
$T(\,\alpha(a + bx + cx^2)\,) = T(\alpha a + \alpha bx + \alpha cx^2)) = \alpha a + \alpha b(x + 1) + \alpha c(x + 1)^2$
$= \alpha(a + b(x + 1) + c(x + 1)^2) = \alpha T(a + bx + cx^2) \Rightarrow T$ is a linear transformation.

10. $T(f + g) = (f + g)(x^2) = f(x^2) + g(x^2) = T(f) + T(g)$ and
$T(\alpha f) = (\alpha f)(x^2) = \alpha f(x^2) = \alpha T(f) \Rightarrow T$ is a linear transformation.

11. $T(f) = (f(x))^2) = (-f(x))^2) = T(-f) \Rightarrow T(-\mathbf{v}) \neq -T(\mathbf{v}) \Rightarrow T$ is *not* linear.

12. Similar to our work in Exercise 10, letting $x^2 = c \Rightarrow T$ is a linear transformation.

13. $S(\,p(x) + q(x)\,) = x(p(x) + q(x)) = xp(x) + xq(x) = S(p(x)) + S(q(x)$ and
$S(\lambda p(x)) = x(\lambda p(x)) = \lambda(xp(x)) = \lambda S(p(x)) \Rightarrow S$ is a linear transformation.

$$T\left(\begin{bmatrix} a \\ b \end{bmatrix} + \begin{bmatrix} a' \\ b' \end{bmatrix}\right) = T\begin{bmatrix} a + a' \\ b + b' \end{bmatrix}$$

$$= (a + a') + ((a + a') + (b + b'))x = (a + (a + b)x) + (a' + (a' + b')x) = T\begin{bmatrix} a \\ b \end{bmatrix} + T\begin{bmatrix} a' \\ b' \end{bmatrix}.$$

$$T\left(\alpha\begin{bmatrix} a \\ b \end{bmatrix}\right) = T\begin{bmatrix} \alpha a \\ \alpha b \end{bmatrix} = \alpha a + (\alpha a + \alpha b)x = \alpha(a + (a + b)x) = \alpha a T\begin{bmatrix} a \\ b \end{bmatrix} \Rightarrow T \text{ is linear.}$$

14. Since $\begin{bmatrix} 5 \\ 2 \end{bmatrix} = 5\begin{bmatrix} 1 \\ 0 \end{bmatrix} + 2\begin{bmatrix} 0 \\ 1 \end{bmatrix}$, we have:

$$T\begin{bmatrix} 5 \\ 2 \end{bmatrix} = 5T\begin{bmatrix} 1 \\ 0 \end{bmatrix} + 2T\begin{bmatrix} 0 \\ 1 \end{bmatrix} = 5\begin{bmatrix} 1 \\ 2 \\ -1 \end{bmatrix} + 2\begin{bmatrix} 3 \\ 0 \\ 4 \end{bmatrix} = \begin{bmatrix} 11 \\ 10 \\ 3 \end{bmatrix}.$$

Since $\begin{bmatrix} a \\ b \end{bmatrix} = a\begin{bmatrix} 1 \\ 0 \end{bmatrix} + b\begin{bmatrix} 0 \\ 1 \end{bmatrix}$, we have:

$$T\begin{bmatrix} a \\ b \end{bmatrix} = aT\begin{bmatrix} 1 \\ 0 \end{bmatrix} + bT\begin{bmatrix} 0 \\ 1 \end{bmatrix} = a\begin{bmatrix} 1 \\ 2 \\ -1 \end{bmatrix} + b\begin{bmatrix} 3 \\ 0 \\ 4 \end{bmatrix} = \begin{bmatrix} a + 3b \\ 2a \\ -a + 4b \end{bmatrix}.$$

15. First, we need to solve $\begin{bmatrix} -7 \\ 9 \end{bmatrix} = c_1 \begin{bmatrix} 1 \\ 1 \end{bmatrix} + c_2 \begin{bmatrix} 3 \\ -1 \end{bmatrix} \Rightarrow \begin{array}{rcr} c_1 + 3c_2 &=& -7 \\ c_1 - c_2 &=& 9 \end{array} \Rightarrow \begin{array}{rcr} c_1 &=& 5 \\ c_2 &=& -4 \end{array}.$

So, since $\begin{bmatrix} -7 \\ 9 \end{bmatrix} = 5 \begin{bmatrix} 1 \\ 1 \end{bmatrix} - 4 \begin{bmatrix} 3 \\ -1 \end{bmatrix}$, we have:

$$T \begin{bmatrix} -7 \\ 9 \end{bmatrix} = 5\,T \begin{bmatrix} 1 \\ 1 \end{bmatrix} - 4\,T \begin{bmatrix} 3 \\ -1 \end{bmatrix} = 5(1-2x) - 4(x+2x^2) = 5 - 14x - 8x^2.$$

Now, we need to solve $\begin{bmatrix} a \\ b \end{bmatrix} = c_1 \begin{bmatrix} 1 \\ 1 \end{bmatrix} + c_2 \begin{bmatrix} 3 \\ -1 \end{bmatrix} \Rightarrow \begin{array}{rcl} c_1 + 3c_2 &=& a \\ c_1 - c_2 &=& b \end{array} \Rightarrow \begin{array}{rcl} c_1 &=& \frac{a+3b}{4} \\ c_2 &=& \frac{a-b}{4} \end{array}.$

So, since $\begin{bmatrix} a \\ b \end{bmatrix} = \frac{a+3b}{4} \begin{bmatrix} 1 \\ 1 \end{bmatrix} + \frac{a-b}{4} \begin{bmatrix} 3 \\ -1 \end{bmatrix}$, we have:

$$T \begin{bmatrix} a \\ b \end{bmatrix} = \frac{a+3b}{4}\,T \begin{bmatrix} 1 \\ 1 \end{bmatrix} + \frac{a-b}{4}\,T \begin{bmatrix} 3 \\ -1 \end{bmatrix} = \frac{a+3b}{4}(1-2x) + \frac{a-b}{4}(x+2x^2) = \frac{a+3b}{4} - \frac{a+7b}{4}\,x + \frac{a-b}{4}\,x^2.$$

16. We need to solve $6 + x - 4x^2 = c_1(1) + c_2(x) + c_3(x^2) \Rightarrow c_1 = 6$, $c_2 = 1$, and $c_3 = -4$.

So, since $6 + x - 4x^2 = 6(1) + 1(x) - 4(x^2)$, we have:

$$\begin{aligned} T(6 + x - 4x^2) &= 6\,T(1) + T(x) - 4\,T(x^2) = 6(3-2x) + (4x - x^2) - 4(2 + 2x^2) \\ &= 10 - 8x - 9x^2. \end{aligned}$$

We need to solve $a + bx + cx^2 = c_1(1) + c_2(x) + c_3(x^2) \Rightarrow c_1 = a$, $c_2 = b$, and $c_3 = c$.

So, since $a + bx + cx^2 = a(1) + b(x) + c(x^2)$, we have:

$$\begin{aligned} T(a + bx + cx^2) &= a\,T(1) + b\,T(x) + c\,T(x^2) = a(3-2x) + b(4x - x^2) + c(2 + 2x^2) \\ &= (3a + 2c) + (4b - 2a)x + (2c - b)x^2. \end{aligned}$$

17. We need to solve $4 - x + 3x^2 = c_1(1+x) + c_2(x+x^2) + c_3(1+x^2)$

$$= (c_1 + c_3) + (c_1 + c_2)x + (c_2 + c_3)x^2 \Rightarrow \begin{array}{rcr} c_1 + c_3 &=& 4 \\ c_1 + c_2 &=& -1 \\ c_2 + c_3 &=& 3 \end{array} \Rightarrow \begin{array}{rcr} c_1 &=& 0 \\ c_2 &=& -1 \\ c_3 &=& 4 \end{array}.$$

So, since $4 - x + 3x^2 = 0(1+x) - (x+x^2) + 4(1+x^2)$, we have:

$$\begin{aligned} T(4 - x + 3x^2) &= 0\,T(1+x) - T(x+x^2) + 4\,T(1+x^2) = 0(1+x^2) - (x - x^2) + 4(1 + x + x^2) \\ &= 4 + 3x + 5x^2. \end{aligned}$$

We need to solve $a + bx + cx^2 = c_1(1+x) + c_2(x+x^2) + c_3(1+x^2)$

$$= (c_1 + c_3) + (c_1 + c_2)x + (c_2 + c_3)x^2 \Rightarrow \begin{array}{rcl} c_1 + c_3 &=& a \\ c_1 + c_2 &=& b \\ c_2 + c_3 &=& c \end{array} \Rightarrow \begin{array}{rcl} c_1 &=& \frac{a+b-c}{2} \\ c_2 &=& \frac{-a+b+c}{2} \\ c_3 &=& \frac{a-b+c}{2} \end{array}.$$

So, since $a + bx + cx^2 = \frac{a+b-c}{2}(1+x) + \frac{-a+b+c}{2}(x+x^2) + \frac{a-b+c}{2}(1+x^2)$, we have:

$$\begin{aligned} T(a + bx + cx^2) &= \frac{a+b-c}{2}\,T(1+x) + \frac{-a+b+c}{2}\,T(x+x^2) + \frac{a-b+c}{2}\,T(1+x^2) \\ &= \frac{a+b-c}{2}(1+x^2) + \frac{-a+b+c}{2}(x-x^2) + \frac{a-b+c}{2}(1+x+x^2) = a + cx + \left(\frac{3a-b-c}{2}\right)x^2. \end{aligned}$$

18. We need to solve $\begin{bmatrix} 1 & 3 \\ 4 & 2 \end{bmatrix} = c_1 \begin{bmatrix} 1 & 0 \\ 0 & 0 \end{bmatrix} + c_2 \begin{bmatrix} 1 & 1 \\ 0 & 0 \end{bmatrix} + c_3 \begin{bmatrix} 1 & 1 \\ 1 & 0 \end{bmatrix} + c_4 \begin{bmatrix} 1 & 1 \\ 1 & 1 \end{bmatrix} \Rightarrow$

$$\begin{aligned} c_1 + c_2 + c_3 + c_4 &= 1 \\ c_2 + c_3 + c_4 &= 3 \\ c_3 + c_4 &= 4 \\ c_4 &= 2 \end{aligned} \Rightarrow \begin{aligned} c_1 &= -2 \\ c_2 &= -1 \\ c_3 &= 2 \\ c_4 &= 2 \end{aligned}.$$

So, since $\begin{bmatrix} 1 & 3 \\ 4 & 2 \end{bmatrix} = -2 \begin{bmatrix} 1 & 0 \\ 0 & 0 \end{bmatrix} - 1 \begin{bmatrix} 1 & 1 \\ 0 & 0 \end{bmatrix} + 2 \begin{bmatrix} 1 & 1 \\ 1 & 0 \end{bmatrix} + 2 \begin{bmatrix} 1 & 1 \\ 1 & 1 \end{bmatrix}$, we have:

$$T\left(\begin{bmatrix} 1 & 3 \\ 4 & 2 \end{bmatrix}\right) = -2T\left(\begin{bmatrix} 1 & 0 \\ 0 & 0 \end{bmatrix}\right) - 1T\left(\begin{bmatrix} 1 & 1 \\ 0 & 0 \end{bmatrix}\right) + 2T\left(\begin{bmatrix} 1 & 1 \\ 1 & 0 \end{bmatrix}\right) + 2T\left(\begin{bmatrix} 1 & 1 \\ 1 & 1 \end{bmatrix}\right)$$

$$= -2(1) - 1(2) + 2(3) + 2(4) = 10.$$

We need to solve $\begin{bmatrix} a & b \\ c & d \end{bmatrix} = c_1 \begin{bmatrix} 1 & 0 \\ 0 & 0 \end{bmatrix} + c_2 \begin{bmatrix} 1 & 1 \\ 0 & 0 \end{bmatrix} + c_3 \begin{bmatrix} 1 & 1 \\ 1 & 0 \end{bmatrix} + c_4 \begin{bmatrix} 1 & 1 \\ 1 & 1 \end{bmatrix} \Rightarrow$

$$\begin{aligned} c_1 + c_2 + c_3 + c_4 &= a \\ c_2 + c_3 + c_4 &= b \\ c_3 + c_4 &= c \\ c_4 &= d \end{aligned} \Rightarrow \begin{aligned} c_1 &= a - b \\ c_2 &= b - c \\ c_3 &= c - d \\ c_4 &= d \end{aligned}.$$

So, since $\begin{bmatrix} a & b \\ c & d \end{bmatrix} = (a-b) \begin{bmatrix} 1 & 0 \\ 0 & 0 \end{bmatrix} + (b-c) \begin{bmatrix} 1 & 1 \\ 0 & 0 \end{bmatrix} + (c-d) \begin{bmatrix} 1 & 1 \\ 1 & 0 \end{bmatrix} + d \begin{bmatrix} 1 & 1 \\ 1 & 1 \end{bmatrix}$, we have:

$$T\left(\begin{bmatrix} a & b \\ c & d \end{bmatrix}\right) = (a-b)T\left(\begin{bmatrix} 1 & 0 \\ 0 & 0 \end{bmatrix}\right) + (b-c)T\left(\begin{bmatrix} 1 & 1 \\ 0 & 0 \end{bmatrix}\right) + (c-d)T\left(\begin{bmatrix} 1 & 1 \\ 1 & 0 \end{bmatrix}\right) + dT\left(\begin{bmatrix} 1 & 1 \\ 1 & 1 \end{bmatrix}\right)$$

$$= (a-b)(1) + (b-c)(2) + (c-d)(3) + (d)(4) = a + b + c + d.$$

19. Let $T(E_{11}) = a$, $T(E_{12}) = b$, $T(E_{21}) = c$, and $T(E_{22}) = d$.

Then note $\begin{bmatrix} w & x \\ y & z \end{bmatrix} = wE_{11} + xE_{12} + yE_{21} + zE_{22} \Rightarrow$

$$T\begin{bmatrix} w & x \\ y & z \end{bmatrix} = T(wE_{11} + xE_{12} + yE_{21} + zE_{22}) = T(wE_{11}) + T(xE_{12}) + T(yE_{21}) + T(zE_{22})$$

$$= aw + bx + cy + dz \text{ as required.}$$

20. If order for T to be a linear transformation, the fact that $\begin{bmatrix} 0 \\ 6 \\ -8 \end{bmatrix} = 6 \begin{bmatrix} 2 \\ 1 \\ 0 \end{bmatrix} - 4 \begin{bmatrix} 3 \\ 0 \\ 2 \end{bmatrix}$ would have to imply that $T \begin{bmatrix} 0 \\ 6 \\ -8 \end{bmatrix} = 6T \begin{bmatrix} 2 \\ 1 \\ 0 \end{bmatrix} - 4T \begin{bmatrix} 3 \\ 0 \\ 2 \end{bmatrix} = 6(1+x) - 4(2-x+x^2) = -2+10x-4x^2$,

but $-2 + 10x - 4x^2 \neq -2 + 2x^2 = T \begin{bmatrix} 0 \\ 6 \\ -8 \end{bmatrix}$. Therefore, T cannot be linear.

21. Recall, by definition, in any vector space V, $\mathbf{v} + (-\mathbf{v}) = \mathbf{0}$. Then Theorem 6.14a $\Rightarrow$
$T(\mathbf{0}) = T(\mathbf{v} + (-\mathbf{v})) = T(\mathbf{v}) + T(-\mathbf{v}) = 0 \Rightarrow T(-\mathbf{v}) = -T(\mathbf{v})$ *by definition.*

22. Let $\mathbf{v} = \sum c_i\mathbf{v}_i$, then $T(\mathbf{v}) = T(\sum c_i\mathbf{v}_i) = \sum c_iT(\mathbf{v}_i) = \sum c_i\mathbf{v}_i = \mathbf{v}$ as required.

23. Let $p(x) = \sum a_kx^k$, then $T(\,p(x)\,) = T(\sum a_kx^k) = \sum a_kT(x^k) = \sum a_kkx^{k-1} = p'(x)$.

24. (a) Must show $\sum c_i\mathbf{v}_i = \mathbf{0} \Rightarrow c_i = 0$ for all i. So, let $\sum c_i\mathbf{v}_i = \mathbf{0} \Rightarrow$
$\mathbf{0} = T(\mathbf{0}) = T(\sum c_i\mathbf{v}_i) = \sum c_iT(\mathbf{v}_i)) \Rightarrow$
$c_i = 0$ for all i because $T(\mathbf{v}_i)$ linearly independent.

(b) Define $T : \mathbb{R}^2 \to \mathbb{R}^2$ as follows: $T\begin{bmatrix} x \\ y \end{bmatrix} = \begin{bmatrix} x \\ 0 \end{bmatrix}$.

Then $\mathbf{e}_1 = \begin{bmatrix} 1 \\ 0 \end{bmatrix}$ and $\mathbf{e}_2 = \begin{bmatrix} 0 \\ 1 \end{bmatrix}$ are linearly independent.

But, $T(\mathbf{e}_1) = \begin{bmatrix} 1 \\ 0 \end{bmatrix}$ and $T(\mathbf{e}_2) = \begin{bmatrix} 0 \\ 0 \end{bmatrix}$ are linearly dependent.

25. $(S \circ T)\begin{bmatrix} 2 \\ 1 \end{bmatrix} = S\left(T\begin{bmatrix} 2 \\ 1 \end{bmatrix}\right) = S\begin{bmatrix} 5 \\ -1 \end{bmatrix} = \begin{bmatrix} 5+(-1) & -1 \\ 0 & 5-(-1) \end{bmatrix} = \begin{bmatrix} 4 & -1 \\ 0 & 6 \end{bmatrix}$.

$(S \circ T)\begin{bmatrix} x \\ y \end{bmatrix} = S\left(T\begin{bmatrix} x \\ y \end{bmatrix}\right) = S\begin{bmatrix} 2x+y \\ -y \end{bmatrix} = \begin{bmatrix} 2x & -y \\ 0 & 2x+2y \end{bmatrix}$.

Domain of $T = \mathbb{R}^2 \neq M_{22} =$ codomain of $S \Rightarrow$ we cannot compute $T \circ S$.

26. $(S \circ T)(3 + 2x - x^2) = S(T(3 + 2x - x^2)) = S(2 - 2x) = 2 + (2 - 2)x + 2(-2)x^2 = 2 - 4x^2$.
$(S \circ T)(a + bx + cx^2) = S(T(a + bx + cx^2)) = S(b + 2cx) = b + (b + 2c)x + 4cx^2$.
Domain of $T = \mathscr{P}_2 =$ codomain of $S \Rightarrow$ we can compute $T \circ S$.
$(T \circ S)(a + bx) = T(S(a + bx)) = T(a + (a + b)x + 2bx^2) = (a + b) + 2(2b)x^2 = (a + b) + 4bx^2$.

27. The Chain Rule states $(p \circ g)'(x) = g'(x)p'(g(x))$. So, $g(x) = x + 1 \Rightarrow g'(x) = 1$.
$(S \circ T)(p(x)) = S(T(p(x))) = S(p'(x)) = p'(x + 1)$.
$(T \circ S)(p(x)) = T(S(p(x))) = T(p(x + 1)) = p'(x + 1)$ (because the derivative of $x + 1$ is 1).

28. $(S \circ T)(p(x)) = S(T(p(x))) = S(x\,p'(x)) = (x + 1)p'(x + 1)$.
$(T \circ S)(p(x)) = T(S(p(x))) = T(p(x + 1)) = (x + 1)p'(x + 1)$.

29. We need to show $S \circ T = I_{\mathbb{R}^2}$ and$T \circ S = I_{\mathbb{R}^2}$.

That is, we must show $S \circ T\begin{bmatrix} x \\ y \end{bmatrix} = \begin{bmatrix} x \\ y \end{bmatrix}$ and $T \circ S\begin{bmatrix} x \\ y \end{bmatrix} = \begin{bmatrix} x \\ y \end{bmatrix}$ for all $\begin{bmatrix} x \\ y \end{bmatrix}$.

$$S \circ T\begin{bmatrix} x \\ y \end{bmatrix} = S\left(T\begin{bmatrix} x \\ y \end{bmatrix}\right) = S\left(\begin{bmatrix} x-y \\ -3x+4y \end{bmatrix}\right) = \begin{bmatrix} 4(x-y)+(-3x+4y) \\ 3(x-y)+(-3x+4y) \end{bmatrix} = \begin{bmatrix} x \\ y \end{bmatrix}.$$

$$T \circ S\begin{bmatrix} x \\ y \end{bmatrix} = T\left(S\begin{bmatrix} x \\ y \end{bmatrix}\right) = T\left(\begin{bmatrix} 4x+y \\ 3x+y \end{bmatrix}\right) = \begin{bmatrix} (4x+y)-(3x+y) \\ -3(4x+y)+4(3x+y) \end{bmatrix} = \begin{bmatrix} x \\ y \end{bmatrix}.$$

30. We need to show $S \circ T = I_{\mathcal{P}_1}$ and $T \circ S = I_{\mathcal{P}_1}$. That is:
we must show $(S \circ T)p(x) = p(x)$ and $(T \circ S)p(x) = p(x)$ for all $p(x) = a + bx$ in $\mathcal{P}_1$.

$$\begin{aligned}(S \circ T)(p(x)) &= S(T(p(x))) = S(T(a+bx)) = S(\tfrac{1}{2}b + (a+2b)x)\\ &= (-4(\tfrac{1}{2}b) + (a+2b)) + 2(\tfrac{1}{2}b)x = a + bx.\\ (T \circ S)(p(x)) &= T(S(p(x))) = T(S(a+bx)) = T((-4a+b) + 2ax)\\ &= \tfrac{1}{2}(2a) + ((-4a+b) + 2(2a))x = a + bx.\end{aligned}$$

31. Let T', T'' be such that $\begin{matrix} T' \circ T = T'' \circ T = I_V \\ T \circ T' = T \circ T'' = I_W \end{matrix}$. Need to show $T'(\mathbf{w}) = T''(\mathbf{w})$ for all $\mathbf{w} \in W$.

$T'(\mathbf{w}) = I_V \circ T'(\mathbf{w}) = T'' \circ T \circ T'(\mathbf{w}) = T'' \circ I_W(\mathbf{w}) = T''(\mathbf{w})$ for all $\mathbf{w} \in W$.

32. (a) Note: $T(\mathbf{v}) = 0 \Rightarrow \{\mathbf{v}, T(\mathbf{v})\}$ linearly dependent since $0\mathbf{v} + T(\mathbf{v}) = \mathbf{0}$.
So assume $T(\mathbf{v}) \neq 0$. Then we have:
$\{\mathbf{v}, T(\mathbf{v})\}$ linearly dependent $\Leftrightarrow$ there exists $c \neq 0$, $c\mathbf{v} + T(\mathbf{v}) = 0 \Rightarrow c = 1 \Leftrightarrow T(\mathbf{v}) = -\mathbf{v}$.

$c\mathbf{v} + T(\mathbf{v}) = \mathbf{0}$ (1) $\Leftrightarrow$ $T(c\mathbf{v} + T(\mathbf{v})) = \mathbf{0}$ $\Leftrightarrow$ $cT(\mathbf{v}) + T \circ T(\mathbf{v}) = \mathbf{0}$

(Now use the fact that $T \circ T = I$.)

$cT(\mathbf{v}) + \mathbf{v} = \mathbf{0}$ (2)

Combining (1) and (2) into a system, we have:

$$\begin{matrix} c\mathbf{v} + T(\mathbf{v}) = \mathbf{0} \\ cT(\mathbf{v}) + \mathbf{v} = \mathbf{0} \end{matrix} \Leftrightarrow (c-1)(\mathbf{v} - T(\mathbf{v})) = \mathbf{0} \Leftrightarrow c = 1 \text{ or } T(\mathbf{v}) = \mathbf{v} \Leftrightarrow T(\mathbf{v}) = \pm\mathbf{v}.$$

(b) Example: $T\begin{bmatrix} x \\ y \end{bmatrix} = \begin{bmatrix} -x \\ -y \end{bmatrix}$.

Then: $T \circ T\begin{bmatrix} x \\ y \end{bmatrix} = T(T\begin{bmatrix} x \\ y \end{bmatrix}) = T\begin{bmatrix} -x \\ -y \end{bmatrix} = \begin{bmatrix} -(-x) \\ -(-y) \end{bmatrix} = \begin{bmatrix} x \\ y \end{bmatrix} \Rightarrow T \circ T = I.$

33. (a) $\{\mathbf{v}, T(\mathbf{v})\}$linearly dependent $\Leftrightarrow$ there exists $c \neq 0$, $c\mathbf{v} + T(\mathbf{v}) = 0 \Rightarrow c = 0 \Leftrightarrow T(\mathbf{v}) = \mathbf{0}$.

$c\mathbf{v} + T(\mathbf{v}) = \mathbf{0}$ (1) $\Leftrightarrow$ $T(c\mathbf{v} + T(\mathbf{v})) = \mathbf{0}$ $\Leftrightarrow$ $cT(\mathbf{v}) + T \circ T(\mathbf{v}) = \mathbf{0} \Leftrightarrow$

(Now use the fact that $T \circ T = T$.)

$cT(\mathbf{v}) + T(\mathbf{v}) = \mathbf{0}$ (2)

Combining (1) and (2) into a system, we have:

$$\begin{matrix} c\mathbf{v} + T(\mathbf{v}) = \mathbf{0} \\ cT(\mathbf{v}) + T(\mathbf{v}) = \mathbf{0} \end{matrix} \Leftrightarrow c(\mathbf{v} - T(\mathbf{v})) = \mathbf{0} \Leftrightarrow c = 0 \text{ or } T(\mathbf{v}) = \mathbf{v} \Leftrightarrow T(\mathbf{v}) = 0 \text{ or } T(\mathbf{v}) = \mathbf{v}.$$

(b) Example: $T\begin{bmatrix} x \\ y \end{bmatrix} = \begin{bmatrix} x \\ 0 \end{bmatrix}$.

Then: $T \circ T\begin{bmatrix} x \\ y \end{bmatrix} = T(T\begin{bmatrix} x \\ 0 \end{bmatrix}) = T\begin{bmatrix} x \\ 0 \end{bmatrix} = \begin{bmatrix} x \\ 0 \end{bmatrix} = T\begin{bmatrix} x \\ y \end{bmatrix} \Rightarrow T \circ T = T.$

34. $$\begin{aligned}(S+T)(\mathbf{v}+\mathbf{w}) &= S(\mathbf{v}+\mathbf{w}) + T(\mathbf{v}+\mathbf{w}) = (S(\mathbf{v}) + S(\mathbf{w})) + (T(\mathbf{v}) + T(\mathbf{w}))\\ &= (S(\mathbf{v}) + T(\mathbf{v})) + (S(\mathbf{w}) + T(\mathbf{w})) = (S+T)(\mathbf{v}) + (S+T)(\mathbf{w}).\end{aligned}$$

$(S+T)(c\mathbf{v}) = S(c\mathbf{v}) + T(c\mathbf{v}) = cS(\mathbf{v}) + cT(\mathbf{v}) = c(S(\mathbf{v}) + T(\mathbf{v})) = c(S+T)(\mathbf{v})$.

35. Below we are intrinsically using the fact that V and W are vector spaces. That tells us that $\mathbf{v}$ in V and $S(\mathbf{v})$ in W satisfy the ten axioms of a vector space.

Note: To prove $S = T$ below, we show $S(\mathbf{v}) = T(\mathbf{v})$ for all $\mathbf{v}$ in V.

1. Exercise 34 $\Rightarrow S, T \in \mathscr{L} \Rightarrow S + T \in \mathscr{L}$.
2. $(S+T)(\mathbf{v}) = S(\mathbf{v}) + T(\mathbf{v}) = T(\mathbf{v}) + S(\mathbf{v}) = (T+S)(\mathbf{v})$.
3. $((R+S)+T)(\mathbf{v}) = (R+S)(\mathbf{v}) + T(\mathbf{v}) = (R(\mathbf{v}) + S(\mathbf{v})) + T(\mathbf{v}) = R(\mathbf{v}) + (S(\mathbf{v}) + T(\mathbf{v}))$
   $= (R + (S+T))(\mathbf{v})$.
4. Set $Z(\mathbf{v}) = \mathbf{0}$ for all $\mathbf{v} \Rightarrow (S+Z)(\mathbf{v}) = S(\mathbf{v}) + Z(\mathbf{v}) = S(\mathbf{v}) + \mathbf{0} = S(\mathbf{v})$.
5. Set $(-S)(\mathbf{v}) = -S(\mathbf{v})$ for all $\mathbf{v} \Rightarrow$
   $(S + (-S))(\mathbf{v}) = S(\mathbf{v}) + (-S)(\mathbf{v}) = S(\mathbf{v}) + (-S(\mathbf{v})) = S(\mathbf{v}) + S(-\mathbf{v})$
   $= S(\mathbf{v} + (-\mathbf{v})) = S(\mathbf{0}) = \mathbf{0}$.
6. Exercise 34 $\Rightarrow S \in \mathscr{L} \Rightarrow cS \in \mathscr{L}$.
7. $c(S+T)(\mathbf{v}) = c(S(\mathbf{v}) + T(\mathbf{v})) = cS(\mathbf{v}) + cT(\mathbf{v})$.
8. $(c+d)S(\mathbf{v}) = cS(\mathbf{v}) + dS(\mathbf{v})$.
9. $(cd)S(\mathbf{v}) = c(dS(\mathbf{v}))$.
10. Set $1(S(\mathbf{v})) = 1\,S(\mathbf{v})$ for all $\mathbf{v}$.

36. (a) $(R \circ (S+T))(\mathbf{v}) = R((S+T)(\mathbf{v}) = R(S(\mathbf{v}) + T(\mathbf{v})) = R(S(\mathbf{v})) + R(T(\mathbf{v}))$
$= (R \circ S)(\mathbf{v}) + (R \circ T)(\mathbf{v})$.

(b) $c(R \circ S)(\mathbf{v}) = c(R(S(\mathbf{v})) = cR(S(\mathbf{v})) = ((cR) \circ S)(\mathbf{v}) = R(cS(\mathbf{v})) = (R \circ (cS))(\mathbf{v})$.

6.5 The Kernel and Range of a Linear Transformation

1. (a) $\begin{bmatrix} a & b \\ c & d \end{bmatrix}$ in $\ker(T) \Leftrightarrow a = d = 0 \Rightarrow$ only (ii) $= \begin{bmatrix} 0 & 4 \\ 2 & 0 \end{bmatrix}$ in $\ker(T)$.

(b) $\begin{bmatrix} a & b \\ c & d \end{bmatrix}$ in $\text{range}(T) \Leftrightarrow b = c = 0 \Rightarrow$ only (iii) $= \begin{bmatrix} 3 & 0 \\ 0 & -3 \end{bmatrix}$ in $\text{range}(T)$.

(c) $\begin{bmatrix} a & b \\ c & d \end{bmatrix}$ in $\ker(T) \Leftrightarrow a = d = 0 \Rightarrow \ker(T) = \left\{ \begin{bmatrix} 0 & b \\ c & 0 \end{bmatrix} \right\}$.

$\begin{bmatrix} a & b \\ c & d \end{bmatrix}$ in $\text{range}(T) \Leftrightarrow b = c = 0 \Rightarrow \text{range}(T) = \left\{ \begin{bmatrix} a & 0 \\ 0 & d \end{bmatrix} \right\}$.

2. Define $A \in M_{22}$ as $A = \begin{bmatrix} a & b \\ c & d \end{bmatrix}$.

(a) A in $\ker(T) \Leftrightarrow a + d = 0 \Rightarrow d = -a$ and b, c free $\Rightarrow$ (ii), (iii) in $\ker(T)$.

(b) $a + d$ in $\text{range}(T) \Leftrightarrow a + d \in \mathbb{R} \Rightarrow$ (i), (ii), (iii) in $\text{range}(T)$.

(c) A in $\ker(T) \Leftrightarrow a + d = 0 \Rightarrow d = -a$, b, c free $\Rightarrow \ker(T) = \left\{ \begin{bmatrix} a & b \\ c & -a \end{bmatrix} \right\}$.

$a + d$ in $\text{range}(T) \Leftrightarrow a + d \in \mathbb{R} \Rightarrow \text{range}(T) = \mathbb{R}$.

3. (a) $p(x)$ in $\ker(T) \Leftrightarrow \begin{matrix} a - b = 0 \\ b + c = 0 \end{matrix} \Leftrightarrow a = b = -c \Rightarrow$ only (iii) in $\ker(T)$.

(b) $p(x)$ in $\text{range}(T) \Leftrightarrow a, b, c \in R \Rightarrow$ (i), (ii), (iii) in $\text{range}(T)$.

(c) $p(x)$ in $\ker(T) \Leftrightarrow a = b = -c \Rightarrow \ker(T) = \{t + tx - tx^2\}$.

$p(x)$ in $\text{range}(T) \Leftrightarrow a, b, c \in \mathbb{R} \Rightarrow \text{range}(T) = \mathbb{R}^2$.

4. (a) $p(x)$ in $\ker(T) \Leftrightarrow xp'(x) = x(b + 2cx) = 0 \Rightarrow b = c = 0 \Rightarrow$ only (ii) $= 2$ in $\ker(T)$.

(b) $p(x)$ in $\text{range}(T) \Leftrightarrow p(x) = x(b + 2cx) = bx + 2cx^2 \Rightarrow a = 0, b$, c free $\Rightarrow$ only (ii) $= x^2$ in $\text{range}(T)$.

(c) $p(x)$ in $\ker(T) \Rightarrow b = c = 0 \Rightarrow \ker(T) = \{a\}$.

$p(x)$ in $\text{range}(T) \Rightarrow a = 0, b$, c free $\Rightarrow \text{range}(T) = \{bx + cx^2\}$.

5. $\ker(T) = \left\{ \begin{bmatrix} 0 & b \\ c & 0 \end{bmatrix} \right\} \Rightarrow \left\{ \begin{bmatrix} 0 & 1 \\ 0 & 0 \end{bmatrix}, \begin{bmatrix} 0 & 0 \\ 1 & 0 \end{bmatrix} \right\}$ is basis $\Rightarrow \text{nullity}(T) = \dim \ker(T) = 2$.

Likewise, $\text{range}(T) = \left\{ \begin{bmatrix} 1 & 0 \\ 0 & 0 \end{bmatrix}, \begin{bmatrix} 0 & 0 \\ 0 & 1 \end{bmatrix} \right\}$ is a basis $\Rightarrow \text{rank}(T) = \dim \text{range}(T) = 2$.

6. $\ker(T) = \left\{ \begin{bmatrix} a & b \\ c & -a \end{bmatrix} \right\} \Rightarrow \left\{ \begin{bmatrix} 1 & 0 \\ 0 & -1 \end{bmatrix}, \begin{bmatrix} 0 & 1 \\ 0 & 0 \end{bmatrix}, \begin{bmatrix} 0 & 0 \\ 1 & 0 \end{bmatrix} \right\}$ is basis $\Rightarrow \text{nullity}(T) = 3$.

$\text{range}(T) = \mathbb{R} \Rightarrow \text{rank}(T) = \dim \text{range}(T) = \dim \mathbb{R} = 1$.

7. $\ker(T) = \{t + tx - tx^2\} \Rightarrow \{1 + x - x^2\}$ is basis $\Rightarrow \text{nullity}(T) = 1$.

$\text{range}(T) = \mathbb{R}^2 \Rightarrow \text{rank}(T) = \dim \text{range}(T) = \dim \mathbb{R}^2 = 2$.

8. $\ker(T) = \{a\} \Rightarrow \{1\}$ is basis $\Rightarrow$ nullity$(T) = 1$.
range$(T) = \{bx + cx^2\} \Rightarrow \{x, x^2\}$ is basis $\Rightarrow$ rank$(T) = 2$.

9. $A \in \ker(T) \Rightarrow = \begin{matrix} a - b = 0 \\ c - d = 0 \end{matrix} \Rightarrow \begin{matrix} a = b \\ c = d \end{matrix} \Rightarrow$

$\ker(T) = \left\{ \begin{bmatrix} a & a \\ c & c \end{bmatrix} \right\} \Rightarrow \ker(T) = \text{span}\left(\begin{bmatrix} 1 & 1 \\ 0 & 0 \end{bmatrix}, \begin{bmatrix} 0 & 0 \\ 1 & 1 \end{bmatrix} \right) \Rightarrow \text{nullity}(T) = \dim \ker(T) = 2.$

Therefore, rank$(T) = \dim M_{22} - \text{nullity}(T) = 4 - 2 = 2$.

10. $T(x) = \begin{bmatrix} p(0) = 0 \\ p(1) = 1 \end{bmatrix}, T(1 - x) = \begin{bmatrix} p(0) = 1 - 0 = 1 \\ p(1) = 1 - 1 = 0 \end{bmatrix} \Rightarrow$

$\text{range}(T) = \text{span}\left(\begin{bmatrix} 0 \\ 1 \end{bmatrix}, \begin{bmatrix} 1 \\ 0 \end{bmatrix} \right) = \mathbb{R}^2 \Rightarrow \text{rank}(T) = \dim \text{range}(T) = 2.$

Therefore, nullity$(T) = \dim \mathcal{P}_2 - \text{rank}(T) = 3 - 2 = 1$.

11. $T(E_{11}) = E_{11}B = \begin{bmatrix} 1 & 0 \\ 0 & 0 \end{bmatrix} \begin{bmatrix} 1 & -1 \\ -1 & 1 \end{bmatrix} = \begin{bmatrix} 1 & -1 \\ 0 & 0 \end{bmatrix}$. Likewise, $T(E_{12}) = E_{12}B = \begin{bmatrix} -1 & 1 \\ 0 & 0 \end{bmatrix}$,

$T(E_{21}) = E_{21}B = \begin{bmatrix} 0 & 0 \\ 1 & -1 \end{bmatrix}, T(E_{22}) = E_{22}B = \begin{bmatrix} 0 & 0 \\ -1 & 1 \end{bmatrix} \Rightarrow$

$\text{range}(T) = \text{span}\left(\begin{bmatrix} 1 & -1 \\ 0 & 0 \end{bmatrix}, \begin{bmatrix} 0 & 0 \\ 1 & -1 \end{bmatrix} \right) \Rightarrow \text{rank}(T) = 2.$

Therefore, nullity$(T) = \dim M_{22} - \text{rank}(T) = 4 - 2 = 2$.

12. $A \in \ker(T) \Rightarrow AB - BA = \begin{bmatrix} c - b & d - a \\ a - d & b - c \end{bmatrix} = 0 \Rightarrow \begin{matrix} b - c = 0 \\ a - d = 0 \end{matrix} \Rightarrow \begin{matrix} c = b \\ d = a \end{matrix} \Rightarrow$

$\ker(T) = \left\{ \begin{bmatrix} a & b \\ b & a \end{bmatrix} \right\} = \text{span}\left(\begin{bmatrix} 1 & 0 \\ 0 & 1 \end{bmatrix}, \begin{bmatrix} 0 & 1 \\ 1 & 0 \end{bmatrix} \right) \Rightarrow \text{nullity}(T) = \dim \ker(T) = 2.$

Therefore, rank$(T) = \dim M_{22} - \text{nullity}(T) = 4 - 2 = 2$.

13. $p(x) \in \ker(T) \Leftrightarrow p'(0) = (b + 2c(0)) = 0 \Rightarrow b = 0$, a, c free $\Rightarrow$
$\ker(T) = \{a + cx^2\} = \text{span}(1, x^2) \Rightarrow \text{nullity}(T) = \dim \ker(T) = 2$.
Therefore, rank$(T) = \dim \mathscr{P}_2 - \text{nullity}(T) = 3 - 2 = 1$.

14. $A \in \ker(T) \Rightarrow A - A^T = 0 \Rightarrow \begin{bmatrix} 0 & a_{12} - a_{21} & a_{13} - a_{31} \\ a_{21} - a_{12} & 0 & a_{23} - a_{32} \\ a_{31} - a_{13} & a_{32} - a_{23} & 0 \end{bmatrix} = 0 \Rightarrow$

$a_{21} = a_{12}$
$a_{31} = a_{13}$ and a_{11}, a_{22}, a_{33} free $\Rightarrow$ nullity$(T) = \dim \ker(T) = 6$.
$a_{32} = a_{23}$

Therefore, rank$(T) = \dim M_{33} - \text{nullity}(T) = 9 - 6 = 3$.

15. (a) $\begin{bmatrix} x \\ y \end{bmatrix} \in \ker(T) \Rightarrow = \begin{matrix} 2x - y = 0 \\ x + 2y = 0 \end{matrix} \Rightarrow x = y = 0 \Rightarrow \ker(T) = \{\mathbf{0}\} \Rightarrow T$ is one-to-one.

(b) Since (a) $\Rightarrow T$ is one-to-one and $V = W = \mathbb{R}^2$, Theorem 6.21 implies T is onto.

16. (a) $\begin{bmatrix} x \\ y \end{bmatrix} \in \ker(T) \Rightarrow = \begin{matrix} x - 2y = 0 \\ 3x + y = 0 \\ x + y = 0 \end{matrix} \Rightarrow x = y = 0 \Rightarrow \ker(T) = \{\mathbf{0}\} \Rightarrow T$ is one-to-one.

(b) (a) $\Rightarrow \operatorname{rank}(T) = \dim \mathbb{R}^2 - \operatorname{nullity}(T) = 2 - 0 = 2 \Rightarrow \dim \operatorname{range}(T) = 2 < \dim \mathbb{R}^3 = 3 \Rightarrow$ T is *not* onto.
Furthermore, $\dim \operatorname{range}(T) \leq 2 < \dim \mathbb{R}^3 = 3 \Rightarrow$ no T can be onto.

17. (a) Note, $T(1) = \begin{bmatrix} 2 \\ 1 \\ -1 \end{bmatrix}$, $T(x) = \begin{bmatrix} -1 \\ 1 \\ 0 \end{bmatrix}$, $T(x^2) = \begin{bmatrix} 0 \\ -3 \\ 1 \end{bmatrix} \Rightarrow$
$-T(1) - 2T(x) = T(x^2) \Rightarrow T(1 + 2x + x^2) = 0 \Rightarrow \ker(T) = \operatorname{span}(1 + 2x + x^2) \neq \{\mathbf{0}\} \Rightarrow$ T is *not* one-to-one.

(b) Since (a) $\Rightarrow T$ is *not* one-to-one and $\dim \mathscr{P}_2 = \dim \mathbb{R}^3$,
Theorem 6.21 implies no T can be onto.

18. (a) Exercise 10 $\Rightarrow \operatorname{nullity}(T) = 1 \Rightarrow \ker(T) \neq \{\mathbf{0}\} \Rightarrow T$ is *not* one-to-one.

(b) Exercise 10 $\operatorname{range}(T) = \operatorname{span}\left(\begin{bmatrix} 0 \\ 1 \end{bmatrix}, \begin{bmatrix} 1 \\ 0 \end{bmatrix}\right) = \mathbb{R}^2 \Rightarrow T$ is onto *by definition.*

19. (a) $\begin{bmatrix} a \\ b \\ c \end{bmatrix} \in \ker(T) \Rightarrow \begin{bmatrix} a-b & b-c \\ a+b & b+c \end{bmatrix} = 0 \Rightarrow \begin{matrix} a = 0 \\ b = 0 \\ c = 0 \end{matrix} \Rightarrow \ker(T) = \{\mathbf{0}\} \Rightarrow T$ is one-to-one.

(b) (a) $\Rightarrow \operatorname{rank}(T) = \dim \mathbb{R}^3 - \operatorname{nullity}(T) = 3 - 0 = 3 \Rightarrow$
$\dim \operatorname{range}(T) = 3 < \dim M_{22} = 4 \Rightarrow T$ is *not* onto.
Furthermore, $\dim \operatorname{range}(T) \leq 3 < \dim M_{22} = 4 \Rightarrow$ no T can be onto.

20. (a) $\begin{bmatrix} a \\ b \\ c \end{bmatrix} \in \ker(T) \Rightarrow \begin{bmatrix} a+b+c & b-2c \\ b-2c & a-c \end{bmatrix} = 0 \Rightarrow a = b = c = 0 \Rightarrow \ker(T) = \{\mathbf{0}\} \Rightarrow$
T is one-to-one.

(b) Since (a) $\Rightarrow T$ is one-to-one and $\dim \mathbb{R}^3 = \dim W = 3$, Theorem 6.21 implies T is onto.
Recall, $\dim W = 3$ since $A = A^T \Rightarrow \begin{bmatrix} a & b \\ c & d \end{bmatrix} = \begin{bmatrix} a & c \\ b & d \end{bmatrix} \Rightarrow \begin{matrix} c = b \\ a,\ d \text{ free} \end{matrix} \Rightarrow$
$\left\{\begin{bmatrix} 1 & 0 \\ 0 & 0 \end{bmatrix}, \begin{bmatrix} 0 & 0 \\ 0 & 1 \end{bmatrix}, \begin{bmatrix} 0 & 1 \\ 1 & 0 \end{bmatrix}\right\}$ is a basis for W.

21. $D_3 = \text{span}(E_{11}, E_{22}, E_{33}) \Rightarrow \dim D_3 = 3 = \dim \mathbb{R}^3 \Rightarrow D_3 \cong \mathbb{R}^3$.

Define $T\begin{bmatrix} x & 0 & 0 \\ 0 & y & 0 \\ 0 & 0 & z \end{bmatrix} = \begin{bmatrix} x \\ y \\ z \end{bmatrix}$. Then $A \in \ker(T) \Rightarrow \begin{bmatrix} x & 0 & 0 \\ 0 & y & 0 \\ 0 & 0 & z \end{bmatrix} = 0 \Rightarrow x = y = z = 0 \Rightarrow$

$\ker(T) = \{\mathbf{0}\} \Rightarrow T$ is one-to-one.

Since T is one-to-one and $\dim D_3 = \dim \mathbb{R}^3$, Theorem 6.21 implies T is onto.

22. $A \in S_3 \Rightarrow A = A^T \Rightarrow\Rightarrow \begin{matrix} a_{21} = a_{12} \\ a_{31} = a_{13} \\ a_{32} = a_{23} \end{matrix}$ and a_{11}, a_{22}, a_{33} free $\Rightarrow \dim S_3 = 6$.

Likewise, $A \in U_3 \Rightarrow \begin{matrix} a_{21} = a_{31} = a_{32} = 0 \\ a_{11}, a_{12}, a_{13}, a_{22}, a_{23}, a_{33} \text{ free} \end{matrix} \Rightarrow \dim U_3 = 6$.

Therefore, $\dim S_3 = \dim U_3 \Rightarrow S_3 \cong U_3$.

Define $T\begin{bmatrix} a_{11} & a_{12} & a_{13} \\ a_{12} & a_{22} & a_{23} \\ a_{13} & a_{23} & a_{33} \end{bmatrix} = \begin{bmatrix} a_{11} & a_{12} & a_{13} \\ 0 & a_{22} & a_{23} \\ 0 & 0 & a_{33} \end{bmatrix}$.

Then $A \in \ker(T) \Rightarrow a_{11} = a_{12} = a_{13} = a_{22} = a_{23} = a_{33} = 0 \Rightarrow \ker(T) = \{\mathbf{0}\} \Rightarrow$
T is one-to-one.

Since T is one-to-one and $\dim S_3 = \dim U_3$, Theorem 6.21 implies T is onto.

23. $A \in S'_3 \Rightarrow A = -A^T \Rightarrow \begin{matrix} a_{21} = a_{12} \\ a_{31} = a_{13} \\ a_{32} = a_{23} \end{matrix}$ and $a_{11} = a_{22} = a_{33} = 0 \Rightarrow \dim S'_3 = 3$.

Exercise 22 $\Rightarrow \dim S_3 = 6 \neq \dim S'_3 = 3 \Rightarrow S_3 \not\cong S'_3$.

24. $\mathscr{P}_2 = \text{span}(1, x, x^2) \Rightarrow \dim \mathscr{P}_2 = 3$.

$p(x) \in W \Rightarrow p(0) = a + b(0) + c(0)^2 + d(0)^3 = 0 \Rightarrow a = 0 \Rightarrow W = \text{span}(x, x^2, x^3) \Rightarrow \dim W = 3$.

Therefore, since $\dim \mathscr{P}_2 = \dim W = 3$, $\mathscr{P}_2 \cong W$.

Define $T(a + bx + x^3) = ax + bx^2 + cx^3$. Then $p(x) \in \ker(T) \Rightarrow ax + bx^2 + cx^3 = 0 \Rightarrow$

$a = b = c = 0 \Rightarrow \ker(T) = \{\mathbf{0}\} \Rightarrow T$ is one-to-one.

Since T is one-to-one and $\dim \mathscr{P}_2 = \dim W$, Theorem 6.21 implies T is onto.

25. $\mathbb{C} = \text{span}(1, i) \Rightarrow \dim \mathbb{C} = 2$. Likewise, $\mathbb{R}^2 = \text{span}\left(\begin{bmatrix} 1 \\ 0 \end{bmatrix}, \begin{bmatrix} 0 \\ 1 \end{bmatrix}\right) \Rightarrow \dim \mathbb{R}^2 = 2$,

Therefore, since $\mathbb{C} = \dim \mathbb{R}^2 = 2$, $\mathbb{C} \cong \mathbb{R}^2$.

Define $T(a + ib) = \begin{bmatrix} a \\ b \end{bmatrix}$.

Then $a + ib \in \ker(T) \Rightarrow a = b = 0 \Rightarrow \ker(T) = \{\mathbf{0}\} \Rightarrow T$ is one-to-one.

Since T is one-to-one and $\dim \mathbb{C} = \dim \mathbb{R}^2$, Theorem 6.21 implies T is onto.

26. Define $A \in M_{22}$ as $A = \begin{bmatrix} a & b \\ c & d \end{bmatrix}$.

$A \in V \Rightarrow \operatorname{tr}(A) = a + d = 0 \Rightarrow a = -d, b,$ c, $free \Rightarrow \dim V = 3 \neq \dim \mathbb{R}^2 = 2 \Rightarrow V \not\cong \mathbb{R}^2$.

27. Need only show $p(x) + p'(x) = 0 \Rightarrow p(x) = 0 \Rightarrow \ker(T) = \{\mathbf{0}\} \Rightarrow T$ is one-to-one.

Let $p(x) = a_0 + a_1 x + \ldots + a_n x^n$ be such that $p(x) + p'(x) = 0 \Rightarrow$

$(a_0 + a_1 x + \ldots + a_n x^n) + (a_1 + 2a_2 x + \ldots + n a_n x^{n-1})$
$= (a_0 + a_1) + (a_1 + 2a_2)x + \ldots + (a_n + (n-1)a_{n-1}x^{n-1}) + a_n x^n = 0 \Rightarrow a_n = 0 \Rightarrow$

$0 + (n-1)a_{n-1} = 0 \Rightarrow a_{n-1} = 0 \ldots\ a_0 = 0 \Rightarrow p(x) = 0.$

28. Need only show $p(x-2) = 0 \Rightarrow p(x) = 0$ since that implies $p(x) \in \ker(T) \Rightarrow p(x) = 0 \Rightarrow$ $\ker(T) = \{\mathbf{0}\} \Rightarrow T$ is one-to-one.

Note, this amounts proving that $\{1, x-2, (x-2)^2, \ldots, (x-2)^n\}$ is a basis for $\mathscr{P}_n$.

We will proceed by induction.

Case: $n = 1$. Then $p(x-2) = a_0 + a_1(x-2) = 0 \Rightarrow a_1 = 0 \Rightarrow a_0 = a_1 = 0 \Rightarrow p(x) = 0$.

Assume $q(x-2) = 0 \Rightarrow q(x) = 0$ for any polynomial of degree $\leq n$.
Then let $p(x) = a_0 + a_1 x + \ldots + a_n x^n$ be such that $p(x-2) = 0$. We will show $p(x) = 0$.

$p(x-2) = 0 \Rightarrow a_n = 0 \Rightarrow$ the degree of $p(x) \leq n \Rightarrow p(x) = 0$ by the induction hypothesis.

Alternative proof using change-of-variable: Need only show $p(x) \in \ker(T) \Rightarrow p(x)$ for all $x \in \mathbb{R}$ since that implies $p(x) = 0 \Rightarrow \ker(T) = \{\mathbf{0}\} \Rightarrow T$ is one-to-one.

$p(x) \in \ker(T) \Rightarrow p(x-2) = 0$ for all $x \in \mathbb{R} \Rightarrow p(y) = 0$ for all y $\in \mathbb{R}$, where $y = x - 2 \Rightarrow$ $p(x) = 0$ for all x $\in \mathbb{R}$.

29. Need only show $x^n p(\frac{1}{x}) = 0 \Rightarrow p(x) = 0$.

Let $p(x) = a_0 + a_1 x + \ldots + a_n x^n$ be such that $x^n p(\frac{1}{x}) = 0$. We will show $p(x) = 0$.

$x^n p(\frac{1}{x}) = a_0 x^n + a_1 x^{n-1} + \ldots + a_n = 0 \Rightarrow a_k = 0$

because x^k are linearly independent $\Rightarrow p(x) = 0$.

30. (a) Define $T(f) = f(x-2)$. Need only show that T is one-to-one and onto.

One-to-one: Need to show $f(x-2) = 0$ for all $x \in [2,3] \Rightarrow f(x) = 0$ for all $x \in [0,1]$.

$f(x-2) = 0$ for all $x \in [2,3] \Rightarrow$

$f(y) = 0$ for all $y \in [0,1]$ where $y = x-2 \Rightarrow$

$f(x) = 0$ for all $x \in [0,1]$.

Onto: Given any $f(x) \in \mathscr{C}[2,3]$, we can define $g \in \mathscr{C}[0,1]$ as $g(x) = f(x+2)$.

Then simply note that $T(g) = f(x+2-2) = f(x)$ which shows T is onto.

(b) Define $T(f) = f(x-a)$. Need only show that T is one-to-one and onto.

One-to-one: $f(x-a) = 0$ for all $x \in [a, a+1] \Rightarrow$

$f(y) = 0$ for all $y \in [0,1]$ where $y = x-a \Rightarrow$

$f(x) = 0$ for all $x \in [0,1]$.

Onto: Given any $f(x) \in \mathscr{C}[a, a+1]$, we can define $g \in \mathscr{C}[0,1]$ as $g(x) = f(x+a)$.

Then simply note that $T(g) = f(x+a-a) = f(x)$ which shows T is onto.

31. Define $T(f) = f(\frac{1}{2}x)$. Need only show that T is one-to-one and onto.

One-to-one: $f(\frac{1}{2}x) = 0$ for all $x \in [0,2] \Rightarrow$

$f(y) = 0$ for all $y \in [0,1]$ where $y = \frac{x}{2} \Rightarrow$

$f(x) = 0$ for all $x \in [0,1]$.

Onto: Given any $f(x) \in \mathscr{C}[0,2]$, we can define $g \in \mathscr{C}[0,1]$ as $g(x) = f(2x)$.

Then simply note that $T(g) = f(2(\frac{1}{2}x)) = f(x)$ which shows T is onto.

32. We want $f(\,t(x)\,)$ to be such $t(c) = a$ and $t(d) = b$.

This amounts to finding the line through (c,a) and $(d,b) \Rightarrow$

$m = \frac{b-a}{d-c} \Rightarrow t(x) = \left(\frac{b-a}{d-c}\right)(x-c) + a$.

One-to-one: $f(\left(\frac{b-a}{d-c}\right)(x-c) + a) = 0$ for all $x \in [c,d] \Rightarrow$

$f(y) = 0$ for all $y \in [c,d]$ where $y = \left(\frac{d-c}{b-a}\right)(x-a) + c \Rightarrow$

$f(x) = 0$ for all $x \in [a,b]$.

Onto: Given any $f(x) \in \mathscr{C}[c,d]$, we can define $g \in \mathscr{C}[a,b]$ as $g(x) = f(\left(\frac{d-c}{b-a}\right)(x-a) + c)$.

Then simply note that $T(g) = f(x)$ which shows T is onto.

33. (a) Recall L is one-to-one if and only if $L(\mathbf{v}) = \mathbf{0} \Leftrightarrow \mathbf{v} = \mathbf{0}$ because $\ker L = \{\mathbf{0}\}$.

So, we need to show $(S \circ T)(\mathbf{u}) = \mathbf{0} \Leftrightarrow \mathbf{u} = 0$.

$(S \circ T)(\mathbf{u}) = \mathbf{0} \Leftrightarrow S(T(\mathbf{u})) = 0 \Leftrightarrow T(\mathbf{u}) = 0 \Leftrightarrow \mathbf{u} = 0$ which shows $S \circ T$ is one-to-one.

(b) Recall L is onto for every $\mathbf{w} \in W$ there exists $\mathbf{v} \in V$ such that $L(\mathbf{v}) = \mathbf{w}$.

So, we need to show for every $\mathbf{w} \in W$ there exists $\mathbf{u} \in U$ such that $(S \circ T)(\mathbf{u}) = \mathbf{w}$.

S onto $\Rightarrow$ for every $\mathbf{w} \in W$ there exists $\mathbf{v} \in V$ such that $S(\mathbf{v}) = \mathbf{w}$.

Furthermore, T onto $\Rightarrow$ there exists $\mathbf{u} \in U$ such that $T(\mathbf{u}) = \mathbf{v} \Rightarrow$

$(S \circ T)(\mathbf{u}) = S(T(\mathbf{u})) = S(\mathbf{v}) = \mathbf{w}$ which shows $S \circ T$ is onto.

34. (a) Need to show $T(\mathbf{u}) = \mathbf{0} \Leftrightarrow \mathbf{u} = \mathbf{0}$.
$T(\mathbf{u}) = \mathbf{0} \Leftrightarrow (S \circ T)(\mathbf{u}) = \mathbf{0} \Leftrightarrow \mathbf{u} = \mathbf{0}$ (because $S \circ T$ is one-to-one.

(b) Need to show for every $\mathbf{w} \in W$ there exists $\mathbf{v} \in V$ such that $S(\mathbf{v}) = \mathbf{w}$.
$S \circ T$ onto $\Rightarrow$ for every $\mathbf{w} \in W$ there exists $\mathbf{u} \in U$ such that $(S \circ T)(\mathbf{u}) = \mathbf{w} \Rightarrow$
$S(T(\mathbf{u})) = \mathbf{w} \Rightarrow$ letting $\mathbf{v} = T(\mathbf{u})$ shows S is onto.

35. (a) Recall Theorem 6.11b from Section 6.2 says if W is a subspace of a finite-dimensional vector space V, then $\dim W = \dim V \Leftrightarrow W = V$.
Note, the contrapositive of this statement is $\dim W \neq \dim V \Leftrightarrow W \neq V$.
Given $T : V \to W$, we will show $\dim V < \dim W \Rightarrow \dim \text{range}(T) \neq \dim W$ which will then imply $\text{range}(T) \neq W$ and therefore T cannot be onto.
Recall, Theorem 6.18b of this section asserts $\text{range}(T)$ is a subspace of W.
Now, Theorem 6.20, therefore implies $\dim \text{range}(T) = \dim V - \text{nullity}(T) \leq \dim V$.
Therefore, $\dim \text{range}(T) < \dim W \Rightarrow T$ cannot be onto.
Need to show $T(\mathbf{u}) = \mathbf{0} \Leftrightarrow \mathbf{u} = \mathbf{0}$.
$T(\mathbf{u}) = \mathbf{0} \Leftrightarrow (S \circ T)(\mathbf{u}) = \mathbf{0} \Leftrightarrow \mathbf{u} = \mathbf{0}$ (because $S \circ T$ is one-to-one).

(b) Since Theorem 6.20 states T is one-to-one $\Leftrightarrow \ker(T) = \{\mathbf{0}\}$,
we need to show $\dim V > \dim W \Rightarrow \ker(T) \neq \{\mathbf{0}\}$.
Recall, Theorem 6.18b of this section asserts
$\text{range}(T)$ is a subspace of $W \Rightarrow \dim W \geq \text{range}(T) = \text{rank}(T)$.
Therefore, $\dim W + \text{nullity}(T) \geq \text{rank}(T) + \text{nullity}(T) = \dim V \Rightarrow$
$\dim \ker(T) = \text{nullity}(T) \geq \dim V - \dim W > 0 \Rightarrow \ker(T) \neq \{\mathbf{0}\}$.

36. $\dim \mathscr{P}_n = n + 1 = \mathbb{R}^n \Rightarrow$ we need only show $\ker(T) = \mathbf{0}$.

In this case, that means we need to show if $T(p(x)) = \mathbf{0}$, then $p(x)$ must be the zero polynomial.

$$T(p(x)) = \begin{bmatrix} p(a_0) \\ p(a_1) \\ \vdots \\ p(a_n) \end{bmatrix} = \begin{bmatrix} 0 \\ 0 \\ \vdots \\ 0 \end{bmatrix} \Rightarrow p(a_i) = 0 \text{ for all } i \Rightarrow p(x) \text{ has n + 1 zeroes.}$$

Therefore, Section 6.2, Exercise 62 $\Rightarrow p(x)$ must be the zero polynomial as required $\Rightarrow$ $\ker(T) = \mathbf{0} \Rightarrow T$ is an isomorphism.

37. **NOTE**: The *null spaces* in the text should be referred to as *kernels.*

Following the hint, we will use the Rank Theorem to show $\ker(T) = \ker(T^2)$.

The Rank Theorem implies $\text{rank}(T) + \text{nullity}(T) = \text{rank}(T^2) + \text{nullity}(T^2) = \dim V$.

So since $\text{rank}(T) = \text{rank}(T^2)$, we have $\text{nullity}(T^2) = \text{nullity}(T)$.

So we will show $\ker(T) \subseteq \ker(T^2)$ and conclude that $\ker(T) = \ker(T^2)$.

Let $\mathbf{v}$ be in $\ker(T)$, then $T^2(\mathbf{v}) = T(T(\mathbf{v})) = T(\mathbf{0}) = \mathbf{0}$ which implies $\ker(T) \subseteq \ker(T^2)$.

So $\ker(T) = \ker(T^2)$ as we were to show.

Now let $\mathbf{v}$ be in $\text{range}(T) \cap \ker(T)$. We want to show $\mathbf{v} = \mathbf{0}$.

Then $\mathbf{v} = T(\mathbf{w})$ and $T(\mathbf{v}) = T(T(\mathbf{w})) = \mathbf{0}$ which implies $\mathbf{w}$ is in $\ker(T^2)$.
But $\ker(T) = \ker(T^2)$, so $\mathbf{w}$ is also in $\ker(T)$.

That is, $\mathbf{v} = T(\mathbf{w}) = \mathbf{0}$ as we were to show.

38. Let U and V be subspaces of a finite-dimensional vector space V with $T(\mathbf{u}, \mathbf{w}) = \mathbf{u} - \mathbf{w}$.

(a) We need to show that T is a linear transformation.

$$\begin{aligned} T(\mathbf{u}_1 + \mathbf{u}_2, \mathbf{w}_1 + \mathbf{w}_2) &= (\mathbf{u}_1 + \mathbf{u}_2) - (\mathbf{w}_1 + \mathbf{w}_2) = (\mathbf{u}_1 - \mathbf{w}_1) + (\mathbf{u}_2 - \mathbf{w}_2) \\ &= T(\mathbf{u}_1, \mathbf{w}_1) + T(\mathbf{u}_2, \mathbf{w}_2). \\ T(c\mathbf{u}, c\mathbf{w}) &= c\mathbf{u} - c\mathbf{w} = c(\mathbf{u} - \mathbf{w}) = cT(\mathbf{u}, \mathbf{w}). \end{aligned}$$

(b) We need to show that $\text{range}(T) = U + W$.

Recall from Exercise 48 of Section 6.1 that $U + W = \{\mathbf{u} + \mathbf{w} : \mathbf{u} \text{ is in } U, \mathbf{w} \text{ is in } W\}$.
Then simply note $T(\mathbf{u}, -\mathbf{w}) = \mathbf{u} - (-\mathbf{w}) = \mathbf{u} + \mathbf{w}$.
Q: Why is this sufficient?
A: Hint: Think about the fact that $-(-\mathbf{w}) = \mathbf{w}$ when substituting into T as well.

(c) We need to show that $\ker(T) \cong U \cap W$.

By Theorem 6.25, since $\ker(T)$ and $U \cap W$ are finite,
we need only show $\dim(\ker(T)) = \text{nullity}(T) = \dim(U \cap W)$.
Let $\mathbf{v} = (\mathbf{u}, \mathbf{w})$ be in $\ker(T)$, then $\mathbf{u} - \mathbf{w} = \mathbf{0} \Rightarrow \mathbf{u} = \mathbf{w} = \mathbf{x}$, where $\mathbf{x}$ is in $U \cap W$.
So, we have $\mathbf{v} = (\mathbf{x}, \mathbf{x})$ where $\mathbf{x}$ is in $U \cap W$.
Therefore, if $\{\mathbf{x}_k\}$ is a basis for $U \cap W$ then $\{(\mathbf{x}_k, \mathbf{x}_k)\}$ is a basis for $\ker(T)$.
So, $\dim(\ker(T)) = \text{nullity}(T) = \dim(U \cap W)$ which implies $\ker(T) \cong U \cap W$.

(d) We need to prove *Grassmann's Identity*: $\dim(U + W) = \dim U + \dim W - \dim(U \cap W)$.

From Exercise 43 in Section 6.2, we have $\dim(U \times W) = \dim U + \dim W$.
From the Rank Theorem we have: $\text{rank}(T) + \text{nullity}(T) = \dim(U \times W) = \dim U + \dim W$.
From parts (a) and (b), we have:
$\text{rank}(T) = \dim \text{range}(T) = \dim(U + W)$ and $\text{nullity}(T) = \dim \ker(T) = \dim(U \cap W)$.

Substituting yields: $\dim(U + W) + \dim(U \cap W) = \dim U + \dim W$ which implies

Grassmann's Identity: $\dim(U + W) = \dim U + \dim W - \dim(U \cap W)$ as we were to show.

6.6 The Matrix of a Linear Transformation

1. Directly, $T(4+2x) = 2-4x$.

$[T(1)]_{\mathcal{C}} = [0-x]_{\mathcal{C}} = \begin{bmatrix} 0 \\ -1 \end{bmatrix}$, $[T(x)]_{\mathcal{C}} = [1-0x]_{\mathcal{C}} = \begin{bmatrix} 1 \\ 0 \end{bmatrix} \Rightarrow [T]_{\mathcal{C}\leftarrow\mathcal{B}} = \begin{bmatrix} 0 & 1 \\ -1 & 0 \end{bmatrix}$.

So, $[T]_{\mathcal{C}\leftarrow\mathcal{B}}\,[4+2x]_{\mathcal{B}} = \begin{bmatrix} 0 & 1 \\ -1 & 0 \end{bmatrix}\begin{bmatrix} 4 \\ 2 \end{bmatrix} = \begin{bmatrix} 2 \\ -4 \end{bmatrix} = [2-4x]_{\mathcal{C}} = [T(4+2x)]_{\mathcal{C}}$.

2. Directly as in Exercise 1, but since $4+2x = 3(1+x)+(1-x)$, we have $[4+2x]_{\mathcal{B}} = \begin{bmatrix} 3 \\ 1 \end{bmatrix}$.

$[T(1+x)]_{\mathcal{C}} = [1-x]_{\mathcal{C}} = \begin{bmatrix} 1 \\ -1 \end{bmatrix}$, $[T(1-x)]_{\mathcal{C}} = [-1-x]_{\mathcal{C}} = \begin{bmatrix} -1 \\ -1 \end{bmatrix} \Rightarrow [T]_{\mathcal{C}\leftarrow\mathcal{B}} = \begin{bmatrix} 1 & -1 \\ -1 & -1 \end{bmatrix}$.

So, $[T]_{\mathcal{C}\leftarrow\mathcal{B}}\,[4+2x]_{\mathcal{B}} = \begin{bmatrix} 1 & -1 \\ -1 & -1 \end{bmatrix}\begin{bmatrix} 3 \\ 1 \end{bmatrix} = \begin{bmatrix} 2 \\ -4 \end{bmatrix} = [2-4x]_{\mathcal{C}} = [T(4+2x)]_{\mathcal{C}}$.

3. Directly, $T(a+bx+cx^2) = a+b(x+2)+c(x+2)^2$.

$[T(1)]_{\mathcal{C}} = [1]_{\mathcal{C}} = \begin{bmatrix} 1 \\ 0 \\ 0 \end{bmatrix}$, $[T(x)]_{\mathcal{C}} = [x+2]_{\mathcal{C}} = \begin{bmatrix} 0 \\ 1 \\ 0 \end{bmatrix}$, $[T(x^2)]_{\mathcal{C}} = [(x+2)^2]_{\mathcal{C}} = \begin{bmatrix} 0 \\ 0 \\ 1 \end{bmatrix} \Rightarrow$

$[T]_{\mathcal{C}\leftarrow\mathcal{B}} = \begin{bmatrix} 1 & 0 & 0 \\ 0 & 1 & 0 \\ 0 & 0 & 1 \end{bmatrix}$. So:

$$[T]_{\mathcal{C}\leftarrow\mathcal{B}}\,[4+2x]_{\mathcal{B}} = \begin{bmatrix} 1 & 0 & 0 \\ 0 & 1 & 0 \\ 0 & 0 & 1 \end{bmatrix}\begin{bmatrix} a \\ b \\ c \end{bmatrix} = \begin{bmatrix} a \\ b \\ c \end{bmatrix} = [a+b(x+2)+c(x+2)^2]_{\mathcal{C}}$$
$$= [T(a+bx+cx^2)]_{\mathcal{C}}.$$

4. Directly as in Exercise 3, but since $a+bx+cx^2 = (a+2b+4c)+(b+4c)(x-2)+c(x-2)^2$, we have $[a+bx+cx^2]_{\mathcal{B}} = \begin{bmatrix} a+2b+4c \\ b+4c \\ c \end{bmatrix}$.

$[T(1)]_{\mathcal{C}} = [1]_{\mathcal{C}} = \begin{bmatrix} 1 \\ 0 \\ 0 \end{bmatrix}$, $[T(x+2)]_{\mathcal{C}} = [x+4]_{\mathcal{C}} = \begin{bmatrix} 4 \\ 1 \\ 0 \end{bmatrix}$,

$[T(\,(x+2)^2\,)]_{\mathcal{C}} = [(x+4)^2]_{\mathcal{C}} = [16+8x+x^2]_{\mathcal{C}} = \begin{bmatrix} 16 \\ 8 \\ 1 \end{bmatrix} \Rightarrow [T]_{\mathcal{C}\leftarrow\mathcal{B}} = \begin{bmatrix} 1 & 4 & 16 \\ 0 & 1 & 8 \\ 0 & 0 & 1 \end{bmatrix}$. So:

$$[T]_{\mathcal{C}\leftarrow\mathcal{B}}\,[4+2x]_{\mathcal{B}} = \begin{bmatrix} 1 & 4 & 16 \\ 0 & 1 & 8 \\ 0 & 0 & 1 \end{bmatrix}\begin{bmatrix} a+2b+4c \\ b+4c \\ c \end{bmatrix} = \begin{bmatrix} a+2b+4c \\ b+4c \\ c \end{bmatrix}_{\mathcal{C}}$$
$$= [(a+2b+4c)+(b+4c)x+cx^2]_{\mathcal{C}} = [a+b(x+2)+c(x+2)^2]_{\mathcal{C}}$$
$$= [T(a+bx+cx^2)]_{\mathcal{C}}.$$

5. Directly, $T(a+bx+cx^2) = \begin{bmatrix} a \\ a+b+c \end{bmatrix}$.

$[T(1)]_{\mathcal{C}} = \begin{bmatrix} 1 \\ 1 \end{bmatrix}_{\mathcal{C}} = \begin{bmatrix} 1 \\ 1 \end{bmatrix}$. Likewise, $[T(x)]_{\mathcal{C}} = [T(x^2)]_{\mathcal{C}} = \begin{bmatrix} 0 \\ 1 \end{bmatrix} \Rightarrow [T]_{\mathcal{C}\leftarrow\mathcal{B}} = \begin{bmatrix} 1 & 0 & 0 \\ 1 & 1 & 1 \end{bmatrix}$.

So, $[T]_{\mathcal{C}\leftarrow\mathcal{B}}\,[a+bx+cx^2]_{\mathcal{B}} = \begin{bmatrix} 1 & 0 & 0 \\ 1 & 1 & 1 \end{bmatrix}\begin{bmatrix} a \\ b \\ c \end{bmatrix} = \begin{bmatrix} a \\ a+b+c \end{bmatrix}_{\mathcal{C}} = [T(a+bx+cx^2)]_{\mathcal{C}}$.

6. Directly as in Exercise 5, but since $\mathcal{B} = \{x^2, x, 1\}$, $[a+bx+cx^2]_{\mathcal{B}} = \begin{bmatrix} c \\ b \\ a \end{bmatrix}$.

$[T(x^2)]_{\mathcal{C}} = [T(x)]_{\mathcal{C}} = \begin{bmatrix} 0 \\ 1 \end{bmatrix}_{\mathcal{C}} = \begin{bmatrix} -1 \\ 1 \end{bmatrix}$ (because $\begin{bmatrix} 0 \\ 1 \end{bmatrix} = -\begin{bmatrix} 1 \\ 0 \end{bmatrix} + \begin{bmatrix} 1 \\ 1 \end{bmatrix}$).

Likewise, $[T(1)]_{\mathcal{C}} = \begin{bmatrix} 1 \\ 1 \end{bmatrix}_{\mathcal{C}} = \begin{bmatrix} 0 \\ 1 \end{bmatrix} \Rightarrow [T]_{\mathcal{C}\leftarrow\mathcal{B}} = \begin{bmatrix} -1 & -1 & 0 \\ 1 & 1 & 1 \end{bmatrix}$. So:

$$[T]_{\mathcal{C}\leftarrow\mathcal{B}}\,[a+bx+cx^2]_{\mathcal{B}} = \begin{bmatrix} -1 & -1 & 0 \\ 1 & 1 & 1 \end{bmatrix}\begin{bmatrix} c \\ b \\ a \end{bmatrix} = \begin{bmatrix} -(b+c) \\ a+b+c \end{bmatrix} = \begin{bmatrix} a \\ a+b+c \end{bmatrix}_{\mathcal{C}}$$
$$= [T(a+bx+cx^2)]_{\mathcal{C}}.$$

Note $\begin{bmatrix} a \\ a+b+c \end{bmatrix} = -(b+c)\begin{bmatrix} 1 \\ 0 \end{bmatrix} + (a+b+c)\begin{bmatrix} 1 \\ 1 \end{bmatrix} \Rightarrow \begin{bmatrix} -(b+c) \\ a+b+c \end{bmatrix} = \begin{bmatrix} a \\ a+b+c \end{bmatrix}_{\mathcal{C}}$.

7. Directly, $T\begin{bmatrix} -7 \\ 7 \end{bmatrix} = \begin{bmatrix} 7 \\ 7 \\ 7 \end{bmatrix}$. Also, note since $\begin{bmatrix} -7 \\ 7 \end{bmatrix} = 2\begin{bmatrix} 1 \\ 2 \end{bmatrix} - 3\begin{bmatrix} 3 \\ -1 \end{bmatrix}$, $\begin{bmatrix} -7 \\ 7 \end{bmatrix}_{\mathcal{B}} = \begin{bmatrix} 2 \\ -3 \end{bmatrix}$.

Likewise, $\left[T\begin{bmatrix} 1 \\ 2 \end{bmatrix}\right]_{\mathcal{C}} = \begin{bmatrix} 5 \\ -1 \\ 2 \end{bmatrix}_{\mathcal{C}} = \begin{bmatrix} 6 \\ -3 \\ 2 \end{bmatrix}$, $\left[T\begin{bmatrix} 3 \\ -1 \end{bmatrix}\right]_{\mathcal{C}} = \begin{bmatrix} 1 \\ -3 \\ -1 \end{bmatrix}_{\mathcal{C}} = \begin{bmatrix} 4 \\ -2 \\ -1 \end{bmatrix} \Rightarrow$

$[T]_{\mathcal{C}\leftarrow\mathcal{B}} = \begin{bmatrix} 6 & 4 \\ -3 & -2 \\ 2 & -1 \end{bmatrix}$.

So, $[T]_{\mathcal{C}\leftarrow\mathcal{B}}\begin{bmatrix} -7 \\ 7 \end{bmatrix}_{\mathcal{B}} = \begin{bmatrix} 6 & 4 \\ -3 & -2 \\ 2 & -1 \end{bmatrix}\begin{bmatrix} 2 \\ -3 \end{bmatrix} = \begin{bmatrix} 0 \\ 0 \\ 7 \end{bmatrix} = \begin{bmatrix} 7 \\ 7 \\ 7 \end{bmatrix}_{\mathcal{C}} = \left[T\begin{bmatrix} -7 \\ 7 \end{bmatrix}\right]_{\mathcal{C}}$.

8. Directly there is nothing to show. On the other hand, note:

$$\begin{bmatrix}1\\0\end{bmatrix} = \tfrac{1}{7}\begin{bmatrix}1\\2\end{bmatrix} + \tfrac{2}{7}\begin{bmatrix}3\\-1\end{bmatrix} \text{ and } \begin{bmatrix}0\\1\end{bmatrix} = \tfrac{3}{7}\begin{bmatrix}1\\2\end{bmatrix} - \tfrac{1}{7}\begin{bmatrix}3\\-1\end{bmatrix}, \begin{bmatrix}a\\b\end{bmatrix}_{\mathcal{B}} = \begin{bmatrix}(a+3b)/7\\(2a-b)/7\end{bmatrix}.$$

$[T]_{\mathcal{C}\leftarrow\mathcal{B}} = \begin{bmatrix}6 & 4\\-3 & -2\\2 & -1\end{bmatrix}$ exactly as in Exercise 7 because the bases have not been changed.

$$\text{So, } [T]_{\mathcal{C}\leftarrow\mathcal{B}}\begin{bmatrix}a\\b\end{bmatrix}_{\mathcal{B}} = \begin{bmatrix}6 & 4\\-3 & -2\\2 & -1\end{bmatrix}\begin{bmatrix}(a+3b)/7\\(2a-b)/7\end{bmatrix} = \begin{bmatrix}2(a+b)\\-(a+b)\\b\end{bmatrix} = \begin{bmatrix}a+2b\\-a\\b\end{bmatrix}_{\mathcal{C}} = \left[T\begin{bmatrix}a\\b\end{bmatrix}\right]_{\mathcal{C}}.$$

$$\text{Note } \begin{bmatrix}2(a+b)\\-(a+b)\\b\end{bmatrix} = 2(a+b)\begin{bmatrix}1\\0\\0\end{bmatrix} - (a+b)\begin{bmatrix}1\\1\\0\end{bmatrix} + b\begin{bmatrix}1\\1\\1\end{bmatrix} = \begin{bmatrix}a+2b\\-a\\b\end{bmatrix}_{\mathcal{C}}.$$

9. Directly $T\begin{bmatrix}a & b\\c & d\end{bmatrix} = \begin{bmatrix}a & b\\c & d\end{bmatrix}^T = \begin{bmatrix}a & c\\b & d\end{bmatrix}.$

$$[T(E_{11})]_{\mathcal{C}} = [E_{11}]_{\mathcal{C}} = \begin{bmatrix}1\\0\\0\\0\end{bmatrix}, \text{ Likewise, } [T(E_{12})]_{\mathcal{C}} = [E_{21}]_{\mathcal{C}} = \begin{bmatrix}0\\0\\1\\0\end{bmatrix},$$

$$[T(E_{21})]_{\mathcal{C}} = [E_{12}]_{\mathcal{C}} = \begin{bmatrix}0\\1\\0\\0\end{bmatrix}, [T(E_{22})]_{\mathcal{C}} = \begin{bmatrix}0\\0\\0\\1\end{bmatrix} \Rightarrow [T]_{\mathcal{C}\leftarrow\mathcal{B}} = \begin{bmatrix}1&0&0&0\\0&0&1&0\\0&1&0&0\\0&0&0&1\end{bmatrix}.$$

$$\text{So, } [T]_{\mathcal{C}\leftarrow\mathcal{B}}[A]_{\mathcal{B}} = \begin{bmatrix}1&0&0&0\\0&0&1&0\\0&1&0&0\\0&0&0&1\end{bmatrix}\begin{bmatrix}a\\b\\c\\d\end{bmatrix} = \begin{bmatrix}a\\c\\b\\d\end{bmatrix} = \left[\begin{bmatrix}a & c\\b & d\end{bmatrix}\right]_{\mathcal{C}} = [T(A)]_{\mathcal{C}}.$$

10. Directly as in Exercise 9, but $[A]_{\mathcal{B}} = \begin{bmatrix} d \\ c \\ b \\ a \end{bmatrix}$ since $\mathcal{B} = \{E_{22}, E_{21}, E_{12}, E_{11}\}$ in that order.

$[T(E_{22})]_{\mathcal{C}} = [E_{22}]_{\mathcal{C}} = \begin{bmatrix} 0 \\ 0 \\ 1 \\ 0 \end{bmatrix}$ also because the *order* of the E_{ij} in $\mathcal{C}$ matter.

Likewise, $[T(E_{21})]_{\mathcal{C}} = [E_{12}]_{\mathcal{C}} == \begin{bmatrix} 1 \\ 0 \\ 0 \\ 0 \end{bmatrix}$, $[T(E_{12})]_{\mathcal{C}} = [E_{21}]_{\mathcal{C}} = \begin{bmatrix} 0 \\ 1 \\ 0 \\ 0 \end{bmatrix}$, $[T(E_{11})]_{\mathcal{C}} = \begin{bmatrix} 0 \\ 0 \\ 0 \\ 1 \end{bmatrix} \Rightarrow$

$$[T]_{\mathcal{C}\leftarrow\mathcal{B}} = \begin{bmatrix} 0 & 1 & 0 & 0 \\ 0 & 0 & 1 & 0 \\ 1 & 0 & 0 & 0 \\ 0 & 0 & 0 & 1 \end{bmatrix}.$$

So, $[T]_{\mathcal{C}\leftarrow\mathcal{B}}\,[A]_{\mathcal{B}} = \begin{bmatrix} 1 & 0 & 0 & 0 \\ 0 & 0 & 1 & 0 \\ 0 & 1 & 0 & 0 \\ 0 & 0 & 0 & 1 \end{bmatrix} \begin{bmatrix} d \\ c \\ b \\ a \end{bmatrix} = \begin{bmatrix} c \\ b \\ d \\ a \end{bmatrix} = \left[\begin{bmatrix} a & c \\ b & d \end{bmatrix}\right]_{\mathcal{C}} = [T(A)]_{\mathcal{C}}.$

Again, $\begin{bmatrix} c \\ b \\ d \\ a \end{bmatrix} = \left[\begin{bmatrix} a & c \\ b & d \end{bmatrix}\right]_{\mathcal{C}}$ because of the order of E_{ij} in $\mathcal{C}$.

11. Directly, $T(A) = AB - BA = \begin{bmatrix} a-b & b-a \\ c-d & d-c \end{bmatrix} - \begin{bmatrix} a-c & b-d \\ c-a & d-b \end{bmatrix} = \begin{bmatrix} c-b & d-a \\ a-d & b-c \end{bmatrix}$.

$$[T(E_{11})]_{\mathcal{C}} = [E_{11}B - BE_{11}]_{\mathcal{C}} = \begin{bmatrix} 0 & -1 \\ 1 & 0 \end{bmatrix}_{\mathcal{C}} = \begin{bmatrix} 0 \\ -1 \\ 1 \\ 0 \end{bmatrix}.$$

Likewise, $[T(E_{12})]_{\mathcal{C}} = \begin{bmatrix} -1 & 0 \\ 0 & 1 \end{bmatrix}_{\mathcal{C}} = \begin{bmatrix} -1 \\ 0 \\ 0 \\ 1 \end{bmatrix}$, $[T(E_{21})]_{\mathcal{C}} = \begin{bmatrix} 1 & 0 \\ 0 & -1 \end{bmatrix}_{\mathcal{C}} = \begin{bmatrix} 1 \\ 0 \\ 0 \\ -1 \end{bmatrix}$,

$[T(E_{22})]_{\mathcal{C}} = \begin{bmatrix} 0 & 1 \\ -1 & 0 \end{bmatrix}_{\mathcal{C}} = \begin{bmatrix} 0 \\ 1 \\ -1 \\ 0 \end{bmatrix} \Rightarrow [T]_{\mathcal{C}\leftarrow\mathcal{B}} = \begin{bmatrix} 0 & -1 & 1 & 0 \\ -1 & 0 & 0 & 1 \\ 1 & 0 & 0 & -1 \\ 0 & 1 & -1 & 0 \end{bmatrix}$. So:

$$[T]_{\mathcal{C}\leftarrow\mathcal{B}}\,[A]_{\mathcal{B}} = \begin{bmatrix} 0 & -1 & 1 & 0 \\ -1 & 0 & 0 & 1 \\ 1 & 0 & 0 & -1 \\ 0 & 1 & -1 & 0 \end{bmatrix} \begin{bmatrix} a \\ b \\ c \\ d \end{bmatrix} = \begin{bmatrix} c-b \\ d-a \\ a-d \\ b-c \end{bmatrix} = \left[\begin{bmatrix} c-b & d-a \\ a-d & b-c \end{bmatrix}\right]_{\mathcal{C}}$$
$$= [AB - BA]_{\mathcal{C}} = [T(A)]_{\mathcal{C}}.$$

12. Directly, $T(A) = A - A^T = \begin{bmatrix} a & b \\ c & d \end{bmatrix} - \begin{bmatrix} a & c \\ b & d \end{bmatrix} = \begin{bmatrix} 0 & b-c \\ c-b & 0 \end{bmatrix}$.

$$[T(E_{11})]_{\mathcal{C}} = [E_{11} - E_{11}]_{\mathcal{C}} = \begin{bmatrix} 0 & 0 \\ 0 & 0 \end{bmatrix}_{\mathcal{C}} = \begin{bmatrix} 0 \\ 0 \\ 0 \\ 0 \end{bmatrix}.$$

Likewise, $[T(E_{12})]_{\mathcal{C}} = \begin{bmatrix} 0 & 1 \\ -1 & 0 \end{bmatrix}_{\mathcal{C}} \begin{bmatrix} 0 \\ 1 \\ -1 \\ 0 \end{bmatrix}$, $[T(E_{21})]_{\mathcal{C}} = \begin{bmatrix} 0 & -1 \\ 1 & 0 \end{bmatrix}_{\mathcal{C}} = \begin{bmatrix} 0 \\ -1 \\ 1 \\ 0 \end{bmatrix}$,

$[T(E_{22})]_{\mathcal{C}} = \begin{bmatrix} 0 & 0 \\ 0 & 0 \end{bmatrix}_{\mathcal{C}} = \begin{bmatrix} 0 \\ 0 \\ 0 \\ 0 \end{bmatrix} \Rightarrow [T]_{\mathcal{C}\leftarrow\mathcal{B}} = \begin{bmatrix} 0 & 0 & 0 & 0 \\ 0 & 1 & -1 & 0 \\ 0 & -1 & 1 & 0 \\ 0 & 0 & 0 & 0 \end{bmatrix}$. So:

$$[T]_{\mathcal{C}\leftarrow\mathcal{B}}\,[A]_{\mathcal{B}} = \begin{bmatrix} 0 & 0 & 0 & 0 \\ 0 & 1 & -1 & 0 \\ 0 & -1 & 1 & 0 \\ 0 & 0 & 0 & 0 \end{bmatrix} \begin{bmatrix} a \\ b \\ c \\ d \end{bmatrix} = \begin{bmatrix} 0 \\ b-c \\ c-b \\ 0 \end{bmatrix} = \left[\begin{bmatrix} 0 & b-c \\ c-b & 0 \end{bmatrix}\right]_{\mathcal{C}}$$
$$= [A - A^T]_{\mathcal{C}} = [T(A)]_{\mathcal{C}}.$$

13. (a) $D(\sin x) = \cos x, D(\cos x) = -\sin x \Rightarrow$
range$(D) = $ span$(\sin x, \cos x) = W$ as was to be shown.

(b) $\mathcal{B} = \{\sin x, \cos x\} \Rightarrow [\sin x]_{\mathcal{B}} = \begin{bmatrix} 1 \\ 0 \end{bmatrix}, [\cos x]_{\mathcal{B}} = \begin{bmatrix} 0 \\ 1 \end{bmatrix} \Rightarrow$

$$[D(\sin x)]_{\mathcal{B}} = \begin{bmatrix} 0 \\ 1 \end{bmatrix}, [D(\cos x)]_{\mathcal{B}} = \begin{bmatrix} -1 \\ 0 \end{bmatrix} \Rightarrow [D]_{\mathcal{B}} = \begin{bmatrix} 0 & -1 \\ 1 & 0 \end{bmatrix}.$$

(c) $[f(x)]_{\mathcal{B}} =]3\sin x - 5\cos x]_{\mathcal{B}} = \begin{bmatrix} 3 \\ -5 \end{bmatrix} \Rightarrow$

$$[D(f(x)]_{\mathcal{B}} = [D]_{\mathcal{B}}[f(x)]_{\mathcal{B}} = \begin{bmatrix} 0 & -1 \\ 1 & 0 \end{bmatrix}\begin{bmatrix} 3 \\ -5 \end{bmatrix} = \begin{bmatrix} 5 \\ 3 \end{bmatrix} \Rightarrow$$

$f'(x) = 5\sin x + 3\cos x$.
Directly, we have $f'(x) = 3(\cos x) - 5(-\sin x) = 5\sin x + 3\cos x$.
This shows the indirect and direct methods give the same answer.

14. (a) $D(e^{2x}) = 2e^{2x}, D(e^{-2x}) = -2e^{-2x} \Rightarrow$
range$(D) = $ span$(e^{2x}, e^{-2x}) = W$ as was to be shown.

(b) $\mathcal{B} = \{e^{2x}, e^{-2x}\} \Rightarrow [e^{2x}]_{\mathcal{B}} = \begin{bmatrix} 1 \\ 0 \end{bmatrix}, [e^{-2x}]_{\mathcal{B}} = \begin{bmatrix} 0 \\ 1 \end{bmatrix} \Rightarrow$

$$[D(e^{2x})]_{\mathcal{B}} = \begin{bmatrix} 2 \\ 0 \end{bmatrix}, [D(e^{-2x})x]_{\mathcal{B}} = \begin{bmatrix} 0 \\ -2 \end{bmatrix} \Rightarrow [D]_{\mathcal{B}} = \begin{bmatrix} 2 & 0 \\ 0 & -2 \end{bmatrix}.$$

(c) $[f(x)]_{\mathcal{B}} = [e^{2x} - 3e^{-2x}]_{\mathcal{B}} = \begin{bmatrix} 1 \\ -3 \end{bmatrix} \Rightarrow$

$$[D(f(x)]_{\mathcal{B}} = [D]_{\mathcal{B}}[f(x)]_{\mathcal{B}} = \begin{bmatrix} 2 & 0 \\ 0 & -2 \end{bmatrix}\begin{bmatrix} 1 \\ -3 \end{bmatrix} = \begin{bmatrix} 2 \\ 6 \end{bmatrix} \Rightarrow f'(x) = 2e^{2x} + 6e^{-2x}.$$

Directly, we have $f'(x) = 2(e^{2x}) - 3(-2e^{-2x}) = 2e^{2x} + 6e^{-2x}$.
This shows the indirect and direct methods give the same answer.

15. (a) $\mathcal{B} = \{e^{2x}, e^{2x}\cos x, e^{2x}\sin x\} \Rightarrow$

$$[D(e^{2x})]_{\mathcal{B}} = [2e^{2x}]_{\mathcal{B}} = \begin{bmatrix} 2 \\ 0 \\ 0 \end{bmatrix}, [D(e^{2x}\cos x)]_{\mathcal{B}} = [2e^{2x}\cos x - e^{2x}\sin x]_{\mathcal{B}} = \begin{bmatrix} 0 \\ 2 \\ -1 \end{bmatrix},$$

$$[D(e^{2x}\sin x)]_{\mathcal{B}} = [e^{2x}\cos x + 2e^{2x}\sin x]_{\mathcal{B}} = \begin{bmatrix} 0 \\ 1 \\ 2 \end{bmatrix} \Rightarrow [D]_{\mathcal{B}} = \begin{bmatrix} 2 & 0 & 0 \\ 0 & 2 & 1 \\ 0 & -1 & 2 \end{bmatrix}.$$

(b) $[f(x)]_{\mathcal{B}} = [3e^{2x} - e^{2x}\cos x + 2e^{2x}\sin x]_{\mathcal{B}} = \begin{bmatrix} 3 \\ -1 \\ 2 \end{bmatrix} \Rightarrow$

$$[D(f(x)]_{\mathcal{B}} = [D]_{\mathcal{B}}[f(x)]_{\mathcal{B}} = \begin{bmatrix} 2 & 0 & 0 \\ 0 & 2 & 1 \\ 0 & -1 & 2 \end{bmatrix}\begin{bmatrix} 3 \\ -1 \\ 2 \end{bmatrix} = \begin{bmatrix} 6 \\ 0 \\ 5 \end{bmatrix} \Rightarrow f'(x) = 6e^{2x} + 5e^{2x}\sin x.$$

Directly, we have:
$f'(x) = 3(2e^{2x}) - (2e^{2x}\cos x - e^{2x}\sin x) + 2(e^{2x}\cos x + 2e^{2x}\sin x) = 6e^{2x} + 5e^{2x}\sin x$.
This shows the indirect and direct methods give the same answer.

16. (a) $\mathcal{B} = \{\cos x, \sin x, x\cos x, x\sin x\} \Rightarrow$

$$[D(\cos x)]_{\mathcal{B}} = [-\sin x]_{\mathcal{B}} = \begin{bmatrix} 0 \\ -1 \\ 0 \\ 0 \end{bmatrix}, [D(\sin x)]_{\mathcal{B}} = [\cos x]_{\mathcal{B}} = \begin{bmatrix} 1 \\ 0 \\ 0 \\ 0 \end{bmatrix},$$

$$[D(x\cos x)]_{\mathcal{B}} = [\cos x - x\sin x]_{\mathcal{B}} = \begin{bmatrix} 1 \\ 0 \\ 0 \\ -1 \end{bmatrix}, [D(x\sin x)]_{\mathcal{B}} = [\sin x + x\cos x]_{\mathcal{B}} = \begin{bmatrix} 0 \\ 1 \\ 1 \\ 0 \end{bmatrix} \Rightarrow$$

$$[D]_{\mathcal{B}} = \begin{bmatrix} 0 & 1 & 1 & 0 \\ -1 & 0 & 0 & 1 \\ 0 & 0 & 0 & 1 \\ 0 & 0 & -1 & 0 \end{bmatrix}.$$

(b) $[f(x)]_{\mathcal{B}} = [\cos x + 2x\cos x]_{\mathcal{B}} = \begin{bmatrix} 1 \\ 0 \\ 2 \\ 0 \end{bmatrix} \Rightarrow$

$$[D(f(x)]_{\mathcal{B}} = [D]_{\mathcal{B}}[f(x)]_{\mathcal{B}} = \begin{bmatrix} 0 & 1 & 1 & 0 \\ -1 & 0 & 0 & 1 \\ 0 & 0 & 0 & 1 \\ 0 & 0 & -1 & 0 \end{bmatrix}\begin{bmatrix} 1 \\ 0 \\ 2 \\ 0 \end{bmatrix} = \begin{bmatrix} 2 \\ -1 \\ 0 \\ -2 \end{bmatrix} \Rightarrow$$

$f'(x) = 2\cos x - \sin x - 2x\sin x$.
Directly, we have $f'(x) = -\sin x + 2(\cos x - x\sin x) = 2\cos x - \sin x - 2x\sin x$.
This shows the indirect and direct methods give the same answer.

17. (a) Let $p(x) = c + dx$, then $T(p(x)) = T(c+dx) = \begin{bmatrix} p(0) \\ p(1) \end{bmatrix} = \begin{bmatrix} c \\ c+d \end{bmatrix} \Rightarrow$

$$(S \circ T)(p(x)) = S(T(p(x)) = S(T(c+dx)) = S\begin{bmatrix} c \\ c+d \end{bmatrix} = \begin{bmatrix} c - 2(c+d) \\ 2c - (c+d) \end{bmatrix} = \begin{bmatrix} -c-2d \\ c-d \end{bmatrix} \Rightarrow$$

$$[(S \circ T)(1)]_{\mathcal{D}\leftarrow\mathcal{B}} = \begin{bmatrix} -1-2(0) \\ 1-0 \end{bmatrix} = \begin{bmatrix} -1 \\ 1 \end{bmatrix} \text{ and}$$

$$[(S \circ T)(x)]_{\mathcal{D}\leftarrow\mathcal{B}} = \begin{bmatrix} -0-2(1) \\ 0-1 \end{bmatrix} = \begin{bmatrix} -2 \\ -1 \end{bmatrix} \Rightarrow [S \circ T]_{\mathcal{D}\leftarrow\mathcal{B}} = \begin{bmatrix} -1 & -2 \\ 1 & -1 \end{bmatrix}.$$

(b) $[T(1)]_{\mathcal{C}\leftarrow\mathcal{B}} = \begin{bmatrix} p(0)=1 \\ p(1)=1 \end{bmatrix} = \begin{bmatrix} 1 \\ 1 \end{bmatrix}$ and $[T(x))]_{\mathcal{C}\leftarrow\mathcal{B}} = \begin{bmatrix} p(0)=0 \\ p(1)=1 \end{bmatrix} = \begin{bmatrix} 0 \\ 1 \end{bmatrix} \Rightarrow$

$$[T]_{\mathcal{C}\leftarrow\mathcal{B}} = \begin{bmatrix} 1 & 0 \\ 1 & 1 \end{bmatrix}.$$

$$[S\begin{bmatrix} 1 \\ 0 \end{bmatrix}]_{\mathcal{D}\leftarrow\mathcal{C}} = \begin{bmatrix} 1-2(0) \\ 2(1)-0 \end{bmatrix} = \begin{bmatrix} 1 \\ 2 \end{bmatrix} \text{ and } S\begin{bmatrix} 0 \\ 1 \end{bmatrix}]_{\mathcal{D}\leftarrow\mathcal{C}} = \begin{bmatrix} 0-2(1) \\ 2(0)-1 \end{bmatrix} = \begin{bmatrix} -2 \\ -1 \end{bmatrix} \Rightarrow$$

$$[S]_{\mathcal{D}\leftarrow\mathcal{C}} = \begin{bmatrix} 1 & -2 \\ 2 & -1 \end{bmatrix}.$$

Therefore, $[S]_{\mathcal{D}\leftarrow\mathcal{B}} = [S]_{\mathcal{D}\leftarrow\mathcal{C}}[T]_{\mathcal{C}\leftarrow\mathcal{B}} = \begin{bmatrix} 1 & -2 \\ 2 & -1 \end{bmatrix}\begin{bmatrix} 1 & 0 \\ 1 & 1 \end{bmatrix} = \begin{bmatrix} -1 & -2 \\ 1 & -1 \end{bmatrix}$ as in (a).

18. (a) Since $(S \circ T)(p(x)) = S(T(p(x)) = S(p(x+1)) = p((x+1)+1) = p(x+2)$, we have:

$$[(S \circ T)(1)]_{\mathcal{D} \leftarrow \mathcal{B}} = [1]_{\mathcal{D}} = \begin{bmatrix} 1 \\ 0 \\ 0 \end{bmatrix} \text{ and } [(S \circ T)(x)]_{\mathcal{D} \leftarrow \mathcal{B}} = [x+2]_{\mathcal{D}} = \begin{bmatrix} 2 \\ 1 \\ 0 \end{bmatrix} \Rightarrow$$

$$[S \circ T]_{\mathcal{D} \leftarrow \mathcal{B}} = \begin{bmatrix} 1 & 2 \\ 0 & 1 \\ 0 & 0 \end{bmatrix}.$$

(b) $$[T(1)]_{\mathcal{C} \leftarrow \mathcal{B}} = [1]_{\mathcal{D}} = \begin{bmatrix} 1 \\ 0 \\ 0 \end{bmatrix} \text{ and } [T(x)]_{\mathcal{C} \leftarrow \mathcal{B}} = [x+1]_{\mathcal{D}} = \begin{bmatrix} 1 \\ 1 \\ 0 \end{bmatrix} \Rightarrow [T]_{\mathcal{C} \leftarrow \mathcal{B}} = \begin{bmatrix} 1 & 1 \\ 0 & 1 \\ 0 & 0 \end{bmatrix}.$$

$$[S(1)]_{\mathcal{D} \leftarrow \mathcal{C}} = [1]_{\mathcal{D}} = \begin{bmatrix} 1 \\ 0 \\ 0 \end{bmatrix}, \ [S(x)]_{\mathcal{D} \leftarrow \mathcal{C}} = [x+1]_{\mathcal{D}} = \begin{bmatrix} 1 \\ 1 \\ 0 \end{bmatrix}, \text{ and}$$

$$[S(x^2)]_{\mathcal{D} \leftarrow \mathcal{C}} = [x^2+2x+1]_{\mathcal{D}} = \begin{bmatrix} 1 \\ 2 \\ 1 \end{bmatrix} \Rightarrow [S]_{\mathcal{D} \leftarrow \mathcal{C}} = \begin{bmatrix} 1 & 1 & 1 \\ 0 & 1 & 2 \\ 0 & 0 & 1 \end{bmatrix}.$$

Therefore, $$[S]_{\mathcal{D} \leftarrow \mathcal{B}} = [S]_{\mathcal{D} \leftarrow \mathcal{C}} [T]_{\mathcal{C} \leftarrow \mathcal{B}} = \begin{bmatrix} 1 & 1 & 1 \\ 0 & 1 & 2 \\ 0 & 0 & 1 \end{bmatrix} \begin{bmatrix} 1 & 1 \\ 0 & 1 \\ 0 & 0 \end{bmatrix} = \begin{bmatrix} 1 & 2 \\ 0 & 1 \\ 0 & 0 \end{bmatrix}$$ as in (a).

19. In Exercise 1, since both bases were already standard, we have:

$[T]_{\mathcal{E}' \leftarrow \mathcal{E}} = \begin{bmatrix} 0 & 1 \\ -1 & 0 \end{bmatrix}$ invertible $\Rightarrow T$ invertible and

$$[T^{-1}(a+bx)]_{\mathcal{E}' \leftarrow \mathcal{E}} = ([T]_{\mathcal{E}' \leftarrow \mathcal{E}})^{-1} \begin{bmatrix} a \\ b \end{bmatrix} = \begin{bmatrix} 0 & 1 \\ -1 & 0 \end{bmatrix}^{-1} \begin{bmatrix} a \\ b \end{bmatrix} = \begin{bmatrix} 0 & -1 \\ 1 & 0 \end{bmatrix} \begin{bmatrix} a \\ b \end{bmatrix} = \begin{bmatrix} -b \\ a \end{bmatrix} \Rightarrow$$

$T^{-1}(a+bx) = -b + ax$.

20. In Exercise 5, we found:

$[T]_{\mathcal{C} \leftarrow \mathcal{B}} = \begin{bmatrix} 1 & 0 & 0 \\ 1 & 1 & 1 \end{bmatrix} \Rightarrow [T]_{\mathcal{C} \leftarrow \mathcal{B}}$ is not square $\Rightarrow [T]_{\mathcal{C} \leftarrow \mathcal{B}}$ is not invertible $\Rightarrow T$ is not invertible.

21. In Exercise 3, $\mathcal{C}$ was not the standard basis, so we need to compute $[T]_{\mathcal{E}'\leftarrow\mathcal{E}}$ before we begin.

$$[T(1)]_{\mathcal{E}} = [1]_{\mathcal{E}} = \begin{bmatrix} 1 \\ 0 \\ 0 \end{bmatrix}, [T(x)]_{\mathcal{E}} = [x+2]_{\mathcal{E}} = \begin{bmatrix} 2 \\ 1 \\ 0 \end{bmatrix}, [T(x^2)]_{\mathcal{E}} = [(x+2)^2]_{\mathcal{E}} = \begin{bmatrix} 4 \\ 4 \\ 1 \end{bmatrix} \Rightarrow$$

$$[T]_{\mathcal{E}'\leftarrow\mathcal{E}} = \begin{bmatrix} 1 & 2 & 4 \\ 0 & 1 & 4 \\ 0 & 0 & 1 \end{bmatrix}.$$

Therefore, $[T]_{\mathcal{E}'\leftarrow\mathcal{E}}$ invertible $\Rightarrow T$ invertible and

$$[T^{-1}(p(x))]_{\mathcal{E}\leftarrow\mathcal{E}'} = [T^{-1}(a+bx+cx^2)]_{\mathcal{E}\leftarrow\mathcal{E}'} = ([T]_{\mathcal{E}'\leftarrow\mathcal{E}})^{-1}[a+bx+cx^2]_{\mathcal{E}'\leftarrow\mathcal{E}}$$

$$= \begin{bmatrix} 1 & 2 & 4 \\ 0 & 1 & 4 \\ 0 & 0 & 1 \end{bmatrix}^{-1} \begin{bmatrix} a \\ b \\ c \end{bmatrix} = \begin{bmatrix} 1 & -2 & 4 \\ 0 & 1 & -4 \\ 0 & 0 & 1 \end{bmatrix} \begin{bmatrix} a \\ b \\ c \end{bmatrix} = \begin{bmatrix} a-2b+4c \\ b-4c \\ c \end{bmatrix} \Rightarrow$$

$T^{-1}(a+bx+cx^2) = (a-2b+4c)+(b-4c)x+cx^2 = a+b(x-2)+c(x-2)^2 \Rightarrow T^{-1}(p(x)) = p(x-2)$.

22. $[T(1)]_{\mathcal{E}} = [0]_{\mathcal{E}} = \begin{bmatrix} 0 \\ 0 \\ 0 \end{bmatrix}, [T(x)]_{\mathcal{E}} = [1]_{\mathcal{E}} = \begin{bmatrix} 1 \\ 0 \\ 0 \end{bmatrix}, [T(x^2)]_{\mathcal{E}} = [2x]_{\mathcal{E}} = \begin{bmatrix} 0 \\ 2 \\ 0 \end{bmatrix} \Rightarrow$

$[T]_{\mathcal{E}'\leftarrow\mathcal{E}} = \begin{bmatrix} 0 & 1 & 0 \\ 0 & 0 & 2 \\ 0 & 0 & 0 \end{bmatrix} \Rightarrow \det([T]_{\mathcal{E}'\leftarrow\mathcal{E}}) = 0 \Rightarrow [T]_{\mathcal{E}'\leftarrow\mathcal{E}}$ is not invertible$\Rightarrow T$ is not invertible.

23. $[T(1)]_{\mathcal{E}} = [1+0]_{\mathcal{E}} = \begin{bmatrix} 1 \\ 0 \\ 0 \end{bmatrix}, [T(x)]_{\mathcal{E}} = [x+1]_{\mathcal{E}} = \begin{bmatrix} 1 \\ 1 \\ 0 \end{bmatrix}, [T(x^2)]_{\mathcal{E}} = [x^2+2x]_{\mathcal{E}} = \begin{bmatrix} 0 \\ 2 \\ 1 \end{bmatrix} \Rightarrow$

$[T]_{\mathcal{E}'\leftarrow\mathcal{E}} = \begin{bmatrix} 1 & 1 & 0 \\ 0 & 1 & 2 \\ 0 & 0 & 1 \end{bmatrix} \Rightarrow [T]_{\mathcal{E}'\leftarrow\mathcal{E}}$ is invertible$\Rightarrow T$ is invertible and

$$[T^{-1}(p(x))]_{\mathcal{E}\leftarrow\mathcal{E}'} = [T^{-1}(a+bx+cx^2)]_{\mathcal{E}\leftarrow\mathcal{E}'} = ([T]_{\mathcal{E}'\leftarrow\mathcal{E}})^{-1}[a+bx+cx^2]_{\mathcal{E}'\leftarrow\mathcal{E}}$$

$$= \begin{bmatrix} 1 & 1 & 0 \\ 0 & 1 & 2 \\ 0 & 0 & 1 \end{bmatrix}^{-1} \begin{bmatrix} a \\ b \\ c \end{bmatrix} = \begin{bmatrix} 1 & -1 & 2 \\ 0 & 1 & -2 \\ 0 & 0 & 1 \end{bmatrix} \begin{bmatrix} a \\ b \\ c \end{bmatrix} = \begin{bmatrix} a-b+2c \\ b-2c \\ c \end{bmatrix} \Rightarrow$$

$T^{-1}(a+bx+cx^2) = (a-b+2c)+(b-2c)x+cx^2 = (a+bx+cx^2)-(b+2cx)+(2c) \Rightarrow$
$T^{-1}(p(x)) = p(x) - p'(x) + p''(x)$.

24. $[T(E_{11})]_{\mathcal{E}} = \begin{bmatrix} 3 \\ 2 \\ 0 \\ 0 \end{bmatrix}$, $[T(E_{12})]_{\mathcal{E}} = \begin{bmatrix} 2 \\ 1 \\ 0 \\ 0 \end{bmatrix}$, $[T(E_{21})]_{\mathcal{E}} = \begin{bmatrix} 0 \\ 0 \\ 3 \\ 2 \end{bmatrix}$, $[T(E_{22})]_{\mathcal{E}} = \begin{bmatrix} 0 \\ 0 \\ 2 \\ 1 \end{bmatrix} \Rightarrow$

$[T]_{\mathcal{E}' \leftarrow \mathcal{E}} = \begin{bmatrix} 3 & 2 & 0 & 0 \\ 2 & 1 & 0 & 0 \\ 0 & 0 & 3 & 2 \\ 0 & 0 & 2 & 1 \end{bmatrix} \Rightarrow [T]_{\mathcal{E}' \leftarrow \mathcal{E}}$ is invertible$\Rightarrow T$ is invertible and

$$[T^{-1}(A)]_{\mathcal{E} \leftarrow \mathcal{E}'} = ([T]_{\mathcal{E}' \leftarrow \mathcal{E}})^{-1}[A]_{\mathcal{E}' \leftarrow \mathcal{E}}$$

$$= \begin{bmatrix} 3 & 2 & 0 & 0 \\ 2 & 1 & 0 & 0 \\ 0 & 0 & 3 & 2 \\ 0 & 0 & 2 & 1 \end{bmatrix}^{-1} \begin{bmatrix} a \\ b \\ c \\ d \end{bmatrix} = \begin{bmatrix} -1 & 2 & 0 & 0 \\ 2 & -3 & 0 & 0 \\ 0 & 0 & -1 & 2 \\ 0 & 0 & 2 & -3 \end{bmatrix} \begin{bmatrix} a \\ b \\ c \\ d \end{bmatrix} = \begin{bmatrix} -a+2b \\ 2a-3b \\ -c+2d \\ 2c-3d \end{bmatrix} \Rightarrow$$

$T^{-1}(A) = \begin{bmatrix} -a+2b & 2a-3b \\ -c+2d & 2c-3d \end{bmatrix} = AB^{-1}$.

25. $\det([T]_{\mathcal{E}' \leftarrow \mathcal{E}}) = 0 \Rightarrow [T]_{\mathcal{E}' \leftarrow \mathcal{E}}$ is not invertible$\Rightarrow T$ is not invertible.

26. $\det([T]_{\mathcal{E}' \leftarrow \mathcal{E}}) = 0 \Rightarrow [T]_{\mathcal{E}' \leftarrow \mathcal{E}}$ is not invertible$\Rightarrow T$ is not invertible.

27. Let $\mathcal{E}' = \mathcal{E} = \{\sin x, \cos x\}$. Then Example 6.83 $\Rightarrow [\int(a\sin x + b\cos x)dx]_{\mathcal{E}'} = [D]_{\mathcal{E}' \leftarrow \mathcal{E}}{}^{-1} \begin{bmatrix} a \\ b \end{bmatrix}$.

Recall from Exercise 13 that $[D]_{\mathcal{E}' \leftarrow \mathcal{E}} = \begin{bmatrix} 0 & -1 \\ 1 & 0 \end{bmatrix} \Rightarrow$

$[\int(\sin x - 3\cos x)dx]_{\mathcal{E}'} = [D]_{\mathcal{E}' \leftarrow \mathcal{E}}{}^{-1} \begin{bmatrix} 1 \\ -3 \end{bmatrix} = \begin{bmatrix} 0 & 1 \\ -1 & 0 \end{bmatrix} \begin{bmatrix} 1 \\ -3 \end{bmatrix} = \begin{bmatrix} -3 \\ -1 \end{bmatrix} \Rightarrow$

$\int(\sin x - 3\cos x)dx = -3\sin x - \cos x + C$.

Directly, we have $(-3\sin x - \cos x)' = -3(\cos x) - (-\sin x) = \sin x - 3\cos x$.

28. Let $\mathcal{E}' = \mathcal{E} = \{e^{2x}, e^{-2x}\}$.

Then Example 6.83 $\Rightarrow [\int 5e^{-2x}\, dx]_{\mathcal{E}'} = [D]_{\mathcal{E}' \leftarrow \mathcal{E}}{}^{-1} \begin{bmatrix} a \\ b \end{bmatrix}$.

Recall from Exercise 14 that $[D]_{\mathcal{E}' \leftarrow \mathcal{E}} = \begin{bmatrix} 2 & 0 \\ 0 & -2 \end{bmatrix} \Rightarrow [D]_{\mathcal{E}' \leftarrow \mathcal{E}}{}^{-1} = \begin{bmatrix} \frac{1}{2} & 0 \\ 0 & -\frac{1}{2} \end{bmatrix} \Rightarrow$

$[\int 5e^{-2x}\, dx]_{\mathcal{E}'} = [D]_{\mathcal{E}' \leftarrow \mathcal{E}}{}^{-1} \begin{bmatrix} 0 \\ 5 \end{bmatrix} = \begin{bmatrix} \frac{1}{2} & 0 \\ 0 & -\frac{1}{2} \end{bmatrix} \begin{bmatrix} 0 \\ 5 \end{bmatrix} = \begin{bmatrix} 0 \\ -\frac{5}{2} \end{bmatrix} \Rightarrow$

$\int 5e^{-2x}\, dx = -\frac{5}{2}e^{-2x}$.

Directly we have $(-\frac{5}{2}e^{-2x})' = -2(-\frac{5}{2}e^{-2x}) = 5e^{-2x}$.

29. Let $\mathcal{E}' = \mathcal{E} = \{e^{2x}, e^{2x}\cos x, e^{2x}\sin x\}$.

Then Example 6.83 $\Rightarrow [\int(a\,e^{2x} + b\,e^{2x}\cos x + c\,e^{2x}\sin x)dx]_{\mathcal{E}'} = [D]_{\mathcal{E}'\leftarrow\mathcal{E}}{}^{-1}\begin{bmatrix} a \\ b \\ c \end{bmatrix}$.

Recall from Exercise 15 that $[D]_{\mathcal{E}'\leftarrow\mathcal{E}} = \begin{bmatrix} 2 & 0 & 0 \\ 0 & 2 & 1 \\ 0 & -1 & 2 \end{bmatrix} \Rightarrow$

$$[\int(e^{2x}\cos x - 2e^{2x}\sin x)dx]_{\mathcal{E}'} = [D]_{\mathcal{E}'\leftarrow\mathcal{E}}{}^{-1}\begin{bmatrix} 0 \\ 1 \\ -2 \end{bmatrix} = \begin{bmatrix} 1/2 & 0 & 0 \\ 0 & 2/5 & -1/5 \\ 0 & 1/5 & 2/5 \end{bmatrix}\begin{bmatrix} 0 \\ 1 \\ -2 \end{bmatrix} = \begin{bmatrix} 0 \\ 4/5 \\ -3/5 \end{bmatrix} \Rightarrow$$

$\int(e^{2x}\cos x - 2e^{2x}\sin x)dx = \frac{4}{5}e^{2x}\cos x - \frac{3}{5}e^{2x}\sin x + C$.

Directly we have

$$(\tfrac{4}{5}e^{2x}\cos x - \tfrac{3}{5}e^{2x}\sin x)' = \tfrac{4}{5}(2e^{2x}\cos x - e^{2x}\sin x) - \tfrac{3}{5}(e^{2x}\cos x + 2e^{2x}\sin x)$$
$$= e^{2x}\cos x - 2e^{2x}\sin x.$$

30. Let $\mathcal{E}' = \mathcal{E} = \{\cos x, \sin x, x\cos x, x\sin x\}$.

Then Example 6.83 $\Rightarrow [\int(a\cos x + b\sin x + cx\cos x + dx\sin x)dx]_{\mathcal{E}'} = [D]_{\mathcal{E}'\leftarrow\mathcal{E}}{}^{-1}\begin{bmatrix} a \\ b \\ c \\ d \end{bmatrix}$.

Recall from Exercise 16 that $[D]_{\mathcal{E}'\leftarrow\mathcal{E}} = \begin{bmatrix} 0 & 1 & 1 & 0 \\ -1 & 0 & 0 & 1 \\ 0 & 0 & 0 & 1 \\ 0 & 0 & -1 & 0 \end{bmatrix} \Rightarrow$

$$[\int(x\cos x + x\sin x)dx]_{\mathcal{E}'} = [D]_{\mathcal{E}'\leftarrow\mathcal{E}}{}^{-1}\begin{bmatrix} 0 \\ 0 \\ 1 \\ 1 \end{bmatrix} = \begin{bmatrix} 0 & -1 & 1 & 0 \\ 1 & 0 & 0 & 1 \\ 0 & 0 & 0 & -1 \\ 0 & 0 & 1 & 0 \end{bmatrix}\begin{bmatrix} 0 \\ 0 \\ 1 \\ 1 \end{bmatrix} = \begin{bmatrix} 1 \\ 1 \\ -1 \\ 1 \end{bmatrix} \Rightarrow$$

$\int(x\cos x + x\sin x)dx = \cos x + \sin x - x\cos x + x\sin x + C$.

Directly we have:

$$(\cos x + \sin x - x\cos x + x\sin x)' = -\sin x + \cos x - (\cos x - x\sin x) + (\sin x + x\cos x)$$
$$= x\cos x + x\sin x.$$

31. Let $\mathcal{E}' = \mathcal{E} = \left\{ \begin{bmatrix} 1 \\ 0 \end{bmatrix}, \begin{bmatrix} 0 \\ 1 \end{bmatrix} \right\}$.

Then $[T\begin{bmatrix} 1 \\ 0 \end{bmatrix}]_{\mathcal{E}} = [\begin{bmatrix} -4(0) \\ 1+5(0) \end{bmatrix}] = \begin{bmatrix} 0 \\ 1 \end{bmatrix}$ and $[T\begin{bmatrix} 0 \\ 1 \end{bmatrix}]_{\mathcal{E}} = [\begin{bmatrix} -4(1) \\ 0+5(1) \end{bmatrix}]_{\mathcal{E}} = \begin{bmatrix} -4 \\ 5 \end{bmatrix} \Rightarrow$

$[T]_{\mathcal{E}} = \begin{bmatrix} 0 & -4 \\ 1 & 5 \end{bmatrix}$.

Example 6.86b $\Rightarrow$ the eigenvectors of $[T]_{\mathcal{E}}$ will give us a basis $\mathcal{C}$.

Section 4.3 $\Rightarrow$ the eigenvalues of $[T]_{\mathcal{E}}$ are 4, 1 with corresponding eigenvectors

$\mathcal{C} = \left\{ \begin{bmatrix} -1 \\ 1 \end{bmatrix}, \begin{bmatrix} -4 \\ 1 \end{bmatrix} \right\}$.

32. Let $\mathcal{E}' = \mathcal{E} = \left\{ \begin{bmatrix} 1 \\ 0 \end{bmatrix}, \begin{bmatrix} 0 \\ 1 \end{bmatrix} \right\}$Then $[T\begin{bmatrix} 1 \\ 0 \end{bmatrix}]_{\mathcal{E}} = [\begin{bmatrix} 1-0 \\ 1+0 \end{bmatrix}] = \begin{bmatrix} 1 \\ 1 \end{bmatrix}$ and

$[T\begin{bmatrix} 0 \\ 1 \end{bmatrix}]_{\mathcal{E}} = [\begin{bmatrix} 0-1 \\ 0+1 \end{bmatrix}]_{\mathcal{E}} = \begin{bmatrix} -1 \\ 1 \end{bmatrix} \Rightarrow [T]_{\mathcal{E}} = \begin{bmatrix} 1 & -1 \\ 1 & 1 \end{bmatrix}$.

Example 6.86b $\Rightarrow$ the eigenvectors of $[T]_{\mathcal{E}}$ will give us a basis $\mathcal{C}$.

Section 4.3 $\Rightarrow$ the eigenvalues of $[T]_{\mathcal{E}}$ are $1+i$, $1-i$.
So the corresponding eigenvectors are not in $\mathbb{R}^2 \Rightarrow T$ is not diagonalizable.

33. Let $\mathcal{E}' = \mathcal{E} = \{1, x\}$.

Then $[T(1)]_{\mathcal{E}} = [(4(1)+2(0)) + (1+3(0))x]_{\mathcal{E}} = [4+x]_{\mathcal{E}} = \begin{bmatrix} 4 \\ 1 \end{bmatrix}$ and

$[T(x)]_{\mathcal{E}} = [(4(0)+2(1)) + (0+3(1))x]_{\mathcal{E}} = [2+3x]_{\mathcal{E}} = \begin{bmatrix} 2 \\ 3 \end{bmatrix} \Rightarrow [T]_{\mathcal{E}} = \begin{bmatrix} 4 & 2 \\ 1 & 3 \end{bmatrix}$.

Example 6.86b $\Rightarrow$ the eigenvectors of $[T]_{\mathcal{E}}$ will give us a basis $\mathcal{C}$.

Section 4.3 $\Rightarrow$ the eigenvalues of $[T]_{\mathcal{E}}$ are 2, 5 with corresponding eigenvectors

$\begin{bmatrix} 1 \\ -1 \end{bmatrix}$ and $\begin{bmatrix} 2 \\ 1 \end{bmatrix} \Rightarrow \mathcal{C} = \{1-x, 2+x\}$.

34. Let $\mathcal{E}' = \mathcal{E} = \{1, x\}$.

Then $[T(1)]_{\mathcal{E}} = [1+(0)(x+1)]_{\mathcal{E}} = [1]_{\mathcal{E}} = \begin{bmatrix} 1 \\ 0 \end{bmatrix}$ and

$[T(x)]_{\mathcal{E}} = [0+(1)(1+x)]_{\mathcal{E}} = [1+x]_{\mathcal{E}} = \begin{bmatrix} 1 \\ 1 \end{bmatrix} \Rightarrow [T]_{\mathcal{E}} = \begin{bmatrix} 1 & 1 \\ 0 & 1 \end{bmatrix} \Rightarrow$

The only eigenvector of $[T]_{\mathcal{E}}$ is $\begin{bmatrix} 1 \\ 0 \end{bmatrix} \Rightarrow$

T is not diagonalizable because it has only 1 linearly independent eigenvector.

35. Let $\mathcal{E}' = \mathcal{E} = \{1, x\}$.

Then $[T(1)]_{\mathcal{E}} = [(1 + (0)]_{\mathcal{E}} = [1]_{\mathcal{E}} = \begin{bmatrix} 1 \\ 0 \end{bmatrix}$ and

$[T(x)]_{\mathcal{E}} = [x + x(1)]_{\mathcal{E}} = [2x]_{\mathcal{E}} = \begin{bmatrix} 0 \\ 2 \end{bmatrix} \Rightarrow [T]_{\mathcal{E}} = \begin{bmatrix} 1 & 0 \\ 0 & 2 \end{bmatrix}$.

$\Rightarrow$ the eigenvalues of $[T]_{\mathcal{E}}$ are 1, 2 with corresponding eigenvectors

$\begin{bmatrix} 1 \\ 0 \end{bmatrix}$ and $\begin{bmatrix} 0 \\ 1 \end{bmatrix} \Rightarrow \mathcal{C} = \{1, x\}$.

36. Let $\mathcal{E}' = \mathcal{E} = \{1, x\}$.

Then $[T(1)]_{\mathcal{E}} = [1]_{\mathcal{E}} = \begin{bmatrix} 1 \\ 0 \\ 0 \end{bmatrix}$, $[T(x)]_{\mathcal{E}} = [3x + 2]_{\mathcal{E}} = \begin{bmatrix} 2 \\ 3 \\ 0 \end{bmatrix}$ and

$[T(x^2)]_{\mathcal{E}} = [(3x + 2)^2]_{\mathcal{E}} = [4 + 12x + 9x^2]_{\mathcal{E}} = \begin{bmatrix} 4 \\ 12 \\ 9 \end{bmatrix} \Rightarrow [T]_{\mathcal{E}} = \begin{bmatrix} 1 & 2 & 4 \\ 0 & 3 & 12 \\ 0 & 0 & 9 \end{bmatrix} \Rightarrow$

The eigenvalues of $[T]_{\mathcal{E}}$ are 1, 3, 9 with corresponding eigenvectors

$\begin{bmatrix} 1 \\ 0 \\ 0 \end{bmatrix}$, $\begin{bmatrix} 1 \\ 1 \\ 0 \end{bmatrix}$ and $\begin{bmatrix} 1 \\ 2 \\ 1 \end{bmatrix} \Rightarrow \mathcal{C} = \{1, 1 + x, 1 + 2x + x^2\} = \{1, x + 1, (x + 1)^2\}$.

37. Let T be reflection in line and let $\mathbf{d} = \begin{bmatrix} d_1 \\ d_2 \end{bmatrix}$ and $\mathbf{d}' = \begin{bmatrix} -d_2 \\ d_1 \end{bmatrix}$ as in Example 6.85.

Then $\mathcal{D} = \{\mathbf{d}, \mathbf{d}'\}$ is a basis for $\mathbb{R}^2$, again as noted in Example 6.85.

In particular, Example 6.85 $\Rightarrow \begin{bmatrix} -d_2 \\ d_1 \end{bmatrix}$ is orthogonal to line $\Rightarrow T \begin{bmatrix} -d_2 \\ d_1 \end{bmatrix} = \begin{bmatrix} d_2 \\ -d_1 \end{bmatrix}$.

Therefore, $[T(\mathbf{d}]_{\mathcal{D}} = \begin{bmatrix} 1 \\ 0 \end{bmatrix}$ and $[T(\mathbf{d}']_{\mathcal{D}} = \begin{bmatrix} 0 \\ -1 \end{bmatrix} \Rightarrow [T]_{\mathcal{D}} = \begin{bmatrix} 1 & 0 \\ 0 & -1 \end{bmatrix}$. Furthermore:

$$P_{\mathcal{E}\leftarrow\mathcal{D}} = \begin{bmatrix} d_1 & d_2 \\ d_2 & -d_1 \end{bmatrix} \Rightarrow P_{\mathcal{D}\leftarrow\mathcal{E}} = (P_{\mathcal{E}\leftarrow\mathcal{D}})^{-1} = \begin{bmatrix} \frac{d_1}{d_1^2+d_2^2} & \frac{d_2}{d_1^2+d_2^2} \\ \frac{d_2}{d_1^2+d_2^2} & -\frac{d_1}{d_1^2+d_2^2} \end{bmatrix} \Rightarrow P_{\mathcal{E}\leftarrow\mathcal{D}} = \begin{bmatrix} d_1 & d_2 \\ d_2 & -d_1 \end{bmatrix} \Rightarrow$$

$$[T]_{\mathcal{E}} = P_{\mathcal{E}\leftarrow\mathcal{D}}[T]_{\mathcal{D}}P_{\mathcal{D}\leftarrow\mathcal{E}} = \begin{bmatrix} d_1 & d_2 \\ d_2 & -d_1 \end{bmatrix} \begin{bmatrix} 1 & 0 \\ 0 & -1 \end{bmatrix} \begin{bmatrix} \frac{d_1}{d_1^2+d_2^2} & \frac{d_2}{d_1^2+d_2^2} \\ \frac{d_2}{d_1^2+d_2^2} & -\frac{d_1}{d_1^2+d_2^2} \end{bmatrix} = \begin{bmatrix} \frac{d_1^2-d_2^2}{d_1^2+d_2^2} & \frac{2d_1 d_2}{d_1^2+d_2^2} \\ \frac{2d_1 d_2}{d_1^2+d_2^2} & \frac{d_2^2-d_1^2}{d_1^2+d_2^2} \end{bmatrix}.$$

38. Let T be projection onto W.

Section 5.2, Example 5.11 $\Rightarrow \mathcal{D} = \left\{ \mathbf{u}_1 = \begin{bmatrix} 1 \\ 1 \\ 0 \end{bmatrix}, \mathbf{u}_2 = \begin{bmatrix} -1 \\ 1 \\ 1 \end{bmatrix}, \mathbf{u}_3 \begin{bmatrix} 1 \\ -1 \\ 2 \end{bmatrix}, \right\}$

is an orthogonal basis for $\mathbb{R}^3$ such that $\mathbf{u}_1, \mathbf{u}_2 \in W$ and $\mathbf{u}_3 \in W^\perp$.

Therefore, $[T(\mathbf{u}_1)]_\mathcal{D} = \begin{bmatrix} 1 \\ 0 \\ 0 \end{bmatrix}$, $[T(\mathbf{u}_2)]_\mathcal{D} = \begin{bmatrix} 0 \\ 1 \\ 0 \end{bmatrix}$ and $[T(\mathbf{u}_3)]_\mathcal{D} = \begin{bmatrix} 0 \\ 0 \\ 0 \end{bmatrix} \Rightarrow [T]_\mathcal{D} = \begin{bmatrix} 1 & 0 & 0 \\ 0 & 1 & 0 \\ 0 & 0 & 0 \end{bmatrix}$.

Furthermore, $P_{\mathcal{E}\leftarrow\mathcal{D}} = \begin{bmatrix} 1 & -1 & 1 \\ 1 & 1 & -1 \\ 0 & 1 & 2 \end{bmatrix} \Rightarrow P_{\mathcal{D}\leftarrow\mathcal{E}} = (P_{\mathcal{E}\leftarrow\mathcal{D}})^{-1} = \begin{bmatrix} 1/2 & 1/2 & 0 \\ -1/3 & 1/3 & 1/3 \\ 1/6 & -1/6 & 1/3 \end{bmatrix} \Rightarrow$

$$[T]_\mathcal{E} = P_{\mathcal{E}\leftarrow\mathcal{D}}[T]_\mathcal{D} P_{\mathcal{D}\leftarrow\mathcal{E}} = \begin{bmatrix} 1 & -1 & 1 \\ 1 & 1 & -1 \\ 0 & 1 & 2 \end{bmatrix} \begin{bmatrix} 1 & 0 & 0 \\ 0 & 1 & 0 \\ 0 & 0 & 0 \end{bmatrix} \begin{bmatrix} 1/2 & 1/2 & 0 \\ -1/3 & 1/3 & 1/3 \\ 1/6 & -1/6 & 1/3 \end{bmatrix}$$

$$= \begin{bmatrix} 5/6 & 1/6 & -1/3 \\ 1/6 & 5/6 & 1/3 \\ -1/3 & 1/3 & 1/3 \end{bmatrix}.$$

Let $\mathbf{v} = \begin{bmatrix} 3 \\ -1 \\ 2 \end{bmatrix}$, then $= \text{proj}_W(\mathbf{v}) = [T]_\mathcal{E}[\mathbf{v}]_\mathcal{E} = \begin{bmatrix} 5/6 & 1/6 & -1/3 \\ 1/6 & 5/6 & 1/3 \\ -1/3 & 1/3 & 1/3 \end{bmatrix} \begin{bmatrix} 3 \\ -1 \\ 2 \end{bmatrix} = \begin{bmatrix} \frac{5}{3} \\ \frac{1}{3} \\ -\frac{2}{3} \end{bmatrix}$

exactly as found in Section 5.2, Example 4.

39. Theorem 6.26 $\Rightarrow [T]_{\mathcal{C}\leftarrow\mathcal{B}} = [\,[T(\mathbf{v}_1)]\,|[T(\mathbf{v}_2)]\,|\ldots|[T(\mathbf{v}_n)\,]$ that is the ith column of $[T]_{\mathcal{C}\leftarrow\mathcal{B}} = [T(\mathbf{v}_i)]$ where $\mathcal{B} = \{\mathbf{v}_1, \ldots, \mathbf{v}_n\}$.

We will show: $A\,[\mathbf{v}]_\mathcal{B} = [T(\mathbf{v})]_\mathcal{C} \Rightarrow \mathbf{a}_i = [T(\mathbf{v}_i)]$.
That is, the ith column of $A = [T(\mathbf{v}_i)] \Rightarrow A = [T]_{\mathcal{C}\leftarrow\mathcal{B}}$.

Note, $[\mathbf{v}_i]_\mathcal{B} = \mathbf{e}_i$ for all i by definition.

$A\,[\mathbf{v}]_\mathcal{B} = [T(\mathbf{v})]_\mathcal{C} \Rightarrow A\,[\mathbf{v}_i]_\mathcal{B} A\,\mathbf{e}_i = \mathbf{a}_i = [T(\mathbf{v}_i)]_\mathcal{C} \Rightarrow$ the ith column of $A = [T(\mathbf{v}_i)] \Rightarrow$ $A = [T]_{\mathcal{C}\leftarrow\mathcal{B}}$ as required.

40. By Exercise 39, $A = [T]_{\mathcal{C}\leftarrow\mathcal{B}} \Leftrightarrow A[\mathbf{b}_i]_\mathcal{B} = A\mathbf{e}_i = \mathbf{a}_i = [T(\mathbf{b}_i)]_\mathcal{C}$.
Also, $\text{nullity}(A) = \dim(\text{null}(A) = \{\mathbf{v} : A\mathbf{v} = \mathbf{0}\})$ and $\text{nullity}(T) = \dim\ker(T) = \{T(\mathbf{v}) : T(\mathbf{v}) = \mathbf{0}\})$.
Will show $\{\mathbf{b}_1, \ldots, \mathbf{b}_m\} \subseteq \mathcal{B}$ is a basis for $\text{null}(A) \Leftrightarrow \{T(\mathbf{b}_1), \ldots, T(\mathbf{b}_m)\}$ is a basis for $\ker(T)$.

Furthermore, $\mathbf{a}_i = A[\mathbf{b}_i]_\mathcal{B} = [T(\mathbf{b}_i)]_\mathcal{C}$ linearly independent $\Leftrightarrow T(\mathbf{b}_i)$ linearly independent.
So we need only show $\text{null}(A) = \text{span}(\mathbf{b}_1, \ldots, \mathbf{b}_m) \Leftrightarrow \ker(T) = \text{span}(T(\mathbf{b}_1), \ldots, T(\mathbf{b}_m))$.

Section 6.2 Theorem 6.7 $\Rightarrow \mathbf{0} = A[\mathbf{b}_i]_\mathcal{B} = [T(\mathbf{b}_i)]_\mathcal{C} \Leftrightarrow T(\mathbf{b}_i) = \mathbf{0} \Rightarrow$
$\text{null}(A) = \text{span}(\mathbf{b}_1, \ldots, \mathbf{b}_m) \Leftrightarrow \ker(T) = \text{span}(T(\mathbf{b}_1), \ldots, T(\mathbf{b}_m)) \Rightarrow \text{nullity}(T) = \text{nullity}(A)$.

41. By Exercise 39, $A = [T]_{\mathcal{C}\leftarrow\mathcal{B}} \Leftrightarrow A[\mathbf{b}_i]_{\mathcal{B}} = A\mathbf{e}_i = \mathbf{a}_i = [T(\mathbf{b}_i)]_{\mathcal{C}}$.
Also, $\text{rank}(A) = \dim \text{col}(A)$, where $\text{col}(A) = \text{span}(\mathbf{a}_i : \mathbf{a}_i \text{ linearly independent})$ and
$\text{rank}(T) = \dim \text{range}(T)$, where $\text{range}(T) = \text{span}(T(\mathbf{b}_i) : T(\mathbf{b}_i) \text{ linearly independent})$.
We will show $\{\mathbf{a}_1, \ldots, \mathbf{a}_m\}$ is a basis for $\text{col}(A) \Leftrightarrow \{T(\mathbf{b}_1), \ldots, T(\mathbf{b}_m)\}$ is a basis for $\text{range}(T)$.

$\mathbf{a}_i = A[\mathbf{b}_i]_{\mathcal{B}} = [T(\mathbf{b}_i)]_{\mathcal{C}}$ linearly independent $\Leftrightarrow T(\mathbf{b}_i)$ linearly independent $\Rightarrow$
$\text{col}(A) = \text{span}(\mathbf{a}_1, \ldots, \mathbf{a}_m) \Leftrightarrow \text{range}(T) = \text{span}(T(\mathbf{b}_1), \ldots, T(\mathbf{b}_m)) \Rightarrow \text{rank}(T) = \text{rank}(A)$.

42. T is diagonalizable $\Leftrightarrow [T]_{\mathcal{C}} = P^{-1}AP \Leftrightarrow A$ is diagonalizable.

43. Section 3.4, Theorem 6.19 $\Rightarrow \dim V = \text{rank}([T]_{\mathcal{C}\leftarrow\mathcal{B}}) + \text{nullity}([T]_{\mathcal{C}\leftarrow\mathcal{B}})$.
So, Exercises 40 and 41 $\Rightarrow \dim V = \text{rank}(T) + \text{nullity}(T)$.

44. We will show $[\mathbf{x}]_{\mathcal{C}'} = P_{\mathcal{C}'\leftarrow\mathcal{C}}[T]_{\mathcal{C}\leftarrow\mathcal{B}}P_{\mathcal{B}\leftarrow\mathcal{B}'}[\mathbf{x}]_{\mathcal{B}'}$ which will imply
$[T]_{\mathcal{C}'\leftarrow\mathcal{B}'} = P_{\mathcal{C}'\leftarrow\mathcal{C}}[T]_{\mathcal{C}\leftarrow\mathcal{B}}P_{\mathcal{B}\leftarrow\mathcal{B}'}$ since this matrix with this property is unique.

$P_{\mathcal{C}'\leftarrow\mathcal{C}}[T]_{\mathcal{C}\leftarrow\mathcal{B}}(P_{\mathcal{B}\leftarrow\mathcal{B}'}[\mathbf{x}]_{\mathcal{B}'}) = P_{\mathcal{C}'\leftarrow\mathcal{C}}([T]_{\mathcal{C}\leftarrow\mathcal{B}}[\mathbf{x}]_{\mathcal{B}}) = P_{\mathcal{C}'\leftarrow\mathcal{C}}[\mathbf{x}]_{\mathcal{C}} = [\mathbf{x}]_{\mathcal{C}'} \Rightarrow$
$[T]_{\mathcal{C}'\leftarrow\mathcal{B}'} = P_{\mathcal{C}'\leftarrow\mathcal{C}}[T]_{\mathcal{C}\leftarrow\mathcal{B}}P_{\mathcal{B}\leftarrow\mathcal{B}'}$ as was to be shown.

45. For $T \in \mathscr{L}(V, W)$, define $\varphi(T) = [T]_{\mathcal{C}\leftarrow\mathcal{B}} \in M_{nn}$. We will show φ is a linear transformation.
Recall, $[T]_{\mathcal{C}\leftarrow\mathcal{B}} = [\,[T\mathbf{b}_i]_{\mathcal{C}}\,]$, where $[\,[T\mathbf{b}_i]_{\mathcal{C}}\,] = [\,[T\mathbf{b}_1]_{\mathcal{C}} \mid \cdots \mid [T\mathbf{b}_n]_{\mathcal{C}}\,]$.

Then, $\varphi(S + T) = [\,[(S + T)\mathbf{b}_i]_{\mathcal{C}}\,] = [\,[S\mathbf{b}_i]_{\mathcal{C}} + [T\mathbf{b}_i]_{\mathcal{C}}\,] = [\,[S\mathbf{b}_i]_{\mathcal{C}}\,] + [\,[T\mathbf{b}_i]_{\mathcal{C}}\,] = \varphi(S) + \varphi(T)$.

Likewise, $\varphi(cT) = [\,[cT\mathbf{b}_i]_{\mathcal{C}}\,] = [\,c[T\mathbf{b}_i]_{\mathcal{C}}\,] = c[\,[T\mathbf{b}_i]_{\mathcal{C}}\,] = c\varphi(T)$.

To show φ is an isomorphism, we need only show that $\ker(\varphi) = \mathbf{0} \Rightarrow \text{nullity}(T) = n$
since the Rank Theorem then implies $\ker(T) = V$, that is $T\mathbf{v} = \mathbf{0}$ for all $\mathbf{v}$
so $T = \mathbf{0}$, the zero linear transformation.

$\varphi(T) = \mathbf{0} \Rightarrow [T]_{\mathcal{C}\leftarrow\mathcal{B}} = [\mathbf{0}]_{nn} \Rightarrow \text{null}(T) = \mathbb{R}^n \Rightarrow \text{nullity}([T]_{\mathcal{C}\leftarrow\mathcal{B}}) = n = \text{nullity}(T)$.

Therefore, φ is one-to-one and hence onto since $\dim \mathscr{L}(V, W) = n^2 = \dim M_{nn} \Rightarrow$
φ is an isomorphism $\Rightarrow \mathscr{L}(V, W) \cong M_{nn}$.

46. This follows from Exercise 45 with $W = \mathbb{R}$ so $\dim \mathbb{R} = 1$ which implies $M_{1n} \cong V$.
Therefore, $V^* = \mathscr{L}(V, W) = \mathscr{L}(V, \mathbb{R}) \cong M_{1n} \cong V$.

Exploration: Tilings, Lattices, and Crystallographic Restriction

1. Let $P =$ the pattern in Figure 6.15.
   We are given $P + \mathbf{u} = P$ and $P + \mathbf{v} = P \Rightarrow P + a\,\mathbf{u} = P$ and $P + b\,\mathbf{v} = P$.
   Therefore, $P + a\,\mathbf{u} + b\,\mathbf{v} = (P + a\,\mathbf{u}) + b\,\mathbf{v} = P + b\,\mathbf{v} = P$.

2. 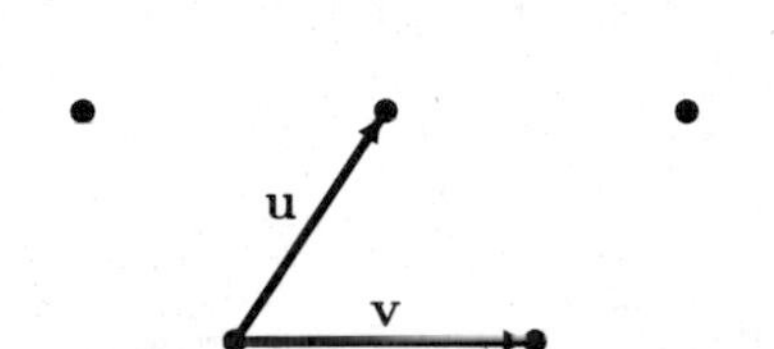

   The lattice points correspond to the centers of the black trifoils.

3. Let $\text{rot}P_O^\theta = P$ rotated by $\theta > 0$ through rotation center O.
   Then $\text{rot}P_O^{n\theta} = \text{rot}P_O^\theta$ n times $= P$ (because $\text{rot}P_O^\theta = P$ every time). Likewise for $-\theta$.
   We need to show that if $0 < \theta \leq 360°$, then $\frac{360}{\theta} = n$ an integer, i.e. $n\theta = 360°$.
   Clearly, $\text{rot}P_O^{360°} = P$ and $\text{rot}P_O^{360°+\varphi} = \text{rot}P_O^\varphi$.
   Let $\theta' = \min\{\varphi : \text{rot}P_O^\varphi = P, 0 < \varphi \leq 360°\}$ and $m = \min\{k : k\theta' \geq 360,\ k \text{ a positive integer}\}$.
   Now let $n\theta' = 360° + \psi$. We will show $\psi < \theta'$ which will imply $\psi = 0°$.
   $m\theta' = 360° + \psi \Rightarrow m\theta' - \psi = 360° > (m-1)\theta'$ (because m is the minimum) $\Rightarrow$
   $m\theta' - \psi > m\theta' - \theta' \Rightarrow -\psi > -\theta' \Rightarrow \psi < \theta'$.
   So, since $P = \text{rot}P_O^{n\theta'} = \text{rot}P_O^{360°+\psi} = \text{rot}P_O^\psi$, we have $\psi = 0°$ (because θ' is the minimum).
   Therefore, $m\theta' = 360° \Rightarrow \frac{360}{\theta'} = m$ an integer.
   Now if θ, $0 < \theta \leq 360°$ and $\text{rot}P_O^\theta = P$ we will show that $\theta = n\theta'$ to complete the proof.
   Let $n = \max\{k : \theta - k\theta' \geq 0,\ k \text{ a positive integer}\}$.
   Now let $\theta - n\theta' = \psi$. We will show $\psi < \theta'$ which will imply $\psi = 0°$.
   $\theta - n\theta' = \psi \Rightarrow \theta - n\theta' - \psi > \theta - (n+1)\theta'$ (because n is the maximum) $\Rightarrow$
   $\theta - n\theta' - \psi > \theta - n\theta' - \theta' \Rightarrow -\psi > -\theta' \Rightarrow \psi < \theta'$.
   So, since $P = \text{rot}P_O^{\theta - n\theta'} = \text{rot}P_O^\psi$, we have $\psi = 0°$ (because θ' is the minimum).
   Therefore, $\theta - n\theta' = 0 \Rightarrow \theta = n\theta' \Rightarrow \frac{\theta}{\theta'} = n \Rightarrow \frac{360}{\theta}$ is an integer.

4. The smallest positive angle of rotational symmetry for the lattice is $60°$. This rotational symmetry is also possessed by the original figure.

Note, as shown in Exercise 3, $120° = 2 \times 60°$.

5. It is impossible to draw a lattice with 8-fold symmetry as we will show below.

6. Since $\theta = 60°$ and $[R_\theta]_{\mathcal{E}} = \begin{bmatrix} \cos\theta & -\sin\theta \\ \sin\theta & \cos\theta \end{bmatrix}$, we have:

$$[R_\theta]_{\mathcal{E}} = \begin{bmatrix} \cos 60° & -\sin 60° \\ \sin 60° & \cos 60° \end{bmatrix} = \begin{bmatrix} \frac{1}{2} & -\frac{\sqrt{3}}{2} \\ \frac{\sqrt{3}}{2} & \frac{1}{2} \end{bmatrix}.$$

With respect to $\mathcal{B} = \{\mathbf{u}, \mathbf{v}\}$, we have:

$$[R_\theta(\mathbf{u})]_{\mathcal{B}} = [\mathbf{v}]_{\mathcal{B}} = \begin{bmatrix} 0 \\ 1 \end{bmatrix} \text{ and } [R_\theta(\mathbf{v})]_{\mathcal{B}} = [-\mathbf{u} + \mathbf{v}]_{\mathcal{B}} = \begin{bmatrix} -1 \\ 1 \end{bmatrix} \Rightarrow [R_\theta]_{\mathcal{B}} = \begin{bmatrix} 0 & -1 \\ 1 & 1 \end{bmatrix}.$$

7. Since the lattice is invariant under R_θ, $\mathbf{u}$ and $\mathbf{v}$ must mapped to lattice points. So:

$$[R_\theta(\mathbf{u})]_{\mathcal{B}} = [a\mathbf{u} + c\mathbf{v}]_{\mathcal{B}} = \begin{bmatrix} a \\ c \end{bmatrix} \text{ and } [R_\theta(\mathbf{v})]_{\mathcal{B}} = [b\mathbf{u} + d\mathbf{v}]_{\mathcal{B}} = \begin{bmatrix} b \\ d \end{bmatrix} \Rightarrow [R_\theta]_{\mathcal{B}} = \begin{bmatrix} a & b \\ c & d \end{bmatrix}.$$

8. Following the hints, we see Exercise 35 in Section 4.4 $\Rightarrow A \sim B \Rightarrow \text{tr}(A) = \text{tr}(B)$. Theorem 6.29 in Section 6.6 $\Rightarrow [T]_{\mathcal{C}} = P^{-1}[T]_{\mathcal{B}}P$, that is $[T]_{\mathcal{C}} \sim [T]_{\mathcal{B}}$.

Now taking $\mathcal{E} = \{\mathbf{e}_1, \mathbf{e}_2\}$ and $\mathcal{B} = \{\mathbf{u}, \mathbf{v}\}$, we have:

Theorem 6.29 in Section 6.6 $\Rightarrow [R_\theta]_{\mathcal{E}} = P^{-1}[R_\theta]_{\mathcal{B}}P$, that is $[R_\theta]_{\mathcal{E}} \sim [R_\theta]_{\mathcal{B}}$.

So, Exercise 35 in Section 4.4 $\Rightarrow \text{tr}([R_\theta]_{\mathcal{E}}) = 2\cos\theta = \text{tr}([R_\theta]_{\mathcal{B}}) =$ an integer.

9. From 8, $2\cos\theta =$ an integer $\Rightarrow \theta = 0°, 60°, 90°, 120°, 180°, 240°, 270°, 300°$, and $360°$.

Now relating these to the integers n, where $n = \frac{360}{\theta}$, we have:

$n = \frac{360}{360} = 1$, $n = \frac{360}{180} = 2$, $n = \frac{360}{120} = 3$, $n = \frac{360}{90} = 4$, and $n = \frac{360}{60} = 6$.

Note the other θ listed above do not give integer results when divided into 360.

10. Here is a website that might serve as a good starting place: http://www.worldofescher.com/gallery/.

6.7 Applications

1. Since $a = -3$, Theorem 6.32 $\Rightarrow \{e^{3t}\}$ is a basis $\Rightarrow$ The solution is of the form $y(t) = c\,e^{3t}$.
So the boundary condition $y(1) = 2 \Rightarrow y(1) = c\,e^3 = 2 \Rightarrow c = \frac{2}{e^3} \Rightarrow y(t) = \frac{2e^{3t}}{e^3}$.

2. Since $a = 1$, Theorem 6.32 $\Rightarrow \{e^{-t}\}$ is a basis $\Rightarrow$ The solution is of the form $x(t) = c\,e^{-t}$.
So the boundary condition $x(1) = 1 \Rightarrow x(1) = c\,e^{-1} = 1 \Rightarrow c = e \Rightarrow x(t) = e^{1-t}$.

3. $y'' - 7y' + 12y = 0$ corresponds to the characteristic equation $\lambda^2 - 7\lambda + 12 = 0$.
Since $\lambda_1 = 3$, $\lambda_2 = 4$, Theorem 6.33 $\Rightarrow$ solution is of the form $y(t) = c_1\,e^{3t} + c_2\,e^{4t}$.

$$\begin{array}{l} y(0)=1 \\ y(1)=1 \end{array} \Rightarrow \begin{array}{l} c_1 + c_2 = 1 \\ c_1\,e^3 + c_2\,e^4 = 1 \end{array} \Rightarrow \begin{array}{l} c_1 = \frac{1-e^4}{e^3-e^4} \\ c_2 = \frac{e^3-1}{e^3-e^4} \end{array} \Rightarrow y(t) = \frac{1}{e^3-e^4}\left((1-e^4)e^{3t} + (e^3-1)e^{4t} \right).$$

4. $x'' + x' - 12x = 0$ corresponds to the characteristic equation $\lambda^2 + \lambda - 12 = 0$.
Since $\lambda_1 = -4$, $\lambda_2 = 3$, Theorem 6.33 $\Rightarrow$ solution is of the form $x(t) = c_1\,e^{-4t} + c_2\,e^{3t}$.

$$\begin{array}{l} x(0)=0 \\ x'(0)=1 \end{array} \Rightarrow \begin{array}{r} c_1 + c_2 = 0 \\ -4c_1 + 3c_2 = 1 \end{array} \Rightarrow \begin{array}{l} c_1 = -\frac{1}{7} \\ c_2 = \frac{1}{7} \end{array} \Rightarrow x(t) = \frac{1}{7}\left(e^{3t} - e^{-4t} \right).$$

5. $f'' - f' - f = 0$ corresponds to the characteristic equation $\lambda^2 + \lambda - 12 = 0$.
Since $\lambda_1 = \frac{1+\sqrt{5}}{2}$, $\lambda_2 = \frac{1-\sqrt{5}}{2}$, Theorem 6.33 $\Rightarrow f(t) = c_1\,e^{(1+\sqrt{5})t/2} + c_2\,e^{(1-\sqrt{5})t/2}$.

$$\begin{array}{l} f(0)=0 \\ f(1)=1 \end{array} \Rightarrow \begin{array}{r} c_1 + c_2 = 0 \\ c_1 e^{(1+\sqrt{5})/2} + c_2 e^{(1-\sqrt{5})/2} = 1 \end{array} \Rightarrow \begin{array}{l} c_1 = \frac{e^{(\sqrt{5}-1)/2}}{e^{\sqrt{5}}-1} \\ c_2 = -\frac{e^{(\sqrt{5}-1)/2}}{e^{\sqrt{5}}-1} \end{array} \Rightarrow$$

$$x(t) = \frac{e^{(\sqrt{5}-1)/2}}{e^{\sqrt{5}}-1}\left(e^{(1+\sqrt{5})t/2} - e^{(1-\sqrt{5})t/2} \right).$$

6. $g'' - 2g = 0$ corresponds to the characteristic equation $\lambda^2 - 2 = 0$.
Since $\lambda_1 = \sqrt{2}$, $\lambda_2 = -\sqrt{2}$, Theorem 6.33 $\Rightarrow g(t) = c_1 e^{\sqrt{2}\,t} + c_2 e^{-\sqrt{2}\,t}$.

$$\begin{array}{l} g(0)=1 \\ g(1)=0 \end{array} \Rightarrow \begin{array}{r} c_1 + c_2 = 1 \\ c_1 e^{\sqrt{2}} + c_2 e^{-\sqrt{2}} = 0 \end{array} \Rightarrow \begin{array}{l} c_1 = \frac{1}{e^{2\sqrt{2}}-1} \\ c_2 = \frac{e^{2\sqrt{2}}}{e^{2\sqrt{2}}-1} \end{array} \Rightarrow$$

$$x(t) = \frac{1}{e^{2\sqrt{2}}-1}\left(e^{\sqrt{2}\,t} + e^{2\sqrt{2}}\,e^{-\sqrt{2}\,t} \right).$$

7. $y'' - 2y' + y = 0$ corresponds to the characteristic equation $\lambda^2 - 2\lambda + 1 = 0$.
Since $\lambda_1 = \lambda_2 = 1$, Theorem 6.33 $\Rightarrow y(t) = c_1 e^t + c_2 t\,e^t$.

$$\begin{array}{l} y(0)=1 \\ y(1)=1 \end{array} \Rightarrow \begin{array}{r} c_1 + c_2 = 1 \\ c_1 e + c_2 e = 1 \end{array} \Rightarrow \begin{array}{l} c_1 = 1 \\ c_2 = -(1-e^{-1}) \end{array} \Rightarrow y(t) = e^t - (1-e^{-1})t\,e^t.$$

8. $x'' + 4x' + 4x = 0$ corresponds to the characteristic equation $\lambda^2 + 4\lambda + 4 = 0$.
Since $\lambda_1 = \lambda_2 = -2$, Theorem 6.33 $\Rightarrow x(t) = c_1 e^{-2t} + c_2 t\,e^{-2t}$.

$$\begin{array}{l} x(0)=1 \\ x'(0)=1 \end{array} \Rightarrow \begin{array}{r} c_1 = 1 \\ -2c_1 + c_2 = 1 \end{array} \Rightarrow \begin{array}{l} c_1 = 1 \\ c_2 = 3 \end{array} \Rightarrow x(t) = e^{-2t} + 3e^{-2t}.$$

9. $y'' - k^2 y = 0$ corresponds to the characteristic equation $\lambda^2 - k^2 = 0$.
Since $\lambda_1 = k$, $\lambda_2 = -k$, Theorem 6.33 $\Rightarrow y(t) = c_1 e^{kt} + c_2\, e^{-kt}$.

$$\begin{matrix} y(0)=1 \\ y'(0)=1 \end{matrix} \Rightarrow \begin{matrix} c_1 + c_2 = 1 \\ kc_1 - kc_2 = 1 \end{matrix} \Rightarrow \begin{matrix} c_1 = \frac{k+1}{2k} \\ c_2 = \frac{k-1}{2k} \end{matrix} \Rightarrow y(t) = \tfrac{1}{2k}(\,(k+1)e^{kt} + (k-1)e^{-kt}\,).$$

10. $y'' - 2ky' + k^2 y = 0$ corresponds to the characteristic equation $\lambda^2 - 2\lambda\, k + k^2 = 0$.
Since $\lambda_1 = \lambda_2 = k$, Theorem 6.33 $\Rightarrow y(t) = c_1 e^{kt} + c_2\, te^{kt}$.

$$\begin{matrix} y(0)=1 \\ y(1)=0 \end{matrix} \Rightarrow \begin{matrix} c_1 = 1 \\ c_1\, e^k + c_2\, e^k = 0 \end{matrix} \Rightarrow \begin{matrix} c_1 = 1 \\ c_2 = -1 \end{matrix} \Rightarrow y(t) = e^{kt} - t\, e^{kt}.$$

11. $f'' - 2f' + 5f = 0$ corresponds to the characteristic equation $\lambda^2 - 2\lambda + 5 = 0$.
Since $\lambda_1 = 1 + 2i$, $\lambda_2 = 1 - 2i$, Theorem 6.33 $\Rightarrow f(t) = c_1\, e^t \cos(2t) + c_2\, e^t \sin(2t)$.

$$\begin{matrix} f(0)=1 \\ f(\frac{\pi}{4})=0 \end{matrix} \Rightarrow \begin{matrix} c_1 = 1 \\ c_2 = 0 \end{matrix} \Rightarrow f(t) = e^t \cos(2t).$$

12. $h'' - 4h' + 5h = 0$ corresponds to the characteristic equation $\lambda^2 - 4\lambda + 5 = 0$.
Since $\lambda_1 = 2 + i$, $\lambda_2 = 2 - i$, Theorem 6.33 $\Rightarrow f(t) = c_1\, e^{2t} \cos t + c_2\, e^{2t} \sin t$.

$$\begin{matrix} h(0)=0 \\ h'(0)=-1 \end{matrix} \Rightarrow \begin{matrix} c_1 = 0 \\ 2c_1 + c_2 = -1 \end{matrix} \Rightarrow \begin{matrix} c_1 = 0 \\ c_2 = -1 \end{matrix} \Rightarrow h(t) = -e^{2t} \sin t.$$

13. (a) $p(t) = c\, e^{kt} \Rightarrow p(0) = c\, e^{0k} = 100 \Rightarrow c = 100 \Rightarrow p(t) = 100e^{kt}$.
So, $p(3) = 100e^{3k} = 1600 \Rightarrow e^{3k} = 16 \Rightarrow k = \frac{\ln 16}{3} \Rightarrow p(t) = 100e^{(\ln 16)t/3}$.

(b) Double $\Rightarrow p(t) = 100e^{(\ln 16)t/3} = 2 \cdot 100 = 200 \Rightarrow e^{(\ln 16)t/3} = 2 \Rightarrow$
$t = \frac{\ln 2}{(\ln 16)/3} = \frac{3 \ln 2}{\ln 16} = \frac{3 \ln 2}{\ln 2^4} = \frac{3 \ln 2}{4 \ln 2} = \frac{3}{4}$ of an hour $= 45$ minutes.

(c) One million $= 10^6 \Rightarrow p(t) = 100e^{(\ln 16)t/3} = 10^6 \Rightarrow e^{(\ln 16)t/3} = 10^4 \Rightarrow$
$t = \frac{\ln 10^4}{(\ln 16)/3} = \frac{12 \ln 10}{4 \ln 2} = \frac{3 \ln 10}{\ln 2} \approx 9.968$ hours.

14. (a) $p(t) = c\, e^{kt} \Rightarrow p(0) = c\, e^{0k} = 76 \Rightarrow c = 76 \Rightarrow p(t) = 76e^{kt}$.
So, $p(1) = 76e^{1k} = 92 \Rightarrow e^{1k} = \frac{92}{76} = \frac{23}{19} \Rightarrow k = \ln \frac{23}{19} \Rightarrow p(t) = 76e^{\ln(23/19)\, t}$.
Year 2000 $\Rightarrow t = 10 \Rightarrow p(10) = 76e^{\ln(23/19) \cdot 10} \approx 513.5 \approx 2 \cdot 281 \Rightarrow$
Our estimate is almost two times too large.
This clearly shows that the population growth is not exponential.

(b) $p(t) = c\, e^{kt} \Rightarrow p(0) = c\, e^{0k} = 203 \Rightarrow c = 203 \Rightarrow p(t) = 203e^{kt}$.
So, $p(1) = 203e^{1k} = 227 \Rightarrow e^{1k} = \frac{227}{203} \Rightarrow k = \ln \frac{227}{203} \Rightarrow p(t) = 76e^{\ln(227/203)\, t}$.
Year 2000 $\Rightarrow t = 3 \Rightarrow p(3) = 203e^{\ln(227/203) \cdot 3} \approx 283.8 \approx 281 \Rightarrow$
Our estimate very nearly agrees with the actual population in Year 2000.

(c) The population growth has been almost exponential since 1970,
but has been cut almost in half compared to the growth rate of 1900-1910.

15. (a) $m(0) = a\,e^{-ct} \Rightarrow m(0) = a\,e^{-0c} = 50 \Rightarrow a = 50 \Rightarrow m(t) = 50e^{-ct}$.
Half-life $= 1590$ years $\Rightarrow m(1590) = 50e^{-1590c} = \frac{1}{2}\,50 = 25 \Rightarrow e^{-1590c} = \frac{1}{2} \Rightarrow$
$c = -\frac{\ln(1/2)}{1590} \Rightarrow m(t) = 50e^{-ct}$ where $c = -\frac{\ln(1/2)}{1590}$.
Therefore, $m(1000) = 50e^{-1000c} \approx 32.33$ mg will be left after 1000 years.

(b) $m(t) = 50e^{-ct} = 10 \Rightarrow e^{-ct} = \frac{1}{5} \Rightarrow t = \frac{\ln(1/5)}{-c} = \frac{\ln(1/5)}{\ln(1/2)/1590} = \frac{1590\ln 5}{\ln 2} \approx 3692$ years.

16. $m(0) = a\,e^{-ct}$ and half-life $= 5730$ years $\Rightarrow m(5730) = a\,e^{-5730c} = \frac{1}{2}\,a \Rightarrow$

$e^{-5730c} = \frac{1}{2} \Rightarrow c = -\frac{\ln(1/2)}{5730} \Rightarrow m(t) = a\,e^{-ct}$ where $c = -\frac{\ln(1/2)}{5730}$.

Therefore, $m(t) = a\,e^{-ct} = 0.45a \Rightarrow e^{-ct} = 0.45 \Rightarrow t = \frac{\ln(0.45)}{-c} = -\frac{5730\ln 0.45}{\ln 2} \approx 6600$ years.

17. The following analysis comes directly from Section 6.7, Example 6.92.

Length $= 10$ when $t = 0 \Rightarrow 10 = x(0) = c_1\cos 0 + c_2\sin 0 = c_1$.
Length $= 5$ when $t = 10 \Rightarrow 5 = x(10) = 10\cos 10\sqrt{K} + c_2\sin 10\sqrt{K} \Rightarrow$

$c_2 = \frac{5-10\cos(10\sqrt{K})}{\sin(10\sqrt{K})} \Rightarrow x(t) = \frac{5-10\cos(10\sqrt{K})}{\sin(10\sqrt{K})}\sin\sqrt{K}\,t + 10\cos\sqrt{K}\,t$.

18. The following comes from Example 6.92. Recall, the period of $\sin bt$ and $\cos bt = \frac{2\pi}{b}$.

Mass $= 50$ and $K = \frac{k}{m} \Rightarrow x(t) = c_1\cos\sqrt{K}\,t + c_1\sin\sqrt{K}\,t = c_1\cos\sqrt{\frac{k}{50}}\,t + c_1\sin\sqrt{\frac{k}{50}}\,t$.

Period $= 10$ and period $= \frac{2\pi}{b} \Rightarrow 10 = \frac{2\pi}{\sqrt{k/50}} \Rightarrow \sqrt{\frac{k}{50}} = \frac{2\pi}{10} = \frac{\pi}{5} \Rightarrow \frac{k}{50} = \frac{\pi^2}{25} \Rightarrow k = 2\pi^2$.

19. The following comes from Example 6.92. Recall, the period of $\sin bt$ and $\cos bt = \frac{2\pi}{b}$.

(a) Since $\theta'' + \frac{g}{L}\,\theta = 0$, Example 6.92 $\Rightarrow K = \frac{g}{L} \Rightarrow$
$\theta(t) = c_1\cos\sqrt{K}\,t + c_1\sin\sqrt{K}\,t = c_1\cos\sqrt{\frac{g}{L}}\,t + c_1\sin\sqrt{\frac{g}{L}}\,t \Rightarrow$
Period, $P = \frac{2\pi}{b} \Rightarrow P = \frac{2\pi}{\sqrt{g/L}} = 2\pi\sqrt{\frac{L}{g}}$.
In this case, since $L = 1$, the period is $2\pi\sqrt{\frac{1}{g}} = \frac{2\pi}{\sqrt{g}}$.

(b) θ_1 does not appear in the formula, therefore, the period does not depend on θ_1.
Furthermore, since g is constant, the period depends only upon the length, L.

20. Section 6.1, Theorem 6.2 $\Rightarrow$ need only show $y_1, y_2 \in W \Rightarrow y_1 + y_2 \in W$ and $y \in W \Rightarrow cy \in W$.
$(y_1 + y_2)'' + a(y_1 + y_2)' + b(y_1 + y_2) = (y_1'' + a\,y_1' + b\,y_1) + (y_2'' + a\,y_2' + b\,y_2) = 0 + 0 = 0 \Rightarrow$
$y_1, y_2 \in W \Rightarrow y_1 + y_2 \in W$.
Likewise, $cy'' + a\,cy' + b\,cy = c(y'' + ay' + by) = c0 = 0 \Rightarrow y \in W \Rightarrow cy \in W$.
Therefore W is a subspace.

21. As stated in the proof of Theorem 6.33a, $\dim S = 2 \Rightarrow$ we need only show $\text{span}(e^{\lambda t}, t\,e^{\lambda t}) \subseteq S$ and $e^{\lambda t}$, $t\,e^{\lambda t}$ linearly independent $\Rightarrow \text{span}(e^{\lambda t}, t\,e^{\lambda t}) = S \Rightarrow \{e^{\lambda t}, t\,e^{\lambda t}\}$ is a basis for S.

$\text{span}(e^{\lambda t}) \subseteq S$ by proof of Section 6.7, Theorem 6.33b. We will show $\text{span}(t\,e^{\lambda t}) \subseteq S$.

Note, λ a solution of $\lambda^2 + a\lambda + b = 0 \Rightarrow \lambda = \frac{-a \pm \sqrt{a^2-4b}}{2}$. Now $\lambda_1 = \lambda_2 = \lambda \Rightarrow$ $a^2 - 4b = 0 \Rightarrow \lambda = -\frac{1}{2}\,a \Rightarrow 2\lambda + a = 0$. We will use this to show $\text{span}(t\,e^{\lambda t}) \subseteq S$.

Let $f(t) = t\,e^{\lambda t} \Rightarrow f' = e^{\lambda t} + \lambda t\,e^{\lambda t}$, and $f'' = 2\lambda\,e^{\lambda t} + \lambda^2 t\,e^{\lambda t} \Rightarrow$
$f'' + af' + b = (2\lambda + a)e^{\lambda t} + (\lambda^2 + a\lambda + b)t\,e^{\lambda t} = 0 + 0 = 0 \Rightarrow \text{span}(t\,e^{\lambda t}) \subseteq S$.

$e^{\lambda t}$, $t\,e^{\lambda t}$ linearly independent since $c_1\,e^{\lambda t} + c_2\,t\,e^{\lambda t} = 0 \Rightarrow$ when $t = 0$, $c_1 = 0 \Rightarrow c_2 = 0$.

Therefore, $\text{span}(e^{\lambda t}, t\,e^{\lambda t}) \subseteq S$ and $\dim S = 2 \Rightarrow \{e^{\lambda t}, t\,e^{\lambda t}\}$ is a basis for S.

22. Need to show $a\,e^{pt}\cos qt + b\,e^{pt}\sin qt = 0 \Rightarrow a = b = 0$. Let $t = 0 \Rightarrow a = 0 \Rightarrow b = 0$.

Note for Exercises 23 to 28. $p = 2 \Rightarrow$ the only scalars are 0 and 1 $\Rightarrow$ the scalar multiples of any vector are $\mathbf{0}$ and $\mathbf{v}$. Note this implies $\mathbf{v} + \mathbf{v} = \mathbf{0}$.

Need only show C is a subspace, i.e., $\mathbf{c}_1$, $\mathbf{c}_2 \in C \Rightarrow \mathbf{c}_1 + \mathbf{c}_2 \in C$ and $\mathbf{c} \in C \Rightarrow a\mathbf{c} \in C$.

If either one of these properties fail for any vector, then C is not a linear code.

23. No, because $\mathbf{c}_1 + \mathbf{c}_2 = \begin{bmatrix}1\\0\end{bmatrix} + \begin{bmatrix}0\\1\end{bmatrix} = \begin{bmatrix}0\\0\end{bmatrix} \notin C$.

24. Yes, since $\mathbf{c}_i + \mathbf{c}_i = \mathbf{0} \in C$, $\mathbf{c}_1 + \mathbf{c}_2 = \mathbf{c}_1 \in C$, and $a\mathbf{c}_i = \mathbf{0}$, $\mathbf{c}_i \in C$.

25. No, because $\mathbf{c}_1 + \mathbf{c}_3 = \begin{bmatrix}1\\0\\1\end{bmatrix} + \begin{bmatrix}1\\1\\0\end{bmatrix} = \begin{bmatrix}0\\1\\1\end{bmatrix} \notin C$.

26. No, because $\mathbf{c}_1 + \mathbf{c}_2 = \begin{bmatrix}1\\0\\0\end{bmatrix} + \begin{bmatrix}0\\1\\0\end{bmatrix} = \begin{bmatrix}1\\1\\0\end{bmatrix} \notin C$.

27. Yes, since $\mathbf{c}_i + \mathbf{c}_i = \mathbf{0} \in C$, $\mathbf{c}_1 + \mathbf{c}_2 = \mathbf{c}_3 \in C$, and $a\mathbf{c}_i = \mathbf{0}$, $\mathbf{c}_i \in C$.

28. No, because $\mathbf{c}_1 + \mathbf{c}_3 = \begin{bmatrix}1\\0\\1\\0\end{bmatrix} + \begin{bmatrix}1\\1\\0\\0\end{bmatrix} = \begin{bmatrix}0\\1\\1\\0\end{bmatrix} \notin C$.

29. Yes, since $w(\mathbf{x} + \mathbf{y}) = w(\mathbf{x}) + w(\mathbf{y}) - 2\,\text{overlap} \Rightarrow w(\mathbf{x} + \mathbf{y})$ is even.
$w(\mathbf{x})$ even $\Rightarrow cw(\mathbf{x})$ is even.

30. No, because $w(\mathbf{x} + \mathbf{y}) = w(\mathbf{x}) + w(\mathbf{y}) - 2\,\text{overlap} \Rightarrow w(\mathbf{x} + \mathbf{y})$ is even $\Rightarrow \mathbf{x} + \mathbf{y} \notin V$.

31. $\mathbf{c}_2 + \mathbf{c}_4 = \mathbf{c}_3 \Rightarrow \text{span}(\mathbf{c}_2, \mathbf{c}_4) = \text{span}(\mathbf{c}_2, \mathbf{c}_3) \Rightarrow \{\mathbf{c}_2, \mathbf{c}_4\}$ is a basis.

$\mathbf{c}_3 + \mathbf{c}_4 = \mathbf{c}_2 \Rightarrow \text{span}(\mathbf{c}_3, \mathbf{c}_4) = \text{span}(\mathbf{c}_2, \mathbf{c}_3) \Rightarrow \{\mathbf{c}_3, \mathbf{c}_4\}$ is a basis.

32. (a) $(9, 4) = (n, k) \Rightarrow P$ is $(n - k) \times n = (9 - 4) \times 9 \Rightarrow P$ is 5×9.

(b) $(n, k) \Rightarrow P$ is $(n - k) \times n$.

33. The proof of Theorem 6.34a $\Rightarrow \dim(C^\perp)^\perp = \dim C$. Therefore, Section 6.2, Theorem 6.11 $\Rightarrow$ We need only show $C \subseteq (C^\perp)^\perp \Rightarrow C = (C^\perp)^\perp$.

$\mathbf{c} \in C$, $\mathbf{c}' \in C^\perp \Rightarrow \mathbf{c} \cdot \mathbf{c}' = 0 \Rightarrow c \in (C^\perp)^\perp$ by definition $\Rightarrow C \subseteq (C^\perp)^\perp \Rightarrow C = (C^\perp)^\perp$.

34. Theorem 6.34*b* $\Rightarrow C^\perp = C \Rightarrow$ the number of vectors in $C =$ the number of vectors in $C^\perp$ $\Rightarrow 2^{n-k} = 2^k \Rightarrow 2^n = 2^{2k} \Rightarrow n = 2k \Rightarrow n$ is even.

35. We will go through the inductive construction of the bases for R_1, R_2, and R_3 from R_0.

$R_0 = \{0, 1\} \Rightarrow \mathbf{1} = 1$ is a basis for $R_0 \Rightarrow$

$$\text{Basis for } R_1 = \left\{ \mathbf{1} = \begin{bmatrix} 1 \\ 1 \end{bmatrix}, \mathbf{u}_{01} = \begin{bmatrix} \mathbf{0} \\ \mathbf{1} \end{bmatrix} = \begin{bmatrix} 0 \\ 1 \end{bmatrix} \right\} \Rightarrow$$

$$\text{Basis for } R_2 = \left\{ \mathbf{1} = \begin{bmatrix} 1 \\ 1 \\ 1 \\ 1 \end{bmatrix}, \mathbf{u}_{01|01} = \begin{bmatrix} \mathbf{u}_{01} \\ \mathbf{u}_{01} \end{bmatrix} = \begin{bmatrix} 0 \\ 1 \\ 0 \\ 1 \end{bmatrix}, \mathbf{u}_{00|11} = \begin{bmatrix} \mathbf{u}_{00} \\ \mathbf{u}_{11} \end{bmatrix} = \begin{bmatrix} 0 \\ 0 \\ 1 \\ 1 \end{bmatrix} \right\} \Rightarrow$$

$$\text{Basis for } R_3 = \left\{ \mathbf{1} = \begin{bmatrix} 1 \\ 1 \\ 1 \\ 1 \\ 1 \\ 1 \\ 1 \\ 1 \end{bmatrix}, \mathbf{u}_{0101|0101} = \begin{bmatrix} 0 \\ 1 \\ 0 \\ 1 \\ 0 \\ 1 \\ 0 \\ 1 \end{bmatrix}, \mathbf{u}_{0011|0011} = \begin{bmatrix} 0 \\ 0 \\ 1 \\ 1 \\ 0 \\ 0 \\ 1 \\ 1 \end{bmatrix}, \mathbf{u}_{0000|1111} = \begin{bmatrix} 0 \\ 0 \\ 0 \\ 0 \\ 1 \\ 1 \\ 1 \\ 1 \end{bmatrix} \right\} \Rightarrow$$

$$R_3 = \left\{ \begin{bmatrix} 0 \\ 0 \\ 0 \\ 0 \\ 0 \\ 0 \\ 0 \\ 0 \end{bmatrix}, \begin{bmatrix} 0 \\ 0 \\ 0 \\ 0 \\ 1 \\ 1 \\ 1 \\ 1 \end{bmatrix}, \begin{bmatrix} 0 \\ 0 \\ 1 \\ 1 \\ 0 \\ 0 \\ 1 \\ 1 \end{bmatrix}, \begin{bmatrix} 0 \\ 0 \\ 1 \\ 1 \\ 1 \\ 1 \\ 0 \\ 0 \end{bmatrix}, \begin{bmatrix} 0 \\ 1 \\ 0 \\ 1 \\ 0 \\ 1 \\ 0 \\ 1 \end{bmatrix}, \begin{bmatrix} 0 \\ 1 \\ 0 \\ 1 \\ 1 \\ 0 \\ 1 \\ 0 \end{bmatrix}, \begin{bmatrix} 0 \\ 1 \\ 1 \\ 0 \\ 0 \\ 1 \\ 1 \\ 0 \end{bmatrix}, \begin{bmatrix} 0 \\ 1 \\ 1 \\ 0 \\ 1 \\ 0 \\ 0 \\ 1 \end{bmatrix}, \begin{bmatrix} 1 \\ 0 \\ 1 \\ 0 \\ 1 \\ 0 \\ 1 \\ 0 \end{bmatrix}, \begin{bmatrix} 1 \\ 0 \\ 1 \\ 0 \\ 0 \\ 1 \\ 0 \\ 1 \end{bmatrix}, \begin{bmatrix} 1 \\ 0 \\ 0 \\ 1 \\ 1 \\ 0 \\ 0 \\ 1 \end{bmatrix}, \begin{bmatrix} 1 \\ 0 \\ 0 \\ 1 \\ 0 \\ 1 \\ 1 \\ 0 \end{bmatrix}, \begin{bmatrix} 1 \\ 1 \\ 0 \\ 0 \\ 1 \\ 1 \\ 0 \\ 0 \end{bmatrix}, \begin{bmatrix} 1 \\ 1 \\ 0 \\ 0 \\ 0 \\ 0 \\ 1 \\ 1 \end{bmatrix}, \begin{bmatrix} 1 \\ 1 \\ 1 \\ 1 \\ 0 \\ 0 \\ 0 \\ 0 \end{bmatrix}, \begin{bmatrix} 1 \\ 1 \\ 1 \\ 1 \\ 1 \\ 1 \\ 1 \\ 1 \end{bmatrix} \right\}$$

36. (a) $G_1 = \begin{bmatrix} 1 & 0 \\ 1 & 1 \end{bmatrix} \Rightarrow G_2 = \left[\begin{array}{c|c} G_1 & \mathbf{0} \\ \hline G_1 & \mathbf{1} \end{array}\right] = \left[\begin{array}{cc|c} 1 & 0 & 0 \\ 1 & 1 & 0 \\ \hline 1 & 0 & 1 \\ 1 & 1 & 1 \end{array}\right] = \begin{bmatrix} 1 & 0 & 0 \\ 1 & 1 & 0 \\ 1 & 0 & 1 \\ 1 & 1 & 1 \end{bmatrix} \Rightarrow$

$$G_3 = \left[\begin{array}{c|c} G_2 & \mathbf{0} \\ \hline G_2 & \mathbf{1} \end{array}\right] = \left[\begin{array}{ccc|c} 1 & 0 & 0 & 0 \\ 1 & 1 & 0 & 0 \\ 1 & 0 & 1 & 0 \\ 1 & 1 & 1 & 0 \\ \hline 1 & 0 & 0 & 1 \\ 1 & 1 & 0 & 1 \\ 1 & 0 & 1 & 1 \\ 1 & 1 & 1 & 1 \end{array}\right] = \begin{bmatrix} 1 & 0 & 0 & 0 \\ 1 & 1 & 0 & 0 \\ 1 & 0 & 1 & 0 \\ 1 & 1 & 1 & 0 \\ 1 & 0 & 0 & 1 \\ 1 & 1 & 0 & 1 \\ 1 & 0 & 1 & 1 \\ 1 & 1 & 1 & 1 \end{bmatrix}.$$

(b) As suggested in the text, we will proceed by induction.

$n = 1$: Part (a) and Exercise 35 $\Rightarrow G_1$ generates R_1.

$n = n - 1$: Assume G_{n-1} generates R_{n-1}. Must show $G_n = \left[\begin{array}{c|c} G_{n-1} & \mathbf{0} \\ \hline G_{n-1} & \mathbf{1} \end{array}\right]$ generates R_n.

In particular, we will show it generates all vectors of the form $\begin{bmatrix} \mathbf{u} \\ \mathbf{u} \end{bmatrix}$ and $\begin{bmatrix} \mathbf{0} \\ \mathbf{1} \end{bmatrix}$,

where $\mathbf{u}$ is a basis vector in R_{n-1} and $\mathbf{0}, \mathbf{1} \in \mathbb{Z}_2^{2n-1}$.

Since G_{n-1} generates R_{n-1}, for every basis vector $\mathbf{u}$ in R_{n-1} there exists a vector $\mathbf{x}$ in $\mathbb{Z}_2^{2n-1}$such that $G_{n-1}\mathbf{x} = \mathbf{u}$.

Therefore, $G_n \begin{bmatrix} \mathbf{x} \\ \mathbf{0} \end{bmatrix} = \left[\begin{array}{c|c} G_{n-1} & \mathbf{0} \\ \hline G_{n-1} & \mathbf{1} \end{array}\right] \begin{bmatrix} \mathbf{x} \\ \mathbf{0} \end{bmatrix} = \begin{bmatrix} \mathbf{u} \\ \mathbf{u} \end{bmatrix}$.

Similarly, $G_n \begin{bmatrix} \mathbf{0} \\ \mathbf{1} \end{bmatrix} = \left[\begin{array}{c|c} G_{n-1} & \mathbf{0} \\ \hline G_{n-1} & \mathbf{1} \end{array}\right] \begin{bmatrix} \mathbf{0} \\ \mathbf{1} \end{bmatrix} = \begin{bmatrix} \mathbf{0} \\ \mathbf{1} \end{bmatrix}$, as we needed to show.

37. Putting G_2 into standard form $\left[\frac{A}{I}\right] \Rightarrow G_2 = \begin{bmatrix} 1 & 0 & 0 \\ 1 & 1 & 0 \\ 1 & 0 & 1 \\ 1 & 1 & 1 \end{bmatrix} \longrightarrow \begin{bmatrix} 1 & 0 & 0 \\ 0 & 1 & 0 \\ 0 & 0 & 1 \\ 1 & 1 & 1 \end{bmatrix} \Rightarrow$

$A = [1\,1\,1] \Rightarrow$ The associated parity check matrix $P = [I \mid A] = [1 \mid 1\,1\,1] = [1\,1\,1\,1]$.

38. Putting G_3 into standard form $\left[\frac{A}{I}\right] \Rightarrow G_3 = \begin{bmatrix} 1 & 0 & 0 & 0 \\ 1 & 1 & 0 & 0 \\ 1 & 0 & 1 & 0 \\ 1 & 1 & 1 & 0 \\ 1 & 0 & 0 & 1 \\ 1 & 1 & 0 & 1 \\ 1 & 0 & 1 & 1 \\ 1 & 1 & 1 & 1 \end{bmatrix} \longrightarrow \begin{bmatrix} 1 & 0 & 0 & 0 \\ 0 & 1 & 0 & 0 \\ 0 & 0 & 1 & 0 \\ 0 & 0 & 0 & 0 \\ 1 & 1 & 1 & 0 \\ 1 & 1 & 0 & 1 \\ 1 & 0 & 1 & 1 \\ 1 & 1 & 1 & 1 \end{bmatrix} \Rightarrow$

$$A = \begin{bmatrix} 1 & 1 & 1 & 0 \\ 1 & 1 & 0 & 1 \\ 1 & 0 & 1 & 1 \\ 1 & 1 & 1 & 1 \end{bmatrix} \Rightarrow \text{parity check matrix } P = [I \mid A] = \left[\begin{array}{cccc|cccc} 1 & 0 & 0 & 0 & 1 & 1 & 1 & 0 \\ 0 & 1 & 0 & 0 & 1 & 1 & 0 & 1 \\ 0 & 0 & 1 & 0 & 1 & 0 & 1 & 1 \\ 0 & 0 & 0 & 1 & 1 & 1 & 1 & 1 \end{array}\right].$$

39. As suggested, we will show $O' = \{\mathbf{c}_0 + \mathbf{e} : \mathbf{e} \in E\} = O$, vectors in C with odd weight.

$w(\mathbf{c}_0 + \mathbf{e}) = w(\mathbf{c}_0) + w(\mathbf{e}) - 2\,\text{overlap} = \text{odd} \Rightarrow \mathbf{c}_0 + \mathbf{e} \in O \Rightarrow O' \subseteq O$.

Recall, for any vector $\mathbf{c} \in C$, $\mathbf{c} + \mathbf{c} = 2\mathbf{c} = \mathbf{0}$. In particular, $\mathbf{c}_0 + \mathbf{c}_0 = \mathbf{0}$.
Also, note $w(\mathbf{c}_0 + \mathbf{o}) = w(\mathbf{c}_0) + w(\mathbf{o}) - 2\,\text{overlap} = \text{even} \Rightarrow \mathbf{c}_0 + \mathbf{o} \in E$.
We will use these facts to show $O \subseteq O' \Rightarrow O' = O$.

$\mathbf{o} \in O \Rightarrow \mathbf{o} = \mathbf{0} + \mathbf{o} = (\mathbf{c}_0 + \mathbf{c}_0) + \mathbf{o} = \mathbf{c}_0 + (\mathbf{c}_0 + \mathbf{o}) \Rightarrow \mathbf{o} \in O' \Rightarrow O \subseteq O' \Rightarrow O' = O$.

This shows either all the code vectors in C have even weight or exactly half of them do.

Chapter 6 Review

1. We will explain and give counter examples to justify our answers below.

   (a) ***False***. See Theorem 6.10 on p.458 of Section 6.2.
   In order for this to be true, we need to know $\dim V = n$ or
   that a basis for V contains n vectors. What other information would be sufficient?
   Q: In $\mathbb{R}^2$, the set $\left\{ \mathbf{v}_1 = \begin{bmatrix} 1 \\ 0 \end{bmatrix}, \mathbf{v}_2 = \begin{bmatrix} 0 \\ 1 \end{bmatrix}, \mathbf{v}_3 = \begin{bmatrix} 1 \\ 1 \end{bmatrix} \right\}$ is a counterexample.
   That is, $\mathbb{R}^2 = \text{span}(\mathbf{v}_1, \mathbf{v}_2, \mathbf{v}_3)$. Prove this.
   Q: Create a counterexample for M_{22}.

   (b) ***True***. See the definition on p.447 in Section 6.2.
   To show $S_a = \{\mathbf{u} + \mathbf{v}, \mathbf{v} + \mathbf{w}, \mathbf{u} + \mathbf{w}\}$ is linearly independent, we need to show
   if $a(\mathbf{u} + \mathbf{v}) + b(\mathbf{v} + \mathbf{w}) + c(\mathbf{u} + \mathbf{w}) = \mathbf{0}$ then $a = b = c = 0$.
   Regrouping, we have $(a + c)\mathbf{u} + (a + b)\mathbf{v} + (b + c)\mathbf{w} = \mathbf{0}$.
   Since $S = \{\mathbf{u}, \mathbf{v}, \mathbf{w}\}$ is linearly independent, $a + c = a + b = b + c = 0 \Rightarrow$
   $b = c \Rightarrow 2b = 0 \Rightarrow b = 0 \Rightarrow c = 0 \Rightarrow a = 0$ as we needed to show.
   Q: What is another way of showing that S_a is linearly independent?
   A: Hint: $\mathbf{u} = \frac{1}{2}[(\mathbf{u} + \mathbf{v}) - (\mathbf{v} + \mathbf{w}) + (\mathbf{u} + \mathbf{w})]$.
   That is, does $\text{span}(S_a) = \text{span}(S)$? Is this sufficient? Is this required?

   (c) ***True***. Is there any reason to suspect this is true for all M_{nn}?
   Let $I = \begin{bmatrix} 1 & 0 \\ 0 & 1 \end{bmatrix}$, $J = \begin{bmatrix} 0 & 1 \\ 1 & 0 \end{bmatrix}$, $L = \begin{bmatrix} 1 & 1 \\ 0 & 1 \end{bmatrix}$, and $R = \begin{bmatrix} 1 & 1 \\ 1 & 0 \end{bmatrix}$.
   Let $\mathcal{E} = \{E_{11}, E_{12}, E_{21}, E_{22}\}$ and $\mathcal{B} = \{I, J, L, R\}$.
   Q: Prove $\mathcal{B}$ is a basis by showing $\text{span}(\mathcal{B}) = \text{span}(\mathcal{E})$.
   Why is this sufficient?

   (d) ***False***. If $W = \{M \in M_{22} : \text{tr}(A) = 0\}$, then $\dim W = 3$. See Exercise 15.
   Q: Why does $\dim W = 3$ imply M_{22} cannot have a basis of matrices all with trace 0?
   A: Because $\dim M_{22} = 4$. Why is this enough?
   Q: Why does the argument above *not* apply to invertible matrices?
   A: Hint: If A and B are invertible, is $A + B$ necessarily invertible?
   But, if A and B have trace zero, does $A + B$ necessarily have trace zero?

1. We will explain and give counter examples to justify our answers below. (continued)

(e) ***False***. The Triangle Inequality asserts $\|\mathbf{x}+\mathbf{y}\| \leq \|\mathbf{x}\| + \|\mathbf{y}\|$.
Q: Which property of a linear transformation does T fails to have?
A: Hint: To be linear, $T(\mathbf{x}+\mathbf{y}) = T(\mathbf{x}) + T(\mathbf{x})$ and $T(c\mathbf{x}) = cT(\mathbf{x})$.
Q: Does $T(\mathbf{x}) = \|\mathbf{x}\|$ at least satisfy $T(c\mathbf{x}) = cT(\mathbf{x})$? Why or why not?

(f) ***True***. This is worth proving directly.
Q: If T is one-to-one, what do we know about the dimension of range(T)?
A: Hint: If $\{\mathbf{u}_i\}$ is a basis for V, is $\{T(\mathbf{u}_i)\}$ a basis for range(T)?
Q: If T is onto, what do we know about a basis for W?
A: Hint: If $\{\mathbf{u}_i\}$ is a basis for V, does $\{T(\mathbf{u}_i)\}$ contain a basis for W?

(g) ***False***. What condition do we need to add to make this statement true?
Q: If $\ker(T) = V$, what does range(T) equal?
Q: If $\ker(T) = V$ and T is onto, what does W equal?
Q: Is $T : \mathbb{R}^2 \to \mathbb{R}^2$ a counterexample to this statement? Why or why not?

(h) ***True***. What Theorem should we cite to prove this?
Q: What is $\dim M_{33}$?
Q: If nullity$(T) = 4$, what is the dimension of range(T)?
Q: Why do the answers to these two questions imply the statement is true?

(i) ***True***. Prove this by finding a basis for V and thereby the dimension of V.
Q: Given $V = \{p(x) \in \mathscr{P}_4 : p(1) = 0\}$, find a basis for V.
A: Hint: is $T : \mathscr{P}_4 \to \mathbb{R}$ defined by $T(p) = p(1)$ a linear transformation?
Q: Given $T : \mathscr{P}_4 \to \mathbb{R}$ defined by $T(p) = p(1)$, is T onto?
Q: What is $\ker(T)$?
Q: What is nullity(T)?
Q: Is $\dim V = 4$ enough to prove that V is isomorphic to $\mathscr{P}_3$?

(j) ***False***. See Example 6.80 in Section 6.6.
If bases $\mathcal{B}$ and $\mathcal{C}$ are different, we get the change-of-basis matrix.

2. As asserted in Theorem 6.2 of Section 6.1, to prove W is a subspace we need only show:

a. $\mathbf{u},\ \mathbf{v} \in W \Rightarrow \mathbf{u}+\mathbf{v} \in W$
b. $\mathbf{u} \in W \Rightarrow c\mathbf{u} \in W$

However, in this case, we note that $W = \left\{ \begin{bmatrix} x \\ y \end{bmatrix} : x^2 + 3y^2 = 0 \right\} = \left\{ \begin{bmatrix} 0 \\ 0 \end{bmatrix} \right\}$. Why?

Because $x^2 + 3y^2 = 0 \Rightarrow x^2 = -3y^2 \Rightarrow x^2 = 0 \Rightarrow x = y = 0$.

Since W is the *zero subspace* of $\mathbb{R}^2$, there is nothing to further to show.

3. As asserted in Theorem 6.2 of Section 6.1, to prove W is a subspace we need only show:

a. $\mathbf{u},\ \mathbf{v} \in W \Rightarrow \mathbf{u}+\mathbf{v} \in W$
b. $\mathbf{u} \in W \Rightarrow c\mathbf{u} \in W$

In this case, we have $W = \left\{ \begin{bmatrix} a & b \\ c & d \end{bmatrix} : a+b = c+d = a+c = b+d \right\}$.

We proceed as in Exercises 46 and 47 of Section 6.1.

1. $A = \begin{bmatrix} a & b \\ c & d \end{bmatrix},\ A' = \begin{bmatrix} a' & b' \\ c' & d' \end{bmatrix} \in W \Rightarrow A + A' = \begin{bmatrix} a+a' & b+b' \\ c+c' & d+d' \end{bmatrix} \in W$
   because $a+b = c+d = a+c = b+d$ and $a'+b' = c'+d' = a'+c' = b'+d'$ imply $(a+a')+(b+b') = (c+c')+(d+d') = (a+a')+(c+c') = (b+b')+(d+d')$.

6. Likewise, $A = \begin{bmatrix} a & b \\ c & d \end{bmatrix} \in W \Rightarrow kA = \begin{bmatrix} ka & kb \\ kc & kd \end{bmatrix} \in W$
   because $a+b = c+d = a+c = b+d$ implies $k(a+b) = k(c+d) = k(a+c) = k(b+d)$ so $ka+kb = kc+kd = ka+kc = kb+kd$.

4. As asserted in Theorem 6.2 of Section 6.1, to prove W is a subspace we need only show:

a. $\mathbf{u},\ \mathbf{v} \in W \Rightarrow \mathbf{u}+\mathbf{v} \in W$
b. $\mathbf{u} \in W \Rightarrow c\mathbf{u} \in W$

In this case, we have $W = \{p(x) \in \mathscr{P}_3 : x^3 p(\frac{1}{x}) = p(x)\}$.

We follow Example 6.13 and proceed as in Exercises 34 and 37 of Section 6.1.

1. $p(x),\ q(x) \in W \Rightarrow (p+q)(x) = p(x) + q(x) \in W$
   because $p(x) + q(x) = x^3 p(\frac{1}{x}) + x^3 q(\frac{1}{x}) = x^3(p(\frac{1}{x}) + q(\frac{1}{x})) = x^3(p+q)(\frac{1}{x})$.

6. Likewise, $p(x) \in W \Rightarrow kp(x) \in W$ because $kx^3 p(\frac{1}{x}) = x^3\left(kp(\frac{1}{x})\right)$.

Q: How does the condition $x^3 p(\frac{1}{x}) = p(x)$ restrict the form of $p(x)$?

A: Hint: If $p(x) = a + bx + cx^2 + dx^3$, then $x^3 p(\frac{1}{x}) = ax^3 + bx^2 + cx + d$.
If $x^3 p(\frac{1}{x}) = p(x)$, what does that tell us about a and d? About b and c?

Q: Is it easier to prove W is a subspace using this insight? Why or why not?

5. As asserted in Theorem 6.2 of Section 6.1, to prove W is a subspace we need only show:

a. $\mathbf{u}, \mathbf{v} \in W \Rightarrow \mathbf{u}+\mathbf{v} \in W$
b. $\mathbf{u} \in W \Rightarrow c\mathbf{u} \in W$

In this case, we have $W = \{f \in \mathscr{F} : f(x+\pi) = f(x) \text{ for all } x\}$.

We follow Example 6.13 and proceed as in Exercises 34 and 37 of Section 6.1.

1. $f(x), g(x) \in W \Rightarrow (f+g)(x) = f(x) + g(x) \in W$
   because $f(x) + g(x) = f(x+\pi) + g(x+\pi) = (f+g)(x+\pi)$.

6. Likewise, $f(x) \in W \Rightarrow kf(x) \in W$ because $kf(x) = kf(x+\pi)$.

6. We apply Example 6.24 and proceed as in Exercises 10, 11, and 14 of Section 6.2.
Recall, to show a set S is linearly dependent, we need only show the following:
A vector in S can be written as a linear combination of the remaining vectors in set.

Q: What theorem supports the assertion made above?
A: Hint: See Section 2.3.

Using this idea, we will show that $S = \{1, \cos 2x, 3\sin^2 x\}$ is a linearly dependent set.

Since $\cos 2x = 1 - 2\sin^2 x$, $S = \{1, \cos 2x, 3\sin^2 x\}$ is a linearly dependent set.

Q: Why does $cos(x+y) = \cos x \cos y - \sin x \sin y$ imply $\cos 2x = 1 - 2\sin^2 x$?

Q: Why does the fact that $\cos 2x = 1 - 2\sin^2 x$ imply that S is linearly dependent?

A: Because $\cos 2x = 1 - \frac{2}{3}\left(3\sin^2 x\right)$.

7. Given A *symmetric* ($A^T = A$)and B *skew-symmetric* ($B^T = -B$), we need to show:
$S = \{A, B\}$ is a linearly independent set.

For further discussion of *symmetric* and *skew-symmetric* see Section 3.2.
Also see Example 6.26 and Exercises 1 through 4 in Section 6.2.
It is important to note that A and B are *nonzero*. Where do we use these facts below?

To prove $\{A, B\}$ is linearly independent, we need only show $cA + dB = O \Rightarrow c = d = 0$.

Given (1) $cA + dB = O$, we take the transpose of both sides to get (2) $cA^T + dB^T = O$.

Now adding equation (1) to equation (2) yields $(cA + cA^T) + (dB + dB^T) = O$.
Since A is symmetric $cA + cA^T = cA + cA = 2cA$.
Likewise, since B is skew-symmetric $cB + cB^T = cB - cB = O$.

So we have $2cA = O \Rightarrow c = 0 \Rightarrow cA + dB = dB = O \Rightarrow d = 0$ as we were required to show.

Q: Why does the fact that $2cA = O$ imply that $c = 0$?
A: Hint: Recall that we are given the fact that A is nonzero.
Does the implication above necessarily follow if A is the zero matrix?

8. We will find a basis for W using the condition on a, b, c, and d that defines it. For further discussion, see Example 6.42 and Exercises 37, 38, and 39 in Section 6.2.

We are given $W = \left\{ \begin{bmatrix} a & b \\ c & d \end{bmatrix} : a + d = b + c \right\}$ which implies $W = \left\{ \begin{bmatrix} a & b \\ c & d \end{bmatrix} : a = b + c - d \right\}$.

That is, W is all 2×2 matrices of the form: $W = \left\{ \begin{bmatrix} b + c - d & b \\ c & d \end{bmatrix} \right\}$.

So, it is obvious that $\mathcal{B} = \left\{ \begin{bmatrix} 1 & 1 \\ 0 & 0 \end{bmatrix}, \begin{bmatrix} 1 & 0 \\ 1 & 0 \end{bmatrix}, \begin{bmatrix} -1 & 0 \\ 0 & 1 \end{bmatrix} \right\}$ is a basis for W and $\dim W = 3$.

Q: Why is it obvious that $\mathcal{B}$ is a basis? Prove this.
A: Hint: Show that matrix vectors in $\mathcal{B}$ are linearly independent and span W.

9. We will find a basis for W using the condition $p(-x) = p(x)$ that defines it. For further discussion, see Example 6.34 and Exercises 34, 35, and 36 in Section 6.2.

Note $W = \{p(x) \in \mathscr{P}_5 : p(-x) = p(x)\}$ implies $W = \{p(x) \in \mathscr{P}_5 :$ degree of $p(x)$ is even $\}$.

That is, W is all polynomials in $\mathscr{P}_5$ of the form: $W = \{p(x) : p(x) = a_0 + a_2x^2 + a_4x^4\}$.

So, it is obvious that $\mathcal{B} = \{1, x^2, x^4\}$ is a basis for W and $\dim W = 3$.

Q: Why is it obvious that the degree of $p(x)$ must be even? Prove this.
A: Hint: Can $p(-x) = p(x)$ if x occurs to an odd power? Like $p(x) = x$, for example?

Q: Why is it obvious that $\mathcal{B}$ is a basis? Prove this.
A: Hint: Show that polynomial vectors in $\mathcal{B}$ are linearly independent and span W.

10. We find $P_{\mathcal{C}\leftarrow\mathcal{B}}$ and $P_{\mathcal{B}\leftarrow\mathcal{C}}$ directly.

Q: Why should we compute these matrices directly instead of indirectly?
A: It is easy to express the vectors in $\mathcal{B}$ in terms of the vectors in $\mathcal{C}$.

Also see Example 6.46, Theorem 6.13, and Exercises 5 through 8 in Section 6.2.

Since $\mathcal{B} = \{1, 1+x, 1+x+x^2\} = \{\mathbf{b}_1, \mathbf{b}_2, \mathbf{b}_3\}$ and $\mathcal{C} = \{1+x, x+x^2, 1+x^2\} = \{\mathbf{c}_1, \mathbf{c}_2, \mathbf{c}_3\}$, $\mathbf{c}_1 = 0\mathbf{b}_1 + 1\mathbf{b}_2 + 0\mathbf{b}_3$, $\mathbf{c}_2 = -1\mathbf{b}_1 + 0\mathbf{b}_2 + 1\mathbf{b}_3$, and $\mathbf{c}_3 = 1\mathbf{b}_1 - 1\mathbf{b}_2 + 1\mathbf{b}_3$.

$$\text{So } P_{\mathcal{B}\leftarrow\mathcal{C}} = \begin{bmatrix} 0 & -1 & 1 \\ 1 & 0 & -1 \\ 0 & 1 & 1 \end{bmatrix} \text{ which implies } P_{\mathcal{C}\leftarrow\mathcal{B}} = (P_{\mathcal{B}\leftarrow\mathcal{C}})^{-1} = \begin{bmatrix} \frac{1}{2} & 1 & \frac{1}{2} \\ -\frac{1}{2} & 0 & \frac{1}{2} \\ \frac{1}{2} & 0 & \frac{1}{2} \end{bmatrix}.$$

Q: Given $P_{\mathcal{B}\leftarrow\mathcal{C}}$, what is the simplest way to compute $P_{\mathcal{C}\leftarrow\mathcal{B}}$?
A: By row reduction. That is, $[\, I \,|\, P_{\mathcal{B}\leftarrow\mathcal{C}} \,] \longrightarrow [\, (P_{\mathcal{B}\leftarrow\mathcal{C}})^{-1} \,|\, I \,]$.

Q: Why is this sufficient?
A: Because $P_{\mathcal{C}\leftarrow\mathcal{B}} = (P_{\mathcal{B}\leftarrow\mathcal{C}})^{-1}$.

Q: How else might we have computed $P_{\mathcal{C}\leftarrow\mathcal{B}}$?
A: Directly as above. That is:

$$\mathbf{b}_1 = \tfrac{1}{2}\mathbf{c}_1 - \tfrac{1}{2}\mathbf{c}_2 + \tfrac{1}{2}\mathbf{c}_3,\ \mathbf{b}_2 = 1\tfrac{1}{2}\mathbf{c}_1 + 0\mathbf{c}_2 + 0\mathbf{c}_3, \text{ and } \mathbf{b}_3 = \tfrac{1}{2}\mathbf{c}_1 + \tfrac{1}{2}\mathbf{c}_2 + \tfrac{1}{2}\mathbf{c}_3.$$

11. NOTE: We need $T(\mathbf{x}) = \mathbf{y}\mathbf{x}^T\mathbf{y}$ otherwise it does not go from $\mathbb{R}^2$ to $\mathbb{R}^2$.

We will show $T(\mathbf{x}) = \mathbf{y}\mathbf{x}^T\mathbf{y}$ is a linear transformation by showing:
$T(\mathbf{x} + c\mathbf{z}) = T(\mathbf{x}) + cT(\mathbf{z})$.

Q: What theorem tells us that it is sufficient to show $T(\mathbf{x} + c\mathbf{z}) = T(\mathbf{x}) + cT(\mathbf{z})$?
A: Hint: See Section 3.5 and Section 6.4.

Since $T(\mathbf{x} + c\mathbf{z}) = \mathbf{y}(\mathbf{x} + c\mathbf{z})^T\mathbf{y} = \mathbf{y}\mathbf{x}^T\mathbf{y} + c\mathbf{y}\mathbf{z}^T\mathbf{y} = T(\mathbf{x}) + cT(\mathbf{z})$ as required.

Q: Is $\mathbf{y}\mathbf{x}^T$ a vector, a scalar, or neither?

Q: Do we need to know the exact value of $\mathbf{y}$ to conclude that T is linear?
A: No. T is linear for any fixed vector $\mathbf{y}$.

Q: Does T need to be a mapping from $\mathbb{R}^2$ to $\mathbb{R}^2$ in order to be linear?
A: No. T can be a mapping from $\mathbb{R}^n$ to $\mathbb{R}^n$ for any n.

Q: Is T still linear if T is a mapping from M_{nn} to M_{nn}? Why or why not?

12. We will show $T(A) = A^TA$ is *not* a linear transformation by showing:
If $A = I$ and $c = -1$, then $T(-1 \cdot I) \neq -1 \cdot T(I)$.

$T(-I) = (-I)^T(-I) = I \neq -I = -1 \cdot (I^TI) = -1 \cdot T(I)$.

Also see Example 6.53 and Exercises 3 and 4 in Section 6.4.

Q: Why is it sufficient to show $T(-1 \cdot I) \neq -1 \cdot T(I)$ to prove T is *not* linear?
A: In order to show T is not linear, we did only show one counterexample.

Q: What is the general pattern underlying this counterexample?
A: $T(cA) = (cA)^T(cA) = c^2(A^TA) = c^2T(A)$ instead of $cT(A)$ as required.

13. We will show $T(p(x)) = p(2x-1)$ is a linear transformation by showing:
$$T(p(x) + cq(x)) = T(p(x)) + cT(q(x)).$$
Also See Exercises 8 and 9 in Section 6.4.

Recall $p(x) + cq(x) = (p + cq)(x)$.

So $T(\ p(x) + cq(x)\ ) = T(\ (p+cq)(x)\ ) = (p+cq)(2x-1)$
$$= p(2x-1) + cq(2x-1) = T(p(x)) + cT(q(x)) \text{ as required.}$$

Q: If $2x-1$ is replaced by $ax+b$ is T still linear?
A: Yes. T is linear for any such substitution.

Q: If $2x-1$ is replaced by ax^2+bx+c is T still linear? Why or why not?

Q: If $2x-1$ is replaced by $a_nx^n + ... + a_0$ is T still linear? Why or why not?

Q: If $2x-1$ is replaced by $a_n\frac{1}{x^n} + ... + a_0$ is T still linear? Why or why not?

14. Compute $T(\ p(x)\ ) = T(5-3x+2x^2)$ by expressing $p(x)$ in terms of the basis vectors in $\mathcal{B}$.

Also see Example 6.55 and Exercises 14 through 18 in Section 6.4.

Since we are given $T(1)$, $T(1+x)$, and $T(1+x+x^2)$,
we let $\mathcal{B} = \{1, 1+x, 1+x+x^2\} = \{\mathbf{b}_1, \mathbf{b}_2, \mathbf{b}_3\}$.

So $p(x) = 8\mathbf{b}_1 - 5\mathbf{b}_1 + 2\mathbf{b}_3 = 8(1) - 5(1+x) + 2(1+x+x^2)$.

Therefore, since T is linear, $T(\ p(x)\ ) = T(5-3x+2x^2) = T(\ 8(1) - 5(1+x) + 2(1+x+x^2)\ )$
$$= 8\cdot T(1) - 5\cdot T(1+x) + 2\cdot T(1+x+x^2) = 8\begin{bmatrix}1 & 0\\0 & 1\end{bmatrix} - 5\begin{bmatrix}1 & 1\\0 & 1\end{bmatrix} + 2\begin{bmatrix}0 & -1\\1 & 0\end{bmatrix} = \begin{bmatrix}3 & -7\\2 & 3\end{bmatrix}.$$

Q: What is $T(x)$? What is $T(x^2)$? Use these to compute $T(\ p(x)\ )$.
Do we get the same matrix as we computed above? Why or why not? Should we?

15. Given $T(A) = \text{tr}(A)$, we will compute nullity(T) using the Rank Theorem, Theorem 6.19.
Recall from Exercises 44 through 46 in Section 3.2, we have $\text{tr}(A) = a_{11} + a_{22} + \cdots + a_{nn}$.
Since $\dim M_{nn} = n^2$ and $\text{tr}(A)$ is a scalar with $\dim \mathbb{R} = 1$, we have nullity$(T) = n^2 - 1$.

Also see Examples 6.64 through 6.67 and Exercise 6 in Section 6.5.

Q: In M_{22}, find a basis $\mathcal{B}$ for ker(T). How many basis vectors must $\mathcal{B}$ have?
A: Since nullity$(T) = n^2 - 1 = 2^2 - 1 = 3$, a basis $\mathcal{B}$ for ker(T) must have 3 vectors.

Since A is in ker(T) means $a_{11} + a_{22} = 0$, we have $a_{11} = -a_{22}$.

So $\mathcal{B} = \left\{\begin{bmatrix}1 & 0\\0 & -1\end{bmatrix}, \begin{bmatrix}0 & 1\\0 & 0\end{bmatrix}, \begin{bmatrix}0 & 0\\1 & 0\end{bmatrix}\right\}$ is a basis for ker(T).

Q: Can we generalize this result to find a basis for T in M_{nn}?

16. We will find: (a) T such that $\ker(T) = W$ and (b) T such that $\text{range}(T) = W$.
See the definitions of $\ker(T)$ (*kernel*) and $\text{range}(T)$ (*range*) given in section 6.5.

Q: Can we find one such T that satisfies both of these requirements at the same time?

Since W is all upper triangular 2×2 matrices, $W = \left\{ \begin{bmatrix} a & b \\ c & d \end{bmatrix} : \begin{bmatrix} a & b \\ 0 & d \end{bmatrix} \right\}$.

Let $A = \begin{bmatrix} a & b \\ c & d \end{bmatrix}$ and basis $\mathcal{B} = \left\{ \begin{bmatrix} 1 & 0 \\ 0 & 0 \end{bmatrix}, \begin{bmatrix} 0 & 1 \\ 0 & 0 \end{bmatrix}, \begin{bmatrix} 0 & 0 \\ 1 & 0 \end{bmatrix}, \begin{bmatrix} 0 & 0 \\ 0 & 1 \end{bmatrix} \right\} = \{\mathbf{b}_1, \mathbf{b}_2, \mathbf{b}_3, \mathbf{b}_4\}$.

(a) We want $\ker(T) = W$. Recall A is in $\ker(T)$ if and only if $T(A) = O$, the zero matrix. Since we want $\ker(T) = W$, T must map a A to O if and only if A is upper triangular. That is, $A \in \ker(T) \Leftrightarrow A \in W$ because W is the set of all upper triangular matrices.

Explicitly, we are looking for a map T such that $T \begin{bmatrix} a & b \\ 0 & d \end{bmatrix} = \begin{bmatrix} 0 & 0 \\ 0 & 0 \end{bmatrix}$.

Q: We claim one such map is $T(A) = c\mathbf{b}_3$. Is T is linear? Does $\ker(T) = W$?
A: Hint: If A is in $\ker(T)$, c must be equal to zero. Why?
Q: Define $T(A) = c\mathbf{b}_1$. Is T linear? Does $\ker(T) = W$? Why or why not?
Q: Define $T(A) = cI$. Is T linear? Does $\ker(T) = W$? Why or why not?
Q: Define $T(A) = kc\mathbf{b}_3$ for a fixed k. Is T linear? Does $\ker(T) = W$? Why or why not?
Q: Define $T(A) = A - A^T$. Is T linear? What is $\ker(T)$?

(b) We want $\text{range}(T) = W$. Recall B is in $\text{range}(T)$ if and only if $B = T(A)$ for some A. That is, $\text{range}(T) = W$, T must map A *onto* B if and only if B is upper triangular. That is, $B \in \text{range}(T) \Leftrightarrow B \in W$ because W is the set of all upper triangular matrices.

Q: We claim one such map is $T(A) = a\mathbf{b}_1 + b\mathbf{b}_2 + d\mathbf{b}_4$. Prove this.
A: Hint: If B is in $\text{range}(T)$, b_{21} must be equal to zero. Why?
Q: Define $T(A) = a\mathbf{b}_1 + b\mathbf{b}_2 + c\mathbf{b}_4$. Does $\text{range}(T) = W$?
Q: Define $T(A) = b\mathbf{b}_1 + c\mathbf{b}_2 + d\mathbf{b}_4$. Does $\text{range}(T) = W$?
Q: Define $T(A) = ka\mathbf{b}_1 + kb\mathbf{b}_2 + kd\mathbf{b}_4$ for fixed k. Does $\text{range}(T) = W$?

17. We will find the change-of-basis matrix $[T]_{\mathcal{C}\leftarrow\mathcal{B}}$,
where $\mathcal{B} = \{1, x, x^2\}$ and $\mathcal{C} = \{E_{11}, E_{12}, E_{21}, E_{22}\}$.

Also see Examples 6.76 and 6.77 and Exercises 1 through 12 in Section 6.6.

We compute $T(\ p(x)\ )$ for the basis vectors in $\mathcal{B}$.
First we express $\{1, x, x^2\}$ in terms of $\{1, 1+x, 1+x+x^2\}$ from Exercise 14.
Then we express the result as a linear combination of the basis vectors in $\mathcal{C}$.

$$T(1) = \begin{bmatrix} 1 & 0 \\ 0 & 1 \end{bmatrix} = 1 \cdot E_{11} + 0 \cdot E_{12} + 0 \cdot E_{21} + 1 \cdot E_{22} \Rightarrow [T(1)]_{\mathcal{C}} = \begin{bmatrix} 1 \\ 0 \\ 0 \\ 1 \end{bmatrix}$$

$$T(x) = T(\ (1+x) - 1\ ) = T(1+x) - T(1)$$

$$= \begin{bmatrix} 1 & 1 \\ 1 & 0 \end{bmatrix} - \begin{bmatrix} 1 & 0 \\ 0 & 1 \end{bmatrix} = \begin{bmatrix} 0 & 1 \\ 0 & 0 \end{bmatrix} = 0 \cdot E_{11} + 1 \cdot E_{12} + 0 \cdot E_{21} + 0 \cdot E_{22} \Rightarrow [T(x)]_{\mathcal{C}} = \begin{bmatrix} 0 \\ 1 \\ 0 \\ 0 \end{bmatrix}$$

$$T(x^2) = T(\ (1+x+x^2) - (1+x)\ ) = T(1+x+x^2) - T(1+x)$$

$$= \begin{bmatrix} 0 & -1 \\ 1 & 0 \end{bmatrix} - \begin{bmatrix} 1 & 1 \\ 0 & 1 \end{bmatrix} = \begin{bmatrix} -1 & -2 \\ 1 & -1 \end{bmatrix} = 1 \cdot E_{11} - 2 \cdot E_{12} + 1 \cdot E_{21} - 1 \cdot E_{22} \Rightarrow \left[T(x^2)\right]_{\mathcal{C}} = \begin{bmatrix} 1 \\ -2 \\ 1 \\ -1 \end{bmatrix}$$

Since these vectors are the columns of $[T]_{\mathcal{C}\leftarrow\mathcal{B}}$, we have: $[T]_{\mathcal{C}\leftarrow\mathcal{B}} = \begin{bmatrix} 1 & 0 & -1 \\ 0 & 1 & -2 \\ 0 & 0 & 1 \\ 1 & 0 & -1 \end{bmatrix}$.

18. To prove S is a basis we need to show: (1) S spans V and (2) S is linearly independent.

Also see Exercise 30 of Section 6.2.

Q: Is this converse of Theorem 6.5. Why or why not?

(1) We must show S spans V.

Q: Every vector in V can be written as a linear combination of the vectors in S.
Does this imply that S spans V? Why or why not?

A: Yes, this is the definition. Symbolically, we write $\text{span}(S) = V$.

(2) We must show S is linearly independent.

Q: Is the following statement true?: S is linearly independent if and only if
there is precisely one way to write $\mathbf{0}$ as a linear combination of the vectors in S.

A: True. S is linearly independent if and only if $c_1\mathbf{v}_1 + \cdots + c_n\mathbf{v}_n = \mathbf{0} \Leftrightarrow$ all the $c_i = 0$.

That means the only way to write $\mathbf{0}$ is $\sum_{i=1}^{n} 0 \cdot \mathbf{v}_i$.

In this case, we know that every vector in V can be written as linear combination
of the vectors in S in exactly one way, including the zero vector, $\mathbf{0}$.

Therefore, S is linearly independent.

19. Given $S \circ T : U \to W$ and range$(T) \subseteq \ker(S)$, we will show $S \circ T$ is the zero map.

See Figure 6.6 in Section 6.4 and Figure 6.8 in Section 6.5
for graphical representations of composition, range, and kernel.

Q: If $\mathbf{v}$ is in $\ker(S)$, what is $S(\mathbf{v})$?

Since range$(T) \subseteq \ker(S)$, if $\mathbf{u}$ in U then $T(\mathbf{u})$ is in $\ker(S)$. Therefore, $S(T(\mathbf{u})) = \mathbf{0}$.

So, $S \circ T : U \to W$ is the zero map. That is, $S \circ T$ maps every vector in U to $\mathbf{0}$ in W.

20. Given $T : V \to V$ and a basis $\{\mathbf{v}_1, \mathbf{v}_2, \ldots, \mathbf{v}_n\}$ for V such that
$\{T(\mathbf{v}_1), T(\mathbf{v}_2), \ldots, T(\mathbf{v}_n)\}$ is also a basis for V.

We will show that T is invertible.

See Corollary 6.23 and Example 6.71 in Section 6.5.
Theorem 6.30 in Section 6.6 provides several ways to prove T is invertible.

Since $\{T(\mathbf{v}_1), T(\mathbf{v}_2), \ldots, T(\mathbf{v}_n)\}$ is a basis for V, range$(T) = V$.
Therefore, by Theorem 6.30t., T is invertible.

Q: Prove that T is invertible using Theorem 6.30s.

Q: Prove that T is invertible by showing that its associated matrix is invertible.

Q: Which of these proofs do you find easier? Why?

Chapter 7

Distance and Approximation

7.1 Inner Product Spaces

1. Example 7.2: $\langle \mathbf{u}, \mathbf{v} \rangle = 2u_1v_1 + 3u_2v_2 \Rightarrow$

(a) $\langle \mathbf{u}, \mathbf{v} \rangle = 2(\,(2)(3)\,) + 3(\,(-1)(4)\,) = 0.$

(b) $\|\mathbf{u}\| = \sqrt{\langle \mathbf{u}, \mathbf{u} \rangle} = \sqrt{2(\,(2)(2)\,) + 3(\,(-1)(-1)\,)} = \sqrt{11}.$

(c) $d(\mathbf{u}, \mathbf{v}) = \|\mathbf{u} - \mathbf{v}\| = \| \begin{bmatrix} -1 \\ -5 \end{bmatrix} \| = \sqrt{2(\,(-1)(-1)\,) + 3(\,(-5)(-5)\,)} = \sqrt{77}.$

2. Example 7.3: $A = \begin{bmatrix} 4 & -2 \\ -2 & 7 \end{bmatrix}$ and $\langle \mathbf{u}, \mathbf{v} \rangle = \mathbf{u}^T A\, \mathbf{v} \Rightarrow$

(a) $\langle \mathbf{u}, \mathbf{v} \rangle = [2 \ \ -1] \begin{bmatrix} 4 & -2 \\ -2 & 7 \end{bmatrix} \begin{bmatrix} 3 \\ 4 \end{bmatrix} = -14.$

(b) $\|\mathbf{u}\| = \sqrt{\mathbf{u}^T A\, \mathbf{u}} = \sqrt{\begin{bmatrix} 2 & -1 \end{bmatrix} \begin{bmatrix} 4 & -2 \\ -2 & 7 \end{bmatrix} \begin{bmatrix} 2 \\ -1 \end{bmatrix}} = \sqrt{31}.$

(c) $d(\mathbf{u}, \mathbf{v}) = \|\mathbf{u} - \mathbf{v}\| = \| \begin{bmatrix} -1 \\ -5 \end{bmatrix} \| = \sqrt{\begin{bmatrix} -1 & -5 \end{bmatrix} \begin{bmatrix} 4 & -2 \\ -2 & 7 \end{bmatrix} \begin{bmatrix} -1 \\ -5 \end{bmatrix}} = \sqrt{159}.$

3. Since $\mathbf{u} = \begin{bmatrix} 2 \\ -1 \end{bmatrix}$ orthogonal means $\langle \mathbf{u}, \mathbf{v} \rangle = 2(2v_1) + 3((-1)v_2) = 0 \Rightarrow$

$4v_1 - 3v_2 = 0 \Rightarrow 4v_1 = 3v_2 \Rightarrow$ any scalar multiple of $\begin{bmatrix} 3 \\ 4 \end{bmatrix}$ is orthogonal to $\mathbf{u}$.

4. Since $\mathbf{u} = \begin{bmatrix} 2 \\ -1 \end{bmatrix}$ orthogonal means $\langle \mathbf{u}, \mathbf{v} \rangle = \begin{bmatrix} 2 & -1 \end{bmatrix} \begin{bmatrix} 4 & -2 \\ -2 & 7 \end{bmatrix} \begin{bmatrix} v_1 \\ v_2 \end{bmatrix} = 0 \Rightarrow$

$10v_1 - 11v_2 = 0 \Rightarrow 10v_1 = 11v_2 \Rightarrow$ any scalar multiple of $\begin{bmatrix} 11 \\ 10 \end{bmatrix}$ is orthogonal to $\mathbf{u}$.

5. Example 7.4: $\langle p(x), q(x) \rangle = a_0\, b_0 + a_1\, b_1 + a_2\, b_2$.
It may be useful to note that this is equivalent to $[p(x)]_{\mathcal{E}}^T \cdot [q(x)]_{\mathcal{E}}$.

(a) $\langle p(x), q(x) \rangle = (2)(1) + (-3)(0) + (1)(-3) = -1.$

(b) $\|p(x)\| = \sqrt{\langle p(x), p(x) \rangle} = \sqrt{(2)(2) + (-3)(-3) + (1)(1)} = \sqrt{14}.$

(c) $d(p(x), q(x)) = \|p(x) - q(x)\| = \|\, 1 - 3x + 4x^2 \,\| = \sqrt{(1)(1) + (-3)(-3) + (4)(4)} = \sqrt{26}.$

6. Example 7.5: $\langle f, g \rangle = \int f(x)g(x)\, dx$ over $\mathscr{P}_2[0,1] \Rightarrow$

(a) $\langle p(x), q(x) \rangle = \int_0^1 p(x)q(x)\, dx = \int_0^1 (1 - 3x^2)(2 - 3x + x^2)\, dx = \frac{29}{60}.$

(b) $\|p(x)\| = \sqrt{\langle p(x), p(x) \rangle} = \sqrt{\int_0^1 (1 - 3x^2)^2\, dx} = \sqrt{\frac{4}{5}}.$

(c) $d(p(x), q(x)) = \|p(x) - q(x)\| = \|\, 1 - 3x + 4x^2 \,\| = \sqrt{\int_0^1 (1 - 3x + 4x^2) dx} = \sqrt{\frac{5}{6}}.$

7. Since $p(x) = 2 - 3x + x^2$ orthogonal means $\langle p(x), r(x)\rangle = 2r_0 - 3r_1 + r_2 = 0 \Rightarrow r_2 = 3r_1 - 2r_0$. Letting $r_0 = 1, r_1 = 0 \Rightarrow r_2 = -2 \Rightarrow r(x) = 1 - 2x^2$ is orthogonal to $p(x)$.

Note, the many choices and the check: $\langle p(x), r(x)\rangle = (2)(1) + (-3)(0) + (1)(-2) = 0$.

8. $\int_0^1 (x^2 - 3x + 2)(a + bx)dx = \frac{b}{4}x^4 + \left(\frac{a-3b}{3}\right)x^3 + \left(\frac{2b-3a}{2}\right)x^2 + 2ax\Big|_0^1 = \frac{b}{4} + \frac{5a}{6} = 0 \Rightarrow 20a = -6b \Rightarrow$ $h(x) = 20x - 6$ is orthogonal to $p(x) = x^2 - 3x + 2$. Can you find one based on $a + bx^2$?

9. Example 7.5: $\langle f, g\rangle = \int f(x)g(x)\,dx$ over $\mathscr{C}[0, 2\pi] \Rightarrow$
The trigonometric identities $\sin^2 x = \frac{1}{2}(1 - \cos 2x)$ and $\cos^2 x = \frac{1}{2}(1 + \cos 2x)$ are useful.

(a) $\langle f, g\rangle = \int_0^{2\pi} \sin x(\sin x + \cos x)\,dx = \int_0^{2\pi} \sin^2 x + \int_0^{2\pi} \sin x \cos x\,dx = \pi$.

(b) $\|f\| = \sqrt{\langle f, f\rangle} = \sqrt{\int_0^{2\pi} \sin^2 x\,dx} = \sqrt{\pi}$.

(c) $d(f, g) = \|f - g\| = \| -\cos x\| = \sqrt{\int_0^{2\pi} \cos^2 x\,dx} = \sqrt{\pi}$.

10. Noting $(x \sin^2 x)' = \sin^2 x + 2\sin x \cos x \Rightarrow$
$\langle f, h\rangle = \int_0^{2\pi} \sin x(\sin x + 2x\cos x)\,dx = \int_0^{2\pi} (\sin^2 x + 2x \sin x \cos x)\,dx = x\sin^2 x]_0^{2\pi} = 0 \Rightarrow$
$h(x) = \sin x + 2\cos x$ is orthogonal to $f(x) = \sin x$.

Can you find an $h(x)$ by differentiating $x^2 \sin^2 x$? $\sin^2 x \cos x$? Can you find a simple $h(x)$ that we have not suggested?
The key is we want $\sin x$ to appear in each term of the derivative. Why?

11. Will show $\langle p(x), q(x)\rangle = p(a)q(a) + p(b)q(b) + p(c)q(c)$ satisfies the inner product axioms. Section 6.2, Exercise 62 $\Rightarrow$ a $p(x)$, degree ≤ 2 with 3 distinct zeroes $\Rightarrow p(x) = \mathbf{0}$.

1. $\langle p(x), q(x)\rangle = p(a)q(a) + p(b)q(b) + p(c)q(c) = q(a)p(a) + q(b)p(b) + q(c)p(c)$
   $= \langle q(x), p(x)\rangle$.
2. $\langle p(x), (q + r)(x)\rangle = p(a)(q + r)(a) + p(b)(q + r)(b) + p(c)(q + r)(c)$
   $= (p(a)q(a) + p(b)q(b) + p(c)q(c)) + (p(a)r(a) + p(b)r(b) + p(c)r(c))$
   $= \langle p(x), q(x)\rangle + \langle p(x), r(x)\rangle$.
3. $\langle p(x), (dq)(x)\rangle = p(a)(dq)(a) + p(b)(dq)(b) + p(c)(dq)(c) = d(p(a)q(a) + p(b)q(b) + p(c)q(c))$
   $= d\langle p(x), q(x)\rangle$.
4. $\langle p(x), p(x)\rangle = p(a)p(a) + p(b)p(b) + p(c)p(c) = p(a)^2 + p(b)^2 + p(c)^2 \geq 0 \Rightarrow \langle p(x), p(x)\rangle \geq 0$ and $p(a)^2 + p(b)^2 + p(c)^2 = 0 \Leftrightarrow p(a) = p(b) = p(c) = 0 \Leftrightarrow p(x)$ has three distinct zeroes $\Leftrightarrow p(x)$ is the zero polynomial, that is $\Rightarrow p(x) = \mathbf{0}$.

12. Exercise 5 with $a = 0$, $b = 1$, $c = 2$, $p(x) = 2 - 3x + x^2$ and $q(x) = 1 - 3x^2$.

(a) $\langle p(x), q(x)\rangle = p(0)q(0) + p(1)q(1) + p(2)q(2) = (2)(1) + (0)(-2) + (0)(-11) = 2$.

(b) $\|p(x)\| = \sqrt{\langle p(x), p(x)\rangle} = \sqrt{p(0)^2 + p(1)^2 + p(2)^2} = \sqrt{2^2 + 0^2 + 0^2} = 2$.

(c) $d(p(x), q(x)) = \sqrt{(p - q)(0)^2 + (p - q)(1)^2 + (p - q)(2)^2} = \sqrt{1^2 + 2^2 + 11^2} = 3\sqrt{14}$.

13. Given $\langle \mathbf{u}, \mathbf{v} \rangle = u_1 v_1$ determine which axioms do not hold and give an example.

Axiom 4 fails: $\mathbf{u} = \begin{bmatrix} 0 \\ u \end{bmatrix} \Rightarrow \langle \mathbf{u}, \mathbf{u} \rangle = (0)(0) = 0$ but $\mathbf{u} \neq \mathbf{0}$.

14. Given $\langle \mathbf{u}, \mathbf{v} \rangle = u_1 v_1 - u_2 v_2$ determine which axioms do not hold and give an example.

Axiom 4 fails: $\mathbf{u} = \begin{bmatrix} u \\ u \end{bmatrix} \Rightarrow \langle \mathbf{u}, \mathbf{u} \rangle = u^2 - u^2 = 0$ but $\mathbf{u} \neq \mathbf{0}$.

15. Given $\langle \mathbf{u}, \mathbf{v} \rangle = u_1 v_2 + u_2 v_1$ determine which axioms do not hold and give an example.

Axiom 4 fails: $\mathbf{u} = \begin{bmatrix} u \\ 0 \end{bmatrix} \Rightarrow \langle \mathbf{u}, \mathbf{u} \rangle = u(0) + (0)u = 0$ but $\mathbf{u} \neq \mathbf{0}$.

Can you come up with another form of $\mathbf{u}$ that shows axiom 4 fails?

16. Given $\langle p(x), q(x) \rangle = p(0)q(0)$ determine which axioms do not hold and give an example.

Axiom 4 fails: $p(x) = ax \Rightarrow \langle p(x), p(x) \rangle = (0)(0) = 0$ but $p(x) \neq \mathbf{0}$.

Can you come up with another form of $p(x)$ that shows axiom 4 fails?

17. Given $\langle p(x), q(x) \rangle = p(1)q(1)$ determine which axioms do not hold and give an example.

Axiom 4 fails: $p(x) = a(1 - x) \Rightarrow \langle p(x), p(x) \rangle = (0)(0) = 0$ but $p(x) \neq \mathbf{0}$.

Can you come up with another form of $p(x)$ that shows axiom 4 fails?

18. Given $\langle A, B \rangle = \det(AB)$ determine which axioms do not hold and give an example.

Axiom 4 fails: $A = \begin{bmatrix} a & a \\ a & a \end{bmatrix} \Rightarrow \langle A, A \rangle = \det A^2 = 4a^4 - 4a^4 = 0$ but $A \neq \mathbf{0}$.

Can you come up with another form of A that shows axiom 4 fails?

19. Given $\langle \mathbf{u}, \mathbf{v} \rangle = 4u_1 v_1 + u_1 v_2 + u_2 v_1 + 4u_2 v_2$ find a symmetric matrix A such that $\langle \mathbf{u}, \mathbf{v} \rangle = \mathbf{u}^T A \mathbf{v}$.

$$\langle \mathbf{u}, \mathbf{v} \rangle = \mathbf{u}^T A \mathbf{v} = \begin{bmatrix} u_1 & u_2 \end{bmatrix} \begin{bmatrix} a & b \\ b & d \end{bmatrix} \begin{bmatrix} v_1 \\ v_2 \end{bmatrix} = a\, u_1 v_1 + b\, u_1 v_2 + b\, u_2 v_1 + d\, u_2 v_2 \Rightarrow$$

So, $\langle \mathbf{u}, \mathbf{v} \rangle = 4u_1 v_1 + u_1 v_2 + u_2 v_1 + 4u_2 v_2 \Rightarrow a = 4,\ b = 1,\ d = 4 \Rightarrow A = \begin{bmatrix} 4 & 1 \\ 1 & 4 \end{bmatrix}$.

20. Given $\langle \mathbf{u}, \mathbf{v} \rangle = u_1 v_1 + 2u_1 v_2 + 2u_2 v_1 + 5u_2 v_2$ find a symmetric matrix A such that $\langle \mathbf{u}, \mathbf{v} \rangle = \mathbf{u}^T A \mathbf{v}$.

$$\langle \mathbf{u}, \mathbf{v} \rangle = \mathbf{u}^T A \mathbf{v} = \begin{bmatrix} u_1 & u_2 \end{bmatrix} \begin{bmatrix} a & b \\ b & d \end{bmatrix} \begin{bmatrix} v_1 \\ v_2 \end{bmatrix} = a\, u_1 v_1 + b\, u_1 v_2 + b\, u_2 v_1 + d\, u_2 v_2 \Rightarrow$$

So, $\langle \mathbf{u}, \mathbf{v} \rangle = u_1 v_1 + 2u_1 v_2 + 2u_2 v_1 + 5u_2 v_2 \Rightarrow a = 1,\ b = 2,\ d = 5 \Rightarrow A = \begin{bmatrix} 1 & 2 \\ 2 & 5 \end{bmatrix}$.

21. Given $\langle \mathbf{u}, \mathbf{v} \rangle = u_1 v_1 + \frac{1}{4} u_2 v_2$ sketch $\|\mathbf{u}\| = \sqrt{\langle \mathbf{u}, \mathbf{u} \rangle} = 1 \Rightarrow \langle \mathbf{u}, \mathbf{u} \rangle = 1$. Let $\mathbf{u} = \begin{bmatrix} x \\ y \end{bmatrix} \Rightarrow$

$x^2 + \frac{1}{4} y^2 = 1 \Rightarrow \frac{x^2}{1} + \frac{y^2}{4} = 1$. Section 5.5, Example 5.30 $\Rightarrow$ ellipse intersecting the x-axis at $(\pm 1, 0)$ intersecting the y-axis at $(0, \pm 2)$.

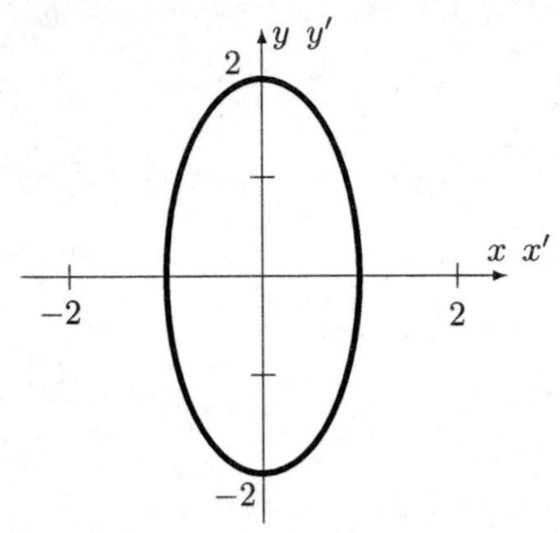

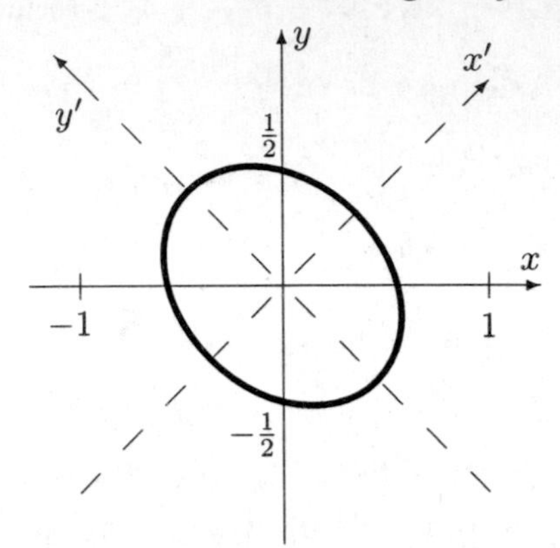

22. Given $\langle \mathbf{u}, \mathbf{v} \rangle = 4u_1 v_1 + u_1 v_2 + u_2 v_1 + 4u_2 v_2$ sketch $\|\mathbf{u}\| = \sqrt{\langle \mathbf{u}, \mathbf{u} \rangle} = 1$. Let $\mathbf{u} = \begin{bmatrix} x \\ y \end{bmatrix} \Rightarrow$

$4x^2 + 2xy + 4y^2 = 1 \Rightarrow \frac{x^2}{1} + \frac{y^2}{1} = 1$. Section 5.5, Example 5.32 $\Rightarrow A = \begin{bmatrix} 4 & 1 \\ 1 & 4 \end{bmatrix}$.

The eigenvectors of $A \Rightarrow \mathbf{e}_1 \longrightarrow \begin{bmatrix} \frac{1}{\sqrt{2}} \\ \frac{1}{\sqrt{2}} \end{bmatrix}$ and $\mathbf{e}_2 \longrightarrow \begin{bmatrix} -\frac{1}{\sqrt{2}} \\ \frac{1}{\sqrt{2}} \end{bmatrix}$.

Furthermore, the eigenvalues of A, namely 5 and 3, $\Rightarrow 5(x')^2 + 3(y')^2 = 1 \Rightarrow \frac{(x')^2}{1/5} + \frac{(y')^2}{1/3} = 1$.

23. Prove Theorem 7.1b. Relevant axioms cited above equal signs below.

$\langle \mathbf{u}, c\mathbf{v} \rangle \stackrel{1}{=} \langle c\mathbf{v}, \mathbf{u} \rangle \stackrel{3}{=} c\langle \mathbf{v}, \mathbf{u} \rangle \stackrel{1}{=} c\langle \mathbf{u}, \mathbf{v} \rangle$.

24. Prove Theorem 7.1c. We will use the fact that $\mathbf{u} - \mathbf{u} = \mathbf{0}$.

$\langle \mathbf{u}, \mathbf{0} \rangle = \langle \mathbf{u}, \mathbf{u} - \mathbf{u} \rangle = \langle \mathbf{u}, \mathbf{u} \rangle - \langle \mathbf{u}, \mathbf{u} \rangle = 0$. Likewise, for $\langle \mathbf{0}, \mathbf{v} \rangle$.

25. $\langle \mathbf{u} + \mathbf{w}, \mathbf{v} - \mathbf{w} \rangle = \langle \mathbf{u}, \mathbf{v} \rangle - \langle \mathbf{u}, \mathbf{w} \rangle + \langle \mathbf{w}, \mathbf{v} \rangle - \langle \mathbf{w}, \mathbf{w} \rangle = 1 - 5 + 0 - (2)^2 = -8$.

26. $\langle 2\mathbf{v} - \mathbf{w}, 3\mathbf{u} + 2\mathbf{w} \rangle = 6\langle \mathbf{v}, \mathbf{u} \rangle + 4\langle \mathbf{v}, \mathbf{w} \rangle - 3\langle \mathbf{w}, \mathbf{u} \rangle - 2\langle \mathbf{w}, \mathbf{w} \rangle = 6(1) + 4(0) - 3(5) - 2(2^2) = -17$.

27. $\|\mathbf{u} + \mathbf{v}\| = \sqrt{\langle \mathbf{u} + \mathbf{v}, \mathbf{u} + \mathbf{v} \rangle} = \sqrt{\langle \mathbf{u}, \mathbf{u} \rangle + 2\langle \mathbf{u}, \mathbf{v} \rangle + \langle \mathbf{v}, \mathbf{v} \rangle} = \sqrt{(1)^2 + 2(1) + (\sqrt{3})^2} = \sqrt{6}$.

28.
$$\begin{aligned} \|2\mathbf{u} - 3\mathbf{v} + \mathbf{w}\| &= \sqrt{\langle 2\mathbf{u} - 3\mathbf{v} + \mathbf{w}, 2\mathbf{u} - 3\mathbf{v} + \mathbf{w} \rangle} \\ &= \sqrt{4\langle \mathbf{u}, \mathbf{u} \rangle - 12\langle \mathbf{u}, \mathbf{v} \rangle + 4\langle \mathbf{u}, \mathbf{w} \rangle + 9\langle \mathbf{v}, \mathbf{v} \rangle - 6\langle \mathbf{v}, \mathbf{w} \rangle + \langle \mathbf{w}, \mathbf{w} \rangle} \\ &= \sqrt{4(1)^2 - 12(1) + 4(5) + 9(\sqrt{3})^2 - 6(0) + (2)^2} = \sqrt{43}. \end{aligned}$$

29. Will show $\mathbf{u}+\mathbf{v}=\mathbf{w}$ by showing $\|\mathbf{u}+\mathbf{v}-\mathbf{w}\|=0$.

$$\begin{aligned}\|\mathbf{u}+\mathbf{v}-\mathbf{w}\| &= \sqrt{\langle \mathbf{u}+\mathbf{v}-\mathbf{w}, \mathbf{u}+\mathbf{v}-\mathbf{w}\rangle} \\ &= \sqrt{\langle \mathbf{u},\mathbf{u}\rangle + 2\langle \mathbf{u},\mathbf{v}\rangle - 2\langle \mathbf{u},\mathbf{w}\rangle + \langle \mathbf{v},\mathbf{v}\rangle - 2\langle \mathbf{v},\mathbf{w}\rangle + \langle \mathbf{w},\mathbf{w}\rangle} \\ &= \sqrt{(1)^2 + 2(1) - 2(5) + (\sqrt{3})^2 - 2(0) + (2)^2} = 0.\end{aligned}$$

30. Note $\|\mathbf{u}\| = \|\mathbf{v}\| = 1 \Rightarrow \langle \mathbf{u},\mathbf{u}\rangle = \langle \mathbf{v},\mathbf{v}\rangle = \|\mathbf{u}\|^2 = \|\mathbf{v}\|^2 = 1$.

$\langle \mathbf{u}+\mathbf{v},\mathbf{u}+\mathbf{v}\rangle = \langle \mathbf{u},\mathbf{u}\rangle + 2\langle \mathbf{u},\mathbf{v}\rangle + \langle \mathbf{v},\mathbf{v}\rangle \geq 0 \Rightarrow 2+2\langle \mathbf{u},\mathbf{v}\rangle \geq 0 \Rightarrow 2\langle \mathbf{u},\mathbf{v}\rangle \geq -2 \Rightarrow \langle \mathbf{u},\mathbf{v}\rangle \geq -1$.

31. $\langle \mathbf{u}+\mathbf{v},\mathbf{u}-\mathbf{v}\rangle = \langle \mathbf{u},\mathbf{u}\rangle + \langle \mathbf{u},\mathbf{v}\rangle - \langle \mathbf{v},\mathbf{u}\rangle + \langle \mathbf{v},\mathbf{v}\rangle = \|\mathbf{u}\|^2 - \|\mathbf{v}\|^2$.

32. $\langle \mathbf{u}+\mathbf{v},\mathbf{u}+\mathbf{v}\rangle = \langle \mathbf{u},\mathbf{u}\rangle + \langle \mathbf{u},\mathbf{v}\rangle + \langle \mathbf{v},\mathbf{u}\rangle + \langle \mathbf{v},\mathbf{v}\rangle = \|\mathbf{u}\|^2 + 2\langle \mathbf{u},\mathbf{v}\rangle + \|\mathbf{v}\|^2$.

33.
$$\begin{array}{rl} \frac{1}{2}\|\mathbf{u}+\mathbf{v}\|^2 & = \frac{1}{2}\|\mathbf{u}\|^2 + \langle \mathbf{u},\mathbf{v}\rangle + \frac{1}{2}\|\mathbf{v}\|^2 \\ +\frac{1}{2}\|\mathbf{u}-\mathbf{v}\|^2 & = \frac{1}{2}\|\mathbf{u}\|^2 - \langle \mathbf{u},\mathbf{v}\rangle + \frac{1}{2}\|\mathbf{v}\|^2 \\ \hline & = \|\mathbf{u}\|^2 + \|\mathbf{v}\|^2 \end{array}$$

34.
$$\begin{array}{rl} \frac{1}{4}\|\mathbf{u}+\mathbf{v}\|^2 & = \frac{1}{4}\|\mathbf{u}\|^2 + \frac{1}{2}\langle \mathbf{u},\mathbf{v}\rangle + \frac{1}{4}\|\mathbf{v}\|^2 \\ -\frac{1}{4}\|\mathbf{u}-\mathbf{v}\|^2 & = -\frac{1}{4}\|\mathbf{u}\|^2 + \frac{1}{2}\langle \mathbf{u},\mathbf{v}\rangle - \frac{1}{4}\|\mathbf{v}\|^2 \\ \hline & = \langle \mathbf{u},\mathbf{v}\rangle \end{array}$$

35. $\|\mathbf{u}+\mathbf{v}\| = \|\mathbf{u}-\mathbf{v}\| \Leftrightarrow \|\mathbf{u}+\mathbf{v}\|^2 = \|\mathbf{u}-\mathbf{v}\|^2 \Leftrightarrow \|\mathbf{u}\|^2+2\langle \mathbf{u},\mathbf{v}\rangle+\|\mathbf{v}\|^2 = \|\mathbf{u}\|^2-2\langle \mathbf{u},\mathbf{v}\rangle+\|\mathbf{v}\|^2 \Leftrightarrow 2\langle \mathbf{u},\mathbf{v}\rangle = -2\langle \mathbf{u},\mathbf{v}\rangle \Leftrightarrow 4\langle \mathbf{u},\mathbf{v}\rangle = 0 \Leftrightarrow \langle \mathbf{u},\mathbf{v}\rangle = 0$ as we needed to show.

Recall, $\mathbf{u}$ and $\mathbf{v}$ are orthogonal means $\langle \mathbf{u},\mathbf{v}\rangle = 0$.

36. $d(\mathbf{u},\mathbf{v}) = \|\mathbf{u}-\mathbf{v}\| = \sqrt{\|\mathbf{u}\|^2+\|\mathbf{v}\|^2} \Leftrightarrow \|\mathbf{u}-\mathbf{v}\|^2 = \|\mathbf{u}\|^2+\|\mathbf{v}\|^2 \Leftrightarrow \|\mathbf{u}\|^2 - 2\langle \mathbf{u},\mathbf{v}\rangle + \|\mathbf{v}\|^2 = \|\mathbf{u}\|^2+\|\mathbf{v}\|^2 \Leftrightarrow -2\langle \mathbf{u},\mathbf{v}\rangle = 0 \Leftrightarrow \langle \mathbf{u},\mathbf{v}\rangle = 0$ as we needed to show.

Recall, $\mathbf{u}$ and $\mathbf{v}$ are orthogonal means $\langle \mathbf{u},\mathbf{v}\rangle = 0$.

37. Note, $\langle \mathbf{v}_1,\mathbf{v}_1\rangle = 2(1)^2+3(0)^2 = 2$ and $\langle \mathbf{v}_1,\mathbf{x}_2\rangle = 2(1)+3(1)(0) = 2 \Rightarrow \frac{\langle \mathbf{v}_1,\mathbf{x}_2\rangle}{\langle \mathbf{v}_1,\mathbf{v}_1\rangle} = \frac{2}{2} = 1$.

$\mathbf{v}_2 = \mathbf{x}_2 - \frac{\langle \mathbf{v}_1,\mathbf{x}_2\rangle}{\langle \mathbf{v}_1,\mathbf{v}_1\rangle}\mathbf{v}_1 = \mathbf{x}_2 - \mathbf{v}_1 = \begin{bmatrix}1\\1\end{bmatrix} - \begin{bmatrix}1\\0\end{bmatrix} = \begin{bmatrix}0\\1\end{bmatrix}$.

38. $\langle \mathbf{v}_1,\mathbf{v}_1\rangle = [1\ \ 0]\begin{bmatrix}4 & -2\\-2 & 7\end{bmatrix}\begin{bmatrix}1\\0\end{bmatrix} = 4$, $\langle \mathbf{v}_1,\mathbf{x}_2\rangle = [1\ \ 0]\begin{bmatrix}4 & -2\\-2 & 7\end{bmatrix}\begin{bmatrix}1\\1\end{bmatrix} = 2 \Rightarrow \frac{\langle \mathbf{v}_1,\mathbf{x}_2\rangle}{\langle \mathbf{v}_1,\mathbf{v}_1\rangle} = \frac{1}{2}$.

$\mathbf{v}_2 = \mathbf{x}_2 - \frac{\langle \mathbf{v}_1,\mathbf{x}_2\rangle}{\langle \mathbf{v}_1,\mathbf{v}_1\rangle}\mathbf{v}_1 = \mathbf{x}_2 - \frac{1}{2}\mathbf{v}_1 = \begin{bmatrix}1\\1\end{bmatrix} - \frac{1}{2}\begin{bmatrix}1\\0\end{bmatrix} = \begin{bmatrix}1/2\\1\end{bmatrix}$.

39. $\mathcal{B} = \{\mathbf{v}_1 = 1, \mathbf{x}_2 = 1+x, \mathbf{x}_3 = 1+x+x^2\}$ and $\langle p(x), q(x)\rangle = \sum a_i b_i \Rightarrow$

$\langle \mathbf{v}_1,\mathbf{v}_1\rangle = 1$, $\langle \mathbf{v}_1,\mathbf{x}_2\rangle = 1 \Rightarrow \frac{\langle \mathbf{v}_1,\mathbf{x}_2\rangle}{\langle \mathbf{v}_1,\mathbf{v}_1\rangle} = 1$.

$\mathbf{v}_2 = \mathbf{x}_2 - \frac{\langle \mathbf{v}_1,\mathbf{x}_2\rangle}{\langle \mathbf{v}_1,\mathbf{v}_1\rangle}\mathbf{v}_1 = \mathbf{x}_2 - \mathbf{v}_1 = (1+x)-1 = x$.

$\langle \mathbf{v}_1,\mathbf{x}_3\rangle = 1$, $\langle \mathbf{v}_2,\mathbf{v}_2\rangle = 1$, $\langle \mathbf{v}_2,\mathbf{x}_3\rangle = 1 \Rightarrow \frac{\langle \mathbf{v}_1,\mathbf{x}_3\rangle}{\langle \mathbf{v}_1,\mathbf{v}_1\rangle} = 1$ and $\frac{\langle \mathbf{v}_2,\mathbf{x}_3\rangle}{\langle \mathbf{v}_2,\mathbf{v}_2\rangle} = 1 \Rightarrow$

$\mathbf{v}_3 = \mathbf{x}_3 - \frac{\langle \mathbf{v}_1,\mathbf{x}_3\rangle}{\langle \mathbf{v}_1,\mathbf{v}_1\rangle}\mathbf{v}_1 - \frac{\langle \mathbf{v}_2,\mathbf{x}_3\rangle}{\langle \mathbf{v}_2,\mathbf{v}_2\rangle}\mathbf{v}_2 = \mathbf{x}_3 - \mathbf{v}_1 - \mathbf{v}_2 = (1+x+x^2)-1-x = x^2 \Rightarrow \mathcal{C} = \{1, x, x^2\}$.

40. $\mathcal{B} = \{1, 1+x, 1+x+x^2\}$ and $\langle p(x), q(x)\rangle = \int pq\,dx$ limits of integration 0 and 1.

$\langle \mathbf{v}_1, \mathbf{v}_1\rangle = \int_0^1 1^2\,dx = 1$, $\langle \mathbf{v}_1, \mathbf{x}_2\rangle = \int 1(1+x)\,dx = \frac{3}{2} \Rightarrow \frac{\langle \mathbf{v}_1, \mathbf{x}_2\rangle}{\langle \mathbf{v}_1, \mathbf{v}_1\rangle} = \frac{3}{2}$.

$\mathbf{v}_2 = \mathbf{x}_2 - \frac{\langle \mathbf{v}_1, \mathbf{x}_2\rangle}{\langle \mathbf{v}_1, \mathbf{v}_1\rangle}\mathbf{v}_1 = \mathbf{x}_2 - \frac{3}{2}\mathbf{v}_1 = (1+x) - \frac{3}{2} = -\frac{1}{2} + x$.

$\langle \mathbf{v}_1, \mathbf{x}_3\rangle = \int_0^1 (1+x+x^2)\,dx = \frac{11}{6}$, $\langle \mathbf{v}_2, \mathbf{v}_2\rangle = \int_0^1 (-\frac{1}{2}+x)^2\,dx = \frac{1}{12}$,

$\langle \mathbf{v}_2, \mathbf{x}_3\rangle = \int (-\frac{1}{2}+x)(1+x+x^2)\,dx = \frac{1}{6} \Rightarrow \frac{\langle \mathbf{v}_1, \mathbf{x}_3\rangle}{\langle \mathbf{v}_1, \mathbf{v}_1\rangle} = \frac{11}{6}$ and $\frac{\langle \mathbf{v}_2, \mathbf{x}_3\rangle}{\langle \mathbf{v}_2, \mathbf{v}_2\rangle} = 2 \Rightarrow$

$\mathbf{v}_3 = \mathbf{x}_3 - \frac{\langle \mathbf{v}_1, \mathbf{x}_3\rangle}{\langle \mathbf{v}_1, \mathbf{v}_1\rangle}\mathbf{v}_1 - \frac{\langle \mathbf{v}_2, \mathbf{x}_3\rangle}{\langle \mathbf{v}_2, \mathbf{v}_2\rangle}\mathbf{v}_2 = (1+x+x^2) - \frac{11}{6}(1) - 2(-\frac{1}{2}+x) = \frac{1}{6} - x + x^2 \Rightarrow$

$\mathcal{C} = \{1, -\frac{1}{2}+x, \frac{1}{6} - x + x^2\}$.

41. Note that the limits of integration are -1 and 1. We will find $l_i(x)$ by dividing by the norms.

(a) $\|1\| = \sqrt{\int_{-1}^1 1^2\,dx} = \sqrt{2} \Rightarrow l_0(x) = \frac{1}{\sqrt{2}}$.

$\|x\| = \sqrt{\int_{-1}^1 x^2\,dx} = \sqrt{\frac{2}{3}} \Rightarrow l_1(x) = \frac{\sqrt{3}\,x}{\sqrt{2}}$.

$\|x^2 - \frac{1}{3}\| = \sqrt{\int_{-1}^1 (x^2 - \frac{1}{3})^2\,dx} = \sqrt{\frac{8}{45}} = \frac{2\sqrt{2}}{3\sqrt{5}} \Rightarrow l_2(x) = \frac{\sqrt{5}}{2\sqrt{2}}(3x^2 - 1)$.

(b) $\langle \mathbf{v}_1, \mathbf{x}_4\rangle = \int_{-1}^1 1(x^3)\,dx = 0$, $\langle \mathbf{v}_2, \mathbf{v}_4\rangle = \int_{-1}^1 x(x^3)\,dx = \frac{2}{5}$, $\langle \mathbf{v}_2, \mathbf{v}_2\rangle = \int_{-1}^1 x^2\,dx = \frac{2}{3}$,

$\langle \mathbf{v}_3, \mathbf{x}_4\rangle = \int_{-1}^1 (x^2 - \frac{1}{3})x^3\,dx = 0 \Rightarrow$

$\mathbf{v}_4 = \mathbf{x}_3 - \frac{\langle \mathbf{v}_2, \mathbf{x}_4\rangle}{\langle \mathbf{v}_2, \mathbf{v}_2\rangle}\mathbf{v}_2 = \mathbf{x}_3 - \frac{2/5}{2/3}\mathbf{v}_2 = \mathbf{x}_3 - \frac{3}{5}\mathbf{v}_2 = x^3 - \frac{3}{5}x \Rightarrow$

$\|x^3 - \frac{3}{5}x\| = \sqrt{\int_{-1}^1 (x^3 - \frac{3}{5}x)^2\,dx} = \sqrt{\frac{8}{175}} = \frac{2\sqrt{2}}{5\sqrt{7}} \Rightarrow l_4(x) = \frac{\sqrt{7}}{2\sqrt{2}}(5x^3 - 3x)$.

42. We use the recurrence relation as stated in the text.

(a) In the following, we use the $l_i(x)$ found in Exercise 41.

$l_0(1) = \frac{1}{\sqrt{2}} \Rightarrow L_0(x) = \frac{\sqrt{2}}{1}\,l_0(x) = 1$.

$l_1(1) = \sqrt{\frac{3}{2}} \Rightarrow L_1(x) = \sqrt{\frac{2}{3}}\,l_1(x) = x$.

$l_2(1) = \frac{\sqrt{5}}{2\sqrt{2}}(3-1) = \sqrt{\frac{5}{2}} \Rightarrow L_2(x) = \sqrt{\frac{2}{5}}\,l_2(x) = \frac{1}{2}(3x^2 - 1) = \frac{3}{2}x^2 - \frac{1}{2}$.

$l_3(1) = \frac{\sqrt{7}}{2\sqrt{2}}(5-3) = \sqrt{\frac{7}{2}} \Rightarrow L_3(x) = \sqrt{\frac{2}{7}}\,l_3(x) = \frac{1}{2}(5x^3 - 3x) = \frac{5}{2}x^3 - \frac{3}{2}x$.

(b) $L_2(x) = \frac{3}{2}L_1(x)\cdot x - \frac{1}{2}L_0(x) = \frac{3}{2}x\cdot x - \frac{1}{2} = \frac{3}{2}x^2 - \frac{1}{2}$.

$L_3(x) = \frac{5}{3}L_2(x)\cdot x - \frac{2}{3}L_1(x) = \frac{5}{3}\left(\frac{3}{2}x^2 - \frac{1}{2}\right)\cdot x - \frac{2}{3}x = \frac{5}{2}x^3 - \frac{3}{2}x$.

$L_4(x) = \frac{7}{4}L_3(x)\cdot x - \frac{3}{4}L_2(x) = \frac{7}{4}\left(\frac{5}{2}x^3 - \frac{3}{2}x\right)\cdot x - \frac{3}{4}\left(\frac{3}{2}x^2 - \frac{1}{2}\right) = \frac{35}{8}x^4 - \frac{15}{4}x^2 + \frac{3}{8}$.

$L_5(x) = \frac{9}{5}L_4(x)\cdot x - \frac{4}{5}L_3(x) = \frac{63}{8}x^5 - \frac{35}{4}x^3 + \frac{33}{5}x$.

43. $\mathcal{B} = \{\mathbf{u}_i\}$ orthogonal basis $\Rightarrow \langle \text{proj}_W(\mathbf{v}), \mathbf{u}_j\rangle = \langle \sum \frac{\langle \mathbf{u}_i, \mathbf{v}\rangle}{\langle \mathbf{u}_i, \mathbf{u}_i\rangle}\mathbf{u}_i, \mathbf{u}_j\rangle = \langle \frac{\langle \mathbf{u}_j, \mathbf{v}\rangle}{\langle \mathbf{u}_j, \mathbf{u}_j\rangle}\mathbf{u}_j, \mathbf{u}_j\rangle = \langle \mathbf{u}_j, \mathbf{v}\rangle \Rightarrow$

$\langle \text{perp}_W(\mathbf{v}), \mathbf{u}_j\rangle = \langle \mathbf{v} - \text{proj}_W(\mathbf{v}), \mathbf{u}_j\rangle = \langle \mathbf{v}, \mathbf{u}_j\rangle - \langle \text{proj}_W(\mathbf{v}), \mathbf{u}_j\rangle = \langle \mathbf{v}, \mathbf{u}_j\rangle - \langle \mathbf{u}_j, \mathbf{v}\rangle = 0 \Rightarrow$

$\text{perp}_W(\mathbf{v})$ is orthogonal to every vector in $\mathcal{B} \Rightarrow \text{perp}_W(\mathbf{v})$ is orthogonal to every $\mathbf{w}$ in W.

44. (a) $\langle t\mathbf{u}+\mathbf{v}, t\mathbf{u}+\mathbf{v}\rangle = t^2\langle \mathbf{u},\mathbf{u}\rangle + 2t\langle \mathbf{u},\mathbf{v}\rangle + \langle \mathbf{v},\mathbf{v}\rangle = \|\mathbf{u}\|^2 t^2 + 2\langle \mathbf{u},\mathbf{v}\rangle t + \|\mathbf{v}\|^2 \Rightarrow$
$a = \|\mathbf{u}\|^2$, $b = 2\langle \mathbf{u},\mathbf{v}\rangle$, $c = \|\mathbf{v}\|^2$.

(b) Since $p(t) = at^2 + bt + c$ is an upward opening parabola ($a = \|u\|^2 > 0$), its vertex is its lowest point and occurs when $t = -\frac{b}{2a} \Rightarrow$
$a\left(-\frac{b}{2a}\right)^2 + b\left(-\frac{b}{2a}\right) + c \geq 0 \Rightarrow \frac{b^2-2b^2+4ac}{4a} \geq 0 \Rightarrow 4ac \geq b^2 \Rightarrow \sqrt{ac} \geq \frac{1}{2}|b|$.

(c) From (a), we have $a = \|\mathbf{u}\|^2$, $b = 2\langle \mathbf{u},\mathbf{v}\rangle$, $c = \|\mathbf{v}\|^2$. Substitution $\Rightarrow$
$\sqrt{\|\mathbf{u}\|^2\,\|\mathbf{v}\|^2} \geq \frac{1}{2}|2\langle \mathbf{u},\mathbf{v}\rangle| \Rightarrow \|\mathbf{u}\|\,\|\mathbf{v}\| \geq |\langle \mathbf{u},\mathbf{v}\rangle|$.

Exploration: Vectors and Matrices with Complex Entries

Before we begin, it may be useful to review Appendix C, *Complex Numbers*, on p.650.

1. We need to show $\|\mathbf{v}\| = \sqrt{|v_1^2| + |v_2^2| + \cdots + |v_n^2|}$.

Since $\mathbf{v} = \begin{bmatrix} v_1 \\ v_2 \\ \vdots \\ v_n \end{bmatrix}$ is in $\mathbb{C}^n$, we have the following:

$$\|\mathbf{v}\| = \sqrt{\mathbf{v}\cdot\mathbf{v}} = \sqrt{\overline{v_1}v_1 + \overline{v_2}v_2 + \cdots + \overline{v_n}v_n} \text{ (as defined in this Exploration)}$$
$$= \sqrt{|v_1^2| + |v_2^2| + \cdots + |v_n^2|} \text{ as we were to show.}$$

Q: What basic fact about complex numbers did we use in the last equality?
A: The fact that $\overline{z}z = |z|^2$. See p.652 in Appendix C, *Complex Numbers*.

Q: Is it obvious that $|z|^2 = |z^2|$. If so, prove it.

Q: Do we need this fact to reach the conclusion above?
A: Yes, so prove that it is true.

2. We use the given $\mathbf{u}$ and $\mathbf{v}$ to compute the values below.

(a) $\mathbf{u}\cdot\mathbf{v} = (-i)\cdot(2-3i) + 1\cdot(1+5i) = -2+3i$.
(b) $\|\mathbf{u}\| = \sqrt{(-i)(i) + (1)(1)} = \sqrt{1+1} = \sqrt{2}$.
(c) $\|\mathbf{v}\| = \sqrt{(2+3i)(2-3i) + (1-5i)(1+5i)} = \sqrt{13+26} = \sqrt{39}$.
(d) $d(\mathbf{u},\mathbf{v}) = \|\mathbf{u}-\mathbf{v}\| = \sqrt{(-2-4i)(-2+4i) + (5i)(-5i)} = \sqrt{20+25} = 3\sqrt{5}$.

3. The key facts are $\mathbf{u}\cdot\mathbf{v} = \sum \overline{u_i}v_i$, $\overline{u_i}v_i = u_i\overline{v_i}$, $\overline{u_i}u_i = \|u_i\|^2$, and $\overline{\overline{u_i}} = u_i$.

(a) $\mathbf{u}\cdot\mathbf{v} = \sum_i u_i v_i = \sum u_i\overline{v_i} = \mathbf{v}\cdot\mathbf{u}$ (because $\overline{\overline{u_i}} = u_i$).
(b) $\mathbf{u}\cdot(\mathbf{v}+\mathbf{w}) = \sum \overline{u_i}(v_i + w_i) = \sum \overline{u_i}v_i + \sum \overline{u_i}w_i = \mathbf{u}\cdot\mathbf{v} + \mathbf{u}\cdot\mathbf{w}$.
(c) $(c\mathbf{u})\cdot\mathbf{v} = \sum \overline{cu_i}v_i = \overline{c}\sum \overline{u_i}v_i = \overline{c}(\mathbf{u}\cdot\mathbf{v})$.
(d) $\mathbf{u}\cdot\mathbf{u} = \sum \overline{u_i}u_i = \sum \|u_i\|^2 \geq 0$ and equals zero if and only if all $u_i = 0$.

4. We simply apply the definition, $A^* = \overline{A^T}$.

(a) $A^* = \overline{A^T} = \begin{bmatrix} -i & i \\ -2i & 3 \end{bmatrix}$

(b) $A^* = \overline{A^T} = \begin{bmatrix} 2 & 5-2i \\ 5+2i & -1 \end{bmatrix}$

(c) $A^* = \overline{A^T} = \begin{bmatrix} 2+i & 4 \\ 1-3i & 0 \\ -2 & 3+4i \end{bmatrix}$

(d) $A^* = \overline{A^T} = \begin{bmatrix} -3i & 1+i & 1-i \\ 0 & 4 & 0 \\ 1-i & -i & i \end{bmatrix}$

5. All of these facts follow easily from the properties of complex numbers.

(a) $\overline{\overline{A}} = \left[\overline{\overline{a_{ij}}}\right] = [a_{ij}] = A.$

(b) $\overline{A+B} = \left[\overline{a_{ij}+b_{ij}}\right] = \left[\overline{a_{ij}}\right] + \left[\overline{b_{ij}}\right] = \overline{A} + \overline{B}.$

The rest of the items are proved similarly and, therefore, left to the reader.

6. All of these facts follow easily from the properties proven in Exercise 5.

(a) $(A^*)^* = \left(\overline{A^T}\right)^* = \overline{\left(\overline{A^T}\right)}^T = \left(\overline{\overline{A}}^T\right)^T = \left(A^T\right)^T = A.$

(b) $(A+B)^* = \overline{(A+B)^T} = \overline{A^T} + \overline{B^T} = A^* + B^*.$

(c) $(cA)^* = \overline{(cA)^T} = \overline{c}\overline{A^T} = \overline{c}A^*.$

(d) $(AB)^* = \overline{(AB)^T} = \overline{(B^TA^T)} = \left(\overline{B^T}\right)\left(\overline{A^T}\right) = B^*A^*.$

7. We need to show that $\mathbf{u}\cdot\mathbf{v} = \mathbf{u}^*\mathbf{v}$.

We have: $\mathbf{u}\cdot\mathbf{v} = \sum \overline{u_i}v_i = \begin{bmatrix} \overline{u_1} & \overline{u_2} & \cdots & \overline{u_n} \end{bmatrix} \begin{bmatrix} v_1 \\ v_2 \\ \vdots \\ v_n \end{bmatrix} = \overline{\mathbf{u}^T}\mathbf{v} = \mathbf{u}^*\mathbf{v}.$

8. We need to show that if $A^* = A$, then a_{ii} is a real number for all i.

Note: If c is a complex number such that $\overline{c} = c$, then c is a real number. Why?
Since $\overline{c} = a - ib = a + ib = c \Rightarrow a = a$ and $b = -b \Rightarrow b = 0 \Rightarrow c$ is a real number.

So, given $A^* = A$, we need only show that $\overline{a_{ii}} = a_{ii}$ for all i.

Since $A^* = \overline{A^T} = [\,\overline{a_{ji}}\,] = [\,a_{ij}\,] = A,$

along the diagonal (when $i = j$) we have $\overline{a_{ii}} = a_{ii}$ as required.

Therefore, the diagonal entries of a Hermitian matrix are real.

9. We will use Exercise 8 and the definition of $A^* = A$ to determine if A is Hermitian.

(a) Since the diagonal entries of A are not real (2 and i), A is *not* Hermitian.

(b) Since $A^* = \begin{bmatrix} -1 & 2+3i \\ 2+3i & 5 \end{bmatrix} \neq \begin{bmatrix} -1 & 2-3i \\ 2-3i & 5 \end{bmatrix} = A$, A is *not* Hermitian.

(c) Since $A^* = \begin{bmatrix} -3 & 1+5i \\ -1-5i & 3 \end{bmatrix} \neq \begin{bmatrix} -3 & -1+5i \\ 1-5i & 3 \end{bmatrix} = A$, A is *not* Hermitian.

(d) Since $A^* = A$, A *is* Hermitian.

(e) Since A has all real entries and $A^T \neq A$, A is *not* Hermitian.

(f) Since A has all real entries and $A^T = A$, A *is* Hermitian.

10. Following the hint, we adapt the proof of Theorem 5.18 on p.398.

Given λ is an eigenvalue of an Hermitian matrix A,
we will show λ is a real number. That is, $\overline{\lambda} = \lambda$.

Let $\mathbf{v}$ be an eigenvector of $A (= A^*)$ with corresponding eigenvalue λ.
Since $\mathbf{v} \neq \mathbf{0}$, by Exercise 7 we have $\mathbf{v}^*\mathbf{v} = \mathbf{v} \cdot \mathbf{v} = \|\mathbf{v}\|^2 \neq 0$.

Note: $\mathbf{v}^*A \overset{A^*=A}{=} \mathbf{v}^*A^* \overset{\text{by 6d.}}{=} (A\mathbf{v})^* = (\lambda\mathbf{v})^* \overset{\text{by 6c.}}{=} \overline{\lambda}\mathbf{v}^*$

So: $\lambda(\mathbf{v}^*\mathbf{v}) = \mathbf{v}^*(\lambda\mathbf{v}) \overset{A\mathbf{v}=\lambda\mathbf{v}}{=} \mathbf{v}^*(A\mathbf{v}) = (\mathbf{v}^*A)\mathbf{v} \overset{\text{by above}}{=} (\overline{\lambda}\mathbf{v}^*)\mathbf{v} = \overline{\lambda}(\mathbf{v}^*\mathbf{v})$

The above implies: $\lambda(\mathbf{v}^*\mathbf{v}) - \overline{\lambda}(\mathbf{v}^*\mathbf{v}) = 0$. That is, $(\lambda - \overline{\lambda})(\mathbf{v}^*\mathbf{v}) = 0$.

Since (as noted above) $\mathbf{v}^*\mathbf{v} \neq 0$, $\lambda - \overline{\lambda} = 0$. This implies $\overline{\lambda} = \lambda$ as we were trying to show.

11. Following the hint, we adapt the proof of Theorem 5.19 on p.399.

Given $\lambda_1 \neq \lambda_2$ are eigenvalues of an Hermitian matrix A corresponding to eigenvectors $\mathbf{v}_1$ and $\mathbf{v}_2$ respectively, we will show $\mathbf{v}_1$ and $\mathbf{v}_2$ are orthogonal. That is, $\mathbf{v}_1 \cdot \mathbf{v}_2 = 0$.

So: $\lambda_1(\mathbf{v}_1 \cdot \mathbf{v}_2) = (\lambda_1\mathbf{v}_1) \cdot \mathbf{v}_2 = (A\mathbf{v}_1) \cdot \mathbf{v}_2 \overset{\text{by 7}}{=} (A\mathbf{v}_1)^*\mathbf{v}_2 \overset{\text{by 6d.}}{=} (\mathbf{v}_1^*A^*)\mathbf{v}_2$

Now we use the fact that A is Hermitian to continue:

$$(\mathbf{v}_1^*A^*)\mathbf{v}_2 \overset{A^*=A}{=} (\mathbf{v}_1^*A)\mathbf{v}_2 = \mathbf{v}_1^*(A\mathbf{v}_2) = \mathbf{v}_1^*(\lambda_2\mathbf{v}_2) = \lambda_2(\mathbf{v}_1^*\mathbf{v}_2) = \lambda_2(\mathbf{v}_1 \cdot \mathbf{v}_2)$$

Therefore: $\lambda_1(\mathbf{v}_1 \cdot \mathbf{v}_2) = \lambda_2(\mathbf{v}_1 \cdot \mathbf{v}_2)$ which implies $(\lambda_1 - \lambda_2)(\mathbf{v}_1 \cdot \mathbf{v}_2) = 0$.

Since $\lambda_1 - \lambda_2 \neq 0$ (because $\lambda_1 \neq \lambda_2$), we have $\mathbf{v}_1 \cdot \mathbf{v}_2 = 0$ as were trying to show.

12. We need only check that $AA^* = I$. Why is this sufficient?

(a) Since $AA^* = I$, the matrix A is unitary with $A^* = \begin{bmatrix} -\frac{i}{\sqrt{2}} & -\frac{i}{\sqrt{2}} \\ \frac{i}{\sqrt{2}} & -\frac{i}{\sqrt{2}} \end{bmatrix}$.

(b) Since $AA^* \neq I$, the matrix A is *not* unitary.

(c) Since $AA^* = I$, the matrix A is unitary with $A^* = \begin{bmatrix} \frac{3}{5} & -\frac{4i}{5} \\ -\frac{4}{5} & -\frac{3i}{5} \end{bmatrix}$.

(d) Since $AA^* = I$, the matrix A is unitary with $A^* = \begin{bmatrix} \frac{1-i}{\sqrt{6}} & 0 & \frac{-1+i}{\sqrt{3}} \\ 0 & 1 & 0 \\ \frac{2}{\sqrt{6}} & 0 & \frac{1}{\sqrt{3}} \end{bmatrix}$.

13. Following the hint, we adapt the proofs of Theorems 5.4 through 5.7 on pp.372 – 373.

$a \Rightarrow e$ Since $U^*U = I$, $U\mathbf{x} \cdot U\mathbf{y} = (U\mathbf{x})^*U\mathbf{y} = \mathbf{x}^*U^*U\mathbf{y} = \mathbf{x}^*I\mathbf{y} = \mathbf{x}^*\mathbf{y} = \mathbf{x} \cdot \mathbf{y}$.

$e \Rightarrow d$ Since $U\mathbf{x} \cdot U\mathbf{y} = \mathbf{x} \cdot \mathbf{y}$, $\|U\mathbf{x}\| = \sqrt{U\mathbf{x} \cdot U\mathbf{x}} = \sqrt{\mathbf{x} \cdot \mathbf{x}} = \|\mathbf{x}\|$.

Since the remaining proofs are equally straightforward adaptations, they are left to the reader.

14. In each case, we simply show that the columns of U do (or do not) form an orthonormal set.

(a) Let $A = \begin{bmatrix} \mathbf{a}_1 & \mathbf{a}_2 \end{bmatrix}$, then simply note the following:

$$\mathbf{a}_1 \cdot \mathbf{a}_1 = \left(-\tfrac{i}{\sqrt{2}}\right)\left(\tfrac{i}{\sqrt{2}}\right) + \left(-\tfrac{i}{\sqrt{2}}\right)\left(\tfrac{i}{\sqrt{2}}\right) = \tfrac{1}{2} + \tfrac{1}{2} = 1$$
$$\mathbf{a}_2 \cdot \mathbf{a}_2 = \left(\tfrac{i}{\sqrt{2}}\right)\left(-\tfrac{i}{\sqrt{2}}\right) + \left(-\tfrac{i}{\sqrt{2}}\right)\left(\tfrac{i}{\sqrt{2}}\right) = \tfrac{1}{2} + \tfrac{1}{2} = 1$$
$$\mathbf{a}_1 \cdot \mathbf{a}_2 = \left(-\tfrac{i}{\sqrt{2}}\right)\left(-\tfrac{i}{\sqrt{2}}\right) + \left(-\tfrac{i}{\sqrt{2}}\right)\left(\tfrac{i}{\sqrt{2}}\right) = -\tfrac{1}{2} + \tfrac{1}{2} = 0$$

Therefore, the columns of A form an orthonormal set.

(b) Let $A = \begin{bmatrix} \mathbf{a}_1 & \mathbf{a}_2 \end{bmatrix}$, then simply note the following:

$\mathbf{a}_1 \cdot \mathbf{a}_1 = (1 - i)(1 + i) + (1 + i)(1 - i) = 2 + 2 = 4$

Since column $\mathbf{a}_1$ is not normalized, A is not a unitary matrix.

Since (b) and (c) are similar to (a), they are left to the reader.

15. In each case, we find the eigenvalues, eigenvectors, and normalize to create U.

(a) Let $A = \begin{bmatrix} 2 & i \\ -i & 2 \end{bmatrix}$, then $\lambda_1 = 1$ with $\mathbf{v}_1 = \begin{bmatrix} \frac{1}{\sqrt{2}} \\ \frac{i}{\sqrt{2}} \end{bmatrix}$ and $\lambda_2 = 3$ with $\mathbf{v}_2 = \begin{bmatrix} \frac{1+i}{2} \\ \frac{1-i}{2} \end{bmatrix}$.

So, we have $U = \begin{bmatrix} \frac{1}{\sqrt{2}} & \frac{1+i}{2} \\ \frac{i}{\sqrt{2}} & \frac{1-i}{2} \end{bmatrix}$ and $U^* = \begin{bmatrix} \frac{1}{\sqrt{2}} & -\frac{i}{\sqrt{2}} \\ \frac{1-i}{2} & \frac{1+i}{2} \end{bmatrix}$.

Verify that $A\mathbf{v}_1 = \mathbf{v}_1$, $A\mathbf{v}_2 = 3\mathbf{v}_2$, and $U^*U = I$.

Finally, verify that $U^*AU = D$, where $D = \begin{bmatrix} 1 & 0 \\ 0 & 3 \end{bmatrix}$.

(b) Let $A = \begin{bmatrix} 0 & -1 \\ 1 & 0 \end{bmatrix}$, then $\lambda_1 = i$ with $\mathbf{v}_1 = \begin{bmatrix} \frac{i}{\sqrt{2}} \\ \frac{1}{\sqrt{2}} \end{bmatrix}$ and $\lambda_2 = -i$ with $\mathbf{v}_2 = \begin{bmatrix} \frac{1+i}{2} \\ \frac{-1+i}{2} \end{bmatrix}$.

So, we have $U = \begin{bmatrix} \frac{i}{\sqrt{2}} & \frac{1+i}{2} \\ \frac{1}{\sqrt{2}} & \frac{-1+i}{2} \end{bmatrix}$ and $U^* = \begin{bmatrix} -\frac{i}{\sqrt{2}} & \frac{1}{\sqrt{2}} \\ \frac{1-i}{2} & \frac{-1-i}{2} \end{bmatrix}$.

Verify that $A\mathbf{v}_1 = i\mathbf{v}_1$, $A\mathbf{v}_2 = -i\mathbf{v}_2$, and $U^*U = I$.

Finally, verify that $U^*AU = D$, where $D = \begin{bmatrix} i & 0 \\ 0 & -i \end{bmatrix}$.

Parts (c) and (d) are left to the reader.

16. We will show every Hermitian, every unitary, and every skew-Hermitian matrix is normal.

Hermitian: $A^* = A$. Then $A^*A = AA = AA^*$.

Unitary: $U^*U = I$. Then $U^*U = I = UU^*$. Why is this obvious?

Skew-Hermitian: $A^* = -A$. Then $A^*A = -AA = A(-A) = AA^*$.

For all the matrices M above, we have shown that $M^*M = MM^*$, that is M is normal.

17. If a square complex matrix A is *unitarily diagonalizable*, then $A^*A = AA^*$. So A is normal.

Exploration: Geometric Inequalities and Optimization Problems

1. Since $\mathbf{u} = \begin{bmatrix} \sqrt{x} \\ \sqrt{y} \end{bmatrix}$ and $\mathbf{v} = \begin{bmatrix} \sqrt{y} \\ \sqrt{x} \end{bmatrix}$, $\|\mathbf{u}\| = \|\mathbf{v}\| = \sqrt{(\sqrt{x})^2 + (\sqrt{y})^2} = \sqrt{x+y}$.

Applying Cauchy-Schwarz, $|\langle \mathbf{u}, \mathbf{v} \rangle| \leq \|\mathbf{u}\| \, \|\mathbf{v}\|$, we have: $\left|\sqrt{xy} + \sqrt{xy}\right| \leq \sqrt{x+y}\,\sqrt{x+y}$.

Recall $x, y \geq 0$, so $\sqrt{x+y}\,\sqrt{x+y} = |x+y| = x+y$.

So, we have: $2\sqrt{xy} \leq x + y \Rightarrow \sqrt{xy} \leq \dfrac{x+y}{2}$.

Now, we show that equality holds $\Leftrightarrow x = y$.

$x = y \Rightarrow x = \sqrt{xx} = \dfrac{x+x}{2} = x \Rightarrow$ equality holds.

$\sqrt{xy} = \dfrac{x+y}{2} \Rightarrow 2\sqrt{xy} = x + y \Rightarrow 4xy = (x+y)^2 = x^2 + 2xy + y^2 \Rightarrow$

$x^2 - 2xy + y^2 = (x-y)^2 = 0 \Rightarrow x - y = 0 \Rightarrow x = y$.

2. (a) $0 \leq (x-y)^2 = x^2 - 2xy + y^2 \Rightarrow 4xy \leq x^2 + 2xy + y^2 = (x+y)^2 \Rightarrow$

$2\sqrt{xy} \leq x + y \Rightarrow \sqrt{xy} \leq \dfrac{x+y}{2}$. Note: $(x-y)^2 = 0 \Leftrightarrow x = y$.

(b) Let $DC = h$. By similar triangles, $\dfrac{y}{h} = \dfrac{h}{x} \Rightarrow xy = h^2 \Rightarrow DC = h = \sqrt{xy}$.

From Figure 7.4, we have: $DC = \sqrt{xy} \leq r = \dfrac{x+y}{2}$ and $DC = r \Leftrightarrow x = y$.

3. Since the area $A = xy = 100$, we have: $10 = \sqrt{xy} \leq \dfrac{x+y}{2} \Rightarrow 40 \leq 2(x+y) = P$

with equality holding $\Leftrightarrow x = y \Rightarrow 40 = 4x \Rightarrow x = y = 10 \Rightarrow$

The square is the rectangle with the smallest perimeter, namely 40.

This follows from the fact that equality holds $\Leftrightarrow x = y$. That is, it is independent of the area.

4. The key is to let $\dfrac{1}{x}$ play the role of y in the Cauchy-Schwarz Inequality. So, we have:

$2\sqrt{x\dfrac{1}{x}} \leq x + \dfrac{1}{x}$ with equality holding $\Leftrightarrow x = \dfrac{1}{x} \Rightarrow$

$x^2 = 1 \Rightarrow x = 1 \Rightarrow f(x) = 2\sqrt{1 \cdot \dfrac{1}{1}} = 2$ is the minimum value of $f(x)$ for $x > 0$.

5. The formula for the volume described is $V = x(10 - 2x)x$. So, we have:

$\sqrt[3]{V} \leq \dfrac{x + (10 - 2x) + x}{3}$ with equality holding $\Leftrightarrow x = 10 - 2x = x \Rightarrow$

$3x = 10 \Rightarrow x = \dfrac{10}{3} \Rightarrow 10 - 2x = \dfrac{10}{3} \Rightarrow$

The dimensions $\dfrac{10}{3} \times \dfrac{10}{3} \times \dfrac{10}{3}$ make the volume of the box as large as possible.

6. Since $f(x,y,z) = (x+y)(y+z)(z+x)$, we have:
$\sqrt[3]{f} \le \dfrac{(x+y)+(y+z)+(z+x)}{3}$ with equality holding $\Leftrightarrow x+y = y+z = z+x \Rightarrow$
$x = y = z$. Therefore, since $xyz = 1$, we have $x = y = z = 1 \Rightarrow$
So, $f(x,y,z) = (1+1)(1+1)(1+1) = 8$ is the minimum value of f when $xyz = 1$.
Note, we could compute this directly from the inequality as follows:
$f = \left(\dfrac{(1+1)+(1+1)+(1+1)}{3}\right)^3 = 2^3 = 8.$

7. We use the substitution $u = x - y \Rightarrow x + \dfrac{8}{y(x-y)} = u + y + \dfrac{8}{uy} \Rightarrow$

We have an AMGM of the form given in Example 7.10, namely: $\sqrt[3]{xyz} \le \dfrac{x+y+z}{3} \Rightarrow$
$\sqrt[3]{uy\,\dfrac{8}{uy}} \le \dfrac{u+y+(8/uy)}{3} \Rightarrow 6 = 3\sqrt[3]{8} \le u + y + \dfrac{8}{uy} = x + \dfrac{8}{y(x-y)}$
with equality holding $\Leftrightarrow u = y = \dfrac{8}{uy} \Rightarrow y = \dfrac{8}{y^2} \Rightarrow y = 2 = u \Rightarrow x = u + y = 4.$
So, the minimum value is 6 and it occurs when $x = 4$ and $y = 2$.

8. Following the method of Example 7.11, we have: $\mathbf{u} = \begin{bmatrix} 1 \\ \sqrt{2} \\ 4 \end{bmatrix}$ and $\mathbf{v} = \begin{bmatrix} x \\ \sqrt{2}\,y \\ z \end{bmatrix}$.

Then the componentwise form of the Cauchy-Schwarz Inequality gives
$(x + 2y + 4z)^2 \le (1^2 + (\sqrt{2})^2 + 4^2)(x^2 + 2y^2 + z^2) = 19$
Thus, the maximum value of our function is $\sqrt{19}$ because the maximum occurs at equality.

9. Following the method of Example 7.11, we have: $\mathbf{u} = \begin{bmatrix} 1 \\ 1 \\ \sqrt{2} \end{bmatrix}$ and $\mathbf{v} = \begin{bmatrix} x \\ y \\ \dfrac{1}{\sqrt{2}}z \end{bmatrix}$.

Then the componentwise form of the Cauchy-Schwarz Inequality gives
$(x + y + z)^2 \le (1^2 + 1^2 + (\sqrt{2})^2)(x^2 + y^2 + \frac{1}{2}z^2)$
So, since $x + y + z = 10$, we have:
$100 \le 4(x^2 + y^2 + \frac{1}{2}z^2) \Rightarrow 25 \le x^2 + y^2 + \frac{1}{2}z^2 \Rightarrow$
The minimum value of our function is 25 because the minimum occurs at equality.

10. Following the method of Example 7.11, we have: $\mathbf{u} = \begin{bmatrix} 1 \\ 1 \end{bmatrix}$ and $\mathbf{v} = \begin{bmatrix} \sin\theta \\ \cos\theta \end{bmatrix}$

because we have the constraint $\sin^2\theta + \cos^2\theta = 1$.

Then the componentwise form of the Cauchy-Schwarz Inequality gives

$(\sin\theta + \cos\theta)^2 \le (1^2 + 1^2)(\sin^2\theta + \cos^2\theta)$

So, since $\sin^2\theta + \cos^2\theta = 1$, we have:

$(\sin\theta + \cos\theta)^2 \le 2 \Rightarrow \sin\theta + \cos\theta \le \sqrt{2} \Rightarrow$

The maximum value of our function is $\sqrt{2}$ because the maximum occurs at equality.

11. Following the method of Example 7.11, we have: $\mathbf{u} = \begin{bmatrix} 1 \\ 2 \end{bmatrix}$ and $\mathbf{v} = \begin{bmatrix} x \\ y \end{bmatrix}$

because we have the constraint $x + 2y = 5$.

The function we want to minimize is $x^2 + y^2$, the distance from the origin.

Then the componentwise form of the Cauchy-Schwarz Inequality gives

$(x + 2y)^2 \le (1^2 + 2^2)(x^2 + y^2)$

So, since $x + 2y = 5 \Rightarrow y = \frac{1}{2}(5 - x)$, we have:

$25 \le 5(x^2 + y^2) \Rightarrow 5 \le x^2 + y^2 \Rightarrow 5 \le x^2 + \frac{1}{4}(5 - x)^2$

Now, since the minimum occurs at equality, we solve the equality:

$4x^2 + x^2 - 10x + 25 = 20 \Rightarrow 5x^2 - 10x + 5 = 0 \Rightarrow x^2 - 2x + 1 = 0 \Rightarrow (x - 1)^2 = 0 \Rightarrow x = 1 \Rightarrow$

The point on the line $x + 2y = 5$ closest to the origin is (1, 2).

12. To show, $\sqrt{xy} \ge \dfrac{2}{1/x + 1/y}$, we will apply AMGM to $x' = 1/x$ and $y' = 1/y \Rightarrow$

$$\frac{1/x + 1/y}{2} \ge \sqrt{\frac{1}{x} \cdot \frac{1}{y}} = \frac{1}{\sqrt{xy}} \Rightarrow \sqrt{xy} \ge \frac{2}{1/x + 1/y}.$$

The last step is an application of the following: When $a,b > 0$, if $a > b$, then $\frac{1}{b} > \frac{1}{a}$.

Since this is an application of AMGM, we have equality $\Leftrightarrow x = y$.

To show, $\sqrt{\dfrac{x^2 + y^2}{2}} \ge \dfrac{x + y}{2}$, we will apply $(x - y)^2 = x^2 - 2xy + y^2 \ge 0 \Rightarrow x^2 + y^2 \ge 2xy$.

We use this fact in Step 2 below.

$$\frac{x^2 + y^2}{2} = \frac{2x^2 + 2y^2}{4} \ge \frac{x^2 + 2xy + y^2}{4} = \left(\frac{x + y}{2}\right)^2 \Rightarrow \sqrt{\frac{x^2 + y^2}{2}} \ge \frac{x + y}{2}.$$

As noted in the list, we do not need to prove the middle inequality AMGM again.

Equality is clear $\Leftrightarrow (x - y)^2 = 0 \Leftrightarrow x = y$.

13. The area of the rectangle is $A = 2xy$, where $x^2 + y^2 = r^2$.

Now from Exercise 12, we have:

$$\sqrt{\frac{x^2+y^2}{2}} \geq \sqrt{xy} \Rightarrow \sqrt{x^2+y^2} \geq \sqrt{2xy} \Rightarrow \sqrt{r^2} = r \geq \sqrt{2xy} = \sqrt{A} \Rightarrow A \leq r^2.$$

So, since the maximum occurs at equality, $A = r^2$ is the maximum area.

14. Following the hint, we have:

$$\frac{(x+y)^2}{xy} = (x+y)(\frac{1}{x} + \frac{1}{y}).$$

So, from Exercise 12, we have:

$$\frac{x+y}{2} \geq \frac{2}{1/x + 1/y} \Rightarrow (x+y)(\frac{1}{x} + \frac{1}{y}) \geq 4.$$

So, since the minimum occurs at equality, 4 is the minimum.

15. Squaring the inequality from Exercise 12, we have:

$$\frac{x^2+y^2}{2} \geq \left(\frac{x+y}{2}\right)^2 \geq \left(\frac{2}{1/x+1/y}\right)^2.$$

So, we have:

$$\frac{\left(x+\frac{1}{x}\right)^2 + \left(y+\frac{1}{y}\right)^2}{2} \geq \left(\frac{\left(x+\frac{1}{x}\right) + \left(y+\frac{1}{y}\right)}{2}\right)^2.$$

Recall, $x + y = 1 \Rightarrow$

$$\left(\frac{\left(x+\frac{1}{x}\right) + \left(y+\frac{1}{y}\right)}{2}\right)^2 = \left(\frac{1}{2} + \frac{\frac{1}{x}+\frac{1}{y}}{2}\right)^2 \geq \left(\frac{1}{2} + \frac{2}{x+y}\right)^2.$$

Again, use the fact that $x + y = 1 \Rightarrow$

$$\left(\frac{1}{2} + \frac{2}{x+y}\right)^2 = \left(\frac{5}{2}\right)^2 = \frac{25}{4} \Rightarrow f(x,y) = \frac{25}{2} \text{ is the minimum when } x+y=1.$$

This equality holds $\Leftrightarrow x = y$. Since $x + y = 1$, the minimum occurs when $x = y = \dfrac{1}{2}$.

7.2 Norms and Distance Functions

1. $\|\mathbf{u}\|_E = \sqrt{(-1)^2 + (4)^2 + (-5)^2} = \sqrt{42}.$
$\|\mathbf{u}\|_s = |-1| + |4| + |-5| = 10.$
$\|\mathbf{u}\|_m = \max\{|-1|, |4|, |-5|\} = 5.$

2. $\|\mathbf{u}\|_E = \sqrt{(2)^2 + (-2)^2 + (0)^2} = \sqrt{8} = 2\sqrt{2}.$
$\|\mathbf{u}\|_s = |2| + |-2| + |0| = 4.$
$\|\mathbf{u}\|_m = \max\{|2|, |-2|, |0|\} = 2.$

3. $\mathrm{d}_E\langle \mathbf{u}, \mathbf{v}\rangle = \sqrt{(-3)^2 + (6)^2 + (-5)^2} = \sqrt{70}.$
$\mathrm{d}_s\langle \mathbf{u}, \mathbf{v}\rangle = |-3| + |6| + |-5| = 14.$
$\mathrm{d}_m\langle \mathbf{u}, \mathbf{v}\rangle = \max\{|-3|, |6|, |-5|\} = 6.$

4. (a) $\mathrm{d}_s\langle \mathbf{u}, \mathbf{v}\rangle$ measures total difference between components.
(b) $\mathrm{d}_m\langle \mathbf{u}, \mathbf{v}\rangle$ measures greatest difference between any pair.

5. $\|\mathbf{u}\|_H = w(\mathbf{u}) =$ number of 1s in $\mathbf{u} = 4.$
$\|\mathbf{v}\|_H = w(\mathbf{v}) =$ number of 1s in $\mathbf{v} = 5.$

6. $\mathrm{d}_H\langle \mathbf{u}, \mathbf{v}\rangle = w(\mathbf{u}-\mathbf{v}) = w(u) + w(\mathbf{v}) - 2$ overlap $= 4 + 5 - 2(2) = 5 =$ number of 1s in $\mathbf{u} - \mathbf{v}$.

7. (a) $\|\mathbf{v}\|_E = \sqrt{\sum v_i^2} = \max\{|v_i|\} \Leftrightarrow v_j = 0, j \neq i, |v_i| = \max \Leftrightarrow$ *non*-max components $= 0$.
(b) $\|\mathbf{v}\|_s = \sum |v_i| = \max\{|v_i|\} \Leftrightarrow v_j = 0, j \neq i, |v_i| = \max \Leftrightarrow$ all *non*-max components $= 0$.
(c) (a) and (b) $\Rightarrow \|\mathbf{v}\|_E = \|\mathbf{v}\|_s = \|\mathbf{v}\|_m = \max\{|v_i|\} \Leftrightarrow$ all *non*-max components $= 0$.

8. (a) $\|\mathbf{u} + \mathbf{v}\|_E = \|\mathbf{u}\|_E + \|\mathbf{v}\|_E \Leftrightarrow \sum(u_i + v_i)^2 = \sum u_i^2 + \sum v_i^2 + 2\sum u_1 v_1 = \sum u_i^2 + \sum v_i^2 \Leftrightarrow 2\sum u_i v_i = 0 \Leftrightarrow \sum u_i v_i = 0 \Leftrightarrow \mathbf{u} \cdot \mathbf{v} = 0.$
(b) $\|\mathbf{u} + \mathbf{v}\|_s = \|\mathbf{u}\|_s + \|\mathbf{v}\|_s \Leftrightarrow \sum |u_i + v_i| = \sum |u_i| + \sum |v_i| \Leftrightarrow u_i v_i \geq 0 \Leftrightarrow \sum u_i v_i \geq 0 \Leftrightarrow \mathbf{u} \cdot \mathbf{v} \geq 0.$
(c) $\|\mathbf{u} + \mathbf{v}\|_m = \|\mathbf{u}\|_m + \|\mathbf{v}\|_m \Leftrightarrow \max\{|u_i + v_i|\} = \max\{|u_i|\} + \max\{|v_i|\} \Leftrightarrow$ there exists k such that $|u_k + v_k| \geq |u_i| + |v_j|$ for all $i, j \Leftrightarrow u_k v_k \geq 0$ for that k.

9. Let $\|\mathbf{v}\|_m = \max\{|v_i|\} = |v_k|$. Then we have: $\|\mathbf{v}\|_E = \sqrt{\sum v_i^2} > \sqrt{v_k^2} = |v_k| = \|\mathbf{v}\|_m$.

10. $\|\mathbf{v}\|_E = \sqrt{\sum v_i^2} \leq \sum \sqrt{v_i^2} = \sum |v_i| = \|\mathbf{v}\|_s$.

11. $\|\mathbf{v}\|_s = \sum |v_i| \leq n \max\{|v_i|\} = n\,\|\mathbf{v}\|_m$.

12. $\|\mathbf{v}\|_E = \sqrt{\sum v_i^2} \leq \sqrt{n \max\{|v_i|^2\}} = \sqrt{n} \max\{|v_i|\} = \sqrt{n}\,\|\mathbf{v}\|_m$.

13. $\|\mathbf{u}\|_s = |\,x\,| + |\,y\,| = 1$ and $\|\mathbf{u}\|_m = \max\{|\,x\,|, |\,y\,|\} = 1 \Rightarrow$ four line segments each.

For $\|\mathbf{u}\|_s = |\,x\,| + |\,y\,| = 1$ we have:

$$\begin{aligned} x + y &= 1; \quad x \geq 0,\ y \geq 0 \\ x - y &= 1; \quad x \geq 0,\ y \leq 0 \\ -x + y &= 1; \quad x \leq 0,\ y \geq 0 \\ -x - y &= 1; \quad x \leq 0,\ y \leq 0 \end{aligned}$$

And for $\|\mathbf{u}\|_m = \max\{|\,x\,|, |\,y\,|\} = 1$ we have:

$$\begin{aligned} x &= 1, \quad |\,y\,| \leq 1 \\ x &= -1, \quad |\,y\,| \leq 1 \\ y &= 1, \quad |\,x\,| \leq 1 \\ y &= -1, \quad |\,x\,| \leq 1 \end{aligned}$$

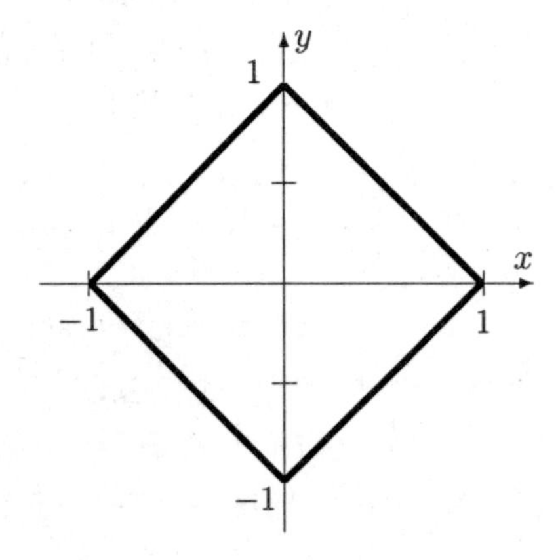

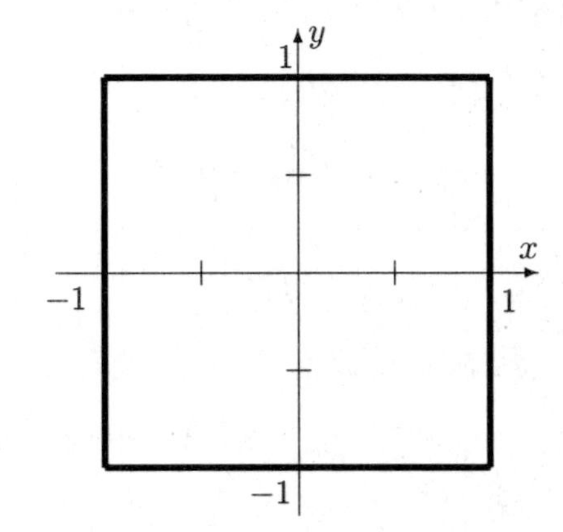

14. Need example that fails $\frac{1}{2}\|\mathbf{u}+\mathbf{v}\|^2 + \frac{1}{2}\|\mathbf{u}-\mathbf{v}\|^2 = \|\mathbf{u}\|^2 + \|\mathbf{v}\|^2$. So, $\mathbf{e}_1 = \begin{bmatrix} 1 \\ 0 \end{bmatrix}$, $\mathbf{e}_2 = \begin{bmatrix} 0 \\ 1 \end{bmatrix}$.

$\frac{1}{2}\|\mathbf{e}_1 + \mathbf{e}_2\|^2 + \frac{1}{2}\|\mathbf{e}_1 - \mathbf{e}_2\|^2 = \frac{1}{2}(2)^2 + \frac{1}{2}(2)^2 = 4 \neq \|\mathbf{e}_1\|^2 + \|\mathbf{e}_2\|^2 = (1)^2 + (1)^2 = 2.$

15.
1. Absolute value in definition $\Rightarrow \geq 0$, $\max\{|2a|, |3b|\} = 0 \Rightarrow a = b = 0 \Rightarrow \mathbf{v} = \mathbf{0}$.
2. $\|c\mathbf{v}\| = \max\{|2ca|, |3cb|\} = |c| \max\{|2a|, |3b|\} = |c|\,\|\mathbf{v}\|$.
3. $\|\mathbf{u}+\mathbf{v}\| = \max\{|2a+2c|, |3b+3d|\} \leq \max\{|2a|+|2c|, |3b|+|3d|\}$
   $\leq \max\{|2a|, |3b|\} + \max\{|2c|, |3d|\} = \|\mathbf{u}\| + \|\mathbf{v}\|$.

16.
1. Absolute value in definition $\Rightarrow \geq 0$, $\max\{|a_{ij}|\} = 0 \Rightarrow a_{ij} = 0 \Rightarrow A = \mathbf{0}$.
2. $\|cA\| = \max\{|ca_{ij}|\} = |c| \max\{|a_{ij}|\} = |c|\,\|A\|$.
3. $\|A+B\| = \max\{|a_{ij}+b_{ij}|\} \leq \max\{|a_{ij}|+|b_{ij}|\}$
   $\leq \max\{|a_{ij}|\} + \max\{|b_{ij}|\} = \|A\| + \|B\|$.

17.
1. Absolute value in definition $\Rightarrow \geq 0$, $\int |f(x)|\,dx = 0 \Rightarrow |f(x)| = 0 \Rightarrow f(x) = \mathbf{0}$.
2. $\|cf\| = \int |cf(x)|\,dx = |c| \int |f(x)|\,dx = |c|\,\|f\|$.
3. $\|f+g\| = \int |f(x)+g(x)|\,dx \leq \int |f(x)|+|g(x)|\,dx = \int |f(x)|\,dx + \int |g(x)|\,dx = \|f\|+\|g\|$.

18.
1. Absolute value in definition $\Rightarrow \geq 0$, $\max|f(x)| = 0 \Rightarrow |f(x)| \leq 0 \Rightarrow f(x) = \mathbf{0}$.
2. $\|cf\| = \max|cf(x)| = |c| \max|f(x)| = |c|\,\|cf\|$.
3. $\|f+g\| = \max|f(x)+g(x)| \leq \max(|f(x)|+|g(x)|) \leq \max|f(x)| + \max|g(x)| = \|f\|+\|g\|$.

19. $\|\mathbf{u}-\mathbf{v}\| = |-1|\,\|\mathbf{u}-\mathbf{v}\| = \|-1(\mathbf{u}-\mathbf{v})\| = \|-\mathbf{u}+\mathbf{v}\| = \|\mathbf{v}-\mathbf{u}\|$.

20. $\|A\|_F = \sqrt{2^2+3^2+4^2+1^2} = \sqrt{30}$.
$\|A\|_1 = \max\{|2|+|4|, |3|+|1|\} = 6$.
$\|A\|_\infty = \max\{|2|+|3|, |4|+|1|\} = 5$.

21. $\|A\|_F = \sqrt{0^2+(-1)^2+(-3)^2+3^2} = \sqrt{19}$.
$\|A\|_1 = \max\{|0|+|-3|, |-1|+|3|\} = 4$.
$\|A\|_\infty = \max\{|0|+|1|, |-3|+|3|\} = 6$.

22. $\|A\|_F = \sqrt{1^2+5^2+2^2+1^2} = \sqrt{31}$.
$\|A\|_1 = \max\{3,6\} = 6$.
$\|A\|_\infty = \max\{6,3\} = 6$.

23. $\|A\|_F = \sqrt{2(2^2)+5(1^2)+2(3^2)} = \sqrt{31}$.
$\|A\|_1 = \max\{4,5,6\} = 6$.
$\|A\|_\infty = \max\{4,6,5\} = 6$.

24. $\|A\|_F = \sqrt{0^2+\ldots+3^2} = \sqrt{89}$.
$\|A\|_1 = \max\{7,10,8\} = 10$.
$\|A\|_\infty = \max\{7,7,11\} = 11$.

25. $\|A\|_F = \sqrt{4^2+\ldots+0^2} = 2\sqrt{11}$.
$\|A\|_1 = \max\{7,6,3\} = 7$.
$\|A\|_\infty = \max\{7,3,6\} = 7$.

26. $\|A\|_1 = \max\|\mathbf{a}_i\|_s = \|\mathbf{a}_1\|_s = \|A\mathbf{x}\|_s$, where $\mathbf{x} = \begin{bmatrix} 1 \\ 0 \end{bmatrix}$ and $\|\mathbf{x}\|_s = 1$ as required.

$\|A\|_\infty = \max\|\mathbf{A}_i\|_s = \|\mathbf{A}_1\|_s = \|A\mathbf{y}\|_m$, where $\mathbf{y} = \begin{bmatrix} 1 \\ 1 \end{bmatrix}$ and $\|\mathbf{y}\|_m = 1$ as required.

27. $\|A\|_1 = \max\|\mathbf{a}_i\|_s = \|\mathbf{a}_2\|_s = \|A\mathbf{x}\|_s$, where $\mathbf{x} = \begin{bmatrix} 0 \\ 1 \end{bmatrix}$ and $\|\mathbf{x}\|_s = 1$ as required.

$\|A\|_\infty = \max\|\mathbf{A}_i\|_s = \|\mathbf{A}_2\|_s = \|A\mathbf{y}\|_m$, where $\mathbf{y} = \begin{bmatrix} -1 \\ 1 \end{bmatrix}$ and $\|\mathbf{y}\|_m = 1$ as required.

28. $\|A\|_1 = \max\|\mathbf{a}_i\|_s = \|\mathbf{a}_2\|_s = \|A\mathbf{x}\|_s$, where $\mathbf{x} = \begin{bmatrix} 0 \\ 1 \end{bmatrix}$ and $\|\mathbf{x}\|_s = 1$ as required.

$\|A\|_\infty = \max\|\mathbf{A}_i\|_s = \|\mathbf{A}_1\|_s = \|A\mathbf{y}\|_m$, where $\mathbf{y} = \begin{bmatrix} 1 \\ 1 \end{bmatrix}$ and $\|\mathbf{y}\|_m = 1$ as required.

29. $\|A\|_1 = \max\|\mathbf{a}_i\|_s = \|\mathbf{a}_3\|_s = \|A\mathbf{x}\|_s$, where $\mathbf{x} = \begin{bmatrix} 0 \\ 0 \\ 1 \end{bmatrix}$ and $\|\mathbf{x}\|_s = 1$ as required.

$\|A\|_\infty = \max\|\mathbf{A}_i\|_s = \|\mathbf{A}_2\|_s = \|A\mathbf{y}\|_m$, where $\mathbf{y} = \begin{bmatrix} 1 \\ 1 \\ 1 \end{bmatrix}$ and $\|\mathbf{y}\|_m = 1$ as required.

30. $\|A\|_1 = \max\|\mathbf{a}_i\|_s = \|\mathbf{a}_2\|_s = \|A\mathbf{x}\|_s$, where $\mathbf{x} = \begin{bmatrix} 0 \\ 1 \\ 0 \end{bmatrix}$ and $\|\mathbf{x}\|_s = 1$ as required.

$\|A\|_\infty = \max\|\mathbf{A}_i\|_s = \|\mathbf{A}_3\|_s = \|A\mathbf{y}\|_m$, where $\mathbf{y} = \begin{bmatrix} -1 \\ -1 \\ 1 \end{bmatrix}$ and $\|\mathbf{y}\|_m = 1$ as required.

31. $\|A\|_1 = \max \|\mathbf{a}_i\|_s = \|\mathbf{a}_1\|_s = \|A\mathbf{x}\|_s$, where $\mathbf{x} = \begin{bmatrix} 1 \\ 0 \\ 0 \end{bmatrix}$ and $\|\mathbf{x}\|_s = 1$ as required.

$\|A\|_\infty = \max \|\mathbf{A}_i\|_s = \|\mathbf{A}_1\|_s = \|A\mathbf{y}\|_m$, where $\mathbf{y} = \begin{bmatrix} 1 \\ -1 \\ -1 \end{bmatrix}$ and $\|\mathbf{y}\|_m = 1$ as required.

32. Let $M = \max\{\|\mathbf{A}_i\|_s\}$ and let $\|\mathbf{x}\|_m = 1$. Then:

$\|A\mathbf{x}\|_m = \max\{\|\mathbf{A}_i\mathbf{x}\|_s\} \le \max\{\|\mathbf{A}_i\|_s\}\max\{|x_i|\} = M\|\mathbf{x}\|_m = M.$

$\|\mathbf{A}_k\|_s = \max\{\|\mathbf{A}_i\|_s\}$, $\mathbf{y} = \begin{cases} y_i = 1 \text{ if } A_{ki} \ge 0 \\ y_i = -1 \text{ if } A_{ki} < 0 \end{cases}$. Then $\|A\mathbf{y}\|_m = \|\mathbf{A}_k\|$ and $\|\mathbf{y}\|_m = 1$.

33. (a) $\|I\| = \max\|I\mathbf{x}\| = \max\|\mathbf{x}\| = 1$ because the max is taken over $\|\mathbf{x}\| = 1$.

(b) $\|I\|_F = \sqrt{\sum 1^2} = \sqrt{n} \Rightarrow$ there is no vector norm that induces the Frobenius norm except when $n = 1$ when it is equivalent to absolute value on the real number line.

34. Let $\mathbf{x}_\lambda$ be an eigenvector for λ, that is $A\mathbf{x}_\lambda = \lambda\mathbf{x}_\lambda$, with $\|\mathbf{x}_\lambda\| = 1$. Then we have:

$\|A\| = \max\|A\mathbf{x}\| \ge \|A\mathbf{x}_\lambda\| = \|\lambda\mathbf{x}_\lambda\| = |\lambda|$. We can choose such an $\mathbf{x}_\lambda$ for every λ.

35. $A^{-1} = \begin{bmatrix} 1 & -\frac{1}{2} \\ -2 & \frac{3}{2} \end{bmatrix} \Rightarrow \begin{matrix} \text{cond}_1(A) = \|A^{-1}\|_1\|A\|_1 = 7 \cdot 3 = 21 \\ \text{cond}_\infty(A) = \|A^{-1}\|_\infty\|A\|_\infty = 6 \cdot \frac{7}{2} = 21 \end{matrix} \Rightarrow A$ is well-conditioned.

36. $\det A = 0 \Rightarrow \text{cond}_1(A) = \text{cond}_\infty(A) = \infty \Rightarrow A$ is ill-conditioned.

37. $A^{-1} = \begin{bmatrix} 100 & -99 \\ -100 & 100 \end{bmatrix} \Rightarrow \begin{matrix} \text{cond}_1(A) = 2 \cdot 200 = 400 \\ \text{cond}_\infty(A) = 2 \cdot 200 = 400 \end{matrix} \Rightarrow A$ is ill-conditioned.

38. $A^{-1} \approx \begin{bmatrix} 40.02 & -2 \\ -30.01 & 1.5 \end{bmatrix} \Rightarrow \begin{matrix} \text{cond}_1(A) \approx 4202 \cdot 70.03 \approx 294,266 \\ \text{cond}_\infty(A) \approx 7003 \cdot 42.02 \approx 294,266 \end{matrix} \Rightarrow A$ is ill-conditioned.

39. $A^{-1} = \begin{bmatrix} 0 & 0 & 1 \\ 6 & -1 & -1 \\ -5 & 1 & 0 \end{bmatrix} \Rightarrow \begin{matrix} \text{cond}_1(A) = 7 \cdot 11 = 77 \\ \text{cond}_\infty(A) = 16 \cdot 8 = 128 \end{matrix} \Rightarrow A$ is moderately ill-conditioned.

40. $A^{-1} = \begin{bmatrix} 9 & -36 & 30 \\ -36 & 192 & -180 \\ 30 & -180 & 180 \end{bmatrix} \Rightarrow \begin{matrix} \text{cond}_1(A) = \frac{11}{6} \cdot 408 = 748 \\ \text{cond}_\infty(A) = \frac{11}{6} \cdot 408 = 748 \end{matrix} \Rightarrow A$ is ill-conditioned.

41. (a) $A^{-1} = \begin{bmatrix} -\frac{1}{k-1} & \frac{k}{k-1} \\ \frac{1}{k-1} & -\frac{1}{k-1} \end{bmatrix} \Rightarrow \text{cond}_\infty(A) = (\max\{|k|+1, 2\})(\max\left\{|\frac{k}{k-1}| + |\frac{1}{k-1}|, |\frac{2}{k-1}|\right\}).$

(b) $k \to 1 \Rightarrow \text{cond}_\infty(A) \to \infty$ as it should since $A \longrightarrow \begin{bmatrix} 1 & 1 \\ 1 & 1 \end{bmatrix}$ which is not invertible.

42. Note $\|AA^{-1}\mathbf{b}\| = \|\mathbf{b}\| \le \|A\|\,\|A^{-1}\mathbf{b}\| \Rightarrow \frac{1}{\|A^{-1}\mathbf{b}\|} \le \frac{\|A\|}{\|\mathbf{b}\|}$.

This fact will be used in the next to the last step of the proof below.

$\frac{\|\Delta\mathbf{x}\|}{\|\mathbf{x}\|} = \frac{\|A^{-1}\Delta\mathbf{b}\|}{\|A^{-1}\mathbf{b}\|} \le \frac{\|A^{-1}\|\,\|\Delta\mathbf{b}\|}{\|A^{-1}\mathbf{b}\|} = \frac{1}{\|A^{-1}\mathbf{b}\|}\|A^{-1}\|\,\|\Delta\mathbf{b}\| \le \frac{\|A\|}{\|\mathbf{b}\|}\|A^{-1}\|\,\|\Delta\mathbf{b}\| \le \text{cond}(A)\,\frac{\|\Delta\mathbf{b}\|}{\|\mathbf{b}\|}.$

43. (a) $A^{-1} = \begin{bmatrix} -\frac{9}{10} & 1 \\ 1 & -1 \end{bmatrix} \Rightarrow \text{cond}_\infty(A) = 20 \cdot 2 = 40.$

(b) $\Delta A = \begin{bmatrix} 10 & 10 \\ 10 & 11 \end{bmatrix} - \begin{bmatrix} 10 & 10 \\ 10 & 9 \end{bmatrix} = \begin{bmatrix} 0 & 0 \\ 0 & 2 \end{bmatrix} \Rightarrow \|\Delta A\| = 2 \Rightarrow$

$\frac{\|\Delta \mathbf{x}\|_m}{\|\mathbf{x}'\|_m} \le \text{cond}_\infty(A)\, \frac{\|\Delta A\|_\infty}{\|A\|_\infty} = 40 \cdot \frac{2}{20} = 4 \Rightarrow$ can produce at most a 400% relative change.

(c) $[\,A\,|\,\mathbf{b}\,] \longrightarrow \left[\begin{array}{cc|c} 1 & 0 & 9 \\ 0 & 1 & 1 \end{array}\right] \Rightarrow \mathbf{x} = \begin{bmatrix} 9 \\ 1 \end{bmatrix}$. $[\,A'\,|\,\mathbf{b}\,] \longrightarrow \left[\begin{array}{cc|c} 1 & 0 & 11 \\ 0 & 1 & -1 \end{array}\right] \Rightarrow \mathbf{x}' = \begin{bmatrix} 11 \\ -1 \end{bmatrix} \Rightarrow$

$\Delta \mathbf{x} = \mathbf{x}' - \mathbf{x} = \begin{bmatrix} 2 \\ 2 \end{bmatrix} \Rightarrow \frac{\|\Delta \mathbf{x}\|_m}{\|\mathbf{x}\|_m} = \frac{2}{9} \approx 22\%$ actual relative error.

(d) $\Delta \mathbf{b} = \begin{bmatrix} 100 \\ 101 \end{bmatrix} - \begin{bmatrix} 100 \\ 99 \end{bmatrix} = \begin{bmatrix} 0 \\ 2 \end{bmatrix} \Rightarrow \|\Delta \mathbf{b}\|_m = 2 \Rightarrow$

$\frac{\|\Delta \mathbf{x}\|_m}{\|\mathbf{x}\|_m} \le \text{cond}_\infty(A)\, \frac{\|\Delta \mathbf{b}\|_m}{\|\mathbf{b}\|_m} = 40 \cdot \frac{2}{100} = 0.8 \Rightarrow$ at most an 80% relative change.

(e) $[\,A\,|\,\mathbf{b}\,] \longrightarrow \left[\begin{array}{cc|c} 1 & 0 & 9 \\ 0 & 1 & 1 \end{array}\right] \Rightarrow \mathbf{x} = \begin{bmatrix} 9 \\ 1 \end{bmatrix}$. $[\,A\,|\,\mathbf{b}'\,] \longrightarrow \left[\begin{array}{cc|c} 1 & 0 & 11 \\ 0 & 1 & -1 \end{array}\right] \Rightarrow \mathbf{x}' = \begin{bmatrix} 11 \\ -1 \end{bmatrix} \Rightarrow$

$\Delta \mathbf{x} = \mathbf{x}' - \mathbf{x} = \begin{bmatrix} 2 \\ 2 \end{bmatrix} \Rightarrow \frac{\|\Delta \mathbf{x}\|_m}{\|\mathbf{x}\|_m} = \frac{2}{9} \approx 22\%$ actual relative error.

44. (a) $A^{-1} = \begin{bmatrix} -10 & 3 & 5 \\ 4 & -1 & -2 \\ 7 & -2 & -3 \end{bmatrix} \Rightarrow \text{cond}_1(A) = 7 \cdot 21 = 147.$

(b) $\Delta A = \begin{bmatrix} 1 & 1 & 1 \\ 1 & 5 & 0 \\ 1 & -1 & 2 \end{bmatrix} - \begin{bmatrix} 1 & 1 & 1 \\ 2 & 5 & 0 \\ 1 & -1 & 2 \end{bmatrix} = \begin{bmatrix} 0 & 0 & 0 \\ -1 & 0 & 0 \\ 0 & 0 & 0 \end{bmatrix} \Rightarrow \|\Delta A\|_1 = 1 \Rightarrow$

$\frac{\|\Delta \mathbf{x}\|_s}{\|\mathbf{x}'\|_s} \le \text{cond}_1(A)\, \frac{\|\Delta A\|_1}{\|A\|_1} = 147 \cdot \frac{1}{7} = 21 \Rightarrow$ can produce at most a 2100% relative change.

(c) $[\,A\,|\,\mathbf{b}\,] \longrightarrow \left[\begin{array}{ccc|c} 1 & 0 & 0 & 11 \\ 0 & 1 & 0 & -4 \\ 0 & 0 & 1 & -6 \end{array}\right] \Rightarrow \mathbf{x} = \begin{bmatrix} 11 \\ -4 \\ -6 \end{bmatrix}$. Likewise, $\mathbf{x}' = \begin{bmatrix} -5.5 \\ 1.5 \\ 5 \end{bmatrix} \Rightarrow$

$\Delta \mathbf{x} = \mathbf{x}' - \mathbf{x} = \begin{bmatrix} -16.5 \\ 5.5 \\ 11 \end{bmatrix} \Rightarrow \frac{\|\Delta \mathbf{x}\|_s}{\|\mathbf{x}\|_s} = \frac{33}{21} \approx 1.57 \Rightarrow 157\%$ actual relative error.

(d) $\Delta \mathbf{b} = \begin{bmatrix} 1 \\ 1 \\ 3 \end{bmatrix} - \begin{bmatrix} 1 \\ 2 \\ 3 \end{bmatrix} = \begin{bmatrix} 0 \\ -1 \\ 0 \end{bmatrix} \Rightarrow \|\Delta \mathbf{b}\|_s = 1 \Rightarrow$

$\frac{\|\Delta \mathbf{x}\|_s}{\|\mathbf{x}\|_s} \le \text{cond}_1(A)\, \frac{\|\Delta \mathbf{b}\|_s}{\|\mathbf{b}\|_s} = 147 \cdot \frac{1}{6} = 24.5 \Rightarrow$ at most a 2450% relative change.

(e) $[\,A\,|\,\mathbf{b}'\,] \longrightarrow \left[\begin{array}{ccc|c} 1 & 0 & 0 & 8 \\ 0 & 1 & 0 & -3 \\ 0 & 0 & 1 & -4 \end{array}\right] \Rightarrow \mathbf{x}' = \begin{bmatrix} 8 \\ -3 \\ -4 \end{bmatrix}$. From above, $\mathbf{x} = \begin{bmatrix} 11 \\ -4 \\ -6 \end{bmatrix} \Rightarrow$

$\Delta \mathbf{x} = \mathbf{x}' - \mathbf{x} = \begin{bmatrix} -3 \\ 1 \\ 2 \end{bmatrix} \Rightarrow \frac{\|\Delta \mathbf{x}\|_s}{\|\mathbf{x}\|_s} = \frac{6}{21} \approx 0.28 \Rightarrow 28\%$ actual relative error.

45. Exercise 33(a) $\Rightarrow 1 = \|I\| = \|A^{-1}A\| \leq \|A^{-1}\|\,\|A\| = \text{cond}(A)$.

46. $\text{cond}(AB) = \|(AB)^{-1}\|\,\|AB\| = \|B^{-1}A^{-1}\|\,\|AB\| \leq \|A^{-1}\|\,\|A\|\,\|B^{-1}\|\,\|B\| = \text{cond}(A)\,\text{cond}(B)$.

47. Section 4.3, Theorem 4.18b $\Rightarrow \lambda$ is an eigenvalue of $A \Leftrightarrow \frac{1}{\lambda}$ is an eigenvalue of A^{-1}.

So, Exercise 34 in Section 4.3 $\Rightarrow \text{cond}(A) = \|A^{-1}\|\,\|A\| \geq |\frac{1}{\lambda_n}|\,|\lambda_1| = \frac{|\lambda_1|}{|\lambda_n|}$.

48. Section 2.5, Exercise 1 and Section 7.2, Example 7.11 $\Rightarrow M = -D^{-1}(L+U)$

$$M = -\begin{bmatrix} 7 & 0 \\ 0 & -5 \end{bmatrix}^{-1} \begin{bmatrix} 0 & -1 \\ 1 & 0 \end{bmatrix} = \begin{bmatrix} -\frac{1}{7} & 0 \\ 0 & \frac{1}{5} \end{bmatrix} \begin{bmatrix} 0 & -1 \\ 1 & 0 \end{bmatrix} = \begin{bmatrix} 0 & \frac{1}{7} \\ \frac{1}{5} & 0 \end{bmatrix} \Rightarrow \|M\|_\infty = \tfrac{1}{5}.$$

$$\mathbf{b} = \begin{bmatrix} 6 \\ -4 \end{bmatrix} \Rightarrow \mathbf{c} = D^{-1}\mathbf{b} = \begin{bmatrix} \frac{1}{7} & 0 \\ 0 & -\frac{1}{5} \end{bmatrix} \begin{bmatrix} 6 \\ -4 \end{bmatrix} = \begin{bmatrix} \frac{6}{7} \\ \frac{4}{5} \end{bmatrix} \Rightarrow \|\mathbf{c}\|_m = \tfrac{6}{7}.$$

Note, $\mathbf{x}_0 = \mathbf{0} \Rightarrow \mathbf{x}_1 = \mathbf{c} \Rightarrow \|M\|_\infty^k\,\|\mathbf{x}_1 - \mathbf{x}_0\|_m = \|M\|_\infty^k\,\|\mathbf{c}\|_m < 0.0005 \Rightarrow k > \frac{4+\log_{10}(\|\mathbf{c}\|_m/5)}{\log_{10}(1/\|M\|_\infty)}$.

So, in this case with $\|M\|_\infty = \frac{1}{5}$ and $\|\mathbf{c}\|_m = \frac{6}{7} \Rightarrow k > \frac{4+\log_{10}(6/35)}{\log_{10} 5} \approx 4.6 \Rightarrow k \geq 5$.

$$\mathbf{x}_{k+1} = M\mathbf{x}_k + \mathbf{c} \Rightarrow \mathbf{x}_5 = \begin{bmatrix} 0.999 \\ 0.999 \end{bmatrix}. \text{ Compare } \mathbf{x} = \mathbf{x}_{actual} = \begin{bmatrix} 1 \\ 1 \end{bmatrix}.$$

49. Section 2.5, Exercise 3 and Section 7.2, Example 7.11 $\Rightarrow M = -D^{-1}(L+U)$

$$M = -\begin{bmatrix} 4.5 & 0 \\ 0 & -3.5 \end{bmatrix}^{-1} \begin{bmatrix} 0 & -0.5 \\ 1 & 0 \end{bmatrix} = \begin{bmatrix} -\frac{2}{9} & 0 \\ 0 & \frac{2}{7} \end{bmatrix} \begin{bmatrix} 0 & -\frac{1}{2} \\ 1 & 0 \end{bmatrix} = \begin{bmatrix} 0 & \frac{1}{9} \\ \frac{2}{7} & 0 \end{bmatrix} \Rightarrow \|M\|_\infty = \tfrac{2}{7}.$$

$$\mathbf{b} = \begin{bmatrix} 1 \\ -1 \end{bmatrix} \Rightarrow \mathbf{c} = D^{-1}\mathbf{b} = \begin{bmatrix} \frac{2}{9} & 0 \\ 0 & -\frac{2}{7} \end{bmatrix} \begin{bmatrix} 1 \\ -1 \end{bmatrix} = \begin{bmatrix} \frac{2}{9} \\ \frac{2}{7} \end{bmatrix} \Rightarrow \|\mathbf{c}\|_m = \tfrac{2}{7}.$$

Note, $\mathbf{x}_0 = \mathbf{0} \Rightarrow \mathbf{x}_1 = \mathbf{c} \Rightarrow \|M\|_\infty^k\,\|\mathbf{x}_1 - \mathbf{x}_0\|_m = \|M\|_\infty^k\,\|\mathbf{c}\|_m < 0.0005 \Rightarrow k > \frac{4+\log_{10}(\|\mathbf{c}\|_m/5)}{\log_{10}(1/\|M\|_\infty)}$.

So, in this case with $\|M\|_\infty = \frac{2}{7}$ and $\|\mathbf{c}\|_m = \frac{2}{7} \Rightarrow k > \frac{4+\log_{10}(2/35)}{\log_{10}(7/2)} \approx 5.06 \Rightarrow k \geq 6$.

$$\mathbf{x}_{k+1} = M\mathbf{x}_k + \mathbf{c} \Rightarrow \mathbf{x}_6 = \begin{bmatrix} 0.2622 \\ 0.3606 \end{bmatrix}. \text{ Compare } \mathbf{x} = \mathbf{x}_{actual} = \begin{bmatrix} 0.2623 \\ 0.3606 \end{bmatrix}.$$

50. $$M = -\begin{bmatrix} 20 & 0 & 0 \\ 0 & -10 & 0 \\ 0 & 0 & 10 \end{bmatrix}^{-1} \begin{bmatrix} 0 & 1 & -1 \\ 1 & 0 & 1 \\ -1 & 1 & 0 \end{bmatrix} = \begin{bmatrix} 0 & -\frac{1}{20} & \frac{1}{20} \\ \frac{1}{10} & 0 & \frac{1}{10} \\ \frac{1}{10} & -\frac{1}{10} & 0 \end{bmatrix} \Rightarrow \|M\|_\infty = \tfrac{2}{10} = \tfrac{1}{5}.$$

$$\mathbf{b} = \begin{bmatrix} 17 \\ 13 \\ 18 \end{bmatrix} \Rightarrow \mathbf{c} = D^{-1}\mathbf{b} = \begin{bmatrix} \frac{2}{9} & 0 & 0 \\ 0 & -\frac{1}{10} & 0 \\ 0 & 0 & \frac{1}{10} \end{bmatrix} \begin{bmatrix} 17 \\ 13 \\ 18 \end{bmatrix} \Rightarrow= \begin{bmatrix} \frac{17}{20} \\ -\frac{13}{20} \\ \frac{9}{5} \end{bmatrix} \Rightarrow \|\mathbf{c}\|_m = \tfrac{9}{5}.$$

In this case with $\|M\|_\infty = \frac{1}{5}$ and $\|\mathbf{c}\|_m = \frac{9}{5} \Rightarrow k > \frac{4+\log_{10}(9/25)}{\log_{10} 5} \approx 5.08 \Rightarrow k \geq 6$.

$$\mathbf{x}_{k+1} = M\mathbf{x}_k + \mathbf{c} \Rightarrow \mathbf{x}_6 = \begin{bmatrix} 0.999 \\ -1.000 \\ 1.999 \end{bmatrix}. \text{ Compare } \mathbf{x} = \mathbf{x}_{actual} = \begin{bmatrix} 1 \\ -1 \\ 2 \end{bmatrix}.$$

51. $M = -\begin{bmatrix} 3 & 0 & 0 \\ 0 & 4 & 0 \\ 0 & 0 & 3 \end{bmatrix}^{-1} \begin{bmatrix} 0 & 1 & 0 \\ 1 & 0 & 1 \\ 0 & 1 & 0 \end{bmatrix} = \begin{bmatrix} 0 & -\frac{1}{3} & 0 \\ -\frac{1}{4} & 0 & -\frac{1}{4} \\ 0 & -\frac{1}{3} & 0 \end{bmatrix} \Rightarrow \|M\|_\infty = \frac{2}{4} = \frac{1}{2}.$

$\mathbf{b} = \begin{bmatrix} 1 \\ 1 \\ 1 \end{bmatrix} \Rightarrow \mathbf{c} = D^{-1}\mathbf{b} = \begin{bmatrix} \frac{1}{3} & 0 & 0 \\ 0 & \frac{1}{4} & 0 \\ 0 & 0 & \frac{1}{3} \end{bmatrix} \begin{bmatrix} 1 \\ 1 \\ 1 \end{bmatrix} = \begin{bmatrix} \frac{1}{3} \\ \frac{1}{4} \\ \frac{1}{3} \end{bmatrix} \Rightarrow \|\mathbf{c}\|_m = \frac{1}{3}.$

In this case with $\|M\|_\infty = \frac{1}{2}$ and $\|\mathbf{c}\|_m = \frac{1}{3} \Rightarrow k > \frac{4+\log_{10}(1/15)}{\log_{10} 2} \approx 9.3 \Rightarrow k \geq 10.$

$\mathbf{x}_{k+1} = M\mathbf{x}_k + \mathbf{c} \Rightarrow \mathbf{x}_{10} = \begin{bmatrix} 0.299 \\ 0.099 \\ 0.299 \end{bmatrix}$. Compare $\mathbf{x} = \mathbf{x}_{actual} = \begin{bmatrix} 0.3 \\ 0.1 \\ 0.3 \end{bmatrix}$.

7.3 Least Squares Approximation

1. $(1,0) \Rightarrow \varepsilon_1 = 0 - (2 \cdot 1 - 2) = 0,$
$(2,1) \Rightarrow \varepsilon_2 = 1 - (2 \cdot 2 - 2) = -1,$
$(3,5) \Rightarrow \varepsilon_3 = 5 - (2 \cdot 3 - 2) = 1 \Rightarrow$
$\|\mathbf{e}\| = \sqrt{0^2 + (-1)^2 + 1^2} = \sqrt{2}.$

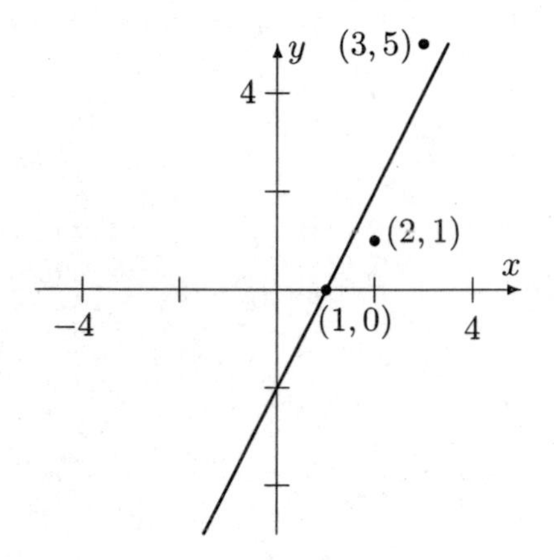

2. $(1,0) \Rightarrow \varepsilon_1 = 0 - (2 \cdot 1 - 3) = 1,$
$(2,1) \Rightarrow \varepsilon_2 = 1 - (2 \cdot 2 - 3) = 0,$
$(3,5) \Rightarrow \varepsilon_3 = 5 - (2 \cdot 3 - 3) = 2 \Rightarrow$
$\|\mathbf{e}\| = \sqrt{1^2 + 0^2 + 2^2} = \sqrt{5}.$

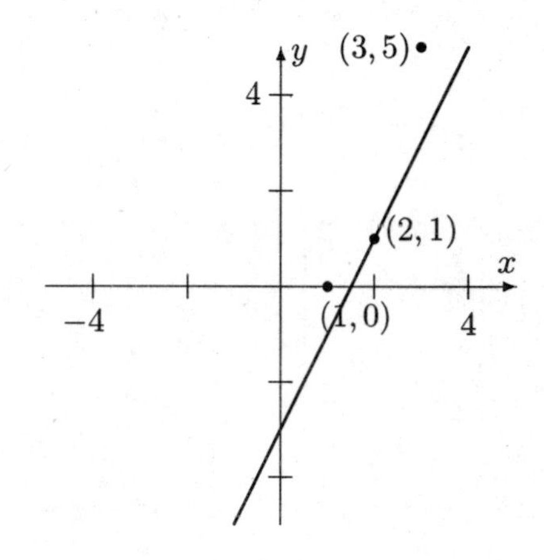

3. $(1,0) \Rightarrow \varepsilon_1 = 0 - (2.5 \cdot 1 - 3) = 0.5,$
$(2,1) \Rightarrow \varepsilon_2 = 1 - (2.5 \cdot 2 - 3) = -1,$
$(3,5) \Rightarrow \varepsilon_3 = 5 - (2.5 \cdot 3 - 3) = 0.5 \Rightarrow$
$\|\mathbf{e}\| = \sqrt{0.5^2 + (-1)^2 + 0.5^2} = \sqrt{6}/2.$

4. $(-5,3) \Rightarrow \varepsilon_1 = 3 - (2 + 5) = -4,$
$(0,3) \Rightarrow \varepsilon_2 = 3 - (2 - 0) = 1,$
$(5,2) \Rightarrow \varepsilon_3 = 2 - (2 - 5) = 5,$
$(10,0) \Rightarrow \varepsilon_4 = 0 - (2 - 10) = 8 \Rightarrow$
$\|\mathbf{e}\| = \sqrt{(-4)^2 + 1^2 + 5^2 + 8^2} = \sqrt{106}.$

5. $(-5,3) \Rightarrow \varepsilon_1 = 3 - 2.5 = 0.5,$
$(0,3) \Rightarrow \varepsilon_2 = 3 - 2.5 = 0.5,$
$(5,2) \Rightarrow \varepsilon_3 = 2 - 2.5 = -0.5,$
$(10,0) \Rightarrow \varepsilon_4 = 0 - 2.5 = 2.5 \Rightarrow$
$\|\mathbf{e}\| = \sqrt{3(0.5)^2 + (2.5)^2} = \sqrt{7}.$

6. $(-5,3) \Rightarrow \varepsilon_1 = 3 - (2 + 0.2 \cdot 5) = 0,$
$(0,3) \Rightarrow \varepsilon_2 = 3 - (2 - 0.2 \cdot 0) = 1,$
$(5,2) \Rightarrow \varepsilon_3 = 2 - (2 - 0.2 \cdot 5) = 1,$
$(10,0) \Rightarrow \varepsilon_4 = 0 - (2 - 0.2 \cdot 10) = 0 \Rightarrow$
$\|\mathbf{e}\| = \sqrt{0^2 + 1^2 + 1^2 + 0^2} = \sqrt{2}.$

7. $\overline{\mathbf{x}} = (A^TA)^{-1}A^T\mathbf{b} = \begin{bmatrix} \frac{7}{3} & -1 \\ -1 & \frac{1}{2} \end{bmatrix} \begin{bmatrix} 1 & 1 & 1 \\ 1 & 2 & 3 \end{bmatrix} \begin{bmatrix} 0 \\ 1 \\ 5 \end{bmatrix} = \begin{bmatrix} -3 \\ \frac{5}{2} \end{bmatrix} \Rightarrow y = -3 + \frac{5}{2}x.$

$$\mathbf{e} = \mathbf{b} - A\overline{\mathbf{x}} = \begin{bmatrix} 0 \\ 1 \\ 5 \end{bmatrix} - \begin{bmatrix} 1 & 1 \\ 1 & 2 \\ 1 & 3 \end{bmatrix} \begin{bmatrix} -3 \\ \frac{5}{2} \end{bmatrix} = \begin{bmatrix} \frac{1}{2} \\ -1 \\ \frac{1}{2} \end{bmatrix} \Rightarrow \|\mathbf{e}\| = \sqrt{(\tfrac{1}{2})^2 + (-1)^2 + (\tfrac{1}{2})^2} = \frac{\sqrt{6}}{2}.$$

8. $\overline{\mathbf{x}} = (A^T A)^{-1} A^T \mathbf{b} = \begin{bmatrix} \frac{7}{3} & -1 \\ -1 & \frac{1}{2} \end{bmatrix} \begin{bmatrix} 1 & 1 & 1 \\ 1 & 2 & 3 \end{bmatrix} \begin{bmatrix} 5 \\ 3 \\ 2 \end{bmatrix} = \begin{bmatrix} \frac{19}{3} \\ -\frac{3}{2} \end{bmatrix} \Rightarrow y = \frac{19}{3} - \frac{3}{2}\,x.$

$$\mathbf{e} = \mathbf{b} - A\overline{\mathbf{x}} = \begin{bmatrix} 5 \\ 3 \\ 2 \end{bmatrix} - \begin{bmatrix} 1 & 1 \\ 1 & 2 \\ 1 & 3 \end{bmatrix} \begin{bmatrix} \frac{19}{3} \\ -\frac{3}{2} \end{bmatrix} = \begin{bmatrix} \frac{1}{6} \\ -\frac{1}{3} \\ \frac{1}{6} \end{bmatrix} \Rightarrow \|\mathbf{e}\| = \sqrt{(\tfrac{1}{6})^2 + (-\tfrac{1}{3})^2 + (\tfrac{1}{6})^2} = \tfrac{\sqrt{6}}{6}.$$

9. $\overline{\mathbf{x}} = (A^T A)^{-1} A^T \mathbf{b} = \begin{bmatrix} \frac{5}{6} & -\frac{1}{2} \\ -\frac{1}{2} & \frac{1}{2} \end{bmatrix} \begin{bmatrix} 0 & 1 & 2 \\ 1 & 1 & 1 \end{bmatrix} \begin{bmatrix} 4 \\ 1 \\ 0 \end{bmatrix} = \begin{bmatrix} \frac{11}{3} \\ -2 \end{bmatrix} \Rightarrow y = \frac{11}{3} - 2x.$

$$\mathbf{e} = \mathbf{b} - A\overline{\mathbf{x}} = \begin{bmatrix} 4 \\ 1 \\ 0 \end{bmatrix} - \begin{bmatrix} 1 & 0 \\ 1 & 1 \\ 1 & 2 \end{bmatrix} \begin{bmatrix} \frac{11}{3} \\ -2 \end{bmatrix} = \begin{bmatrix} \frac{1}{3} \\ -\frac{2}{3} \\ \frac{1}{3} \end{bmatrix} \Rightarrow \|\mathbf{e}\| = \sqrt{(\tfrac{1}{3})^2 + (-\tfrac{2}{3})^2 + (\tfrac{1}{3})^2} = \tfrac{\sqrt{6}}{3}.$$

10. $\overline{\mathbf{x}} = (A^T A)^{-1} A^T \mathbf{b} = \begin{bmatrix} \frac{5}{6} & -\frac{1}{2} \\ -\frac{1}{2} & \frac{1}{2} \end{bmatrix} \begin{bmatrix} 0 & 1 & 2 \\ 1 & 1 & 1 \end{bmatrix} \begin{bmatrix} 2 \\ 2 \\ 5 \end{bmatrix} = \begin{bmatrix} \frac{3}{2} \\ \frac{3}{2} \end{bmatrix} \Rightarrow y = \frac{3}{2} + \frac{3}{2}\,x.$

$$\mathbf{e} = \mathbf{b} - A\overline{\mathbf{x}} = \begin{bmatrix} 2 \\ 2 \\ 5 \end{bmatrix} - \begin{bmatrix} 1 & 0 \\ 1 & 1 \\ 1 & 2 \end{bmatrix} \begin{bmatrix} \frac{3}{2} \\ \frac{3}{2} \end{bmatrix} = \begin{bmatrix} \frac{1}{2} \\ -1 \\ \frac{1}{2} \end{bmatrix} \Rightarrow \|\mathbf{e}\| = \sqrt{(\tfrac{1}{2})^2 + (-1)^2 + (\tfrac{1}{2})^2} = \tfrac{\sqrt{6}}{2}.$$

11. $\overline{\mathbf{x}} = (A^T A)^{-1} A^T \mathbf{b} = \begin{bmatrix} \frac{3}{10} & -\frac{1}{50} \\ -\frac{1}{50} & \frac{1}{125} \end{bmatrix} \begin{bmatrix} 1 & 1 & 1 & 1 \\ -5 & 0 & 5 & 10 \end{bmatrix} \begin{bmatrix} -1 \\ 1 \\ 2 \\ 4 \end{bmatrix} = \begin{bmatrix} \frac{7}{10} \\ \frac{8}{25} \end{bmatrix} \Rightarrow y = \frac{7}{10} + \frac{8}{25}\,x.$

$$\mathbf{e} = \begin{bmatrix} -1 \\ 1 \\ 2 \\ 4 \end{bmatrix} - \begin{bmatrix} 1 & -5 \\ 1 & 0 \\ 1 & 5 \\ 1 & 10 \end{bmatrix} \begin{bmatrix} \frac{7}{10} \\ \frac{8}{25} \end{bmatrix} = \begin{bmatrix} -\frac{1}{10} \\ \frac{3}{10} \\ -\frac{3}{10} \\ \frac{1}{10} \end{bmatrix} \Rightarrow \|\mathbf{e}\| = \sqrt{(-\tfrac{1}{10})^2 + (\tfrac{3}{10})^2 + (-\tfrac{3}{10})^2 + (\tfrac{1}{10})^2} = \tfrac{\sqrt{5}}{5}.$$

12. $\overline{\mathbf{x}} = (A^T A)^{-1} A^T \mathbf{b} = \begin{bmatrix} \frac{3}{10} & -\frac{1}{50} \\ -\frac{1}{50} & \frac{1}{125} \end{bmatrix} \begin{bmatrix} 1 & 1 & 1 & 1 \\ -5 & 0 & 5 & 10 \end{bmatrix} \begin{bmatrix} 3 \\ 3 \\ 2 \\ 0 \end{bmatrix} = \begin{bmatrix} \frac{5}{2} \\ -\frac{1}{5} \end{bmatrix} \Rightarrow y = \frac{5}{2} - \frac{1}{5}\,x.$

$$\mathbf{e} = \begin{bmatrix} 3 \\ 3 \\ 2 \\ 0 \end{bmatrix} - \begin{bmatrix} 1 & -5 \\ 1 & 0 \\ 1 & 5 \\ 1 & 10 \end{bmatrix} \begin{bmatrix} \frac{5}{2} \\ -\frac{1}{5} \end{bmatrix} = \begin{bmatrix} -\frac{1}{2} \\ \frac{1}{2} \\ \frac{1}{2} \\ -\frac{1}{2} \end{bmatrix} \Rightarrow \|\mathbf{e}\| = \sqrt{(-\tfrac{1}{2})^2 + (\tfrac{1}{2})^2 + (\tfrac{1}{2})^2 + (-\tfrac{1}{2})^2} = 1.$$

13. $\overline{\mathbf{x}} = (A^TA)^{-1}A^T\mathbf{b} = \begin{bmatrix} \frac{11}{10} & -\frac{3}{10} \\ -\frac{3}{10} & \frac{1}{10} \end{bmatrix} \begin{bmatrix} 1 & 1 & 1 & 1 & 1 \\ 1 & 2 & 3 & 4 & 5 \end{bmatrix} \begin{bmatrix} 1 \\ 3 \\ 4 \\ 5 \\ 7 \end{bmatrix} = \begin{bmatrix} -\frac{1}{5} \\ \frac{7}{5} \end{bmatrix} \Rightarrow y = -\frac{1}{5} + \frac{7}{5}\,x.$

$$\mathbf{e} = \begin{bmatrix} 1 \\ 3 \\ 4 \\ 5 \\ 7 \end{bmatrix} - \begin{bmatrix} 1 & 1 \\ 1 & 2 \\ 1 & 3 \\ 1 & 4 \\ 1 & 5 \end{bmatrix} \begin{bmatrix} -\frac{1}{5} \\ \frac{7}{5} \end{bmatrix} = \begin{bmatrix} -\frac{1}{5} \\ \frac{2}{5} \\ 0 \\ -\frac{2}{5} \\ \frac{1}{5} \end{bmatrix} \Rightarrow \|\mathbf{e}\| = \sqrt{(-\tfrac{1}{5})^2 + (\tfrac{2}{5})^2 + 0^2 + (-\tfrac{2}{5})^2 + (\tfrac{1}{5})^2} = \tfrac{\sqrt{10}}{5}.$$

14. $\overline{\mathbf{x}} = (A^TA)^{-1}A^T\mathbf{b} = \begin{bmatrix} \frac{11}{10} & -\frac{3}{10} \\ -\frac{3}{10} & \frac{1}{10} \end{bmatrix} \begin{bmatrix} 1 & 1 & 1 & 1 & 1 \\ 1 & 2 & 3 & 4 & 5 \end{bmatrix} \begin{bmatrix} 10 \\ 8 \\ 5 \\ 3 \\ 0 \end{bmatrix} = \begin{bmatrix} \frac{127}{10} \\ -\frac{5}{2} \end{bmatrix} \Rightarrow y = \frac{127}{10} - \frac{5}{2}\,x.$

$$\mathbf{e} = \begin{bmatrix} 10 \\ 8 \\ 5 \\ 3 \\ 0 \end{bmatrix} - \begin{bmatrix} 1 & 1 \\ 1 & 2 \\ 1 & 3 \\ 1 & 4 \\ 1 & 5 \end{bmatrix} \begin{bmatrix} \frac{127}{10} \\ -\frac{5}{2} \end{bmatrix} = \begin{bmatrix} -\frac{1}{5} \\ \frac{3}{10} \\ -\frac{1}{5} \\ \frac{3}{10} \\ -\frac{1}{5} \end{bmatrix} \Rightarrow \|\mathbf{e}\| = \sqrt{3\cdot(-\tfrac{1}{5})^2 + 2\cdot(\tfrac{3}{10})^2} = \tfrac{\sqrt{30}}{10}.$$

15. $A^TA = \begin{bmatrix} 1 & 1 & 1 & 1 \\ 1 & 2 & 3 & 4 \\ 1 & 4 & 9 & 16 \end{bmatrix} \begin{bmatrix} 1 & 1 & 1 \\ 1 & 2 & 4 \\ 1 & 3 & 9 \\ 1 & 4 & 16 \end{bmatrix} = \begin{bmatrix} 40 & 10 & 30 \\ 10 & 30 & 100 \\ 30 & 100 & 354 \end{bmatrix},\ A^T\mathbf{b} = \begin{bmatrix} 1 & 1 & 1 & 1 \\ 1 & 2 & 3 & 4 \\ 1 & 4 & 9 & 16 \end{bmatrix} \begin{bmatrix} 1 \\ -2 \\ 3 \\ 4 \end{bmatrix} = \begin{bmatrix} 6 \\ 22 \\ 84 \end{bmatrix} \Rightarrow$

$$[\,A^TA\,|\,A^T\mathbf{b}\,] \longrightarrow [\,I\,|\,\overline{\mathbf{x}}\,] \Rightarrow \overline{\mathbf{x}} = \begin{bmatrix} 3 \\ -\frac{8}{15} \\ 1 \end{bmatrix} \Rightarrow y = 3 - \tfrac{18}{5}\,x + x^2.$$

16. $A^TA = \begin{bmatrix} 1 & 1 & 1 & 1 \\ 1 & 2 & 3 & 4 \\ 1 & 4 & 9 & 16 \end{bmatrix} \begin{bmatrix} 1 & 1 & 1 \\ 1 & 2 & 4 \\ 1 & 3 & 9 \\ 1 & 4 & 16 \end{bmatrix} = \begin{bmatrix} 40 & 10 & 30 \\ 10 & 30 & 100 \\ 30 & 100 & 354 \end{bmatrix},\ A^T\mathbf{b} = \begin{bmatrix} 1 & 1 & 1 & 1 \\ 1 & 2 & 3 & 4 \\ 1 & 4 & 9 & 16 \end{bmatrix} \begin{bmatrix} 8 \\ 7 \\ 5 \\ 2 \end{bmatrix} = \begin{bmatrix} 22 \\ 45 \\ 113 \end{bmatrix} \Rightarrow$

$$[\,A^TA\,|\,A^T\mathbf{b}\,] \longrightarrow [\,I\,|\,\overline{\mathbf{x}}\,] \Rightarrow \overline{\mathbf{x}} = \begin{bmatrix} 8 \\ \frac{1}{2} \\ -\frac{1}{2} \end{bmatrix} \Rightarrow y = 8 + \tfrac{1}{2}\,x - \tfrac{1}{2}\,x^2.$$

17. $A^TA = \begin{bmatrix} 1 & 1 & 1 & 1 & 1 \\ -2 & -1 & 0 & 1 & 2 \\ 4 & 1 & 0 & 1 & 4 \end{bmatrix} \begin{bmatrix} 1 & -2 & 4 \\ 1 & -1 & 1 \\ 1 & 0 & 0 \\ 1 & 1 & 1 \\ 1 & 2 & 4 \end{bmatrix} = \begin{bmatrix} 5 & 0 & 10 \\ 0 & 10 & 0 \\ 10 & 0 & 34 \end{bmatrix}$, $A^T\mathbf{b} = \begin{bmatrix} 13 \\ -17 \\ 19 \end{bmatrix} \Rightarrow$

$$[\,A^TA\,|\,A^T\mathbf{b}\,] \longrightarrow [\,I\,|\,\overline{\mathbf{x}}\,] \Rightarrow \overline{\mathbf{x}} = \begin{bmatrix} \frac{18}{5} \\ -\frac{17}{10} \\ -\frac{1}{2} \end{bmatrix} \Rightarrow y = \tfrac{18}{5} - \tfrac{17}{10}\,x - \tfrac{1}{2}\,x^2.$$

18. $A^TA = \begin{bmatrix} 1 & 1 & 1 & 1 & 1 \\ -2 & -1 & 0 & 1 & 2 \\ 4 & 1 & 0 & 1 & 4 \end{bmatrix} \begin{bmatrix} 1 & -2 & 4 \\ 1 & -1 & 1 \\ 1 & 0 & 0 \\ 1 & 1 & 1 \\ 1 & 2 & 4 \end{bmatrix} = \begin{bmatrix} 5 & 0 & 10 \\ 0 & 10 & 0 \\ 10 & 0 & 34 \end{bmatrix}$, $A^T\mathbf{b} = \begin{bmatrix} 22 \\ 18 \\ 12 \end{bmatrix} \Rightarrow$

$$[\,A^TA\,|\,A^T\mathbf{b}\,] \longrightarrow [\,I\,|\,\overline{\mathbf{x}}\,] \Rightarrow \overline{\mathbf{x}} = \begin{bmatrix} -\frac{62}{5} \\ \frac{9}{5} \\ 4 \end{bmatrix} \Rightarrow y = -\tfrac{62}{5} + \tfrac{9}{5}\,x + 4x^2.$$

19. $[\,A^TA\,|\,A^T\mathbf{b}\,] \longrightarrow [\,I\,|\,\overline{\mathbf{x}}\,] \Rightarrow \left[\begin{array}{cc|c} 11 & 6 & 5 \\ 6 & 6 & 4 \end{array}\right] \longrightarrow \left[\begin{array}{cc|c} 1 & 0 & \frac{1}{5} \\ 0 & 1 & \frac{7}{15} \end{array}\right] \Rightarrow \overline{\mathbf{x}} = \begin{bmatrix} \frac{1}{5} \\ \frac{7}{15} \end{bmatrix}.$

20. $[\,A^TA\,|\,A^T\mathbf{b}\,] \longrightarrow [\,I\,|\,\overline{\mathbf{x}}\,] \Rightarrow \left[\begin{array}{cc|c} 14 & -6 & 6 \\ -6 & 9 & -3 \end{array}\right] \longrightarrow \left[\begin{array}{cc|c} 1 & 0 & \frac{2}{5} \\ 0 & 1 & -\frac{1}{15} \end{array}\right] \Rightarrow \overline{\mathbf{x}} = \begin{bmatrix} \frac{2}{5} \\ -\frac{1}{15} \end{bmatrix}.$

21. $[\,A^TA\,|\,A^T\mathbf{b}\,] \longrightarrow [\,I\,|\,\overline{\mathbf{x}}\,] \Rightarrow \left[\begin{array}{cc|c} 14 & 8 & 12 \\ 8 & 38 & -21 \end{array}\right] \longrightarrow \left[\begin{array}{cc|c} 1 & 0 & \frac{4}{3} \\ 0 & 1 & -\frac{5}{6} \end{array}\right] \Rightarrow \overline{\mathbf{x}} = \begin{bmatrix} \frac{4}{3} \\ -\frac{5}{6} \end{bmatrix}.$

22. $[\,A^TA\,|\,A^T\mathbf{b}\,] \longrightarrow [\,I\,|\,\overline{\mathbf{x}}\,] \Rightarrow \left[\begin{array}{cc|c} 15 & 0 & 5 \\ 0 & 6 & 4 \end{array}\right] \longrightarrow \left[\begin{array}{cc|c} 1 & 0 & \frac{1}{3} \\ 0 & 1 & \frac{2}{3} \end{array}\right] \Rightarrow \overline{\mathbf{x}} = \begin{bmatrix} \frac{1}{3} \\ \frac{2}{3} \end{bmatrix}.$

23. $[A^TA|A^T\mathbf{b}] \longrightarrow [I|\overline{\mathbf{x}}] \Rightarrow \left[\begin{array}{cccc|c} 3 & 0 & 2 & 1 & 2 \\ 0 & 3 & -2 & -1 & -5 \\ 2 & -2 & 3 & 2 & 3 \\ 1 & -1 & 2 & 2 & -1 \end{array}\right] \longrightarrow \left[\begin{array}{cccc|c} 1 & 0 & 0 & -1 & 4 \\ 0 & 1 & 0 & 1 & -5 \\ 0 & 0 & 1 & 2 & -5 \\ 0 & 0 & 0 & 0 & 0 \end{array}\right] \Rightarrow \overline{\mathbf{x}} = \begin{bmatrix} 4+t \\ -5-t \\ -5-2t \\ t \end{bmatrix}.$

24. $[\,A^TA\,|\,A^T\mathbf{b}\,] \longrightarrow [\,I\,|\,\overline{\mathbf{x}}\,] \Rightarrow \left[\begin{array}{cccc|c} 3 & 0 & 3 & 0 & 3 \\ 0 & 3 & 1 & 2 & 8 \\ 3 & 1 & 4 & 0 & 8 \\ 0 & 2 & 0 & 2 & -2 \end{array}\right] \longrightarrow \left[\begin{array}{cccc|c} 1 & 0 & 0 & 1 & -5 \\ 0 & 1 & 0 & 1 & -1 \\ 0 & 0 & 1 & -1 & 6 \\ 0 & 0 & 0 & 0 & 0 \end{array}\right] \Rightarrow \overline{\mathbf{x}} = \begin{bmatrix} -5-t \\ -1-t \\ 6+t \\ t \end{bmatrix}.$

25. $[\,A^TA\,|\,A^T\mathbf{b}\,] \longrightarrow [\,I\,|\,\overline{\mathbf{x}}\,] \Rightarrow \left[\begin{array}{ccc|c} 11 & 7 & -5 & 35 \\ 7 & 6 & -5 & 18 \\ -5 & -5 & 7 & -1 \end{array}\right] \longrightarrow \left[\begin{array}{ccc|c} 1 & 0 & 0 & \frac{42}{11} \\ 0 & 1 & 0 & \frac{19}{11} \\ 0 & 0 & 1 & \frac{42}{11} \end{array}\right] \Rightarrow \overline{\mathbf{x}} = \begin{bmatrix} \frac{42}{11} \\ \frac{19}{11} \\ \frac{42}{11} \end{bmatrix}.$

26. $\left[\,A^TA\,|\,A^T\mathbf{b}\,\right] \longrightarrow \left[\,I\,|\,\overline{\mathbf{x}}\,\right] \Rightarrow \left[\begin{array}{ccc|c} 6 & 6 & 4 & 35 \\ 6 & 15 & 5 & 84 \\ 4 & 5 & 4 & 14 \end{array}\right] \longrightarrow \left[\begin{array}{ccc|c} 1 & 0 & 0 & \frac{469}{66} \\ 0 & 1 & 0 & \frac{224}{33} \\ 0 & 0 & 1 & -\frac{133}{11} \end{array}\right] \Rightarrow \overline{\mathbf{x}} = \left[\begin{array}{c} \frac{469}{66} \\ \frac{224}{33} \\ -\frac{133}{11} \end{array}\right].$

27. $\overline{\mathbf{x}} = R^{-1}Q^T\mathbf{b} = \left[\begin{array}{cc} \frac{1}{3} & -\frac{1}{3} \\ 0 & 1 \end{array}\right] \left[\begin{array}{ccc} \frac{2}{3} & \frac{2}{3} & \frac{1}{3} \\ \frac{1}{3} & -\frac{2}{3} & \frac{2}{3} \end{array}\right] \left[\begin{array}{c} 2 \\ 3 \\ 1 \end{array}\right] = \left[\begin{array}{c} \frac{5}{3} \\ -2 \end{array}\right].$

28. $\overline{\mathbf{x}} = R^{-1}Q^T\mathbf{b} = \left[\begin{array}{cc} \frac{1}{\sqrt{6}} & \frac{1}{\sqrt{2}} \\ 0 & \sqrt{2} \end{array}\right] \left[\begin{array}{ccc} \frac{1}{\sqrt{6}} & \frac{2}{\sqrt{6}} & -\frac{1}{\sqrt{6}} \\ \frac{1}{\sqrt{2}} & 0 & \frac{1}{\sqrt{2}} \end{array}\right] \left[\begin{array}{c} 1 \\ 1 \\ 1 \end{array}\right] = \left[\begin{array}{c} \frac{4}{3} \\ 2 \end{array}\right].$

29. $A^TA = \left[\begin{array}{cccccc} 1 & 1 & 1 & 1 & 1 & 1 \\ 20 & 40 & 48 & 60 & 80 & 100 \end{array}\right] \left[\begin{array}{cc} 1 & 20 \\ 1 & 40 \\ 1 & 48 \\ 1 & 60 \\ 1 & 80 \\ 1 & 100 \end{array}\right] = \left[\begin{array}{cc} 6 & 348 \\ 348 & 24304 \end{array}\right],\ A^T\mathbf{b} = \left[\begin{array}{c} 259.5 \\ 18058 \end{array}\right] \Rightarrow$

$\left[\,A^TA\,|\,A^T\mathbf{b}\,\right] \longrightarrow \left[\,I\,|\,\overline{\mathbf{x}}\,\right] \Rightarrow \overline{\mathbf{x}} \approx \left[\begin{array}{c} 0.92 \\ 0.73 \end{array}\right] \Rightarrow y = 0.92 + 0.73\,x.$

30. (a) $A^TA = \left[\begin{array}{cccc} 1 & 1 & 1 & 1 \\ 2 & 4 & 6 & 8 \end{array}\right] \left[\begin{array}{cc} 1 & 2 \\ 1 & 4 \\ 1 & 6 \\ 1 & 8 \end{array}\right] = \left[\begin{array}{cc} 4 & 20 \\ 20 & 120 \end{array}\right],\ A^T\mathbf{b} = \left[\begin{array}{c} 42.1 \\ 231 \end{array}\right] \Rightarrow$

$\left[\,A^TA\,|\,A^T\mathbf{b}\,\right] \longrightarrow \left[\,I\,|\,\overline{\mathbf{x}}\,\right] \Rightarrow \overline{\mathbf{x}} \approx \left[\begin{array}{c} 5.4 \\ 1.025 \end{array}\right] \Rightarrow L(F) = 5.4 + 1.025\,F.$

a represents the length of the spring hanging freely with no force being applied.

(b) $L(5) = 5.4 + 1.025\,(5) = 10.525$ in.

31. (a) $A^TA = \left[\begin{array}{cc} 8 & 28 \\ 28 & 140 \end{array}\right],\ A^T\mathbf{b} = \left[\begin{array}{c} 534.5 \\ 1992.9 \end{array}\right] \Rightarrow \left[\,A^TA\,|\,A^T\mathbf{b}\,\right] \longrightarrow \left[\,I\,|\,\overline{\mathbf{x}}\,\right] \Rightarrow$

$\overline{\mathbf{x}} \approx \left[\begin{array}{c} 56.6 \\ 0.29 \end{array}\right] \Rightarrow s(t) = 56.6 + 0.29\,t \Rightarrow$

The life expectancy of someone born in 2000 is $s(80) = 56.6 + 0.29\,(80) = 79.8$ years.

(b) $\mathbf{e} = \mathbf{b} - A\overline{\mathbf{x}} \Rightarrow \|\mathbf{e}\| \approx 29.2 \Rightarrow$ Error almost half the output $\Rightarrow$ model is not good.

32. (a) $A^TA = \begin{bmatrix} 5 & 8 & 16.5 \\ 8 & 16.5 & 39.5 \\ 16.5 & 39.5 & 103.125 \end{bmatrix}$, $A^T\mathbf{b} = \begin{bmatrix} 90 \\ 154 \\ 321 \end{bmatrix} \Rightarrow [\, A^TA \,|\, A^T\mathbf{b} \,] \longrightarrow [\, I \,|\, \overline{\mathbf{x}} \,] \Rightarrow$

$\overline{\mathbf{x}} \approx \begin{bmatrix} 1.92 \\ 20.31 \\ -4.97 \end{bmatrix} \Rightarrow s(t) = 1.92 + 20.31\,t - 4.97\,t^2 \Rightarrow$

(b) Height at which the object was released $= s_0 =$ the constant $= 1.92$ m.
Initial velocity $= v_0 =$ the coefficient of $t = 20.31$ m/s.
Acceleration due to gravity $(g) =$ twice the coefficient of $t^2 = 2(-4.97) = -9.94$ m/s.

(c) Hitting the ground $\Rightarrow$ height is zero $\Rightarrow s(t) = 0 \Rightarrow 1.92 + 20.31t - 4.97t^2 = 0$.
So, the Quadratic Formula $\Rightarrow t = \frac{-20.31 \pm \sqrt{(20.31)^2 - 4(-4.97)(1.92)}}{2(-4.97)} \approx 4.18$ seconds.

33. (a) $p(t) = c\,e^{kt} \Rightarrow p(0) = 150 \Rightarrow p = 150e^{kt} \Rightarrow \ln p = \ln(150e^{kt}) = \ln 150 + kt \Rightarrow$
$kt = \ln p - \ln 150 \Rightarrow kt = \ln \frac{p}{150}$.

So, Example 7.29 $\Rightarrow A = \begin{bmatrix} t_1 \\ t_2 \\ t_3 \\ t_4 \\ t_5 \end{bmatrix} = \begin{bmatrix} 1 \\ 2 \\ 3 \\ 4 \\ 5 \end{bmatrix}$ and $\mathbf{b} = \begin{bmatrix} kt_1 \\ kt_2 \\ kt_3 \\ kt_4 \\ kt_5 \end{bmatrix} = \begin{bmatrix} \ln \frac{179}{150} \\ \ln \frac{203}{150} \\ \ln \frac{227}{150} \\ \ln \frac{250}{150} \\ \ln \frac{281}{150} \end{bmatrix} \Rightarrow$

$A^TA = [55]$ and $A^T\mathbf{b} \approx [7.21] \Rightarrow \overline{\mathbf{x}} \approx \left[\frac{7.21}{55}\right] = 0.131 = k \Rightarrow p(t) = 150e^{0.131t}$.

(b) The population in 2010 will be $p(6) = 150e^{0.131(6)} \approx 329$ million.

34. (a) $A = \begin{bmatrix} 1 & 0 & 0 \\ 0 & 5 & 25 \\ 1 & 10 & 100 \\ 1 & 15 & 225 \\ 1 & 20 & 400 \\ 1 & 25 & 625 \\ 1 & 30 & 900 \end{bmatrix} \Rightarrow A^T A = \begin{bmatrix} 7 & 105 & 2275 \\ 105 & 2275 & 55125 \\ 2275 & 55125 & 1421875 \end{bmatrix}, \; A^T\mathbf{b} = \begin{bmatrix} 4193.3 \\ 103823.5 \\ 2738397.5 \end{bmatrix} \Rightarrow$

$$[\, A^T A \,|\, A^T\mathbf{b} \,] \longrightarrow [\, I \,|\, \overline{\mathbf{x}} \,] \Rightarrow \overline{\mathbf{x}} \approx \begin{bmatrix} 74.14 \\ -26.03 \\ 2.82 \end{bmatrix} \Rightarrow y = 74.14 - 26.03\,t + 2.82\,t^2.$$

(b) $p(t) = c\,e^{kt} \Rightarrow p(0) = 150 \Rightarrow p = 150e^{kt} \Rightarrow \ln p = \ln(150e^{kt}) = \ln 150 + kt \Rightarrow$
$kt = \ln p - \ln 150 \Rightarrow kt = \ln \frac{p}{150}$.

$$\text{So, Example 7.29} \Rightarrow A = \begin{bmatrix} t_1 \\ t_2 \\ t_3 \\ t_4 \\ t_5 \\ t_6 \end{bmatrix} = \begin{bmatrix} 1 \\ 2 \\ 3 \\ 4 \\ 5 \\ 6 \end{bmatrix} \text{ and } \mathbf{b} = \begin{bmatrix} kt_1 \\ kt_2 \\ kt_3 \\ kt_4 \\ kt_5 \\ kt_6 \end{bmatrix} = \begin{bmatrix} \ln \frac{44.7}{29.3} \\ \ln \frac{143.8}{29.3} \\ \ln \frac{371.6}{29.3} \\ \ln \frac{597.5}{29.3} \\ \ln \frac{1110.8}{29.3} \\ \ln \frac{1895.6}{29.3} \end{bmatrix} \Rightarrow$$

$A^T A = [91]$ and $A^T\mathbf{b} \approx [66.480] \Rightarrow \overline{\mathbf{x}} \approx \left[\frac{66.480}{91}\right] = 0.731 = k \Rightarrow p(t) = 29.3\,e^{0.731t}$.

(c) $\mathbf{e} = \mathbf{b} - A\overline{\mathbf{x}} \Rightarrow \|\mathbf{e}_y\| \approx 158, \; \|\mathbf{e}_p\| \approx 468 \Rightarrow$
Quadratic gives better approximation because exponential is growing too quickly.

(d) The average in 2010 will be $y(40) = 74.14 - 26.03(40) + 2.82(40)^2 \approx 3,545$ thousand.
Compare to salary given by exponential: $p(8) = 29.3\,e^{0.731(8)} = 10,154$ thousand.

35. $$\text{Example 7.29} \Rightarrow A = \begin{bmatrix} t_1 \\ t_2 \\ t_3 \end{bmatrix} = \begin{bmatrix} 30 \\ 60 \\ 90 \end{bmatrix} \text{ and } \mathbf{b} = \begin{bmatrix} kt_1 \\ kt_2 \\ kt_3 \end{bmatrix} = \begin{bmatrix} \ln \frac{172}{200} \\ \ln \frac{148}{200} \\ \ln \frac{128}{200} \end{bmatrix} \Rightarrow$$

$A^T A = [12600]$ and $A^T\mathbf{b} \approx [-62.8] \Rightarrow \overline{\mathbf{x}} \approx \left[-\frac{62.8}{12600}\right] = -0.005 = k \Rightarrow p(t) = c\,e^{-0.005t}$.
So, Section 6.7, Example 6.88 $\Rightarrow$ we can find the half-life as follows: $\frac{1}{2}c = c\,e^{-0.005t} \Rightarrow$

$e^{-0.005t} = \frac{1}{2} \Rightarrow -0.005t = \ln(1/2) \Rightarrow t = -\frac{\ln(1/2)}{0.005} \approx 139$ days is the half life.

36. $$A = \begin{bmatrix} 1 & 0 & -4 \\ 1 & 5 & 0 \\ 1 & 4 & -1 \\ 1 & 1 & -3 \\ 1 & -1 & -5 \end{bmatrix}, \; \mathbf{b} = \begin{bmatrix} 0 \\ 0 \\ 1 \\ 1 \\ -2 \end{bmatrix} \Rightarrow A^T A = \begin{bmatrix} 5 & 9 & -13 \\ 9 & 43 & -2 \\ -13 & -2 & 51 \end{bmatrix}, \; A^T\mathbf{b} = \begin{bmatrix} 0 \\ 7 \\ 6 \end{bmatrix} \Rightarrow$$

$$[\, A^T A \,|\, A^T\mathbf{b} \,] \longrightarrow [\, I \,|\, \overline{\mathbf{x}} \,] \Rightarrow \overline{\mathbf{x}} \approx \begin{bmatrix} 14.33 \\ -2.66 \\ 3.66 \end{bmatrix} \Rightarrow z = 14.33 - 2.66\,x + 3.66\,y.$$

Note, $\mathbf{e} = \mathbf{b} - A\overline{\mathbf{x}} \Rightarrow \|\mathbf{e}\| \approx 1.63 \Rightarrow$ plane $z = 14.33 - 2.66\,x + 3.66\,y$ is a good fit to the data.

37. $A = \begin{bmatrix} 1 \\ 1 \end{bmatrix} \Rightarrow A^T A = \begin{bmatrix} 2 \end{bmatrix} \Rightarrow (A^T A)^{-1} = \begin{bmatrix} \frac{1}{2} \end{bmatrix}$. So, with $P =$ the standard matrix we have:

$$P = A(A^T A)^{-1} A^T = \begin{bmatrix} 1 \\ 1 \end{bmatrix} \begin{bmatrix} \frac{1}{2} \end{bmatrix} \begin{bmatrix} 1 & 1 \end{bmatrix} = \begin{bmatrix} \frac{1}{2} & \frac{1}{2} \\ \frac{1}{2} & \frac{1}{2} \end{bmatrix} \Rightarrow \text{proj}_W(\mathbf{v}) = P\mathbf{v} = \begin{bmatrix} \frac{1}{2} & \frac{1}{2} \\ \frac{1}{2} & \frac{1}{2} \end{bmatrix} \begin{bmatrix} 3 \\ 4 \end{bmatrix} = \begin{bmatrix} \frac{7}{2} \\ \frac{7}{2} \end{bmatrix}.$$

38. $A = \begin{bmatrix} 1 \\ -2 \end{bmatrix} \Rightarrow A^T A = \begin{bmatrix} 5 \end{bmatrix} \Rightarrow (A^T A)^{-1} = \begin{bmatrix} \frac{1}{5} \end{bmatrix}$. So, $P = A(A^T A)^{-1} A^T \Rightarrow$

$$P = \begin{bmatrix} 1 \\ -2 \end{bmatrix} \begin{bmatrix} \frac{1}{5} \end{bmatrix} \begin{bmatrix} 1 & -2 \end{bmatrix} = \begin{bmatrix} \frac{1}{5} & -\frac{2}{5} \\ -\frac{2}{5} & \frac{4}{5} \end{bmatrix} \Rightarrow \text{proj}_W(\mathbf{v}) = P\mathbf{v} = \begin{bmatrix} \frac{1}{5} & -\frac{2}{5} \\ -\frac{2}{5} & \frac{4}{5} \end{bmatrix} \begin{bmatrix} 1 \\ 1 \end{bmatrix} = \begin{bmatrix} -\frac{1}{5} \\ \frac{2}{5} \end{bmatrix}.$$

39. $A = \begin{bmatrix} 1 \\ 1 \\ 1 \end{bmatrix} \Rightarrow A^T A = \begin{bmatrix} 3 \end{bmatrix} \Rightarrow (A^T A)^{-1} = \begin{bmatrix} \frac{1}{3} \end{bmatrix}$. So, $P = A(A^T A)^{-1} A^T \Rightarrow$

$$P = \begin{bmatrix} 1 \\ 1 \\ 1 \end{bmatrix} \begin{bmatrix} \frac{1}{3} \end{bmatrix} \begin{bmatrix} 1 & 1 & 1 \end{bmatrix} = \begin{bmatrix} \frac{1}{3} & \frac{1}{3} & \frac{1}{3} \\ \frac{1}{3} & \frac{1}{3} & \frac{1}{3} \\ \frac{1}{3} & \frac{1}{3} & \frac{1}{3} \end{bmatrix} \Rightarrow \text{proj}_W(\mathbf{v}) = P\mathbf{v} = \begin{bmatrix} \frac{1}{3} & \frac{1}{3} & \frac{1}{3} \\ \frac{1}{3} & \frac{1}{3} & \frac{1}{3} \\ \frac{1}{3} & \frac{1}{3} & \frac{1}{3} \end{bmatrix} \begin{bmatrix} 1 \\ 2 \\ 3 \end{bmatrix} = \begin{bmatrix} 2 \\ 2 \\ 2 \end{bmatrix}.$$

40. $A = \begin{bmatrix} 2 \\ 2 \\ -1 \end{bmatrix} \Rightarrow A^T A = \begin{bmatrix} 9 \end{bmatrix} \Rightarrow (A^T A)^{-1} = \begin{bmatrix} \frac{1}{9} \end{bmatrix}$. So, $P = A(A^T A)^{-1} A^T \Rightarrow$

$$P = \begin{bmatrix} 2 \\ 2 \\ -1 \end{bmatrix} \begin{bmatrix} \frac{1}{9} \end{bmatrix} \begin{bmatrix} 2 & 2 & -1 \end{bmatrix} = \begin{bmatrix} \frac{4}{9} & \frac{4}{9} & -\frac{2}{9} \\ \frac{4}{9} & \frac{4}{9} & -\frac{2}{9} \\ -\frac{2}{9} & -\frac{2}{9} & \frac{1}{9} \end{bmatrix} \Rightarrow \text{proj}_W(\mathbf{v}) = P\mathbf{v} = P \begin{bmatrix} 1 \\ 0 \\ 0 \end{bmatrix} = \begin{bmatrix} \frac{4}{9} \\ \frac{4}{9} \\ -\frac{2}{9} \end{bmatrix}.$$

41. $A = \begin{bmatrix} 1 & 1 \\ 0 & 1 \\ -1 & 1 \end{bmatrix} \Rightarrow A^T A = \begin{bmatrix} 2 & 0 \\ 0 & 3 \end{bmatrix} \Rightarrow (A^T A)^{-1} = \begin{bmatrix} \frac{1}{2} & 0 \\ 0 & \frac{1}{3} \end{bmatrix} \Rightarrow$. So, $P = A(A^T A)^{-1} A^T \Rightarrow$

$$P = \begin{bmatrix} 1 & 1 \\ 0 & 1 \\ -1 & 1 \end{bmatrix} \begin{bmatrix} \frac{1}{2} & 0 \\ 0 & \frac{1}{3} \end{bmatrix} \begin{bmatrix} 1 & 0 & -1 \\ 1 & 1 & 1 \end{bmatrix} = \begin{bmatrix} \frac{5}{6} & \frac{1}{3} & -\frac{1}{6} \\ \frac{1}{3} & \frac{1}{3} & \frac{1}{3} \\ -\frac{1}{6} & \frac{1}{3} & \frac{5}{6} \end{bmatrix} \Rightarrow \text{proj}_W(\mathbf{v}) = P\mathbf{v} = P \begin{bmatrix} 1 \\ 0 \\ 0 \end{bmatrix} = \begin{bmatrix} \frac{5}{6} \\ \frac{1}{3} \\ -\frac{1}{6} \end{bmatrix}.$$

42. $A = \begin{bmatrix} 1 & 1 \\ -2 & 0 \\ 1 & -1 \end{bmatrix} \Rightarrow A^T A = \begin{bmatrix} 6 & 0 \\ 0 & 2 \end{bmatrix} \Rightarrow (A^T A)^{-1} = \begin{bmatrix} \frac{1}{6} & 0 \\ 0 & \frac{1}{2} \end{bmatrix} \Rightarrow$. So, $P = A(A^T A)^{-1} A^T \Rightarrow$

$$P = \begin{bmatrix} 1 & 1 \\ -2 & 0 \\ 1 & -1 \end{bmatrix} \begin{bmatrix} \frac{1}{6} & 0 \\ 0 & \frac{1}{2} \end{bmatrix} \begin{bmatrix} 1 & -2 & 1 \\ 1 & 0 & -1 \end{bmatrix} = \begin{bmatrix} \frac{2}{3} & -\frac{1}{3} & -\frac{1}{3} \\ -\frac{1}{3} & \frac{2}{3} & -\frac{1}{3} \\ -\frac{1}{3} & -\frac{1}{3} & \frac{2}{3} \end{bmatrix} \Rightarrow \text{proj}_W(\mathbf{v}) = P \begin{bmatrix} 1 \\ 2 \\ 3 \end{bmatrix} = \begin{bmatrix} -1 \\ 0 \\ 1 \end{bmatrix}.$$

43. $A = \begin{bmatrix} 1 & 1 \\ 1 & 3 \\ 0 & 1 \end{bmatrix} \Rightarrow A^TA = \begin{bmatrix} 2 & 4 \\ 4 & 11 \end{bmatrix} \Rightarrow (A^TA)^{-1} = \begin{bmatrix} \frac{11}{6} & -\frac{2}{3} \\ -\frac{2}{3} & \frac{1}{3} \end{bmatrix} \Rightarrow$. So, $P = A(A^TA)^{-1}A^T \Rightarrow$

$P = \begin{bmatrix} 1 & 1 \\ 1 & 3 \\ 0 & 1 \end{bmatrix} \begin{bmatrix} \frac{11}{6} & -\frac{2}{3} \\ -\frac{2}{3} & \frac{1}{3} \end{bmatrix} \begin{bmatrix} 1 & 1 & 0 \\ 1 & 3 & 1 \end{bmatrix} = \begin{bmatrix} \frac{5}{6} & \frac{1}{6} & -\frac{1}{3} \\ \frac{1}{6} & \frac{5}{6} & \frac{1}{3} \\ -\frac{1}{3} & \frac{1}{3} & \frac{1}{3} \end{bmatrix}$ as found in Example 7.31.

44. (a) Section 3.3, Theorem 3.9d $\Rightarrow (A^{-1})^T = (A^T)^{-1} \Rightarrow ((A^TA)^{-1})^T = ((A^TA)^T)^{-1}$.
We will use this fact in the third step. We need to show $P^T = P = A(A^TA)^{-1}A^T$.
$P^T = (A(A^TA)^{-1}A^T)^T = (A^T)^T((A^TA)^{-1})^TA^T = A((A^TA)^T)^{-1}A^T = A(A^TA)^{-1}A^T$.

(b) We need to show $PP = P = A(A^TA)^{-1}A^T$.
$PP = (A(A^TA)^{-1}A^T)(A(A^TA)^{-1}A^T) = A((A^TA)^{-1}(A^TA))(A^TA)^{-1}A^T = A(A^TA)^{-1}A^T$.

45. $A^TA = [\,5\,] \Rightarrow (A^TA)^{-1} = [\,\frac{1}{5}\,] \Rightarrow A^+ = (A^TA)^{-1}A^T = [\,\frac{1}{5}\,][\,1\ \ 2\,] = [\,\frac{1}{5}\ \ \frac{2}{5}\,]$.

46. $A^TA = [\,6\,] \Rightarrow (A^TA)^{-1} = [\,\frac{1}{6}\,] \Rightarrow A^+ = (A^TA)^{-1}A^T = [\,\frac{1}{6}\,][\,1\ \ {-1}\ \ 2\,] = [\,\frac{1}{6}\ \ -\frac{1}{6}\ \ \frac{1}{3}\,]$.

47. $A^TA = \begin{bmatrix} 2 & 2 \\ 2 & 14 \end{bmatrix} \Rightarrow (A^TA)^{-1} = \begin{bmatrix} \frac{7}{12} & -\frac{1}{12} \\ -\frac{1}{12} & \frac{1}{12} \end{bmatrix} \Rightarrow A^+ = (A^TA)^{-1}A^T = \begin{bmatrix} \frac{1}{3} & -\frac{2}{3} & -\frac{1}{6} \\ \frac{1}{6} & \frac{1}{6} & \frac{1}{6} \end{bmatrix}$.

48. $A^TA = \begin{bmatrix} 14 & 10 \\ 10 & 14 \end{bmatrix} \Rightarrow (A^TA)^{-1} = \begin{bmatrix} \frac{7}{48} & -\frac{5}{48} \\ -\frac{5}{48} & \frac{7}{48} \end{bmatrix} \Rightarrow A^+ = (A^TA)^{-1}A^T = \begin{bmatrix} -\frac{1}{6} & \frac{1}{3} & \frac{1}{12} \\ \frac{1}{3} & -\frac{1}{6} & \frac{1}{12} \end{bmatrix}$.

49. $A^TA = \begin{bmatrix} 1 & 1 \\ 1 & 2 \end{bmatrix} \Rightarrow (A^TA)^{-1} = \begin{bmatrix} 2 & -1 \\ -1 & 1 \end{bmatrix} \Rightarrow A^+ = (A^TA)^{-1}A^T = \begin{bmatrix} 1 & -1 \\ 0 & 1 \end{bmatrix}$.

50. $A^TA = \begin{bmatrix} 10 & 14 \\ 14 & 20 \end{bmatrix} \Rightarrow (A^TA)^{-1} = \begin{bmatrix} 5 & -\frac{7}{2} \\ -\frac{7}{2} & \frac{5}{2} \end{bmatrix} \Rightarrow A^+ = (A^TA)^{-1}A^T = \begin{bmatrix} -2 & 1 \\ \frac{3}{2} & -\frac{1}{2} \end{bmatrix}$.

51. $A^TA = \begin{bmatrix} 3 & 1 & 2 \\ 1 & 2 & 2 \\ 2 & 2 & 3 \end{bmatrix} \Rightarrow (A^TA)^{-1} = \begin{bmatrix} \frac{2}{3} & \frac{1}{3} & -\frac{2}{3} \\ \frac{1}{3} & \frac{5}{3} & -\frac{4}{3} \\ -\frac{2}{3} & -\frac{4}{3} & \frac{5}{3} \end{bmatrix} \Rightarrow A^+ = (A^TA)^{-1}A^T = \begin{bmatrix} \frac{2}{3} & 0 & -\frac{1}{3} & \frac{1}{3} \\ \frac{1}{3} & -1 & \frac{1}{3} & \frac{2}{3} \\ -\frac{2}{3} & 1 & \frac{1}{3} & -\frac{1}{3} \end{bmatrix}$.

52. $A^TA = \begin{bmatrix} 2 & 3 & -2 \\ 3 & 6 & -3 \\ -2 & -3 & 9 \end{bmatrix} \Rightarrow (A^TA)^{-1} = \begin{bmatrix} \frac{15}{7} & -1 & \frac{1}{7} \\ -1 & \frac{2}{3} & 0 \\ \frac{1}{7} & 0 & \frac{1}{7} \end{bmatrix} \Rightarrow A^+ = (A^TA)^{-1}A^T = \begin{bmatrix} \frac{1}{7} & -\frac{8}{7} & \frac{6}{7} & \frac{2}{7} \\ \frac{1}{3} & \frac{2}{3} & -\frac{1}{3} & 0 \\ \frac{1}{7} & -\frac{1}{7} & -\frac{1}{7} & \frac{2}{7} \end{bmatrix}$.

53. (a) $A^+ = (A^TA)^{-1}A^T = A^{-1}(A^T)^{-1}A^T = A^{-1}$.

(b) Recall, A has orthonormal columns $\mathbf{q}_i$ means $\mathbf{q}_i \cdot \mathbf{q}_j = \begin{cases} 0 \text{ if } i \neq j \\ 1 \text{ if } i = j \end{cases}$.

So, $A^TA = \begin{bmatrix} \mathbf{q}_1 \\ \vdots \\ \mathbf{q}_n \end{bmatrix} [\,\mathbf{q}_1 | \cdots | \mathbf{q}_n\,] = I_n \Rightarrow A^+ = (A^TA)^{-1}A^T = I_nA^T = A^T$.

Note, we should have expected this since if A were an $n \times n$ matrix, $A^{-1} = A^T$.

54. $A^+AA^+ = ((A^TA)^{-1}A^T)\,A\,A^+ = (A^TA)^{-1}(A^TA)A^+ = A^+.$

55. Will show $A^+A = I$, which is symmetric. $A^+A = (A^TA)^{-1}A^TA = (A^TA)^{-1}(A^TA) = I.$

56. (a) We will show $(cA)^+ = \frac{1}{c}\,A^+$, as should be expected since $(cA)^{-1} = \frac{1}{c}\,A^{-1}$.
$(cA)^+ = (cA^TcA)^{-1}cA^T = \frac{1}{c^2}\,(A^TA)^{-1}cA^T = \frac{1}{c}\,(A^TA)^{-1}A^T = \frac{1}{c}\,A^+.$

(b) Exercise 53(a) $\Rightarrow$ A square $(A^{-1})^+ = (A^{-1})^{-1}$. So $(A^+)^+ = (A^{-1})^{-1} = A$.

(c) Exercise 53(a) $\Rightarrow$ A square $\Rightarrow$ $(A^T)^+ = (A^T)^{-1}$ and Section 3.3, Theorem 3.9d $\Rightarrow$ $(A^T)^{-1} = (A^{-1})^T$. So $(A^T)^+ = (A^T)^{-1} = (A^{-1})^T = (A^+)^T$.

57. We will construct A as in Example 7.26, then show the columns of A are linearly independent.

Theorem 7.9 $\Rightarrow A\mathbf{x} = \mathbf{b} = \begin{bmatrix} y_1 \\ \vdots \\ y_n \end{bmatrix} \in \mathbb{R}^n$ has unique least squares solution $\overline{\mathbf{x}} = \begin{bmatrix} a_0 \\ a_1 \end{bmatrix} \in \mathbb{R}^2 \Rightarrow$
$p(x) = a_0 + a_1 x$ is the unique least squares approximating line.

So, let $A = \begin{cases} a_{1i} = 1 & \text{for all } i \\ a_{2i} = x_i & \text{for all } i \end{cases}$, that is $A = \begin{bmatrix} 1 & x_1 \\ \vdots & \vdots \\ 1 & x_n \end{bmatrix}$.

The fact that all the points do not lie on the same vertical line $\Rightarrow$ there exists $x_j \neq x_k$.
We will use this fact to show that the columns of A, that is $\mathbf{a}_i$, are linearly independent.

$a_0\mathbf{a}_1 + a_1\mathbf{a}_2 = 0 \Leftrightarrow a_0 + a_1 x_i = 0$ for all $i \Rightarrow p(x) = a_0 + a_1 x$ has at least two distinct zeroes (since $x_j \neq x_k$). Therefore, Section 6.2, Exercise 62 $\Rightarrow p(x)$ is the zero polynomial $\Rightarrow a_0 = a_1 = 0 \Rightarrow$ the columns of A, that is $\mathbf{a}_i$, are linearly independent.

58. We will construct A as in Example 7.26, then show the columns of A are linearly independent.

Theorem 7.9 $\Rightarrow A\mathbf{x} = \mathbf{b} = \begin{bmatrix} y_1 \\ \vdots \\ y_n \end{bmatrix} \in \mathbb{R}^n$ has unique least squares solution $\overline{\mathbf{x}} = \begin{bmatrix} a_0 \\ a_1 \\ \vdots \\ a_k \end{bmatrix} \in$
$\mathbb{R}^{k+1} \Rightarrow$
$p(x) = a_0 + a_1 x + \ldots + a_k x^k$ is the unique least squares approximating polynomial.

So, let $A = \begin{cases} a_{1j} = 1 & \text{for all } j \\ a_{ij} = x_j^{i-1} & \text{for } i \geq 2 \text{ and all } j \end{cases}$, that is $A = \begin{bmatrix} 1 & x_1 & x_1^2 & \ldots & x_1^k \\ \vdots & \vdots & \vdots & \ldots & \vdots \\ 1 & x_n & x_n^2 & \ldots & x_n^k \end{bmatrix}$.

$a_0\mathbf{a}_1 + a_1\mathbf{a}_2 + \ldots + a_k\mathbf{a}_{k+1} = 0 \Leftrightarrow a_0 x_i + a_1 x_i^2 + \ldots + a_k x_i^k = 0$ for all $i \Rightarrow$
$p(x) = a_0 + a_1 x + \ldots + a_k x^k$ has at least $k+1$ distinct zeroes ($k+1$ of x_i are distinct).
So, Section 6.2, Exercise 62 $\Rightarrow p(x)$ is the zero polynomial $\Rightarrow a_0 = a_1 = \ldots = a_k = 0 \Rightarrow$
the columns of A, that is $\mathbf{a}_i$, are linearly independent.

7.4 The Singular Value Decomposition

1. $A = \begin{bmatrix} 2 & 0 \\ 0 & 3 \end{bmatrix} \Rightarrow A^T A = \begin{bmatrix} 4 & 0 \\ 0 & 9 \end{bmatrix} \Rightarrow$ the eigenvalues of $A^T A$ are $\lambda_1 = 9$, $\lambda_2 = 4 \Rightarrow$ the singular values of A are $\sigma_1 = \sqrt{9} = 3$, $\sigma_2 = \sqrt{4} = 2$.

2. $A = \begin{bmatrix} 3 & 1 \\ 1 & 3 \end{bmatrix} \Rightarrow A^T A = \begin{bmatrix} 10 & 6 \\ 6 & 10 \end{bmatrix} \Rightarrow$ the eigenvalues of $A^T A$ are $\lambda_1 = 16$, $\lambda_2 = 4 \Rightarrow$ the singular values of A are $\sigma_1 = \sqrt{16} = 4$, $\sigma_2 = \sqrt{4} = 2$.

3. $A^T A = \begin{bmatrix} 1 & 1 \\ 1 & 1 \end{bmatrix} \Rightarrow$ eigenvalues $2, 0 \Rightarrow$ singular values of A are $\sigma_1 = \sqrt{2}$, $\sigma_2 = 0$.

4. $A^T A = \begin{bmatrix} 2 & \sqrt{2} \\ \sqrt{2} & 3 \end{bmatrix} \Rightarrow$ eigenvalues $4, 1 \Rightarrow$ singular values of A are $\sigma_1 = 2$, $\sigma_2 = 1$.

5. $A^T A = \begin{bmatrix} 25 \end{bmatrix} \Rightarrow$ eigenvalue $25 \Rightarrow$ singular value of A is $\sigma_1 = 5$.

6. $A^T A = \begin{bmatrix} 9 & 12 \\ 12 & 16 \end{bmatrix} \Rightarrow$ eigenvalues $25, 0 \Rightarrow$ singular values of A are $\sigma_1 = 5$, $\sigma_2 = 0$.

7. $A^T A = \begin{bmatrix} 4 & 0 \\ 0 & 9 \end{bmatrix} \Rightarrow$ eigenvalues $9, 4 \Rightarrow$ singular values of A are $\sigma_1 = 3$, $\sigma_2 = 2$.

8. $A^T A = \begin{bmatrix} 5 & -4 \\ -4 & 5 \end{bmatrix} \Rightarrow$ eigenvalues $9, 1 \Rightarrow$ singular values of A are $\sigma_1 = 3$, $\sigma_2 = 1$.

9. $A^T A = \begin{bmatrix} 4 & 0 & 2 \\ 0 & 4 & 0 \\ 2 & 0 & 1 \end{bmatrix} \Rightarrow$ eigenvalues $5, 4, 0 \Rightarrow$ singular values of A are $\sigma_1 = \sqrt{5}$, $\sigma_2 = 2$, $\sigma_3 = 0$.

10. $A^T A = \begin{bmatrix} 2 & 0 & 2 \\ 0 & 9 & 0 \\ 2 & 0 & 2 \end{bmatrix} \Rightarrow$ eigenvalues $9, 4, 0 \Rightarrow$ singular values of A $\sigma_1 = 3$, $\sigma_2 = 2$, $\sigma_3 = 0$.

11. $A^T A = \begin{bmatrix} 1 & 1 \\ 1 & 1 \end{bmatrix} \Rightarrow$ eigenvectors of $A^T A$ are $\begin{bmatrix} \frac{1}{\sqrt{2}} \\ \frac{1}{\sqrt{2}} \end{bmatrix}, \begin{bmatrix} \frac{1}{\sqrt{2}} \\ -\frac{1}{\sqrt{2}} \end{bmatrix} \Rightarrow$

$\mathbf{u}_1 = \frac{1}{\sigma_1} A\mathbf{v}_1 = \frac{1}{\sqrt{2}} \begin{bmatrix} 1 & 1 \\ 0 & 0 \end{bmatrix} \begin{bmatrix} \frac{1}{\sqrt{2}} \\ \frac{1}{\sqrt{2}} \end{bmatrix} = \begin{bmatrix} 1 \\ 0 \end{bmatrix}$. $\sigma_2 = 0 \Rightarrow$ extend with $\mathbf{u}_2 = \begin{bmatrix} 0 \\ 1 \end{bmatrix} \Rightarrow$

$A = U\Sigma V^T = \begin{bmatrix} 1 & 0 \\ 0 & 1 \end{bmatrix} \begin{bmatrix} \sqrt{2} & 0 \\ 0 & 0 \end{bmatrix} \begin{bmatrix} \frac{1}{\sqrt{2}} & \frac{1}{\sqrt{2}} \\ \frac{1}{\sqrt{2}} & -\frac{1}{\sqrt{2}} \end{bmatrix} = \begin{bmatrix} 1 & 1 \\ 0 & 0 \end{bmatrix}$.

12. $A^TA = \begin{bmatrix} 4 & 0 \\ 0 & 0 \end{bmatrix} \Rightarrow$ eigenvectors of A^TA are $\begin{bmatrix} -1 \\ 0 \end{bmatrix}, \begin{bmatrix} 0 \\ 1 \end{bmatrix} \Rightarrow$

$\mathbf{u}_1 = \frac{1}{\sigma_1} A\mathbf{v}_1 = \frac{1}{2}\begin{bmatrix} -2 & 0 \\ 0 & 0 \end{bmatrix}\begin{bmatrix} -1 \\ 0 \end{bmatrix} = \begin{bmatrix} 1 \\ 0 \end{bmatrix}$. $\sigma_2 = 0 \Rightarrow$ extend with $\mathbf{u}_2 = \begin{bmatrix} 0 \\ 1 \end{bmatrix} \Rightarrow$

$A = U\Sigma V^T = \begin{bmatrix} 1 & 0 \\ 0 & 1 \end{bmatrix}\begin{bmatrix} 2 & 0 \\ 0 & 0 \end{bmatrix}\begin{bmatrix} -1 & 0 \\ 0 & 1 \end{bmatrix} = \begin{bmatrix} -2 & 0 \\ 0 & 0 \end{bmatrix}$.

13. $A^TA = \begin{bmatrix} 9 & 0 \\ 0 & 4 \end{bmatrix} \Rightarrow$ eigenvectors of A^TA are $\begin{bmatrix} -1 \\ 0 \end{bmatrix}, \begin{bmatrix} 0 \\ -1 \end{bmatrix} \Rightarrow$

$\mathbf{u}_1 = \frac{1}{\sigma_1} A\mathbf{v}_1 = \frac{1}{3}\begin{bmatrix} 0 & -2 \\ -3 & 0 \end{bmatrix}\begin{bmatrix} -1 \\ 0 \end{bmatrix} = \begin{bmatrix} 0 \\ 1 \end{bmatrix}$. $\mathbf{u}_2 = \frac{1}{\sigma_2} A\mathbf{v}_2 = \frac{1}{2}\begin{bmatrix} 0 & -2 \\ -3 & 0 \end{bmatrix}\begin{bmatrix} 0 \\ -1 \end{bmatrix} = \begin{bmatrix} 1 \\ 0 \end{bmatrix}$.

$A = U\Sigma V^T = \begin{bmatrix} 0 & 1 \\ 1 & 0 \end{bmatrix}\begin{bmatrix} 3 & 0 \\ 0 & 2 \end{bmatrix}\begin{bmatrix} -1 & 0 \\ 0 & -1 \end{bmatrix} = \begin{bmatrix} 0 & -2 \\ -3 & 0 \end{bmatrix}$.

14. $A^TA = \begin{bmatrix} 2 & 0 \\ 0 & 2 \end{bmatrix} \Rightarrow$ eigenvectors of A^TA are $\begin{bmatrix} 1 \\ 0 \end{bmatrix}, \begin{bmatrix} 0 \\ 1 \end{bmatrix} \Rightarrow$

$\mathbf{u}_1 = \frac{1}{\sigma_1} A\mathbf{v}_1 = \frac{1}{\sqrt{2}}\begin{bmatrix} 1 & -1 \\ 1 & 1 \end{bmatrix}\begin{bmatrix} 1 \\ 0 \end{bmatrix} = \begin{bmatrix} \frac{1}{\sqrt{2}} \\ \frac{1}{\sqrt{2}} \end{bmatrix}$. $\mathbf{u}_2 = \frac{1}{\sigma_2} A\mathbf{v}_2 = \frac{1}{\sqrt{2}}\begin{bmatrix} 1 & -1 \\ 1 & 1 \end{bmatrix}\begin{bmatrix} 0 \\ 1 \end{bmatrix} = \begin{bmatrix} -\frac{1}{\sqrt{2}} \\ \frac{1}{\sqrt{2}} \end{bmatrix} \Rightarrow$

$A = U\Sigma V^T = \begin{bmatrix} \frac{1}{\sqrt{2}} & -\frac{1}{\sqrt{2}} \\ \frac{1}{\sqrt{2}} & \frac{1}{\sqrt{2}} \end{bmatrix}\begin{bmatrix} \sqrt{2} & 0 \\ 0 & \sqrt{2} \end{bmatrix}\begin{bmatrix} 1 & 0 \\ 0 & 1 \end{bmatrix} = \begin{bmatrix} 1 & -1 \\ 1 & 1 \end{bmatrix}$.

15. $A^TA = \begin{bmatrix} 25 \end{bmatrix} \Rightarrow$ eigenvector of A^TA is $\begin{bmatrix} 1 \end{bmatrix} \Rightarrow$

$\mathbf{u}_1 = \frac{1}{\sigma_1} A\mathbf{v}_1 = \frac{1}{5}\begin{bmatrix} 3 \\ 4 \end{bmatrix}\begin{bmatrix} 1 \end{bmatrix} = \begin{bmatrix} \frac{3}{5} \\ \frac{4}{5} \end{bmatrix}$. There is no $\sigma_2 \Rightarrow$ extend with $\mathbf{u}_2 = \begin{bmatrix} -\frac{4}{5} \\ \frac{3}{5} \end{bmatrix} \Rightarrow$

$A = U\Sigma V^T = \begin{bmatrix} \frac{3}{5} & -\frac{4}{5} \\ \frac{4}{5} & \frac{3}{5} \end{bmatrix}\begin{bmatrix} 5 \\ 0 \end{bmatrix}\begin{bmatrix} 1 \end{bmatrix} = \begin{bmatrix} 3 \\ 4 \end{bmatrix}$.

16. $A^TA = \begin{bmatrix} 9 & 12 \\ 12 & 16 \end{bmatrix} \Rightarrow$ eigenvectors of A^TA are $\begin{bmatrix} \frac{3}{5} \\ \frac{4}{5} \end{bmatrix}, \begin{bmatrix} \frac{4}{5} \\ -\frac{3}{5} \end{bmatrix} \Rightarrow$

$\mathbf{u}_1 = \frac{1}{\sigma_1} A\mathbf{v}_1 = \frac{1}{5}\begin{bmatrix} 3 & 4 \end{bmatrix}\begin{bmatrix} \frac{3}{5} \\ \frac{4}{5} \end{bmatrix} = \begin{bmatrix} 1 \end{bmatrix} \Rightarrow A = U\Sigma V^T = \begin{bmatrix} 1 \end{bmatrix}\begin{bmatrix} 5 & 0 \end{bmatrix}\begin{bmatrix} \frac{3}{5} & \frac{4}{5} \\ \frac{4}{5} & -\frac{3}{5} \end{bmatrix} = \begin{bmatrix} 3 & 4 \end{bmatrix}$.

17. $A^TA = \begin{bmatrix} 4 & 0 \\ 0 & 9 \end{bmatrix} \Rightarrow$ eigenvectors of A^TA are $\begin{bmatrix} 0 \\ 1 \end{bmatrix}, \begin{bmatrix} 1 \\ 0 \end{bmatrix} \Rightarrow$

$\mathbf{u}_1 = \frac{1}{\sigma_1} A\mathbf{v}_1 = \frac{1}{3}\begin{bmatrix} 0 & 0 \\ 0 & 3 \\ -2 & 0 \end{bmatrix}\begin{bmatrix} 0 \\ 1 \end{bmatrix} = \begin{bmatrix} 0 \\ 1 \\ 0 \end{bmatrix}$. $\mathbf{u}_2 = \frac{1}{2} A\mathbf{v}_2 = \begin{bmatrix} 0 \\ 0 \\ -1 \end{bmatrix}$.

Extend $\mathbf{u}_3 = \begin{bmatrix} 1 \\ 0 \\ 0 \end{bmatrix} \Rightarrow A = U\Sigma V^T = \begin{bmatrix} 0 & 0 & 1 \\ 1 & 0 & 0 \\ 0 & -1 & 0 \end{bmatrix}\begin{bmatrix} 3 & 0 \\ 0 & 2 \\ 0 & 0 \end{bmatrix}\begin{bmatrix} 0 & 1 \\ 1 & 0 \end{bmatrix} = \begin{bmatrix} 0 & 0 \\ 0 & 3 \\ -2 & 0 \end{bmatrix}$.

18. $A^TA = \begin{bmatrix} 5 & -4 \\ -4 & 5 \end{bmatrix} \Rightarrow$ eigenvectors of A^TA are $\begin{bmatrix} \frac{1}{\sqrt{2}} \\ -\frac{1}{\sqrt{2}} \end{bmatrix}, \begin{bmatrix} \frac{1}{\sqrt{2}} \\ \frac{1}{\sqrt{2}} \end{bmatrix} \Rightarrow$

$\mathbf{u}_1 = \frac{1}{\sigma_1} A\mathbf{v}_1 = \frac{1}{3} \begin{bmatrix} 1 & 0 \\ 0 & 1 \\ -2 & 2 \end{bmatrix} \begin{bmatrix} \frac{1}{\sqrt{2}} \\ -\frac{1}{\sqrt{2}} \end{bmatrix} = \begin{bmatrix} \frac{1}{3\sqrt{2}} \\ -\frac{1}{3\sqrt{2}} \\ -\frac{4}{3\sqrt{2}} \end{bmatrix}$. $\mathbf{u}_2 = \frac{1}{\sigma_2} A\mathbf{v}_2 = \begin{bmatrix} \frac{1}{\sqrt{2}} \\ \frac{1}{\sqrt{2}} \\ 0 \end{bmatrix}$.

Extend $\mathbf{u}_3 = \begin{bmatrix} \frac{2}{3} \\ -\frac{2}{3} \\ \frac{1}{3} \end{bmatrix} \Rightarrow A = U\Sigma V^T = \begin{bmatrix} \frac{1}{3\sqrt{2}} & \frac{1}{\sqrt{2}} & \frac{2}{3} \\ -\frac{1}{3\sqrt{2}} & \frac{1}{\sqrt{2}} & -\frac{2}{3} \\ -\frac{4}{3\sqrt{2}} & 0 & \frac{1}{3} \end{bmatrix} \begin{bmatrix} 3 & 0 \\ 0 & 1 \\ 0 & 0 \end{bmatrix} \begin{bmatrix} \frac{1}{\sqrt{2}} & -\frac{1}{\sqrt{2}} \\ \frac{1}{\sqrt{2}} & \frac{1}{\sqrt{2}} \end{bmatrix} = \begin{bmatrix} 1 & 0 \\ 0 & 1 \\ -2 & 2 \end{bmatrix}$.

19. $A^TA = \begin{bmatrix} 4 & 0 & 2 \\ 0 & 4 & 0 \\ 2 & 0 & 1 \end{bmatrix} \Rightarrow$ eigenvectors of A^TA are $\begin{bmatrix} \frac{2}{\sqrt{5}} \\ 0 \\ \frac{1}{\sqrt{5}} \end{bmatrix}, \begin{bmatrix} 0 \\ 1 \\ 0 \end{bmatrix}, \begin{bmatrix} \frac{1}{\sqrt{5}} \\ 0 \\ -\frac{2}{\sqrt{5}} \end{bmatrix} \Rightarrow$

$\mathbf{u}_1 = \frac{1}{\sigma_1} A\mathbf{v}_1 = \frac{1}{\sqrt{5}} \begin{bmatrix} 2 & 0 & 1 \\ 0 & 2 & 0 \end{bmatrix} \begin{bmatrix} \frac{2}{\sqrt{5}} \\ 0 \\ \frac{1}{\sqrt{5}} \end{bmatrix} = \begin{bmatrix} 1 \\ 0 \end{bmatrix}$. $\mathbf{u}_2 = \frac{1}{2} A\mathbf{v}_2 = \begin{bmatrix} 0 \\ 1 \end{bmatrix}$.

$A = U\Sigma V^T = \begin{bmatrix} 1 & 0 \\ 0 & 1 \end{bmatrix} \begin{bmatrix} \sqrt{5} & 0 & 0 \\ 0 & 2 & 0 \end{bmatrix} \begin{bmatrix} \frac{2}{\sqrt{5}} & 0 & \frac{1}{\sqrt{5}} \\ 0 & 1 & 0 \\ \frac{1}{\sqrt{5}} & 0 & -\frac{2}{\sqrt{5}} \end{bmatrix} = \begin{bmatrix} 2 & 0 & 1 \\ 0 & 2 & 0 \end{bmatrix}$.

20. $A^TA = \begin{bmatrix} 2 & 2 & 2 \\ 2 & 2 & 2 \\ 2 & 2 & 2 \end{bmatrix} \Rightarrow$ eigenvectors of A^TA are $\begin{bmatrix} \frac{1}{\sqrt{3}} \\ \frac{1}{\sqrt{3}} \\ \frac{1}{\sqrt{3}} \end{bmatrix}, \begin{bmatrix} \frac{2}{\sqrt{6}} \\ -\frac{1}{\sqrt{6}} \\ -\frac{1}{\sqrt{6}} \end{bmatrix}, \begin{bmatrix} 0 \\ \frac{1}{\sqrt{2}} \\ -\frac{1}{\sqrt{2}} \end{bmatrix} \Rightarrow$

$\mathbf{u}_1 = \frac{1}{\sigma_1} A\mathbf{v}_1 = \frac{1}{\sqrt{6}} \begin{bmatrix} 1 & 1 & 1 \\ 1 & 1 & 1 \end{bmatrix} \begin{bmatrix} \frac{1}{\sqrt{3}} \\ \frac{1}{\sqrt{3}} \\ \frac{1}{\sqrt{3}} \end{bmatrix} = \begin{bmatrix} \frac{1}{\sqrt{2}} \\ \frac{1}{\sqrt{2}} \end{bmatrix}$. Extend with $\mathbf{u}_2 \begin{bmatrix} \frac{1}{\sqrt{2}} \\ -\frac{1}{\sqrt{2}} \end{bmatrix}$.

$A = U\Sigma V^T = \begin{bmatrix} \frac{1}{\sqrt{2}} & \frac{1}{\sqrt{2}} \\ \frac{1}{\sqrt{2}} & -\frac{1}{\sqrt{2}} \end{bmatrix} \begin{bmatrix} \sqrt{6} & 0 & 0 \\ 0 & 0 & 0 \end{bmatrix} \begin{bmatrix} \frac{1}{\sqrt{3}} & \frac{1}{\sqrt{3}} & \frac{1}{\sqrt{3}} \\ \frac{2}{\sqrt{6}} & -\frac{1}{\sqrt{6}} & -\frac{1}{\sqrt{6}} \\ 0 & \frac{1}{\sqrt{2}} & -\frac{1}{\sqrt{2}} \end{bmatrix} = \begin{bmatrix} 1 & 1 & 1 \\ 1 & 1 & 1 \end{bmatrix}$.

21. Exercise 11 $\Rightarrow \sigma_1 = \sqrt{2}$, $\sigma_2 = 0$, $\mathbf{v}_1 = \begin{bmatrix} \frac{1}{\sqrt{2}} \\ \frac{1}{\sqrt{2}} \end{bmatrix}$, $\mathbf{v}_2 = \begin{bmatrix} \frac{1}{\sqrt{2}} \\ -\frac{1}{\sqrt{2}} \end{bmatrix}$, $\mathbf{u}_1 = \begin{bmatrix} 1 \\ 0 \end{bmatrix}$, $\mathbf{u}_2 = \begin{bmatrix} 0 \\ 1 \end{bmatrix} \Rightarrow$

outer product form of SVD is $A = \sigma_1\mathbf{u}_1\mathbf{v}_1^T + \sigma_2\mathbf{u}_2\mathbf{v}_2^T = \sqrt{2}\begin{bmatrix} 1 \\ 0 \end{bmatrix} \begin{bmatrix} \frac{1}{\sqrt{2}} & \frac{1}{\sqrt{2}} \end{bmatrix} + 0 \begin{bmatrix} 0 \\ 1 \end{bmatrix} \begin{bmatrix} \frac{1}{\sqrt{2}} & -\frac{1}{\sqrt{2}} \end{bmatrix}$.

22. Exercise 14 $\Rightarrow \sigma_1 = \sqrt{2}$, $\sigma_2 = \sqrt{2}$, $\mathbf{v}_1 = \begin{bmatrix} 1 \\ 0 \end{bmatrix}$, $\mathbf{v}_2 = \begin{bmatrix} 0 \\ 1 \end{bmatrix}$, $\mathbf{u}_1 = \begin{bmatrix} \frac{1}{\sqrt{2}} \\ \frac{1}{\sqrt{2}} \end{bmatrix}$, $\mathbf{u}_2 = \begin{bmatrix} -\frac{1}{\sqrt{2}} \\ \frac{1}{\sqrt{2}} \end{bmatrix} \Rightarrow$ outer product form of SVD is $A = \sigma_1 \mathbf{u}_1 \mathbf{v}_1^T + \sigma_2 \mathbf{u}_2 \mathbf{v}_2^T = \sqrt{2} \begin{bmatrix} \frac{1}{\sqrt{2}} \\ \frac{1}{\sqrt{2}} \end{bmatrix} \begin{bmatrix} 1 & 0 \end{bmatrix} + \sqrt{2} \begin{bmatrix} -\frac{1}{\sqrt{2}} \\ \frac{1}{\sqrt{2}} \end{bmatrix} \begin{bmatrix} 0 & 1 \end{bmatrix}$.

23. Exercise 17 $\Rightarrow A = \sigma_1 \mathbf{u}_1 \mathbf{v}_1^T + \sigma_2 \mathbf{u}_2 \mathbf{v}_2^T = 3 \begin{bmatrix} 0 \\ 1 \\ 0 \end{bmatrix} \begin{bmatrix} 0 & 1 \end{bmatrix} + 2 \begin{bmatrix} 0 \\ 0 \\ -1 \end{bmatrix} \begin{bmatrix} 1 & 0 \end{bmatrix}$.

24. Exercise 19 $\Rightarrow A = \sigma_1 \mathbf{u}_1 \mathbf{v}_1^T + \sigma_2 \mathbf{u}_2 \mathbf{v}_2^T + \sigma_3 \mathbf{u}_3 \mathbf{v}_3^T = \sqrt{5} \begin{bmatrix} 1 \\ 0 \end{bmatrix} \begin{bmatrix} \frac{2}{\sqrt{5}} & 0 & \frac{1}{\sqrt{5}} \end{bmatrix} + 2 \begin{bmatrix} 0 \\ 1 \end{bmatrix} \begin{bmatrix} 0 & 1 & 0 \end{bmatrix}$.

25. Counter examples: In Exercise 11, we could extend with $\mathbf{u}_2 = \begin{bmatrix} 0 \\ -1 \end{bmatrix} \Rightarrow$

$$A = U\Sigma V^T = \begin{bmatrix} 1 & 0 \\ 0 & -1 \end{bmatrix} \begin{bmatrix} \sqrt{2} & 0 \\ 0 & 0 \end{bmatrix} \begin{bmatrix} \frac{1}{\sqrt{2}} & \frac{1}{\sqrt{2}} \\ \frac{1}{\sqrt{2}} & -\frac{1}{\sqrt{2}} \end{bmatrix} = \begin{bmatrix} 1 & 1 \\ 0 & 0 \end{bmatrix}.$$

In Exercise 14, we could take eigenvectors $\mathbf{v}_1 = \begin{bmatrix} -1 \\ 0 \end{bmatrix}$, $\mathbf{v}_2 = \begin{bmatrix} 0 \\ -1 \end{bmatrix} \Rightarrow U = \begin{bmatrix} -\frac{1}{\sqrt{2}} & \frac{1}{\sqrt{2}} \\ -\frac{1}{\sqrt{2}} & -\frac{1}{\sqrt{2}} \end{bmatrix} \Rightarrow$

$$A = U\Sigma V^T = \begin{bmatrix} -\frac{1}{\sqrt{2}} & \frac{1}{\sqrt{2}} \\ -\frac{1}{\sqrt{2}} & -\frac{1}{\sqrt{2}} \end{bmatrix} \begin{bmatrix} \sqrt{2} & 0 \\ 0 & \sqrt{2} \end{bmatrix} \begin{bmatrix} -1 & 0 \\ 0 & -1 \end{bmatrix} = \begin{bmatrix} 1 & -1 \\ 1 & 1 \end{bmatrix}.$$

26. (a) A symmetric $\Rightarrow A^TA = A^2$, so Section 4.3, Theorem 4.18c $\Rightarrow$ the eigenvalues of $A^TA = \lambda^2$,
where λ is an eigenvalue of $A \Rightarrow$ the singular values $\sigma = \sqrt{\lambda^2} = |\lambda|$.

(b) A positive definite $\Rightarrow A$ symmetric and $\lambda > 0$, that is $|\lambda| = \lambda$, so part (a) $\Rightarrow$ the singular values $\sigma = \sqrt{\lambda^2} = |\lambda| = \lambda$.

27. (a) Recall U, V orthogonal $\Rightarrow U^T = U^{-1}$, $V^T = V^{-1}$. So, $A = U\Sigma V^T \Rightarrow$ $U^TAV = \Sigma$, where Σ is diagonal. Therefore, we only need to show $Q = U = V$.
From Exercise 27(b), A positive definite $\Rightarrow \sigma_i = \lambda_i$, where λ_i is an eigenvalue of A.
Also, recall $\mathbf{v}_i$ is the eigenvector of A corresponding to λ_i that is $A\mathbf{v}_i = \lambda_i \mathbf{v}_i$.
So, $\mathbf{u}_i = \frac{1}{\sigma_i} A\mathbf{v}_i = \frac{1}{\lambda_i} A\mathbf{v}_i = \frac{1}{\lambda_i}\lambda_i \mathbf{v}_i = \mathbf{v}_i \Rightarrow Q = U = V$ as we needed to show.

(b) $A = \sigma_1 \mathbf{u}_1 \mathbf{v}_1^T + \ldots + \sigma_n \mathbf{u}_n \mathbf{v}_n^T = \lambda_1 \mathbf{q}_1 \mathbf{q}_1^T + \ldots + \lambda_n \mathbf{q}_n \mathbf{q}_n^T$ the spectral decomposition because A positive definite $\Rightarrow \sigma_i = \lambda_i$ and part (a) $\Rightarrow Q = U = V$.

28. A invertible and $A = U\Sigma V^T \Rightarrow \Sigma = U^TAV \Rightarrow \Sigma$ is invertible because U^T, A, and V are. Furthermore, $A^{-1} = (U\Sigma V^T)^{-1} = V\Sigma^{-1}U^T$.
In the last step, we have again used U, V orthogonal $\Rightarrow U^T = U^{-1}$, $V^T = V^{-1}$.

29. We need to show the columns of U (the left singular vectors) are the eigenvectors of AA^T.
$AA^T = (U\Sigma V^T)(U\Sigma V^T)^T = U\Sigma(V^TV)\Sigma U^T = U\Sigma^2U^T \Rightarrow U^TAA^TU = \Sigma^2$ (diagonal).
Therefore, by Section 4.4, Theorem 4.23, the columns of U are the eigenvectors of AA^T.

30. We need to show the eigenvalues of $A^TA =$ the eigenvalues of $(A^T)^TA^T = AA^T$.
From Exercise 29, $AA^T = U\Sigma^2U^T$. Likewise, $A^TA = V\Sigma^2V^T \Rightarrow$
eigenvalues of $A^TA =$ eigenvalues of $AA^T = (A^T)^TA^T$ (diagonal entries of Σ^2) $\Rightarrow$
singular values of $A =$ positive square roots of eigenvalues of $A^TA =$ singular values of A^T.

31. We need to show the eigenvalues of A^TA are the same as the eigenvalues of $(QA)^T(QA)$.
In fact, we will use the fact that Q is orthogonal to show that $(QA)^T(QA) = A^TA$.

So, $(QA)^T(QA) = A^TQ^T(QA) = A^T(Q^TQ)A = A^TA$.

Q: What key property of orthogonal matrices did we use in the equation above?
A: When Q is orthogonal $Q^{-1} = Q^T$, therefore, $Q^TQ = I$.

32. We know $\{\mathbf{v}_1, \ldots, \mathbf{v}_r\}$ is an orthonormal set $\Rightarrow$ it is linearly independent.

Now $A = U\Sigma V^T \Rightarrow A^T = (U\Sigma V^T)^T = V\Sigma^TU^T \Rightarrow V\Sigma^T = A^TU \Rightarrow$
$\mathbf{v}_i\sigma_i = A^T\mathbf{u}_i \Rightarrow \mathbf{v}_i = \frac{1}{\sigma_i}A^T\mathbf{u}_i \Rightarrow \mathbf{v}_i \in \text{row}(A)$. Furthermore, $r = \text{rank}(A) = \dim(\text{row}(A))$.

Therefore, $\{\mathbf{v}_1, \ldots, \mathbf{v}_r\}$ is a basis for row(A).

33. Exercises 3 and 11 $\Rightarrow \text{rank}(A) = 1 < 2$ (from $\mathbb{R}^2$), so the proof of Theorem 7.16 $\Rightarrow y_1 \le 1$.

Furthermore, since y_1 corresponds to $\mathbf{e}_1 = \begin{bmatrix} 1 \\ 0 \end{bmatrix} \Rightarrow y_1 \le 1$ is the interval $[-1, 1]$.

34. Exercises 7 and 17 $\Rightarrow \text{rank}(A) = 2$ (from $\mathbb{R}^2$), so the proof of Theorem 7.16 $\Rightarrow$
$(\frac{y_1}{3})^2 + (\frac{y_2}{2})^2 = 1$ (because $\sigma_1 = 3$ and $\sigma_2 = 2$). After simplifying, we have $\frac{y_1^2}{9} + \frac{y_2^2}{4} = 1$.

35. Exercises 9 and 19 $\Rightarrow \text{rank}(A) = 2 < 3$ (from $\mathbb{R}^3$), so the proof of Theorem 7.16 $\Rightarrow$
$(\frac{y_1}{\sqrt{5}})^2 + (\frac{y_2}{2})^2 \le 1$ (because $\sigma_1 = \sqrt{5}$ and $\sigma_2 = 2$). After simplifying, we have $\frac{y_1^2}{5} + \frac{y_2^2}{4} \le 1$.

36. Exercise 10 $\Rightarrow \text{rank}(A) = 2 < 3$ (from $\mathbb{R}^3$), so the proof of Theorem 7.16 $\Rightarrow$
$(\frac{y_1}{3})^2 + (\frac{y_2}{2})^2 \le 1$. So, we have $\frac{y_1^2}{9} + \frac{y_2^2}{4} \le 1$.

37. Exercise 3 $\Rightarrow$ (a) $\|A\|_2 = \sigma_1 = \sqrt{2}$. (b) $\text{cond}_2(A) = \frac{\sigma_1}{\sigma_2} = \frac{\sqrt{2}}{0} = \infty$.

38. Exercise 8 $\Rightarrow$ (a) $\|A\|_2 = \sigma_1 = 3$. (b) $\text{cond}_2(A) = \frac{\sigma_1}{\sigma_2} = \frac{3}{1} = 3$.

39. $A^TA = \begin{bmatrix} 2 & 1.9 \\ 1.9 & 1.81 \end{bmatrix} \Rightarrow$ the eigenvalues of A^TA are $\lambda_1 \approx 3.807$, $\lambda_2 \approx 0.0026 \Rightarrow$

(a) $\|A\|_2 = \sigma_1 = \sqrt{\lambda_1} \approx \sqrt{3.807} \approx 1.95$. (b) $\text{cond}_2(A) = \frac{\sigma_1}{\sigma_2} \approx \frac{\sqrt{\lambda_1}}{\sqrt{\lambda_2}} \approx \frac{\sqrt{3.807}}{\sqrt{0.0026}} \approx 38.27$.

40. $A^TA = \begin{bmatrix} 10100 & 10100 & 100 \\ 10100 & 10100 & 100 \\ 100 & 100 & 1 \end{bmatrix} \Rightarrow$ the eigenvalues of A^TA are $\lambda_1 \approx 20200$, $\lambda_2 \approx 0.0099 \Rightarrow$

(a) $\|A\|_2 = \sigma_1 \approx \sqrt{20200} \approx 142.13$. (b) $\text{cond}_2(A) = \frac{\sigma_1}{\sigma_2} \approx \frac{\sqrt{20200}}{\sqrt{0.0099}} \approx 1428.43$.

41. Exercise 11 $\Rightarrow A^+ = V\Sigma^+U^T = \begin{bmatrix} \frac{1}{\sqrt{2}} & \frac{1}{\sqrt{2}} \\ \frac{1}{\sqrt{2}} & -\frac{1}{\sqrt{2}} \end{bmatrix} \begin{bmatrix} \frac{1}{\sqrt{2}} & 0 \\ 0 & 0 \end{bmatrix} \begin{bmatrix} 1 & 0 \\ 0 & 1 \end{bmatrix} = \begin{bmatrix} \frac{1}{2} & 0 \\ \frac{1}{2} & 0 \end{bmatrix}$.

42. Exercise 18 $\Rightarrow A^+ = V\Sigma^+U^T = \begin{bmatrix} \frac{1}{\sqrt{2}} & \frac{1}{\sqrt{2}} \\ -\frac{1}{\sqrt{2}} & \frac{1}{\sqrt{2}} \end{bmatrix} \begin{bmatrix} \frac{1}{3} & 0 & 0 \\ 0 & 1 & 0 \end{bmatrix} \begin{bmatrix} \frac{1}{3\sqrt{2}} & -\frac{1}{3\sqrt{2}} & -\frac{4}{3\sqrt{2}} \\ \frac{1}{\sqrt{2}} & \frac{1}{\sqrt{2}} & 0 \\ \frac{2}{3} & -\frac{2}{3} & \frac{1}{3} \end{bmatrix} = \begin{bmatrix} \frac{5}{9} & \frac{4}{9} & -\frac{2}{9} \\ \frac{4}{9} & \frac{5}{9} & \frac{2}{9} \end{bmatrix}$.

43. Exercise 19 $\Rightarrow A^+ = V\Sigma^+U^T = \begin{bmatrix} \frac{2}{\sqrt{5}} & 0 & \frac{1}{\sqrt{5}} \\ 0 & 1 & 0 \\ \frac{1}{\sqrt{5}} & 0 & -\frac{2}{\sqrt{5}} \end{bmatrix} \begin{bmatrix} \frac{1}{\sqrt{5}} & 0 \\ 0 & \frac{1}{2} \\ 0 & 0 \end{bmatrix} \begin{bmatrix} 1 & 0 \\ 0 & 1 \end{bmatrix} = \begin{bmatrix} \frac{2}{5} & 0 \\ 0 & \frac{1}{2} \\ \frac{1}{5} & 0 \end{bmatrix}$.

44. Exercise 10 $\Rightarrow A^TA = \begin{bmatrix} 2 & 0 & 2 \\ 0 & 9 & 0 \\ 2 & 0 & 2 \end{bmatrix} \Rightarrow$ singular values of A $\sigma_1 = 3$, $\sigma_2 = 2$, $\sigma_3 = 0 \Rightarrow$

$$V = \begin{bmatrix} 0 & \frac{1}{\sqrt{2}} & \frac{1}{\sqrt{2}} \\ 1 & 0 & 0 \\ 0 & \frac{1}{\sqrt{2}} & -\frac{1}{\sqrt{2}} \end{bmatrix},\ \Sigma = \begin{bmatrix} 3 & 0 & 0 \\ 0 & 1 & 0 \\ 0 & 0 & 0 \end{bmatrix},\ U = \begin{bmatrix} 0 & \frac{1}{\sqrt{2}} & \frac{1}{\sqrt{2}} \\ -1 & 0 & 0 \\ 0 & \frac{1}{\sqrt{2}} & -\frac{1}{\sqrt{2}} \end{bmatrix} \Rightarrow$$

$$A^+ = V\Sigma^+U^T = \begin{bmatrix} 0 & \frac{1}{\sqrt{2}} & \frac{1}{\sqrt{2}} \\ 1 & 0 & 0 \\ 0 & \frac{1}{\sqrt{2}} & -\frac{1}{\sqrt{2}} \end{bmatrix} \begin{bmatrix} \frac{1}{3} & 0 & 0 \\ 0 & 1 & 0 \\ 0 & 0 & 0 \end{bmatrix} \begin{bmatrix} 0 & -1 & 0 \\ \frac{1}{\sqrt{2}} & 0 & \frac{1}{\sqrt{2}} \\ \frac{1}{\sqrt{2}} & 0 & -\frac{1}{\sqrt{2}} \end{bmatrix} = \begin{bmatrix} \frac{1}{2} & 0 & \frac{1}{2} \\ 0 & -\frac{1}{3} & 0 \\ \frac{1}{2} & 0 & \frac{1}{2} \end{bmatrix}.$$

45. $A^TA = \begin{bmatrix} 5 & 10 \\ 10 & 20 \end{bmatrix} \Rightarrow \sigma_i = 5, 0 \Rightarrow V = \begin{bmatrix} \frac{1}{\sqrt{5}} & \frac{2}{\sqrt{5}} \\ \frac{2}{\sqrt{5}} & -\frac{1}{\sqrt{5}} \end{bmatrix}$, $\Sigma = \begin{bmatrix} 5 & 0 \\ 0 & 0 \end{bmatrix}$, $U = \begin{bmatrix} \frac{1}{\sqrt{5}} & \frac{2}{\sqrt{5}} \\ \frac{2}{\sqrt{5}} & -\frac{1}{\sqrt{5}} \end{bmatrix} \Rightarrow$

$$A^+ = \begin{bmatrix} \frac{1}{\sqrt{5}} & \frac{2}{\sqrt{5}} \\ \frac{2}{\sqrt{5}} & -\frac{1}{\sqrt{5}} \end{bmatrix} \begin{bmatrix} \frac{1}{5} & 0 \\ 0 & 0 \end{bmatrix} \begin{bmatrix} \frac{1}{\sqrt{5}} & \frac{2}{\sqrt{5}} \\ \frac{2}{\sqrt{5}} & -\frac{1}{\sqrt{5}} \end{bmatrix} = \begin{bmatrix} \frac{1}{25} & \frac{2}{25} \\ \frac{2}{25} & \frac{4}{25} \end{bmatrix} \Rightarrow \overline{\mathbf{x}} = \begin{bmatrix} \frac{1}{25} & \frac{2}{25} \\ \frac{2}{25} & \frac{4}{25} \end{bmatrix} \begin{bmatrix} 3 \\ 5 \end{bmatrix} = \begin{bmatrix} 0.52 \\ 1.04 \end{bmatrix}.$$

46. $A^TA = \begin{bmatrix} 9 & 0 & 0 \\ 0 & 0 & 0 \\ 0 & 0 & 4 \end{bmatrix} \Rightarrow \sigma_i = 3, 2, 0 \Rightarrow V = \begin{bmatrix} 1 & 0 & 0 \\ 0 & 0 & 1 \\ 0 & 1 & 0 \end{bmatrix}$ and $U = \begin{bmatrix} 1 & 0 \\ 0 & 1 \end{bmatrix} \Rightarrow$

$$A^+ = \begin{bmatrix} 1 & 0 & 0 \\ 0 & 0 & 1 \\ 0 & 1 & 0 \end{bmatrix} \begin{bmatrix} \frac{1}{3} & 0 \\ 0 & \frac{1}{2} \\ 0 & 0 \end{bmatrix} \begin{bmatrix} 1 & 0 \\ 0 & 1 \end{bmatrix} = \begin{bmatrix} \frac{1}{3} & 0 \\ 0 & 0 \\ 0 & \frac{1}{2} \end{bmatrix} \Rightarrow \overline{\mathbf{x}} = \begin{bmatrix} \frac{1}{3} & 0 \\ 0 & 0 \\ 0 & \frac{1}{2} \end{bmatrix} \begin{bmatrix} 3 \\ 0 \end{bmatrix} = \begin{bmatrix} 1 \\ 0 \\ 0 \end{bmatrix}.$$

47. $A^TA = \begin{bmatrix} 3 & 3 \\ 3 & 3 \end{bmatrix} \Rightarrow \sigma_i = \sqrt{6}, 0 \Rightarrow V = \begin{bmatrix} \frac{1}{\sqrt{2}} & \frac{1}{\sqrt{2}} \\ \frac{1}{\sqrt{2}} & -\frac{1}{\sqrt{2}} \end{bmatrix}$ and $U = \begin{bmatrix} \frac{1}{\sqrt{3}} & \frac{2}{\sqrt{6}} & 0 \\ \frac{1}{\sqrt{3}} & -\frac{1}{\sqrt{6}} & \frac{1}{\sqrt{2}} \\ \frac{1}{\sqrt{3}} & -\frac{1}{\sqrt{6}} & \frac{1}{\sqrt{2}} \end{bmatrix} \Rightarrow$

$$A^+ = \begin{bmatrix} \frac{1}{\sqrt{2}} & \frac{1}{\sqrt{2}} \\ \frac{1}{\sqrt{2}} & -\frac{1}{\sqrt{2}} \end{bmatrix} \begin{bmatrix} \frac{1}{\sqrt{6}} & 0 & 0 \\ & 0 & 0 & 0 \end{bmatrix} \begin{bmatrix} \frac{1}{\sqrt{3}} & \frac{1}{\sqrt{3}} & \frac{1}{\sqrt{3}} \\ \frac{2}{\sqrt{6}} & -\frac{1}{\sqrt{6}} & -\frac{1}{\sqrt{6}} \\ 0 & \frac{1}{\sqrt{2}} & \frac{1}{\sqrt{2}} \end{bmatrix} = \begin{bmatrix} \frac{1}{6} & \frac{1}{6} & \frac{1}{6} \\ \frac{1}{6} & \frac{1}{6} & \frac{1}{6} \end{bmatrix} \Rightarrow \overline{\mathbf{x}} = \begin{bmatrix} \frac{1}{6} & \frac{1}{6} & \frac{1}{6} \\ \frac{1}{6} & \frac{1}{6} & \frac{1}{6} \end{bmatrix} \begin{bmatrix} 1 \\ 2 \\ 3 \end{bmatrix} = \begin{bmatrix} 1 \\ 1 \end{bmatrix}.$$

48. $A^TA = \begin{bmatrix} 2&0&2 \\ 0&1&0 \\ 2&0&2 \end{bmatrix} \Rightarrow \sigma_i = 2, 1, 0 \Rightarrow V = \begin{bmatrix} \frac{1}{\sqrt{2}} & 0 & \frac{1}{\sqrt{2}} \\ 0 & 1 & 0 \\ \frac{1}{\sqrt{2}} & 0 & -\frac{1}{\sqrt{2}} \end{bmatrix}$ and $U = \begin{bmatrix} \frac{1}{\sqrt{2}} & 0 & \frac{1}{\sqrt{2}} \\ 0 & 1 & 0 \\ \frac{1}{\sqrt{2}} & 0 & -\frac{1}{\sqrt{2}} \end{bmatrix} \Rightarrow$

$$A^+ = \begin{bmatrix} \frac{1}{\sqrt{2}} & 0 & \frac{1}{\sqrt{2}} \\ 0 & 1 & 0 \\ \frac{1}{\sqrt{2}} & 0 & -\frac{1}{\sqrt{2}} \end{bmatrix} \begin{bmatrix} \frac{1}{2}&0&0 \\ 0&1&0 \\ 0&0&0 \end{bmatrix} \begin{bmatrix} \frac{1}{\sqrt{2}} & 0 & \frac{1}{\sqrt{2}} \\ 0 & 1 & 0 \\ \frac{1}{\sqrt{2}} & 0 & -\frac{1}{\sqrt{2}} \end{bmatrix} = \begin{bmatrix} \frac{1}{4} & 0 & \frac{1}{4} \\ 0&1&0 \\ \frac{1}{4} & 0 & \frac{1}{4} \end{bmatrix} \Rightarrow \overline{\mathbf{x}} = A^+ \begin{bmatrix} 1\\1\\1 \end{bmatrix} = \begin{bmatrix} \frac{1}{2} \\ 1 \\ \frac{1}{2} \end{bmatrix}.$$

49. (a) Normal equations: $A^TA\overline{\mathbf{x}} = A^T\mathbf{b}$. So, $A^TA = \begin{bmatrix} 2&2 \\ 2&2 \end{bmatrix}$ and $A^T\mathbf{b} = \begin{bmatrix} 1\\1 \end{bmatrix}$.

So, since $\left[\begin{array}{cc|c} 2&2&1 \\ 2&2&1 \end{array}\right] \longrightarrow \left[\begin{array}{cc|c} 1&1&\frac{1}{2} \\ 0&0&0 \end{array}\right] \Rightarrow \overline{\mathbf{x}} = \begin{bmatrix} \frac{1}{2} \\ 0 \end{bmatrix} \Rightarrow x + y = \frac{1}{2}$.

(b) Following the method of Section 1.3, Example 1.20, we have:

$$\mathbf{x} = \mathbf{p} + t\mathbf{d} \Rightarrow \begin{bmatrix} x\\y \end{bmatrix} = \begin{bmatrix} \frac{1}{2} \\ 0 \end{bmatrix} + t\begin{bmatrix} 1 \\ -1 \end{bmatrix} = \begin{bmatrix} \frac{1}{2}+t \\ t \end{bmatrix}.$$

(c) We need to minimize $\left\| \begin{bmatrix} \frac{1}{2}+t \\ t \end{bmatrix} \right\|^2 = (1+t)^2 + (-t)^2 = 2t^2 + t + \frac{1}{4}$.

This is an upward opening parabola, so its minimum occurs at $t = -\frac{b}{2a} \Rightarrow$

$t = -\frac{1}{2(2)} = -\frac{1}{4} \Rightarrow \overline{\mathbf{x}}$ of minimum length$= \begin{bmatrix} \frac{1}{2} + (-\frac{1}{4}) \\ -(-\frac{1}{4}) \end{bmatrix} = \begin{bmatrix} \frac{1}{4} \\ \frac{1}{4} \end{bmatrix}$ as found in Example 7.40.

50. As suggested by Exercise 28, A invertible $\Rightarrow \Sigma^+ = \Sigma^{-1}$, $\Sigma^T = \Sigma$. With $A = U\Sigma V^T$, we have:
$A^T = (U\Sigma V^T)^T = V\Sigma U^T \Rightarrow A^TA = (V\Sigma U^T)(U\Sigma V^T) = V\Sigma^2V^T \Rightarrow (A^TA)^{-1} = V(\Sigma^{-1})^2V^T \Rightarrow$
(7.3) $A^+ = (A^TA)^{-1}A^T = (V(\Sigma^{-1})^2V^T)(V\Sigma U^T) = V\Sigma^{-1}U^T = V\Sigma^+U^T = A^+$ (7.4).

51. Note, $\Sigma^+_{n\times m}\Sigma_{m\times n} = \begin{bmatrix} I_{r\times r} & 0 \\ 0 & 0 \end{bmatrix}_{n\times n} \Rightarrow \Sigma\Sigma^+\Sigma = \Sigma(\Sigma^+\Sigma) = \begin{bmatrix} D_{r\times r} & 0 \\ 0 & 0 \end{bmatrix} \begin{bmatrix} I_{r\times r} & 0 \\ 0 & 0 \end{bmatrix}_{n\times n} = \Sigma.$

(a) Will show $AA^+A = A$ using the fact that $\Sigma\Sigma^+\Sigma = \Sigma$ (shown above) in step 4.
$AA^+A = A(V\Sigma^+U^T)A = (U\Sigma V^T)(V\Sigma^+U^T)(U\Sigma V^T) = U\Sigma(V^TV)\Sigma^+(U^TU)\Sigma V^T$
$= U(\Sigma\Sigma^+\Sigma)V^T = U\Sigma V^T = A.$
We are also using the fact (again) that U, V orthogonal $\Rightarrow V^TV = U^TU = I$.

(b) Will show $A^+AA^+ = A^+$ using the fact that $\Sigma^+\Sigma\Sigma^+ = \Sigma^+$ (see note) in step 4.
$A^+AA^+ = (V\Sigma^+U^T)A(V\Sigma^+U^T) = (V\Sigma^+U^T)(U\Sigma V^T)(V\Sigma^+U^T)$
$= V\Sigma^+(U^TU)\Sigma(V^TV)\Sigma^+V^T = V(\Sigma^+\Sigma\Sigma^+)U^T = V\Sigma^+U^T = A^+.$

(c) Will show $(AA^+)^T = AA^+$ using the fact that $(\Sigma^+)^T(\Sigma)^T = \Sigma\Sigma^+$ (see note) in step 5.
$(AA^+)^T = (AV\Sigma^+U^T)^T = U(\Sigma^+)^TV^TA^T = U(\Sigma^+)^T(V^TV)(\Sigma)^TU^T$
$= U(\Sigma^+)^T(\Sigma)^TU^T = U\Sigma\Sigma^+U^T = (U\Sigma V^T)(V\Sigma^+U^T) = AA^+.$
Likewise, $(\Sigma)^T(\Sigma^+)^T = \Sigma^+\Sigma \Rightarrow (A^+A)^T = A^+A$.

52. As suggested in the hint, we will show $A' = A'AA^+$ and $A^+ = A'AA^+$.
Will cite Penrose conditions $(a)', (b)', (c)'$ and $(a)^+, (b)^+, (c)^+$ as they are used for A' and A^+.

$$\begin{aligned} A' &\overset{(b)'}{=} A'AA' = ((A')^T A^T (A')^T)^T &&\overset{(a)^+}{=} ((A')^T (AA^+A)^T (A')^T)^T \\ &= ((A')^T (A^T (AA^+)^T)(A')^T)^T &&= (((A')^T A^T)(AA^+)^T)(A')^T)^T \\ &\overset{(c)'}{=} ((AA')(AA^+)^T)(A')^T)^T &&\overset{(c)^+}{=} ((AA')(AA^+)(A')^T)^T \\ &= ((AA'A)A^+(A')^T)^T &&\overset{(a)'}{=} (AA^+(A')^T)^T = A'(A^+)^T A^T \overset{(c)^+}{=} A'AA^+. \end{aligned}$$

Likewise, $A^+ = A'AA^+ \Rightarrow A' = A^+ \Rightarrow A^+$ is unique.

53. We will show A satisfies the Penrose conditions for A^+, so by Exercise 52 $A = (A^+)^+$.
Comparing the Penrose conditions for A^+ and A, we have:

$$\begin{array}{llll} (a)^+ & A^+(A^+)^+A^+ = A^+ & (a) & AA^+A = A \\ (b)^+ & (A^+)^+A^+(A^+)^+ = (A^+)^+ & (b) & A^+AA^+ = A^+ \\ (c)^+ & i.\ (A^+(A^+)^+)^T = A^+(A^+)^+ & (c) & i.\ (AA^+)^T = AA^+ \\ & ii.\ ((A^+)^+A^+)^T = (A^+)^+A^+ & & ii.\ (A^+A)^T = A^+A \end{array}$$

Since A satisfies conditions $(a), (b), (c)$, will show A satisfies conditions $(a)^+, (b)^+, (c)^+$.

$(a)^+$ A satisfies $(b) \Rightarrow A^+AA^+ = A^+ \Rightarrow A$ satisfies $(a)^+$, that is $A^+(A^+)^+A^+ = A^+$.

$(b)^+$ A satisfies $(a) \Rightarrow AA^+A = A \Rightarrow A$ satisfies $(b)^+$, that is $(A^+)^+A^+(A^+)^+ = (A^+)^+$.

$(c)^+$ A satisfies $(c)\, ii. \Rightarrow (A^+A)^T = A^+A \Rightarrow A$ satisfies $(c)^+ i.$,
that is $(A^+(A^+)^+)^T = A^+(A^+)^+$. Likewise, for $(c)^+ ii. \Rightarrow$

A satisfies the Penrose conditions for $A^+ \Rightarrow A = (A^+)^+$.

54. We will show $(A^+)^T$ satisfies the Penrose conditions for A^T, so by Exercise 52 $(A^+)^T = (A^T)^+$.
Comparing Penrose conditions for A^T and the transpose of conditions for A, we have:

$$\begin{array}{llll} (a)^T & A^T(A^T)^+A^T = A^T & (a) & A^T(A^+)^TA^T = A^T \\ (b)^T & (A^T)^+A^T(A^T)^+ = (A^T)^+ & (b) & (A^+)^TA^T(A^+)^T = (A^+)^T \\ (c)^T & i.\ (A^T(A^T)^+)^T = A^T(A^T)^+ & (c) & i.\ ((A^+)^TA^T)^T = (AA^+)^T = (A^+)^TA^T \\ & ii.\ ((A^T)^+A^T)^T = (A^T)^+A^T & & ii.\ (A^T(A^+)^T)^T = (A^+A)^T = A^T(A^+)^T \end{array}$$

Since $(A^+)^T$ satisfies conditions $(a), (b), (c)$ will show A satisfies conditions $(a)^T, (b)^T, (c)^T$.

$(a)^T$ $(A^+)^T$ satisfies $(a) \Rightarrow A^T(A^+)^TA^T = A^T \Rightarrow (A^+)^T$ satisfies $(a)^T$.

$(b)^T$ $(A^+)^T$ satisfies $(b) \Rightarrow (A^+)^TA^T(A^+)^T = (A^+)^T \Rightarrow (A^+)^T$ satisfies $(b)^T$.

$(c)^T$ $(A^+)^T$ satisfies $(c)\, ii. \Rightarrow (A^+)^T$ satisfies $(c)^T i.$ Likewise, for $(c)^T ii. \Rightarrow$

$(A^+)^T$ satisfies the Penrose conditions for $A^T \Rightarrow (A^+)^T = (A^T)^+$.

55. We will show A satisfies the Penrose conditions for A, so by Exercise 52 $A = A^+$.
Comparing Penrose conditions for A and conditions derived from $A = A^T = A^2$, we have:

$(a)\ AA^+A = A$	$(a)'\ AAA = A^2A = A^2 = A$
$(b)\ A^+AA^+ = A^+$	$(b)'\ AAA = A^2A = A^2 = A$
$(c)\ i.\ (AA^+)^T = AA^+$	$(c)'\ i.\ (AA)^T = A^TA^T = AA$
$ii.\ (A^+A)^T = A^+A$	$ii.\ (AA)^T = A^TA^T = AA$

Since A satisfies conditions $(a)', (b)', (c)'$ will show A satisfies conditions $(a), (b), (c)$.

(a) A satisfies $(a)' \Rightarrow AAA = A \Rightarrow A$ satisfies (a), that is $AA^+A = A$.

(b) A satisfies $(b)' \Rightarrow AAA = A \Rightarrow A$ satisfies (b), that is $A^+AA^+ = A^+$.

(c) A satisfies $(c)'\,i. \Rightarrow A$ satisfies $(c)\,i.$ Likewise, for $(c)\,ii. \Rightarrow$

A satisfies the Penrose conditions for $A \Rightarrow A = A^+$.

56. We need only show $(QA)^+ = A^+Q^T$ satisfies the Penrose conditions for QA.

(a) $(QA)(A^+Q^T)(QA) = Q(AA^+)(Q^TQ)A = Q(AA^+A) = QA$

(b) $(A^+Q^T)(QA)(A^+Q^T) = A^+(Q^TQ)AA^+Q^T) = (A^+AA^+)Q^T = A^+Q^T$

(c) (i) $\left((QA)(A^+Q^T)\right)^T = (A^+Q^T)^T(QA)^T = Q((A^+)^TA^T)Q^T = Q(AA^+)^TQ^T$

$$= Q(AA^+)Q^T = (QA)(A^+Q^T)$$

(ii) $\left((A^+Q^T)(QA)\right)^T = (QA)^T(A^+Q^T)^T = A^T(Q^TQ)(A^+)^T = A^T(A^+)^T = (A^+A^T)^T$

$$= A^+A = A^+(Q^TQ)A = (A^+Q^T)(QA)$$

A^+Q^T satisfies the Penrose conditions for $QA \Rightarrow A^+Q^T = (QA)^+$.

57. From Section 5.5, p.415, we know a positive definite matrix A is *symmetric*. So, $A^T = A$.
Therefore, $A^T = (U\Sigma V^T)^T = V\Sigma U^T = U\Sigma V^T = A$.

Q: Is this enough to prove that $U = V$? Why or why not?
If not, what additional implication of being positive definite do we need to evoke?

58. Before we begin, let us review some implications of a diagonal matrix A.

Q: In a diagonal matrix, the maximum value for every *column* occurs on the diagonal. Why?
A: The only nonzero entry occurs on the diagonal. In particular, $\max\{|a_{ij}|\} = |a_{ii}|$.

Q: Since a_{ii} is the only nonzero entry, what can we conclude about $\|\mathbf{a}_i\|$?
A: We can conclude that $\|\mathbf{a}_i\| = |a_{ii}|$.

Q: Likewise, what can we conclude about $|a_{i1}| + \cdots + |a_{in}|$?
A: We can conclude that $|a_{i1}| + \cdots + |a_{in}| = |a_{ii}|$.

Q: Do the statements above also hold for the *rows* of a diagonal matrix A?
A: Yes.

Recall the statement of Theorem 7.7 on p.569 of Section 7.2:

Thm 7.7 p.569 7.2 Let A have column vectors $\mathbf{a}_i$ and row vectors $\mathbf{A}_i$ then:

$$\text{a. } \|A\|_1 = \max\{\|\mathbf{a}_i\|_s\} = \max\left\{\sum_{i=1}^{n} |a_{ij}|\right\}$$

$$\text{b. } \|A\|_\infty = \max\{\|\mathbf{A}_i\|_s\} = \max\left\{\sum_{j=1}^{n} |a_{ij}|\right\}$$

So for a diagonal matrix A, $\|A\|_1 = \|A\|_\infty$.

Q: Recall $\|A\|_2$ is based upon $\|\mathbf{v}\|_2 = \|\mathbf{v}\|_E = \sqrt{|v_1|^2 + \cdots + |v_n|^2}$.
Why does this imply $\|A\|_1 = \|A\|_2$ for a diagonal matrix A?

Q: How might we state these observations in words?
A: In a diagonal matrix, the maximum value of each column and row occurs on the diagonal. Furthermore, the only nonzero entry in each column is on the diagonal.

59. We follow the hint to prove $\|A\|_2^2 \le \|A\|_1 \|A\|_\infty$.

Let σ be the largest singular value of A, then as noted in the hint: $\|A\|_2^2 = \sigma^2 = |\lambda|$.

Now since $\|A^T\|_1 = \|A\|_\infty$, $\|A^TA\|_1 = \|A^T\|_1\|A\|_1 = \|A^T\|_1\|A\|_1 = \|A\|_\infty\|A\|_1$.

By Exercise 34 of Section 7.2, $\|A\|_1\|A\|_\infty \ge |\lambda| = \sigma^2 = \|A\|_2^2$ as required.

60. Note, A square $\Rightarrow (\Sigma_{n\times n})^T = \begin{bmatrix} D_{r\times r} & 0 \\ 0 & 0 \end{bmatrix}^T_{n\times n} = \begin{bmatrix} D_{r\times r} & 0 \\ 0 & 0 \end{bmatrix}_{n\times n} = \Sigma_{n\times n} = \Sigma$.

Also, U orthogonal $\Rightarrow U^TU = I_n \Rightarrow \Sigma U^TU = \Sigma(U^TU) = \Sigma_{n\times n} I_n = \Sigma$.
Therefore, $A = U\Sigma V^T = U\Sigma(U^TU)V^T = (U\Sigma U^T)(UV^T) = RQ$.

Now, $R = U\Sigma U^T$ has eigenvalues $\sigma_i \ge 0$ and $R^T = (U\Sigma U^T)^T = U\Sigma^T U^T = U\Sigma U^T = R \Rightarrow$
R is symmetric with non-negative eigenvalues $\Rightarrow R$ is positive semidefinite by definition.

U, V^T orthogonal $\Rightarrow Q = UV^T$ orthogonal (by Section 5.1, Theorem 8d).

Therefore, $A = RQ$ is a polar decomposition of the square matrix A.

61. Recall, from Exercises 3 and 11, $U = I_2$, so we have:

$$A = RQ = (U\Sigma U^T)(UV^T) = (I_2\,\Sigma\, I_2^T)(I_2\, V^T) = \Sigma V^T = \begin{bmatrix} \sqrt{2} & 0 \\ 0 & 0 \end{bmatrix}\begin{bmatrix} \frac{1}{\sqrt{2}} & \frac{1}{\sqrt{2}} \\ \frac{1}{\sqrt{2}} & -\frac{1}{\sqrt{2}} \end{bmatrix}.$$

62. Recall, from Exercise 14, $V = V^T = I_2$ and $\Sigma = \sqrt{2}\,I_2$, so we have:

$$A = RQ = (\sqrt{2}\,I_2 UU^T)(U) = \sqrt{2}\,I_2 U = \begin{bmatrix} \sqrt{2} & 0 \\ 0 & \sqrt{2} \end{bmatrix} \begin{bmatrix} \frac{1}{\sqrt{2}} & -\frac{1}{\sqrt{2}} \\ \frac{1}{\sqrt{2}} & \frac{1}{\sqrt{2}} \end{bmatrix}.$$

63. Though it is possible to solve this problem exactly, you may wish to solve it using a CAS. However, your answers should be exact as the following analysis shows.

$$A = \begin{bmatrix} 1 & 2 \\ -3 & -1 \end{bmatrix} \Rightarrow A^T A = \begin{bmatrix} 10 & 5 \\ 5 & 5 \end{bmatrix} \Rightarrow \begin{vmatrix} 10-\lambda & 5 \\ 5 & 5-\lambda \end{vmatrix} = 0 \Rightarrow$$

$$\lambda_1 = \frac{15+5\sqrt{5}}{2},\ \mathbf{v}_1 = \begin{bmatrix} \frac{1+\sqrt{5}}{2} \\ 1 \end{bmatrix} \text{ and } \lambda_2 = \frac{15-5\sqrt{5}}{2},\ \mathbf{v}_2 = \begin{bmatrix} \frac{1-\sqrt{5}}{2} \\ 1 \end{bmatrix}.$$

Normalizing these vectors, we create V and then use the columns of V to create U:

$$V = \begin{bmatrix} \sqrt{\frac{2}{5-\sqrt{5}}} & -\sqrt{\frac{2}{5+\sqrt{5}}} \\ \sqrt{\frac{2}{5+\sqrt{5}}} & \sqrt{\frac{2}{5-\sqrt{5}}} \end{bmatrix} \text{ so } V \text{ has the form } V = \begin{bmatrix} v_{11} & -v_{21} \\ v_{21} & v_{11} \end{bmatrix} \Rightarrow$$

$$\mathbf{u}_1 = \frac{1}{\sigma_1} A\mathbf{v}_1 = \sqrt{\tfrac{1}{\lambda_1}} \begin{bmatrix} 1 & 2 \\ -3 & -1 \end{bmatrix} \begin{bmatrix} v_{11} \\ v_{21} \end{bmatrix} = \begin{bmatrix} \sqrt{\frac{1}{\lambda_1}}(v_{11}+2v_{21}) \\ -\sqrt{\frac{1}{\lambda_1}}(3v_{11}+v_{21}) \end{bmatrix} = \begin{bmatrix} v_{21} \\ -v_{11} \end{bmatrix}.$$

This last equality and ones like it are the reason this solution is exact. We show one below:

$\sqrt{\frac{1}{\lambda_1}}(v_{11}+2v_{21}) = v_{21} \Leftrightarrow \sqrt{\frac{1}{\lambda_1}}(\frac{v_{11}}{v_{21}}+2)v_{21} = v_{21} \Leftrightarrow \frac{v_{11}}{v_{21}} + 2 = \sqrt{\lambda_1}$ that is:

$$\frac{\sqrt{\frac{2}{5-\sqrt{5}}}}{\sqrt{\frac{2}{5+\sqrt{5}}}} + 2 = \sqrt{\frac{15+5\sqrt{5}}{2}} \Rightarrow \sqrt{\frac{5+\sqrt{5}}{5-\sqrt{5}}} + 2 = \frac{5+\sqrt{5}}{\sqrt{20}} + \frac{2\sqrt{20}}{\sqrt{20}} = \frac{5+5\sqrt{5}}{\sqrt{20}}$$

$$= \sqrt{\frac{(5+5\sqrt{5})^2}{20}} = \sqrt{\frac{150+50\sqrt{5}}{20}} = \sqrt{\frac{15+5\sqrt{5}}{2}} = \sqrt{\lambda_1}.$$

Likewise, $\mathbf{u}_2 = \frac{1}{\sigma_2} A\mathbf{v}_2 = \begin{bmatrix} v_{11} \\ v_{21} \end{bmatrix} \Rightarrow U = \begin{bmatrix} v_{21} & v_{11} \\ -v_{11} & v_{21} \end{bmatrix}$. Therefore, $Q = UV^T$

$$= \begin{bmatrix} v_{21} & v_{11} \\ -v_{11} & v_{21} \end{bmatrix} \begin{bmatrix} v_{11} & v_{21} \\ -v_{21} & v_{11} \end{bmatrix} = \begin{bmatrix} v_{11}v_{21} - v_{11}v_{21} = 0 & v_{11}^2 + v_{21}^2 = 1 \\ -(v_{11}^2 + v_{21}^2) = -1 & v_{11}v_{21} - v_{11}v_{21} = 0 \end{bmatrix} = \begin{bmatrix} 0 & 1 \\ -1 & 0 \end{bmatrix}.$$

Similarly, we find $R = U\Sigma U^T$

$$= \begin{bmatrix} v_{21} & v_{11} \\ -v_{11} & v_{21} \end{bmatrix} \begin{bmatrix} \sqrt{\lambda_1} & 0 \\ 0 & \sqrt{\lambda_2} \end{bmatrix} \begin{bmatrix} v_{11} & v_{21} \\ -v_{21} & v_{11} \end{bmatrix}$$

$$= \begin{bmatrix} \sqrt{\lambda_2}v_{11}^2 + \sqrt{\lambda_1}v_{21}^2 = 2 & -\sqrt{\lambda_1}v_{11}v_{21} + \sqrt{\lambda_2}v_{11}v_{21} = -1 \\ -\sqrt{\lambda_1}v_{11}v_{21} + \sqrt{\lambda_2}v_{11}v_{21} = -1 & \sqrt{\lambda_1}v_{11}^2 + \sqrt{\lambda_2}v_{21}^2 = 3 \end{bmatrix} = \begin{bmatrix} 2 & -1 \\ -1 & 3 \end{bmatrix}.$$

So the polar decomposition of A is $A = RQ = \begin{bmatrix} 2 & -1 \\ -1 & 3 \end{bmatrix} \begin{bmatrix} 0 & 1 \\ -1 & 0 \end{bmatrix} = \begin{bmatrix} 1 & 2 \\ -3 & -1 \end{bmatrix}.$

64. $A^TA = \begin{bmatrix} 36 & 0 & 0 \\ 0 & 9 & 0 \\ 0 & 0 & 81 \end{bmatrix} \Rightarrow \sigma_i = 9, 6, 3 \Rightarrow V = \begin{bmatrix} 0 & 1 & 0 \\ 0 & 0 & 1 \\ 1 & 0 & 0 \end{bmatrix}, \Sigma = \begin{bmatrix} 9 & 0 & 0 \\ 0 & 6 & 0 \\ 0 & 0 & 3 \end{bmatrix}, U = \begin{bmatrix} -\frac{1}{3} & \frac{2}{3} & \frac{2}{3} \\ \frac{2}{3} & -\frac{1}{3} & \frac{2}{3} \\ \frac{2}{3} & \frac{2}{3} & -\frac{1}{3} \end{bmatrix} \Rightarrow$

$$A = RQ = (U\Sigma U^T)(UV^T) = \begin{bmatrix} 5 & -2 & 0 \\ -2 & 6 & 2 \\ 0 & 2 & 7 \end{bmatrix} \begin{bmatrix} \frac{2}{3} & \frac{2}{3} & -\frac{1}{3} \\ -\frac{1}{3} & \frac{2}{3} & \frac{2}{3} \\ \frac{2}{3} & -\frac{1}{3} & \frac{2}{3} \end{bmatrix}.$$

7.5 Applications

1. From Example 7.41, we have $g(x) = \text{proj}_W(x^2) = \frac{\langle 1,x^2\rangle}{\langle 1,1\rangle}\,1 + \frac{\langle x,x^2\rangle}{\langle x,x\rangle}\,x$, so we compute:

$$\langle 1,x^2\rangle = \int_{-1}^{1} x^2\,dx = \tfrac{1}{3}x^3\Big|_{-1}^{1} = \tfrac{2}{3} \qquad \langle x,x^2\rangle = \int_{-1}^{1} x^3\,dx = \tfrac{1}{4}x^4\Big|_{-1}^{1} = 0$$

$$\langle 1,1\rangle = \int_{-1}^{1} 1\,dx = x\Big|_{-1}^{1} = 2 \qquad \langle x,x\rangle = \int_{-1}^{1} x^2\,dx = \tfrac{1}{3}x^3\Big|_{-1}^{1} = \tfrac{2}{3}$$

$g(x) = \text{proj}_W(x^2) = \frac{\langle 1,x^2\rangle}{\langle 1,1\rangle}\,1 + \frac{\langle x,x^2\rangle}{\langle x,x\rangle}\,x = \frac{2/3}{2}\,1 + \frac{0}{2/3}\,x = \frac{1}{3}$ is the best linear approximation.

2. Recall 1 and x are orthogonal in $\mathscr{P}_1[-1,1]$, that is $\langle 1,x\rangle = 0$.
So, building on our work in Exercise 1, we compute:

$$\langle 1,x^2+2x\rangle = \langle 1,x^2\rangle + 2\langle 1,x\rangle = \tfrac{2}{3} \qquad \langle x,x^2+2x\rangle = \langle x,x^2\rangle + 2\langle x,x\rangle = \tfrac{4}{3}$$

$g(x) = \frac{\langle 1,x^2+2x\rangle}{\langle 1,1\rangle}\,1 + \frac{\langle x,x^2+2x\rangle}{\langle x,x\rangle}\,x = \frac{2/3}{2}\,1 + \frac{4/3}{2/3}\,x = \frac{1}{3} + 2x$ is the best linear approximation.

3. Building on our work in Exercises 1 and 2, we compute:

$$\langle 1,x^3\rangle = \int_{-1}^{1} x^3\,dx = \langle x,x^2\rangle = 0 \qquad \langle x,x^3\rangle = \int_{-1}^{1} x^4\,dx = \tfrac{1}{5}x^5\Big|_{-1}^{1} = \tfrac{2}{5}$$

$g(x) = \frac{\langle 1,x^3\rangle}{\langle 1,1\rangle}\,1 + \frac{\langle x,x^3\rangle}{\langle x,x\rangle}\,x = \frac{0}{2}\,1 + \frac{2/5}{2/3}\,x = \frac{3}{5}x$ is the best linear approximation.

4. Recall $\cos(-kx) = \cos(kx)$ and $\sin(-kx) = -\sin(kx)$. So, we have:

$$\int_{-c}^{0} f(x)\,dx = \int_{c}^{0} f(-x)\,(-dx) = -\int_{c}^{0} f(-x)\,dx = \int_{0}^{c} f(-x)\,dx \Rightarrow$$

$$\int_{-a}^{0}\cos(kx)\,dx = \int_{0}^{a}\cos(kx)\,dx \quad\text{and}\quad \int_{-a}^{0}\sin(kx)\,dx = -\int_{0}^{a}\sin(kx)\,dx$$

$$\begin{aligned}
\langle 1,\sin\left(\tfrac{\pi x}{2}\right)\rangle &= \int_{-1}^{1}\sin\left(\tfrac{\pi x}{2}\right)dx = \int_{0}^{1}\sin\left(\tfrac{\pi x}{2}\right)dx - \int_{0}^{1}\sin\left(\tfrac{\pi x}{2}\right)dx = 0\\
\langle x,\sin\left(\tfrac{\pi x}{2}\right)\rangle &= \int_{-1}^{1} x\sin\left(\tfrac{\pi x}{2}\right)dx = \int_{0}^{1} x\sin\left(\tfrac{\pi x}{2}\right)dx - \int_{0}^{1}(-x)\sin\left(\tfrac{\pi x}{2}\right)dx\\
&= \int_{0}^{1} x\sin\left(\tfrac{\pi x}{2}\right)dx + \int_{0}^{1} x\sin\left(\tfrac{\pi x}{2}\right)dx\\
&= 2\int_{0}^{1} x\sin\left(\tfrac{\pi x}{2}\right)dx = 2\left(-\tfrac{2x}{\pi}\cos\left(\tfrac{\pi x}{2}\right) + \tfrac{4}{\pi^2}\sin\left(\tfrac{\pi x}{2}\right)\right)\Big|_{0}^{1} = \tfrac{8}{\pi^2} \Rightarrow
\end{aligned}$$

$$g(x) = \frac{\langle 1,\sin\left(\frac{\pi x}{2}\right)\rangle}{\langle 1,1\rangle}\,1 + \frac{\langle x,\sin\left(\frac{\pi x}{2}\right)\rangle}{\langle x,x\rangle}\,x = \frac{0}{2}\,1 + \frac{8/\pi^2}{2/3}\,x = \frac{12}{\pi^2}x.$$

5. From Example 7.42, we have the orthogonal basis $\left\{1, x, x^2 - \frac{1}{3}\right\} \Rightarrow$

$$\langle 1,|x|\rangle = \int_{0}^{1} x\,dx - \int_{-1}^{0} x\,dx = 1 \qquad \langle x^2-\tfrac{1}{3},|x|\rangle = \int_{0}^{1}(x^3-\tfrac{1}{3}x)\,dx - \int_{-1}^{0}(x^3-\tfrac{1}{3}x)\,dx = \tfrac{1}{6}$$

$$\langle x,|x|\rangle = \int_{0}^{1} x^2\,dx - \int_{-1}^{0} x^2\,dx = 0 \qquad \langle x^2-\tfrac{1}{3},x^2-\tfrac{1}{3}\rangle = \int_{-1}^{1}(x^2-\tfrac{1}{3})^2\,dx = \tfrac{8}{45} \Rightarrow$$

$$g(x) = \frac{\langle 1,|x|\rangle}{\langle 1,1\rangle}\,1 + \frac{\langle x,|x|\rangle}{\langle x,x\rangle}\,x + \frac{\langle x^2-\frac{1}{3},|x|\rangle}{\langle x^2-\frac{1}{3},x^2-\frac{1}{3}\rangle}\,x = \frac{1}{2}\,1 + \frac{0}{2/3}\,x + \frac{1/6}{8/45}(x^2-\tfrac{1}{3}) = \frac{3}{16} + \frac{15}{16}x^2.$$

6. Building on our work in Exercise 4 and using the basis $\{1, x, x^2 - \frac{1}{3}\} \Rightarrow$, we have:

$$\begin{aligned}
\langle 1, \cos(\tfrac{\pi x}{2})\rangle &= \int_{-1}^{1} \cos(\tfrac{\pi x}{2})\,dx &&= \int_0^1 \cos(\tfrac{\pi x}{2})\,dx + \int_0^1 \cos(\tfrac{\pi x}{2})\,dx \\
&= 2\int_0^1 \cos(\tfrac{\pi x}{2})\,dx &&= 2\left(\tfrac{2}{\pi}\sin(\tfrac{\pi x}{2})\right)\Big|_0^1 = \tfrac{4}{\pi} \\
\langle x, \cos(\tfrac{\pi x}{2})\rangle &= \int_{-1}^{1} x\cos(\tfrac{\pi x}{2})\,dx &&= \int_0^1 x\cos(\tfrac{\pi x}{2})\,dx + \int_0^1 (-x)\cos(\tfrac{\pi x}{2})\,dx \\
& &&= \int_0^1 x\cos(\tfrac{\pi x}{2})\,dx - \int_0^1 x\cos(\tfrac{\pi x}{2})\,dx \quad = 0
\end{aligned}$$

$$\begin{aligned}
\langle x^2 - \tfrac{1}{3}, \cos(\tfrac{\pi x}{2})\rangle &= \langle x^2, \cos(\tfrac{\pi x}{2})\rangle - \tfrac{1}{3}\langle 1, \cos(\tfrac{\pi x}{2})\rangle \\
&= 2\int_0^1 x^2\cos(\tfrac{\pi x}{2})\,dx - \tfrac{1}{3}\left(\tfrac{4}{\pi}\right) \\
&= 2\left(\tfrac{2x^2}{\pi}\sin(\tfrac{\pi x}{2}) + \tfrac{8x}{\pi^2}\cos(\tfrac{\pi x}{2}) - \tfrac{16}{\pi^3}\sin(\tfrac{\pi x}{2})\right)\Big|_0^1 - \tfrac{4}{3\pi} \\
&= 2\left(\tfrac{2}{\pi} - \tfrac{16}{\pi^3}\right) - \tfrac{4}{3\pi} = \tfrac{8\pi^2 - 96}{3\pi^3} \Rightarrow
\end{aligned}$$

$$\begin{aligned}
g(x) &= \tfrac{\langle 1, \cos(\frac{\pi x}{2})\rangle}{\langle 1,1\rangle}\,1 + \tfrac{\langle x, \cos(\frac{\pi x}{2})\rangle}{\langle x,x\rangle}\,x + \tfrac{\langle x^2 - \frac{1}{3}, \cos(\frac{\pi x}{2})\rangle}{\langle x^2 - \frac{1}{3}, x^2 - \frac{1}{3}\rangle}\,(x^2 - \tfrac{1}{3}) \\
&= \tfrac{4/\pi}{2}\,1 + \tfrac{0}{2/3}\,x + \tfrac{(8\pi^2 - 96)/3\pi^3}{8/45}\,(x^2 - \tfrac{1}{3}) \\
&= \left[\tfrac{2}{\pi} - \tfrac{1}{3}\tfrac{15\pi^2 - 180}{\pi^3}\right] + \tfrac{15\pi^2 - 180}{\pi^3}\,x^2 = \tfrac{60 - 3\pi^2}{\pi^3} + \tfrac{15\pi^2 - 180}{\pi^3}\,x^2.
\end{aligned}$$

7. We have $\mathbf{x}_2 = x$ and $\mathbf{v}_1 = \mathbf{x}_1 = 1 \Rightarrow$

$\langle 1, 1\rangle = \int_0^1 1\,dx = x\big|_0^1 = 1$ and $\langle 1, x\rangle = \int_0^1 x\,dx = \frac{1}{2}x^2\big|_0^1 = \frac{1}{2} \Rightarrow$

$\mathbf{v}_2 = x - \frac{\langle 1,x\rangle}{\langle 1,1\rangle}\,1 = x - \frac{1/2}{1}\,1 = x - \frac{1}{2} \Rightarrow$ an orthogonal basis for $\mathscr{P}_1[0,1]$ is $\mathcal{B} = \{1, x - \frac{1}{2}\}$.

8. Building on our work in Exercise 7, we compute:

$\langle 1, x^2\rangle = \int_0^1 x^2\,dx = \langle x, x\rangle = \frac{1}{3}$.

$\langle x - \frac{1}{2}, x^2\rangle = \langle x, x^2\rangle - \frac{1}{2}\langle 1, x^2\rangle = \int_0^1 x^3\,dx - \frac{1}{2}\left(\frac{1}{3}\right) = \frac{1}{4}x^4\big|_0^1 - \frac{1}{6} = \frac{1}{4} - \frac{1}{6} = \frac{1}{12}$.

$\langle x - \frac{1}{2}, x - \frac{1}{2}\rangle = \langle x, x\rangle - \langle 1, x\rangle + \frac{1}{4}\langle 1, 1\rangle = 1\left(\frac{1}{3}\right) - 1\left(\frac{1}{2}\right) + \frac{1}{4}(1) = \frac{1}{12}$.

$\mathbf{v}_3 = x^2 - \frac{\langle 1,x^2\rangle}{\langle 1,1\rangle}\,1 - \frac{\langle x-\frac{1}{2},x^2\rangle}{\langle x-\frac{1}{2},x-\frac{1}{2}\rangle}(x - \frac{1}{2}) = x^2 - \frac{1/3}{1}\,1 - \frac{1/12}{1/12}(x - \frac{1}{2}) = x^2 - \frac{1}{3} - (x - \frac{1}{2}) = x^2 - x + \frac{1}{6} \Rightarrow$

An orthogonal basis for $\mathscr{P}_2[0,1]$ is $\mathcal{B} = \{1, x - \frac{1}{2}, x^2 - x + \frac{1}{6}\}$.

9. Building on our work in Exercise 8, we have:

$g(x) = \frac{\langle 1,x^2\rangle}{\langle 1,1\rangle}\,1 + \frac{\langle x-\frac{1}{2},x^2\rangle}{\langle x-\frac{1}{2},x-\frac{1}{2}\rangle}(x - \frac{1}{2}) = \frac{1/3}{1}\,1 + \frac{1/12}{1/12}(x - \frac{1}{2}) = x - \frac{1}{6}$.

10. Similar to our work in Exercise 7 and building on our work in Exercise 8, we have:

$\langle 1, \sqrt{x}\rangle = \int_0^1 \sqrt{x}\,dx = \frac{2}{3}x^{\frac{3}{2}}\big|_0^1 = \frac{2}{3}$ and

$\langle x - \frac{1}{2}, \sqrt{x}\rangle = \langle x, \sqrt{x}\rangle - \frac{1}{2}\langle 1, \sqrt{x}\rangle = \int_0^1 x\sqrt{x}\,dx - \frac{1}{2}\left(\frac{2}{3}\right) = \frac{2}{5}x^{\frac{5}{2}}\big|_0^1 - \frac{1}{3} = \frac{2}{5} - \frac{1}{3} = \frac{1}{15}$.

$g(x) = \frac{\langle 1,\sqrt{x}\rangle}{\langle 1,1\rangle}\,1 + \frac{\langle x-\frac{1}{2},\sqrt{x}\rangle}{\langle x-\frac{1}{2},x-\frac{1}{2}\rangle}(x - \frac{1}{2}) = \frac{2/3}{1}\,1 + \frac{1/15}{1/12}(x - \frac{1}{2}) = \frac{2}{3} + \frac{4}{5}(x - \frac{1}{2}) = \frac{4}{5}x + \frac{4}{15}$.

11. $\langle x - \frac{1}{2}, e^x\rangle = \int_0^1 (x\,e^x - \frac{1}{2}\,e^x)\,dx = x\,e^x - \frac{3}{2}\,e^x\big|_0^1 = -\frac{1}{2}\,e + \frac{3}{2} \Rightarrow$

$g(x) = \frac{\langle 1,e^x\rangle}{\langle 1,1\rangle}\,1 + \frac{\langle x-\frac{1}{2},e^x\rangle}{\langle x-\frac{1}{2},x-\frac{1}{2}\rangle}(x - \frac{1}{2}) = \frac{e-1}{1}\,1 + \frac{-e/2+3/2}{1/12}(x - \frac{1}{2}) = (4e - 10) + (18 - 6e)x$.

12. Similar to our work in Exercise 6 and building on our work in Exercise 8, we have:

$\langle 1, \sin\left(\frac{\pi x}{2}\right)\rangle = \int_0^1 \sin\left(\frac{\pi x}{2}\right)\,dx = -\frac{2}{\pi}\cos\left(\frac{\pi x}{2}\right)\Big|_0^1 = \frac{2}{\pi}$ and

$\langle x - \frac{1}{2}, \sin\left(\frac{\pi x}{2}\right)\rangle = \langle x, \sin\left(\frac{\pi x}{2}\right)\rangle - \frac{1}{2}\langle 1, \sin\left(\frac{\pi x}{2}\right)\rangle = \int_0^1 x\sin\left(\frac{\pi x}{2}\right)\,dx - \frac{1}{2}\left(\frac{2}{\pi}\right)$

$= -\frac{2x}{\pi}\cos\left(\frac{\pi x}{2}\right) + \frac{4}{\pi^2}\sin\left(\frac{\pi x}{2}\right)\Big|_0^1 - \frac{1}{\pi} = \frac{4}{\pi^2} - \frac{1}{\pi} = \frac{4-\pi}{\pi^2}.$

$g(x) = \frac{\langle 1, \sin\left(\frac{\pi x}{2}\right)\rangle}{\langle 1,1\rangle}\,1 + \frac{\langle x-\frac{1}{2}, \sin\left(\frac{\pi x}{2}\right)\rangle}{\langle x-\frac{1}{2}, x-\frac{1}{2}\rangle}(x - \frac{1}{2}) = \frac{2/\pi}{1}\,1 + \frac{(4-\pi)/\pi^2}{1/12}(x - \frac{1}{2}) = \frac{8\pi-24}{\pi^2} + \frac{48-12\pi}{\pi^2}\,x.$

13. Similar to our work in Exercise 7 and building on our work in Exercise 8, we have:

$\langle 1, x^3\rangle = \int_0^1 x^3\,dx = \frac{1}{4}x^4\Big|_0^1 = \frac{1}{4}$ and

$\langle x - \frac{1}{2}, x^3\rangle = \langle x, x^3\rangle - \frac{1}{2}\langle 1, x^3\rangle = \int_0^1 x^4\,dx - \frac{1}{2}\left(\frac{1}{4}\right) = \frac{1}{5}x^5\Big|_0^1 - \frac{1}{8} = \frac{1}{5} - \frac{1}{8} = \frac{3}{40}.$

$\langle x^2 - x + \frac{1}{6}, x^3\rangle = \langle x^2, x^3\rangle - \langle x, x^3\rangle + \frac{1}{6}\langle 1, x^3\rangle = \int_0^1 x^5\,dx - 1\left(\frac{1}{5}\right) + \frac{1}{6}\left(\frac{1}{4}\right).$

$= \frac{1}{6}x^6\Big|_0^1 - \frac{1}{5} + \frac{1}{24} = \frac{1}{6} - \frac{19}{120} = \frac{1}{120}.$

$\langle x^2 - x + \frac{1}{6}, x^2 - x + \frac{1}{6}\rangle = \langle x^2, x^2\rangle - 2\langle x, x^2\rangle + \frac{4}{3}\langle x, x\rangle - \frac{1}{3}\langle 1, x\rangle + \frac{1}{36}\langle 1, 1\rangle$

$= 1\left(\frac{1}{5}\right) - 2\left(\frac{1}{4}\right) + \frac{4}{3}\left(\frac{1}{3}\right) - \frac{1}{3}\left(\frac{1}{2}\right) + \frac{1}{36}(1) = \frac{1}{180}.$

$g(x) = \frac{\langle 1, x^3\rangle}{\langle 1,1\rangle}\,1 + \frac{\langle x-\frac{1}{2}, x^3\rangle}{\langle x-\frac{1}{2}, x-\frac{1}{2}\rangle}(x - \frac{1}{2}) + \frac{\langle x^2-x+\frac{1}{6}, x^3\rangle}{\langle x^2-x+\frac{1}{6}, x^2-x+\frac{1}{6}\rangle}(x^2 - x + \frac{1}{6})$

$= \frac{1/4}{1}\,1 + \frac{3/40}{1/12}(x - \frac{1}{2}) + \frac{1/120}{1/180}(x^2 - x + \frac{1}{6}) = \frac{1}{20} - \frac{3}{5}x + \frac{3}{2}x^2.$

14. Building on our work in Exercise 10, we have:

$\langle x^2 - x + \frac{1}{6}, \sqrt{x}\rangle = \langle x^2, \sqrt{x}\rangle - \langle x, \sqrt{x}\rangle + \frac{1}{6}\langle 1, \sqrt{x}\rangle$

$= \int_0^1 x^2\sqrt{x}\,dx - 1\left(\frac{2}{5}\right) + \frac{1}{6}\left(\frac{2}{3}\right) = \frac{2}{7}x^{\frac{7}{2}}\Big|_0^1 - \frac{2}{5} + \frac{1}{9} = \frac{2}{7} - \frac{13}{45} = -\frac{1}{315}.$

$g(x) = \frac{\langle 1, \sqrt{x}\rangle}{\langle 1,1\rangle}\,1 + \frac{\langle x-\frac{1}{2}, \sqrt{x}\rangle}{\langle x-\frac{1}{2}, x-\frac{1}{2}\rangle}(x - \frac{1}{2}) + \frac{\langle x^2-x+\frac{1}{6}, \sqrt{x}\rangle}{\langle x^2-x+\frac{1}{6}, x^2-x+\frac{1}{6}\rangle}(x^2 - x + \frac{1}{6})$

$= \left(\frac{4}{15} + \frac{4}{5}x\right) - \frac{1/315}{1/180}(x^2 - x + \frac{1}{6}) = \frac{6}{35} + \frac{48}{35}x - \frac{4}{7}x^2.$

15. Building on our work in Exercise 11, we have:

$\langle x^2 - x + \frac{1}{6}, e^x\rangle = \langle x^2, e^x\rangle - \langle x, e^x\rangle + \frac{1}{6}\langle 1, e^x\rangle$

$= \int_0^1 x^2 e^x\,dx - 1(1) + \frac{1}{6}(e - 1) = x^2\,e^x - 2x\,e^x + 2e^x\Big|_0^1 + \left(\frac{1}{6}\,e - \frac{7}{6}\right)$

$= (e - 2) + \left(\frac{1}{6}\,e - \frac{7}{6}\right) = \frac{7}{6}\,e - \frac{19}{6}$

$g(x) = \frac{\langle 1, e^x\rangle}{\langle 1,1\rangle}\,1 + \frac{\langle x-\frac{1}{2}, e^x\rangle}{\langle x-\frac{1}{2}, x-\frac{1}{2}\rangle}(x - \frac{1}{2}) + \frac{\langle x^2-x+\frac{1}{6}, e^x\rangle}{\langle x^2-x+\frac{1}{6}, x^2-x+\frac{1}{6}\rangle}(x^2 - x + \frac{1}{6})$

$= \left[\,(4e - 10) + (18 - 6e)x\,\right] + \frac{(7e-19)/6}{1/180}(x^2 - x + \frac{1}{6})$

$= (39e - 105) + (588 - 216e)x + (210e - 570)x^2.$

16. Building on our work in Exercise 12, we have:

$$\begin{aligned}\langle x^2 - x + \tfrac{1}{6}, \sin\left(\tfrac{\pi x}{2}\right)\rangle &= \langle x^2, \sin\left(\tfrac{\pi x}{2}\right)\rangle - \langle x, \sin\left(\tfrac{\pi x}{2}\right)\rangle + \tfrac{1}{6}\langle 1, \sin\left(\tfrac{\pi x}{2}\right)\rangle \\ &= \int_0^1 x^2 \sin\left(\tfrac{\pi x}{2}\right)\, dx - 1\left(\tfrac{4}{\pi^2}\right) + \tfrac{1}{6}\left(\tfrac{2}{\pi}\right) \\ &= -\tfrac{2x^2}{\pi}\cos\left(\tfrac{\pi x}{2}\right) + \tfrac{8x}{\pi^2}\sin\left(\tfrac{\pi x}{2}\right) + \tfrac{16}{\pi^3}\cos\left(\tfrac{\pi x}{2}\right)\Big|_0^1 + \left(\tfrac{\pi - 12}{3\pi^2}\right) \\ &= \left(\tfrac{8\pi - 16}{\pi^3}\right) + \left(\tfrac{\pi - 12}{3\pi^2}\right) = \tfrac{\pi^2 + 12\pi - 48}{3\pi^3}\end{aligned}$$

$$\begin{aligned}g(x) &= \tfrac{\langle 1, \sin\left(\frac{\pi x}{2}\right)\rangle}{\langle 1, 1\rangle} 1 + \tfrac{\langle x - \frac{1}{2}, \sin\left(\frac{\pi x}{2}\right)\rangle}{\langle x - \frac{1}{2}, x - \frac{1}{2}\rangle}\left(x - \tfrac{1}{2}\right) + \tfrac{\langle x^2 - x + \frac{1}{6}, \sin\left(\frac{\pi x}{2}\right)\rangle}{\langle x^2 - x + \frac{1}{6}, x^2 - x + \frac{1}{6}\rangle}\left(x^2 - x + \tfrac{1}{6}\right) \\ &= \left(\tfrac{18\pi^2 + 96\pi - 480}{\pi^3}\right) + \left(\tfrac{2880 - 672\pi - 72\pi^2}{\pi^3}\right) x + \left(\tfrac{60\pi^2 + 720\pi - 2880}{\pi^3}\right) x^2.\end{aligned}$$

17. We need to show $\langle 1, \cos(kx)\rangle = 0$ and $\langle 1, \sin(kx)\rangle = 0$ in $\mathscr{C}[-\pi, \pi]$.

$\langle 1, \cos(kx)\rangle = \int_{-\pi}^{\pi} \cos(kx)\, dx = \frac{1}{k}\sin(kx)\Big|_{-\pi}^{\pi} = \frac{2}{k}\sin(k\pi) = \frac{2}{k}\cdot 0 = 0.$

$\langle 1, \sin kx\rangle = \int_{-\pi}^{\pi} \sin kx\, dx = -\frac{1}{k}\cos(kx)\Big|_{-\pi}^{\pi} = -\frac{1}{k}(\cos(k\pi) - \cos(k\pi)) = -\frac{1}{k}\cdot 0 = 0.$

18. We need to show $\langle \cos(jx), \cos(kx)\rangle = 0$ in $\mathscr{C}[-\pi, \pi]$, provided $j \neq k$.

$$\begin{aligned}\langle \cos(jx), \cos(kx)\rangle &= \int_{-\pi}^{\pi} \cos(jx)\cos(kx)\, dx = \tfrac{1}{2}\int_{-\pi}^{\pi} \cos(j+k)x \,+\, \cos(j-k)x\, dx \\ &= \tfrac{1}{2}\left(\tfrac{\sin(j+k)x}{j+k} + \tfrac{\sin(j-k)x}{j-k}\right)\Big|_{-\pi}^{\pi} = 0.\end{aligned}$$

Note $j \neq k \Rightarrow$ we can divide by $j - k \neq 0$, as we did in the last step.

19. We need to show $\langle \sin(jx), \sin(kx)\rangle = 0$ in $\mathscr{C}[-\pi, \pi]$, provided $j \neq k$.

$$\begin{aligned}\langle \sin(jx), \sin(kx)\rangle &= \int_{-\pi}^{\pi} \sin(jx)\sin(kx)\, dx = \tfrac{1}{2}\int_{-\pi}^{\pi} \cos(j-k)x \,-\, \cos(j+k)x\, dx \\ &= \tfrac{1}{2}\left(\tfrac{\sin(j-k)x}{j-k} - \tfrac{\sin(j+k)x}{j+k}\right)\Big|_{-\pi}^{\pi} = 0.\end{aligned}$$

Note $j \neq k \Rightarrow$ we can divide by $j - k \neq 0$, as we did in the last step.

20. $\|1\|^2 = \langle 1, 1\rangle = \int_{-\pi}^{\pi} 1^2\, dx = x\Big|_{-\pi}^{\pi} = 2\pi.$

$$\begin{aligned}\|\cos(kx)\|^2 &= \langle \cos(kx), \cos(kx)\rangle = \int_{-\pi}^{\pi} \cos^2(kx)\, dx \\ &= \int_{-\pi}^{\pi}\left(\tfrac{1}{2} + \tfrac{1}{2}\cos(2kx)\right) dx = \tfrac{1}{2}x + \tfrac{\sin(2kx)}{4k}\Big|_{-\pi}^{\pi} = \pi.\end{aligned}$$

21. $a_0 = \frac{1}{2\pi}\int_{-\pi}^{\pi} |x|\, dx = \frac{1}{2\pi}\left(\int_0^{\pi} x\, dx + \int_0^{\pi} x\, dx\right) = \frac{1}{\pi}\int_0^{\pi} x\, dx = \frac{1}{\pi}\frac{x^2}{2}\Big|_0^{\pi} = \frac{\pi}{2}.$

$$\begin{aligned}a_1 &= \tfrac{1}{\pi}\int_{-\pi}^{\pi} |x|\cos x\, dx = \tfrac{1}{\pi}\left(\int_0^{\pi} x\cos x\, dx + \int_0^{\pi} x\cos x\, dx\right) = \tfrac{2}{\pi}\int_0^{\pi} x\cos x\, dx \\ &= \tfrac{2}{\pi}(x\sin x + \cos x)\Big|_0^{\pi} = -\tfrac{4}{\pi}.\end{aligned}$$

$a_2 = \frac{2}{\pi}\left(\frac{1}{2}x\sin(2x) + \frac{1}{4}\cos(2x)\right)\Big|_0^{\pi} = \frac{2}{\pi}\left(\frac{1}{4} - \frac{1}{4}\right) = 0.$

$a_3 = \frac{2}{\pi}\left(\frac{1}{3}x\sin(3x) + \frac{1}{9}\cos(3x)\right)\Big|_0^{\pi} = \frac{2}{\pi}\left(-\frac{1}{9} - \frac{1}{9}\right) = -\frac{4}{9\pi}.$

$b_k = \frac{1}{\pi}\int_{-\pi}^{\pi} |x|\sin(kx)\, dx = \frac{1}{\pi}\left(\int_0^{\pi} x\sin(kx)\, dx - \int_0^{\pi} x\sin(kx)\, dx\right) = 0.$

So, the third-order Fourier approximation of $f(x) = |x|$ on $[-\pi, \pi]$ is:

$$\begin{aligned}\mathrm{proj}_W(|x|) &= a_0 + a_1\cos x + a_2\cos(2x) + a_3\cos(3x) \\ &= \tfrac{\pi}{2} - \tfrac{4}{\pi}\cos x + 0\cos(2x) - \tfrac{4}{9\pi}\cos(3x) = \tfrac{\pi}{2} - \tfrac{4}{\pi}\left(\cos x + \tfrac{\cos(3x)}{9}\right).\end{aligned}$$

22. $a_0 = \frac{1}{2\pi}\int_{-\pi}^{\pi} x^2\,dx = \frac{1}{2\pi}\left(\int_0^{\pi} x^2\,dx + \int_0^{\pi} x^2\,dx\right) = \frac{1}{\pi}\int_0^{\pi} x^2\,dx = \frac{1}{\pi}\frac{x^3}{3}\Big|_0^{\pi} = \frac{\pi^2}{3}.$

$a_1 = \frac{1}{\pi}\int_{-\pi}^{\pi} x^2\cos x\,dx = \frac{1}{\pi}\left(\int_0^{\pi} x^2\cos x\,dx + \int_0^{\pi} x^2\cos x\,dx\right) = \frac{2}{\pi}\int_0^{\pi} x^2\cos x\,dx$

$= \frac{2}{\pi}(x^2\sin x + 2x\cos x - 2\sin x)\Big|_0^{\pi} = \frac{2}{\pi}(-2\pi) = -4.$

$a_2 = \frac{2}{\pi}\left(\frac{1}{2}x^2\sin(2x) + \frac{1}{2}x\cos(2x) - \frac{1}{4}\sin(2x)\right)\Big|_0^{\pi} = \frac{2}{\pi}\left(\frac{1}{2}\pi\right) = 1.$

$a_3 = \frac{2}{\pi}\left(\frac{1}{3}x^2\sin(3x) + \frac{2}{9}x\cos(3x) - \frac{2}{27}\sin(3x)\right)\Big|_0^{\pi} = \frac{2}{\pi}\left(\frac{1}{9}(-2\pi)\right) = -\frac{4}{9}.$

$b_k = \frac{1}{\pi}\int_{-\pi}^{\pi} x^2\sin(kx)\,dx = \frac{1}{\pi}\left(\int_0^{\pi} x^2\sin(kx)\,dx - \int_0^{\pi} x^2\sin(kx)\,dx\right) = 0.$

So, the third-order Fourier approximation of $f(x) = x^2$ on $[-\pi, \pi]$ is:

$\text{proj}_W(x^2) = a_0 + a_1\cos x + a_2\cos(2x) + a_3\cos(3x)$

$= \frac{\pi^2}{3} - 4\cos x + \cos(2x) - \frac{4}{9}\cos(3x).$

23. $a_0 = \frac{1}{2\pi}\int_{-\pi}^{\pi} f(x)\,dx = \frac{1}{2\pi}\left(\int_0^{\pi} 1\,dx + \int_0^{\pi} 0\,dx\right) = \frac{1}{2\pi}\int_0^{\pi} 1\,dx = \frac{1}{2\pi}x\Big|_0^{\pi} = \frac{1}{2}.$

$a_k = \frac{1}{\pi}\int_{-\pi}^{\pi} f(x)\cos(kx)\,dx = \frac{1}{\pi}\int_0^{\pi}\cos(kx)\,dx = \frac{1}{\pi}\left(\frac{\sin(kx)}{k}\right)\Big|_0^{\pi} = 0.$

$b_k = \frac{1}{\pi}\int_{-\pi}^{\pi} f(x)\sin(kx)\,dx = \frac{1}{\pi}\int_0^{\pi}\sin(kx)\,dx = \frac{1}{\pi}\left(-\frac{\cos(kx)}{k}\right)\Big|_0^{\pi} = \frac{1-(-1)^k}{\pi k}.$

24. $a_0 = \frac{1}{2\pi}\int_{-\pi}^{\pi} f(x)\,dx = \frac{1}{2\pi}\left(\int_0^{\pi} 1\,dx - \int_0^{\pi} 1\,dx\right) = 0.$

$a_k = \frac{1}{\pi}\int_{-\pi}^{\pi} f(x)\cos(kx)\,dx = \frac{1}{\pi}\left(\int_0^{\pi}\cos(kx)\,dx - \cos(kx)\,dx\right) = 0.$

$b_k = \frac{1}{\pi}\int_{-\pi}^{\pi} f(x)\sin(kx)\,dx = \frac{1}{\pi}\left(\int_0^{\pi}\sin(kx)\,dx + \sin(kx)\,dx\right) = \frac{2}{\pi}\int_0^{\pi}\sin(kx)\,dx$

$= \frac{2}{\pi}\left(-\frac{\cos(kx)}{k}\right)\Big|_0^{\pi} = \frac{2(1-(-1)^k)}{\pi k}.$

25. $a_0 = \frac{1}{\pi}\int_{-\pi}^{\pi}(\pi - x)\,dx = \frac{1}{\pi}\left(\pi x - \frac{1}{2}x^2\right)\Big|_{-\pi}^{\pi} = \pi.$

$a_k = \langle \pi - x, \cos(kx)\rangle = \pi\langle 1, \cos(kx)\rangle - \langle x, \cos(kx)\rangle = \frac{1}{2}\int_{-\pi}^{\pi}\cos(kx)\,dx - \frac{1}{\pi}\int_{-\pi}^{\pi} x\cos(kx)\,dx$

$= \frac{1}{2}\left(\frac{\sin(kx)}{k}\right)\Big|_{-\pi}^{\pi} - \frac{1}{\pi}\left(\frac{x\sin(kx)}{k} + \frac{\cos(kx)}{k^2}\right)\Big|_{-\pi}^{\pi} = \frac{1}{2}(0) - \frac{1}{2\pi}\left(\frac{\cos(\pi k)}{k^2} - \frac{\cos(\pi k)}{k^2}\right) = 0.$

$b_k = \langle \pi - x, \sin(kx)\rangle = \pi\langle 1, \sin(kx)\rangle - \langle x, \sin(kx)\rangle = \frac{1}{2}\int_{-\pi}^{\pi}\sin(kx)\,dx - \frac{1}{\pi}\int_{-\pi}^{\pi} x\sin(kx)\,dx$

$= \frac{1}{2}\left(-\frac{\cos(kx)}{k}\right)\Big|_{-\pi}^{\pi} - \frac{1}{\pi}\left(\frac{-x\cos(kx)}{k} + \frac{\sin(kx)}{k^2}\right)\Big|_{-\pi}^{\pi}$

$= \frac{1}{2}\left(-\frac{\cos(\pi k)}{k} + \frac{\cos(\pi k)}{k}\right) - \frac{1}{\pi}\left(-\frac{\pi\cos(\pi k)}{k} - \frac{\pi\cos(\pi k)}{k}\right) = \frac{1}{2}(0) - \frac{1}{\pi}\left(\frac{-2\pi(-1)^k}{k}\right) = \frac{2(-1)^k}{k}.$

26. $a_0 = \frac{1}{2\pi}\int_{-\pi}^{\pi} |x|\,dx = \frac{1}{2\pi}\left(\int_0^{\pi} x\,dx + \int_0^{\pi} x\,dx\right) = \frac{1}{\pi}\int_0^{\pi} x\,dx = \frac{1}{\pi}\frac{x^2}{2}\Big|_0^{\pi} = \frac{\pi}{2}.$

$a_k = \frac{1}{\pi}\int_{-\pi}^{\pi} |x|\cos(kx)\,dx = \frac{1}{\pi}\left(\int_0^{\pi} x\cos(kx)\,dx + \int_0^{\pi} x\cos(kx)\,dx\right) = \frac{2}{\pi}\int_0^{\pi} x\cos(kx)\,dx$

$= \frac{2}{\pi}\left(\frac{x\sin(kx)}{k} + \frac{\cos(kx)}{k^2}\right)\Big|_0^{\pi} = \frac{2}{\pi}\left(\frac{(-1)^k - 1}{k^2}\right).$

$b_k = \frac{1}{\pi}\int_{-\pi}^{\pi} |x|\sin(kx)\,dx = \frac{1}{\pi}\left(\int_0^{\pi} x\sin(kx)\,dx - \int_0^{\pi} x\sin(kx)\,dx\right) = 0.$

27. (a) $\int_{-\pi}^{\pi} f(x)\,dx = \int_0^{\pi} f(x)\,dx + \int_0^{\pi} f(-x)\,dx = \int_0^{\pi} f(x)\,dx - \int_0^{\pi} f(x)\,dx = 0.$

(b) $a_k = \int_{-\pi}^{\pi} f(x)\cos(kx)\,dx = \int_0^{\pi} f(x)\cos(kx)\,dx + \int_0^{\pi} f(-x)\cos(-kx)\,dx$
$= \int_0^{\pi} f(x)\cos(kx) - \int_0^{\pi} f(x)\cos(kx)\,dx = 0.$

28. (a) $\int_{-\pi}^{\pi} f(x)\,dx = \int_0^{\pi} f(x)\,dx + \int_0^{\pi} f(-x)\,dx = \int_0^{\pi} f(x)\,dx + \int_0^{\pi} f(x)\,dx = 2\int_0^{\pi} f(x)\,dx.$

(b) $b_k = \int_{-\pi}^{\pi} f(x)\sin(kx)\,dx = \int_0^{\pi} f(x)\sin(kx)\,dx + \int_0^{\pi} f(-x)\sin(-kx)\,dx$
$= \int_0^{\pi} f(x)\sin(kx) - \int_0^{\pi} f(x)\sin(kx)\,dx = 0.$

29. $a - b = a + b$ in $\mathbb{Z}_2 \Rightarrow \mathrm{d}_H(x,y) = \|x-y\|_H = \|x+y\|_H = w(x+y) \Rightarrow$
$w(\mathbf{0}+\mathbf{v}_1) = w(\mathbf{v}_1) = 1,\ w(\mathbf{0}+\mathbf{v}_2) = w(\mathbf{v}_2) = 2,\ w(\mathbf{v}_1+\mathbf{v}_2) = w(\mathbf{e}_1) = 1 \Rightarrow.$
Therefore, since $\mathrm{d}(C) = \min\{\mathrm{d}_H(x,y) : x \neq y \in C\}$ we have $\mathrm{d}(C) = \min\{1,2,1\} = 1.$

30. With $C = \{\mathbf{c}_1, \mathbf{c}_2, \mathbf{c}_3, \mathbf{c}_4\}$, we have:

$w(\mathbf{c}_1+\mathbf{c}_2) = 4,\quad w(\mathbf{c}_1+\mathbf{c}_3) = 2,\quad w(\mathbf{c}_1+\mathbf{c}_4) = 2$
$w(\mathbf{c}_2+\mathbf{c}_3) = 2,\quad w(\mathbf{c}_2+\mathbf{c}_4) = 2,\quad w(\mathbf{c}_3+\mathbf{c}_4) = 4 \Rightarrow\ \mathrm{d}(C) = \min\{4,2\} = 2.$

31. Note, $\mathbf{t} = \begin{bmatrix} 1 \\ 1 \\ 0 \\ \vdots \\ 0 \end{bmatrix} \in E_n \Rightarrow w(\mathbf{0}+\mathbf{t}) = w(\mathbf{t}) = 2 \Rightarrow \mathrm{d}(C) \le 2.$

We will show $\mathbf{x} \neq \mathbf{0} \in E_n \Rightarrow \mathrm{d}_H(x,y) \ge 2$ which will imply $\mathrm{d}(C) = 2$.

Let $j \ge k$ ($j \ge 1$ i.e. $\mathbf{x} \neq \mathbf{0}$) with $w(\mathbf{x}) = 2j$, $w(\mathbf{y}) = 2k$, $i =$ overlap (so $i \le k$) $\Rightarrow$
$\mathrm{d}_H(x,y) = w(\mathbf{x}) + w(\mathbf{y}) - 2\,\text{overlap} = 2j + 2k - 2i \ge 2j + 2k - 2k = 2j \ge 2 \Rightarrow \mathrm{d}(C) = 2.$

32. Recall, $\mathrm{Rep}_n = \{\mathbf{0}, \mathbf{1}\} \Rightarrow \mathrm{d}(C) = w(\mathbf{0}+\mathbf{1}) = n.$

33. Recall in $\mathbb{Z}_2$, $\mathbf{a}_k + \mathbf{a}_j = 0 \Leftrightarrow \mathbf{a}_k - \mathbf{a}_j = 0 \Leftrightarrow \mathbf{a}_k = \mathbf{a}_j$.
Let $A = \begin{bmatrix} \mathbf{a}_1 & \mathbf{a}_2 & \mathbf{a}_3 & \mathbf{a}_4 & \mathbf{a}_5 & \mathbf{a}_6 & \mathbf{a}_7 \end{bmatrix}$. Since none of the $\mathbf{a}_i$ are identical, $\mathrm{d}(C) > 2$.
On the other hand, $\mathbf{a}_1 + \mathbf{a}_2 + \mathbf{a}_3 = 0 \Rightarrow \mathrm{d}(C) \le 3 \Rightarrow \mathrm{d}(C) = 3.$

34. $P = \begin{bmatrix} 1 & 1 & 0 & 0 \\ 1 & 1 & 1 & 1 \\ 1 & 0 & 0 & 1 \end{bmatrix} \longrightarrow \begin{bmatrix} 1 & 0 & 0 & 0 \\ 0 & 1 & 0 & 0 \\ 0 & 0 & 1 & 1 \end{bmatrix} \Rightarrow \mathrm{d}(C) = 4.$ Or, rows of P linearly independent $\Rightarrow$ $\mathrm{d}(C) = 3 + 1 = 4.$

35. With $C = \{\mathbf{c}_1, \mathbf{c}_2, \mathbf{c}_3, \mathbf{c}_4\}$, we have:

$\mathrm{d}_H(\mathbf{c}_1,\mathbf{c}_2) = 3,\quad \mathrm{d}_H(\mathbf{c}_1,\mathbf{c}_3) = 4,\quad \mathrm{d}_H(\mathbf{c}_1,\mathbf{c}_4) = 3$
$\mathrm{d}_H(\mathbf{c}_2,\mathbf{c}_3) = 3,\quad \mathrm{d}_H(\mathbf{c}_2,\mathbf{c}_4) = 4,\quad \mathrm{d}_H(\mathbf{c}_3,\mathbf{c}_4) = 3 \Rightarrow\ \mathrm{d}(C) = \min\{3,4\} = 3.$

$\mathrm{d}_H(\mathbf{c}_1,\mathbf{u}) = 2,\quad \mathrm{d}_H(\mathbf{c}_2,\mathbf{u}) = 1,\quad \mathrm{d}_H(\mathbf{c}_3,\mathbf{u}) = 4,\quad \mathrm{d}_H(\mathbf{c}_4,\mathbf{u}) = 3$

Since $\min\{\mathrm{d}_H(\mathbf{c}_i,\mathbf{u})\} = 1$ occurs only for $\mathbf{c}_2 \Rightarrow \mathbf{u}$ decodes as $\mathbf{c}_2$.

$\mathrm{d}_H(\mathbf{c}_1,\mathbf{v}) = 3,\quad \mathrm{d}_H(\mathbf{c}_2,\mathbf{v}) = 2,\quad \mathrm{d}_H(\mathbf{c}_3,\mathbf{v}) = 3,\quad \mathrm{d}_H(\mathbf{c}_4,\mathbf{v}) = 2$

Since $\min\{\mathrm{d}_H(\mathbf{c}_i,\mathbf{v})\} = 2$ occurs for $\mathbf{c}_2$ and $\mathbf{c}_4$, $\mathbf{v}$ cannot be decoded.

$\mathrm{d}_H(\mathbf{c}_1,\mathbf{w}) = 3,\quad \mathrm{d}_H(\mathbf{c}_2,\mathbf{w}) = 2,\quad \mathrm{d}_H(\mathbf{c}_3,\mathbf{w}) = 1,\quad \mathrm{d}_H(\mathbf{c}_4,\mathbf{w}) = 4$

Since $\min\{\mathrm{d}_H(\mathbf{c}_i,\mathbf{w})\} = 1$ occurs only for $\mathbf{c}_3 \Rightarrow \mathbf{w}$ decodes as $\mathbf{c}_3$.

36. G generates the following code:

$$C = \left\{ \begin{bmatrix}0\\0\\0\\0\\0\\0\\0\end{bmatrix}, \begin{bmatrix}1\\0\\0\\1\\1\\0\\1\end{bmatrix}, \begin{bmatrix}0\\1\\0\\1\\0\\1\\1\end{bmatrix}, \begin{bmatrix}0\\0\\1\\0\\1\\1\\1\end{bmatrix}, \begin{bmatrix}1\\1\\0\\0\\1\\1\\0\end{bmatrix}, \begin{bmatrix}1\\0\\1\\1\\0\\1\\0\end{bmatrix}, \begin{bmatrix}0\\1\\1\\1\\1\\0\\0\end{bmatrix}, \begin{bmatrix}1\\1\\1\\0\\0\\0\\1\end{bmatrix} \right\}.$$

So, letting $C = \{\mathbf{c}_1, \mathbf{c}_2, \mathbf{c}_3, \mathbf{c}_4, \mathbf{c}_5, \mathbf{c}_6, \mathbf{c}_7, \mathbf{c}_8\}$, we have:

Since $\min\{d_H(\mathbf{c}_i, \mathbf{u})\} = 1$ occurs only for $\mathbf{c}_5 \Rightarrow \mathbf{u}$ decodes as $\mathbf{c}_5$.

Since $\min\{d_H(\mathbf{c}_i, \mathbf{v})\} = 1$ occurs only for $\mathbf{c}_8 \Rightarrow \mathbf{v}$ decodes as $\mathbf{c}_8$.

Since $\min\{d_H(\mathbf{c}_i, \mathbf{w})\} = 1$ occurs only for $\mathbf{c}_4 \Rightarrow \mathbf{w}$ decodes as $\mathbf{c}_4$.

37. $\text{Rep}_8 = \{\mathbf{0}, \mathbf{1}\}$ in $\mathbb{Z}_2^8$ works since we saw in Exercise 32 that $d(\text{Rep}_n) = n$, so $d(\text{Rep}_8) = 8$.

38. A parity check matrix P for such a code is $(8-2) \times 8 = 6 \times 8 \Rightarrow$
the number of rows $\le 6 \Rightarrow$ the number of linearly independent rows $\le 6 \Rightarrow$
the number of linearly independent columns $\le 6 \Rightarrow$
the number of linearly dependent columns $\le 6 + 1 = 7 \Rightarrow$
$d(C) \le 7 < 8 \Rightarrow$ No linear $(8, 2, 8)$ code exists.

39. A parity check matrix P for such a code is$(8-5) \times 8 = 3 \times 8 \Rightarrow$
the number of rows $\le 3 \Rightarrow$ the number of linearly independent rows $\le 3 \Rightarrow$
the number of linearly independent columns $\le 3 \Rightarrow$
the number of linearly dependent columns $\le 3 + 1 = 4 \Rightarrow$
$d = d(C) \le 4 < 5 \Rightarrow$ No linear $(8, 5, 5)$ code exists.

40. A $(8, 4, 4)$ linear code *does* exist. Proof is left to the reader.

41.

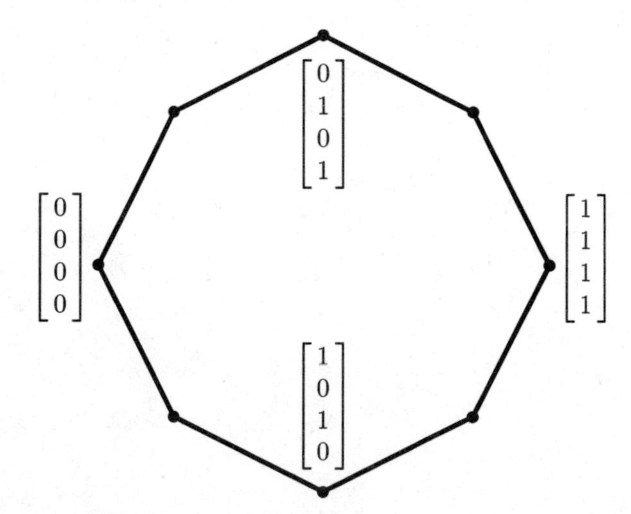

42. C a linear code $\Rightarrow C$ is a subspace (definition Section 6.7) $\Rightarrow \mathbf{c}(\neq \mathbf{0}), \mathbf{d} \in C \Rightarrow \mathbf{c}+\mathbf{d} \in C$, so $d = \min\{d_H(\mathbf{c},\mathbf{d}) : \mathbf{c},\mathbf{d} \in C\} = \min\{w(\mathbf{c}+\mathbf{d}) : \mathbf{c},\mathbf{d} \in C\} = \min\{w(\mathbf{c}) : \mathbf{c}(\neq \mathbf{0}) \in C\}$.

43. A parity check matrix P for such a code is $(n-k) \times n \Rightarrow$ the number of rows $\leq n-k \Rightarrow$
the number of linearly independent rows $\leq n-k \Rightarrow$
the number of linearly independent columns $\leq n-k \Rightarrow$
the number of linearly dependent columns $\leq n-k+1 \Rightarrow$
$d =$ the minimum number of linearly dependent columns $\leq n-k+1 \Rightarrow$
$d-1 \leq n-k$.

44. From 43, we have $d \leq n-k+1$.
Now if every $n-k$ columns of P are linearly independent $\Rightarrow n-k < d \Rightarrow d = n-k+1$.

If $d = n-k+1$, then Theorem 7.21 $\Rightarrow n-k+1$ is the smallest integer for which P has $n-k+1$ linearly dependent columns $\Rightarrow$ every $n-k$ columns of P are linearly independent.

Chapter 7 Review

1. We will explain and give counter examples to justify our answers below.

(a) ***True***. See definition on p.540 of Section 7.1.
Simply verify that $\langle \mathbf{u}, \mathbf{v} \rangle = u_1v_1 + \pi\, u_2v_2$ satisfies the definition.
Q: Does $\langle \mathbf{u}, \mathbf{v} \rangle = u_1v_1 + c\, u_2v_2$ work for any scalar c?
What about $\langle \mathbf{u}, \mathbf{v} \rangle = u_1v_2 + c\, u_2v_1$?

(b) ***True***. Check the properties carefully.

(c) ***False***. Which property or properties should we expect to fail?
Consider $A = \begin{bmatrix} 0 & 1 \\ 1 & 0 \end{bmatrix}$, $\langle A, A \rangle = \text{tr}(A) + \text{tr}(A) = 0 + 0 = 0$, but $A \neq 0$.
Q: Which property does this violate? That is, which property fails?

(d) ***True***. Simply perform the calculation implied by the definitions.
$\|\mathbf{u} + \mathbf{v}\| = \sqrt{\langle \mathbf{u} + \mathbf{v}, \mathbf{u} + \mathbf{v} \rangle} = \sqrt{\|\mathbf{u}\|^2 + 2\langle \mathbf{u}, \mathbf{v} \rangle + \|\mathbf{v}\|^2}$
$\sqrt{4^2 + 2 \cdot 2 + (\sqrt{5})^2} = \sqrt{16 + 4 + 5} = \sqrt{25} = 5$

(e) ***True***. Below we apply the definitions when $n = 1$.

sum norm	p.561	7.2	$\|\mathbf{v}\|_s = \|v_1\|$
max norm	p.562	7.2	$\|\mathbf{v}\|_m = \max\{\|v_1\|\} = \|v_1\|$
Euclidean norm	p.562	7.2	$\|\mathbf{v}\|_E = \sqrt{\|v_1\|^2} = \|v_1\|$

(f) ***False***. See definition on p.570 in Section 7.2 and the remarks following.
Since the definition uses an inequality, the converse may not hold.

(g) ***True***. See definition on p.570 in Section 7.2 and the remarks following.

(h) ***False***. It is not necessarily *unique*. See Theorem 7.9 on p.584 of Section 7.3.
Q: Under what conditions is the least squares solution of a linear system unique?
A: Hint: See Theorem 7.10 on p.591 of Section 7.3.

(i) ***True***. See Theorem 7.11 on p.592 of Section 7.3.
Q: What does the fact that A has orthonormal columns tell us about A?
A: Hint: Consider the effect upon the product A^TA.

(j) ***False***. We construct a simple counterexample below.
Consider $A = \begin{bmatrix} 2 & 0 \\ 0 & 2 \end{bmatrix}$. A is symmetric with eigenvalue 2 and singular value $\sqrt{2}$.

2. See the definition of an *inner product* on p.540 and Example 7.2 in Section 7.1.

This is obviously *not* an inner product. Why not?
Consider Property 4: $\langle \mathbf{u}, \mathbf{u} \rangle = 0$ if and only $\mathbf{u} = \mathbf{0}$. Is this true?

Let $\mathbf{u} = u(x) = x$, then $\langle \mathbf{u}, \mathbf{u} \rangle = u(0)u(1) + u(1)u(0) = 2u(0)u(1) = 0$ but $\mathbf{u} \neq \mathbf{0} = z(x) = 0$.

Q: Can we *fix* the definition so this is an *inner product*?
A: Hint: Would it help if $\langle p(x), q(x) \rangle = p(1)q(-1) + p(-1)q(1)$? Why or why not?

3. We will show that $\langle A, B \rangle = \text{tr}(A^T B)$ for A and B in M_{22} is an inner product by showing that it satisfies all four properties of the definition.

Let $\mathbf{u} = A = \begin{bmatrix} a & b \\ c & d \end{bmatrix}$ and $\mathbf{v} = B = \begin{bmatrix} e & f \\ g & h \end{bmatrix}$.

1. $\langle A, B \rangle = \text{tr}(A^T B) = ae + cg + bf + dh = \text{tr}(B^T A) = \langle B, A \rangle$.
2. $\langle A, B + C \rangle = \text{tr}(A^T(B + C)) = \text{tr}(A^T B) + \text{tr}(A^T C) = \langle A, B \rangle + \langle A, C \rangle$.
3. $\langle cA, B \rangle = \text{tr}(cA^T B) = c\text{tr}(A^T B) = c\langle A, B \rangle$.
4. $\langle A, A \rangle = \text{tr}(A^T A) = a^2 + b^2 + c^2 + d^2 \geq 0$.
   Also, $a^2 + b^2 + c^2 + d^2 = 0$ if and only if $a = b = c = d = 0$ that is, $A = O$.

Therefore, $\langle A, B \rangle = \text{tr}(A^T B)$ for A and B in M_{22} is an inner product.

4. See the definition of an *inner product* on p.540 and Example 7.2 in Section 7.1.

This is *not* an inner product. Why not?
Consider Property 2: $\langle \mathbf{u}, \mathbf{v} + \mathbf{w} \rangle = \langle \mathbf{u}, \mathbf{v} \rangle + \langle \mathbf{u}, \mathbf{w} \rangle$.

Let $\mathbf{u} = f(x) = x$, $\mathbf{v} = g(x) = x$, and $\mathbf{w} = h(x) = -x$. Does Property 2 hold?
Note that $(g + h)(x) = 0$ but g maximizes when $x = 1$. Why does this cause a problem?

5. We need only recall that $\|\mathbf{v}\| = \sqrt{\langle \mathbf{v}, \mathbf{v} \rangle}$ on p.544 in Section 7.1.

So $\|1 + x + x^2\| = \sqrt{\langle 1 + x + x^2, 1 + x + x^2 \rangle} = \sqrt{1 \cdot 1 + 1 \cdot 1 + 1 \cdot 1} = \sqrt{3}$.

6. We need only recall that $\text{d}(\mathbf{u}, \mathbf{v}) = \|\mathbf{u} - \mathbf{v}\|$ on p.544 in Section 7.1.

So $\text{d}(x, x^2) = \sqrt{\langle x - x^2, x - x^2 \rangle} = \sqrt{\int_0^1 (x - x^2)^2\, dx} = \sqrt{\int_0^1 x^2 - 2x^3 + x^4\, dx}$

$= \sqrt{\frac{x^3}{3} - \frac{x^4}{2} + \frac{x^5}{5}\Big|_0^1} = \sqrt{\frac{1}{3} - \frac{1}{2} + \frac{1}{5}} = \sqrt{\frac{1}{30}}$.

7. See Example 7.8 and Exercises 37 through 39 in Section 7.1.

$\mathcal{B} = \left\{ \mathbf{v}_1 = \begin{bmatrix} 1 \\ 1 \end{bmatrix}, \mathbf{x}_2 = \begin{bmatrix} 1 \\ 2 \end{bmatrix} \right\}$ and $\langle \mathbf{u}, \mathbf{v} \rangle = \mathbf{u}^T A \mathbf{v}$, $A = \begin{bmatrix} 6 & 4 \\ 4 & 6 \end{bmatrix}$.

$$\langle \mathbf{v}_1, \mathbf{v}_1 \rangle = \mathbf{v}_1^T A \mathbf{v}_1 = \begin{bmatrix} 1 & 1 \end{bmatrix} \begin{bmatrix} 6 & 4 \\ 4 & 6 \end{bmatrix} \begin{bmatrix} 1 & 1 \end{bmatrix} = 20,$$

$$\langle \mathbf{v}_1, \mathbf{x}_2 \rangle = \mathbf{v}_1^T A \mathbf{x}_2 = \begin{bmatrix} 1 & 1 \end{bmatrix} \begin{bmatrix} 6 & 4 \\ 4 & 6 \end{bmatrix} \begin{bmatrix} 1 & 2 \end{bmatrix} = 30,$$

So $\frac{\langle \mathbf{v}_1, \mathbf{x}_2 \rangle}{\langle \mathbf{v}_1, \mathbf{v}_1 \rangle} = \frac{30}{20} = \frac{3}{2}$.

Therefore, $\mathbf{v}_2 = \mathbf{x}_2 - \frac{\langle \mathbf{v}_1, \mathbf{x}_2 \rangle}{\langle \mathbf{v}_1, \mathbf{v}_1 \rangle} \mathbf{v}_1 = \mathbf{x}_2 - \frac{3}{2}\mathbf{v}_1 = \begin{bmatrix} 1 \\ 2 \end{bmatrix} - \frac{3}{2} \begin{bmatrix} 1 \\ 1 \end{bmatrix} = \begin{bmatrix} -\frac{1}{2} \\ \frac{1}{2} \end{bmatrix}$.

Therefore, the orthogonal set is $\mathcal{C} = \left\{ \begin{bmatrix} 1 \\ 1 \end{bmatrix}, \begin{bmatrix} -\frac{1}{2} \\ \frac{1}{2} \end{bmatrix} \right\}$.

Verify that $\langle \mathbf{v}_1, \mathbf{v}_2 \rangle = 0$.

8. See Example 7.8 and Exercises 37 through 39 in Section 7.1.

$\mathcal{B} = \{\mathbf{v}_1 = 1, \mathbf{x}_2 = x, \mathbf{x}_3 = x^2\}$ and $\langle p(x), q(x) \rangle = \int_0^1 p(x)q(x)\,dx$.

$$\langle \mathbf{v}_1, \mathbf{v}_1 \rangle = \langle 1, 1 \rangle = \int_0^1 1\,dx = x\Big|_0^1 = 1,$$

$$\langle \mathbf{v}_1, \mathbf{x}_2 \rangle = \langle 1, x \rangle = \int_0^1 x\,dx = \frac{x^2}{2}\Big|_0^1 = \frac{1}{2}.$$

So $\frac{\langle \mathbf{v}_1, \mathbf{x}_2 \rangle}{\langle \mathbf{v}_1, \mathbf{v}_1 \rangle} = \frac{1/2}{1} = \frac{1}{2}$.

Therefore, $\mathbf{v}_2 = \mathbf{x}_2 - \frac{\langle \mathbf{v}_1, \mathbf{x}_2 \rangle}{\langle \mathbf{v}_1, \mathbf{v}_1 \rangle} \mathbf{v}_1 = \mathbf{x}_2 - \frac{1}{2}\mathbf{v}_1 = x - \frac{1}{2}(1) = x - \frac{1}{2}$.

$$\langle \mathbf{v}_1, \mathbf{x}_3 \rangle = \langle 1, x^2 \rangle = \int_0^1 x^2\,dx = \frac{x^3}{3}\Big|_0^1 = \frac{1}{3}.$$

So $\frac{\langle \mathbf{v}_1, \mathbf{x}_3 \rangle}{\langle \mathbf{v}_1, \mathbf{v}_1 \rangle} = \frac{1/3}{1} = \frac{1}{3}$.

$$\langle \mathbf{v}_2, \mathbf{v}_2 \rangle = \langle x - \tfrac{1}{2}, x - \tfrac{1}{2} \rangle = \int_0^1 x^2 - x + \tfrac{1}{4}\,dx = \frac{x^3}{3} - \frac{x^2}{2} + \frac{1}{4}x\Big|_0^1 = \frac{1}{3} - \frac{1}{2} + \frac{1}{4} = \frac{1}{12}.$$

$$\langle \mathbf{v}_2, \mathbf{x}_3 \rangle = \langle x - \tfrac{1}{2}, x^2 \rangle = \int_0^1 x^3 - \frac{x^2}{2}\,dx = \frac{x^4}{4} - \frac{x^3}{6}\Big|_0^1 = \frac{1}{4} - \frac{1}{6} = \frac{1}{12}.$$

So $\frac{\langle \mathbf{v}_2, \mathbf{x}_3 \rangle}{\langle \mathbf{v}_2, \mathbf{v}_2 \rangle} = \frac{1/12}{1/12} = 1$.

$$\mathbf{v}_3 = \mathbf{x}_3 - \frac{\langle \mathbf{v}_1, \mathbf{x}_3 \rangle}{\langle \mathbf{v}_1, \mathbf{v}_1 \rangle}\mathbf{v}_1 - \frac{\langle \mathbf{v}_2, \mathbf{x}_3 \rangle}{\langle \mathbf{v}_2, \mathbf{v}_2 \rangle}\mathbf{v}_2 = \mathbf{x}_3 - \tfrac{1}{3}\mathbf{v}_1 - \mathbf{v}_2 = x^2 - \tfrac{1}{3} \cdot 1 - \left(x - \tfrac{1}{2}\right)$$
$$= x^2 - \tfrac{1}{3} \cdot 1 - \left(x - \tfrac{1}{2}\right) = x^2 - x + \tfrac{1}{6}.$$

Therefore, the orthogonal set is $\mathcal{C} = \{1, x - \frac{1}{2}, x^2 - x + \frac{1}{6}\}$.

Verify that $\langle \mathbf{v}_1, \mathbf{v}_2 \rangle = \langle \mathbf{v}_1, \mathbf{v}_3 \rangle = \langle \mathbf{v}_2, \mathbf{v}_3 \rangle = 0$.

9. See the definition of *norm* and Example 7.12 on p.561 in Section 7.2.

We will show that $\|\mathbf{v}\| = \mathbf{v}^T\mathbf{v}$ is *not* a norm because it fails to satisfy Property 2.

2. $\|c\mathbf{v}\| = c\mathbf{v}^T c\mathbf{v} = c^2\mathbf{v}^T\mathbf{v} = c^2\|\mathbf{v}\| \neq c\|\mathbf{v}\|$.

Therefore, $\|\mathbf{v}\| = \mathbf{v}^T\mathbf{v}$ is *not* a norm.

10. See the definition of *norm* and Example 7.12 on p.561 in Section 7.2.

We will show that $\|p(x)\| = |p(0)| + |p(1) - p(0)|$ is a norm
because it satisfies Properties 1 through 3.

Let $p(x) = a + bx$, so $p(0) = a$ and $p(1) = a + b$.

1. Absolute value in definition implies the norm is greater than or equal to zero.
   $\|p(x)\| = |p(0)| + |p(1) - p(0)| = |a| + |a + b - a| = |a| + |b| = 0$
   $\Leftrightarrow a = b = 0 \Leftrightarrow p(x) = 0$, as required.

2. $\|cp(x)\| = |cp(0)| + |cp(1) - cp(0)| = |c||p(0)| + |c||p(1) - p(0)| = c\|p(x)\|$, as required.

3. $$\begin{aligned}\|(p+q)(x)\| &= |(p+q)(0)| + |(p+q)(1) - (p+q)(0)| \\ &\leq (|p(0)| + |p(1) - p(0)|) + (|q(0)| + |q(1) - q(0)|) \\ &= \|p(x)\| + \|q(x)\|, \text{ as required.}\end{aligned}$$

Therefore, $\|p(x)\| = |p(0)| + |p(1) - p(0)|$ is a norm.

11. See the definition of *ill-conditioned* and Example 7.20 on p.570 in Section 7.2.

We find A^{-1} then compute $\|A\|\|A^{-1}\|$, the condition number(see p.571).

$$A^{-1} = \begin{bmatrix} 1 & -11 & 10 \\ -11 & -990 & 1000 \\ 10 & 1000 & -1000 \end{bmatrix} \Rightarrow \begin{array}{l} \text{cond}_1(A) = 1.21 \cdot 2010 = 2432.1 \\ \text{cond}_\infty(A) = 1.21 \cdot 2010 = 2432.1 \end{array} \Rightarrow A \text{ is ill-conditioned.}$$

12. See the definition of the *Frobenius norm* on p.565 in Section 7.2, namely $\|Q\|_F = \sqrt{\sum\limits_{i,j=1}^{n} q_{ij}^2}$.

Let $Q = \begin{bmatrix} \mathbf{q}_1 & \mathbf{q}_2 & \cdots & \mathbf{q}_n \end{bmatrix}$ be orthogonal.
Recall, that all of the $\mathbf{q}_i$ are unit vectors. What does that tell us?
For each $\mathbf{q}_i$, $\|\mathbf{q}_i\| = 1$ so $\|\mathbf{q}_i\|^2 = 1$.

In terms of the entries of Q, we have: $\|\mathbf{q}_i\| = \sqrt{\sum\limits_{j=1}^{n} q_{ij}^2} = \sqrt{1}$.

Therefore, $\|Q\|_F = \sqrt{\sum\limits_{i,j=1}^{n} q_{ij}^2} = \sqrt{1+1+\cdots+1} = \sqrt{n}$.

Q: How might we describe the fact that $\|Q\|_F = \sqrt{n}$ in words?
A: Since Q is orthogonal, every column is a unit vector.
Therefore, the sum of the squares of each column is 1.
Since there are n columns, $\|Q\|_F = \sqrt{1+1+\cdots+1} = \sqrt{n}$.

13. See Example 7.26 and Exercises 7 through 14 in Section 7.3.

The given points imply $A = \begin{bmatrix} 1 & 1 \\ 1 & 2 \\ 1 & 3 \\ 1 & 4 \end{bmatrix}$.

Therefore, $A^TA = \begin{bmatrix} 4 & 10 \\ 10 & 30 \end{bmatrix}$ which implies $(A^TA)^{-1} = \begin{bmatrix} \frac{3}{2} & -\frac{1}{2} \\ -\frac{1}{2} & \frac{1}{5} \end{bmatrix}$. So:

$$\overline{\mathbf{x}} = (A^TA)^{-1}A^T\mathbf{b} = \begin{bmatrix} \frac{3}{2} & -\frac{1}{2} \\ -\frac{1}{2} & \frac{1}{5} \end{bmatrix} \begin{bmatrix} 1 & 1 & 1 & 1 \\ 1 & 2 & 3 & 4 \end{bmatrix} \begin{bmatrix} 2 \\ 3 \\ 5 \\ 7 \end{bmatrix} = \begin{bmatrix} 0 \\ \frac{17}{10} \end{bmatrix} \Rightarrow y = 0 + \tfrac{17}{10}\,x.$$

14. See Example 7.30 and Exercises 19 through 22 in Section 7.3.

We begin by finding a QR factorization of A.
See Example 5.15 on p.391 in Section 5.3.

We use the Gram-Schmidt Process to find an orthonormal basis for col(A).
Since $\mathbf{a}_1 \cdot \mathbf{a}_1 = 6$ and $\mathbf{a}_2 \cdot \mathbf{a}_1 = 0$, we need only normalize the columns of A.

Since $A = \begin{bmatrix} 1 & 2 \\ 1 & 0 \\ 2 & -1 \\ 0 & 5 \end{bmatrix}$, we have: $Q = \begin{bmatrix} \frac{1}{\sqrt{6}} & \frac{2}{\sqrt{30}} \\ \frac{1}{\sqrt{6}} & 0 \\ \frac{1}{\sqrt{6}} & -\frac{1}{\sqrt{30}} \\ 0 & \frac{5}{\sqrt{30}} \end{bmatrix}$.

Now let $R = Q^T A = \begin{bmatrix} \frac{1}{\sqrt{6}} & \frac{1}{\sqrt{6}} & \frac{1}{\sqrt{6}} & 0 \\ \frac{2}{\sqrt{30}} & 0 & -\frac{1}{\sqrt{30}} & \frac{5}{\sqrt{30}} \end{bmatrix} \begin{bmatrix} 1 & 2 \\ 1 & 0 \\ 2 & -1 \\ 0 & 5 \end{bmatrix} = \begin{bmatrix} \sqrt{6} & 0 \\ 0 & \sqrt{30} \end{bmatrix}$.

Verify that $A = QR$.

Now we follow Example 7.30 in Section 7.3.

We have $Q^T\mathbf{b} = \begin{bmatrix} \frac{1}{\sqrt{6}} & \frac{1}{\sqrt{6}} & \frac{1}{\sqrt{6}} & 0 \\ \frac{2}{\sqrt{30}} & 0 & -\frac{1}{\sqrt{30}} & \frac{5}{\sqrt{30}} \end{bmatrix} \begin{bmatrix} 1 \\ 0 \\ -1 \\ 3 \end{bmatrix} = \begin{bmatrix} -\frac{1}{\sqrt{6}} \\ \frac{18}{\sqrt{30}} \end{bmatrix}$.

So, $R\overline{\mathbf{x}} = Q^T\mathbf{b}$. That is: $\begin{bmatrix} \sqrt{6} & 0 \\ 0 & \sqrt{30} \end{bmatrix} \overline{\mathbf{x}} = \begin{bmatrix} -\frac{1}{\sqrt{6}} \\ \frac{18}{\sqrt{30}} \end{bmatrix}$.

Back substitution quickly yields: $\overline{\mathbf{x}} = \begin{bmatrix} -\frac{1}{6} \\ \frac{3}{5} \end{bmatrix}$.

15. See Example 7.31 and Exercises 37 through 42 in Section 7.3.

Since $A = \begin{bmatrix} 1 & 1 \\ 0 & 1 \\ 1 & 0 \end{bmatrix}$, $A^T A = \begin{bmatrix} 1 & 0 & 1 \\ 1 & 1 & 0 \end{bmatrix} \begin{bmatrix} 1 & 1 \\ 0 & 1 \\ 1 & 0 \end{bmatrix} = \begin{bmatrix} 2 & 1 \\ 1 & 2 \end{bmatrix}$. So, $(A^T A)^{-1} = \begin{bmatrix} \frac{2}{3} & -\frac{1}{3} \\ -\frac{1}{3} & \frac{2}{3} \end{bmatrix}$.

Now we take $P = A(A^T A)^{-1} A^T = \begin{bmatrix} 1 & 0 & 1 \\ 1 & 1 & 0 \end{bmatrix} \begin{bmatrix} \frac{2}{3} & -\frac{1}{3} \\ -\frac{1}{3} & \frac{2}{3} \end{bmatrix} \begin{bmatrix} 1 & 1 \\ 0 & 1 \\ 1 & 0 \end{bmatrix} = \begin{bmatrix} \frac{2}{3} & \frac{1}{3} & \frac{1}{3} \\ \frac{1}{3} & \frac{2}{3} & -\frac{1}{3} \\ \frac{1}{3} & -\frac{1}{3} & \frac{2}{3} \end{bmatrix}$.

So the projection is: $\text{proj}_W(\mathbf{v}) = P\mathbf{v} = \begin{bmatrix} \frac{2}{3} & \frac{1}{3} & \frac{1}{3} \\ \frac{1}{3} & \frac{2}{3} & -\frac{1}{3} \\ \frac{1}{3} & -\frac{1}{3} & \frac{2}{3} \end{bmatrix} \begin{bmatrix} 1 \\ 2 \\ 3 \end{bmatrix} = \begin{bmatrix} \frac{7}{3} \\ \frac{2}{3} \\ \frac{5}{3} \end{bmatrix}$.

16. The key is to let $A = \begin{bmatrix} \mathbf{u} & \mathbf{v} \end{bmatrix}$.

Since $\mathbf{u}, \mathbf{v}$ are orthonormal, $\|\mathbf{u}\| = \sqrt{\mathbf{u} \cdot \mathbf{u}} = \|\mathbf{v}\| = \sqrt{\mathbf{v} \cdot \mathbf{v}} = 1$ and $\mathbf{u} \cdot \mathbf{v} = 0$.

Furthermore, $\mathbf{u}^T\mathbf{v} = \mathbf{u} \cdot \mathbf{v} = 0$, $\mathbf{v}^T\mathbf{u} = \mathbf{v} \cdot \mathbf{u} = 0$, $\mathbf{u}^T\mathbf{u} = \mathbf{u} \cdot \mathbf{u} = 1$, and $\mathbf{v}^T\mathbf{v} = \mathbf{v} \cdot \mathbf{v} = 1$.

Then $A^T A = \begin{bmatrix} \mathbf{u}^T \\ \mathbf{v}^T \end{bmatrix} \begin{bmatrix} \mathbf{u} & \mathbf{v} \end{bmatrix} = \begin{bmatrix} \mathbf{u}^T\mathbf{u} & \mathbf{u}^T\mathbf{v} \\ \mathbf{v}^T\mathbf{u} & \mathbf{v}^T\mathbf{v} \end{bmatrix} = \begin{bmatrix} 1 & 0 \\ 0 & 1 \end{bmatrix}$. Thus, $(A^T A)^{-1} = \begin{bmatrix} 1 & 0 \\ 0 & 1 \end{bmatrix}$.

So, we have $P = A(A^T A)^{-1} A^T = \begin{bmatrix} \mathbf{u} & \mathbf{v} \end{bmatrix} \begin{bmatrix} 1 & 0 \\ 0 & 1 \end{bmatrix} \begin{bmatrix} \mathbf{u}^T \\ \mathbf{v}^T \end{bmatrix} = \mathbf{u}\mathbf{u}^T + \mathbf{v}\mathbf{v}^T$ as required.

Q: Why is this sufficient?

A: Hint: See Theorem 7.11 on p.593 in Section 7.3.

17. See Examples 7.33, 7.34, and 7.39 in Section 7.4.

(a) We follow Example 7.33 to find the singular values of A.

The matrix $A^TA = \begin{bmatrix} 1 & 0 & 1 \\ 1 & 0 & -1 \end{bmatrix} \begin{bmatrix} 1 & 1 \\ 0 & 0 \\ 1 & -1 \end{bmatrix} = \begin{bmatrix} 2 & 0 \\ 0 & 2 \end{bmatrix}$ has eigenvalues $\lambda_1 = \lambda_2 = 2$.

So, the singular values of A are $\sigma_1 = \sigma_2 = \sqrt{\lambda_1} = \sqrt{\lambda_2} = \sqrt{2}$.

(b) We follow Example 7.34 to find the singular value decomposition of A.

The corresponding eigenvectors of A^TA are $\mathbf{v}_1 = \begin{bmatrix} 1 \\ 0 \end{bmatrix}$ and $\mathbf{v}_2 = \begin{bmatrix} 0 \\ 1 \end{bmatrix}$.

So, $V = \begin{bmatrix} 1 & 0 \\ 0 & 1 \end{bmatrix}$ and $\Sigma = \begin{bmatrix} \sqrt{2} & 0 \\ 0 & \sqrt{2} \\ 0 & 0 \end{bmatrix}$. To find U, we compute $\mathbf{u}_1$, $\mathbf{u}_2$.

$$\mathbf{u}_1 = \frac{1}{\sigma_1} A\mathbf{v}_1 = \frac{1}{\sqrt{2}} \begin{bmatrix} 1 & 1 \\ 0 & 0 \\ 1 & -1 \end{bmatrix} \begin{bmatrix} 1 \\ 0 \end{bmatrix} = \begin{bmatrix} \frac{1}{\sqrt{2}} \\ 0 \\ \frac{1}{\sqrt{2}} \end{bmatrix} \text{ and } \mathbf{u}_2 = \frac{1}{\sigma_2} A\mathbf{v}_2 = \begin{bmatrix} \frac{1}{\sqrt{2}} \\ 0 \\ -\frac{1}{\sqrt{2}} \end{bmatrix}.$$

Verify that $A = U\Sigma V^T = \begin{bmatrix} \frac{1}{\sqrt{2}} & \frac{1}{\sqrt{2}} & 0 \\ 0 & 0 & 1 \\ \frac{1}{\sqrt{2}} & -\frac{1}{\sqrt{2}} & 0 \end{bmatrix} \begin{bmatrix} \sqrt{2} & 0 \\ 0 & \sqrt{2} \\ 0 & 0 \end{bmatrix} \begin{bmatrix} 1 & 0 \\ 0 & 1 \end{bmatrix} = \begin{bmatrix} 1 & 1 \\ 0 & 0 \\ 1 & -1 \end{bmatrix}$.

(c) We follow Example 7.39 to find the pseudoinverse of A, that is A^+.

Given $\Sigma = \begin{bmatrix} D & O \\ O & O \end{bmatrix}$, we have $\Sigma^+ = \begin{bmatrix} D^{-1} & O \\ O & O \end{bmatrix}$.

So in this case, $\Sigma^+ = \begin{bmatrix} \frac{1}{\sqrt{2}} & 0 & 0 \\ 0 & \frac{1}{\sqrt{2}} & 0 \end{bmatrix}$.

Then $A^+ = V\Sigma^+U^T = \begin{bmatrix} 1 & 0 \\ 0 & 1 \end{bmatrix} \begin{bmatrix} \frac{1}{\sqrt{2}} & 0 & 0 \\ 0 & \frac{1}{\sqrt{2}} & 0 \end{bmatrix} \begin{bmatrix} \frac{1}{\sqrt{2}} & 0 & \frac{1}{\sqrt{2}} \\ \frac{1}{\sqrt{2}} & 0 & -\frac{1}{\sqrt{2}} \\ 0 & 1 & 0 \end{bmatrix} = \begin{bmatrix} \frac{1}{2} & 0 & \frac{1}{2} \\ \frac{1}{2} & 0 & -\frac{1}{2} \end{bmatrix}$.

18. See Examples 7.33, 7.34, and 7.39 in Section 7.4.

(a) We follow Example 7.33 to find the singular values of A.

$$A^TA = \begin{bmatrix} 1 & 1 \\ 1 & 1 \\ -1 & -1 \end{bmatrix} \begin{bmatrix} 1 & 1 & -1 \\ 1 & 1 & -1 \end{bmatrix} = \begin{bmatrix} 2 & 2 & -2 \\ 2 & 2 & -2 \\ -2 & -2 & 2 \end{bmatrix} \text{ implies } \lambda_1 = 0, \lambda_2 = 6, \lambda_3 = 0.$$

So, the singular values of A are $\sigma_1 = \sqrt{\lambda_1} = 0$, $\sigma_2 = \sqrt{\lambda_2} = \sqrt{6}$, and $\sigma_3 = \sqrt{\lambda_3} = 0$.

(b) We follow Example 7.34 to find the singular value decomposition of A.
The normalized eigenvectors of A^TA are:

$$\mathbf{v}_1 = \begin{bmatrix} \frac{2}{\sqrt{6}} \\ -\frac{1}{\sqrt{6}} \\ \frac{1}{\sqrt{6}} \end{bmatrix}, \mathbf{v}_2 = \begin{bmatrix} \frac{1}{\sqrt{3}} \\ \frac{1}{\sqrt{3}} \\ -\frac{1}{\sqrt{3}} \end{bmatrix}, \mathbf{v}_3 = \begin{bmatrix} 0 \\ \frac{1}{\sqrt{2}} \\ -\frac{1}{\sqrt{2}} \end{bmatrix}.$$

So, $V = \begin{bmatrix} \frac{2}{\sqrt{6}} & \frac{1}{\sqrt{3}} & 0 \\ -\frac{1}{\sqrt{6}} & \frac{1}{\sqrt{3}} & \frac{1}{\sqrt{2}} \\ \frac{1}{\sqrt{6}} & -\frac{1}{\sqrt{3}} & \frac{1}{\sqrt{2}} \end{bmatrix}$ and $\Sigma = \begin{bmatrix} 0 & 0 & 0 \\ 0 & \sqrt{6} & 0 \end{bmatrix}$. To find U, we compute $\mathbf{u}_2$.

$$\mathbf{u}_2 = \frac{1}{\sigma_2}A\mathbf{v}_2 = \frac{1}{\sqrt{6}} \begin{bmatrix} 1 & 1 & -1 \\ 1 & 1 & -1 \end{bmatrix} \begin{bmatrix} \frac{1}{\sqrt{3}} \\ \frac{1}{\sqrt{3}} \\ -\frac{1}{\sqrt{3}} \end{bmatrix} = \begin{bmatrix} \frac{1}{\sqrt{2}} \\ \frac{1}{\sqrt{2}} \end{bmatrix}.$$

Verify that $A = U\Sigma V^T = \begin{bmatrix} \frac{1}{\sqrt{2}} & \frac{1}{\sqrt{2}} \\ -\frac{1}{\sqrt{2}} & \frac{1}{\sqrt{2}} \end{bmatrix} \begin{bmatrix} 0 & 0 & 0 \\ 0 & \sqrt{6} & 0 \end{bmatrix} \begin{bmatrix} \frac{2}{\sqrt{6}} & -\frac{1}{\sqrt{6}} & \frac{1}{\sqrt{6}} \\ \frac{1}{\sqrt{3}} & \frac{1}{\sqrt{3}} & -\frac{1}{\sqrt{3}} \\ 0 & \frac{1}{\sqrt{2}} & \frac{1}{\sqrt{2}} \end{bmatrix} = \begin{bmatrix} 1 & 1 & -1 \\ 1 & 1 & -1 \end{bmatrix}.$

(c) We follow Example 7.39 to find the pseudoinverse of A, that is A^+.

Given $\Sigma = \begin{bmatrix} D & O \\ O & O \end{bmatrix}$, we have $\Sigma^+ = \begin{bmatrix} D^{-1} & O \\ O & O \end{bmatrix}$.

So in this case, $\Sigma^+ = \begin{bmatrix} 0 & 0 \\ 0 & \frac{1}{\sqrt{6}} \\ 0 & 0 \end{bmatrix}$.

Then $A^+ = V\Sigma^+U^T = \begin{bmatrix} \frac{2}{\sqrt{6}} & \frac{1}{\sqrt{3}} & 0 \\ -\frac{1}{\sqrt{6}} & \frac{1}{\sqrt{3}} & \frac{1}{\sqrt{2}} \\ \frac{1}{\sqrt{6}} & -\frac{1}{\sqrt{3}} & \frac{1}{\sqrt{2}} \end{bmatrix} \begin{bmatrix} 0 & 0 \\ 0 & \frac{1}{\sqrt{6}} \\ 0 & 0 \end{bmatrix} \begin{bmatrix} \frac{1}{\sqrt{2}} & -\frac{1}{\sqrt{2}} \\ \frac{1}{\sqrt{2}} & \frac{1}{\sqrt{2}} \end{bmatrix} = \begin{bmatrix} \frac{1}{6} & \frac{1}{6} \\ \frac{1}{6} & \frac{1}{6} \\ -\frac{1}{6} & -\frac{1}{6} \end{bmatrix}.$

19. See Exercise 31 in Section 7.4.

We need to show the eigenvalues of $A^TA =$ the eigenvalues of $(PAQ)^T(PAQ)$.
$(PAQ)^T(PAQ) = Q^TA^T(P^TP)AQ = Q^T(A^TA)Q$. Therefore, by definition
A^TA and $(PAQ)^T(PAQ)$ are similar. So, Section 4.4, Theorem 4.22 $\Rightarrow$
eigenvalues of $A^TA =$ eigenvalues of $(PAQ)^T(PAQ) \Rightarrow$
singular values of $A =$ singular values of PAQ.

20. Since A is square, $(A^2)^T = (A^T)^2 = A^TA^T = O^T = O$.

By the Penrose conditions, we have:

$$(A^+)^2 = (A^+AA^+)(A^+AA^+) = A^+(AA^+)(A^+A)A^+ = A^+(AA^+)^T(A^+A)^TA^+$$
$$= A^+(A^+)^T(A^TA^T)(A^+)^TA^+ = A^+(A^+)^TO(A^+)^TA^+ = O \text{ as required.}$$

Appendix I: Key Definitions and Concepts

This list includes most but not all of the definitions listed at the end of each section. We begin the list with key symbols and symbol-based definitions in chapter order.

Chapter 1

$\mathbf{v} = \mathbf{w}$	p.3	1.1	... if and only if *corresponding* components are equal
$\mathbf{u} + \mathbf{v}$	p.5	1.1	$\mathbf{u} + \mathbf{v} = [u_1 + v_1, u_2 + v_2]$ *vector addition*
$\mathbf{u} = \sum c_i \mathbf{v}_i$	p.12	1.1	the c_i are scalars (*linear combination*)
$\mathbf{u} - \mathbf{v}$	p.8	1.1	$\mathbf{u} - \mathbf{v} = \mathbf{u} + (-\mathbf{v})$ *vector subtraction*
$\mathbf{u} \parallel \mathbf{v}$	p.8	1.1	$\mathbf{u} \parallel \mathbf{v} \Leftrightarrow \mathbf{v} = c\mathbf{u}$ if and only if scalar multiples (*parallel*)
$\mathbf{v}$	p.3	1.1	vectors denoted by a single, boldface, lowercase letter
$\parallel$	p.8	1.1	parallel (word is used not symbol)
$\Rightarrow$	p.4	1.1	implies (**if** vectors have the same direction, **then** they are parallel)
$\Leftrightarrow$	p.4	1.1	if and only if (phrase is used on p. 4 of Section 1.1)
$\sum c_i \mathbf{v}_i$	p.12	1.1	shorthand for $c_1\mathbf{v}_1 + c_2\mathbf{v}_2 + \cdots + c_n\mathbf{v}_n$
$\overrightarrow{AB}$	p.3	1.1	an overhead arrow (vector $\overrightarrow{AB}$, differs from line segment $\overline{AB}$)

Chapter 2

$R_i \leftrightarrow R_j$	p.70	2.2	interchange two rows
kR_i	p.70	2.2	multiply a row by a nonzero constant
$R_i + kR_j$	p.70	2.2	add a multiple of a row to another row
A_c		2.3	Created by taking vectors $\mathbf{v}_i$ as its columns
A_r		2.3	Created by taking vectors $\mathbf{v}_i$ as its rows
$\mathbf{b} = \sum c_i \mathbf{v}_i$	p.638	*A*.5	$\mathbf{b}$ is a linear combination of $\mathbf{v}_i$ (see Appendix A)
$\mathbf{a}_i$	p.90	2.3	rows or columns of A written as vectors
$[A \mid \mathbf{b}]$	p.91	2.3	augmented matrix of a linear system
$[A \mid \mathbf{0}]$	p.97	2.3	augmented matrix of a homogenous linear system

Chapter 3

A^T	p.149	3.1	A-transpose, created by switching rows and columns
A^{-1}	p.162	3.3	A-inverse: satisfies $AA^{-1} = I$ and is unique
A^{-n}	p.167	3.3	This is defined to be $(A^{-1})^n = (A^n)^{-1}$ and is unique
$\mathbf{e}_i$	p.142	3.1	standard $1 \times m$ unit vector
$\mathbf{e}_j$	p.142	3.1	standard $n \times 1$ unit vector
E_i	p.169	3.3	Matrix created by an elementary row operation on I
L	p.179	3.4	Often used to stand for a unit lower triangular matrix
O	p.152	3.2	O is commonly used to stand for the zero matrix
P	p.185	3.4	Often used to stand for a permutation matrix
U	p.179	3.4	Often used to stand for an upper triangular matrix
LU **factorization**	p.179	3.4	$A = LU$ with L unit lower triangular and and U upper triangular
LDU **factorization**	p.189	3.4	$A = LDU$ same as LU except D is diagonal and U *unit* upper triangular
$P^T LU$ **factorization**	p.184	3.4	Adjustment to the LU factorization when we have to permute the rows of A
$[\mathbf{v}]_{\mathcal{B}}$	p.198	3.5	coordinate vector with respect to $\mathcal{B}$
col(A)	p.193	3.5	column space of A spanned by the columns of A
dim(S)	p.201	3.5	the number of vectors in a basis called its *dimension*
null(A)	p.195	3.5	null space of A, $\mathbf{x}$ such that $A\mathbf{x} = \mathbf{0}$
nullity(A)	p.203	3.5	the dimension of null(A)
row(A)	p.193	3.5	row space of A spanned by the rows of A
det A	p.163	3.3	Determinant of A for 2×2 matrices $\det A = ad - bc$

Chapter 4

λ	p.253	4.1	λ often used to denote an eigenvalue of A
E_λ	p.255	4.1	$E_\lambda = \text{null}(A - \lambda I) = \{\text{eigenvectors of } \lambda\} \cup \{\mathbf{0}\}$
$\sum_{i=1}^{n} a_{ij}C_{ij}$	p.263	4.2	$\sum_{j=1}^{n} a_{ij}C_{ij} = a_{i1}C_{i1} + a_{i2}C_{i2} + \cdots + a_{in}C_{in}$
$\prod_{i=1}^{n} A_i$	p.263	4.2	$\prod_{i=1}^{n} A_i = A_1A_2\cdots A_n$ Shorthand for products like $\sum_{j=1}^{n} a_{ij}C_{ij}$ is for sums
A_{ij}	p.263	4.2	submatrix of A obtained by deleting row i and column j
$\det A_{ij}$	p.263	4.2	$\det A_{ij}$, the (i, j)-minor of A
$A_i(\mathbf{b})$	p.273	4.2	$A_i(\mathbf{b})$ obtained by replacing column i of A by $\mathbf{b}$ That is, $A_i(\mathbf{b}) = \begin{bmatrix} \mathbf{a}_1 & \cdots & \mathbf{b} & \cdots & \mathbf{a}_n \end{bmatrix}$ (see Cramer's Rule)
$\text{adj}\, A$	p.275	4.2	$\text{adj}\, A = [C_{ij}]^T$, the adjoint of A
$\det A$	p.264	4.2	$\det A = \sum_{j=1}^{n} a_{1j}(-1)^{i+1} \det A_{1j}$ or $\det A = \sum_{j=1}^{n} a_{1j}C_{1j}$
C_{ij}	p.265	4.2	$C_{ij} = (-1)^{i+j} \det A_{ij}$, the (i, j)-cofactor of A
$c_A(\lambda)$	p.289	4.3	The *characteristic* polynomial is $c_A(\lambda) = \det(A - \lambda I)$. It is important to note that $c_A(\lambda)$ is a polynomial in λ. That is, λ is *not* a fixed eigenvalue, but a variable like x.
$\prod_{i=1}^{n} (\lambda_i - \lambda)$	p.289	4.3	$c_A(\lambda)$ using the eigenvalues of A, λ_i.
$\prod_{i=1}^{m} (\lambda_i - \lambda)^{k_i}$	p.289	4.3	$c_A(\lambda)$ emphasizing the algebraic multiplicity of λ_i, k_i.
$A \sim B$	p.298	4.4	$A \sim B$ (A is *similar* to B) if $P^{-1}AP = B$, P invertible
λ_1	p.308	4.5	*dominant eigenvalue*, $\lvert\lambda_1\rvert > \lvert\lambda_i\rvert$
$\mathbf{v}_1$	p.308	4.5	*dominant eigenvector*, associated with dominant eigenvalue

Chapter 5

$\mathcal{B}$	p.369	5.1	$\mathcal{B}$ often used to denote a set of basis vectors
$\mathbf{q}_i$	p.369	5.1	$\mathbf{q}_i$ often used to denote a vector in an *orthonormal* basis
Q	p.371	5.1	Q often used to denote an *orthogonal* matrix
$\mathbf{w}^\perp$	p.379	5.2	component of $\mathbf{v}$ orthogonal to W, $\mathbf{v} - \text{proj}_W(\mathbf{v})$
$W^\perp$	p.375	5.2	$W^\perp = \{\mathbf{v} \text{ in } \mathbb{R}^n : \mathbf{v} \cdot \mathbf{w} = 0 \quad \text{for all } \mathbf{w} \text{ in } W\}$
$\text{perp}_W(\mathbf{v})$	p.379	5.2	component of $\mathbf{v}$ orthogonal to W, $\mathbf{v} - \text{proj}_W(\mathbf{v})$
$\text{proj}_W(\mathbf{v})$	p.379	5.2	*orthogonal projection* of $\mathbf{v}$ onto W, $\sum_{k=1}^{n} \left(\frac{\mathbf{u}_k \cdot \mathbf{v}}{\mathbf{u}_k \cdot \mathbf{u}_k}\right) \mathbf{u}_k$
$\mathbf{1}$	p.411	5.5	$\mathbf{1}$ (the vector in $\mathbb{Z}_2^n$ all of whose entries are 1)
C	p.408	5.5	C (a set of code vectors in $\mathbb{Z}_2^n$)
$C^\perp$	p.408	5.5	$C^\perp = \{\mathbf{x} \text{ in } \mathbb{Z}_2^n : \mathbf{c} \cdot \mathbf{x} = 0 \quad \text{for all } \mathbf{c} \text{ in } C\}$
G	p.405	5.5	$G = \left[\frac{I_m}{A}\right]$ (standard generator matrix for a code)
P	p.405	5.5	$G = [\, B \mid I_{n-m} \,]$ (standard parity check matrix for a code)
$w(\mathbf{x})$	p.411	5.5	$w(\mathbf{x})$ (*weight* of $\mathbf{x}$), the number of 1s in $\mathbf{x}$ (in $\mathbb{Z}_2^n$)
$f(\mathbf{x})$	p.412	5.5	$f(\mathbf{x}) = \mathbf{x}^T A\mathbf{x}$ (*quadratic form*), (A, the matrix associated with f, is symmetric)
$\mathbf{x}^T A\mathbf{x}$	p.412	5.5	$\mathbf{x}^T A\mathbf{x} = a_{11}x_1^2 + a_{22}x_2^2 + \cdots + a_{nn}x_n^2 + \sum_{i<j} 2a_{ij}x_ix_j$

Chapter 6

V	p.433	6.1	V often used to denote a *vector space*
W	p.438	6.1	W often used to denote a *subspace*
$\text{span}(S)$	p.442	6.1	If $S = \{\mathbf{v}_i\}$, then $\text{span}(S)$ equals all linear combinations of $\mathbf{v}_i$.
$\{\mathbf{e}_i\}$	p.451	6.2	$\{\mathbf{e}_i\}$, standard basis for $\mathbb{R}^n$ (see Example 6.29)
$\{x^i\}$	p.451	6.2	$\{x^i\}$, standard basis for $\mathscr{P}_n$ (see Example 6.30)
$\{E_{ij}\}$	p.451	6.2	$\mathcal{E} = \{E_{ij}\}$, standard basis for M_{mn} (see Example 6.31)
$[\mathbf{v}]_{\mathcal{B}}$	p.453	6.2	If $\mathcal{B} = \{\mathbf{u}_i\}$ is a basis and $\mathbf{x} = \sum_{i=1}^{n} c_i\mathbf{u}_i$, then $[\mathbf{v}]_{\mathcal{B}} = \begin{bmatrix} c_1 \\ \vdots \\ c_n \end{bmatrix}$.
$P_{\mathcal{C}\leftarrow\mathcal{B}}$	p.469	6.3	*change-of-basis* matrix: $P_{\mathcal{C}\leftarrow\mathcal{B}} = \begin{bmatrix} [\mathbf{u}_1]_{\mathcal{C}} & \cdots & [\mathbf{u}_n]_{\mathcal{C}} \end{bmatrix}$
$\mathscr{C}[a,b]$	p.477	6.4	The space of continuous functions (See Example 6.52).
$\int_a^b f(x)\,dx$	p.477	6.4	$S : \mathscr{C}[a,b] \to \mathbb{R}$, $S(f) = \int_a^b f(x)\,dx$ (See Example 6.52).
I	p.478	6.4	$I : V \to W$ maps every vector to itself. That is, $I(\mathbf{v}) = \mathbf{v}$ (See Example 6.54).
$S \circ T$	p.481	6.4	If $T : U \to V$ and $S : V \to W$ then the *composition of S with T* is $(S \circ T)(\mathbf{u}) = S(T(\mathbf{u}))$.
T_0	p.478	6.4	$T_0 : V \to W$ maps every vector to the zero vector. That is, $T_0(\mathbf{v}) = \mathbf{0}$ (See Example 6.54).
T^{-1}	p.483	6.4	T^{-1}: $T^{-1} \circ T = I_V$, $T \circ T^{-1} = I_W$ (See *invertible*).

Chapter 6, continued

$\ker(T)$	p.486	6.5	$\ker(T) = \{\mathbf{v} \text{ in } V : T(\mathbf{v}) = \mathbf{0}\}$ (See *kernel.*)
nullity(T)	p.488	6.5	nullity$(T) = \dim(\ker(T))$
range(T)	p.486	6.5	range$(T) = \{\mathbf{w} \text{ in } W : \mathbf{w} = T(\mathbf{v}) \text{ for some } \mathbf{v} \text{ in } V\}$ (See *range.*)
rank(T)	p.488	6.5	rank$(T) = \dim(\text{range}(T))$ (See *rank.*)
$V \cong W$	p.497	6.5	If $T : V \to W$ where T is an *isomorphism.* Then: V and W are *isomorphic.* This is written $V \cong W$.
$[T]_{\mathcal{C}\leftarrow\mathcal{B}}$	p.502	6.6	The matrix of T with respect to bases $\mathcal{B}$ and $\mathcal{C}$. By Thm 6.26 it satisfies: $[T]_{\mathcal{C}\leftarrow\mathcal{B}}[\mathbf{v}]_{\mathcal{B}} = [T(\mathbf{v})]_{\mathcal{C}}$ See remarks following Thm 6.26.
$[T]_{\mathcal{B}}$	p.502	6.6	Special case when $V = W$ and $\mathcal{B} = \mathcal{C}$. By Thm 6.26 it satisfies: $[T]_{\mathcal{B}}[\mathbf{v}]_{\mathcal{B}} = [T(\mathbf{v})]_{\mathcal{B}}$ See remarks following Thm 6.26.
$[T(\mathbf{v})]_{\mathcal{C}}$	p.502	6.6	$A[\mathbf{v}]_{\mathcal{B}} = [T(\mathbf{v})]_{\mathcal{C}}$ We are asked to show this in Exercise 39. See remarks following Thm 6.26.

Chapter 7

$\|\mathbf{v}\|$	p.544	7.1	The *length* (or *norm*) of $\mathbf{v}$ is $\|\mathbf{v}\| = \sqrt{\langle \mathbf{v}, \mathbf{v} \rangle}$
$\mathrm{d}(\mathbf{u}, \mathbf{v})$	p.544	7.1	The *distance* between $\mathbf{u}$ and $\mathbf{v}$ is $\mathrm{d}(\mathbf{u}, \mathbf{v}) = \|\mathbf{u} - \mathbf{v}\|$
$\langle \mathbf{u}, \mathbf{v} \rangle = 0$	p.544	7.1	$\mathbf{u}$ and $\mathbf{v}$ are orthogonal if $\langle \mathbf{u}, \mathbf{v} \rangle = 0$
$\mathrm{proj}_W(\mathbf{v})$	p.547	7.1	$\mathrm{proj}_W(\mathbf{v}) = \sum_{k=1}^{n} \frac{\langle \mathbf{u}_k \cdot \mathbf{v} \rangle}{\langle \mathbf{u}_k \cdot \mathbf{u}_k \rangle} \mathbf{u}_k$ (*orthogonal projection*)
$\mathrm{perp}_W(\mathbf{v})$	p.547	7.1	$\mathrm{perp}_W(\mathbf{v}) = \mathbf{v} - \mathrm{proj}_W(\mathbf{v})$ This is the *component of* $\mathbf{v}$ *orthogonal to* W)
$\|\mathbf{v}\|_1 = \|\mathbf{v}\|_s$	p.562	7.2	$\|\mathbf{v}\|_1 = \|\mathbf{v}\|_s = \|v_1\| + \cdots + \|v_n\|$
$\|\mathbf{v}\|_\infty = \|\mathbf{v}\|_m$	p.562	7.2	$\|\mathbf{v}\|_\infty = \|\mathbf{v}\|_m = \max\{\|v_1\|, \ldots, \|v_n\|\}$
$\|\mathbf{v}\|_p$	p.562	7.2	$\|\mathbf{v}\|_p = (\|v_1\|^p + \cdots + \|v_n\|^p)^{1/p}$
$\|\mathbf{v}\|_2 = \|\mathbf{v}\|_E$	p.562	7.2	$\|\mathbf{v}\|_2 = \|\mathbf{v}\|_E = \sqrt{\|v_1\|^2 + \cdots + \|v_n\|^2}$
$\|\mathbf{v}\|_H$	p.563	7.2	$\|\mathbf{v}\|_H = w(\mathbf{v})$ (weight). See Example 7.15.
$\overline{\mathbf{v}}$	p.579	7.3	A vector $\overline{\mathbf{v}}$ is *best approximation* to $\mathbf{v}$ in W if $\|\mathbf{v} - \overline{\mathbf{v}}\| < \|\mathbf{v} - \mathbf{w}\|$ for all $\mathbf{w} \neq \overline{\mathbf{v}}$
$\|\mathbf{e}\|$	p.581	7.3	$\|\mathbf{e}\| = \sqrt{\varepsilon_1^2 + \varepsilon_2^2 + \varepsilon_3^2}$ is the *least squares error* of the approximation
A^+	p.594	7.3	If A is a matrix with linearly independent columns, then $A^+ = (A^T A)^{-1} A^T$ is the *pseudoinverse.*
σ_i	p.599	7.4	If A is an $m \times n$ matrix, the *singular values* of A are the square roots of the eigenvalues of $A^T A$. They are denoted by $\sigma_1, \ldots, \sigma_n$. They are usually arranged $\sigma_1 \geq \sigma_1 \geq \ldots \geq \sigma_n$.
Σ	p.601	7.4	Σ is created from the σ_i (detailed description p.601)
$A = U \Sigma V^T$	p.601	7.4	A *singular value decomposition* (SVD) of $A = U \Sigma V^T$. where U and and V are orthogonal matrices. Σ is constructed using the σ_i.
$A^+ = V \Sigma^+ U^T$	p.611	7.4	The *pseudoinverse* (*Moore-Penrose*) is $A^+ = V \Sigma^+ U^T$.

A

adjacency matrix	p.236	3.7	$A(G)$ where $a_{ij} = 1$ if edge, 0 otherwise
adjoint	p.275	4.2	$\text{adj}\, A = [C_{ij}]^T$, the adjoint of A
algebraic multiplicity	p.291	4.3	The multiplicity of λ as a root of $\det(A - \lambda I) = 0$. So, if $c_A(\lambda) = \prod_{i=1}^{m} (\lambda_i - \lambda)^{k_i}$ where λ_i are distinct, then the algebraic multiplicity of λ_i is k_i. So, the algebraic multiplicity of an eigenvalue is equal to the exponent of its associated factor in the characteristic equation.
angle	p.21	1.2	$\cos\theta = \frac{\mathbf{u}\cdot\mathbf{v}}{\|\mathbf{u}\|\,\|\mathbf{v}\|}$
attractor	p.347	4.6	See definition on p.347 and Example 4.48.
augmented matrix	p.62	2.1	coefficient matrix augmented by the constants

B

back substitution	p.62	2.1	procedure used to solve Example 2.5 on p. 62
basis	p.196	3.5	linearly independent set of vectors that spans S
basis	p.450	6.2	A subset $\mathcal{B}$ of a vector space V is a *basis* for V if 1. $\mathcal{B}$ spans V, that is $\text{span}(\mathcal{B}) = V$ 2. $\mathcal{B}$ is linearly independent
best approximation	p.579	7.3	A vector $\overline{\mathbf{v}}$ is *best approximation* to $\mathbf{v}$ in W if $\|\mathbf{v} - \overline{\mathbf{v}}\| < \|\mathbf{v} - \mathbf{w}\|$ for all $\mathbf{w} \neq \overline{\mathbf{v}}$
best approximation	p.620	7.5	If $\{\mathbf{u}_1, \ldots, \mathbf{u}_k\}$ is an orthogonal basis for W then $\text{proj}_W(f) = \frac{\langle \mathbf{u}_1, f\rangle}{\langle \mathbf{u}_1, \mathbf{u}_1\rangle}\mathbf{u}_1 + \cdots + + \frac{\langle \mathbf{u}_k, f\rangle}{\langle \mathbf{u}_k, \mathbf{u}_k\rangle}\mathbf{u}_k$
binary	p.47	1.4	see the text discussion of arithmetic in $\mathbb{Z}_2$
branch	p.104	2.4	a directed edge of a network

C

change-of-basis	p.469	6.3	*change-of-basis* matrix: $P_{\mathcal{C}\leftarrow\mathcal{B}} = \left[\, [\mathbf{u}_1]_\mathcal{C} \ \cdots \ [\mathbf{u}_n]_\mathcal{C} \,\right]$
characteristic equation	p.289	4.3	The *characteristic* equation is $\det(A - \lambda I) = 0$. Solutions of $\det(A - \lambda I) = 0$ are the *eigenvalues* of A.
characteristic polynomial	p.289	4.3	The *characteristic* polynomial is $c_A(\lambda) = \det(A - \lambda I)$. Solutions of $\det(A - \lambda I) = 0$ are the *eigenvalues* of A.
check digit	p.49	1.4	component added to vector to make *parity* even
circuit	p.237	3.7	a closed *path* (ends and begins at the same *vertex*
code	p.48	1.4	a set of vectors of the same *length* (m-ary)
code	p.240	3.7	See discussion prior to Example 3.69 and Section 1.4
coefficient matrix	p.62	2.1	a matrix of coefficients taken from a linear system
coefficients	p.59	2.1	the a_i in $a_1x_1 + a_2x_2 + \cdots + a_nx_n = b$
cofactor	p.265	4.2	$C_{ij} = (-1)^{i+j} \det A_{ij}$, the (i, j)-cofactor of A
column space	p.193	3.5	$\text{col}(A)$ spanned by the columns of A
commutative diagram	p.503	6.6	See detailed remarks following Thm 6.26. The diagram referred to is on p.502.
compatible	p.565	7.2	A matrix norm $\lVert A \rVert$ is compatible with a vector norm $\lVert \mathbf{x} \rVert$ if $\lVert A\mathbf{x} \rVert \leq \lVert A \rVert \lVert \mathbf{x} \rVert$
components	p.3	1.1	individual coordinates of a vector like 3, 2 of $[3, 2]$
composition of S with T	p.481	6.4	If $T: U \to V$ and $S: V \to W$ then the *composition of S with T* is the mapping $S \circ T$: $(S \circ T)(\mathbf{u}) = S(T(\mathbf{u}))$.

C, continued

conditioned	p.570	7.2	For description of *ill-* and *well-conditioned*, see text.
condition number	p.571	7.4	Condition number of a square matrix is $\|A^{-1}\|\|A\|$. If A is not invertible, we define $\text{cond}(A) = \infty$.
conditions	p.333	4.6	*initial conditions.* See definition p. 333.
conservation of flow	p.104	2.4	at each node, the flow in equals the flow out
consistent	p.61	2.1	system of equations with at least one solution
constant term	p.59	2.1	b in $a_1x_1 + a_2x_2 + \cdots + a_nx_n = b$
converges	p.123	2.5	when iterates approach a solution
coordinate vector	p.453	6.2	If $\mathcal{B} = \{\mathbf{u}_i\}$ is a basis and $\mathbf{x} = \sum_{i=1}^{n} c_i\mathbf{u}_i$, then $[\mathbf{v}]_{\mathcal{B}} = \begin{bmatrix} c_1 \\ \vdots \\ c_n \end{bmatrix}$.
correction	p.627	7.5	See text for description of *correcting k errors.*
Current Law (nodes)	p.106	2.4	sum flowing into a node equals sum out

D

decoding	p.48	1.4	converting code vectors into a message
decoding	p.627	7.5	See text for full description of *nearest neighbor decoding.*
definite	p.415	5.5	A quadratic form $f(\mathbf{x}) = \mathbf{x}^T A\mathbf{x}$ is classified as follows: 1. *positive definite* $f(\mathbf{x}) > 0$ for all $\mathbf{x} \neq \mathbf{0}$ 2. *positive semidefinite* $f(\mathbf{x}) \geq 0$ for all $\mathbf{x}$ 3. *negative definite* $f(\mathbf{x}) < 0$ for all $\mathbf{x} \neq \mathbf{0}$ 4. *negative semidefinite* $f(\mathbf{x}) \leq 0$ for all $\mathbf{x}$ 5. *indefinite* if $f(\mathbf{x})$, both positive and negative The associated matrix A is classified in the same way.
detection	p.627	7.5	See text for full description of *detecting k errors.*
determinant	p.163	3.3	Determinant of A for 2×2 matrices $\det A = ad - bc$
determinant	p.264	4.2	$\det A = \sum_{j=1}^{n} a_{1j}(-1)^{i+1} \det A_{1j}$ or $\det A = \sum_{j=1}^{n} a_{1j}C_{1j}$

D, continued

Term	Page	Section	Definition
diagonalizable	p.300	4.4	$A \sim D$ if $P^{-1}AP = D$ where D is diagonal
diagonalizable	p.513	6.6	Let V be a finite dimensional vector space. Let $T : V \to V$ be a linear transformation. Then: T is called *diagonalizable* if there is a basis $\mathcal{C}$ such that $[T]_{\mathcal{C}}$ is diagonal matrix.
differential equation	p.522	6.7	See detailed remarks before Thm 6.31. Here we show the equations for *first* and *second* order. *first-order*: $y' + ay = 0$ (homogeneous, because $= 0$). Solution: $y = e^{-at}$ (see Thm 6.31) *second-order*: $y'' + ay' + by = 0$ (homogeneous). Solution: $y = c_1 e^{\lambda_1 t} + c_2 e^{\lambda_2 t}$ (see Thm 6.31) Where: λ_1 and λ_2 are solutions of $\lambda^2 + a\lambda + b = 0$
differential operator	p.477	6.4	The *differential operator* is defined as follows: $D : \mathscr{D} \to \mathscr{F}$, $D(f) = f'$ (See Example 6.51).
dimension	p.201	3.5	$\dim(S)$, the number of vectors in a basis
dimensional **finite-dim** **infinite-dim** **zero-dim**	p.457	6.2	A vector space is *blank*-dimensional if: A basis has finitely many vectors A basis has infinitely many vectors The dimension of the subspace $\mathbf{0}$ is defined to be 0
direction vector	p.32	1.3	$\mathbf{d}$, parallel to any vector on line ℓ
distance	p.20	1.2	$d(\mathbf{u}, \mathbf{v}) = \|\mathbf{u} - \mathbf{v}\|$
distance	p.544	7.1	The *distance* between $\mathbf{u}$ and $\mathbf{v}$ is $d(\mathbf{u}, \mathbf{v}) = \|\mathbf{u} - \mathbf{v}\|$
distance	p.563	7.2	See Example 7.16 for $d_s(\mathbf{u}, \mathbf{v})$, $d_E(\mathbf{u}, \mathbf{v})$, $d_H(\mathbf{u}, \mathbf{v})$ ...
divergence	p.125	2.5	when iterates do not approach a solution
dominant eigenvalue	p.308	4.5	$\lvert\lambda_1\rvert > \lvert\lambda_i\rvert$, where λ_i are the *other* eigenvalues of A
dominant eigenvector	p.308	4.5	The eigenvector associated with $\lvert\lambda_1\rvert > \lvert\lambda_i\rvert$, that is, $A\mathbf{x}_1 = \lambda_1 \mathbf{x}_1$, where λ_1 is dominant
dot product	p.15	1.2	$\mathbf{u} \cdot \mathbf{v} = u_1 v_1 + u_2 v_2 + \ldots + u_n v_n$

E

eigenspace	p.255	4.1	$E_\lambda = \text{null}(A - \lambda I) = \{\text{eigenvectors of } \lambda\} \cup \{\mathbf{0}\}$ All vectors such that $(A - \lambda I)\mathbf{x} = \mathbf{0}$ or $A\mathbf{x} = \lambda\mathbf{x}$
eigenvalue	p.253	4.1	if $A\mathbf{x} = \lambda\mathbf{x}$, then A has eigenvalue λ
eigenvalue	p.289	4.3	The solutions of $\det(A - \lambda I) = 0$.
eigenvector	p.253	4.1	if $A\mathbf{x} = \lambda\mathbf{x}$, then $\mathbf{x}$ is an eigenvector of A
elementary matrix	p.169	3.3	Any matrix that can be obtained by performing an elementary row operation on an identity matrix
elementary row operations	p.70	2.2	EROs: $R_i \leftrightarrow R_j,\ kR_i,\ R_i + kR_j$
encoding	p.48	1.4	converting a message into code vectors
equivalent	p.61	2.1	linear systems that have the same solution set
Euclidean norm	p.562	7.2	$\lVert\mathbf{v}\rVert_E = \lVert\mathbf{v}\rVert_2 = \sqrt{\lvert v_1\rvert^2 + \cdots + \lvert v_n\rvert^2}$ (same as 2-norm)
expansion	p.265	4.2	$\det A = \sum_{j=1}^{n} a_{ij}C_{ij}$ along the ith row $\det A = \sum_{i=1}^{n} a_{ij}C_{ij}$ along the jth column

F

Fourier approximation	p.624	7.5	The best approximation using the *Fourier coefficients.* Given below and used in a *trigonometric polynomial.*
Fourier coefficients	p.624	7.5	These coefficients are used in the *Fourier approximation*: $a_0 = \frac{\langle 1,f\rangle}{\langle 1,1\rangle} = \frac{1}{2\pi}\int_{-\pi}^{\pi} f(x)\,dx$ $a_k = \frac{\langle \cos kx, f\rangle}{\langle \cos kx, \cos kx\rangle} = \frac{1}{\pi}\int_{-\pi}^{\pi} f(x)\cos kx\,dx$ $b_k = \frac{\langle \sin kx, f\rangle}{\langle \sin kx, \sin kx\rangle} = \frac{1}{\pi}\int_{-\pi}^{\pi} f(x)\sin kx\,dx$
Fourier series	p.624	7.5	When *Fourier trigonometric polynomial* becomes infinite, the result is called the *Fourier series.* $f(x) = a_0 + \sum_{k=1}^{\infty}(a_k\cos kx + b_k\sin kx) \quad \text{on } [-\pi, \pi]$

F, continued

Term	Page	Section	Definition
free variable	p.75	2.2	a variable free to take on any value
Frobenius norm	p.565	7.2	$\|A\|_F = \sqrt{\sum_{i,j=1}^{n} a_{ij}^2}$. See Example 7.18.
fundamental subspaces	p.377	5.2	There are four *fundamental subspaces* (two pair): $\text{row}(A), \text{null}(A)$ and $\text{col}(A), \text{null}(A^T)$

G

Term	Page	Section	Definition
Gauss-Jordan Elimination	p.76	2.2	see procedure described in box on p. 77
Gauss-Seidel method	p.122	2.5	This process is applied in Example 2.35
Gaussian Elimination	p.72	2.2	see procedure described in box on p. 72
general form	p.33	1.3	$ax + by = c$ where $\mathbf{n} = \begin{bmatrix} a \\ b \end{bmatrix}$ is normal for ℓ
generator matrix	p.240	3.7	G: generates a *code*. See definition before Thm 3.34
generator matrix	p.406	5.5	For $n > m$, an $n \times m$ matrix G and an $(n-k) \times n$ matrix P (with entries in $\mathbb{Z}_2$) are a *generator matrix* and a *parity check matrix* respectively for an (n, k) binary code C if: 1. The columns of G are linearly independent. 2. The rows of P are linearly independent. 3. $PG = O$.
geometric multiplicity	p.291	4.3	Dimension of eigenspace associated with λ, $\dim E_\lambda$, where $\dim E_\lambda$ is the number of vectors in a basis. That is, *geometric multiplicity* $= \dim E_\lambda$.
Gerschgorin disk	p.316	4.5	$D_i = \{z \text{ in } \mathbb{C} : \|z - a_{ii}\| \leq r_i\}$ where $r_i = \sum_{j \neq i} \|a_{ij}\|$
graph	p.236	3.7	finite set of *points* and *edges*

H

Hamming code	p.242	3.7	See Example 3.70 for construction
Hamming norm	p.563	7.2	$\|\mathbf{v}\|_H = w(\mathbf{v})$ (weight). See Example 7.15.
homogeneous	p.79	2.2	system in which each constant term is 0

I

identity transformation	p.478	6.4	$I : V \to W$ maps every vector to itself. That is, $I(\mathbf{v}) = \mathbf{v}$ (See Example 6.54).
if and only if	p.17	1.2	signals a *double implication*
inconsistent	p.61	2.1	a system of linear equations with no solutions
induction	p.147	3.1	See Appendix B, *Mathematical Induction*
inner product	p.15	1.2	generalized notion of the dot product (Ch. 7)
inner product	p.540	7.1	An *inner product* on a vector space V with $\langle \mathbf{u}, \mathbf{v} \rangle$ has the following properties: 1. $\langle \mathbf{u}, \mathbf{v} \rangle = \langle \mathbf{v}, \mathbf{u} \rangle$ 2. $\langle \mathbf{u}, \mathbf{v} + \mathbf{w} \rangle = \langle \mathbf{u}, \mathbf{v} \rangle + \langle \mathbf{u}, \mathbf{w} \rangle$ 3. $\langle c\mathbf{u}, \mathbf{v} \rangle = c\langle \mathbf{u}, \mathbf{v} \rangle$ 4. $\langle \mathbf{u}, \mathbf{u} \rangle \geq 0$ and $\langle \mathbf{u}, \mathbf{u} \rangle = 0$ if and only if $\mathbf{0} = 0$
inner product space	p.540	7.1	A vector space with an *inner product* is called an *inner product space*
inverse	p.161	3.3	A-inverse: satisfies $AA^{-1} = I$ and is unique
inverse	p.483	6.4	T^{-1}: $T^{-1} \circ T = I_V$, $T \circ T^{-1} = I_W$ (See *invertible*).
inverse transformation	p.219	3.6	$S \circ T = I_n$ and $T \circ S = I_n$ $S(T(\mathbf{v})) = [S][T]\mathbf{v} = \mathbf{v} \Rightarrow [T] = [S]^{-1}$

I, continued

invertible	p.161	3.3	If A^{-1} exists, A is called *invertible.*
invertible	p.482	6.4	A linear transformation $T : V \to W$ is *invertible* if there is a linear transformation $T' : W \to V$ such that: $T' \circ T = I_V$ and $T \circ T' = I_W$ In this case, T' is called the *inverse* for T.
isomorphic	p.497	6.5	If $T : V \to W$ where T is an *isomorphism.* Then: V and W are *isomorphic.* This is written $V \cong W$.
isomorphism	p.497	6.5	$T : V \to W$ is an *isomorphism* if T is *one-to-one* and *onto.*
iterates	p.123	2.5	vectors found through the iterative process

J

Jacobi's method	p.122	2.5	This process is applied in Example 2.35

K

kernel (of T)	p.486	6.5	Let $T : V \to W$ be a linear transformation. *Kernel* of T is the set of all vectors that T maps to zero. That is, $\ker(T) = \{\mathbf{v} \text{ in } V : T(\mathbf{v}) = \mathbf{0}\}$.
Kirchoff's Laws	p.106	2.4	Current Law (nodes), Voltage Law (circuits)

L

leading entry	p.68	2.2	the first nonzero entry in a row of a matrix
least squares error	p.581	7.3	$\|\mathbf{e}\| = \sqrt{\varepsilon_1^2 + \varepsilon_2^2 + \varepsilon_3^2}$ is the *least squares error* of the approximation
least squares solution	p.583	7.3	A *least squares solution* of $A\mathbf{x} = \mathbf{b}$ is $\overline{\mathbf{x}}$ such that $\|\mathbf{b} - A\overline{\mathbf{x}}\| \le \|\mathbf{b} - A\mathbf{x}\|$ for all $\mathbf{x}$ in $\mathbb{R}$
left singular vectors	p.602	7.4	Given a singular value decomposition (SVD) of $A = U\Sigma V^T$. The columns of U (an orthogonal matrix) are called the *left singular values* of A.
length	p.17	1.2	$\|\mathbf{v}\| = \sqrt{\mathbf{v} \cdot \mathbf{v}}$ (Means the same thing as *norm*)
length	p.237	3.7	number of edges a *path* contains
length	p.544	7.1	The *length* (or *norm*) of $\mathbf{v}$ is $\|\mathbf{v}\| = \sqrt{\langle \mathbf{v}, \mathbf{v} \rangle}$
length (m-ary)	p.51	1.4	the number of components in in a vector
Leslie matrix	p.234	3.7	Population matrix L described after Example 3.66
linear code	p.529	6.7	A p-ary *linear code* is a subspace of C of $\mathbb{Z}_p^n$.
linear combination	p.12	1.1	$\mathbf{u} = c_1\mathbf{v}_1 + c_2\mathbf{v}_2 + \cdots + c_n\mathbf{v}_n$ where c_i are scalars
linear combination	p.95	2.3	$\mathbf{v} = c_1\mathbf{v}_1 + \cdots + c_n\mathbf{v}_n$ $\mathbf{v} = \sum c_i\mathbf{v}_i$
linear combination	p.152	3.2	$\mathbf{B} = c_1\mathbf{A}_1 + \cdots + c_n\mathbf{A}_n$ $\mathbf{B} = \sum c_i\mathbf{A}_i$
linearly dependent	p.155	3.2	one *can* be written as linear combination of others $c_1\mathbf{A}_1 + \cdots + c_n\mathbf{A}_n = O$ with at least one $c_i \neq 0$
linearly dependent	p.447	6.2	vector can be written as linear combination of others $c_1\mathbf{v}_1 + \cdots + c_n\mathbf{v}_n = \mathbf{0}$ with at least one $c_i \neq 0$
linearly dependent set of vectors	p.94	2.3	vector can be written as linear combination of others $c_1\mathbf{v}_1 + \cdots + c_n\mathbf{v}_n = 0$ with at least one $c_i \neq 0$ $\sum c_i\mathbf{v}_i = 0$ with at least one $c_i \neq 0$

L, continued

linear equation	p.59	2.1	$a_1x_1 + a_2x_2 + \cdots + a_nx_n = b$
linearly independent	p.155	3.2	matrices are *not* linear combinations of each other $c_1\mathbf{A}_1 + \cdots + c_n\mathbf{A}_n = O \Leftrightarrow$ all the $c_i = 0$
linearly independent	p.447	6.2	no vector can be written as linear combination of others or $c_1\mathbf{v}_1 + \cdots + c_n\mathbf{v}_n = \mathbf{0} \Leftrightarrow$ all the $c_i = 0$
linearly independent set of vectors	p.94	2.3	no vector in the set can be written as a llnear combination of the other vectors in the set $c_1\mathbf{v}_1 + \cdots + c_n\mathbf{v}_n = 0 \Leftrightarrow$ all the $c_i = 0$ $\sum c_i\mathbf{v}_i = 0 \Leftrightarrow$ all the $c_i = 0$
linear system	p.59	2.1	a set of linear equations with the same variables
linear transformation	p.211	3.6	$T(c\mathbf{v}) = cT(\mathbf{v})$ and $T(\mathbf{u}+\mathbf{v}) = T(\mathbf{u}+\mathbf{v})$ If T is linear, then $T(\mathbf{v}) = A\mathbf{v}$ for some matrix A.
linear transformation	p.476	6.4	A *linear transformation* from a vector space V to a vector space W is a mapping $T : V \to W$ such that: 1. $T(\mathbf{u}+\mathbf{v}) = T(\mathbf{u}+\mathbf{v})$ 2. $T(c\mathbf{u}) = cT(\mathbf{u})$ or $T(c_1\mathbf{v}_1 + \cdots + c_k\mathbf{v}_k) = c_1T(\mathbf{v}_1) + \cdots + c_kT(\mathbf{v}_k)$

M

m-ary	p.51	1.4	see the text discussion of arithmetic in $\mathbb{Z}_m$
Markov chain	p.228	3.7	See description given prior to Example 3.64
matrix	p.62	2.1	a rectangular array of numbers in rows and columns
matrix norm	p.565	7.2	A *matrix norm* on M_{nn} has the following properties: 1. $\|A\| \geq 0$, and $\|A\| = 0$ if and only if $A = O$ 2. $\|cA\| = |c|\|A\|$ 3. $\|A+B\| \leq \|A\| + \|B\|$ 4. $\|AB\| \leq \|A\|\|B\|$

M, continued

matrix product	p.139	3.1	$C = AB \Rightarrow c_{ij} = a_{i1}b_{1j} + a_{i2}b_{2j} + \cdots + a_{in}b_{nj}$
max norm	p.562	7.2	$\|\mathbf{v}\|_m = \|\mathbf{v}\|_\infty = \max\{\lvert v_1\rvert, \ldots, \lvert v_n\rvert\}$ (also ∞-norm)
minimum distance	p.626	7.5	The smallest distance between any two distinct vectors in C: $\mathrm{d}(C) = \min\{\mathrm{d}_H(\mathbf{x}, \mathbf{y}) : \mathbf{x} \neq \mathbf{y} \text{ in } C\}$
minor	p.263	4.2	$\det A_{ij}$, the (i, j)-minor of A
modular	p.50	1.4	see the text development of Modular Arithmetic
multiplier	p.181	3.4	the scalar k in $R_i - kR_j$

N

network	p.104	2.4	nodes connected by a series of branches
norm	p.17	1.2	$\|\mathbf{v}\| = \sqrt{\mathbf{v} \cdot \mathbf{v}}$ (using the dot product)
norm	p.544	7.1	The *length* (or *norm*) of $\mathbf{v}$ is $\|\mathbf{v}\| = \sqrt{\langle \mathbf{v}, \mathbf{v} \rangle}$
norm	p.561	7.2	A *norm* on a vector space V satisfies:
normal equations	p.584	7.3	$A^T A\overline{\mathbf{x}} = A^T \mathbf{b}$
normal form	p.33	1.3	$\mathbf{n} \cdot (\mathbf{x} - \mathbf{p}) = 0$ or $\mathbf{n} \cdot \mathbf{x} = \mathbf{n} \cdot \mathbf{p}$
normalizing	p.18	1.2	a unit vector in same direction ($\mathbf{u} = \frac{1}{\|\mathbf{v}\|}\mathbf{v}$)
normal vector	p.31	1.3	$\mathbf{n}$, orthogonal to any vector on line ℓ or plane $\mathscr{P}$
normed linear space			1. $\|\mathbf{v}\| \geq 0$, and $\|\mathbf{v}\| = 0$ if and only if $\mathbf{v} = \mathbf{0}$ 2. $\|c\mathbf{v}\| = \lvert c\rvert \|\mathbf{v}\|$ 3. $\|\mathbf{u} + \mathbf{v}\| \leq \|\mathbf{u}\| + \|\mathbf{v}\|$
nullity	p.203	3.5	the dimension of null(A)
nullity (of T)	p.488	6.5	Let $T : V \to W$ be a linear transformation. The *nullity* of T is the dimension of the kernel of T. That is, nullity(T) = dim(ker(T)).
null space	p.195	3.5	null(A), $\mathbf{x}$ such that $A\mathbf{x} = \mathbf{0}$

O

Ohm's Law	p.106	2.4	force = resistance × current, $E = RI$
one-to-one	p.492	6.5	T maps distinct vectors in V to distinct vectors in W. That is, $\mathbf{v} \neq \mathbf{u}$ implies $T(\mathbf{v}) \neq T(\mathbf{u})$, or $T(\mathbf{v}) = T(\mathbf{u})$ implies $\mathbf{v} = \mathbf{u}$.
onto	p.492	6.5	If range$(T) = W$, then T is called *onto*. For all $\mathbf{w}$, there is a $\mathbf{v}$ such that $\mathbf{w} = T(\mathbf{v})$.
operator norm	p.568	7.2	$\|A\| = \max\|A\mathbf{x}\|$ where $\|\mathbf{x}\| = 1$. See Theorem 7.6. $\|A\|_1 = \max\|A\mathbf{x}\|_s$, $\|A\|_2 = \max\|A\mathbf{x}\|_E$, $\|A\|_\infty = \max\|A\mathbf{x}\|_m$
ordered	p.3	1.1	$[3, 2]$ ***vs.*** $[2, 3]$: these are *not* the same vector
orthogonal	p.23	1.2	$\mathbf{u}$ and $\mathbf{v}$ are orthogonal if $\mathbf{u} \cdot \mathbf{v} = 0$
orthogonal	p.544	7.1	$\mathbf{u}$ and $\mathbf{v}$ are orthogonal if $\langle \mathbf{u}, \mathbf{v} \rangle = 0$
orthogonal set	p.546	7.1	$\{\mathbf{v}_1, \ldots, \mathbf{v}_k\}$ such that $\langle \mathbf{v}_i, \mathbf{v}_j \rangle = 0$ when $\mathbf{v}_i \neq \mathbf{v}_j$
orthogonal		5.1	This adjective is used for sets, bases, and matrices:
set	p.366		$\mathbf{v}_i \cdot \mathbf{v}_j = 0$ when $i \neq j$
basis	p.367		a basis $\mathcal{B}$ that is an orthogonal set *square* matrix
matrix	p.371		whose columns form an orthonormal set That is, Q where columns $\{\mathbf{q}_i\}$ are orthonormal Matrix is ortho*gonal* when columns are ortho*normal*
orthogonal		5.2	This adjective applies to *complements* and *projections*:
complement	p.375		$W^\perp = \{\mathbf{v} \text{ in } \mathbb{R}^n : \mathbf{v} \cdot \mathbf{w} = 0 \text{ for all } \mathbf{w} \text{ in } W\}$
component	p.379		$\text{perp}_W(\mathbf{v}) = \mathbf{w}^\perp = \mathbf{v} - \text{proj}_W(\mathbf{v})$
projection	p.379		$\text{proj}_W(\mathbf{v}) = \sum_{k=1}^{n} \left(\frac{\mathbf{u}_k \cdot \mathbf{v}}{\mathbf{u}_k \cdot \mathbf{u}_k}\right) \mathbf{u}_k = \sum_{k=1}^{n} \text{proj}_{\mathbf{u}_k}(\mathbf{v})$
orthogonally diagonalizable	p.397	5.4	If there exists an orthogonal Q and diagonal D such that $Q^T A Q = D$, A is *orthogonally diagonalizable.*
orthonormal		5.1	This adjective is used for sets and bases:
	p.369		For a set to be orthonormal it must satisfy: $\mathbf{q}_i \cdot \mathbf{q}_j = \begin{cases} 0 & \text{if } i \neq j \text{ Property 1} \\ 1 & \text{if } i = j \text{ Property 2} \end{cases}$
set	p.369		an orthogonal set of *unit* vectors
basis	p.369		a basis $\mathcal{B}$ that is an orthonormal set
orthonormal set	p.546	7.1	An orthogonal set of unit vectors, that is $\|\mathbf{v}_i\| = 1$
outer product	p.145	3.1	see description of process after Example 3.10

P

Term	Page	Section	Definition
p-norm	p.562	7.2	$\|\mathbf{v}\|_p = (\lvert v_1\rvert^p + \cdots + \lvert v_n\rvert^p)^{1/p}$
parallel	p.8	1.1	if and only if scalar multiples of each other
parametric equations	p.33	1.3	component equations from $\mathbf{x} = \mathbf{p} + t\mathbf{d}$
parity	p.49	1.4	the number of 1s in a code vector
parity check	p.406	5.5	See *generator matrix* for details: the definitions and conditions are associated.
parity matrix	p.240	3.7	P: checks a *code*. See definition before Thm 3.34
path	p.237	3.7	set of *edges* that connects one *vertex* to another
Penrose conditions	p.595	7.3	A^+ (pseudoinverse) satisfies the *Penrose conditions*. a. $AA^+A = A$ b. $A^+AA^+ = A^+$ c. AA^+ and A^+A are symmetric
permutation matrix	p.185	3.4	$P = P_k \cdots P_2P_1$ where multiplication by P_i performs a row interchange like $R_i \leftrightarrow R_j$
Power Method	p.312	4.5	This two-step iterative method is described in detail on p. 312 and illustrated in Example 4.31
position vector	p.3	1.1	vector with tail at the origin O, i.e. $\overrightarrow{OA}$
pivoting	p.70	2.2	see explanation in solution to Example 2.9 p. 70
probability vector	p.229	3.7	vector with nonnegative components that add to 1
projection	p.24	1.2	$\text{proj}_{\mathbf{u}}(\mathbf{v}) = \left(\frac{\mathbf{u}\cdot\mathbf{v}}{\mathbf{u}\cdot\mathbf{u}}\right)\mathbf{u}$
projection	p.547	7.1	$\text{proj}_W(\mathbf{v}) = \sum_{k=1}^{n} \frac{\langle \mathbf{u}_k\cdot\mathbf{v}\rangle}{\langle \mathbf{u}_k\cdot\mathbf{u}_k\rangle}\mathbf{u}_k$ (*orthogonal projection*)
projection form	p.402	5.4	If A is a real symmetric matrix with $A = QDQ^T$, then the *projection form of the Spectral Theorem* is $A = \lambda_1\mathbf{q}_1\mathbf{q}_1^T + \cdots + \lambda_n\mathbf{q}_n\mathbf{q}_n^T = \sum_{i=1}^{n} \lambda_i\mathbf{q}_i\mathbf{q}_i^T$.
pseudoinverse	p.594	7.3	If A is a matrix with linearly independent columns, then $A^+ = (A^TA)^{-1}A^T$ is the *pseudoinverse*.
pseudoinverse	p.611	7.4	The *pseudoinverse* (*Moore-Penrose*) is $A^+ = V\Sigma^+U^T$.

Q

Term	Page	Section	Definition
quadratic form	p.412	5.5	$f(\mathbf{x}) = \mathbf{x}^T A\mathbf{x}$ (*quadratic form*), (A, the matrix associated with f, is symmetric)

R

Term	Page	Section	Definition
range (of T)	p.486	6.5	Let $T : V \to W$ be a linear transformation. The *range* of T is the set of all vectors that are images under T. So, range$(T) = \{\mathbf{w} \text{ in } W : \mathbf{w} = T(\mathbf{v}) \text{ for some } \mathbf{v} \text{ in } V\}$.
rank	p.75	2.2	number of nonzero rows in the REF of a matrix
rank	p.202	3.5	dimension of row(A) or col(A) (they are the same)
rank (of T)	p.488	6.5	Let $T : V \to W$ be a linear transformation. The *rank* of T is the dimension of the range of T. That is, rank(T) = dim(range(T)).
recurrence	p.333	4.6	*linear recurrence.* See definition p. 333.
reduced row echelon form	p.76	2.2	RREF: REF; leading entries, 1; all else, 0s
Reed-Muller	p.532	6.7	The (first order) *Reed-Muller codes* R_n (see p.532).
repeller	p.349	4.6	See definition on p.349 and Example 4.50.
right singular vectors	p.602	7.4	Given a singular value decomposition (SVD) of $A = U\Sigma V^T$. The columns of V (an orthogonal matrix) are called the *right singular values* of A.
root mean square error	p.621	7.5	This is the root means square error: $\|f - g\| = \sqrt{\int_{-1}^{1}(f(x) - g(x))^2\, dx}$
row echelon form	p.68	2.2	REF: zero rows, bottom; leading entries, left
row equivalent	p.72	2.2	A can be reduced to B using EROs
row reduction	p.70	2.2	applying EROs to bring a matrix into REF
row space	p.193	3.5	row(A) spanned by the rows of A

S

saddle point	p.349	4.6	See definition on p.349 and Example 4.50.
scalar	p.8	1.1	a real number c (that is, c is ***not*** a vector)
scalar product	p.15	1.2	another name for dot product (result is a scalar)
similar	p.298	4.4	$A \sim B$ if $P^{-1}AP = B$ where P is invertible
simple path	p.237	3.7	a *path* that does not contain the same *edge* twice
singular values	p.599	7.4	The square roots of the eigenvalues of A^TA. Usually denoted by $\sigma_1, \ldots, \sigma_n$.
singular value decomposition	p.601	7.4	A *singular value decomposition* (SVD) of $A = U\Sigma V^T$. where U and and V are orthogonal matrices. Σ is constructed using the σ_i.
skew-symmetric	p.160	3.2	$A^T = -A$
solution	p.60	2.1	$[s_1, s_2, \ldots, s_n]$ where $a_1s_1 + a_2s_2 + \cdots + a_ns_n = b$
span(S)	p.92	2.3	all linear combinations of $S = \{\mathbf{s}_1, \mathbf{s}_2, \ldots, \mathbf{s}_n\}$
spanning set for $\mathbb{R}^n$	p.92	2.3	set S such that span$(S) = \mathbb{R}^n$, that is set S such that $\mathbf{v}$ in $\mathbb{R}^n \Rightarrow \mathbf{v} = \sum c_i\mathbf{s}_i$
spectral decomposition	p.402	5.4	If A is a real symmetric matrix with $A = QDQ^T$, then $A = \lambda_1\mathbf{q}_1\mathbf{q}_1^T + \cdots + \lambda_n\mathbf{q}_n\mathbf{q}_n^T = \sum_{i=1}^{n} \lambda_i\mathbf{q}_i\mathbf{q}_i^T$.
spectrum	p.400	5.4	Eigenvalues of an *orthogonally diagonalizable* matrix.
standard basis	p.19	1.2	$\mathbf{e}_i$ with a 1 in the ith component and 0s elsewhere
state vector	p.229	3.7	$\mathbf{x}_k$: $\mathbf{x}_k = P^k\mathbf{x}_0$ in a Markov Chain
steady state	p.231	3.7	$\mathbf{x}$: $\mathbf{x} = P\mathbf{x}$ in a Markov Chain
strictly diagonally dominant	p.126	2.5	$\lvert a_{11}\rvert > \lvert a_{12}\rvert + \lvert a_{13}\rvert + \cdots + \lvert a_{1n}\rvert$ $\lvert a_{22}\rvert > \lvert a_{21}\rvert + \lvert a_{23}\rvert + \cdots + \lvert a_{2n}\rvert$ and $\lvert a_{nn}\rvert > \lvert a_{n1}\rvert + \lvert a_{n2}\rvert + \cdots + \lvert a_{n,n-1}\rvert$
subspace	p.190	3.5	$\mathbf{0}$, $\mathbf{u} + \mathbf{v}$, and $c\mathbf{v}$ are in S
subspace	p.438	6.1	$W \subseteq V$, where W is a vector space
submatrix	p.263	4.2	obtained from A by deleting row i and column j
sum norm	p.561	7.2	$\lVert\mathbf{v}\rVert_s = \lVert\mathbf{v}\rVert_1 = \lvert v_1\rvert + \cdots + \lvert v_n\rvert$ (same as 1-norm)
symmetric	p.149	3.1	$A^T = A$
symmetric	p.149	3.1	A matrix A is *symmetric* if and only if $A^T = A$.

T

ternary	p.50	1.4	see the text discussion of arithmetic in $\mathbb{Z}_3$
trace	p.160	3.2	$\text{tr}(A) = a_{11} + a_{22} + \cdots + a_{nn}$ So, the *trace* is the sum of the diagonal entries
trajectory	p.346	4.6	See definition on p.346 prior to Example 4.48.
transition matrix	p.229	3.7	P: $\mathbf{x}_k = P^k \mathbf{x}_0$ in a Markov Chain
transpose	p.148	3.1	A^T, create by switching rows and columns
transposition	p.52	1.4	the interchange of two adjacent components
trigonometric polynomial	p.623	7.5	A function of this form is a *trigonometric polynomial.* $p(x) = a_0 + a_1 \cos x + a_2 \cos 2x + \cdots + a_n \cos nx +$ $b_1 \sin x + b_2 \sin 2x + \cdots + b_n \sin nx.$
trivial subspaces	p.441	6.1	The subspaces $\{\mathbf{0}\}$ and V are called the *trivial subspaces* of V. Therefore, all other subspaces are nontrivial.

U

unit lower diagonal	p.179	3.4	L with all the entries above the diagonal are 0 and all the entries on the diagonal are 1. That is, $i > j \Rightarrow a_{ij} = 0$ and $a_{ii} = 1$.
unit vector	p.18	1.2	vector of length 1
upper diagonal	p.179	3.4	U where all the entries below the diagonal are 0. That is, $i < j \Rightarrow a_{ij} = 0$.
upper triangular	p.160	3.2	all entries below the main diagonal are zero

V

vector	p.3	1.1	*directed* line segment with *length* and *direction*
vector addition	p.5	1.1	$\mathbf{u}+\mathbf{v}=[u_1+v_1, u_2+v_2]$
vector form	p.33	1.3	$\mathbf{x}=\mathbf{p}+t\mathbf{d}$ where $\mathbf{d}$ is a direction vector for ℓ
vector space	p.433	6.1	If the following axioms hold, then V is a *vector space*: 1. $\mathbf{u}+\mathbf{v}$ is in V (*Closure under addition*) 2. $\mathbf{u}+\mathbf{v}=\mathbf{v}+\mathbf{u}$ (*Commutativity*) 3. $(\mathbf{u}+\mathbf{v})+\mathbf{w}=\mathbf{u}+(\mathbf{v}+\mathbf{w})$ (*Associativity*) 4. $\mathbf{u}+\mathbf{0}=\mathbf{u}$ (*Zero vector*) 5. $\mathbf{u}+(-\mathbf{u})=\mathbf{0}$ 6. $c\mathbf{u}$ is in V (*Closure under scalar multiplication*) 7. $c(\mathbf{u}+\mathbf{v})=c\mathbf{u}+c\mathbf{v}$ (*Distributivity*) 8. $(c+d)\mathbf{u}=c\mathbf{u}+d\mathbf{u}$ (*Distributivity*) 9. $c(d\mathbf{u})=(cd)\mathbf{u}$ 10. $1\mathbf{u}=\mathbf{u}$
Voltage Law	p.106	2.4	voltage *drops* equal total voltage

W

weight	p.411	5.5	$w(\mathbf{x})$ (*weight* of $\mathbf{x}$), the number of 1s in $\mathbf{x}$ (in $\mathbb{Z}_2^n$)
weighted dot product	p.541	7.1	An *inner product* defined on $\mathbb{R}^n$ such that $\langle\mathbf{u},\mathbf{v}\rangle=w_1u_1v_1+\ldots+w_nu_nv_n$ with $w_i\geq 0$ (see p.541 for details)

X Y Z

zero transformation	p.478	6.4	$T_0 : V \to W$ maps every vector to the zero vector. That is, $T_0(\mathbf{v})=\mathbf{0}$ (See Example 6.54).
zero vector	p.4	1.1	$\mathbf{0}$, *all* components are 0, so length is 0
zero vector	p.23	1.2	$\mathbf{0}$. Note: $\mathbf{0}\cdot\mathbf{v}=0$ for *every* vector $\mathbf{v}$ in $\mathbb{R}^n$

Appendix II: Theorems

Theorems

In the summary, we will occasionally list only the central result of the theorem.
For the complete statement of the theorem, refer to the text.
Theorems with names are listed in alphabetical order at the end of this section.

Theorems, Chapter 1

Thm 1.1	p.10	1.1	Algebraic Properties of Vectors in $\mathbb{R}^n$
Thm 1.2	p.16	1.2	Properties of the dot product ($\mathbf{u} \cdot \mathbf{v} = \mathbf{v} \cdot \mathbf{u}$ ...)
Thm 1.3	p.17	1.2	Properties of the norm ($\|\mathbf{v}\| = 0 \Leftrightarrow \mathbf{v} = \mathbf{0}$ ...)
Thm 1.4	p.19	1.2	Cauchy-Schwarz: $|\mathbf{u} \cdot \mathbf{v}| \leq \|\mathbf{u}\| \|\mathbf{v}\|$
Thm 1.5	p.19	1.2	Triangle Inequality: $\|\mathbf{u} + \mathbf{v}\| \leq \|\mathbf{u}\| + \|\mathbf{v}\|$
Thm 1.6	p.23	1.2	Pythagoras: $\|\mathbf{u} + \mathbf{v}\|^2 = \|\mathbf{u}\|^2 + \|\mathbf{v}\|^2 \Leftrightarrow \mathbf{u} \cdot \mathbf{v} = 0$

Theorems, Chapter 2

Thm 2.1	p.72	2.2	A and B are row equivalent $\Leftrightarrow$ they reduce to same REF
Thm 2.2	p.74	2.2	**Rank Thm:** number of free variables $= n - \text{rank}(A)$
Thm 2.3	p.80	2.2	$[A\,|\,\mathbf{0}]$: m equations $< n$ variables $\Rightarrow$ infinitely many solutions
Thm 2.4	p.91	2.3	$[A\,|\,\mathbf{b}]$ is consistent $\Leftrightarrow \mathbf{b} = \sum c_i \mathbf{a}_i$
Thm 2.5	p.95	2.3	linearly dependent set $\Leftrightarrow$ linear combination of the others
Thm 2.6	p.97	2.3	linearly dependent set $\Leftrightarrow [A_c\,|\,0]$ has nontrivial solution
Thm 2.7	p.98	2.3	linearly dependent set $\Leftrightarrow \text{rank}(A_r) < m$ where A_r is $m \times n$
Thm 2.8	p.99	2.3	m vectors in $\mathbb{R}^n$ are linearly dependent if $m > n$
Thm 2.9	p.124	2.5	A strictly diagonally dominant $\Rightarrow$ iterates converge
Thm 2.10	p.124	2.5	methods converge $\Rightarrow$ they converge to the solution

Theorems, Chapter 3

Thm 3.1	p.142	3.1	$\mathbf{e}_i A$ is row of A, $A\mathbf{e}_j$ is column of A
Thm 3.2	p.152	3.2	Properties of Matrix Addition and Scalar Multiplication
Thm 3.3	p.156	3.2	Properties of Matrix Multiplication
Thm 3.4	p.157	3.2	Properties of the Transpose, A^T
Thm 3.5	p.159	3.2	a. $A + A^T$ is symmetric if A is square AA^T and A^TA b. AA^T and A^TA are always symmetric
Thm 3.6	p.162	3.3	If A is invertible, then its inverse is unique. If $AA^{-1} = I$, then A^{-1} is unique.
Thm 3.7	p.162	3.3	$A\mathbf{x} = \mathbf{b}$ has unique solution $\mathbf{x} = A^{-1}\mathbf{b}$ The inverse A^{-1} exists if and only if $\det A = ad - bc \neq 0$
Thm 3.8	p.163	3.3	Formula for A^{-1} for 2×2 matrices using $\det A$
Thm 3.9	p.166	3.3	Key Properties of A^{-1}, for example $(A^{-1})^{-1} = I$
Thm 3.10	p.170	3.3	Perform row operation E on A is equivalent to EA
Thm 3.11	p.171	3.3	E^{-1} is created by undoing the operation that created E
Thm 3.12	p.171	3.3	Five conditions that are equivalent to A being invertible
Thm 3.13	p.173	3.3	If $AB = I$ or $BA = I$, then $B = A^{-1}$
Thm 3.14	p.173	3.3	If $E_nE_{n-1}\cdots E_1A = I$, then $A^{-1} = E_nE_{n-1}\cdots E_1$
Thm 3.15	p.180	3.4	If $A \longrightarrow$ REF without $R_i \leftrightarrow Rj$, then $A = LU$ exists.
Thm 3.16	p.184	3.4	If A is invertible, then we have the following: If $A = L_1U_1$ and $A = L_2U_2$ then $L_1 = L_2$ and $U_1 = U_2$. That is, if an LU factorization exists, it is unique.
Thm 3.17	p.185	3.4	If P is a permutation matrix, then $P^{-1} = P^T$.
Thm 3.18	p.186	3.4	If A is square, then $A = P^TLU$ exists.
Thm 3.19	p.190	3.5	$\text{span}(\mathbf{v}_1, \mathbf{v}_2, \ldots, \mathbf{v}_n)$ is a subspace
Thm 3.20	p.194	3.5	If $A \longrightarrow B$, then $\text{row}(A) = \text{row}(B)$
Thm 3.21	p.194	3.5	$\text{null}(A)$, $\mathbf{x}$ such that $A\mathbf{x} = \mathbf{0}$, is a subspace
Thm 3.22	p.195	3.5	$A\mathbf{x} = \mathbf{b}$ has exactly none, one, or infinitely many solutions
Thm 3.23	p.200	3.5	Any two bases have the same number of vectors (*dimension*)
Thm 3.24	p.202	3.5	$\text{row}(A)$ and $\text{col}(A)$ have the same dimension
Thm 3.25	p.202	3.5	$\text{rank}(A) = \text{rank}(A^T)$
Thm 3.26	p.203	3.5	**Rank Theorem**: $\text{rank}(A) + \text{nullity}(A) = n$
Thm 3.27	p.204	3.5	**Fundamental Theorem: Version 2**
Thm 3.28	p.205	3.5	$\text{rank}(A^TA) = \text{rank}(A)$ and A^TA invertible $\Leftrightarrow \text{rank}(A) = n$
Thm 3.29	p.206	3.5	If $\{\mathbf{v}_1, \ldots, \mathbf{v}_k\}$ is a basis, then $\mathbf{v} = \sum c_i\mathbf{v}_i$ is unique
Thm 3.30	p.212	3.6	Given A, then $T_A(\mathbf{v}) = A\mathbf{v}$ is linear
Thm 3.31	p.214	3.6	$[T] = \left[\, T(\mathbf{e}_1)\ \ T(\mathbf{e}_2)\ \ \ldots\ \ T(\mathbf{e}_n) \,\right]$
Thm 3.32	p.218	3.6	Given $S \circ T$, $S(T(\mathbf{v})) = [S][T]\mathbf{v}$
Thm 3.33	p.220	3.6	$S \circ T = I_n$, $S(T(\mathbf{v})) = [S][T]\mathbf{v} = \mathbf{v} \Rightarrow [T] = [S]^{-1}$
Thm 3.34	p.241	3.7	Gives conditions on G and P for a given *code*

Theorems, Chapter 4

Thm 4.1	p.265	4.2	$\det A = \sum_{j=1}^{n} a_{ij}C_{ij}$ (any row) or $\sum_{i=1}^{n} a_{ij}C_{ij}$ (any column)
Thm 4.2	p.268	4.2	If A is triangular, then $\det A = a_{11}a_{22}\cdots a_{nn}$
Thm 4.3	p.268	4.2	a. through f. detail row (column) operations effects on $\det A$ a. $\mathbf{A}_i = \mathbf{0} \Rightarrow \det A = 0$ b. $A \overset{R_i \leftrightarrow R_j}{\longrightarrow} B \Rightarrow \det B = -\det A$ c. $\mathbf{A}_i = \mathbf{A}_j \Rightarrow \det A = 0$ d. $A \overset{kR_i}{\longrightarrow} B \Rightarrow \det B = k \det A$ e. $\mathbf{C}_i = \mathbf{A}_i + \mathbf{B}_i \Rightarrow \det C = \det A + \det B$ f. $A \overset{R_i + kR_j}{\longrightarrow} B \Rightarrow \det B = \det A$
Thm 4.4	p.270	4.2	a. through c. detail row (column) operations effects on $\det E$ a. $I \overset{R_i \leftrightarrow R_j}{\longrightarrow} E \Rightarrow \det E = -1$ b. $I \overset{kR_i}{\longrightarrow} E \Rightarrow \det E = k$ c. $I \overset{R_i + kR_j}{\longrightarrow} E \Rightarrow \det E = 1$
Lem 4.5	p.271	4.2	E, elementary, then $\det(EB) = (\det E)(\det B)$
Thm 4.6	p.271	4.2	A is invertible if and only if $\det A \neq 0$
Thm 4.7	p.271	4.2	$\det(kA) = k^n \det A$
Thm 4.8	p.272	4.2	$\det(AB) = (\det A)(\det B)$
Thm 4.9	p.273	4.2	$\det(A^{-1}) = \frac{1}{\det A}$
Thm 4.10	p.273	4.2	$\det A = \det(A^T)$
Thm 4.11	p.274	4.2	If $A\mathbf{x} = \mathbf{b}$, then $x_i = \frac{\det(A_i(\mathbf{b}))}{\det A}$ (Cramer's Rule)
Thm 4.12	p.276	4.2	$A^{-1} = \frac{1}{\det A} \text{adj } A$
Lem 4.13	p.276	4.2	(row 1) $\det A = \sum_{j=1}^{n} a_{1j}C_{1j} = \sum_{i=1}^{n} a_{i1}C_{i1}$ (column 1)
Lem 4.14	p.276	4.2	$A \overset{R_i \leftrightarrow R_j}{\longrightarrow} B \Rightarrow \det B = -\det A$ (see Thm 4.3(f))

Thm 4.15	p.292	4.3	If A is triangular, then its eigenvalues are its diagonal entries.
Thm 4.16	p.292	4.3	A is invertible if and only if all $\lambda \neq 0$.
Thm 4.17	p.293	4.3	a. through o. give equivalent conditions for A to be invertible.
Thm 4.18	p.293	4.3	a. through c. relate eigenvalues, A^n and A^{-1}: For any integer n if $A\mathbf{x} = \lambda\mathbf{x}$, then $A^n\mathbf{x} = \lambda^n\mathbf{x}$.
Thm 4.19	p.294	4.3	If $\mathbf{x} \in \text{span}(\mathbf{v}_i)$, then $A^k\mathbf{x} = c_1\lambda_1^k\mathbf{v}_1 + c_2\lambda_2^k\mathbf{v}_2 + \cdots + c_m\lambda_m^k\mathbf{v}_m$ where $\mathbf{v}_i$ is the eigenvector corresponding to λ_i.
Thm 4.20	p.294	4.3	If λ_i are distinct, then $\{\mathbf{v}_i\}$ are linearly independent.
Thm 4.21	p.291	4.4	a. $A \sim A$, b. $A \sim B \Rightarrow B \sim A$, c. $A \sim B$, $B \sim C \Rightarrow A \sim C$
Thm 4.22	p.299	4.4	parts a. through e. list implications of $A \sim B$ a. $\det A = \det B$ b. A is invertible if and only if B is invertible c. $\text{rank}(A) = \text{rank}(B)$ d. $\det(A - \lambda I) = \det(B - \lambda I)$ e. λ is an eigenvalue of A if and only if λ is an eigenvalue of B
Thm 4.23	p.300	4.4	$P^{-1}AP = D \Leftrightarrow [D]_{ii} = \lambda_i$ and $\mathbf{p}_i = \mathbf{v}_i$ (the eigenvectors)
Thm 4.24	p.302	4.4	If λ_i are distinct, their basis vectors are linearly independent.
Thm 4.25	p.303	4.4	If all the λ_i are distinct, then A is diagonalizable.
Lem 4.26	p.303	4.4	geometric multiplicity $\leq$ algebraic multiplicity: That is, $\dim(E_{\lambda_i}) \leq k_i$ where $c_A(\lambda) = \prod_{i=1}^{m} (\lambda_i - \lambda)^{k_i}$.
Thm 4.27	p.304	4.4	parts a. through c. list equivalences to $A = P^{-1}DP$ a. A is diagonalizable, $A = P^{-1}DP$ b. bases of the eigenvectors contain n vectors c. geometric multiplicity = algebraic multiplicity, $\dim(E_{\lambda_i}) = k_i$
Thm 4.28	p.309	4.5	If A is diagonalizable with dominant eigenvalue λ_1, then $\mathbf{x}_1 = A\mathbf{x}_0$, $\mathbf{x}_2 = A\mathbf{x}_1$, $\mathbf{x}_k = A\mathbf{x}_{k-1}$ where the $\mathbf{x}_k$ are approaching the dominant eigenvector of A
Thm 4.29	p.318	4.5	Every eigenvalue of A is contained in a Gerschgorin Disk

Thm 4.30	p.322	4.6	If P is the $n \times n$ transition matrix of a Markov chain, then 1 is an eigenvalue of P
Thm 4.31	p.322	4.6	Let P be an $n \times n$ transition matrix with eigenvalue λ. a. $\lvert\lambda\rvert \leq 1$ b. If P is regular and $\lambda \neq 1$, then $\lvert\lambda\rvert < 1$.
Lem 4.32	p.324	4.6	Let P be an $n \times n$ transition matrix. If P is diagonalizable, then the dominant eigenvalue $\lambda_1 = 1$ has algebraic multiplicity 1.
Thm 4.33	p.325	4.6	Let P be an $n \times n$ transition matrix. Then as $k \longrightarrow \infty$, P^k approaches an $n \times n$ matrix L whose columns are $\mathbf{x}$, the steady state probability vector for P.
Thm 4.34	p.326	4.6	Let P be an $n \times n$ transition matrix ... as in Theorem 4.33. Then for any probability vector $\mathbf{x}_0$, $\mathbf{x}_k$ approaches $\mathbf{x}$.
Thm 4.35	p.328	4.6	Every Leslie matrix has a unique positive eigenvalue and a corresponding eigenvector with positive components.
Thm 4.36	p.330	4.6	Let A be a positive $n \times n$ matrix. A has a real eigenvalue λ_1: a. $\lambda_1 > 0$ b. λ_1 has a corresponding positive eigenvector. c. If λ is any other eigenvalue of A, then $\lvert\lambda\rvert \leq \lambda_1$.
Thm 4.37	p.332	4.6	Let A be an irreducible nonnegative $n \times n$ matrix. Then: a. $\lambda_1 > 0$ b. λ_1 has a corresponding positive eigenvector. c. If λ is any other eigenvalue of A, then $\lvert\lambda\rvert \leq \lambda_1$. If A is primitive, then this inequality is strict. d. If λ is an eigenvalue of A such that $\lvert\lambda\rvert = \lambda_1$, then λ is a (complex) root of the equation $\lambda^n - \lambda_1^n = 0$. e. λ_1 has algebraic multiplicity 1.
Thm 4.38	p.336	4.6	See statement of Theorem on p. 336.
Thm 4.39	p.337	4.6	See statement of Theorem on p. 337.
Thm 4.40	p.339	4.6	See statement of Theorem on p. 339.
Thm 4.41	p.344	4.6	See statement of Theorem on p. 344.
Thm 4.42	p.349	4.6	See statement of Theorem on p. 349.

Theorems, Chapter 5

Thm 5.1	p.366	5.1	$\mathcal{B} = \{\mathbf{v}_1, \mathbf{v}_2, \ldots, \mathbf{v}_n\}$ orthogonal $\Rightarrow \mathcal{B}$ is linearly independent
Thm 5.2	p.368	5.1	If $\mathcal{B} = \{\mathbf{v}_1, \mathbf{v}_2, \ldots, \mathbf{v}_n\}$ is orthogonal and $\mathbf{w} = \sum_{i=1}^{n} c_i\mathbf{v}_i$, then $c_i = \frac{\mathbf{w}\cdot\mathbf{v}_i}{\mathbf{v}_i\cdot\mathbf{v}_i}$.
Thm 5.3	p.370	5.1	$\mathbf{w} = \sum_{i=1}^{n}(\mathbf{w}\cdot\mathbf{q}_1)\mathbf{q}_i$ (this representation is unique)
Thm 5.4	p.371	5.1	$Q^TQ = I$ if and only if $\{\mathbf{q}_i\}$ is an orthonormal set.
Thm 5.5	p.371	5.1	Q is orthogonal if and only if $Q^{-1} = Q^T$
Thm 5.6	p.372	5.1	a. through c. give equivalent conditions for Q to be orthogonal: a. Q is orthogonal b. $\lVert Q\mathbf{x}\rVert = \lVert\mathbf{x}\rVert$ for every $\mathbf{x}$ in $\mathbb{R}^n$ c. $Q\mathbf{x}\cdot Q\mathbf{x} = \mathbf{x}\cdot\mathbf{y}$ for every $\mathbf{x}$, $\mathbf{y}$ in $\mathbb{R}^n$
Thm 5.7	p.373	5.1	Q is orthogonal, then its *rows* form an orthonormal set.
Thm 5.8	p.372	5.1	a. through d. list implications of Q being orthogonal: a. Q^{-1} is orthogonal b. $\det Q = \pm 1$ c. If λ is an eigenvalue of Q, then $\lvert\lambda\rvert = 1$. d. If Q_1 and Q_2 are orthogonal $n \times n$ matrices, then so is Q_1Q_2.
Thm 5.9	p.376	5.2	a. through d. give properties of $W^\perp$: a. $W^\perp$ is a subspace of $\mathbb{R}^n$ b. $\left(W^\perp\right)^\perp = W$ c. $W \cap W^\perp = \{\mathbf{0}\}$ d. If $W = \text{span}(\mathbf{w}_1, \ldots, \mathbf{w}_k)$, then $\mathbf{v}$ is in $W^\perp$ if and only if $\mathbf{v}\cdot\mathbf{w}_i = 0$ for all i
Thm 5.10	p.376	5.2	$(\text{row}(A))^\perp = \text{null}(A)$ and $(\text{col}(A))^\perp = \text{null}(A^T)$
Thm 5.11	p.381	5.2	**The Orthogonal Decomposition Theorem**: $\mathbf{v} = \mathbf{w} + \mathbf{w}^\perp$
Cor 5.12	p.382	5.2	$(W^\perp)^\perp = W$
Thm 5.13	p.383	5.2	$\dim W + \dim W^\perp = n$
Cor 5.14	p.381	5.2	**The Rank Theorem**: Given A is $m \times n$ then ... $\text{rank}(A) + \text{nullity}(A) = n$ and $\text{rank}(A) + \text{nullity}(A^T) = m$

Thm 5.15 p.389 5.3 **The Gram-Schmidt Process:**
Let $\{\mathbf{x}_1, \ldots, \mathbf{w}_k\}$ be a basis for W then ...
$\mathbf{v}_1 = \mathbf{x}_1$, $W_1 = \text{span}(\mathbf{x}_1)$
$\mathbf{v}_2 = \mathbf{x}_2 - \left(\frac{\mathbf{v}_1 \cdot \mathbf{x}_2}{\mathbf{v}_1 \cdot \mathbf{v}_1}\right)\mathbf{v}_1$, $W_2 = \text{span}(\mathbf{x}_1, \mathbf{x}_2)$
$\mathbf{v}_3 = \mathbf{x}_3 - \left(\frac{\mathbf{v}_1 \cdot \mathbf{x}_3}{\mathbf{v}_1 \cdot \mathbf{v}_1}\right)\mathbf{v}_1 - \left(\frac{\mathbf{v}_2 \cdot \mathbf{x}_3}{\mathbf{v}_2 \cdot \mathbf{v}_2}\right)\mathbf{v}_2$, $W_3 = \text{span}(\mathbf{x}_1, \mathbf{x}_2, \mathbf{x}_3)$
...
$\mathbf{v}_k = \mathbf{x}_k - \sum_{i=1}^{k-1} \frac{\mathbf{v}_i \cdot \mathbf{x}_k}{\mathbf{v}_i \cdot \mathbf{v}_i}$, $W_k = \text{span}(\{\mathbf{x}_k\})$
... $\{\mathbf{v}_1, \ldots, \mathbf{v}_k\}$ is an orthogonal basis for W.

Thm 5.16 p.390 5.3 **The QR Factorization:**
Let A be an $m \times n$ matrix with linearly independent columns.
Then A can be factored as $A = QR$, where
Q is an $m \times n$ matrix with orthonormal columns and
R is an invertible upper triangular matrix.

Thm 5.17 p.398 5.4 If A is *orthogonally diagonalizable*,
then A is *symmetric.*

Thm 5.18 p.398 5.4 If A is a *real* symmetric matrix,
then the eigenvalues of A are *real.*

Thm 5.19 p.399 5.4 If A is a *symmetric* matrix, then any two eigenvectors
corresponding to distinct eigenvalues of A are *orthogonal.*

Thm 5.20 p.400 5.4 **The Spectral Theorem:**
Let A be an $n \times n$ *real* matrix. Then A is *symmetric*
if and only if A is *orthogonally diagonalizable.*

Thm 5.21 p.409 5.5 If C is an (n, k) binary code with generator matrix G and parity check matrix P, then $C^\perp$ is an $(n, n-k)$ binary code such that
a. G^T is parity check matrix for $C^\perp$.
b. P^T is generator matrix for $C^\perp$.

Thm 5.22 p.411 5.5 If C is a self dual code, then
a. Every vector in C has even weight.
b. **1** is in C.

Thm 5.23 p.411 5.5 **The Principal Axes Theorem**:
Every quadratic form can be diagonalized:
$\mathbf{x}^T A\mathbf{x} = \mathbf{y}^T D\mathbf{y} = \lambda_1 y_1^2 + \cdots + \lambda_n y_n^2$
(see statement of Theorem in the text for details)

Thm 5.24 p.416 5.5 The quadratic form $f(\mathbf{x}) = \mathbf{x}^T A\mathbf{x}$ is:
a. *positive definite* $\Leftrightarrow \lambda_i > 0$
b. *positive semidefinite* $\Leftrightarrow \lambda_i \geq 0$
c. *negative definite* $\Leftrightarrow \lambda_i < 0$
d. *negative semidefinite* $\Leftrightarrow \lambda_i \leq 0$
e. *indefinite* $\Leftrightarrow \lambda_i$ are both positive and negative

Thm 5.25 p.417 5.5 Given $f(\mathbf{x}) = \mathbf{x}^T A\mathbf{x}$,
$\lambda_1 \geq \lambda_2 \geq \cdots \geq \lambda_n$, and $\|\mathbf{x}\| = 1$:
a. $\lambda_1 \geq f(\mathbf{x}) \geq \lambda_n$
b. max $f(\mathbf{x}) = \lambda_1$ occurs for a unit eigenvector
c. min $f(\mathbf{x}) = \lambda_n$ occurs for a unit eigenvector

Theorems, Chapter 6

Thm 6.1	p.437	6.1	Let V be a vector space, $\mathbf{u}$ a vector, c a scalar: a. $0\mathbf{u} = \mathbf{0}$ b. $c\mathbf{0} = \mathbf{0}$ c. $(-1)\mathbf{u} = -\mathbf{u}$ d. If $c\mathbf{u} = \mathbf{0}$, then $c = 0$ or $\mathbf{u} = \mathbf{0}$.
Thm 6.2	p.438	6.1	Let $W(\neq \emptyset) \subseteq V$, then W is a subspace if: a. $\mathbf{u}, \mathbf{v} \in W \Rightarrow \mathbf{u} + \mathbf{v} \in W$ b. $\mathbf{u} \in W \Rightarrow c\mathbf{u} \in W$
Thm 6.3	p.445	6.1	Let $\mathbf{v}_1, \mathbf{v}_2, \ldots, \mathbf{v}_k$ be vectors in a vector space V. a. $\text{span}(\mathbf{v}_1, \mathbf{v}_2, \ldots, \mathbf{v}_k)$ is a subspace of V b. $\text{span}(\mathbf{v}_1, \mathbf{v}_2, \ldots, \mathbf{v}_k)$, smallest subspace with $\mathbf{v}_1, \mathbf{v}_2, \ldots, \mathbf{v}_k$
Thm 6.4	p.448	6.2	vector can be written as linear combination of others $\Leftrightarrow$ *linearly dependent*
Thm 6.5	p.452	6.2	If $\mathcal{B} = \{\mathbf{u}_i\}$ is a basis and $\mathbf{x} = \sum_{i=1}^{n} c_i\mathbf{u}_i$, then c_i are unique.
Thm 6.6	p.455	6.2	Let $\mathcal{B} = \{\mathbf{u}_i\}$ be a basis for V then: a. $[\mathbf{u} + \mathbf{v}]_{\mathcal{B}} = [\mathbf{u}]_{\mathcal{B}} + [\mathbf{v}]_{\mathcal{B}}$ b. $[c\mathbf{u}]_{\mathcal{B}} = c[\mathbf{u}]_{\mathcal{B}}$
Thm 6.7	p.456	6.2	Let $\mathcal{B} = \{\mathbf{u}_i\}$ be a basis for V then: $\{\mathbf{u}_i\}$ are linearly independent in V if and only if $[\mathbf{u}_i]_{\mathcal{B}}$ are linearly independent in $\mathbb{R}^n$
Thm 6.8	p.456	6.2	Let $\mathcal{B} = \{\mathbf{u}_i\}$ be a basis for V then: a. Any set of more than n vectors in V must be linearly dependent b. Any set of fewer than n vectors cannot span V
Thm 6.9	p.457	6.2	Every basis has exactly the same number of vectors
Thm 6.10	p.458	6.2	Let V be a vector space with $\dim V = n$. Then: a. Any linearly independent set in V contains at most n vectors. b. Any spanning set for V contains at least n vectors. c. Any lin indpt set of exactly n vectors in V is a basis for V. d. Any spanning set for V of exactly n vectors is a basis for V. e. Any lin indpt set in V can be extended to a basis for V. f. Any spanning set for V can be reduced to a basis for V.
Thm 6.11	p.460	6.2	Let W be a subspace of a finite-dimensional vector space V. a. W is finite-dimensional and $\dim W \leq \dim V$. b. $\dim W = \dim V$ if and only if $W = V$.

Thm 6.12 p.469 6.3 Let $\mathcal{B} = \{\mathbf{u}_1, \ldots, \mathbf{u}_n\}$, $\mathcal{C} = \{\mathbf{v}_1, \ldots, \mathbf{v}_n\}$ be bases for V.
let $P_{\mathcal{C}\leftarrow\mathcal{B}}$ be a *change-of-basis* matrix from $\mathcal{B}$ to $\mathcal{C}$. Then:
a. $P_{\mathcal{C}\leftarrow\mathcal{B}}[\mathbf{x}]_{\mathcal{B}} = [\mathbf{x}]_{\mathcal{C}}$ for all $\mathbf{x}$ in V.
b. $P_{\mathcal{C}\leftarrow\mathcal{B}}$ is the unique matrix P with property (a).
c. $P_{\mathcal{C}\leftarrow\mathcal{B}}$ is invertible and $(P_{\mathcal{C}\leftarrow\mathcal{B}})^{-1} = P_{\mathcal{B}\leftarrow\mathcal{C}}$.

Thm 6.13 p.474 6.3 Let $\mathcal{B} = \{\mathbf{u}_1, \ldots, \mathbf{u}_n\}$, $\mathcal{C} = \{\mathbf{v}_1, \ldots, \mathbf{v}_n\}$ be bases for V.
Let $B = \left[\, [\mathbf{u}_1]_{\mathcal{E}} \;\cdots\; [\mathbf{u}_n]_{\mathcal{E}} \,\right]$, $C = \left[\, [\mathbf{v}_1]_{\mathcal{E}} \;\cdots\; [\mathbf{v}_n]_{\mathcal{E}} \,\right]$.
Then: $\left[\, C \,|\, B \,\right] \longrightarrow \left[\, I \,|\, P_{\mathcal{C}\leftarrow\mathcal{B}} \,\right]$.

Thm 6.14 p.479 6.4 Let $T : V \to W$ be a linear transformation. Then:
a. $T(\mathbf{0}) = \mathbf{0}$.
b. $T(-\mathbf{v}) = -T(\mathbf{v})$ for all $\mathbf{v}$ in V.
c. $T(\mathbf{u} - \mathbf{v}) = T(\mathbf{u}) - T(\mathbf{v})$ for all $\mathbf{u}$, $\mathbf{v}$ in V.

Thm 6.15 p.480 6.4 Let $T : V \to W$ be a linear transformation.
Let $\mathcal{B} = \{\mathbf{v}_1, \ldots, \mathbf{v}_n\}$ be a spanning set for V. Then:
$T(\mathcal{B}) = \{T(\mathbf{v}_1), \ldots, T(\mathbf{v}_n)\}$ is a spanning set for the range of T.

Thm 6.16 p.481 6.4 If $T : U \to V$ and $S : V \to W$ then
$S \circ T : U \to W$, $(S \circ T)(\mathbf{u}) = S(T(\mathbf{u}))$, is a linear transformation.

Thm 6.17 p.483 6.4 If T is an *invertible* linear transformation, then its *inverse* is unique.
If $R \circ T = I_V$, $T \circ R = I_W$ and $S \circ T = I_V$, $T \circ S = I_W$, then $R = S$.

Thm 6.18 p.488 6.5 Let $T : V \to W$ be a linear transformation. Then:
a. The *kernel* of T is a subspace of V
b. The *range* of T is a subspace of W

Thm 6.19 p.490 6.5 Let $T : V \to W$ be a linear transformation. Then:
$\operatorname{rank}(T) + \operatorname{null}(T) = \dim V$

Thm 6.20 p.494 6.5 Let $T : V \to W$ be a linear transformation. Then:
T is *one-to-one* if and only if $\ker(T) = \{\mathbf{0}\}$.

Thm 6.21 p.494 6.5 Let $\dim V = \dim W = n$. Then:
$T : V \to W$ is *one-to-one* if and only if T is *onto*.

Thm 6.22 p.495 6.5 Let $T : V \to W$ be a *one-to-one* linear transformation.
If $S = \{\mathbf{v}_1, \ldots, \mathbf{v}_k\}$ is a linearly independent set in V. Then:
$T(S) = \{T(\mathbf{v}_1), \ldots, T(\mathbf{v}_k)\}$ is a linearly independent set in W.

Cor 6.23 p.495 6.5 Let $\dim V = \dim W = n$. Then:
If T is *one-to-one*, then T maps a basis for V to a basis for W.

Thm 6.24 p.495 6.5 Let $T : V \to W$ be a linear transformation. Then:
T is *invertible* if and only if T is *one-to-one* and *onto*.

Thm 6.25 p.498 6.5 Let V and W be two finite dimensional vector spaces. Then:
V is *isomorphic* to W ($V \cong W$) if and only if $\dim V = \dim W$.

Thm 6.26 p.502 6.6 Let V and W be two finite dimensional vector spaces
with bases $\mathcal{B}$ and $\mathcal{C}$ respectively, where $\mathcal{B} = \{\mathbf{v}_1, \ldots, \mathbf{v}_n\}$.
Let $T : V \to W$ be a linear transformation, then
the $n \times n$ matrix $A = \left[\, [T(\mathbf{v}_1)]_{\mathcal{C}} \,|\, [T(\mathbf{v}_2)]_{\mathcal{C}} \,|\cdots|\, [T(\mathbf{v}_n)]_{\mathcal{C}} \,\right]$
satisfies $A[\mathbf{v}]_{\mathcal{B}} = [T(\mathbf{v})]_{\mathcal{C}}$ for every vector $\mathbf{v}$ in V.

Thm 6.27 p.508 6.6 Let U, V, and W be finite dimensional vector spaces
with bases $\mathcal{B}$, $\mathcal{C}$, and $\mathcal{D}$ respectively.
Let $T : U \to V$ and $S : V \to W$ be linear transformations. Then:
$[S \circ T]_{\mathcal{D} \leftarrow \mathcal{B}} = [S]_{\mathcal{D} \leftarrow \mathcal{C}} [T]_{\mathcal{C} \leftarrow \mathcal{B}}$

Thm 6.28 p.509 6.6 Let V and W be finite dimensional vector spaces
with bases $\mathcal{B}$ and $\mathcal{C}$, respectively.
Let $T : V \to W$ be a linear transformation. Then:
T is invertible if and only if the matrix $[T]_{\mathcal{C} \leftarrow \mathcal{B}}$ is invertible.
In this case, $([T]_{\mathcal{C} \leftarrow \mathcal{B}})^{-1} = [T^{-1}]_{\mathcal{C} \leftarrow \mathcal{B}}$

Thm 6.29 p.512 6.6 Let V be a finite dimensional vector space
with bases $\mathcal{B}$ and $\mathcal{C}$.
Let $T : V \to V$ be a linear transformation. Then:
$[T]_{\mathcal{C}} = P^{-1}[T]_{\mathcal{B}} P$
where P is the change-of-basis matrix from $\mathcal{C}$ to $\mathcal{B}$.

Thm 6.30 p.516 6.6 *Fundamental Theorem of Invertible Matrices*: Version 4
Let A be an $n \times n$ matrix and
let $T : V \to W$ be a linear transformation
whose matrix ($[T]_{\mathcal{C} \leftarrow \mathcal{B}}$) with respect to bases
$\mathcal{B}$ and $\mathcal{C}$ of V and W respectively, is A.
Statements a. through t. are equivalent (see p.516).

p. T is invertible
q. T is one-to-one
r. T is onto
s. $\ker(T) = \mathbf{0}$
t. $\text{range}(T) = W$

Thm 6.31	p.522	6.7	The set S of all solutions to $y' + ay = 0$ is a subspace of $\mathscr{F}$.
Thm 6.32	p.523	6.7	If S is the solution space of $y' + ay = 0$, then: dim $= 1$ and $\{e^{-at}\}$ is a basis for S.
Thm 6.33	p.526	6.7	Let S be the solution space of $y'' + ay' + by = 0$ and let λ_1 and λ_2 be the roots of $\lambda^2 + a\lambda + b = 0$. a. If $\lambda_1 \neq \lambda_2$, then $\{e^{\lambda_1 t}, e^{\lambda_2 t}\}$ is a basis for S. b. If $\lambda_1 = \lambda_2$, then $\{e^{\lambda_1 t}, te^{\lambda_1 t}\}$ is a basis for S.
Thm 6.34	p.531	6.7	Let C be an (n, k) linear code. a. The dual code $C^{\perp}$ is an $(n, n-k)$ linear code. b. C contains 2^k vectors, and $C^{\perp}$ contains 2^{n-k} vectors.
Thm 6.35	p.533	6.7	For $n \geq 1$, the *Reed-Muller code* R_n is: a $(2^n, n+1)$ linear code in which every vector (except $\mathbf{0}$ and $\mathbf{1}$) has weight 2^{n-1}.

Theorems, Chapter 7

Thm 7.1 p.544 7.1
Let $\mathbf{u}$, $\mathbf{v}$, and $\mathbf{w}$ be vectors in an inner product space V and c a scalar:

a. $\langle \mathbf{u}+\mathbf{v}, \mathbf{w}\rangle = \langle \mathbf{u}, \mathbf{v}\rangle + \langle \mathbf{u}, \mathbf{w}\rangle$
b. $\langle \mathbf{u}, c\mathbf{v}\rangle = c\langle \mathbf{u}, \mathbf{v}\rangle$
c. $\langle \mathbf{u}, \mathbf{0}\rangle = \langle \mathbf{0}, \mathbf{v}\rangle = 0$

Thm 7.2 p.546 7.1
Let $\mathbf{u}$ and $\mathbf{v}$ be vectors in an inner product space V.
Then $\mathbf{u}$ and $\mathbf{v}$ are orthogonal if and only if $\|\mathbf{u}+\mathbf{v}\|^2 = \|\mathbf{u}\|^2 + \|\mathbf{v}\|^2$

Thm 7.3 p.548 7.1
Let $\mathbf{u}$ and $\mathbf{v}$ be vectors in an inner product space V.
Then $|\langle \mathbf{u}, \mathbf{v}\rangle| \leq \|\mathbf{u}\|\|\mathbf{v}\|$ with equality holding if and only if $\mathbf{u}$ and $\mathbf{v}$ are scalar multiples of each other

Thm 7.4 p.549 7.1
Let $\mathbf{u}$ and $\mathbf{v}$ be vectors in an inner product space V.
Then $\|\mathbf{u}+\mathbf{v}\| \leq \|\mathbf{u}\| + \|\mathbf{v}\|$

Thm 7.5 p.544 7.2
Let d be a distance function defined on a normed linear space V.
The following properties hold for all vectors $\mathbf{u}$, $\mathbf{v}$, and $\mathbf{w}$ in V.

a. $\mathrm{d}(\mathbf{u}, \mathbf{v}) \geq 0$, and $\mathrm{d}(\mathbf{u}, \mathbf{v}) = 0$ if and only if $\mathbf{u} = \mathbf{v}$
b. $\mathrm{d}(\mathbf{u}, \mathbf{v}) = \mathrm{d}(\mathbf{v}, \mathbf{u})$
c. $\mathrm{d}(\mathbf{u}, \mathbf{w}) \leq \mathrm{d}(\mathbf{u}, \mathbf{v}) + \mathrm{d}(\mathbf{v}, \mathbf{w})$

Thm 7.6 p.567 7.2
If $\|\mathbf{x}\|$ is a vector norm on $\mathbb{R}$,
then $\|A\| = \max\|A\mathbf{x}\|$ where $\|\mathbf{x}\| = 1$
defines a matrix norm on M_{nn} that is compatible with the vector norm that induces it.

Thm 7.7 p.569 7.2
Let A have column vectors $\mathbf{a}_i$ and row vectors $\mathbf{A}_i$ then:

a. $\|A\|_1 = \max\{\|\mathbf{a}_i\|_s\} = \max\left\{\sum_{i=1}^{n} |a_{ij}|\right\}$
b. $\|A\|_\infty = \max\{\|\mathbf{A}_i\|_s\} = \max\left\{\sum_{j=1}^{n} |a_{ij}|\right\}$

Thm 7.8 p.579 7.3
If W is a finite-dimensional subspace of an inner product space V and if $\mathbf{v}$ is a vector in V,
then $\mathrm{proj}_W(\mathbf{v})$ is the *best approximation* to $\mathbf{v}$ in W.

Thm 7.9 p.584 7.3
$A\mathbf{x} = \mathbf{b}$ has at least one *least squares solution*, $\overline{\mathbf{x}}$.

a. $\overline{\mathbf{x}}$ is a *least squares solution* of $A\mathbf{x} = \mathbf{b}$ if and only if $\overline{\mathbf{x}}$ is a solution of the normal equations $A^T A\overline{\mathbf{x}} = A^T\mathbf{b}$.
b. A has linearly independent columns $\Leftrightarrow$ $A^T A$ is invertible. Then $\overline{\mathbf{x}} = (A^T A)^{-1}A^T\mathbf{b}$ is the *unique* solution.

Thm 7.10 p.591 7.3 Let A have linearly independent columns with QR factorization $A = QR$ (see Theorem 5.16 below). Then $\overline{\mathbf{x}} = R^{-1}Q^T\mathbf{b}$ is the *unique* solution of $A\mathbf{x} = \mathbf{b}$.

Thm 7.11 p.592 7.3 Let the columns of A be a basis for W. Then $P : \mathbb{R}^m \to \mathbb{R}^n$ that projects $\mathbb{R}^m$ onto W has matrix $A(A^TA)^{-1}A^T$ and $\text{proj}_W(\mathbf{v}) = A(A^TA)^{-1}A^T\mathbf{v}$.

Thm 7.12 p.595 7.3 Let A have linearly independent columns. Then A^+ the pseudoinverse of A satisfies the *Penrose conditions*.

a. $AA^+A = A$
b. $A^+AA^+ = A^+$
c. AA^+ and A^+A are symmetric

Thm 7.13 p.602 7.4 **The Singular Value Decomposition Theorem**
Let A be an $m \times n$ matrix with singular values $\sigma_1 \geq \sigma_1 \geq \ldots \geq \sigma_r > 0$ and $\sigma_{r+1} = \sigma_{r+2} = \ldots = \sigma_n = 0$. Then there exist an $m \times m$ orthogonal matrix U and an $n \times n$ orthogonal matrix V and an $m \times n$ matrix Σ (see p.601) such that $A = U\Sigma V^T$.

Thm 7.14 p.605 7.4 Let A be an $m \times n$ matrix with singular values $\sigma_1 \geq \sigma_1 \geq \ldots \geq \sigma_r > 0$ and $\sigma_{r+1} = \sigma_{r+2} = \ldots = \sigma_n = 0$. Let $\mathbf{u}_1, \ldots, \mathbf{u}_r$ be left singular values and let $\mathbf{v}_1, \ldots, \mathbf{v}_r$ be left singular values. Then: $A = \sigma_1\mathbf{u}_1\mathbf{v}_1^T + \cdots + \sigma_r\mathbf{u}_r\mathbf{v}_r^T$.

Thm 7.15 p.606 7.4 Let $A = U\Sigma V^T$ be an *SVD* of an $m \times m$ matrix A with nonzero singular values $\sigma_1, \ldots, \sigma_r$.

a. The rank of A is r.
b. $\{\mathbf{u}_1, \ldots, \mathbf{u}_r\}$ is an orthonormal basis for col(A).
c. $\{\mathbf{u}_{r+1}, \ldots, \mathbf{u}_m\}$ is an orthonormal basis for null(A^T).
d. $\{\mathbf{v}_1, \ldots, \mathbf{v}_r\}$ is an orthonormal basis for row(A).
e. $\{\mathbf{v}_{r+1}, \ldots, \mathbf{v}_m\}$ is an orthonormal basis for null(A).

Thm 7.16 p.607 7.4 Let $A = U\Sigma V^T$ be an *SVD* of an $m \times n$ matrix A with rank r. Then the image of the unit sphere in $\mathbb{R}^n$ under the matrix transformation that maps $\mathbf{x}$ to $A\mathbf{x}$ is

a. The surface of an ellipsoid in $\mathbb{R}^m$ if $r = n$.
b. A solid ellipsoid in $\mathbb{R}^m$ if $r < n$.

Thm 7.17	p.610	7.4	Let A be an $m \times n$ matrix and let $\sigma_1, \ldots, \sigma_r$ be all the nonzero singular values of A. Then: $\|A\|_F = \sqrt{\sigma_1^2 + \cdots + \sigma_r^2}$
Thm 7.18	p.613	7.4	The least squares problem $A\mathbf{x} = \mathbf{b}$ has a unique least squares solution $\overline{\mathbf{x}}$ of minimal length that is given by $\overline{\mathbf{x}} = A^{+}\mathbf{b}$.
Thm 7.20	p.628	7.5	Let C be a (binary) code with minimum distance d. a. C detects k errors if and only if $d \geq k + 1$. b. C detects k errors if and only if $d \geq 2k + 1$.
Thm 7.21	p.628	7.5	Let C be a (n, k) with parity check matrix P. Then the minimum distance of C is the smallest integer d for which there are d linearly dependent columns.

Theorems, Chapter 7, continued

The Fundamental Theorem of Invertible Matrices: Final Version

Thm 7.19 p.614 7.4 a. through u. give equivalent conditions for A to be invertible:

a. A is invertible.
b. $A\mathbf{x} = \mathbf{b}$ has a unique solution for every $\mathbf{b}$ in $\mathbb{R}^n$
c. $A\mathbf{x} = \mathbf{0}$ has only the trivial solution.
d. The reduced row echelon form of A is I_n.
e. A is the product of elementary matrices.
f. $\text{rank}(A) = n$
g. $\text{nullity}(A) = 0$
h. The column vectors of A are linearly independent.
i. The column vectors of A span $\mathbb{R}^n$.
j. The column vectors of A form a basis for $\mathbb{R}^n$.
k. The row vectors of A are linearly independent.
l. The row vectors of A span $\mathbb{R}^n$.
m. The row vectors of A for a basis for $\mathbb{R}^n$.
n. $\det(A) \neq 0$
o. 0 is not an eigenvalue of A
p. T is invertible
q. T is one-to-one
r. T is onto
s. $\ker(T) = \{\mathbf{0}\}$
t. $\text{range}(T) = W$
u. 0 is not a singular value of A

Thm 7.8	p.579	7.3	**The Best Approximation Theorem** If W is a finite-dimensional subspace of an inner product space V and if $\mathbf{v}$ is a vector in V, then $\text{proj}_W(\mathbf{v})$ is the *best approximation* to $\mathbf{v}$ in W.
Thm 1.4	p.19	1.2	**The Cauchy-Schwarz Inequality** $\lvert\mathbf{u}\cdot\mathbf{v}\rvert \le \lVert\mathbf{u}\rVert\,\lVert\mathbf{v}\rVert$
Thm 7.3	p.548	7.1	**The Cauchy-Schwarz Inequality** Let $\mathbf{u}$ and $\mathbf{v}$ be vectors in an inner product space V. Then $\lvert\langle\mathbf{u},\mathbf{v}\rangle\rvert \le \lVert\mathbf{u}\rVert\lVert\mathbf{v}\rVert$ with equality holding if and only if $\mathbf{u}$ and $\mathbf{v}$ are scalar multiples of each other
Thm 4.11	p.274	4.2	**Cramer's Rule** If $A\mathbf{x} = \mathbf{b}$, then $x_i = \frac{\det(A_i(\mathbf{b}))}{\det A}$
Thm 5.15	p.386	5.3	**The Gram-Schmidt Process** Let $\{\mathbf{x}_1, \ldots, \mathbf{w}_k\}$ be a basis for W then ... $\mathbf{v}_1 = \mathbf{x}_1,\ W_1 = \text{span}(\mathbf{x}_1)$ $\mathbf{v}_2 = \mathbf{x}_2 - \left(\frac{\mathbf{v}_1\cdot\mathbf{x}_2}{\mathbf{v}_1\cdot\mathbf{v}_1}\right)\mathbf{v}_1,\ W_2 = \text{span}(\mathbf{x}_1, \mathbf{x}_2)$ $\mathbf{v}_3 = \mathbf{x}_3 - \left(\frac{\mathbf{v}_1\cdot\mathbf{x}_3}{\mathbf{v}_1\cdot\mathbf{v}_1}\right)\mathbf{v}_1 - \left(\frac{\mathbf{v}_2\cdot\mathbf{x}_3}{\mathbf{v}_2\cdot\mathbf{v}_2}\right)\mathbf{v}_2,\ W_3 = \text{span}(\mathbf{x}_1, \mathbf{x}_2, \mathbf{x}_3)$ ... $\mathbf{v}_k = \mathbf{x}_k - \sum_{i=1}^{k-1} \frac{\mathbf{v}_i\cdot\mathbf{x}_k}{\mathbf{v}_i\cdot\mathbf{v}_i},\ W_k = \text{span}(\{\mathbf{x}_k\})$ ... $\{\mathbf{v}_1, \ldots, \mathbf{v}_k\}$ is an orthogonal basis for W.
Thm 4.1	p.265	4.2	**The Laplace Expansion Theorem** $\det A = \sum_{j=1}^{n} a_{ij}C_{ij}$ (any row) or $\sum_{i=1}^{n} a_{ij}C_{ij}$ (any column)
Thm 7.9	p.584	7.3	**The Least Squares Theorem** $A\mathbf{x} = \mathbf{b}$ has at least one *least squares solution*, $\overline{\mathbf{x}}$. a. $\overline{\mathbf{x}}$ is a *least squares solution* of $A\mathbf{x} = \mathbf{b}$ if and only if $\overline{\mathbf{x}}$ is a solution of the normal equations $A^TA\overline{\mathbf{x}} = A^T\mathbf{b}$. b. A has linearly independent columns $\Leftrightarrow$ A^TA is invertible. Then $\overline{\mathbf{x}} = (A^TA)^{-1}A^T\mathbf{b}$ is the *unique* solution.
Thm 5.11	p.381	5.2	**The Orthogonal Decomposition Theorem** $\mathbf{v} = \mathbf{w} + \mathbf{w}^{\perp}$
Thm 4.36	p.330	4.6	**Perron's Theorem** See Theorem 4.36.
Thm 4.37	p.332	4.6	**Perron-Frobenius Theorem** See Theorem 4.37.

Theorems by name, continued

Thm 5.23	p.411	5.5	**The Principal Axes Theorem** Every quadratic form can be diagonalized: $\mathbf{x}^T A\mathbf{x} = \mathbf{y}^T D\mathbf{y} = \lambda_1 y_1^2 + \cdots + \lambda_n y_n^2$ (see statement of Theorem in the text for details)
Thm 1.6	p.23	1.2	**Pythagoras' Theorem** $\|\mathbf{u}+\mathbf{v}\|^2 = \|\mathbf{u}\|^2 + \|\mathbf{v}\|^2 \Leftrightarrow \mathbf{u}\cdot\mathbf{v} = 0$
Thm 7.2	p.546	7.1	**Pythagoras' Theorem** Let $\mathbf{u}$ and $\mathbf{v}$ be vectors in an inner product space V. Then $\mathbf{u}$ and $\mathbf{v}$ are orthogonal $\Leftrightarrow \|\mathbf{u}+\mathbf{v}\|^2 = \|\mathbf{u}\|^2 + \|\mathbf{v}\|^2$
Thm 5.16	p.390	5.3	**The QR Factorization** Let A be an $m \times n$ matrix with linearly independent columns. Then A can be factored as $A = QR$, where Q is an $m \times n$ matrix with orthonormal columns and R is an invertible upper triangular matrix.
Thm 2.2	p.75	2.2	**The Rank Theorem**: number of free variables $= n - \text{rank}(A)$
Thm 3.26	p.203	3.5	**The Rank Theorem**: $\text{rank}(A) + \text{nullity}(A) = n$
Cor 5.14	p.381	5.2	**The Rank Theorem** Given A is $m \times n$ then ... $\text{rank}(A) + \text{nullity}(A) = n$ and $\text{rank}(A) + \text{nullity}(A^T) = m$
Thm 6.19	p.490	6.5	**The Rank Theorem** Let $T : V \to W$ be a linear transformation. Then: $\text{rank}(T) + \text{null}(T) = \dim V$
Thm 5.20	p.400	5.4	**The Spectral Theorem** Let A be an $n \times n$ *real* matrix. Then A is *symmetric* if and only if A is *orthogonally diagonalizable.*
Thm 1.5	p.19	1.2	**The Triangle Inequality**: $\|\mathbf{u}+\mathbf{v}\| \leq \|\mathbf{u}\| + \|\mathbf{v}\|$
Thm 7.4	p.549	7.1	**The Triangle Inequality** Let $\mathbf{u}$ and $\mathbf{v}$ be vectors in an inner product space V. Then $\|\mathbf{u}+\mathbf{v}\| \leq \|\mathbf{u}\| + \|\mathbf{v}\|$
Thm 7.13	p.602	7.4	**The Singular Value Decomposition Theorem** Let A be an $m \times n$ matrix with singular values $\sigma_1 \geq \sigma_1 \geq \ldots \geq \sigma_r > 0$ and $\sigma_{r+1} = \sigma_{r+2} = \ldots = \sigma_n = 0$. Then there exist an $m \times m$ orthogonal matrix U and an $n \times n$ orthogonal matrix V and an $m \times n$ matrix Σ (see p.601) such that $A = U\Sigma V^T$.